CW01529408

FOODBORNE PATHOGENS

AN ILLUSTRATED TEXT

FOODBORNE PATHOGENS

AN ILLUSTRATED TEXT

A. H. Varnam
BSc, PhD
Consultant Microbiologist
Southern Biological, Reading

M. G. Evans
Photographer

Wolfe Publishing Ltd
1991

Copyright © A.H. Varnam, M.G. Evans, 1991
Published by Wolfe Publishing Ltd, 1991
Printed by BPCC Hazell Books, Aylesbury, England
ISBN 0 7234 1521 8

A CIP catalogue record for this book is available from
the British Library.

For full details of all Wolfe titles please write to Wolfe
Publishing Ltd, Brook House, 2–16 Torrington Place,
London WC1E 7LT, England.

Contents

Preface   6

Acknowledgements   7

1   Food poisoning: medical and microbiological overview   9
2   Food poisoning: economic, social and personal aspects   21
3   Foodborne pathogenic micro-organisms: a perspective   27
4   *Salmonella*   51
5   *Shigella*   87
6   *Escherichia coli*   101
7   *Yersinia*   129
8   *Vibrio*   157
9   *Aeromonas*   185
10   *Plesiomonas shigelloides*   201
11   *Campylobacter*   209
12   *Staphylococcus aureus*   235
13   *Bacillus*   267
14   *Clostridium botulinum*   289
15   *Clostridium perfringens*   312
16   *Listeria monocytogenes*   327
17   Other bacterial agents of foodborne disease   355
18   Foodborne viral infection   363
19   Protozoa   373
20   Control of pathogenic micro-organisms in food: management aspects   387
21   Control of pathogenic micro-organisms in meat   427
22   Control of pathogenic micro-organisms in milk   447
23   Control of pathogenic micro-organisms in eggs   459
24   Control of pathogenic micro-organisms in fish and shellfish   463
25   Control of pathogenic micro-organisms in miscellaneous foods and processes   473

References   485

Further reading   549

Index   550

Preface

Foodborne Pathogens: An Illustrated Text is intended to be used as a text and reference book for those, including advanced undergraduates and graduate students, who have a professional concern or interest in the problems posed by microbial food poisoning and the organisms responsible. The book is also intended to serve as a practical guide for people involved in laboratory examination for food poisoning micro-organisms, and to provide assistance to those responsible for the safe production and handling of food.

It is the author's firm belief that the various different aspects of food poisoning cannot be studied in isolation, but that an overall appreciation of the subject is required, even where the reader is working in a highly specialised field. The book is designed, therefore, to give a fully comprehensive coverage, this being achieved by dividing the contents into three complementary sections.

The first section serves to place the problems presented by food poisoning into a true perspective with respect to the medical and economic consequences, and also to provide an overview of specialist aspects of microbiology which relate to the study of the causative organisms. The second section provides a detailed account of the causative organisms themselves and is concerned with both established pathogens, such as *Salmonella*, and 'emerging pathogens', including other bacteria, viruses and protozoa. Particular attention has been given to recent findings concerning the properties of foodborne pathogens, and also to the laboratory methods used for detection and isolation of the organisms and for determination of their characteristics. The third section is concerned with the prevention of food poisoning. It integrates management responsibilities with technical control systems, such as the hazard analysis critical control point (HACCP) system, and the practical measures required to ensure safety during food manufacture and handling.

The text is highly illustrated, with photographs and line and colour artwork chosen to reflect the scope and contents of the book. However, it is recognised that not all aspects of the subject material are amenable to an illustrative treatment, and effort has therefore been concentrated on those areas where illustrations may be used most effectively.

Microbiological food poisoning is a large and complex subject involving many aspects of microbiology, medicine and food technology. However, in all cases, reduction in the morbidity due to food poisoning must be seen as the ultimate common objective, an objective which represents the underlying rationale of this book. Responsibility is no less real for being shared amongst many persons, and failure to accept responsibility, whether as employee or manager, as microbiologist or food scientist, cannot be excused. Those who choose not to be part of the solution inevitably become part of the problem.

Acknowledgements

In writing a text of this nature it is, inevitably, necessary to obtain the help and advice of a large number of people and organisations and the authors are grateful to all who have so freely offered their assistance and encouragement. More specifically, we would like to offer particular thanks to those below.

API-bioMérieux (UK) Ltd., Bioenterprises Pty Ltd., Biomet Laboratories Ltd., Difco Laboratories Ltd., Rhône-Poulenc Diagnostics Ltd., Roche Products Ltd. and Unipath Ltd. for the gifts of media and other laboratory materials. In this context particular assistance was given by Dr J. Broughall of Difco Laboratories, Mr R. Fielder of Rhône-Poulenc, Mr D. Post of Unipath and Dr B. Thomas of API-bioMérieux.

Dr M. Arnott, Reading Scientific Services Ltd., Dr I. Bousfield, National Collections of Industrial and Marine Bacteria Ltd., Dr D. Cope, Public Health Laboratory, Leicester; Dr C.E.R. Dodd, School of Agriculture, University of Nottingham; Dr J.M. Grainger, Department of Microbiology, University of Reading; Dr R.N. Peel, Public Health Laboratory, Leeds; and Dr J. Sutherland, J. Sainsbury plc: our thanks for supplying cultures.

Nikon Instruments (UK) for the loan of a microscope. Debbie Andrews for producing all of the colour and the black and white drawings. Dr T.K. Sawyer, Rescon Associates Inc., for advice concerning free-living amoebae. Computer Care South Ltd. for assistance concerning the installation and operation of computer equipment and software. AFRC Institute of Food Research, Reading Laboratory and J. Sainsbury plc for valuable assistance in a number of areas. The library of the University of Reading for allowing access to books and periodicals. The Editor of *Communicable Disease Report* for permission to make use of material from that publication, and the Secretary of Campden Food Research Association for permission to quote from research memoranda. Aspells Ltd., Clearwater Fish Farms Ltd. and Cliffords Dairies plc for permission to photograph their operations.

Although most of the photographs used were taken specially for this book, a number were supplied by other microbiologists. The sources are acknowledged individually in the accompanying captions, but in addition we are particularly grateful for the help given by Dr E.O. Caul, Public Health Laboratory, Bristol; Mr P.T. Feldsine, BioControl Systems Inc.; Dr S. Knutton, Institute of Child Health, University of Birmingham; Dr J.M. Kramer, Central Public Health Laboratory, Colindale; Dr M.P. Osborne, Dr J. Stephen and Dr T.S. Wallis, School of Biological Sciences, University of Birmingham; and Professor T. Yamamoto, Department of Bacteriology, Juntendo University.

1. Food poisoning: medical and microbiological overview

Introduction

In recent years, food poisoning has become a major topic of both public and scientific debate. On the one hand a public, increasingly concerned with 'green' issues and anxious over the safety and quality of food, has been subjected to a barrage of information, and indeed misinformation, concerning a seemingly endless number of malignant micro-organisms. On the other hand those professionals concerned with public health must face the reality of an increasing incidence of food transmitted disease. In reviewing the current situation, Archer and Young (1988) noted:

'Food has been shown to be a vector of bacteria previously thought to infect humans only by other routes, and newly discovered causes of disease have been associated with particular foods. Some long-recognised pathogens have appeared in foods once believed to be incapable of supporting their growth. The finding that some of these pathogens are far more resistant to long standing food processing and storage techniques than expected has caused alarm within the food industry'.

Despite the high level of concern, confusion remains as to the nature of food poisoning. Among the general public this is hardly surprising given the complexities of the subject and the quality of information available. However, the terms 'food poisoning' and the rather broader term 'foodborne disease' may also be confusing to professional microbiologists.

The bulk of 'foreign' material that humans encounter in their daily lives is food, and it is therefore inevitable that it can serve as a vehicle for the transmission of agents of disease. These may be intrinsic agents, including allergens and toxic compounds naturally present, or extrinsic agents, which include micro-organisms and toxic compounds present in the food as contaminants. In the case of food poisoning due to micro-organisms, the food may serve as either an active vehicle, in which multiplication occurs, or as a passive vehicle, in which no growth takes place.

Further confusion may result from the range of symptoms that are produced by foodborne micro-organisms. Classical food poisoning symptoms are those of acute gastrointestinal disturbances; diarrhoea, vomiting, or both, together with related symptoms such as abdominal pain and discomfort. However, enteric pathogens may also cause acute and possibly life threatening extraintestinal illness, including respiratory, renal and perinatal disease. These symptoms may occur independently or accompany 'classic' intestinal symptoms. In addition to acute intestinal and extraintestinal disease, there is increasing recognition of the role of foodborne pathogens in chronic diseases, including joint disease and autoimmune thyroid disease.

Agents of foodborne disease are surveyed in **Table 1**.

It is regrettable that there are still attempts to produce a narrow, artificial and misleading definition of food poisoning by excluding, for example, agents that do not produce 'classic' intestinal symptoms or that do not multiply in foods. Arguments resulting from these narrow definitions are as futile as arguments concerning the number of angels able to stand on the head of a pin, and have no place in modern food microbiology. Indeed, it is considered that the most satisfactory approach is to view foodborne disease in the context of total disease. Foodborne disease is a subset of enteric disease, which in turn is a major category of the disease total (Archer and Young, 1988).

Table 1. Survey of food poisoning agents.

Micro-organisms	Bacteria
	Rickettsia (one species)
	Viruses
	Protozoa
	Microfungi (moulds)
Chemicals	Intrinsic
	Extrinsic
	direct contamination
	environmental
Parasites	Animal parasites
	Fish parasites

Food poisoning: a disease of our times?

An inevitable consequence of the increased incidence of food poisoning observed in recent years is the suspicion thrown on modern food processing and handling techniques. Such suspicion is not wholly justified, but does raise legitimate questions as to why morbidity due to food poisoning is increasing at a time when, in the more developed world at least, a decline in the number of cases might reasonably be expected.

To some extent the increase in cases may be explained by more efficient reporting. In cases of abortion, physicians are more likely to consider *Listeria monocytogenes* as a potential cause, while illness following the consumption of egg dishes will inevitably arouse suspicion of infection with *Salmonella enteritidis* phage type 4. Food poisoning remains, however, a vastly under-reported illness and it is not possible to attribute the entire increased morbidity to improved reporting.

In discussing the increased incidence of food poisoning, a number of possible contributory factors emerge. However, before examining these, it is necessary to place current problems in a historical perspective.

Microbiology is a relatively young science and, indeed, the relationship between bacteria such as *Salmonella* and *Clostridium botulinum* became apparent only between 1880 and 1900 (Gilbert, 1983). That we should be recognising 'new pathogens' more than 80 years later must reflect developments within the science as well as new patterns of food handling and of micro-organism – host relationships.

Despite this, there is a perception among many that food poisoning is a 'new' problem. It is certainly not uncommon to hear otherwise well-informed persons comment on food poisoning in this way and yet, as always, a cursory examination of history is sufficient to demonstrate the illusion of a past 'golden age'. The first *Adulteration of Food and Drink Act*, for example, was passed in Britain in 1860 in response to the massive adulteration of foods which was then current practice. A significant factor in the successful parliamentary passage of this Act was a major outbreak of food poisoning which affected several hundred people and caused 21 deaths as a result of eating peppermint lozenges, which had been mistakenly adulterated with arsenic instead of the commonly used Plaster of Paris (Turner, 1980). As late as 1900 a mysterious illness among heavy beer drinkers in the north and midlands of England, which affected at least 6,000 people and caused 70 deaths, was attributed to arsenic contamination (Burnett, 1966).

During the twentieth century micro-organisms have replaced adulterants as the major recognised agent of food poisoning, although significant outbreaks of chemical food poisoning still occur (see page 16). It is necessary only to recall the significant diseases associated with milk before the introduction of wide-scale pasteurisation and other measures to improve its hygienic status (**Table 2**) to realise that food poisoning is by no means a new problem, nor is it a problem uniquely associated with modern production and processing methods.

Table 2. Illness previously associated with consumption of milk.*

Illness	Causative organism
Brucellosis	*Brucella abortus*
Tuberculosis	*Mycobacterium tuberculosis*
Typhoid/paratyphoid	*Salmonella typhi/paratyphi*
Dysentery	*Shigella* species
Diptheria	*Corynebacterium diptheriae*
Various streptococcal infections including post streptococcal nephritis	*Streptococcus* species
Q fever	*Coxiella burnetii*
Listeriosis	***Listeria monocytogenes***
Gastroenteritis	***Salmonella* (serovars other than *typhi/paratyphi*)** ***Eschericia coli*** ***Yersinia enterocolitica*** ***Campylobacter* species**

* **Bold** type indicates illnesses and micro-organisms that are a current as well as a historic problem.

However, it must be recognised that we should be less concerned with debunking the myths of a golden past than facing the problems of today. Advances in many areas, including milk hygiene, must be set against problems resulting from modern practices. Intensive agriculture is the most obvious example and many microbiologists are unhappy with the standards attained and the procedures employed. Less obvious, but nonetheless presenting a significant problem, is the ever increasing storage life of chilled food and the associated risk from low temperature pathogens, including *Listeria*. The greatest cause for concern may not, however, lie with 'new' problems, but with the way in which well-established tenets of good hygienic practice continue to be ignored. For example, during the summer of 1989, 888 known cases of salmonellosis, in four unrelated outbreaks, occurred in the UK following the consumption of traditional cooked-meat products. In each case, the cause was attributed not to modern food production and processing methods, nor to a 'new' or unusual pathogen but, quite simply, to ignorance of well-established principles.

A question that has inevitably been asked in recent years is: Are there any enteric pathogens which may accurately be described as 'new'? It is not possible to attempt to answer this without engaging in speculation, but it seems likely that most newly recognised pathogens are not truly new. In some cases, apparently new organisms may be linked with newly recognised clinical conditions, but in most cases the organism has probably existed for some time. For example, haemorrhagic colitis was described accurately in 1963, but was not associated with the major causative organism *Escherichia coli* 0157:H7 until 19 years later (Hoskin and Lior, 1985). In other cases, the organism itself may have changed. The AIDS virus is believed to have crossed a species barrier to infect man and,

with more relevance to food, there is speculation of a similar process in bovine spongiform encephalopathy. Micro-organisms may also acquire virulence factors through genetic exchange. Heat-stable enterotoxin production has been detected in a number of genera of the *Enterobacteriaceae*, which have not generally been considered as enteric pathogens (Archer and Young, 1988). It seems likely that certain plasmidborne virulence genes are transferred readily between certain genera, possibly along with antibiotic resistance (Rosenberg *et al.*, 1974).

Virulence among micro-organisms cannot be considered in isolation of the host response. A number of newly recognised enteric pathogens primarily affect people with underlying symptoms, such as cancer or an impaired immune function. The number of people with such conditions in the community is much greater than in previous years. A further consideration is that the proportion of elderly people in the total population is increasing. The elderly are at a greater general risk of food poisoning due to general debility and, if institutionalised, may also be at greater specific risk.

Food poisoning: the scale of the problem

On a global basis food poisoning is a major disease. It is estimated that 1,000 million cases of acute diarrhoea occur annually in children under the age of 5 years in the countries of Africa, Asia (not including China) and Latin America. Some 5 million of these cases are fatal (Snyder and Merson, 1982).

The situation in such countries is very different to that found in the richer industrialised countries. Childhood diarrhoea in the so-called Third World results not only from contaminated food and water, but also from living conditions in which adequate standards of hygiene are impossible to maintain. These factors are exacerbated by extreme poverty and malnutrition. Further, the technological solutions that may be sought in industrialised nations are often inappropriate elsewhere. Morbidity and mortality due to acute diarrhoea in developing nations can be reduced significantly by the provision of easily operated and maintained, cheap supplies of clean water, the breast feeding of infants and avoidance of expensive milk substitutes, and the provision of kits for the rapid rehydration of diarrhoea victims.

In the richer industrialised nations, the scale of the problem is less although still of considerable significance. It is difficult to assess the true scale due to the very considerable degree to which food poisoning is under-reported. It is estimated that the actual incidence is 10 to 100 times (or even greater) that actually

reported. Therefore, it is perhaps more useful to relate food poisoning to total disease. Exercises of this nature have been carried out in the USA where enteric disease ranks second to respiratory diseases (Guerrant *et al.*, 1985), and it has been estimated that, on average, every citizen suffers at least one bout of enteric disease each year. One-third of these cases are conservatively estimated to involve food as a vehicle of infection, a total number in excess of 80 million cases.

Overall figures obviously cannot illustrate the number of cases that may arise from either a single source of infection, or from a given type of food. The numbers involved in each instance vary enormously, from family outbreaks involving only a small number of people, to national and international outbreaks involving very large numbers. In this context, the largest single source outbreak to date in the USA involved over 15,000 known cases and was due to contamination of pasteurised milk with an antibiotic resistant *Salmonella* (Ryan *et al.*, 1987). It has also been estimated that as many as 2 million people in the USA may have suffered salmonellosis following consumption of egg dishes.

The full impact of food poisoning cannot be appreciated solely on the basis of the numbers affected. It is also necessary to be aware of both the human suffering and the economic consequences. These aspects are discussed in Chapter 2, pages 21–26.

Food poisoning: relative importance of different types of agent

A number of attempts have been made to assess and quantify the risks arising from various food safety hazards. Differences may result from the criteria employed, but almost invariably foodborne disease heads any list (Wheelock, 1988). One of the more widely quoted assessments is that of Truswell *et al.* (1978), who found that the risk from 'microbial contamination' was 1,000 times greater than that from 'environmental pollution' and 100,000 times greater than that from 'pesticides' (**Table 3**).

Some caution may be necessary in the interpretation of such assessments. For example, it may be argued that while the greatest risk of acute food poisoning indubitably lies with micro-organisms, a much higher number of people may be at risk from longterm effects caused by environmental pollutants entering the food chain, and even from pesticides and other agricultural chemicals.

Table 3. Relative risk associated with different aspects of food safety (Truswell *et al.*, 1978).

Aspect	Relative risk
Microbial contamination	1,000,000
Nutritional imbalance	1,000,000
Environmental pollution	100
Pesticides	1
Food additives	1

It will be appreciated that micro-organisms themselves form a broad category. Within this category there is no question that bacteria are of prime importance with respect both to the number of cases and types of food poisoning.

Food poisoning: causes and effects

Micro-organisms

Micro-organisms are responsible for a wide range of types of food poisoning. The involvement of individual genera and species in food poisoning is discussed fully in the appropriate Chapter but is also summarised below.

Classical food poisoning

Bacteria

Classical food poisoning (enteric symptoms including vomiting and diarrhoea) caused by bacteria is summarised in **Table 4**. Bacteria are the most important micro-organisms associated with food poisoning, and most processing and handling precautions are primarily directed against them. In many cases, food poisoning bacteria are able to grow in foods, which thus have an active role in the transmission of enteric disease.

In addition to food poisoning being caused by specific pathogens, it may also result from the growth of spoilage micro-organisms in foods. Production of histamine from histidine is one part of the pathogenic process, and is of potential importance in two types of food, fish (usually scombroid) and mature cheese. The problem is discussed in greater detail on pages 465–466 and 458 respectively.

Viruses

Compared to bacteria, the role of viruses in foodborne disease is less well defined. In part, this is due to the technical problems involved in the detection and cultivation of viruses. This requires special facilities which are not available in the majority of food microbiology laboratories.

Despite these difficulties, rotavirus is now well established as a cause of diarrhoea in infants, while a group of structurally similar, but antigenically distinct viruses, referred to as Norwalk-like viruses in the USA and small, round, structured viruses in the UK, are increasingly implicated in outbreaks of gastroenteritis. In the case of small, round, structured viruses, illness is usually associated with consumption of seafood.

Viruses are not able to grow in foods, and suggestions to the contrary are founded in ignorance and should be discounted.

Viruses known to be associated with gastroenteritis and the symptoms produced are summarised in **Table 5**.

Protozoa

Although *Entamoeba histolytica* has been recognised for many years as a water and, to a lesser extent, a foodborne pathogen, the role of other protozoa as agents of diarrhoea has, until recently, been obscure.

Both *Giardia lamblia* and *Cryptosporidium parvum* were originally associated with homosexuals and AIDS patients, but both are known to be common causes of diarrhoea in the general population. Food and water are recognised vehicles of infection (Casemore, 1989).

Protozoa are unable to grow in foods, but cysts are able to survive for prolonged periods.

Relatively few food microbiology laboratories have expertise with protozoa, and it is probable that in the food industry the possible involvement of the organism in outbreaks of diarrhoea is only rarely considered.

The role of protozoa in gastroenteritis is summarised in **Table 6**.

Table 4. Symptoms of classical bacterial food poisoning.

Organism	Symptoms
Salmonella	Diarrhoea, vomiting, fever and malaise, usually 12 to 16 hours after ingestion
Shigella	Diarrhoea with mucoid, bloody stools, 12 to 50 hours after ingestion
Campylobacter spp.	Prodromal fever and malaise, 2 to 11 days after ingestion followed by abdominal pain and profuse diarrhoea
Escherichia coli	Symptoms vary according to type of *E. coli* infection (see Chapter 6)
Yersinia enterocolitica	Abdominal pain, fever, headache, diarrhoea and possibly vomiting and malaise, 24 to 48 hours (and possibly longer) after ingestion
Vibrio parahaemolyticus	Profuse diarrhoea which leads to dehydration, vomiting and fever, 2 to 48 hours after ingestion. Symptoms due to other species of *Vibrio* including *V. cholerae* and *V. vulnificus* often differ
Aeromonas	Diarrhoea in normal patients, disseminating infections in the immunocompromised
Plesiomonas shigelloides	Predominantly diarrhoea accompanied possibly by abdominal pain, nausea and vomiting. Symptoms appear within 48 hours of ingestion
Staphylococcus aureus	Vomiting, abdominal pain and diarrhoea on some occasions, 2 to 6 hours after ingestion. Severe dehydration may result in collapse
Bacillus cereus	Two distinct syndromes: diarrhoeal and emetic (see Chapter 13)
Clostridium botulinum	Fatigue, lassitude, dizziness and effects on the central nervous system including speech difficulties and visual disturbances. Onset is 24 to 72 hours after ingestion
Clostridium perfringens	Abdominal pain, nausea and vomiting (occasional diarrhoea), 8 to 22 hours after ingestion
Listeria monocytogenes	Meningitis in neonates, abortion in pregnant females, septicaemia. Usually extended period between ingestion and appearance of symptoms

Note: The above symptoms and incubation periods are summaries and represent those most commonly encountered. Many enteric micro-organisms produce a spectrum of symptoms and these are discussed fully in the relevant chapters.

Table 5. Gastroenteritis and other foodborne disease due to viruses.

Illness	Virus
Gastroenteritis	Small, round, structured viruses including Norwalk, Snow Mountain and Ditchling agents Astrovirus Calicivirus Parvovirus Rotavirus
Hepatitis	Hepatitis A virus Hepatitis E virus (enterically transmitted non-A, non-B hepatitis)
Spongiform encephalopathies??	Scrapie-like agent
Other syndromes	Echovirus Coxsackievirus

Table 6. Gastroenteritis and other foodborne disease due to protozoa.

Illness	Organism
Gastroenteritis	*Giardia lamblia* *Cryptosporidium parvum* *Entamoeba histolytica* *Isopora belli*[1] *Balantidium coli*
Respiratory symptoms	*Cryptosporidium parvum*
Other symptoms	Free living pathogenic amoeba??

[1]These usually only occur in AIDS patients or the immunocompromised.

Algae

Algae do not cause food poisoning directly. However, algal toxins may enter the flesh of fish either through fish feeding on dinoflagellates, or indirectly through large carnivorous fish eating smaller herbivores. Examples are paralytic shellfish poisoning and ciguatera poisoning. These are discussed in further detail on pages 468–470 and 464–465 respectively.

Acute extraintestinal disease

Foodborne micro-organisms are responsible for a range of acute extraintestinal diseases. Indeed, infections with enteric pathogens have produced symptoms in virtually all organ systems (Archer and Young, 1988).

The role of enteric pathogens in respiratory disease is of particular note, since in many countries this type of disease is the most common. Respiratory symptoms have been associated with a number of foodborne micro-organisms including enteric viruses, cryptosporidia and bacteria. In a number of cases, the portal of infection is not known, but in some instances appears likely to be via the gastrointestinal tract. Lung infections due to *Campylobacter jejuni*, for example, probably originate in the gastrointestinal tract, although this is not yet proven (Blaser *et al.*, 1986).

Acute extraintestinal infections and associated enteric micro-organisms are summarised in **Table 7**.

Table 7. Examples of acute extraintestinal disease due to foodborne micro-organisms.

Organs affected	Micro-organism
Lungs and respiratory tract	*Campylobacter jejuni* *Salmonella* spp. *Cryptosporidium* spp. Enteric viruses
Kidneys	*Escherichia coli* O157:H7 *Salmonella* spp. *Shigella* spp. *Campylobacter* spp.
Heart	*Yersinia enterocolitica* *Salmonella* spp.
Central nervous system	*Salmonella* spp. *Listeria monocytogenes* Enterovirus
Foetal tissue	*Listeria monocytogenes*
Skin and soft tissue	*Campylobacter* spp. *Salmonella* spp. *Yersinia enterocolitica*

Chronic disease

There is growing evidence that foodborne pathogens directly or indirectly cause, or predispose, man to chronic diseases usually by acting as environmental triggers. Two microbial mechanisms appear to be of particular importance, antigenic heterogeneity and molecular mimicry.

The triggering of chronic disease involves complex interactions between the micro-organism or its products and the host. The host immune system is commonly involved, and it is pertinent to note that the intestinal immune system is in many ways distinct from the systemic immune system. Its importance may be measured from the fact that, in the mouse, 80% of all immunoglobulin secreting cells are located in the small intestine (Van der Heijden *et al.*, 1987).

The role of foodborne micro-organisms in triggering chronic disease is summarised in **Table 8**.

Table 8. Examples of foodborne micro-organisms as triggers of chronic disease.

Disease type	Micro-organism
Joint diseases	*Yersinia enterocolitica* *Campylobacter jejuni* *Salmonella* spp. *Shigella* spp. *Escherichia coli* *Staphylococcus aureus*
Autoimmune thyroid diseases	*Yersinia enterocolitica*
Neural and neuromuscular disorders	*Campylobacter jejuni* *Yersinia enterocolitica* *Escherichia coli*
Long-term heart and vascular disease	*Salmonella* Enteric viruses

Foodborne micro-organisms as reflectors and indicators of cancer

People suffering from cancer or who are immunocompromised for other reasons are predisposed to bacteraemia (Bodey, 1985). The occurrence in the bloodstream of either *S. typhimurium*, *Aeromonas hydrophila*, *C. fetus* or *L. monocytogenes* is a reflector and indicator of cancer (Beebe, 1986). Each organism has a predilection for people with specific types of tumour (**Table 9**). This relationship is reliable and the isolation of these organisms from the blood of apparently normal people may well be indicative of the presence of undetected cancer.

Table 9. Relationship between bacteraemia due to enteric micro-organisms and specific types of neoplasm.

Organism in bloodstream	Neoplasm
Salmonella typhimurium ⎤ ⎥ Listeria monocytogenes ⎦	⎡ Hodgkin's disease ⎢ Leukaemia ⎣ Lymphoma
Campylobacter fetus	Leukaemia
Aeromonas hydrophila	⎰ Leukaemia ⎱ Lymphoma

Travellers' diarrhoea

Travellers' diarrhoea is a syndrome characterised by a twofold, or greater, increase in the frequency of unformed stools. Other symptoms including abdominal cramps, nausea and malaise are frequently associated with the diarrhoea (Anon, 1986). Travellers' diarrhoea has many vernacular descriptions including 'Montezuma's revenge', 'Mexican quick step', or simply 'the trots'.

Episodes of travellers' diarrhoea are typically of abrupt onset, develop during travel or shortly after return home and are self-limiting. Travellers at risk are those from a developed country visiting a country, or region, where there is an increased risk of contracting diarrhoeal disease. Attack rates of 20 to 50% are commonly reported, the risk varying from region to region. High-risk countries include most of the developing nations of Latin America, Africa, the Middle East and Asia, while medium-risk areas include most southern European countries and some of the Caribbean islands.

The micro-organisms responsible for travellers' diarrhoea comprise a long 'list of villains', which includes four viruses, at least five and possibly nine bacteria, three important and three less important protozoa, several metazoa and some fungi (Kean, 1986).

Although travellers' diarrhoea is usually of limited duration, a number of patients develop chronic diarrhoea, which may last many months (Gianella, 1986). This syndrome is not fully understood at present.

Travellers' diarrhoea is often discussed primarily in economic terms because of its deterrent effect on tourism (see page 24), or in social terms because of its effect on individual happiness or achievement (see page 26). Important though these aspects are, it is salutary to remember that, in overall medical terms, travellers' diarrhoea is but a pale reflection of the severe problem of diarrhoeal disease in developing countries.

Fungal toxins

Many food-associated moulds are able to produce toxic metabolites (mycotoxins). These comprise a diverse group of chemical compounds ranging from simple aromatic compounds to complex polycyclics and toxic peptides (Jarvis et al., 1982). In many cases, potential problems involve the possibility of cancer or delayed organ damage due to repeated ingestion of subacute levels of mycotoxins. Aflatoxins produced by members of the *Aspergillus flavus* group, have been associated with hepatitis, cirrhosis, Reye's syndrome, udorn encephalopathy and hepatoma in man, food animals such as pigs, and in other animals. In some cases, the toxicity to man is unknown.

In addition to long-term illness, acute symptoms can also result from consumption of mould contaminated foods. For example, trichothecenes produced by species of *Fusarium* cause haemorrhagic enteritis as well as other symptoms such as skin irritation and desquamation. A serious condition, alimentary toxic aleukia, is due to tricothecene production by *Fusarium* species growing on grain crops before harvest, but this condition is rare today.

Mycotoxins can enter the diet directly, by consumption of contaminated foods or, in some cases, indirectly, by consumption of meat from animals fed on contaminated feed.

Problems with mycotoxins are greatest in less developed countries where storage facilities for grains etc are inadequate and where food shortages dictate the consumption of mould contaminated food. However there may also be significant problems in developed nations. For example, farmers in the midwest of the USA suffered major losses due to mycotoxin contamination of crops during the 1989 growing season.

Mention should also be made of ergotism, a disease associated with the consumption of bread made from rye or, less commonly, other cereals infected by the pathogen *Claviceps purpurea*. Ergotism is due to the production by the sclerotia of the pharmacologically active compound ergotamine and takes two basic forms, gangrenous and convulsive. The disease is rare today, the last major European outbreak being in France in the early 1950s.

Examples of mycotoxigenic moulds and foods commonly contaminated are listed in **Table 10**.

Table 10. Summary of mycotoxin producing moulds and examples of foods commonly contaminated (Jarvis et al., 1982).

Mycotoxin	Mould	Foods at risk
Aflatoxin	*Aspergillus flavus*	Nuts, rice, coffee, cocoa, soya, corn and corn products, milk and milk products
Citrinin	*Penicillium citrinum* *P. viridicatum*	Wheat, oats, rye, rice, cheese
Ochratoxins	*A. ochraceus* *P. viridicatum*	Cereal crops, nuts, citrus fruits, coffee, cocoa, soya, cheese
Penicillic acid	*P. cyclopium* *P. martensii*	Corn, beans, fruit
Sterigmatocystin and derivatives	*A. versicolor*	Grains, coffee, miscellaneous foods
Trichothecenes	*Fusarium graminearum* *F. tricinctum*	Corn and cereal crops
Zearalenone	*F. graminearum*	Corn and cereal crops

Note: Only the most important mycotoxin producing species are listed.

Chemicals

Chemical food poisoning may be either intrinsic or extrinsic. In the former case, the agent may either be a component of an accepted food such as the haemagglutinins present in raw red kidney beans, *Phaeseolus vulgaris* (Noah *et al.*, 1980), or a toxic compound present in products consumed in error. A common example of the latter includes confusion of edible with non-edible fungi.

Food may be subject to extrinsic contamination from a wide range of sources. The adulteration of olive oil with contaminated rape oil, which affected more than 15,000 people in Spain with more than 200 deaths (Ross, 1981), is evidence that adulteration is not entirely an historic problem. More recently there have been an increasing number of cases of the deliberate addition of toxic substances and other hazardous material, such as broken glass, to food for political or criminal purposes.

In its broadest sense, chemical food poisoning embraces not only those contaminants which provoke an acute response, but also the consequences of long-term exposure to hazards which have no detectable short-term effect. Such hazards include environmental pollutants such as sub-critical levels of industrial chemicals, examples being heavy metals and polychlorinated biphenyls, radionucleides, agricultural chemicals such as pesticides and fungicides, and some food ingredients or their derivatives such as nitrosamines.

The role of foods and food ingredients may also be discussed in relation to 'allergic' symptoms. However, some caution is required to avoid linguistic confusion since some authors restrict the term 'allergy' to con-

Table 11. Summary of types of chemical food poisoning.

Intrinsic	
Plants	Food plant improperly prepared or processed
	Poisonous plant eaten in error
	Poisonous plant contaminating food plants
Fungi	Poisonous fungi eaten in error
Fish	Food fish improperly prepared
	Poisonous fish eaten in error
All foods	Allergic response
	Food intolerance?
Extrinsic	
Direct contamination	Toxic substances added in error
	Accidental contamination
	Sabotage
	Excess of normal ingredient
	Permitted additive, pesticide or fungicide
	Toxic substances from packaging, utensils etc
Environmental contamination	Agricultural chemicals
	Industrial chemicals, radionucleides etc
All sources	Allergic response
	Food intolerance?

ditions in which the immune system is demonstrably involved, whereas others include 'delayed' or 'masked' allergies. The latter conditions may also be referred to as food intolerance and this term is used in the present work.

Irrespective of linguistic confusion, it is necessary to differentiate between the classic type of food allergy and food intolerance. In the former, response to the allergen is rapid and the food responsible – shellfish or eggs, for example – is usually easily recognised. Symptoms are usually well defined and often include vomiting. Food intolerance is less easily recognised and, indeed, some physicians are unwilling to accept that such conditions exist (Gamlin, 1989) blaming psychosomatic disorders. No typical set of symptoms exists in food intolerances, the typical patient suffering from 'thick-note syndrome' and visiting the doctor frequently with a wide range of symptoms. More than 40 possible symptoms have been reported ranging from headaches, depression, asthma, constipation and recurrent mouth ulcers, to aching joints, water retention, stomach ulcers and a constant runny nose (Gamlin, 1989). It is this plethora of symptoms which makes recognition of the condition difficult and leads to the assumption that the underlying causes are psychosomatic.

The various types of chemical food poisoning are summarised in **Table 11**. Further discussion is beyond the scope of this book.

Parasites of animals and fish

Raw and undercooked foods are well known as vehicles of parasitic infections. The overall problem is greatest in developing countries where a high proportion of food animals are infected with parasites, where cooking facilities are poor and where economic circumstances necessitate the consumption of infected meat. In developed countries, problems may arise from the conscious consumption of raw meat or fish. For example, the increased popularity of 'sushi' restaurants in recent years has led to increased numbers of people infected by piscine parasites. In this context, it is noteworthy that species previously thought to be harmless to man such as the common cod worm (*Phocanema* spp) is now recognised as a cause of human parasitic disease.

Although parasitic infections are usually associated with consumption of meat or fish, liver fluke (*Fasciola hepatica*) may also be transmitted by drinking water containing the cecarial stage of the fluke or by eating plants such as watercress grown in contaminated water.

Infections of man by parasites of animals and fish are summarised in **Table 12**. The organisms themselves are not discussed further but precautions against parasitic infections are briefly discussed on pages 433, 437–438 (meat), 464 (fish) and 476 (watercress).

Table 12. Foodborne disease due to parasites of animals and fish.

Animal parasites	
Tapeworms	*Taenia saginata* (beef)
	T. solium (pork)
Nematodes	*Trichinella spiralis*
Liver flukes	*Fasciola hepatica*
Protozoa	*Toxoplasma gondii*
Sarcocysts[1]	*Sarcocystis bovihominis* (beef)
	S. suihominis (pork)
Fish parasites	
Nematodes	*Anisakis simplex*
	Other *Anisakis* species
	Terranova species
	Phocanema species[2]

[1]Sarcocysts are often considered to be non-infective to man, but digestive complaints have been noted after consumption of heavily-affected meat (Heydoorn, 1977).
[2]*Phocanema* is of low infectivity to man.

Food poisoning: where and why?

Location of outbreaks of food poisoning

The widespread publicity given to outbreaks of food poisoning in the UK during the late 1980s was followed by arguments and counter arguments concerning responsibility. On the one hand the food industry was accused of placing profit above safety, while on the other the housewife was accused of having lost the essential common sense necessary for safe food handling in the home. While it is not the intention to become involved in this controversy, with its inevitable political overtones, comment is required on some of the issues.

It is recognised that most outbreaks of food poisoning occur in catering institutions or in private households. The need for improved practices in the catering industry and for improved education of the general public is well recognised and each topic is discussed in detail elsewhere (see pages 424–425 and 406).

Table 13. Relationship between type of food and pathogenic micro-organisms.

Food type	Micro-organism
Meat or milk eaten raw or undercooked	Zoonoses such as *Salmonella* and *Campylobacter*; animal parasites
Fish eaten raw or undercooked	Pathogens of marine origin such as *Vibrio*; foodborne viruses; fish parasites
Cooked food contaminated by handler	*Staphylococcus aureus*, less commonly *Salmonella*, *Shigella* etc
Cooked food contaminated by raw materials	Zoonoses such as *Salmonella* and *Campylobacter*
Cooked food held warm	*Clostridium perfringens*; *Bacillus cereus*
Cooked food held for extended periods at low temperature	*Listeria monocytogenes*; *Yersinia enterocolitica*
Cooked food held for extended periods under anaerobiosis	*Clostridium botulinum*
Retorted food* underprocessed	*Clostridium botulinum*
Fermented food incomplete fermentation	*Staphylococcus aureus*, possibly *Clostridium botulinum*

*Food processed at 121°C or equivalent to ensure 'commercial sterility'.

Table 14. Factors contributing to outbreaks of food poisoning: England and Wales.

Factor	Per cent outbreaks			
	Salmonella spp.	*Clostridium perfringens*	*Staphylococcus aureus*	*Bacillus cereus*
Preparation too far in advance	42	88	48	86
Storage at ambient temperature	30	53	45	62
Inadequate cooling	22	60	7	27
Inadequate reheating	13	52	3	52
Contaminated processed food	19	4	16	6
Contaminated canned food	0	1	25	2
Undercooking	25	14	1	2
Inadequate thawing	11	6	0	0
Cross contamination	15	2	1	0
Raw food eaten	15	0	1	0
Incorrect warm handling	3	10	0	13
Infected food handlers	2	0	30	0
Use of left-overs	4	5	7	2
Excessive amounts prepared	5	3	1	0

Notes:
1 Several factors were often linked to one outbreak and thus the sum of values for each organism exceeds 100.
2 Figures are based on a study of 1,320 outbreaks occurring between 1970 and 1982 (Roberts, 1985).

However, it must be accepted that with the exception on some occasions of *Staphylococcus aureus*, food poisoning micro-organisms are usually derived not from the home or catering environments but from raw food. While in the case of pathogens such as *Clostridium perfringens* it is probably unrealistic to attempt to reduce the incidence in raw food, a reduction in the incidence of other micro-organisms such as *Salmonella* and *Campylobacter* would seem feasible providing the will exists, and would result in a reduction in morbidity due to these organisms. Although this may be achieved by irradiation, this is not seen as the answer in the absence of other moves toward more acceptable practices.

It must also be appreciated that the domestic kitchen can never be an ideal place for food preparation. It is not realistic to expect separate refrigerators for the storage of raw and cooked foods; also many kitchens simply do not have enough space to adequately separate raw and cooked preparation areas. Indeed, it may be asked cynically and perhaps simplistically why, if the food industry with its multimillion pound investments and trained technical staff cannot prevent the contamination of cooked food, the housewife should be expected to do better?

Types of food involved

Although many types of food can cause food poisoning, there are distinct trends which relate certain types of food to certain pathogens. These are the result of the natural contamination of the different foods and the processing or cooking customarily applied (**Table 13**).

Poor practice and food poisoning

There are many stages at which poor practice may lead to food poisoning. A comprehensive study of the relative importance of poor practice at various stages has been made in the UK (Roberts, 1985) and similar, if rather smaller, studies have been made in the USA (Bryan, 1978) and Canada (Todd, 1983). The results of these studies are summarised in **Tables 14** and **15**.

It is not possible to compare directly the results of the different studies, since the categories chosen do not exactly coincide. However, there are common trends.

Poor temperature control during the cooling and refrigerated storage of food is an obvious factor leading to illness, and is a major contributory factor to problems related to preparing food too far in advance of consumption. Other major factors include undercooking, inadequate reheating and the use of contaminated processed foods. Infected food handlers are not a significant cause of food poisoning, except in

Table 15. Factors contributing to outbreaks of food poisoning: Canada and USA.

Factor	Per cent outbreaks	
	Canada	USA
Storage at ambient temperature	61	NA
Inadequate refrigeration	5	63
Food left in car	5	NA
Preparation too far in advance	NA	29
Warming devices at ideal growth temperature	NA	27
Infected food handlers	NA	26
Inadequate reheating	NA	25
Cross contamination	NA	15
Use of left-overs	NA	7
Inadequate cooking	NA	5
Others	29	10

Notes:
1 Figures for Canada were based on a study of 212 outbreaks occurring between 1973 and 1977 (Todd, 1983).
2 Figures for the USA were based on a study of outbreaks occurring between 1973 and 1976 (Bryan, 1978). In this study, several factors were often linked to one outbreak and thus the sum of values exceeds 100.
3 The definition of categories used in the two studies referred to above and that made in England and Wales (**Table 14**) may not exactly correspond, and comparisons between the three studies require caution in interpretation.
4 NA = Category not applicable.

the case of *Staph. aureus* and while infected handlers are found in outbreaks of *Salmonella* food poisoning, in most cases they are victims rather than originators.

Since the publication of the above reports, a greater emphasis has been placed on 'low temperature' pathogens such as *L. monocytogenes* and *Yersinia enterocolitica*. There would seem to be little doubt that the extremely long storage lives applied to many refrigerated foods are a major contributory factor in cases of food poisoning caused by bacteria capable of growth at temperatures below 6 to 7°C.

It will be appreciated that in many cases there are underlying factors which influence more immediate technical causes of food poisoning. For example, the very long storage lives of refrigerated foods are largely a consequence of the commercial requirements of the large supermarket chains. It is also ironic that moves to prevent consumers storing food for longer than recommended periods by improving labelling instructions are to some extent confounded by the pattern of shopping associated with large edge of town superstores, which encourage bulk shopping at weekly or longer intervals.

2 Food poisoning: economic, social and personal aspects

Introduction

Most people involved with public health and the prevention and control of food poisoning are concerned primarily with medical microbiological or toxicological aspects. This is natural and correct since it is in these areas that personal interest and expertise lies. However, it is necessary to have an awareness of the broader issues of food poisoning; the economic loss to the individuals affected, the companies involved and national economies; the social loss of the ruined holiday or other special occasion; and, last but by no means least, personal loss. In extreme circumstances this involves death, but even 'mild' food poisoning is likely to involve an extremely unpleasant experience for the patient.

Economic loss resulting from food poisoning

Food poisoning causes considerable economic loss to all involved. It is usually difficult, or impossible, to estimate total direct and indirect costs, but examples of direct costs to manufacturers involved in large outbreaks (**Table 16**) illustrate the considerable financial impact.

In determining the cost of a single outbreak of food poisoning, it is necessary to take account of a number of factors including medical care, loss of earnings and cost of investigating the outbreak, as well as direct costs to the manufacturer. These are summarised in **Table 17** and discussed in further detail below.

Table 16. The cost of foodborne disease: examples of large outbreaks.

Year	Illness	Food	Cost
1964	Typhoid	Canned corned beef	£25m
1974	Salmonellosis	Cold roast pork	£350,000
1978	Botulism (UK)	Canned salmon	£2m
1979	Staphylococcal intoxication	Corned beef	£1m
1985	Salmonellosis	Dried baby milk	£22m*

*Difference in estimated value of company before outbreak and price realised when sold.

Affected persons

In most instances financial loss to people affected by food poisoning is related to loss of earnings and medical expenses. In each case, circumstances may vary widely. With earnings loss, for example, much will depend on the arrangements made by employers as well as the national support system. However, even where earnings loss is not total, there may be loss of bonus and overtime payments. In extreme cases where job security is not legally ensured, even short absences through illness may result in the loss of employment. This can have further implications in the food industry where food handlers may be unwilling to report enteric illness for fear of dismissal (see pages 409–411).

Cost of medical treatment will obviously depend on the severity and duration of the illness and whether or not hospital in-treatment is required. The proportion

Table 17. Direct and indirect costs in outbreaks of food poisoning.

Affected people	Loss of earnings and productivity
	Cost of medical treatment
	Cost of death
Company involved	Destruction of stock
	Loss of production
	Cleaning and renovation
	In-house investigation
	Staff retraining
	Loss of sales and brand rehabilitation
	Compensation
	Legal costs and fines
National costs	Cost of investigation

of the cost borne by the patient will vary from country to country. Where there is a national health service, as in the UK and several other European countries, most or all of the cost will be met by the state. In other countries such as the USA the patient, or his insurers, is likely to have to meet all the cost.

Except in certain circumstances, food poisoning is rarely fatal. Where death does occur there is an obvious financial loss to dependents in terms of loss of potential income as well as immediate costs such as burial expenses.

In some cases attempts have been made to put a value on human life *per se*. There is a long-standing precedence for this; the English social reformer Edwin Chadwick estimating in 1845 that a 30-year-old railway labourer represented an investment of £350 (Coleman, 1965). Many people find the concept of valuing human life distasteful and certainly few would wish to take such a stark, arithmetic approach as Chadwick. However, concepts of this type are used in attempts to determine the national costs of food poisoning and payments of compensation (see below).

Company involved

The major short-term costs for the company involved derive from the need to withdraw and ultimately destroy stock and from loss of production and hence income. In either case, the loss of income at a time when high costs are incurred are likely to lead to severe, and possibly insuperable, cash flow problems. In the case of perishable foods where stock levels are usually relatively small, loss of production is likely to be the major problem, but with non-perishables, particularly canned goods produced on a seasonal basis, the value of stock may be very high indeed.

The cost of cleaning and renovation may also be high and may well be necessary before production can restart. The purchasers of a British baby milk manufacturer whose products had caused *Salmonella ealing* infections (Rowe *et al.*, 1987) immediately replaced the existing spray drier with new equipment at a cost of £8 million. Not all manufacturing operations require such expensive equipment, but with small companies the relative cost of re-equipping may be proportionately greater.

The cost of in-house investigations is often small in relation to the total costs of a food poisoning outbreak. Staff retraining is often necessary and should be undertaken before production restarts. In situations where poor staff training is considered to be a contributory cause of the outbreak, a commitment must be made to training, not merely in direct response to an emergency, but on a continued basis. This also requires a share of increasingly limited resources.

The short-term response to food poisoning must obviously relate to immediate problems posed by withdrawal of stock and loss of production. However, in the longer term public loss of confidence in the affected brand is likely to prove as great, or even a greater, problem. In some cases where multiple brand names are owned, it may be possible to switch production to another label or to develop an entirely new brand identity. This latter strategy was adopted by a manufacturer of luxury chocolates following a loss of public confidence following sabotage. The strategy is expensive, and it is necessary to abandon the considerable investment involved in building the original brand.

A further consideration involving loss of brand confidence following an outbreak of food poisoning, is the accounting practice adopted by some large companies of incorporating the estimated value of a brand name into the company balance sheet as an asset. A fall in the value of the brand name as a result of food poisoning must therefore lead to increased financial vulnerability.

Compensation to be paid, either to victims or to organisations indirectly affected, is difficult to assess. Where possible companies seek to agree compensation without court hearings – both to minimise further adverse publicity and, where settlements are substantial, to avoid a large number of copycat claims. In the UK, it is rare for claims for compensation following food poisoning to come to court, although in most cases claims are settled fairly. The situation in the USA is rather different, as litigation against corporations and individuals is much more commonplace and where, due to lawyers' fees being based on the amount awarded, amounts claimed are often very high (**Table 18**).

Table 18. Examples of legal proceedings resulting from outbreaks of food poisoning.

Organism	Total settlement	Average per person
Clostridium botulinum	$603,410	$5,640
Clostridium botulinum	$5,856,032	$132,222
Salmonella	$38,400	$3,488
Salmonella	$227,548	$3,937
Staphylococcus aureus	>$1,719,829	$343,966 (?)
Virus	$106,483	$1,183

Notes:
1 Settlements are in US dollars corrected to 1985 values.
2 Figures from Todd (1987a) where further examples and outbreak information is available.

The whole question of legal liability and its economic impact on the food industry has been reviewed, with particular emphasis on the USA and Canada, by Todd (1987) and this paper should be consulted for a full discussion.

In many countries, fines can be imposed if a distinct breach of regulations has occurred. In many cases, the fine payable is likely to be small in relation to the overall cost of the outbreak and considerably less than legal expenses if the case is defended.

Punitive measures may also be directed at individuals and can include probation, community service and prison sentences. For example, the owner of a US company involved in a criminal case concerned with *Salmonella* contamination of frog legs paid a personal fine of $10,000 (in addition to a company fine of $15,000) and was placed on probation for one year. The vice-president of a company which produced a Mexican style cheese contaminated with *Listeria monocytogenes* and which caused up to 80 deaths, was imprisoned for 60 days with a further two years probation and fined $9,300.

Justice may not always be seen to be done. Rioting followed the light sentences and acquittals of senior officials involved in the adulteration of olive oil with industrial rape seed oil in Spain, which led to some 200 deaths out of more than 15,000 affected.

National costs

In countries where a national health scheme operates, the greatest direct cost is likely to be that of providing medical care. In most cases, the major part of the cost of investigation is met by the state. This can be significant where large, nation-wide outbreaks are involved. Under some circumstances, it may be necessary to pay compensation to food producers affected by precautions taken to limit food poisoning. For example, in the UK compensation was paid to egg producers forced to destroy birds as a result in fall of demand following publicity concerning *S. enteritidis* PT 4 in eggs.

It is important that the direct national costs of individual outbreaks of food poisoning should not be confused with the overall cost to a national economy. This is discussed in pages 24–26.

Industry wide repercussions of food poisoning

The impact of a major outbreak of food poisoning is not restricted to the brands directly affected, but can have a major effect on sales throughout the industry. Following an outbreak of botulism in the UK involving hazelnut yoghurt, sales collapsed not only of all types and brands of yoghurt, but also of flavoured soft cheeses and other products which are similar in nature and sold in similar packaging from adjacent sections of display cabinets. This public response is similar to that noted earlier when food poisoning outbreaks have received significant publicity. Manufacturers and, to a lesser extent, retailers are therefore faced with a substantial loss of income through loss of sales, and may face expenditure by attempting to reassure the public as to the safety of their product. Thus, during the 1988 publicity concerning *S. enteritidis* in eggs, the word *Salmonella* appeared in such unusual and surprising places as supermarket windows, press advertisements and even a pizza house menu!

The measures taken were only partly successful, and a fall in profits of £30 million suffered by a major UK food manufacturer during 1988–89 was attributed to the public reaction to adverse publicity concerning food safety, even though the manufacturer had not been directly involved.

The impact of food poisoning may go beyond the manufacturers of foods which have been directly involved, and threaten the income of whole sectors of the food industry because of an overall lack of confidence. As a whole, retailers are particularly at risk since while it is certainly true that man must eat, a distrust of highly prepared, high-profit margin foods such as cook-chill 'ready meals' could result in a return to more basic foods of lower profit margin. In what should perhaps be termed 'a spirit of enlightened self-interest', major UK retailers set up a telephone-based food safety advisory service to advise and reassure customers. The cost of establishing and operating this service is not known, although it is obviously small in relation to the turnover of the companies involved. Despite this, the very existence of the service and its attendant costs, however small, serves to demonstrate the impact food poisoning can have across even a financially stable and sophisticated food industry.

Impact of food poisoning on national economies

In addition to the financial impact of individual outbreaks of food poisoning, it is necessary to discuss briefly the overall cost to national economies. In the current context, this refers to money which is no longer available for investment or expenditure elsewhere in the economy, irrespective of whether direct loss is to individuals, companies, or the exchequer.

The national economy may be affected in a number of ways, including direct costs such as medical treatment which have been discussed above (page 22), as well as indirect costs such as loss of exports and long-term costs, such as those of research into prevention of food poisoning. These costs are summarised in **Table 19**. Although the economies of all nations are affected by the costs of food poisoning, there are differences between industrialised and non-industrialised nations, and the two situations are discussed separately.

Non-industrialised nations

Although the cost of food poisoning to the national economies of non-industrialised nations is lower in absolute terms than in industrialised countries, the relative cost to the national product is considerably greater. In the case of infant diarrhoea, Todd (1987b) estimates a cost of $50 per case (including the value of death), which results in a total world-wide cost of $50,000 million from this illness alone. In addition to the further costs of treating adults, there are the indirect effects associated with excessive load on health services.

Loss of productivity due to illness or death may well have a disproportionately great effect due to the labour intensive nature of agricultural and manufacturing operations. A more serious potential problem is the loss of foreign exchange earnings.

The export economies of many non-industrialised nations are vulnerable to the impact of food poisoning in two ways. First, because of their dependence on a limited number of export commodities, the market for which may collapse in the event of a food poisoning 'scare' in consumer nations. Second, because of their dependence, in some cases, on tourism and the loss of income associated with food poisoning among tourists.

The first aspect may be illustrated by reference to the 1964 outbreak of typhoid in Aberdeen, Scotland, caused by corned beef canned in South America by a US owned company. The dramatic fall in sales of canned corned beef had a serious adverse effect not only on the economy of the producing country, but on other beef exporting nations, including Tanzania.

A secondary effect in some instances of this nature occurs when locally owned companies are involved. Where these are unable to meet financial commit-

Table 19. Summarised costs of food poisoning to national economies.

Value of deaths	
Medical treatment	Medication
	Hospitalisation
	Treatment of sequellae
Welfare payments	Income supplements
	Dependent support
Loss of productivity	Loss of output
	Reduction in spending power
Loss of export income	Loss of export markets
	Loss of tourist income
Preventative measures	Cost of infrastructure
	Cost of control of production
	Cost of research
Outbreak costs	Value of food lost
	Cost of investigation

Note: Not all costs are applicable in all economies.

ments, local control may be lost to overseas companies with a consequent net loss to the national economy.

The total international market for tourism is enormous and while there are considerable concentrations of resorts in areas such as the Mediterranean, cheap air fares and growing affluence among certain sectors of the population of industrialised nations means that holidays are more commonly taken in 'exotic' destinations. Problems of poor hygiene and gastroenteritis are not restricted to resorts in developing nations, but the incidence of enteric illness in tourists from countries with good hygienic standards to those with low standards is high. Problems may also be exacerbated by local culinary customs such as the use of fish prone to ciguatera toxicity in the West Indies.

In the short term, the effect on the economy is the reduction of tourist spending. A person suffering from acute diarrhoea is unlikely to wish to eat, visit a discotheque or buy souvenirs. There may be some increased expenditure on medicines but this is unlikely to offset the overall reduction. The long-term effect is likely to be more serious since if the risk of illness becomes well known, tourists will simply not visit the area. While the young and healthy may accept the risk of 'Montezuma's revenge' or 'Delhi belly' as a small price to pay for an adventurous holiday, the majority will not.

It is not possible to calculate the exact cost of food poisoning to tourism for specific nations or regions. However, it has been estimated, that the cost to world tourism is no less than $20,000 million (Todd, 1987b). It is likely that the relative cost is greatest for non-industrialised nations.

Industrialised nations

Although the relative cost of food poisoning is less in industrialised nations it is still very considerable. Accurate figures are not available but Todd (1987b) estimates costs for Canada of $2,200 million and for the USA of $12,700 million. Figures have subsequently been calculated for the loss to the US economy caused by each of the major aetiological agents of food poisoning (Todd, 1989; **Table 20**). There is no reason to believe that the USA and Canada represent unusual situations, and while it is possible to dispute these figures, it seems certain that the order of magnitude is correct. The message therefore is clear, the economies of industrialised countries are losing thousands of millions of dollars (or its equivalent in local currency) per year.

Although the relative number of cases of food poisoning is lower in industrialised countries, medical costs, where treatment is necessary, are probably greater due to the higher standard of care routinely available. Under these circumstances, long-term sequelae such as arthritis and heart disease may become significant in terms of medical costs. The actual costs are not known, but in the case of congenital toxoplasmosis, at least half of which is attributed to consumption of contaminated pork by pregnant women, the 3,300 affected children born each year in the USA require a special service cost for life of $430 million (Roberts, 1985; Todd, 1987b).

Food poisoning is a significant cause of loss of productivity and output in industrialised countries. The number of working days lost is not known, but is probably second only to respiratory disease and certainly greater than industrial action, which is much more publicised.

Although the loss of an export market through concerns over food poisoning is less significant in a diversified economy, there may still be a marked effect on trade balances and localised hardship. Food production is obviously centred on rural areas where unemployment is high and wages low. Even a reduction of wages through loss of overtime working, bonuses etc, is likely to result in a significant loss of spending power with secondary effects on local retailers and service industries.

Industrialised nations are able to spend considerably greater sums in prevention of food poisoning. Indeed, in the UK the historic force behind public health reform is the assertion that disease is cheaper to prevent than to cure. A major expense lies in providing an infrastructure which includes water purification and waste treatment, and provision of refrigerated transport and storage facilities. Secondary costs lie in the control of food producing and processing activities and in research into the causes and prevention of food poisoning. In many countries, Government expenditure on prevention of food poisoning is small in relation to the magnitude of the problem, and considerably less than expenditure in other areas such as defence. In the UK, Government expenditure on research into prevention of food poisoning has become an important political issue, highlighted by the announced closure in 1989–90 of a major food research laboratory.

In recent years, there has been increasing interest in quantifying the economic benefits obtained by preventative measures against food poisoning. One of the most convincing demonstrations of the benefits of intervention on a national scale is that of the introduction of mandatory milk pasteurisation to Scotland. The introduction of pasteurisation is estimated to

Table 20. Cost to USA economy of the major aetiological agents of food poisoning (Todd, 1989).

Aetiological agent	Total cost US$
Bacteria	
Salmonella (excluding S. typhi)	3,991,000,000
Staph. aureus	1,516,000,000
L. monocytogenes	313,000,000
Campylobacter	156,000,000
E. coli (excluding O157:H7)	139,000,000
Cl. perfringens	123,000,000
Y. enterocolitica	109,000,000
Cl. botulinum	87,000,000
E. coli O157:H7	84,000,000
Shigella	63,000,000
V. vulnificus	37,000,000
B. cereus	36,000,000
Other bacteria	123,000,000
Total bacteria	6,777,000,000
Viral	
Hepatitis A virus	176,000,000
Norwalk virus (SRSV)	161,000,000
Total viral	337,000,000
Total parasitic	625,000,000
Total seafood toxins	125,000,000
Total chemical	33,000,000

have prevented 257 people becoming ill each year, the benefit of which outweighed the cost of purchase of pasteurisation equipment (Cohen *et al.*, 1983). An alternative approach is to consider the benefit of preventing, or limiting, a single large outbreak. Applying such an analysis to an outbreak of food poisoning due to chocolate bars containing *S. napoli*, it was conservatively estimated that the saving achieved by successful intervention was £1,670,000. However, prevention of the outbreak altogether would have saved some £2,136,780. The ratio of benefit to cost was high, even allowing for the likelihood of a significant underestimate of savings (Roberts *et al.*, 1989).

Social impact of food poisoning

Food poisoning often has a significant social impact in terms of disrupted lives, lost leisure time, disturbed personal relations and other factors. The most disruptive factor is a death where the personal effects on immediate family members is usually far more significant than purely financial considerations such as loss of family income. The death of a parent of a young child, or of a child or infant, is likely to have particularly long-lasting personal effects, which frequently need psychological guidance to resolve.

Fortunately, death is rare in food poisoning, but considerable disruption and distress may still result from cases which, in the medical sense, are mild. In the case of large outbreaks centred on a single location, there may be considerable disruption, not merely to individuals, but to the community as a whole. This may be demonstrated by reference to the situation in Aberdeen, Scotland during the 1964 typhoid epidemic. During the earlier stages of the epidemic, the streets were sprayed with disinfectant, students at one college were advised not to use public transport, and holidaymaking Aberdonians were not only made to feel unwelcome at hotels but even turned away. This reaction was extreme, possibly due to the severe nature of typhoid and the mistaken perception that the disease is highly communicable (Howie, 1968), but similar reactions have been noted elsewhere.

The disruptive effects of food poisoning are usually most marked when special occasions, such as holidays and wedding receptions, are involved. Ironically, the risks associated with travel to warm climates and with large-scale catering mean that such occasions are all to often marred by food poisoning. A particular case is that of a marathon race at Honolulu, Hawaii, where a large number of entrants contracted salmonellosis at the pre-race pasta party, the pasta being prepared under totally unsuitable conditions. In another incident, an outbreak of salmonellosis at a European summit conference in the Netherlands affected 600 to 700 people (Beckers *et al.*, 1985). The politicians themselves were not affected!

Personal relations may be seriously affected when the person responsible for preparing the incriminated food is a friend or relation. Thus a lady whose chicken meal had resulted in *Clostridium perfringens* food poisoning for a large number of friends and neighbours, not only had to endure herself the short-term symptoms, but also the long-term stigma of the new nickname 'Lucretia Borgia' (C.F. Monty, personal communication).

People responsible for preparing food implicated in food poisoning may feel considerable guilt, even if not directly involved with the victims. Others not directly involved may also be affected in this way, the vice-president of a Japanese airline whose passengers suffered staphylococcal food poisoning during a flight being sufficiently affected to commit suicide.

Personal impact of food poisoning

In discussing food poisoning it is easy to list symptoms, financial consequences and social disruption, while forgetting the impact on the person in the context of the illness suffered. Food poisoning is an extremely unpleasant and debilitating experience even when symptoms are, from a medical viewpoint, 'mild'.

It is not possible to describe exact feelings and to attempt to do so would serve no purpose. Merely to remember the personally unpleasant nature of food poisoning serves to place problems posed and solutions offered within a human context, and enables the illness to be placed in a wider context than the purely technical or, indeed, financial.

3. Foodborne pathogenic micro-organisms: a perspective

Introduction

Foodborne pathogenic bacteria are, from both a taxonomic and a physiological viewpoint, a diverse set of micro-organisms. It must be appreciated that in each case there are common factors which relate both to the organisms themselves and to their study. To avoid repetition and to provide a broader perspective, these factors, where of particular relevance, are discussed briefly below. Emphasis is placed on bacteria, but many of the comments are also applicable to other foodborne micro-organisms to which reference is made when appropriate.

The science of microbiology

Microbiology is a broad-based discipline which, in addition to its own specific subject material, incorporates aspects of other disciplines including biochemistry, molecular biology and mathematics. The application of microbiology in the food industry or elsewhere requires properly trained and experienced people working in properly equipped laboratories. It is an unfortunate consequence of the apparent simplicity of techniques and a misunderstanding of the complexities of the science, that the need for properly qualified personnel is sometimes not recognised. It is further regrettable that some organisations should promulgate this view by offering short training courses, which are sometimes of only a few days duration, to enable totally inexperienced people to function as microbiologists.

Isolation of foodborne micro-organisms

With clinical specimens it is often possible to isolate pathogenic micro-organisms by direct plating onto blood agar media followed by confirmatory tests on isolates. In foods, pathogenic micro-organisms are often present in relatively low numbers, outnumbered by spoilage micro-organisms and possibly damaged by processing operations. Isolation is a more complex and time-consuming procedure often involving three stages: resuscitation or pre-enrichment, enrichment and selective plating. Although these are well established procedures, there remains a considerable amount of confusion concerning both concepts and practice.

Resuscitation

Stressed cells are of enhanced sensitivity both to selective agents and to inhibitors which naturally occur in some solid media. For this reason a resuscitation stage is included in the isoloation protocol for many foodborne micro-organisms. Resuscitation may involve preliminary incubation in nonselective media or, alternatively the addition of protective compounds such as pyruvate or egg yolk to solid selective media. Stressed cells are often particularly susceptible to reactive oxygen metabolites and the addition of free radical quenching agents or anaerobic incubation may be highly effective in resuscitation. Many workers believe however that with organisms such as *Salmonella*, a resuscitation stage should always be used (van Leusden *et al.*, 1982).

Enrichment

Enrichment is applied where pathogenic micro-organisms are present only in small numbers, particularly if large numbers of spoilage micro-organisms are present. It is not normally applied where the organism is a common contaminant of foods in low (and harmless) numbers, such as *Bacillus cereus* in dried foods. If used in these circumstances, results must be interpreted with caution.

Liquid enrichment media tend to select the micro-organisms of highest growth rate among all members of the population, which are able to grow under the conditions prevailing. Growth dynamics of *Salmonella* in Rappaport-Vassiliadis medium are illustrated in **Figure 1**.

Most enrichment media used for isolation of food poisoning micro-organisms are nutritionally complex,

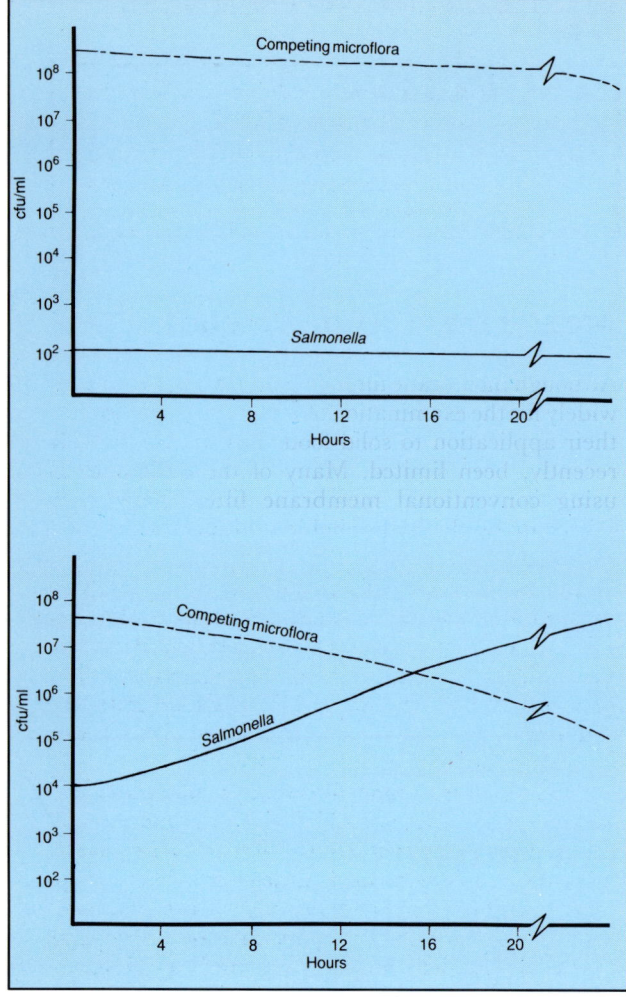

1. Dynamics of *Salmonella* isolation in selective enrichment. Two of several possible circumstances are illustrated. In the first, in selenite broth, the enrichment is unsuccessful, *Salmonella* fails to multiply and the competing microflora remains numerically dominant. Failure of the enrichment was attributed to the low initial number and the relative sensitivity of the *Salmonella*, a strain of *S. typhimurium*; to selenite. In the second illustration, a successful enrichment of the same strain of *S. typhimurium* is made in Rappaport–Vassiliadis broth. Initial numbers are higher and, within 12 h increase to the level necessary for detection on a streak plate (10^5–10^6/ml). The competing microflora declines and *Salmonella* is numerically dominant in the later stages of enrichment.

and employ the principle of back selection in which inhibitors reduce or prevent the growth of competitors. However, there are exceptions – low temperature enrichment of *Yersinia enterocolitica* is based on the ability of that organism to grow at low temperatures, while minerals modified glutamate medium used for *Eschericha coli* is a chemically defined medium which is elective rather than selective, for example.

Although in many cases enrichment media are used in conjunction with selective plating media, they may be used to enumerate specific micro-organisms, or groups of micro-organisms using the most probable number technique. In such circumstances, it is necessary to incorporate a diagnostic (differential) reaction to distinguish the organism being enumerated from other organisms, which are able to grow under the imposed conditions. Examples are lactose fermentation in media for 'coliforms' and *E. coli* and the 'sulphide' reaction in media for clostridia. Further confirmatory tests are required where reactions are positive.

Selective plating

Selective plating media are solid media formulated to select for a specific micro-organism, or group of micro-organisms. With food poisoning micro-organisms, complex media containing selective agents are used. The principle is similar to that of liquid enrichment cultures, although the spatial separation of organisms on the surface of the agar means that competition between different types is much reduced.

A diagnostic system is incorporated to differentiate the organism it is intended to isolate from others able to form colonies on the selective medium. However, such differentiation must be regarded as presumptive and confirmatory tests are required (see below).

Selective plating media may either be streaked with a loopful of enrichment broth or, if numbers are high, be used for direct plating. In the latter case, enumeration may be made if required.

When using selective plating media it is a common error to define organisms in terms of their ability to grow on specific selective media. It must be stressed that the ability to grow on a selective medium does not mean that the organism growing is that for which the medium was developed. Conversely, failure of an organism to grow on a selective medium does not mean that the organism is not that for which the medium was developed. In the former case it is good practice, even where colonies have a typical appearance, to make some simple confirmatory examinations before drawing conclusions or even proceeding to more extensive testing. The value of examining a Gram-stained smear is frequently overlooked, while simple biochemical tests such as the catalase, oxidase and coagulase reactions, correctly applied, can prevent much confusion and wastage of time. Some caution is required: cells grown on selective media may not have typical microscopic appearances, while Gram-positive bacteria may appear Gram-variable or Gram-negative after prolonged incubation. Similarly false-negative catalase and oxidase reactions may occur when colonies are taken direct from isolation plates, while the coagulase reaction should not be determined directly if high NaCl media are used.

The performance of both enrichment and plating media containing selective inhibitors may be defined in terms of sensitivity and selectivity (**Table 21**). The requirement for both to be as high as possible is contradictory, since a medium that excludes all competitors is likely to inhibit a number of strains of the organism being tested for, while a medium which recovers all strains of the organism being tested for may permit a disproportionate number of competitors to grow. To overcome this problem, many workers recommend the use of two media, one of low and one of high selectivity. While indubitably a logical solution, this can result in serious problems where large numbers of samples are to be examined, since two types of both enrichment and selective plating media must be used.

Quite apart from logistic considerations, many less experienced microbiologists (and, not infrequently, many more experienced microbiologists) are confused and bewildered by the sheer number of media available for the isolation of specific micro-organisms. Development of new media, modification of existing media and continuing modification of modifications, is a favourite activity of many microbiologists and one which many equally feel is unfruitful. Writing with reference to *Salmonella*, Mossel (1983) noted:

'For over 35 years a deluge of papers on the quantification of salmonellas in food, feeds and water has appeared and increasingly sophisticated, although still far from perfect methods have been developed to that end . . . All these efforts were notably unsuccessful in reducing salmonellosis morbidity'.

There is certainly no doubt that Mossel and many other workers are correct in their assertion that intervention at manufacturing level is a far more effective means of solving problems of food poisoning than laboratory examination (see page 387 *et seq*); but at the same time the need for the most efficient possible means of isolation remains. The practising microbiologist must therefore steer a course between the Scylla and Charibdis of, on the one hand, continuing to use inefficient procedures and, on the other, the whirlpool of continuing modifications, comparisons and comparisons of comparisons. Choice of media should not be a 'lucky dip', but should be made logically, taking into consideration factors such as ease of use and availability of ingredients, as well as selectivity and sensitivity. Factors affecting choice of media are summarised in **Table 22**.

Table 21. Sensitivity and selectivity of isolation media.

Sensitivity	Ability to recover the bacteria for which the medium was designed
Selectivity	Ability to prevent the growth of bacteria other than those for which the medium was designed

Membrane filter techniques

Although membrane filter techniques have been used widely for the examination of water and other liquids, their application to solid foods has, until relatively recently, been limited. Many of the difficulties of using conventional membrane filters have been overcome by the hydrophobic grid membrane filter technique.

The hydrophobic grid membrane filter (HGMF) uses non-toxic water repellent materials to define a grid, which limits physically the size and degree of spreading of bacterial colonies on a membrane filter (Sharpe and Michaud, 1974). This permits the counting of up to 10^4 colonies on a single filter (Sharpe and Michaud, 1975).

Although membrane filtration is an established technique for the microbiological examination of water, application to foods was limited in the past by difficulties in filtering food suspensions. These problems have been largely overcome by pre-treatment with a surfactant or by enzymic digestion. Problems due to food particles may be overcome by preliminary filtration through nylon or stainless steel mesh prefilters (Entis, 1986).

Table 22. Factors affecting choice of selective media.

Performance	Sensitivity Selectivity Reliability of diagnostic system
Type of food	Interference from food components 'Target' organisms from specific foods may be unusually sensitive to selective agents Competing micro-organisms from specific foods may be unusually resistant to selective agents
Ease of use	Ingredients readily available Difficulty of preparation Stability of prepared media Requirement for special equipment Ease of recognition of 'target' micro-organisms
Expense	Should be a secondary consideration only

HGMF protocols have been developed for a number of foodborne pathogens including *Salmonella*, *E. coli*, *Staphylococcus aureus*, *Vibrio parahaemolyticus* and *Y. enterocolitica* (Entis, 1986). In subsequent chapters HGMF is discussed in conjunction with Rapid Cultural Techniques.

Membrane filtration is also involved in the direct epifluorescent technique, which is discussed below.

Rapid methods for detection of foodborne pathogenic micro-organisms

Traditional methods for the isolation of micro-organisms are dependent on the growth of micro-organisms, to the extent of being able either to form visible colonies on solid media, or to produce a detectable chemical change in liquid media. The process of microbial growth in selective media is inevitably variable and failure to recover organisms is a continuing problem.

Conventional methods have further practical drawbacks in being labour intensive – although requiring relatively little capital expenditure – and in requiring an extended period before even a presumptive isolation is made. This is a major difficulty in food microbiology where early recognition of an actual or potential problem is necessary.

For many years, there appeared to be no prospect of rapid detection methods for specific micro-organisms, although rapid methods based on dye reduction were widely used to assess the hygienic quality of dairy products. In recent years, this situation has been changed by major developments in three separate areas – electronics, immunology and DNA technology – as well as less-radical developments in more conventional methodology. A number of methods for rapid detection of specific foodborne pathogens are now available (**Table 23**), the principles of these are briefly discussed below.

Rapid cultural techniques

Rapid cultural techniques are attractive to many microbiologists since they form a tangible link with conventional methodology, require no new interpretational skills and no major capital expenditure. However, the scope for the development of such techniques is restricted and only two, the Salmonella Rapid Test and the BioControl 1–2 Test© (immunoimmobilisation) are in common use. These tests are discussed in greater detail on pages 78–80.

Microscopic techniques

Microscopic techniques offer a truly rapid means of enumerating bacteria, and an early method, the Breed smear, was widely used with milk despite a number of drawbacks. The fluorescent antibody technique which also involves microscopy, is discussed on page 32. A more recent development, the Direct Epifluorescent Filter Technique (DEFT; Pettipher, 1983), overcomes a number of the problems formerly associated with microscopic techniques and also distinguishes between live and dead cells. As originally devised, DEFT does not differentiate between morphologically similar micro-organisms and had no application for the detection of foodborne pathogens. Attempts to widen the application of DEFT to specific bacteria have involved the growth of microcolonies on the surface of membrane filters placed on selective agar media, followed by staining of the colonies and

Table 23. Rapid methods for detection of foodborne pathogens.

Rapid cultural techniques	Salmonella rapid test Immunoimmobilisation
Microscopic techniques	Direct epifluorescent filter technique
Electrical techniques	Impedance measurement Non-impediometric techniques
Assay of adenosine triphosphate	
Serological techniques	Fluorescent antibody technique Latex co-agglutination Enzyme linked immunosorbent assays
Genetic techniques	DNA hybridisation
Labelled bacteriophages	

examination by epifluorescent microscopy (Rodrigues and Kroll, 1988). This technique has been applied on an experimental basis to *Staph. aureus* (**Figure 2**) and *Salmonella*. The lower limit of detection is in the order of 10^3 cfu/g, the use of a laser-light pulse counting method offering the promise of improved sensitivity (Kroll *et al.*, 1989). Further work is required to validate the technique and to extend application to other micro-organisms.

Electrical techniques

Impedance measurement

Impedance measurement offers relatively rapid detection of micro-organisms by monitoring metabolic activities through the mediation of electrical impedance. The number of bacteria initially present in the sample may be related to the time taken to produce an acceleration of the impedance curve, providing that other factors such as temperature remain constant. The physical and mathematical background to the technique have been discussed in detail elsewhere (Firstenberg-Eden and Eden, 1984; Firstenberg-Eden, 1986), and reference should be made to these publications if further theoretical information is required.

Impedance measurement forms the basis of a number of types of commercially available equipment including the Bactometer©, Malthus© and Rabit© instruments. Each of these is capable of handling a large number of samples simultaneously and represents a significant capital investment.

Impedance measurement is used successfully for detection of a number of foodborne pathogens including *Salmonella*, *E. coli*, *Staph. aureus* and *Listeria monocytogenes*. Detection times vary according to the food and the organism, but with the detection of *Salmonella* in confectionery, for example, results are available 3 days earlier than with conventional methodology (Bullock and Frodsham, 1989).

Detection of specific micro-organisms by impedance monitoring requires growth in selective media similar to those used in conventional methodology. It is therefore possible to modify published protocols to individual requirements, although this may lead to excessive time being devoted to media engineering.

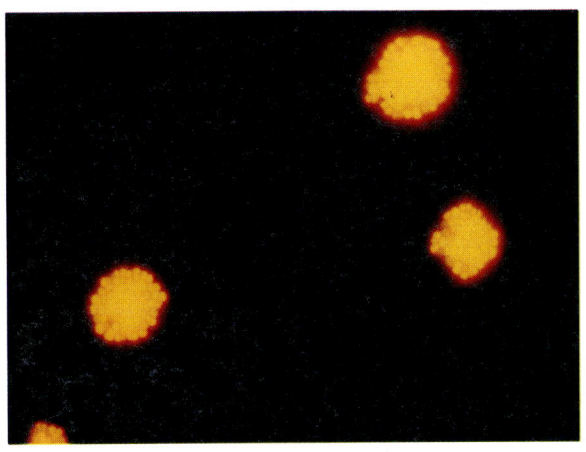

2. Selective enumeration of bacteria using the direct epifluorescent technique (DEFT). Microcolonies are grown on polycarbonate membrane filters overlaid on selective agar medium, stained with acridine orange and examined by epifluorescent microscopy. The colonies illustrated are *Micrococcus* sp., stained after growth on azide blood agar for 4 h at 30°C. (Reproduced with permission from Rodrigues, U.M. and Kroll, R.G. *Journal of Applied Bacteriology* **64**: 65–78. © 1988 Blackwell Scientific Publications Ltd.)

Non-impediometric techniques

An alternative electrical method operating on the principle of a simple battery has been described (Ackland *et al.*, 1984; Ackland and Manvell, 1987). The method is based on the measurement of the voltage generated between two electrodes of dissimilar metals (aluminium and gold) placed in an electrolyte (growth medium).

Although the technique is by no means fully developed, it offers a potential alternative to impediometry. Unlike impedance, voltage is not sensitive to small changes in temperature and thus the need for sophisticated temperature control equipment is avoided.

Assay of adenosine triphosphate

Assay of microbial adenosine triphosphate (ATP) has attracted much interest as a means of assessing the hygienic status of foods (Stannard and Wood, 1983), although there are continuing problems with

3. Detection of bacteria by the fluorescent antibody technique. Although much effort has been expended in developing and improving conventional fluorescent antibody technology, the technique has not been widely used in food microbiology. The major problem is false-positive reactions and there are lesser, but still significant, problems such as operator fatigue. Novel approaches to the technique may, however, lead to increased interest and wider application (see pages 82, 350). The figure illustrates *Salmonella* inoculated into sausage meat and detected by the fluorescent antibody technique using transmitted illumination. (Reproduced by courtesy of J. Sainsbury plc.)

interference from somatic ATP. Selective lysis of pathogens using either lytic bacteriophages or, in the case of *Staph. aureus*, lysostaphin, is required for the detection of specific micro-organisms.

The detection limit of ATP based on the sensitive firefly luciferin: luciferase assay and current instrumentation is equivalent to 10^5 cfu/g. In most cases, therefore, an enrichment stage would be required.

Serological techniques

Serological methods have attracted attention for a number of years as a highly specific and sensitive means of detecting micro-organisms in foods. However, until recently, it has not been possible to overcome the problem of false positive reactions due to non-specific cross reactions. A very considerable amount of effort, for example, was expended in efforts to develop a fluorescent antibody assay for *Salmonella* (Insalata *et al.*, 1973; Thomason, 1981), but despite being approved for use in the USA, the technique was never widely adopted due to the number of false positives and other difficulties (**Figure 3**; Swaminathan and Konger, 1986).

The situation has changed in recent years due to the development of techniques which permit the production of highly specific monoclonal antibodies, and of new detection systems such as latex

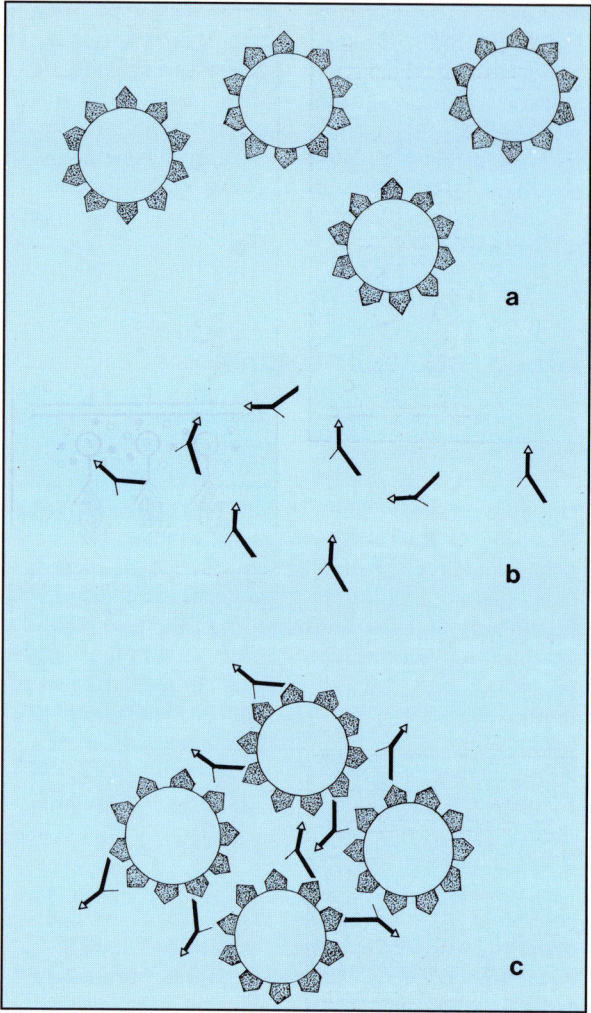

4. Detection of antibodies by latex co-agglutination. Latex co-agglutination uses latex particles to amplify antigen: antibody agglutination reactions. Latex particles are coated with antigens (**a**); addition of antibody (**b**) causes the particles to agglutinate (**c**).

In the configuration illustrated, latex co-agglutination is used in the serological diagnosis of infection. In food microbiology, the reverse configuration is more common in which the latex particle is coated with antibody for the detection of an antigen such as a toxin or a component of a bacterial cell. The principle, however, is the same in each case.

Applications of latex co-agglutination are illustrated with respect to Salmonella, *page 83*, E. coli O157:H7, *page 124 and* Staph. aureus, *page 259.*

co-agglutination (Essers and Radebold, 1980) and the enzyme linked immunosorbent assay (ELISA; Engvall and Perlmann, 1971).

The principle of latex co-agglutination is illustrated in **Figure 4**. The technique is straightforward providing appropriate antibody sensitised latex is commercially available. A commercial kit is available for

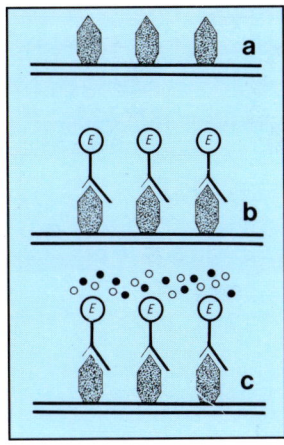

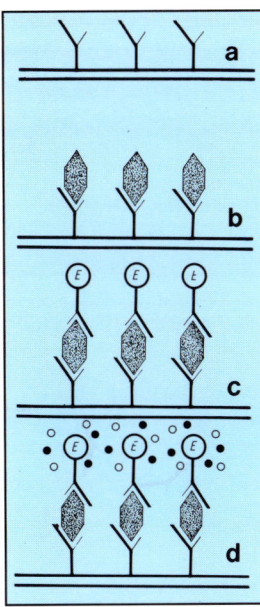

5. Enzyme linked immunosorbent assay: direct configuration. The 'target' antigen is bound onto the solid phase (**a**) and washed. The enzyme labelled antibody is added (**b**), a proportion of which binds to the antigen and the remainder is removed by washing. A substrate is added to the bound antigen:labelled antibody complex (**c**). After a pre-determined time, the reaction is stopped. The quantity of antigen present is equal to the enzyme product. (Reproduced with permission from Swaminathan, B., Aleixo, J.A.G. and Minnich, S.A. *Food Technology* **39**(3): 83–89. © 1985 Institute of Food Technologists.)

6. Enzyme linked immunosorbent assay: sandwich configuration. The antibody is absorbed onto the solid phase (**a**) and washed. The test solution containing the 'target' antigen is added (**b**); the antibody is in excess and all of the antigen is complexed. After washing, the enzyme labelled antibody is added (**c**). A proportion binds to the antigen and the remainder is removed by washing. An enzyme substrate is added (**d**) and reacted for a predetermined time. Quantity of antigen = enzyme product. (Reproduced with permission from Swaminathan, B., Aleixo, J.A.G. and Minnich, S.A. *Food Technology* **39**(3): 83–89. © 1985 Institute of Food Technologists.)

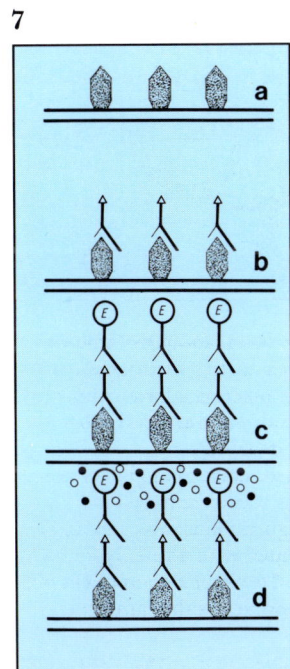

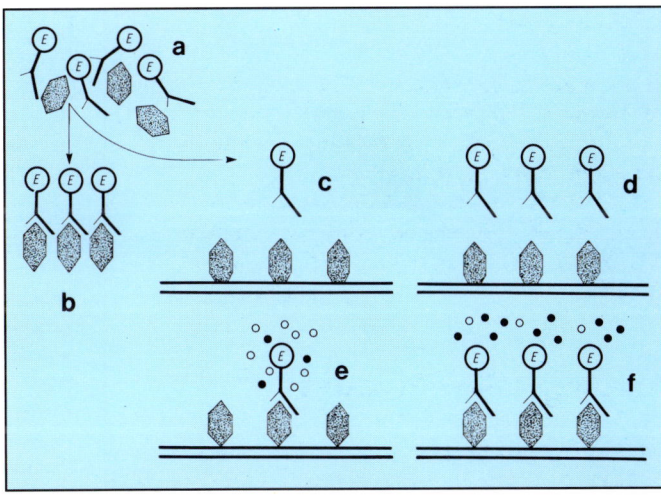

7. Enzyme linked immunosorbent assay: indirect configuration. The 'target' antigen is bound to the solid phase (**a**). After washing specific, but unlabelled, antibody A is added (**b**), antibody A is in excess and all of the 'target' antigen is complexed. The system is washed and an enzyme labelled antibody to A is added (**c**). The labelled antibody binds to the antigen:antibody A complex, the excess is removed by washing and the enzyme substrate added (**d**) and reacted for a pre-determined time. Quantity of antigen = enzyme product. (Reproduced with permission from Swaminathan, B., Aleixo, J.A.G. and Minnich, S.A. *Food Technology* **39**(3): 83–89. © 1985 Institute of Food Technologists.)

8. Enzyme linked immunosorbent assay: competitive configuration. In contrast to other ELISAs, the first stage of the competitive configuration is carried out with both the 'target' antigen and the labelled antibody in solution. The two components are mixed (**a**) and the resulting antigen:antibody complex removed by centrifugation (**b**); the antibody is in excess and thus some will remain uncomplexed. At the same time, two sets of immobilised antigen are prepared; the free labelled antibody remaining in solution after complexing is added to the first immobilised antigen (**c**) and a known quantity of labelled antibody added to the second (control) immobilised antigen (**d**). After washing, an enzyme substrate is added to each system (**e**, **f**) and allowed to react for a pre-determined time. Quantity of antigen = (enzyme product in control) − (enzyme product in test). (Reproduced with permission from Swaminathan, B., Aleixo, J.A.G. and Minnich, S.A. *Food Technology* **39**(3): 83–89. © 1985 Institute of Food Technologists.)

Table 24. Advantages and disadvantages of different ELISA assays (Swaminathan and Konger, 1986).

Assay	Advantages/disadvantages
Competitive	Rapid, requiring only one incubation and one washing stage Purified antigen required Prone to interference from proteases etc Antigen should be of relatively low molecular weight
Non-competitive and indirect competitive	No problem with interference Possible amplification effect Single conjugate detects a variety of antigens (indirect) At least two incubations and several washings required Antigen should be of relatively low molecular weight (indirect)
Sandwich	Less sensitive particularly if monoclonal antibodies are used Most suitable configuration when testing food samples and other mixtures of proteins

Table 25. Design considerations for ELISA.

Factor	Example
Configuration	Direct competitive Indirect competitive Double antibody sandwich Direct
Matrix	Microtiter plate Nylon beads Polystyrene coated ferro-magnetic beads
Antibody	Polyclonal Monoclonal
Enzyme	Peroxidase Alkaline phosphatase β-galactosidase
Conjugation	One step glutaraldehyde Two step glutaraldehyde Periodate oxidation Cross linking with: N, N'-o-phenylenedimaleimide m-maleimidobenzoyl-N-hydroxysuccinimide ester N-succinimidyl 3-(2-pyridyldithio) propionate
Substrate	
Peroxidase	p-nitrophenyl phosphate 4-methyl umbelliferyl phosphate[*]
Alkaline phosphatase	3-3' diaminobenzidine + H_2O_2 o-toluidine + H_2O_2 3-(p-hydroxyphenyl) propionic acid[*]
β-galactosidase	o-nitrophenyl β-D-galactopyranoside 4-methyl umbelliferyl β-D-galactopyranosidase[*]

[*] Fluorogenic substrate.

Note: See Swaminathan and Konger (1986) for a full discussion of these factors.

detection of *Salmonella* in foods (Spectate©; Clark *et al.*, 1989). The sensitivity is such that enrichment is required.

ELISA assays may be divided into two major types; competitive and non-competitive. Both types have been used for detection of bacteria in foods including *Salmonella*, *L. monocytogenes* and *Campylobacter*. Several assay configurations have been devised, four of which are illustrated in **Figures 5–8**. Each has its advantages and disadvantages and these are listed in **Table 24**.

In designing an ELISA assay of any type or configuration it is necessary to take account of a number of factors (**Table 25**). In most cases, the design of an assay is beyond the scope of non-specialist laboratories and it is usually necessary to rely on commercial kits. It must be appreciated that while such kits provide all of the required materials in a convenient form, ELISA assays still require rigorous technique and a considerable amount of manipulative time. Automation is available for some functions including washing and plate reading and indeed, is essential where large numbers of samples are processed. However, in many cases automation offers only a partial solution.

Despite attempts to develop an ELISA assay sufficiently sensitive to permit direct detection of micro-organisms in food samples, current levels of sensitivity require enrichment.

One of the major difficulties affecting the acceptability and application of ELISA and co-agglutination techniques is the unfamiliar nature of the manipulations and the lack of a relationship, apart from the enrichment stage, to conventional methodology. To some extent this difficulty lies with expertise: a manipulation considered simple by an experienced immunologist may present difficulties to a person who has no prior experience with serological techniques or whose experience is limited to slide agglutination reactions for *Salmonella*. In other cases, the difficulty lies in established thought patterns. For example, rapid methods for *Listeria*, where there are few ingrained prejudices concerning isolation methods, are likely to be more readily accepted than those for *Salmonella*.

Both ELISA and co-agglutination techniques are widely used for detection of toxins, and co-agglutination is used increasingly as a means of confirming the identity of a micro-organism.

Genetic methods

Genetic methods for the detection of micro-organisms have resulted from earlier fundamental studies concerning the structure and function of DNA. Such methods are used widely in clinical microbiology, but as yet application to routine food microbiology has been limited. It is likely that in the long term genetic methods will largely replace immunological methods (Sharpe, 1986).

DNA Hybridisation

The principle of DNA hybridisation is illustrated in **Flow diagram 1**. There are a number of possible approaches to the design of DNA hybridisation assays and the main factors are listed in **Table 26**. As with serological methods, the development of assays is beyond the scope of most food industry laboratories and it is necessary to rely on commercial kits. Such kits are available for *Salmonella*, *Campylobacter* and *Listeria*.

Enrichment is required for detection of microorganisms by DNA hybridisation. An alternative

Flow diagram 1. Detection of micro-organisms by DNA hybridisation.

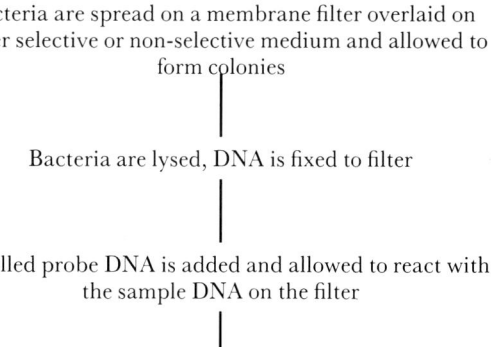

Bacteria are spread on a membrane filter overlaid on either selective or non-selective medium and allowed to form colonies

↓

Bacteria are lysed, DNA is fixed to filter

↓

Labelled probe DNA is added and allowed to react with the sample DNA on the filter

↓

Non-hybridised labelled DNA is removed by washing. Hybridised DNA is detected by means of label

approach with some organisms is to develop probes for genes coding for a virulence factor such as the heat labile toxin (LT) of enterotoxigenic *E. coli*. Such an assay may be combined with conventional selective plating to permit direct isolation of enterotoxigenic *E. coli* from foods (Romick *et al.*, 1987). In such an application time saving lies not in the cultivation stage but in obviating the need to screen a large number of cultures for toxin production. Detection of a gene controlling virulence cannot, of course, be taken as definitive proof that an organism is virulent, since the property involved may not be expressed. However, a knowledge even of potential virulence is very valuable under many circumstances. A further advantage of DNA probes is that the gene is detected in other genera if intergeneric transfer has occurred.

DNA hybridisation techniques require a considerable amount of manipulative time and problems of acceptability may occur due to unfamiliarity with both concepts and methods. A major obstacle to acceptability is the use in many systems of radiolabelled probes. The use of radiolabelled material is acceptable in clinical laboratories where facilities and expertise for safe handling have existed for many years. Such facilities, however, are generally lacking in the food industry and the introduction of radiolabelled material into industry-based laboratories is considered unacceptable.

An alternative and more widely acceptable method of labelling probes, enzyme labelling, has been investigated by a number of workers using biotin or haptens as reporter molecules or by direct labelling of the DNA probe (Fitts, 1986).

Table 26. Design considerations for DNA hybridisation assays.

Sample preparation	Nature of solid support
	Extraction of DNA
	Attachment of DNA to solid support
Probe sequence	Nature of target (genus, species or specific strain)
	Presence of closely related organisms in environment
Probe labelling and detection	Nick-translation (isotopic/non-isotopic)
	Oligonucleotide synthesis (isotopic/non-isotopic)
	RNA probe (isotopic/non-isotopic)
Hybridisation stage	Specificity required

Note: See Fitts (1986) for a full discussion of these factors.

The polymerase chain reaction is a gene amplification procedure and a highly sensitive means of detecting specific micro-organisms by detection of a segment of their DNA (Saiki *et al.*, 1985,1988). DNA amplification involves two oligonucleotide primers that flank the DNA segment to be amplified and repeated cycles of heat denaturation of the DNA, annealing of the primers to their complementary sequences and extension of the annealed primers with thermostable DNA polymerase from the thermophile *Thermus aquaticus*.

In principle PCR is a quantitative technique, each cycle doubling the amount of DNA synthesised in the previous cycle and resulting in an exponential accumulation of the specific target segment of DNA. The number of amplification cycles can therefore be preselected to yield a predetermined level of sensitivity, the accumulated DNA being detected either using a probe or by gel electrophoresis.

The PCR technique is capable of detecting ca 10 cells of the target micro-organism in a 10g food sample without the need for a cultural stage and in the presence of a large number of other micro-organisms. The method has particular advantages where plasmid instability means that gene probes are an unreliable means of detection. A disadvantage (under most circumstances) is that dead cells are detected although this problem may be overcome by a brief cultural stage (3 to 5 cell doublings).

PCR technology has been applied to a number of micro-organisms (see pages 99–100, 116, 125, 386) and wider scale application may be anticipated.

Labelled bacteriophage detection

Labelled bacteriophages offer a highly specific means of detecting micro-organisms. As with sophisticated serological and genetic assays, development is beyond the scope of non-specialist laboratories, and it is usually necessary to make use of commercial kits or instrumentation. Systems based on labelled phage are available for *Salmonella* and *Listeria*.

Flow diagram 2. Detection of bacteria using enzyme labelled bacteriophage.

Filter sample containing 'target' bacteria through a membrane filter

Add labelled bacteriophage specific to 'target' bacteria

Bacteriophage attach to 'target' bacteria, but not to other micro-organisms present. Non-attached phage is removed by washing

Enzyme substrate added. Quantity of enzyme product may be directly related to number of 'target' bacteria present

Enzymes are used to label the phage and either colourimetric (Vitek©) or fluorogenic (Lumophage©) detection may be used. In either case, an enrichment stage is necessary. The Vitek© system is automated and a high sample throughput is required. The principle of this system is illustrated in **Flow diagram 2**.

An alternative to labelling phage with an enzyme is the incorporation of the *lux* gene from the marine luminescent bacterium *V. fischeri* into the phage DNA. Target bacteria produce light after taking up the *lux* gene and this may be measured using a suitable photometer. This method is not yet fully developed but initial tests suggest that as few as 10 cells/ml may be detected within one hour (Waites, 1988).

Criteria for acceptability of rapid methods

Minimum criteria must be met before any rapid method can be adapted. These are listed in **Table 27**.

Table 27. Criteria for the acceptability of new methods of isolation and detection (Andrews, 1985).

Sensitivity	Must be equivalent to, but not necessarily greater than, the accepted cultural method
Applicability	Suitable for all foods of interest
Rate of incorrect reactions	False positives and false negatives should not exceed *ca* 10%
Overall acceptability	Method should be as simple, rapid and inexpensive as possible without compromising reliability

Identification of foodborne pathogenic micro-organisms

For some purposes a presumptive identification based on colonial appearance on differential, selective isolation media may be considered adequate, while in other cases only a small number of tests are required for confirmation of identity. Typical colonial appearance on Baird-Parker medium, cell shape and arrangement, Gram-reaction and a positive coagulase test, for example, would usually be acceptable as identification criteria for *Staph. aureus*. However, in other circumstances a significantly larger number of tests must be performed.

Although most tests used in the identification of bacteria are based on biochemical reactions, serological confirmation of identity is of increasing importance and, in the longer term, a general application of genetic techniques is likely.

Most identification requires the use of pure cultures of micro-organisms. Obtaining pure cultures by the streak plate technique is a fundamental part of microbiological technique, but despite this it is the authors' experience that false identifications may often be attributed to failure to obtain pure cultures. This usually results from attempting purification on selective media that do not permit contaminating micro-organisms to be detected, or through failure to obtain isolated colonies on the streak plate.

There is, in fact, no single 'correct' means of streaking a plate and many methods are used. For general purposes, the methods described by Harrigan and McCance (1976) are recommended.

It is appreciated that miniaturised identification systems are usually inoculated direct from isolation plates and that purification introduces a delay of at least 18 to 24 hours. Direct inoculation from an isolation plate is permissible, providing that colonies are well separated and the plate has been checked by an experienced microbiologist. It is strongly recommended that a streak plate of non-selective media is inoculated concurrently with the identification system, to serve both as a retrospective check of purity and to provide a culture for any further testing such as serotyping.

Biochemical tests for identification

Biochemical tests have been applied to bacterial identification since the early days of microbiology. Some, such as the production of acids from carbohydrates, remain important tests today. Others, the determination of specific enzymes such as β-galactosidase (ONPG) or lysine decarboxylase, for example, have been developed more recently.

The choice of tests for identification varies according to the type of bacteria, and tables listing tests for identification to both genus and, where appropriate, species level are included in Chapters 4 to 16 concerned with specific types of pathogen. It must be appreciated that tests which are able to discriminate between species of one genus may very well be useless with another genus.

Conventional biochemical tests are generally easy to use, although inoculation may be time consuming and a considerable amount of work is involved in preparing and dispensing the media. In many cases, the composition of media for determining the same reaction varies from one family of bacteria to another. For example, conventional Hugh and Leifson medium, used for the determination of oxidation or fermentation of glucose by Gram-negative, rod-shaped bacteria, is not suitable for determining the same reactions by *Staphylococcus/Micrococcus*, and a modified version must be used. Furthermore, most biochemical tests are discriminatory only when used under defined conditions of incubation, including atmosphere, temperature and time. Failure to abide by these conditions is likely to lead to false results. Incubation of Thornley's arginine dihydrolase medium under strictly anaerobic conditions in an anaerobic jar, rather than under partially anaerobic conditions under a layer of liquid paraffin, for example, will often lead to false-negative reactions.

In most situations, the microbiologist is concerned only with the identification of the 'target' organism. Thus during examinations for *Salmonella*, the identity of any non-salmonellas is usually of no importance. Therefore, it is usually possible to restrict the number of biochemical tests to a relatively small number of key criteria. With the *Enterobacteriaceae*, this concept was extended to multi-test media, which incorporate two or more media into a single tube and which permit identification at a generic level of the more important members of the family. Such media include lysine iron agar (**Figure 9**), Kligler iron agar (**Figure 10**), triple sugar-iron agar and Kohn's two tube medium. A significant saving of both preparation time and inoculation time was obtained and media of this type are still in use although to a large extent superseded by miniaturised identification systems.

 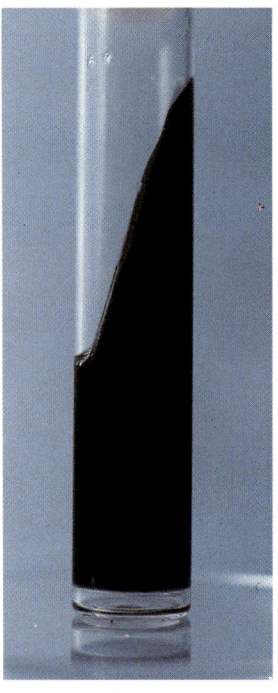

9 & 10. Use of multi-test media for identification of bacteria. Both lysine-iron agar (**9**) and Kligler iron agar (**10**) have been widely used for the presumptive identification of *Salmonella* and other members of the *Enterobacteriaceae*. *Salmonella* is recognised in lysine-iron agar by the decarboxylation of lysine and the production of hydrogen sulphide. In the example illustrated, production of lysine decarboxylase by *S. typhimurium* is detected by the purple (alkaline) colouration of the medium, but hydrogen sulphide production is very weak. Differentiation from other bacteria such as *Klebsiella* would not be possible. In contrast, the same strain of *S. typhimurium* produces such large quantities of hydrogen sulphide on Kligler iron medium that other reactions are masked and differentiation from *Proteus* and some other bacteria would not be possible. (*Both media: 21 h at 37°C.*)

Miniaturised identification systems

Since the early 1970s, the use of miniaturised identification systems has become widespread in both clinical and food microbiology laboratories. Such systems have many advantages over conventional media in terms of economy, ease of use and reproducibility. Their use permits a greater number of tests to be performed during routine identification and this, together with associated computer-based probabilistic identification enhances the likelihood that identifications are correct.

The first systems developed were designed for identification of the *Enterobacteriaceae*, a family that is both of obvious medical importance and biochemically active. A number of approaches were made to miniaturisation and two contrasting systems are illustrated: the API 20E© (**Figure 11**) and the Enterotube© (**Figure 12**).

The earliest miniaturised systems essentially reproduced sets of conventional tests, and were largely based on the taxonomic schemes of Edwards and Ewing (1972). Subsequently the range of kits available for different families, or physiological groups, has increased considerably, while tests based on determination of specific pre-formed enzymes rather than miniaturised conventional tests, have become of increasing importance. Use has also been made of carbon source utilisation tests. At the same time, two further types of kit have evolved: those which will fully identify all members of a family, or physiological group; and those which will confirm the identity of a narrow range of 'target' organisms, but give no indication as to the identity of organisms outside that range. The various types of kit available are summarised in **Table 28** and examples are illustrated in **Figures 13–16**.

A large number of comparisons of the performance of kits have been published. It is not intended to

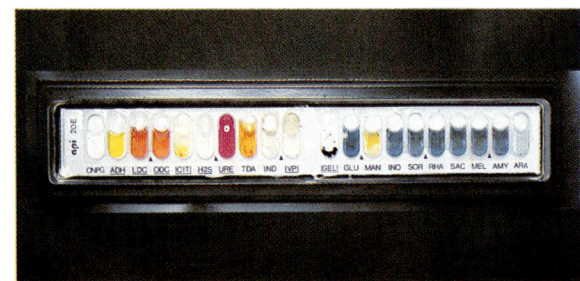

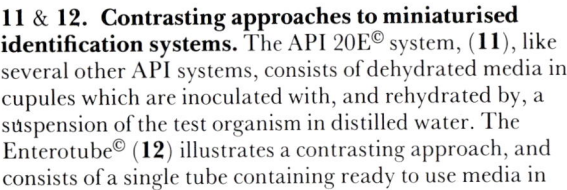

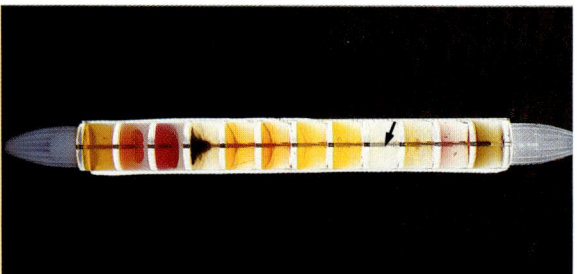

11 & 12. Contrasting approaches to miniaturised identification systems. The API 20E© system, (**11**), like several other API systems, consists of dehydrated media in cupules which are inoculated with, and rehydrated by, a suspension of the test organism in distilled water. The Enterotube© (**12**) illustrates a contrasting approach, and consists of a single tube containing ready to use media in discrete compartments. The tube is inoculated in a single operation using a built-in needle (arrowed).

A number of other approaches have been taken to miniaturisation. These include the use of paper discs impregnated with dehydrated media and dehydrated media in microtitre plates.

13

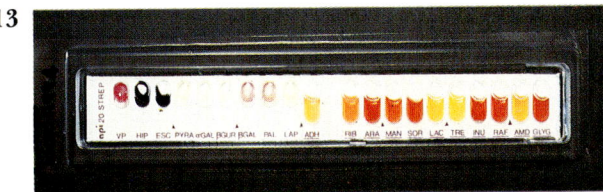

14

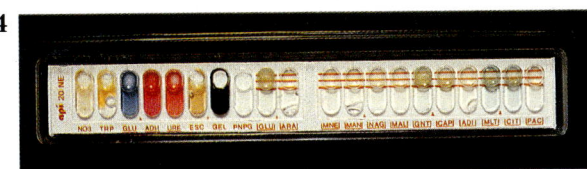

15

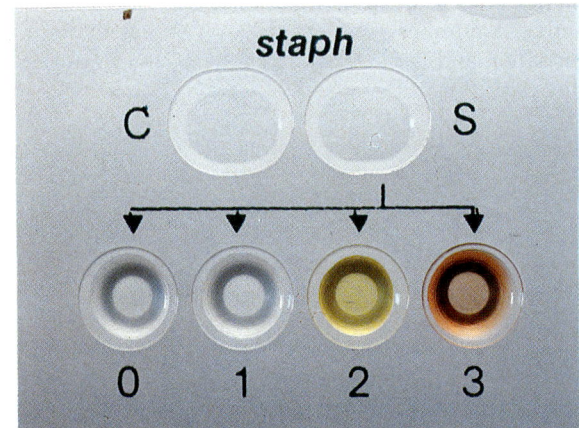

16

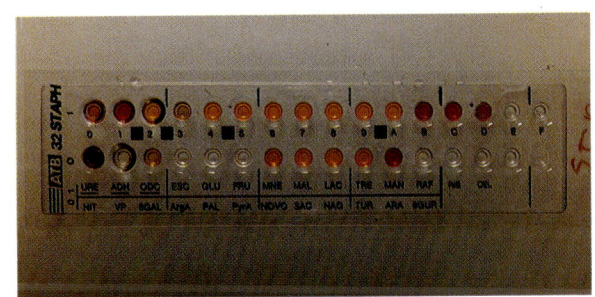

Table 28. Examples of miniaturised identification kits.

Full identification within a given family or group of bacteria

Based on conventional biochemical tests

API 10E; 20E; Rapid 20E©	*Enterobacteriaceae*[1]
Enterotube©	*Enterobacteriaceae*[1]
Minitek©	*Enterobacteriaceae*[2]
Oxi/Ferm©	Gram-negative, oxidase positive, rod-shaped bacteria
API 20A©	Anaerobes including clostridia
API 50CHB©	*Bacillus* spp.[3]

Based on conventional biochemical tests and pre-formed enzymes

API 20 Strep©	*Streptococcus*[4]
API ATB Staph©	*Staphylococcus*[5]

Based on pre-formed enzymes

API ATB 32A©	Anaerobes including clostridia
API ATB 32E©	*Enterobacteriaceae*[1]

Based on carbon source utilisation

Biolog GN MicroPlate©	*Enterobacteriaceae*

Based on conventional biochemical tests and carbon source utilisation

API 20NE©	Non-enteric, Gram-negative rods

Confirmation of identity of a limited number of bacteria[6]

API Z©	*Salmonella, Shigella, Yersinia*
API Rapidec Staph©	Main clinical species of *Staphylococcus*

[1] Will also identify some other Gram-negative, rod-shaped bacteria.
[2] System may be configured for many types of bacteria including anaerobes.
[3] Same system may be used for other bacteria depending on the suspension medium used.
[4] Will also identify *Listeria, Gemella* and *Gardnerella*.
[5] Will also identify *Micrococcus, Stomatococcus* and *Aerococcus*.
[6] Based on preformed enzymes.

13–16. Examples of further types of miniaturised identification system. In addition to large scale use for the identification of the *Enterobacteriaceae*, miniaturised systems are available for a wide range of micro-organisms, including many that are of significance with respect to food poisoning.

The API-strep© (**13**) was originally designed for use with *Streptococcus*, but is widely used in food microbiology for the identification of *Listeria* (but not for differentiation between species of *Listeria*; see page 352). The system uses a combination of conventional biochemical tests and the determination of specific enzymes. (*API-strep: 24h at 37°C.*)

The API 20NE© (**14**) is designed to identify Gram-negative bacteria other than the *Enterobacteriaceae*. Such bacteria are often inactive in conventional biochemical tests, and so the system includes carbon source utilisation tests to permit identification to be based on a reactive test pattern. The system may be used to identify oxidase-positive, fermentative bacteria implicated in food poisoning such as *V. parahaemolyticus, A. hydrophila* (see page 199) and *P. shigelloides*, each of which may also be identified by many of the systems designed for the *Enterobacteriaceae*. (*API 20NE: 24h at 30°C.*)

The Rapidec Staph© (**15**) is designed as a simple means of identifying *Staph. aureus* and a limited number of other species of *Staphylococcus* which are of medical or veterinary importance. The system is based on the detection of pre-formed enzymes and a result is obtained in two hours (see page 261). (*Rapidec Staph: 2h at 37°C.*)

The API ATB 32 Staph© (**16**) also detects pre-formed enzymes but provides a full identification of members of the genus as well as members of other genera of Gram-positive cocci such as *Micrococcus* (see pages 260–261). (*API 32 Staph: 24h at 37°C.*)

attempt any such comparison here, except that to note that identifications are typically greater than 90% correct, and that different atypical isolates are likely to cause confusion to different kits. Databases are largely prepared using the characteristics of isolates from medical sources and, in general terms, a higher proportion of unidentifiable isolates may be expected from foods. Choice of system not infrequently comes down to personal preference or prejudice; ease of use and clarity of reactions must therefore be seen as important factors in choice.

Although miniaturised systems are easier to use than conventional one, the need for good practice and rigorous technique is no less. Conditions and length of incubation may be more critical than with conventional media and arbitrary variations should be avoided.

Automated identification systems

A number of automated identification systems have been developed, although not all have been commercially successful. Most automated systems are based on conventional biochemical tests or on determination of specific enzymes, but alternative technology has been employed. The Hewlett Packard HP 5898A Microbial Identification System, for example, used high resolution gas chromatography of cellular fatty acids as a means of identification. This instrument is no longer available.

The major interest in automated identification lies in medical laboratories where sample throughput is often very high. They are of only limited application in the food industry and, therefore, are not discussed further.

Typing of foodborne pathogenic micro-organisms

Although identification alone is adequate for many safety assurance purposes, it provides insufficient discrimination for epidemiological purposes. In this situation one, or more, typing schemes must be used. Typing may be based on any character, or set of characters, which permit epidemiologically valid discrimination among a single genus or species. Large-scale typing schemes based on serotyping and phage typing have been developed over the past years, while biotyping and various other methods are less commonly used, but can still be very useful. More recently, techniques based on genetic analysis have been developed including plasmid profiling and bacterial restriction endonuclease analysis (BRENDA). Both of these methods offer a high degree of discrimination, but their full potential has not yet been realised.

In the case of many pathogenic micro-organisms, serotyping and phage typing schemes are operated on a national basis, behind which lies a considerable degree of international co-operation and standardisation. In some cases, such as *B. cereus*, it is necessary to refer the culture to the appropriate national laboratory – in the UK, the Central Public Health Laboratory, Colindale, for example.

Although large and sophisticated typing schemes are a necessity, the value of observations made at the laboratory bench should not be overlooked. The most obvious of these is colony morphology, but unusual biochemical properties may also provide a useful epidemiological marker (Olds, 1975).

Serotyping

Serotyping is the most widely used form of typing and schemes exist for most foodborne pathogens. Serotyping is based on the immunological analysis of the surface structures of the cell. In the case of the *Enterobacteriaceae*, which have probably been most extensively studied, these are the polysaccharide component of the lipopolysaccharides in the outer membrane of the cell (O antigen); capsular polysaccharides (K antigen) and flagellar protein (H antigen). In the example of *Salmonella*, serotyping is based on analysis of the O and H antigenic structure (**Figure 17**) while with *E. coli* the K antigen is also used.

Antigens are usually detected by simple slide or tube (tray) agglutination reactions using whole cells.

17

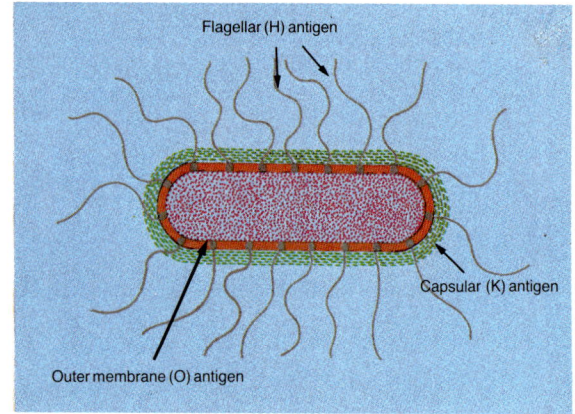

Flagellar (H) antigen

Capsular (K) antigen

Outer membrane (O) antigen

17. Antigenic structure of bacteria.

Care is needed during the cultivation of some bacteria for typing, to ensure that surface structures such as capsules or flagella are present.

In some cases, such as *B. cereus*, serotyping alone may permit strain discrimination at a level suitable for epidemiological purposes, while in others, notably *Salmonella*, further typing is required. In the case of *Staph. aureus*, it has not been possible to develop a serotyping scheme although its enterotoxins are serologically distinct.

Epidemiological patterns change with time, and it is necessary to update typing antisera in accord with these changes. This may be illustrated by reference to pathogenic strains of *E. coli*, where typing sera originally used are now of little use.

Phage typing

Phage typing schemes may be developed where an organism harbours phages which are reasonably specific in host range, and which may therefore distinguish between epidemiologically unrelated strains. Phage typing may be used to discriminate within serovars as with *Salmonella*, or in place of serotyping as with *Staph. aureus*.

Routine typing involves preparation of a lawn plate of the organism being typed, onto which suspensions of the typing phages are spotted. Following incubation, lysis is usually easily detected by clearing in the lawn plate. A generalised protocol is shown in **Flow diagram 3**.

Although most phages are fairly easily propagated, great care is necessary to ensure that only those of correct specificity are produced. For this reason, the propagation of typing phages is usually the responsibility of a single specialist centre in each country.

Bacteriocin typing

Bacteriocins are produced by a wide range of bacteria. They consist of plasmid-encoded proteinaceous toxins, which are active only against closely related strains of bacteria. In many cases, the activity is sufficiently specific for application to typing, and schemes have been developed for a number of pathogens including *E. coli*, *Salmonella* and *Shigella*.

A generalised protocol for a bacteriocin typing scheme is illustrated in **Flow diagram 4**.

Flow diagram 3. General protocol for phage typing.

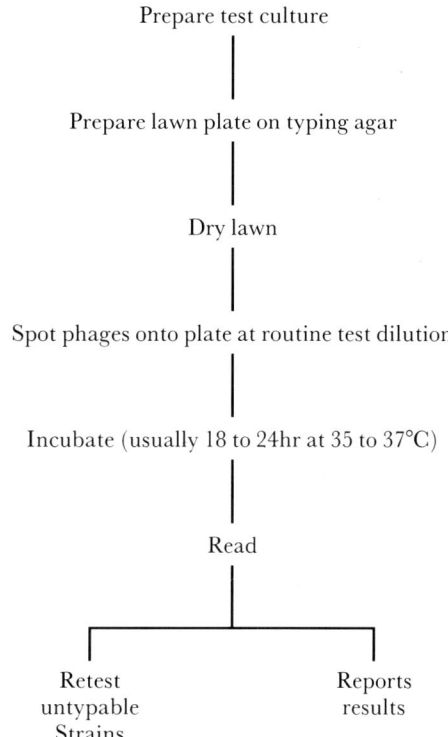

Prepare test culture

Prepare lawn plate on typing agar

Dry lawn

Spot phages onto plate at routine test dilution

Incubate (usually 18 to 24hr at 35 to 37°C)

Read

Retest untypable Strains   Reports results

Lysis may be recorded in several ways but the criteria of Anderson and Williams (1956) is widely used.

Plaque size: L = large; N = normal; S = small; m = minute (visible only with hand lens); μ = micro.

Number: 0–5; ± = 6–20; + = 21–40; + ± = 41–60; + + = 61–80 + + ± = 81–120; + + + = >120.

Lysis: SCL = semiconfluent; CL = confluent; <SCL, <CL = intermediate degrees; OL = confluent opaque lysis (opacity due to heavy secondary growth).

Flow diagram 4. General protocol for bacteriocin typing.

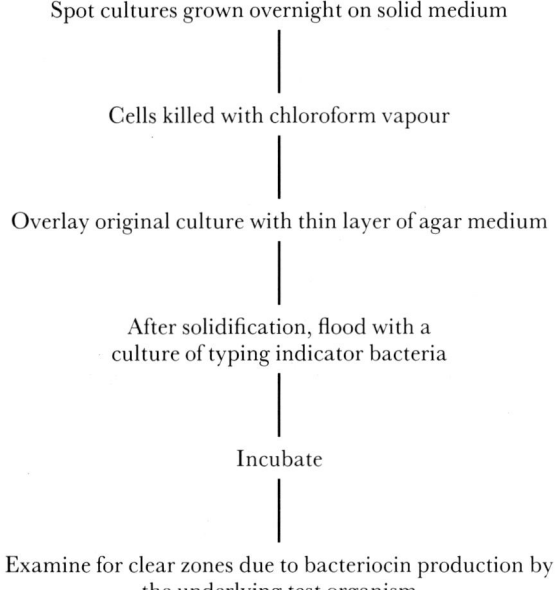

Overlay method

Spot cultures grown overnight on solid medium

Cells killed with chloroform vapour

Overlay original culture with thin layer of agar medium

After solidification, flood with a culture of typing indicator bacteria

Incubate

Examine for clear zones due to bacteriocin production by the underlying test organism

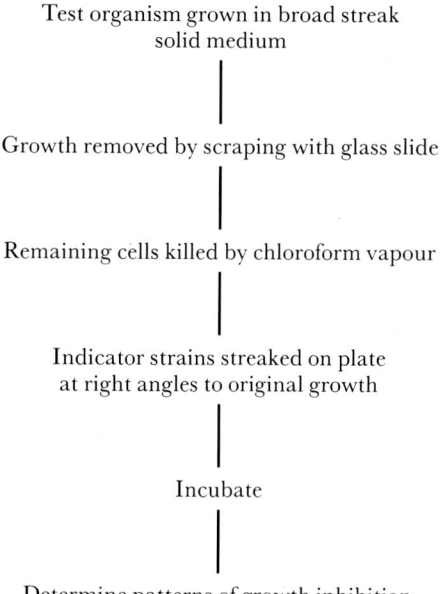

Streak method

Test organism grown in broad streak solid medium

Growth removed by scraping with glass slide

Remaining cells killed by chloroform vapour

Indicator strains streaked on plate at right angles to original growth

Incubate

Determine patterns of growth inhibition

Biotyping

Biovars are defined by differences in biochemical characteristics and may be used either to subdivide a species or to discriminate within a serovar as with the xylose$^+$: xylose$^-$ characteristic of *S. typhi*. Biotyping can be of value both epidemiologically and for determining the origin of micro-organisms during ecological studies concerning pathogens in foods.

Genetic methods

Genetic methods of typing bacteria are of relatively recent development, and their application is currently restricted to a small number of laboratories. Two methods, plasmid profiling and bacterial restriction endonuclease analysis, are briefly described below.

Plasmid profiling

Plasmid profiling involves the isolation of plasmid DNA and its separation by gel electrophoresis. The method is rapid and avoids the necessity for prior preparation and maintenance of specific antibodies and typing phages. The method is discriminatory in the absence of a prior knowledge of an organism's characteristics (Dodd, 1988) and has now been used successfully in epidemiological investigations concerning a wide range of foodborne pathogens.

A generalised protocol for plasmid profiling is illustrated in **Flow diagram 5**.

Bacterial restriction endonuclease analysis

Bacterial restriction endonuclease analysis (BRENDA) offers a means of 'fingerprinting' bacteria at a high level of discrimination. The application of BRENDA does not depend on the presence of plasmids, and the method has been successfully applied, in conjunction with plasmid profiling, in investigations concerning *E. coli* and *V. cholerae*.

Flow diagram 5. General protocol for plasmid profiling. (Dodd, 1988)

Suspend cells in isoosmotic buffer containing **tris/EDTA** and **lysozyme**. Incubate

Tris/EDTA removes the outer coating of cells such as those of E. coli. Lysozyme causes partial digestion of peptidoglycan of cell wall. Alternative methods such as lysostaphin for Staph. aureus must be used for lysozyme resistant bacteria

Add detergent such as **sodium dodecyl sulphate** (SDS)
SDS or other suitable detergent gently disrupts the cell membrane

Cell lysis

High speed centrifugation
Removes cell debris and high molecular weight chromosomal DNA. Cleared lysate contains plasmid DNA, RNA, carbohydrate, protein and residual lower molecular weight DNA

Precipitate with 70% **ethanol** at −20°C

Resuspend **pellet containing nucleic acids** in Tris/EDTA buffer

Treat with **ribonuclease**
Removes RNA which may interfere with detection of low molecular weight plasmid DNA

Add **loading buffer** and load onto **agarose gel** containing **ethidium bromide**
Loading buffer contains dye and a 'heavy molecule' such as ficoll or sucrose

Electrophoresis, usually overnight

View under **ultra-violet** light

Note: Key procedures or reagents are printed in **bold**.

Control of foodborne pathogenic micro-organisms

Control of foodborne pathogenic micro-organisms relies on four aspects of good practice: minimising contamination of raw materials; processing for safety; prevention of re-contamination of cooked or processed foods; and prevention of growth of bacteria in foods, whether cooked or raw. Each of these aspects is discussed fully in relation both to individual foods and the various micro-organisms, but it is considered that some brief comments are appropriate at this stage concerning those factors which control the growth of micro-organisms in foods.

Over the years, a large number of publications have attempted to define the parameters which permit growth and, where appropriate, toxin production by pathogenic micro-organisms in foods. Parameters which have been investigated include temperature, pH value, a_w level, redox potential, as well as the presence of various inhibitors such as NaCl and nitrite.

It is necessary to place these studies in a realistic perspective to the actual situation in foods. In some cases, studies of this nature do no more than indicate the ability of certain strains of an organism in pure culture to grow under defined conditions in an artificial medium. This may bear little relationship to the likely behaviour of that organism in food, and should be used only as an approximate guide.

Inoculation experiments offer a possibly more realistic alternative, but can neither reproduce the true physiological condition of bacteria in food at the point of growth initiation, nor truly predict the ability of bacteria to grow in foods under even marginally altered conditions.

It is therefore considered that the only valid approach to predicting the growth, or non-growth, of bacteria in foods is by using multi-factorial experiments, and using the results of these to develop mathematical models for prediction of the likely behaviour of bacteria under any likely combination of circumstances. This approach is described in greater detail on pages 304–305, 393–394.

Mathematical modelling is a relatively new technique, and to date has been concerned with pure culture studies. A major difficulty with foods is heterogeneity, which can result in niches which are favourable to the growth of pathogenic micro-organisms when the overall environment is inhibitory.

Mechanisms of pathogenicity

In recent years, there has been a considerable advance in the understanding of the mechanisms of microbial pathogenicity, although in most cases understanding is far from complete. It has also become appreciated that the pathogenicity, or non-pathogenicity of an organism, cannot be viewed in isolation from the host (Archer and Young, 1988), and a holistic approach must be taken to gain a full appreciation of the host/ parasite relationship (Isenberg, 1988). The body has at its disposal a wide range of host defences and barriers to pathogenic micro-organisms which, in turn, have a wide range of evasive mechanisms. Both host defences and microbial evasion mechanisms are summarised in **Table 29** and discussed below.

Genetic control of virulence

The genetic control of virulence is often complex and can involve either plasmid or chromosomal genes, or both. Bacteriophages are also involved in virulence in some foodborne pathogens, such as *Shigella*. External factors may also play an important role in the expression of virulence. For example, temperature regulation of virulence genes is a common phenomenon, and may be a general strategy for human pathogens to ensure that virulence genes are expressed only when the organism is in its host niche (Maurelli, 1989).

Microbial ecology of the gastrointestinal tract

With the exception of organisms such as *Staph. aureus*, which produces a pre-formed toxin during growth in foods, it is necessary for acute foodborne pathogens either to colonise the epithelial surface at some point, or to invade the epithelium without colonisation. Examples of each type of behaviour may be drawn from a single species, *E. coli*, where enterotoxigenic strains colonise the epithelium in contrast to entero-invasive strains which invade.

Enteric pathogens are obviously not present in the gastrointestinal tract in isolation and account must be taken of the autochthonous (indigenous) micro-flora. The total number of prokaryotic cells in the gastrointestinal tract is in the order of 10^{14} (Luckey, 1972). In most cases, the population is relatively stable, and autochthonous micro-organisms must have the ability to cope with changes caused by cell turnover, mucus flow, peristalsis and nutrient availability (Savage, 1983). Conditions vary from section to section of the gastrointestinal tract, and differences in the microenvironment of pH value, oxygen tension and nutrients suggest that each section has its own autochthonous flora.

Under most circumstances, all ecological niches in the gastrointestinal tract are occupied by autochthonous organisms and allochthonous microbes are usually able to become established only under exceptional conditions. Such conditions include long-term antibiotic therapy or, in infants, at weaning.

Successful foodborne pathogens posses a variety of mechanisms, virulence factors, which permit colonisation or invasion (see below). However, the normal autochthonous flora still present formidable difficulties to allochthonous pathogens.

In the normal host, colonisation by pathogens is a temporary phenomenon, which will ultimately be reversed as the autochthonous organisms re-colonise.

The role of autochthonous micro-organisms as pathogens is not clear, but in some cases plasmid-borne virulence factors may be acquired through genetic exchange with allochthonous organisms (Archer and Young, 1988).

Table 29. Summary of host defences and the evasionary strategies of foodborne micro-organisms (Gotschlich, 1983).

Host defence	Mechanism of evasion
Hydrodynamic flow	Attachment
Mucus barrier	Attachment Penetration
Deprivation of essential nutrients	High-affinity uptake
Lysozyme	Resistance to lysis
Surface immunoglobulin	Absent or low immunogenicity Antigenic heterogeneity Masking of antigen
Intact tissue surface	Penetration
Serum defences	
Recognition by antibody	Antigenic heterogeneity Masking of antigen Antigenic variation Destruction of antibody?
Complement system	Failure to activate alternative pathway Inactivation of components Resistance to lysis
Phagocytosis	Inhibition of attachment and ingestion Inhibition of metabolic burst Resistance to oxidative attack

Stomach acidity, bile and proteolytic enzymes

The conditions inside the gastrointestinal tract are generally hostile to allochthonous bacteria. Food-borne bacteria are generally resistant to bile and proteolytic enzymes and enteroviruses are also notably acid resistant. Most bacteria are favoured by a slightly alkaline environment, and stomach acidity can be an important host defence mechanism. This may be illustrated by the infective dose of less than 10^3 cells for *Shigella* compared with 10^9 cells for enteroinvasive *E. coli*, which is otherwise similar biochemically, antigenically and in the nature of the disease caused (Williams *et al.*, 1988). This difference is probably due to the greater sensitivity of *E. coli* to gastric acidity.

Food is an important factor in reducing the effect of gastric acidity: particles of any food offer protection but the effect is most marked with high-fat foods such as chocolate and hard cheese.

Achlorhydria, impaired production of acid by the host, is an important predisposing factor for gastrointestinal infection, especially in the case of bacteria such as *V. cholerae*, which are particularly acid sensitive. It has also been suggested that dietary modification or habits may predispose individuals to food poisoning by reducing stomach acidity. People taking large quantities of antacid preparations, for example, may be at increased risk.

Hydrodynamic flow

The continuous downward flow through the gastro-intestinal tract, together with the constantly changing contours of the tract surface, present problems to both autochthonous and allochthonous intestinal micro-organisms. Foodborne pathogens must therefore have the ability either to attach to the epithelial surface, or to invade the epithelium very rapidly. Mechanisms are discussed in greater detail below.

The mucus layer

Relatively few micro-organisms are able to colonise the mucus layer (Savage, 1980) secreted by the eponymous mucosal epithelium. The extent to which the mucus layer, consisting of proteins and polysaccharides, acts as a physical barrier is debatable (Mims, 1982), but it certainly provides a hostile environment in which bacteria can be trapped. For example, mucus contains IgA antibodies that protect immune persons against infection, as well as iron binding proteins, which reduce the iron available to micro-organisms.

Many enteric pathogens are motile, and it is likely that in many cases motility is an important virulence factor and there is evidence that non-motile variants of *V. cholerae* are less virulent (Jones *et al.*, 1976). *V. cholerae* also produces a mucinase, which may enable a more rapid transit through the mucus.

Passage through the mucus layer to the intestinal epithelium does not depend on a random directional motility, but on movement along a chemical gradient (chemotaxis). With *V. cholerae*, and possibly *Campylobacter*, taxis towards mucin is an important factor (see pages 167, 218).

Non-motile enteric pathogens, such as *Shigella* must rely on random and passive transport through the mucus layer.

Adherence

Adherence is the penultimate stage in the pathogenic process (before toxin production) for non-invasive pathogens, but only a prelude to penetration for invasive pathogens. Bacteria adhere to specific receptors by means of adhesins, which are secreted protein or carbohydrate structures (Christensen *et al.*, 1985). Classically, protein adhesins have been divided into two classes, fimbriae, which can be discerned by electron microscope as hair-like fibres a few micrometers long and *ca* 2-8 nm in diameter (Duguid and Olds, 1980), and non-fimbrial adhesins, which lack a structure discernible by electron microscope. This distinction has become blurred by more recent findings. In some cases, adherence by fimbriae is mediated by distinct non-fimbrial proteins carried on the fimbriae (Minion *et al.*, 1986), while fine fimbrial structures have been discerned in some adhesins previously classified as non-fimbrial (Hinson *et al.*, 1987).

Fimbriae may be divided into two types on the basis of inhibition of haemagglutination by mannosides. Where haemagglutination is inhibited, fimbriae are referred to as mannose-sensitive (MS; type 1); and where not inhibited, as mannose-resistant (MR). MS fimbriae constitute a group of more or less related antigens, whereas MR fimbriae are antigenically diverse. MR fimbriae often function as virulence factors, and can be both species specific and organ specific in adhesive character (Orskov, 1984).

Carbohydrate adhesins are associated with the capsules (glycocalyces) produced by freshly isolated members of the *Enterobacteriaceae* Williams *et al.*, 1988. A major role for some capsules is in the serum resistance of the organism (see pages 48, 171), but in other cases capsules may be involved in permitting adherent microcolonies to form on mucosal surfaces (Costerton *et al.*, 1985).

Control of synthesis of most, if not all, adhesins involves the process of phase variation, whereby bacteria reversibly switch between expression and non-expression (Williams *et al.*, 1988). Phase variation is of considerable importance in microbial pathogenicity and is apparently a random process.

In the specific case of adhesin synthesis, suspension may permit cells to be released from 'old' epithelial tissue, which is being sloughed off, to re-adhere to newly exposed tissue when adhesin synthesis is resumed. Expression of a range of adhesins of different tissue specificity also permits colonisation of a range of sites. This is certainly of importance in *E. coli* infections of the urinary tract (Williams *et al.*, 1984), but its role in enteric infection is not known.

Although much is known of the adhesins of non-invasive micro-organisms, especially enterotoxigenic *E. coli*, relatively little is known of the mechanisms by which invasive organisms attach – although it seems probable that adhesion to epithelial cells must involve much closer contact (Williams *et al.*, 1988).

Although work on mechanisms of adherence have concentrated on bacteria, it should be appreciated that other types of intestinal pathogen must also have a means of adhering to the epithelial surface. For example, *Giardia lamblia*, attaches to the upper small intestine of man by a sucking disc (Mims, 1982).

Invasion of intestinal cells

The initial stages in the invasive process tend to be similar over a wide range of enteric pathogens, although the site at which invasion occurs varies. For example, *Salmonella* and *Yersinia*, invade the ileum, in contrast to *Shigella* and enteroinvasive *E. coli*, which invade the colon.

Receptor-mediated endocytosis

Receptor-mediated endocytosis (RME) is an important mechanism by which bacteria enter cells, and is a virtually identical process in a number of genera. Host cell engulfment of bacteria is triggered by the sequential binding of receptors on the host cell membrane to ligands on the bacterial surface. The nature of the components which induce RME are not known, and probably vary from one organism to another.

Intracellular proliferation

The site of intracellular proliferation varies according to the organism. Using the four invasive species of the *Enterobacteriaceae* as examples, *Salmonella* multiplies in

the *lamina propria*, *Shigella* and enteroinvasive *E. coli* in the cytoplasm, and *Y. enterocolitica* in lymphoid tissues. The host response and hence symptoms and pathological processes vary accordingly.

Systemic infection

In many cases progression of foodborne invasive pathogens beyond the gastric epithelium and adjacent tissues is rare and occurs only when exceptional circumstances affect the host. These usually involve extremes of age, an underlying illness such as cancer, or an immunocompromised condition. However, in other cases, systemic infection is a normal part of the progress of the illness and causative organisms, which include bacteria e.g. *Listeria*, rickettsia e.g. *Coxiella burnettii*, protozoan parasites e.g. *Toxoplasma* and viruses e.g. Hepatitis A, may effect no intestinal pathology discernible by light microscopy (Wells *et al.*, 1988).

Translocation

Translocation is a mechanism by which enteric pathogens are able to escape the intestine and cause systemic disease (Wells *et al.*, 1988). These authors have postulated four major roles for motile phagocytes.

1 Ingestion of intestinal bacteria.
2 Transportation of intestinal bacteria to extra-intestinal sites.
3 Failure to accomplish intracellular killing.
4 Liberation of the bacteria at the intracellular site.

This hypothesis is supported by a large number of observations made in the literature, which include the observation that agents that modulate immune (including phagocytic) function can alter the rate of translocation. Furthermore, the intestinal bacteria which can most readily translocate out of the intestine are those classified as facultative intracellular parasites (see below) and include *Salmonella*, *Shigella* and *Listeria* (Brubaker, 1985).

Survival and proliferation

Once established in the host body and tissues, micro-organisms must survive and ultimately proliferate in the hostile environment created by the host defences. It should be appreciated that these defences are similar to those at the mucosal surfaces, and thus are part of a continuing problem for the invading organism.

Facultative intracellular parasites

Facultative intracellular parasites are of two types. Specialised types are defined as those which are capable of penetrating and growing within non-professional phagocytes, but which are unable to exist within macrophages (Brubaker, 1985). An example is *Shigella*, which 'specialises' in parasitising intestinal epithelial cells. Generalised types are capable of growth within normal macrophages, and may also be able to survive within neutrophils and monocytes. In most cases, non-professional phagocytes are also penetrated. An example is *Salmonella*.

Phagocytosis

Phagocytosis represents the most effective non-specific host defence mechanism (Stendahl, 1983), although phagocytes play a role in the translocation of bacteria from the intestine.

The major professional phagocytes of the host defences are neutrophils (polymorphonuclear leukocytes, PMNs), monocytes and macrophages (**Flow diagram 6**). Neutrophils and monocytes are of greatest importance in responding to early stages of infection.

Phagocytes are attracted to invading bacteria by various chemotaxins including peptides containing N-formylmethionine and early components of the complement cascade. Invasive bacteria vary in their attractiveness to phagocytes and *in vitro*, at least, may actively inhibit chemotaxis. However, such bacteria are not necessarily of high virulence.

Intracellular killing by phagocytes is by two main processes. In the first, the membrane bound vacuole, the phagosome, which contains the engulfed bacterium, fuses with a lysosome producing a new vacuole, a phagolysosome. A number of hydrolytic enzymes are contributed to the new vacuole by the lysosome, including lysozyme, phospholipase and several proteases, and non-resistant micro-organisms are killed rapidly.

The second process leads to the synthesis of the highly lethal radical superoxide (O_2^-) and the rather less lethal hydrogen peroxide (H_2O_2). The mechanism is illustrated in **Figure 18**.

Invasive intestinal pathogens have mechanisms of resistance to these two processes. In the first case, inhibition of the activation of the complement cascade by mechanisms such as phase and antigenic variation may be of importance, while lipopolysaccharide antigens and the various acidic polysaccharide antigens of *Salmonella* and *E. coli* have been implicated in resistance both to phagocytosis and serum killing (Smith, 1983). The increase in surface hydrophobicity due to lipopolysaccharides may be a factor in resistance, while surface structures such as capsules may also be involved.

The resistance of *S. typhi* and *S. typhimurium* to reactive oxygen metabolites has recently been confirmed (Ishibashi and Arai, 1989) and, in these and other bacteria, may be due to either a direct resistance to these compounds, or to a failure to trigger the respiratory burst of macrophages.

Flow diagram 6. Summary of professional phagocytes.

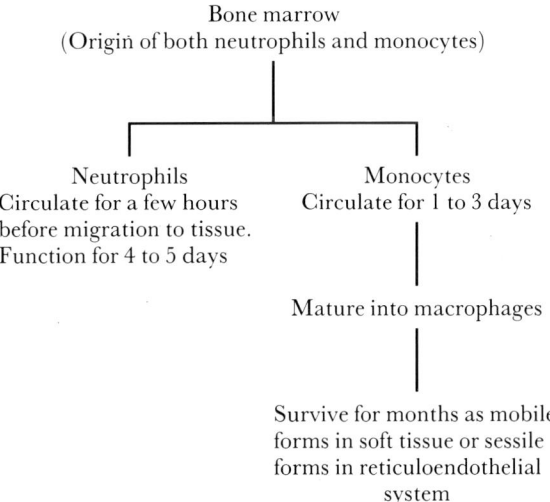

Bone marrow
(Origin of both neutrophils and monocytes)

Neutrophils
Circulate for a few hours before migration to tissue. Function for 4 to 5 days

Monocytes
Circulate for 1 to 3 days

Mature into macrophages

Survive for months as mobile forms in soft tissue or sessile forms in reticuloendothelial system

Note: Neutrophils and monocytes are the primary scavengers invading bacteria and respond to the early stages of infection.

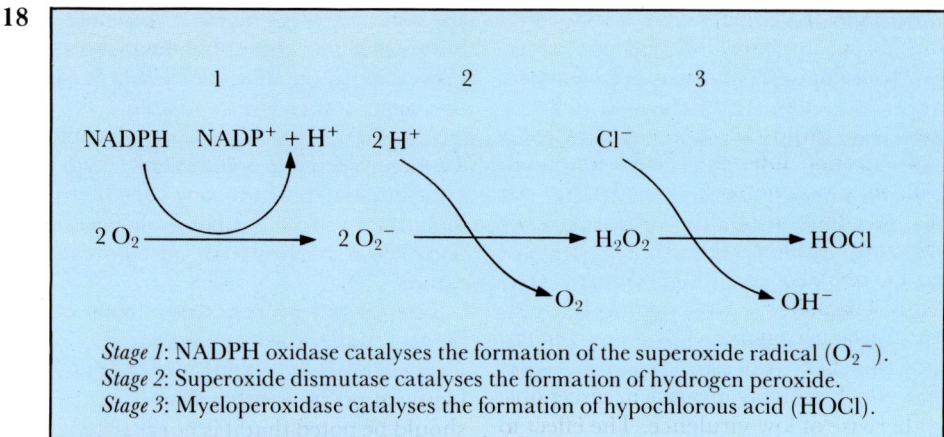

Stage 1: NADPH oxidase catalyses the formation of the superoxide radical (O_2^-).
Stage 2: Superoxide dismutase catalyses the formation of hydrogen peroxide.
Stage 3: Myeloperoxidase catalyses the formation of hypochlorous acid (HOCl).

18. Myeloperoxidase pathway of neutrophils and some macrophages.

Bactericidal effect of normal serum

The bactericidal effect of normal serum is a primary defence mechanism against bacterial invasion (Taylor, 1983). Cells of many species are killed when the five terminal proteins of the complement cascade, organised as a macromolecular complex described as the membrane attack complex (MAC), are first deposited and then inserted into the bacterial outer membrane. The integrity of the outer membrane is destroyed and it is unable to maintain an effective barrier, thus increasing the susceptibility of the peptidoglycan layer to lysozyme. In most cases, degradation of the peptidoglycan layer leads to cell lysis. The most important pathway for the activation of complement is the specific interaction of antibodies with antigenic determinants at the cell surface. These include lipopolysaccharides and outer membrane proteins.

Invasive micro-organisms have a high level of resistance to the bactericidal effects of serum. In the invasive members of the *Enterobacteriaceae*, and possibly in other bacteria, the basis for the resistance is multi-factorial, the overall effect being a reduction of the MAC activity (Williams *et al.*, 1988). Structural features such as lipopolysaccharide O antigen may be involved. These structures reduce membrane fluidity and may thus block the integration of MACs into the membrane (Taylor, 1983). Evidence for this lies in the fact that 'rough' variants of *E. coli*, *Salmonella* and *Shigella*, which lack all or part of their lipopolysaccharide structure, are markedly sensitive to serum. The basis of serum resistance cannot entirely be explained in this way, since there is considerable variation in susceptibility among isolates with a normal lipopolysaccharide structure.

Although outer membrane proteins may also have a role in resistance to serum, this appears to be marginal and there is no clear link between synthesis of outer membrane proteins, resistance to serum or virulence (Williams, 1988).

Capsules which are produced by many strains of the *Enterobacteriaceae* have for many years been considered to impart resistance to serum. For example, some capsules are poorly antigenic and may mask stronger antigens from the immune system. The role of capsules in serum resistance may well have been overstated, since there is considerable variation among capsulated strains. Further members of the family which do not produce capsules, such as most salmonellas, may well have a higher level of resistance than capsulated members.

Iron deprivation

Iron is required for all life forms including bacteria. It has a role in a variety of enzymatic reactions in which the cation serves as a cofactor *per se* or as part of a prosthetic group, primarily haem. Bacteriostasis is achieved when iron deprivation is sufficient to inhibit these reactions, but extreme deficiency can cause lethal lesions at the level of protein synthesis (Brubaker, 1983). Under normal conditions, the host body contains an abundance of iron, but most is located intracellularly where it is relatively inaccessible to micro-organisms (Finkelstein *et al.*, 1983). The small amounts present in body fluids are complexed with host iron binding and transport proteins, such as transferrin in plasma and lactoferrin in mucosal secretions.

In the normal host, bacterial invasion prompts a rapid reduction in plasma iron levels by a process known as the hypoferraemia of infection. This results from the suppression of intestinal iron assimilation and increased iron storage in the liver induced by a low molecular weight protein, interleukin 1. Fever may further restrict the amount of iron available for micro-organisms (Williams, 1988). The ability to reduce the available iron has been described as nutritional immunity, and may be reversed experimentally by supplying exogenous iron (Kochan, 1983).

Iron overload is an important factor in predisposing persons to systemic infection. This condition, hyperferraemia, may be due to liver or haemolytic diseases, or to introduction of excess iron, possibly through incorrect medication. Hypotransferrinaemia resulting from kwashiorkor, also increases susceptibility to invasion. The invasive ability of a wide range of foodborne bacteria are enhanced by iron overload in the host, but the effect is possibly most dramatic with *V. vulnificus*, which is highly invasive in susceptible persons, but otherwise of low virulence. The effect to the host of excess iron may be further enhanced by interference with phagocytic activity (Brock, 1986).

Micro-organisms have responded to the problems of iron deprivation by producing powerful iron binding factors of their own. These are known generically as siderophores, and those identified belong to two chemically distinct classes: the phenolates (catechols) and the hydroxamates. Production of both types is genetically determined and iron responsive. A general mode of action is that the siderophores are synthesised and exported, iron is then captured followed by internalisation through the mediation of specific receptors. These are outer membrane proteins and are also genetically determined and iron responsive (Finkelstein *et al.*, 1983).

In addition to a general role in both pathogenic and non-pathogenic bacteria, siderophores may be important virulence determinants. Aerobactin, for example, is a virulence determinant in many strains of enteroinvasive members of the *Enterobacteriaceae* (Roberts *et al.*, 1986) as well as members of that family involved in extraintestinal infections (Linggood *et al.*, 1987)

Phase and antigenic variation

The role of phase variation in synthesis of adhesins has been discussed earlier (pages 45–46). Both phase and antigenic variation play a wider role in permitting organisms to change their antigenic profile. The mechanism has been demonstrated in *V. cholerae* and *C. jejuni*, and in the latter case may be a significant factor in persistent diarrhoea (Blaser *et al.*, 1979). It should be noted that it is not strictly correct to use the terms 'phase variation' and 'antigenic variation' synonymously, since in the former single antigens are expressed or repressed independently, while in the latter alternative antigens are synthesised by a co-ordinated expression system. Similar mechanisms of genetic control may, however, be involved (Williams *et al.*, 1988).

Mention should also be made of molecular mimicry, a concept in which the major antigens of a particular micro-organism cross-react with human tissue antigens. This has been described in the context of the specific disease, ankylosing spondylitis, as the cross tolerance hypothesis (Ebringer, 1983). This hypothesis predicts that the sharing of some antigens permit host and micro-organism to coexist (Archer and Young, 1988). In some people, the shared antigens, or shared epitope(s) may lead to problems of chronic disease. However, molecular mimicry is not the only predisposing factor involved.

Microbial toxins

Microbial toxins may be defined as:
'Any of various microbial products or components which, when present in low concentrations in cells or tissues of a higher (multicellular) organism, can cause injury by interfering with the structural or functional integrity of those cells or tissues' (Singleton and Sainsbury, 1987).

These authors also point out that the basis of classification is arbitrary and thus lactic acid and hydrogen sulphide are not classified as toxins despite their poisonous properties. Types of toxin are defined in **Table 30**.

Table 30. Toxins: definition of terms.

Endotoxin	1 A generic term for the lipopolysaccharide of Gram-negative bacteria 2 Any microbial toxin which is released only on cell lysis
Exotoxin	1 An extracellular toxin 2 A toxin secreted by a living cell
Enterotoxin	Any toxin which either, when ingested or produced by an organism within the intestine, acts on the intestinal mucosa in an abnormal way
Cytotoxin	Any toxin that kills host cells by enzymatic attack
Neurotoxin	Any toxin that interferes with the normal transmission of nerve impulses
Haemolysin (haemolytic toxin)	Any of the various products which cause haemolysis by damaging the cytoplasmic membrane of red blood cells

Note: These definitions are derived from Singleton and Sainsbury (1987) with additions.

A wide range of toxins are produced by enteric pathogens, examples of which are listed in **Table 31**. In some cases the role in the pathogenic process is not understood, while others have been extensively studied. The situation with respect to chemical structure of the toxins is similar.

It is striking that there is a close relationship between toxins produced not only by different groups of bacteria, but even by different micro-organisms. The similarity between toxins produced by *V. cholerae* and some strains of *E. coli* has been known for some years, and a related toxin may also be produced by *Salmonella* (Scotland, 1988). A cholera-like toxin is also produced by some strains of *C. jejuni* (Blaser *et al.*, 1979) and, most remarkably, by the protozoan pathogen *G. lamblia* (Archer and Young, 1988).

The production of similar toxins by different micro-organisms does not necessarily mean a similar role in pathogenicity.

In most cases toxin production is only a part of the pathogenic process. However, with three foodborne pathogens, *Staph. aureus*, *B. cereus* and, under most circumstances, *Clostridium botulinum*, toxins are pre-formed in the food and the presence of viable bacteria is not required in the disease process.

Table 31. Examples of toxins produced by foodborne pathogenic micro-organisms.

Toxin	Micro-organisms
Cholera and related toxins	*Vibrio cholerae, Escherichia coli*[1], *Campylobacter* spp., *Salmonella, Aeromonas* spp., *Giardia lamblia*
Shiga and related toxins	*Shigella dysenteriae* 1, other *Shigella* spp., *E. coli*[2]
E. coli heat stable toxin	*E. coli*[3], *Citrobacter* spp.
Tetrodotoxin	*Vibrio* spp., *Aeromonas* spp.
B. cereus toxins Emetic Diarrhoeal	*Bacillus cereus*, other *Bacillus* spp?
Botulinum toxin	*Clostridium botulinum*, strains of *Cl. butyricum* and *Cl. baratii*, other *Clostridium* spp?
Staphylococcal enterotoxin	*Staphylococcus aureus*, other *Staphylococcus* spp?

[1] Enterotoxigenic strains producing labile toxin.
[2] Enteropathogenic and enterohaemorrhagic strains.
[3] Enterotoxigenic strains producing stable toxin.

Notes: A large number of other putative toxins are produced by the various enteric pathogens. In many cases, their role in pathogenicity is not known.

4. *Salmonella*

Introduction

The genus *Salmonella* is a typical member of the family *Enterobacteriaceae* and consists of Gram-negative, oxidase-negative, straight-sided, rod-shaped bacteria which are catalase-positive and have both a respiratory and a fermentative metabolism of carbohydrates. Most salmonellas are motile by peritrichous flagella.

Members of the genus are responsible for diseases of man and animals. The degree of host adaptation varies and affects the pathogenicity for man in three ways:

1 Serovars adapted to man, such as *S. typhi*, *S. paratyphi* A and *S. sendai*, usually cause grave diseases with septicaemic-typhoidic syndrome (enteric fever). These serovars are not usually pathogenic to animals.
2 Ubiquitous serovars such as *S. typhimurium*, which affect both man and a range of animals, cause gastrointestinal infections of varying severity (but usually less severe than enteric fever). In addition to 'classical' food poisoning, these serovars are involved in infantile and travellers' diarrhoea.
3 Serovars which are highly adapted to an animal host, such as *S. abortovis* (sheep) and *S. gallinarum* (poultry) usually produce no or very mild symptoms in man. However, *S. choleraesuis* which has the pig as primary host, also causes a severe systemic illness (NRC, 1969).

Salmonellas have been recognised as causes of enteric disease for many years, and methods of control are well established. Despite this, *Salmonella* remains the most important reported cause of food poisoning and recent years have seen both massive outbreaks and a major new vehicle of infection, hens eggs, emerge.

S. typhi and other human-adapted salmonellas are less commonly transmitted by food than ubiquitous serovars with waterborne and person-to-person transmission being more important. For this reason, the human-adapted serovars are often excluded from discussion of *Salmonella* infection. This appears to be an artificial restriction:

'. . . *there appears little justification to consider the human-adapted species and* S. choleraesuis . . . *differently from other* Salmonella' (NRC 1985).

Isolation and detection of *Salmonella*

Conventional cultural isolation techniques

In the case of clinical samples, isolation can usually be made by streaking directly onto a suitable selective medium or, in some cases, a relatively non-selective blood agar. With foods and other non-clinical materials, salmonellas are usually present only in small numbers and are greatly outnumbered by other bacteria. Some cells may be stressed by processing or environmental conditions. It is therefore customary to use a three-stage process – resuscitation (pre-enrichment), enrichment, selective plating.

Resuscitation

The need for resuscitation is now widely accepted for all types of food, and not merely those which have been dried or frozen.

Non-selective media such as buffered peptone water and lactose broth are most widely used for resuscitation.

Selective enrichment

Over the years a wide range of media has been developed for enrichment. Of these various formulations of selenite and tetrathionate broths have been widely used, although in recent years there has been increasing use of the malachite green-based Rappaport-Vassiliadis (RV) broth.

Selective plating

As with enrichment broths, a wide range of media has been devised for selective plating. Where competition from other bacteria is insignificant, a general-purpose medium for the *Enterobacteriaceae* such as one of the MacConkey agars may be used. In many cases, greater selectivity is required and it is necessary to use a medium devised specifically for *Salmonella*, such as Brilliant green agar, Salmonella-shigella agar, DCLS agar and Hektoen enteric agar.

Rapid cultural methods

The full cultural method for isolation of salmonellas is time consuming, taking up to 5 days to obtain a presumptive result. Attempts have been made to shorten incubation periods at various stages, but this carries an unacceptably high risk of false negative results.

The hydrophobic grid membrane filter technique has been used to reduce the time taken, and two rapid cultural techniques are available as commercial kits. The first of these, the Salmonella Rapid Test (Oxoid©) is based on conventional methodology, while the second is based on the immobilisation of salmonellas by flagellar antibodies (immunoimmobilisation).

Rapid detection methods

The importance of *Salmonella* means that, as with development of conventional media, a wide range of rapid detection methods is available for the organism. These include serological, genetic and electrical methods.

Methods for the isolation and detection of Salmonella *are discussed in greater detail with respect to choice of method and laboratory practice in* **Laboratory Methods**, *on pages 71–83.*

Taxonomy of *Salmonella*

Differentiation of *Salmonella* from other genera

Salmonella may be differentiated from closely-related genera by biochemical tests. Commonly used differentiating criteria are listed in **Table 32**. In common with other members of the *Enterobacteriaceae*, *Salmonella* may harbour 'foreign' replicons (temperate phages or plasmids), which may code for metabolic characters used in identification such as lactose and sucrose fermentation (Le Minor, 1984). Particular difficulties arise when two or more differential characters are affected simultaneously. Differentiation from other species may also be complicated by pleiotrophic mutations (Le Minor *et al.*, 1969) which, in the instance described, simultaneously affected nitrate, tetrathionate and thiosulphate reduction as well as hydrogenlyase activity.

The importance of *Salmonella* in medical microbiology means that there is a constant demand for faster and less onerous means of differentiating the organism. Multi-test media (see Chapter 3) have been in use for *Salmonella* since 1911, and while displaced to a large extent by miniaturised identification kits, are still in use, particularly in the food industry. Miniaturised kits have found their widest application with *Salmonella* and other members of the *Enterobacteriaceae* and a wide variety is available (see pages 38–39, 84–85). Considerable interest has also been shown in a fluorogenic means of identification, the MUCAP test (Humbert *et al.*, 1989). Although bio-

chemical test methods permit presumptive identification of *Salmonella* to a high level, the similarity of the genus to some other members of the *Enterobacteriaceae*, together with possible additional problems due to atypical strains, means that biochemical identification should be supported by more definitive criteria.

Serology is usually used to provide confirmation of biochemical identification. In the past, Spicer-Edwards polyvalent antiflagellar antiserum (anti-H) has been used in a slide agglutination test, but this has to some extent been superseded by latex co-agglutination tests based on detection of somatic antigens (Clark *et al.*, 1989). Some of these tests, which are available as commercial kits, allow a preliminary serogrouping and can also be used as part of a rapid detection system (see pages 82–83).

Although serological methods are most widely used to confirm identification of *Salmonella*, some workers have proposed the use of the 0-1 bacteriophage (Felix and Callow, 1943). More than 98% of strains isolated from routine clinical sources were susceptible to this phage (Bockemuhl, 1972), and susceptibility among other members of the *Enterobacteriaceae* is uncommon. Use of the 0-1 phage in this context is rare, due to difficulties in maintaining phage stocks and the lack of a clear advantage over serological methods.

Table 32. Differentiation of *Salmonella* from other members of the *Enterobacteriaceae*.

	Salmonella	*Shigella*	*Citrobacter*	*Edwardsiella*
β-galactosidase	−[2]	+/−	+	−
Arginine dihydrolase	+/−	−	+/−	−
Lysine decarboxylase	+	−	−	+
Ornithine decarboxylase	+	−[3]	+/−	+
Simmon's citrate	+	−	+	−
H₂S production	+	−	−/+	+
Acid from				
Lactose	−[2]	−[4]	+/−	−
Dulcitol	+/−	−	−/+	−
Melibiose	+	−/+	−	−
Sorbitol	+	−/+	+	−
Xylose	+	−	+	−
Motility	+	−	+	+

[1] Reactions for *Salmonella* are based on those of ubiquitous serovars of importance in food poisoning. Reactions of *S. typhi* and other host adapted serovars may differ.
[2] *S. arizonae* and strains of some other serovars are positive.
[3] *Sh. sonnei* is usually positive.
[4] Delayed fermentation is a feature of some strains of *Sh. sonnei*.

Differentiation within *Salmonella*

The usual rules of nomenclature are not followed in the case of *Salmonella*. If the principle is followed that bacteria which are related by 70% or more on the basis of DNA:DNA hybridisation experiments belong to the same genospecies (Le Minor, 1984) then the genus *Salmonella* must be considered to be a single species. This may be further divided into 'subgenera'. Originally, four subgenera were proposed by Kauffman (1966) but this number was increased to five by Le Minor (1984). Genetically, the level of the subgenera is that of subspecies but the earlier terminology is likely to remain. Most recently, Le Minor and Popoff (1987) have proposed that *S. enterica* be used as the species name instead of *S. choleraesuis* and that there should be seven subspecies, *enterica*, *salamae*, *arizonae*, *diarizonae*, *houtenae*, *bongori* and *indica*. Under this system *S. typhimurium* would be reported as *S. enterica* subspecies *enterica* serovar typhimurium.

Despite these changes in nomenclature, it seems likely that the species-like epithets currently applied to serovars commonly associated with food poisoning will continue to be used. However, it is, important that these species-like epithets given to serovars of *Salmonella* are not confused with the true species names given to the species within other genera or to the subspecies of *Salmonella*. The use of species-like epithets for serovars is a practice unique to *Salmonella*, and stems from the early recognition of the importance of the organism in pathology. Since 1966 serovars, other than those of 'subgenus', have been designated purely on the basis of antigenic formulae, a practice which has a number of advantages.

It will be appreciated that the situation regarding the taxonomy and nomenclature of *Salmonella* can hardly be described as satisfactory. The International Subcommittee on *Enterobacteriaceae* consider, however, that the diagnostic importance of the Kauffman-White serotyping scheme is of over-riding importance and that current practices should continue.

Methods used in differentiating Salmonella *from other genera and for determining subgenera are discussed in more detail with respect to choice of method and laboratory practice in* **Laboratory Methods**, *pages 83–85.*

Typing of *Salmonella*

Serotyping

Serotyping is a routine procedure for *Salmonella*, and typing sera for common serovars are commercially available in many countries. The Kauffman-White (Kauffman, 1966) typing scheme in which organisms are represented by the numbers and letters given to the different O (somatic), Vi (capsular) and H (flagellar) antigens is commonly used. However, this formula is not a complete record of the antigens present, but indicates only those of primary diagnostic value. The scheme now covers all subgenera (Le Minor, 1984).

In the case of uncommon serovars, no further subdivision may be necessary, but with those commonly encountered such as *S. typhimurium* and *S. enteritidis*, serotyping alone is not sufficiently discriminatory for epidemiological investigation. A means of subdivision is therefore required for common serovars.

Serotyping of Salmonella *is described in greater detail in* **Laboratory Methods**, *page 85.*

Phage typing

Phage typing has been applied to *Salmonella* for some years, typing of *S. typhi* and other salmonellas which possess the Vi antigen being based on a set of phages adapted from the Vi-II phage of Craigie and Yen (1938). Similar principles are applied to typing of all serovars, but each requires a specific set of typing phages.

Phage typing has been used successfully with many serovars in addition to *S. typhi*. These include *S. typhimurium* (Scholtens, 1962; Anderson, 1964), *S. enteritidis* (Ward *et al.*, 1987), *S. virchow* (Chambers *et al.*, 1987) and *S. bareilly* (Jayasheela *et al.*, 1987). It has not been possible to develop phage typing schemes which are sufficiently discriminatory for some serovars. For example, 81 different phagovars of *S. montevideo* have been demonstrated, but it was not possible to demonstrate any relationship between phagovar and source (Vieu *et al.*, 1981).

Acquisition of a drug resistance plasmid may change the phagovar of a salmonella. This phenomenon has previously been observed with *S. typhi* and *S. typhimurium*, and more recently with *S. enteritidis* (Frost *et al.*, 1989). In the latter example, acquisition of a resistance plasmid converted *S. enteritidis* PT 4 to PT 24. The apparent increase in PT 24 infections during 1989 should therefore be seen as a part of the larger increase due to PT 4.

Biotyping

Biotyping has been used successfully to subdivide serovars in a number of epidemiological investigations. A model for biotyping schemes is that developed by Duguid *et al.* (1975) for application to *S. typhimurium*. This scheme is highly discriminatory and recognises 184 full biovars. Modified versions have been developed which are suitable for other salmonellas such as *S. agona* (Barker *et al.*, 1982) and *S. montevideo* (Old *et al.*, 1985). An evaluation of the discriminatory ability of biotyping schemes for *S. typhimurium* and *S. paratyphi* B showed that while discrimination was high when used alone, the schemes were most effectively used in conjunction with phage typing (Old and Barker, 1989).

Bacteriocin typing

Bacteriocin typing of salmonellas has been discussed by Gillies (1978). An overlay technique is used (see page 42) and various additional tests may be used to characterise the bacteriocins (D'Aoust, 1989).

Bacteriocin typing has found only limited application, but on some occasions has been used to separate isolates of the same phagovar and biovar. On other occasions, bacteriocin typing has failed to produce any useful information.

Genetic and molecular typing methods

Genetic typing methods are becoming used increasingly in epidemiological investigations, although the necessary expertise currently exists in relatively few laboratories. Of the available methods (see pages 42–43), plasmid profiling has been most widely used. The method has been applied successfully in a number of epidemiological investigations including, in the UK, a major hospital outbreak of *S. typhimurium* food poisoning and an outbreak involving contamination of pâté with *S. gold-coast* (Threlfall *et al.*, 1986a,b). A notable investigation in the USA involved a nationwide outbreak of salmonellosis due to contamination of marijuana with *S. muenchen* where phage and other conventional typing methods were unsuitable (Taylor *et al.*, 1982).

With some serovars plasmid profiling is most

effectively used as an adjunct to phage typing, rather than for primary profiling (Threlfall et al., 1989) and the technique has been succesfully applied to the subdivision of phage-type 49 of *S. typhimurium*, the most common phage type in the UK (Threlfall et al., 1990).

Resistance plasmid analysis can also be of value and has been used to differentiate the epidemic strain in an outbreak involving *S. newport* (Holmberg et al., 1984).

Symptoms of *Salmonella* infections

Enteric fever

Symptoms of enteric fever due to *S. typhi* usually appear within 10 to 14 days of infection, although the incubation period may be up to 5 weeks or less than 1 week depending on the dose. Initial symptoms include anorexia, headache, a non-productive cough and abdominal pain. These symptoms are accompanied by an increasing remitted fever which reaches as high as 40°C. Constipation is more common than diarrhoea in the early stages of the disease.

In most cases, enteric fever increases in severity during the second week. A rash of 'rose spots' is a useful diagnostic feature, and the patient is likely to have bradycardia and splenomegaly. Polymorphonuclear leucopenea is common. In untreated patients, the fever declines during the third and fourth week, although a mortality of 10% may be expected.

Although complete recovery may be anticipated, 10 to 20% of patients suffer a relapse within the first few months of the original illness. This phenomenon may be strain dependent, and may also be linked to the presence of a 24 or 38 kilobase plasmid (Gotuzzo et al., 1987).

Enteric fever due to other human-adapted serovars such as *S. paratyphi* A is less severe. The onset of the disease is more rapid and early onset of diarrhoea is more common.

Salmonella enteritis

The symptoms of *Salmonella* enteritis (salmonellosis) usually develop within 12 to 36 hours of infection. An extremely short incubation period of 3 hours was noted in an outbreak of *S. enteritidis PT 4* infection (Stevens et al., 1989) while at the other extreme, periods of up to 72 hours are not uncommon. Extended periods of 1 to 12 days have been reported for *S. newport*, 1 to 8 days for *S. heidelberg* and 3 to 7 days for *S. typhimurium*. Such extended periods lead to obvious difficulties in epidemiological investigation. Typical symptoms include abdominal pain, diarrhoea, vomiting and fever (Pinegar and Suffield, 1982). The severity of the symptoms can vary considerably, particularly in newborn babies and infants, from a grave typhoidlike illness to transitory diarrhoea or asymptomatic infection. Mild relapses may occur in the months following recovery, particularly if the initial symptoms were severe.

Although bloody diarrhoea is not usually associated with salmonellosis, 42% of affected people reported blood in stools as a symptom of *S. typhimurium* DT 124 infection, associated with salami sticks (Cowden et al., 1989).

Unusual prodromal symptoms have occasionally been reported. In an outbreak caused by *S. enteritidis* PT 4, diarrhoea was preceded in the previous 24 to 48 hours by severe headache, myalgia, fever, neck stiffness and photophobia (Drs Y. K. Lau and A. C. Maddocks, personal communication).

A possible association, based on anecdotal evidence, has also been suggested between *Salmonella* infection and ulcerative colitis (Taylor-Robinson et al., 1989).

Extraintestinal infection

The major symptoms of infection with *S. typhi* and other human-adapted serovars are largely extraintestinal. It may be said that there is no tissue that has not at some time been affected by such infections.

Extraintestinal infections with non-human adapted salmonellas are well chronicled (Cohen et al., 1987; Wilkins and Roberts, 1988). In addition to typhoidlike and septicaemic illnesses, these include other acute and chronic diseases. Some serovars of nonhuman adapted salmonellas have a greater potential for invasiveness, serovars identified in this way are *S. choleraesuis*, *S. dublin*, *S. virchow*, *S. panama* and *S. london* (Wilkins and Roberts, 1988).

Extraintestinal disease caused by serovars of *Salmonella* is summarised in **Table 33**.

Table 33. Summary of extraintestinal infection due to *Salmonella*.

Septicaemia	
Haemolytic uraemic syndrome	
Erythema nodosum	
Focal infections	Meningitis
	Osteomyelitis
	Septic arthritis
	Pneumonia
	Cholecystitis
	Peritonitis
	Pyelonephritis
	Cystitis
	Abscesses in various parts of the body
	Endocarditis
	Pericarditis
	Vasculitis

Persons at particular risk from *Salmonella* infection

Susceptibility

Within the community, susceptibility to *Salmonella* infections varies. Individual susceptibility varies according to a number of factors and is partly genetically controlled. High susceptibility is common in the very young and old, for whom symptoms are often more severe, and those with underlying illnesses including malignancies. *Salmonella* infections may be associated with Reiter's syndrome, which has a high predilection for young males. *Salmonella* should also be considered among the pathogens associated with human acquired immunodeficiency virus (AIDS; Celum *et al.*, 1987), and a recurrent *Salmonella* bacteraemia may be the first indication of the syndrome (Sperber and Schleupner, 1987). Food counselling has been advised for AIDS sufferers to enable them to avoid foods which are of high risk for *Salmonella* (Griffin and Tauxe, 1988).

In some cases, susceptibility may be increased by host factors directly enhancing virulence. For example, iron overload may increase the virulence of *S. typhimurium* (Madden *et al.*, 1986).

An association between antibiotic therapy and increased susceptibility to infection by antibiotic resistant strains of *Salmonella* has been recognised for some years. In a massive milk-associated outbreak in Illinois due to an antibiotic resistant strain of *S. typhimurium*, 16% of victims had received antibiotics in the month before infection, and it is estimated that the use of antibiotics increased the risk of symptomatic infection more than fivefold (Ryan *et al.*, 1987). It was postulated that the antimicrobials increased host susceptibility by depleting the normal intestinal microflora and creating favourable conditions for the resistant salmonellas. However a similar increase in susceptibility following antibiotic therapy also has been described in infection with sensitive strains of *S. havana* (Pavia *et al.*, 1990).

A protein deficient diet increases the susceptibility to salmonellas as well as to other enteric pathogens. Work with rats has shown that while the febrile response is normal in cases of protein deficiency, there is evidence of increased invasion of the liver and spleen (Bradley and Kauffman, 1988).

Exposure

People may be at greater risk from *Salmonella* because of greater than average exposure rather than individual susceptibility. People working with animals, handlers of unprocessed meat and poultry, and some other workers such as sewer workers, face a higher risk of exposure particularly if personal hygiene is poor or preventive measures inadequate.

The risk of acquiring *Salmonella* infections is obviously greater where hygiene standards are low and the climate is warm. These conditions apply in many areas popular with tourists, and visitors from countries where hygienic standards are high are at particular risk. This has been identified as an increasing problem with *S. typhi* infections in the USA (Ryan *et al.*, 1989).

Localised conditions can also lead to increased risk of exposure. Hygiene standards may be difficult to maintain, for example, in prisons, asylums and refugee camps where problems are exacerbated by overcrowding and a lack of basic sanitation. Schools can, to a lesser extent, be a problem, especially where young children are involved, while hospitals are always at particular risk.

Dietary habits may put individuals at increased risk of exposure to *Salmonella*, and people who consciously consume raw meat, milk or eggs are at higher risk than the general population. Apart from common sense precautions, no virtue is seen in undertaking large-scale dietary modification in attempts to avoid infection by *Salmonella*. The risk of dietary inbalance resulting from such attempts is likely to be greater than the risk of salmonellosis or any other form of food poisoning.

Treatment of *Salmonella* infections

Enteric fever

Enteric fever, and other severe systemic infections due to human-adapted serovars of *Salmonella*, are usually treated by oral antibiotic therapy. Typical regimes for treatment of *S. typhi* infections include amoxycillin, cotrimoxazole and chloramphenicol. More recently, ceftriaxone has been suggested as a superior alternative to chloramphenicol (Islam *et al.*, 1988). Treatment with amoxycillin and cotrimoxazole is usually continued for 14 to 21 days, with chloramphenicol for 10 days and with ceftriaxone for 7 days. Continued excretion of *S. typhi* usually requires the treatment to be repeated.

Intravenous ampicillin at a high dose rate has been shown to be an effective and economic therapy for treatment of typhoid fever in developing countries (Finkelstein and Markel, 1990).

Usually, antibiotic therapy is not considered effective in the treatment of chronic carriers, although success has been claimed for ampicillin or amoxycillin continued for 90 days. In many cases, the only effective treatment of the chronic carrier is the rather drastic step of surgical excision of the gallbladder or other infected organs. Less drastically, there have been reports that acidophilus milk shortens the duration of carriage.

Antibiotic resistance is common among salmonellas, and multiple resistance is reported with increasing frequency – a strain of *S. typhi* resistant to chloramphenicol, trimethoprim, ampicillin and sulphonamides, has been isolated, for example, (Anon, 1987). In theory, choice of antibiotic should be guided by the outcome of sensitivity tests, but in practice it is necessary to commence treatment before the results of sensitivity tests are available. In this context, Rowe *et al.* (1987a) considered that chloramphenicol remains the drug of choice for 'front line' treatment of typhoid in the UK, despite a growing number of reports of chloramphenicol resistance elsewhere.

Salmonella enteritis

Treatment with antibiotics is not desirable in enteritis since it is ineffective, it increases the likelihood of a carrier state developing, and may promulgate the further spread of antibiotic resistance. The situation is aptly summarised by Duguid *et al.* (1978):

'Thus the practitioner, although he may have a feeling of futility in not being able to offer a specific treatment, should refrain from exhibiting what are in cases of salmonella food poisoning useless and potentially harmful drugs.'

However, antibiotic therapy is advocated in *Salmonella* enteritis where there is associated septicaemia. Clinical determination of septicaemia is difficult in the early stages of the illness when antibiotic treatment is most successful. In elderly patients, delay may lead to severe dehydration and often fatal illness. Early use of antibiotic therapy may, therefore, be advisable in elderly patients with severe diarrhoea (Mandal and Brennand, 1988). Chloramphenicol, ampicillin and cotrimoxazole are most commonly used.

In all types of *Salmonella* infection, rehydration and restoration of the electrolyte balance may be required if diarrhoea is particularly severe.

Mechanisms of pathogenicity

19. Invasion by *S. typhimurium*. a. Individual organisms, visualised by immunofluoresence microscopy invading the lateral region of a villus, 2h after inoculation.
b. A similar preparation, taken 2h after inoculation and examined by transmission electron microscopy (TEM). Note only local disturbance of the brush border (arrowed **a**) and invasion of enterocytes. Organisms were not seen penetrating through tight-cell junctions (arrowed **b**).
c. A fluorescent-antibody-stained section of *S. typhimurium* infected ileal loop, taken 12h after inoculation. Note the villus-tip enterocyte, laden with bacteria, rounded up and extruded. The mucosa (arrowed) appears to become rapidly resealed.
d. A similar preparation taken 18h after incubation. The number of bacteria in the bases of the villi has increased considerably. (Reproduced with permission from Wallis, T.S., Starkey, W.G., Stephen, J. *et al.*, *Journal of Medical Microbiology* **22:** 39–49. © 1986 The Pathological Society of Great Britain and Ireland.)

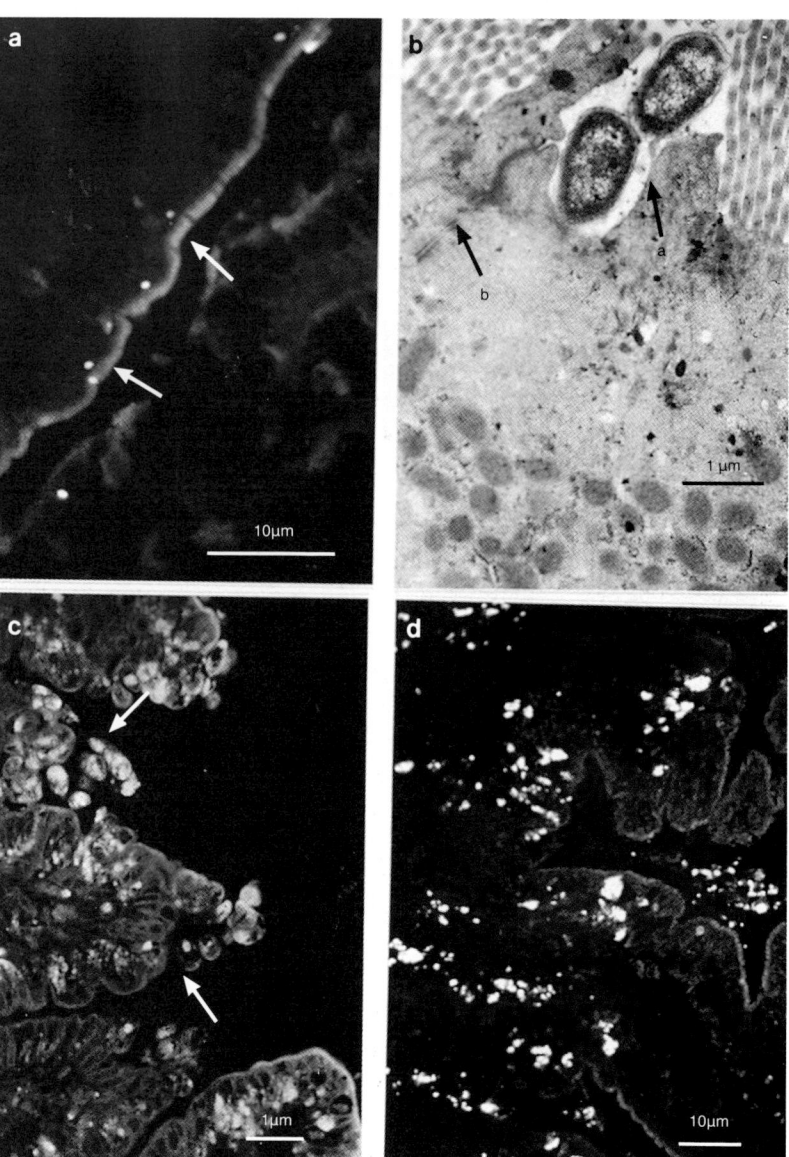

Infective dose

An infective dose in healthy people varies according to serovar as well as the type of food involved. Human adapted serovars, such as *S. typhi*, have been considered for many years to have lower infective doses than non-adapted serovars such as *S. typhimurium*.

In the case of *S. typhi*, infection at levels of less than 10^3 organisms ingested has occurred in a high proportion of outbreaks (Blaser and Newman, 1982), confirming the generally held view of the organism's infectivity. In the case of non-human adapted serovars, there is evidence that volunteer feeding experiments used to determine infective doses have given a false impression of the numbers needed to initiate infection. Such studies have consistently indicated that a dose of 10^5 to 10^7 organisms are required to initiate infection, whereas in actual outbreaks the numbers

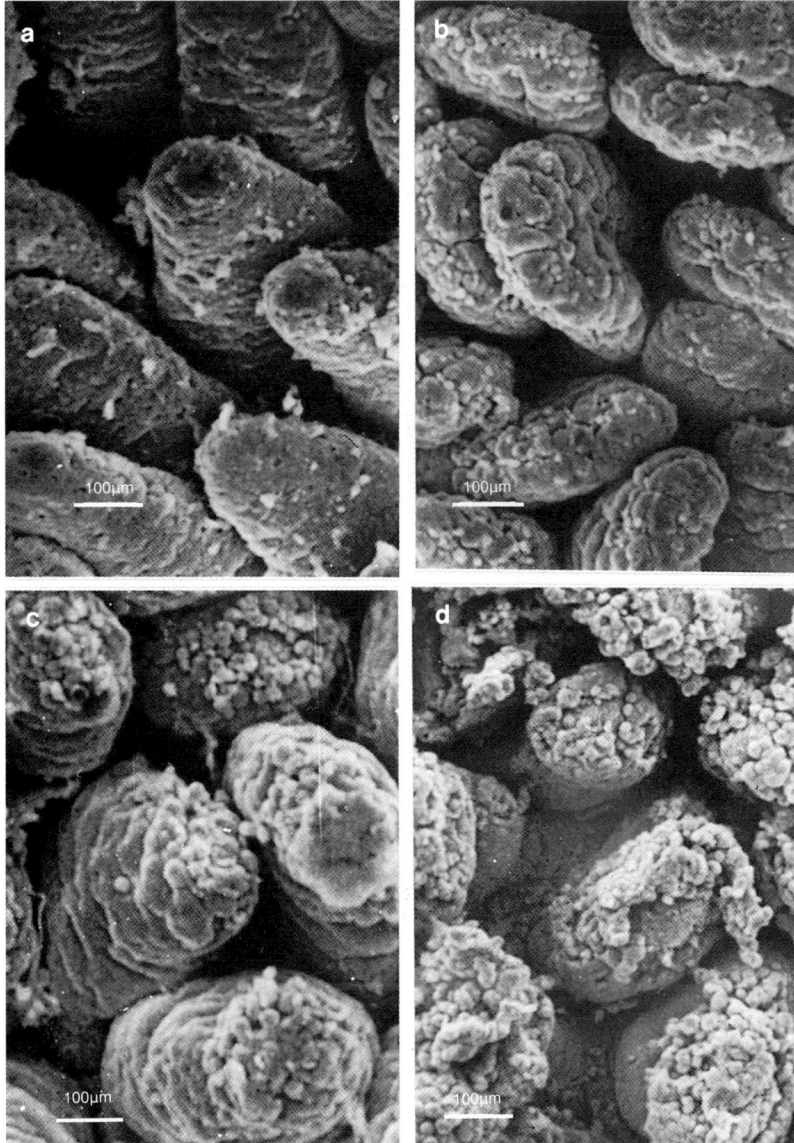

20. Scanning electron microscope (SEM) analysis of *S. typhimurium*-induced changes in villus architecture. Biopsies were taken from separate rabbits 4, 8, 12, and 16 h after inoculation of rabbit ileal loops, processed and analysed by SEM. The appearance of the villi range through normal (**a**), through surface wrinkling (**b**), to progressive shedding of cell tips (**c** and **d**). It is useful to compare the appearance of **20c** viewed by SEM with that of **19c** viewed by TEM.

It should be particularly noted that the extent and timing of the damage illustrated in this figure varied, and did not occur when rabbits of a different source and breed were used. It follows therefore that this kind of gross architectural damage is not a prerequisite for the initiation of salmonella-induced fluid secretion.

20 *and* **21** *comprise representative regions of mucosa selected from a larger field at 400× magnification and containing around 50 villi.* (Reproduced with permission from Wallis, T.S., Starkey, W.G., Stephen, J. *et al.*, *Journal of Medical Microbiology* **22:** 39–49. © 1986 The Pathological Society of Great Britain and Ireland.)

ingested have been very much lower. Retrospective analysis has shown infective doses of less than 10^3 to have caused illness when ingested in a number of foods (**Table 34**) and water.

It has been suggested that the infective dose is lower in foods of high fat or protein content, due to protection of cells from gastric acidity (Fontaine, *et al.*, 1980; Blaser and Newman, 1982). This may be true in the majority of cases, but an obvious exception is a waterborne outbreak where the infective dose of *S. typhimurium* was calculated to be 1.7×10^1 cells (Boring *et al.*, 1971).

Table 34. Examples of *Salmonella* infections where the number of cells ingested was less than a thousand (Blaser and Newman, 1982).

Serovar	Estimated dose	Vehicle
S. typhimurium	1.7×10^1	Water
S. newport	$6–23 \times 10^1$	Hamburger
S. eastbourne	$1–2.5 \times 10^1$	Chocolate
S. heidelberg	$1–5 \times 10^2$	Cheese

Note: Estimated dose in infections with *S. typhi* is frequently less than 10^3.

21. SEM analysis of *S. typhimurium*-induced changes in villus architecture. Biopsies were taken from the same rabbits at 18 h: **a** after inoculation of rabbit ileal loop with *S. typhimurium*, and **b** from a control loop that received sterile medium. Note that there is evidence of recovery of villi, which, although shortened, show less tip derangement and a return to a more normal surface (arrowed). Ligation for 18 h *per se* does not induce changes in villus surface structure and the picture shown was virtually indistinguishable from control biopsies taken at 0 h. (Reproduced with permission from Wallis, T.S., Starkey, W.G., Stephen, J. *et al.*, *Journal of Medical Microbiology* **22:** 39–49. © 1986 The Pathological Society of Great Britain and Ireland.)

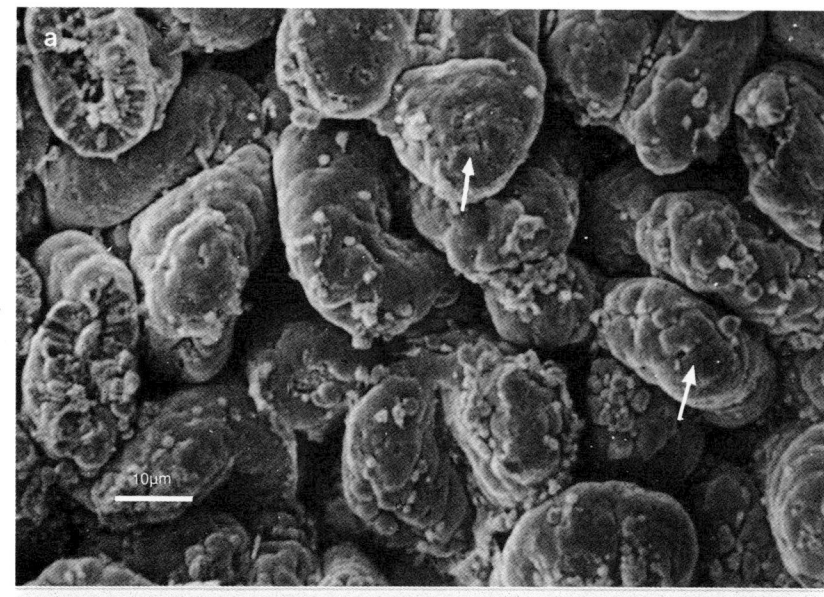

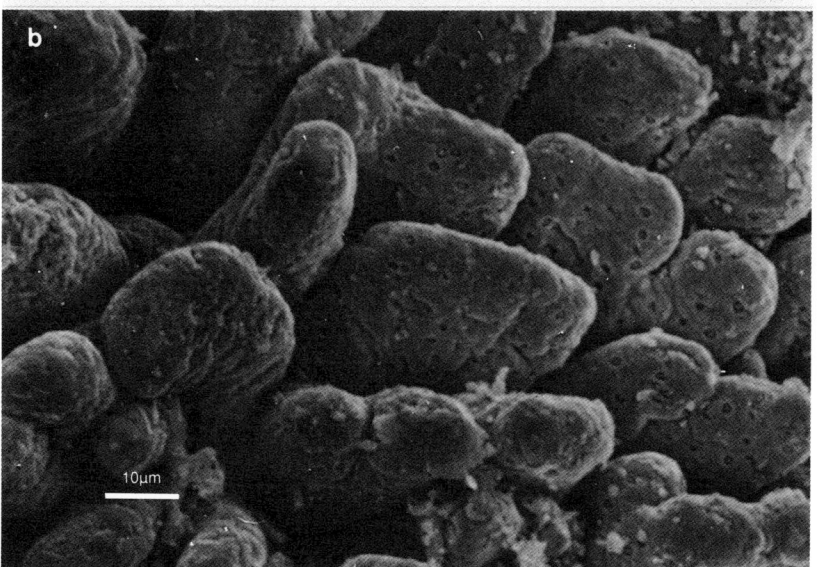

Genetic control of virulence

Although the genetic control of many aspects of *Salmonella* virulence has not been elucidated, both plasmid borne and chromosomal genes appear to be involved.

Salmonellas, with the exception of *S. typhi*, carry a 50 to 60 Mda plasmid which has been associated with virulence in a number of serovars. The primary virulence plasmid phenotype is in the ability to spread beyond the initial site of infection (Gulig, 1990), and plasmid-linked genes have been associated with systemic infection by many serovars, including *S. dublin*, *S. enteritidis*, *S. choleraesuis* and *S. paratyphi* C (Barrow *et al.*, 1989).

Chromosomal genes are also important in determining virulence, and appear to play a major role in determining the ability to survive and multiply in cells of the reticulo-endothelial system (Barrow *et al.*, 1989). The *phoP* gene has been identified as being necessary for virulence and survival, and may be a regulatory sequence necessary for the expression of a number of virulence factors (Fields *et al.*, 1989).

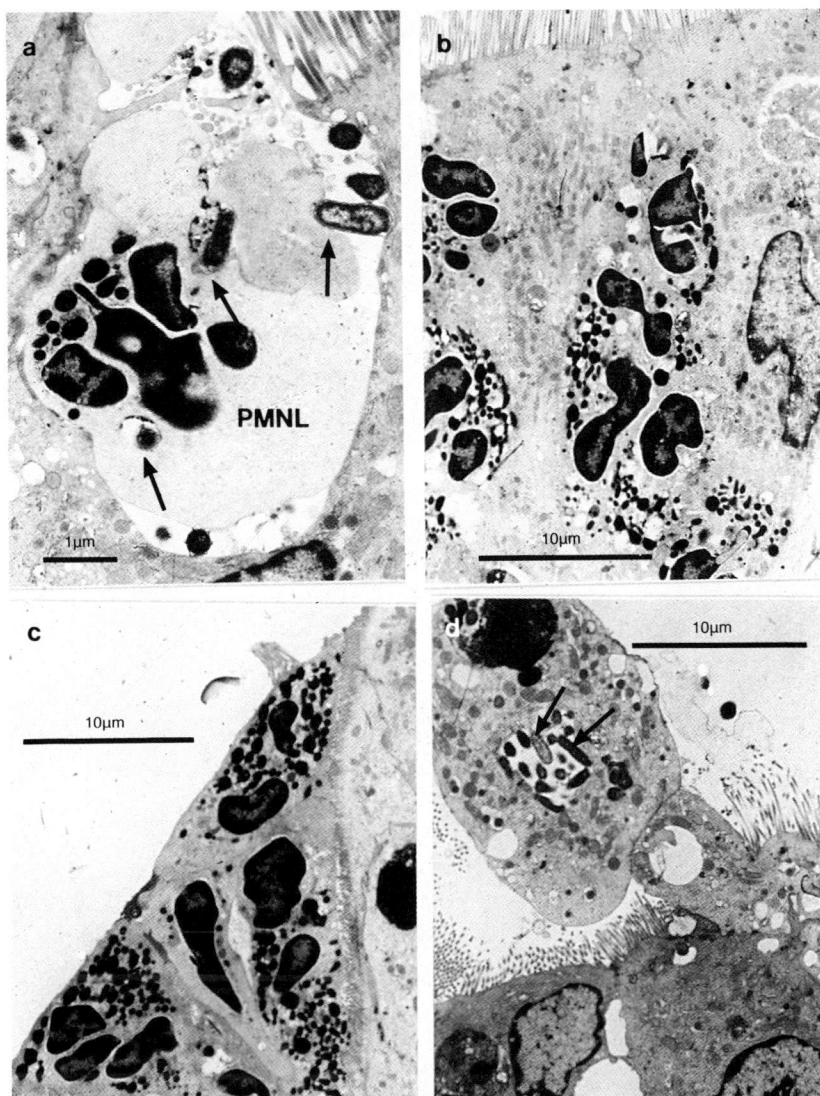

22

22. TEM analysis of tissues infected with S. typhimurium.
Biopsies were taken at 6, 8, 8 and 14h (**a**, **b**, **c** and **d** respectively) after inoculation of rabbit ileal loops from three different animals.
a. Polymorphonuclear leukocyte (PMNL) interacting with bacteria (arrowed) at the surface of the enterocyte layer.
b. Massive filtration of PMNLs into the mucosa between adjacent enterocytes.
c. PMNLs apparently being extruded into the lumen.
d. Extrusion of rounded-up, bacteria-containing enterocyte (arrowed; *cf.* **19c**, **20c**, **20d**). The cell–cell junctions appear to remain intact until the final expulsion of the infected enterocyte.

Note that the massive influx of PMNLs always occurred and was *not* dependent on a prior induction of damage illustrated in **20**.
(Reproduced with permission from Wallis, T.S., Starkey, W.G., Stephen, J. *et al.*, *Journal of Medical Microbiology* **22**: 39–49. © 1986 The Pathological Society of Great Britain and Ireland.)

Motility and chemotaxis

The role of motility in virulence is not well defined. In studies with an avirulent strain of *S. typhimurium*, it was demonstrated that *in vitro* motility and chemotaxis play no major role in intestinal colonisation, and that growth in mucus inhibits motility by this serovar (McCormick *et al.*, 1988; Lockman and Curtis, 1990). However, in a separate study with *S. typhi* it was concluded that intact motility is an invasion-related factor (Liu *et al.*, 1988); while Worton *et al.* (1989), using a specially built organ culture apparatus, were able to

show that motility almost certainly has a role in association of *S. typhimurium* with the mucus layer. These findings support earlier work by Jones *et al.* (1981), who found that in invasion of HeLa cells by *S. typhimurium*, motility was an important factor in maintaining close contact between the bacterium and the HeLa cells in the stage before adherence. Chemotaxis was also of importance as the HeLa cells released a diffusible attractant which increased the collision frequency and hence the attachment of chemotactic *S. typhimurium*

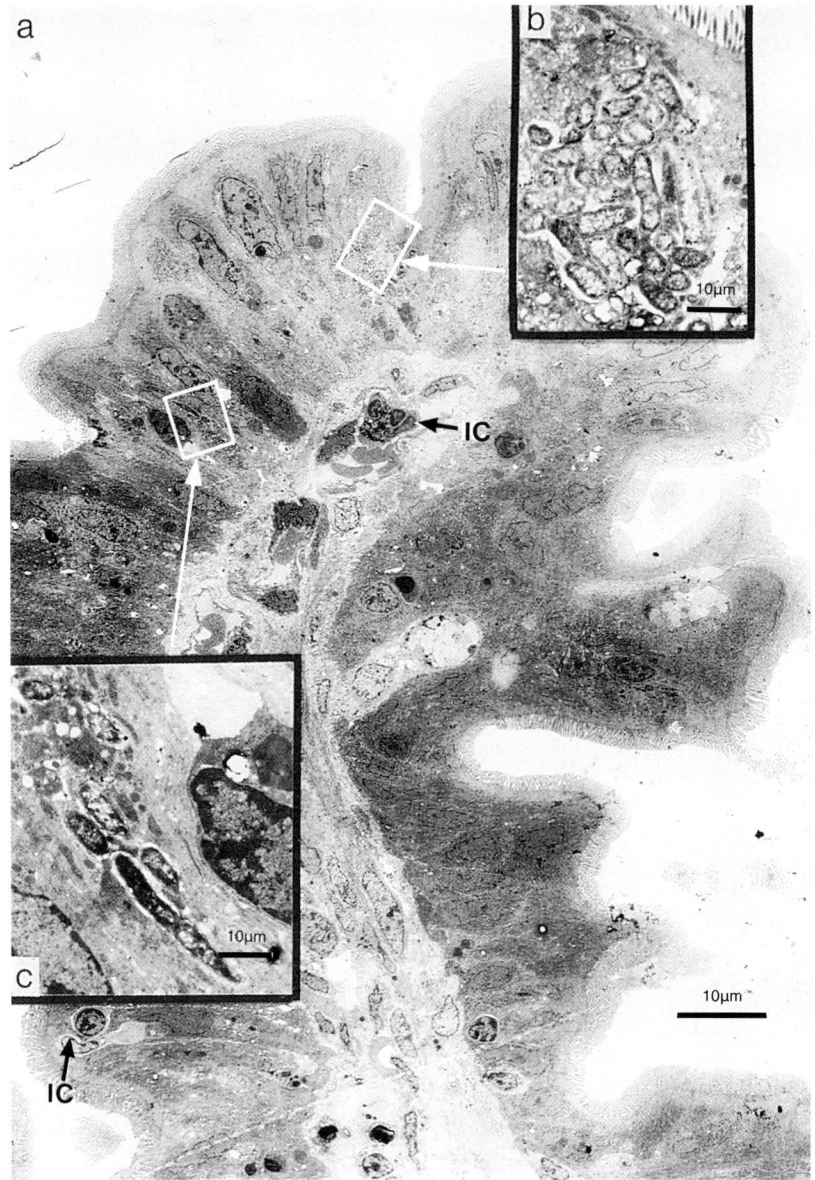

23. TEM analysis of *S. typhimurium*-infected villus.
Biopsy was taken 8 h after inoculation of a rabbit ileal loop. Note the highly invaginated surface of the enterocyte layer (**a**; *cf*. **20b**) and large numbers of invading (**b**) and dividing (**c**) bacteria. There is also evidence for infiltrating cells of the immune system (IC; arrowed). (Reproduced with permission from Wallis, T.S., Starkey, W.G., Stephen, J. *et al.*, *Journal of Medical Microbiology* **22:** 39–49. © 1986 The Pathological Society of Great Britain and Ireland.)

Adherence and penetration

Salmonella adheres to the epithelial cells in the ileum by means of mannose-resistant fimbriae. It is necessary, however, to distinguish between adherence of *Salmonella* and the 'anchor-like' attachment of *Vibrio cholerae* and enterotoxigenic *E. coli*. With *S. typhimurium*, adherence should be viewed as a first step in a biological continuum which leads to penetration of enterocytes and ultimately to invasion of deeper tissues.

Penetration of the epithelial cells is by receptor-mediated endocytosis (RME), although salmonellas may also penetrate the epithelium at the boundary between adjacent cells. The bacteria enter individually, the vacuoles containing them later appearing to coalesce. The nature of the components which induce RME are not currently known (Williams *et al.*, 1988); anaerobiosis, however, has been shown to increase internalisation and may be a controlling factor (Ernst *et al.*, 1990).

Although the microvilli are disrupted by penetration of the bacteria, they reform and there is virtually no cell leakage resulting from RME. Entry of *Salmonella* into the epithelium and its consequences are illustrated in **Figures 19–23**.

Intracellular multiplication was previously thought to be limited in the epithelial cells, and invasion of these cells was seen as being merely an initial, or even incidental, stage (Small *et al.*, 1987). Subsequently, however, rapid multiplication in the membrane-bound vacuoles has been shown to follow a lag phase, a generation time of 40 to 50 minutes being observed during optimum growth (Finlay and Falkow, 1989).

At the *lamina propria*, salmonellas are extruded and elicit an inflammatory response. This varies according to serovar, *S. typhi* attracts macrophages, whereas serovars causing enteritis stimulate a response from neutrophils. Salmonellas also induce production of interferon-γ (IFN-γ), which enhances killing by macrophages during early stages of infection (Muotiala and Makela, 1990). IFN-γ is produced by fibroblasts as well as by lymphoid and myeloid cells (Hess *et al.*, 1989).

It has been postulated that *S. typhi* may 'hide away' in cells such as the epithelium or impaired mononuclear phagocytes, where they are protected against host defence mechanisms and injected drugs (Smith, 1984a). This may be of importance in the carrier state of typhoid (Smith, 1984b).

Systemic spread

Food poisoning serovars of *Salmonella* usually progress no further than the submucosa, and thus systemic infection does not occur although the possibility of a transitory bacteraemia cannot be entirely discounted. Systemic spread by non-typhoid salmonellas may occur depending on host factors and serovar (see page 55).

With *S. typhi* and other host-adapted serovars, systemic spread is the norm and involves two possible routes – invasion of and multiplication in the lymphatic system, from which the organism spreads to all parts of the body, and translocation in motile phagocytes.

Survival and proliferation in host tissue

Resistance to phagocytosis

As noted above, *Salmonella* has for several years been considered to be a generalised intracellular parasite able to survive and multiply within many phagocytes. The situation is complicated by the finding that phagocytic cells in the peritoneal cavity rapidly kill *Salmonella*, whereas those in the spleen and liver do not (Saxen, 1984), and host factors may also be important in determining survival (Vladoianu *et al.*, 1990). In a more radical appraisal, Lin *et al.* (1989) have ques-

tioned the role of salmonellas as intracellular parasites, finding phagocytes to be active against *Salmonella* within 5 hours of infection and to remain active over the course of the disease. Multiplication of *Salmonella* was predominantly extracellular.

Cell wall components are of major importance in resistance to both oxidative and nonoxidative phagocytic killing, the latter being of greatest importance. Lipopolysaccharide (LPS) has a major role in resistance to nonoxidative killing, the resistance of *S. typhimurium* being directly proportional to its length and complexity (Stinavage *et al.*, 1989). Loss of LPS during conversion of *S. enteritidis* phage type 4 to phage type 7 is also correlated with loss of virulence (Chart *et al.*, 1989).

Outer membrane proteins (OMPs) are involved in resistance to oxidative killing, Stinavage *et al.* (1990) describing a 59 Kda OMP, production of which is mediated by the *phoP* gene. Porins, a major type of OMP, have been shown to interact with the macrophage membrane, decreasing the oxidative burst and hydrophobicity (Tufano *et al.*, 1989). The underlying mechanism appears to be activation of the adenylate cyclase system.

Effect of iron deprivation

Salmonellas require iron for growth and produce siderophores. In common with most members of the *Enterobacteriaceae*, the cathecol siderophore, enterochelin, is produced. The role of enterochelin as a virulence factor is in doubt, as it is produced by a wide range of both pathogenic and non-pathogenic bacteria (Williams *et al.*, 1988). A second siderophore, aerobactin, is produced only by pathogenic strains of *Salmonella* (Colonna *et al.*, 1985) and other members of the *Enterobacteriaceae*, and is considered to be a virulence determinant. More recently, Fernandez-Beros *et al.* (1989) found that iron deprived *S. typhi* produced enterochelin, but not aerobactin, as well as 'new' high-molecular weight protein compounds which were immunogenic in infected people.

Toxin production

It has been postulated that in salmonellas, invasive ability is essential for the development of diarrhoea and other enteric symptoms, since only invasive strains cause enteritis and fluid secretion. However, invasion cannot be solely responsible for the observed symptoms, since some enteroinvasive strains do not cause fluid secretion (Glatz, 1986). Another factor, or factors, must therefore be involved and it has now been established that in addition to an endotoxin, *Salmonella* produces at least three enterotoxins and a cytotoxin.

Endotoxin

The toxicity of *Salmonella* endotoxin is derived from the lipid portion of the lipopolysaccharide molecule, lipid A. When endotoxin is released from the bacterial cell wall as in enteric fever '. . . *profound and often lethal effects are produced*' (Stephen and Pietrowski, 1981). The role of endotoxin may, however, be rather more subtle and indeed devious, in that it acts as an antibody stimulator, and directs the host response towards antibody production rather than to the more effective cell mediated immunity (Mims, 1987).

Enterotoxins

The first of three known enterotoxins was originally demonstrated in *S. typhimurium* and *S. enteritidis* by Copal and Deibel (1975). This toxin, which produces fluid loss in infant mice, is a protein associated with the cell wall or outer cell membrane, but which is distinct from the endotoxin.

The other two enterotoxins are both detected by an increase in vascular permeability of rabbit skins (Sandefur and Peterson, 1976). The first is heat stable and rapid in action, producing a response in not more than 1 hour after injection. The second is heat labile, and has a delayed response of *ca* 18 hours. This toxin is commonly referred to as the delayed permeability factor and in several, but not all studies, a relationship between this toxin and cholera toxin or *E. coli* labile toxin has been demonstrated (Scotland, 1988).

This toxin further resembles cholera toxin in its mode of action and interferes with the control of adenyl cyclase activity, thus causing over-production of cyclic adenosine monophosphate (cAMP) Caprioli *et al.*, 1981; Jiwa, 1984). However, in contrast to cholera toxin, *Salmonella* requires an influx of neutrophils, and there is a correlation between virulence of *S. typhimurium* and the ability to evoke such an influx. It has been postulated that *S. typhimurium* of the correct virulent genotype induce an influx of leukocytes, mainly neutrophils, during invasion of the intestinal mucosa. The neutrophils then react with either luminal or invading bacteria (or both) which, if of the correct phenotype, release a processed cholera 'toxin-like' toxin (Wallis *et al.*, 1990).

Cytotoxin

Salmonellas produce at least one cytotoxin. This acts by inhibition of protein synthesis (Koo and Peterson, 1982) and may be at least partly responsible for damage to the intestinal mucosa (Koo *et al.*, 1984). In some cases, salmonella cytotoxin is closely related to Shiga toxin, but this is not an invariable property (Scotland, 1988).

Detection of Salmonella *toxins*

Tests for the heat labile enterotoxin (delayed permeability factor) are based on antisera against cholera toxin or *E. coli* labile toxin. An ELISA method was described by Hariharan *et al.* (1986), and an immuno-dot-blot assay and a latex co-agglutination test by Panigrahi *et al.* (1987). These tests are not currently suitable for general application.

Human carriage of *Salmonella*

Carriage of salmonellas following infection with human adapted serovars is a common phenomenon, and asymptomatic excretion may continue for weeks, months or years, an extreme example being of typhoid carriage for 52 years. Excretion of salmonellas by carriers is usually in faeces but urinary excretion also occurs. An outbreak of typhoid in Aberystwyth, Wales, due to ice cream, was attributed to contamination by the manufacturer who was an active urinary excretor, and who had suffered typhoid 9 years earlier, but been declared free of *S. typhi* after recovery.

Long-term carriage of non-human adapted serovars of *Salmonella* is rare, although not unknown. A survey based on 2,814 patients showed that persistent excretion for more than 1 year occurred in less than 1% of subjects (Buchwald and Blaser, 1984). Convalescent carriage is common, with a median duration of excretion of 5 weeks. It seems likely that most 'carriers' among food handlers are actually convalescent, or actively suffering the disease (Roberts, 1982).

Excretion after infection is more prolonged in children aged less than 5 years, people with symptomatic infections, and those infected with serovars other than *S. typhimurium*. However, long-term, chronic carriers are more frequently female and aged over 60 years.

Precautions against *Salmonella* contamination of food by carrier handlers is part of the larger question of medical examinations and the health of handlers, and is discussed in this context on pages 408–413.

Salmonella in the environment

Salmonellas inhabit the intestinal tracts of a wide range of animals, birds and reptiles. They may be isolated from sewage, and fresh water and estuarine water contaminated with sewage or with faeces. Survival in water depends on many factors, but may be relatively prolonged in both estuarine waters (Rhodes and Kator, 1988) and in tropical freshwater (Jiminez, 1989). In common with many other bacteria, *Salmonella* may enter a viable, nonculturable phase in water and require special techniques for detection (Desmonts *et al.*, 1990). In many situations, *E. coli* is likely to die faster than *Salmonella* and may have no validity as an index organism.

Salmonella may also be isolated from land irrigated with contaminated water or fertilised with slurry. However, there is little evidence of animal infection resulting from this practice (Ayanwale *et al.*, 1980).

Salmonella may become established in the food processing environment, and persist for significant periods. For example, the source of infection in an outbreak of *S. ealing* associated with infant dried milk was the insulating material surrounding the chamber of the spray drier.

Foods commonly involved in transmission of *Salmonella*

Meat and meat products

Raw meat and poultry is commonly contaminated with *Salmonella*. Numerous surveys have been made on the incidence in different species, but it is considered that no useful purpose is served by discussing these in detail. In summary, the situation is that the incidence varies according to species being highest among poultry, and also varies according to agricultural practice (see pages 427–430), and hygienic standards during slaughter and subsequent handling (see pages 430–433). In extreme cases, contamination of intensively-reared poultry may approach 100%.

Infections with salmonellas are rarely acquired directly from raw meats, although occasional cases result from conscious consumption of dishes such as steak tartare or, less commonly, raw sausage. In most outbreaks, the meat is either undercooked or has been contaminated after cooking. It should be noted that the thawing of frozen poultry with the production of large quantities of drip is a particular hazard. Cooked meats may be contaminated from sources other than raw meats including food handlers and ingredients added after cooking – gelatin added to pork pies and used to glaze some pâtés is notorious.

The continuing problems with cooked meats are illustrated by four major outbreaks which occurred during the summer of 1989 in the UK (**Table 35**; Communicable Disease Surveillance Centre, 1989). In each case, the underlying cause was the neglect of basic principles of hygiene.

Fermented meats such as salami are made from raw meat, and salmonellas are known to survive the fermentation process (NRC, 1985). An outbreak involving a salami snack stick occurred in the UK during 1988; the causative serovar was *S. typhimurium* DT 124, and 71 cases were confirmed, of whom 55 were children under 16 years of age (Cowden *et al.*, 1989). Earlier outbreaks due to salami had occurred in Australia (*S. newport*; Taplin, 1982) and in Italy (*S. choleraesuis*), but *Salmonella* is usually secondary to *Staphylococcus aureus* as a hazard in fermented meats.

Table 35. *Salmonella* **food poisoning and cooked meat products: UK, Summer 1989 (Communicable Disease Surveillance Centre, 1989).**

Product	Serovar	Number affected
Various cold meats	*S. typhimurium* DT 12	538
Roast turkey	*S. kedougou*	61
Roast pork	*S. typhimurium* DT 193	206
Ham and other cold meats	*S. falkensee*	83

Milk and milk products

Raw milk is inevitably contaminated with *Salmonella*, and consumption of raw milk is a recognised risk factor for salmonellosis, as well as other enteric infections. The commercial distribution of raw milk in California led to continuing outbreaks caused by *S. dublin*, while in Scotland, where milkborne salmonellosis was particularly prevalent, the introduction of mandatory pasteurisation eliminated the problem (Sharp, 1986). It is recognised that mandatory pasteurisation dramatically reduces morbidity due to *Salmonella*, and it is regrettable that a vociferous and ill-informed minority should propagandise for continuing sale of unpasteurised milk (see page 448).

In recent years, there have been major outbreaks due to pasteurised milk. The largest of these occurred in 1985 in Chicago, USA, and involved 16,284 known cases (Ryan *et al.*, 1987). This was the largest outbreak of salmonellosis ever recorded, and the actual number of people affected may have been as high as 250,000 (Archer and Young, 1988). The underlying causes of this and a smaller outbreak in Cambridge, UK, which affected 54 people (Rampling *et al.*, 1987) were similar, involving poor plant design and incorrect operation (see pages 396–398, 448–449).

Milk products are a diverse group of commodities which undergo a number of different processes. In all cases, safety is dependent on heat processing the raw material and prevention of re-contamination.

Dried milk

Infant dried milk was the cause of a significant outbreak involving *S. ealing* in the UK during 1985. The source of contamination was the insulation surrounding the drying chamber. This cause was similar to that in an earlier outbreak involving infant feeds in Australia, where *S. bredeney* was the causative organism (Craven, 1978). Surveys suggest that *Salmonella* contamination in the environment of spray drying plants may be fairly common (Rowe *et al.*, 1987b).

Cheese

Salmonellas are destroyed or inactivated during the fermentation of high-acid products (lactic acid *ca* 1%,

pH value less than 4.55) such as yoghurt and soft cheese (Chapman and Sharpe, 1981; Robinson and Tamime, 1981), although the effect is less in cheese due to protection by casein (Rubin, 1985). In contrast, growth of salmonellas may occur in the curd of low-acid cheese (pH value greater than 4.95).

The number of salmonellas decline during ripening, the effect being greatest at higher temperatures (Goepfort, 1968). Small numbers may persist for significant periods, and the practice of ageing raw milk cheese for 60 days at not less than 4.4°C (ICMSF, 1986) is therefore unlikely to serve as an effective control measure (D'Aoust *et al.*, 1987).

Although cheeseborne salmonellosis is often associated with raw-milk cheese, a large outbreak which occurred in Canada in 1984 was attributed to improper pasteurisation (Ratnam *et al.*, 1984). An estimated 10,000 people were affected (Ratnam and Marsh, 1986), the causative serovar being *S. typhimurium* PT 10. It should also be appreciated that some salmonellas may survive the reduced heat treatments (thermisation) applied to cheese milk by some producers.

Contamination of cheese by *Salmonella* may occur during handling at any stage. An outbreak of salmonellosis, involving the now notorious Swiss cheese vacherin Mont d'Or, due to *S. typhimurium*, was attributed to contamination of partly or fully ripened cheeses by piglets kept in a sty adjacent to the factory.

Other milk products

Salmonella infections have occasionally been associated with other dairy products. Ice cream gained notoriety following its involvement in the 1947 Aberystwyth typhoid epidemic, but despite notably poor standards of hygiene among many vendors, outbreaks today have involved home-made ice cream.

It seems likely that raw eggs were the source of the salmonellas in these cases and in others involving home-made frozen desserts, such as cassata (Bryan, 1983).

Cream-filled pastries are known to constitute a foodborne disease problem (ICMSF, 1989). *Staph. aureus* is involved in most cases, but in a survey of 439 outbreaks in the USA, 12.5% were attributed to *Salmonella* (Bryan, 1976).

Eggs and egg products

Between 1986 and 1988, the incidence of *Salmonella* food poisoning due to consumption of raw, or lightly cooked, hens eggs increased dramatically to become, in the UK, the commonest source of outbreaks in which an attributable food was identified (Anon, 1989). Virtually all of the infections were due to *S. enteritidis* PT 4. Problems involving this organism have been reported from other western European countries, especially Spain (Perales and Audicana, 1989), while other phage types of *S. enteritidis* (PT 8, PT 13a) have been involved in a significant rise in infections in the USA.

Although contamination of egg products with salmonellas is a long-standing problem (see below), this has been attributed either to the use of damaged eggs or to contamination at, or after, breaking. The situation with *S. enteritidis* of phage types 4, 8 and 13a is different, in that the organism can invade the ovaries and oviducts of chickens, with the result that the contents of the intact egg may be infected by transovarian infection. The true incidence is not known, but is recognised as being very low (Anon, 1989). However, the very large number of eggs eaten means that eggborne infections are a serious public health problem.

Most egg-associated infections have involved either consumption of raw eggs or of products made with raw egg, such as home-made mayonnaise, ice cream or other desserts (Humphrey *et al.*, 1989).

Lightly cooked eggs including 'hard' boiled eggs have also been implicated. It should be appreciated that *S. enteritidis* is *not* unusually heat resistant, but that when boiled the yolks of eggs are unlikely to reach a sufficiently high temperature to inactivate the organism until after about 7 minutes' cooking. From a practical viewpoint, this means that domestically cooked eggs with a 'runny' yolk may contain viable *S. enteritidis* (Humphrey *et al.*, 1989).

Historically, egg products including liquid and frozen egg have been an important source of human salmonellosis. In most countries, the situation has improved dramatically since the mid-1960s, and problems with egg products are now rare. Liquid egg was, however, considered to be the origin of salmonellas in outbreaks involving frozen pasta (Anon, 1986), and has been implicated in outbreaks of salmonellosis in the USA during 1989.

Fish

Shellfish

Contamination of shellfish with salmonellas due to growth in sewage-polluted waters has been a continuing problem in many parts of the world, and has led to a number of outbreaks of *Salmonella* food poisoning. Depuration provides some protection, but primary precautions involve ensuring that shellfish for human consumption are grown only in pollution-free waters.

Other fish

Salmonella food poisoning is only infrequently associated with fish other than shellfish, although isolations of the organism from fish have occasionally been reported.

Fresh salmon has been an occasional cause of *Salmonella* food poisoning in the UK in recent years (Watson, 1985; Cartwright and Evans, 1988). In outbreaks described in detail, the fish was probably contaminated from chicken under circumstances of poor kitchen hygiene and practice.

Confectionery

Chocolate

Chocolate (cocoa beans) are subject to contamination at source, and the heating during manufacture is insufficient to kill the organism. The low a_w level prevents growth in the finished product, but survival may be prolonged (Tamminga *et al.*, 1977).

Two large international outbreaks of food poisoning involving *S. napoli* (Gill *et al.*, 1983) and *S. eastbourne* (Craven *et al.*, 1975; D'Aoust *et al.*, 1975) have occurred in which chocolate was the vehicle of infection. A notable feature of these outbreaks was the low infective dose of salmonellas ingested in chocolate (Greenwood and Hooper, 1983). Subsequently, a large outbreak occurred in Scandinavia in which *S. typhimurium* was implicated (Kapperud *et al.*, 1989).

Cocoa powder has also been implicated in *Salmonella* food poisoning (Gastrin *et al.*, 1972).

Desiccated coconut

Desiccated coconut is used as an ingredient of many types of confectionery, and represents a *Salmonella* hazard. Contamination occurs in the shell through leakage, the milk supporting growth of salmonellas. Serovars involved include the lactose fermenting *S. ferlac*.

Pasta

Some serovars of *Salmonella* survive the drying of pasta and persist in the final product (Hsieh *et al.*, 1976). The organism would normally be killed by cooking, but pasta has been implicated in a number of outbreaks of salmonellosis (Anon, 1986).

Salads

Two major types of risk are presented by salads. The first involves contamination of salad vegetables either during growth or subsequent handling. The risk is greatest where nightsoil or animal manures are used as fertiliser, or where polluted water is used for irrigation. Salad vegetables are not usually cooked and protection lies in thorough washing.

The second risk involves the dressings applied to some types of salad. Mayonnaise is the main hazard, especially where raw eggs are used, or where the acetic acid content and acidity are too low. Very large outbreaks of food poisoning have resulted from mayonnaise-based salads. Mayonnaise from a salad factory caused approximately 10,000 cases in Denmark during 1975 (NRC, 1985), while in 1976, outbreaks of salmonellosis among passengers on four charter flights from Las Palmas to Helsinki affected several hundred people and caused six deaths. The vehicle of infection was mayonnaise contaminated with *S. typhimurium* (Davies and Wahba, 1976). The number of cases involving mayonnaise have increased in recent years, reflecting higher levels of contamination of shell eggs. It is noteworthy that despite widescale publicity concerning the hazards of raw eggs, outbreaks of food poisoning due to the use of raw eggs in mayonnaise continues. However, not all recent outbreaks have been due to *S. enteritidis*, and a large UK outbreak during 1988 was attributed to *S. typhimurium* DT 49. The source of contamination was probably egg but the problem was exacerbated by poor handling (Mitchell *et al.*, 1989).

An unusual outbreak due to *S. saint-paul* which occurred in the UK during 1988, was attributed to eating bean sprouts either as salad ingredients or lightly fried. The organism may have originated from the seeds and growth may have occurred during the sprouting process.

Miscellaneous foods

Outbreaks of *Salmonella* food poisoning have been caused by a wide range of foods and other commodities. Examples are listed in **Table 36**.

Table 36. Miscellaneous products involved in *Salmonella* food poisoning.

Product	Further information
Marijuana	Large, multi-state outbreak of unusual epidemiology
Rattlesnake capsules	Rattlesnake capsules are used by the Hispanic community in the US for a variety of chronic complaints ranging from acne to arthritis. Causative organism was *Salmonella subgroup 3 (S. arizonae)*, which is common in snakes and other reptiles. All patients had an underlying illness such as AIDS or cancer
Peanut butter	*Salmonella* has been isolated from peanut butter, or confectionery made with peanut butter, on several occasions. No case of salmonellosis has been attributed to this product
Frogs' legs	The level of contamination of frogs' legs with salmonellas remains high despite improvements in handling. Irradiation has been proposed as a means of control
Dried yeast	Dried yeast has been responsible for food poisoning when used as an ingredient in vegetarian foods that are often eaten unheated. Yeast-based flavouring was suspected of being the vehicle of infection in outbreaks involving savoury snacks in the UK during 1989

Factors affecting the growth and survival of *Salmonella* in foods

Temperature

As a generalisation, salmonellas grow over the range 5°C to 45 to 47°C (Simonsen *et al.*, 1987), with an optimum of *ca* 37°C. There have been anecdotal reports of growth at lower temperatures, but these have not been substantiated. Ability to grow at temperatures below 7°C is serovar and strain dependent, lowest growth temperatures being for strains of *S. bredeney*, *S. typhimurium* and *S. virchow* which are able to grow at 5 to 6°C and *S. agona*, *S. senftenberg* and *S. hadar* which are able to grow above 6°C (Alcock, 1985).

Growth rates at temperatures below 10°C are very low, but the ability to grow at relatively low temperatures may be significant where refrigeration is poor and shelf life prolonged.

Salmonellas are destroyed in high a_w foods by heating at a minimum temperature of 74°C (Simonsen *et al.*, 1987). This usually permits a margin of safety; D_{60} values for most serovars are in the range 0.2 to 6.5 minutes, while Z values are commonly 4 to 5 minutes (Roberts, 1982). *S. senftenberg* is considerably more resistant than other serovars, but this is probably of limited consequence since the organism appears to be of low pathogenicity to man.

There is interaction between thermal resistance and pH value, and several serovars of *Salmonella* are unusual in having a greater heat resistance at pH 5.5 than pH 8.5. The heat resistance also increases markedly at low a_w levels, particularly in foods such as chocolate, which also have a high fat content (ICMSF, 1980). In a study with milk chocolate (Goepfort and Biggie, 1968) the D_{70} value for a strain of *S. senftenberg* was 6 to 8 hours, and for a strain of *S. typhimurium* 12 to 18 hours (at high a_w levels the relative heat resistance of the two serovars is reversed). It should also be appreciated that it is not possible to predict the heat resistance of salmonellas in foods of low a_w levels and D and z values must be determined experimentally.

There is also evidence that successive sub-lethal heat treatments increases the resistance of *S. typhimurium* (**Figure 24**).

Salmonellas decline in numbers during frozen storage, the rate being greater at temperatures around the freezing point of meat (-2 to -5°C) than at the commercially used -23°C. Commercial conditions could be adjusted to discourage the survival of salmonellas (Mead, 1982), but it is unlikely that the organism could be eliminated totally in this way.

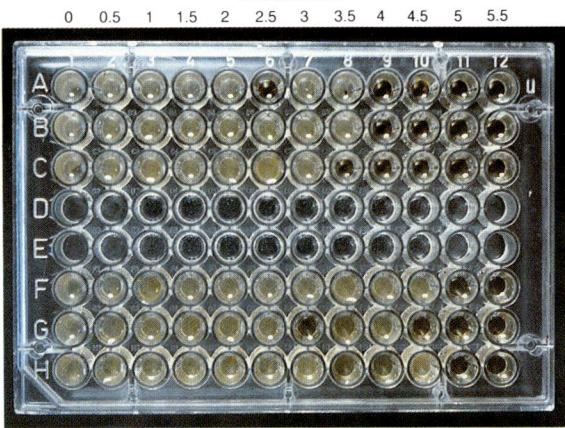

24. Increase in thermoresistance of *S. typhimurium* after mild heat shock. Cells which had not been subject to heat shocking at 48°C for 30 min survived for only 3.5 min at 60°C (ABC). In contrast, cells which had been heat shocked survived for up to 4.5 min (FGH). There are obvious implications for safety in processes which involve a sub-lethal preheating followed by a main heat treatment which is marginal with respect to safety (Mackey and Derrick, 1987a). Similar increases in heat resistance occur when cells are heated slowly as in some forms of cooking (Mackey and Derrick, 1987b). *One drop of a cell suspension which had been heated in nutrient broth at 60°C for each of the times shown was transferred to a well of a microtitre plate containing fresh nutrient broth and incubated at 30°C for 36 h. The experiment was conducted in this manner for the purposes of illustration.*

Irradiation

Irradiation has been proposed as a means of controlling salmonellas in poultry, shellfish and frogs' legs. In Canada a dose of 3 Kgy is recommended for poultry at 'on ice' temperature (Ouwerkerk, 1981) and a similar dose is used in the Netherlands for treatment of shellfish (Ley, 1983).

Work with liquid egg suggests that the efficiency of irradiation can be increased by combining irradiation with a heat treatment (thermoradiation). Lower pH values also lead to a higher rate of inactivation.

pH value

Salmonellas grow over the range of pH values 4.5 to 9.0, with an optimum of pH 6.5 to 7.5. At pH values below 4.1, inactivation and death will occur. There is

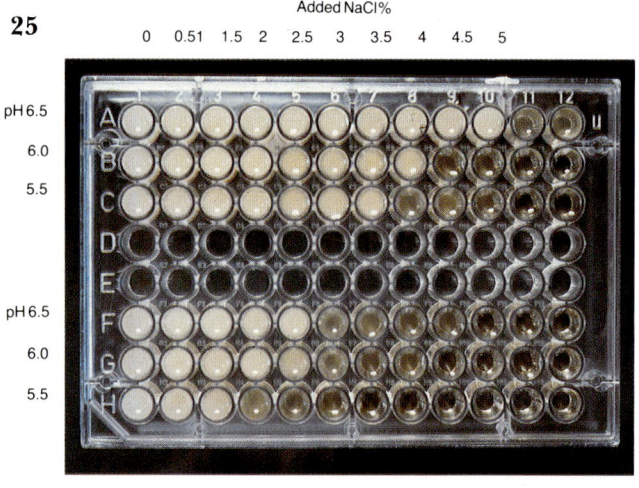

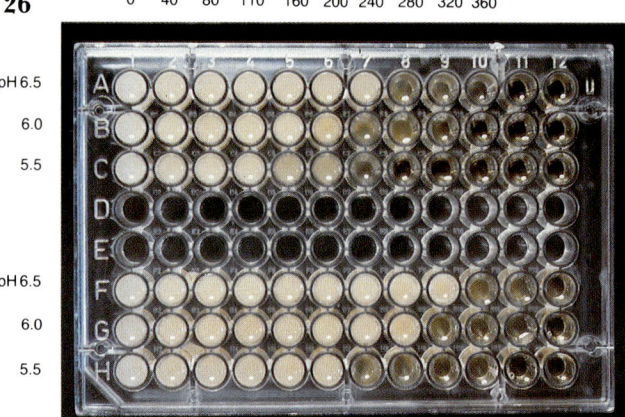

25 & 26. Variation in resistance to curing ingredients between strains of *S. typhimurium*. Considerable variation in resistance to curing ingredients can exist between strains of a serovar of *Salmonella* as well as between serovars. The figures illustrate the differing resistance of two strains of *S.typhimurium* to NaCl (**25**) and NaNO₂ (**26**) at each of three different pH values. Strain A is of greater resistance to NaCl at each pH value, but strain B has greater resistance to NaNO₂. Such variation can make prediction of safe formulation difficult, although differences in resistance to curing ingredients are less when used in combination. (*Nutrient broth plus NaCl or NaNO₂: 48h at 30° C.*)

interaction with other factors with respect to growth, survival and death. The nature of the acidulant can be of particular importance. Thus in the manufacture of mayonnaise, acetic acid may ensure safety under conditions where citric, or hydrochloric acids will not.

In acidified milk products, casein exerts a protective effect (Rubin, 1985), and similar protection may occur in other products.

Water activity

The lower limit for growth is a_w 0.93. Higher limiting values will apply when other inhibitory factors are present.

Oxidation-reduction potential

Growth is inhibited below E_h 30 mV. Higher limiting values will apply when other inhibitory factors are present.

Curing ingredients

When other growth conditions are optimal, salmonellas are able to grow in *ca* 4% NaCl and *ca* 350 mg/l NaNO₂ depending on serovar and strain. These levels are markedly lower when the two inhibitors are used in combination and there is a strong interaction with pH value and temperature (*see* **Figures 25 & 26**). Growth of salmonellas may occur on mildly cured, cooked meats, but in raw, cured meats further factors such as competition from other bacteria become significant (see below).

Salmonella can survive to the point of consumption in some fermented meats, although actual outbreaks of salmonellosis from such products have been rare (see pages 53, 437). In an outbreak in the UK due to a salami snack, it has been postulated that survival of *S. typhimurium* DT 124 was enhanced by the presence of large chunks of fat, and by the use of low-temperature instead of ambient temperature storage (Cowden *et al.*, 1989).

Organic acids

The use of acetic acid to control salmonellas in mayonnaise has been noted in relation to pH value (see above). In addition, the use of 'generally recognised as safe' (GRAS) chemicals to kill salmonellas on poultry carcasses, has been investigated with particular attention being paid to succinic acid, which has been found to be highly effective against this organism (Mountney and O'Malley, 1965). In later work, Thompson *et al.* (1976) were able to eliminate salmonellas from chicken carcasses using a hot dip in a solution containing 1% succinic acid and 50 mg/l chlorine. Application is limited by skin discoloration. In a contrasting approach, Olson *et al.* (1981) combined freezing and treatment with succinic acid, while a model triple stress system devised by Obafemi and Davies (1986) consisted of cold shock exposure at 5°C to 1% succinic acid and 5 mg/l free chlorine at pH 2.5. This treatment sensitised *S. typhimurium* to subsequent freezing, frozen storage and thawing.

Formic acid treatment of poultry feed has been suggested as an alternative to heat treatment for elimination of salmonellas (Humphrey and Lanning, 1988).

Competition from other micro-organisms

The competitive ability of *Salmonella* varies according to strain. Many compete poorly with food spoilage micro-organisms, particularly the lactic acid bacteria. This is of importance in food fermentations and may be due to a specific inhibitory mechanism as well as to acidification (Ashenafi and Busse, 1989).

Laboratory methods

Conventional cultural isolation methods

Conventional cultural isolation techniques remain in general use in many laboratories. This involves resuscitation (pre-enrichment), selective enrichment and selective plating followed by biochemical and/or serological confirmation of the identity of suspect colonies.

Resuscitation

A large number of media have been proposed for the resuscitation of salmonellas, although it seems unlikely that there are major differences in performance. Lactose broth and buffered peptone water (BPW) are most widely used, the latter being recommended for routine purposes (van Leusden, 1982; Fricker, 1987).

Although non-selective resuscitation broths are suitable for most foods, problems may occur where large numbers of Gram-positive bacteria such as *Streptococcus lactis* (*Lactococcus lactis*) are present. This problem may be overcome by the addition of low levels of dyes such as brilliant green (0.002% w/v) or crystal violet (0.004% w/v) to the resuscitation medium (North, 1960). More recently, van Schothorst and Renaud (1985) proposed the addition of 0.01% malachite green to BPW for recovery of salmonellas from milk powder, and this is recommended in all situations where the sample material is heavily contaminated with bacteria other than members of the *Enterobacteriaceae*.

Although incubation at 43°C has been used occasionally, 37°C is recommended for resuscitation of salmonellas.

The use of a resuscitation stage increases the time taken to obtain a result, and there have been attempts to reduce this by using a short (*ca* 6 hours) incubation period. However, there is an obvious potential for an extended lag phase when cells are damaged and an incubation period of 24 hours is recommended.

Selective enrichment

Many types of inhibitors have been proposed for the selective enrichment of salmonellas, the most widely used of which are bile, tetrathionate, selenite and dyes including brilliant green and malachite green. These inhibitors have been incorporated, either singly or in combination, into an even wider range of media, the most widely used of which are summarised in **Table 37**.

With the exception of RV medium, enrichment media in use today are well established, and based on modifications of earlier media. RV medium is itself an evolutionary development of Rappaport broth (Rappaport, 1956) with the malachite green content reduced to 0.004% w/v to avoid problems of over selectivity. A number of studies have been made com-

Table 37. Examples of commonly used media for the selective enrichment of salmonellas.

Medium	Inhibitors	Applications and limitations
Tetrathionate broth (Muller-Kauffman)[1]	Tetrathionate, brilliant green, ox-bile	Not suitable for host-adapted serovars
Selenite F broth[2]	Selenite	Suitable for most serovars including *S. typhi*, *S. dublin* and *S. choleraesuis*
Selenite-cystine broth[3]	Selenite	Cystine enhances *Salmonella* growth. Widely used for foods
Brilliant green-MacConkey broth[4]	Brilliant green, bile salts	Very effective with *S. choleraesius* but not widely used
RV broth[5]	Malachite green, magnesium chloride, 'low' pH value	Medium of choice for foods. May fail to recover *S. typhi* and *S. dublin*. May also be over selective for other serovars

[1] Kauffman (1935). This medium may also be known as Tetrathionate-brilliant green broth and should not be confused with standard tetrathionate broth.
[2] Leifson (1936).
[3] North and Bartram (1953).
[4] McCullough and Byrne (1952).
[5] Vassiliadis *et al.* (1976).

paring RV broth with the two most widely used traditional media tetrathionate broth and selenite F broth. Predictably, findings have been variable, but most have found RV to be superior for isolation of salmonellas from human and animal food (Harvey and Price, 1982; Vassiliadis, 1983). RV broth is known to be over selective with respect to *S. typhi* and *S. dublin*, and for this reason should not be used with human or animal faeces. However, RV may be inhibitory to a wider range of serovars with consequent risk of under recovery (Patil and Parhad, 1986), and some unpublished findings suggest that under recovery may be a significant problem with some types of food.

If RV medium is to be used it is important that the correct ratio of sample to broth is used. With selenite and tetrathionate broths, an inoculation ratio of 1:10 is most common, but with RV a much smaller inoculum is necessary. Inoculation ratios in the range 1:100 to 1:2,000 give good results (Harvey and Price, 1980), and 1:100 is generally recommended (Fricker *et al.*, 1985; Vassiliadis *et al.*, 1985).

Selenite and tetrathionate broths are not sensitive to the amount of sample added, and as much as 80g animal faeces may be added to 100 ml broth (Harvey and Price, 1982).

Traditionally, enrichment broths have been incubated at 35 to 37°C, with loopfuls removed for streaking at 18 to 24 hours and 42 to 48 hours. A temperature of 41°C is required when using RV broth to obtain the necessary selectivity, and this temperature has also become widely used with selenite F and tetrathionate broths. Incubation at 41°C should not be used with selenite cystine broth.

It is accepted that, where applicable, incubation at the higher temperature increases the general efficiency

of enrichment for salmonellas (D'Aoust, 1984). However, *S. typhi* is not recovered by enrichment at this temperature, and strains of other serovars may also be unable to grow. The occurrence of such strains is probably infrequent on an overall basis, but may be more common in an individual food.

The control of temperature is more critical at 41°C than at 37°C, since too low a temperature may permit the growth of competing organisms, while too high a temperature is likely to inhibit growth of salmonellas. For this reason, incubators capable of maintaining 41 +/− 0.1°C throughout the cabinet must be used. Waterbaths can be used for this purpose but this practice may lead to cross-contamination.

Although a single incubation period of 18 to 24 hours is sometimes used for enrichment cultures, this may not be adequate and a second streaking at 48 hours is required. Extending the incubation beyond 48 hours is not recommended.

The possibility of using short incubation periods has been investigated, with results being equivocal or contradictory. Enrichment for 6 hours (Andrews *et al.*, 1985) or 8 hours (Rappold *et al.*, 1984) was successful in some circumstances, but earlier Price (1983) had shown that incubation of enrichment cultures for 6 hours is likely to result in significant under recovery of *Salmonella*. This conclusion was supported by work on the dynamics of enrichment for salmonellas (van Schothorst and Renaud, 1983; Beckers *et al.*, 1987).

Although further work is required before the use of a long resuscitation and a short enrichment can be recommended, van Schothorst and Renaud were able to suggest a more modest time saving in that the substitution of soya peptone for tryptone in RV medium made continuation of incubation beyond 24 hours unnecessary.

The use of additional manipulations to increase enrichment efficiency has also been investigated, and Truscott and Lamerding (1987) combined membrane filtration of the resuscitation culture with the use of RV enrichment broth and claimed satisfactory results after 6 hours enrichment.

Various attempts have been made to combine the resuscitation and enrichment stages to save both overall time and manipulations (Sveum and Kraft, 1981). Addition of selenite to selenite cystine broth was delayed until 4 hours after commencing incubation (Alford and Knight, 1969), while alternatively selective agents may be encapsulated in a wax coating and released after several hours incubation (Lanz and Hartmann, 1976). None of these methods is used on a wide scale.

Selective plating media

As with enrichment broths, most of the selective plating media in use today are modifications of earlier media. Selective agents used are similar in each case with bile salts, or sodium deoxycholate, and brilliant green being used widely.

Selective plating media for *Salmonella* all contain a diagnostic system to permit differentiation of the organism from non-salmonellas. This is commonly based on the inability of most salmonellas to ferment lactose and, in some cases, other carbohydrates such as sucrose and salicin. Bile containing media often employ a second diagnostic system based on the ability of *Salmonella* to produce hydrogen sulphide. This increases their usefulness, particularly when dealing with materials which frequently contain lactose fermenting salmonellas, although lactose-positive, H_2S-negative salmonellas may also be isolated (see **Figure 34**).

Mention should be made of a further medium originally formulated by Wilson and Blair (1927), which is based on different selective and diagnostic principles. In this medium, the selective agents are bismuth sulphite, sodium sulphite and brilliant green and salmonellas are recognised by their ability to reduce sulphite to sulphide when growing in the presence of a fermentable carbohydrate. Sulphide is detected by blackening of the medium in the presence of iron (Fricker, 1987). There is disagreement over the performance of this medium with foods (Cox *et al.*, 1972; Erdman, 1974), but it is widely used where lactose fermenting salmonellas are common and also recommended for *S. typhi*.

Commonly used selective plating media for *Salmonella* are summarised in **Table 38** and their use illustrated in **Figures 27–42**.

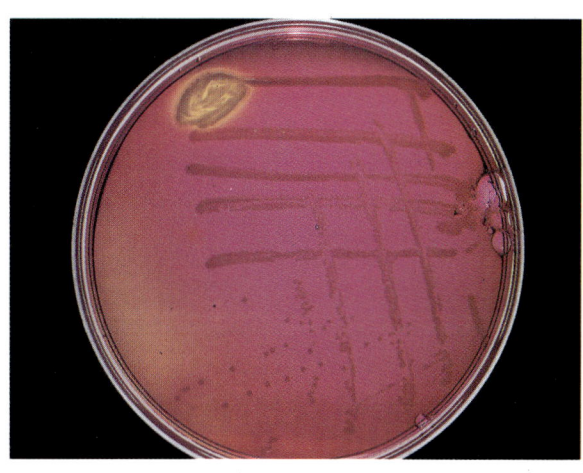

27

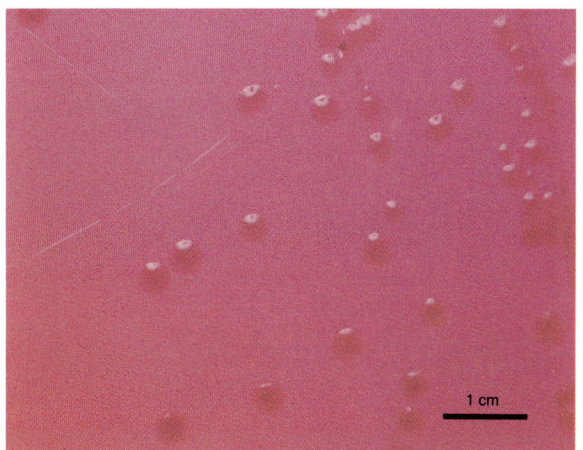

28

27 & 28. Isolation of *Salmonella* on brilliant green agar medium. Brilliant green agar is an effective medium for the isolation of a wide range of salmonellas from many environments, and is the medium of choice in many circumstances. The distinctive bright red colour of the medium when *Salmonella* is present (**27**) provides an easy recognition of positive cultures. The colonies themselves are red on the red background (**28**). A number of modifications have been made to improve the selectivity of brilliant green medium. These include the addition of mandelic acid and sodium sulphacetamide (sulphamandelate; Watson and Walker, 1978) which is recommended when competition from other bacteria is a particular problem (see **39**). (*Brilliant green medium: 24 h at 37°C.*)

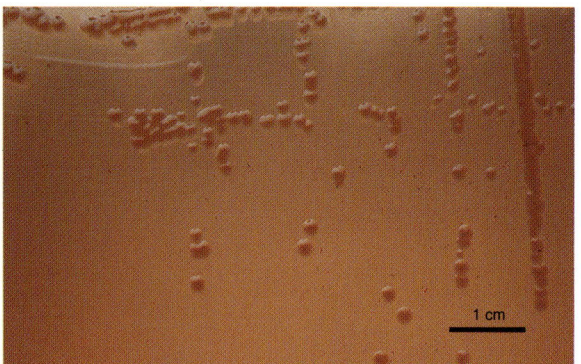
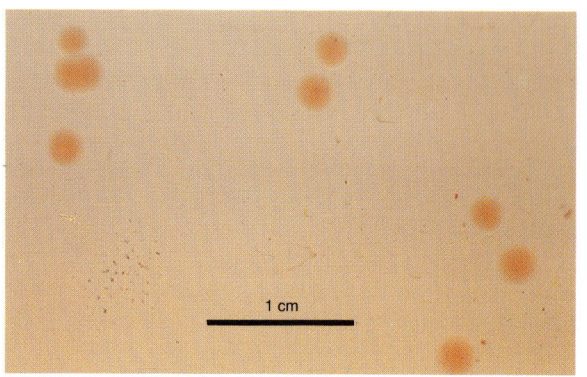

29

Table 38. Examples of commonly used media for the selective plating of salmonellas.

Brilliant green agar[1]
Inhibitors	Brilliant green
Diagnostic system	Fermentation of lacrose and sucrose
Applications and limitations	Widely used in food industry. Not suitable for *S. typhi*

Brilliant green–sulphamandelate agar[2]
Inhibitors	Brilliant green, Na sulphacetamide, Na mandelate
Diagnostic system	Fermentation of lactose and sucrose
Applications and limitations	Effective where *Proteus* and *Pseudomonas* are present in large numbers. Restricts growth of some salmonellas

Deoxycholate–citrate agar[3]
Inhibitors	Deoxycholate
Diagnostic system	Lactose fermentation, H_2S production
Applications and limitations	Relatively low selectivity

DCLS agar
Inhibitors	Deoxycholate
Diagnostic system	Lactose and sucrose fermentation, H_2S production
Applications and limitations	Improved differentiation of *Proteus*

Salmonella-shigella agar
Inhibitors	Brilliant green, bile salts
Diagnostic system	Lactose fermentation, H_2S production
Applications and limitations	Effective with many foods

Xylose–lysine–deoxycholate agar[4]
Inhibitors	Deoxycholate
Diagnostic system	Lactose, xylose and sucrose fermentation. Decarboxylation of lysine. H_2S production
Applications and limitations	Relatively low selectivity

Hektoen enteric agar[5]
Inhibitors	Bile salts
Diagnostic system	Lactose, salicin and sucrose fermentation, H_2S production
Applications and limitations	Good differentiation, relatively low selectivity

Bismuth sulphite agar[6]
Inhibitors	Bismuth sulphite, Na sulphite, brilliant green
Diagnostic system	Reduction of sulphite to sulphide in the presence of fermentable carbohydrate
Applications and limitations	Often recommended for *S. typhi*, effective with lactose-positive salmonellas. Performance with foods may be variable

[1] van Leusden *et al.* (1982).
[2] Watson and Walker (1978).
[8] Hynes (1942).
[4] Taylor (1965).
[5] King and Metzger (1968).
[6] Wilson and Blair (1927).

29 & 30. Isolation of *Salmonella* on DCLS agar medium. DCLS agar is a derivative of deoxycholate-citrate medium and contains sucrose as well as lactose to improve the differentiation of salmonellas from *Proteus* and other lactose-negative, sucrose-positive bacteria. The selective inhibitor is deoxycholate. The growth of salmonellas is less easily recognised than on brilliant green agar, with no colour change occurring in the medium (**29**), the colonies being pale, translucent or colourless (**30**). (*DCLS medium: 24 h at 37°C.*)

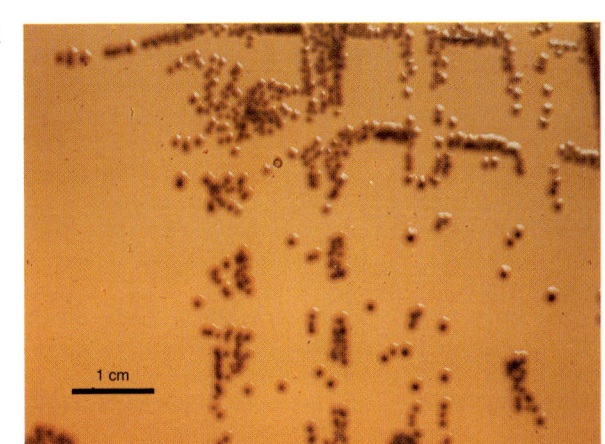

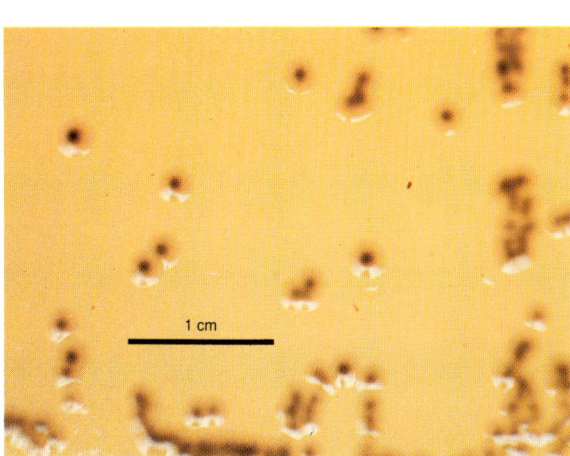

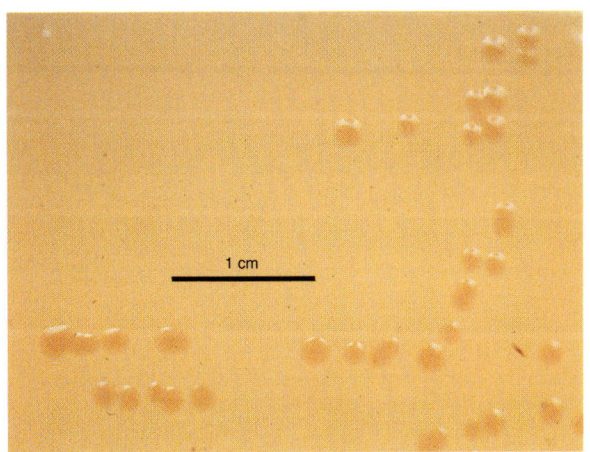

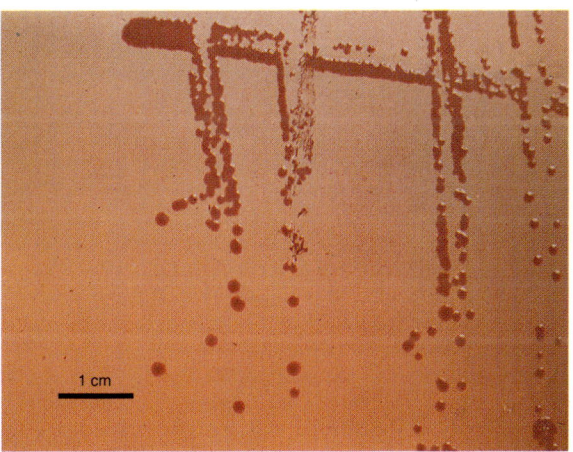

31–34. Isolation of *Salmonella* on Salmonella-shigella agar medium. Salmonella-shigella agar is a widely used medium which employs both bile salts and brilliant green as selective inhibitors. The blackening of colonies due to hydrogen sulphide production provides a second differential system to lactose fermentation. Typical H_2S-positive, lactose-negative salmonellas are easily recognised (**31 & 32**), the colonies being transparent with a black centre. Paradoxically, the ease with which typical salmonellas are recognised can mean that H_2S-negative strains, which lack the black centre, are likely to be overlooked by the less experienced (**33**). Lactose fermenting organisms have pink colonies, lactose-positive strains of *Salmonella* usually being differentiated by H_2S production. Atypical salmonellas which are lactose-positive and H_2S-negative such as some strains of *S. london* exist and may be relatively common in some foods such as dried milk powder (**34**). Colonies of these salmonellas resemble those of lactose fermenting bacteria such as most strains of *E. coli* but it may be possible to recognise salmonellas by faster growth rate and larger colony size (*cf.* **41 & 42**). (*Salmonella-shigella agar: 24 h at 37°C.*)

35

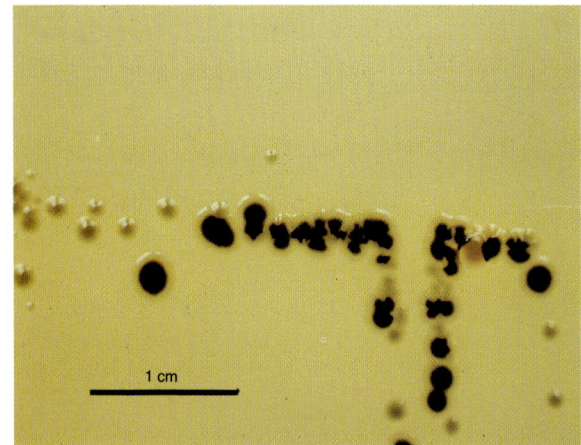

1 cm

36

1 cm

Incubation of selective plating media should be at 35 to 37°C. It is recommended that incubation should be for a full 24 hours since shorter periods may give false-negative results, particularly where strongly selective media are used. Where necessary, incubation may be extended to 48 hours, but there is no value in extending the period of time further.

Design of a complete protocol

The choice of media for each stage of the isolation should be made according to logical principles, such as those outlined on page 29. Standard methods, such as that of the International Organisation for Standardisation, are available (van Leusden, 1982), but it must be remembered that while inter-laboratory standardisation is an admirable ideal, the development of such methods inevitably involves compromise, and may not meet the specific needs of individual laboratories. Similarly, it is necessary to ensure that the conclusions of published comparisons of media are relevant in other situations.

It must be appreciated that while it is possible to recommend a single resuscitation medium, no single enrichment or selective plating medium is equally effective for recovery of all strains or serovars of *Salmonella*. To recover the widest possible range of salmonellas, it is therefore necessary to use two enrichment media, each of which is streaked onto two selective plating media after both 18 to 24 hours and 48 hours incubation.

35 & 36. Isolation of *Salmonella* on Hektoen enteric agar medium. Hektoen enteric agar contains bile salts as a selective inhibitor and bromthymol blue as an indicator. Differentiation from other species is based on fermentation of salicin as well as lactose and sucrose, and an H_2S indicator is also included in the medium. Hektoen enteric agar is of relatively low selectivity, but typical lactose-negative, H_2S-positive salmonellas are readily differentiated even in the presence of large numbers of other bacteria (**35**). Blackening due to H_2S production is pronounced and the centre is typically surrounded by only a narrow green or green-blue margin. As with Salmonella-shigella agar recognition and differentiation of H_2S negative strains is more difficult. (*Hektoen enteric agar: 24h at 37°C.*)

37–42. Selectivity of plating media for *Salmonella*. Selectivity varies considerably but none of the media are fully selective. Non-lactose fermenting members of the *Enterobacteriaceae* can be a particular problem. *Proteus*, for example, grows well on Hektoen enteric medium (**37**) although it is readily differentiated from typical salmonellas. Brilliant green agar is usually more selective than Hektoen enteric agar but on some occasions *Proteus* (**38**) may overgrow *Salmonella*. This may be controlled by the addition of sulphamandelate supplement (**39**). Problems may also be caused by non-*Enterobacteriaceae* including *Aeromonas* (see page 197) and *Pseudomonas* (**40**). In this example, *Pseudomonas*, growing on Salmonella-shigella agar, may be recognised by the large quantities of water diffusible pigment in the medium. However, pigment production is not a property of all strains of *Pseudomonas* and is less common in Europe than in the US.

It is a common misconception that lactose non-fermenting strains of *E. coli* are a major problem in the examination of some foods. On media such as Salmonella-shigella agar, colonies of lactose-negative *E. coli* (**41**) are similar in appearance to those of H_2S-negative *Salmonella* (**42**), but develop much more slowly and may thus be distinguished by an experienced microbiologist. (*All media: 24h at 37°C.*)

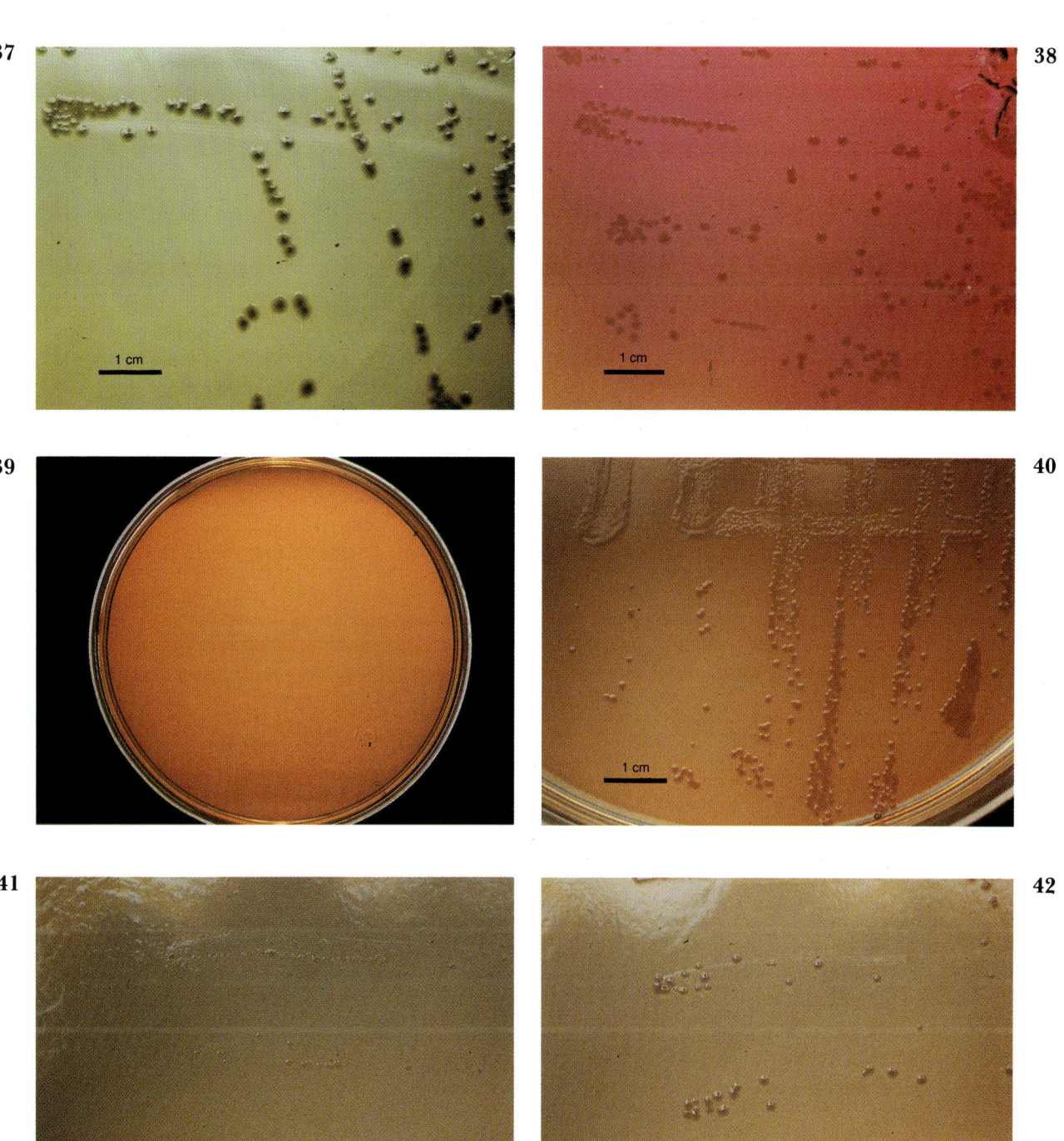

Flow diagram 7. Sample protocol for the recovery of salmonellas by conventional methods.

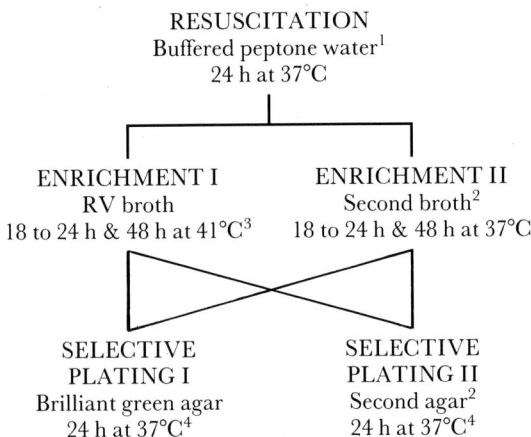

RESUSCITATION
Buffered peptone water[1]
24 h at 37°C

ENRICHMENT I
RV broth
18 to 24 h & 48 h at 41°C[3]

ENRICHMENT II
Second broth[2]
18 to 24 h & 48 h at 37°C

SELECTIVE
PLATING I
Brilliant green agar
24 h at 37°C[4]

SELECTIVE
PLATING II
Second agar[2]
24 h at 37°C[4]

[1] Suitable for most foods. Lactose broth may be preferred for dairy products.
[2] Any of the media in **Tables 37** & **38** according to choice.
[3] RV broth at 41°C may not be suitable for dairy products. Selenite broth at 35 to 37°C may be preferred.
[4] If necessary incubation may be extended to 48 h.

The use of two media in this way requires a considerable number of manipulations, and uses considerable quantities of media. This may place a considerable strain on the resources of laboratories, and in many cases single enrichment and plating media are used. It is considered that the use of two media is desirable, and should always be applied when investigating complaints of food poisoning. However, for routine purposes it must be an individual decision based on knowledge of the food sampled, serovars usually isolated, and the purpose of the examination.

A sample protocol for isolation of salmonellas from foods is illustrated in **Flow diagram 7**.

Most probable number enumeration

A most probable number technique for enumeration of salmonellas has been described by Morgan-Jones (1982). Lysine-iron-cystine-neutral red broth (LICNR) is used (Hargrove *et al.*, 1971), salmonellas being detected by an alkaline reaction due to decarboxylation of lysine, and a black precipitate due to hydrogen sulphide production. LICNR is non-selective, and may be used without resuscitation, although the addition of novobiocin reduces the overgrowth of salmonellas by competing organisms.

In use, LICNR is set up in a 3, 5 or 10 tube configuration, and incubated for 48 hours at 37°C. The colour change in the medium gives only a presumptive indication of the presence of *Salmonella*, and streaking onto selective agar followed by biochemical and serological confirmation is required in the case of presumptive positive tubes. The advantage of this technique lies primarily in reducing the amount of confirmatory work required to give an accurate estimation of numbers by early elimination of negative enrichments. A further saving in effort, as well as in time to obtain a result, might be obtained by replacing conventional confirmatory testing with the latex co-agglutination test or the MUCAP test (see pages 82–83; 85).

Rapid cultural methods

Hydrophobic grid membrane filtration (HGMF)

A method for the isolation of *Salmonella* using HGMF has been described by Entis (1985). This is based on conventional methodology, but uses concentration of bacteria by filtration to permit a short (6 hour) enrichment, and exploits the properties of the HGMF to produce isolated colonies of *Salmonella* and avoid the risk of overgrowth when incubated on selective media. Tetrathionate-brilliant green broth was used for enrichment but, theoretically, other enrichment broths could be used. The short enrichment permits a presumptive result to be obtained in 48 to 72 hours, and sensitivity is equal to that of conventional cultural techniques.

Salmonella Rapid Test

The Salmonella Rapid Test (SRT; Oxoid©) is based on selective motility enrichment methodology. The system and its use is illustrated in **Figures 43–46**. The SRT has been evaluated by Holbrook *et al.* (1989) and found to be satisfactory for a range of foods. Identification is presumptive and confirmation may be either by conventional tests or, more conveniently, by latex co-agglutination. In the latter case, a result is available in 42 hours. The SRT will detect only motile salmonellas. On an overall basis, the percentage of non-motile strains of serovars commonly involved in food poisoning is small, but such strains may be more common in specific circumstances. Flagella may also be damaged by processing, but de Smedt *et al.* (1986) claimed that repair of damage occurs during resuscitation.

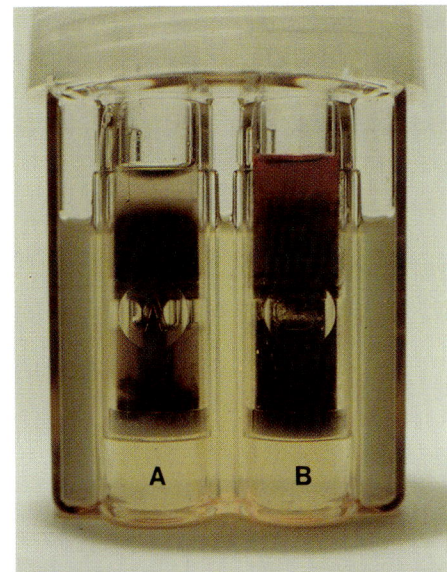

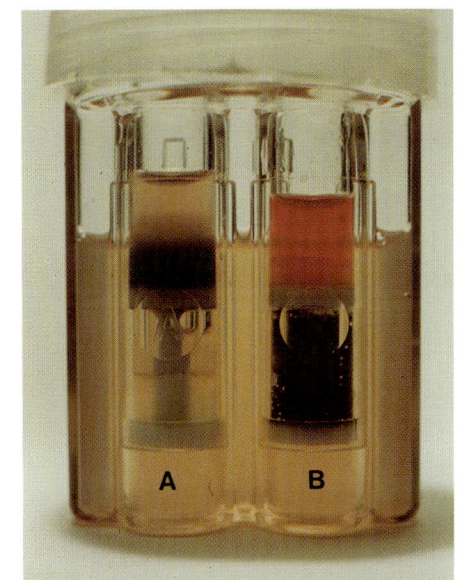

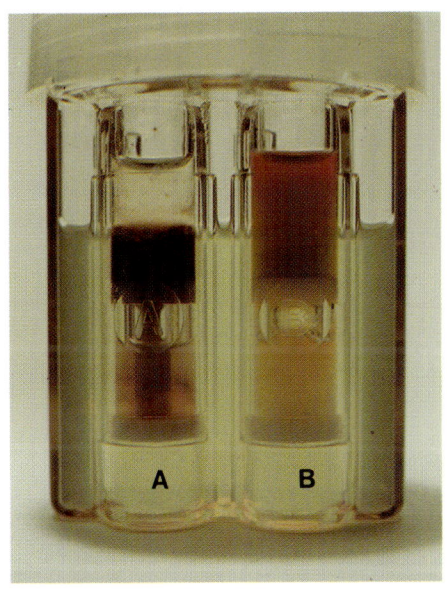

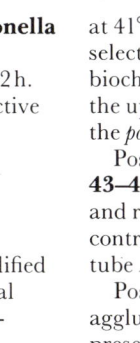

43–46. Detection of salmonellas using the Salmonella Rapid Test. The Salmonella Rapid Test (Oxoid©) permits the detection of salmonellas in foods within 42 h. The test consists of a culture vessel containing an elective medium for salmonellas and two tubes. Each tube contains a lower compartment containing a selective medium, and an upper compartment containing a differential medium. The two compartments are separated by a porous partition. Tube A contains modified RV broth as the selective medium, and modified lysine-iron-cystine-neutral red broth as the differential medium, while Tube B contains modified lysine-iron-deoxcholate broth and modified brilliant green broth.

Food samples are resuscitated in buffered peptone water (20 h at 37°C), and 1 ml of this culture is added to the elective medium in the culture vessel. During incubation at 41°C for 24 h, salmonellas migrate actively through the selective medium to the differential medium where their biochemical activities lead to distinctive colour changes in the upper compartment. A change in *either* tube indicates the *possible* presence of *Salmonella*.

Positive reactions of three salmonellas are shown in **43–45**. Varying degrees of blackening are seen in tube A, and red, or red and black, coloration in tube B. In contrast, a strain of *Proteus* (**46**) produced no blackening in tube A and a yellow-green colouration in tube B.

Positive tubes should be tested further using latex agglutination. Positive results indicate the *presumptive* presence of *Salmonella*, which should be confirmed by further biochemical and serological tests. (*Salmonella Rapid Test (Oxoid©): 24 h at 41°C.*)

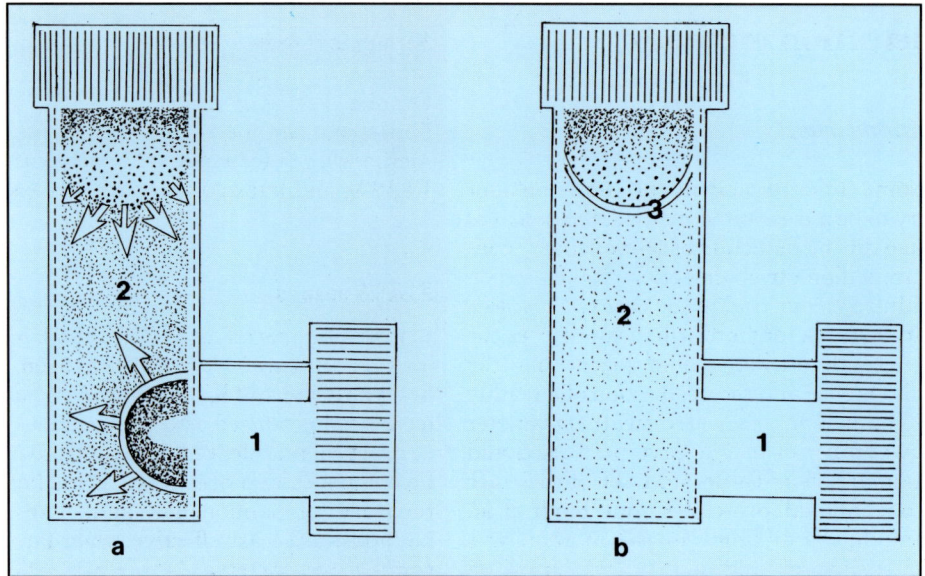

47. Detection of salmonellas using the 1–2 Test© immunoimmobilisation system. The 1–2 Test immunoimmobilisation system consists of a specially-designed device consisting of an inoculation chamber (**1**) and a larger motility chamber (**2**). When ready for use, the inoculation chamber contains brilliant green-tetrathionate broth of high L-serine concentration, and the motility chamber contains a non-selective motility medium. The device is inoculated with an enrichment culture and at the same time a suspension of flagellar antibodies is added to the top of the motility chamber.

During the first 4 h of incubation at 35 to 37°C, the antibodies diffuse radially downward into the motility medium while any salmonellas present multiply in the enrichment medium. L-serine is a chemoattractant, and at this stage its concentration in the enrichment medium prevents migration into the motility chamber.

Between 4–8 h of incubation (**a**), the antibody continues to diffuse downwards forming an antibody concentration gradient. Meanwhile, the L-serine concentration in the inoculation chamber falls due to bacterial metabolism and the motile salmonellas migrate into the motility medium. Migration proceeds upwards in response to a gradient of L-serine resulting from continued metabolism of the amino-acid. As the salmonellas move into the antibody concentration gradient, they are progressively immobilised by the antibody binding of the flagella.

At the end-point (**b**), the cells are completely immobilised and form a visible and permanent immunoband (**3**), the U-shape of which reflects the radial diffusion pattern of the antibodies. (Reproduced by courtesy of BioControl Systems Inc.)

Immunoimmobilisation (1-2 Test©)

The immunoimmobilisation technique for detection of salmonellas was first described by Greenwood and Swaminathan (1981), and subsequently developed by other workers (de Smedt *et al.*, 1986; de Smedt and Bolderdijk, 1987). The 1-2 Test represents the commercial development of the technique into an easily used format, which permits detection of salmonellas in 32 to 38 hours. The principles of the test are illustrated in **Figure 47**.

As originally devised, the 1-2 Test was used as a means of detecting *Salmonella*, but subsequently it has been used as part of a most probable number examination (Dickson, 1989).

An evaluation by D'Aoust and Sewell (1988) of a prototype 1-2 Test system showed a large number of false negative reactions. This was attributed to the low selectivity of cultural conditions used by these authors and the consequent inability of the test to detect salmonellas in the presence of a large competing microflora. At a later date, Nath *et al.* (1989) showed the performance of the test to be markedly improved by incorporating enrichment in tetra-thionate brilliant green broth for 24 hours into the protocol, this period having subsequently been reduced to 7 hours (Oggel *et al.*, 1990). The 1-2 Test has been shown to compare favourably with ELISA and DNA hybridisation techniques (St Clair and Klenk, 1990) and has received AOAC approval.

Potential problems with non-motile strains of *Salmonella* are the same as those for the SRT.

Rapid detection methods

Impediometric methods

Although some microbiologists do not consider impediometry to be a truly rapid method, the potential for time saving during detection of salmonellas is considerable. Impediometric detection depends on growth in culture media, and in the case of *Salmonella* it is necessary to use a medium which suppresses the growth of non-salmonellas, while permitting the growth of salmonellas and the production of a readily detectable change in the electrical parameter measured. A number of media have been used and these are summarised in **Table 39**. Each of these are based on conventional principles, but Pugh *et al.* (1988) found that confidence in the method was increased by using, in addition, a test based on specific bacteriophage.

Impediometric determination of salmonellas has potential for application in all areas of the food industry, but to date has been of particular value with confectionery where positive release systems are applied before use of many ingredients. The use of impediometric detection in combination with a 24 hour conventional resuscitation, permits negative samples to be released in 48 hours. The method has the additional advantage of detecting lactose-positive, hydrogen sulphide-negative isolates (*see* **Figure 34**), which are likely to be missed by conventional techniques.

Serological techniques

Despite an occasional resurgence of interest in the fluorescent antibody technique, most serological methods for detection of salmonellas are of two types, ELISAs and latex co-agglutination assays.

ELISA assays

A large number of ELISA assays have been developed, and earlier types have been reviewed by Swaminathan and Konger (1986). A number of these have been produced as commercial kits.

As a general rule, ELISA assays have been found to be comparable to conventional methods in terms of recovery of salmonellas and, like conventional methods, are least effective when large numbers of competitive bacteria are present. There are differences in performance between the different assays, but these are relatively minor and in many situations choice is likely to depend on convenience and ease of use.

The Bio-EnzaBead© assay is one of the more widely used, and has been the subject of a number of evaluations using pure cultures (D'Aoust and Sewell, 1986) or either artificially or naturally contaminated foods (Beckers *et al.*, 1986,1988; Flowers *et al.*, 1987, Todd *et al.*, 1987). Evaluations were largely favourable, although the presence of *Citrobacter* may result in false-positives (D'Aoust and Sewell, 1986). There are also indications that different foods affect the perform-

Table 39. Media for the detection of salmonellas by impediometry.

Medium	Applications and limitations
Selenite–cystine–trimethylamine oxide–dulcitol[1]	Based on a modification of selenite F broth. Enzymes necessary for dulcitol fermentation and TMAO reduction are inducible, and resuscitation should be in a broth containing these ingredients
Selenite–cystine–trimethylamine oxide–mannitol[2]	Designed to overcome problems with dulcitol negative strains. Resuscitation should be in a broth containing mannitol and dimethylsulphoxide
Lactalbumin hydrolysate–glucose–lysine–selenite[3]	Used in conjunction with above medium to overcome false-positive reactions due to some strains of *Citrobacter*. Resuscitation should be in buffered peptone water containing glucose and lysine
Modified lysine decarboxylase broth[4]	Also used in conjunction with media containing TMAO. Contains an indicator system to detect H_2S production by the blackening of positive wells by salmonellas and thus will eliminate false-positive reactions due to *Klebsiella pneumoniae*

[1] Easter and Gibson (1985).
[2] Gibson (1987). A similar medium was described by Ogden and Cann (1987).
[3] Ogden (1988).
[4] Arnott *et al.* (1988).

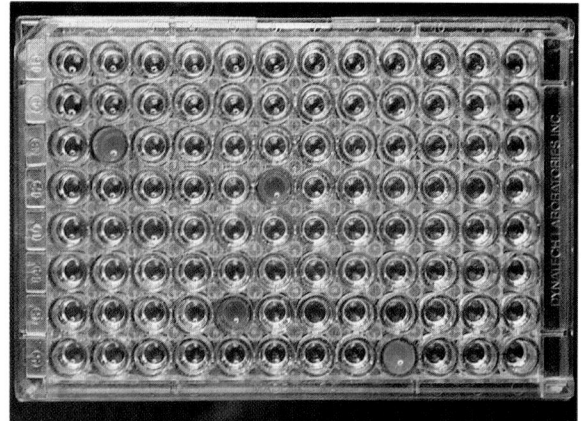

48. Detection of Salmonellas using the TECRA© ELISA assay. Positive samples are indicated by the blue-green colour in the well of the microtitre plate. Visual reading obviates the need for an expensive photometer-based plate reader. However, the ELISA assay involves a number of manipulative steps and automation is necessary where sample throughput is high.

ance of the assay, and that the criterion recommended by the manufacturer as a limit for all foods should be adjusted to suit the type of food under investigation (Beckers *et al.*, 1986).

A further drawback to the Bio-EnzaBead is the need for centrifugation, while the incubation periods recommended were inconvenient for obtaining a result in 42 hours. A modification eliminates the need for centrifugation and permits a result to be obtained in 40 hours (Flowers *et al.*, 1988a).

The ELISA Screening Kit© is less well known than the Bio-EnzaBead, and has been less fully evaluated. In operation, the kit has the potential advantage of using time-lapse sampling in the determination of enzyme activity, which may avoid the need to adjust the baseline for different foods. However, this kit gave a higher number of false-negatives than the Bio-EnzaBead, but it is still a useful alternative to conventional procedures (Beckers *et al.*, 1988).

The TECRA© Salmonella Visual Immunoassay is of particular interest, in that readings of enzyme activity may be made visually without the need for expensive instrumentation (**Figure 48**). Evaluations have been favourable (Blackburn *et al.*, 1988; Flowers *et al.*, 1988b; Hughes *et al.*, 1987), initial problems of cross-reactions with *Citrobacter* having been successfully overcome.

Although ELISA assays permit a significant time saving, an increase in sensitivity of detection could further reduce the time taken to obtain a result, by eliminating or shortening some cultural stages. A number of assays which will detect salmonellas after pre-enrichment or short enrichment periods have been developed (Minnich *et al.*, 1982; Prusak-Sochaczewski and Luong, 1989; Lee *et al.*, 1990), although further development is required in each case. The method of Lee *et al.* is considered to be of particular importance, since it enables *S. typhimurium* to be detected in 19 hours even when outnumbered by 10^6 to 1. Detection is made by a microtitre plate-antibody capture assay using a monoclonal antibody, following enrichment in a non-selective chemically defined medium. Further development to enable all salmonellas to be detected is required.

Latex co-agglutination

In addition to its use in confirming the identity of *Salmonella*, latex co-agglutination may be used as the basis for a rapid detection assay. A novel latex test has been described (Hadfield *et al.*, 1987), and a commercial kit based on this principle, the Spectate© (**Figures 49 & 50**), has been developed (Clark *et al.*, 1989). The test may be applied at any stage of the conventional procedure, and with raw foods detection is possible after enrichment. Selenite broth enrichments may be applied direct to the latex reagents, but enrichment broths such as RV and tetrathionate require either filtration or subculture into a nutrient broth and a short incubation to avoid interference from solid material or from carry over of colour.

The Spectate© test may also be used in conjunction with impedance methods to confirm the presence of salmonellas in selective broths showing a positive response.

The specificity of latex co-agglutination is high, and there are few reported problems with either false-positives or false-negatives. However, some strains of *Proteus* and *Citrobacter* may cross-react.

Fluorescent antibody-Microcolony technique

The combination of the fluorescent antibody technique with growth of microcolonies on nuclepore membranes overcomes many of the problems associated with the conventional fluorescent antibody method. *Salmonella*s were detected within 24 hours at a sensitivity equal to that of cultural methods (Rodrigues and Kroll, 1990), but validation of the technique using a wide range of foods is required.

DNA:DNA hybridisation

Salmonella presented particular problems in development of a DNA hybridisation assay in that it has no virulence or toxin genes equivalent to those in *E. coli* or *Y. enterocolitica*, nor is it possible to find a widely distributed plasmid, or set of plasmid genes that can

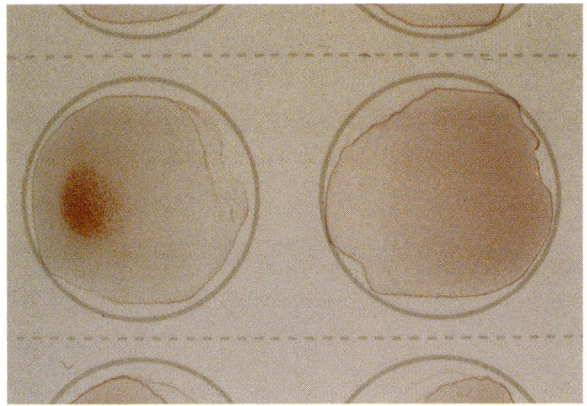

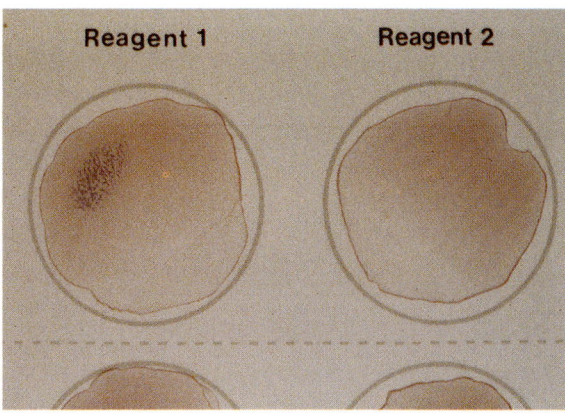

**49 & 50. Detection of salmonellas using the Spectate©
latex co-agglutination test.** Latex co-agglutination tests
are widely used in diagnostic microbiology (see pages 32,
34). The principle has been refined in the Spectate and
similar tests by the use of a mixture of red, blue or green
latex particles, each colour of which is sensitised with a
different serogroup specific antibody. Two reagents are
used to cover groups A, B, C, D, E and G, as well as the
presence of Vi antigen.

The agglutination is carried out on a special card, the
group being identified by the colour of the agglutination.
S. agona, group B, for example, agglutinates the red latex of

reagent 1 (**49**), and *S. montevideo*, group C, agglutinates the
blue latex of reagent 1 (**50**). The reaction with reagent 2 is
negative in each case.

Although most common foodborne salmonellas are
included in the serogroups covered by the test, *S. ealing*
which was implicated in a large outbreak of salmonellosis
due to infant dried milk (see page 66), is a group O
Salmonella and would not therefore have agglutinated with
either reagent. Failure to detect unusual serovars such as
S. ealing may be of little importance in many applications,
but this example illustrates the hazards of over reliance on
any single detection system.

detect all salmonellas (Fitts, 1985). However, a suitable probe has been constructed (Fitts *et al.*, 1983;
Fitts, 1986), and used as an assay method for salmonellas in foods.

A commercial DNA hybridisation assay for salmonellas (GeneTrak©) is available. The sensitivity
approximates to that of ELISA assays, and thus
enrichment is necessary. The performance of the test
is generally considered satisfactory, there being less
than 5% false-positives, and a negligible percentage of
false-negatives (Emswiler-Rose *et al.*, 1987). The major
obstacle to wider use is the use of isotopic labelling.

Phage based detection systems

Two phage-based systems are available for detection
of salmonellas, the Vitek© based on enzyme-labelled
phage, and Lumiphage© based on phage labelled
with a luminescent precursor. Both of these systems
detect *Salmonella* directly after resuscitation, and are
thus faster than methods such as ELISA, which require enrichment stages.

There is little published work concerning the performance of these systems and many laboratories are
probably unwilling to purchase the instrumentation
required for a system using unfamiliar technology.

Identification of salmonellas

There is considerable choice available in the identification of salmonellas, ranging from full characterisation and identification by conventional biochemical
tests, to the use of rapid enzyme systems which indicate
only that an isolate is likely to be *Salmonella*.

Conventional methods

The media and methods recommended are those of
Cowan and Steel (1974). Full characterisation may
be made according to the reactions listed in **Table 32**,
but is not required for routine purposes. Sufficient
tests must, however, be made to enable atypical
strains to be identified. Conventional methods have
largely been replaced by multi-test media and miniaturised kits, but a knowledge of the methodology is
necessary if additional tests are required.

Multi-test media

Multi-test media have themselves been largely
replaced by miniaturised kits, although they are still
used in a number of laboratories. At best, such media

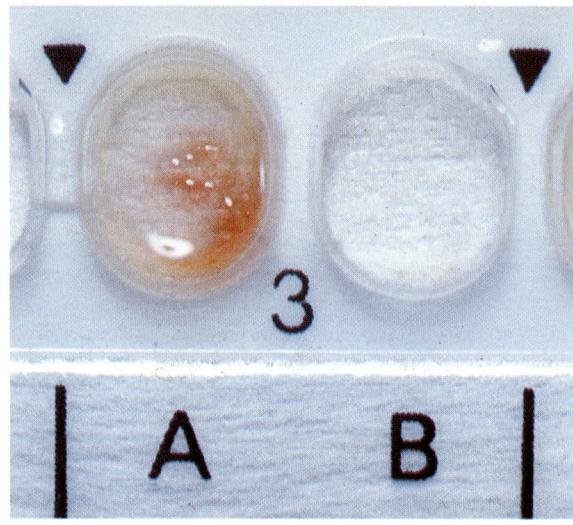 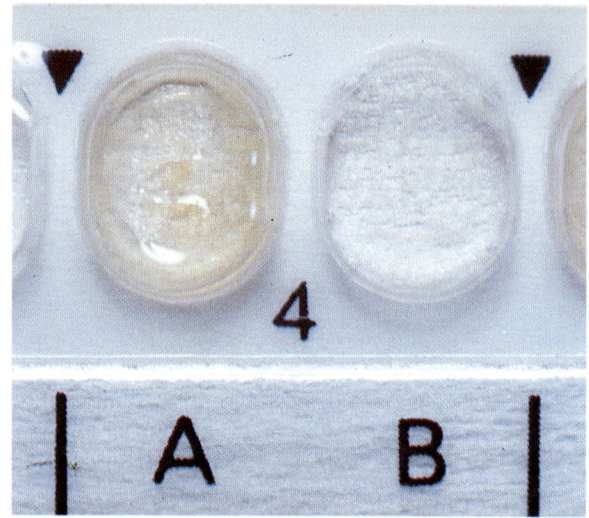

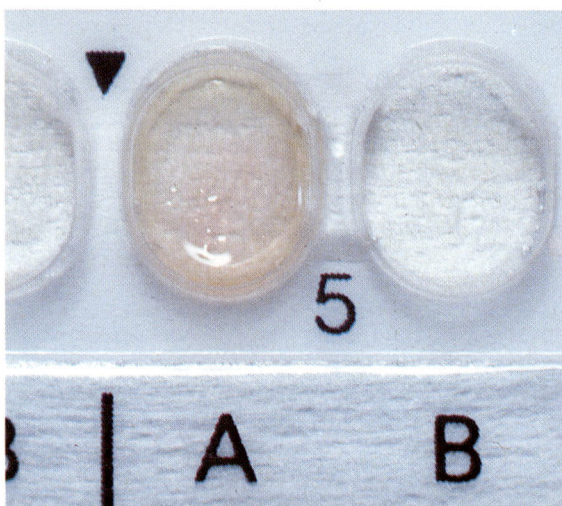 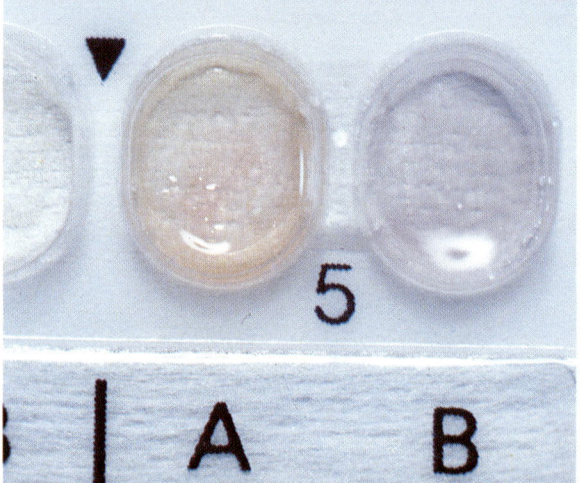

51–54. Use of the API Z© system for screening suspect *Salmonella* colonies. The API Z test can be particularly useful for screening when the appearance of colonies on selective media is equivocal. A red coloration after the addition of the reagent fast violet blue is due to the presence of a specific esterase, and is strongly suggestive of Salmonella (**51**), while a pale beige colour is less strongly suggestive of either *Salmonella*, *Shigella* or *Y. enterocolitica* (**52**). In either case, it is necessary to proceed to further testing.

Problems are more common with oxidase-positive bacteria such as *Pseudomonas* (**53**) than with members of the *Enterobacteriaceae*. For this reason, it is essential to carry out an oxidase test (**54**, cupule B). (*API Z: 2h at 37°C.*)

Flow diagram 8. Serotyping of salmonellas.

NB Serotyping of salmonellas involves the use of live cultures and full aseptic precautions must be taken. Slides used must be sterilised before cleaning. Serotyping should be undertaken only by experienced personnel.

Inoculate moist agar slope with test culture. Incubate overnight at 37°C

|

Test against polyvalent O, polyvalent H (specific and non-specific) and Vi by slide agglutination. (Place drop of saline solution on clean microscope slide, emulsify culture in drop, check for autoagglutination, add undiluted antiserum, mix. Observe carefully for agglutination against a dark background)

|

If there is agglutination only with Vi, heat a cell suspension for 1 h at 100°C to remove Vi, centrifuge, resuspend and retest

|

O group may usually be determined by slide agglutination followed by H group. If culture fails to agglutinate with non-specific H antiserum, determine phase I antigens by further tests with single factor antisera. If culture agglutinates with non-specific antiserum, use Craigie tube method to change the H phase before determining phase II antigens

|

Confirmation of slide agglutination by tube agglutination using formaldehyde treated broth culture is recommended

suffer from general problems of poorly defined colour reactions, and interference between different tests (see page 38). Further, such media are based largely on sugar fermentation and the production of hydrogen sulphide, and are unable to recognise strains which are atypical in these reactions.

Media of this type in current use include Kligler iron medium, Kohn's two-tube medium, lysine-iron medium and triple sugar-iron medium.

The MUCAP test

The MUCAP test is a rapid means of identifying salmonellas from selective media by detection of a specific C_8 esterase. A methylumbelliferyl conjugated substrate is hydrolysed by the esterase releasing the fluorescent umbelliferone. Evaluations have shown that the reaction is of high sensitivity but low specificity, since many genera capable of growth on selective media for salmonellas give positive reactions (Humbert *et al.*, 1989; Aguine *et al.*, 1990), particular problems being caused by *Pseudomonas*. The predictive value is very low when H_2S^+ strains are present.

Miniaturised identification kits

A wide range of kits is available for *Salmonella*. Most are designed to identify the isolate to either genus or species level, and are able to correctly identify atypical strains of *Salmonella*. It is not intended to attempt comparative evaluations of such kits. Many such studies have been published, and in general indicate that kits are comparable in performance, but may differ in their ability to recognise certain organisms (usually non-salmonellas).

A full identification is not always required and the API Z© system permits rapid recognition of *Salmonella*, *Shigella* and *Yersinia*, but does not necessarily distinguish between these species (**Figures 51–54**).

Serological confirmation of identity

In most cases, serological confirmation of identity is required. Polyvalent O antisera may be used in slide agglutination tests, or a co-agglutination test such as Spectate, which also determines the serogroup, may be applied.

Typing of salmonellas

Serotyping

Serotyping of salmonellas is carried out according to the protocol in **Flow diagram 8**.

5. *Shigella*

Introduction

Shigella is a Gram-negative, catalase-positive (with exceptions in one species), rod-shaped bacterium which shares the general properties of the *Enterobacteriaceae* (**Figure 55**). Gas is not usually produced during the fermentation of carbohydrates and, shigellas are non-motile (**Figure 56**; Rowe and Gross, 1984).

Shigella is closely related to *Escherichia coli* and on the basis of their DNA homology form a single species, Brenner (1984) commenting that: '*Shigellae are actually metabolically inactive biogroups of E. coli*'. According to Brenner, the two species remain separate because of the ease of communication provided for medical microbiology, and because of the resistance and confusion that would be caused by reclassification. More recently, Goullet and Picard (1987), using esterase electrophoretic polymorphism were able to distinguish *Shigella* from *E. coli*, and thus substantiate phenotypic differences at a molecular level.

Currently, the genus *Shigella* consists of four species: *Sh. dysenteriae* (subgroup A), *Sh. flexneri* (subgroup B), *Sh. boydii* (subgroup C) and *Sh. sonnei* (subgroup D). Separate species status has been proposed for *Sh. boydii* 13.

All species of *Shigella* are pathogenic to man, causing bacillary dysentery of varying severity. Shigellas are host adapted and cause illness in other primates but, with the exception of rare infections in dogs, are pathogenic to no other animal except under artificially imposed laboratory conditions.

Although infections due to *Sh. dysenteriae* are common in countries where the organism is endemic, the species is rarely found in countries where hygiene standards are high and where the most common species involved in diarrhoea is *Sh. sonnei* (Smith, 1987).

There has been a tendency among food microbiologists to ignore *Shigella* as a foodborne pathogen. It is indubitably true that person-to-person transmission via the oral-faecal route or waterborne transmission are of considerable importance, especially where hygiene standards are low. But Smith (1987) considered that foodborne outbreaks are a significant, if not the major cause of shigellosis in the USA. Earlier, Mossel (1982) stated that while the reported incidence of shigellosis is lower than that of salmonellosis, the problem of foodborne *Shigella* infections is probably greater than generally realised.

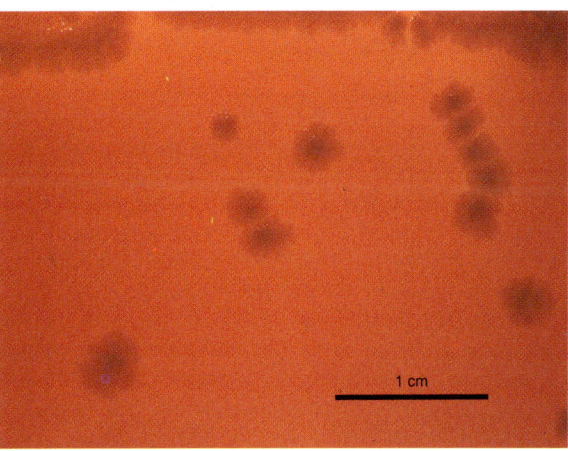

55

55. Growth of *Shigella* species on blood agar. Colonies are non-haemolytic and have a 'fried egg' morphology with a dark centre surrounded by a translucent perimeter. In some cases, the entire colony is translucent and can be very difficult to see. (*Blood agar: 24h at 37°C.*)

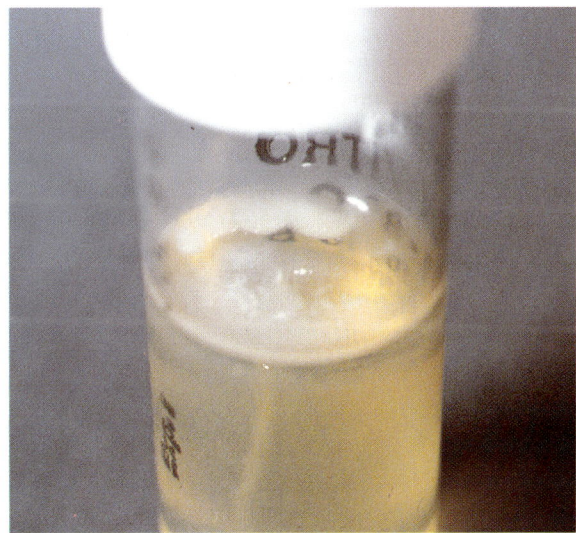

56

56. Lack of motility by *Shigella* species. All species of *Shigella* are non-motile. Although this may be a useful differential feature, it should not be accorded undue weight since non-motile variants of normally motile bacteria including *Salmonella* and *E. coli* are relatively common. Non-motile strains of *E. coli* occur particularly frequently among biochemically inactive strains which closely resemble *Shigella*. (*Motility medium (API-bioMerieux©): 24h at 37°C.*)

Isolation and detection of *Shigella*

Conventional cultural isolation methods

One of the difficulties in assessing the importance of *Shigella* in foods is that there are no reliable and effective enrichment procedures. A three-stage procedure consisting of resuscitation, selective enrichment and selective plating has been used, but there is no agreement concerning choice of media and methods. The situation is complicated by there being no media specifically designed for shigellas, and while modifications to media such as Salmonella-shigella and DCLS have been made to improve recovery, their performance remains variable at best. Recovery of heat-stressed cells presents particular difficulties.

Rapid detection methods

The situation concerning development of rapid detection methods for *Shigella* also reflects the lack of interest in the organism as a foodborne pathogen. However, a small number of rapid detection methods based on both genetic and serological principles have been developed.

Methods for the isolation and detection of Shigella *are discussed in greater detail with respect to choice of method and laboratory practice on pages 97–100.*

Taxonomy of *Shigella*

Differentiation of *Shigella* from other genera

Shigella may be differentiated from other genera on the basis of biochemical reactions, although serological confirmation of identity is required. Either conventional methods, or miniaturised identification kits, may be used. Shigellas are relatively inactive biochemically, and the greatest difficulties arise in differentiation from other biochemically inactive members of the *Enterobacteriaceae*. In this context, the greatest difficulty lies in differentiation between shigellas and anaerogenic, non-motile biovars of *E. coli*, the so-called 'Alkalescens-Dispar' group (**Figure 57**).

The major biochemical characteristics used in differentiation of *Shigella* from closely related genera are listed in **Table 40**.

Serological confirmation is carried out by slide agglutination using polyvalent antisera.

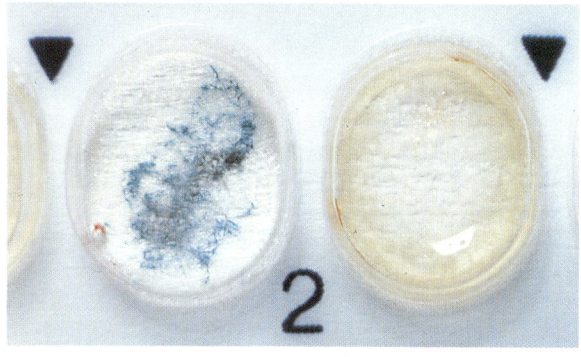

57. Production of β-galactosidase by *Sh. sonnei*. Most strains of *Sh. sonnei* and some of *Sh. dysenteriae* and (*Sh. boydii*) produce β-galactosidase (ONPG positive; left-hand cupule). The enzyme is not present in *Sh. flexneri*. (*API Z: 2h at 37°C.*)

Differentiation within *Shigella*

Shigella is divided into species on the basis of biochemical reactions (**Table 41**). It should be noted that some miniaturised identification kits are unable to differentiate between the species of *Shigella*, and in these cases conventional methods must be used.

Methods used for differentiation of Shigella *from other genera and for differentiation within the genus are discussed in greater detail with respect to choice of method and laboratory practice in* **Laboratory Methods**, *page 100.*

Table 40. Differentiation of *Shigella* from other members of the *Enterobacteriaceae*.

	Shigella	Escherichia	Hafnia	Providencia
β-galactosidase	+/−	+[1]	+	−
Lysine decarboxylase	−	+/−	+	−
Ornithine decarboxylase	−[2]	+/−	+	−
Indole	−/+	+	−	+
Urease	−	−	−	+/−
Phenylalanine deaminase	−	−	−	+
Gas from glucose	−	+[3]	+	+
Acid from				
Lactose	−[4]	+[1]	−	−
Melibiose	−/+	+/−	−	−
Rhamnose	−/+[2]	+/−	+	−/+
Sorbitol	−/+	+[5]	−	−
Xylose	−	+	+	−
Motility	−	+/−[1]	+	+

[1] 'Inactive' strains are usually negative.
[2] *Sh. sonnei* is usually positive.
[3] Occasional strains are negative.
[4] Delayed fermentation is a feature of some strains of *Sh. sonnei*.
[5] Serogroups O517 and occasional strains of other serogroups are negative.

Table 41. Differentiation of species of *Shigella* (Rowe and Gross, 1984; Smith, 1987).

	Shigella			
	dysenteriae	flexneri	boydii	sonnei
Gas from glucose	−	−	−[1]	−[1]
β-galactosidase	+/−[2]	−	+/−	+
Ornithine decarboxylase	−	−	−[3]	+
Indole	+/−[4]	+/−[4]	+/−	−
Acid from				
Dulcitol	−[5]	−[5]	−	−
Lactose	−	−[6]	−[6]	+[7]
Mannitol	−	+	+	+
Raffinose	−	+/−	−	+[7]
Sucrose	−	−	−	+[7]
Xylose	−	−	+/−	−

[1] Gas produced only by certain biovars of *Sh. flexneri* 6 and *Sh. boydii* 13 and 14 strains.
[2] *Sh. dysenteriae* 1 strains are always positive, some other serovars are occasionally positive.
[3] *Sh. boydii* 13 strains are always positive.
[4] *Sh. dysenteriae* 1 and *Sh. flexneri* 6 strains are always negative. *Sh. dysenteriae* 2 strains are positive.
[5] *Sh. dysenteriae* 5 and *Sh. flexneriae* 6 strains may be positive.
[6] *Sh. flexneri* 2a and *Sh. boydii* 9 strains may be positive.
[7] Positive reaction delayed for more than 24 hours.

Typing of *Shigella*

Despite its importance as an enteric pathogen, epidemiological investigations of *Shigella* have been relatively few. However, typing systems analogous to those of *Salmonella* and other bacteria have been devised.

Serotyping

Serotyping schemes based on the O antigen have been successfully devised for *Sh. dysenteriae*, *Sh. flexneri* and *Sh. boydii*. However, *Sh. sonnei* contains only one serovar which exists in two 'phases', I and II, each of which has a distinctive antigen. Recognised serovars of *Shigella* species are listed in **Table 42**.

Slide agglutination is used for the typing of *Shigella*, and typing antisera are commercially available in many countries.

Table 42. Serotyping of *Shigella*.

Species	Serovar	Sub-serovar
Sh. dysenteriae (Sub-group A)	1, 2, 3, 4, 5, 6, 7, 8, 9, 10	
Sh. flexneri (Sub-group B)	1, 2, 3	1a, 1b, 2a, 2b, 3a, 3b, 3c
	4, 5, 6[1]	
Sh. boydii (Sub-group C)	1, 2, 3, 4, 5, 6, 7, 8, 9, 10, 11, 12, 13, 14, 15	
Sh. sonnei (Sub-group D)		

[1] Strains of *Sh. flexneri* 6 resemble *Sh. boydii* immunochemically, and reclassification as *Sh. boydii* has been proposed (Petrovskaya and Bondarenko, 1977).

Phage typing

A number of phage typing schemes for species of *Shigella* have been devised. Interest has been primarily with *Sh. flexneri* and *Sh. sonnei* (Bergan, 1979).

Colicin typing

Colicin typing has been found to be of greatest value with *Sh. sonnei*, where serotyping is not possible. The scheme currently used is based on that devised by Abbott and Shannon (1958) and distinguishes 14 types using 15 indicator strains.

Plasmid profiles

Plasmid profiles of antibiotic-resistant strains of *Sh. dysenteriae* type 1 have successfully been used as part of epidemiological investigations (Haider *et al.*, 1988). With this organism, plasmid profiles may be used to monitor possible pandemic strains as well as individual epidemic strains.

Symptoms of *Shigella* infections

Classical bacillary dysentery is characterised by the frequent passing of liquid stools, which contain blood, mucus and inflammatory cells. Associated symptoms include fever, severe abdominal cramps, tenesmus, ulceration of the gastric mucosa and fluid and electrolyte loss into the intestinal lumen. Symptoms usually appear within 12 to 50 hours of infection, and typically last for 3 to 4 days. In severe cases, overt symptoms may continue for 10 to 14 days or, rarely, longer. In developed countries, death is rare except with patients at the extremes of age, or where the illness is complicated by pre-existing conditions. In tropical developing countries, where standards of nutrition are low, shigella diarrhoea accounts for the death of at least 500,000 young children every year (Guerrant, 1985).

Sh. dysenteriae almost invariably produces symptoms of classic dysentery, which are more severe than those of other species. Classic symptoms may also be produced by *Sh. boydii* and *Sh. flexneri*, but in other cases, symptoms are restricted to relatively mild diarrhoea without ulceration of the mucosa and the resultant bloody stools. Symptoms of *Sh. sonnei* infections usually involve diarrhoea only.

With young children, symptoms of *Shigella* infections often vary according to age. In neonates, among whom infection is rare (see page 93), shigellosis resembles invasive septicaemia (Mobassaleh 1988), while in slightly older babies, up to the age one, vomiting, watery stools and dehydration are common. In children older than one year, symptoms tend to be those of classical dysentery, with bloody stools and abdominal pain. Convulsions may occur in children under 4 years of age, and are common in severe cases which require hospitalisation. In such cases, there may also be headaches, delirium and lethargy.

Systemic complications are relatively rare with *Shigella* infection. A small, but significant, number of cases do progress towards the development of disseminated vascular complications and haemolytic uraemic syndrome (HUS; Obrig *et al.*, 1988). Progression to HUS can be associated with inappropriate treatment with antimicrobial drugs (*cf.* enterohaemorrhagic *E. coli*). Bacteraemia is most common in young people under 16 years of age, and most cases involve patients aged less than 5 years. Bacteraemia is intermittent requiring serial blood cultures for detection and the death rate may approach 50% (Wachsmuth and Morris, 1989).

Shigella infections have also been associated with chronic rheumatoid diseases including Reiter's syndrome and reactive arthritis (Archer, 1985). Other, less specific, long-term effects include retarded height gains among affected children in developing countries (Anon, 1985).

Persons at particular risk from *Shigella* infection

Susceptibility

Greater than normal susceptibility to *Shigella* infections is found among the old, the young (with the exception of neonates) and those with an underlying disease. A gradual increase in the average age of males infected with *Sh. flexneri* in the USA (5 to 24 years between 1968 and 1984) suggested increased male homosexual transmission (Wachsmuth and Morris, 1989).

Exposure

Exposure to shigellas is greatest in countries where hygiene is poor, ambient temperatures are high and the disease is endemic. Travellers from countries of higher hygienic standards are at greatest risk, and *Shigella* has been reported to be the second most common cause of 'travellers' diarrhoea' for US citizens (DuPont *et al.*, 1982).

As with salmonellas and other food poisoning organisms, local breakdown in hygiene in developed nations can put groups of people at risk. Schools, asylums and childrens' homes are of relatively high risk and sporadic, large-scale outbreaks have been associated with military barracks, summer camps and religious communities. One of the largest out-breaks of this type involved 1,328 culture-confirmed cases of *Sh. sonnei* gastroenteritis in New York during 1986-87. Most of the victims were tradition-observant Jews, and it is likely that as many as 13,000 people were affected in total. Smaller, related outbreaks occurred in other parts of the USA. In cases such as this, person-to-person spread is of greatest import-ance, and control lies in personal hygiene, although measures based on this are difficult to impose where young children are involved.

The urban poor, migrant workers and native peoples in the USA and Australia are at a continuing high risk of exposure.

Treatment of *Shigella* infections

In most cases, shigellosis is a self-limiting disease, and no treatment is required except in severe cases where fluid replacement is seen as the basis for success-ful therapy (Berkow, 1982). In milder cases, where fluid replacement is not necessary, a non-specific treatment such as kaolin may be useful.

Antibiotic therapy may be used if symptoms are particularly severe or, more commonly, to decrease morbidity and limit the secondary spread of infection (Weissman *et al.*, 1974). Ampicillin and cotrimoxazole are recommended, but the situation has been com-plicated by the increasing incidence of multiple drug resistance among shigellas. Since multiple, transfer-able drug resistance was first noted in Japan by Ochiai *et al.* (1959), there has been a dramatic increase in drug resistance in the USA (Neu *et al.*, 1975), England and Wales (Thomas and Tillett, 1973; Gross *et al.*, 1981) and elsewhere. For this reason, the decision to use antimicrobials in mild cases should be balanced against the risk of producing resistant strains (Anon, 1988a).

A more serious risk associated with resistant strains is that inappropriate administration of antimicrobials is a risk factor for the development of haemolytic uraemic syndrome (*cf.* enterohaemorrhagic *E. coli*, page 109; Butler *et al.*, 1987). This complication has arisen where resistant strains of *Shigella* have been treated with ampicillin or cotrimoxazole (Parsonnet *et al.*, 1989).

Infections caused by multiply resistant strains of *Shigella* that require antimicrobial therapy may be treated with nalidixic acid or norfloxacin but care must be taken with children because of possible side effects (Anon, 1988b).

In developing countries, management of shigellosis in patients at high risk of fatality is difficult, even where relatively intensive patient care is possible (Bennish *et al.*, 1990). However, early antibiotic therapy may be helpful.

Mechanisms of pathogenicity

Infective dose

The infective dose of *Shigella* is low and considered to be from 10^1 to 10^4 cells (Bryan, 1979; Morris, 1986) depending on the virulence of the infecting strain and on host factors.

Genetic control of virulence

At least three separate genetic determinants essential for full expression of virulence have been identified on the *Shigella* chromosome, but the genetic deter-minants for the first stage in infection, invasion, are located on a large (120,000 to 140,000 dalton) plasmid, which is present in all virulent shigellas and in entero-invasive *E. coli* (Sansonetti *et al.*, 1982,1983; Adler *et al.*, 1989).

Although the large 'virulence' plasmid is associated with invasive ability, its functions vary from species to species. For example, in *Sh. sonnei* the plasmid also codes for synthesis of the specific O side chains of the lipopolysaccharide (Sansonetti *et al.*, 1981). Loss of the plasmid leads both to a loss of virulence and a change from the Form I cells, which produce smooth colonies to the Form II which produce rough colonies. In contrast, in *Sh. flexneri* the large plasmid has no role in production of the O antigen, and loss of the plasmid leads to loss of virulence, but not to a change from a smooth to a rough colonial morphology (Hale *et al.*, 1983). *Sh. flexneri* also carries a plasmid, pWR100, which specifies the cell-bound haemolysin involved in early release of the organism from endocytic vacuoles during invasion (see below).

Virulent strains of *Sh. dysenteriae* contain both the large 140,000 dalton plasmid and a small plasmid (6,000 daltons), which encodes for virulence and for production of the O antigen (Watanabe and Timmis, 1984). In contrast to enteroinvasive *E. coli*, the small plasmid is stable in *Sh. dysenteriae*, stability being associated with the histidine region of the chromosome, which is unique to that species of *Shigella* (Aqua *et al.*, 1988). This plasmid is not required for cell invasion.

In addition to invasion, plasmid genes are involved in the early blockage of respiration in infected cells and their subsequent death, but the mechanism is not known (Scotland, 1988).

The plasmid-controlled invasion process in shigellas is temperature dependent, and virulence is expressed at 37°C, and not at 30°C (Maurelli *et al.*, 1984). Loss of virulence is not due to loss of the plasmid.

As well as the association with the stability of the small plasmid of *Sh. dysenteriae*, chromosomally encoded functions have further roles in the full virulence of shigellas, but the nature of these is poorly defined (Kopecko *et al.*, 1985). The production of Shiga toxin and related cytotoxins also appears to be under chromosomal rather than plasmid control.

Motility

Shigellas are non-motile, and it is assumed that movement through the intestinal mucosa is passive. Production of a mucinase has been reported (Prizant and Reed, 1980), and degradation of the glycoproteins of the mucosa would permit direct access to the surface of the epithelial cells.

Adherence and penetration

Shigella adheres to the colon, the adhesins being outer membrane proteins. Entry into the epithelial cells is by a directed phagocytosis, involving a local and transient polymerisation of actin. Clathrin is also involved in invasion, possibly by supporting a membrane turnover process (Clerc and Sansonetti, 1989). The phagocytic signal is produced only by virulent strains (Hale *et al.*, 1979), and may be associated with a plasmid-mediated 101 Kda protein which binds haem and may thus disguise the bacteria as a desirable molecule for intestinal epithelial cells (Stagard *et al.*, 1989). Immediately after penetration, shigellas are enclosed in membrane bound vacuoles, but escape into the cytoplasm is rapid. Escape is promoted by a cell-bound haemolysin (Sansonetti *et al.*, 1986).

Multiplication occurs in the cytoplasm, *Shigella* being a specialised intracellular pathogen. The organism spreads laterally, invading adjacent cells and initiating an acute inflammatory response involving neutrophils in the *lamina propria*. A 120 Kda outer membrane protein plays an important role in lateral spread by allowing interaction with F-actin microfilaments within the cell (Bernadini *et al.*, 1989). Cell damage results in ulcerative lesions, which vary in severity up to a diffuse ulceration. Damage is similar to that caused by enteroinvasive *E. coli*, which is illustrated on page 114.

Invasive ability is seen as a major virulence factor in shigellas. Invasive strains produced an inflammatory response and structural changes in guinea-pig intestine, irrespective of toxin production, while non-invasive strains had no effect (Formal *et al.*, 1972). These experiments were substantiated by feeding tests with human volunteers (Stephen and Pietrowski, 1981).

Systemic spread

Infection by shigellas is usually limited to the outer layers of the colon. In cases other than those involving neonates, further spread usually only occurs in the immunocompromised (Struelens *et al.*, 1985).

Survival and proliferation in host tissue

Resistance to phagocytosis

Professional phagocytes, primarily macrophages, are killed by invading shigellas following their uptake of the organism. Superoxide dismutase plays a major role in protecting the organism from the respiratory burst (Frorgren *et al.*, 1990), while the cell-bound haemolysin, which promotes early release from endocytic vacuoles, may be involved in the killing of the macrophages (Clerc *et al.*, 1987).

Shigella resembles *Salmonella* in inducing interferon-γ production, but invading cells must be metabolically active (Hess *et al.*, 1990).

Effects of iron deprivation

Aerobactin is typically produced by clinical strains of *Sh. boydii* and *Sh. flexneri*, and enterochelin by *Sh. dysenteriae* and *Sh. sonnei*. Some strains produce both siderophores.

The ability to scavenge iron by the aerobactin system is likely to be important in the initial stages of cell penetration rather than after invasion (Williams and Roberts, 1985), although aerobactin synthesis does give a selective advantage during intracellular growth. Aerobactin does not appear to be a major virulence factor (Payne, 1988).

Toxin production

Shiga toxin

Invasive strains of *Sh. dysenteriae* type 1 produce high levels of a cytotoxin usually referred to as Shiga toxin. Other serovars of *Sh. dysenteriae*, as well as other species of *Shigella*, produce low levels of cytotoxic activity. Despite extensive studies, the role of Shiga toxin is thus not well defined (Neill *et al.*, 1988).

Classic Shiga toxin is a protein exotoxin (Keusch *et al.*, 1972) that, in addition to cytotoxicity, has enterotoxic, neurotoxic and other biological activities (**Table 43**).

Shiga toxin is usually produced during the exponential phase of growth. During this phase toxin is secreted into the periplasmic space of the cell, but during the subsequent stationery phase large quantities appear in the culture medium (Donohue-Rolfe and Keusch, 1983; Keusch *et al.*, 1985). The toxin consists of two polypeptide subunits, A and B, which in the intact toxin are combined in the ratio 1:5-7. Subunit A is of molecular weight 30,500 to 32,000 daltons, and B of molecular weight 5,000 to 6,500 daltons. Subunit A may be dissociated into A_1 and A_2 subunits, with molecular weight 28,000 to 29,000 and 3,000 respectively, the two subunits being linked by a disulphide bridge. Structure-function analysis of Shiga and Shiga-like toxins has shown the A subunits to possess enzymatic activity identical to that of the

plant toxin ricin, and to be unlike any previously characterised bacterial toxin (Jackson, 1990).

Shiga toxin is synthesised as a zymogen and requires activation before biological activity can be demonstrated. During *Shigella* infections, activation by cellular proteases probably occurs after the organism has penetrated the epithelial cells (Smith, 1987). HeLa cells have two receptors for Shiga toxin, globotriaosylceramide (Gb_3), which is of high affinity but low capacity, and a second receptor of high capacity but low affinity (Jacewicz *et al.*, 1989). The mechanism of the events which follow binding, and which lead to inhibition of protein synthesis and the observed effects listed in **Table 43**, are illustrated and summarised in **Figure 58**.

Sensitivity of cells to Shiga toxin is proportional to the number of high affinity sites present. The host cell receptors are unusual in being developmentally regulated, and their absence in neonates is thought to explain both the low incidence and unusual symptoms of shigellosis in that age group (Mobassaleh *et al.*, 1988).

A number of attempts have been made to evaluate the role of Shiga toxin in diarrhoeal disease by using mutants of *Sh. dysenteriae* type 1, which are unable to synthesize the toxin. The finding that some atoxigenic mutants were still capable of causing diarrhoea in man suggested a limited role for Shiga toxin. However, a chlorate resistant derivative of *Sh. dysenteriae* associated with a loss of high-level toxin production has been found to produce low levels of a cytotoxin that is neutralisable with antiserum to Shiga toxin, but which is determined by different genes to those coding for synthesis of classic Shiga toxin (Neill *et al.*, 1988). It is suggested that this low level of cytotoxic activity is sufficient to play a role in disease. Low-level cytotoxic activity in other shigellas is due to a toxin similar to that produced by the chlorate resistant *Sh. dysenteriae* mutant.

Table 43. Biological activities of purified Shiga toxin (Keusch, *et al.*, (1985).

Activity	Effect
Cytotoxicity	Death of cells in tissue culture
Enterotoxicity	Fluid accumulation in rabbit ileal loops
Neurotoxicity	Paralysis and lethality in animals after parenteral injection with toxin
Neuronotoxicity	Cytotoxic effect on rat neurons
Inhibition of protein synthesis	Inactivation of 60S ribosomes with resultant cessation of peptide chain elongation

58. Mechanism of Shiga toxin. Shiga toxin enters the cell by receptor mediated endocytosis, the B subunit of the toxin molecule binding to the host cell receptor. The clathrin coated pit is pinched off to form a vesicle, is acidified and may fuse with lysosomes.

In the cytosol, the A_1 fragment is probably generated by proteoloytic nicking and reduction of S:S bonds of the A subunit. The A_1 subunit binds to 60S ribosomes, protein synthesis is blocked followed by cell death. (Reproduced with permission from O'Brien, A.D. and Holmes, R.K. *Microbiological Reviews* **51**: 206–220. © 1987 The American Society for Microbiology.)

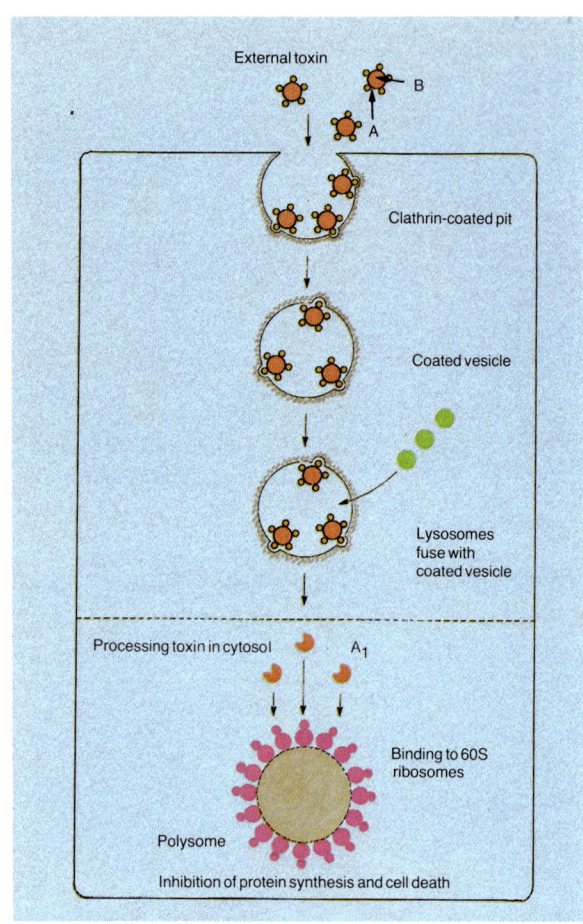

Other toxins produced by Shigella

A number of studies have suggested that shigellas may produce further toxins, which are distinct from classical Shiga toxin and which, unlike the cytotoxins produced by shigellas other than *Sh. dysenteriae*, are not neutralised by antiserum to Shiga toxin. Many of these toxins appear to be cell-bound and their in vivo significance requires assessment, although their importance may lie in stages of infection after invasion of epithelial cells (Scotland, 1988). Other possible *Shigella* toxins are summarised in **Table 44**.

Table 44. Other toxins produced by shigellas.

Toxin	Activity
Cytotoxin	Cytotoxic to HeLa cells, in most strains not neutralised by anti-Shiga toxin. Partially neutralised in others (Bartlett, *et al.*, 1986). Other cytotoxins have been isolated, including a 100–125 Kda protein toxin produced at low levels (Ashkenazi *et al.*, 1990). The significance of these toxins and their relationships remain unresolved.
Cell-free lysate	Induced alterations in myoelectric activity of rabbit ileal segments, also caused severe to moderate inflammation of the mucosa and enterocyte necrosis (Fernandez *et al.*, 1984).
CHO elongation factor	Caused elongation of Chinese hampster ovary cells. Activity low compared with *E. coli* LT Taleda *et al.*, 1979).
Heat stable enterotoxin	Active against infant mice (Ketyi, 1985).

In-vitro assessment of virulence in shigellas

Methodology of the congo red binding and salt aggregation tests is described in greater detail in **Laboratory Methods**, *page 100.*

Both the ELISA and the DNA hybridisation assays developed for the detection of shigellas (see pages 98–99) detect only virulent strains, and therefore may also be used for virulence assessment. Where animal tests are required the Sereny test, a whole animal test that measures the ability of invasive *Shigella* species to cause a keratoconjunctivitis in adult guinea-pigs or rabbits, correlates well with virulence. Many workers prefer to avoid animal tests and a relatively simple tissue-culture test has been developed (Oaks *et al.*, 1985) using HeLa cells which may be suitable for use in food microbiology laboratories (Madden *et al.*, 1986).

A number of workers have employed the ability of wild-type strains of *Shigella* to bind congo red dye as a marker for virulence. A direct relationship exists between dye binding and the existence of the large virulence plasmid in *Sh. flexneri* and *Sh. boydii* but not in *Sh. dysenteriae* type 1 or *Sh. sonnei* (Qadri *et al.*, 1988). Dye binding is a reliable indicator of virulence in smooth strains, but not in rough.

An alternative test to binding of congo red dye is the salt aggregation test, which determines cell hydrophobicity and which is also a reliable indicator of virulence in smooth strains.

Detection of Shiga toxin

The conventional means of detecting Shiga toxin is the HeLa or Vero cell tissue culture assay. Inhibition of protein synthesis by the toxin leads to sloughing of the cell culture monolayer (Wachsmuth and Morris, 1989).

Immunoassays for Shiga toxin based on ELISA and immunoblotting techniques have been available, although not on a commercial basis, for some years, and are sufficiently sensitive to detect toxin in faeces. Attention has recently been paid to ELISA assays in which Gb_3, the natural receptor, is used to bind the toxin. Such assays are completed within one day, and are able to detect all cytotoxins which bind to Gb_3 (Ashkenazi *et al.*, 1990).

A number of gene probes have been developed for the genes encoding toxin production. Obviously, these do not determine the actual presence of toxin, but an indication of the potential of an isolate to produce toxin can be of great value in epidemiological and other studies. To date, these probes have largely been used for research purposes.

Human carriage of *Shigella*

Asymptomatic carriage of *Shigella* may persist for several months following recovery from symptoms, and is a factor in the spread of shigellosis (Bryan, 1978). Carriage of shigellas has been studied less systematically than that of salmonellas, and the true incidence is not known. Similarly, it is likely that at least some outbreaks where contamination of food has been attributed to a carrier food handler, were in fact due to a food handler who was actually suffering clinical symptoms of shigellosis.

In recent years, there have been anecdotal accounts of transmission of shigellas to children from primates such as monkeys. Particular attention has been paid to monkeys kept by itinerant photographers at Mediterranean holiday resorts, which pose with children. While there is no doubt that such monkeys are frequently poorly treated and unhealthy, there is little evidence of primate: human infection in this way. The only well-documented case involved a case of *Sh. flexneri* infection acquired from a monkey which touched a child's ice cream in the pet's corner of a large store (Rothwell, 1981).

Shigella in the environment

Shigellas are found only in the intestinal tract of man and other primates, or in materials contaminated with the faeces of man and primates. It is thus possible to isolate shigellas from sewage and sewage-contaminated river and estuarine water, particularly in areas where shigellosis is endemic. Shigellas are also present in sewage sludge and may also be isolated from soil fertilised with sludge or with nightsoil.

It has been generally assumed that *Shigella* will survive for only a very short time in the environment. This assumption appears to be based on the concept of shigellas as being particularly delicate micro-organisms (see below), and may not always be justified. Few studies have been published, but in one case shigellas survived for more than 7 months in anaerobically digested sewage sludge applied to land (Hyde, 1976).

Isolation of *Shigella* from uncooked broiler chickens has been reported (Bok *et al.*, 1986). This finding could not be explained in terms of contamination by carrier food handlers and the source of the organism is not known.

Foods commonly involved in transmission of *Shigella*

Except on very rare occasions, (see page 95), *Shigella* is acquired from man. In the case of foodborne transmission, this means contamination of food which is not to be cooked before eating either from sewage contaminated environmental sources such as water or soil, or from an infected, or carrier, food handler.

In countries where hygiene is poor, both sources of contamination are important. Salads, which may be grown in contaminated soil, washed in contaminated water or prepared by an infected food handler, have long been considered high-risk items along with drinking water and ice cream. Any food is at risk of contamination, and an outbreak of dysentery in the Netherlands due to *Sh. flexneri* 2, which caused 14 deaths among *ca* 100 at risk was attributed to frozen shrimps imported from the Far East (Kayser and Mossel, 1984).

In countries where the standard of hygiene is high, foodborne shigellosis almost invariably involves contamination by a food handler (except, of course, where imported food contaminated at source is the vehicle). In the majority of cases, contamination occurs either in a family or a food service operation. In the latter situation, this has been attributed to the employment of young and inexperienced food handlers and the failure to provide adequate hygiene training. In some cases, low wages and lack of job security place pressure on handlers to work while unwell. In many cases, inadequate refrigeration of prepared foods has been an important secondary factor in the transmission of shigellas.

It should be appreciated that any food, which is not heated directly before consumption, handled by a *Shigella* infected person with poor personal hygiene, is a potential vehicle of foodborne shigellosis (Smith, 1987). Examples of recent outbreaks of foodborne shigellosis are summarised in **Table 45**.

Table 45. Examples of foodborne outbreaks of *Shigella* infection.

Food	Circumstances
Potato, tuna and egg salads, coleslaw[1]	Vehicle was dressing common to all salads which was probably contaminated by an asymptomatic carrier of *Sh. sonnei*. Foods were kept for long periods at room temperature during preparation and serving. 248 people were affected
Spaghetti[1]	12 of 26 food handlers had diarrhoea including 1 handler whose stools yielded *Sh. boydii*, the causative organism. Both sauce and pasta were prepared several hours in advance, but were said to have been heated before serving. 176 people were affected
Tuna and other salads[1]	Handler preparing salad had diarrhoea on first day of outbreak and was stool positive for the causative organism, *Sh. sonnei*. Handler had been exposed to a child with severe diarrhoea. 282 people were affected.
Shrimp cocktail[2]	Food prepared from frozen, cooked shrimps which had been contaminated after cooking in the country of origin. The causative organism was *Sh. flexneri*. Up to 173 people were affected with 14 deaths among the elderly patients
Lettuce?[3]	Several hundred airline passengers including members of the Minnesota Vikings football team were affected on about 140 internal air journeys. The cause has not been fully identified

[1] Smith (1987).
[2] Kayser and Mossel (1984).
[3] Anon (1988a).

Factors affecting the growth and survival of *Shigella* in foods

Although shigellas are often considered to be fragile and to survive for only limited periods outside the host, there is evidence that, under some circumstances at least, survival is prolonged (Bryan, 1978). It is further incorrect to assume that shigellas are unable to grow in foods, as growth will take place in many foods if refrigeration is poor. However, studies of growth and survival of *Shigella* have been extremely limited in comparison with other foodborne pathogens.

Temperature

All species of *Shigella* have an optimum for growth at 37°C, but there is species variation in maximum and minimum growth temperatures. *Sh. sonnei* will grow over the range 7 to 46°C, and *Sh. flexneri* over the range 10 to 44°C (Fehlhaber, 1981).

Shigellas are readily destroyed by heating, most strains being killed by heating at 63°C for 5 minutes.

Irradiation

Irradiation has been proposed as a means of controlling *Shigella* in foods imported from areas where the organism is endemic (Kayser and Mossel, 1984).

Sensitivity of shigellas is similar to that of other members of the *Enterobacteriaceae*, a dose of 3 Kgy ensuring inactivation.

pH value

Shigellas have been demonstrated to grow over the pH range 5.5 to 7.0 (Hentges, 1967), although growth at higher or lower values do not appear to have been investigated. At acid pH values there is interaction with organic acids, and in the presence of formic or acetic acids there is no growth below pH 6.0.

Survival of shigellas in low pH value environments is poor, both *Sh. flexneri* and *Sh. sonnei* being inactivated within 30 minutes at pH 3.5 and within 4 hours at pH 4.0 and 4.5 (Hentges, 1967).

Curing ingredients

The general relationship of *Shigella* to curing ingredients is similar to that for *Salmonella* (page 70). There is species variation with respect to nitrite sensitivity, *Sh. sonnei* being inhibited by 700 mg/l and *Sh. flexneri* by 450 mg/l at pH 5.5 (Nakamura *et al.*, 1964).

Laboratory methods

Conventional cultural isolation techniques

It has been noted above that in contrast to *Salmonella* and, indeed, pathogens of considerably less importance, little effort has been expended in developing media for *Shigella*. It seems likely that improvement of this situation will lie not in conventional methods, but in newer techniques such as DNA hybridisation. However, even here, a lack of interest in the organism means that commercial kits are unavailable, and most laboratories still rely on conventional three-stage isolation.

Resuscitation (pre-enrichment)

Although the phenomenon of sub-lethal damage has been recognised in shigellas (Tollison and Johnson, 1985), no generally accepted resuscitation procedures have been devised. However, Tollison and Johnson recommended the plating of food samples onto the non-selective tryptic-phytone-glucose agar medium and incubating for 8 hours at 35°C before overlaying with selective media.

59

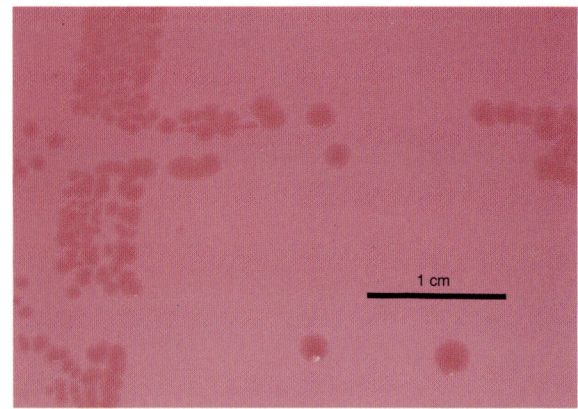

1 cm

60

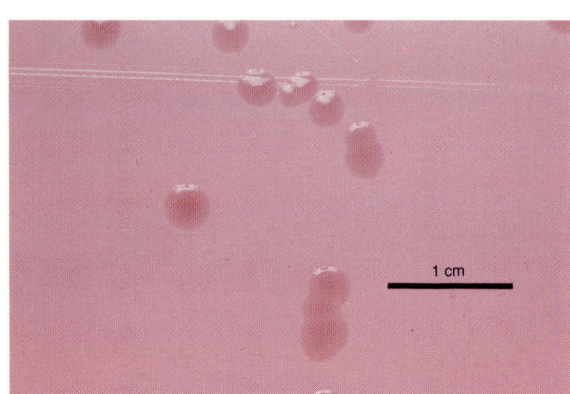

1 cm

61

1 cm

59–61. Isolation of *Shigella* species on general-purpose selective media for the *Enterobacteriaceae*.
Media of relatively low selectivity are often preferred for the isolation of *Shigella*, although this is not possible if competing bacteria are present in significant numbers. Colonies are usually lactose-negative, and while boring is a descriptive adjective rarely applied to colonial morphology, it is apt with respect to the appearance of *Shigella* on both MacConkey-purple (**59**) and Drigalski (**60**) agars. The appearance may be of diagnostic value for the experienced microbiologist, but differentiation from *Proteus* (**61**) may be difficult on MacConkey agars. (*All media: 24 h at 30°C.*)

Selective enrichment

Gram-negative broth (Hajna, 1955) is generally considered to be 'useful' for enrichment of foodborne shigellas, and may also be of value in examination of stool samples (Rowe and Gross, 1984). The performance of Gram-negative broth is improved by adjusting the pH value to 6.0-7.0 after addition of the food sample (Morris, 1984). Selenite-cystine broth, in parallel with Gram-negative broth, has also been recommended (Twedt, 1978).

Incubation of both Gram-negative and selenite-cystine broths should be for 18 to 24 hours at 35°C.

Selective plating media

Media for the selective plating of shigellas are either weakly selective, general-purpose media for recovery of the *Enterobacteriaceae*, or more strongly selective media devised primarily for recovery of *Salmonella*. Some strains of *Shigella* are sensitive to bile, and it is recommended that two media of different selectivity should be used in parallel. The use of each type of media is illustrated in **Figures 59–65**.

A number of selective media have been evaluated from the viewpoint of recovery of heat-damaged shigellas (Smith and Dell, 1990). Most media were unsuitable for this purpose, although some improvement in performance could be achieved by addition of 1% sodium pyruvate.

Rapid detection methods

Serological techniques

Two serological techniques have been used, ELISA and latex agglutination. An ELISA assay, which detects the virulence marker antigen (VMA) of virulent shigellas was described by Pai *et al.* (1985). Virulent strains of all four species are detected by this technique, but avirulent strains are not. Virulent strains of enteroinvasive *E. coli* also possess the VMA, and hence are also detected by the assay.

A commercial kit system, Bactigen©, has been developed, which detects both *Salmonella* and *Shigella*. An evaluation showed the kit to be satisafactory for detection of *Salmonella*, but its relative efficiency for *Shigella* is not known.

DNA:DNA hybridisation

In contrast to *Salmonella*, the existence of virulence associated genes in *Shigella* means that construction of

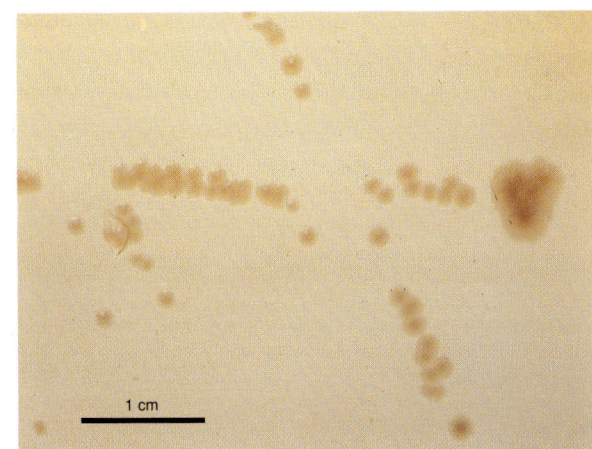

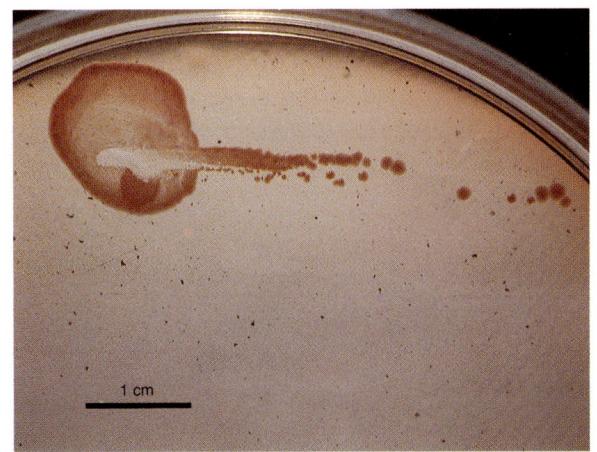

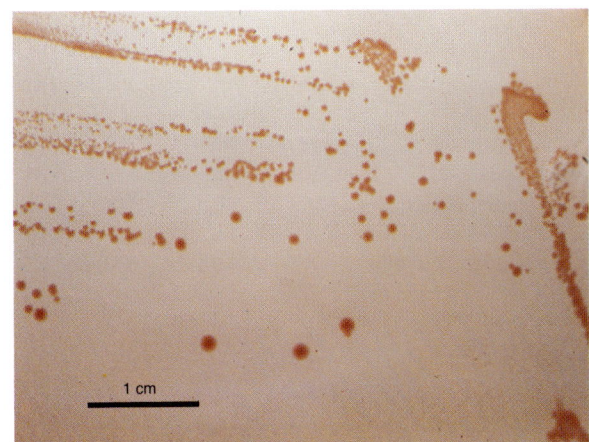

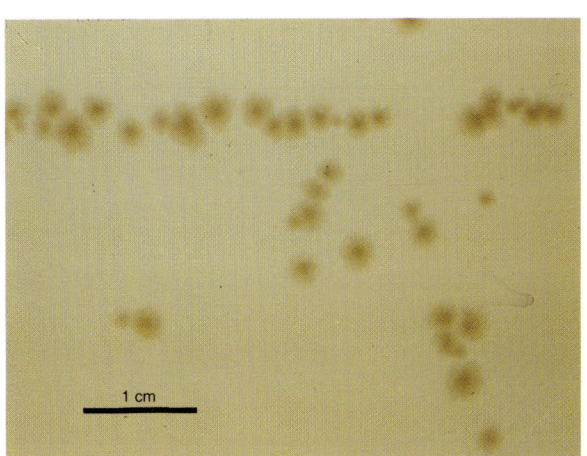

62–65. Isolation of *Shigella* species on specially-designed or modified media. The recovery of *Shigella* remains difficult even on media designed, or modified, for the purpose. Salmonella-shigella agar (**62**) supports only moderate growth of many shigellas, despite modifications to improve their recovery. Colonies are readily confused with those of H_2S-negative strains of *Citrobacter* and *Proteus*. Brilliant green agar (**63**) was not designed for the recovery of shigellas and growth is often very poor. DCLS agar (**64**) recovers most shigellas, although colony size is often small.

Typical shigellas form colourless colonies, but some strains of *Sh. sonnei* form pale pink colonies due to late lactose fermentation. These colonies are usually readily distinguished from those of *E. coli* and other strongly lactose fermenting genera which produce bright red colonies. Hektoen enteric agar (**65**) supports relatively good growth of many strains of *Shigella*. The pale green colouration of colonies permits differentiation from *Salmonella* and from many other members of the *Enterobacteriaceae*. (*All media: 24 h at 30°C.*)

specific probes is relatively straightforward.

Assays have been developed by a number of workers, including Sethabutyr *et al*. (1985) and Venkatesan *et al*. (1988,1989). Most assays have used invasion essential gene segments as probes and thus detect virulent strains only. However, problems arise due to the loss of the virulence plasmid and selective deletion of invasion associated genes, during the cultivation and storage of isolates and the *ipaH* gene sequence, which is present both on the chromosome, and the virulence plasmid was found to be more effective in the detection of both virulent and

avirulent strains (Venkatesan *et al*., 1990).

Both isotopic and non-isotopic labelling has been used for probes for *Shigella*, the latter having obvious advantages in food laboratories.

The polymerase chain reaction

The polymerase chain reaction (PCR) has been applied to the detection of invasive strains of *Sh. flexneri* in foods, and offers a sensitive method which can detect low numbers of cells in less than one day, even when the numbers of non-shigellas present is high (Lampel *et al.*, 1990). Problems associated with the instability of the virulence plasmid are avoided, and damaged cells which cannot be recovered by conventional methods are detectable. However, dead cells may be detected, although this problem can be overcome by incorporating a brief growth phase (3–5 cell divisions) followed by dilution.

Identification of shigellas

Differentiation from other genera

Shigella may be differentiated from other members of the *Enterobacteriaceae* by conventional test methods according to the reactions in **Table 40**, page 89 or using miniaturised identification kits.

Particular difficulties may be encountered in differentiating shigellas from biochemically inactive strains of *E. coli*, and there may also be difficulties in differentiation of aerogenic biovars of *Sh. flexneri* 6 (Newcastle and Manchester biovars) and *Sh. boydii* 13 and 14, or slow lactose fermenting strains of *Sh. sonnei* with more typical strains of *E. coli*. Miniaturised identification kits may misidentify lactose fermenting strains of *Sh. sonnei* as *E. coli*.

Care must also be taken to avoid confusing *Shigella* with some strains of *Hafnia alvei*, *Providencia* and *Aeromonas* (Rowe and Gross, 1984). The oxidase test is useful for differentiating *Aeromonas* from *Shigella*, but false-negative reactions may occur with colonies taken directly from selective isolation plates.

Flow diagram 9. Congo red test for the determination of virulence of shigellas.

Streak onto Congo red agar
(Tryptone-soy broth + 0.6%
yeast extract, 1.5% agar
and 0.003% Congo red)

|

Incubate 18 h at 37°C

|

Examine colonies for uptake of Congo red

|

Test positive colonies for
rough strains by autoagglutination

Flow diagram 10. Salt aggregation test for the determination of virulence of shigellas.

Grow cells on tryptone-soy agar: 18 h at 37°C

|

Test for presence of rough strains by autoagglutination

|

If rough strains absent, test for aggregation in ammonium sulphate solution in 0.02 M phosphate buffer in concentrations ranging from 0.5 to 0.3 M

Serological confirmation of biochemical identification of *Shigella* is required. Confirmation should be made by slide agglutination using polyvalent O antisera.

Differentiation of species of shigella

Differentiation should be made according to the characteristics listed in **Table 41**, page 89. Identification kits may be unable to reliably differentiate between species and, especially with species other than *Sh. sonnei*, conventional media and methods as described by Cowan and Steel (1974) should be used.

Typing of shigellas

Serotyping

The basic serotyping of shigellas is carried out as for salmonellas (see page 85), but only O antisera are used. Untreated cell suspensions can normally be used but, if agglutination is poor, the suspension should be heated for 15 to 30 minutes at 100°C and then re-tested.

Determination of virulence using the congo red and salt aggregation tests

Both the Congo red and salt agglutination tests offer a simple and effective means of determining virulence in shigellas. These tests are particularly useful in laboratories with limited facilities. The methodology of the Congo red test and the salt aggregation test is described in **Flow diagrams 9** and **10** respectively.

6. *Escherichia coli*

Introduction

Escherichia is the type genus of the family *Enterobacteriaceae*, and shares the common characteristics of that family. In contrast to *Salmonella* and *Shigella*, most escherichias are able to ferment lactose with production of acid and gas. However, this and other features may be readily lost by mutation, while 'forbidden' traits may be acquired (**Figures 67–70**).

The genus *Escherichia* contains six species, *E. coli*, the type species, *E. adecarboxylata*, *E. fergusonii* (Farmer *et al.*, 1985b), *E. hermanii* (Brenner *et al.*, 1982a), *E. vulneris* (Brenner *et al.*, 1982b) and *E. blattae* (Burgess *et al.*, 1973). The reclassification of *E. adecarboxylata* in a new genus *Leclercia* has been proposed (Tamura *et al.*, 1986), but the status of this genus is uncertain (Jones, 1988). With the exception of *E. blattae*, a denizen of the hind gut of cockroaches, all of the species of *Escherichia* have been isolated from clinical specimens. *E. coli* is closely related to *Shigella*, and on some criteria the two form a single species (see page 87).

E. coli is the most common aerobic organism in the gastrointestinal tract of man and many other animals. In this connection, the organism has a traditional, and often misunderstood, role in food and water microbiology as an index of faecal pollution.

An association between *E. coli* and animal disease was made as early as the 1890s, while its association with disease in children was confirmed in the late 1940s, following investigations into outbreaks in hospital nurseries. Despite this, *E. coli* has largely been regarded as an opportunistic pathogen, and in the food industry interest in the organism has largely been restricted to its role as an index organism.

However, in recent years *E. coli* has been recognised as a specific pathogen in both intestinal and extra-intestinal disease. Five types of diarrhoea producing *E. coli* are known:

- Enteropathogenic (EPEC), Class I and II (includes strains referred to as 'attaching-effacing *E. coli*')
- Enterotoxigenic (ETEC)
- Enteroinvasive (EIEC; *Shigella*-like *E. coli*'; 'Dysentery-like *E. coli*')
- Enterohaemorrhagic (EHEC; Verocytoxin-producing *E. coli* (VTEC); 'colohaemorrhagic *E. coli*')
- Enteroadherent-aggregative (EA-AggEC)

The four main types, EPEC, ETEC, EIEC and EHEC (**Table 46**), each have a distinct clinical pattern and there are also differences in epidemiology, pathogenicity and O:H serovar. Common properties include plasmid-encoded virulence factors, character-

66

67

66 & 67. Delayed lactose fermentation by *E. coli*.
Although lactose fermentation is an important diagnostic feature for *E. coli*, strains which fail to ferment the sugar, or in which fermentation is delayed, are not uncommon. This feature is characteristic of a particular biovar, previously described as the Alkalescens-Dispar group, but is also found in the general *E. coli* population. In the example illustrated, lactose was not fermented after 24 h incubation at 35°C (**66**), but acid was produced after 72 h incubation at the same temperature (**67**). The situation was complicated by the fact that while late fermentation occurred in the identification medium which contained no bile salts, lactose was not fermented, irrespective of the length of incubation, on media containing bile salt. This may be due to alterations in the cell wall permeability in the presence of bile salts, the delayed fermentation in bile-free media suggesting lack of a permease. The strain isolated was, in fact, isolated on Salmonella-shigella medium, and initially mistaken for *Salmonella*. In other strains, particularly those isolated from dairy products, fermentation of lactose and other sugars may be temperature dependent, the property being expressed at 30°C but not at 37°C (*cf. Y. enterocolitica*, pages 131–132). (*Enterotube*©: 24 h/72 h at 35°C.)

Note: Extended incubation of identification test media may lead to loss of discrimination with some tests or, in other cases, atypical reactions due to reversion of indicators.

68. An anaerogenic strain of *E. coli*. Failure to produce detectable gas from carbohydrates is not restricted to any single biotype, and the characteristic is found in the general population of *E. coli*. Gas production from carbohydrates is of differential value among members of the *Enterobacteriaceae*, but gas production reflects the possession of only a single enzyme; formic hydrogenlyase. This enzyme is not essential for fermentative metabolism, and may be lost by mutation without affecting the fermentative capacity. Undue weight should not, therefore, be placed on this characteristic. (*Hugh and Leifson medium: 48h at 37°C.*)

istic interaction with the intestinal mucosa, and synthesis of various enterotoxins or cytotoxins (Levine, 1987).

The position of enteroadherent-aggregative strains is less well defined, and has been the cause of some confusion. These strains have a characteristic aggregative pattern in the HEp-2 assay, are negative by tests with DNA probes for enteropathogenic, enterotoxigenic, enteroinvasive and enterohaemorrhagic *E. coli*, and do not fit these categories by serovar (Vial *et al.*, 1988). A statistical likelihood of involvement of enteroadherent-aggregative strains in persistent diarrhoea among Indian children has been demonstrated by Bhan *et al.* (1989). However, it has been claimed that the characteristic 'stacked brick' appearance of these strains in HEp-2 assay is a consequence of a particular assay system, and that other assay protocols demonstrate a localised or diffuse adherence typical of EPEC strains (Matthewson and Caviato, 1989).

Mention should also be made of non-toxigenic, colonising strains of *E. coli*, distinct from the recognised pathogenic types which, like enteroadherent-aggregative strains, are involved in persistent diarrhoea (Schlager *et al.*, 1990). It is possible that with the emergence of persistent diarrhoea as a major cause of illness (see pages 106, 108), further types of diarrhoeagenic *E. coli* may be recognised as important human pathogens.

69. Failure of *E. coli* to produce indole. Production of indole from tryptophane is an important diagnostic feature for *E. coli*, and is considered by some authors to be a more reliable characteristic than lactose fermentation for use in selective media (see **84 & 85**). Although in excess of 90% of strains are indole-positive, negative strains may be dominant in certain foods. For example, *E. coli* strains isolated from an English soft cheese were consistently indole-negative although typical in other key biochemical characteristics. (*Enterotube©: 24h at 30°C.*)

70. Utilisation of citrate by a strain of *E. coli*. Citrate utilisation by *E. coli* was first described by Washington and Timm (1976) and is one of a number of 'forbidden' phenotypic traits. Other 'forbidden' traits include H_2S and urease production which, like citrate utilisation, are plasmid determined in contrast to most of the classical characteristics of *E. coli*, which are chromosomally determined. The strain illustrated was isolated from an environmental swab taken at a poultry packing plant and in other respects was typical of *E. coli*. (*Simmons citrate medium: 36h at 37°C.*)

Table 46. Summary of diarrhoeaogenic strains of *E. coli*.

Category	Serogroups	Virulence factors
Enteropathogenic		
Class I	O55, O86, O111, O119, O125, O126, O127, O128ab, O142	Shiga toxin production? Fimbrial or outer membrane protein adhesion in some strains
Class II	O18, O44, O112, O114	
Enterotoxigenic	O6, O8, O15, O20, O25, O27, O63, O78, O80, O85, O115, O128ac, O139, O148, O153, O159, O167	Heat-labile and heat-stable toxins. Adhesive factors
Enteroinvasive	O28ac, O29, O124, O136, O143, O144, O152, O164, O167	Epithelial invasion
Enterohaemorrhagic	O26[1], O157, O111	High levels of Shiga toxin. Fimbrial adhesion
Enteroadherent-aggregative	not defined	Epithelial adherence. Toxin production?

[1] Serogroup 026 was originally categorised as enteropathogenic, but is now considered to be enterohaemorrhagic.

Isolation and detection of *E. coli*

Conventional cultural techniques

Resuscitation

Although the need for resuscitation is well recognised in determining *E. coli* in water (Hartmann *et al.*, 1986) relatively little attention has been paid to resuscitation of *E. coli* in foods. However, a number of enumeration protocols do incorporate a resuscitation stage (Fishbein *et al.*, 1976; Holbrook *et al.*, 1980).

Selective enrichment

Various broths have been used for enrichment of *E. coli*. Most common are EE broth and lauryl-tryptose broth.

Selective plating

Media used for the selective plating of *E. coli* are largely traditional lactose-based media, such as the MacConkey and violet red-bile agars. An alternative approach is the use of tryptone-bile agar while, more recently, the incorporation of the fluorogenic substrate 4-methylumbelliferyl-β-D-glucuronide for detection of β-glucuronidase, has also been used as a differential agent for *E. coli*. None of these media will differentiate pathogenic serogroups from non-pathogenic serogroups, although specific media and methods are available for serogroup O157.

Serological and genetic techniques for detection of specific type of diarrhoeagenic *E. coli*

Serological and/or genetic methods have been developed for all recognised diarrhoeagenic types of *E. coli*, including enteroadherent-aggregative strains. Methods based on the polymerase chain reaction have also been devised for enteroinvasive and enterohaemorrhagic strains.

Further details of isolation methods for pathogenic types of E. coli, *with respect to choice of media and laboratory practice, are discussed in* **Laboratory Methods**, *pages 120–127.*

Taxonomy of *Escherichia*

Differentiation of *Escherichia* from other genera

Differentiation is made on the basis of biochemical tests (**Table 47**). Conventional test media or, more conveniently, miniaturised identification kits may be used. Multi-test media should be used for preliminary screening only, and would seem to have little value.

Differentiation of *E. coli* from other species of *Escherichia*

E. coli may be differentiated from other species by the biochemical tests listed in **Table 48**. Conventional test media may be used, but some miniaturised kits require supplementation by additional tests.

Specific characters of diarrhoeagenic strains

Certain properties are more common among diarrhoeagenic strains than among the general population of *E. coli*. With the exception of certain properties of serovar O157:H7, it is not possible to use these properties as phenotypic markers, although they provide valuable information to the experienced microbiologist. At the same time, difficulties can be caused in the isolation and recognition of pathogenic strains by their tendency to atypical properties.

All types

The most common atypical property shared by all types of diarrhoeal *E. coli* is the failure to ferment lactose. It is generally accepted that *ca* 10% of all strains of *E. coli* fail to ferment lactose (Ewing, 1972) and there is a disproportionately high percentage among pathogenic strains. No less than 49% of enteropathogenic serovars isolated from children with diarrhoea were lactose-negative (Sakazaki *et al.*, 1967), while the corresponding figure for enteroinvasive strains is 35% (Taylor *et al.*, 1988).

Failure to ferment lactose may not be an absolute phenomenon. In some cases, fermentation is delayed and occurs too late to be detected during the incubation periods usually employed (**Figures 66 & 67**) while in some cases *E. coli* is able to ferment lactose at 37°C but not at 44°C.

The practical implications of failure of *E. coli* to ferment lactose may be demonstrated by the initial failure to detect an enteropathogenic strain responsible for food poisoning from soft cheese (Marier *et al.*, 1973). More recently, strains of enteropathogenic *E. coli* serogroup O4, which is lactose-negative and of epidemiological importance in Somalia, have been overlooked for the same reason (Nicoletti *et al.*, 1988).

Production of indole from tryptophane at 44°C is often considered a more reliable property than lactose fermentation (Anderson and Baird-Parker, 1975). Although some 99% of all strains of *E. coli* produce indole, the percentage of strains which are negative (**Figure 69**) is higher among pathogenic types (Hartmann *et al.*, 1986). However, the percentage of indole-negative strains would seem to be significantly lower than that of lactose-negative strains. Indole production may also be temperature dependent in that some strains are positive at 37°C but not at 44°C. This trait tends to be associated with *E. coli* O157:H7, but also occurs among other pathogenic types.

The inability to grow at temperatures in the region of 44°C is again a trait associated with serovar O157: H7. 'Temperature sensitive' strains of other diarrhoeaogenic types may also be isolated although at a lower frequency.

Enteroinvasive *E. coli*

Strains of EIEC show a number of atypical properties in addition to those common to all pathogenic types. In many cases, these are properties in common with the closely related *Shigella*. A high proportion of EIEC strains are non-motile, and many also fail to produce gas during sugar fermentation. This latter property is not unique to EIEC, and may be found in strains of both pathogenic and non-pathogenic types (**Figure 68**).

Enterohaemorrhagic *E. coli* *(O157:H7)*

Isolates of EHEC serovar O157:H7 have three distinct properties which may be used both in the isolation and differentiation of the serovar. Failure to ferment sorbitol is a very common property in serovar O157: H7, but relatively rare among other strains of *E. coli*. Strains of serovar O157:H7 also fail to produce β-glucuronidase, and both of these characteristics may be exploited as differential features in selective isolation media (pages 121, 124).

A less well-known property of this serovar is sensitivity to bromthymol blue at high incubation temperatures, and this has been suggested as the basis of a simple confirmatory test (Gubash *et al.*, 1988). Other possible biochemical markers for serovar O157:H7 are failure to produce lysine and ornithine decarboxylase (Haldane *et al.*, 1986), and fermentation of raffinose and dulcitol (Ratnam *et al.*, 1988).

It must be appreciated that the above discussion refers specifically to EHEC serovar O157:H7 and not to any other EHEC serovar.

Table 47. Differentiation of *Escherichia* from related genera.

	Escherichia	*Salmonella*[1]	*Klebsiella*	*Enterobacter*
β-galactosidase	+[2]	−[3]	+/−	+
Lysine decarboxylase	+/−	+	+/−	−/+
Ornithine decarboxylase	+/−	+	−/+	+
Indole	+	−	−/+	−
Simmons' citrate	−	+	+/−	+
Voges-Proskauer	−	−	+/−	+
Aesculin hydrolysis	−/+	−	+/−	+
Acid from				
Lactose	+[2]	−[3]	+/−	+
Raffinose	−	−	+	+
Salicin	−/+	−	+	+
Sucrose	−/+	−	+/−	+/−

[1] Reactions for *Salmonella* are based on those of ubiquitous serovars of importance in food poisoning. Reactions of *S. typhi* and other host adapted serovars may differ.
[2] 'Inactive' strains are usually negative.
[3] *S. arizonae* and strains of some other serovars are positive.

Table 48. Differentiation of *E. coli* from other species of *Escherichia*.

	Escherichia			
	coli	*adecarboxylata*	*hermanii*	*vulneris*
Arginine dihydrolase	−	−	−	+/−
Lysine decarboxylase	+/−	−	−	+/−
Ornithine decarboxylase	+/−	−	+	−
Indole	+	+	+	−
Aesculin hydrolysis	−	+	−	−
Malonate	−	+	−	+/−
Acid from				
Adonitol	−	+	−	−
Amygdalin	−	+	+	+/−
Cellobiose	−	+	+	+
Sorbitol	+[1]	−	−	−

[1] Serogroup O157 and strains of some other serogroups are negative.

Typing of *E. coli*

Serotyping

Serotyping of *E. coli* is a well-established procedure, and may be used either for detection of known pathogenic serogroups or in epidemiological investigations.

Serotyping schemes are analogous to those for *Salmonella*, but three types of antigen are determined; somatic (O), capsular (K) and flagellar (H). In the Kauffman-Knipschildt-Vahlne scheme the primary subdivision is made according to the antigenicity of the cell wall lipopolysaccharide O antigen. A serogroup is thus defined by the presence of a characteristic O antigen, and may then be subdivided into serovars on the basis of the K and H antigens. The need to proceed to the determination of the K and H antigens depends on the serogroup (Robins-Browne, 1987). For example, serogroup O55 is rarely found in healthy persons, and should be regarded as pathogenic if isolated from children with diarrhoea. In this case, full serotyping is thus unnecessary. In contrast, serogroup O86 is commonly found in healthy people and, if isolated, serotyping is necessary since only serovar O86:H34 is pathogenic.

Changes in the designation of the K antigens have taken place, and some care is necessary to avoid confusion.

Practical aspects of serotyping are described in greater detail in **Laboratory Methods**, *page 127.*

Phage typing

Phage typing procedures are well established for *E. coli* although not widely used. Most of the earlier schemes were devised for EPEC, or for serovars involved in extraintestinal disease. More recent applications include the development of a phage typing scheme for serogroup O157 in Canada (Ahmed *et al.*, 1987), which has been applied successfully to strains isolated in the UK (Frost *et al.*, 1989), and a scheme for tracing labile-toxin producing strains of enterotoxigenic *E. coli* (Monsur *et al.*, 1989a).

Genetic methods

Restriction endonuclease digest patterns have been used successfully in epidemiological investigations where discrimination was not possible by other methods, including plasmid profiling. These include an outbreak of diarrhoea due to enterotoxigenic *E. coli* on a cruise ship, and in two separate outbreaks involving *E. coli* O157:H7.

Symptoms of *E. coli* infections

Enteropathogenic *E. coli*

Enteropathogenic strains of *E. coli* (EPEC) are responsible for many cases of infantile diarrhoea. In such cases, the severity of symptoms varies from extremely mild, which may pass undetected, to severe and possibly life threatening. In typical cases, symptoms appear within 12 to 36 hours of infection. Watery stools are passed which occasionally contain mucus and, rarely, blood. Vomiting and low-grade fever usually accompany the diarrhoea. Symptoms may be prolonged and dehydration together with acidosis and shock is a common complication. Infections with Class I (adherent) EPEC strains may result in a persistent, life threatening, secretory diarrhoea, with severe small intestine enteropathy. Colonisation by multiple strains is common among neonates, and the likelihood of death correlates with the number of EPEC adherence factor positive strains present (Senerwa *et al.*, 1989).

Enterotoxigenic *E. coli*

The symptoms of enterotoxigenic *E. coli* (ETEC) tend to reflect the toxin or toxins produced. Strains producing the heat labile, LT toxin, which resembles cholera toxin (see pages 167–170), produce symptoms which at their most severe closely resemble those of

Vibrio cholerae 01, both in type and severity. A sudden and explosive onset is common, although symptoms may appear gradually. Diarrhoea consisting of watery stools, almost invariably without blood or mucus, is the predominant symptom, but is often accompanied by abdominal pain and vomiting. There is little or no fever and the temperature may be subnormal. Dehydration is common and may result in death if treatment is delayed.

Symptoms of such severity due to ETEC are rare and are usually reported only in areas where cholera itself is endemic (Pickering, 1979). In other areas, the severity varies from the mild and transitory, to something approaching the full cholera-like disease. Symptoms usually appear 12 to 36 hours after infection but the period may be shorter. These involve watery diarrhoea accompanied by mild abdominal cramps. Vomiting may occur in the early stages of the illness, and low-grade fever is an occasional feature with children but not adults. The duration of symptoms is usually no more than 1 to 2 days but may persist, gradually reducing in severity, over several days.

Strains of ETEC that produce only heat stable, ST toxin, are relatively uncommon, and have received less attention as pathogens. The first reported case was described by Sack *et al.*, (1975) and involved symptoms similar to those of LT producing strains: watery diarrhoea without blood or mucus, possibly accompanied by nausea, general malaise, fever, abdominal cramps and vomiting. The onset of illness due to ST producing strains is generally more rapid than that due to LT producing strains, and may be as little as 4 hours or, occasionally, less.

There is evidence of a high incidence of multiple infections in ETEC infections with three different serovars and/or phagovars being isolated from the stools of some patients (Monsur *et al.*, 1989b).

Enteroinvasive *E. coli*

Symptoms of infections due to enteroinvasive *E. coli* (EIEC) are essentially those of shigellosis. Patients pass bloody, mucoid stools which may contain neutrophils, there is urging and tenesmus and fever and colitis are usually present. The severity of symptoms can vary from a mild form resembling *Sh. sonnei* infection, to an extreme form resembling classical dysentery.

Enterohaemorrhagic *E. coli*

Enterohaemorrhagic *E. coli* (EHEC) produces two types of illness, haemorrhagic colitis and haemolytic uraemic syndrome.

Haemorrhagic colitis

Haemorrhagic colitis results from colonic mucosal oedema, erosion and haemorrhage. The incubation period in community-wide outbreaks is 3 to 4 days (Riley, 1987), but an average period of 8 days was noted in an outbreak in a nursing home (Ryan *et al.*, 1986). In an outbreak among Canadian children, incubation periods between 1 and 14 days were recorded (Anon, 1987).

The onset of symptoms is marked by a sudden severe pain followed by watery diarrhoea. Nausea and vomiting occur in the early stages of the illness, and there is abdominal distention. The severity of the pain has been described as 'worse than appendicitis' or, in female patients, as 'like labour pains', and may be a distinguishing feature of the disease (Riley, 1987). However, similar comments have been made concerning infections with other micro-organisms including *Campylobacter* and *Yersinia enterocolitica*.

After the onset, the disease progresses over 1 to 2 days to bloody diarrhoea. Temperature is usually normal, but a high fever may be present in the elderly. Prognosis in cases where a high fever develops late in the course of the disease is poor. Recovery is usually uneventful, but death may occur among the elderly especially where suffering from pre-existing conditions. Death rates in some reported outbreaks have been as high as 36%, while in others no deaths have occurred.

Although bloody diarrhoea is considered a characteristic of EHEC infections, it is usually associated with serovar O157:H7, and in disease caused by non-O157 EHEC serogroups, stools may remain watery throughout. This is common in western Canada where the incidence of EHEC infections is high, equalling *Salmonella* and *Campylobacter* in importance (Kaper, 1987).

Haemolytic uraemic syndrome

Haemolytic uraemic syndrome (HUS) usually, but not invariably, commences with diarrhoea followed by a triad of microangiopathic haemolytic anaemia, thrombocytopenia and acute renal failure (Riley, 1987). Less commonly a related syndrome, thrombotic, thrombocytopenic purpura, may occur (Morrison *et al.*, 1986). This is an extension of HUS and includes fever and neurological symptoms. Long-term kidney damage may result and death can occur several years after the abatement of acute symptoms.

The highest risk of systemic complications is associated with the Shiga-like toxin II producing genotype (Ostroff *et al.*, 1989). Both bloody stools and a high temperature in O157:H7 have been considered predictive of the development of HUS.

A number of less common complications of EHEC

infections have been reported. These include haemor-rhagic cystitis and balanitis (Grandsen *et al.*, 1985), convulsions, sepsis with another organism, anaemia and iatrogenic upper gastrointestinal tract bleeding resulting from attempts to relieve abdominal disten-sion by nasogastric suction (Ryan, *et al.*, 1986).

Enteroadherent-aggregative *E. coli*

Enteroadherent-aggregative *E. coli* has been incrimi-nated as a cause of diarrhoea on epidemiological grounds only, and full details of symptoms are not known. In cases attributed to these strains, symptoms have been prolonged, diarrhoea persisting for at least 14 days, with bloody stools occurring in *ca* 11% of cases (Bhan, 1989).

Other diarrhoeal symptoms attributed to *E. coli*

Heavy colonisation with non-toxigenic *E. coli* can induce diarrhoea in 10 to 60% of hosts. Diarrhoea is prolonged, and impaired intestinal absorption may have nutritional consequences (Schlager *et al.*, 1990).

Persons at particular risk from *E. coli* infection

Susceptibility

Susceptibility to *E. coli* infections follows the usual pattern of the old, the young and those with under-lying infections being at greatest risk. Infections due to both EPEC and ETEC strains are particularly common in infants, and classical EPEC serovars are usually associated only with infections in children aged less than 18 months. Within this age group, there is evidence that males are more susceptible than females (Antai and Anozie, 1987).

Infections due to ETEC are only rarely found in areas where EPEC infections are present. This is probably due to the two organisms competing for the same intestinal binding sites.

Although the epidemiology of EHEC strains is not fully established, it would appear that those at extremes of age, children aged under 5 years and geriatrics, are most susceptible (Pai *et al.*, 1988. Haemolytic uraemic syndrome is most common in children under 16 years of age and in the elderly. This may reflect host adaptation by Shiga-like toxin II producing strains, which are rare in patients aged 17 to 61 years (Ostroff *et al.*, 1989).

Risk of EHEC infections is also increased by gaster-ectomy, prior antibiotic therapy and H_2 blocking agents (Edelman *et al.*, 1988). Antibiotic therapy during the course of the disease can lead to haemolytic uraemic syndrome (see below).

Exposure

With the exception of EHEC strains, the prevalence of pathogenic *E. coli* is greater in countries where general standards of hygiene are poor. Thus ETEC strains are a common, and in many cases the major, cause of travellers' diarrhoea.

Infantile diarrhoea due to EPEC was previously a problem in hospital nurseries where transmission by staff, or from child to child, can be important vehicles. Outbreaks associated with nurseries and day care centres still occur, but in general, control of hygiene means that EPEC is of diminishing import-ance in developed nations. Elsewhere, however, the organism remains a major cause of infant mortality. The high incidence of EPEC means that children are exposed directly after weaning when most cases occur. Children fed on artificial formula foods are at a signi-ficantly greater risk than those who are breast fed. This is largely due to the lack of suitable water for reconstituting dried formulations, although the failure to develop a protective gut microflora may also be involved. There is no doubt that in developing countries 'breast is best' (Duguid *et al.*, 1978), but the efforts of health educators often make little impact against the seductive marketing of the infant-food manufacturers.

EIEC is rare in the developed world, and people most at risk are those in areas where hygiene is poor and other disease, particularly cholera, is endemic. Infections with EIEC strains are not associated specifically with travellers' diarrhoea, but people visiting areas where the incidence is high are obviously at greater risk.

There is evidence of considerable geographical variation in the incidence of EHEC infections. For example, the incidence is high in the Calgary area of western Canada. There may also be marked variation within a country. In Sweden, the incidence in Stockholm is 1.3 cases per million of the population, compared with 15 cases per million of the population in Lund (Edelman et al., 1988).

There is also evidence that people in institutions are at a higher than average risk from EHEC. To some extent this may be attributed to the greater susceptibility of the young and old, but it is also possible that nursing staff, infected as a result of repeated exposure, play a part in the spread of infection.

Cattle may be an important reservoir of EHEC, and consumption of raw milk, or raw or undercooked meat has been defined as a risk factor, a particular association being considered to exist with the consumption of hamburgers from fast food outlets. However, the thesis that EHEC is a foodborne zoonosis has been challenged by Walmer-Touws (1990), who considers that insufficient evidence exists to implicate cattle as the origin of the organism in human disease. Other contemporary work has supported the concept that cattle serve as a reservoir (Bryant et al., 1989; Montenegro et al., 1990), but not that the risk is associated with any specific type of food. A general association exists, however, between EHEC infections and improper cooking and handling of foods (Bryant et al., 1989).

Handling of foods rather than consumption may be a risk factor. A community-wide outbreak in eastern England, which involved 24 cases with one death, was attributed to infection acquired when handling raw potatoes which, it is surmised, had been contaminated with manure.

Treatment of *E. coli* infections

In most cases infections with diarrhoeagenic *E. coli* are self limiting, and treatment is usually restricted to general care, although in severe cases rehydration and restoration of the fluid balance is necessary. Antibiotic therapy is normally neither effective nor desirable, but is indicated when infection with adherent EPEC strains leads to persistent and life-threatening diarrhoea, and severe small intestine enteropathy (Hill et al., 1988). Intravenous administration of antibiotics includes gentamicin and trimoxazole (Thoren et al., 1980).

The use of anti-motility agents is not recommended in cases of infection with EHEC or ETEC, since intestinal stasis may promote absorption of the toxins. In the case of ETEC infections, the use of blocking agents to prevent the binding of toxin to receptor sites is of long-term interest for the control of LT producing strains (Donta et al., 1988; Greenhough et al., 1988).

The need for careful monitoring of the patients' condition is of particular importance with EHEC infections, where severe complications are relatively common (Riley, 1987). Treatment with antimicrobial drugs is not necessary, and treatment with cotrimoxazole can lead to the development of haemolytic uraemic syndrome (cf. *Shigella*, page 91; Pavia et al., 1988). This may be related to the *in vitro* stimulation of both intracellular and extracellular Shiga-like toxin production by sub-inhibitory levels of this drug.

Mechanisms of pathogenicity

Enteropathogenic *E. coli*

Although EPEC is well recognised as a pathogen, the mechanism of diarrhoea production is yet to be established. However, in recent years a number of advances have been made which, it is hoped, point the way to a full understanding of the factors involved.

Genetic control

Adherence has been shown to be a two-stage process (Kaper, 1987; **Figure 71**). The first stage (initial adherence) is thought to be under plasmid control, but the second stage (late adherence), has been considered to be mediated by chromosomal genes. How-

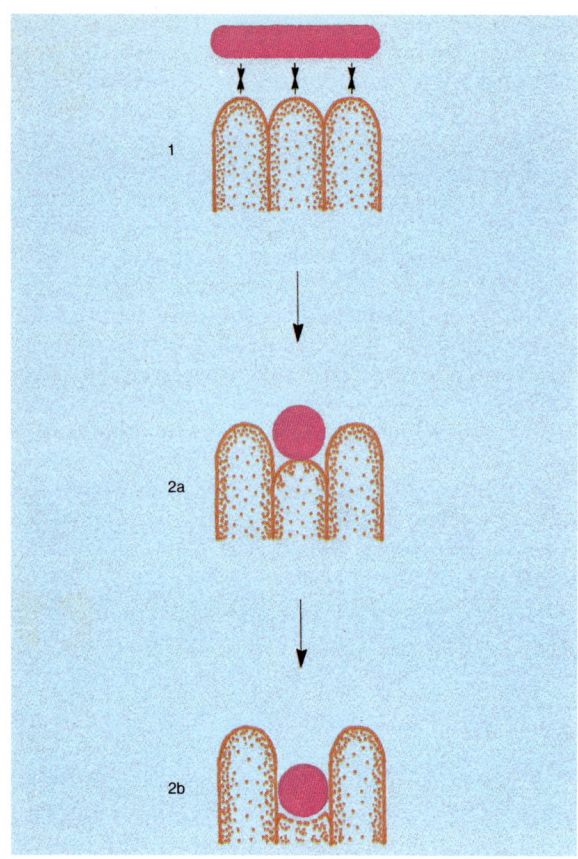

71. Adherence of enteropathogenic *E. coli* and destruction of microvilli: diagrammatic summary.
The process involves two distinct stages. In the first, the cell of *E. coli* is attached to the intact brush border (**1**). In the second, the cell becomes much more intimately attached to the brush border and destruction of the microvilli commences (**2a**). Destruction procedes by a process of membrane vesiculation resulting in the characteristic pedestal on which the bacterial cell sits (**2b**).

A mechanism for the shedding of the microvilli, which draws on the similarity with calcium induced shedding, has been proposed (Knutton *et al.*, 1987b). At calcium concentrations below 10^{-6} M (normal in the cytosol), the microvillous core cytoskeleton is a stable, rigid structure. At concentrations above 10^{-6} M the actin filaments of the core are broken into short filaments resulting in a thermodynamically unstable microvillous membrane that fragments into vesicles. The residual enterocyte plasma membrane lacks an organised cytoskeleton and would allow enteropathogenic *E. coli* to bind strongly over a large surface to produce pedestals.

Shedding of the microvilli is thought to be a defence mechanism which, in the case of infections with enteropathogenic *E. coli*, is ineffective. (Reproduced with permission from Knutton, S., Lloyd, D.R. and McNeish, A.S. *Infection and Immunity* **55**: 69–77. © 1987 The American Society for Microbiology.)

ever, colonisation is enhanced when the plasmid is present (Knutton *et al.*, 1987a), and it now seems likely that at least one of the genes involved in damage to the brush border during late adherence is carried on a plasmid (Fletcher *et al.*, 1990).

Invasion by EPEC strains also involves genes situated in both plasmid and chromosome (Donnenberg *et al.*, 1990).

Production of a Shiga-like toxin is not plasmid mediated, but in at least some strains, may be governed by a bacteriophage (O'Brien *et al.*, 1982).

Adherence

Adherence is known to be important in the pathogenesis of EPEC infections, although almost certainly other factors are involved. Two distinct patterns of adherence by EPEC have been described (Scaletsky *et al.*, 1984): localised adherence (LA), in which organisms attach to one or two small areas of the cell surface, and diffuse adherence (DA), in which organisms adhere to the whole cell surface or not at all. This is commonly referred to as Class I and Class II adherence respectively.

Adherence in Class I serovars, which include those commonly incriminated in diarrhoea, has been associated with the presence of a *ca* 55 Mdalton plasmid, which codes for production of a 94 Kdalton adhesin, the EPEC adhesive factor (EAF; Levine *et al.*, 1985). There is some controversy over the nature of this adhesin, which was previously thought to be an outer membrane protein (Scotland *et al.*, 1983; Levine *et al.*, 1985), but which was later stated to be fimbrial Knutton *et al.*, 1987b). An explanation for this disparity may lie in the suggestion that only a few EPEC cells in a culture are fimbriate, and fimbriae are thus difficult to detect (Law, 1988).

Little is known of receptors for EAF, but a glycoprotein, fibronectin, is believed to be involved. Work suggests that two or more receptors may be involved.

Possession of EAF is not essential for pathogenicity, and EAF$^-$ strains involved in diarrhoeal disease probably produce an as yet unidentified mediator of adherence (Bowen *et al.*, 1989).

The nature of the adhesin involved in diffuse adherence remains largely unknown, although it is mediated by a plasmid distinct from that which mediates localised adherence (Nataro *et al.*, 1985a). There is also evidence that adherence is assisted by hydrophobic properties conferred by the DA adhesin (Nataro *et al.*, 1985a). Doubt has been cast on the role of this class of EPEC in disease (Levine *et al.*, 1988), although the virulence of some strains has been demonstrated in feeding trials (Levine *et al.*, 1985. More recently, an association has been demonstrated with childhood diarrhoea in New Zealand (Gunzburg *et al.*, 1990).

Some workers consider that Class II EPEC strains may form a category distinct both from other EPEC strains and the other types of pathogenic *E. coli*. This situation may, however, be further confused by the finding that some strains, which were considered to be non-adherent, were found to adhere weakly when a more sensitive fluorescent staining technique was applied (Knutton *et al.*, 1989).

In common with other pathogenic as well as non-pathogenic types of *E. coli*, 40 to 80% of EPEC strains possess type I fimbriae. Their role in the pathogenicity of EPEC infections is at best uncertain, and it has been postulated that they may even act as 'negative virulence factors' by mediating attachment of EPEC cells to the mucus layer, and thus both hindering their attachment to the epithelium and facilitating removal by peristalsis (Evans and Evans, 1983).

Adherence by Class I serovars results in characteristic lesions in the brush border. The possible mechanism is summarised in **Figure 71** and the damage to the brush border illustrated in **Figures 72–75**.

Invasion

Epithelial cell invasion by EPEC has been an overlooked property in the past. Adherence is a prerequisite for invasion, the bacteria entering by an endocytic process (Andrade *et al.*, 1989). Only metabolically active cells are internalised, the cells penetrating the

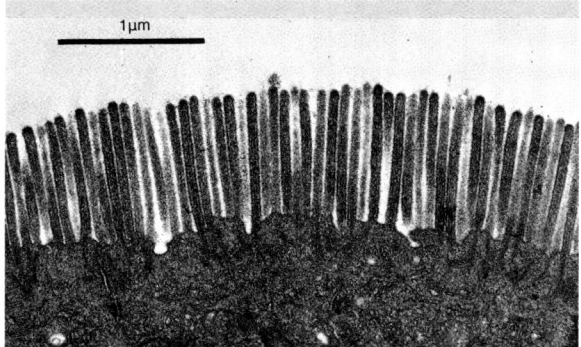

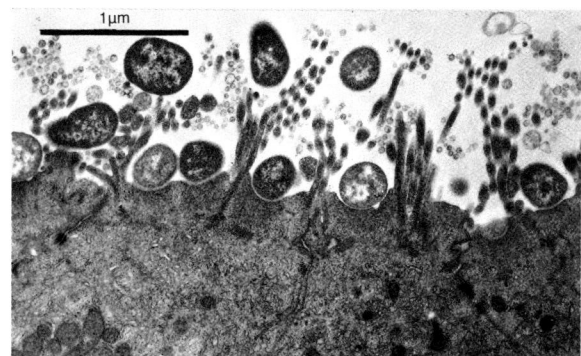

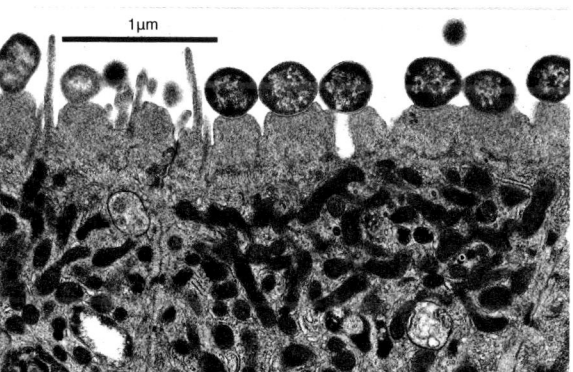

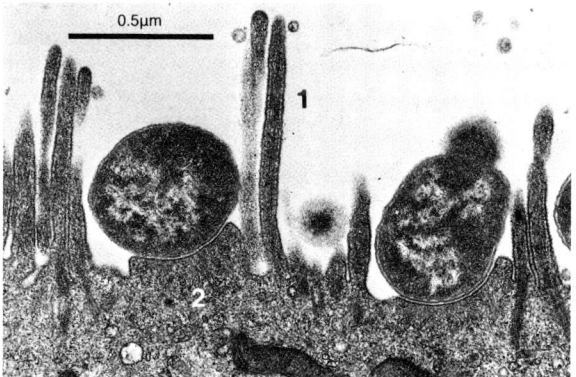

72–75. Destruction of microvilli during formation of lesions by enteropathogenic *E. coli*. The normal brush border of the human duodenal mucosa is illustrated in (**72**). Vesicles formed during breakdown of the microvilli are localised at sites of attachment of enteropathogenic *E. coli* and brush border destruction (**73**). The vesicles have membrane profiles identical to those of the microvilli, but lack any organised cytoskeleton. A further characteristic is the dramatic elongation of some microvilli that usually precedes vesiculation. Microvilli of up to 3 μm in length may be observed in contrast to a normal length of <1 μm.

Microvilli are totally destroyed at the site of attachment, and the bacteria become intimately attached to the mucosal surface (**74**). The lesion is seen at higher magnification in (**75**); the bacteria may be seen 'sitting' in the cup-like pedestals, projections from the apical enterocyte membrane, with only *ca* 10 nm seperating bacteria and enterocyte membranes. An intact microvillus remains (**1**), but elsewhere these have been destroyed and the associated cytoskeletal elements disrupted.

Concentrations of short filaments are present in the apical cytoplasm beneath the adherent bacteria (**2**). (Reproduced with permission from Knutton, S., Lloyd, D.R. and McNeish, A.S. *Infection and Immunity* **55**: 69–77. © 1987 The American Society for Microbiology.)

host cell cytoplasm enclosed in membrane-bound vacuoles, which underlie localised sites of adherence. Adhesins are the endocytic trigger. Intracellular multiplication occurs, internalised EPEC cells being observed both enclosed in vacuoles and in the cytoplasm (Donnenberg *et al.*, 1989). The invasive capabilities of some EPEC strains are considered to be equal to, or greater than, those of enteroinvasive *E. coli*. However, only a relatively small proportion of adhering bacteria is able to invade host cells, and factors other than adherence must be involved (Donnenberg *et al.*, 1990).

Toxin production

Although there have been reports of both LT and ST production by EPEC (see discussion of enterotoxigenic *E. coli* below), many of these were probably due to incomplete characterisation of the EPEC serovar or serogroup.

The well-substantiated reports of Shiga-like toxin (SLT) by EPEC (O'Brien and LaVeck, 1983) are of considerably greater significance. Two SLTs are produced which are probably related, but only one of which, SLTI, is neutralisable by Shiga toxin antibody.

The role of SLT in the pathogenesis of EPEC diarrhoea is not clear. EPEC strains are both more likely to produce SLT and to produce larger quantities than non-pathogenic strains from healthy individuals (Cleary *et al.*, 1985; O'Brien *et al.*, 1982), although patterns of production may be blurred by a lack of sensitivity in the assay technique. Apart from serogroups O26 (now classified as EHEC) and O128, most EPEC strains produce undetectable amounts of toxin unless grown in an iron deficient medium. In volunteer experiments, no direct relationship exists between the amount of toxin produced and the ability to cause disease (O'Brien *et al.*, 1982). *In vivo* toxin production may be greater (Law, 1988) and the possibility also exists that low levels of SLT may damage host cells if produced by strains that adhered avidly (O'Brien and Holmes, 1987).

Classical Shiga toxin is discussed in greater detail in Chapter 5, *Shigella* and Shiga-like toxin production by enterohaemorrhagic *E. coli* in the present Chapter, pages 115–116.

The possibility of toxins other than SLT or LT and ST being produced by EPEC strains has been considered by a number of workers, and it has been suggested that endotoxins may be involved (Mullaney and Cantey, 1984). Nicoletti *et al.* (1988) subsequently considered that haemolysin was a virulence factor in serogroup O4 strains, and hypothesised that diarrhoea was induced by adherence to erythrocytes and mucosal injury.

Protein kinase C activity

An important role has been proposed for protein kinase C in the pathogenesis of EPEC infection. EPEC induces specific changes in host protein phosphorylation patterns, similar to those induced by known activities of protein kinase C (Baldwin *et al.*, 1990). This results in extreme phosphorylation of host cell membrane proteins and ultimately in diarrhoea (see page 168).

In vitro determination of virulence

Apart from feeding trials using human volunteers, tissue culture assays using HEp-2 or HeLa cells have been used to determine patterns of adherence by putative enteropathogenic strains of *E. coli*. Such assays are obviously unsuitable for routine use in food industry laboratories. A radiolabelled DNA probe has been developed (Nataro *et al.*, 1985b), which hybridises with DNA from EPEC strains showing localised adherence. This is a highly sensitive and specific probe but is not yet commercially available. Serovar-specific variation in the association of *E. coli* strains with a localised adherence pattern means that the probe should be used in conjunction with serotyping (Tardelli Gomes *et al.*, 1989). However, in many clinical laboratories O-serogrouping remains the only method for routine detection of EPEC.

Enterotoxigenic *E. coli*

Many serovars of *E. coli* can be toxigenic, but a limited number of O:H serovars occur repeatedly throughout the world and account for a majority of ETEC strains. These recurrent O:H serovars appear to be successful ETEC clones that have spread far and wide (Levine, 1987). Classical ETEC serovars also appear to have received enterotoxin plasmids long ago, while rare serovars acquired plasmids recently by conjugation, but have not spread to any degree (Danbara *et al.*, 1988). Serovars commonly associated with diarrhoea usually elaborate both types of toxin (see below) and possess fimbrial colonisation factors.

Genetic control

Plasmids encode for fimbrial adhesins and for production of both labile toxin I (LTI) and stable toxin (ST). The same plasmid almost always encodes for both adhesins, and ST and often LTI as well.

Strains of ETEC which produce both LTI and ST

usually carry genes for both toxins on a single plasmid molecule of 55 to 61 Mdaltons. This plasmid is constant in size and in physical and genetic characteristics from one strain to the next. However, in some strains that produce both toxin types, the ST and LTI genes are carried on separate plasmids (Glatz, 1986).

In strains that produce ST only, plasmids are diverse in molecular weight it appears likely that plasmids in strains which only produce LTI are also heterogeneous. However, the LTI gene has also been found on a phage and the possibility of its introduction into non-toxigenic strains by lysogenic conversion exists (Glatz, 1986).

A second, antigenically distinct, labile toxin, LTII, is produced by some strains, but production is under chromosomal rather than plasmid control.

Adherence

Attachment or colonisation factors have been identified in all of the main ETEC O serogroups (Levine, 1987). All have been characterised as fimbriae, which share a number of common features (**Table 49**).

Four groups of colonisation factor fimbriae are recognised: colonisation factor antigen I (CFAI), CFA/II, CFA/III (putative colonisation factor 8775; E8775) and PCFO159. The situation is confused by the fact that both CFA/II and CFA/III are composed of three structurally and antigenically distinct components (CS1, 2, 3 and 4, 5, 6 respectively). Colonisation factor fimbriae and their relation to O serogroups are summarised in **Table 50**.

Relatively little is known of the host cell receptor sites for fimbriae, but a glycoprotein containing important sialic acid moieties has been identified on human erythrocytes (Pieroni *et al.*, 1988). Collagen-binding sites have been identified on ETEC and binding of collagen and fibronectin may be a mechanism for binding to the small intestine (Vsai *et al.*, 1990).

Toxins

Heat-labile enterotoxins

Heat-labile toxins of ETEC are inactivated at 60°C for 30 minutes. LTI closely resembles cholera toxin (CT) in structure, antigenicity and activity (see pages 167–170, for a full discussion of cholera toxin). LTI binds to the same ganglioside receptors on mammalian cell surfaces (Eidels *et al.*, 1983), although there appear to be additional receptors, identified as galactoprotein in the rabbit intestine (Griffiths *et al.*, 1986). Like CT, LTI stimulates adenylate cyclase activity (Stephen and Pietrowski, 1981), which ultimately leads to water and electrolyte outflow into the lumen of the small intestine and consequent watery diarrhoea.

Table 49. Common features of fimbrial colonisation factors of human strains of enterotoxigenic *E. coli* (Levine, 1987).

1 Consist either of rigid structures 6 to 7 nm in diameter, or of wiry, flexible structures 2 to 3 nm in diameter.
2 Plasmid encoded.
3 Made up of protein subunits 14 to 22 Kdaltons in size.
4 Most are mannose resistant.
5 Fimbrial colonisation factors are expressed at 37°C but not at 18°C.
6 Particular colonisation fimbriae are usually restricted to certain O:H servovars.

In contrast to CT which is secreted, LTI is located in the periplasmic space and only a small amount is released into the culture fluid (Scotland, 1988).

The second labile toxin, LTII, has similar biological properties to LTI, but is antigenically distinct. The toxin is composed of A and B subunits with molecular weights of 28,000 and 11,800 respectively. The A subunit is cleaved by trypsin to A1 and A2 subunits of molecular weights 21,000 and 7,000 (Scotland, 1988). The role of LTII in illness is not known. ETEC strains producing LTII appear to be relatively uncommon, and in a survey by Seriwatana *et al.*, (1988) were shown to be isolated primarily from cows, buffaloes and beef, fewer isolations being made from children.

Table 50. Summary of colonisation factor fimbriae produced by enterotoxogenic *E. coli*.

Fimbrial antigen	Type	O serogroups
CFA/I	Rigid	15, 25, 63, 78, 128, 153
CFA/II		
CS1	Rigid	6, 139
CS2	Rigid	6
CS3	Wiry	6, 8, 80, 85, 139
CFA/III (PCF 8775)		
CS4	Rigid	25
CS5	Rigid	92, 115
CS6	Wiry	25, 27, 92, 115, 148, 169
PCFO159	?	159

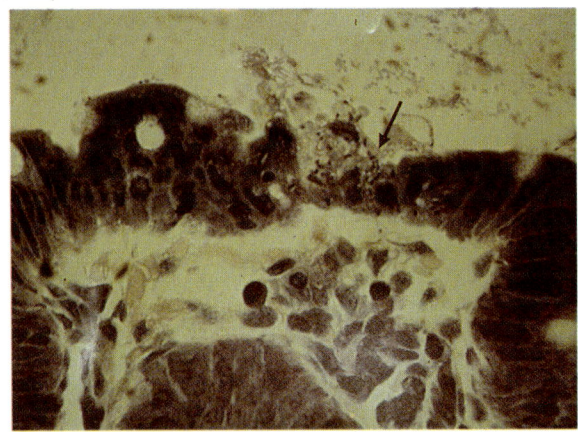

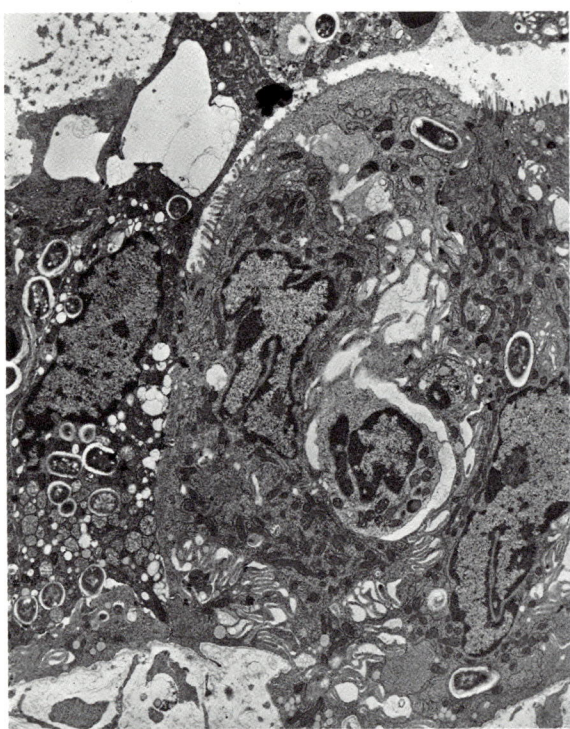

76 & 77. Infection of cultured human colonic mucosa with enteroinvasive *E. coli*. Like *Shigella*, enteropathogenic *E. coli* enters cells by receptor mediated endocytosis. Heavy infection of colonic mucosal cells is seen in (**76**) (arrowed). An electron micrograph of an area similar to that arrowed (**77**) shows the heavily infected cell on the left being extruded from the mucosa, the brush border is still apparent in the lightly infected cells on the right. *In vivo* effects are apparently identical to those in the cultured cells. (Reproduced by courtesy of Dr S. Knutton, Institute of Child Health, University of Birmingham.)

Heat-stable enterotoxins

Like LT, ST consists of at least two toxic products, ST_A and ST_B. In contrast to LT, both of these toxins are of low molecular weight and are poorly immunogenic. ST_A acts by stimulation of the guanylate cyclase system, leading to enhanced levels of cyclic guanosine monophosphate and activation of a cGMP-dependent protein kinase (Hiroyama *et al.*, 1989). ST_A has two precursors, the second of which, the 53 amino-acid pro-ST_A, is cleaved extracellularly to form the mature toxin (Rasheed, 1990).

ST_A may be distinguished from ST_B by the failure of the latter to cause fluid secretion in the intestines of suckling mice, and by it being insoluble in methanol. ST_B is only rarely found in human strains, and its role in human disease is not known.

Detection of LT and ST toxins

In addition to tissue culture tests on cells such as CHO, Vero and Y1, a number of immunological methods have been developed for LTI, or derived from existing methods for CT. Various methods including ELISA, reverse passive latex agglutination and gel diffusion assays have been described. Some methods are available as commercial kits. STs are poorly immunogenic and require either coupling to bovine-serum albumen or polymerisation with glutaraldehyde for raising antibodies. Radioimmunoasays and ELISAs have been developed for both ST_A and ST_B which will detect the toxin in faecal specimens (Svennerholm *et al.*, 1986), and an ELISA kit is now commercially available.

Gene probes have been developed for the genes controlling synthesis of each type of toxin, although these are not commercially available. A method based on the polymerase chain reaction has also been developed.

Further details of methods for the detection of LT and ST may be found in **Laboratory Methods**, *pages 127–128.*

Enteroinvasive *E. coli*

The mechanism of invasion and the intracellular lifestyle of EIEC is similar to that of *Shigella* (see pages 91–93), which it also resembles biochemically and antigenically. The main difference is that EIEC is much less efficient as a pathogen, with 10^9 cells required to cause illness, compared with fewer than 10^3 for *Shigella*. This is probably due to a lower resistance to gastric acidity resulting from differences in outer membrane proteins (Small and Falkow, 1988).

There is no evidence of production of Shiga toxin (or any other toxin) by EIEC (Cleary and Murry, 1988), and virulence is entirely associated with invasiveness and the response provoked in host tissue (**Figures 76** & **77**).

DNA probes constructed for detection of *Shigella*, as well as methods based on the polymerase chain reaction, will also detect EIEC.

Enterohaemorrhagic *E. coli*

Haemorrhagic colitis and its complications have received a considerable amount of attention in recent years. Most work has concerned the first recognised EHEC serovar, O157:H7, which remains the most common, but it should be appreciated that there may be differences between O157:H7 and the other recognised serogroups O26 and O111.

Genetic control

Virulent strains of O157:H7 possess a 60 Mdalton plasmid which encodes for production of fimbrial adhesins. Strains of serovar O26:H11 also possess a 60 Mdalton plasmid, and homology has been demonstrated between that plasmid and the plasmid possessed by O157:H7 strains (Levine, 1987).

Production of large quantities of Shiga-like toxin is an important property of O157:H7 strains and production is mediated by converting phages (O'Brien *et al.*, 1982; Strockbine, 1986). A widely studied EHEC strain, 933, has been shown to harbour two converting phages, and it is likely that a broad family exists, the distribution and degree of relatedness of which has not yet been established.

The mechanism by which phages control production of SLT is not known. Production continues at a low level in the absence of phage, and the toxin structural gene, or genes, probably resides on the genome.

Adherence

Human strains of EHEC, in possible contrast to animal strains, are not invasive (Robins-Browne, 1987), but adhere to epithelial cells. Attachment is mediated by a unique fimbrial antigen (Wells *et al.*, 1983; Karch *et al.*, 1987), but outer membrane constituents may also be involved (Sherman and Soni, 1988). A tissue culture assay with Henle 407 cells showed a characteristic pattern of adherence by O157:H7 containing plasmids, small numbers of bacteria (usually two to four) attaching in a central location. Plasmid-cured derivatives fail to adhere (Karch *et al.*, 1987). Adherence involves a two-stage process resembling that of EPEC strains (Toth *et al.*, 1990). However, while lesions produced by EHEC were originally considered to closely resemble those of classical EPEC serovars, the two types can be clearly differentiated in gnotobiotic piglets (**Table 51**; Levine, 1987).

Table 51. Comparison of intestinal lesions produced by enterohaemorrhagic and enteropathogenic *E. coli*.

	EHEC	ETEC
Site of involvement	Caecum and colon	Whole intestine
Severity	+++	++
Leukocytes	No infiltration	Some infiltration

Toxins

A distinctive feature of EHEC is the production of large quantities of Shiga-like toxin. This may also be referred to as verocytotoxin (verotoxin; VT), as these toxins are almost identical (O'Brien *et al.*, 1982).

Two distinct SLTs are produced, the first, SLTI (VT1), is neutralisable by anti-Shiga toxin, while the second SLTII (VT2) is not (Scotland *et al.*, 1985). Small but significant differences have been demonstrated between SLTII and VT2 (Head *et al.*, 1988), and there may be greater heterogeneity between toxins of this type than previously supposed. Most strains of O157:H7 produce either SLTI and SLTII, or SLTII only, but a strain producing SLTI only has also been described (Smith *et al.*, 1987). SLTII producing strains predominate in North America, the toxin produced being related, but not identical, to the SLTII produced by UK isolates (Dickie *et al.*, 1989).

SLTI is structurally very similar to Shiga toxin, and has the same biological activity (see page 93). By analogy with Shiga toxin, the molecular basis of activity of SLTI is considered to be the catalytic inactivation of 60S ribosomes in toxin-sensitive cells (O'Brien and Holmes, 1987), although these authors note that the analogy has not been tested.

SLTII differs not only immunologically but also in amino acid composition and physicochemical properties. Like Shiga toxin and SLTI, SLTII consists of A and B subunits, but these are larger than those of SLTI with molecular weights of 35,000 and 10,700 respectively (Yutsudo *et al.*, 1987). However, SLTII does share 60% amino acid sequence homology with Shiga toxin (Oku *et al.*, 1989).

A toxin immunologically related to SLTII has been described, which is neutralised by anti-SLTII serum and which has strong cytotoxic activity against Vero cells, but much less activity against HeLa cells. This toxin, SLTII variant (Oku *et al.*, 1989), may be related to a toxin which is involved in pig oedema disease.

A possible fourth toxin has been isolated from a strain of serovar 0157:H7 (Padyhe *et al.*, 1986). This toxin also differs immunologically from SLTI, but does not appear to be composed of subunits. The molecular weight is in the order of 64,000.

The role of Shiga-like and other possible toxins in EHEC infections, is not entirely clear although circumstantial evidence is strong. A number of investigations of the pathogenicity of EHEC have involved animal challenge experiments with live bacteria, where it is not possible to discriminate between the manifestations due to adherence and those due to toxins. Results of studies with purified toxins indicate that there are colonic, renal and neurological effects associated with each type (Scotland, 1988) and that, despite other differences, the biological activity is the same.

A new type of *E. coli* haemolysin, enterohaemolysin, is produced by *ca* 90% of EHEC strains (Beutin *et al.*, 1989). The significance of enterohaemolysin in pathogenesis is not known, but its production may be a useful epidemiological marker.

Some animal EHEC strains produce ST or LT in addition to SLT (Smith *et al.*, 1988), but production of ST or LT has never been reported by a human strain.

Detection of Shiga-like toxins

The classic means of detecting SLTs is by tissue culture. A number of ELISA assays have also been developed, and advanced techniques such as receptor-specified ELISA (Basta *et al.*, 1989) offer an extremely high level of specificity. However, in recent years most attention has been paid to detection of the genes mediating toxin production, using either probes or the polymerase chain reaction (see page 125).

Enteroadherent-aggregative *E. coli*

Little is known of the mechanisms of pathogenicity of this type of pathogenic *E. coli*. Most strains tested by Vial *et al.* (1988) had a 55 to 65 Mdalton plasmid, transfer of which, in one strain, was accompanied by transfer of smooth polysaccharide, expression of fimbriae and aggregative properties.

Distinct histopathological lesions are produced, the enterocytes of the tips and sides of villi being destroyed, and the villi severely blunted. A denuded connective tissue core with a haemorrhagic surface remains (Bhan *et al.*, 1989). The involvement of a toxin, possibly Shiga-like, has been postulated but evidence is lacking.

Human carriage of *E. coli*

In discussing human carriage of *E. coli*, it is important that a distinction should be made between the non-pathogenic strains which are present in the gut of most of the population, and strains which are recognised enteric pathogens. A number of the more sensational reports concerning food hygiene which appeared during the late 1980s failed to recognise this distinction and resulted in misleading conclusions.

It seems likely that in many cases 'carriers' are, in fact, suffering unrecognised mild or asymptomatic infections. Adults are generally relatively resistant to EPEC infections, and cases are known where children have been infected by an apparently healthy adult. This may occur within the family, particularly where housing conditions are poor but is of particular importance in nurseries, etc.

Asymptomatic carriage also exists with strains of ETEC which produce the labile toxin. The rate of carriage is obviously higher where infections by the organism are common, and was found to be 6% of 223 healthy students attending a Mexican university (Pickering *et al.*, 1977). Carriage is transient but, nevertheless, may be important in the spread of infection.

With ETEC the carrier rate varies according to serogroup. Differences in the geographical distribution of serogroups may, therefore, lead to variation in carrier rate from locality to locality. For example, Serogroup O126 which has been isolated only in Hong Kong and Guangzhou, has a low rate of asymptomatic carriage which is reflected in the low rates of carriage in these localities (Yam *et al.*, 1988).

The extent of human carriage of EIEC strains is not known, but the situation is likely to resemble that with *Shigella* (page 95).

The possible importance of human carriage in the spread of EHEC infections by nurses in retirement homes and similar institutions has been noted above (page 109).

E. coli in the environment

Although *E. coli* is present in the gut of man and other animals, it is often assumed that diarrhoeagenic strains, with the exception of enterohaemorrhagic strains, are of human origin. In some cases, isolations have been made from animals and it is suggested, on the basis of limited data, that some ETEC strains responsible for human disease may be carried in the gut of animals (Doyle and Padyhe, 1989).

The organism is excreted in large numbers and thus enters water courses and, where hygiene is poor, the general environment via sewage. There is evidence that *E. coli* persists in freshwater and estuarine waters for shorter periods than *Salmonella*, but in many situations contamination is continuous.

Cattle are thought to be a reservoir for *E. coli* O157:H7, and zoonotic transmission of pathogenic strains from animals to man is thought to occur (Dorn *et al.*, 1989). The serovar may be more common in mature dairy cattle than in younger beef animals, and in the USA there has been concern that the increased amount of cow beef marketed may lead to an increased number of EHEC infections.

It seems likely that other animals and birds are also reservoirs of serovar O157:H7. The organism can readily colonise the ceca of chickens and may be excreted for several months (Beery *et al.*, 1985). The isolation of the organism from retail pork and lamb also suggests that pigs and sheep are reservoirs.

There has been only one reported isolation of O157:H7 from a sick animal, a calf (Orskov *et al.*, 1987). Other EHEC serogroups are common causes of diarrhoea in cattle (Orskov *et al.*, 1987) and diarrhoea and oedema disease in pigs (Smith *et al.*, 1988). These serogroups are not pathogenic to man.

A comparison of non-O157:H7 strains isolated from ill people with strains of the same serogroups isolated from cattle and lambs showed differences with respect to VT production, plasmid profiles and adhesive properties (Dorn *et al.*, 1989).

Isolation of EHEC serovar O157:H7 has been reported from untreated reservoir water in the USA (McGowan *et al.*, 1989). There was no evidence of contamination from human or bovine sources, but it was considered possible that wild deer, which had access to the shores of the reservoir, were involved.

Foods commonly involved in transmission of E. coli

In areas where diarrhoeagenic *E. coli* is endemic, it is usually not possible to associate an infection with specific foods, although a link between diarrhoea in infants and formula feeds made up with contaminated water is well established. It may also be difficult to establish an association between specific foods and *E. coli* infections in developed countries, even where a large number of cases is involved. In part this may be due to the relatively rare occurrence of *E. coli* infections, or to difficulties in isolating pathogenic serovars. However, in some cases, food handlers may be important rather than specific foods.

Enteropathogenic *E. coli*

Although EPEC is a well established cause of diarrhoea among infants, particularly in developing countries, it is rarely considered as a cause in adults, and there is consequently relatively little knowledge of foods which have been associated with EPEC infections.

During the early 1970s, a large number of outbreaks of food poisoning due to EPEC (along with ETEC and EIEC) were attributed to soft cheese imported into the USA (Marier *et al.*, 1973). There is strong evidence that EPEC is able to grow to hazardous levels in soft cheese if is present in the early stages of manufacture (Park *et al.*, 1978; see page 457).

Water has been associated with large outbreaks of EPEC infection in the USA and Sweden. In the USA outbreak at a conference centre near Washington DC where upwards of 170 were affected, incorrect construction of wells permitted contamination of water with sewage.

Meat products were implicated in two separate UK outbreaks involving cold pork and meat pies, and a rather unusual outbreak in Romania involved coffee substitute (Doyle and Padyhe, 1989).

There have been few publications concerning the incidence of EPEC in market foods, but four EPEC strains were found among 60 *E. coli* isolates from cheese in Iraq. EPEC was also isolated from Egyptian butter (Abbar, 1988).

Enterotoxigenic *E. coli*

ETEC has been involved in food poisoning due to soft cheese imported into the USA. The serovar implicated, O27:H20, produced only ST. A single manufacturing plant was involved, and the pattern of infection suggested intermittent contamination over a period of time but the source of contamination was not known (MacDonald *et al.*, 1986).

Water has been implicated as a vehicle of ETEC infection on several occasions, including an outbreak which involved more than 2,000 people at Crater Lake National Park, Oregon, USA. This was attributed to the contamination of the water supply with raw sewage. Waterborne outbreaks have also occurred in Japan and on cruise ships sailing out of the USA and the UK (Doyle and Padyhe, 1989). Cruise ships have an unfortunate association with ETEC and *E. coli* O27 was isolated from 55%, 61% and 20% of travellers who suffered diarrhoea on three successive voyages of the same ship. On these occasions, the infection was probably foodborne (Hobbs *et al.*, 1976).

Some large outbreaks of ETEC have occurred where no association was made with a food. In some cases, contamination of a number of foods by infected handlers was likely. Food handlers are also likely to be an important vehicle of transmission of ETEC in travellers' diarrhoea, since person-to-person spread among adults is uncommon.

Information concerning the incidence of ETEC in foods is inconclusive. ETEC was isolated from up to 10% of meat products obtained from a supermarket in Sao Paulo, Brazil, but none of the isolates was of a serogroup associated with human diarrhoea in the city (Reis *et al.*, 1980). A survey of milk products in West Germany detected four ETEC strains among 77 isolates from Camembert cheese, and one ETEC strain among four ETEC isolates from yoghurt (Franke *et al.*, 1984). Eight percent of *E. coli* strains isolated from a variety of foods in the USA were found to be EPEC, but the most common serogroup was O149, which is associated with illness in piglets rather than in humans (Doyle and Padyhe, 1989).

Enteroinvasive *E. Coli*

EIEC infections are rare in developed countries and many outbreaks are sporadic, the source of infection frequently remaining unknown. Serogroup O124 has been implicated in a number of outbreaks, including one of the first to be fully identified where canned salmon was the suspected vehicle of infection. EIEC serogroup O124 was also involved in outbreaks of gastroenteritis involving soft cheese, and more than 380 people were known to have been affected. The source of contamination was improperly treated river water used for cleaning at the producing plant. Serogroup O124 has also been associated with a waterborne outbreak in Hungary, a hospital outbreak in the UK, and an outbreak at a school for the mentally retarded in the USA. This outbreak was the first recorded instance of person-to-person transmission of EIEC, and presumably involved the oral-faecal route (Doyle and Padyhe, 1989).

A cruise ship was again the setting for a second major food-associated outbreak in the USA, caused by a nontypable strain of EIEC. Potato salad served at a cold buffet was incriminated.

Enterohaemorrhagic *E. coli*

EHEC infections have been linked epidemiologically with minced beef for hamburgers (Riley, 1987). Evidence was first obtained from outbreaks in Oregon and Michigan, which involved 26 and 21 cases respectively (Riley *et al.*, 1983). More recently an outbreak which affected 31 school children in Minnesota was attributed to beef patties. In all cases, the products were undercooked, but in the latter case undercooking occurred at the manufacturing plant rather than at the food service outlet.

Cold, cooked meats, probably a turkey roll, were implicated as cause of a point source outbreak following a christening party (Salmon *et al.*, 1989). Although the possibility of mishandling in the home (where the buffet meal implicated was prepared) could not be totally discounted, the meat was probably contaminated before pre-packing.

Haemorrhagic colitis and its complications must be added to the infections associated with consumption of raw milk. *E. coli* O157:H7 has been isolated from cattle and a bulk milk tank on a farm where two people became infected with the organism after drinking raw milk (Martin *et al.*, 1986). In a later study the incidence of EHEC in milk filters in south-western Ontario was determined using a vero-cell culture assay (Clarke *et al.*, 1989). Recovery rates varied from 0.44 to 0.99%, which was considered to be an underestimate. None of the isolates was of serogroup O157, but two isolates, serovars O26:H7 and O155:H25, were known human pathogenic strains. The pathogenicity to humans or animals of other isolates is not known. The dangers of concentrating solely on serovar O157:H7 are highlighted by this work.

Relatively little information is available concerning the incidence of EHEC in foods at market level. Studies in the USA and Canada have shown serovar O157:H7 to be present in 3.7% of ground beef, 1.5% of pork, 1.5% of poultry and 2% of lamb samples. The incidence of the organism was high (31% of beef samples) in Calgary, western Canada, and this correlates with the high incidence of infection in the area (Doyle and Schoeni, 1987).

There is no information concerning non-O157 serogroups of EHEC.

Factors affecting the growth and survival of *E. coli* in foods

The general behaviour of *E. coli* in foods is similar to that of *Salmonella* and other members of the *Enterobacteriaceae*, and discussion is therefore concentrated on factors where *E. coli* differs. It should be noted that in a number of cases, limiting conditions are not specifically those of pathogenic strains of *E. coli*. Caution is therefore required since the behaviour of pathogenic strains may differ.

Temperature

The general pattern of growth-temperature relationships for *E. coli* is to grow over the range 7 to 48°C with an optimum temperature of 37°C. However, there are important exceptions which involve pathogenic strains.

Enterotoxigenic strains of *E. coli* are recognised as low-temperature pathogens (Palumbo, 1986) and some, if not all, strains will grow at 4 to 5°C, and possibly at lower temperatures. Circumstantial evidence suggests that some strains of other pathogenic types, including *E. coli* O157:H7, are also able to grow below 5°C, but definite evidence is lacking.

The maximum growth temperature for *E. coli* O157: H7 is lower than for other serovars, and several strains are unable to grow at 44°C, a temperature used widely during the isolation of *E. coli*. However, strains unable to grow at 44°C may be more common than is appreciated, particularly with isolates from dairy products.

Although all pathogenic strains of *E. coli* are capable of growth at 37°C, the optimum growth temperature for some may be as low as 30°C. This trait is again particularly common in isolates from dairy products, but does not necessarily involve diarrhoeal strains.

The heat resistance of *E. coli* is of the same order as for most serovars of *Salmonella*, and D values are typically 5 minutes at 55°C and 0.1 minutes at 60°C. Resistance varies according to the nature of the heating menstruum, and there is strain variation. There is no evidence whatsoever that any strain of *E. coli* is sufficiently heat resistant to survive correctly applied pasteurisation or cooking. 'Thermal resistant' or even 'spore forming' *E. coli*, which occasionally appear in anecdotal accounts within the food industry, are nothing more than convenient inventions to conceal the shortcomings of process operations or sanitation.

E. coli is capable of prolonged survival in frozen foods. For example, serovar O157:H7 survives for up to 9 months at −20°C in ground beef (Doyle and Schoeni, 1987).

Irradiation

The sensitivity of *E. coli* to irradiation is similar to that of *Salmonella*, and a dose of 3 Kgy proposed for control of that organism would be effective against *E. coli*.

pH value

E. coli is capable of growth over the pH range 4.4 to 9.0 (ICMSF, 1980). There is interaction with other factors including temperature, a_w and the nature of the acidulant.

Water activity

The minimum a_w level for growth where other conditions are optimal is 0.95.

Curing ingredients

The effect of combinations of curing ingredients on a cocktail of 'enteropathogenic' strains of *E. coli* has been studied by Gibson and Roberts (1986). The upper NaCl concentration for growth was 6%, but growth in this concentration occurred only at pH values between 5.6 and 6.8 and at temperatures between 15 and 35°C. At pH 5.6, growth at 10 to 15°C was inhibited by 200 mg/l $NaNO_2$ and at 20°C by 400 mg/l $NaNO_2$. Growth occurred in any combination of 0 to 4% NaCl and 0 to 400 mg/l $NaNO_2$ at pH values between 6.2 and 6.8 and at temperatures between 15 and 35°C. *E. coli* is thus somewhat more resistant than *Salmonella*.

Competition from other micro-organisms

E. coli is a more effective competitor with spoilage micro-organisms than *Salmonella*, and in food fermentations may reach significant numbers before competition from lactic acid bacteria becomes significant.

However, at low temperatures *E. coli* is likely to be overgrown by psychrotrophic spoilage organisms such as *Pseudomonas*, even if capable of growing below 5°C in the absence of competition.

Laboratory methods

Conventional cultural isolation methods

Resuscitation

Resuscitation of *E. coli* has received little attention compared with *Salmonella*. Non-selective broths such as nutrient, or tryptone-soy, are commonly used. Solid media, nutrient agar or the elective, minerals modified glutamate agar are used with the filter method of Holbrook *et al.* (1980). Resuscitation periods are short, being usually no longer than 4 to 6 hours at 35 to 37°C. Longer periods may be required where cells are severely damaged.

Selective enrichment

Commonly used enrichment broths for *E. coli* are EE broth, which contains brilliant green and bile as selective agents, and lauryl-tryptose broth, which contains sodium lauryl sulphate. Other media that may be used are MacConkey broth and GN broth.

In a number of methods, elevated incubation temperatures of 41 to 45°C are advised in order to increase selectivity and reduce the incubation period. *E. coli* O157:H7 as well as strains of other serovars are often unable to grow at such temperatures, or may lose virulence plasmids, and if elevated temperature incubation is used, a second enrichment broth incubated at 35 to 37°C is often recommended. This is of particular importance with strains from dairy sources, but is likely to lead to problems due to reduced selectivity. In order to overcome this dilemma, Mehlmann and Romero (1982) developed two broths, tryptone-phosphate and glucose-lactose-Tween 80, which permit the growth of pathogenic *E. coli*, other than serovar O157:H7, at 44°C. Tryptone-phosphate broth is recommended for enrichment (Mehlmann and Lovett, 1984).

Selective plating

Various selective agents are used in selective plating media for *E. coli*, although bile salts are most common. For the purpose of discussing the media classification is best made on the basis of the differential system used.

Lactose fermentation is the traditional means of differentiating *E. coli* from non-lactose fermenting members of the *Enterobacteriaceae*. The most serious objection to the use of lactose fermentation as a differential system is the high proportion of pathogenic strains of *E. coli* which fail to ferment the sugar (see pages 104–105). In situations where conventional media of this type are used, it is prudent to select lactose non-fermenting colonies as well as lactose fermenting colonies for further testing. However, this may involve testing a very large number of colonies. An alternative approach is to replace all, or some, of the lactose with another sugar, such as glucose or arabinose, but testing of a large number of colonies may still be necessary.

The use of sugar fermentation as a differential characteristic is avoided entirely in tryptone-bile medium, which differentiates *E. coli* on its ability to produce indole from tryptophane when grown on a cellulose acetate membrane at 44°C. This reaction is unique to *E. coli* and no biochemical confirmation is required. This medium is widely used in the food industry, but it is considered that the medium and associated methodology suffer conceptual and practical shortcomings with respect to the isolation of pathogenic serogroups.

Detection of β-glucuronidase has been used as a differential system for *E. coli* since the widespread availability of fluorogenic substrates such as 4-methylumbelliferyl-β-D-glucuronide (MUG) (Feng and Hartmann, 1982). This substrate is usually combined with a conventional medium. It must be appreciated that β-glucuronidase activity is not unique to *E. coli*, and that the enzyme is also produced by some salmonellas and shigellas and a few yersinia. This may be of little practical significance, but false positive reactions may also be produced by spoilage bacteria which may be present in large numbers including enterococci, pseudomonads and other Gram-negative bacteria. It should also be appreciated that β-glucuronidase-negative strains of *E. coli* other than serovar O157:H7 (see below) exist, and that these may be more widespread than previously appreciated. Up to 34% of *E. coli* from the faeces of healthy people were found to be negative by Chang *et al.*, (1989), although this is probably of greater consequence when the test is used for determining *E. coli* as an index organism. Thus while the incorporation of

MUG can improve the precision of selective plating media for *E. coli*, the need for caution when interpreting results remains.

The replacement of MUG with a novel substrate 5-bromo-4-chloro-3-indoxyl-β-D-glucuronide has been suggested (Frampton *et al.*, 1988; Watkins *et al.*, 1988). This is chromogenic rather than fluorogenic but, while more convenient, is likely to suffer the same problems as MUG.

Selective plating media are summarised in **Table 52** and their use illustrated in **Figures 78–85**.

Recovery of serovar O157:H7

As noted above, serovar O157:H7 has phenotypic characteristics which may be exploited in the development of media for its selective isolation. MacConkey-sorbitol agar (March and Ratnam, 1986) has been developed for the isolation and presumptive identification of this serovar. Its use, and that of a variant which incorporates MUG as a second differential feature (Szabo *et al.*, 1986), is illustrated in **Figures 86–90**.

Design of a complete protocol

Few food microbiology laboratories examine for pathogenic serogroups of *E. coli* on a regular basis and, if examination is required, they develop an *ad hoc* method based on procedures for determining *E. coli* as an index organism. A sample protocol which is designed to overcome the problems associated with the different types of media is shown in **Flow diagram 11**. This protocol is designed for use with all diarrhoeagenic *E. coli*, and alternatives may be employed where a specific type is to be isolated. For example, a procedure for EHEC O157:H7 has been devised (Okrend *et al.*, 1990). This comprises enrichment in EC broth of reduced bile salt concentration containing novobiocin, incubated at 37°C for 6 hours (shaken), or at 35°C for 24 hours (static). Spread dilutions of the enrichment culture are then plated onto MacConkey-sorbitol agar incubated at 42°C overnight. Suspect colonies are subcultured for confirmatory testing. It should be appreciated that some strains of EHEC O157:H7 may be unable to grow at 42°C.

Table 52. Examples of selective plating media for isolation of *E. coli*.

MacConkey agar (various formulations)
Inhibitors	Bile sales, crystal violet[1]
Diagnostic system	Lactose fermentation[2]

Violet red–bile agar (various formulations)
Inhibitors	Bile salts, crystal violet
Diagnostic system	Lactose fermentation[2]

Drigalski agar
Inhibitors	Bile salts, crystal violet
Diagnostic system	Lactose fermentation

MUG agar
MUG agar is based on one of the above media and provides an additional diagnostic test for *E. coli* by determining the production of β-glucuronidase.

Eosin–methylene blue agar
Inhibitors	Methylene blue
Diagnostic system	Lactose fermentation

Endo agar
Inhibitors	Sodium sulphite, basic fuchsin
Diagnostic system	Lactose fermentation

Tryptone–bile agar
Inhibitors	Bile salts (at 44°C)
Diagnostic system	Indole production (at 44°C)

[1] Crystal violet is not part of most formulations.
[2] Other carbohydrates may be used with, or instead, of lactose.

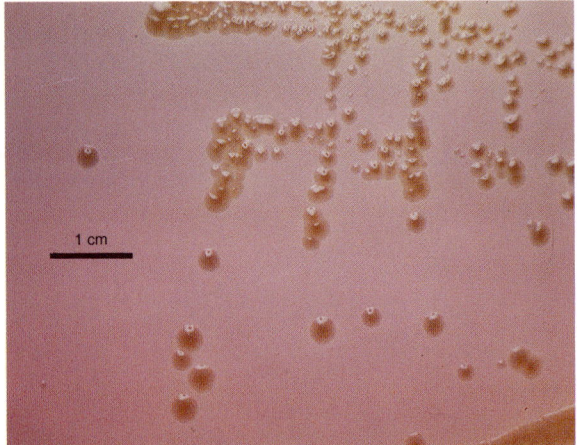

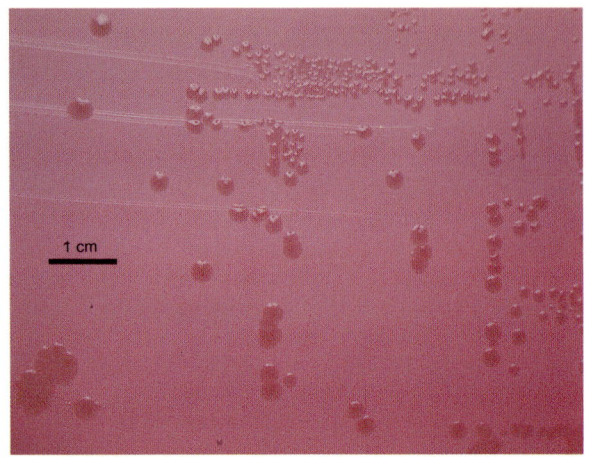

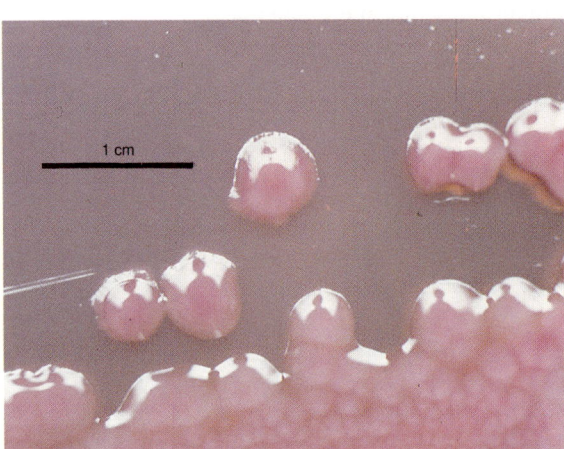

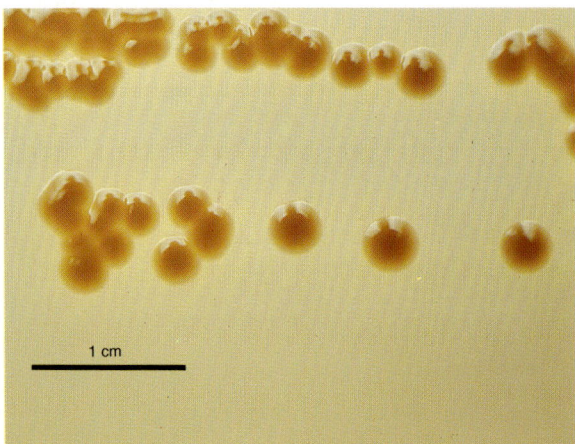

78–81. Isolation of *E. coli* on selective agar media containing bile and lactose. Media containing bile as the selective agent and using acid production from lactose as the diagnostic system have been used for the isolation of *E. coli* and other members of the *Enterobacteriaceae* for many years. Many variations have been proposed, but the basic principles remain the same.

MacConkey-purple agar (**78** & **79**) is a modification of the original formulation and contains bromocresol purple as the indicator of acid production in place of neutral red, bromocresol purple being less inhibitory and providing a more sensitive indicator of acid production. Lactose fermenting strains of *E. coli* (**78**) are readily distinguished from lactose-negative members of the *Enterobacteriaceae* such as *Salmonella* (**79**) by the formation of yellow colonies. However, it is not possible to distinguish between *E. coli* and other lactose fermenting members of the family.

MacConkey agars are of low selectivity and support the growth of non-*Enterobacteriaceae*, including enterococci and staphylococci as well as other Gram-negative bacteria such as *Aeromonas*. Violet red-bile-lactose agar (**80**) is a MacConkey-type medium which is often preferred for the isolation of 'coliforms' and *E. coli* from foods. Crystal violet is used as a selective agent in addition to bile and neutral red is the indicator. Lactose fermenting colonies are pink and there is no differentiation between *E. coli* and other lactose-positive bacteria. A modification, violet red-bile-glucose agar, is used for the determination of 'total' *Enterobacteriaceae* (lactose fermenting and non-fermenting species).

Drigalski agar (**81**) resembles violet red-bile-lactose agar in containing both crystal violet and bile as selective agents, but the ratio of crystal violet to bile salts is higher in Drigalski medium. Bromothymol blue is used as the indicator and gives a clearer differentiation between acid-producing bacteria and non-acid producing than neutral red. (*All media: 24h at 37°C.*)

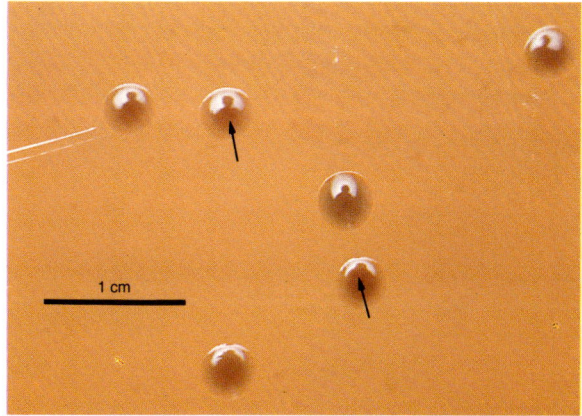

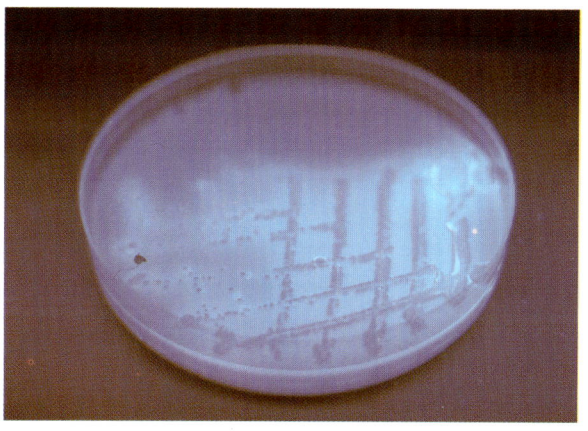

82. Isolation of *E. coli* on eosin-methylene blue agar.
Eosin-methylene blue agar is widely used in the USA for the isolation of *E. coli* from water and may also be used with foods. The medium contains lactose as a fermentable carbohydrate, but methylene blue and eosin act as both selective and diagnostic agents. Colonies of *E. coli* may be differentiated from those of other lactose fermenting bacteria by the dark purple centres (arrowed) visible under transmitted light. Colonies may also have a greenish metallic sheen when viewed under reflected light, but this property is not consistent. Colonies of *E. coli* are, however, very highly reflective. (*Eosin-methylene blue agar medium: 24h at 37°C.*)

83. Incorporation of 4-methylumbelliferyl-β-D-glucuronide into MacConkey agar. 4-methylumbelliferyl-β-D-glucuronide (MUG) serves as a means of differentiating *E. coli* from other lactose fermenting bacteria that are able to develop on MacConkey and similar media. MUG may also be incorporated into liquid media used in most probable number determinations. In use, *E. coli* hydrolyses MUG by the enzyme β-glucuronidase and positive colonies are detected by fluorescence under long wavelength ultra-violet illumination. Although a useful method, there are shortcomings (see page 121) and some laboratories have experienced poor results as a consequence of uncritical acceptance of the method as a solution to all problems. (*MacConkey agar plus MUG: 24h at 37°C.*)

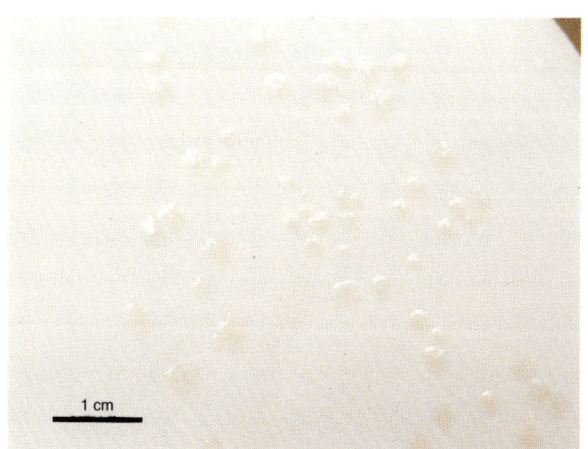

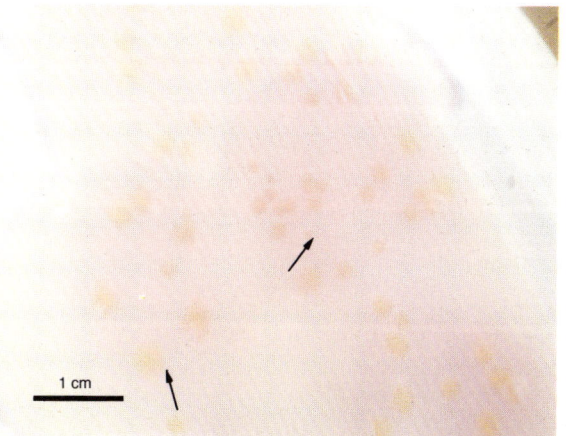

84 & 85. Isolation of *E. coli* on tryptone-bile agar.
Colonies develop on membrane filters overlaying the medium (**84**) and the indole reagent is added to detect the indole-positive colonies of *E. coli* (**85**). Although the dependence on lactose fermentation as diagnostic character is avoided, false-negatives occur due to indole-negative strains of *E. coli*, which may be more prevalent among pathogenic strains. A pure culture of *E. coli* was used in the preparation of the illustrations, but indole-negative colonies are present (arrowed) among the pink indole-positive.

From a practical viewpoint, it may be difficult to determine which colonies are indole-positive when a large number of non-*E. coli* are present. A more serious problem is the lethal effect of the indole reagent, which necessitates replica plating to a second medium before the addition of the reagent if confirmatory testing is required. Where a significant background microflora is present, it often proves impossible to recognise the indole-positive colonies on the replicate plate.

Trypyone-bile agar incubated at 44°C may also be over-selective to some strains of *E. coli* from long shelf-life chilled foods, even when a resuscitation step is included. (*Tryptone-bile agar: 24h at 44°C.*)

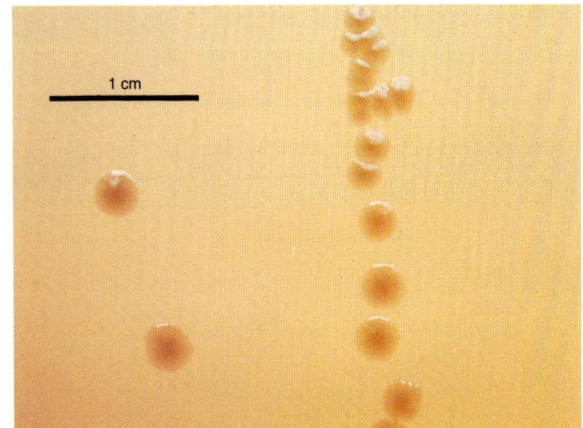

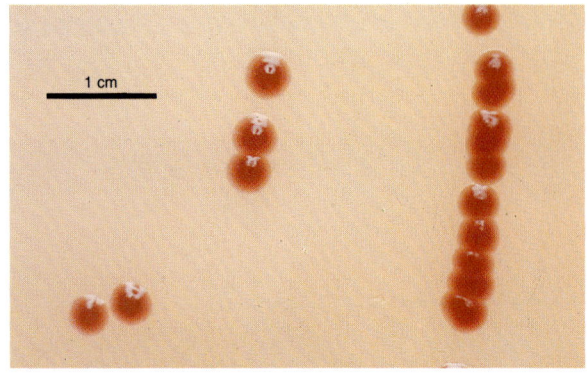

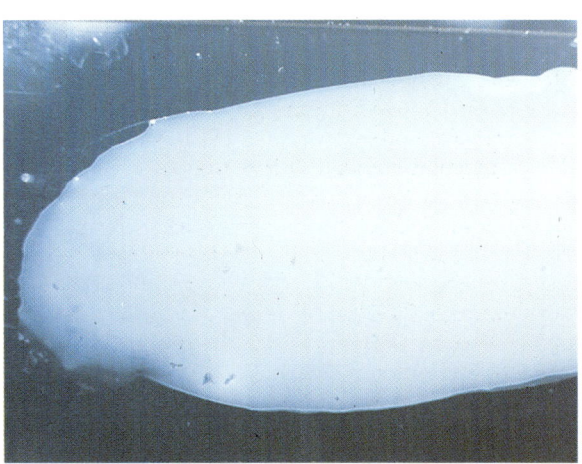

86–90. Isolation of *E. coli* O157:H7. MacConkey-sorbitol agar in which sorbitol replaces lactose as a fermentable carbohydrate is a simple and, within limits, effective means of isolating and differentiating *E. coli* O157:H7. Most strains of this serovar are unable to ferment sorbitol, and colonies developing on MacConkey-sorbitol agar are either colourless or pale pink (**86**). In contrast, most other isolates of *E. coli* as well as many other bacteria able to grow on the medium ferment sorbitol, and colonies are deep pink or red (**87**). MacConkey-sorbitol agar, however, permits only presumptive isolation and differentiation since the inability to ferment sorbitol is by no means a unique characteristic of *E. coli* O157:H7. Specificity is increased by incorporating MUG into the medium since this serovar also fails to produce β-glucuronidase and thus gives a negative (non-fluorescent) reaction (**88**). In all cases, confirmation using a latex co-agglutination test (Unipath) is recommended. The positive reaction illustrated (**89**) is of a presumptive isolate from beef mince. The isolate giving a negative reaction (**90**) was from the same source and was biochemically identified as a species of *Aeromonas*.

Particular problems may be encountered with some strains of *E. hermanii*, which will not only agglutinate with commercial O157 antiserum but also fail to ferment sorbitol or hydrolyse MUG. *E. hermanii* is rare in foods, but in cases of doubt a full biochemical differentiation is required.

E. coli serotypes other than O157:H7 may cause haemorrhagic colitis and its complications, and in such cases the above procedures are of no value. (*MacConkey-sorbitol agar: 24h at 35°C.*)

Flow diagram 11. Sample protocol for the recovery of pathogenic serogroups of *E. coli* from foods

RESUSCITATION I
Nutrient broth:
6 h at 37°C

RESUSCITATION II
Membrane overlaid on minerals
modified glutamate agar: 4 h at 37°C

ENRICHMENT
EE broth: 18 h: 37°C and 44°C

SELECTIVE
PLATING I
MacConkey –
arabinose agar:
24 h at 37°C

SELECTIVE
PLATING II
MacConkey –
sorbitol agar:
24 h at 37°C

SELECTIVE
PLATING III
Transfer
membrane to
TBA: 18 to 24 h at 44°C

Serological and genetic methods for the detection of specific types of diarrhoeagenic *E. coli*

Enterotoxigenic E. coli

A number of colony hybridisation assays have been developed for detection of ETEC. A novel visual assay designed for detection of LT producing ETEC in foods (Romick *et al.*, 1987) was subsequently evaluated and considered to have considerable potential, despite problems with backgound 'noise' from other bacteria and food constituents (Romick *et al.*, 1989). The protocol is shown in **Flow diagram 12**. Further developments have included the use of disoxigenin labelling in place of biotinylation, such probes being cheap, simple in use, and having fewer problems with background 'noise' (Riley and Caffrey, 1990), and the development of a RNA probe capable of detecting LT and ST genes simultaneously (Suez-Llorens, 1989).

A method based on the polymerase chain reaction has been devised for LTI producing ETEC (Furrer *et al.*, 1990). This is adaptable for examination of food or environmental samples.

Enterohaemorrhagic E. coli

Both colony hybridisation and immunoblotting techniques have been applied to the detection of EHEC by gene probes. When used in conjunction with enrichment, results are available within 48 hours with immunoblotting, but 3 to 4 days are required for colony hybridisation (Samadpour *et al.*, 1990). In the latter case, however, colonies are available for confirmatory testing.

A sophisticated method of isolating EHEC serovar O157:H7, which combined hydrophobic grid membrane filtration with immunoblotting, was developed by Doyle and Schoeni (1987). This technique was used successfully in a survey of the incidence of serovar O157:H7 in retail meats and poultry, but was not considered amenable to routine testing.

More recently, Todd *et al.* (1988) have described a rapid method which combines HGMF techniques with an enzyme labelled antibody procedure which yielded results within 24 hours and detected *E. coli* O157:H7 in 95% of artificially contaminated meat samples at a sensitivity of 10/g. A highly specific monoclonal antibody was used, which largely resolved the problems of false-positive reactions experienced by Doyle and Schoeni (1987). The method detects all members of serogroup O157 and thus specific typing for H7 is required for confirmation of serovar O157:H7. Serogroups other than O157 are not detected. The protocol is shown in **Flow diagram 13**.

The polymerase chain reaction forms the basis of a technique for detecting any SLT genes (Karch and Meyer, 1989). The technique is capable of detecting 10^2 EHEC cells in 10^7 competitors, but although developed for use with clinical samples, it cannot be used direct on stools due to interference. Adaptation for use with foods is considered to be feasible.

Enteroadherent-aggregative E. coli

A biotin labelled probe has been developed for use with clinical materials (Baudry *et al.*, 1990).

Flow diagram 12. Protocol of colony hybridisation assay for detection of LT producing enterotoxigenic _E. coli_ (Romick _et al._, 1987)

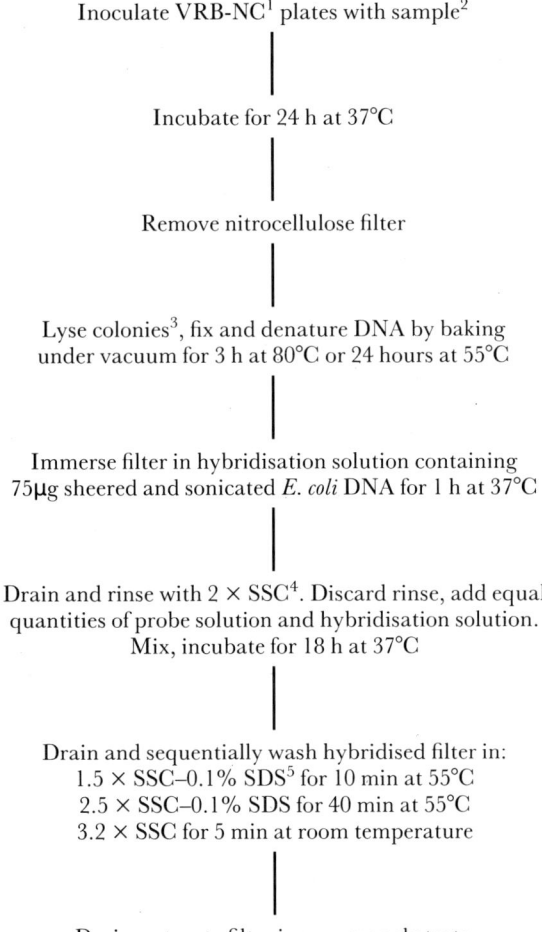

Inoculate VRB-NC[1] plates with sample[2]

|

Incubate for 24 h at 37°C

|

Remove nitrocellulose filter

|

Lyse colonies[3], fix and denature DNA by baking under vacuum for 3 h at 80°C or 24 hours at 55°C

|

Immerse filter in hybridisation solution containing 75μg sheered and sonicated _E. coli_ DNA for 1 h at 37°C

|

Drain and rinse with 2 × SSC[4]. Discard rinse, add equal quantities of probe solution and hybridisation solution. Mix, incubate for 18 h at 37°C

|

Drain and sequentially wash hybridised filter in:
1.5 × SSC–0.1% SDS[5] for 10 min at 55°C
2.5 × SSC–0.1% SDS for 40 min at 55°C
3.2 × SSC for 5 min at room temperature

|

Drain, saturate filter in enzyme substrate. Incubate, in dark, for 2 h at 25°C

|

Positive colonies appear brown when compared with negative.

[1] Violet red-bile agar overlaid with nitrocellulose filter.
[2] In practice enrichment may be required.
[3] Method of Moseley _et al._ (1980).
[4] Sodium chloride, 0.015 mol/l; sodium citrate, 0.0015 mol/l; sodium dodecyl sulphate.
[5] Sodium dodecyl sulphate.

Flow diagram 13. Protocol of HGMF-enzyme labelled antibody procedure for detection of _E. coli_ 0157

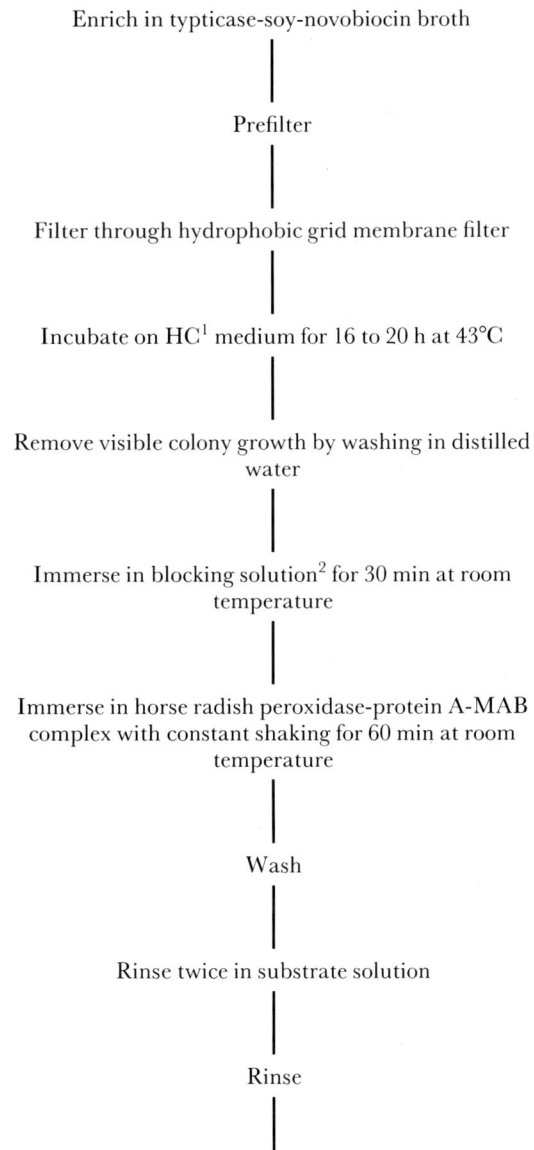

Enrich in typticase-soy-novobiocin broth

|

Prefilter

|

Filter through hydrophobic grid membrane filter

|

Incubate on HC[1] medium for 16 to 20 h at 43°C

|

Remove visible colony growth by washing in distilled water

|

Immerse in blocking solution[2] for 30 min at room temperature

|

Immerse in horse radish peroxidase-protein A-MAB complex with constant shaking for 60 min at room temperature

|

Wash

|

Rinse twice in substrate solution

|

Rinse

|

Positive colonies appear purple when compared with negative

[1] Medium of Szabo (1986)
[2] Gelatin (3%) in tris buffered saline.

Identification of diarrhoeagenic _E. coli_

Differentiation from other genera

Differentiation is made according to the criteria in **Table 47**, page 104. There are usually few difficulties with the exception of possible confusion with _Shigella_, although according to Tamura _et al._ (1986) many food-derived bacteria identified as _E. coli_ are probably _Leclercia adecarboxylata_ (_E. adecarboxylata_), which is allegedly common in foods. Confusion would seem most likely with EHEC which, like _L. adecarboxylata_, is negative in lysine and ornithine decarboxylase reactions, and fails to ferment sorbitol. Therefore, where biochemical testing is used, a full range of reactions should be determined.

Differentiation from other species of Escherichia

Differentiation is made according to the criteria in **Table 48**, page 105. In practice, species other than *E. coli* are usually rare in foods, although *E. hermanii* can be a particular problem with respect to EHEC O157: H7 since it is sorbitol-negative, β-glucuronidase-negative and may agglutinate in commercial latex agglutination tests for serogroup O157. A full range of tests must be made to discriminate between the species and commercial identification kits based on a small number of tests are not suitable.

Typing of *E. coli*

Serotyping

Serotyping may be either by slide or, more commonly, tube agglutination. It must be appreciated that fimbrial antigens will take part in agglutination reactions when cells are in the fimbrial phase (Cruickshank *et al.*, 1975). Such antigens are widely shared between strains of different serovars, and thus misleading cross-reactions may occur. It is therefore necessary either to use cells from agar cultures which are non-fimbriate, or cells which have been defrimbiated by heating for 1 hour at 100°C.

Co-agglutination tests are often used for confirming the presence of EHEC serogroup O157:H7. Commercial systems commonly use latex sensitised by a polyclonal antibody, but problems with false-positive reactions may be largely eliminated by use of a monoclonal antibody. A specific and sensitive assay described by Perry *et al.* (1988) used cells of *Staphylococcus aureus* Cowan Group 1, sensitised by a monoclonal antibody in a slide agglutination test.

Detection of toxins

Of the various toxins produced by *E. coli*, assay methods suitable for routine application are available only for the LT and ST toxins of enterotoxigenic strains.

Detection of LT

LT is closely related to cholera toxin and it is often possible to use the same methods to detect both toxins.

Methods to date are applicable to the detection of LTI only. However, the importance of LTII in human illness is probably limited.

Selection of colonies

False-negatives have occurred in toxin tests due to simple failure to ensure that colonies chosen for testing have consisted of enterotoxin producing cells. This can be a particular hazard when selecting colonies from a mixed culture (Wachsmuth, 1986). Further problems may occur due to loss of toxin production through loss of plasmids encoding that property. This is a well-recognised phenomenon that may correlate with serovar (Evans *et al.*, 1977).

Production of LT

Only a small quantity of LT is released into the culture medium and considerable attention must be paid to ensuring adequate yield. Techniques commonly used are summarised in **Flow diagram 14**.

Detection methods

A number of serological methods have been described but two, ELISA and reverse passive latex agglutination methods, are most widely used. Of particular interest with respect to ELISA are those assays which use the GM_1 ganglioside, the physiological target for LT, as a ligand for the toxin. This approach was first described by Svennerholm and Holmgren (1978), their assay being modified by Back *et al.* (1979) and Sack *et al.* (1980) for use with microtitre plates. Subsequent modifications have reduced the time necessary for the assay to 8 hours and increased the sensitivity to equal that of the CHO cell assay (Svennerholm and Wiklund, 1983; Bongaerts *et al.*, 1985). However, Bongaerts *et al.* noted that the increasing sensitivity of the GM_1 ELISA may make the interpretation of results more difficult. Alternative procedures include the nitrocellulose-ELISA (Beutin *et al.*, 1984) and the immunospot assays (Czerkinsky and Svennerholm, 1983).

Although the major interest has been with ELISAs, reverse passive latex agglutination (RPLA) has been developed as an alternative and is available as a commercial kit (Denka Seiken; Unipath). The assay is less sensitive than ELISA, but is technically simpler and suitable for use in the smaller quality assurance laboratory.

In addition to the more sophisticated techniques, gel diffusion techniques have been described. Gel diffusion techniques have a number of inherent disadvantages, including low sensitivity, but require no specialised equipment or sophisticated reagents and may therefore be suitable for application in laboratories with limited facilities or in field epidemiological studies.

Flow diagram 14. Cultural conditions for production of *E. coli* heat labile enterotoxin.

Isolates producing high levels of toxin

Medium[1]

Casein hydrolysate	200 g
Yeast extract	6 g
NaCl	2.5 g
K_2HPO_4	8.7 g
Trace salts mix	1 ml
Distilled water	1000 ml pH 8.5

(Trace salts mix: $MgSO_4$ 5%; $MnCl_2$ 0.5%; $FeCl_3$ 0.5% dissolved in 0.001N H_2SO_4).

Incubate inoculated medium for 18 h at 37°C in an orbital incubator at 150 rpm.

Isolates producing low levels of toxin

Medium

Brain-heart infusion agar supplemented with 90 μg/ml lincomycin

|

Cultivate on slope for 18 h at 37°C

|

Suspend cells in 1 ml 0.085% NaCl containing 10,000 units/ml polymyxin B[2]

|

Shake suspension for 30 min at 37°C

|

Centrifuge for 20 min at 3,000 rpm

|

Remove supernatant and use as test sample

[1] Medium of Evans *et al.* (1974).
[2] Bongaerts *et al.* (1985) considered use of polymyxin B unnecessary.

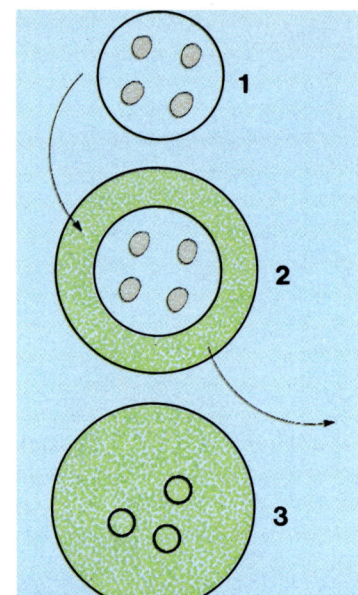

91. Membrane filter method for the detection of *E. coli* LT enterotoxin. The membrane filter method is a simple method which is considered to be worth further development for use where facilities are limited. When anti-cholera toxin was used as the antiserum, the sensitivity of the method was 84% (compared with combined results of an ELISA and a DNA hybridisation assay) and the specificity 95%.

Subcultures were spot-inoculated onto 0.45 μm cellulose acetate membrane filters and incubated overnight at 44°C on pads soaked in membrane lauryl-sulphate broth (**1**). The pads were transferred to Biken agar medium containing 2% unpurified anti-cholera toxin and incubated at 44°C for a further 48 h (**2**). The membrane filter was then removed and the underlying agar examined for zones of precipitation (**3**).

Two gel diffusion techniques were evaluated (Vadivelu *et al.*, 1987), the Biken plate (Honda *et al.*, 1981) and a membrane filter assay (**Figure 91**; Vadivelu *et al.*, 1987), the latter being of significantly greater sensitivity.

Detection of ST

A competitive ELISA has been developed which correlates extremely closely with the infant mouse asssay. The ELISA assay uses a synthetic peptide toxin analogue and a monoclonal antibody, which results in a high level of specificity when used for the detection of ST in culture fluid. Only ST_A is detected but ST_B, like LTII, appears to be of little importance in human disease. The assay is available as a commercial kit (Denka Seiken; Unipath).

7. *Yersinia*

Introduction

The genus *Yersinia* is a typical member of the *Enterobacteriaceae*, although it has a number of distinctive features, including the size and morphology of colonies (**Figures 92–94**). Cells of *Yersinia* are small and have a coccoid morphology, which resembles that of the *Pasteurellaceae* rather than the *Enterobacteriaceae* (Bercovier and Mollaret, 1984). Depending on the medium and temperature of incubation, pleomorphism may be observed, a mixture of coccoid cells and rods occurring singly or in short chains being present.

Yersinias are 'low temperature' pathogens and are able to grow at 4°C. A striking feature of the genus is that many of its characteristics are temperature dependent, being expressed at 28 to 30°C but not at 37°C (see page 131). Incubation at 37°C also results in loss of the virulence plasmid and thus of plasmid mediated properties (see page 139).

The genus *Yersinia* contains three species which are recognised pathogens of humans; *Y. pestis, Y. pseudotuberculosis* and *Y. enterocolitica.*

Y. pestis is well known, both historically and in modern medicine, as the causative organism of the plague (the 'Black Death' of medieval Europe). In its most common form, bubonic plague is spread by fleas, the organism invading the lymph nodes corresponding to the bite and then colonising deep organs via the blood stream (Baker, 1983). If untreated, death usually occurs within 3 to 5 days of infection.

Y. pseudotuberculosis has been known in the past as a pathogen of animals, but may also infect man via the oral route, and the frequency of infection may be considerably underestimated (Attwood *et al.*, 1987). Food has been implicated as a vehicle of infection in Japan (Shiozawa *et al.*, 1988).

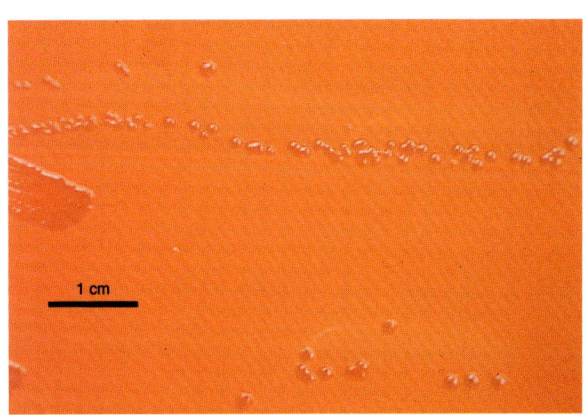

92

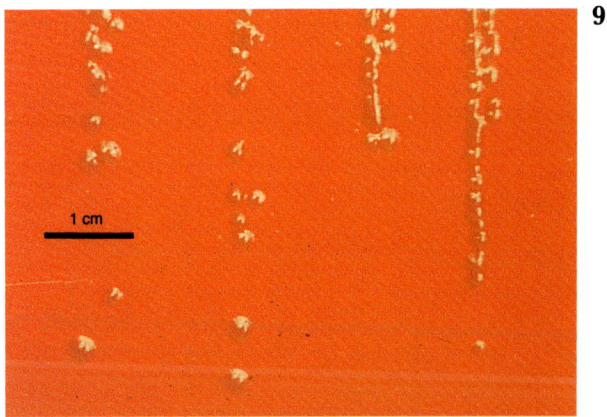

93

92 & 93. Growth of *Y. enterocolitica* on blood agar medium. *Yersinia* species (**92**) have a smaller colony diameter than other members of the *Enterobacteriaceae* such as *E. coli* (**93**). Blood and other nutritional supplements such as yeast extract have little or no effect on colony size. (*Blood agar: 24 h at 28°C.*)

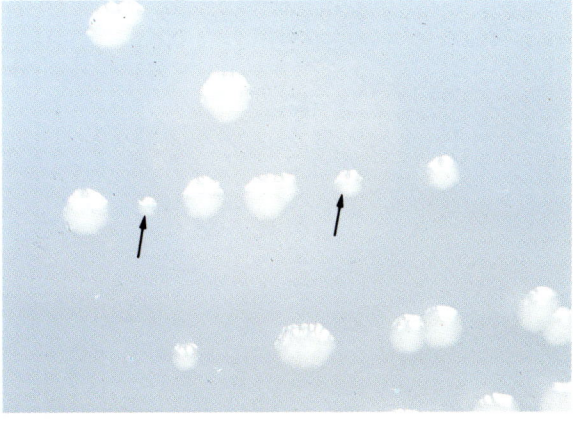

94

94. Dissociation of colonies of *Y. enterocolitica*. In common with all other species of *Yersinia*, *Y. enterocolitica* dissociates into small (*ca* 0.5 mm) and large (*ca* 2 mm) colonies after *ca* 48 h incubation. This can be a valuable feature in presumptive recognition of the genus, but appears to be dependent on the medium used (Bercovier *et al.*, 1979). Some colonies also have a 'chinese hat' morphology (arrowed) after incubation for this period. (*Nutrient agar medium: 48 h at 25°C.*)

Y. pestis and *Y. pseudotuberculosis* are closely related, and it has been proposed, on the basis of DNA relatedness, that the two should constitute a single species containing two subspecies, *Y. pseudotuberculosis subsp. pseudotuberculosis* and *subsp. pestis* (Bercovier *et al.*, 1980a,b). This was rejected (Anon, 1987) following opposition by Williams (1984) on the grounds that change in nomenclature could lead to dangerous confusion in public health laboratories.

Y. enterocolitica is currently considered to be of greatest importance as a foodborne pathogen. Symptoms are primarily those of gastroenteritis, but a number of complicating or secondary symptoms may also occur.

A number of other species of *Yersinia* are recognised, some of which were previously described as '*Yersinia enterocolitica*-like organisms'. These may be considered to be environmental organisms, and while there have been occasional reports of association with illness, their role has been opportunistic.

Isolation of *Yersinia*

Following the recognition of *Y. enterocolitica* as a foodborne pathogen, a considerable effort has been expended in developing methods for its isolation. In comparison, the isolation of *Y. pseudotuberculosis* from foods has received little attention.

Conventional cultural techniques

Enrichment

Two approaches have been taken to the enrichment of *Y. enterocolitica*. The first involves enrichment at a low temperature, usually in a non-selective medium. The second involves the use of selective enrichment broths incubated at 25°C. A number of broths have been used, but only a modified Rappaport broth and bile–oxalate–sorbose broths appear suitable.

Cold enrichment is also a suitable means of enriching for *Y. pseudotuberculosis*.

Alkali tolerance

Both *Y. enterocolitica* and *Y. pseudotuberculosis* are more tolerant of alkali conditions than most other bacteria, and this has been exploited in a pre-plating treatment of enrichment cultures to selectively reduce the level of competing micro-organisms.

Selective plating media

Traditional general purpose media for the *Enterobacteriaceae* or media designed for *Salmonella* were used initially as selective plating media for *Y. enterocolitica*, and some, such as MacConkey agar, remain in wide use for clinical samples and occasionally for foods. However, media devised specifically for *Y. enterocolitica*, such as cefsulodin–irgasan–novobiocin (CIN), are generally preferred.

CIN medium is not suitable for recovery of *Y. pseudotuberculosis*, but this organism grows well on most plating media for enteric organisms.

Rapid detection of *Yersinia*

In comparison with many other enteric pathogens, little attention has been given to rapid detection methods for yersinias. However, colony hybridisation techniques have been developed, which do not require pre-enrichment.

Methods for the isolation and detection of Yersinia *are discussed in greater detail, with respect to choice of method and laboratory practice, in* **Laboratory Methods**, *pages 146–152.*

Taxonomy of *Yersinia*

Effect of incubation temperature on biochemical characteristics

A distinctive feature of *Yersinia* is the temperature dependence of a number of biochemical reactions. A number of reactions are positive at 25°C, but negative at 37°C. The transition point is usually *ca* 30°C, and to avoid confusion an incubation temperature of 25°C is recommended for biochemical tests. The biochemical reactions of yersinias which are known to be temperature dependent are listed in **Table 53**, and some examples are illustrated in **Figures 95–99**.

Differentiation of *Yersinia* from other genera

Differentiation of *Yersinia* from related genera may be made according to the criteria in **Table 54**. Conventional test media, miniaturised identification kits and, under some circumstances, multi-test media may be used. Some properties of *Yersinia* are illustrated in **Figures 100–103**.

Table 53. Temperature dependent biochemical characteristics of *Yersinia*.

Motility
ONPG hydrolysis
Ornithine decarboxlase
Indole
Voges–Proskauer[*]
Acid from
Cellobiose
Raffinose
Sorbitol[*]

Note: The reactions listed are examples only and other biochemical reactions may also be temperature sensitive. Examples are from Bercovier and Mollaret (1984) except for those marked [*], which are from the Authors' unpublished observations.

Differentiation within *Yersinia*

In addition to the three recognised human pathogenic species of *Yersinia*, there are a number of further species, some of which are common in food. Three of these species; *Y. intermedia* (Brenner *et al.*, 1980a), *Y. frederiksenii* (Ursing *et al.*, 1980b) and *Y. kristensenii* (Bercovier, 1980a) were those previously referred to as '*Yersinia enterocolitica*-like organisms'. *Y. aldovae* (Bercovier *et al.*, 1984) is comprised of strains previously described as 'Group X' (Brenner, 1980b) and appears to be associated with aquatic habitats, while *Y. rohdei* (Aleksic *et al.*, 1987) has been isolated from surface waters and faeces. Most recently it has been proposed to transfer biovars 3A and 3B of *Y. enterocolitica* to two new species, *Y. mollaretii* and *Y. bercovieri* (Wauters *et al.*, 1988).

Criteria for differentiation of *Y. enterocolitica*, *Y. pseudotuberculosis* and other foodborne yersinias are listed in **Table 55**.

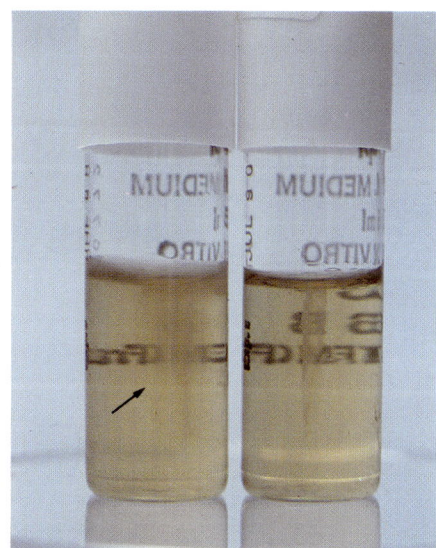

95

95. Influence of incubation temperature on the motility of *Y. enterocolitica*. At 28°C, *Y. enterocolitica* is motile and is seen (arrowed) spreading away from the stab inoculum during growth in semi-solid medium. At 37°C, the organism is non-motile and growth is restricted to the line of inoculation. (*Motility medium (API-bioMerieux©): 48h at 28/37°C.*)

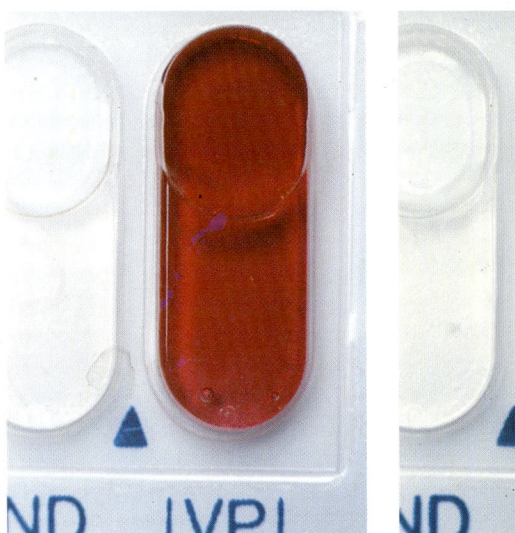
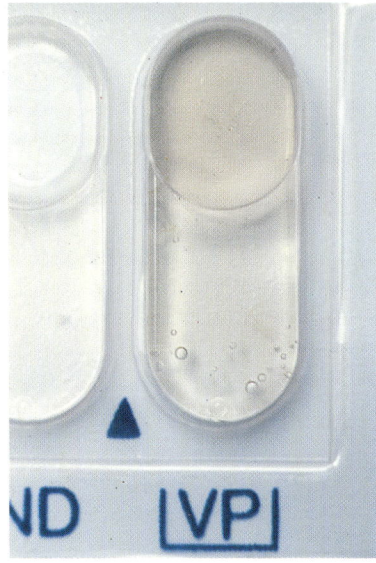

96–99. Temperature dependent biochemical characteristics of *Y. enterocolitica*. Temperature dependency of biochemical characteristics is a feature of all species of *Yersinia*. Within any species there are strain differences, although temperature dependency is more common with some characteristics than others. Two examples of temperature dependency in a virulent strain of *Y. enterocolitica* biovar 4 are illustrated, the Voges-Proskauer reaction (positive, **96**; negative, **97**), and the fermentation of sorbitol (positive, **98**; negative, **99**). Each reaction was positive at 25°C but negative at 37°C. (*API 20E©: 24 h at 25/37°C.*)

Table 54. Differentiation of *Yersinia* from related genera.

	Yersinia	*Escherichia*	*Klebsiella*	*Proteus*	*Hafnia*
Colony size[1]	<1 mm	>1 mm	>1 mm	>1 mm	>1 mm
Lysine decarboxylase	−[2]	+/−	+/−	−	+
Ornithine decarboylase	+/−	+/−	−/+	−/+	+
Simmon's citrate (37°C)	−	−	+/−	−/+	−
Voges–Proskauer (25°C)	+/−	−	+/−	+/−	+
Urease	+	−	+/−	+	−
Phenylalanine deaminase	−	−	−	+	−
Gas from glucose	−[3]	+[4]	−	+	+
Motility					
25°C	+	+[5]	−	+	+
37°C	−	+[5]	−	+	+

[1] Nutrient agar, 24 hours at 37°C.
[2] *Y. ruckeri* is usually positive.
[3] A few bubbles may be produced; see **Figure 100**.
[4] Occasional strains are negative.
[5] 'Inactive' strains are usually negative.

Table 55. Differentiation of main foodborne species of *Yersinia*.

	Y. pseudotuberculosis	*Y. enterocolitica*	*Y. frederiksenii*	*Y. intermedia*	*Y. kristensenii*	*Y. mollaretii*	*Y. bercovieri*
Ornithine decarboxylase[1]	−	−	+	+	+	−	−
Simmons' citrate (25°C)	−[2]	−	+/−	+	−	−	−
Voges–Proskaur (25°C)	−	+	+	+	−	−	−
Indole production	−	+/−	+	+	+/−	−	−
Acid from							
Cellobiose	−	+	+	+	+	+	+
α–methyl–D–glucoside	−	−	−	+	−	−	−
Melibiose	+	−	−	+	−	−	−
Mucate	−	−	+/−	+/−	−	+	+
Raffinose	+/−	−	−	+	−	−	−
Rhamnose	+	−	+	+	−	−	−
Sorbitol	−	+	+	+	+	+	+
Sorbose	−	+	+	+	+	+	−
Sucrose	−	+	+	+	−	+	+
Pyrazinamidase	−[3]	+/−	+	+	+	+	+

[1] Moller's method.
[2] Serogroup IV strains are positive.
[3] Occasional strains are weakly positive.

133

100. Weak gas production by a strain of Y. enterocolitica biovar 3. Although *Yersinia* is usually considered to ferment carbohydrates without gas production, species other than *Y. pestis* and *Y. pseudotuberculosis* may produce one or two bubbles after 2 to 3 days incubation at temperatures of 28° to 30°C. (*Hugh and Leifson medium: 3 days at 28°C.*)

101 & 102. ONPG activity of Y. enterocolitica. In contrast to the ONPG activity of other members of the *Enterobacteriaceae*, that of yersinias corresponds to an ONPG-ase and not to a true β-galactosidase. This does not, however, affect the value of the test in identification. The reaction is often very weak (**101**) and should always be compared with a negative control (**102**). ONPG activity is often temperature dependent and not expressed at 37°C. (*API20E©: 20 h at 28°C.*)

103. Production of a brown pigment by a strain of Y. frederikensii. Pigmentation is not usually associated with *Yersinia*, but strains of both *Y. frederikensii* and an avirulent strain of *Y. enterocolitica* biovar 1 have been found to produce a pigment after extended incubation (>5 days) in semi-solid medium. The pigment is water soluble and non-fluorescent; production does not appear to be temperature dependent in the range 25° to 37°C. (*Motility medium (API-bioMerieux©): 7 days at 25°C.*)

Typing of *Yersinia*

Biotyping

Biotyping of *Y. enterocolitica* is of considerable importance, and in some circumstances has ranked with serotyping as a means of discriminating within the species. Various schemes have been proposed in the past, and confusion may result from failure to specify which scheme has been used. It should also be appreciated that some biovars specified in typing schemes are now recognised to be distinct species of *Yersinia*.

The biotyping scheme of Bercovier and Mollaret (1984) is widely used. This originally contained five biovars, but biovar 3 was subsequently distinguished from 3A and 3B. As noted above, these two biovars are now classified as *Y. mollaretii* and *Y. bercovieri* respectively. A further subdivison was proposed by Cornelis (1987), who separated biovar 1 into the non-pathogenic environmental strains which occur worldwide, and the highly virulent American strains which usually occur only in North America.

The biochemical criteria for biotyping *Y. enterocolitica* are listed in **Table 56**. The use of the pyrazinamidase test (**Figures 104** & **105**) is of particular interest, since the enzyme is only very rarely found in potentially pathogenic biovars (Kandolo and Wauters, 1985; see page 141).

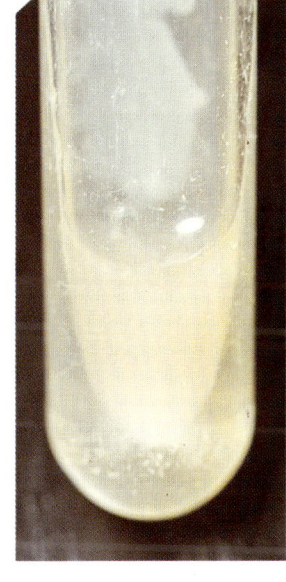

104 & **105**. **Determination of pyrazinamidase activity in *Y. enterocolitica*.** The pyrazinamidase test is a useful means of distinguishing potentially pathogenic biovars of *Y. enterocolitica* from non-pathogenic biovars, although the test does not correlate directly with the presence or absence of a virulence plasmid. Pyrazinamidase activity is determined on tryptic-soy agar containing pyrazinecarboxamide (Merck or Sigma). After incubation, the slope is flooded with ferrous ammonium sulphate which reacts with the end product, pyrazinoic acid, to produce a pink-brown coloration (Kandolo and Wauters, 1985).

104 illustrates the positive reaction of *Y. enterocolitica* biovar 1 ('European'), serovar O5 which is non-pathogenic. In contrast, *Y. enterocolitica* biovar 1 ('American'), serovar O8 which is pathogenic to man has a colourless, negative reaction (**105**). (*Pyrazinamidase medium: 48h at 28°C.*)

Table 56. Biotyping of *Yersinia enterocolitica*.

	Biovar					
	1e	*1a*	*2*	*3*	*4*	*5*
Hydrolysis of						
Tween	+	+	−	−	−	−
Aesculin	+/−	+	−	−	−	−
Indole production	+	+	+[*]	−	−	−
Acid from xylose	+	+	+	+	−	+/−
Ornithine decarboxylase Acid from trehalose Nitrate reduction	+	+	+	+	+	−
Pyrazinamidase	+	−	−	−	−	−
Deoxyribonuclease	−	−	−	−	+	+

[*] Positive reaction but usually delayed.

Serotyping

A serotyping scheme based on O antigens has been developed for *Y. enterocolitica*. First described by Winblad (1967), only eight antigenic factors were originally involved. The scheme has since been extended on a number of occasions, and now involves 57 antigenic factors (Wauters, 1981). A number of the antigenic factors in Wauters' scheme were associated solely with '*Yersinia enterocolitica*-like organisms' (Walker, 1987), and a revised and simplified scheme containing 18 serovars with 20 O antigens has been proposed by Aleksic and Bockemuhl (1984).

A marked relationship has been demonstrated between the different biovars and serovars of *Y. enterocolitica*, geographical distribution and pathogenicity

Table 57. Distribution and pathogenicity of bioserovars of _Y. enterocolitica_.

Potentially pathogenic bioserovars

Biovar 1 ('American')
Serovars: O4: O8: O13a: 13b: O18: O20: O21 etc
Ecology: Humans (North America), possibly rats and
rodent fleas (O21)
Pathogenicity: Humans

Biovar 2
Serovars: O9: O5, 27
Ecology: Humans (Europe, Japan), pigs (healthy),
possibly cattle and sheep (O5, 27)
Pathogenicity: Humans

Biovar 3
Serovars: O1, 2, 3: O5, 27
Ecology: Chinchillas (Europe)
Pathogenicity: Chinchillas, occasionally humans

Biovar 4
Serovars: O3
Ecology: Humans (Europe, Japan, Southern Africa),
pigs (healthy)
Pathogenicity: Humans

Biovar 5
Serovars: O2, 3
Ecology: Hares, goat (Europe)
Pathogenicity: Hares, goat

Environmental bioserovars

Biovar 1
Serovars: Numerous
Ecology: Water, soil, food, faeces, humans and animals
Pathogenicity: None

Table 58. Geographical distribution of phage types of _Y. enterocolitica_ bioserovar 4/O3.

Origin of isolates	Phage type
Europe	VIII
Japan	VIII
Canada	IVB
South Africa	IVA

(**Table 57**). However, over an 11-year period in the USA, over 40% of _Yersinia_ infections were caused by serogroups generally thought to be non-pathogenic (Bisset _et al._, 1990), suggesting a changing and expanding spectrum of _Y. enterocolitica_ serogroups associated with infection.

A serotyping scheme for _Y. pseudotuberculosis_ also exists. The organism has been classified into six sero-groups (I to VI) on the basis of heat-stable O antigens. Groups I, II, IV and V may each be divided into two subgroups, A and B. Serogroup I is most commonly associated with human infection, although a cluster of 19 cases involving serogroup III was described in Finland (Schiemann, 1989).

Phage typing

Two phage typing schemes, the 'French' and the 'Swedish', have been proposed for 'European' strains of _Y. enterocolitica_. The 'French' scheme, which is based on a primary set of 12 phages, is most widely used, and is able to discriminate, for example, between strains of the pathogenic biovar 4, serovar O3 isolated in geographically distinct locations (Mollaret _et al._, 1979; **Table 58**).

Phages used in the European schemes are of relatively low activity against North American strains of serovar O8, and a set of 24 lytic phages from sewage that are active against _Y. enterocolitica_ (including O8) and also _Y. intermedia_, _Y. frederikensii_ and _Y. kristensenii_ have been used in a typing scheme. They were considered useful for epidemiological purposes despite a variety of phage types within a species and different patterns within a serogroup being observed (Schiemann, 1989).

Symptoms of _Yersinia_ infections

Yersinia enterocolitica

The symptoms of infections due to _Y. enterocolitica_ may vary considerably according to strain, the dose ingested, host genetic factors and the age and health of the host (Bottone, 1977; Larson, 1979). Indeed, the variability of the clinical symptoms may be the reason for past failures to report human yersiniosis (Walker, 1987).

In most cases, the predominant symptoms are

those of gastroenteritis, which appear within two to three days of infection, characterised by abdominal pain and diarrhoea, which may vary in severity from a few loose stools to a fulminant enterocolitis with ulcerative lesions (Cornelis *et al.*, 1987; Mingrone *et al.*, 1987). The temperature is usually raised and there may also be vomiting, although not usually as a primary symptom. The pain may evoke appendicitis as a result of the acute terminal ileitis and severe inflammation of the mesenteric lymph nodes. In infants and young children, typical gastrointestinal symptoms prevail, but in teenaged children the pain may be confined to the right fossa iliaca (lower right quadrant of the trunk). This syndrome, right iliac fossa syndrome, is particularly likely to be confused with acute appendicitis and has led to unnecessary surgical intervention.

In adults, complications may occur some weeks after the initial infection. These usually affect the skin and connective tissue, the most striking being erythema nodosum – an eruption of the skin leading to painful red swellings, which is most common in women aged between 40 and 70 years (Larson, 1979).

Connective tissue disorders commonly include painful joints, arthralgia and various forms of arthritis. It should be noted that some workers have refuted a role for *Y. enterocolitica* in arthritis (Schiemann, 1989). Less common complications may arise independently, or in conjunction with arthritis, and include inflammation of the iris, ciliary body and choroid of the eye, the muscular walls of the heart or the pericardium.

In a minority of cases, further long-term complications appear. These are manifested in a number of forms, important among which are: rheumatoid arthritis, Sjøgren's syndrome (a condition usually affecting middle-aged women, which involves inflammation of the cornea and conjunctiva of the eye and of the pharynx, enlargement of glands, dry throat and polyarthritis), arterial inflammation and damage; inflammatory dermatitis; hardening and shrinkage of connective tissue; and autoimmune thyroid disease, particularly Graves' disease.

Only one serovar, O3, appears to be involved in autoimmune disease. A strong association has been observed between autoimmune disease and antibodies against plasmid-encoded proteins of *Y. enterocolitica* O3 (Wenzel *et al.*, 1988) while Heyma *et al.* (1986) reported that thyroid-stimulating hormone binding sites were present on *Y. enterocolitica*, and were recognised by immunoglobulins from Graves' disease patients. A role for molecular mimicry in thyroid disease thus seems probable, although a suppressor cell dysfunction may also be involved in the pathology of the disease (Davies and Platzer, 1986).

Although septicaemia has been considered rare, many cases have been reported. Furthermore, these have not been restricted to the elderly with underlying symptoms, but have also occurred in otherwise healthy children. However, septicaemia is probably most common in the elderly, the immunocompromised and those with an underlying condition (Wormser and Keusch, 1981; Kapperud and Bergan, 1984). In the elderly and the immunocompromised, the fatality rate may exceed 50%.

Manifestations of yersiniosis vary to some extent according to the causative serovar. For example, serovar O3 is often associated with arthritis and erythema nodosum, and these symptoms are particularly common in Scandinavia where that serovar is predominant (Winblad, 1981). In contrast, the more highly invasive serovar O8 is predominant in North America, and the incidence of systemic infections including soft tissue infections, conjunctivitis and osteomyelitis is accordingly higher (Wormser and Keusch, 1981).

Extraintestinal infections are not invariably preceded by gastrointestinal symptoms. Sore throats have been reported in the absence of abdominal pain or diarrhoea and exudative pharyngitis (which was fatal), has been suspected of occurring under the same circumstances (Rose *et al.*, 1987). Observations made during outbreaks associated with pasteurised milk in the USA and a cooked pork product in Hungary, suggest that extraintestinal symptoms in the absence of gastroenteritis are relatively common among adults, and in some outbreaks may be the rule rather than the exception (see page 143).

Yersinia pseudotuberculosis

Historically, *Y. pseudotuberculosis* infections in man have been associated with a typhoid-like septicaemia, which is usually fatal. Less severe gastrointestinal disease is now recognised, the most commonly reported symptoms being abdominal pain, fever, anorexia, nausea and vomiting. Diarrhoea is unusual, and the symptoms are readily confused with appendicitis. Abdominal pain results from inflammation of the mesenteric lymph nodes and terminal ileitis. Terminal ileitis associated with *Y. pseudotuberculosis* infection has also been confused with Crohn's disease.

More recently, *Y. pseudotuberculosis* has been recognised as a relatively uncommon cause of haemolytic uraemic syndrome.

In Japan, secondary complications of *Y. pseudotuberculosis* infections include erythema nodosum, arthritis and renal deficiency. These symptoms are readily confused with Kawasaki syndrome (mucocutaneous lymph node syndrome), an important syndrome among Japanese children.

Persons at particular risk from *Yersinia* infections

Susceptibility

Among non-compromised individuals, children are at greatest risk with infants aged less than one year being most susceptible (Vandepitte and Wauters, 1979). Susceptibility remains high, until about 14 years of age.

Compromised adults include the immunosuppressed, the immunodeficient and those with a predisposing condition. Of particular significance is cirrhosis, or some other liver disorder, and iron overload.

Conditions such as liver malfunction and iron overload also increase susceptibility to the more severe forms of *Yersinia* infection, such as systemic infections with *Y. enterocolitica* and typhoid-like septicaemia with *Y. pseudotuberculosis*. Iron overload is of interest in that it may result from dietary and cultural practices, as in the example of South African adult males who habitually consume large quantities of Bantu beer, containing significant amounts of ionisable iron. This results in the deposition of haemosiderin in the viscera which, in turn, increases the susceptibility to generalised *Y. enterocolitica* infections (Robins-Browne *et al.*, 1979). There have also been cases of generalised *Yersinia* infections associated with oral overdose of iron tablets (Melby *et al.*, 1982) and the use of the siderophore desferrioxamine B (Desferal©) in the treatment of iron overload can potentiate patients to systemic *Yersinia* infection (Williams *et al.*, 1988).

Exposure

With respect to dietary habits, it is inevitably those who habitually consume unpasteurised milk, under-cooked meat (especially pork), and unchlorinated water who are at enhanced risk through increased exposure.

Contact with dogs and possibly, to a lesser extent, other domestic animals such as cats, is a possible risk factor with respect to exposure to *Y. enterocolitica*. Dogs were first implicated as direct sources of human yersiniosis by Gutman *et al.* (1975) but cultural proof was lacking. A proven association was described by Wilson *et al.* (1976), who isolated the same serovar from a sick infant and from dog faeces obtained in the infant's home. The role of the dog in transmission of yersiniosis is not universally accepted, and Swaminathan *et al.* (1982) quoted a communication which suggested that either humans and dogs acquired *Y. enterocolitica* from a common source, or that dogs were infected by humans.

It has been hypothesised that the highly virulent O8 and O21 serovars found in North America may be transmitted from a rodent reservoir by fleas (Schiemann, 1989; see page 142). If correct, this would seem to imply that those living in poor and overcrowded conditions are at risk, not only because of poor food hygiene, but through increased risk of flea bites. Occupational groups such as sewer workers, pest-control operators and agricultural workers would also appear to be at high risk of exposure.

Y. enterocolitica may also be important as a nosocomial pathogen. Hospitalised patients are often of increased susceptibility, and strains of the organism that cannot be identified in the laboratory as pathogens may cause disease in these circumstances (Schiemann, 1989).

Y. pseudotuberculosis is common among animals of all types, and contact with animals is a well-recognised risk factor (Daniels, 1965).

Treatment of *Yersinia* infections

In most cases, infections with *Y. enterocolitica* or *Y. pseudotuberculosis* are self-limiting, and treatment with antimicrobial drugs is neither necessary nor desirable.

Antimicrobial therapy may be required for systemic infections or where symptoms are prolonged. Tetracyclines, aminoglycosides, sulphonamides and chloramphenicol are effective against most isolates, but penicillins and cephalosporins are not recommended because of β-lactamase production by many strains (Bottone, 1977). However, ampicillin and carbenicillin may be effective against strains of *Y. enterocolitica* serovar O8.

Mechanisms of pathogenicity

Yersinia enterocolitica shows a distinct pattern of pathogenicity which it shares with *Y. pseudotuberculosis*, and the apparently very different *Y. pestis*. The situation was effectively summarised by Cornelis (1987):

*'Although these three micro-organisms infect the host via different routes and cause diseases of very different severity, they share a marked tropism for lymphoid tissue and a remarkable ability to resist the primary immune system of the host. Moreover, they share the common feature of invasion via mucosal interaction be it in the gut (*Y. enterocolitica, Y. pseudotuberculosis*), or in the lung (pneumonic plague).'*

Genetic control of virulence

In common with other members of the *Enterobacteriaceae*, both chromosomal and plasmid genes are involved in the control of virulence. Chromosomal control appears to be of prime importance in mediating the initial stages of invasion, although plasmids may be involved in adherence but not invasion *per se*. *Y. pseudotuberculosis* contains the chromosomal *ivv* gene, while two chromosomal loci have been identified in *Y. enterocolitica* (Miller and Falkow, 1988). The *inv* locus, which is homologous with the *ivv* gene of *Y. pseudotuberculosis*, allows a uniformly high level of invasion irrespective of tissue type, while the *ail* locus is tissue specific.

All three pathogenic species of *Yersinia* harbour 40 to 48 Mdalton plasmids (Gemski *et al.*, 1980; Portnoy *et al.*, 1981; Zinc *et al.*, 1980). A common designation has been adopted for these plasmids (Portnoy and Falkow, 1981). The general designation for *Yersinia* virulence plasmids is pVY, while pVYe refers specifically to that of *Y. enterocolitica*.

The virulence plasmids of yersinias are involved with a number of temperature-regulated phenotypes, including the expression of certain outer membrane proteins (Yops; Bolin *et al.*, 1985; 1988), autoagglutination (Laird and Cavanaugh, 1988), serum resistance (Martinez, 1983), cytotoxicity for macrophages (Goguen *et al.*, 1986), production of V and W antigens (Perry and Brubaker, 1983) and the low calcium response (LCR; Gemski *et al.*, 1980, Brubaker, 1983). The LCR is manifested as growth restriction and production of Yops when the temperature is raised from 26°C to 37°C in the absence of calcium. Yops appear about 1 hour after the shift to the higher temperature, and synthesis continues until cell growth ceases, usually after one to two generations (Brubaker, 1983). Yops are known to have an important role in virulence (see page 140).

Although temperature regulation of virulence genes is a common factor among many enteric pathogens, the situation with *Yersinia* is complicated in that some virulence-associated factors are turned on at 26°C and off at 37°C, while for others the reverse is true (**Table 59**). It has been suggested that the organism is primed for virulence and invasion during growth outside the mammalian body at a lower temperature, and that only an initial burst of this activity is required on entry to the host in order to establish an infection (Maurelli, 1989). Subsequent steps in pathogenesis may then require the down-regulation of the genes for adherence and internalisation.

Yersinias differ further from other enteric pathogens in that calcium also provides a stimulus during regulation. The nutritional requirement for calcium ions at 37°C is in turn potentiated by magnesium ions, and while the significance of this phenotype is not fully understood (Cornelis *et al.*, 1987) it may be correlated with the levels of calcium and magnesium in host intracellular and extracellular fluids. The concentration of calcium in a eukaryotic cell is very low, but *ca* 2.5 mM is present in the extracellular fluid. In contrast, magnesium is an important intracellular ion with a concentration of *ca* 20 mM in leukocytes. Calcium dependence may therefore be viewed as a response to the intracellular and extracellular environment (Brubaker, 1979), and inhibition of intracellular growth correlated with the synthesis of elements involved in pathogenesis (Cornelis, 1976).

Table 59. Temperature regulated virulence genes of *Y. enterocolitica* and *Y. pseudotuberculosis* (Maurelli, 1989).

	Gene/phenotype	26°C	37°C
Y. enterocolitica	Adherence to HeLa cells	++++	+
	Invasion of HeLa cells	++++	+
	Internalisation by PMNs	++++	++
	Serum resistance	−	++++
	yop84, yop25, yop51, (Yop2b)	+	++++
	virABCD	+	++++
Y. pseudotuberculosis	Mouse lethality	++++	+
	Adherence to HeLa cells	++++	+
	Internalisation by macrophages	++++	++
	Serum resistance	−	++++
	yopE (Yop5)	+	++++
	yopH (Yop2b)	+	++++
	inv (invasin)	++++	+

Adherence and penetration

Y. enterocolitica adheres to the cells of the ileum, adherence being mediated by outer membrane proteins which are able to recognise several receptors (Isberg, 1989a). Internalisation is primarily by receptor-mediated endocytosis, indistinguishable from that of *Salmonella* or *Shigella* although, in contrast to these organisms, non-viable cells will stimulate endocytosis by HeLa cells. A further contrast with *Shigella* is that yersinias do not multiply intracellularly in HeLa cells after release from the endocytic vacuole.

Although functional *inv*-homologous sequences are usually required for invasion (Pierson and Falkow, 1990), some *inv*⁻ yersinias are capable of low level entry into HEp-2 cells in the presence of the virulence plasmid (Isberg, 1989b).

From the epithelial cells, both *Y. enterocolitica* and *Y. pseudotuberculosis* progress to Peyer's patches. A study of serovar O8 showed Peyer's patches to be 1,000 times more heavily colonised than other tissue. *Y. enterocolitica* multiplies in the follicles and then spreads to the lamina propria. At either site, multiplication was mostly extracellular, with only a small percentage multiplying within phagocytes (Hanski *et al.*, 1989). *Y. enterocolitica* could not pass the intact basement membrane, which is thus likely to play a role in determining the route of entry and the limit of spread.

Systemic spread

Systemic spread of *Y. enterocolitica* and *Y. pseudotuberculosis* is rare, and usually occurs only in persons with elevated serum iron levels. This is probably due to the inability of yersinias to capture transferrin-bound iron (see below). Systemic spread commonly involves *Y. enterocolitica* serovar O8, which has a relatively low requirement for iron. Systemic spread probably involves both translocation and, more significantly, the lymphatic system.

Survival and proliferation in host tissue

Production of V and W antigens

Virulent strains of *Y. pestis* and *Y. pseudotuberculosis* both produce two soluble antigens, the V and W antigens, while *Y. enterocolitica* produces only the V antigen (Perry and Brubaker, 1983). Calcium dependence and production of V and W antigens were previously considered to be linked characters, but it is now recog-

nised that the restricted state *per se*, rather than the ability to produce the antigens, is responsible for pathogenicity in yersinias.

Phagocytosis

Protection of yersinias from phagocytosis is considered to be one of the major roles of the plasmid encoded, calcium regulated, outer membrane proteins (Yops). For example, non-pathogenic strains of *Y. enterocolitica* are readily phagocytised by macrophages and readily killed. Pathogenic strains are also readily phagocytised and an initial decrease in numbers occurs. This is followed some hours later by a marked increase in numbers which may be attributed to induction of Yops (Une, 1977). For example, the cytotoxic protein Yop E, acting in concert with Yop H, has been shown to have a major function in obstructing the antibacterial activities of the professional phagocytes of the regional lymph nodes (Rosqvist *et al.*, 1990). Yops have also been shown to reduce the luminol-enhanced chemiluminescence of zymosan stimulated neutrophils either by inhibiting phagocytosis or by blocking the respiratory burst (Williams *et al*,. 1988).

Bactericidal effect of normal serum

Virulent strains of *Y. enterocolitica* show a high level of resistance to the complement component of human serum when grown at 37°C in the absence of calcium. Avirulent strains that do not contain the VY plasmid are sensitive, as are virulent strains grown at 28°C (Perry and Brubaker, 1983). Yops, particularly Y1, play a major role in resistance (Martinez, 1983; Balligund *et al.*, 1985).

The mechanism of serum resistance differs in *Y. pseudotuberculosis* in being temperature dependent, but not mediated by the virulence plasmid.

Effects of iron deprivation

Yersinias are unusual in that siderophore production has never been demonstrated, and the organisms are thus unable to capture transferrin-bound iron. Haemin may be used as a source of iron, and it is possible that cells of yersinias have receptors for siderophores such as aerobactin and enterochelin produced by other bacteria, and for desferrioxamine B produced by fungi (Perry and Brubaker, 1979). Haemin binding may be plasmid mediated, but pVY is probably not involved in the utilisation of desferrioxamine B (Prpic *et al.*, 1983).

Highly pathogenic strains of *Yersinia* synthesise two high molecular weight proteins in response to iron starvation. In strains of low pathogenicity the gene coding for synthesis of these proteins is absent (Corniel et al., 1989).

Toxin production

Production of a heat-stable enterotoxin is a feature of both virulent and avirulent strains of *Y. enterocolitica*, as well as of other non-pathogenic yersinias particularly *Y. kristensenii*. The toxin, Y-ST, resembles the ST_A toxin of *E. coli* in biological activity, and the two toxins are similar, but not identical, immunologically (Okamoto et al., 1983). For example, Y-ST was not detected by a competitive GM_1-ELISA when either monoclonal or polyclonal anti-ST_A sera were used (Svennerholm et al., 1986). Production of Y-ST is under chromosomal control, and is greatest at 26°C and usually ceases at 30°C. Unusual biovars and serovars may produce toxin at 37°C (Schiemann, 1989), as may strains of *Y. kristensenii* (Kapperud and Langeland, 1981).

The ability of some strains of *Y. enterocolitica* to produce toxin in artificially inoculated milk (Francis et al., 1980; Walker, 1986) and meat (Velin, 1984), has been demonstrated at 25°C, but not at refrigeration temperatures. This has led to speculation that toxin pre-formed in foods may cause illness (Kapperud, 1982), but evidence is entirely lacking (Walker, 1987).

More significantly, there is doubt that enterotoxin plays any role in *Y. enterocolitica* infections whatsoever. This results from the facts that toxin production usually ceases at 30°C and above; that the toxin is produced by non-pathogenic as well as pathogenic strains; and that production cannot be demonstrated *in-vivo* (Pai et al., 1980; Swaminathan et al., 1982; Scotland, 1988). In contrast, Cornelis et al., do not totally preclude a role for enterotoxin in yersiniosis, and point out that toxin production by virulent strains is qualitatively greater than by avirulent strains. The situation remains unclear.

Other virulence associated factors

A high level of lipolytic activity has been postulated as an additional factor in the greater virulence of *Y. enterocolitica* serovar O8 (Aulisio et al., 1983). It is suggested that a lipase acts on lipid barriers, permitting the bacterium to penetrate and infect neighbouring tissues more easily.

Y. enterocolitica serovars O3 and O9 have a cytotoxic activity on cells in tissue culture (Portnoy et al., 1981, Vesikari et al., 1983), which results in rounding and partial detachment of cells within 1.5 hours. This activity is distinct from changes attributable to adher-

ence and appears to be plasmid mediated. Its significance is not known.

Markers of virulence

A number of characteristics have been used as markers of virulence in *Y. enterocolitica* (**Table 60**). Several of these correlate directly with the presence of the pVYe plasmid, and may be determined in laboratories without specialist facilities. *Determination of virulence markers is discussed in greater detail and illustrated in* **Laboratory Methods**, *pages 153–155.*

More recently, DNA hybridisation and enzyme immunoassay techniques have been developed for determining pathogenicity of both *Y. enterocolitica* and *Y. pseudotuberculosis*. The DNA hybridisation assay employs a probe for the virulence plasmid of *Y. enterocolitica* serovar O8, biovar 1 ('American') in a colony hybridisation assay that will detect pathogenic strains of *Y. pestis* as well as *Y. enterocolitica* and *Y. pseudotuberculosis* (Robins-Browne et al., 1989). The enzyme immunoassay employs antiserum against plasmid-encoded proteins of *Y. enterocolitica* serovar O3, and will detect plasmid containing strains of *Y. enterocolitica* and *Y. pseudotuberculosis* (Kaneko and Maruyama, 1989). Each of these assays is highly specific for virulent strains, and more reproducible than older methods which largely depend on phenotypic traits. DNA hybridisation and ELISA assays are not currently suitable for use in laboratories with limited facilities.

Table 60. Virulence markers in *Y. enterocolitica*.

Marker	Plasmid mediated
Calcium dependency 37°C	+
Autoagglutination 37°C	+
Colony morphology 37°C	+
V and W antigens 37°C	+
Congo red uptake 25 or 37°C	+
Unique outer membrane proteins 37°C	+
Mannose-resistant haemmagglutination	+
Hydrophobicity 37°C	+
25°C	−
Pyrazinamidase	−
Serum resistance 37°C	?

Detection of toxin

The only current means of detecting *Y. enterocolitica* enterotoxin is the infant mouse test, and simple but sensitive means of detecting the toxin are required.

Human carriage of *Yersinia*

Although carriage of *Y. enterocolitica* by apparently healthy people is a well-documented phenomenon, the extent and epidemiological significance remains a matter of controversy. In Japan, reported rates of carriage range from over 10% (Kanazawa and Ikemura, 1979) to less than 1% (Asakawa *et al.*, 1979). More recently, in the UK, Greenwood and Hooper (1987) reported a carriage rate of 3.5% over one year, a rate higher than that of *Salmonella*, and attributed reports of lower carriage rates to inadequate methodology. In most cases the serovars of *Y. enterocolitica* isolated from healthy carriers have not been those associated with disease.

In addition to the true carrier state, it appears that weak and non-diagnosed, or asymptomatic infections, are relatively common. In a survey of 62 apparently healthy blood donors, 39% were diagnosed using an immunoblotting technique to have, or to have had, a *Y. enterocolitica* infection.

Yersinia in the environment

Y. enterocolitica

The animal kingdom appears to be a significant resevoir of *Y. enterocolitica* (Swaminathan *et al.*, 1982). In many cases, bioserovars that are pathogenic to animals, are only rarely pathogenic, or non-pathogenic, to man. For example, biovar 5 is pathogenic only to hares and goats, and isolation of other non-human pathogenic strains has been made from a wide range of creatures including cattle, sheep, deer, foxes, mink, racoons, frogs and toads, snails, flies and fleas (Mollaret, 1979; Swaminathan *et al.*, 1982). A pathogen of chinchillas, serovar O1, causes human disease occasionally.

The only food animal known to harbour human pathogenic serovars at a high rate of carriage is the pig. Young pigs are readily infected and become long-term carriers while remaining symptom free. Infection is usually from other pigs, contaminated pens or faeces, and the variation in the incidence from herd to herd in the same geographical area may be related to piggery hygiene. Differences in agricultural practice may be involved in determining regional variations in carriage. In Denmark, where more than 80% of pigs carry *Y. enterocolitica*, high rates are associated with the practice of 'buying in' pigs to the farm (Tauxe *et al.*, 1987). Carriage may be either intestinal or, more commonly, pharyngeal, and isolation rates from the throat often exceed 50%. Biovar 2, serovar O:5,27 and Biovar 4, serovar O:3, are usually involved in porcine carriage. Strains of Biovar 1 ('American') serovars O:8 and O:21 are rarely, if ever, isolated from pigs. Control by vaccination of pigs is a possibility.

Cattle and sheep may occasionally be involved in the carriage of human pathogenic strains of *Y. enterocolitica*, and biovar 2, serovar O:5,27 has been isolated from cattle faeces, as well as from aborted ovine and bovine foetuses.

The possible role of dogs, other domestic animals and rats as possible reservoirs of human serovars of *Y. enterocolitica* has been noted above. The isolation of serovar O:21 from wild rodent fleas lends support to the hypothesised role for rats in human yersiniosis. However, wild rodents may also be involved in contamination of water (see below).

Although yersinias may be readily isolated from the environment, including water, these are usually members of species other than *Y. enterocolitica*. The widespread carriage of *Y. enterocolitica* by animals means that contamination both of water supplies and of soil and vegetation is inevitable. Surface run-off from piggeries has been considered to be an important source of contamination, although rats may also be of importance in the contamination of water as well as the environment. The organism survives well in soil, but its longevity in water depends on the activity of eukaryotic predators (Choo *et al.*, 1988).

Y. pseudotuberculosis

Y. pseudotuberculosis is found in numerous species of animals, birds and reptiles. As with *Y. enterocolitica*, the carriage rate among pigs is high, but the serogroup usually involved, III, is rare in human infections. In addition to direct infection from animals to humans, contamination of the environment is inevitable.

Foods commonly involved in transmission of *Yersinia*

Y. enterocolitica

Although *Y. enterocolitica* is well established as an enteric pathogen, the number of cases which can be associated with a specific food remains small. Most of these have been in North America, and have involved serovar O8. Indeed, in many countries where yersiniosis is endemic and the common serovars are O3 and O9, a specific food has never been identified as being responsible for an outbreak.

Milk

Milk has been responsible for three large outbreaks of yersiniosis in the USA, including the first-known food-borne outbreak in Oneida County, New York (Black, 1978; Anon, 1979). Chocolate milk was involved, the organism being introduced with the chocolate syrup, which was added after the milk had been pasteurised. Appendectomies were performed on no less than 18 of 38 known victims.

The largest outbreak to date involved pasteurised milk and affected several thousand people in Memphis and other areas. The rare O13a and 13b serovars were involved, and the course of infection was unusual in that, in adults, the predominant symptom tended to be pharyngitis with positive throat cultures. The origin of the organism was pigs, the milk being contaminated after processing by an unusual route involving milk crates (**Flow diagram 15**; CDC, 1982).

Two milkborne outbreaks have occurred among hospitalised children in the UK (Barrett, 1986). Pasteurised milk was involved in each outbreak and in one, which probably continued for several months, *Y. fredericksenii* was isolated from some patients as well as *Y. enterocolitica*.

Although *Y. enterocolitica* has been isolated from raw milk, the bioserovars are not those usually associated with human illness. However, consumption of raw milk is suspected of being a common factor in sporadic outbreaks of yersiniosis in Europe, the ultimate origin of infection being pigs.

Pork

The high rate of carriage of some types of *Y. enterocolitica*

Flow diagram 15. Transmission of *Y. enterocolitica* from pigs to milk.

Waste milk sent from dairy to pig farm, crates contaminated with pig manure at farm, then returned to dairy.

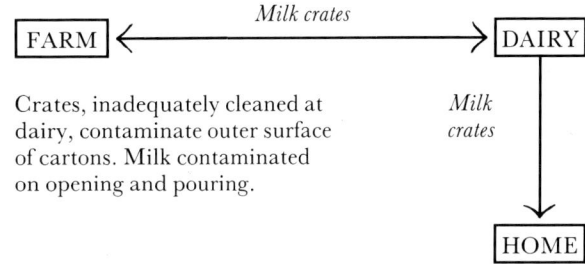

Crates, inadequately cleaned at dairy, contaminate outer surface of cartons. Milk contaminated on opening and pouring.

by pigs suggests that pork at a retail level would be commonly contaminated with the organism, and in turn that pork products should be associated with yersiniosis. In practice, isolation of pathogenic serovars, usually O3, can be made consistently from fresh pork tongue, but not from other pork products. However, serovar O3 has been isolated from ground pork in Japan (Fukushima, 1985) and from pork in Hungary (Marjai *et al.*, 1987). In Norway, the incidence of this serovar on cut surfaces of carcasses and in the slaughterhouse environment have been attributed to modern meat cutting and meat inspection practices (Nesbakken, 1988). The failure to isolate *Y. enterocolitica* in other circumstances may be due not to the absence of the organism, but to difficulties in isolating it in the presence of a competitive microflora (Schiemann, 1989).

Only one case of yersiniosis attributed directly to a pork product is known – a family outbreak due to *Y. enterocolitica* serovar O3 in Hungary. The food implicated was 'disznosajt' (literally pork cheese, a type of cooked brawn). Symptoms in adults involved sore throats and fever rather than gastrointestinal symptoms, and an initial diagnosis of influenza was made. In Belgium, which has the highest-known incidence of yersiniosis, a strong epidemiological link has been established with the national habit of consuming raw pork mince (Tauxe *et al.*, 1987). It seems likely that slaughterhouse practice permits the mince to become contaminated with tonsil.

Tofu

An outbreak of yersiniosis involving 87 people in the USA was attributed to Tofu (a soya product). Serovars O8 and O21 (O:Tacoma) were implicated, the source of contamination being water used to wash the product (Tacket *et al.*, 1985).

Water

Although not conclusively proven, water has been implicated in *Y. enterocolitica* infections in Scandinavia (Lassen, 1972) and in the USA (Eden *et al.*, 1977; Carberry *et al.*, 1984).

Other foods

Isolation of *Y. enterocolitica* has been made from a wide range of foods including meat products, poultry, fish and shellfish and fruits and vegetables. In raw foods, the organism may be isolated from as many as 33% of samples examined (Delmas and Vidon, 1985). These figures should be treated with caution when assessing the potential importance of foods as vehicles of yersiniosis since many earlier reports failed to distinguish adequately between *Y. enterocolitica* and 'environmental' yersinias such as *Y. intermedia*. Even where *Y. enterocolitica sensu stricto* was isolated, the serovars involved, with the exception of pork tongues, were non-pathogenic and included those organisms now classified as *Y. mollaretii* and *Y. bercovieri*. The mere isolation of *Y. enterocolitica* from a food is, in any case, insufficient to ascribe any pathogenic significance to such isolation (Swaminathan *et al.*, 1982).

Y. pseudotuberculosis

Infections due to animal contact remain of prime importance in human infections due to *Y. pseudotuberculosis*, but a role for food and waterborne transmission has also been postulated.

A proven link exists between water and enteric *Y. pseudotuberculosis* infection in Japan (Sato, 1987) but the link with food is more tenuous. However, since 1981 there have been a number of large Japanese outbreaks in which food has been strongly implicated (Inoue *et al.*, 1984; Sanbe *et al.*, 1987).

As with *Y. enterocolitica*, pigs and their meat may play an important role (Shiozawa *et al.*, 1988) and potentially pathogenic strains of *Y. pseudotuberculosis* have been isolated from 2% of pigs tongues (Fukushima, 1985) and 0.8% of ground pork (Shiozawa *et al.*, 1987). Other suspect foods include sandwiches and vegetable juice (Tsubokura *et al.*, 1989).

Factors affecting the growth and survival of *Yersinia* in foods

Temperature

A minimum growth temperature of 4°C is generally accepted for *Y. enterocolitica* (Bercovier and Mollaret, 1984), but other reports suggest that the organism may continue to multiply at temperatures as low as 1°C in meat (Hanna *et al.*, 1977) and 0.5°C in broth media (Heim *et al.*, 1984). The growth rate is much reduced at low temperatures (Buckeridge *et al.*, 1980), although in some circumstances this is partially offset by reduced competition from spoilage microorganisms (Schiemann and Olson, 1984).

The much reduced growth rate at low temperatures means that refrigeration is still an effective means of control of *Y. enterocolitica* under many circumstances. Difficulties for the food industry are likely to occur only where storage lives are of excessive length or where refrigeration is inefficient and permits relatively rapid growth of *Y. enterocolitica*. In either case, problems will be exacerbated by the absence of secondary controlling factors, such as a low pH value or preservatives.

The optimum growth temperature of *Y. enterocolitica* is 28 to 29°C and the maximum *ca* 42°C, these temperatures being lower than normal for the *Enterobacteriaceae*.

The growth: temperature relationships of *Y. pseudotuberculosis* are broadly the same as those of *Y. enterocolitica*.

The repeated isolation of *Y. enterocolitica* from commercially pasteurised milk and cream (Hughes, 1979) gave rise to initial suggestions that the organism possessed a high degree of heat resistance. It has subsequently been shown that its presence in pasteurised products reflects improper heat treatment, and while there is considerable strain variation in thermal tolerance the organism will not survive properly-applied pasteurisation. D values in whole milk at 62.8°C ranged from 0.7 to 17.0 seconds when 21 strains were tested (Francis *et al.*, 1980) although Lovett *et al.* (1982) found a higher resistance with three strains having D values ranging from 0.24 to 0.96 minutes under the same conditions. The latter workers also reported z values to range from 5.11 to 5.78°C.

The introduction of a pasteurisation stage has been specifically recommended as a means of increasing the safety of soya products such as tofu (van Kooij and deBoer, 1985).

Irradiation

Specific information on the irradiation tolerance of *Y. enterocolitica* is limited, but El-Zawahry and Rowley (1979) considered the organism to be among the most radiation sensitive of food-borne pathogens. In trypticase-soy broth, strains were inactivated by doses in the range 0.7 to 1.2 Kgy, values being approximately doubled in ground beef at 25°C. *Y. enterocolitica* would therefore be controlled by the dose of 3 Kgy proposed for the elimination of *Salmonella*.

pH value and acidity

Y. enterocolitica is generally considered to be sensitive to low pH conditions, and acidity has been proposed as a means of controlling its growth (Moustafa *et al.*, 1983; Neilsen and Zeuthen, 1985). In pure culture studies, Stern *et al.* (1980) found the pH range for growth to be 4.4 to 9.0, while Brackett (1986) found that at 5°C, *Y. enterocolitica* remained viable for 21 days at pH 4.0.

Experience with foodstuffs is limited and superficially, at least, contradictory. For example, Aldova *et al.* (1975) found *Y. enterocolitica* to survive in tartare sauce, but in contrast Brackett (1986) was unable to recover the organism from artificially contaminated tartare sauce or spoonable salad dressing at any time, or from mayonnaise after 48 hours.

In a detailed appraisal of the situation in high acid foods, Brackett (1987) noted that the effect of different acids had not been considered. Statistical analysis showed that pH value alone had relatively little effect on survival, and that the relative activity of acids against *Y. enterocolitica* differed with different types of compared data.

Other inhibitors

Relatively few studies have been made concerning the effect of other inhibitors used in foods on the survival and growth of *Y. enterocolitica*. In general terms, the resistance of the organism is similar to that of other members of the *Enterobacteriaceae*. *Y. enterocolitica* is able to grow (in broth medium) in 5% NaCl (Stern *et al.*, 1980), but concentrations above 7% are bactericidal as is $NaNO_2$ at a concentration of *ca* 200 mg/l (Raccach and Henningen, 1984).

Y. enterocolitica may also be controlled by 0.1 to 0.2% sorbate at pH 5.5 (Restaino *et al.*, 1982).

Laboratory methods

Conventional cultural isolation techniques

Although methodology for isolation of *Y. enterocolitica* has been much improved in recent years, difficulties persist with the isolation of the organism from heavily contaminated foods. Difficulties are also caused by the fact that different bioserogroups respond differently to the selective agents used in both liquid and solid media.

Enrichment

Although direct isolation of *Y. enterocolitica* may be possible from clinical samples and even from meats under some circumstances (Fukushima *et al.*, 1985; see below), enrichment is usually necessary for foods and environmental samples and advantageous for clinical material.

Many enrichment procedures have exploited the ability of *Y. enterocolitica* to grow at low temperatures, and cold enrichments in a non-selective medium such as phosphate buffered saline (PBS; Feeley *et al.*, 1976) have been widely used for faecal samples (Weisfield and Sonnenwirth, 1980) and foods (Leistner *et al.*, 1975). However, cold enrichment in PBS does have a number of disadvantages. From an applications viewpoint, the long incubation period required severely limits its use in the diagnosis of human disease and in the quality assurance of most foods (Walker, 1987). Furthermore the method may fail to recover *Y. enterocolitica* due either to competition from other psychrotrophic organisms present in foods, or to nutritional limitations (Mehlman *et al.*, 1978). In other circumstances, however, low temperature enrichment is superior (**Figures 106** & **107**).

Attempts to improve the performance of the cold enrichment method were made by introducing selective enrichment media such as bile–oxalate–sorbose broth (BOS; Mehlman and Aulisio, 1978; Mehlman *et al.*, 1978). Subsequently, strongly selective media were found to be unnecessary for low temperature enrichment and high protein media without carbohydrates, such as trypticase-soy broth or PBS supplemented with 2% peptone, are now preferred particularly in the USA. Performance may be improved further by adjustment of the pH value to 8.0 to 8.3.

An incubation temperature of 4°C is commonly used for a period of 21 to 28 days. This temperature does not appear to have been chosen on logical grounds, but because of the ready availability of refrigerators running at approximately 4°C. Some workers have claimed that the same result could be obtained with incubation at 15°C for 2 days (Schiemann, 1989), but this is not supported by experience, at least in the UK. An incubation temperature of 9°C was recommended by Greenwood and Hooper (1988) for isolation of *Y. enterocolitica* from faeces. More than 75% of strains were recovered after 7 days, the remainder requiring 10 to 14 days.

A number of selective enrichment broths have been investigated for use at incubation temperatures of *ca* 25°C for up to 5 days. Attention has largely been concentrated on two of these, bile–oxalate–sorbose broth and a modified Rappaport broth. On an overall basis, both Vidon and Delmas (1981) and Walker and Gilmour (1986a) found BOS to be the most suitable. BOS is particularly suitable for the recovery of the virulent American strains of serovar O8, but some strains of serovar O5,27 grow only poorly, if at all. This inhibition may be overcome by the addition of 2.5% NaCl to the broth (the mechanism of this phenomenon is not known), but this addition inhibits strains of some other serovars including O3. It is recommended that BOS should be used both with and without the addition of NaCl to ensure the optimal recovery of the different serovars which may be present in foods (Doyle, 1986).

Modified Rappaport broth containing irgasan, ticarcillin and potassium chlorate is preferred to BOS for recovery of serovars O3 and possibly O9 by some workers (Wauters, 1988). Carbenicillin is included in some formulations of Rappaport broth. This is inhibitory to some strains of O3, but may be omitted without reducing the selectivity of the medium.

A number of two-stage procedures have been devised. Many of these involve cold and selective enrichment stages, but in a comprehensive study of 26 enrichment procedures (Walker and Gilmour, 1986a), the most effective two-stage enrichment involved a first-stage, non-selective enrichment in trypticase-soy broth (24 hours at 22°C) followed by a second-stage selective enrichment in BOS (5 days at 22°C).

Cold enrichment is commonly used for *Y. pseudotuberculosis*. Simple media such as PBS are used and the possibility of using high protein media does not appear to have been explored.

Alkali tolerance

Exploitation of the high level of tolerance to alkaline conditions exhibited by both *Y. enterocolitica* and *Y.*

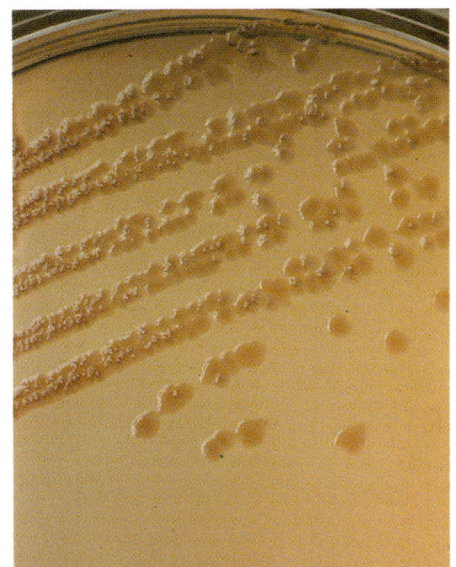

106 & 107. Comparison of enrichment techniques for *Y. enterocolitica*. The relative efficiency of low and high temperature enrichment techniques is affected by various factors, including the nature of the competing microflora. In this example the food sample, a dairy cream dessert, contained relatively large numbers of *Morganella morganii* as well as a small number of *Y. enterocolitica*. *M. morganii* was able to grow during enrichment in modified Rappaport broth at 25°C, and overgrew *Y. enterocolitica* on the solid selective (CIN) medium (**106**). When low temperature enrichment in phosphate buffered saline at 4°C was used, *M. morganii* was unable to grow and *Y. enterocolitica* was readily isolated from the selective agar (**107**). (*Modified Rappaport broth: 3 days at 25°C; phosphate buffered saline: 21 days at 4°C; CIN agar: 2 days at 30°C.*)

pseudotuberculosis was first proposed by Aulisio *et al.* (1980). Treatment is usually applied to enrichment broths, but has also been used with a meat homogenate prior to direct plating on selective media (**Table 61**).

Alkali treatment has been widely recommended, but its use has met with a variable response in practice. Serovar O8 is known to be sensitive to alkali, but recovery may also be variable with resistant serovars. It is possible that some of the less than enthusiastic reports of the value of alkali treatment were the consequence of an excessively long (5 minutes) exposure to alkaline conditions (Schiemann, 1983), and the short treatment of no more than 15 seconds has been found to give improved recoveries by many workers (Doyle and Hugdahl, 1983; Doyle, 1986; de Boer and Seldon, 1987). In contrast, alkaline treatment was not recommended by Walker and Gilmour (1986a), and even after improvements to the technique, was inferior to selective enrichment in BOS for recovery of *Y. enterocolitica* from inoculated foods.

Alkali treatment is not recommended for general use, but should be considered where specific problems with competing micro-organisms occur (see **Figures 106 & 107**).

Selective plating media

Selective plating media for *Y. enterocolitica* are of three types: media devised for general purpose use with the *Enterobacteriaceae* or with *Salmonella*; modifications of some of these media by the addition of higher levels of surface active agents, such as bile salts or tween; media specifically developed for *Y. enterocolitica*. These are summarised in **Table 62** and the use of examples of each type are illustrated in **Figures 108–119**.

For most purposes, cefsulodin–irgasan–novobiocin medium (CIN; *Yersinia* medium) is recommended. Strains of biovar 3 may be inhibited and the use of a second medium is advised. MacConkey agar was used alongside CIN by Doyle and Hugdahl (1983), and is generally considered to be useful for the isolation of *Y. enterocolitica*, despite its low selectivity. Salmonella-shigella medium with added deoxycholate and CaCl$_2$ has also been recommended for the recovery of sero-

Table 61. Alkali treatments for the isolation of *Y. enterocolitica*.

Post-enrichment treatment
Mix 0.5 ml enrichment broth with 4.5 ml dilute aqueous alkali solution (0.5% KOH: 0.5% NaCl) for 15 seconds and plate immediately (Aulisio *et al.*, 1980).

Pre-direct plating treatment
Mix 0.5 ml sample homogenate with 0.5 ml dilute aqueous alkaline solution (0.72% KOH: 0.5% NaCl) for 30 seconds and plate immediately onto CIN or MacConkey agar (Fukushima, 1985).

108

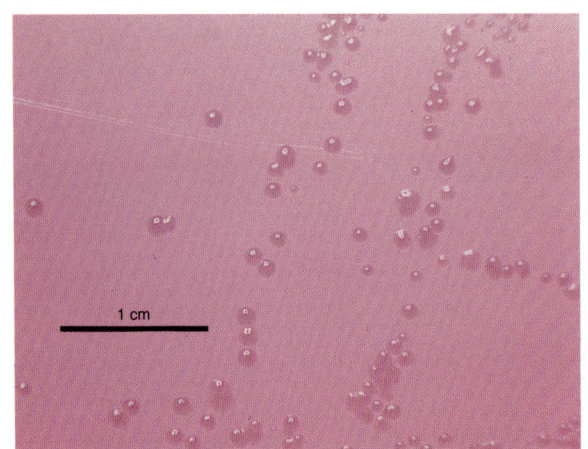

1 cm

108–110. Isolation of _Y. enterocolitica_ on general-purpose selective media for the _Enterobacteriaceae._
Y. enterocolitica grows well on all commonly used media of this type, including the various MacConkey agars, violet red-bile agars and Drigalski agar. These media, however, are of relatively low selectivity and the colonial appearance of yersinias is similar to that of other genera. The plates illustrate _Y. enterocolitica_ on MacConkey-purple agar (**108**) and on Drigalski agar (**109 & 110**). The colony diameter on the latter medium is larger than on most other media. (_Both media: 24h at 28°C._)

109

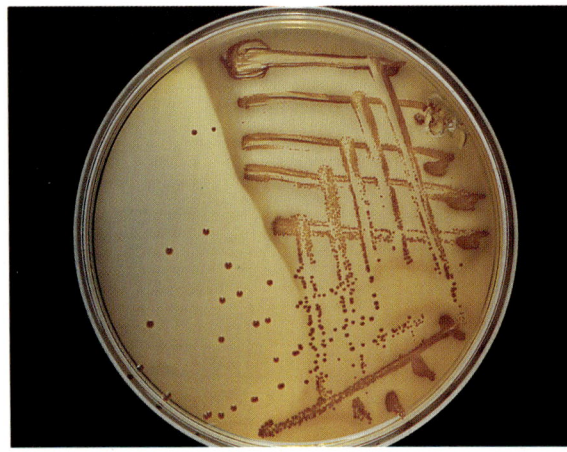

110

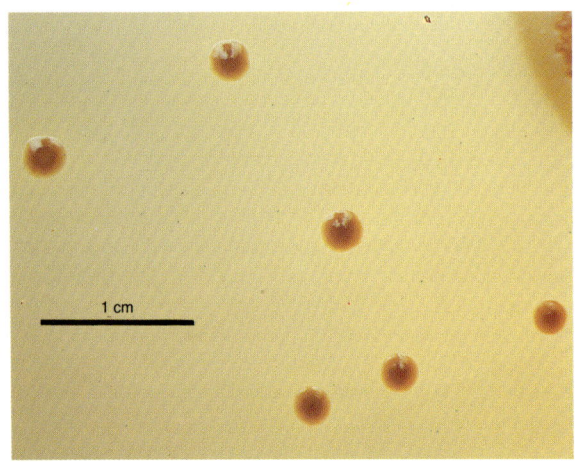

1 cm

Table 62. Selective plating media for _Y. enterocolitica._

Media designed for the Enterobacteriaceae _or_ Salmonella
MacConkey
Drigalski
Salmonella–shigella
DCLS
Hektoen enteric
Modifications to above media: MacConkey–tween
Salmonella–shigella–deoxycholate

Media specifically designed for Y. enterocolitica
Cefsulodin-irgasan-novobiocin
    Inhibitors: Deoxycholate, crystal violet, cefsulodin, irgasan, novobiocin
    Diagnostic system: Mannitol fermentation

VYE medium
    Inhibitors: Irgasan, josamycin, cefsulodin, oleandomycin
    Diagnostic system: Aesculin hydrolysis

111–113. Isolation of *Y. enterocolitica* on selective media devised for *Salmonella* and *Shigella*. Media devised for *Salmonella* and *Shigella* are generally more selective for yersinias than media such as MacConkey agar. Differentiation of *Y. enterocolitica* from other members of the *Enterobacteriaceae* is often difficult, and the appearance of colonies often differs from published descriptions. (*All media: 24 h at 28°C.*)

111. Salmonella-shigella agar. Many strains of *Y. enterocolitica* fail to grow on Salmonella-shigella agar in either its standard formulation, or modified by addition of 1% sodium deoxycholate (Wauters, 1973). Strains that are able to grow produce small, white or pale pink colonies which are difficult to distinguish when other bacteria are present.

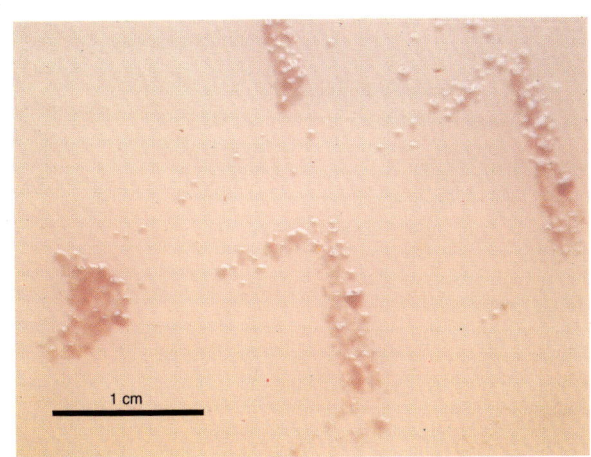

112. DCLS agar. *Y. enterocolitica* grows well on DCLS but the colonial appearance is similar to that of many other members of the *Enterobacteriaceae*.

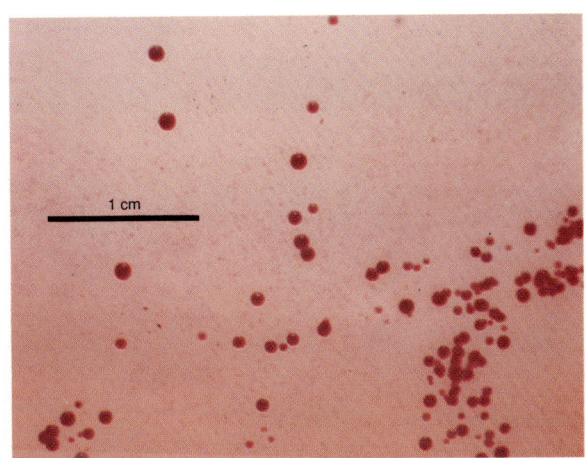

113. Hektoen enteric agar. Growth on Hektoen enteric agar is usually good, although some strains produce very small colonies (*ca* 0.5 mm). Colonial appearance is similar to that of several other genera within the *Enterobacteriaceae*.

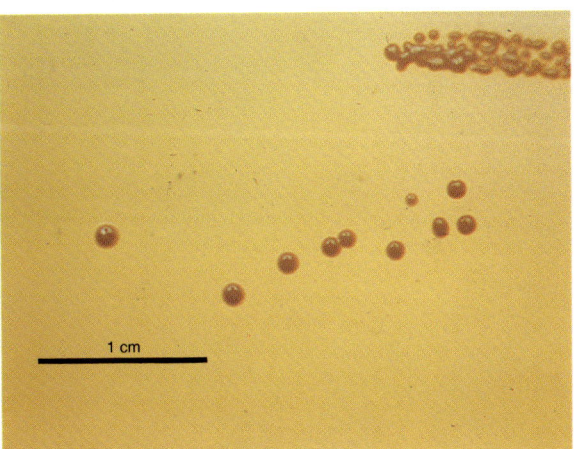

114

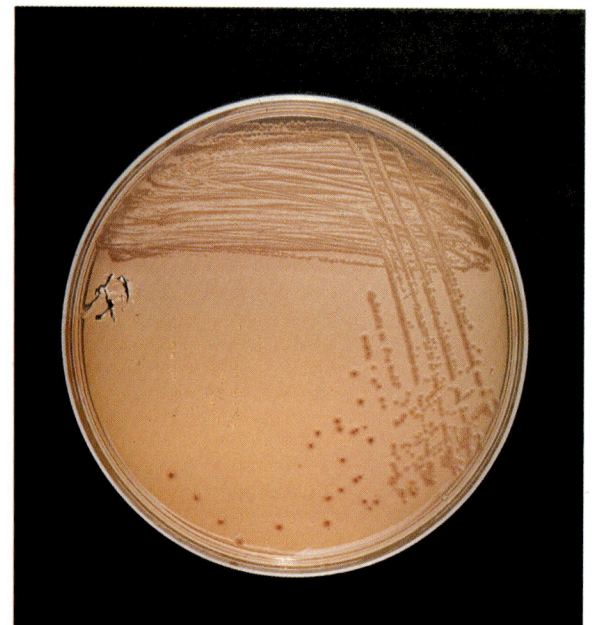

114–116. Isolation of *Y. enterocolitica* on Cefsulodin-Irgasan-Novobiocin (CIN) agar medium.

CIN agar is widely used for the selective isolation of *Y. enterocolitica*. Used with novobiocin at a concentration of 2.5 mg/l, the medium supports good growth (**114**) of all strains of *Y. enterocolitica*, including strains of serovar 08, which are inhibited by the concentration of novobiocin (15 mg/l) used in the original formulation. Typical colonies develop as a dark red 'bullseye' surrounded by a transparent border (**115**), but there is considerable variation in colony size, roughness and width of the border. With some strains, no 'bullseye' is produced and the border is absent (**116**). Environmental species of *Yersinia* are also recovered on CIN medium and resemble *Y. enterocolitica*. (*CIN medium: 24 h at 28°C.*)

115

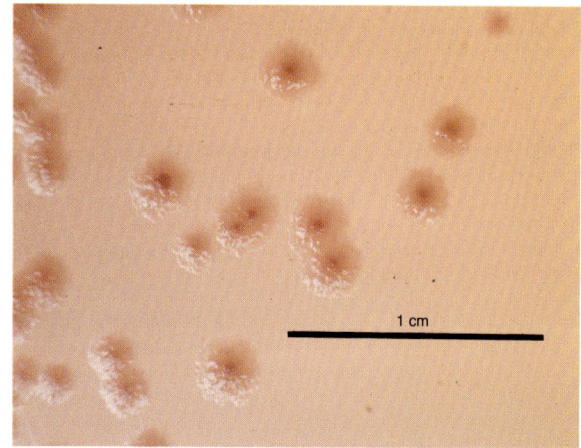

1 cm

116

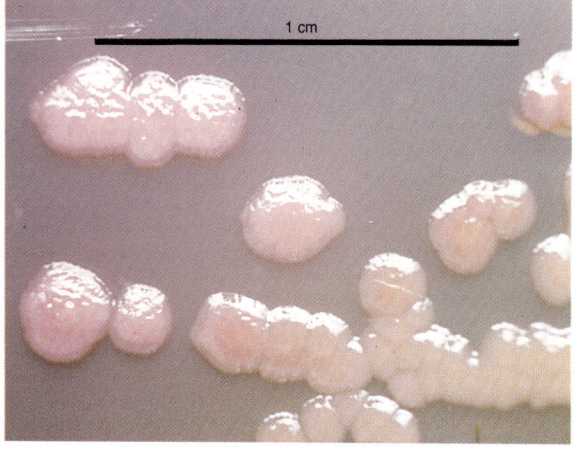

1 cm

117–119. Selectivity of CIN medium. CIN medium is not totally selective for yersinias and permits the growth of some other members of the *Enterobacteriaceae* as well as *Aeromonas*. In some cases, differentiation between *Y. enterocolitica* and other bacteria growing on CIN medium is difficult.

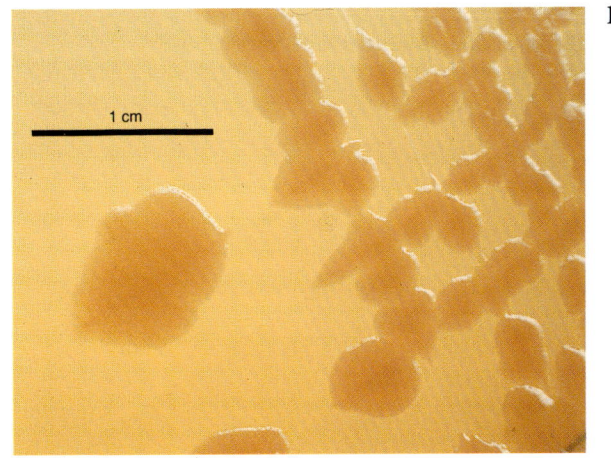

117

117. *Morganella morganii*. Colonies can be readily distinguished from those of *Y. enterocolitica* but the large size and rapid growth rate of *M. morganii* can lead to problems of overgrowth (see **106**).

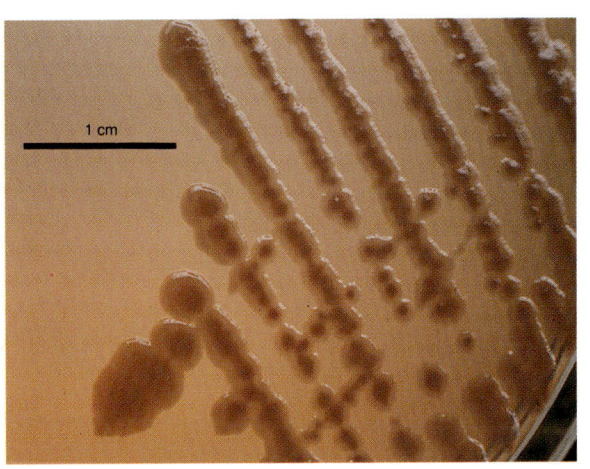

118

118. *Aeromonas hydrophila*. *A. hydrophila* grows extremely well on CIN medium which can be used for its selective isolation (see page 197). The colonies are sufficiently different from those of *Y. enterocolitica* to at least raise suspicion. Care should be taken if using the oxidase reaction as a means of differentiation, as *Aeromonas* may give a false-negative reaction if tested direct from the isolation plate.

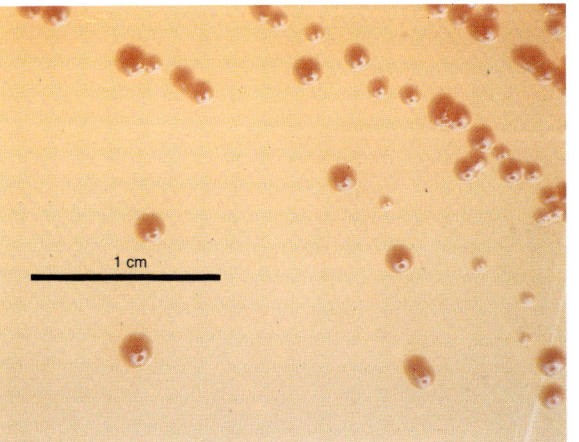

119

119. *Serratia liquefaciens*. Colonies of *S. liquefaciens* closely resemble those of some strains of *Y. enterocolitica*. *Citrobacter freundii* and *Enterobacter agglomerans* may also have a similar colonial appearance on CIN. In these cases, biochemical testing is required for confirmation of identity. (*CIN medium: 24h at 28°C.*)

Table 63. Summary of factors affecting the choice of media and methods for the isolation of *Y. enterocolitica*.

> *Type of sample*
> Stools
> Food
>   Processed or unprocessed
>   Level of competing microflora and any known
>   problems
>   Bioserovars likely to be present
>
> *Information required*
> All yersinias or specific bioserovars
> Time limit for results

Flow diagram 16. Model protocol for the isolation of *Y. enterocolitica* from foods.

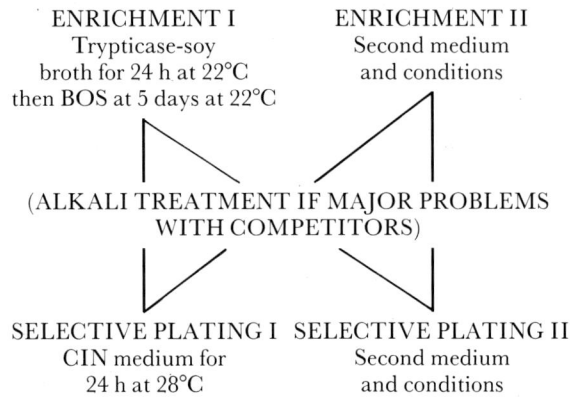

ENRICHMENT I	ENRICHMENT II
Trypticase-soy broth for 24 h at 22°C then BOS at 5 days at 22°C	Second medium and conditions

(ALKALI TREATMENT IF MAJOR PROBLEMS WITH COMPETITORS)

SELECTIVE PLATING I	SELECTIVE PLATING II
CIN medium for 24 h at 28°C	Second medium and conditions

var 3 (Wauters, 1988). This medium is not suitable for use in combination with alkaline treatment.

The VYE medium of Fukushima *et al.*, (1987) has attracted considerable interest, although in the UK application appears to have been very limited. The medium is reportedly highly selective, containing irgasan, josamycin, cefsulodin and oleandomycin as selective agents, but is not available in a dehydrated form and is consequently time consuming to prepare. Hydrolysis of aesculin is used as a differential system and permits differentiation between pathogenic biovars (usually aesculin-negative) and non-pathogenic biovars (usually aesculin-positive). Caution is required when using aesculin hydrolysis in this context and confirmatory testing is required.

Little attention has been given to selective plating media for *Y. pseudotuberculosis*. CIN medium is known to be unsuitable and should not be employed. *Y. pseudotuberculosis* may be recovered on any of the commonly used media for *Enterobacteriaceae* with the exception of bismuth sulphite agar. For many purposes, MacConkey medium is preferred.

Complete protocol for the isolation of Yersinia

A protocol for the isolation of *Y. enterocolitica* is shown in **Flow diagram 16**. In accordance with the recommendation of many authors (Stern, 1982; Doyle and Hugdahl, 1983; Feeley and Schiemann, 1984; de Boer and Seldom, 1987; Wauters *et al.*, 1988), the use of two enrichment and selective plating media is considered essential where maximum recovery of all serovars is required. Beyond that it must be stressed that the successful recovery of *Y. enterocolitica* requires a considerable level of skill and experience in selecting the most appropriate combination of media. Factors which must be considered are summarised in **Table 63**.

Detection by DNA hybridisation

DNA hybridisation techniques may be particularly valuable with *Y. enterocolitica* since it is the only organism in which presence of a plasmid, as determined by plasmid–specific DNA probes, totally reflects the pathogenicity of the organism as determined by animal testing (Aulisio *et al.*, 1983). Therefore, it is rather surprising that relatively little interest has been shown in developing probes for the detection of yersinias in foods. However, colony hybridisation techniques suitable for use with foods have been developed (Meliotis *et al.*, 1989; Kapperud *et al.*, 1990). Although common pathogenic serogroups may be detected without enrichment, the possibility exists that serogroups which are not recognised pathogens will be missed. A risk of false negatives due to loss of the virulence plasmid during cultivation also exists although this is minimised by cultivation at 26°C.

Identification of *Yersinia*

Differentiation of Yersinia from other genera

Differentiation from other species requires only a small number of tests. A convenient procedure adopted by Walker and Gilmour (1986b) uses only three media (**Table 64**). An alternative is the use of miniaturised identification kits. Most systems are suitable, but incubation should be at 25°C for 36 to 48 hours rather than at 37°C for 18 to 24 hours as recommended by the manufacturer. This would seem to be of particular importance with the API© systems.

Where full characterisation is required using conventional methodology, the criteria in **Table 54**, (page 133), should be used. Tests should be carried out according to Leininger (1976).

Systems employing a larger number of tests usually permit identification direct to species level without additional tests being required.

Differentiation of Y. enterocolitica and Y. pseudotuberculosis from other species

Where conventional test media are used, differentiation may be made according to **Table 55**, (page 133). Tests should be carried out according to Leininger (1976).

Typing of *Y. enterocolitica*

Biotyping

Biotyping may be carried out according to the tests listed in **Table 56** (page 135). It should be noted that

Table 64. Differentiation of *Yersinia* from other genera by the three media scheme (Walker and Gilmour, 1986b).

Medium	Typical reaction
Kliger iron agar (24 hours at 35°C)	Acid butt: alkaline slope No gas or H_2S production
Urea agar (24 hours at 28°C)	Positive
Mannitol–peptone water (24 hours at 28°C	Acid with very little or no gas

Note: Urease-negative strains have been reported but none are known to be pathogenic.

strains previously typed as biovars 3A and 3B are not included (see page 131). The methods of Leininger (1976) should be used for all tests except for pyrazinamidase activity, which is determined using the media and methods of Kandolo and Wauters (1985; **Figures 104** & **105**, page 135).

Serotyping

The scheme of Aleksic and Bockemuhl (1984) is recommended. O antigens may be determined directly from Kliger iron media if used for confirmation of identity or, preferably, plate count agar. H antigens must be grown in soft (semisolid) agar. In each case, slide agglutination is used with living organisms.

Typing antisera for *Y. enterocolitica* are not commercially available, and thus serotyping is not possible in most food industry laboratories.

In vitro determination of virulence of *Y. enterocolitica*

A number of methods exist (**Table 60**, page 141), many of which correlate directly with the presence of the VYe plasmid. Two methods, calcium dependence at 37°C and colony morphology at 37°C, which are suitable for use on a routine basis are described and illustrated in **Figures 120–125**.

Autoagglutination is also suitable for routine use, and has been described as the most useful test for identifying plasmid containing pathogenic strains of *Y. enterocolitica*. However, autoagglutination has been observed in some biochemically atypical non-pathogenic strains (Schiemann, 1989). A fourth method, congo red uptake at 25 or 37°C, is also technically straightforward, but results are often very variable.

The pyrazinamidase test may also be applied at this stage, but many workers prefer to use it earlier, either as part of biotyping, or to screen for presumptive pathogenicity.

120

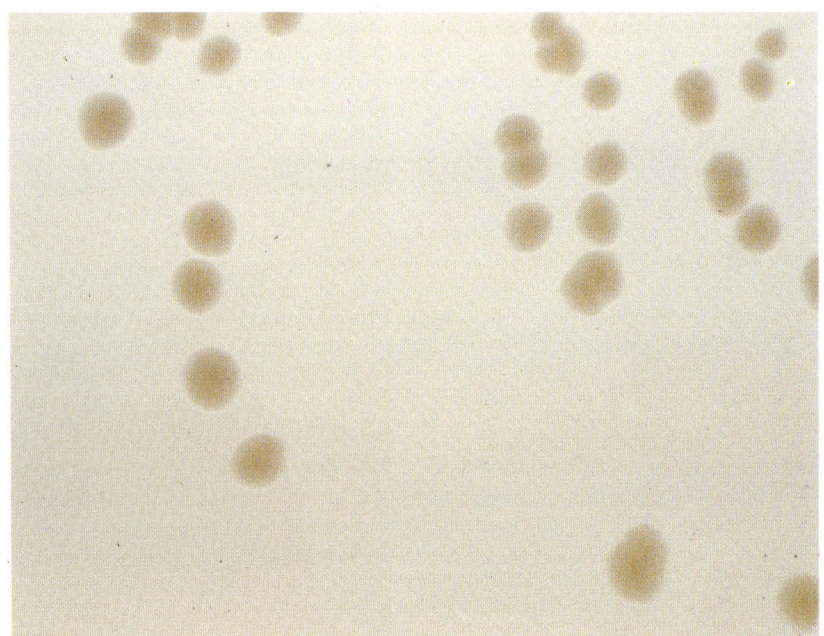

121

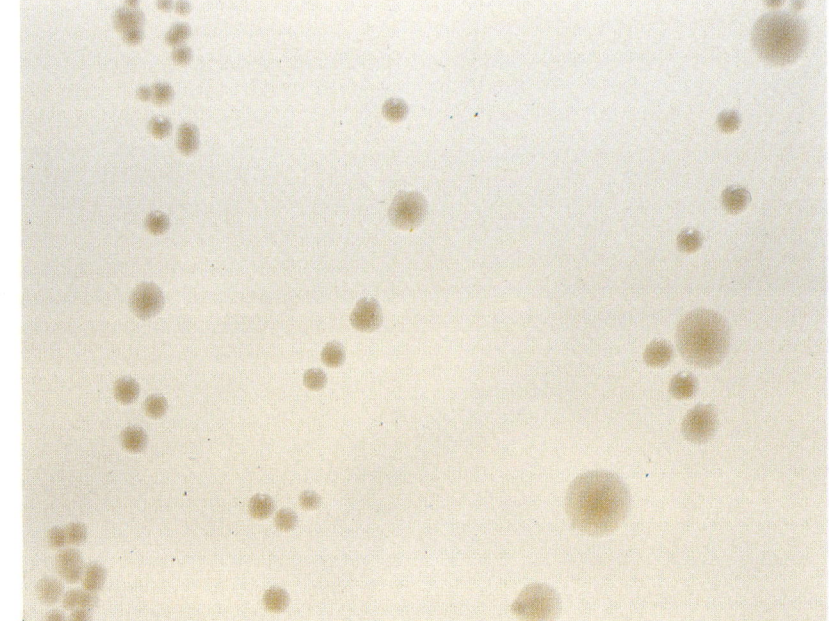

120 & 121. Determination of virulence in *Y. enterocolitica*: colonial morphology on trypticase-soy agar at 37°C. Strains oof avirulent *Y. enterocolitica* such as biovar 1 ('European'), serovar O5 have larger and more regular colonies (**120**) when grown on trypticase-soy agar at 37°C than strains of virulent *Y. enterocolitica* such as biovar 4, serovar O3 (**121**). Interpretation is highly subjective and the test should be used with caution (Schiemann, 1989). The test is, however, technically simple and trypticase-soy agar is commercially available. (*Trypticase-soy agar: 24 h at 37°C.*)

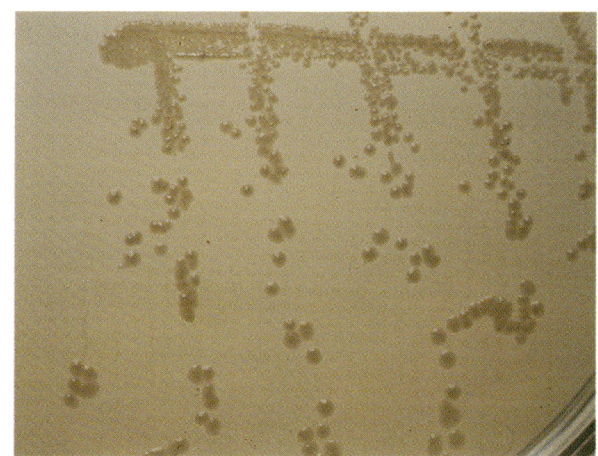 

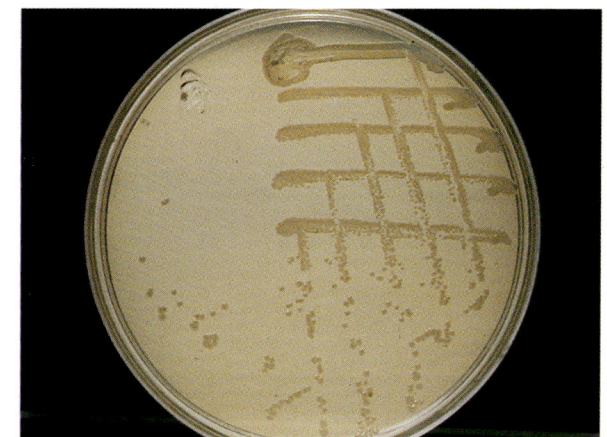 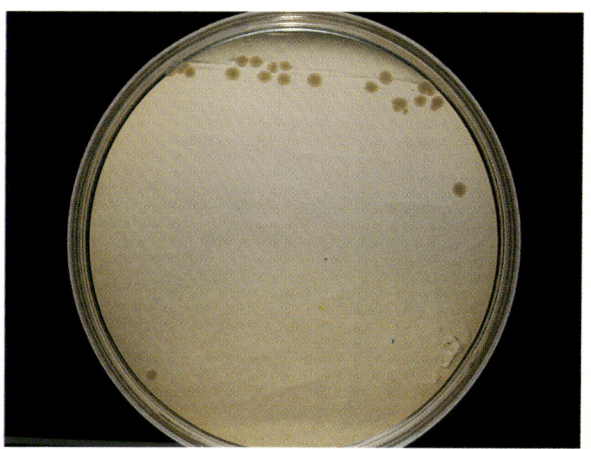

122–125. Determination of virulence in *Y. enterocolitica*: calcium dependence at 37°C. Virulent strains of *Y. enterocolitica* are dependent on Ca^{++} ion for growth at 37°C (Gemski *et al.*, 1980). Tests are normally made on magnesium oxalate agar but calcium-free blood agar base may give better differentiation (Marjai *et al.*, 1987). Growth, at 37°C, of a virulent strain of *Y. enterocolitica* serovar O8 on blood agar base supplemented with calcium (**122**), may be compared with growth of the same strain on unsupplemented blood agar base (**123**). The formation of a relatively large number of pin-point colonies in the absence of calcium tends to be a characteristic of serovar O8, and dependency is often more marked with other virulent serovars such as O3. Growth of serovar O3 is normal on calcium supplemented medium (**124**), but largely restricted to the initial line of inoculum

when calcium is absent. Colonies which do develop are of normal size and may be have developed from cells which have lost the plasmid (**125**).

Calcium dependency is a widely used means of determining probable virulence and is generally reliable. Interpretational difficulties may, however, arise with some strains of serovar O8, or with avirulent strains which grow poorly at 37°C under any circumstances (Schiemann, 1989). In some circumstances, interpretation may be aided by comparing numbers of colonies developing on media with and without calcium supplementation. It should be appreciated that all virulent strains are calcium dependent at 37°C, but that not all calcium dependent strains are virulent (Prpic *et al.*, 1985). (*Calcium-free blood agar base with or without 252.5mM calcium: 48h at 37°C.*)

8. *Vibrio*

Introduction

The genus *Vibrio* is the type genus of the family *Vibrionaceae* and comprises straight or curved rod-shaped bacteria that are Gram-negative, oxidase-positive and capable of both respiratory and fermentative metabolism. Motility is usually by polar flagella, although lateral flagella may be formed under certain cultural conditions such as on solid media (Baumann and Schubert, 1984). The polar flagella of *Vibrio* are unusual in being enclosed in a sheath which is continuous with the cell wall outer membrane (Baumann *et al.*, 1984). In addition to flagella, some marine strains possess tubular appendages of unknown function (Allen and Baumann, 1971).

Many vibrios are of marine origin and growth of all species is stimulated by concentrations of Na^+ ion greater than those of unsupplemented laboratory media (**Figures 126 & 127**) and, with the exception of *V. cholerae*, an obligate Na^+ requirement exists. Levels of Na^+ for optimal growth vary widely, from 5 to 15 mM for *V. cholerae* and *V. metschnikovii*, to 600 to 700 mM for halophilic species (Kushner, 1978; Baumann *et al.*, 1980).

In recent years, it has been found that that vibrios are able to respond to adverse environmental conditions by entering a viable, but non-culturable, phase. This phase, which may be part of a 'life cycle' has similarities to endospore formation by Gram-positive bacteria (Colwell, 1986).

The genus *Vibrio* contains a number of species which are pathogenic to man as well as to marine vertebrates and invertebrates. Of those pathogenic to man, eight species have been associated with foodborne disease, although in some cases the number of instances is very small. *Vibrios* are also associated with extraintestinal infections, primarily wound and ear infections and septicaemia. Human infections by vibrios are summarised in **Table 65**.

Three species of *Vibrio*, *V. cholerae*, *V. parahaemolyticus* and *V. vulnificus*, are responsible for most foodborne infections. The classical biovar of *V. cholerae* serovar O1, the 'King Cholera' of the nineteenth century pandemics and the causative organism of Asiatic Cholera, is today restricted in occurrence to parts of Asia, such as Bangladesh, and most cholera is caused by the el tor biovar. Water remains of major importance in the spread of cholera in areas where the disease is pandemic, but food is of greater importance elsewhere. In addition to cholera, strains of *V. cholerae* produce diarrhoeal disease varying in severity from the mild to that approaching classic cholera (Morris and Black, 1985). It is not known why *V. cholerae* O1

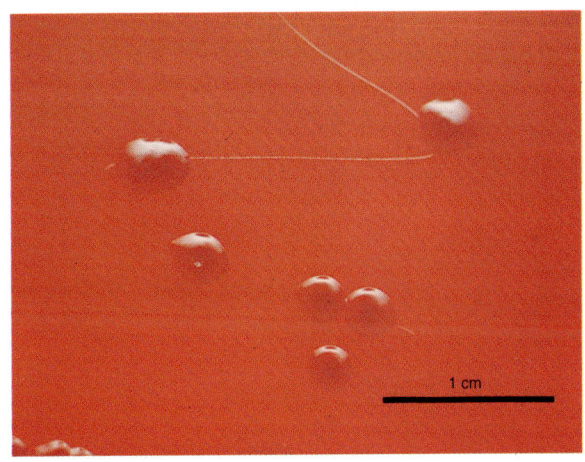

126

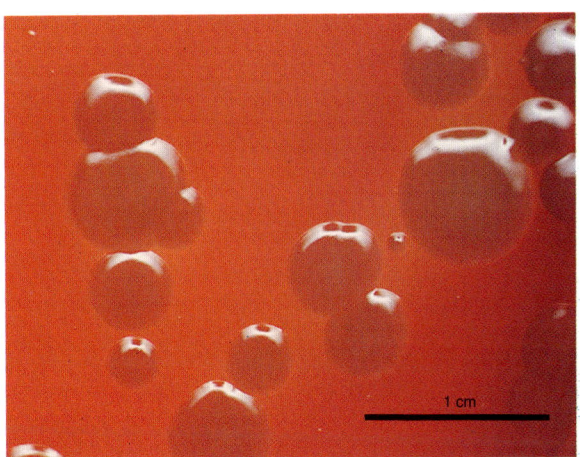

127

126 & 127. Growth of enteropathogenic vibrios on blood agar medium. Enteropathogenic vibrios including *V. parahaemolyticus* (**126**) and *V. vulnificus* (**127**) grow well on normal laboratory media. This continues to cause confusion among those who misunderstand the concepts of halophilic life and who confuse enteropathogenic vibrios with bacteria which have extremely high Na^+ requirements for growth, for example, members of the Archaebacteria such as *Halobacter*. The highest Na^+ requirments in the genus *Vibrio* are those of *V. costicola*, an organism associated with bacon curing environments. (*Blood agar: 24 h at 37°C.*)

Table 65. Summary of human infections due to *Vibrio* species.

Vibrio	Gastroenteritis	Wound infection	Ear/eye infection	Septicaemia
alginolyticus	−	+	(+)	(+)
cholerae O1	+	−	−	−
non O1	+	(+)	(+)	(+)
cincinnatiensis	−	−	−	+
damsela	−	+	−	−
fluvialis	+	−	−	−
furnisii	+	−	−	−
hollisae	+	−	−	−
metschnikovii	(+)	−	−	?
mimicus	+	−	(+)*	−
parahaemolyticucs	+	+	(+)	(+)
vulnificus	+	+	−	+

Key: + well established causal relationship.
    (+) reported on only a few occasions.
    ? probable but not proven cause.
    − not reported.
    * non-toxigenic strains only.

causes epidemic and pandemic disease, whereas non-O1 serovars do not.

V. parahaemolyticus was first isolated in Japan in 1951 during investigations into an outbreak of seafood associated enteritis. The organism is of particular significance in that country, and accounts for 40 to 70% of the nation's food poisoning (Hackney and Dicharry, 1988). Diarrhoea is the predominant symptom of *V. parahaemolyticus* infections.

V. vulnificus was first isolated from blood cultures by Hollis *et al.* (1976). The organism is one of the most invasive species to be described, and although gastro-intestinal symptoms may be present, infections are characterised by primary septicaemia. Victims almost invariably have an underlying predisposing condition (see page 164), and the high mortality rate has led to the organism being described as the '*new terror of the deep*'.

Diarrhoea is the predominant symptom of gastro-intestinal infections due to other species of *Vibrio*. Symptoms are usually mild and victims often have predisposing illnesses.

Isolation of *Vibrio*

Conventional cultural methods

In general, the same media and methods may be used for the isolation of all species of enteropathogenic vibrios, although *V. hollisae* will not grow on the widely used thiosulphate–bile–citrate–sucrose (TCBS) medium. Direct plating on TCBS may be used for faeces and other clinical samples and, possibly, for heavily contaminated foods, but food samples usually require a two-stage isolation procedure involving enrichment and selective plating.

Enrichment

There are two approaches to enrichment: the use of weakly selective media such as alkaline peptone water supplemented with 3% NaCl; or the use of more strongly selective media such as sodium–gelatin–phosphate broth for *V. cholerae* or glucose–salt–teepol broth for *V. parahaemolyticus*.

Selective plating media

TCBS agar medium is currently widely used for the isolation of all enteropathogenic species of *Vibrio* with the exception of *V. hollisae*, for which sucrose–tellurite–teepol agar is recommended. A number of attempts have been made to develop a medium specifically for the isolation of *V. vulnificus*, but none are fully satisfactory.

Rapid detection methods

The application of rapid methods to the detection of vibrios in foods has been limited, although gene probes have been developed for detection of *V. cholerae* non-O1 and *V. vulnificus*.

A novel method for detection of *V. parahaemolyticus*, based on a fluorogenic assay for trypsin-like activity, has also been described.

Methods for the isolation and detection of vibrios in foods are discussed in greater detail, with respect to choice of media and laboratory practice, in **Laboratory Methods**, *pages 178–180.*

Taxonomy of *Vibrio*

Effect of NaCl concentration and temperature on biochemical characteristics

The biochemical characteristics of *V. parahaemolyticus* and probably of other enteropathogenic vibrios, may be markedly affected by small changes in NaCl concentration (**Figures 128–130**). A wide range of biochemical characteristics may be affected and there is considerable variation from strain to strain. In general, the greatest number of biochemical characteristics are expressed at, or just below the optimal NaCl concentration for growth. Temperature dependence appears to be secondary to NaCl concentration in that expression of characteristics at 30°C, but not at 37°C, occurs at NaCl concentrations above and below the growth optimum but not at the optimum concentration itself.

128–130. Effect of NaCl concentration on the biochemical reactions of a strain of *V. parahaemolyticus.* NaCl concentration often has a marked effect on the biochemical activity of enteropathogenic vibrios. The widest spectrum of biochemical activity may usually be demonstrated at, or just below, the optimum NaCl concentration for growth, in this example, 3% (**128**) and is reduced during growth in both 2% (**129**) and 4% NaCl (**130**). It should be noted that, quantitatively, growth at these concentrations is very similar to that at the optimum of 3%.

Both carbohydrate fermentations and decarboxylase reactions are inhibited at NaCl concentrations above, or below, the optimum, although fermentation of amygdalin is inhibited at 2% NaCl but not at 4%.

The response to NaCl among enteropathogenic vibrios tends to form a continuous spectrum and considerable strain variation is likely. (*API 20E© with bacteria suspended in 3%, 2% and 4% NaCl respectively: 24h at 37°C.*)

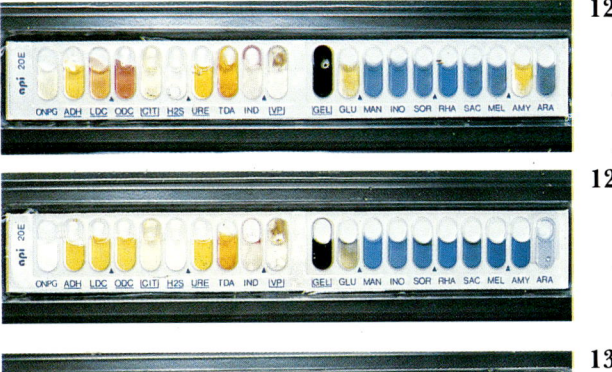

128

129

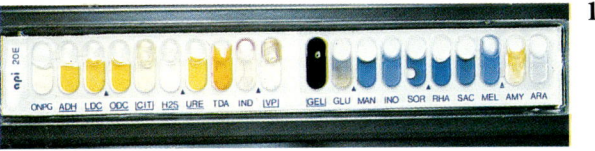

130

Differentiation of *Vibrio* from other genera

Vibrio may be differentiated from species with similar biochemical characteristics according to the criteria in **Table 66**. Either conventional media or miniaturised systems may be used, but in either case the media should contain an enhanced concentration of NaCl (usually 2-3%; Taylor *et al.*, 1982; Twedt *et al.*, 1984).

Differentiation of enteropathogenic species from other *Vibrios*

Enteropathogenic species of *Vibrio* may be differentiated from other species of the genus according to the criteria in **Table 67**. Either conventional media or miniaturised systems may be used (supplemented with NaCl as above), but in the latter case additional tests may be required.

The taxonomy of Vibrio *is discussed in greater detail, with respect to laboratory practice, in* **Laboratory Methods**, *pages 180–182.*

Table 66. Differentiation of *Vibrio* from other genera.

	Vibrio	Aeromonas	Plesiomonas
Na$^+$ requirement	+	−	−
Inhibition by			
vibriostat 0/129 (150 µg)	+	−	+
methylene blue (1%)	+	−	−
Gas from glucose	−[2]	+/−	−
Ornithine decarboxylase	+/−	−[3]	+
Acid from inositol	−[4]	−	+
Acid from mannitol	+/−	+/−	−

[1] Requiring, or stimulated by, levels greater than those in standard laboratory media.
[2] Gas is produced by *V. furnissii*.
[3] *A. veronii* is positive.
[4] Some strains of *V. metschnikovii* are positive.

Table 67. Differentiation of the main enteropathogenic species of *Vibrio* from similar, non-enteropathogenic species.

	V. alginolyticus	V. cholerae	V. metschnikovi	V. mimicus	V. parahaemolyticus	V. vulnificus
Oxidase	+	+	+	+	+	+
Thornley's arginine	−	−	+	−	−	−
Lysine decarboxylase	+	+	+/−	+	+	+
Ornithine decarboxylase	+	+	−	+	+	+
Voges–Proskauer[1]	+	+/−	+	−	−	−
β-galactosidase[1]	−	+	+/−	+	−	+
Acid from						
Lactose	−	−	+/−	−	−	+
Sucrose	+	+	+	−	−	−
Nitrate reduction	+	+	−	+	+	+
Growth in NaCl						
0%	−	+	+/−	+	−	−
6%	+	−	+	−	+	+
8%	+	−	+/−	−	+	−
Growth at 43°C	+	+	+/−	+	+	+
Inhibition by vibriostat[2]						
10 µg	−	+	+	+	−	+
150 µg	+	+	+	+	+	+

[1] After 24 hours incubation.
[2] 2,4-diamino-6,7-diisopropylpteridine phosphate.

Typing of *Vibrio*

V. cholerae

Serotyping

V. cholerae is divided into two serogroups on the basis of its O antigens. *V. cholerae* O group 1 (O1) is the causative organism of cholera; all other serogroups, which fail to agglutinate in O1 antisera, being commonly referred to as *V. cholerae* non-O1 or, erroneously, as non-cholera vibrio (NCV) or non-agglutinating vibrio (NAG).

V. cholerae O1 is not usually serotyped further until after the determination of the biovar (see below). Serogroup O1 may then be subdivided into three serovars; Ogawa and Inaba, which agglutinate only with their specific antisera, and Hikojima, which agglutinates with antisera to both Ogawa and Inaba.

Biotyping

V. cholerae O1 is divided into two biovars, classic and el tor. In the past, distinction between the two was made entirely on the basis of haemolysis of sheep erythrocytes, the originally described el tor serovars being strongly haemolytic. More recently, isolated strains have been either weakly haemolytic or non-haemolytic, while non-toxigenic strains which usually do not identify strongly with either biovar, may also be strongly haemolytic (Baumann *et al.*, 1984). Differentiation between classical and el tor biovars is currently made on the basis of physiological, or bacteriophage sensitivity, patterns (**Table 68**).

Other typing schemes

A number of phage typing schemes have been proposed for *V. cholerae* (Mukerjee, 1978). The most widely used for classic serovars is that of Mukerjee *et al.*, (1957), and for el tor that of Basu and Mukerjee (1968). The latter scheme is limited since it only recognises six types, and the more recent scheme of Lee and Furniss (1981) would seem to be of greater epidemiological value. In this scheme, over 1,000 isolates of *V. cholerae* el tor fell into 24 phage types although three types were still dominant (Baumann *et al.*, 1984).

Plasmids are of a relatively rare occurrence in *V. cholerae* and plasmid typing fails to differentiate between biovars. It has been suggested that the mere

Table 68. Differentiation between classic and el tor biovars of *Vibrio cholerae*.

	Biovar	
	Classic	el tor
Haemolysis	−	+/−[1]
Voges–Proskauer	−	+
Haemagglutination	−	+
Sensitivity to		
Polymyxin (50 iμ)	+	−
Classic phage IV	+	−
El Tor phage V	−	+

Note: The originally described el tor isolates were all haemolytic but those isolated in recent years were either only weakly haemolytic or non-haemolytic.

presence of plasmids may be a valuable epidemiological feature (Bartowsky and Manning, 1988), but further work is required to validate this suggestion.

Typing schemes for both *V. cholerae* O1 and non-O1 have been devised, but have not been applied to any significant extent. The topic has been reviewed by Brandis (1978).

V. parahaemolyticus

Serotyping

Serotyping of *V. parahaemolyticus* is based on both the O and K antigens. The scheme involves 11 O groups and 65 K antigens (**Table 69**). Most clinical strains are typable although others, particularly from the environment, are not (Twedt, 1989). It is generally considered that new serovars should not be added to the scheme unless they represent unusual clinical isolates.

Biotyping

No biotyping scheme exists for *V. parahaemolyticus* which, according to many criteria, form a homogeneous species. However, a subdivision may be made on the basis of the Kanagawa reaction. This reaction determines the presence of a thermostable direct haemolysin (Vp-TDH), which is thought to be closely related to pathogenicity (Nishibuchi *et al.*, 1985). Some Kanagawa-negative strains have been

161

shown to produce a second haemolysin, distinct from, but immunologically related to, Vp-TDH (Honda *et al.*, 1988a). Such strains have also been implicated as causes of gastroenteritis (Honda *et al.*, 1987; see page 171).

Table 69. Serotyping of *Vibrio parahaemolyticus*.

O group	K types
1	1, 25, 26, 32, 38, 41, 56, 58, 64, 69
2	3, 28
3	4, 5, 6, 7, 29, 30, 31, 33, 43, 45, 48, 54, 57, 58, 59, 65
4	4, 8, 9, 10, 11, 12, 13, 34, 42, 49, 53, 55, 63, 67
5	15, 17, 30, 47, 60, 61, 68
6	18, 46
7	19
8	20, 21, 22, 39, 70
9	2, 3, 44
10	19, 24, 52, 66, 71
11	36, 40, 50, 51, 61

Symptoms of *Vibrio* infections

V. cholerae

Symptoms of *V. cholerae* O1 infections range from mild diarrhoea to severe and life threatening illness. The most severe form (*cholera gravis*) is caused only by strains of serovar O1, which produce cholera toxin. Symptoms appear after an incubation period which varies from six hours to five days. Initial stools are brown with faecal material, but the classic 'rice water' stools soon appear. Passage of stools results in an enormous fluid loss, which may exceed one litre per hour and which results in dehydration and circulatory collapse. In turn, concurrent loss of potassium and bicarbonate in the stools leads to acid-base and electrolyte disturbances. Renal function is impaired and secondary symptoms such as chills, leg cramps, severe thirst, hoarse speech, rapid pulse and weakness appear. In severe cases, the body may be visibly dehydrated. When treatment is not available, death may be rapid and the observation, in 1832, of the French physician, Professor Majendie, that *'cholera is a disease that begins where others end: in death'* is still appropriate today.

The death rate varies from virtually zero in developed nations, where appropriate treatment is given rapidly, to *ca* 30% in underdeveloped areas where treatment is delayed (Tauxe *et al.*, 1988). Death is more common in infection by classic strains than by el tor, el tor infections being generally less severe.

Toxin production is not an absolute necessity for infection, and non-toxigenic strains of *V. cholerae* O1 also cause illness. This usually involves diarrhoea and symptoms vary from the mild and transitory, to an illness approaching the severity of *cholera gravis*. Infections with non-toxigenic *V. cholerae* O1 may be of increasing prevalence (Honda *et al.*, 1988b).

V. cholerae non-O1 is also associated with diarrhoeal illness. In the past it has been thought that both toxigenic and non-toxigenic strains can be involved, but Morris *et al.* (1990) consider that toxin production is a prerequisite for pathogenicity. Non-toxigenic strains have been associated with illness in the USA, but Morris *et al.* have suggested that some such strains are not truly pathogenic. However, it is accepted that other pathogenic determinants may be present. Incubation periods range from six hours to three days with a period of *ca* 12 hours being most common. Abdominal cramps and fever are common symptoms, and in *ca* 25% of cases bloody stools are passed. Nausea and vomiting occur but are less common. The illness due to non-O1 strains is generally milder than that due to O1 strains, but occasionally diarrhoea is severe (Morris and Black, 1985). Toxin synthesising strains may cause symptoms identical with *cholera gravis* (Nair *et al.*, 1988).

In contrast to O1 serovars, *V. cholerae* non-O1 has

been involved in wound and ear infections and is capable of systemic infection. Septicaemia may be secondary to another infection or may occur without an obvious focus (Madden *et al.*, 1989).

V. parahaemolyticus

The incubation period in *V. parahaemolyticus* infections is usually 9 to 24 hours, although symptoms may appear as soon as 2 hours after ingesting contaminated food (Pinegar and Suffield, 1982) or be delayed for up to 96 hours. Diarrhoea is the predominant symptom, and is usually accompanied by abdominal cramps and nausea. Vomiting, headache, low-grade fever and chills are less common, but significant symptoms (Fujiro *et al.*, 1974).

A second, dysenteric, form of the illness has been described in Bangladesh, India and, in a small number of cases the USA (Twedt, 1989). This form, which is more severe, is characterised by bloody, or mucoid, stools.

Although a number of other symptoms have been attributed to *V. parahaemolyticus* (Blake *et al.*, 1980a), these reports should be treated with caution and may actually refer to mis-identified *V. vulnificus* (West *et al.*, 1986). However, *V. parahaemolyticus* is recognised as a cause of wound, eye and ear infections and has an invasive ability.

Death is rare in infections with *V. parahaemolyticus*; no deaths have been reported in the USA and the annual rate in Japan is 0.04% (Hackney and Dicharry, 1988).

V. vulnificus

Symptoms of foodborne *V. vulnificus* infections usually appear within 16 to 48 hours of infection. The organism is highly invasive and the most common symptoms are those of primary septicaemia. The onset is marked by a general malaise followed by chills, fever and prostration. Hypotension is a feature in *ca* 33% of cases. Septicaemic *V. vulnificus* infections progress rapidly towards death, which often occurs in only a few days. Patients dying within 2 to 24 hours of admission to hospital have been noted in many studies, although the time to death may be as long as 6 weeks (Oliver, 1989). The overall mortality rate is 40 to 60%.

Diarrhoea is a relatively uncommon symptom in cases of primary septicaemia (Blake *et al.*, 1979) and occurred in only 40% of cases reviewed by Tacket *et al.*, (1984). It has been suggested that gastrointestinal symptoms result from a concurrent infection with a second micro-organism, but this seems improbable and support for an alternative hypothesis, that gastrointestinal symptoms represent a separate type of *V. vulnificus* infection with distinct pathogenic determinants, comes from reports of gastrointestinal symptoms, in the absence of septicaemia, in both immuno-compromised and normal people.

V. vulnificus is an important cause of wound infections and other extraintestinal infections including meningitis, myositis and pneumonia (Okada *et al.*, 1987a).

Other species of *Vibrio*

Foodborne infections due to other species of *Vibrio* tend to resemble those due to *V. cholerae* non-O1. Infections are often transitory, lasting no more than one day, but symptoms may persist for as long as six days. Bloody stools are more common in infections with *V. mimicus* than with other species, and terminal ileitis has been reported in infection with both *V. mimicus* and *V. fluvialis* (Hoge *et al.*, 1989).

There is some evidence that *V. hollisae* shares with *V. vulnificus* a predilection for bloodstream invasion in people with predisposing factors (Tilton and Ryan, 1987; Rank *et al.*, 1988), although infections are less severe.

Persons at particular risk from *Vibrio* infection

Susceptibility

V. cholerae

Endemic *cholerae* is historically an illness of the poor and undernourished. To some extent, this is due to low standards of hygiene and other factors which increase the risk of exposure to *V. cholerae* O1, but it is also a consequence of poor nutrition and the presence of underlying illnesses.

In developed countries, poor nutrition is unlikely to be a predisposing factor to *V. cholerae* infection. In any circumstances, low gastric acidity, whether due to diet, consumption of antacids or other medication, or to clinical factors, markedly increases the suscepti-bility to infection (Blake *et al.*, 1988a).

Genetically determined host factors are also of importance, and blood group O persons have a greater susceptibility to *V. cholerae* infections.

The general situation with *V. cholerae* non-O1 is similar to that with serovar O1. There are indications that, in common with many other enteropathogenic vibrios, underlying liver disease may be an important host factor for increased risk (Hackney and Dicharry, 1988).

V. parahaemolyticus

V. parahaemolyticus resembles *V. cholerae* in that suscepti-bility is markedly higher in individuals with reduced stomach acidity. Victims are usually adults but there is no discernible pattern of age group or sex.

V. vulnificus

V. vulnificus infections are rare in healthy people, and almost invariably restricted to gastroenteritis. People suffering from a number of underlying conditions are at high risk of infection and the development of primary septicaemia. Such conditions include liver disease, iron overload, chronic alcohol abuse, cancer, chronic renal malfunction, gastric disease including inflam-matory bowel disease and achlorhydria, steroid dependency and immunodeficiency (Chart and Griffiths, 1985; Johnston *et al.*, 1985). The effect of iron was quantitatively demonstrated by Wright *et al.*, (1981), who showed that in a mouse model the addition of ferric ammonium citrate reduced the LD_{50} from 10^6 cells to 1.1.

Cases of *V. vulnificus* infections have been reported in children as young as 7 and 8 years of age, but the majority of cases involve people over the age of 50 years. Males are markedly more susceptible and account for over 80% of *V. vulnificus* infections. The reason for this is not known (Oliver, 1989).

Other species

Factors determining susceptibility to other entero-pathogenic vibrios have not been defined. Underlying liver disease and iron overload do appear to be involved in invasive disease caused by *V. hollisae*, and many reported cases of infections with other species have involved similar predisposing conditions.

Exposure

V. cholerae

The risk of exposure to *V. cholerae* O1 is obviously greatest in areas where cholera is either endemic or epidemic. It should be appreciated that endemic regions are not defined by any immutable law, and it was notable that during the seventh pandemic, cholera became established in countries, largely in the sub-Saharan region of Africa, where it had previously been rare. To some extent, this was attributable to the more robust nature of the now dominant el tor biovar, and to its higher infectivity:virulence ratio, but a further major factor was drought in the Sahel regions.

Cholera epidemics are well known to be associated with drought (Fleurat, 1986). Major causative factors, apart from increased susceptibility due to malnutri-tion, are the rapid spread of infection from region to region as a result of increased travel in search of food, and insanitary and overcrowded conditions at relief camps. There are other and less obvious factors which provide an illustration of the interplay of apparently unrelated factors in increasing the exposure of whole populations to *V. cholerae* O1 and thus to risk of cholera (**Flow diagram 17**).

Breast-fed children rarely contract cholera even where the disease is epidemic but, as with other enteric pathogens, the substitution of formula foods made up with contaminated water is a major risk factor (Gunn *et al.*, 1979).

Flow diagram 17. Drought associated factors increasing exposure to *V. cholerae* in sub-Saharan Africa.

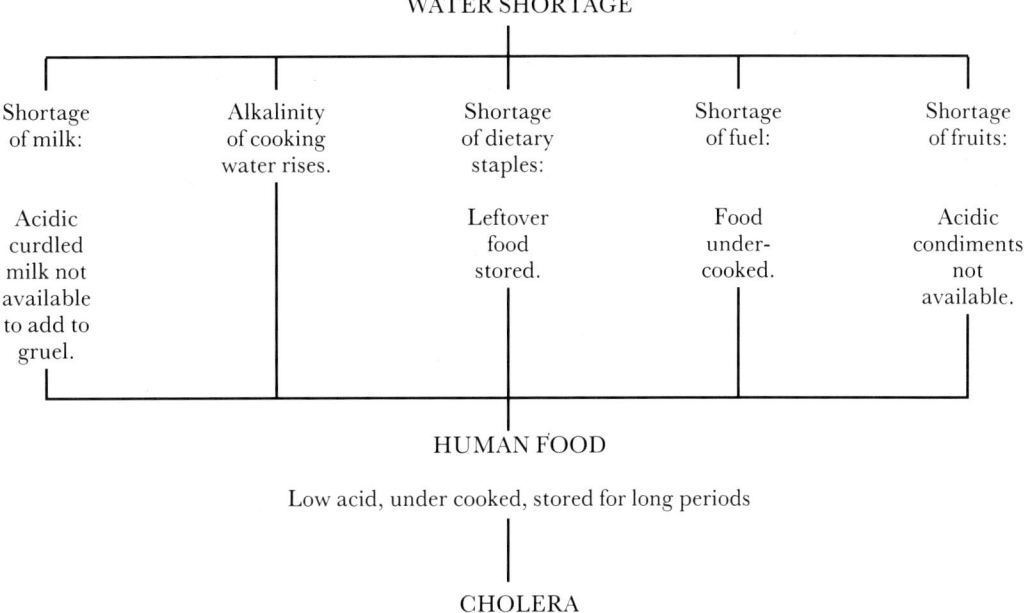

Note: Information from Fleurat (1986); Tauxe *et al.* (1988).

The profuse and watery nature of the stools means that direct transmission, or transmission via contaminated bedding is of importance where high standards of hygiene cannot be enforced (Mhalu *et al.*, 1984). Local customs such as corpse-washing are also important factors in increasing exposure.

Apart from the immediate contacts of returning travellers who have contacted cholera abroad, exposure to both O1 and non-O1 serovars in developed countries is restricted to the consumption of shellfish. Risk of infection is greatest in warm months, which may be due both to the presence of greater numbers of organisms and to increased consumption of raw shellfish.

Other species

The major risk of exposure to other enteropathogenic vibrios involves consumption of raw fish or shellfish. In Japan, *V. parahaemolyticus* infections tend to be associated with raw whitefish, while in the USA shellfish are more commonly involved. The high mortality of *V. vulnificus* infections in susceptible persons and its association with raw seafood has led to proposals in the USA for 'targeted education' for those at risk, including the possibility of 'health warnings' on packaging (Anon, 1988).

Occupational exposure in groups, such as fishermen, can be an important factor in the acquisition of *Vibrio* infections, and both intestinal and extra-intestinal disease may be involved.

Treatment of *Vibrio* infections

V. cholerae

The basic treatment for *cholera gravis* consists of intravenous rehydration using, preferably, Ringer's lactate solution together with careful correction of acid:base and electrolyte disturbances (Greenhough, 1985). In less severe, cases oral rehydration solution may be used.

Anti-microbial therapy reduces the volume and duration of diarrhoea, lessens the need for rehydration, and shortens the post-infection period of shedding of *V. cholerae*. Tetracycline or, less commonly, furoziladene are used. Strains may be resistant to a wide spectrum of anti-microbial drugs, and striking differences have been noted between classic and el tor strains responsible for simultaneous epidemics in Bangladesh, classic strains being largely resistant to tetracycline and sensitive to ampicillin, while a reversed pattern was usual for el tor strains (Siddique, 1989).

Fluid replacement is also required in the treatment of severe diarrhoea due to *V. cholerae* non-O1, and the use of tetracycline is considered reasonable (Madden *et al.*, 1989). Systemic infections should be treated with tetracycline or chloramphenicol, and these may be combined with an aminoglycoside if required.

Treatment of cholera with blocking agents, which block toxin receptor sites, has been evaluated (*cf E. coli*, page 109) but field trials were not fully successful. However, this approach is considered to hold considerable promise (Greenhough *et al.*, 1988).

V. parahaemolyticus

Infections with *V. parahaemolyticus* usually require no treatment other than supportive therapy, including rehydration where diarrhoea is severe. Tetracycline is beneficial where symptoms are prolonged.

V. vulnificus

Tetracycline is the most effective antibiotic against *V. vulnificus*, although in cases involving septicaemia many deaths have occurred despite extensive antibiotic therapy. Surviving patients generally require several weeks hospitalisation.

Other species

The general approach to treatment of infections due to other enteropathogenic vibrios involves supportive therapy with rehydration where required. Tetracycline is the drug of choice where antibiotic therapy is considered beneficial.

Mechanisms of pathogenicity

V. cholerae

V. cholerae is generally considered to be non-invasive, although reports of septicaemia due to non-O1 serovars suggest a limited invasive ability by these strains.

V. cholerae is well known for the production of cholera toxin (CT) by many strains. It is important not to overlook the role of other toxins in pathogenicity, or virulence factors involved in colonisation (**Table 70**).

Genetic control

In contrast to the production of LT by *Escherichia coli*, plasmids are not involved in the production of CT, which is under chromosomal control (Glatz, 1986). Both structural and regulatory genes are situated on the chromosome. Structural genes are not closely linked to regulatory genes, and regulatory genes may be found in different, unlinked, locations.

V. cholerae provides an example of co-ordinate production of virulence factors. TOXR protein (coded by the *toxR* gene) is required for the expression of several virulence genes, including those encoding production of CT and fimbrial adhesion. TOXR responds to environmental signals. These are complex but the CT and other genes on the regulon are induced by low osmolarity and by asparagine (Pugsley, 1988).

Mechanisms for the control of virulence appear to be the same in both classic and el tor biovars.

Motility and chemotaxis

The motility of *V. cholerae* aids in the penetration of the intestinal mucosa, and is a virulence factor for both O1 and non-O1 strains. There is also evidence that chemotaxis towards mucus increases virulence (**Figure 131**).

Mucus presents a physical barrier to bacteria and *V. cholerae* produces a mucase which aids penetration by reducing viscosity. However, neither motility nor mucase production are essential for pathogenicity (Mims, 1987).

Adherence

Despite considerable current interest, relatively little is known about the mechanism of adherence (Honda *et al.*, 1988c). Work with formalised human small intestines has shown that the mucus coat is the primary adhesion target, and that cell associated haemagglutinins have an important role in adherence (**Figures 132 & 133**; Yamamoto and Yokota, 1988) of *V. cholerae* non-O1. It has further been demonstrated that while a direct correlation exists between the titre of cell-associated haemagglutinins and adherence, similar levels are present in both clinical and environmental isolates, suggesting that intestinal adherence is an essential, but not a sufficient prerequisite for pathogenicity (Datta-Roy *et al.*, 1989).

A study of the outer membrane proteins (OMPs) of *V. cholerae* showed that while those produced by O1 strains were strikingly similar, there were differences with non-O1 strains. However, OMPs critical for adherence are shared by *V. cholerae* irrespective of biovar, serovar etc (Sengupta *et al.*, 1989). A possible role for non-OMPs was also postulated.

Fimbriae have been observed on both *V. cholerae* O1 (Ehara *et al.*, 1986) and non-O1 (Honda *et al.*, 1988b). The latter authors have postulated a role for fimbriae as colonisation factors based an analogy to *E. coli*.

Toxins

Cholera toxin

CT is a potent enterotoxin and as little as 5 µg administered orally caused diarrhoea in human volunteers (Levine *et al.*, 1983). The toxin has been studied extensively and detailed reviews of its structure and function

Table 70. Virulence factors of *Vibrio cholerae*.

	Colonisation	Toxins
O1 toxigenic	Motility Mucase Adhesins?	Classic cholera toxin; one or more toxins distinct from cholera toxin
O1 non-toxigenic	Motility Mucase Adhesins?	One or more toxins distinct from cholera toxin
Non-O1 toxigenic	Motility Mucase Adhesins?	Classic cholera toxin; one or more toxins distinct from cholera toxin
Non-O1 non-toxigenic	Motility Mucase Adhesins?	One or more toxins distinct from cholera toxin

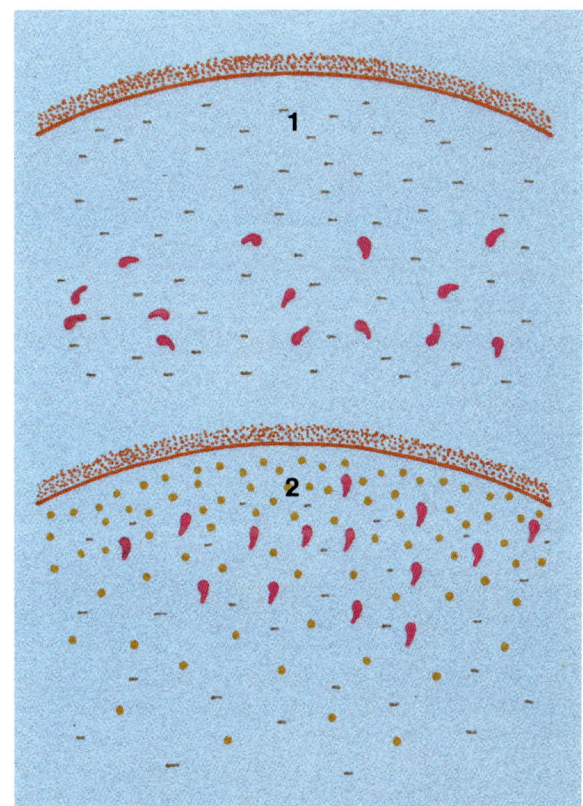

131

131. Role of chemotaxis in virulence of *V. cholerae*. In the absence of a chemoattractant, the movement of cells is random and penetration through the mucus towards the intestinal epithelium surface is a chance event (**1**). When a chemoattractant is present, cells are orientated and move along the gradient towards the higher concentrations of chemoattractant and thus the epithelial surface (**2**).

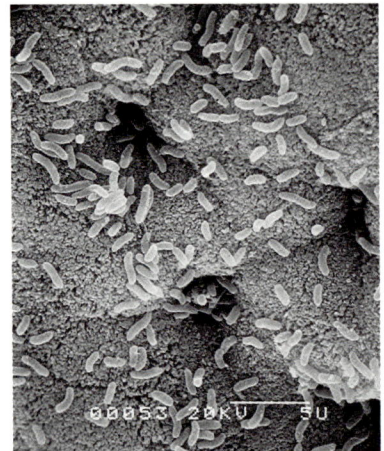

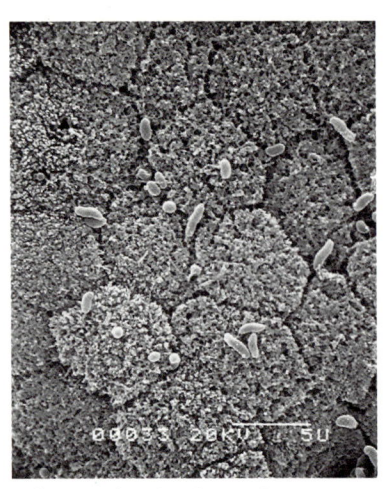

132 & 133. Role of haemagglutinins in the adherence of *V. cholerae* non-O1 to mucus. Adherence to the mucus layer of human ileal villi by a strain of *V. cholerae* non-O1 which produces a high level of cell-associated haemagglutinin (**132**) is contrasted with a variant of the same strain, which produces only a low level of haemagglutinin (**133**). The illustrations show the relative degree of adherence, which indicates at least a partial role for haemagglutinins. (Reproduced with permission from Yamamoto, T. and Yokota, T. *Infection and Immunity* **26:** 2018–2024. © 1988 The American Society for Microbiology.)

134

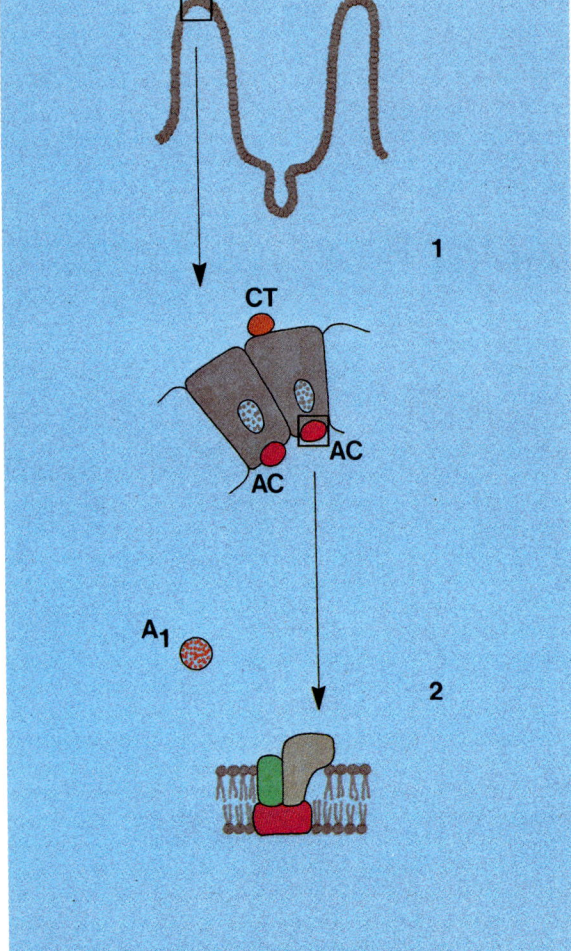

134. Cholera toxin and diarrhoea. In the first stage (**1**), the toxin binds to the mucosal cells of the ileal epithelium, which is followed by the entry of fragment A_1.

A_1 interferes with the adenylate cyclase complex of the basal membrane (**2**) and inhibits the GTPase switch-off of the adenylate cyclase. Inhibition probably involves the ADP-ribosylation of the GTP dependent regulator protein. This results in increased levels of cyclic AMP, activation of cAMP dependent protein kinase, and phosphorylation of some intestinal cell membrane proteins, leading to a biochemical reaction cascade causing diarrhoea. (Reproduced with permission from Stephen, J. and Pietrowski, R.A. *Bacterial Toxins.* © 1981 Chapman and Hall Ltd.)

135. Mechanism for the regulation of adenylate cyclase activity and site of action of cholera toxin. (Reproduced with permission from Stephen, J. and Pietrowski, R.A. *Bacterial Toxins.* © 1981 Chapman and Hall Ltd.)

13

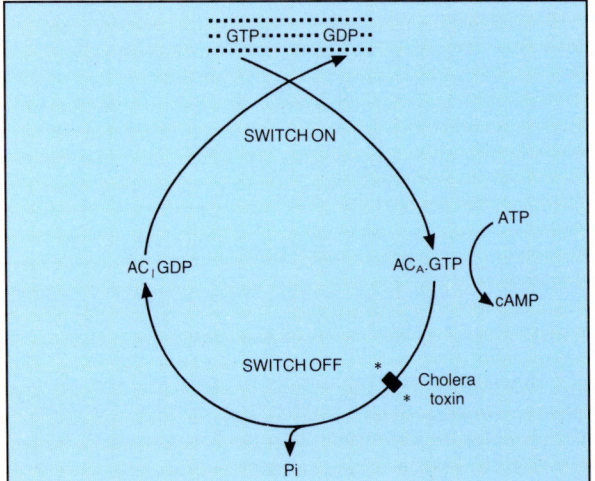

have been prepared by a number of authors including, van Heyningen (1977); Field (1979) and Enomoto and Gill (1980).

CT is an oligomeric protein of molecular weight 84,000, which is composed of A_1, A_2 and five B subunits. The B subunits (MW 11,000 each) bind to the receptor GM_1 ganglioside in the cell membrane. Blood group O people, who are more susceptible to *V. cholerae* infection, are thought to lack compounds present in persons with AB blood groups, which react with CT and prevent receptor binding (Bennun *et al.*, 1989). A_1 subunits (MW 24,000) gain entry to the cytoplasm of the enterocyte and irreversibly activates adenylate cyclase, resulting in increased intracellular levels of cyclic adenosine monophoshate (cAMP) and hypersecretion of salt and water. A_2 subunits (MW 5,000) are thought to mediate the movement of toxin into the host cell. The mechanism of action of cholera toxin and its entry into the cell is illustrated in **Figures 134–136**.

Both O1 and non-O1 serovars of *V. cholerae* produce identical proteases, which have both haemagglutinating and proteolytic activity. The proteases enhance CT activity by eroding the mucus layer, thus permitting closer contact between the receptors and the toxin and by 'nicking' the toxin (Honda *et al.*, 1989a).

Other toxins

V. cholerae produces a number of toxins in addition to CT. Neither the full significance of the toxins, nor the relative frequency with which they occur, is known, but their diverse nature may account for the range of symptoms observed in *V. cholerae* infections.

Toxins described are:

1 A toxin produced by *V. cholerae* non-O1 which is similar, but not identical, to cholera toxin.
2 A toxin which is produced by both CT-positive (Ogawa and Inaba serogroups) and CT-negative strains of *V. cholerae* O1 (Sanyal *et al.*, 1983; Saha and Sanyal, 1988; 1990). This toxin caused diarrhoea in infant rabbits, fluid accumulation in rabbit ileal loops, and increased the permeability of rabbit skin. It differs from CT in its antigenic nature, receptor site, mode of action and genetic homology. Toxins from different strains are immunobiologically similar (Setarunnahar and Sanyal, 1989). These toxins may be referred to as new cholera toxins (NCTs).
3 A haemolysin (NAG – rTDH) produced by some non-O1 strains that is identical biochemically and immunologically to *V. parahaemolyticus* thermostable direct haemolysin (Honda *et al.*, 1985).
4 A heat stable toxin (NAG – ST) similar to the ST toxin of enterotoxigenic *E. coli* (Honda *et al.*, 1985).
5 A sodium-channel inhibitor identified with, or similar to, the neurotoxin, tetrodotoxin, produced by the puffer fish (Tamplin *et al.*, 1987). The

significance is not known, but large quantities are produced by a CT-negative mutant which has retained mild pathogenic properties. Tetrodotoxin is also produced by species of *Vibrio* which, as far as is known, are non-pathogenic.
6 A toxin resembling the Shiga toxin of *Shigella dysenteriae* 1, which is produced by both O1 and non-O1 strains (O'Brien *et al.*, 1984).

In addition to the above toxins, there has been renewed speculation concerning a possible role for endotoxin (Ashan and Ciznar, 1987), which was previously thought to be responsible for the symptoms of cholera.

V. cholerae O1 releases endotoxin during the late logarithmic phase of growth. CT-negative strains release approximately four times as much as CT-positive strains. It is recognised that endotoxin cannot cross the normal intestinal membrane, although it binds strongly to that membrane. It is postulated that, if released in the form of blebs, it may function as a vehicle to deliver other biologically active products direct to the cell membrane underlying the adhering cell of *V. cholerae*, and thus increase the pathogenic potential.

V. parahaemolyticus

V. parahaemolyticus differs from *V. cholerae*, both in being invasive, and in failing to produce a classical enterotoxin. There has been much interest in the direct thermostable haemolysin (Vp – TDH) and its role, but other aspects of pathogenicity tend to have been neglected.

Genetic control

The presence of plasmids in *V. parahaemolyticus* has been known for some years, and it has been previously suggested that haemolysin production and pathogenicity were plasmid mediated (Barker *et al.*, 1975). It now appears unlikely that there is any involvement of plasmids, pathogenic mechanisms being under chromosomal control (Glatz, 1986).

Adherence and penetration

V. parahaemolyticus is able to adhere to epithelial cells, and flagella appear to be involved: sheared, aflagellate or heat-treated cells failing to adhere (Twedt, 1989). A role has also been demonstrated for cell-associated haemagglutinins, adherence being largely to the lymphoid follicle epithelium rather than to the mucus coat (*cf V. cholerae*, page 168). Some studies have shown differences in adherence between Kanagawa-positive and Kanagawa-negative strains,

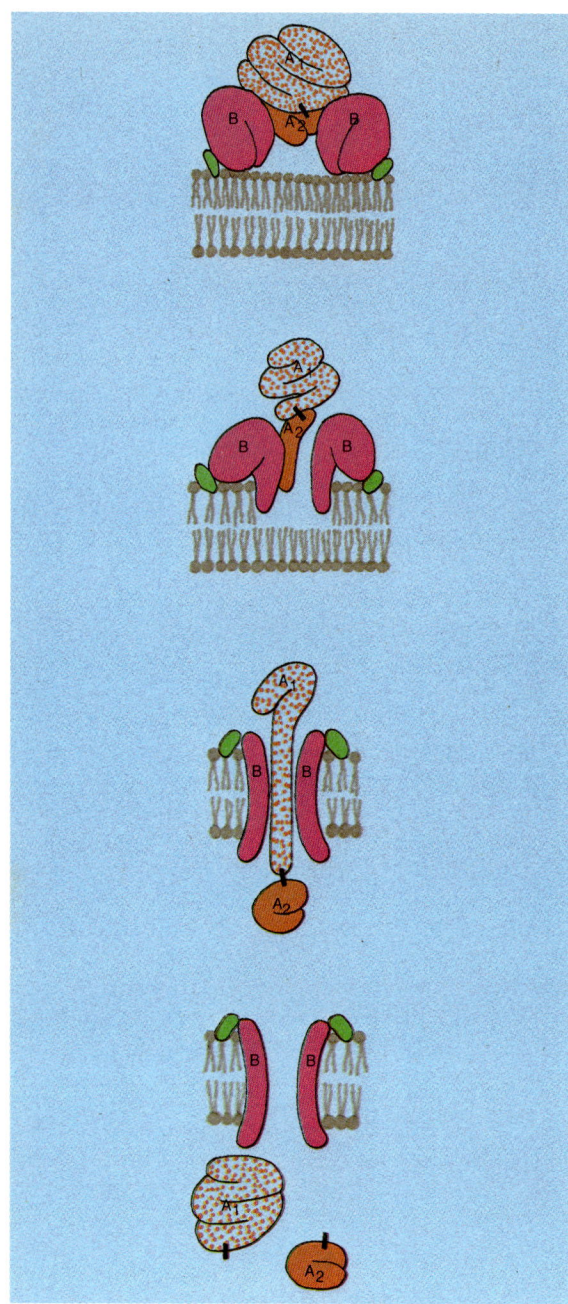

136. Proposed mechanism of entry of cholera toxin subunit A₁ into the cell. The mechanism of entry is a three-stage process followed by separation of the A_1 and A_2 subunits.

1 B subunits bind to the GM_1 ganglioside receptors.

2 The subunits undergo a change in conformation and are inserted into the cell membrane.

3 Insertion produces a hydrophilic channel and the linked A_1 and A_2 subunits diffuse into the cytoplasm. Movement is mediated by the A_2 subunit.

4 The S-S bridge linking the A_1 and A_2 subunits is reduced and the two subunits seperate. The A_1 subunit activates adenylate cyclase. (Reproduced with permission from Gill, D.M. *Biochemistry* **15:** 1242–1248. © 1976 American Chemical Society.)

but it appears more probable that epithelial adherence is a property of most strains.

V. parahaemolyticus enters the epithelium and penetrates to the *lamina propria*. The mechanism for this has not been fully elucidated. Invasiveness would appear to be a characteristic of most strains of *V. parahaemolyticus* and not restricted to Kanagawa-positive.

Systemic spread

In animal studies, *V. parahaemolyticus* has been shown to spread from the *lamina propria* to the heart, spleen, pancreas and liver, thus indicating circulatory spread.

Thermostable direct haemolysin

For a number of years, Vp–TDH has been recognised as the principle virulence factor. It is also responsible for the Kanagawa reaction.

Vp-TDH is a heat stable, trypsin-sensitive protein of MW 44,000. It is composed of two identical monomeric subunits of MW 22,000. Production of the haemolysin is controlled by pH value and occurs only when between pH 5.5 and 6.5. Vp–TDH is lethal, cytotoxic and cardiotoxic, and has a minimum haemolytic dose (purified) of 0.1 μg. Despite the known properties of Vp-TDH there is no direct evidence of involvement in production of diarrhoea (Takeda, 1983).

In addition to Vp–TDH, *V. parahaemolyticus* produces four further haemolytic constituents: a heat labile direct haemolysin, phospholipase A, lysophospholipase and glycerylphosphorylcholine esterase. In contrast to Vp–TDH, the heat labile haemolysin is distributed randomly among *V. parahaemolyticus*.

Although pathogenic strains of *V. parahaemolyticus* were previously thought to be Kanagawa-positive, diarrhoeal illness due to Kanagawa-negative strains is now well documented and indeed such strains are the common cause of infection outside tropical countries (Kelly and Stroh, 1989). A new haemolysin, Vp-TDH related haemolysin (Vp-TRH), was purified from a Kanagawa-negative isolate of *V. parahaemolyticus* serovar O3:K6 (Honda *et al.*, 1987). This haemolysin is immunologically related to, but distinct from, Vp-TDH and has identical biological activity. The same haemolysin has been purified from a clinical isolate of *V. parahaemolyticus* O6:K46 (Honda *et al.*, 1988a; 1989b).

Many attempts have been made to isolate an alternative cell-free enterotoxin from *V. parahaemolyticus*, but none has been successful (Twedt, 1989).

Tetrodotoxin and the related anhydrotetrodotoxin (see page 169) are also produced by some strains of *V. parahaemolyticus* and, in common with *V. cholerae*, a Shiga-like toxin has been reported (O'Brien *et al.*, 1984).

V. vulnificus

The high fatality rate associated with *V. vulnificus* infections in susceptible people, and the very high level of invasiveness, means that the mechanisms of pathogenicity have received considerable attention in recent years. Despite this, many features remain unresolved.

Genetic control

Plasmids are found in only 12% of *V. vulnificus*, and there is no correlation with any putative virulence factor. Virulence may, therefore, be assumed to be under chromosomal control.

Adherence and penetration

Adherence is an important virulence factor for *V. vulnificus*, clinical isolates having a much greater capacity for adherence than environmental isolates (Reyes *et al.*, 1985).

Penetration of the epithelial tissue is rapid and accompanied by extensive tissue damage. When the rabbit ligated ileal loop model is used, *V. vulnificus* appears in the animal's bloodstream within 5 hours of its introduction. Loss of intestinal epithelium and swelling of the sub-mucosa may be detected microscopically as early as 1 to 2 hours after its introduction. Tissue damage continues until necrosis extends to the muscularis, and there is breakdown of the endothelial cells of blood vessels. It seems likely, therefore, that *V. vulnificus* produces one, or more, toxins which promote access to the circulatory system through tissue damage (Oliver, 1989).

Survival and proliferation in host tissue

Resistance to phagocytosis

Virulence of *V. vulnificus* is associated, at least in part, with resistance to phagocytosis (Yoshida *et al.*, 1985). Resistance appears to be due to an antiphagocytic surface component that may be referred to as a capsule. The antiphagocytic compound is thought to be an acidic polysaccharide, but an involvement of membrane proteins cannot be discounted.

Resistance to normal human serum

Many pathogenic strains of *V. vulnificus* have a marked resistance to normal human serum. A role has been postulated for the capsule but evidence is lacking (Wright *et al.*, 1990).

Some strains of *V. vulnificus* are not susceptible to the activity of serum by either the classic or alternative pathways of complement activation, while a second group fully activate the classic pathway, but only partly activate the alternative pathway. Individuals suffering liver cirrhosis are largely dependent on the alternative pathway activation for protection against Gram-negative bacteria and this, together with iron overload (see below), explains the high level of susceptibility in such individuals.

Effect of iron deprivation

Although iron enhancement of pathogenicity is a common feature of Gram-negative pathogens, the relationship between iron levels and the virulence of *V. vulnificus* is striking. In mice, dramatic decreases in the size of inoculum required for mortality are observed when the animals are iron overloaded. *V. vulnificus*, which is killed rapidly in normal human serum, is able to survive and then multiply as the iron content increases. Iron may thus be considered as a major limiting factor for survival and growth in the human body.

Production of siderophores has been noted in pathogenic vibrios generally (Andrews *et al.*, 1983), and in *V. vulnificus* in particular, (Simpson and Oliver, 1983). *V. vulnificus* is unique among vibrios in producing both a hydroxamate and a phenolate siderophore. Of the two, the hydroxamate siderophore has greater chelating activity.

Virulent strains of *V. vulnificus* are able to obtain iron from iron-binding serum proteins, such as transferrin and lactoferrin only when the proteins are saturated with iron. Avirulent strains are unable to obtain iron from iron-binding proteins even when fully saturated.

V. vulnificus differs from other bacteria in being able to obtain iron from haemoglobin when it is complexed with haptoglobin (Helms *et al.*, 1984). Haptoglobin, a component of normal human serum, binds specifically and irreversibly to haemoglobin and thus acts as a bacteriostat for most pathogens (Eaton *et al.*, 1982). In a later study, Zakaria-Meehan *et al.* (1988) showed that *V. vulnificus* was able to obtain iron from haemoglobin bound by haptoglobin phenotypes 1 and 2, but not by phenotype 2-1. These authors considered that the attribute may be a particularly powerful virulence factor.

Toxins

Haemolysin-cytotoxin

V. vulnificus synthesises at least two haemolysins. The most important of these, which is probably identical to the 'lethal toxin' in the culture supernatant of *V. vulnificus* strain FCC (Okada *et al*., 1987b) is a 56 Kda heat labile protein. The binding step is inhibited by cholesterol (Yamanaka *et al*., 1987), but *V. vulnificus* haemolysin differs from other cholesterol inhibited haemolysins.

V. vulnificus haemolysin is lethal for mice, and possesses cytotoxic activity against CHO cells, haemolytic activity against mammalian erythrocytes, and vascular permeability factor activity in guinea-pig skin. The haemolysin is produced *in vivo* during the course of infections.

V. vulnificus haemolysin is produced by both clinical and environmental strains. In a study using mutants derived from a virulent strain, similar quantities were produced by both the virulent parent strain and avirulent isogenic mutants, while the virulence of non-haemolytic mutants was as great as that of the parent strain (Massad *et al*., 1988). Thus while probably contributing to virulence, haemolysin is not an essential determinant.

A second 50 Kda haemolysin has been shown to act on the membrane of mast cells, inducing cytolysis and histamine release. This haemolysin also has a cytolytic effect on leucocytes (Yamanaka *et al*., 1990)

Enterotoxin

Virulent strains of *V. vulnificus* produce a soluble enterotoxin which is distinct from both haemolysins and proteases (Stelma *et al*., 1988). This enterotoxin is involved in diarrhoeal disease, although other pathogenic determinants are necessary.

Other products

V. vulnificus produces a variety of other substances which may have a role in pathogenicity: these include two phospholipases (Testa *et al*., 1984); one or more proteases which have activity against collagen (Smith and Menkel, 1982), elastin and albumin (Kothary and Kreger, 1985); deoxyribonuclease; lipase; mucinase; chondroitin sulphatase; hyaluronidase; fibrinolysin and alkyl sulphatase (Oliver, 1989). The significance of most of these substances is not known, but proteases active against elastin and albumin are likely to have a role in pathogenesis (Kothary and Kreger, 1985), which may be of greater importance in wound infections.

Other enteropathogenic species

The mechanisms of pathogenicity of other enteropathogenic vibrios has not been studied in detail. In general terms, availability of iron is probably of importance to invasive species such as *V. mimicus* (Andrews *et al*., 1983; Tilton and Ryan, 1987), and it also seems likely that a variety of extracellular products are involved in virulence. For example, *V. metschnikovii* produces a cytolysin distinct from other vibrio cytolysins (Miyake *et al*., 1988), while *V. hollisae* produces a haemolysin immunologically related to, but distinct from, Vp-TDH and Vp-TRH of *V. parahaemolyticus* (Honda *et al*., 1989c). *V. fluvialis* has at least four toxic activities (Joseph *et al*., 1982) and V. mimicus produces Mimicus-ST, a toxin closely related to the NAG-ST produced by non-O1 *V. cholerae* (Morris *et al*., 1990).

In vitro determination of virulence

V. cholerae

At present, factors such as adherence and colonisation can be determined only by use of whole animal models of which the removable intestinal tie-adult rabbit model (RITARD) is probably most effective. Such models cannot be applied on a large scale, or in the vast majority of food industry laboratories.

Demonstration of cholera toxin (CT) is probably the easiest and most meaningful assay for the demonstration of potential virulence. Tissue culture assays such as CY_1-adrenal cells (Sack and Sack, 1975; Madden *et al*., 1984) and chinese hamster ovary cells (Guerrant *et al*., 1974) have been widely used, but for routine use are now being replaced by serological assays for the toxin or by probes for the genes controlling synthesis. The assays used are largely those employed for *E. coli* LT, and are discussed in greater detail in pages 127–128.

V. parahaemolyticus

The Kanagawa reaction correlates well with the production of a direct haemolysin by *V. parahaemolyticus* and in the past has generally been accepted as a means of distinguishing pathogenic strains. However, the common occurrence of pathogenic strains that are Kanagawa-negative inevitably casts doubt on the validity of the Kanagawa reaction, and other assays for direct haemolysin, as an indicator of virulence.

The determination of virulence of V. parahaemolyticus *is discussed in greater detail with respect to practical aspects in* **Laboratory Methods**, *page 183.*

V. vulnificus

Virulence testing of *V. vulnificus* usually involves animal testing – routine monitoring using normal adult mice. However, it is possible to relate colonial morphology to the possession of the antiphagocytic surface antigen. This is discussed in greater detail in **Laboratory Methods**, page 183.

The haemolysin of *V. vulnificus* cannot form the basis of a virulence test, since it is produced by both virulent and avirulent strains.

Human carriage of *Vibrio*

V. cholerae

Human carriage of *V. cholerae* is of obvious importance in the spread of cholera in epidemic periods, a large number of those involved in carriage being asymptomatic or having only very mild symptoms. Long-term carriage for months or years, as occurs with *Salmonella typhi*, is not a problem with *V. cholerae*.

It is not known whether humans or the environment are the prime resevoir of *V. cholerae* during the periods between seasonal epidemics in Asian countries. However, studies in Bangladesh suggest year-round human carriage (Madden *et al.*, 1989).

V. parahaemolyticus

In Japan, *V. parahaemolyticus* can be isolated from the stools of asymptomatic persons, as well as from individuals with, or recovering from, actual illness, during the warm months of June to October. Carriage rates of 0.3% among healthy individuals and 2.5% among sushi chefs have been reported. Duration of carriage is short: 3 to 7 days in stools of healthy individuals and 10 to 16 days in those recuperating from diarrhoea.

Other species

The extent of human carriage among other species is not known. In most cases the situation probably resembles that with *V. parahaemolyticus*.

Vibrio in the environment

V. cholerae

In endemic and epidemic areas *V. cholerae* enters surface water systems, including those used as sources of potable water, from sewage, and from there contaminates the entire aquatic environment. The aquatic environment also forms the natural habitat of *V. cholerae* in non-endemic or epidemic areas and, in the USA, Australia and possibly elsewhere, the critical reservoir appears to be estuarine waters where the organism is a part of the normal microflora. The incidence varies according to water temperature, and numbers increase with increasing water temperature. This correlates with the higher incidence of *V. cholerae* infections in warm weather, and may be related not only to temperature but to the greater availability of chitin, which *V. cholerae* is able to utilise, from marine crustaceans.

Survival of *V. cholerae* in estuarine waters at low temperatures is greater than anticipated, and this in turn has been attributed to the protective effect of chitin and to a lesser extent some amino acids and peptides (Huq *et al.*, 1983; Amako *et al.*, 1987).

Of greater significance in understanding the long-term survival of *V. cholerae* in the environment is the finding that the organism can enter a non-culturable, but viable, phase in response to unfavourable environmental conditions. This phase may be part of a life cycle and helps to explain the observed seasonality of infections with *V. cholerae* and other enteropathogenic vibrios.

V. cholerae is not exclusively of estuarine origin and is thought to be indigenous to tropical fresh waters (Peres-Rosa and Hazen, 1989). The organism is also found in freshwater habitats in temperate climates, and the distribution of non-O1 strains in non-saline environments is probably much wider than thought previously (Pitrak and Gindorf, 1989).

Strains of *V. cholerae* isolated from the environment tend to be non-O1 serovars, and both O1 and non-O1 isolates are usually non-toxigenic.

The role of land animals and birds in the spread of *V. cholerae* has been examined in some depth. There is no evidence that animals act as a reservoir of *V. cholerae* O1, but non-O1 serovars have been isolated from domestic animals and wildlife. There is increasing evidence that aquatic birds may spread *V. cholerae* across areas where cholera is not endemic (West, 1989). With the exception of *V. fluvialis*, other species of enteropathogenic vibrios do not appear to be harboured by animals or birds.

In common with other species, *V. cholerae* may be found in association with the roots of aquatic plants, which are involved in transmission of the organism in the environment (Chowdbury *et al.*, 1989). In this context, the fast growing water hyacinth may be of particular importance in Asian countries such as Bangladesh.

V. parahaemolyticus

The natural habitat of *V. parahaemolyticus* is similar to that of *V. cholerae*, the organism being widely distributed in estuarine and coastal waters, but not in the open sea (Colwell, 1984). There is also evidence that *V. parahaemolyticus* is associated with the digestive tract of shellfish, such as clams (Greenberg *et al.*, 1982). This supports the suggestion of Baumann and Baumann (1977) that marine vibrios may be considered to be marine enteric bacteria.

Occasional isolations of *V. parahaemolyticus* have been made from freshwater or non-marine fish. In these cases, it is assumed that salinity has been raised locally.

It is likely that a viable, non-culturable stage similar to that of *V. cholerae* exists, and this would largely explain the persistence of the organism under conditions of no growth.

In recent years, a restricted bioserovar of Kanagawa-negative *V. parahaemolyticus* (O4:K12, urease[+]) has become dominant in coastal waters of the western United States and Mexico, and is also the major cause of *V. parahaemolyticus* infection (Abbott *et al.*, 1989; Kaysner *et al.*, 1990). Dominance may reflect a more effective response to environmental pressures.

Other species

Other enteropathogenic vibrios resemble *V. cholerae* and *V. parahaemolyticus* in habitat, and are widely distributed in estuarine and marine habitats. The sodium requirement of the different species varies, and thus their relative frequency in different locations often reflects salinity. For example, *V. vulnificus* is of widespread distribution in estuarine waters (Tilton and Ryan, 1987), especially if the salinity is low (Kelly, 1982), while *V. mimicus* has been isolated from fresh water (Hackney and Dicharry, 1988), and *V. fluvialis* from the faeces of cattle, pigs and rabbits (Lee, *et al.*, 1981). Little is known of the natural habitat of *V. hollisae*.

Foods commonly associated with *Vibrio* infections

V. cholerae

Historically, *cholera gravis* is regarded as a water-borne disease and there is no doubt of the importance of this route in areas where the disease is endemic and a continuing cycle of drinking water pollution and consumption exists (Levine and Nalin, 1976). Even under such circumstances, food may be an important, and possibly underestimated, vehicle of transmission. For example, during an investigation of a cholera epidemic in Mali, two means of transmission were identified, a well supplying drinking water and leftover millet gruel (Tauxe *et al.*, 1988).

It may be postulated that water used in preparation of food, or for washing utensils, is a common source of contamination, although both seafood and vegetables may be contaminated at source. Contamination of food by handlers is also likely to be of importance. A wide variety of foods have been implicated in the transmission of cholera, including soft drinks, fruit and vegetables, milk, seafood, locally brewed beer, as well as millet gruel. Although in some circumstances food may serve only as a passive vehicle of infection, in other circumstances growth in foods is of major importance.

Waterborne cholera is usually considered to be restricted to the developing nations of Africa and Asia. However, bottled, non-carbonated mineral water was a vehicle of cholera in Portugal in 1974 (Blake *et al.*, 1977).

In many industrial countries, such as the UK, cholera is almost invariably an imported disease (Galbraith *et al.*, 1987). In other countries, food is likely to be a major vehicle of both serovar O1 and non-O1. The estuarine habitat of *V. cholerae* is reflected in the relatively high level of contamination of seafood harvested from estuaries (Blake *et al.*, 1980b; Desmarchelier and Reichelt, 1982; Madden *et al.*, 1982; Blake, 1983; Eyles and Davey, 1984). During 1986 a continuing epidemic of *V. cholerae* O1 Inaba, in the US, claimed 12 victims in Louisiana and 1 in Florida (CDC, 1986). Although these numbers are insignificant compared with *Salmonella* and other major pathogens, the USA, which had remained free of indigenous cholera from 1911 to 1973, has suffered 44 cases between 1973 and 1986. In several of these cases, cooked shellfish were involved, the fish having either been inadequately cooked or contaminated after cooking.

In the USA, infection due to non-O1 serovars is more common than with O1 serovars, and this reflects the distribution of the two types in seafoods.

Other foods have been identified in small-scale outbreaks of *V. cholerae* infections in developed countries. These include raw vegetables, egg and asparagus salad, potatoes (Hackney and Dicharry, 1988) and frog legs (Sang *et al.*, 1987). Only in the case of frog legs is any intrinsic hazard likely to exist.

V. parahaemolyticus

The primary source of *V. parahaemolyticus* is fish or seafood. In Japan, raw finfish is the most common vehicle of infection, while in the USA cooked crustaceans have usually been involved although raw oysters (NRC, 1985), and crabs (Blake, 1980a) have also been implicated. The common involvement of cooked food reflects poor standards of hygiene.

A seasonal variation in numbers reflecting the numbers in the environment is apparent with oysters harvested in Louisiana (Paile *et al.*, 1987), but not in the Gulf of Mexico (Thompson and Vanderzant, 1976).

Other species

Although in some cases full evidence is lacking, other species of enteropathogenic vibrios appear to be associated solely with fish and seafood. Various types of seafood have been implicated, but risk is generally greatest with raw oysters. *V. hollisae* infections have been associated with the consumption of fried fish (Lowry *et al.*, 1986) and smoked dried fish (Rank *et al.*, 1988), as well as raw fish, and may withstand some processing better than other species (West, 1989). The epidemiology of *V. furnissii* remains obscure, although the organism has been involved in food poisoning on board aircraft.

Factors affecting the growth and survival of *Vibrio* in foods

Vibrios are readily destroyed by cooking, but this means of control is not available with seafoods that are traditionally eaten raw. Much emphasis has been placed on the control of growing beds to minimise faecal pollution and depuration (purification) before consumption. While no doubt contributing to a reduced overall morbidity, such measures, cannot guarantee the absence of enteropathogenic species of *Vibrio*, and a finite risk of illness from these and other agents of food poisoning such as some viruses remains. Many consumers of seafood seem prepared to accept the risk, and short of a total embargo on the sale of raw shellfish during the high-risk months, which would in any case be difficult to enforce, there seems little that can be done from a regulatory viewpoint except to attempt an extensive and ongoing public education exercise. Such an exercise is of particular importance with respect to *V. vulnificus* and people with predisposing conditions.

Temperature

Enteropathogenic vibrios grow over the range 5 to 43°C, with an optimum of *ca* 37°C. There is variation both between species and among strains within a species, and in general terms *V. cholerae* will grow at lower temperatures than *V. parahaemolyticus*.

At suitable temperatures, growth of *V. parahaemolyticus* can be very rapid indeed, generation times as short as 8 to 9 minutes having been observed during broth culture under optimal conditions. In foods, generation times of 13 to 18 minutes and 12 minutes at 30°C were observed in raw horse mackerel and boiled octopus respectively. Growth rates at lower temperatures are naturally lower, but counts increased from 10^2 to 10^8 cfu/g after 24 hours storage at 25°C in homogenised shrimp, and from 5×10^3 to 5×10^8 cfu/g after 7 days storage at 12°C in homogenised oysters (Twedt, 1989). Temperature control in the seafood industry is generally poor, and the public health implications of such rapid growth are obvious.

Enteropathogenic vibrios are not heat resistant and are readily destroyed by cooking. Resistance depends on several factors, including heating menstruum and physiological condition. There is also considerable variation between species, and while information is neither complete, nor, in some cases, fully reliable, *V. cholerae* would seem to have a higher level of heat resistance than *V. parahaemolyticus* (**Table 71**).

Care must be taken in translating D values into safe cooking procedures especially, with seafood such as crab, which often vary considerably in size. Many commonly used procedures were developed empirically, not for safety but for organoleptic reasons, or to facilitate opening. Accordingly, there is some disagreement concerning the safety of processes in common use. Thus while Boutin *et al.* (1982) found traditional recipes to effectively eliminate 10^5 cfu/g *V. cholerae* O1 from naturally contaminated oysters, Blake *et al.*, (1980) found *V. cholerae* O1 to survive boiling for up to 8 minutes and steaming for up to 25 minutes in crabs inoculated by being placed in seawater containing 10^4 cells/ml. A 'cooked' appearance was not a satisfactory indicator of product safety.

The much lower level of heat resistance of *V. parahaemolyticus* means that traditional processes are often adequate. The commercial practice of heat shocking oysters in boiling water to facilitate opening reduced counts of *V. parahaemolyticus* and other non *V. cholerae* vibrios to 'undetectable' levels (Hackney *et al.*, 1980). However, steaming to facilitate opening is not adequate to ensure safety.

The isolation of *V. hollisae* from fried fish in the absence of evidence of post-process contamination (Lowry *et al.*, 1986) suggests the possibility of a higher level of heat resistance than other species.

V. parahaemolyticus is particularly sensitive to temperatures below the lower limit for growth, and numbers may fall during refrigerated storage at temperatures between 0 and 5°C. At these temperatures, a fall in numbers may be due not to the direct effect of temperature but to the loss of ecological niches to organisms such as *Aeromonas*. Death occurs more rapidly at temperatures below 0°C, although NaCl exerts a certain protective effect.

Low-temperature storage of seafood has been proposed as a means of eliminating *V. parahaemolyticus*. This method is not of sufficient reliability for

Table 71. Thermal resistance of *Vibrio cholerae* and *Vibrio parahaemolyticus*.

	Temperature (°C)	D value (mins)
V. cholerae[1]	49	8.15
	60	2.65
	71	0.3
V. parahaemolyticus[2]	49	0.7
	51	0.54
	53	0.31
	55	0.24

[1] Shultz *et al.* (1984); heating menstruum crab slurry.
[2] Delmore and Crisley (1979); heating menstruum clam slurry.

commercial application, particularly as imponderables remain concerning the most suitable storage temperature.

V. cholerae is more robust than *V. parahaemolyticus* with respect to low temperatures and, at temperatures above freezing, will persist beyond the point of visible spoilage (Riley and Hackney, 1985), while evidence concerning *V. vulnificus* is contradictory. The organism has been reported to be unable to survive in whole oysters at temperatures close to freezing point (Oliver, 1981), but in a controlled study the organism survived for up to two weeks in oyster shellstock at 2°C (Kaysner *et al.*, 1989). Discrepancies may be due to the organism becoming non-culturable rather than non-viable, Oliver and Wanucho (1990) showing that while cells rapidly lose culturability at 5 or 10°C, a significant proportion remain viable and metabolically active.

Irradiation

In general, *Vibrio* species are sensitive to ionising irradiation, and doses of 1 Kgy have been shown to be lethal to *V. cholerae*, *V. parahaemolyticus* and *V. vulnificus* in shellstock oysters, with *V. cholerae* being slightly more sensitive than the other species (Kilgen *et al.*, 1987). Other enteropathogenic species were thought to have a similar level of resistance to *V. parahaemolyticus* and *V. vulnificus*. Similar levels of sensitivity were recorded for *V. parahaemolyticus* in shrimp by Bandekar *et al.* (1987), who suggested that a dose of 1 Kgy would form the basis of an effective commercial process with no organoleptic side effects. A commercial process was also proposed for the elimination of *V. cholerae* from frog legs using a dose of 0.5 to 1 Kgy (Sang *et al.*, 1987).

Ultraviolet irradiation is used indirectly in the control of vibrios by purification of water in depuration plants.

pH value

Although *V. parahaemolyticus* has been reported as growing at pH 4.8 (ICMSF, 1980), vibrios are generally sensitive to pH values below 7.0. There is a strong interaction with temperature and, at temperatures approaching the minimum for growth, the lower limiting pH value in fish is likely to be in the region of 7.0 or above.

There is also a strong interaction with NaCl concentration, the lower limiting pH value tending to rise as the NaCl concentration increases.

All species of enteropathogenic vibrios grow well at alkaline pH values, the upper limiting values being pH 10-11.

Composition of atmosphere

Controlled atmosphere packaging is widely used for extending the storage life of fish and other seafood. Contrary to unsubstantiated and irresponsible claims made in some sectors of the packaging industry, there is no evidence that the gas mixtures used significantly reduce the growth rate of enteropathogenic vibrios, let alone inactivate the organisms.

NaCl concentration

All enteropathogenic vibrios require NaCl for growth, but the limiting levels, while variable, are low and the requirement is met readily by the concentrations naturally present in fish, etc.

Most species grow well over the range between the lower and upper growth limits, growth reducing markedly towards the upper limit. The optimum for growth in pure culture is *ca* 3%, but may be higher in mixed cultures due to the inhibition of competing micro-organisms at higher concentrations.

The upper growth limit varies markedly from species to species. For example, *V. cholerae* and *V. mimicus* are unable to grow in concentrations of 6%, while *V. fluvialis* and *V. parahaemolyticus* are able to grow at 8% and, in some cases, at 10%. In practice, the upper limit is of little relevance since all species are able to grow in marine foods with the exception of acid-preserved or heavily salted and dried fish. *V. hollisae* is able to survive in dried salt fish (Rank *et al.*, 1988).

Competition from other micro-organisms

Enteropathogenic species of *Vibrio* are generally poor competitors, and are likely to be overgrown by a range of organisms, including non-pathogenic vibrios.

Laboratory methods

During any laboratory work with enteropathogenic vibrios, it is necessary to be aware of two factors. The first is the organism's extreme sensitivity to cold, which can lead to inactivation of the organism through cold shock or through storage of samples etc at low temperatures. The second is the osmotic fragility of many species if they are transferred to a medium of low ionic strength, and the effect of the NaCl concentration of the medium on their growth and biochemical reactions. All people working with enteropathogenic vibrios should be aware of these factors, and the measures required to avoid problems should be incorporated into the laboratory working procedures.

Conventional cultural isolation techniques

Enrichment

In many laboratories, the use of relatively non-selective enrichment media, such as alkaline peptone water + 3% NaCl (APWS) or trypticase–soy broth + 3% NaCl (TSBS), is preferred. These have the advantage of being easy to prepare and suitable for any known enteropathogenic species. APWS is the medium of choice for *V. cholerae* (Baumann *et al.*, 1984; Twedt *et al.*, 1984; Hoover, 1985). Incubation should be at 37°C for 6 to 8 hours to prevent loss of selectivity and overgrowth of *V. cholerae* by other bacteria. This time period may cause logistical problems in some laboratories, in which case an incubation temperature of 20°C maintains the selective effect for a longer period (Taylor *et al.*, 1982). More recently, the use of an incubation temperature of 42°C has been found to give significantly higher recoveries of *V. cholerae* and to minimise problems with competing micro-organisms (De Paola *et al.*, 1988).

The use of more strongly selective enrichment media reduces problems caused by competing bacteria, but such media are usually suitable only for isolation of a specific species. An exception is thiosulphate–citrate–bile–salts–sucrose broth (TCBSB; the broth equivalent of TCBS agar), which is suitable for all enteropathogenic vibrios except *V. hollisae*.

A considerable amount of attention has been paid to *V. parahaemolyticus*, and several enrichment broths are available for this species. Examples are glucose–salt–teepol broth (GSTB; Beuchat, 1977), salt–colistin broth (SCB; Sakazaki and Balows, 1981) and salt-polymyxin broth (SPB). GSTB, incubated overnight at 37°C, is most widely used, and while SCB and SPB are more selective, each may be inhibitory to some strains of *V. parahaemolyticus* (Richard and Lhullier, 1977; Oscroft, 1987).

The need to use strongly selective media for the enrichment of *V. parahaemolyticus* is not universally accepted, and APWS has been found to be most effective for the isolation of the organism from both oysters (Eyles *et al.*, 1985) and prawn homogenate (Oscroft, 1987). As with *V. cholerae*, it is necessary to use a short incubation period of 6 to 8 hours at 37°C to prevent overgrowth.

APWS is widely used as enrichment medium for *V. vulnificus*, although there is some doubt concerning its efficiency (Oliver, 1989). An alternative broth medium, starch–gelatin–polymyxin broth, has been found effective in an environmental study, but requires further evaluation.

Recovery of *V. vulnificus* from oysters may present difficulties due to a combined effect of cold shock and an inhibitor present in the oysters (Joseph, 1982). This may be overcome by careful tempering of diluents and the enrichment medium (good practice at all times), and by increasing the ratio of enrichment medium to sample.

Selective plating media

The thiosulphate–citrate–bile–salts–sucrose agar (TCBS) of Kobayashi *et al.* (1963) is widely used for all enteropathogenic vibrios except *V. hollisae*, and is accepted as the medium of choice for *V. cholerae* and *V. parahaemolyticus*. The use of the medium is illustrated in **Figures 137** & **138**.

The use of TCBS has become so generally accepted that non-specialists may never even consider other media. The use of a second selective medium is, as ever, considered good practice, and *Vibrio* agar (Tamura *et al.*, 1972) was recommended by Sakazaki and Balows (1981) for use in parallel with TCBS. Other workers, prefer to use a non-selective second medium, such as gelatin agar or gelatin–phosphate–salt agar (GPSA; Doyle, 1986).

A number of media other than TCBS have been devised for *V. parahaemolyticus*, including the eponymous *Vibrio parahaemolyticus* agar (De *et al.*, 1977) and bromothymol blue salt teepol agar (Sakazaki, 1965). None of these media offer any clear advantage over TCBS and none are available as complete dehydrated media.

Although TCBS is used for isolation of *V. vulnificus*, there have been a number of attempts to devise a selective medium to differentiate between this organism and other sucrose non-fermenting vibrios such as

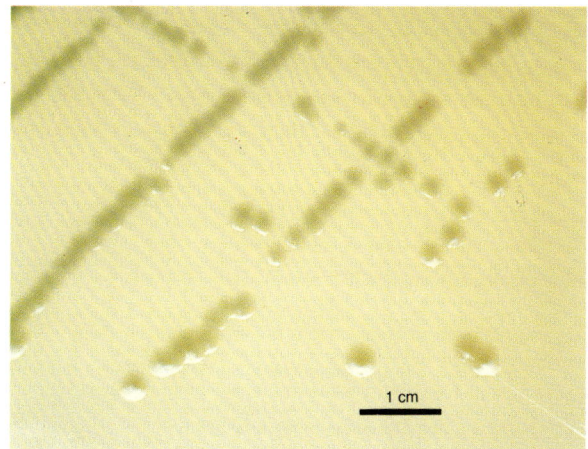

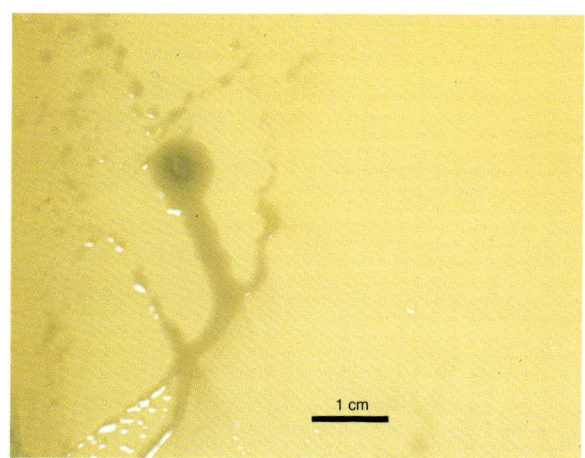

137 & 138. Isolation of enteropathogenic vibrios on TCBS medium. Species such as *V. parahaemolyticus* (**137**) which do not ferment sucrose are readily recognised by the dull green colonies. In contrast, sucrose fermenting species such as *V. cholerae* produce yellow colonies which cannot be differentiated from those of non-enteropathogenic vibrios such as *V. alginolyticus*. Problems due to such species may, however, be reduced by enrichment at 42°C where possible.

TCBS is recognised as being unsuitable for *V. hollisae*, and occasional strains of species other than *V. cholerae* and *V. parahaemolyticus* may also grow poorly. A strain of *V. vulnificus*, for example, failed to grow significantly after the initial streak where mucoid and ill-defined colonies developed (**138**). TCBS is not fully selective and *Aeromonas* can be a problem. Strains of *Proteus*, *Enterobacter* and *Enterococcus* may also grow on the medium and may, superficially, resemble vibrios. (*TCBS: 24h at 37°C*.)

V. parahaemolyticus. Such media include colistin–polymyxin–cellobiose agar (CPC), sodium dodecyl sulphate–polymyxin–sucrose agar (SPS), and *Vibrio vulnificus* agar (VV). Although VV appears to be of limited value, both SPS and CPC have considerable promise as a selective and differential medium for *V. vulnificus* (Oliver, 1989).

No selective medium exists for *V. hollisae* but sucrose–tellurite–teepol agar (STT; Chatterjee *et al.*, 1977a,b) may be used in lieu of the unsuitable TCBS.

Protocol for the isolation of Vibrio

A model protocol for the isolation of vibrios from foods is shown in **Flow diagram 18**.

Membrane filter techniques

A membrane filter technique was first described by Watkins *et al.*, (1976) for the direct enumeration of *V. parahaemolyticus*. This technique used a conventional membrane filter and was modified by Entis and Boleszczyk (1983). In this latter technique, a hydrophobic grid membrane filter (HGMF) was used, and the protocol incorporated a 4 to 5 hour resuscitation on non-selective agar at 35°C, followed by overnight

Flow diagram 18. Model protocol for the isolation of *Vibrio* from foods.

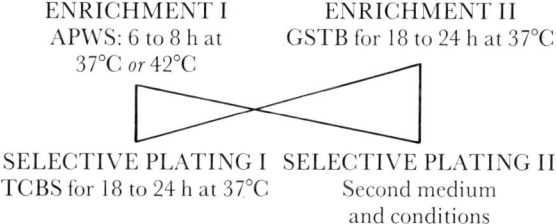

ENRICHMENT I
APWS: 6 to 8 h at
37°C *or* 42°C

ENRICHMENT II
GSTB for 18 to 24 h at 37°C

SELECTIVE PLATING I
TCBS for 18 to 24 h at 37°C

SELECTIVE PLATING II
Second medium
and conditions

Notes: 1 The use of two media for enrichment and selective plating may not be considered necessary for routine purposes.
2 TCBS is not suitable for *V. hollisae*.

incubation on a specially-formulated *Vibrio parahaemolyticus*–sucrose agar (VpS). No attempt was made to carry out confirmatory testing on colonies apart from the built-in, non-fermentation of sucrose. Entis and Boleszczyk were fully aware of the dangers of relying on a single diagnostic feature, but rightly pointed out that the situation is hardly unique in microbiology, and that such a reliance is acceptable in a screening test providing that users are aware of the limitations. Support for this method came later in a comparison of methods which showed it not only to be the better of two membrane filtration methods, but superior to most probable number techniques (De Paola *et al.*, 1988).

Rapid detection of *Vibrio*

V. cholerae

A fluorescent antibody technique for the detection and enumeration of *V. cholerae* in water has been developed (Brayton *et al.*, 1987). This technique is claimed to be more sensitive than MPN techniques, and is also able to detect viable, but non-culturable, cells. The suitability of the technique for foods is not known.

Genetic techniques have not been widely applied to *Vibrio*, but an oligonucleotide probe for detection of non-O1 *V. cholerae* carrying the gene for heat stable enterotoxin (NAG-ST) has been developed (Hoge *et al.*, 1989).

V. parahaemolyticus

A method for the rapid detection of *V. parahaemolyticus* by means of a fluorogenic assay for trypsin-like activity has been devised by Miwamoto *et al.* (1990). Samples are enriched in arabinose-gluconate medium for 6 to 8 hours at 37°C, centrifuged, and the trypsin-like activity of the microorganisms in the centrifugate is assayed using the fluorogenic substrate benzoyl-L-arginine-7-aminomethyl-coumarin. It is claimed that the assay can detect as few as 10 cells within 10 hours.

V. vulnificus

A DNA probe for the cytotoxin haemolysin gene of *V. vulnificus* has been developed by Morris *et al.* (1987). The method employs colony hybridisation and was used by Kaysner *et al.* (1987) in a study of *V. vulnificus* in estuarine waters on the west coast of the USA.

Identification

Quality control of media

Vibrios are particularly sensitive to minor changes in the composition of both isolation and biochemical test media. In practice, this usually leads to false negative reactions rather than to false positives. To avoid misidentification of isolates it is necessary to set up positive control cultures in parallel with cultures of unknown identity (Twedt, 1984). Strict quality control is of particular importance where media is 'home-made' from basic ingredients, or in laboratories where enteropathogenic *Vibrio* species are encountered only infrequently.

Differentiation from other genera

Vibrio may be differentiated from biochemically similar genera according to the criteria in **Table 66**, page 160. Conventional media, multi-test media or some types of miniaturised kits, may be used. Media used for the *Enterobacteriaceae* (supplemented with 3% NaCl) are generally suitable where conventional media are used. An andrade indicator should be used in sugar fermentation tests to overcome possible problems due to indicator reduction.

Triple sugar-iron medium and sulphide-indole-motility medium each provide a useful means of screening isolates although, as with the *Enterobacteriaceae*, the use of miniaturised kits is often preferred. Kits such as the API© systems, which can be supplemented with additional NaCl, are suitable although some difficulties in interpretation may be encountered. For example, *V. hollisae* produces a bright lime green colouration in the API tests for fermentation of glucose and arabinose.

Particular care should be taken in the application of the tests for 0/129 sensitivity (**Figures 139–141**) and Na$^+$ dependence. Sensitivity to 0/129 must be determined at 150µg, and at this level small zones of inhibition may be noted with some members of the *Enterobacteriaceae*. The test must be controlled carefully with known positive and negative strains.

Determination of Na$^+$ dependence is complicated by the fact that some species of enteropathogenic *Vibrio*, such as *V. cholerae*, grow well in standard laboratory media. Further, Na$^+$ ions are derived not only from added NaCl, but from ingredients such as peptones and from acids and alkalis used in the adjustment of pH value. Unless extremely rigorous steps are taken, the level of Na$^+$ ions in media with 'no added' NaCl can support good growth of some strains of vibrios. It is possible to determine a response to Na$^+$ by comparing growth in a medium with minimal levels, with that in a medium with optimal levels, but the test must be carefully controlled and supervised by experienced people.

Differentiation between species

The criteria for differentiation of enteropathogenic species of *Vibrio* are listed in **Table 67**, page 160. Particular care must be taken over the use of lactose fermentation in the identification of *V. vulnificus*. In the past, lactose fermentation has been considered to be of particular importance in the identification of this species, the organism having been referred to as 'lactose fermenting vibrio'. Lactose fermentation is now recognised as a common trait among marine and estuarine vibrios (West *et al.*, 1986) and only *ca* 20% of lactose fermenting vibrios from marine environ-

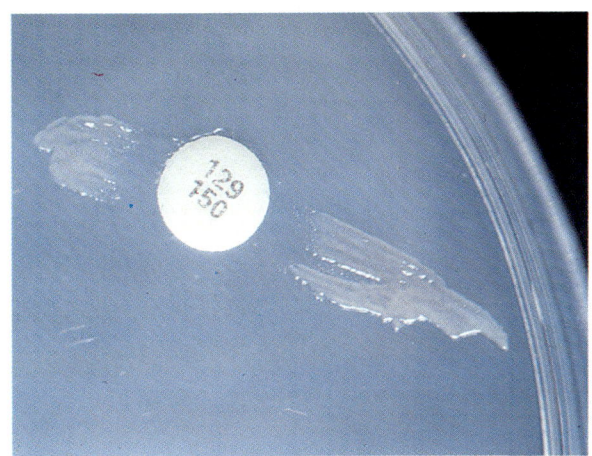

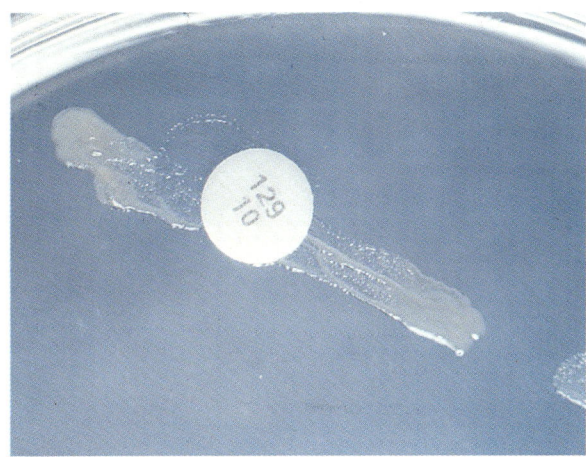

139–141. Sensitivity of *Vibrio* species to the vibriostat 0/129. Sensitivity to the vibriostat 0/129, 2,4-diamino-6,7-diisopropylpteridine phosphate, may be used both to differentiate *Vibrio* from other species and for differentiation of genera of *Vibrio*. All vibrios of clinical importance are inhibited by 0/129 at a concentration of 150 μg (**139**). *V. vulnificus*, like *V. cholerae*, is also inhibited to some extent at 10 μg (**140**), while *V. parahaemolyticus* and several other enteropathogenic species are resistant (**141**).

Sensitivity to 0/129 is not unique to *Vibrio*, *P. shigelloides* is also inhibited at 150 μg and, in some cases, at 10 μg, while occasional strains of members of the *Enterobacteriaceae* are inhibited at the higher concentration.

Although sensitivity to 0/129 is simple to determine using commercially available sensitivity discs (Unipath), sensitivity can be affected by several factors, and positive and negative control cultures must always be included. (*Tryptone-soy agar + 3% NaCl: 24 h at 37°C.*)

ments are identified with *V. vulnificus* (Oliver *et al.*, 1983). Further lactose fermentation is not a feature of wild-type *V. vulnificus*, but results from lactose fermenting mutants which emerge during laboratory culture (Baumann *et al.*, 1981). For these reasons, lactose fermentation should be regarded with caution as a diagnostic feature for *V. vulnificus*, and over-reliance may lead to the species being misidentified as *V. parahaemolyticus* (West *et al.*, 1986).

Either conventional media or suitable miniaturised kits may be used for species differentiation, although the API 20E© system is not suitable for identifying *V. mimicus* without additional tests, such as NaCl tolerance (Chowdbury, 1989). In either case, care should be taken in the interpretation of the arginine dihydrolase reaction, and it may be necessary to use a modification of the Thornley (1960) method (Baumann and Baumann, 1981). With some strains of *V. metschnikovii*, a positive reaction is due to a mechanism other than arginine dihydrolase (Baumann *et al.*, 1984). How-

ever, this is of no significance in identification.

A multi-test medium, VC medium, has been developed specifically for rapid, presumptive identification of *V. cholerae* from environmental sources. This medium simultaneously determines acid and gas production from glucose, fermentation of inositol or arabinose, arginine dihydrolase activity and the production of indole and hydrogen sulphide.

Rapid tests for V. cholerae

Two rather unusual tests have been described for rapid identification of *V. cholerae*. It is likely that tests of this nature will be superseded largely by sensitive and specific serological (see below) and possibly genetic tests. Neither of these tests requires sophisticated reagents, and each may therefore be of continuing utility where facilities are limited.

142–144. Rapid Visual Test for the differentiation of epidemic strains of _V. cholerae_ O1. The Rapid Visual Test (RVT) is based on the ability of lysates of epidemic strains of _V. cholerae_ O1 to reverse the discolouration of a redox indicator. The mechanism of the test is not known, but there appears to be no direct correlation between a positive reaction in the RVT and production of cholera toxin.

Use of the test involves lysis of cells with Triton X-100, followed by the addition of an assay solution containing nicotinamide adenine dinucleotide phosphate, glucose-6-phosphate and the redox indicator, dichlorophenol indophenol. In the first phase of the reaction, the redox indicator is decolourised from blue (**142**) to colourless (**143**) during incubation at room temperature for 30 min. At the end of this period, tubes are transferred to incubation at 5°C and held overnight. During the second phase, in positive tubes, the colour change is reversed (**144**) and the blue colouration reappears (Salles and Momen, 1981).

The 'string' test

The 'string' test (Smith, 1970) has been found to be a useful test for presumptive identification of both O1 and non-O1 serovars of _V. cholerae_. A single large colony from T_1N_1 agar is emulsified in a large drop of 5% sodium deoxycholate in 0.85% saline solution. _V. cholerae_ forms a mucoid mass which 'strings' when a loopful is lifted 2 to 3 cm.

The 'rapid visual' test

This test permits the differentiation of epidemic strains of _V. cholerae_ O1 from non-epidemic strains, and from _V. cholerae_ non-O1 strains (Salles and Momen, 1981). The rapid visual test is illustrated in **Figures 142–144**.

Serological confirmation of identity

Serological confirmation of _V. cholerae_ O1 is made simply using slide agglutination with polyvalent O1 antisera (Twedt, 1984). However, this technique is not suitable for direct application to colonies from commonly used isolation media such as TCBS. More recently, a co-agglutination test using sensitised cells of Cowan Group I _Staphylococcus aureus_ has been developed, which may be applied to colonies on primary isolation (Rayman _et al._, 1989) and which, when further refined, may obviate the need for biochemical testing.

The rapid progression of _V. vulnificus_ infections and the high mortality rate, mean that a rapid means of confirming the identity of the organism is highly desirable. This is not possible using biochemical testing, and work has been concentrated on a serological test based on species specific anti-flagellar anti-sera (Tassin _et al._, 1983). A microimmunodiffusion technique was devised by Nisibuchi and Seidler (1985), while Kaysner *et al.* (1987) found an agglutination technique a valuable aid to the recognition of *V. vulnificus* from the environment. The most promising development, uses anti-flagellar monoclonal antibodies in a co-agglutination assay with sensitised *Staph. aureus* cells or latex beads (Simonson and Siebeling, 1988). This test may also be used for the identification of *V. cholerae* and *V. mimicus*.

Determination of virulence

V. parahaemolyticus

The Kanagawa reaction remains the most widely used means of determining virulence. It is essential that the test is made strictly according to accepted methods (**Figure 145; Table 72**). In the past, the test has often been read incorrectly or modified, and some published results are of limited value only (Taylor *et al.*, 1982).

Serological methods have been proposed to detect the haemolysin. The simplest is a modified ELEK test where the haemolysin is detected by the formation of lines of precipitation between the colonies of *V. parahaemolyticus* and anti-haemolysin sera incorporated in the agar. This test was no more sensitive than the classic procedure using Wagatsuma agar but is more consistent and not affected by variations in the blood used (Twedt, 1989). More recently, highly sensitive

145. The Kanagawa reaction of *V. parahaemolyticus*. The Kanagawa reaction is valid only when correct procedures (**Table 72**) are followed. Clear zones of β-haemolysis around the culture under test indicate a positive reaction. Haemolysis, or lack of haemolysis, on standard blood agars are *not* indicative of the Kanagawa reaction (*cf.* **126** & **127**). (*Wagatsuma agar: 20h at 37°C.*)

ELISA and reverse passive latex agglutination assays have been developed that are able to detect nanogram quantities of haemolysin per ml.

The most sensitive means of detecting haemolysin related virulence is a DNA probe for the chromosomal gene controlling synthesis. Use of this probe has so far been restricted to research purposes, but work to date indicates that the correlation is better between the gene probe and serological assays than between the probe and the classical Kanagawa assay.

V. vulnificus

In most laboratories, colonial morphology is the only means of determining virulence of *V. vulnificus*. Virulent strains carry a heavy surface layer of the anti-phagocytic surface antigen, and have an opaque colonial morphology, while isogenic mutants, which lack the material, have translucent colonies. Virulent strains of *V. vulnificus* produce, therefore, a mixture of translucent and opaque colonies, while avirulent strains produce only translucent colonies (Oliver, 1989).

Colonial morphology also correlates with the ability to use transferrin-bound iron, this property being restricted to strains producing opaque colonies.

Table 72. Determination of the Kanagawa reaction (Taylor *et al.*, 1982).

Wagatsuma medium

Yeast extract	5g	Trypticase	10g
Sodium chloride	70g	Mannitol	5g
Crystal violet	0.001g	Dist. H$_2$O	1l

pH 7.5

Add 10ml of a 20% suspension of fresh human red blood cells to 100ml of molten base at 50°C.

Procedure

1 Spot or stab inoculate plates of Wagatsuma medium with overnight broth cultures. Incorporate a positive control with the test cultures.
2 Read after 18 to 24 hours incubation at 37°C. Incubation must not be continued beyond 24 hours.
3 Clear zones of β-haemolysis indicate positive reactions. Discoloration, or α-haemolysis must not be recorded as positive.

9. *Aeromonas*

Introduction

The genus *Aeromonas* comprises Gram-negative, catalase-positive, oxidase-positive, rod-shaped bacteria that are capable of both respiratory and fermentative metabolism of glucose (**Figures 146 & 147**). The genus has been classified with the family *Vibrionaceae* (Baumann and Schubert, 1984), but more recently Colwell *et al.*, (1986), noting the low level of DNA relatedness of *Vibrio* and *Aeromonas*, have proposed a new family, the *Aeromonadaceae*.

Aeromonas has a distinctive cellular morphology in that there is considerable variation in the shape and size of cells, with some strains appearing as short rods, while others appear thin and filamentous. Curved cells may also be present, but these are readily distinguished from those of *Vibrio*. Motility is by a single, unsheathed flagellum, although many strains produce lateral flagella in young culture.

The presence of glucose in a growth medium is associated with a suicide phenomenon (**Figure 148**). The basis of this phenomenon is the suppression of the tricarboxylic acid cycle by supplied glucose, which leads to accumulation of acetate and cell death (Namdari and Cabelli, 1989).

The genus *Aeromonas* contains two well separated groups (Popoff, 1984). The first group consists of a single psychrotrophic, non-motile species, *A. salmonicida*, which is a strict parasite under natural conditions. *A. salmonicida* is the causative organism of the economically important furunculosis disease of salmon, and trout, but is not a pathogen of man.

The second group consists of mesophilic motile strains. Three species were described by Popoff *et al.*, (1981): *A. hydrophila, A. sobria* and *A. caviae*. Subsequent DNA hybridisation studies have shown there to be at least 11 species of mesophilic aeromonads, seven of which have taxonomic standing. Most human patho-

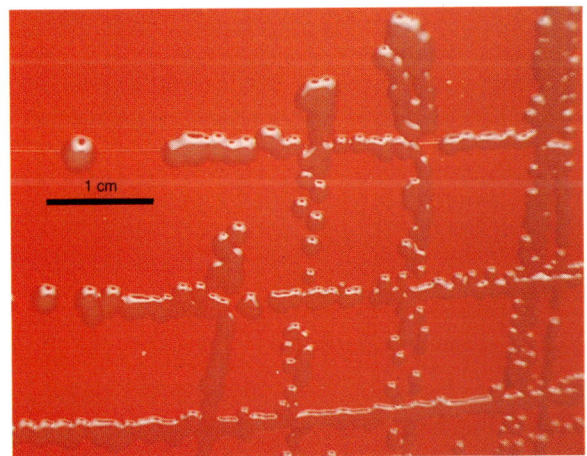

146 & 147. Growth of *A. hydrophila* and *A. sobria* on blood agar. Motile strains of *Aeromonas* grow well on a wide range of media including nutrient and blood agars. Weak haemolysis is exhibited by *A. hydrophila* (**146**) but not *A. sobria* (**147**). Some strains of each species produce a strong odour, but this character is variable. (*Blood agar: 24 h at 28°C.*)

148. The suicide phenomenon in *A. sobria*. 'Suicide' is detected by the sedimentation of cells following growth in glucose containing medium. The phenomenon is most common among strains of *A. caviae* and rare among strains of *A. hydrophila*.

genic strains fall into hybridisation groups 1 (*A. hydrophila*), 4 (*A. caviae*) and 8 (*A. sobria*). Other species have also been associated with diarrhoeal disease, including *A. veronii* (hybridisation group X) and *A. schubertii* (Hickman-Brenner *et al.*, 1988).

Although motile aeromonads may be clearly differentiated into groups by DNA hybridisation techniques, there is no simple means of differentiating these groups by biochemical reactions. Furthermore, species other than *A. hydrophila* may only be differentiated by means of a large number of tests, and for this reason many laboratories continue to group all motile aeromonads in the general category *A. hydrophila* group or complex (Hickman-Brenner *et al.*, 1987).

A. hydrophila, and other motile aeromonads, are recognised pathogens of reptiles (Marcus, 1971), amphibians (Shotts *et al.*, 1972), fish (Trust and Sparrow, 1974; Heuschmann-Brunner, 1978) and cattle (Wohlgemuth *et al.*, 1972). In humans the organisms are recognised as the cause of disseminating infections in the immunocompromised (Davis *et al.*, 1978; Wolff *et al.*, 1980; Ellison and Mostow, 1984), and are also wound pathogens (Davis *et al.*, 1978). In recent years, *Aeromonas* has received increasing attention as an agent of foodborne diarrhoeal disease in otherwise healthy people (Champsaur *et al.*, 1982; Gracey *et al.*, 1982; Janda *et al.*, 1983; Agger *et al.*, 1985; Palumbo *et al.*, 1985a). However, the role of *Aeromonas* as an enteric pathogen is not fully clarified, and the genus is best described as a 'putative enteropathogen'.

The basic difficulty in relating *Aeromonas* to diarrhoeal disease lies in the fact that while it is fairly easy to establish statistical and epidemiological links (Holmberg, 1986; Morgan and Ward, 1988), feeding studies with human volunteers have failed to confirm a pathogenic role. Furthermore, there are only a few cases where a causal link between a food contaminated with *Aeromonas* and cases of enteric disease exists.

The results of seven epidemiological investigations, each made in a different country, into the relationship between *Aeromonas* and diarrhoeal disease has been reviewed by Morgan and Ward (1988). A statistically significant relationship was established in six, the exception being an Italian study (Figura *et al.*, 1986). However, in work not covered by the review, Millership *et al.* (1983) also failed to establish a relationship between *Aeromonas* and diarrhoea in

Table 73. Incidence of *Aeromonas* and other infective agents in childhood gastroenteritis (Burke *et al.*, 1983).

Agent	Isolation rates	
	% diarrhoeaic (n = 975)	% non-diarrhoeaic (n = 975)
E. coli (ETEC)	1.9	0.1
Salmonella	5.7	0.9
Shigella	1.3	0.0
Campylobacter	7.4	0.6
Aeromonas	10.8	0.7
Rotavirus	12.7	11.1

England, while in Thailand, a statistical relationship existed among visitors but not among native Thais (Pitarangsi, 1982). Of the studies indicating a positive relationship between *Aeromonas* and diarrhoea, the most persuasive was a very large, age-matched control study (Burke *et al.*, 1983), which found that *Aeromonas* was the most commonly isolated bacterial pathogen from children suffering diarrhoea (**Table 73**).

Although the results of feeding experiments appear to contradict the epidemiological evidence, Morgan and Wood (1988) believe that the administration of the challenge culture in bicarbonate solution, a standard procedure with other enteropathogens, is inappropriate for *Aeromonas*. It should also be appreciated that the recovery of any micro-organism from diarrhoeal stools does not prove a causative role and that *Aeromonas*, like other organisms, may have a purely commensal role. It is likely that confusion arises from the fact that some *Aeromonas* strains have a primary causative role in diarrhoea while others are, indeed, commensals (Stelma, 1989). However, the paradigm is not complete and, while the contention of several authors that the recognition of more virulence determinants will permit the enteropathogenicity of the organism to be fully understood is almost certainly correct, the status of *Aeromonas* as merely a putative pathogen cannot currently be challenged.

Isolation of *Aeromonas*

Isolation of *Aeromonas* from foods may either be by direct plating onto a selective medium, or by a two-stage process involving selective enrichment followed by streaking onto selective agar. Tryptone–soy broth supplemented with ampicillin provides optimal recovery (Okrend *et al.*, 1987). This medium may also be used in a most probable number determination of *Aeromonas* (von Gravenitz and Bucher, 1983).

A wide range of media have been used as selective plating media for *Aeromonas*. These include media designed for members of the *Enterobacteriaceae*, media containing ampicillin as a selective agent such as starch-ampicillin medium and CIN medium, originally developed for *Yersinia enterocolitica*.

The isolation of Aeromonas *is discussed in greater detail with respect to choice of media and laboratory practice in* **Laboratory Methods**, *pages 195–199.*

Taxonomy of *Aeromonas*

Differentiation from other genera

Aeromonas may be differentiated from other genera by the criteria listed in **Table 74**. Either conventional test media, or suitable miniaturised identification kits such as API 20E©, may be used. A multi-test medium has also been devised (Kaper *et al.*, 1979). When the API© system is used, additional tests, such as sensitivity to the vibriostat 0/129, may be required to distinguish *Aeromonas* species from *V. fluvialis* and *A. veronii* from *V. cholerae* and *V. mimicus*.

Differentiation between species

Differentiation between species may be made according to the criteria listed in **Table 75**. The suitability of miniaturised systems varies, the API 20NE© and ATB 32E© are able to differentiate *A. hydrophila* from *A. sobria* but not from *A. caviae*, while the API 20E cannot distinguish between *A. hydrophila* and the other species. Even where full identification is not possible using the database, these systems are convenient means of determining many reactions. The GN Microplate©, a novel identification system based on carbon substrate oxidation profiles, has been found to give good discrimination between the species of *Aeromonas* (Carnahan *et al.*, 1989).

A number of workers have sought a simplified means of differentiating between *Aeromonas* species. Tests for a CAMP-like factor have been used to differentiate between *A. hydrophila*, *A. caviae* and *A. sobria* (Figura and Guglielmetti, 1987), while cephalothin sensitivity has been proposed as a marker for *A. sobria*. Neither of these approaches has been fully successful, and subsequently Namdari and Bottone (1989) have devised a simple means of differentiation based on the 'suicide phenomenon' (see page 200), aerogenicity and aesculin hydrolysis. The pyrazinamidase reaction (*cf Y. enterocolitica*, page 135) also has a role in distinguishing *A. sobria* (-ve) from *A. hydrophila* and *A. caviae* (Carnahan *et al.*, 1990).

Taxonomy of Aeromonas *is discussed in greater detail with respect to choice of method and laboratory practice in* **Laboratory Methods**, *pages 199–200.*

Table 74. Differentiation of *Aeromonas* from other genera.

	Aeromonas	*Pseudomonas*	*Shigella*
Oxidase	+	+	−
Fermentation of glucose	+	−	+
Gas from glucose	+/−	−	−
Simmons' citrate	−/+	+/−	−
Acid from mannitol	+	−	+/−
Motility	+	+	−

Note: Aeromonas may be differentiated from Vibrio and Plesiomonas according to the criteria listed in Chapter 8, Vibrio, page 160.

Table 75. Differentiation between species of *Aeromonas*.

	Aeromonas			
	hydrophila	*caviae*	*sobria*	*veronii*
Ornithine decarboxylase	−	−	−	+
Aesculin hydrolysis	+	+	−	+
Growth in KCN	+	+	−	+/−
Gas from glucose	+	−	+	+
Voges–Proskauer	+	−	+/−	+
H₂S from cysteine	+	−	+	−
Utilisation of L-histidine and L-arginine	+	+	+	−
L-arabinose	+	+	−	−
Acid from salicin	+	+	−	+

Notes:
1 Anaerogenic strains may be present in gas-producing species.
2 The Voges–Proskauer reaction may be temperature dependent.

Typing of *Aeromonas*

In comparison with many other enteropathogenic bacteria, the typing of *Aeromonas* is in its infancy. It is considered that the development of a simple, suitably discriminatory typing scheme would be of very considerable value in enhancing the understanding of the epidemiology of *Aeromonas* and its role in human disease.

Serotyping

Although serotyping has been applied to *A. hydrophila* pathogenic to fish, relatively little attention has been paid to human pathogenic strains. A serotyping system for *A. hydrophila*, *A. sobria* and *A. caviae* based on lipopolysaccharide antigens was developed by Fricker (1987) and, using 16 antisera, was able to type 46% of strains isolated from human faeces. Subsequently, an extended scheme has been developed (Thomas *et al.*, 1990), which recognises 52 provisional new serogroups, and which should permit the typing of mesophilic aeromonads from a wide range of sources.

Other typing schemes

Difficulties with serotyping led Elbashir and Millership (1989) to devise a typing scheme based on the haemagglutinating activity of *Aeromonas* from different sources, but this scheme was of little value. A more promising approach was the use of sodium dodecyl sulphate–polyacrylamide gel electrophoresis to produce protein fingerprints. Radiolabelling was used originally to visualise the proteins (Stephenson *et al.*, 1987), but silver staining was found to be quicker and more convenient (Millership and Want, 1989). Each strain of *Aeromonas* tested appeared to have a unique 'fingerprint' and the potential value in epidemiological studies is considerable. Protein fingerprinting is not suitable for use in non-specialist laboratories even where the use of radioisotopes is avoided.

Molecular typing methods are now being more widely applied to *Aeromonas* and both DNA restriction endonuclease analysis and polyacrylamide gel electrophoresis of cell envelopes have been used (Kuijper *et al.*, 1989).

Symptoms of *Aeromonas* infections

Pathogenic species of *Aeromonas* resemble *V. vulnificus* in that while symptoms in healthy people usually involve gastroenteritis, disseminating infections are the norm among the immunocompromised and other people of high susceptibility.

Gastrointestinal infections are of two distinct types. The first type, which accounts for *ca* 75% of cases, is a cholera-like illness characterised by watery stools with a mild or absent fever. There may be vomiting in children aged less than two years and, in patients of any age, diarrhoea may be accompanied by abdominal pain or cramps. The second type of infection is dysentery-like, characterised by bloody, mucoid stools. Vomiting is rare, but there may be abdominal pain.

Gastroenteritis due to *Aeromonas* species is usually mild and self-limiting (Holmberg and Farmer, 1984) but severe and life-threatening cases of both types of infection have been reported (Rhamon and Willoughby, 1980; Champsaur *et al.*, 1982). In a severe case of cholera-like infection, the symptoms resemble those of *cholera gravis*, and differential diagnosis is not possible in the absence of the bacteriological examination of stools. Such infections may be fatal.

The duration of symptoms in severe cases of dysentery-like infection may be prolonged for a month or more, although complete recovery is expected after appropriate antibiotic therapy.

Aeromonas can produce a wide range of extraintestinal infections in susceptible persons. Septicaemia and meningitis are most common (Davis *et al.*, 1978; Ellison and Mostow, 1984), and are often accompanied by hypotension, abdominal pain and, less commonly, nausea, vomiting and skin disturbances. Other, unusual manifestations of extraintestinal *Aeromonas* infections include eye infections, pneumonia, urinary tract infections, osteomyelitis (Turnbull *et al.*, 1984) and endocarditis (Morgan and Ward, 1988). In some of these cases, the causative role of *Aeromonas* has not been fully established.

The mortality rate due to extraintestinal *Aeromonas* infections in the immunocompromised may exceed 60% (Abeyta and Wekell, 1988).

The route of infection in disseminating infections is probably via the gastrointestinal tract following the ingestion of contaminated food (Harris *et al.*, 1985; Beebe, 1986).

Persons at particular risk from *Aeromonas* infection

Susceptibility

Among healthy people children are at greatest risk (Travis and Washington, 1986; Megraud, 1986). Most cases occur between six months and two years of age and the frequency falls markedly among children older than five years. In adults, *Aeromonas* infections are most common in people over 60 years of age.

Formula fed children may be at enhanced risk of infection with *A. caviae*, this organism being favoured by the elevated intestinal pH value (greater than 7.5) found in such children (Namdari and Bottone, 1990).

Disseminating infection is largely a disease of the immunocompromised, particularly in individuals suffering from leukaemia or cirrhosis. Sickle-cell anaemia and haemodialysis have also been associated with septicaemia due to *Aeromonas*. Males account for over 80% of people with underlying cancer who suffer *Aeromonas* infections (Harris *et al.*, 1985).

Exposure

In common with many other Gram-negative entero-pathogens, illness due to *Aeromonas* is more common where general standards of hygiene are poor. *Aeromonas* has been associated with travellers' diarrhoea in Asia (Echeverria *et al.*, 1984; Gracey *et al.*, 1984 and may also be a cause of the syndrome in Latin America and Africa (Black, 1986).

Treatment of *Aeromonas* infections

Most *Aeromonas* infections are self-limiting, and specific treatment is not required although rehydration and supportive therapy may be necessary where diarrhoea is severe. Anti-microbial therapy is indicated where symptoms are prolonged and in systemic infections. The drugs of choice are chloramphenicol, tetracycline and cotrimoxazole (Janda and Duffey, 1988). Environmental isolates of *Aeromonas* rarely show significant drug resistance, but human isolates are not infrequently resistant to chloramphenicol and tetracycline.

Mechanisms of pathogenicity

Although there has been considerable interest in the virulence factors of *Aeromonas*, there remain many aspects of the pathogenicity of the organism which are unknown, or only poorly understood.

Genetic control of virulence

There is no evidence of an involvement of plasmids in the control of virulence. Production of the cholera toxin-like cytotonic enterotoxin is known to be under chromosomal control, but firm information is lacking for other virulence factors (Chakraborty *et al.*, 1984).

Temperature-dependent alterations in growth kinetics and protein profiles have been reported (Shattner *et al.*, 1988), but the significance is uncertain.

Adherence

Although mannose-resistant haemagglutination, which is often considered to be a reliable indicator of the presence of pili (Archer and Kvenberg, 1988), does not appear to correlate with diarrhoea and adherence/colonisation in *Aeromonas* (Stelma, 1988), a role for pili has been proposed. Direct evidence has been presented by Corrello *et al.*, (1988), who were able to demonstrate an association between diarrhoea and high level adherence. Type I pili are probably involved.

Systemic spread

Significant systemic spread usually occurs only in the immunocompromised. Very little is known of the pathogenicity of such infections (Stelma, 1988), but it is possible to demonstrate marked differences in the behaviour of the three species, *A. sobria*, *A. hydrophila* and *A. caviae*. *A. sobria* is considered to be highly virulent with respect to bacteraemia; *A. hydrophila* to be of intermediate virulence; and *A. caviae* to be avirulent and to be involved only in rare polymicrobic infections (Janda *et al.*, 1983).

Despite these observed differences, it has not been possible to correlate virulence in disseminating infections with enterotoxin production, haemolysis, cytotoxin production or resistance to normal serum (Janda *et al.*, 1984) and the only factor that may be related to differences between species is the generally greater invasiveness of *A. sobria* (Watson *et al.*, 1985). Later work with *A. hydrophila* has raised the possibility that an extra surface protein layer, the S-layer, is involved in systemic infections (Dooley *et al.*, 1988; Murray *et al.*, 1988). This layer may also be involved in resistance to normal serum.

Effect of iron deprivation

In addition to enterobactin production by *A. hydrophila*, a novel siderophore amonabactin is produced by strains of the three main species. Amonabactin is a phenolate siderophore, which exists in two biologically-active forms containing either tryptophane or phenylalanine (Borghcathi *et al.*, 1989). Amonabactin is produced by 76% of *A. hydrophila* strains, 19% of *A. sobria* strains and all three strains of *A. caviae* tested.

Toxins

Species of *Aeromonas* produce a wide range of putative toxins, but the relationship of these products, either singly or in combination, to pathogenicity in man remains largely unresolved. There is no evidence that toxins pre-formed in foods play any role, and while it is possible that toxins are produced *in-vivo*, it has been postulated that their role in the pathogenicity of *Aeromonas* differs from that in other organisms, and that rather than being direct causes of enteritis, they facilitate the establishment of the organism in the intestine and aid adherence and/or invasion (Todd *et al.*, 1989).

A further study which may be of considerable importance to the understanding of *Aeromonas* pathogenicity is that of Knochel (1989). This showed that only 9% of aeromonads from 'cold' sources (refrigerated foods, an environment in a cool climate) were able to produce large quantities of haemolysins during growth at 37°C, compared with 65% from 'warm' sources (clinical samples, warm water aquaculture). It is not known whether production of other toxins is similarly affected, but it seems likely that isolates from low temperature environments have become adapted to, or selected for, life at low temperatures, and are thus less likely to be pathogenic or even to be able to grow at 37°C (Knochel, 1990).

Enterotoxins

The situation with respect to the enterotoxins of *Aeromonas* can be confusing, and care must be taken to ensure that correct nomenclature is used.

Cytotonic enterotoxin

A cytotonic enterotoxin produced by *Aeromonas* was first described fully by Ljungh *et al.* (1981, 1982). The toxin is of MW 50,000 and is stable to heating for 10 minutes at 56°C. The toxin was considered initially to be antigenically distinct from cholera toxin (Ljungh *et al.*, 1982; Chakraborty *et al.*, 1984), but this contention was not supported by Potomski *et al.*, (1987a,b), who described a cytotonic enterotoxin which cross-reacted with anti-CT serum. The relationship between the *Aeromonas* cytotonic enterotoxin and cholera toxin was defined by use of a synthetic oligonucleotide probe, to show relatedness between the two toxins (Schultz and McCardell, 1988).

Evidence currently available suggests that the cytotonic enterotoxin fulfils requirements for a role in diarrhoeal disease. However, the toxin has only been isolated from a low percentage of *Aeromonas* strains (Stelma, 1988).

A further 44 Kda cytotonic enterotoxin was described by Chopra and Houston (1989). Although this enterotoxin was not cross-reactive with cholera toxin, it induced elevated levels of cyclic AMP in chinese hamster ovary cells, and thus may act by a similar mechanism to cholera toxin and *Escherichia coli* labile toxin.

Cytotoxic enterotoxin

Indirect evidence for the production of a cytotoxic enterotoxin was presented by Cumbernatch *et al.* (1979) and Johnson and Lior (1981). The first direct evidence was that of Asao *et al.* (1984), who showed that the toxin to be a protein of MW 50,000, which was inactivated by heating for 5 minutes at 56°C. There was a strong correlation between the cytotoxin and haemolysin activity (Burke *et al.*, 1983), the haemolysin subsequently being shown to be a β-haemolysin.

Although some workers, such as Stelma *et al.* (1988), contend that the β-haemolysin alone can cause diarrhoea, others have found no correlation between cytotoxin production and gastroenteritis.

The toxin is unrelated to either cholera toxin (Timmis *et al.*, 1984) or Shiga toxin (Kindschuch *et al.*, 1987).

Cholera toxin–cross–reactive cytolytic enterotoxin (Aerolysin)

A small number of strains of *A. hydrophila* produce a cytolytic enterotoxin that cross-reacts with anti-CT (Shimoda *et al.*, 1984; Campbell and Houston, 1985; Honda *et al.*, 1985). The toxin does not have any cholera toxin-like activity, and it has been suggested that the serological relatedness may be weak, and to involve only one common antigenic site (Stelma, 1988).

Aerolysin is a β-haemolysin and consists of a single polypeptide of MW 52,000 that possesses haemolytic, enterotoxic and cytotoxic activity (Rose *et al.*, 1987; 1989a,b). The toxic mechanism is similar to that of the α-toxin of *Staphylococcus aureus*. Two precursor forms bind to the eukaryotic cells and aggregate to form holes of *ca* 3 nm diameter, which lead to destruction of the membrane permeability barrier and osmotic lysis (Howard *et al.*, 1987).

Although aerolysins produced by different isolates are all biologically similar, significant chemical and immunological differences may exist (Rose *et al.*, 1989a). In all cases the cytolytic and enterotoxic activities are likely to contribute significantly to the pathogenesis of *Aeromonas* infections (Rose *et al.*, 1989b).

Proteases

A. hydrophila produces two endoproteases, proteinase I and II, as well as an aminopeptidase (Pansone *et al.*, 1986). A role as extracellular virulence factors has been proposed (Ljungh and Wadstrom, 1983), but proteases, while possibly contributing to pathogenicity, are unlikely to play a major role.

Sodium channel inhibitor

A tetrodotoxin-like sodium channel inhibitor has been isolated from strains of *A. hydrophila* (Tamplin *et al.*, 1987). It has been postulated that this may have an important, and hitherto unrecognised, role in the pathogenicity of *Aeromonas* infections (Stelma, 1988)

but, as with *Vibrio* (see page 169), the full significance of such compounds in bacteria is not known.

Determination of virulence

The determination of virulence in aeromonads is usually by animal testing, although the cytotonic enterotoxin may be detected by most methods used for the detection of cholera toxin or *E. coli* LT. A number of phenotypic characters have been proposed as markers of enteropathogenicity. The most important of these are haemolysis, lysine decarboxylase activity (**Figure 149**), voges-proskauer reaction (Kirov *et al.*, 1986; Callister and Agger, 1987) and sorbitol fermentation (Callister and Agger, 1987). The validity of these characters as markers of virulence has been questioned by other workers (Figura *et al.*, 1986; Kindschuch *et al.*, 1987). The autoagglutination of cells of *Aeromonas* after boiling has been correlated with pathogenicity to mice, and this phenomenon has also been proposed as a virulence marker (**Figures 150 & 151**; Janda *et al.*, 1987), while lack of pyrazinamidase activity may correlate with virulence in *A. sobria* (Carnahan *et al.*, 1990; *cf Yersinia*, page 135).

149

149. Lysine decarboxylase reaction of *A. hydrophila*.
The lysine decarboxylase reaction in *Aeromonas* species is variable and may correlate with pathogenicity. The reaction illustrated is weak and differences in methodology may account for discrepancies between the results of different workers (Kindschuch *et al.*, 1987). (*API 20E*©: *18h at 28°C.*)

150 & 151. Autoagglutination of *Aeromonas* as a marker of virulence. Autoagglutination is detected simply by observing broth cultures for sedimentation before and after boiling (glucose containing media should not be used to avoid confusion with the suicide phenomenon; (see **148**)). Results with two diarrhoea-associated strains of *A. hydrophila* were variable; the first (**150**) showed clear sedimentation after boiling but the second (**151**) was equivocal. (*Tryptone-soy broth: 24h at 28°C.*)

Human carriage of *Aeromonas*

Aeromonas is usually only a transient inhabitant of the intestinal tract, but carriage may be extended among the immunocompromised. Isolation rates from human faeces are highest in tropical regions such as Asia, South America and Australia, and lowest in the US and Europe.

No special precautions are necessary with respect to carriage of *Aeromonas* and food handling.

Aeromonas in the environment

Aeromonads are ubiquitous in freshwater environments. Numbers in rivers and lakes range from less than 1 cfu/ml to greater than 10^4 cfu/ml, and members of the genus may also be isolated from mineral springs, the watershed between rivers (Abeyta and Wekell, 1988), and drilled wells. Water from a drilled well in Sweden was found to contain up to 640 cfu *A. hydrophila*/ml and aeromonads were considered to be a more accurate indicator of water quality than the number of coliforms or *E. coli* (Krovacek, 1989).

The number of *Aeromonas* present in rivers and lakes depends on the extent of sewage pollution, the trophic state of the water, and the ambient temperature. Aeromonads probably receive at least some of their nutrients from phytoplankton, and positive correlations have been found between the numbers of *Aeromonas* and concentrations of chlorophyll a and phosphorus (van der Kooij, 1988). For this reason, the number of *Aeromonas* is a good indicator of the trophic state of fresh water (Rippey and Cabrelli, 1980).

Aeromonas is able to grow in sewage systems and numbers may attain 10^7 to 10^8 cfu/ml. The organism will also grow in waste water drainage systems and in the traps of baths and wash basins, and other domestic plumbing such as dish washer outlets. However, the contribution from these sources is insufficient to account for the numbers of aeromonads present in the environment.

Although *Aeromonas* is often thought of as being restricted to fresh water, the organism may also be isolated from saline and estuarine waters (Williams and LaRoche, 1985; Abeyta *et al.*, 1986). In these environments, the organism is not indigenous but transient (Abeyta and Wekell, 1988).

In addition to human carriage, *Aeromonas* may be isolated from the faeces of healthy animals. Reported isolation rates vary from *ca* 4% (Stern *et al.*, 1987) to 11.8% (Gray, 1984); while Morse and Hind (1984) were able to isolate the organism from 5.7% of pig lymph nodes. A high level of carriage may occur in poultry.

Foods commonly associated with *Aeromonas* infections

Although *Aeromonas* may be isolated from a wide range of food at the retail level and from potable water, there are only a small number of cases where a direct causal relationship has been established between contaminated food or water and gastroenteritis. There is some disagreement over the most important source of human infection; Holmberg (1986) considered water to be the major source, while Agger and Callister (1987) contended that foods were more commonly involved.

Water

A. hydrophila has been isolated from both chlorinated and unchlorinated piped water supplies (LeChevalier *et al.*, 1982, Burke *et al.*, 1984). The organism has also been isolated from bored well water (Krovacek, 1989) and from bottled spring water (Slade *et al.*, 1986). *Aeromonas* is able to grow in water distribution systems, and is able to utilise nutrients derived from materials used in plumbing, such as oleates derived from soft soap (van der Kooij, 1988).

The isolation of *A. hydrophila* from chlorinated water has led to the widespread assumption that the organism is particularly resistant to chlorine. This has been refuted by Cattabiani (1986). An increase in numbers of *A. hydrophila* in the drinking water of The Hague, Netherlands, was attributed to poor performance of a slow sand filter and was successfully controlled by chlorination (van der Kooij, 1988).

A statistical link has been demonstrated between contaminated drinking water and *Aeromonas* infections (Burke *et al.*, 1984).

More recently, long-lasting diarrhoea suffered by a child of one and half years was attributed to *Aeromonas* contamination of the family water supply (Krovacek *et al.*, 1989).

In The Netherlands, an 'indicative maximum value' for aeromonads of 20 cfu/100 ml has been proposed for drinking water at the waterworks, and 200 cfu/100 ml for water during distribution (van der Kooij, 1988). It is considered that standards are required for aeromonads in bottled spring water. Current publicity concerning the quality of mains water supplies is likely to increase the use of bottled spring water for making up infant feeds and for consumption by the immunocompromised, and standards applied should reflect these high-risk consumers.

Meat

Aeromonas is well established as a component of the

152. Gelatin hydrolysis by *A. hydrophila*. Motile species of *Aeromonas* produce powerful proteolytic enzymes, and the importance of the organism in the spoilage of meat and dairy products was known before the pathogenic role was recognised. (*API 20E*©: *18h at 28°C.*)

spoilage microflora of raw meat (**Figure 152**; Dainty *et al.*, 1983), and appears to be of particular importance in vacuum packs (Simard *et al.*, 1984), and in meat packed in carbon dioxide or nitrogen (Enfors *et al.*, 1979; Grau *et al.*, 1985). No distinction between toxigenic and non-toxigenic strains were made in early studies, but more recently Okrend *et al.* (1987) noted a high incidence of toxigenic *Aeromonas* species in beef and pork.

With the exception of poultry, intestinal carriage cannot explain the incidence of *Aeromonas* on meats, and water used in washing etc has been considered the most likely source (Stern *et al.*, 1987). Water may also be a source of contamination of poultry, but the evisceration stage is a critical stage for contamination during processing (Anon, 1988).

Fish and seafood

Aeromonads may be readily isolated from freshwater and estuarine fin fish (Gram *et al.*, 1987), and from shellfish and other seafood (Palumbo *et al.*, 1985a). A direct link has been established between the consumption of oysters and outbreaks of *Aeromonas* gastroenteritis in the USA (Herrington, 1984; Abeyta *et al.*, 1986). In the largest of these outbreaks, over 470 cases were reported. In the UK, imported cooked prawns have been implicated in three outbreaks, the largest of which involved more than 20 known cases.

Milk

Aeromonas is readily isolated from raw milk and may occur in up to 50% of samples (FDA, 1985). Multiplication probably occurs during storage in refrigerated bulk tanks and silos.

Produce

The potential importance of produce as a source of *Aeromonas* has been demonstrated by Callister and Agger (1987), who isolated the organism from all of 12 varieties of produce (including parsley, spinach, celery, alfalfa sprouts, broccoli and lettuce) examined after retail purchase. Ninety-two percent of the 12 varieties yielded cytotoxic strains. The widespread occurrence of *Aeromonas* in water means that total elimination from produce is unlikely to be feasible. Growth of the organism on cold stored produce may be anticipated, although multiplication rates are lower than on meat (Palumbo and Buchanan, 1988). Potential problems with *Aeromonas* in produce may be exacerbated by the use of modified atmosphere packaging to obtain very long storage lives at low temperatures.

Other products

Isolation of *Aeromonas* has been reported from a number of products including eggs, frog legs and snails. A small outbreak of diarrhoea has been attributed to the consumption of snails that were shown to be contaminated with *Aeromonas* (Abeonglabor, 1982).

A. caviae was isolated from 4.7% of ice cream samples in Wales (Hunter and Burge, 1987). The significance of the organism was considered to be uncertain.

Cases of *Aeromonas* infections associated with soup and starchy broths have been reported in Hungary (Janossy and Torjan, 1980), but details are not known.

Control of *Aeromonas*

Temperature

Clinical isolates of *A. hydrophila* have been shown to be capable of significant growth at 4°C (Palumbo *et al.*, 1985b), while strains isolated from foods have been shown to have even lower minimum growth temperatures of −0.1 to +1.2°C (Walker and Stringer, 1987). Optimum growth of mesophilic aeromonads is at *ca* 28°C and maximum in the range 38 to 41°C. Isolates from 'cold' environments may be unable to grow at 37°C (Knochel, 1990).

Aeromonas is eliminated by heat treatments designed for the elimination of *Salmonella*. However, survivor curves show a diphasic response (**Figure 153**) due to the presence of a relatively heat resistant sub-population (Palumbo *et al.*, 1987). The significance of this phenomenon to food safety is not known.

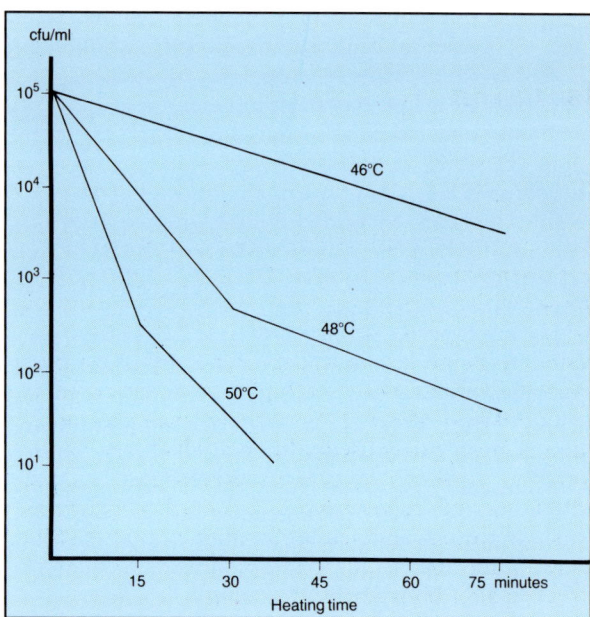

153. Diphasic survivor curves of A. hydophila.
Diphasic survivor curves after heating are relatively common with both vegetative bacteria and endospores (*cf.* page 282). The effect is seen strongly with *A. hydrophila*, but the underlying reason for the existence of the relatively heat resistant sub-population is not known.

Irradiation

Aeromonas is relatively sensitive to ionising irradiation with *D* values in the range 1.4 to 2.2 Kgy (Palumbo *et al.*, 1986). A pasteurising dose of 3 Kgy would be sufficient to eliminate the organism from fresh foods.

Ultra-violet irradiation of spring water prior to bottling is common practice, but despite reports of *Aeromonas* in bottled water there is no evidence of unusually high resistance to this treatment.

pH value

Aeromonas is sensitive to pH values below 6.0 (Palumbo and Buchanan, 1988), and it is unlikely to be a problem in naturally acidic or acidified foods or in carbonated spring water. The decline in numbers of *Aeromonas* sometimes observed during the storage of raw oysters may be due to the fermentation of the relatively large quantities of glucose present to lactic acid with a concomitant fall in pH value.

NaCl

Aeromonas species are often considered to be sensitive to NaCl and cannot grow in broth media containing 5% NaCl (Popoff, 1984). There is a strong interaction between pH value and NaCl tolerance, with small differences in pH value resulting in relatively large changes in NaCl tolerance. In practice, growth is unlikely to be a problem where the pH value is less than 6.5 and the NaCl content greater than 3.0%.

Composition of atmosphere

Carbon dioxide has a specific inhibitory effect on *Aeromonas hydrophila*, extending the lag phase at 30°C and causing a decline in the population at 5°C. In contrast, nitrogen enhanced growth of both injured and uninjured cells at 5°, but not 30°C. This was attributed to impaired functioning of the oxygen protective system at 5°C, permitting a protective role for nitrogen (Golden *et al.*, 1989).

Laboratory methods

Isolation

Enrichment

Although direct plating may be used when *Aeromonas* is present in large numbers in foods, recovery is improved by enrichment. A number of enrichment media have been used including alkaline peptone water, tryptone–soy broth, tryptone–soy broth plus ampicillin, tryptone–soy broth plus NaCl and tryptone broth (Okrend, 1987). Tryptone–soy broth plus ampicillin is most widely used, although alkaline peptone water is preferred by some workers and has the advantage of recovering occasional ampicillin-sensitive strains. An incubation temperature of 28°C is recommended, and the enrichment should be streaked after both 6 and 24 hours.

Selective plating media

A wide range of selective plating media have been used for the recovery of *Aeromonas* from foods (Joseph *et al.*, 1988). Most attention has been paid to the recovery of *A. hydrophila* and, as a general rule, media are less effective with *A. sobria* and *A. caviae*. Most strains of *Aeromonas* will grow on many of the selective media devised for members of the *Enterobacteriaceae*, such as MacConkey agar, Drigalski agar, Salmonella–shigella agar and cefsulodin–irgasan–novobiocin (CIN) agar. In some cases, *Aeromonas* has been recovered from media of this type when its presence in food was not expected. Their use is illustrated in **Figures 154–160**.

A number of modifications have been made to media for the *Enterobacteriaceae* to improve their performance for the recovery of *Aeromonas*. Such media include Rimler-Shotts agar (Shotts and Rimler, 1973), xylose–deoxycholate–citrate agar (Shread *et al.*, 1981), and bile salts–brilliant green agar plus starch (Nishikawa and Kishi, 1987). MacConkey agar has also been modified by the addition of ampicillin (see below).

Bile salts based media have the advantage of recovering ampicillin-sensitive strains of *Aeromonas*, but these are uncommon and ampicillin is currently the most widely used selective agent. Ampicillin-

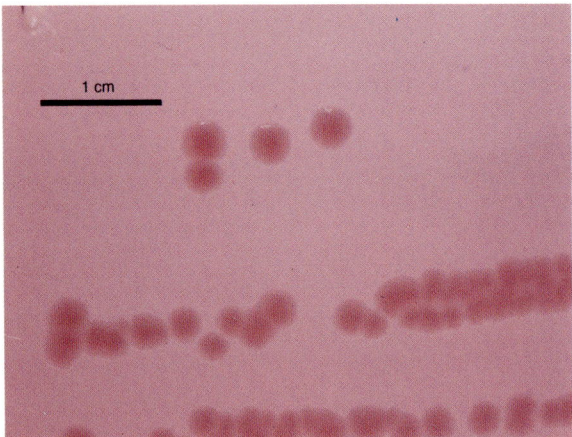

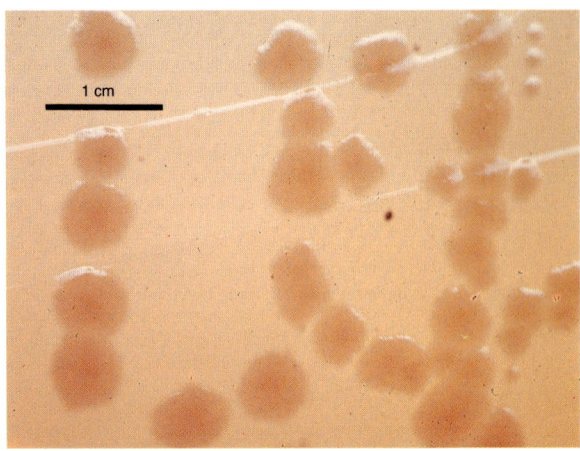

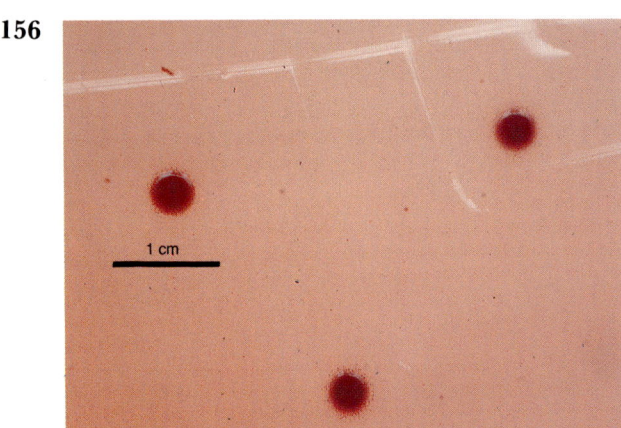

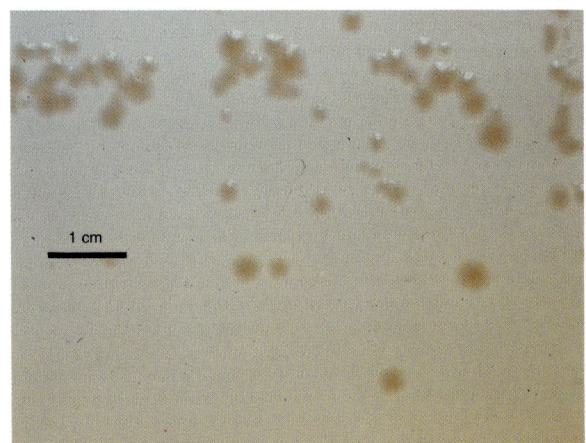

154–157. Isolation of motile *Aeromonas* species on general-purpose selective media for the *Enterobacteriaceae*. *Aeromonas* grows well on media of this type, including MacConkey-purple agar (**154**), violet red-bile agar (**155** & **156**) and Drigalski agar (**157**).

Colonies are usually lactose-negative and flattened in appearance (**154**, **155**, **157**). In a small number of cases, colonies are lactose-positive and resemble those of lactose fermenting members of the *Enterobacteriaceae* (**156**). (*All media: 24h at 28°C.*)

based selective plating media used for foods are summarised in **Table 76**.

Blood agar-ampicillin is a productive medium, but effectively non-differential since haemolysis by species other than *A. hydrophila* is often weak, and at least 10% of isolates will be overlooked if this criterion is used to differentiate *Aeromonas* from other species (Kelly *et al.*, 1988). Blood agar-ampicillin is widely used for the isolation of aeromonads from stools, and in this context is most effective when it is used in tandem with CIN medium. Brain-heart infusion agar should be used as base.

Starch-ampicillin medium has been the medium of choice for a number of workers (Callister and Agger,

1987; Okrend *et al.*, 1987; Stern *et al.*, 1987). This medium is generally satisfactory with respect to both sensitivity and selectivity, although with fish and seafood major problems with starch fermenting vibrios may be encountered.

Of the various MacConkey-based media, MacConkey–mannitol–ampicillin has been reported to be as effective as starch-ampicillin after enrichment, although less effective for direct plating (Okrend *et al.*, 1987). Other variants were less effective.

Aeromonas medium is a commercially developed medium which has become widely used in the food industry. The use of the medium is illustrated in **Figures 161–163**.

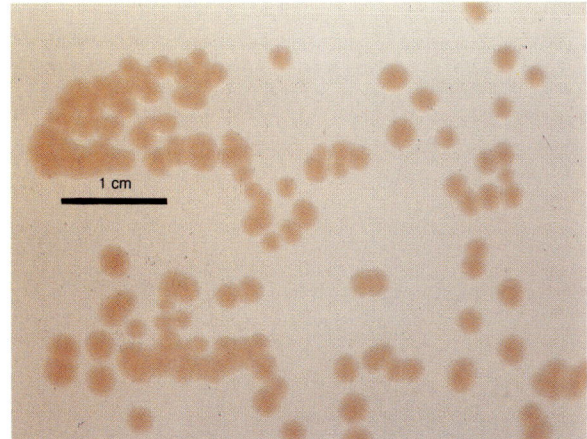

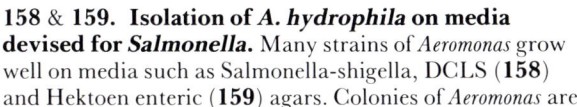

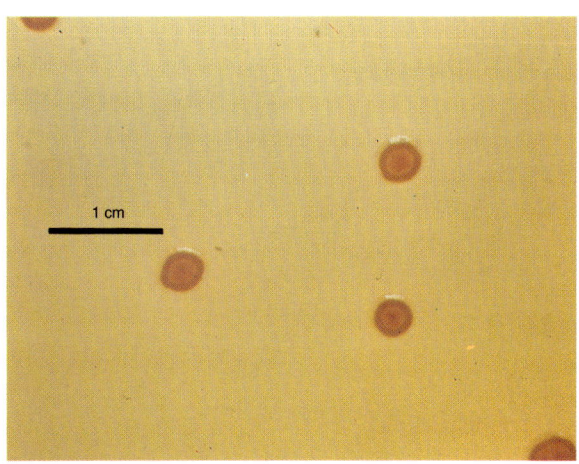

158 & 159. Isolation of *A. hydrophila* on media devised for *Salmonella*. Many strains of *Aeromonas* grow well on media such as Salmonella-shigella, DCLS (**158**) and Hektoen enteric (**159**) agars. Colonies of *Aeromonas* are easily differentiated from those of *Salmonella* (see pages 74–76), but on DCLS resemble, superficially at least, some strains of *Shigella*. (*Both media: 24h at 30°C.*)

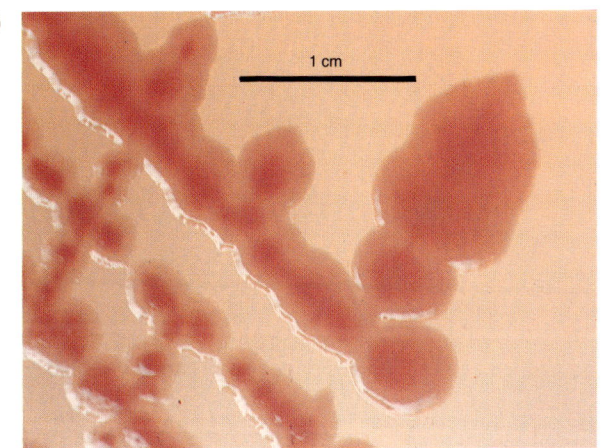

160. Isolation of *A. hydrophila* on CIN medium. Although CIN medium was designed for the selective isolation of *Yersinia enterocolitica*, it supports good growth of *A. hydrophila* and moderate growth of other motile species, and has been used for the selective isolation of that organism. Colonies are usually large, but there can be considerable variation in appearance. (*CIN medium: 24h at 28°C.*)

Table 76. Ampicillin-based selective plating media for *Aeromonas*.

Blood agar-ampicillin
Starch-ampicillin agar
MacConkey-ampicillin agar
MacConkey-tween-ampicillin agar
MacConkey-mannitol-ampicillin agar
MacConkey-xylose-ampicillin agar
Aeromonas agar

161

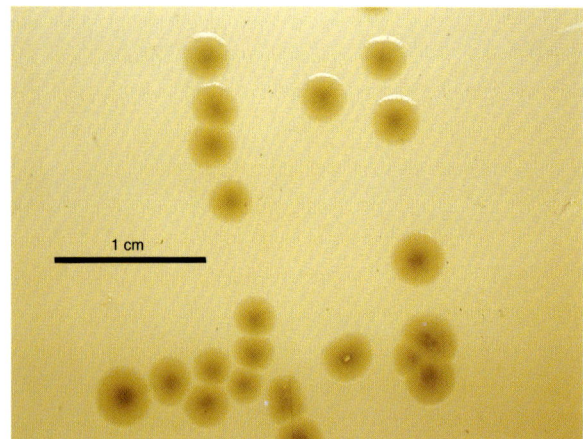

161. Isolation of *A. hydrophila* on Aeromonas medium. Colonies of *A. hydrophila* have a typically dark green centre surrounded by a yellow-green zone. In some cases, the outer zone is absent and the entire colony is dark green. Although developed for *A. hydrophila*, other motile species grow reasonably well on the medium and have a similar colonial appearance. (*Aeromonas medium (Unipath): 24 h at 28°C.*)

162

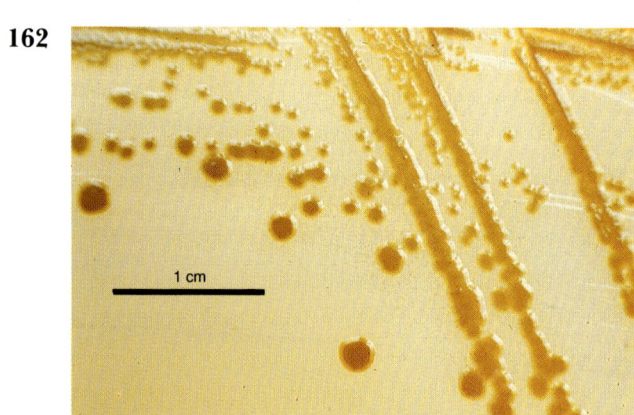

163

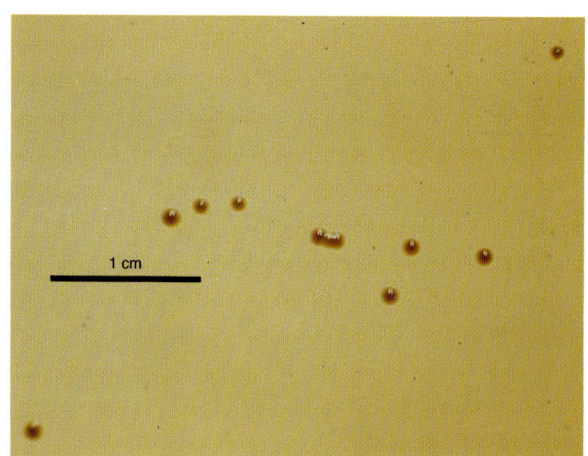

162 & 163. Selectivity of Aeromonas medium. In addition to motile species of *Aeromonas*, Aeromonas medium supports the growth of some members of the *Enterobacteriaceae* as well as *Pseudomonas* species. *E. coli* (**162**) grows well, although it is easily differentiated from *Aeromonas* by colonial appearance. Most strains of *Pseudomonas* produce only small colonies at 28°C (**163**), but in products such as fresh meat they make detection of *Aeromonas* difficult due to their presence in greater numbers. (*Aeromonas medium (Unipath©): 24 h at 28°C.*)

Identification of *Aeromonas*

Effect of incubation temperature

Although the extent to which temperature-sensitive reactions occur in aeromonads has not been fully investigated, some reactions including arginine dihydrolase may be positive at 28 to 30°C, but not at higher temperatures.

Differentiation from other genera

Aeromonas may be differentiated from other genera according to the criteria listed in **Table 74**, page 187. Either conventional media, or miniaturised kits may be used, but little value is seen in the multi-test medium described by Kaper (1979). Where conventional methodology is used, the media and methods described for the *Enterobacteriaceae* by Edwards and Ewing (1972) are suitable.

Although differentiation of *Aeromonas* from other genera is usually relatively straightforward, care must be taken with some reactions.

Oxidase test

The oxidase reaction of *Aeromonas* species should be determined only with young cultures of not more than 24 hours old grown at 28°C on carbohydrate-free medium. Failure to observe these precautions may result in false-negative reactions and confusion with the *Enterobacteriaceae*.

O/129 sensitivity

This test is discussed in further detail in Chapter 8, *Vibrio*, pages 180–181.

Differentiation between species

Differentiation between species of *Aeromonas* may be made according to the criteria listed in **Table 75**,

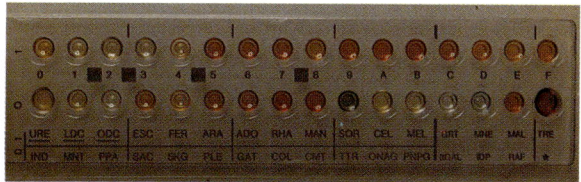

164 & 165. Identification of *Aeromonas hydrophila* using miniaturised systems. Oxidase-positive, fermentative, Gram-negative rods such as *Aeromonas* and *Vibrio* may be identified using systems designed both for the *Enterobacteriaceae* and for non-fermentative bacteria. The API ATB 32E© (**164**) and the API 2ONE© (**165**) both differentiate between *A. hydrophila* and *A. sobria* but not *A. caviae*. (*Both systems: 24 h at 28°C.*)

page 187. Where conventional methodology is used, the media and methods of Edwards and Ewing (1972) are suitable. Miniaturised identification systems may be used, subject to limitations (**Figures 164 & 165**).

Care must be taken to avoid over dependence on gas production from glucose when differentiating *A. hydrophila* from *A. caviae*, since while *A. hydrophila* is usually aerogenic, non-aerogenic strains are not uncommon.

The 'suicide phenomenon' has been proposed as a simple means of differentiating between *A. caviae*, *A. hydrophila* and *A. sobria* (**Figures 166 & 167; Flow diagram 19**). The 'suicide phenomenon' may also be of use in conjunction with miniaturised systems.

166 & 167. The use of the suicide phenomenon in differentiation of *A. hydrophila* from *A. sobria* and *A. caviae*. The three main species of motile aeromonads may be differentiated on the basis of the suicide phenomenon (page 185), together with gas production from glucose and aesculin hydrolysis (see **Flow diagram 19**). *A. hydrophila* may thus be differentiated by the absence of the suicide phenomenon (**166**) and the hydrolysis of aesculin (**167**). The determination of gas production from glucose is not required in this example.

Flow diagram 19. The 'suicide phenomenon' in the differentiation of *Aeromonas* species.

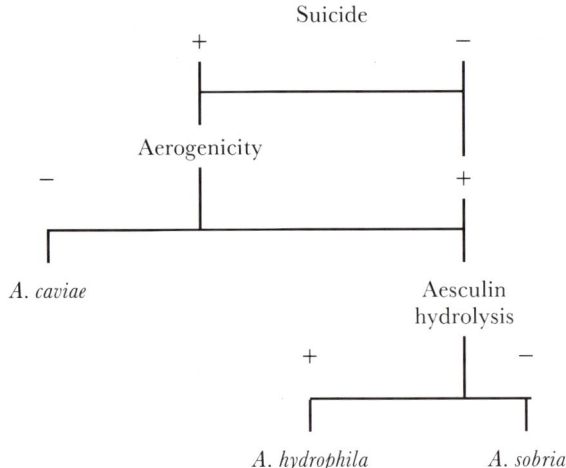

10. *Plesiomonas shigelloides*

Introduction

Plesiomonas shigelloides is a Gram-negative, oxidase-positive, rod-shaped bacterium which, in common with *Vibrio* and *Aeromonas*, has both a fermentative and a respiratory metabolism of glucose. The organism contains the *Enterobacteriaceae* common antigen (LeMinor *et al.*, 1972).

Cells of *P. shigelloides* are long, straight rods which sometimes appear filamentous. Most strains are motile by polar flagella, and lateral flagella are also produced by young cultures. Flagellate but non-motile and aflagellate strains also occur.

Although *P. shigelloides* does not synthesise poly-β-hydroxybutyrate, inclusion bodies have been described (Pastian and Bromel, 1984). Earlier reports that these bodies were electron dense and predictably bipolar (Miller and Koburger, 1985) have been disputed, the location in the cell depending on the phase of the growth cycle (Brendan *et al.*, 1988), but it has been postulated that the bodies may play an important role in the survival of *P. shigelloides* as minicells.

P. shigelloides is a taxonomic mendicant having been, at various times, assigned to the family *Enterobacteriaceae*, the genus *Pseudomonas* and, more recently, the genus *Aeromonas*. The current genus *Plesiomonas* was created by Habs and Schubert (1962) in recognition of the fact that there are essential differences with either *Vibrio* or *Aeromonas*. *Plesiomonas* was placed in the family *Vibrionaceae* by Schubert (1984).

Subsequent work has shown *P. shigelloides* to be highly homogeneous (Brendan *et al.*, 1988), but 5S rRNA studies have shown it to be more closely related to *Proteus vulgaris* and *Pr. mirabilis* than to members of the *Vibrionaceae* (MacDonell and Colwell, 1985), and its transfer to the genus *Proteus* as *Pr. shigelloides* has been proposed (Basu, 1985; MacDonell *et al.*, 1986). Such a transfer would require a redefinition of the genus *Proteus* (Jones, 1988).

P. shigelloides is not a normal inhabitant of the human intestinal tract, but the organism is increasingly associated with diarrhoea and, less commonly, extraintestinal infections. Infections are usually opportunistic, involving the very young, the aged or people with a predisposing condition (see page 204).

Despite evidence concerning the role of *Plesiomonas* as a causative agent of diarrhoea, the situation is analogous to that with mesophilic aeromonads in that a causal link is not definitely proven. Evidence for such a role is of three types; reports of individual cases, reports of outbreaks involving several cases and general epidemiological considerations (**Table 77**, Brendan *et al.*, 1988).

The most comprehensive evidence for a causal role for *P. shigelloides* was presented by Holmberg *et al.*, (1980), who matched 31 patients, 28 of whom had no recognised cause of gastroenteritis other than *P. shigelloides*, with 62 controls. In contrast, not all evidence for a causal role is as comprehensive and some is best described as weak. More significant may be the failure of volunteer feeding studies to demonstrate causal links. This situation is again analogous to that with *Aeromonas* (see page 186) and may result from the failure to protect *Plesiomonas* from the acidic conditions in the gut. Until further information is available concerning the role of *Plesiomonas*, it is probably prudent to regard the organism as a putative pathogen.

Table 77. Evidence for a causal role for *Plesiomonas shigelloides* in diarrhoea (Brendan *et al.*, 1988).

A Case reports
1 No other enteric pathogen found.
2 Usually predominant micro-organism in cases of gastroenteritis.
3 Isolation in high numbers from multiple gastrointestinal sites.
4 Resolution of symptoms concurrent with disappearance of *Plesiomonas* after therapy.

B Outbreak reports
1 Isolation from multiple patients associated with same outbreak.
2 Isolation of a single predominant serovar from each outbreak.

C Epidemiology
1 Low carriage rate in healthy people.
2 Significant association of infection with consumption of raw shellfish and foreign travel.

Isolation of *P. shigelloides*

Enrichment

Although direct plating may be used for clinical samples such as diarrhoeal stools from which *P. shigelloides* may be isolated as predominant organism, enrichment is recommended for its isolation from foods, particularly if the organism is stressed (Miller and Koburger, 1986a). Two media have been widely used, alkaline peptone water and tetrathionate broth.

Selective plating media

In the past, media designed for *Salmonella* and other members of the *Enterobacteriaceae* were in general use for the isolation of *P. shigelloides*. Three specific media have subsequently been developed, inositol-brilliant green-bile salts, *Plesiomonas* medium and a modification of Rimler-Shotts medium.

Isolation of P. shigelloides *is discussed in greater detail, with respect to practical aspects and choice of media in* **Laboratory Methods**, *pages 207–208.*

Taxonomy of *P. shigelloides*

P. shigelloides is currently considered to be a single species genus (Schubert, 1984), and while attempts have been made to further classify the organism (Miller and Koburger, 1985) these have not been widely recognised.

The main biochemical characteristics of *P. shigelloides* are listed in **Table 78**. These may be determined using either conventional techniques or suitable miniaturised systems such as the API©. Alternatively, a combination of triple sugar-iron and inositol-gelatin media may be used as a screen.

The taxonomy of P. shigelloides *is discussed in greater detail with respect to practical aspects in* **Laboratory Methods**, *pages 207–208.*

Table 78. Biochemical characteristics of *Plesiomonas shigelloides* (Von Graevenitz, 1980).

Motility	+[1]
Catalase	+
Oxidase	+
Carbohydrate metabolism	Fermentative
Gas from glucose	−
Inositol fermentation	+
Arginine dihydrolase	+
Ornithine and lysine decarboxylase	+
Proteinase	−
Lipase	−
Voges–Proskauer	−
Indole	+
0/129 sensitivity	+[2]
β-haemolysis	−

[1] Some strains are non-motile.
[2] Some strains carry a plasmid coding resistance.
Note: Characteristics in **bold** are minimum identification criteria.

Typing of *P. shigelloides*

Serotyping is the only established means of typing *P. shigelloides*. A typing scheme based on 50 O-serogroups and 17 H-antigens has been described (Shimada and Sakazaki, 1978). Some serovars are ubiquitous, whereas others are isolated only in specific regions (Aldova, 1985).

Although it was previously thought that all strains of *P. shigelloides* share a common O-antigen with *Shigella*, this has been shown to be a minority characteristic (Schubert, 1984).

Symptoms of *P. shigelloides* infections

Gastrointestinal

Gastrointestinal symptoms usually appear within 48 hours of ingestion. Diarrhoea is the predominant symptom occurring in 94% of cases. Accompanying symptoms vary, but abdominal pain, nausea, vomiting, headaches, chills and dehydration are most common. The severity of symptoms varies, and the duration is usually seven days or longer.

Diarrhoeal symptoms of *Plesiomonas* infections are of three types, secretory, shigella-like and cholera-like (Brendan *et al.*, 1988).

The secretory form is the most commonly reported to date, and varies in severity from a mild illness of short duration, to severe diarrhoea with the passage of up to 30 stools per day at its peak. Symptoms may be of a prolonged duration, and in severe cases persist for an average of 11 days and as long as 21 days. In many cases, the organism is isolated from the stools in virtual pure culture.

The shigella-like form is less common, but is usually severe, being characterised by abdominal pain and bloody, mucoid stools which contain polymorphonuclear leukocytes. A case has been reported (van Loon *et al.*, 1989) involving a healthy person with no risk factors in which bloody diarrhoea persisted for 6 months. Pseudomembranous colitis was also involved.

Cholera-like symptoms have been reported in only one case, a co-infection with *A. sobria*. Symptoms were prolonged and the patient finally succumbed (Sawle *et al.*, 1986).

Extraintestinal infections

Extraintestinal infections due to *P. shigelloides* are relatively rare, and most common in people with solid malignancies or other predisposing conditions (Saraswathi *et al.*, 1983; see page 204). The portal of entry is difficult to determine, but is likely to be via the gut in at least some cases. In one reported case involving an immunocompetent patient, consumption of a seafood meal was associated with a one day history of gastrointestinal symptoms followed by a blood stream infection (Ingram *et al.*, 1987) and bacteraemia has also been reported in a healthy girl with mild gastroenteritis (Paul *et al.*, 1990).

Extraintestinal infections take a number of forms of which meningitis is most common. Septicaemia and osteomyelitis are less common. Meningitis has a mortality rate of *ca* 80% while, in people with predisposing conditions, blood-stream infections have been invariably fatal (Humphreys *et al.*, 1985; Nolte *et al.*, 1988). However, isolation of *P. shigelloides* from the blood-stream of healthy people does not necesarily lead to fatality or to fulminant disease (Paul *et al.*, 1990).

Persons at particular risk from *P. shigelloides* infections

Susceptibility

P. shigelloides infections occur in both children and adults (McNeeley *et al.*, 1984; Holmberg and Farmer, 1984). The very young, the aged and the infirm are generally considered to be at greatest risk (Freund *et al.*, 1988) although Brendan *et al.*, (1988) considers infection to be most common among adults.

Superimposed on age-linked differences in susceptibility is the effect of underlying conditions. Malignancies such as leukaemia are often involved, and cirrhosis is also important. Low gastric acidity resulting from malnutrition, gastric surgery or a malignancy may also increase susceptibility to a *Plesiomonas* infection (Huq and Islam, 1983; Janda, 1987), and for this reason a minority of people taking antacids as medication may also be at a heightened risk (Martin and Gustafsen, 1985; Janda, 1987).

Preliminary evidence has been presented (Janda, 1987) that normal gut inhabitants such as *Enterococcus faecium* (*Streptococcus faecium*) and *Pseudomonas aeruginosa* produce compounds which have a specific antimicrobial effect on *Plesiomonas*. Disturbance of the normal gut microflora by treatments such as long-term antibiotic therapy may also increase susceptibility to infection.

Exposure

Plesiomonas infections are most widely reported in warm climates, cases reported in countries in temperate climates often being imported diseases. Infection in industrialised countries of the northern hemisphere are often associated with overseas travel.

The organism has been isolated from the stools of US Peace Corps volunteers with travellers' diarrhoea in Thailand, and the prevalence of *P. shigelloides* infection is also high in visitors to other parts of Asia, including Japan and to Africa. However, in general *P. shigelloides* is a less important cause of travellers' diarrhoea than other bacteria, such as enteropathogenic *Escherichia coli*, *Salmonella* and *Campylobacter jejuni* (Koburger, 1989).

Consumption of raw seafood, particularly oysters, is a well-established risk factor for *P. shigelloides* infections. Habitual consumers in warm climates are at markedly enhanced risk of exposure.

Increasing incidence of P. shigelloides *infections*

In several countries, the incidence of infection with *P. shigelloides* is increasing. To some extent this is probably attributable to greater recognition of the organism as a potential enteropathogen, but host associated factors are also likely to be involved (see **Table 79**).

Table 79. Host-associated factors leading to an increasing incidence of *Plesiomonas shigelloides* infections.

A Increased life expectancy of immunocompromised people.

B Heightened exposure of general population to infection.
1 Increased foreign travel.
2 Increasing consumption of raw shellfish.
3 Increased recreational use of aquatic facilities.

Treatment of *P. shigelloides* infections

Plesiomonas infections may be mild and self limiting, requiring no treatment, or fluid replacement only. However, *ca* 60% of known cases have required hospitalisation and/or antibiotic therapy because of the severity or the prolonged duration of symptoms (Brendan *et al.*, 1988).

When antibiotic therapy is indicated, chloramphenicol, gentamycin and tetracycline are usually effective. Multiple drug resistance does not currently appear to be a problem, although resistance to penicillin, ampicillin and carbenicillin, possibly due to β-lactamase production may be increasing.

Mechanisms of pathogenicity

Relatively little is known of the mechanisms of pathogenicity in *P. shigelloides*, and controversy remains concerning the enteropathogenicity of the organism. A number of putative virulence factors have been described (**Table 80**) although in most cases the significance and relative importance is not known.

Table 80. Putative virulence factors of *P. shigelloides*.

Adhesin(s)	
In vitro effect:	Agglutination of erythrocytes; fimbriae
Site of action:	Gastrointestinal tract
Role *in vivo*:	Adherence to mucosa
Invasiveness	
In vitro effect:	Invasion of HeLa cells
Site of action:	Gastrointestinal tract
Role *in vivo*:	Invasion of mucosal lining
Enterotoxin(s)	
In vitro effect:	Suckling mouse, rabbit ileal loop positive
Site of action:	Gastrointestinal tract
Role *in vivo*:	Fluid accumulation
Endotoxin	
In vitro effect:	Toxicity to mice and rabbits, pyrogenicity
Site of action:	Reticulo-endothelial system
Role *in vivo*:	Complement activation
Elastase	
In vitro effect:	Digestion of elastin fibres
Site of action:	Connective tissue
Role *in vivo*:	Tissue degradation?

Motility

Motility would appear to be a virulence factor, with invasive ability being demonstrated only in motile strains (Bimms *et al.*, 1984). *P. shigelloides* is rapidly inactivated in the high acid conditions of the gut (Janda, 1987), and motility may assist the organism to by-pass the acidity barrier.

Adherence

Adherence has been described, although the mechanism is poorly understood (Ljungh and Wadstrom, 1985). A glycocalyx outer layer may be involved.

Invasion

Invasiveness is dependent on the strain and cannot be demonstrated in all models. The Sereny test is invariably negative, but invasive ability has been demonstrated using HeLa cells (Bimms *et al.*, 1984).

There is no evidence of *Shigella*-like invasiveness (Olsvik *et al.*, 1985), or for a correlation between invasiveness and the possession of the *Sh. sonnei* antigen (Bimms *et al.*, 1984).

The potential for invasion is probably restricted to a small number of strains, and may explain the occasional dysentery-like symptoms associated with *Plesiomonas* infections.

Toxins

Enterotoxin

The evidence for *P. shigelloides* enterotoxin is conflicting. Negative results have been reported in a number of studies (Johnson and Lior, 1981; Sanyal *et al.*, 1980; Olsvik *et al.*, 1985) while, in contrast, evidence for the existence of an enterotoxin was presented as early as 1977 (Gurwith and Williams, 1977). This evidence was supported by that of Huq and Islam (1983) and Saraswathi *et al.* (1983), the latter work suggesting the production of an enterotoxin effective in mediating cyclic AMP. Studies using partially purified material showed that two toxins were present, a heat stable toxin which was active in both rabbit ileal loop and suckling mouse assays, and a heat labile toxin active only in ileal loop assays.

Endotoxin

An endotoxin similar to the classic Gram-negative lipopolysaccharide endotoxin has been isolated from *P. shigelloides* (Foster and Rao, 1976). This endotoxin is lethal to both mice and rabbits and pyrogenic in rabbits. It was noted by Brendan *et al.*, (1988) that there has been no speculation concerning a possible role in pathogenicity.

Elastase

Elastase production by *P. shigelloides* was noted by Ljungh and Wadstrom (1985). Elastase is a potential virulence factor, but there is no knowledge of any specific role in *Plesiomonas*.

Human carriage of *P. shigelloides*

P. shigelloides is not a part of the normal gut flora of man, and the rate of carriage among healthy people is generally low, a range of 0.0078 to 0.26% being reported by Arai *et al.*, (1980). The rate of carriage obviously varies considerably with the location, and a rate of 6.2% has been reported in Thailand (Pitarangsi *et al.*, 1982). In this case, a progressive increase in carriage rate with age was demonstrated, carriage increasing from 2 to 3% in children aged 0 to 2 years to 24% in adults.

P. shigelloides in the environment

Water is a major natural habitat for *P. shigelloides*, and probably the most important source of the organism in human infections (Arai *et al.*, 1980; Schubert, 1981). The organism is usually found in fresh or estuarine waters, but may also be isolated from sea water during the warm months (Brendan *et al.*, 1988).

In temperate climates, there is a marked seasonal variation in the numbers of *P. shigelloides* isolated from water, which may be related directly to the inability of the organism to grow at temperatures below *ca* 8°C.

Carriage in animals and the intestines of fish is relatively common, and isolation of *P. shigelloides* has been reported from food animals (pigs, cows, poultry, sheep and goats); pet animals (cats and dogs); pri-mates (monkeys); finfish and shellfish; amphibia (newts); and snakes (Davis *et al.*, 1978; Arai *et al.*, 1980; Vandepitte *et al.*, 1980). In the USA the organism forms part of the intestinal microflora of wild turkey vultures, and these birds may be important reservoirs (Winson *et al.*, 1981). It seems likely that the primary reservoir of *P. shigelloides* is fish, which would account for the ease of isolation from the aquatic environment (Koburger, 1989).

The possibility of direct infection from animals exists, and in a well-documented case a zoo employee apparently acquired infection from a diseased boa constrictor.

Foods commonly associated with *P. shigelloides* infections

Water

Water has been implicated in a number of outbreaks of diarrhoea. In one example *P. shigelloides* was the only suspect organism isolated both from the implicated water and from stool specimens. Some outbreaks have been large and involved several hundred people (Koburger, 1989). There have been no reports of *Plesiomonas* infections associated with chlorinated water.

In addition to transmission of *P. shigelloides* by drinking water, a number of cases have been associated with recreational use of water.

Fish

As noted above, contamination of fish with *P. shigelloides* is widespread, and an incidence of 95% has been found in a survey of cultured, wild and retail samples of various finfish including bream, catfish, perch, carp and bass (Koburger, 1989). Isolations have also been made from a range of shellfish.

Fish and shellfish which have been associated with *Plesiomonas* infections include crab, (Claesson *et al.*, 1984), shrimp (Holmberg *et al.*, 1980), oysters (Lieh, 1983; Rutala *et al.*, 1983; Koburger and Miller, 1984), salt mackerel (Hori *et al.*, 1966) and cuttle fish salad (Miller and Koburger, 1985). In the US, ray oysters appear to be the most common vehicle of infection (Koburger and Miller, 1984; Hackney and Dicharry, 1988).

Other foods

The relatively high rate of carriage of *P. shigelloides* by food animals would suggest a similarly high level of contamination at a retail level, and an association with food poisoning through consumption of under-cooked or recontaminated foods. In fact, there is only one known case associated with food other than fish, chicken being implicated (Newsome and Gallais, 1982).

Factors affecting the growth and survival of *P. shigelloides*

Temperature

The minimum growth temperature of *P. shigelloides* is widely accepted to be 8°C (Schubert, 1981). However, Miller and Koburger (1986b), found only *ca* 22% of isolates to be able to grow at that temperature, although all could grow at 10°C. In contrast, at least one strain has been reported to grow at 0°C (Hackney and Dicharry, 1988).

Optimum growth is between 37 and 38°C, and the maximum growth temperature is between 40 and 45°C.

P. shigelloides is not heat resistant and is destroyed by heating at 60°C for 30 minutes (Miller and Koburger, 1986b).

pH value

P. shigelloides has a greater tolerance of low pH values than many Gram-negative bacteria, and *ca* 60% of strains are able to initiate growth at pH 4.0. All are able to grow at pH 4.5 and 85% at pH 9.0 (Miller and Koburger, 1986b). However, the organism is rapidly inactivated at pH values of less than 4.0 (Janda, 1987).

NaCl concentration

The limiting NaCl concentration for the growth of *P. shigelloides* is dependent on the nutritional status of the growth medium. In tryptone broth, the organism is able to grow in 3% but not in 4% NaCl, while in the more nutritious trypticase–soy broth, all strains grow in 4%, and 65% of strains in 5% NaCl (Miller and Koburger, 1986b).

Laboratory methods

Isolation

Although *P. shigelloides* has received increasing attention in recent years, most media development has been directed towards clinical and environmental samples. With the exception of fish and other seafood little, or no, attention has been paid to the isolation of the organism from foods, and care must therefore be taken when choosing media, since conclusions valid

for faeces and environmental samples may not be valid for foods.

Enrichment

Although enrichment is desirable in principle, there is evidence that neither of the two widely-used enrichment media, alkaline peptone water and tetrathionate broth are particularly effective. This is probably due to the limited competitive ability and somewhat demanding nutritional requirements of *Plesiomonas* (Koburger, 1989).

Comparisons of alkaline peptone water and tetrathionate broth have produced conflicting results. Alkaline peptone water was found to be superior by von Graevenitz and Bucher (1983), while Millership and Chattopadhja (1984) obtained the best results using tetrathionate broth. Tetrathionate broth (without the addition of iodine) was also preferred by Freund *et al.* (1988).

Incubation of enrichment cultures is usually at 35 to 37°C, but Freund *et al.* (1988) used a temperature of 40°C to reduce competition from other microorganisms. The higher temperature is close to the maximum growth temperature of some strains.

Selective plating media

In the past, media designed for *Salmonella* and other members of the *Enterobacteriaceae* have been widely used for the isolation of *P. shigelloides*, and their use has continued, if only for convenience. However, the selective agents used in these media are inhibitory, to a greater or lesser extent, to *Plesiomonas* (Schubert, 1977; Freund *et al.*, 1988), and their use is less than ideal. Of these media, the most satisfactory results are likely to be obtained using Salmonella-shigella agar or MacConkey agar.

Three selective plating media have been developed for the isolation of *P. shigelloides*; inositol-brilliant green-bile salts (IBB; Schubert, 1984), *Plesiomonas* agar (PA; Miller and Koburger, 1986a), and a modified Rimler-Shotts agar (mRS; Sakata and Todaka, 1987). In a comparative study of IBB and PA media, IBB was found to be superior on an overall basis and when numbers of *Plesiomonas* were small, but PA was more effective for the recovery of heat and cold stressed cells (Miller and Koburger, 1986a). Ideally, both media should be used in parallel, but where this is not possible, the choice should depend on the nature of the food and the processing received.

Both PA and mRS media support the growth of *Pseudomonas* species and the screening of colonies is necessary. Anaerobic incubation of mRS has been proposed to overcome this problem (Sakata and Todaka, 1987), and this solution may also be applicable to PA. Incubation of media for the selective plating of *P. shigelloides* should be at 35 to 37°C for up to 48 hours.

Identification of *P. shigelloides*

Differentiation of *P. shigelloides* presents little difficulty whether conventional (see **Table 78**, page 202) or miniaturised methodology is used. In the former case, media used for the *Enterobacteriaceae* are suitable. Acid production from carbohydrates is often weak, irrespective of the type of media used and incubation should be continued to 48 hours at 37°C if necessary.

Although the use of triple sugar-iron medium together with inositol gelatin medium is suitable for screening colonies, the API© systems are preferred since identification is based on a larger number of characters.

11. *Campylobacter*

Introduction

The genus *Campylobacter* comprises slender, spiral, Gram-negative rod-shaped bacteria, which have a characteristic 'corkscrew-like' motility produced by a single polar flagellum at one, or both, ends of the cell (Smibert, 1984).

Campylobacter is a microaerophilic organism with a respiratory metabolism. Oxygen is required for growth at concentrations of 3 to 15%, and carbon dioxide at concentrations of 3 to 5%. In some cases, growth occurs in aerobic conditions of 20% oxygen. Although oxygen is required for growth in most circumstances, some campylobacters are able to grow anaerobically in the presence of certain substrates such as nitrate and fumarate, and some require the presence of hydrogen.

The genus *Campylobacter* has been the subject of a major taxonomic study based on 16S ribosomal RNA sequence analysis (Thompson *et al.*, 1988). This divided campylobacters into three homology groups, each of which represents a species. Group I consists of the 'true' campylobacters and contains two subgroups: the misleadingly named thermophilic group comprising *C. coli*, *C. jejuni*, *C. laridis* (proposed name *C. lari*) and '*C. upsaliensis*'; and the classical group comprising *C. fetus*, *C. hyointestalinis*, *C. sputorum*, *C. conciscus* and *C. mucosalis*. Group II contains *C. pylori*, *C. cinaedi* and *C. fennelliae* together with *Wolinella succinoides*, while Group III contains *C. cryoaerophila* and *C. nitrofigilis*. These groupings were broadly supported by further work involving analysis of isoprenoid quinones (Moss *et al.*, 1990), but Skirrow (1989) noted that while groupings I and III are in accord with phenotypic characteristics and logic, the proposed grouping of *C. pylori*, *C. cinaedi* and *C. fennelliae* with *Wolinella* should be regarded as provisional. The current situation is that a new genus, *Helicobacter*, is proposed for *C. pylori*, but that all other species remain classified with *Campylobacter*.

Campylobacters are responsible for a range of disease in animals and man (**Table 81**). In man, enteritis is the predominant symptom, and *Campylobacter* has risen rapidly in importance from obscurity to a significance approaching, or even exceeding, that of *Salmonella*. Most cases of enteritis are caused by the 'thermophilic species', *C. jejuni*, *C. coli* and *C. laridis*, with *C. jejuni* being the most important in most, but not all, regions. However, disease due to other species may be underestimated.

Campylobacter and *Helicobacter* represent but two regions of a spectrum of microaerophilic microorganisms some of which, at least, play further roles in disease in humans. Such bacteria may include the 'unclassified microaerophilic bacterium' (Romero *et al.*, 1988), which was associated with gastroenteritis

Table 81. Diseases of animals and man associated with species of *Campylobacter*.

Species	Host	Disease
C. jejuni	Man	**Enteritis** Systemic infection, abortion, perinatal disease
	Sheep	Abortion
C. coli	Man	**Enteritis**
C. laridis	Man	Enteritis
C. hyointestinalis	Man	Enteritis Proctitis
	Pigs	Proliferative adenitis Adenomatosis
'*C. upsaliensis*'	Cats and dogs	Diarrhoea
	Man	Enteritis?
C. cinaedi	Man	Enteritis
C. fennelliae	Man	Enteritis
C. conciscus	Man	Periodontal disease?
C. fetus subsp. *fetus*	Man	Enteritis? Systemic infection, abortion, perinatal sepsis and meningitis Enteritis
	Cattle and sheep	Sporadic abortion
subsp. *venerealis*	Cattle	Abortion, infertility
C. mucosalis	Pigs	Intestinal adenomatosis, necrotic enteritis, regional ileitis, proliferative haemorrhagic enteropathy
C. cryoaerophila	Man	Enteritis?
	Cattle and pigs	Abortion
H. pylori	Man	**Gastritis** **Peptic ulcers**

Notes:
1 Major diseases of man are printed in **bold**.
2 *H. pylori* was previously *C. pylori* and is included for comparison.

and which, despite some biochemical similarities with *H. pylori*, is probably distinct from that organism and from *Campylobacter*. The spectrum may indeed extend to include spiral rods with a strictly anaerobic metabolism, such as *Anaerobiospirillium*, which has recently been implicated in diarrhoeal disease (Malnick *et al.*, 1990).

Isolation and detection of *Campylobacter*

The recognition of *Campylobacter* as a major cause of human and animal disease led to a rapid development of media and techniques for the isolation of the organism. Although those currently available are satisfactory for the recovery of the major species, such as *C. coli* and *C. jejuni*, some species are not recovered by generally used methods, and as a consequence remain poorly understood.

There are two basic approaches to the isolation of campylobacters. The first is based on the physical separation of campylobacter from other organisms by filtration through a 0.65µm membrane filter, followed by plating of the filtrate on non-selective media. The second follows a similar protocol to that for other food poisoning micro-organisms and involves selective enrichment and plating stages, preceded, in some cases, by resuscitation.

Resuscitation

The situation regarding resuscitation is not clear and while recommended by some workers (Humphrey, 1986a,b), others have found no advantage (Beuchat, 1987). Currently, resuscitation is not widely used.

Enrichment

Enrichment is widely used for the isolation of *Campylobacter* from foods, although the value of enrichment for clinical samples is not universally accepted. Enrichment media are based on a nutrient rich broth containing combinations of anti-microbial agents, such as polymyxin, trimethoprim and vancomycin. Blood is added to most enrichment media.

Selective plating

A range of media of differing selectivity has been developed, commonly-used examples being Skirrow's medium, Butzler's medium, Blaser-Wang medium and Preston medium. A blood-free selective medium has also been developed. Such media may not be suitable for some lesser known species.

Rapid detection

Rapid detection of *Campylobacter* has received little attention, although a latex co-agglutination technique is available.

The isolation and detection of Campylobacter *is discussed in greater detail, with respect to choice of media and laboratory methods, in* **Laboratory Methods**, *pages 225–230.*

Taxonomy of *Campylobacter*

In comparison with the longer established pathogens such as *Salmonella*, the taxonomy of *Campylobacter* and related bacteria is in a state of continuing evolution, with the recognition of several 'new' species and the reassignment of an existing species, *C. pylori* to the new genus *Helicobacter*. The difficulties that such changes inevitably bring the practising microbiologist are compounded by problems which relate to the inherent nature of *Campylobacter*. For example, members of the genus are relatively inert biochemically, and the number of easily determined differential characteristics is relatively small. Furthermore, species often differ more in the degree to which a characteristic is demonstrable than in the number of different characters (Penner, 1988). For these reasons, the use of DNA-hybridisation to define species is widespread with *Campylobacter*, but only a few laboratories are able to perform these genetic tests.

Differentiation of *Campylobacter* from other genera

Differentiation of *Campylobacter* from other genera is relatively straightforward in comparison with differentiation of species within the genus. In practice, isolates from foods, or from people suffering enteritis, may be identified by cell morphology, the organisms having a characteristic spiral or S-shaped appearance. Examination of a wet film for the characteristic darting in corkscrew fashion is a valuable means of confirming the identity of campylobacters.

Further testing is not usually necessary in routine work, but in cases of doubt *Campylobacter* may be recognised by the criteria listed in **Table 82**. Alternatively serological and genetic methods are available some of which may be applied on a routine basis.

Table 82. Characteristics of *Campylobacter*.

Oxidase positive
Neither ferment nor oxidise carbohydrates
Reduce nitrate[1]

[1] *C. fennelliae* does not reduce nitrate.

Table 83. Differentiation of the major foodborne species of *Campylobacter*.

	C. jejuni	*C. coli*	*C. laridis*
Growth at			
30.5°C	−[1]	+	+
45.5°C	+/−	+	+
Hippurate hydrolysis	+	−	−
H$_2$S in iron medium[2]	+/−	−	+
Resistance to nalidixic acid[4]	−	−	+[5]
Tolerance of tetrazolium trichloride	+/−	+	−
Anaerobic growth with Aspartate Fumarate TMAO[6]	−	−	+

[1] Some strains of *C. jejuni* may grow at this and lower temperatures.
[2] Method of Skirrow and Benjamin (1982).
[3] Biovar 1 is negative; biovar 2 is positive.
[4] 30 μg
[5] Nalidixic acid sensitive strains, which are also urease-positive exist.
[6] Trimethylamine *N*-oxide hydrochloride.
Note: Minimum reactions for identification are printed in **bold**.

Differentiation of species of *Campylobacter*

Differential reactions for *Campylobacter* are listed in **Table 83**. Differentiation is often made on the basis of a small number of phenotypic characteristics, and distinctions can be blurred by the existence of 'intermediate' strains. Furthermore, the recognition of newly described species has tended to reduce the value of some traditionally used tests. For example, the catalase test has been widely used in the past to differentiate the catalase-positive pathogenic species from the catalase-negative commensals but, more recently, the non-pathogenic but catalase-positive species, *C. nitrofigilis*, has been described, while the putative pathogen '*C. upsaliensis*' is usually catalase-negative.

Enteropathogenic species

The major human enteropathogens are all members of the thermophilic (more correctly thermotolerant) group of campylobacters. Each species is closely related, but can be differentiated on the basis of biochemical reactions.

C. jejuni and *C. coli*

C. jejuni and *C. coli* were the first campylobacters to be recognised as important human pathogens. For several years, there was no reliable means of differentiating between the two species, a situation which led to considerable confusion, but which was resolved by the finding that *C. jejuni* was capable of hydrolysing hippurate, whereas *C. coli* was not (Harvey, 1980). The hippurate test is of major importance in the taxonomy of the thermophilic campylobacters, and in most cases permits reliable differentiation between these two important species. However, the test is not, infallible and a small number of strains of *C. jejuni* are hippurate-negative (Hebert *et al.*, 1984). The existence of hippurate-positive strains of *C. coli* is not proven.

The situation with *C. jejuni* may resemble that which previously applied to *Yersinia enterocolitica* in that biochemically 'atypical' strains isolated from the environment have no role in disease (Maurer, 1988a, b). Currently, such strains are readily confused with pathogenic strains isolated from man.

Nitrate-negative campylobacters have been assigned to a subspecies of *C. jejuni*, *C. jejuni* subsp *doylei* (Steele and Owen, 1988). Methods used for recovery of conventional *C. jejuni* are unsuitable for many strains of

C. jejuni subsp *doylei*. The role of this subspecies in disease is undefined.

Conventional strains of *C. jejuni* should strictly be referred to as *C. jejuni* subsp *jejuni*.

C. laridis

Strains of *C. laridis* were originally referred to as nalidixic acid-resistant thermophilic campylobacters, the species name *C. laridis* being proposed by Benjamin *et al.* (1983) and currently subject to a change to *C. lari*. Although resistance to nalidixic acid remains a differential feature of the species, it is not absolute, as nalidixic acid-resistant strains of *C. coli* and *C. jejuni* have also been isolated. Anaerobic growth using trimethylamine oxide as a terminal hydrogen acceptor is a definitive characteristic of *C. laridis*, but the property is also present in *C. hyointestalinis* and two of the biovars of *C. sputorum*. *C. laridis* also produces H_2S, but sensitive methods are required for its detection.

Variants of *C. laridis* which produce urease and which are sensitive to nalidixic acid have been isolated from environmental and human sources (Bolton *et al.*, 1985), and may also be a cause of human illness (Megraud *et al.*, 1988). The position of these bacteria as biovars of *C. laridis* has been confirmed using electrophoretic protein patterns (Owen *et al.*, 1988).

'C. upsaliensis'

'C. upsaliensis' is a putative pathogen and a causative role in diarrhoea is not yet fully established. The organism has been isolated from blood cultures from paediatric patients and may be an opportunistic pathogen in young children (Lastovica *et al.*, 1989). The species is catalase-negative, or weakly positive, and sensitive to cephalothin; at least 4% hydrogen is required in the atmosphere for cultivation. Phenotypic characteristics may not form a suitable basis for identification (Lastovica *et al.*, 1989).

C. cinaedi and C. fennelliae

Originally referred to as 'campylobacter-like organisms' (CLOs) 1 and 2, these species were isolated from homosexual males together with a third isolate CLO-3, the taxonomic status of which remains obscure (Totten *et al.*, 1985). *C. cinaedi* infections probably occur more widely than usually assumed and are not restricted to homosexual or bisexual males (Vandamme *et al.*, 1990). Both species grow under hydrogen-containing microaerophilic conditions at 37°C, but not at 25 or 42°C. *C. cinaedi* may be differentiated from *C. fennelliae* on the basis of its ability to reduce nitrate and intermediate sensitivity to cephalothin. A further unusual characteristic of *C. fennelliae* is the smell of hypochlorite produced by cultures.

C. cinaedi may be divided into two subgroups by means of DNA probes, but these groups cannot be differentiated by biochemical tests.

C. hyointestalinis

Although *C. hyointestalinis* was first considered to be a pathogen of pigs, the organism is now known to be an occasional human pathogen. *C. hyointestalinis* may be differentiated from *C. fetus* by the production of H_2S in the presence of hydrogen, and by anaerobic growth using trimethylamine oxide as a terminal hydrogen acceptor. Reported outbreaks have involved homosexuals.

C. fetus

C. fetus has two subspecies, *C. fetus* subsp. *fetus* and subsp. *venerealis*. Both are primarily associated with animals, but *C. fetus* subsp. *fetus*, although better known as a cause of sporadic abortion in cattle and sheep, is also a rare cause of human disease. *C. fetus* subsp. *fetus* is sensitive to cephalothin, but while most strains are unable to grow at 42°C, a number of atypical strains capable of growth at this temperature have been isolated from cases of diarrhoea.

C. cryoaerophila

C. cryoaerophila differs from other campylobacters in being capable of growth in air, and in having an optimum temperature for growth of 30°C. The organism has been associated with bovine and porcine abortions, and there is also a report of isolation from a homosexual male suffering intermittent diarrhoea (Tee *et al.*, 1988). *C. cryoaerophila* grows best on media for leptospiras, and would not usually be recovered on the media, or at the incubation temperatures, commonly used.

C. conciscus

C. conciscus is catalase-negative and was previously thought to be non-pathogenic, despite its presence in large numbers in periodontal disease. A group of campylobacters associated with gastroenteritis, EF Group 22, have been identified as *C. conciscus* (Vandamme *et al.*, 1989), and the species is now considered to be a putative enteropathogen.

The taxonomy of Campylobacter *is discussed in greater detail, with respect of choice of methods and laboratory practice, in* **Laboratory Methods**, *pages 230–232.*

Typing of *Campylobacter*

Serotyping

Both *C. jejuni* and *C. coli* are antigenically hetero-geneous, and several serotyping schemes have been devised (Abbott *et al.*, 1980; Kosunen *et al.*, 1980; Penner and Hennessy, 1980; Lauwers *et al.*, 1981; Lior *et al.*, 1982; Penner *et al.*, 1983; Lior, 1984a). These are of two types:

1. Schemes based on the detection of heat stable anti-gens by passive haemagglutination. The widely used scheme of Penner *et al.*, (1983) recognises 42 serovars of *C. jejuni* and 18 serovars of *C. coli*. Most North American isolates can be typed using this scheme, but many from Europe are untypable. New serovars identified among untypable isolates are being incorporated into an extended scheme, while the similar typing scheme of Lauwers *et al.*, (1981) is being integrated into the Penner *et al.*, (1983) scheme (Penner, 1988).

2. Schemes based on the detection of heat labile anti-gens by slide agglutination. The most widely used scheme of this type is that of Lior *et al.*, (1982), which recognises 68 serovars of *C. jejuni*.

Typing by heat stable antigens requires the extrac-tion of antigens and passive haemagglutination titra-tions, and is thus seen as being labour intensive. However, the technique was critically appraised by Fricker *et al.*, (1987), who developed a more rapid passive haemagglutination technique. Subsequently, a rapid serotyping technique was developed using pass-ive haemagglutination combined with co-agglutination (Illingworth and Fricker, 1987).

The slide agglutination technique used for the detection of heat labile antigens is, in principle, much simpler, and for this reason has been recommended for use in clinical laboratories, especially in develop-ing nations (Lior and Butzler, 1986). However, it is necessary to use absorbed antisera and there may be problems due to autoagglutination.

Although serotyping has been of great value in a large number of epidemiological investigations, some caution is necessary. Serological grouping does not necessarily indicate a definite identity among isolates of the same serogroup (Karmali *et al.*, 1983), while heterogeneity may exist among similar strains (Hutchinson *et al.*, 1987). Thus strains which are of the same serovar in both the Penner and Lior typing schemes, may be of different biovar, while strains of the same biovar may all be of the same serovar by one typing scheme, but of different serovars by the other. Hutchinson *et al.*, concluded that at least two typing methods are required when working with epidemio-logically related strains, a combination of a serotyping scheme with an extended biotyping scheme being most effective.

Although serotyping schemes are of epidemiologi-cal value they are not considered useful for identifying potentially pathogenic campylobacters (Johnson and Lior, 1984; Fricker and Park, 1989).

Serotyping of Campylobacter *is discussed in greater detail, with respect to laboratory practice, in* **Laboratory Methods**, *pages 232–234.*

Phage typing

Despite its potential value, phage typing has not been widely applied to *Campylobacter* in the past (Penner, 1988). A scheme has, however, been devised which may be used alone or, preferably, in conjunction with Penner serotyping (Saloma *et al.*, 1990). In use with isolates from outbreaks, phage typing subdivided serovar 2 and serogroups 4. 13, 16 and 50, but strains of phage group 39 from three different outbreaks were different Penner serovars.

Biotyping

A number of biotyping schemes have been proposed, ranging in complexity from that of Skirrow and Benjamin (1982), which defines two biovars of *C. jejuni*, to the more complex schemes such as that of Lior (1984b), which defined four biovars of *C. jejuni* and two each of *C. coli* and *C. laridis*.

The lack of readily determined phenotypic charac-teristics has led to a number of alternative approaches to biotyping. These include the 'Preston' scheme (Bolton *et al.*, 1984) , which is essentially a resisto-typing method, carbon substrate utilisation tests (Elharif and Megraud, 1986), and the use of lectins (Wong *et al.*, 1986).

Biotyping has been found to be of value in epidemio-logical investigations, and is most effective when combined with a serotyping scheme. However, bio-typing is only of epidemiological value and has no clinical significance (Hebert *et al.*, 1982).

Practical aspects of the biotyping scheme of Skirrow and Benjamin (1982) are discussed in greater detail in **Labora-tory Methods**, *page 234.*

Other typing schemes

Protein profiles

A number of workers have employed protein profiling for the typing of *Campylobacter*. Methodology has involved either polyacrylamide gel electrophoresis (Morris and Park, 1973; Costas and Owen, 1983), or isoelectric focusing (Abrams, 1985). The technique is not suitable for routine use.

Bacterial restriction endonuclease analysis (BRENDA)

BRENDA has been successfully used to type *C. jejuni* and *C. coli* from animals and man (Kakoyiannis *et al.*, 1988). The potential of this type of technique is greatest at the infrasubspecific level, but performance is dependent on method of digestion and data analysis. A potential application exists where serotyping is not possible (Owen *et al.*, 1989).

Symptoms of *Campylobacter* infections

C. jejuni and *C. coli*

C. jejuni and *C. coli* are the two most common causes of campylobacter infections. Symptoms may vary considerably both in severity and in the relative importance of the various manifestations of the disease. Superimposed on this variation are differences between infections in adults and infants.

The incubation period varies between 2 and 11 days (Pinegar and Suffield, 1982), with 3 to 5 days being most common. Reports of excessively long incubation periods are spurious (Skirrow, 1984), but a long incubation period is a general characteristic of *Campylobacter* infections, the shortest recorded being 20 hours.

Enteritis in adults

The onset of clinical symptoms in adults tends to be abrupt. In some cases the early symptoms are intestinal, but in others an influenza-like prodromal period occurs with associated malaise, headache, shivering, dizziness and generalised myalgia. The prodrome usually lasts for one day, but ranges in duration from a few hours to two to three days. Enteric symptoms are usually more severe when following a prodrome.

Initial intestinal symptoms may either be abdominal pain or nausea and vomiting. In a small number of cases, a temperature of *ca* 40°C develops, which can be associated with delirium.

Diarrhoea, which is of two distinct types, is an important symptom in most, but not all cases. The severity varies from the severe and prostrating to a few loose stools, but is generally less than that of *Salmonella* infections.

The two types of diarrhoeal symptom may be the result of different mechanisms of pathogenicity predominating in different strains, or may be representative of the two ends of the clinical spectrum arising from translocation (Walker *et al.*, 1986). The first type is secretory, with profuse watery stools which are usually bile stained and of foul odour. The second type resembles dysentery, stools containing inflammatory cells and occult, or overt, blood. Colonic infection is a common feature, and in some cases symptoms are predominantly those of colitis (Skirrow, 1984).

Abdominal pain is a major feature of *Campylobacter* enteritis, and is usually of greater severity than that due to *Salmonella* and other enteric pathogens. Pain is initially colicky, but later becomes continuous and may move to the right iliac fossa. Temporary relief is gained by flatus and the passage of stools. Campylobacteriosis may be confused with acute appendicitis, although the clinical distinction between the two conditions is fairly straightforward (Skirrow, 1987). However, *Campylobacter* has been associated with 'true' appendicitis (Pearson *et al.*, 1983).

Vomiting is not a major symptom of *Campylobacter* enteritis and, despite its early appearance in some cases, rarely occurs more than once or twice during the course of an outbreak.

Campylobacter infections are usually self-limiting, symptoms lasting less than one week in healthy adults, and recovery is usually uneventful. The patient has a persistent weakness during recovery, and abdominal pain and discomfort may persist for several days after other symptoms disappear. Recurrence of symptoms is a feature of the infection in *ca* 25% of cases (Blaser *et* al., 1983), and is probably due to the ability of *Campylobacter* to undergo rapid phase variation (see page 49). Recurrence may also be associated with stomach overloading as a result of too rapid a return to normal dietary habits. Death is rare even among those with an underlying condition.

Symptoms due to *C. jejuni* and *C. coli* are essentially the same, although those due to *C. coli* are generally less severe and bloody stools are less common. The predominance of *C. jejuni* makes direct inter-species comparisons of symptoms difficult. However, two direct comparisons produced directly conflicting results. The first, conducted in Yugoslavia (Popovic-Uroic *et al.*, 1988), confirmed the more serious nature of *C. jejuni* enteritis, while the second, conducted in Italy (Figura and Guglielmetti, 1988), reached the opposite conclusion (**Table 84**). There was no evidence of hypervirulent strains of *C. coli* distorting the findings in the Italian survey, but discrepancies may be due to differences in the environmental circulation of *C. coli*, or to differences in the age of patients.

Table 84. Clinical comparison of *Campylobacter jejuni* and *Campylobacter coli* enteritis.

| | Per cent of cases | | |
	C. jejuni	C. coli	Significance
Yugoslavia			
Watery stools	73.6	68.8	NS
Occult blood	35.5	21.2	<0.01
Fever >38°C	71.4	53.4	<0.01
Peak temperature	38.8°C	38.7°C	NS
Maximum number of stools/day	7.4	7.1	NS
Anti-microbial therapy needed	48.6	39.8	<0.05
Italy			
Mucoid stools	89	88	NS
Bloody stools	60	53	NS
Diarrhoea lasted >7 days	33	67	<0.05
Abdominal pain	29	47	NS
Vomiting	18	56	<0.01
Dehydration	6	6	NS
Fever	56	57	NS
White cell count >11,000/µl	21	53	<0.01

Notes:
1 NS = Not significant.
2 Based on Popovic-Uroic *et al.* (1988); Figura and Guglielmetti (1988).

Enteritis in children

In addition to differences in the frequency of some symptoms between adults and children, there are also significant differences between symptoms observed in industrialised and developing nations. Symptoms, particularly in developing nations, are mild and, in contrast to other forms of infantile diarrhoea, severe dehydration is rare. However, symptoms may be prolonged and the consequent malabsorption of nutrients has been associated with a failure to thrive (Pignata *et al.*, 1984).

Vomiting is relatively common being recorded in up to 50% of cases, but few patients have abdominal pain although the incidence of pseudo-appendicitis is higher among children than among adults. Fever is uncommon in children aged less than one year, but older children may develop a high fever accompanied by convulsions, hallucinations and meningism. Bloody stools are a common feature in industrialised nations but, elsewhere, watery stools predominate.

Differences in symptoms between industrialised and developing nations are probably due to hyperendemic exposure to *Campylobacter* in the latter (Molbak *et al.*, 1988; Taylor *et al.*, 1988).

Complications and extraintestinal infections

Campylobacter enteritis is infrequently accompanied by complications. These may be classified as being of early or late onset and are summarised in (**Table 85**).

Table 85. Complications associated with *Campylobacter* enteritis.

Early onset	Late onset
Haemolytic uraemic syndrome	Reactive arthritis
Gastrointestinal haemorrhage	Reiter's syndrome
Biliary tract infection	Guillain-Barré syndrome
Cholecystitis	Erythema nodosum
Hepatitis	
Pancreatitis	
Appendicitis	
Pancreatitis	
Facial rash	

Campylobacter has been established as one of the causative organisms of haemolytic uraemic syndrome, while the role in Guillain-Barré syndrome is of particular note. In a retrospective study, C. jejuni was considered to be the most common single, identifiable pathogen associated with the syndrome, which is more serious among people with campylobacteriosis (Kaldor and Speed, 1984). Although Guillain-Barré syndrome is considered rare, it is one of the most common causes of hospital admission for a neurological cause, and Campylobacter has been associated with both adult and paediatric cases (Archer and Young, 1988).

Both C. jejuni and C. coli have been associated with extraintestinal infections where the portal of entry was believed to be the intestine, although links with specific foods were not proven. For example, C. jejuni has been associated with lung infections (Blaser et al., 1986), while both C. coli and C. jejuni are causes of abortion and perinatal sepsis. In the case of abortion, infection of the foetus is probably due to transplacental passage rather than to ascending transcervical infection (Simon et al., 1986) although mild infections of neonates are probably acquired at birth.

C. laridis

The number of reported cases of illness due to C. laridis is small, although the importance of the organism may be underestimated. In most cases, the symptoms have resembled those of C. jejuni (Penner, 1988), although a case of bacteraemia in an elderly man with multiple myeloma has also been reported (Nachamkin et al., 1984).

C. cinaedi and C. fennelliae

Symptoms associated with these species are typically those of enteritis, but bacteraemia has also been reported, including successive bacteraemias due to C. cinaedi followed by C. fennelliae (Ng et al., 1987).

C. hyointestilinis

A small number of human infections due to C. hyointestalinis have been reported (Fennell et al., 1986; Edmonds et al., 1987), and it is possible that the organism is under-reported. Symptoms primarily involve diarrhoea, but proctitis has also been reported.

C. fetus subsp. fetus

Human infections due to C. fetus subsp. fetus are rare, and have been largely restricted to people with a predisposing condition. Symptoms are usually those of a non-specific, febrile illness, which is typically subacute and relapsing with spontaneous recovery occurring even in the compromised individual. Mortality may be high, but this is due primarily to the underlying symptoms.

In a minority of cases, focal infections develop. Meningitis is most common, but other infections include pericarditis, spontaneous peritonitis, salpingitis, pleural infection, septic arthritis, hepatitis and cellulitis.

The tropism for foetal tissue exhibited by C. fetus subsp. fetus in cattle and sheep, may also be manifest in humans, and abortion and perinatal sepsis have occasionally been reported. Meningitis may also occur in neonates.

Although C. fetus subsp. fetus was previously thought not to be associated with gastroenteritis, the organism has been isolated from the faeces of people suffering the disease.

Persons at particular risk from *Campylobacter* infection

Susceptibility

In industrialised countries, Campylobacter infections show a bimodal distribution with respect to age. In the UK, for example, the incidence is highest in children aged one to four years with a frequency of attack of ca 5.4%/year. A second peak of infection occurs with young adults aged 15 to 24 years, when the attack rate is ca 2%/year. Children in the intervening years suffer an attack rate of only 0.3%/year, and the incidence is also low in adults aged over 21 years (Skirrow, 1984; 1987). The severity of the disease is greatest in young adults and hospitalisation is most common in this age group.

A high level of environmental exposure means that Campylobacter enteritis tends to be a disease of infancy in developing countries.

In adults over 21 years, there is no difference in the incidence of infection between males and females, but in young adults and children, 1.5 times as many males as females are affected. Attempts have been made to explain this in terms of exposure rather than susceptibility. For example, Skirrow (1987) postulated that males may be at greater risk through occupational exposure, handling chickens etc, while the higher incidence among male American college students than female students was attributed to the culinary incompetence of the former, leading to consumption of undercooked chickens (Tauxe et al., 1985). Neither of these explanations appear convincing.

As a group, male homosexuals are highly susceptible both to *Campylobacter* enteritis and systemic infections. Indeed, *C. cinaedi* and *C. fennelliae* were initially uniquely associated with homosexuals. Susceptibility among homosexuals is a general phenomenon distinct from AIDS, but appears to be related to the immunosuppressed state common among those practising anal intercourse. In some cases *Campylobacter* would seem to be sexually transmitted, a fact which was probably generally recognised by homosexuals before the 'straight' members of the medical profession:

'Sexually transmitted diseases were common, especially among those who had many partners. Lice, scabies, urethritis, proctitis . . . herpes, worms, shigella, campylobacter and those familiar vaudevillians syphilis and gonorrhoea.' (Black, 1986).

Systemic *Campylobacter* infections are almost always in patients who have an underlying predisposing condition. Such conditions include malignant disease such as leukaemia, immunosuppression, alcoholism, liver cirrhosis, diabetes, chronic renal failure and gastrectomony (Skirrow, 1984). Pregnancy is also a risk factor, probably due to the suppression of cell-mediated immunity (Weinberg, 1984). With the obvious exception of pregnancy, systemic *Campylobacter* infections usually involve males aged more than 45 years.

AIDS has not been considered to be of particular importance as a predisposing factor for campylobacteriosis. Infections may be overlooked due to the unusual symptoms; chronic or recurrent infections of both a septicaemic and a digestive nature; and it is likely that AIDS is of greater importance than previously thought. Risk of *Campylobacter* infection is increased by additional digestive tract infections such as cytomegalovirus or candidosis (Bernard et al., 1989).

A wide range of conditions may predispose children to infection by '*C. upsaliensis*' including kwashiorkor, foetal alcohol syndrome, anaemia and burns (Lastovica, 1989).

Exposure

Exposure to *Campylobacter* is greatest in developing countries where standards of hygiene are poor. *Campylobacter* is associated with travellers' diarrhoea and, in Sweden, between 50 and 75% of infections with the organism are associated with overseas travel. Multiple infections with two, or more, serovars is common in developing nations such as Mexico, but does not occur in industrialised nations (Sjogren et al., 1989).

Occupational exposure is also an important risk factor in the acquisition of *Campylobacter* infection (Mancinelli et al., 1987). Farm workers, veterinary surgeons, slaughterhouse and poultry plant workers, and butchers are all at enhanced risk of infection. This risk may extend to *H. pylori* (*C. pylori*), which is possibly of zoonotic origin (Vaira et al., 1988). Contact with domestic animals, primarily puppies, but also cats and kittens, is also a risk factor in *Campylobacter* infections (see page 221). Zoonotic acquisition of *C. jejuni* infection from a coyote has also been reported (Fox et al., 1989).

Dietary habits may increase the risk of exposure to campylobacters with, as ever, the conscious consumption of raw milk and meat being significant risk factors.

Treatment of *Campylobacter* infections

Campylobacter infections are often self limiting, and no treatment is required. In severe cases, dehydration may occur and a glucose electrolyte mix may be prescribed. Anti-diarrhoeal agents such as codeine, diphenoxylate (Lomotil©) or loperamide (Imodium©) can be effective but should not be administered to children (Skirrow, 1984).

Chemotherapy is indicated in severe cases or where complications are present, and erythromycin is gener-

ally accepted as the drug of choice (Anders et al., 1982; Pai et al., 1983). However, erythromycin resistance may be a problem (Bernard et al., 1989). Tetracyclines, particularly minocycline, are also effective but the high level of resistant strains means that laboratory control is essential.

Nalidixic acid is effective against *C. coli* and *C. jejuni*, but the emergence of resistant strains is causing increasing problems (Altwegg, 1987).

Mechanisms of pathogenicity

Infective dose

Campylobacter is unable to grow in foods and, in some cases at least, infective doses are low. As few as 500 organisms were estimated to have successfully established infection in both a waterborne (Pearson, 1986) and a foodborne (Robinson, 1981) outbreak. However, there is considerable variation in the infective dose, and in a milkborne outbreak 10^6 organisms were estimated to have been required for infection (Steele and McDermott, 1978).

Variation in the infective dose is probably due to differences in the virulence of different *Campylobacter* strains (Coid *et al.*, 1987), differences in host susceptibility, or to a combination of these factors. Evidence from milkborne outbreaks suggests the possibility of acquired immunity. In an outbreak at a Kansas bible school due to raw milk, the attack rate was only 1 in 17 in children who drank raw milk at home, but seven times greater in children who had not previously been exposed to the organism (Anon, 1989). Asymptomatic reinfection among Mexican patients who had previously suffered campylobacteriosis due to multiple serovars, is also evidence of acquired immunity (Sjogren *et al.*, 1989).

Motility and chemotaxis

The ability of *Campylobacter* to colonise the gastric mucosa is considered to be a major determinant of pathogenicity (Lee *et al.*, 1986). Motility and chemotaxis are the two main factors involved in successful colonisation.

The role postulated for motility is similar to that in *Vibrio cholerae*. The characteristic darting movement of *Campylobacter*, together with the corkscrew shape of the cells, is suited to movement in a viscous environment and may confer a selective advantage in the intestinal mucosa. Furthermore, the mechanism of motility can change to maintain efficiency in different viscosities (Ferrero and Lee, 1988).

Chemotactic direction is likely to greatly enhance the role of motility as a determinant of pathogenicity.

An affinity for mucus has been demonstrated (Lee *et al.*, 1983), mucin functioning as a chemoattractant primarily because of taxis towards L-fucose (Hugdahl and Doyle, 1985; Hugdahl *et al.*, 1988). Taxis towards L-fucose is sufficiently powerful to overcome repulsion by bile.

Adherence and penetration

Adherence may be demonstrated *in vitro* using cell cultures (Walker *et al.*, 1986) and porcine brush borders (Naess *et al.*, 1983). Both adherence and invasion require active participation by the host cell and two distinct antigens are involved, one which initiates interaction between the bacteria and the host cells, and one which initiates phagocytosis (Konkel and Joens, 1989). The role of surface exposed outer membrane proteins has been explored by DeMelo and Pechere (1990), who described a 32 Kda protein, which was directly correlated with *in vitro* binding to eukaryotic cells and invasion. It was suggested that this protein might form part of a microcapsule.

Campylobacter cell-free supernatant has been associated with disturbance of myoelectric activity in isolated segments of rabbit ileum (Sninsky *et al.*, 1985). The possible *in vivo* consequence is the discontinuation of peristalsis and the creation of conditions favourable for adherence and proliferation.

Penetration of the intestinal epithelium occurs in the terminal ileum and colon, although the secretory form of diarrhoea is considered to be a disease of the small bowel.

Proliferation of *Campylobacter* may occur in the intestinal epithelium, resulting in lesions which resemble those of *Shigella*. Alternatively, penetration may involve minimal damage to the intestinal mucosa, which serves as a passage to the *lamina propria* and the mesenteric lymph nodes, which are the sites of proliferation. In this circumstance, the pathogenic mechanism of *Campylobacter* resembles that of *Salmonella*.

Systemic spread

Systemic spread is rare in infections of otherwise healthy individuals by *C. jejuni*, *C. coli* and *C. laridis*, although in susceptible people the organisms are able to spread from the intestine to colonise internal organs.

In contrast, systemic spread is a feature of *C. fetus* subsp. *fetus* infections in man. The difference between the species appears to lie not in differences in invasion determining properties, but in differences in resistance to serum.

Survival and proliferation in host tissue

Resistance to normal serum

C. jejuni, *C. coli* and *C. laridis* are generally sensitive to normal serum, although there is strain variation, and with *C. jejuni* a higher level of resistance to serum is found in bloodstream isolates than in stool isolates (Walker, 1986). A significant level of resistance is found in some strains of *C. fetus* subsp. *fetus* (Blaser *et al.*, 1988) but again there is strain variation and Coid *et al.*, (1987) noted that two of four strains investigated were readily killed.

A capsule of MW 100,000 may be responsible for serum resistance and also have a role in resistance to phagocytic killing in *C. fetus* subsp. *fetus* (Blaser *et al.*, 1988).

Resistance to phagocytosis

The liver has been shown to play an important role in the elimination of *C. jejuni* during systemic infection. Cells are phagocytosed, but not killed, by granulocytes, the 'infected' granulocytes being cleared in the liver (Bar *et al.*, 1989). It must therefore be assumed that resistant strains of *C. fetus* subsp. *fetus* must be resistant to killing both within the phagocyte, and during clearing in the liver. In addition to a possible role for the capsule in resistance to phagocytosis, a surface array protein (S-layer) may also be of importance (Dubreuil *et al.*, 1988; Zei and Blaser, 1990). Possession of a S-layer appears to be restricted to *C. fetus* subsp. *fetus*, and none has been demonstrated in other human pathogenic campylobacters.

The ability to survive within phagocytes seems likely to be an important means by which *C. fetus* subsp. *fetus* is distributed throughout the host reticulo-endothelial system (Kiehlbauch *et al.*, 1985).

Toxin production

Both *C. jejuni* and *C. coli* are known to produce at least two toxins, a heat-labile enterotoxin which has many features in common with cholera toxin (CT), and a cytotoxin (Johnson and Lior, 1984). The role of these toxins in human disease is not yet fully established.

Properties of the two toxins are summarised in **Table 86**.

Table 86. Properties of *Campylobacter* toxins.

	Enterotoxin	Cytotoxin
Molecular weight	60 to 70,000	>30,000
Heat inactivation	56°C/1 h 96°C/10 mins	100°C/30 mins
pH inactivation	pH 4 (partial) pH 2.8 (full)	Not known Not known
Trypsin sensitive	Not known	Sensitive

Note: Based on data from McCardell *et al.* (1984); Klipstein and Engert (1984); Guerrant *et al.* (1985).

Enterotoxin

The production of a CT-related enterotoxin by some strains of enteropathogenic campylobacters is well established (Ruiz-Palacios *et al.*, 1983; McCardell *et al.*, 1984). *Campylobacter* enterotoxin shares functional and immunological properties with CT, and secretory diarrhoea is induced by stimulation of adenylate cyclase activity in the intestinal mucosa and by disrupting normal ion transport in the erythrocytes (Ruiz-Palacios *et al.*, 1983).

Iron appears to play an important role in the production of enterotoxin by *Campylobacter* species. Production was stimulated by iron at levels above those required for growth, and strains previously thought to be non-enterotoxigenic were able to produce toxin when grown in a high iron medium. Such strains reverted to non-production when grown in a medium of low iron content (McCardell and Madden, 1985).

Cytotoxin

Production of a cytotoxin has been demonstrated by both *C. jejuni* and *C. coli* (Yeen *et al.*, 1983; Guerrant *et al.*, 1985). There is evidence that cytotoxin production is inducible, and thus may not be produced by isolates from a hostile environment until a correct stimulus is received (Fricker and Park, 1989).

Role of toxins in Campylobacter infections

As noted above, the role of toxins in *Campylobacter* infections is not yet fully established. Enterotoxin producing strains appear to be associated with the secretory form of *Campylobacter* diarrhoea, and cytotoxin producing strains with the dysentery-like type but, in a large survey, neither toxin could be isolated from campylobacters from patients with diarrhoea. This may have been due to the excretion of more than one serovar of *Campylobacter* and the selection of a non-toxigenic serovar for testing (Fricker and Park, 1989).

There has been more doubt concerning a role for the cytotoxin than for the enterotoxin, and Klipstein *et al.* (1985) considered that cytotoxin played no part in the pathophysiology of campylobacteriosis. Although this view was later challenged by a number of workers including Pang *et al.* (1987), who considered the views of Klipstein *et al.* to be 'premature', further evidence that the role of the cytotoxin is, at most, limited was presented by Perez-Perez *et al.* (1989). In this study, cytotoxin production was found to be rare, or at low level only, while a host immune response was absent.

Determination of virulence

Several models have been proposed for the determination of virulence including newly hatched chicks, chicken embryos, hamsters and rabbits. The use of such models is restrictive, and *in vitro* tissue culture methods have been developed for adherence (Newell *et al.*, 1985), penetration of eukaryotic cell membranes (Newell and Pearson, 1984), enterotoxins (Klipstein *et al.*, 1985) and cytotoxin (Yeen *et al.*, 1983). Tissue culture techniques cannot be applied in most food industry laboratories, and the development of serological tests is thus of considerable interest. ELISA assays similar to those used to detect CT have been developed for *Campylobacter* enterotoxin (McCardell *et al.*, 1984; Fricker and Park, 1989). However, assays are not commercially available.

Human carriage of *Campylobacter*

Human carriage has been implicated in the spread of *Campylobacter* via foods, but its involvement is rare (Blaser, 1985). Exclusion of people from work as food handlers who have *fully* recovered is not usually required providing adequate precautions are taken (see pages 411–413). However, an outbreak of *C. jejuni* enteritis at an Israeli army base was attributed to a temporary food handler (Cohen *et al.*, 1984). The handler had suffered acute enteritis prior to the outbreak, and may not have been fully recovered before returning to work.

Asymptomatic infection is common and as many as 25% of infected people fail to develop overt symptoms (Porter and Reid, 1980; Walker *et al.*, 1986). Asymptomatic infection is particularly common among children aged 5 to 14 years, and this may be an important factor in sporadic outbreaks of campylobacteriosis. There may also be implications where catering operations are conducted from domestic premises, and where children assist in food preparation.

Person-to-person transmission of *Campylobacter* is known, but is usually of limited importance requiring close contact between people. The spread is usually from a child to its mother and/or siblings (Skirrow, 1984), although in six cases of person-to-person transmission studied by Oosterom *et al.* (1984) only one case involved a child. Mother to child transmission is of importance in perinatal campylobacteriosis (Simon *et al.*, 1986).

Campylobacter in the environment

Enteropathogenic campylobacters are of zoonotic origin. In many cases, food animals are the source of infection in man, but there is evidence that the primary natural reservoir is small rodents. *Campylobacters* may also be readily isolated from fresh and estuarine water, but are thought to be transient contaminants.

Food animals and birds

Campylobacters have been isolated from all common food animals and birds (Blaser *et al.*, 1984). The organism is usually present as a commensal, although *C. jejuni* has been associated with abortion in sheep. The incidence of the organism in sheep is lower than in other food animals (Franco, 1988).

C. jejuni is the most common campylobacter in food animals, although the incidence of *C. coli* is relatively high in pigs.

Serotyping studies have shown that with *C. jejuni*, strains from food animals and birds tended to be of common serogroup with strains from humans (Munroe *et al.*, 1983; Vogt *et al.*, 1984; Rogol *et al.*, 1985). These findings were broadly confirmed by Kakoyiannis *et al.*, (1988), who used BRENDA to show that 61% of human strains could be matched to food animals.

Studies of the serological relatedness of animal and human strains of *C. coli* have produced variable results (Lior *et al.*, 1982; Munroe *et al.*, 1983; Rogol and Sechter, 1987), but Kakoyiannis *et al.* (1988) found that pig and human strains usually differed.

Domestic animals

Dogs, and to a lesser extent, cats have been implicated as a reservoir of *Campylobacter*. The incidence is higher in immature animals than in adult animals, and higher also in kennel populations and among strays than in household pets (Franco, 1988).

Animal to man infection is well documented, the classic scenario being the new puppy which develops diarrhoea and subsequently infects children. A community outbreak involving people all infected by the same litter of puppies has been described (Hutchinson *et al.*, 1987).

Wild animals and birds

Wild animals, especially rodents, have been suspected of being primary reservoirs of *Campylobacter* for a number of years (Fernie and Park, 1977). There is evidence that small rodents, in particular voles, are of importance in areas devoid of human habitation (Pacha *et al.*, 1987), while rats may be of importance in infecting food animals.

*Campylobacter*s have also been isolated from a variety of wild birds including pigeons, sea gulls, crows and ravens (Franco, 1988). In many cases, isolates from avian sources differ from those from man (Kakoyiannis *et al.*, 1988; Whelan *et al.*, 1988), although contamination of water supplies by bird droppings has been responsible for outbreaks of human campylobacteriosis (see page 223).

Other environmental sources

Although campylobacters are not normally considered to be free-living, they may be readily isolated from environmental sources of which water is most important, isolation rates from stream and river water sometimes exceeding 50%. In a systematic study of a river system in north west England (Bolton *et al.*, 1987), sewage works were found to be a major source of campylobacters despite the fact that up to 99.9% are removed during treatment (Arimi *et al.*, 1986). Surface runoff from farmland was also an important source of contamination especially after heavy rain. A separate study made in the Reading area of southern England showed similar results (Fricker and Park, 1989). In this study, it was shown that serovars present in sewage were almost entirely those common in human infections, but in a river receiving a major sewage effluent discharge other serovars, presumably derived from sources such as farm runoff, were also present.

Although the incidence of contamination of rivers is lowest in samples taken from rural sites and fast-flowing stretches of water (Bolton *et al.*, 1987), *Campylobacter* has also been isolated from remote mountain streams in wilderness areas. Taylor *et al.* (1983) commented that; *'No water should be regarded as free from the organism, including pristine mountain streams.'*

Survival of campylobacters in water is greatest at lower temperatures (Blaser *et al.*, 1980), and the incidence in natural water is highest in the autumn and winter (Bolton *et al.*, 1987; Carter *et al.*, 1987).

Although most surveys have been concerned with fresh water, *Campylobacter* has also been recovered from sea water (Franco, 1988), where it is presumably derived from rivers and sewage outfalls.

Foods commonly associated with *Campylobacter* infection

Outbreaks of *Campylobacter* food poisoning are of two types: large outbreaks, often in the community, in which a cause can be determined, and sporadic outbreaks. The epidemiology of sporadic outbreaks is not understood, but in many instances food appears to be involved.

Milk

Milk has been the vehicle of infection in a large number of outbreaks of *Campylobacter* food poisoning ranging in size from incidents involving a small number of people in a family to many hundreds of people.

The vast majority of outbreaks involve unpasteurised milk. Examples of large outbreaks include 50 children at a campsite in New Zealand, 616 people (two outbreaks) in a rural area of Scotland; and more than 500 people who had competed in a road race in Switzerland. Cows' milk has usually been involved, but a single case of *C. coli* infection due to goat's milk has been reported (Hutchinson *et al.*, 1985a).

Pasteurised milk has been involved in a small number of outbreaks, but in each case improper pasteurisation was involved. In a large outbreak, which affected more than 2,500 school children in two English towns, raw milk was thought to have by-passed the pasteurisation plant, although evidence was lacking. In a smaller outbreak, *C. jejuni*, survived the batch pasteurisation applied to milk privately pasteurised for a US boarding school. In this incidence, the severity of the process had been arbitrarily lessened to reduce 'burnt flavours' in the milk.

Despite the importance of milk as a vehicle of infection for *Campylobacter*, the reported incidence of the organism in raw milk is low, ranging from 0.4% (McManus and Lanier, 1987) to 8.1% (Humphrey and Beckett, 1987). In the latter survey, the milk was taken from herds which were known to be infected by *Campylobacter*.

Although there may be some under recovery of *Campylobacter* from raw milk, the low incidence probably reflects the sporadic nature of contamination, and may explain the variation in the number of milk-borne outbreaks from year to year. The contamination of milk probably occurs from faeces (Waterman *et al.*, 1984; Humphrey and Beckett, 1987), and while some support exists for the theory that *Campylobacter* is a cause of mastitis (Sandstedt, 1982, Hutchinson *et al.*, 1985b), this route seems unlikely.

Campylobacteriosis has been associated only with liquid milk and not with dairy products. The organism is sensitive to lactic acid (Skirrow *et al.*, 1982), and thus is unlikely to be a risk in fermented products such as yoghurt even if made from raw milk (Cuk *et al.*, 1987).

Poultry

Poultry, especially chickens, is an important cause of *Campylobacter* infections. In cases where a direct food link is known, the poultry has usually been undercooked. Cross-contamination from raw poultry to foods which are not cooked before eating, however, probably plays a large, if not a major, role in sporadic cases of campylobacteriosis (Dawkins *et al.*, 1984; Coates *et al.*, 1987). In an investigation by Pearson *et al.*, (1985), the route of transmission from chickens to salads and cooked meats via the hands of inexperienced chefs, which were contaminated during evisceration, was clearly demonstrated. Chickens from a single producer were also shown to be responsible for a large number of sporadic cases throughout southern England.

The incidence of *Campylobacter* may be 100% on freshly packed chickens (Hartog *et al.*, 1983) and while death of campylobacters occurs during storage, the incidence of contamination on fresh chill birds may exceed 50% at a retail level (Franco, 1988). Over 50% of *Campylobacter* isolates from chickens are serovars that are common in human disease, and under these conditions the likely consequence of cross-contamination from raw birds is obvious. Problems are exacerbated by the protective effect exerted on campylobacters by chicken drip (Coates *et al.*, 1987).

C. jejuni is the usual species isolated from poultry, but a high incidence of *C. coli* has been reported from the Zagreb region of Yugoslavia (Popovic-Uroic, 1989) and from France (Marinescu *et al.*, 1987).

Red meat

Red meat is believed to be an important vehicle of infection, although relatively few outbreaks have been directly attributed to such foods. Examples of outbreaks include undercooked beefburger, sliced roast beef and lamb kebabs; there have also been anecdotal accounts of infection acquired by the conscious consumption of raw meat. In most cases, the importance of red meat lies in sporadic outbreaks where it is probably secondary to poultry.

Beef and lamb are of major importance in most

countries of western Europe and the US. The incidence of infection is generally lower than with poultry, usually being *ca* 20% (Stern *et al.*, 1985; Lammerding *et al.*, 1988) although an incidence as high as 43.1% was reported in veal at the retail level (Lammerding *et al.*, 1988). *C. jejuni* is predominant and a high proportion of serovars isolated are those of importance in human disease (Fricker and Park, 1989).

Although there has been little interest in species other than *C. jejuni*, *C. coli* and *C. laridis* in foods, Grau (1988) reported high numbers of *C. hyointestalinis* in the rumen content of cattle, especially lot fed cattle, and a corresponding contamination of the carcass with the organism. The significance of this in relation to human disease is not known.

The situation with pork is less clear than that of beef and lamb. Slaughterhouse methods mean a high level of carcass contamination, but the incidence at retail level is usually markedly lower than that of beef and lamb, usually being less than 10%. This is attributed to the marked dehydration of pork carcasses during chilling (Hudson and Roberts, 1982; Oosterom, 1983). Furthermore, most isolates are of *C. coli* and of serovars not usually associated with human disease. Therefore, in countries such as the UK, pork is not considered to be of importance as a vehicle of *Campylobacter*. Care must be taken in extrapolating this conclusion to other countries since pork is a major cause of campylobacteriosis in parts of Africa (Central African Republic) and the rural areas surrounding Zagreb, Yugoslavia (Popovic-Uroic *et al.*, 1988). Pork has also been implicated as a secondary risk factor in the Netherlands (Oosterom *et al.*, 1983). Predictably, infections from pork usually involve *C. coli*.

Most work has concerned raw meat from which *Campylobacter* can be eliminated by correct cooking. However, *Campylobacter* has been isolated from commercially prepared cooked meats (Fricker and Park, 1989), and while the incidence was low (*ca* 2%), such products are of obvious potential importance as vehicles of infection.

Shellfish

Raw clams were considered to be the vehicle of infection in an outbreak in New Jersey, USA, the underlying cause probably being sewage pollution of growing beds (Blaser *et al.*, 1982). *Campylobacter* has also been isolated from oysters. Work with artificially contaminated oysters showed that commercially applied depuration could not guarantee safety (Arumugaswamy *et al.*, 1988).

Salads

Salads were implicated in a large outbreak in British Columbia, Canada (Anon, 1985). The salads were probably contaminated at the point of preparation or sale, and surveys have indicated that salads probably do not present an inherent risk.

Water

Water has been responsible for both large outbreaks and sporadic cases of *Campylobacter* enteritis.

Contamination of stored drinking water by wild birds was responsible for two outbreaks, the first involving contamination of a community supply in Florida, USA, which affected 871 people (CDC, 1983), the second involving the tank supplying an English boys' school (Palmer *et al.*, 1983).

Community outbreaks in the US have also been attributed to untreated spring water forming the supply to remote settlements (CDC, 1978, Mentzing, 1981).

Sporadic outbreaks have usually involved water not intended for human consumption and many of those affected have been walkers who drank from mountain springs and streams (Taylor *et al.*, 1983; Pacha *et al.*, 1987). *Campylobacter* is eliminated by halide-based water disinfection tablets, but these are eschewed by those walkers who wish to return to nature and who consequently may gain not only a true wilderness experience, but an unpleasant bout of 'trekkers' trots'.

Factors affecting the survival of *Campylobacter*

Predictions that campylobacters would be able to grow in foods under some conditions, such as a controlled atmosphere and vacuum packs, have proved unfounded. As far as is known, there are no conditions which permit the growth of the organism in foods and the microbiologist is concerned solely with its survival.

Temperature

Campylobacter is readily destroyed by temperatures used in pasteurisation and cooking. *D* values range from 7.3 minutes at 50°C to 1.1 minutes at 55°C and *z* values from 2.8–5.3°C with a mean of 3.7°C (Skirrow *et al.*, 1982). Values obtained by other workers were similar, but there can be considerable strain variation (Waterman, 1982; Sorqvist, 1989).

Campylobacter survives poorly during storage at room temperature although many factors interact to determine the actual period of survival. Lowering the storage temperature enhances survival, which in most cases is greatest at *ca* 3°C. Survival in many foods is prolonged over the range 1 to 10°C, which is the temperature range of commercial refrigeration.

Survival in frozen foods varies considerably and the physiological condition of the organism at the time of freezing may be an important determinant. Freezing cannot be relied upon to eliminate the organism even when storage is prolonged.

Irradiation

Campylobacter is less resistant to ionising radiation than *Salmonella*, and is eliminated by a dose of 1 Kgy.

Drying

Campylobacter is unusually sensitive to drying. The organism is therefore eliminated rapidly from foods of only slightly reduced a_w. The marked fall in numbers on air-chilled meat is attributed to the drying of the carcass and is particularly significant with pork. It is possible also that air-chilled poultry has lower levels of contamination than wet-chilled poultry, although there appears to be no comparative information available.

The sensitivity of campylobacters to drying has wider consequences for food handling. Adequate *drying* as well as washing, is of considerable importance in preventing the transfer of *Campylobacter* on the hands of food handlers (Coates *et al.*, 1987). Drying would also seem to be of importance following the cleaning of equipment and utensils.

pH value and organic acids

Campylobacter is generally recognised as being sensitive to low pH values and survival is optimal in the range 6.5 to 7.5. The organism is particularly sensitive to lactic acid (see page 222) as well as other organic acids, including acetic acid and ascorbic and isoascorbic acids. In the case of ascorbic and isoascorbic acids, sensitivity is attributed to oxidation products rather than to the acids *per se* (Juven *et al.*, 1988).

Composition of atmosphere

The survival of *Campylobacter* is greater in vacuum, or controlled atmosphere, packs than in air (Blankenship and Craven, 1982, Hanninen *et al.*, 1984). A contributory factor may be the maintenance of moist conditions on the surface of the food in vacuum and controlled atmosphere packs.

Competition from other organisms

Survival of *Campylobacter* is reduced by the presence of other micro-organisms. In some cases, this may be attributed to a reduction of the pH value and the production of organic acids. Some caution is required since in some cases the apparent reduction in survival is an artefact due to the greater difficulty of recovering *Campylobacter* when large numbers of competitive bacteria are present (authors' unpublished observation).

Laboratory methods

Cultivation

Sample handling

The death of campylobacters during transportation is a serious problem. The underlying principles must be to reduce delays to the minimum, and to refrigerate samples during transport. Where mechanical refrigeration is not available, samples should be pre-cooled to 4°C and packed in a properly constructed insulated container. Freezing should be avoided.

Survival is greatest in an oxygen-free atmosphere. Provision of such an atmosphere is often not possible when samples are collected by inexperienced people. However, for liquid samples, the addition of 0.15% sodium thioglycollate is an acceptable, and practical alternative (Koidis and Doyle, 1984). Survival is further enhanced by the addition of 0.01% sodium bisulphate.

With meat and other solid samples, transport and storage in an equal volume of Cary-Blair medium with reduced agar concentration (Luechtefeld et al., 1981) is an acceptable alternative to holding in nitrogen (Stern and Kotula, 1982). If necessary, a liquid enrichment medium (see below) may be used in place of Cary-Blair medium.

Cultivation media

The basic requirements are the same for both selective and non-selective cultivation media.

Nutrient requirements

In the authors' experience, difficulties with general cultivation of campylobacters have often resulted from the failure to meet the organism's nutritional requirements. However, while complex media are required for routine cultivation, the requirements are readily met by widely used media such as blood agar and *Brucella* broth.

Oxygen scavengers

Oxygen scavengers are required in solid media and in broth media for large-scale cultivation. Although not necessary in broth media for small-scale cultivation, some workers prefer to use oxygen scavengers at all times.

Blood is commonly added to solid media and standard formulations of blood agar are recommended for routine, non-selective cultivation of campylobacters.

Blood is also widely used in enrichment media but it is inconvenient to use. Adequate supplies of sterile blood are not available in some countries and an effective and convenient alternative is FBP supplement (Hoffman et al., 1979), which consists of 0.05% each of ferrous sulphate, sodium metabisulphite and sodium pyruvate. Nutrient broth supplemented with FBP may be used for the cultivation of campylobacters as well as for the determination of H_2S production (see page 231), and FBP may also replace blood in solid media.

Composition of atmosphere

The most convenient method of obtaining a suitable atmosphere for non-hydrogen requiring campylobacters is by using commercially prepared gas generator kits such as those produced by Becton Dickinson, bioMerieux and Unipath Ltd. These may be used with suitable anaerobic jars fitted with an active catalyst or with anaerobic pouches such as Generbag© (bioMerieux) and CampyPouch© (Becton Dickinson). Pouches are particularly convenient where samples are plated out away from a laboratory.

In countries where convenient and suitable generator kits are unavailable, conventional anaerobic jars may be used by removing the catalyst, extracting two-thirds of the air by means of a vacuum pump, and restoring atmospheric pressure with a suitable gas mixture. Mixtures which have been used successfully include $95\% N_2 : 5\% CO_2$ and $85\% H_2 : 15\% CO_2$. For safety reasons, the mixture containing nitrogen is preferable.

Hydrogen:carbon dioxide generator kits may be used in a two-jar method described by Simmons (1977).

Laboratory facilities in developing countries may be very limited, and it may be necessary to use less effective means of obtaining suitable atmospheres. Candle extinction jars, while not recommended (Bolton and Coates, 1983), may be the only means available. If the candle extinction jar is used, media should contain FBP supplement as well as blood. An alternative is to use the Fortner principle, which exploits the ability of a rapidly growing facultative anaerobe to reduce the oxygen tension in a closed system. This has been used successfully for the isolation of campylobacters from faeces (Karmali and Fleming, 1979).

Several species of *Campylobacter*, *C. cinaedi*, *C. fennel-*

liae, *C. conciscus* and '*C. upsaliensis*' require a hydrogen containing atmosphere for growth. A minimum of 4% H_2 is required by most strains, and a suitable atmosphere comprises 4% CO_2: 5% O_2: 7.5% H_2: 83.5% N_2. This may be obtained by replacing 75% of the air in an anaerobic jar (without catalyst) with a gas mixture containing 5% CO_2: 10% H_2 and 85% N_2 (Vandamme *et al.*, 1990).

Isolation of *Campylobacter*

Filtration

Although differential filtration has been little used for routine purposes, it is of considerable value in investigations of those less well-known strains and species of *Campylobacter*, which are sensitive to the inhibitors used in selective media. In this context, Steele and McDermott (1984) used the technique to isolate strains of *C. jejuni* from faeces which were unable to grow on selective media, and the technique is recommended for *C. jejuni* subsp *doylei*, '*C. upsaliensis*' and other 'atypical' species. Filtration is often considered to be too insensitive for general use with foods, but has been used as a post-enrichment procedure when enrichment has been only partly successful (Park *et al.*, 1984), while Megraud (1987) described a combined enrichment-filtration technique that was devised for the isolation of *Campylobacter* from pigeon faeces, but which may have applications with foods.

Combining filtration with enrichment introduces selectivity, and thus at least partly negates the major advantage of filtration techniques. However, the use of only mildly selective enrichments is possible.

Filtration techniques are included in the protocols for the isolation of *Campylobacter* in **Flow diagram 21**, page 230.

Resuscitation

The basis for the inclusion of a resuscitation stage is the sensitivity of stressed campylobacters to selective agents, including sodium deoxycholate (Humphrey and Cruickshank, 1985) and the antibiotics rifampicin and polymyxin (Humphrey and Cruickshank, 1985; Humphrey 1986a,b; Ray and Johnston, 1984). An appraisal of the efficacy of resuscitation (Humphrey, 1989), indicated that recovery of *Campylobacter* in liquid culture was improved by 4 hours incubation at 37°C before further incubation at 43°C, but that with solid selective media the inclusion of agents which quench toxic derivatives of oxygen was more important.

Enrichment

A number of antibiotic containing enrichment media have been developed and the more widely used are summarised in **Table 87**. There have been relatively few comparative studies, but for the isolation of campylobacters from foods both Preston enrichment broth and Doyle and Roman enrichment broth are considered satisfactory (Fricker, 1984; 1987).

Enrichment cultures must be incubated under microaerobic conditions, and for most purposes an incubation temperature of 42 to 43°C is recommended, subcultures onto selective plating media being made at 24 and 48 hours. Some species such as *C. fetus* subsp. *fetus* do not grow at 42°C, and in such cases

Table 87. Commonly used enrichment media for *Campylobacter*.

Media	Selective agents
Preston broth[1]	Polymyxin B, rifampicin, trimethoprim, actidione
Doyle and Roman broth[2]	Vancomycin, trimethoprim, polymyxin B, cycloheximide
Rosef broth[3]	Vancomycin, trimethoprim lactate, polymyxin B
VTP brucella-FBP broth[4]	Vancomycin, trimethoprim, polymyxin B

[1] Bolton and Robertson (1982); Bolton *et al.* (1983).
[2] Doyle and Roman (1982).
[3] Rosef (1983).
[4] Park *et al.* (1983; 1984).

incubation should be at 37°C. Some strains of *C. jejuni* are also inhibited at 42°C, and for this reason incubation of enrichment broths in parallel at the two temperatures is sometimes recommended (Tee *et al.*, 1987). Incubation at 37°C was found to give a significantly higher recovery than at 43°C in the isolation of *C. jejuni* from water and artificially contaminated milk, but not from chicken skin (Humphrey and Muscat, 1989). Incubation of enrichment broths at 37°C obviates the need to resuscitate injured cells.

Enrichment of campylobacters from raw milk may be unsuccessful due to the inhibitory effect of the lactoperoxidase system (LPS). Recovery of the organism is therefore enhanced if the LPS is inactivated by raising the pH value of the milk to 7.5 by the addition of 1 mol/l NaOH (Beumer *et al.*, 1988).

Attempts have been made to improve the recovery of campylobacters by combining enrichment with physical concentration. Concentration by centrifugation is recommended for their isolation from water, and the same procedure may be adopted for milk and other liquid samples. A technique for pieces of meat and poultry has also been developed (Park *et al.*, 1984), although swabbing an area of 25 to 100 cm^2 is more effective for the carcasses of large animals.

Concentration is necessary for the isolation of campylobacters from water, and is an effective means of increasing the sensitivity of isolation from foods. However, the procedures used are time consuming and are not considered necessary for routine use. Concentration procedures are summarised in **Flow diagram 20**.

Selective plating

A wide range of media is available for the recovery of the three main species of *Campylobacter* — *C. jejuni*, *C. coli* and *C. laridis* — from foods. The principle of each is similar: media consisting of a nutrient-rich, and usually blood-containing base, with selectivity provided by a mixture of anti-microbial agents.

The first medium to gain widespread acceptance was that of Skirrow (Skirrow, 1977), but other widely used media include Butzler (Butzler and Skirrow, 1979), Blaser-Wang (Campy-BAP; Blaser *et al.*, 1979), and Preston (Bolton and Robertson, 1982). All of these media contain blood; successful blood-free alternatives are CCD medium (blood-free campylobacter medium; Bolton *et al.*, 1984) and blood-free Preston medium (Goossens *et al.*, 1986).

Incubation of selective plating media is at 42°C for 48 hours under microaerobic conditions. Incubation at 37°C should be used in parallel if required.

The sensitivity of campylobacters to dehydration means that the agar surface must be moist and a freshly prepared medium is recommended (Fricker, 1985). The use of commercially prepared, prepoured plates, while convenient and satisfactory for most purposes, may result in lower recoveries (Laughan *et al.*, 1988).

Flow diagram 20. Concentration techniques for *Campylobacter* in food and water (Park *et al.* 1984).

A Liquid Sample

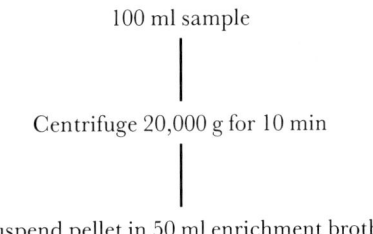

100 ml sample

|

Centrifuge 20,000 g for 10 min

|

Suspend pellet in 50 ml enrichment broth

B Solid Samples

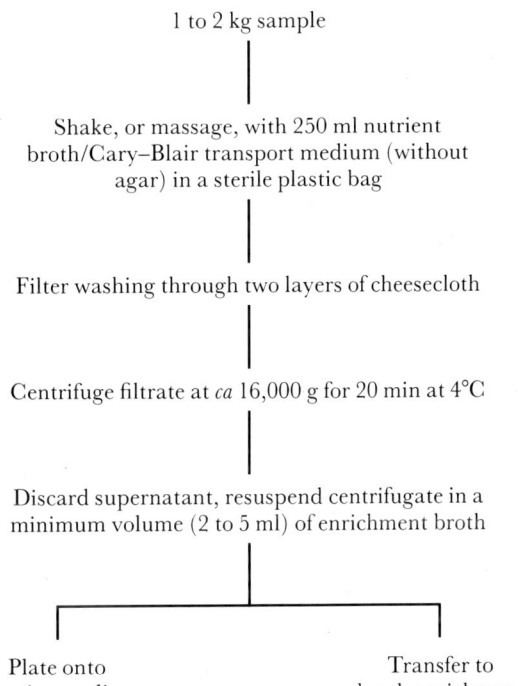

1 to 2 kg sample

|

Shake, or massage, with 250 ml nutrient broth/Cary–Blair transport medium (without agar) in a sterile plastic bag

|

Filter washing through two layers of cheesecloth

|

Centrifuge filtrate at *ca* 16,000 g for 20 min at 4°C

|

Discard supernatant, resuspend centrifugate in a minimum volume (2 to 5 ml) of enrichment broth

Plate onto selective medium   Transfer to broth enrichment

Media in general use for the isolation of campylo-bacters from foods are summarised in **Table 88** and illustrated in **Figures 168–173**.

168–173. Selective media for the isolation of *Campylobacter* species.
Skirrow medium, the first selective medium devised for the isolation of campylobacters, remains widely used alongside alternative media. Blaser-Wang medium, for example, is more selective than Skirrow medium but, while suitable for most strains of *C. jejuni* (**168** & **169**) and other common campylobacters, contains cephalothin and is therefore not suitable for the isolation of strains sensitive to that antibiotic.

Preston medium does not contain cephalothin and is often regarded as the medium of choice, being both highly sensitive and selective. *C. jejuni* (**170**) and *C. coli* (**171**) both form discrete colonies which are of similar appearance.

Blood-free Preston medium is convenient in use and has a performance similar to that of other media. *C. fetus* subsp. *fetus* produces larger colonies (**172**) than *C. coli* (**173**) or *C. jejuni* but requires incubation at 37°C. (*All media: 4h at 37°C: O_2 5%/ CO_2 10%/ N_2 85%.*)

Table 88. Commonly used selective plating media for *Campylobacter*.

Media	Selective agents
Skirrow agar[1]	Vancomycin, polymyxin B, trimethoprim
Butzler agar[2]	Novobiocin, bacitracin, acitdione, colistin, cephalothin
Blaser-Wang agar[3]	Vancomycin, trimethoprim, polymyxin B, amphotericin, cephalothin
Preston agar[4]	Polymyxin B, rifampicin, trimethoprim lactate, actidione
Blood-free Preston agar[5]	Cefoperazone
CCD agar[6]	Cefazolin

[1] Skirrow *et al.* (1977).
[2] Butzler and Skirrow (1979).
[3] Blaser *et al.* (1979).
[4] Bolton and Robertson (1982).
[5] Goossens *et al.* (1986).
[6] Bolton *et al.* (1984).

168
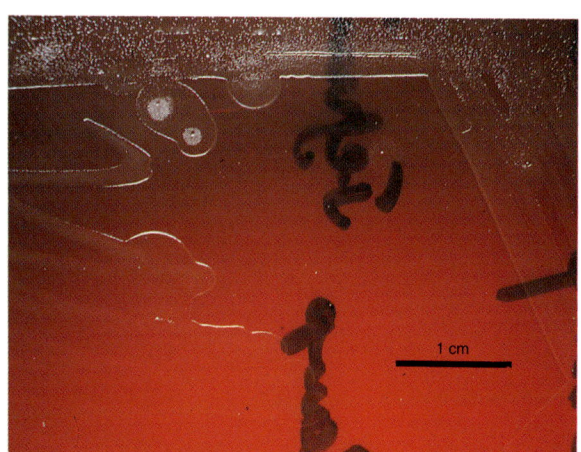

169
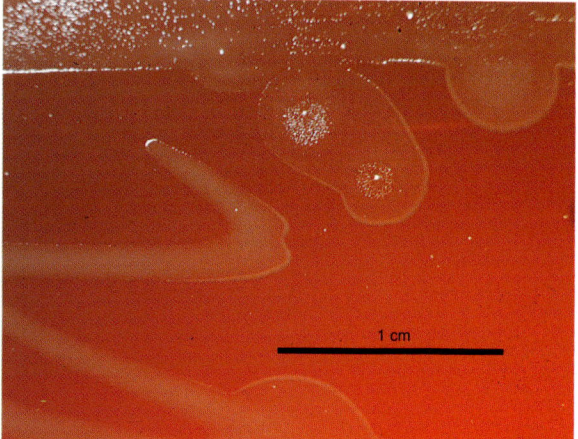

170
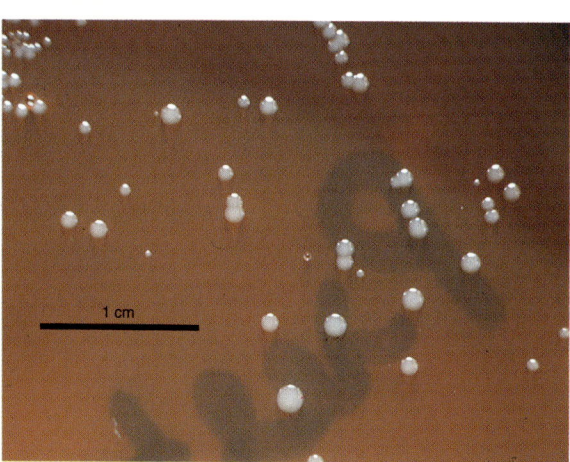

171
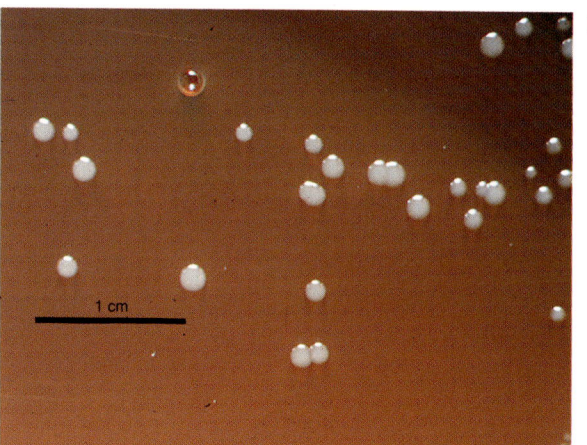

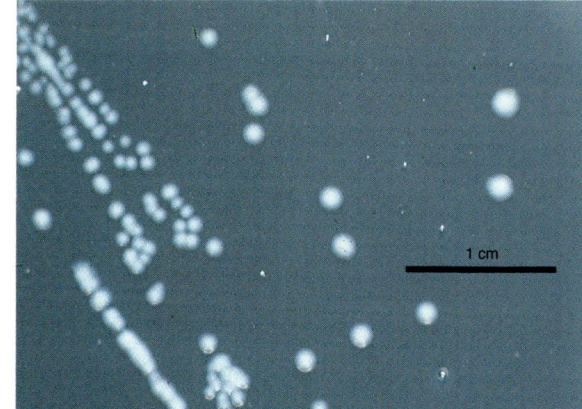

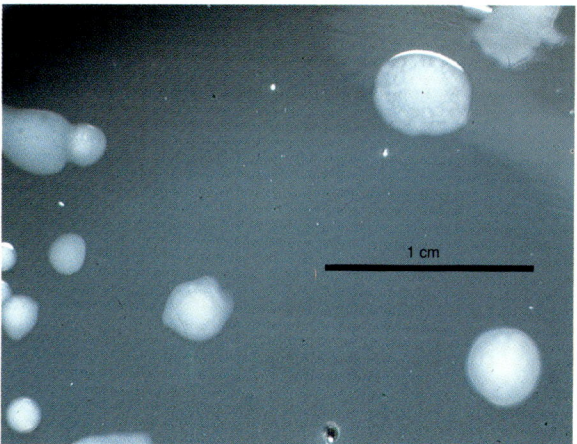

Media designed primarily for *C. jejuni*, although suitable for *C. coli* and *C. laridis*, may fail to recover other species. This is probably of little practical significance with foods, but may be of importance in clinical laboratories. Particular problems may be caused by the sensitivity of species including *C. fetus* subsp. *fetus*, *C. cinaedi*, *C. fennelliae*, and *C. hyointestinalis* to cephalothin, which is used as a selective agent in Butzler and Blaser-Wang selective agars. These species should therefore be isolated on cephalothin-free medium incubated at 35 to 37°C instead of 42°C, if necessary. Alternatively, the filtration method of Steele and McDermott (1984) used in conjunction with a non-selective medium is often preferred (see above). A blood-free, charcoal-containing medium, which employs cefoperazone, vancomycin and cyclo-heximide has been developed for isolation of '*C. upsaliensis*' from faeces (Walmsley and Karmali, 1989), but Goosens and Butzler (1989) are strongly opposed to the use of antibiotic-containing media for '*C. upsaliensis*' and recommend the use of filtration to maximise recovery. *C. cryoaerophila*, despite growing optimally on leptospiral media, was recovered by Tee *et al.* (1988) on conventional media used for campylobacters.

Complete isolation procedures

Three protocols for the isolation of campylobacters from foods are illustrated in **Flow diagram 21**. These comprise the standard enrichment:selective streaking procedure; and two combined enrichment:filtration procedures. The standard procedure is most commonly used and is satisfactory for most purposes.

Flow diagram 21. Protocols for the isolation of *Campylobacter* **from foods.**

A Conventional two-stage procedure (Fricker, 1984, 1987)

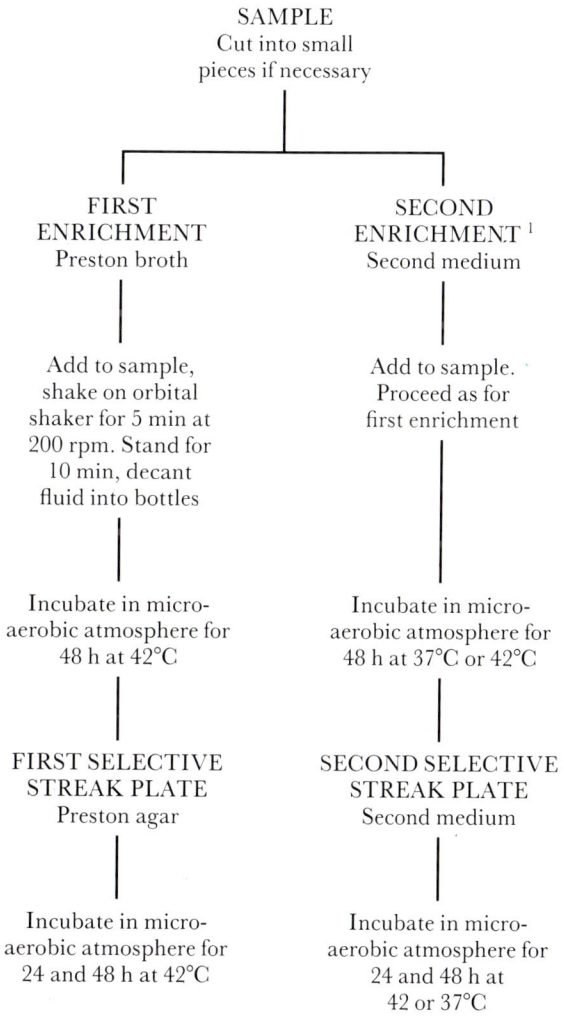

SAMPLE
Cut into small pieces if necessary

FIRST ENRICHMENT
Preston broth

SECOND ENRICHMENT [1]
Second medium

Add to sample, shake on orbital shaker for 5 min at 200 rpm. Stand for 10 min, decant fluid into bottles

Add to sample. Proceed as for first enrichment

Incubate in micro-aerobic atmosphere for 48 h at 42°C

Incubate in micro-aerobic atmosphere for 48 h at 37°C or 42°C

FIRST SELECTIVE STREAK PLATE
Preston agar

SECOND SELECTIVE STREAK PLATE
Second medium

Incubate in micro-aerobic atmosphere for 24 and 48 h at 42°C

Incubate in micro-aerobic atmosphere for 24 and 48 h at 42 or 37°C

Flow diagram 21. Protocols for the isolation of *Campylobacter* from foods.

B Use of filtration to enhance enrichment (Park *et al.* 1984)

C Combined enrichment: Filtration Procedure

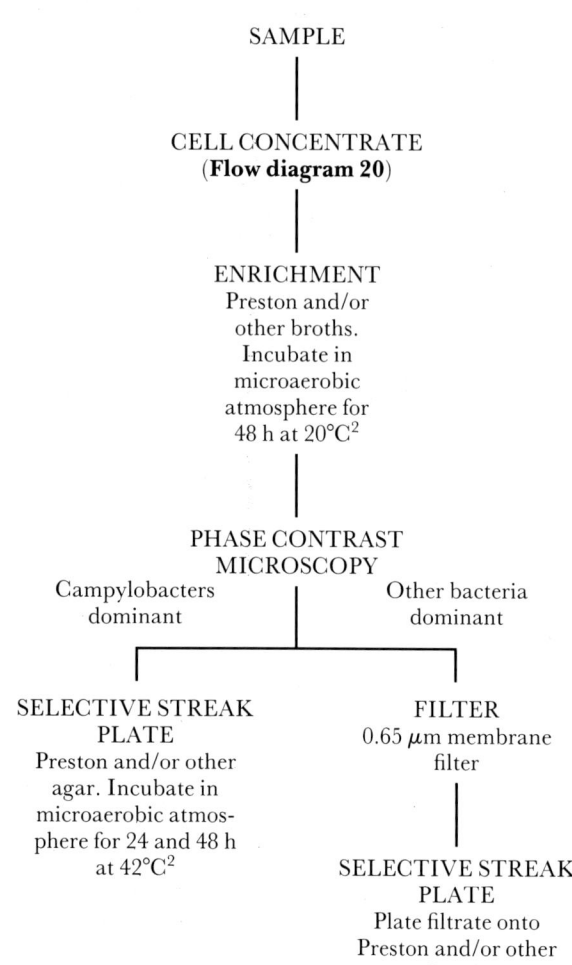

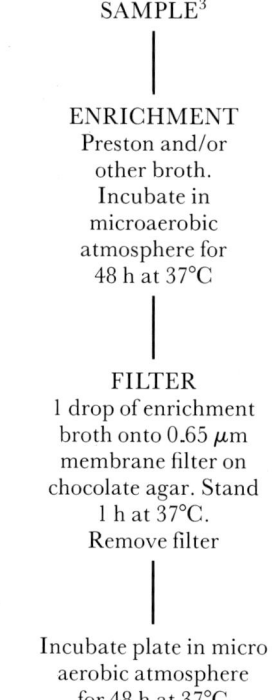

[1] Second medium is not normally used unless isolation of 'atypical' campylobacters is intended. In such a case, neither enrichment nor plating media should contain cephalothin and incubation should be at 37°C.
[2] Incubation should be at 37°C for 'atypical' campylobacters.
[3] Solid samples should be prepared as in Protocol A.

Detection of *Campylobacter*

Direct detection of *Campylobacter* in faeces is possible using latex co-agglutination (Fricker and Park, 1987). A commercial kit based on this principle is available (Mercia Diagnostics), which may also be used to detect foodborne campylobacters direct from the enrichment broth. This kit should not be confused with other types that are suitable only for confirming the identity of colonies on isolation media.

Identification of *Campylobacter*

Differentiation from other species

Differentiation of *Campylobacter* from other species is based on a positive oxidase-reaction, cell morphology and Gram-reaction, and the characteristic darting motility determined preferably by phase-contrast microscopy. These tests permit differentiation from other species able to develop on selective media for *Campylobacter* and few precautions are required. There are, however, anecdotal reports of weak, or negative, oxidase reactions when colonies have been tested direct from selective isolation plates. Furthermore the darting motility may be restricted in cells from solid media which tend to become 'anchored' together. In cases of difficulty, identity may be confirmed either by latex co-agglutination or by DNA probe. In either case, only the more common species, *C. jejuni*, *C. coli* and *C. laridis* are detected, and differentiation between the species is not possible. Latex co-agglutination is simple, but problems occur with some makes of kit due to *Pseudomonas aeruginosa* (Hodinka and Gilligan, 1988). The AccuProbe© Assay System (Gene-Probe Inc) has been used as a culture confirmation assay for thermophilic campylobacters, and is notable in being a simple homogenous, chemiluminescent probe, in which all reactions occur in a single tube (Tenover *et al.*, 1990).

Differentiation between species

For routine purposes, *C. jejuni*, *C. coli* and *C. laridis* may be differentiated by three tests: hippurate hydrolysis, production of H_2S in FBP broth, and resistance to nalidixic acid (Fricker and Park, 1989). The application of these tests to identification of species is illustrated in **Figures 174–178**.

If further confirmation of identity is required, the additional tests listed in **Table 83**, page 211, should be used. The recommended methodology is that of Skirrow *et al.* (1982).

174 & 175. Differentiation of *C. jejuni* from *C. coli* by hippurate hydrolysis. Hippurate hydrolysis is determined by suspending a loopful of growth in a solution of sodium hippurate in phosphate buffer and incubating for 2 h at 37°C. Hippurate hydrolysing cultures, such as those of *C. jejuni*, produce glycine which may be detected by the addition of ninhydrin solution to give a blue colour (**174**). Other campylobacters such as, *C. coli*, do not hydrolyse hippurate and the assay mixture remains colourless (**175**).

176–178. Hydrogen sulphide production as a differential test for campylobacters. H_2S production is a means both of differentiating between species of *Campylobacter* and between the Skirrow biovars 1 and 2 of *C. jejuni*. The use of FBP medium is simple, rapid and effective. A large lump of culture is suspended in FBP and incubated for 4 h at 37°C. A positive reaction is indicated by blackening of the medium. In many cases, blackening is restricted to the area of the lump of culture, but in the example illustrated *C. jejuni* biotype 2 (**176**), sufficient cells were distributed at the top of the medium during suspension to give a more general blackening. *C. jejuni* biotype 1 does not produce H_2S and no blackening occurs around the lump of culture (**177**).

FBP is a highly sensitive medium for the detection of H_2S, and traditional media such as triple sugar-iron (TSI) will not detect H_2S production by *C. jejuni* biotype 2 or *C. laridis*. TSI is, however, suitable for use with the more strongly H_2S producing species such as *C. hyointestalis* (**178**). The reaction illustrated is exceptionally strong.

Methods involving the use of lead acetate paper are unreliable and not recommended for use with campylobacters. (*FBP medium: 4 h at 37°C; TSI medium: 30 h at 37°C.*)

176 **177** **178**

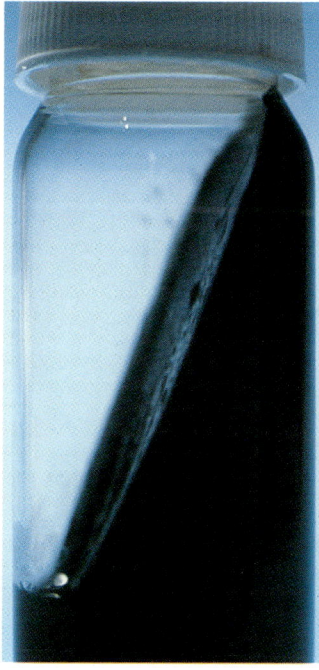

'4

5

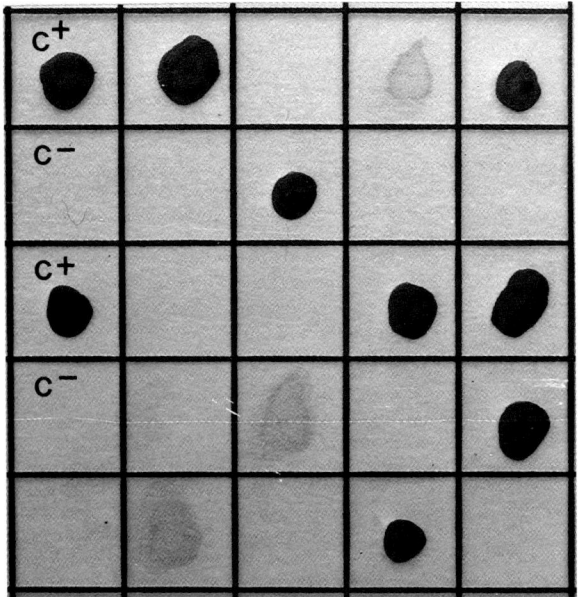

179. Identification of *C. cinaedi* by DNA hybridisation. Isolates are cultured, from standardised inocula, on nitrocellulose filters. The colony hybridisation method (see page 126) is then applied using a ^{32}P-labelled probe. The extent of homology of each test strain with the probe DNA is indicated by the intensity of darkening of X-ray film compared with the controls (C).

The subspecies of *C. jejuni* may be differentiated on the basis of nitrate reduction and the ability to grow at 42°C.

C. fetus subsp. *fetus* and other species may be differentiated according to Penner (1988). However, differentiation based on phenotypic characteristics is inadequate in some cases, and while a number of studies have been made using DNA hybridisation techniques (Steele *et al.*, 1985; Totten *et al.*, 1985, 1987; Korolik *et al.*, 1988), these are not yet suitable for general application. The taxonomic spot blot test, developed during studies of 'campylobacter-like organisms' isolated from homosexuals, is seen as a significant development towards more widespread application of DNA hybridisation by clinical and, ultimately, food laboratories (**Figure 179**; Totten *et al.*, 1985,1987). The use of isotopic labels is not acceptable in many food industry laboratories, but the use of radio-isotopes was avoided by Chevrier *et al.*, (1988) who used 2-acetylaminofluorine as label.

Typing of *Campylobacter*

Serotyping

Of the two most widely used typing schemes, that of Penner *et al.*, (1983) is considered definitive despite its being more difficult to apply. Both the Penner scheme and that of Lior *et al.*, (1982) are in use, and it is necessary to distinguish between the schemes used. Serovars defined by the Penner scheme are designated as, for example, PEN 2, and those in the Lior scheme as, for example, LIO 4.

As noted earlier, the major drawback to the Penner scheme is the technical complexity involved in extracting the heat-stable antigen, and in performing the necessary passive haemagglutination titrations. For this reason, the combined co-agglutination:passive haemagglutination technique of Illingworth and Fricker (1987) is recommended for screening purposes. This technique is not only simpler, but more specific than the passive haemagglutination originally used. Where required, serovar may be confirmed using the modified passive haemagglutination method of Fricker *et al.*, (1987). Each of these techniques is summarised in **Flow diagram 22**.

Despite improvements to techniques, many laboratories rely on a central facility for serotyping. Strains may lose viability during transport, and to overcome this difficulty antigenic extracts may be sent for typing rather than live bacteria (Fricker, 1986).

Typing antisera for the determination of heat-stable antigens are not commercially available and, if required, should be raised as described by Fricker *et al.*, (1986). Cross-reactions can cause serious problems and may be attributed to antigenic variation (Preston and Penner, 1989).

Flow diagram 22. Penner serotyping scheme for *Campylobacter.*

A Passive haemagglutination (Fricker *et al.*, 1987)

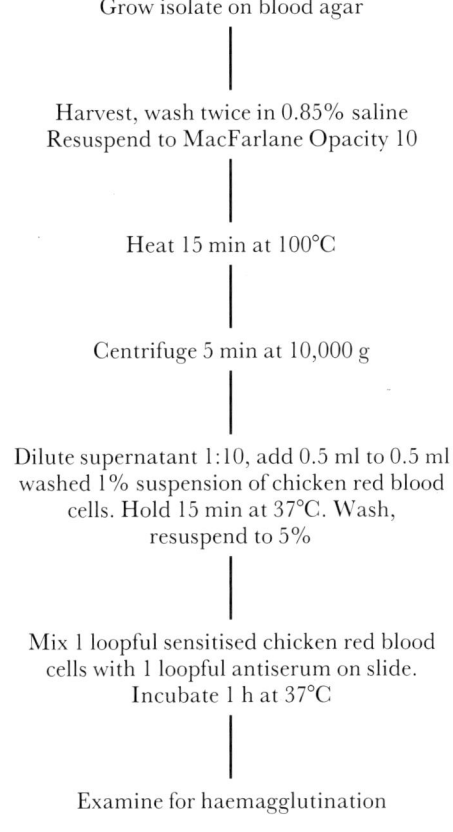

Grow isolate on blood agar

Harvest, wash twice in 0.85% saline
Resuspend to MacFarlane Opacity 10

Heat 15 min at 100°C

Centrifuge 5 min at 10,000 g

Dilute supernatant 1:10, add 0.5 ml to 0.5 ml
washed 1% suspension of chicken red blood
cells. Hold 15 min at 37°C. Wash,
resuspend to 5%

Mix 1 loopful sensitised chicken red blood
cells with 1 loopful antiserum on slide.
Incubate 1 h at 37°C

Examine for haemagglutination

B Combined co-agglutination and haemagglutination
(Illingworth and Fricker, 1987)

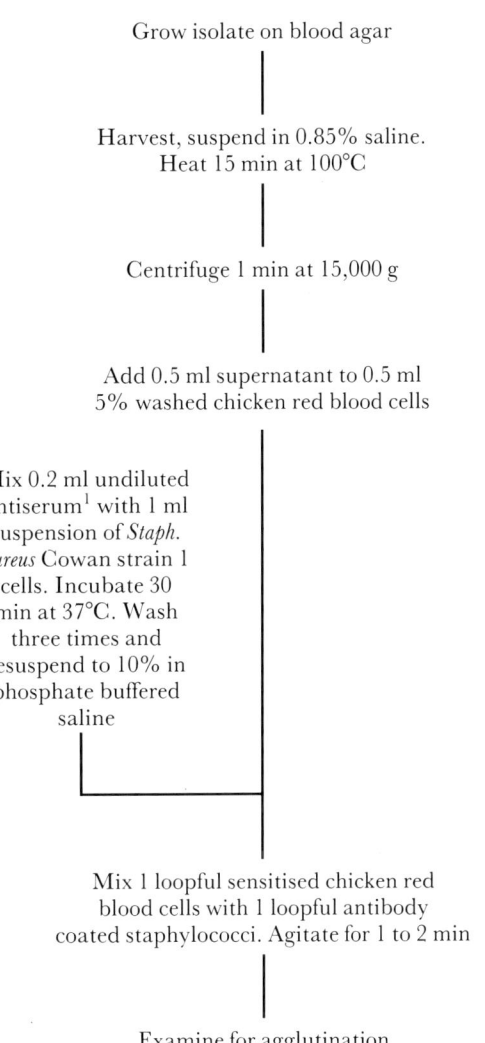

Grow isolate on blood agar

Harvest, suspend in 0.85% saline.
Heat 15 min at 100°C

Centrifuge 1 min at 15,000 g

Add 0.5 ml supernatant to 0.5 ml
5% washed chicken red blood cells

Mix 0.2 ml undiluted
antiserum[1] with 1 ml
suspension of *Staph.
aureus* Cowan strain 1
cells. Incubate 30
min at 37°C. Wash
three times and
resuspend to 10% in
phosphate buffered
saline

Mix 1 loopful sensitised chicken red
blood cells with 1 loopful antibody
coated staphylococci. Agitate for 1 to 2 min

Examine for agglutination

[1] For a preliminary grouping, 'pooled' reagents consisting
of staphylococci coated with four antisera against four
Campylobacter O groups may be used.

The determination of serovar according to the Lior scheme, which is based on heat-labile antigens, is summarised in **Flow diagram 23**. Although the procedure is simpler, autoagglutination is sometimes a major problem that may require repeated subculture of test isolates to overcome. Alternatively cell suspensions that autoagglutinate may be treated with deoxyribonuclease to remove the extruded DNA which is responsible for autoagglutination. Typing antisera are commercially available (Sopar-Biochem).

Flow diagram 23. Lior serotyping scheme for *Campylobacter* (Lior *et al.*, 1982).

Grow isolate on fresh, moist
Mueller–Hinton agar plus blood

|

Remove small loopful, emulsify in small
drop of phosphate buffered saline. Add
drop diluted antiserum. Hold for 30 to 45 sec

|

Examine for agglutination

|

Repeat positive tests *without* antiserum to
avoid false-positives due to auto-agglutination

Biotyping

Despite the existence of more complex schemes, the simple scheme of Skirrow and Benjamin (1980), which recognises two biovars of *C. jejuni*, is of considerable practical value. The basis of differentiation is shown in **Table 89**.

Skirrow biovars 1 and 2 may also be differentiated by a single test for the enzyme gamma-glutamyl aminopeptidase, which is absent in biovar 1 strains but present in biovar 2 strains (McNulty and Dent, 1987).

Table 89. The Skirrow biotyping scheme for *Campylobacter* (Skirrow and Benjamin, 1980).

	Biotype 1	Biotype 2
H_2S in iron medium	−	+
Growth at 45.5°C	+/−	+
Tolerance of TTC*	−	+/−

* Filter paper strip soaked in 4% tri-phenyltetrazolium chloride and dried.

12. *Staphylococcus aureus*

Introduction

The genus *Staphylococcus* is a member of the family *Micrococcaceae* and consists of Gram-positive, catalase-positive cocci, which usually have both an oxidative and a fermentative metabolism of glucose (**Figures 180** & **181**). Cells of *Staphylococcus* have a characteristic arrangement (**Figures 182** & **183**) and a number

180

181

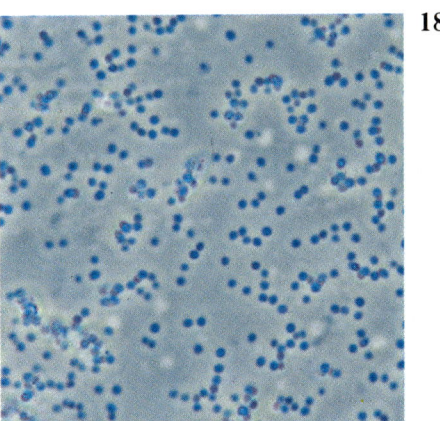

182

183

180 & **181. Failure of *Staphylococcus* to ferment glucose.** Fermentation of glucose (**180**) is considered to be a generic characteristic of *Staphylococcus* and an important feature in the differentiation of that genus from *Micrococcus*. The property can, however, be variable and over 18% of *Staphylococcus* spp. from milk are afermentative (**181**; Harvey and Gilmour, 1985). Failure to ferment glucose is most common in strains of *Staph. saprophyticus*, but occurs in strains of other species including *Staph. aureus*. Conversely, some strains of *Micrococcus* may produce small quantities of acid from glucose when incubated under anaerobic conditions. (*API 50CHB©: 48h at 37°C.*)

182 & **183. Cellular morphology *Staph. aureus.*** The characteristic arrangement of cells in clusters is not an artefact but a consequence of the mode of division of *Staph. aureus*. When growing in a cured meat slurry, clusters can be seen, by phase contrast microscopy, forming within 4 h at 25°C (**182**). Under some conditions, however, cluster formation is limited, a high percentage of cells occurring singly, or in short chains (**183**). Cultural conditions (plate count agar: 24 h at 37°C) and Gram-staining procedure were normal and the reason for this cell arrangement is not known. The culture resembles 'old' (7-day) cultures of some coryneform bacteria.

of members of the genus produce yellow, or golden carotenoid pigments (**Figures 184** & **185**; Kloos and Schleifer, 1986).

The genus *Staphylococcus* currently contains more than 20 species, many of which have been implicated as causative agents of disease in man and animals. Of these species, *Staph. aureus* is a major cause of food poisoning in man as well as of a range of extraintestinal infections (**Table 90**).

Staph. aureus produces a number of extracellular compounds including haemolysins, staphylococcal enterotoxins, coagulases, nucleases and lipases. Enterotoxins are responsible for the symptoms of staphylococcal food poisoning and may have a role in the pathogenicity of some other staphylococcal diseases, while coagulase and, to a lesser extent, thermonuclease are used both as markers for enterotoxin production and as a means of identification. Staphylococcal enterotoxins are discussed in detail on pages 242–243 and the significance of coagulase and thermonuclease on pages 243–244.

For many years, enterotoxin production was uniquely associated with *Staph. aureus*. More recently, enterotoxin production has been demonstrated by strains of *Staph. capitis*, *Staph. caprae*, *Staph. chromogenes*,

Table 90. Common extraintestinal infections of man caused by *Staph. aureus*.

Wounds and burn infections	Scalded skin syndrome
Boils	Bacteraemia
Impetigo	Meningitis
Pneumonia	Toxic shock syndrome
Pseudomembranous colitis	

Staph. cohnii, *Staph. epidermidis*, *Staph. haemolyticus*, *Staph. hysicus* (Hoover *et al.*, 1983), *Staph. intermedius* (Hirooka *et al.*, 1983), *Staph. lentus*, *Staph. sciuri*, *Staph. warneri* (Valle *et al.*, 1990) and *Staph. xylosus* (Bautista *et al.*, 1988)., as well as *Micrococcus luteus* (Genigeorgis, 1989). In the absence of information to the contrary, these species must be considered to be potential agents of food poisoning, although no cases have been reported.

Species of *Staphylococcus* are generally considered to be undesirable in processed foods, but one, *Staph. carnosus*, which has no known pathogenicity, is used as a starter organism in fermented sausages.

184

185

184 & 185. Pigmentation of *Staphyloccus aureus*.
The eponymous golden carotenoid pigment produced by *Staph. aureus* (**184**) is often the best-known feature of the organism to the less knowledgeable. It should be appreciated, however, that not only are there considerable strain differences, with depth of pigmentation varying markedly, but that the ability to form pigments may be lost by mutation. Expression of pigmentation may also be affected by temperature of incubation and the nature of the growth medium.

Carotenoid pigments are also produced by strains of other species such as *Staph. warneri* (**185**). These pigments tend to be yellow rather than gold, but reliable differentiation between *Staph. aureus* and other species cannot be made on the basis of pigmentation alone. (*Plate count agar medium: 36 h at 37°C plus 36 h at room temperature in reduced light.*)

Isolation of *Staph. aureus*

Conventional cultural techniques

Staph. aureus may be isolated from foods by direct plating where numbers are sufficiently high. High-performance media, such as the egg yolk-tellurite-pyruvate-glycine agar of Baird-Parker (1962) and its later derivatives, and the KRANEP agar of Sinell and Baumgart (1967), are now in general use.

Enrichment is not normally required for the isolation of *Staph. aureus* and, where used, the results must be treated with caution since *Staph. aureus* is present in many foods in small numbers, and is of little significance unless prior growth has occurred. However, detection of small numbers of *Staph. aureus* by enrichment may, in conjunction with enterotoxin assays, yield valuable epidemiological information during the investigation of food poisoning outbreaks. The broth medium of Giolitti and Cantoni (1966) as modified by Van Doone *et al.* (1981) is most suitable.

Isolation of species of *Staphylococcus* other than *Staph. aureus* usually requires different selective media. SK agar (Schleifer and Kramer, 1980) has been used successfully for the isolation of a range of species of *Staphylococcus* from raw milk (Harvey and Gilmour, 1985), while *Staphylococcus* 110 agar (Chapman, 1945) was found to be most suitable for coagulase-negative staphylococci.

Rapid detection methods

Efforts to develop rapid and novel methodology for *Staph. aureus* have been concentrated on the detection of enterotoxins (see pages 261–265). A method using gene probes for the detection of enterotoxigenic strains has been developed (Notermans *et al.*, 1988; Neill *et al.*, 1990), and a protocol for the determination of *Staph. aureus* using impediometric monitoring has also been published (Firstenberg-Eden, 1986).

On a longer term basis, Tuncan and Martin (1987) have discussed the potential of a technique involving the differential extraction of adenosine triphosphate from *Staph. aureus* using lysostaphin and its assay by the firefly luminescence method.

The isolation of Staph. aureus *is discussed in greater detail, with respect to choice of media and laboratory practice,* in **Laboratory Methods**, *pages 255–258.*

Taxonomy of *Staph. aureus*

Differentiation of *Staphylococcus* from other genera

In routine screening of foods for *Staph. aureus*, it is common practice to proceed direct to screening tests for the species (see below). These should include the minimum criteria for differentiation of *Staphylococcus* from other genera likely to grow on the selective medium used, Gram reaction, cell morphology and catalase reaction.

Where species other than *Staph. aureus* are involved, or where isolation has been made on non-selective, or weakly selective media, differentiation from other

Table 91. Differentiation of *Staphylococcus* from similar genera.

	Staphylococcus	*Micrococcus*	*Streptococcus*[1]
Catalase	+	+	−
Fermentation of glucose	+[2]	−[3]	+
Cells arranged in irregular clusters[4]	+	+	−
Sensitivity to			
lysostaphin	+	−	*
lysozyme	−	+	*
Acid from glycerol in presence of erthyromycin	+	−	*

[1] Criteria also applicable to *Enterococcus*, *Lactococcus*, *Pediococcus* and *Leuconostoc*.
[2] A small number of strains are afermentative (see page 235).
[3] A small number of strains are weakly fermentative.
[4] See page 235, smears should be made from young (24 hour) cultures to avoid possible confusion with coryneform bacteria.
* = Test not applicable.

genera is required before further testing. This is usually possible using the three tests listed above, together with the determination of the fermentative metabolism of glucose (**Table 91**). However, the possibility of non-fermentative strains of *Staphylococcus* and of weakly fermentative strains of *Micrococcus* mean that additional tests may be required to differentiate these two genera. Effective tests for *Staphylococcus* are sensitivity to lysostaphin, resistance to lysozyme and acid production from glycerol in the presence of erythromycin.

Differentiation of *Staph. aureus* from other species of *Staphylococcus*

For routine purposes, *Staph. aureus* is commonly differentiated from other species on the basis of the coagulase reaction using either the tube coagulase test or a rapid agglutination test (see page 259) together with, if required, the thermonuclease test. The latter test is particularly useful for differentiating coagulase-negative strains of *Staph. aureus*, or strains of some other species that give a positive coagulase reaction in the tests commonly used. It is necessary to be aware of the possibility of strains of *Staph. aureus* that are both coagulase and thermonuclease negative.

A full differentiation between strains may be made using biochemical tests. This is necessary with species other than *Staph. aureus*, or with some atypical strains of *Staph. aureus*. The definitive scheme is that of Kloos and Schleifer (1986), although new species have been recognised since that publication. The number of species likely to be encountered is relatively small, and a more limited scheme, such as that of Harvey and Gilmour (1985), forms a suitable basis for identification (**Table 92**).

In addition to identification based on conventional methodology, miniaturised identification kits are available. These are of two types:

1. A rapid means (two hours) of identifying *Staph. aureus* and a small number of other medically important species (API Rapidec Staph©).

2. Full identification of currently described species of *Staphylococcus* based on either conventional tests or conventional tests and the determination of specific enzymes (API Staph©; ATB 32 Staph©).

Identification of Staph. aureus *is discussed in greater detail, with respect to methodology and laboratory practice, in* **Laboratory Methods**, *pages 258–261. The relationship between the coagulase and thermonuclease reactions and enterotoxin production is discussed in greater detail in* **Mechanisms of Pathogenicity**, *pages 243–244.*

Table 92. Differentiation between the main foodborne species of *Staphylococcus* (**Harvey and Gilmour, 1985**).

	Staphylococcus							
	aureus	*hysicus* subsp. *hysicus*	*hysicus* subsp. *chromogenes*	*simulans*	*intermedius*	*epidermidis*	*capitis*	*hominis*
Coagulase	+	−	−	−	+	−	−	−
Thermonuclease	+	+	+/−	−	+	−	−	−
Haemolysis	+	−	−	+/−	+	+/−	−	+/−
Acetoin	+	−	−	−	−	+	+/−	+/−
Pigment	+	−	+	+/−	−	−	−	+/−
Acid (aerobic)								
Sucrose	+	+	+	+	+	+	+	+
Trehalose	+	+/−	+	+/−	+	−	−	+
Mannitol	+	−	+/−	+	+	−	+	−
Cellobiose	−	−	−	−	−	−	−	−
Maltose	+	−	+/−	+/−	−	+	−	+
Mannose	+	+	+	+/−	+	+	+	−
Xylose	−	−	−	−	−	−	−	−
Phosphatase	+	+	+	+/−	+	+	+/−	−
Novobiocin (1.6 µg/ml)	S	S	S	S	S	S	S	S

Table 92. Continued.

	Staphylococcus						
	warneri	*haemolyticus*	*cohnii*	*saprophyticus*	*xylosus*	*sciuri* subsp. *sciuri*	*sciuri* subsp. *lentus*
Coagulase	−	−	−	−	−	−	−
Thermonuclease	−	−	−	−	−	−	−
Haemolysis	+/−	+	+/−	−	−	−	−
Acetoin	+	+	+	+	+/−	−	−
Pigment	+	+/−	+/−	+/−	+/−	+	−
Acid (aerobic)							
Sucrose	+	+	−	+	+	+	+
Trehalose	+	+	+	+	+	+	+
Mannitol	+/−	+/−	+	+	+	+	+
Cellobiose	−	−	−	−	−	+	+
Maltose	+	+	+	+	+	+/−	+/−
Mannose	−	−	+/−	−	+	+/−	+
Xylose	−	−	−	−	+	−	+/−
Phosphatase	+/−	+/−	−	−	+	+	+/−
Novobiocin (1.6 µg/ml)	S	S	R	R	R	R	R

Notes:
1 Coagulase reaction is that using human plasma.
2 Haemolysis is that of sheep blood.

Typing of *Staph. aureus*

Serotyping

The complex nature of the antigens means that serotyping is of no practical application to *Staph. aureus*. However enterotoxins, are antigenically distinct and typing is of considerable value in epidemiological studies.

Phage typing

Phage typing is a well-established technique with *Staph. aureus*, the method having been developed from the work of Fisk (1942). Typing laboratories have been established in a large number of countries and the continuing development of the scheme is controlled by the International Subcommittee on Phage-typing of Staphylococci.

The typing set comprises 23 phages and is updated as required on a four-yearly basis. The typing system used differs from that used with most other organisms, such as *Salmonella*, in that lysis with a specific phage is uncommon, strains being identified on the basis of patterns of lysis.

Phage typing of Staph. aureus *is discussed in greater detail, with respect to methodology, in* **Laboratory Methods**, page 261.

Biotyping

Distinct biovars (ecovars) of *Staph. aureus* exist according to their origin (animal host). These may be defined on the basis of phage sensitivity, antigenic cell wall components, plasma coagulation patterns, haemolysis and fibrinolysin activity. Some of these methods are not suitable for routine purposes, and a number of simpler schemes have been described. Among the more widely used are those of Hajek and Marsalek (1973) and Devriese (1984). These schemes are summarised in **Table 93**. More recently, Mead *et al.* (1989) employed a relatively large number of tests and a sophisticated statistical technique, cluster analysis (Gower, 1966) to differentiate strains of *Staph. aureus* 'endemic' in a poultry packing plant from those derived from freshly killed poultry.

Although biotyping has been used successfully in a number of of studies (Gibbs *et al.*, 1978; Harvey *et al.*, 1982; Devriese *et al.*, 1985), the technique has limitations and in some situations is inadequate to establish the origin of *Staph. aureus* (Notermans *et al.*, 1983).

Table 93. A comparison of the biotyping schemes of Hajek and Marsalek and Devriese.

	Biovars	Criteria
Hajek and Marsalek (1973)	A (Human) B–F (Animal)	Fibrinolysin, pigmentation, coagulase (human and bovine plasma), haemolysins, crystal violet reaction
Devriese (1984)	(Ecovars) Human Poultry Sheep/goat Cattle	Fibrinolysin, ß-haemolysin, coagulase (bovine plasma), crystal violet reaction
	5 additional non-host specific serovars	

Other typing methods

Plasmid profiles

Plasmid profiling of *Staph. aureus* has been used in investigations of nosocomial infections (McGowan *et al.*, 1979) and in studies of bovine mastitis (Baumgartner *et al.*, 1984). Application of the technique to *Staph. aureus* in foods has been restricted to the typing of isolates from poultry packing plants (Dodd *et al.*, 1987,1988). In this application, plasmid profiling is considered to be a more rapid and sensitive means of differentiating between isolates than biotyping or phage typing.

Symptoms of *Staph. aureus* intoxication

The symptoms of staphylococcal intoxication are well known, and have been described on numerous occasions. Despite this a retrospective survey in the USA of 131 outbreaks that involved 7,000 individual cases (Holmberg and Blake, 1984) revealed a number of commonly held misconceptions concerning both the common symptoms associated with the intoxication and various other aspects.

Symptoms usually appear within 2 to 4 hours of consumption of contaminated food (Bergdoll, 1979). Incubation periods as short as 30 minutes or in excess of 8 hours have been reported, but are poorly documented and largely anecdotal. Commonly reported symptoms include nausea, retching, vomiting and, less frequently, diarrhoea. Headache, dizziness and weakness are reported in a minority of cases and there have been rare, and unsubstantiated, complaints of double vision and other visual disturbances. Diarrhoea does not appear to occur in the absence of vomiting.

Temperature is usually considered to be subnormal and, indeed, the presence of subjective fever is sometimes used to rule out the possibility of staphylococcal food poisoning (Holmberg and Blake, 1984). However, these authors found low-grade fever to be a symptom in *ca* 16% of cases, accompanied by, on some occasions, chills and perspiration.

Staphylococcal intoxication is self-limiting and symptoms usually persist for no more than 24 hours. In severe cases, dehydration leads to shock and collapse, accompanied by a weak pulse and shallow breathing. It is noteworthy that, for a disease commonly described as 'mild', as many as 10% of those affected seek hospital care (Holmberg and Blake, 1984).

Death is rare and usually occurs only when the patient is elderly, very young, or suffering from a debilitating disease. Death has, however, been reported occasionally in normal adults (Currier, 1973).

There is some evidence that both the pattern and severity of symptoms are affected by the age of the patient. In an investigation of an outbreak where chocolate-flavoured milk was the vehicle of infection, 89% of children aged 5 to 9 years suffered vomiting, in contrast to less than 65% of children aged 10 to 19 years (Evenson *et al.*, 1988).

The symptoms of staphylococcal intoxication are readily confused with those of *Bacillus cereus* emetic syndrome (Newsome, 1988). Particular care is needed to avoid confusion in cases where both organisms are isolated from the same sample of suspect food.

In addition to staphylococcal food poisoning, *Staph. aureus* has been implicated as a cause of pseudomembranous colitis in people who have received oral administration of broad-spectrum antibiotics. In these situations, overgrowth by antibiotic resistant, enterotoxin producing strains of *Staph. aureus* may occur (Jawetz *et al.*, 1982; Ciborowski and Jeljaszewicz, 1985). Symptoms are abdominal cramps, severe diarrhoea, dehydration and electrolyte imbalance. These clinical manifestations, the isolation of antibiotic resistant *Staph. aureus* from stools in pure culture and necrosis of the intestinal tract, serve to differentiate this syndrome from *Staph. aureus* food poisoning.

Persons at particular risk from *Staph. aureus* intoxication

Susceptibility

There is considerable variation in susceptibility to *Staph. aureus* enterotoxin among normal adults. The basis for this is not known, although prior exposure to enterotoxins may confer a degree of immunity or tolerance.

Susceptibility is greatest in young children and the elderly. There is a marked difference between children aged 5 to 9 years and those aged 10 to 19 years (Evenson *et al.*, 1988). When exposed to an average of 144 ng enterotoxin A, the attack rate among the former age group was 42% in contrast to 28.6% in the latter.

Although unhealthy persons are of generally higher susceptibility, there is no specific complaint which predisposes to staphylococcal intoxication.

Exposure

Although *Staph. aureus* food poisoning is associated with well-defined types of food, there is no relationship with particular dietary habits and thus no cultural or other group at particular risk.

Staph. aureus food poisoning is sometimes considered to be an illness of the summer months but, in the USA at least, this would seem to be erroneous (Holmberg and Blake, 1984).

Treatment of *Staph. aureus* intoxication

In most cases, no treatment is required and complete recovery follows quickly after cessation of symptoms.

In severe cases, rehydration and treatment for shock is necessary and hospitalisation may be required.

Mechanisms of pathogenicity

The symptoms of staphylococcal food poisoning are due entirely to the action of the pre-formed enterotoxin and under most conditions no growth occurs in the intestine. However, growth and enterotoxin production does occur *in vivo* in cases of pseudomembranous colitis and toxic shock syndrome in which enterotoxin B as well as toxic shock toxin is implicated.

Despite a wide range of biological properties, staphylococcal enterotoxins have not been considered for a role in disease other than food poisoning, pseudomembranous colitis and toxic shock syndrome. This has been challenged by Humphreys *et al.* (1989) who suggest a wider role in the pathogenesis of staphylococcal disease. However, supportive evidence is currently lacking.

Nature of enterotoxins

Staph. aureus enterotoxins (SEs) are a heterogeneous group of heat-stable, water-soluble, single-chain globular proteins with a molecular weight of between 28,000 and 35,000 daltons. Seven immunologically distinct enterotoxins are produced: A, B, C_1, C_2, D, E and toxic shock toxin (TST). Further unidentified enterotoxins may exist (Tatini *et al.*, 1984). All of the known toxins, with the exception of TST, have been associated with food poisoning. Enterotoxins A and D, singly or in combination, are most frequently involved (Bennett, 1986).

Determination of the nucleotide sequence of SEA demonstrated a degree of amino acid sequence relatedness with SEB, SEC_1 and TST. A natural DNA probe to the *ent A* gene hybridised to an internal fragment of the *ent E* gene and an oligonucleotide derived from the *ent A* gene sequence hybridised with strains producing SEA, SEB, SEC_1, and SED. These results suggest the occurrence of a considerable sequence divergence within the family of enterotoxins (Betley and Mekalanos, 1988).

Synthesis of enterotoxins

Staphylococcal enterotoxins, in common with other secreted compounds, are produced for the sole purpose of cell export (Tweten and Iandolo, 1983). Genetic control is not fully understood. SEA production appears to be under chromosomal control, the *ent A* gene probably residing at more than one locus and possibly carried on a temperate phage integrated into the chromosome (Genigeorgis, 1989). The situation with SEB and SEC is less clear, but it appears

likely that the structural gene is chromosomal, while a regulator gene, or genes, is carried on a plasmid. SED synthesis also appears to be plasmid controlled. The nature of genetic control of production of other staphylococcal enterotoxins is not known.

Enterotoxin production is strain-specific, although a single strain of *Staph. aureus* can synthesise multiple toxin serovars. Synthesis of enterotoxins is selective in that SEA and SED are produced during the logarithmic phase of growth, while SEB and SEC are produced during the late logarithmic and early stationery

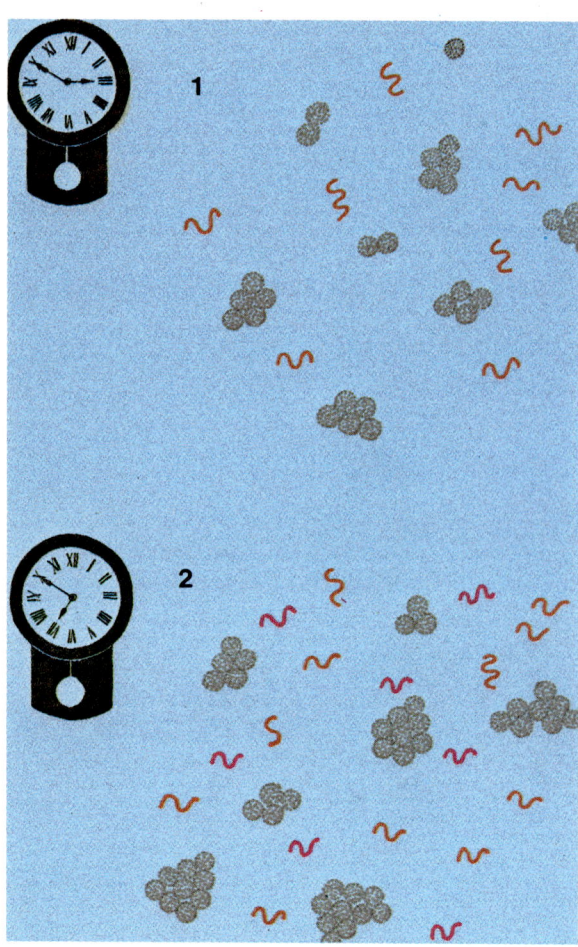

186. Selective synthesis of staphylococcal enterotoxins. The results of pure culture investigations into the synthesis of enterotoxins by a strain of *Staph. aureus* implicated in food poisoning in which pasta was the vehicle are summarised in the illustration. During growth in brain-heart infusion broth, enterotoxin A appeared first (**1**), with production of enterotoxin C delayed by 4 h (**2**; Bryant *et al.*, 1988). This phenomenon causes obvious problems during epidemiological investigations and may also partially explain the higher frequency of toxin types A and D in food poisoning.

phases (**Figure 186**; Bennett and Amos, 1982,1983). It appears that very large numbers of *Staph. aureus* (greater than 2.8×10^8) must be present for the production of detectable amounts of toxin (Otero *et al.*, 1987).

Enterotoxin production is affected by a range of extracellular factors (**Table 94**). Many of these factors are important determinants of toxin production in foods, and are discussed in greater detail in **Factors affecting growth, survival and enterotoxin production**, pages 250–255.

Table 94. Summary of extracellular factors affecting enterotoxin production.

Nutritional factors	Atmospheric factors
pH value	Sodium chloride
Temperature	Other inhibitors
Water activity	Other micro-organisms

Action of staphylococcal enterotoxins

Staphylococcal enterotoxins do not appear to act directly on intestinal cells, and thus are not considered to be classical enterotoxins such as cholera toxin (Halpin-Dohnalek and Marth, 1989). The site of the emetic action lies in the abdominal viscera where receptors activated by the enterotoxin generate impulses which reach the subcortical vomiting centre of the brain via the vagus nerve and sympathetic afferents (Sugiyama *et al.*, 1961). Staphylococcal enterotoxins are thus more correctly thought of as neurotoxins (Stephen and Pietrowski, 1981).

There is evidence that the action of SEB differs from that of the other enterotoxins in that the emetic effect resembles a pseudoallergic reaction (Scheuber *et al.*, 1987). An intradermal injection of SEB produced a skin reaction in monkeys similar to a histamine-induced response (Scheuber *et al.*, 1985). Both the skin reaction and the emetic response of SEB could be expunged by oral administration of a selective inhibitor of leukotrienes D_4 and E_4 (Scheuber *et al.*, 1987).

The potency of staphylococcal enterotoxins differs according to the serovar, SEA followed by SED being generally considered to be the most potent. In most cases, the quantity of enterotoxin required to produce an illness has been estimated on the basis of the quantity recovered from foods which had been implicated in food poisoning. On this basis the minimum quantity required has been estimated at 1µg or less (Bergdoll, 1972; Bennett; 1986).

An exceptional case occurred among school children where circumstances permitted the dose received to be calculated with a significantly higher degree of accuracy than in other cases. In this case, the quantity of SEA required to produce symptoms when ingested in chocolate milk ranged from 94 to 184ng with a mean of 144ng. This figure was considerably lower than previous estimates, although it is recognised that larger quantities of other enterotoxins would probably be needed.

The quantity of toxin required to produce symptoms does not necessarily correlate with the severity of symptoms and in a large study, 46% of patients affected by SEB received hospital attention, compared with only 5% of those affected by SEA (Holmberg and Blake, 1984). These figures should be treated with some caution, as only two outbreaks involved SEB.

Relationship between enterotoxin, coagulase and thermonuclease production

It is not currently possible to assay for staphylococcal enterotoxins on a fully routine basis, and over the years a number of biochemical properties have been proposed as markers of enterotoxigenicity. The most widely used property is the coagulase test, while thermonuclease (TNase) production has been proposed as an ancillary test to coagulase (Lachica *et al.*, 1972). This test is not intended to be a substitute for the coagulase test (Bennett, 1986), but to support the coagulase test where reactions are equivocal (ICMSF, 1978).

Coagulase and TNase are produced by most strains of *Staph. aureus*, and there appears to be no significant difference in patterns of production between enterotoxigenic and non-enterotoxigenic strains (Sperber and Tatini, 1975; Bennett *et al.*, 1986; Castro *et al.*, 1986; Umoh *et al.*, 1988). According to the test method used (see page 259), other species of *Staphylococcus* produce a positive coagulase reaction, while a thermonuclease is produced by *Staph. intermedius*, *Staph. hysicus*, *Staph. simulans* and some enterococci. However, *Staph. aureus* thermonuclease may be differentiated serologically (see page 260).

Further difficulties are caused by the existence of enterotoxigenic strains of *Staph. aureus* which are coagulase and/or TNase negative (Bennett *et al.*, 1986; Bautista *et al.*, 1988). Such strains have been implicated in food poisoning (Breckenridge and Bergdoll, 1971).

In contrast to coagulase, TNase may be detected in food extracts. The enzyme persists after the death of the producing cells and assay has been proposed as a means of determining the likely presence of entero-toxin in fermented foods such as cheese or sausage (NRC, 1985; ICMSF, 1986).

Although it is accepted that TNase may be present in foods in the absence of an enterotoxin, the question of the presence of an enterotoxin in the absence of TNase is not fully resolved. For example, in cheese, Todd *et al.* (1981) noted that cheese containing an enterotoxin did not always contain TNase, while Ibrahim and Baldock (1981) detected the enzyme whenever counts of *Staph. aureus* reached *ca* 10^6/g. It is possible that this discrepancy was due to differences in the sensitivity of the assay methods used (ICMSF, 1986). More recently, a lack of correlation between TNase activity and enterotoxin production was noted in canned fish (Stersky *et al.*, 1986), while enterotoxin was produced in cooked egg noodles in the absence of TNase (Gockler *et al.*, 1988). In associative growth with *B. subtilis* and *Enterococcus (Streptococcus) faecalis* var. *faecalis*, TNase was a reliable indicator of *Staph. aureus* growth and enterotoxin production but could not be used to predict either the number of cells present or the quantity of enterotoxin (Daoud and Debevere, 1985).

It is concluded that while TNase may have a role in the screening of foods for the likely presence of entero-toxin, results must be interpreted with care. The further development of enterotoxin assays to permit fully routine use is seen as the most valuable long-term approach.

Assay methods for coagulase and TNase are discussed in **Laboratory Methods**, *pages 258–260.*

Detection of enterotoxins

Although human volunteers have been used in the detection of staphylococcal enterotoxins (Dangerfield *et al.*, 1973), most early work depended on animal models such as kittens and, less commonly, monkeys. Kittens are still used for the determination of possible unidentified enterotoxins, and for certain other purposes (Bennett, 1986), but, with these exceptions, serological methods are in almost universal use. Gel diffusion methods, especially double gel diffusion, were the methods of choice for several years, but more sensitive methods such as ELISA and co-agglutination are now more widely used. However, methods for the extraction and concentration of enterotoxins, remain less than fully satisfactory for many foods.

Methods for the detection of enterotoxins are discussed in greater detail in **Laboratory Methods**, *pages 261–265.*

Human carriage of *Staph. aureus*

The high rate of human carriage of *Staph. aureus* is an important feature of the organism with respect to its role as a foodborne pathogen. Indeed, staphylococcal intoxication is the only major form of food poisoning in which food handlers play a significant role (Gilbert, 1983).

Staph. aureus is an important pathogen of the skin and thus is present in large numbers in boils, infected cuts and other skin lesions. It must be emphasised that human carriage is not restricted to people with overt infections, but is common among healthy people.

Staph. aureus may be present at a number of sites on the human body including the skin, nose, throat, and hair, and may even be present in stools (Holmberg and Blake, 1984). Of these sites, the nose is considered to be most important with respect to the contamination of foods (Bergdoll, 1979). The rate of nasal carriage is estimated at 10 to 40% (Noble, 1981), and the numbers in the nasal fossae usually exceed 10^3 cfu/swab (Garcia *et al.*, 1986).

The significance of human carriage of *Staph. aureus* to food safety is generally well understood, but many of the precautions taken remain based on the misconception that carriage is related to overt lesions. Action, which involves exclusion from food handling, is obviously necessary where there is overt infection but such action should not be permitted to divert attention from the need to take precautions against the healthy carrier, or against other sources of *Staph. aureus* such as raw materials or environmental colonisation.

Human carriage of staphylococci other than *Staph. aureus* may occur occasionally. Animal strains (novo-biocin resistant) of *Staph. saprophyticus* and *Staph. hysicus* subsp. *chromogenes*, for example, have been isolated from the nasal fossae of workers in abattoirs and butchers' shops (Garcia *et al.*, 1986).

Staph. aureus in the environment

The natural reservoirs of *Staphylococcus* are man and other warm-blooded animals. *Staph. aureus* is found both in healthy people and in people with skin disorders, infected cuts, etc (see above). The organism may also be isolated from domestic and food animals including dogs, poultry, sheep, cattle, pigs and goats. *Staph. aureus* in animals may be associated with disease, especially mastitis, in cattle, goats and sheep.

It is usually possible to distinguish between strains of *Staph. aureus* of human and animal origin by biotyping or other methods, and distinct ecovars (biovars) related to animal species exist (see page 240).

Enterotoxin production by animal strains of *Staph. aureus* is less common than by human strains.

Staph. aureus is highly resistant to stress and is able to adapt to survival outside the animal host. The organism can be a long-term pollutant of water (Borrego *et al.*, 1987), and may also colonise the food processing environment and plant. The most striking example of plant colonisation is that of poultry packing plants, especially of the defeathering equipment (**Figures 187–190**; Mead and Adams, 1986; Bolton *et al.*, 1988).

187–189. Colonisation of poultry packing equipment by *Staph. aureus*. *Staph. aureus* is able to colonise equipment in poultry packing plants, the rubber fingers of feather pluckers being particularly vulnerable. Colonisation leads to high numbers of *Staph. aureus* on the packaged poultry, and while this is of little direct significance on raw birds, the risk of cross-contamination to cooked products is increased.

The origin of strains of *Staph. aureus* which are able to succesfully colonise the equipment, the endemic strains, has been investigated using plasmid profiling (Dodd *et al.*, 1988) and appears to be the incoming poultry.

Apart from differences in plasmid profiles, endemic strains have a number of characteristic features, some of which are directly related to their ability to colonise equipment.

187 & 188. Clumping phenotype. Endemic strains produce clumps when grown in liquid culture and have a characteristic 'stickiness' on solid culture medium. Clumping growth in tryptone–soya broth is illustrated in **187**, and is easily distinguished from growth of non-endemic strains (**188**). Clumping probably aids the attachment to plucker fingers and increases the resistance to chlorine (Bolton *et al.*, 1988).

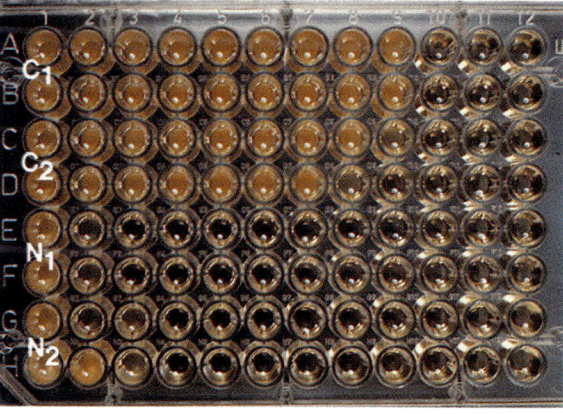

189. Chlorine resistance of endemic strains of *Staph. aureus*. Endemic strains (C_1, C_2) are up to eight times more resistant to sodium hypochlorite, a widely used disinfectant in the poultry processing industry, than non-endemic strains (N_1, N_2).

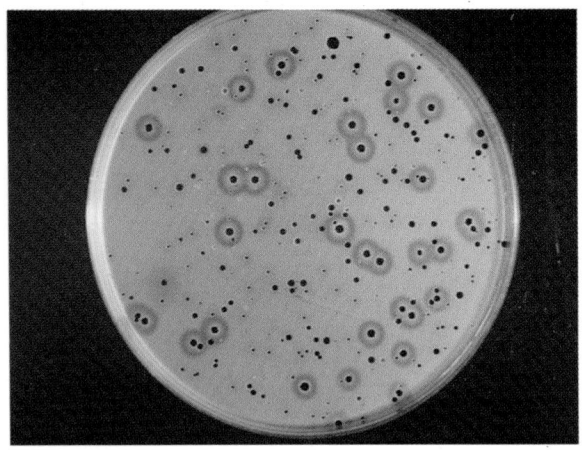

190. Aerial contamination of a poultry packing plant by *Staph. aureus*. The extent to which *Staph. aureus* can become established in poultry packing plants is indicated by the large number of colonies developing on Baird-Parker selective medium after exposure of the agar surface to plant air for 10 min. (Reproduced by courtesy of Dr T.A. Roberts, AFRC Institute of Food Research, Reading Laboratory.)

Species other than *Staph. aureus* also originate in man and warm-blooded animals. Some species, for example, *Staph. caprae* in goats and *Staph. gallinarum* in poultry, have a strong association with a given animal, while others such as *Staph. epidermidis* and *Staph. hysicus* have a much wider distribution (Kloos and Schleifer, 1986).

Foods commonly associated with *Staph. aureus*

Significance of *Staph. aureus* in foods and role as an indicator organism

With most food poisoning micro-organisms such as *Salmonella*, the presence in a food of a virulent strain indicates a definite hazard to consumers. However, the situation is not the same with *Staph. aureus*, since safety is defined by the presence of enterotoxin and not that of the producing organism. For this reason, it has been suggested on a number of occasions that there is no value in examining foods for *Staph. aureus* and that quality assurance should be based on the determination of enterotoxin.

In foods such as salamis where enterotoxin is usually present in the absence of viable cells of *Staph. aureus*, this is indeed a valid approach, providing that assay methods suitable for truly routine application can be developed. Caution is necessary since as many as 5% of food poisoning outbreaks are caused by 'unknown' toxins, which are not detectable by current serological methods (Kokan and Bergdoll, 1987).

In foods where numbers of *Staph. aureus* are not significantly reduced by processing, the situation is different in that concentration on assay for enterotoxin ignores the valuable role of *Staph. aureus* as an indicator organism. The presence of excessive numbers of *Staph. aureus* on, for example, sliced cooked meat, is indicative of a failure to adequately control critical process points (contamination and temperature), *irrespective* of the presence or absence of enterotoxin. Correctly applied and interpreted, viable counts for *Staph. aureus* provide information which cannot be obtained by enterotoxin assay or by examination for other indicator or index organisms.

Raw foods of animal origin

Meat and poultry

Staph. aureus is inevitably present on raw meat and poultry. Many of the strains present are of animal origin, although human strains are also likely to be present especially if a high level of handling is involved during butchery and hygiene is poor. Numbers of *Staph. aureus* on raw poultry may be high as a result of the colonisation of the processing equipment by particular strains of the organism (see above).

The presence of *Staph. aureus* in raw meat is of little direct significance since the organism is unable to grow, and since in these circumstances it has no role as an indicator organism. However, high levels of a pathogen in a raw product can lead to high morbidity in the processed product due to an increased risk of cross contamination. For example, high numbers of *Staph. aureus* in raw poultry is likely to lead to the poultry being rejected for further processing (Dodd *et al.*, 1987).

Milk

Milk from most species including cows (Harvey and Gilmour, 1985), goats (Harvey and Gilmour, 1988) and sheep (Hajek, 1978) contains small numbers of *Staph. aureus*. Greater numbers may be present if mastitic milk is present, although this effect is less marked with *Staph. aureus* than with another common mastitis pathogen, *Strep. uberis* (Bramley *et al.*, 1984).

In general, the presence of small numbers of the organism in raw milk is of little consequence although, in the case of the mastitic udder, a high percentage of strains may be enterotoxigenic and thus of greater potential public health significance (Lombai *et al.*, 1980). There have, been reports of *Staph. aureus* growing and synthesising enterotoxin in raw milk before heat treatment. Examples include dried milk, Indian milk sweets and chocolate milk. The latter outbreak, in which the milk was subsequently pasteurised, involved at least 500 school children in California, USA (Holmberg and Blake, 1984).

There have also been unsubstantiated suggestions that staphylococcal food poisoning of infants has been caused by the use of raw cows' milk to make up dried formula foods, the relatively low water activity of the made up food favouring the growth of *Staph. aureus* during holding at room temperature.

Unpasteurised cream is thought to have been the origin of *Staph. aureus* in an outbreak involving mushroom sauce (Anon, 1984).

Heat processed (pasteurised) food

Meat products

Staph. aureus is frequently found in small numbers (*ca* 10^2 cfu/g) on cooked meats. Significantly higher numbers of the organism (10^4–10^5 cfu/g) are indicative of an unacceptable level of contamination during handling and/or poor temperature control. Cold cooked meats such as ham, roast beef and cooked chicken are common vehicles of staphylococcal food poisoning. Small outbreaks may occur in the home, but those of a larger scale are usually associated with catering (**Table 95**). The traditional use of cold meats in buffets combined with the catering practice of preparation well in advance of eating, means that social functions such as weddings and christening

parties are often involved. In many cases, the outbreak could be avoided by attention to temperature control, but too often socialising takes precedence over refrigeration (Holmberg and Blake, 1984).

Staphylococcal food poisoning due to sliced, vacuum-packed cooked meats is relatively rare, although cases are known. This has been attributed to the suppression of enterotoxin production at low oxygen tensions, but the inhibition of *Staph. aureus* by lactic acid bacteria, which inevitably reach large numbers in commercial packs, may be of greater importance.

The growth of *Staph. aureus* may occur during the common catering practice of 'warm holding'. A large outbreak occurred in a factory canteen where beefburgers were held at 35 to 40°C to avoid the dehydration which occurred at higher temperatures. The source of infection was the canteen manageress, who had an infected cut and who, despite being warned not to handle food, insisted on 'helping out'.

The production of some cooked meats involves a two-stage cooking process, and cases are known where the growth of, and the production of enterotoxin by, *Staph. aureus* has occurred in the interval between the two heat treatments. A chopped, shaped, burger-type product made with lamb was responsible for a common source outbreak in which six family clusters of cases were reported over a 5-week period. Enterotoxin A was isolated from the product in the absence of viable *Staph. aureus*. Food handlers were believed to be the source of the organism.

Although most outbreaks attributed to cooked meats have involved strains of human origin, poultry was considered to be the source of an enterotoxin A producing strain of *Staph. aureus* that was implicated in an outbreak involving cooked chicken. Cross contamination from the raw birds occurred via trays, working surfaces and hands.

Staph. aureus may be introduced to cooked products from ingredients added after cooking. As with *Sal-*

Table 95. Examples of outbreaks of staphylococcal food poisoning associated with catering.

Circumstances	Food	Number affected	Source
Aircraft charter flight[1]	Ham omelette	197	Chef
Wedding reception (food provided by guests)[2]	Not known	>50	?
Company social events (3 events in 24 h period)[3]	Ham sandwich	93	Food handler
Camp site[4]	Roast beef	40	?
Christening party[5]	Beef sausage (also rice salad)[6]	18	Food handler

[1] Board (1983).
[2] Pinegar and Suffield (1982).
[3] Anon (1984).
[4] Communicable Disease Surveillance Centre (1988).
[5] Communicable Disease Surveillance Centre (1987).
[6] *Bacillus cereus* was isolated from the rice salad.

monella, gelatin is a major hazard, and has been the vehicle of infection for staphylococcal food poisoning when injected into pork pies and used to glaze pâté.

Milk and milk products

Pasteurised liquid milk is rarely involved in staphylococcal food poisoning, the major exception being the outbreak noted above involving chocolate milk where growth occurred before pasteurisation. Cream packaged for domestic use is also rarely involved, although an outbreak of staphylococcal food poisoning which followed a Christmas meal attended by members of an extended family was attributed to the cream served with the Christmas pudding. This had been mixed with whisky the previous night, and deliberately left in a warm place to mingle!

Growth of, and enterotoxin production by, *Staph. aureus* in cream is reputed to be poor (Ikram and Luedecke, 1977). Despite this, cakes and other confections containing whipped cream are notorious as a vehicle of staphylococcal food poisoning. The underlying cause is poor hygiene in bakeries, some machines used for whipping cream being reported to be virtually impossible to clean and sanitise correctly. The number of cases has, in fact, decreased in recent years due to the more widespread use of refrigerated display cabinets, although sporadic outbreaks continue. A recent large outbreak involved 215 of 715 passengers on a cruise ship (Waterman *et al.*, 1987). The source of contamination was human carriage, and temperature abuse was a major contributory factor.

Fish and fish products

Occasional outbreaks of staphylococcal food poisoning have been attributed to cooked shellfish (Bryan, 1980). There have also been unconfirmed reports of outbreaks due to fried finfish that were held 'warm' before consumption.

Eggs and egg products

Hard-boiled eggs were the vehicle of infection in two outbreaks involving an enterotoxin B producing strain of *Staph. aureus* in the US (Harbrecht and Bergdoll, 1980). Hard-boiled egg sandwiches were also suspected of involvement in an outbreak in the UK.

Liquid egg is believed to have been the source of

Staph. aureus in outbreaks involving both fresh and dried pasta (see below; Woolaway, 1986).

Vegetables

Staph. aureus grows well on several types of vegetable, although on uncooked vegetables the organism is usually inhibited by competition from other microorganisms. There have been occasional reports of staphylococcal food poisoning due to its growth on cooked vegetables that had been kept warm before serving; and anecdotal reports of growth and enterotoxin production on blanched peas before both freezing and canning. French fried potatoes (chips) have been implicated in well-substantiated outbreaks that may have been due to the contamination and subsequent growth of *Staph. aureus* in the interval between the two frying stages commonly used in restaurant catering.

Fermented foods

Meat products

Staph. aureus can be a major problem in salamis and other fermented meats if the initial fermentation is inadequate and over the years a number of cases of food poisoning have occurred. *Staph. aureus* dies fairly rapidly in fermented sausages during maturation and thus the absence of the organism, or its presence in small numbers, in the finished product is not indicative of safety. The thermonuclease test may be applied to the finished product to determine the *likely* presence of enterotoxin, but criteria based on counts of viable *Staph. aureus* should be applied immediately after fermentation, when numbers should not exceed 10^4 cfu/g.

Milk products

Staph. aureus can be isolated in small numbers from a wide range of fermented dairy products, although staphylococcal food poisoning is usually only associated with certain types of hard cheese. The presence of *Staph. aureus* in significant numbers at the end of fermentation is attributed to failure of the starter culture. The greatest number of the organism in the finished product usually occurs in hard cheeses such as cheddar or similar varieties, where the drained whey is held at a high temperature for some time before salting.

Cheese milk may be contaminated either from human sources after pasteurisation or, in the case of cheese made from unpasteurised milk, from the producing animal. There is some disagreement concerning the importance of *Staph. aureus* strains from milk producing animals. Enterotoxin producing strains of coagulase-negative *Staph. aureus* have been reported in cheese made with raw milk, especially sheep milk (Bautista *et al.*, 1988). In other studies, coagulase and thermonuclease-negative strains from raw milk cheese have been found to be non-enterotoxigenic (Castro *et al.*, 1986). Sheep milk cheese has, however, been implicated in staphylococcal food poisoning, the organism being derived from mastitic udders (Bone *et al.*, 1989).

Reported cases of food poisoning have been restricted to cheddar and similar varieties of cheese (ICMSF, 1980, 1986) and Swiss-type cheese (Todd *et al.*, 1981). However, a wider range of varieties of cheese have been implicated in outbreaks which have not been officially investigated including the semi-hard Stilton (non-mould ripened), and both soft ripened and unripened varieties.

The number of cases of staphylococcal food poisoning involving cheese has fallen in recent years, although the problem persists (NRC, 1985).

Vegetables

There have been a number of unconfirmed reports of staphylococcal food poisoning resulting from the consumption of home-made, salt-pickled vegetables including sauerkraut. There have also been a number of cases resulting from the consumption of vinegar-pickled onions and cucumber with dill. *Staph. aureus* is resistant to the anti-bacterial effect of onion components, and the organism probably grew on the peeled onions before immersion in the brine, or during the early stages of pickling.

Concentrated and dried products

Staph. aureus is present in small numbers in a wide range of dried and concentrated products. Normally this is of no significance. However, in some circumstances the selective advantage imparted to *Staph. aureus* by the lowered water activity level during the concentration, or drying, process permits the growth of the organism and the production of significant quantities of enterotoxin

Spray-dried milk was associated with a number of outbreaks of food poisoning in the mid-1950s (Anderson and Stone, 1955), and occasional cases are still reported (El-Dairouty, 1989).

Concentrated milk itself does not appear to be a problem in Western nations, but the Indian concentrated milk, Khou, used as a base for Indian sweets, has been responsible for a number of outbreaks of staphylococcal food poisoning (Varadaraj and Nambudriped, 1986).

In the mid-1980s, dried pasta (egg lasagne) was responsible for an international outbreak of staphylococcal food poisoning with cases reported in three countries: Italy (the country of production), Luxembourg and the UK. In the largest UK outbreak, 30 girls (of 270 at risk) at a girls' boarding school were affected. The source of the organism is believed to have been the egg used in the lasagne — *Staph. aureus* grew in the fresh dough and while it was drying, and survived in the final product (Woolaway, 1986).

Bakery products

Although a well-recognised problem exists with cream confections, other bakery goods have been implicated in staphylococcal food poisoning. In the UK, products cooked on a hot-plate including potato cakes, drop scones and scotch pancakes have been involved. In most cases, the organism survived the very short cooking and was able to grow in the finished product during storage at room temperature. In other cases, the growth of *Staph. aureus* occurred in the batter mix before cooking.

There have also been unconfirmed reports of staphylococcal intoxication following the consumption of bread. These cases were not fully investigated and no enterotoxin determinations were made, but *Staph. aureus* was isolated in small numbers from suspect material. Contamination probably occurred while the bread was being sliced.

Although there is some awareness of potential problems with bakery goods (Halpin-Dohnalek and Marth, 1989), these products are often overlooked as a vehicle for *Staph. aureus*. The potential for the growth of the organism in some types of dough before baking is seen as an important factor which is often ignored.

Canned goods

Staph. aureus has been involved in a number of cases of food poisoning involving post-process leakage of cans. In some cases, this has been attributed to handling of the warm cans, but in others the contents were probably contaminated from can runways or other equipment. During 1989, there has been a continuing problem in the USA due to canned mushrooms imported from China. A number of factories have been involved and the problem appears to be industry-wide. During the same year, a number of outbreaks in the UK have been attributed to pasteurised ham imported from Romania in catering size cans.

Further examples include corned beef, (Gilbert, 1983), fish (Stersky *et al.*, 1986), peas (Bashford *et al.*, 1960) and ox-tongue (Communicable Disease Surveillance Centre, 1989).

Miscellaneous products

The ubiquitous nature of *Staph. aureus* means that isolations have been made from a wide range of foods. In most cases, numbers are very small and the probability of growth is very low. Less common products that have been associated with staphylococcal food poisoning include whipped butter (CDC, 1970,1977), garlic butter (Anon, 1984), margarine (Anon, 1984) and potato salad (Anon, 1984).

Species other than *Staph. aureus* in foods

Species other than *Staph. aureus* may be readily isolated from raw milk and meat, as well as from products such as raw milk cheese and salami. The species isolated tend to reflect the producing animal, but in cheese and salami fermentation introduces selective pressures. In the case of salami, most isolates have a high NaCl tolerance and a probable role in fermentation. A study of fermented sausages imported into the UK (Seager *et al.*, 1986) showed the predominant species to be *Staph. xylosus* (38%), *Staph. saprophyticus* (29%) and *Staph. warnerii* (10%). The predominant species may be determined by the site of production.

Staphylococci, other than *Staph. aureus*, are uncommon in foods not of animal origin, an unusual example being the isolation of *Staph. warnerii* from lemonade (*ca* 10^3 cfu/ml). This appeared to be associated with plant contamination.

There have been no known outbreaks of food poisoning due to staphylococci other than *Staph. aureus*, but where enterotoxin producing species are present the possibility cannot be discounted.

Factors affecting the growth and survival of *Staph. aureus* in foods and the production and stability of its enterotoxins

Enterotoxin production is not an inevitable consequence of the growth of *Staph. aureus*, and growth may occur in the absence of enterotoxin synthesis. In this context, there is a marked difference between SEA and SED producing strains whose growth and enterotoxin production correlate closely, and SEB and SEC producing strains whose growth in the absence of enterotoxin production is a common occurrence.

It must be appreciated when discussing the conditions that limit the production of enterotoxin that some of the widely accepted criteria were determined using methods of detection considerably less sensitive than those in use today. In some cases, enterotoxin production may continue at a low level, under more stringent conditions than those published. It must further be appreciated, when discussing the inactivation of enterotoxins, that inactivation as determined by serological methods may not correlate with a loss of biological activity. Circumstantial evidence suggests that a correlation exists (Notermans *et al.*, 1988), but this has not been directly proven.

Temperature

According to Tatini (1973), the cardinal temperatures for the growth of, and the enterotoxin production by, *Staph. aureus* are:

Growth Range 7 to 47.8°C; optimum 37°C
Enterotoxin production Range 10 to 46°C; optimum 40 to 45°C.

These temperatures were determined using pure cultures cultivated in broths, and should be used only as a guide to the organism's behaviour in foods where other factors affect growth:temperature relationships. For example, no growth was recorded on turkey meat at 10°C (Yang *et al.*, 1988), whereas growth occurred at 11°C in corned beef (Whiting *et al.*, 1985), and at 12°C in a nitrite-free salted pork product (Post *et al.*, 1988).

Although it is not possible to predict the relationship between the organism's growth and enterotoxin production in foods, generalised guidelines have been prepared (Smith *et al.*, 1983):

1 In any given food, the optimum temperature for enterotoxin production is a few degrees higher than that for growth.
2 Temperature changes affect enterotoxin synthesis much more than growth.

In dairy products, SEA was produced in milk at 20°C, but not at 10°C (Donnelly *et al.*, 1968), while Scheusner *et al.* (1973) found SEB but not other enterotoxins to be produced during growth at 13°C in brain-heart infusion broth.

There have been a number of anecdotal accounts of the organism's growth and enterotoxin production at temperatures below 5°C. There is no properly documented evidence to support these contentions, which appear to be based on an unproven and unjustified faith in the efficiency of commercial refrigeration equipment.

Cells of *Staph. aureus* are destroyed at temperatures commonly used in food processing. There is considerable strain variation in resistance, but this is unlikely to have any consequence in processing. Heat resistance is modified by a number of factors including heating menstruum and the age of cells, resistance is also greater in cells grown at 46°C than at 37°C. Survivor curves show 'tailing', and there is a small subpopulation of considerably greater heat resistance.

Published D_{60} values range from 0.43 to 7.9 minutes, and z values from 4.5°C to 10°C (Roberts, 1982).

Heat-stressed cultures of *Staph. aureus* may lose the ability to synthesise enterotoxin as well as coagulase and thermonuclease (Batish *et al.*, 1989). Strains of SEB producing *Staph. aureus* appear to be most heat resistant, and least prone to the loss of enterotoxin production. The lost characteristics are regained after four to five subcultures.

Enterotoxins are recognised as having a considerably greater level of heat resistance than *Staph. aureus* cells, and to survive cooking and most commercially applied heat-treatment. In a systematic investigation using levels of SEA, SEB and SEC, similar to those in foods, Tibora *et al.* (1987) confirmed that normal time:temperature combinations used in cooking were 'unlikely' to destroy toxins.

However, inactivation of toxins is not an all or nothing phenomenon. For example, SEB lost 60 to 70% of its activity during the first few minutes of heating at 80 and 100°C (Reichert and Fung, 1976), although the remaining 30 to 40% of activity was lost only slowly. The effect of heating is complicated by reactivation, which may occur during storage, during continued heating and even during reheating to a higher temperature (Reichert and Fung, 1976).

The stability of the enterotoxin molecule is affected by various physical factors and there is evidence that thermal inactivation can occur at temperatures lower than generally accepted. For example, the heat resistance of both SEA and SEB is much lower in soya bean ferment than in brain-heart infusion broth (Nout *et al.*, 1988). Cooking in soya bean ferment for one minute at 80°C completely inactivated SEA and inactivated SEB by 92%.

Staph. aureus is a robust bacterium and can survive for long periods at temperatures below those which permit growth. Freezing has little effect on its viability at pH values of most frozen foods even when repeated freeze-thaw cycles are applied (Demchick *et al.*, 1982).

Enterotoxins are also stable at low temperatures and while a loss in titre may be observed during the refrigerated storage of foods such as cheese (Otero *et al.*, 1988), this is probably either due to other factors or is, possibly, an artefact caused by the binding of enterotoxin to food proteins.

Irradiation

Published results show a wide range of sensitivity of cells of *Staph. aureus* to ionising radiation, doses ranging from less than 0.5 Kgy (Niven, 1958) to 1 to 2 (Quinn *et al.*, 1967). However, most studies suggest that the organism is inactivated at doses of 0.3 to 0.5 Kgy. Planned radiation pasteurisation processes are largely directed at *Salmonella*, and *Staph. aureus* is not of major concern. However, the potential for the rapid growth of *Staph. aureus* in the absence of a spoilage microflora is considerable, and the organism must be seen as one of the potential hazards of irradiation treatment.

Staphylococcal enterotoxins are not inactivated by irradiation at dose levels acceptable in foods.

pH value

Under otherwise optimal conditions, the pH value:growth relationships of *Staph. aureus* are: range 4.0 to 9.8, optimum 6.0 to 7.0. However, the pH range for its growth and enterotoxin production is strongly modified by other factors, especially anaerobiosis.

In aerobic conditions, Genigeorgis (1976) was able to demonstrate growth and toxin production by the organism at pH 4.0, whereas under anaerobic conditions, limiting values were pH 4.6 and 5.3 respectively. Some variation according to toxin serovar is likely (Barber and Deibel, 1972). In aerobic conditions, minimum values were: SEA 4.9; SEB 5.0; SEC_2 4.9; SEE 4.8; while corresponding values under anaerobic conditions were SEA, SEB, SEC_2: 5.7; SEE 6.0. Care must be taken in extrapolating specific results to the general situation, as considerable strain to strain variation exists. For example, minimum pH value for SEA production, varied from 4.9 to 5.7 under aerobic conditions, and from 5.7 to 7.0 under anaerobic.

The nature of the acidulant is also of considerable importance in determining the minimum pH value for *Staph. aureus* growth and enterotoxin production. The limiting pH value for SEA was 4.5 when hydrochloric acid was used as acidulant, but 5.0 when lactic acid was used (Tatini *et al.*, 1971). This factor is of considerable importance when formulating preservative systems.

Water activity level

The ability of *Staph. aureus* to grow and produce enterotoxins at relatively low a_w is of considerable importance in the formulation of intermediate moisture foods. The growth of *Staph. aureus* occurs over the range 0.83 to greater than 0.99, while corresponding values for enterotoxin production are 0.86 to greater than 0.99. In each case, the optimum is a value of greater than 0.99 (Tatini, 1973). The minimum a_w level for both growth and enterotoxin production is likely to be raised by other factors, including temperature and pH value.

At low a_w levels there are marked differences between SEA and SEB production. Reducing the a_w level from 0.996 to 0.91 had only a limited effect on the growth of a SEA producing strain of *Staph. aureus* but reduced toxin production fivefold. In contrast, SEB production was reduced 43 fold although again there was only a marginal effect on the growth of the producing strain (Troller and Stinson, 1978). Similar effects were found by Ewald and Notermans (1988),

the growth and production of SEA and SED continuing, when other conditions were optimal, at a_w 0.86, while the production of SEB and SEC ceased at a_w 0.96 to 0.93.

The minimum a_w level for both growth and enterotoxin production varies according to the humectant used. Thus for SEB production, the limiting a_w was 0.98 to 0.99 when glycerol was used as humectant, 0.90 to 0.92 with NaCl and less than 0.90 with a mixture of NaCl and Na_2SO_4 (Troller, 1971). The effect of less commonly used humectants has also been assessed (Vaamande *et al.*, 1982). The lower limit for the *growth* of *Staph. aureus* was a_w 0.90 when sodium lactate was used as humectant, but with the two polyols, erythritol and xylitol, inhibition occurred at levels as a_w high as 0.94 and 0.92 respectively.

Composition of the atmosphere

Although the importance of aerobic conditions in promoting enterotoxin production may be readily demonstrated in pure culture (Dietrich *et al.*, 1972; Woodburn *et al.*, 1973), the situation in foods is more complex.

As noted above, enterotoxin production is affected by anaerobic conditions to a greater extent than the growth of the organism (Smith *et al.*, 1983). In vacuum packed ham, SEB production is variable and unpredictable even when numbers of *Staph. aureus* are high (Genigeorgis *et al.*, 1969), while enterotoxin production was inhibited when a nitrite-free salt pork product was either vacuum, or nitrogen, packed (Post, 1988).

Curing ingredients

NaCl

Staph. aureus is a halotolerant organism with a growth range, in artificial media, of 0 to 10-20% added NaCl. The growth range in foods is similar, although the upper limit may be lower due to the presence of other inhibitory factors. In many foods, there is little reduction in its growth at NaCl concentrations up to 5–7%, and in fermented products containing less than 5% NaCl growth of *Staph. aureus* may be enhanced at the higher level. However, this is an indirect effect due to the partial inhibition of starter bacteria at the higher levels reducing the competitive pressure on *Staph. aureus* (Marcy *et al.*, 1985).

Under optimal conditions, the upper limit for

enterotoxin production is *ca* 10%, but there are again marked differences between SEA and SEB. SEA production relative to growth remains relatively constant, while the effect on SEB production is considerably greater than on growth (Smith *et al.*, 1983).

In recent years, concern over the possible adverse health effects of dietary sodium has led to moves to replace NaCl with KCl, $CaCl_2$, $MgCl_2$, or a combination of these salts. Relatively few studies have been made of the implications of such a substitution on *Staph. aureus*, but Morita *et al.* (1979) showed that the magnesium ion, in contrast to the sodium ion, stimulated enterotoxin production.

Nitrite

Staph. aureus is generally considered to be resistant to nitrite (Newsome, 1988), although few quantitative studies have been made. An early study of the effect of the salt on enterotoxin production showed 200 mg/l nitrite to have no effect on SEB production, although a small effect was noted in combination with 2% NaCl (McLean *et al.*, 1968). The effect of nitrite is enhanced by anaerobic conditions. Slight sensitivity to 100 mg/l nitrite was noted at a pH value as high as 7.3 under strictly anaerobic conditions, while at pH 6.0 inhibition was marked. No inhibition by 100 mg/l nitrite, pH 7.3 was observed under aerobic conditions although there was a slight inhibitory effect at 500 mg/l, pH 7.3 (Fang *et al.*, 1985). An anti-microbial effect of nitrite at pH 7.3 is itself unusual since there is effectively no undissociated nitrous acid present; nitrous acid being considered responsible for the inhibitory effects of nitrite.

Other curing ingredients

Although information is limited, $NaNO_3$ and curing adjuncts appear to have no significant effect on the growth of *Staph. aureus* or the production of enterotoxin.

Other preservatives

Staph. aureus is resistant to many preservatives used in foods, potassium sorbate (0.25%), for example, had no effect on growth of *Staph. aureus* in either a simulated processed cheese (Parada *et al.*, 1982) or pumpkin pie (Wyatt and Guy, 1981). However, sorbate was effective at both 0.13 and 0.26% against *Staph. aureus* in bacon (Pierson *et al.*, 1979). The inhibitory effect of sorbate is enhanced by modified atmospheres containing high concentrations of carbon dioxide (Elliott *et al.*, 1982).

The antioxidants, butylated hydroxyanisole (BHA) and butylated hydroxytoluene (BHT), have activity against *Staph. aureus* at levels of 100 mg/l upwards (Ayaz *et al.*, 1980). There is synergism with sorbate (Parada *et al.*, 1982) and activity is also enhanced in fermented foods by the presence of starter organisms (Metaxopoulos *et al.*, 1981).

Type of food

In addition to well-defined parameters which affect staphylococcal growth and enterotoxin production, variation exists according to the type of food. The underlying causes are not fully understood, although nutritional factors and the repression of enterotoxin formation by some carbohydrates (Jarvis *et al.*, 1975) may be involved.

For example, growth on peas is often very good, but enterotoxin is produced only in small quantities (less than 4µg/10g) after prolonged incubation, while staphylococcal growth to an equivalent extent on pork or chicken commonly yields more than 100µg/g (Notermans and von Olterdijk, 1985). Cooked turkey meat has been found to be a poor medium for enterotoxin production (Yang *et al.*, 1988).

Inhibition by other micro-organisms

Staph. aureus is usually considered to compete poorly with other micro-organisms and to rarely reach significant numbers, or produce significant amounts of enterotoxin when a spoilage microflora is present. Noleto *et al.*, (1987) showed that enterotoxin was produced only when *Staph. aureus* significantly outnumbered competitors (*B. cereus*, *Ent. faecalis*, *Pseudomonas aeruginosa*) when growing in broth cultures. Inhibition by spoilage organisms is less easy to demonstrate on foods than in broths due, probably, to the spatial separation of the different organisms in foods. In many circumstances, toxin production is more influenced by temperature than by the presence of competitors (Herten *et al.*, 1989).

The inhibitory effect of lactic acid bacteria in starter cultures is of particular interest because of the potential for the growth of *Staph. aureus* in improperly fermented foods (**Figures 191–193**). For several years inhibition was thought to be caused by production of lactic acid. It is now recognised that other factors are also involved including hydrogen peroxide formation, reduction of E_h and bacteriocins.

Bacteriocin production by lactic acid bacteria was first described by Upredi and Hinsdill (1975), and is a feature both of *Lactobacillus* and *Pediococcus*. Activity is acid mediated, being greater at low pH values (Anderson, 1986; Attaie *et al.*, 1987).

Bacteriocin production appears to be under plasmid control (Gecs *et al.*, 1983), and the possibility therefore exists of using genetic engineering techniques to enhance the specific inhibitory effects of lactic acid bacteria against *Staph. aureus* and other pathogens such as *Listeria monocytogenes*. This approach may be limited in value by the ability of *Staph. aureus* to adapt to the inhibitory effect of bacteriocins (Anderson, 1986).

Not all of the interactions between *Staph. aureus* and other micro-organisms lead to inhibition. In cheddar cheese, surface growth by *Penicillium* raised the pH value in the subsurface sections of the cheese and permitted staphylococcal growth and SED production (Duitschaever and Irvine, 1971).

191–193. Patterns of inhibition of growth and enterotoxin production by *Staph. aureus* in the presence of starter cultures of lactic acid bacteria.
Three patterns of inhibition have been described:

191

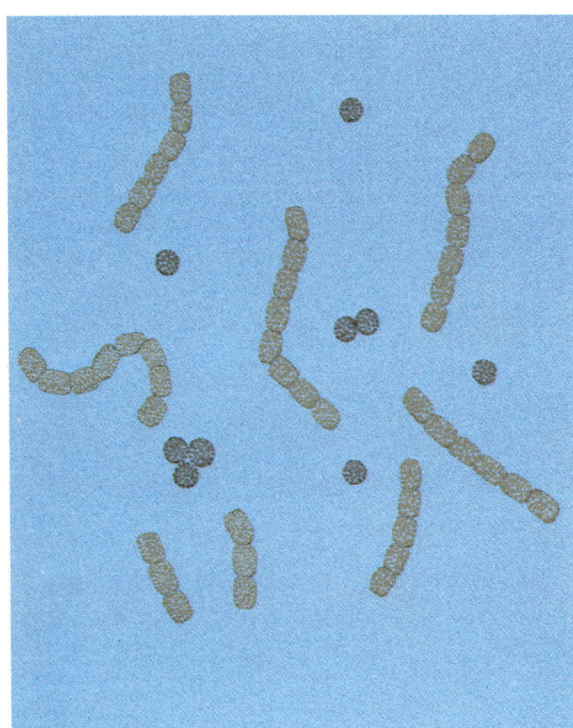

192

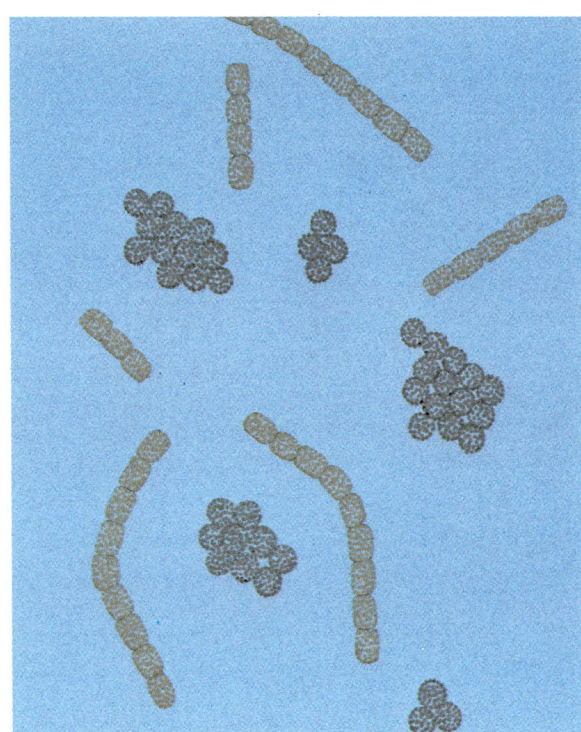

191. Inhibition of growth of *Staph. aureus* during growth of lactic acid bacteria (Attaie *et al.*, 1987). Under some conditions, an existing population of *Staph. aureus* may be reduced in number (Raccach, 1981).

192. No inhibition of growth of *Staph. aureus*, but inhibition of enterotoxin synthesis (Metaxopoulos *et al.*, 1981).

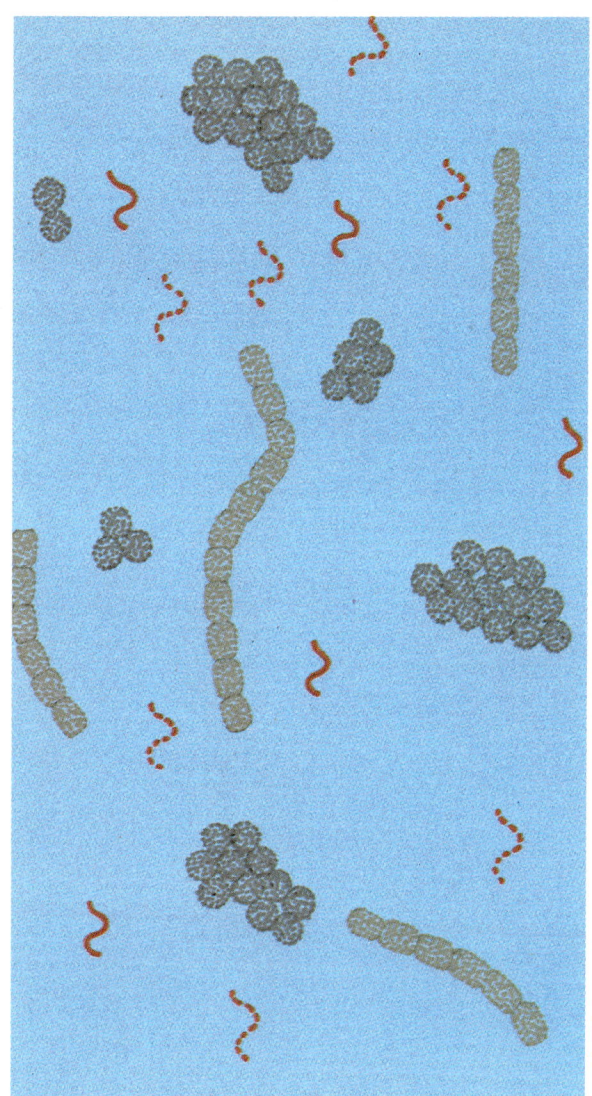

193. Destruction of preformed enterotoxin during growth of lactic acid bacteria (Otero *et al.*, 1988). This is probably an uncommon phenomenon.

Note: According to Niskanen and Nurmi (1976), lactic acid bacteria *alone* are unable to inhibit growth and enterotoxin synthesis, but are able to act synergistically with other factors especially temperature.

Laboratory methods

Isolation of *Staph. aureus*

Enrichment

Although hypertonic media such as salt meat broth are still widely used for the enrichment of *Staph. aureus*, they are not recommended. Major disadvantages are the poor recovery of stressed cells of *Staph. aureus* and the inability to prevent the growth of *Bacillus*. A suitable alternative is GC broth (Giolitti and Cantoni, 1966) as modified by Van Doorne *et al.*, (1981). Cultures showing blackening after 24 hours at

37°C are considered to be presumptive *Staph. aureus*, although confirmatory testing is required (Bouwer-Hertzberger *et al.*, 1982). GC broth is also suitable for use in most probable number determinations (AOAC, 1984).

Selective plating media

Media containing high levels of NaCl have also been widely used as selective plating media for *Staph. aureus* (Chapman, 1945). Some media of this type such as *Staphylococcus* Medium No. 110 and mannitol-salt

194

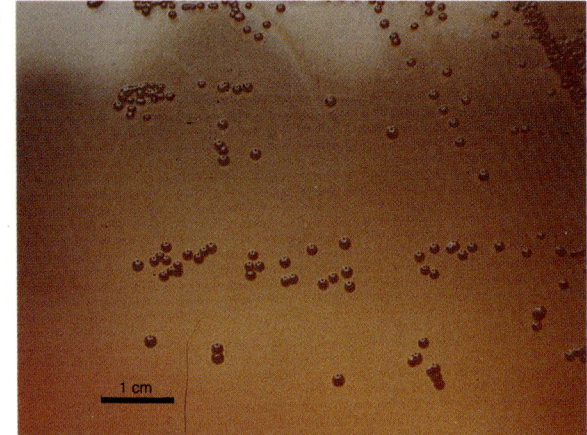

195

194 & 195. Isolation of *Staph. aureus* on mannitol–salt agar medium. Mannitol–salt agar is unsatisfactory, both because of the inhibitory effect of high NaCl concentrations on stressed cells of *Staph. aureus*, and because of its low selectivity to other NaCl tolerant micro-organisms. Mannitol–salt agar, however, is simple to prepare, contains readily available ingredients and is cheap. Its use may therefore be indicated in situations where preparation of technically superior but more complicated and expensive media such as Baird–Parker causes difficulties. (*Mannitol-salt agar: 36 h at 37°C.*)

agar (**Figures 194–195**) remain in use, but neither these, nor later media such as the polymyxin containing phenolphthalein diphosphate agar, can be recommended where the use of superior, but more complicated, media such as Baird-Parker agar is possible.

Baird-Parker agar (egg yolk-glycine-tellurite-pyruvate agar; Baird-Parker, 1962) represented a major breakthrough in media for *Staph. aureus*, and is a striking example of the logical application of the principles of selective and diagnostic media.

Baird-Parker agar combines a high degree of selectivity with the ability to recover the stressed cells of *Staph. aureus*, and is particularly effective for recovery of the organism in the presence of large numbers of competing micro-organisms (Rayman *et al.*, 1988).

The functions of the active ingredients of the medium are listed in **Table 96** and its use illustrated in **Figures 196–200**.

Colonial appearance on Baird-Parker medium gives, at best, a presumptive identification and two modifications have been made to reduce the need for further testing. The first of these incorporates the direct determination of coagulase activity to replace egg yolk clearing activity as a diagnostic feature. In the first medium of this type, pig plasma was used (Stadhouders *et al.*, 1976), bovine fibrinogen subsequently being incorporated to improve the clarity of the coagulase reaction (Hauschild *et al.*, 1979). A similar medium of higher specificity containing rabbit plasma in place of pig plasma (RPF agar) was developed by Beckers *et al.* (1984), but was over-selective to some strains of *Staph. aureus* as a result of the toxicity of tellurite in the absence of egg yolk. This problem was overcome by reducing the tellurite content of the medium fourfold (Sawhney, 1986).

Table 96. Role of functional ingredients in Baird–Parker and KRANEP media for the isolation of *Staph. aureus*.

	Baird-Parker	KRANEP
Selective inhibitors	Tellurite Lithium chloride[*]	Actidione Sodium azide Potassium thiocyanate Lithium chloride[*]
Selective growth stimulators	Pyruvate Glycine	Pyruvate
Diagnostic system	Egg yolk Tellurite reduction	Egg yolk Pigmentation

[*] Limited inhibitory effect only.

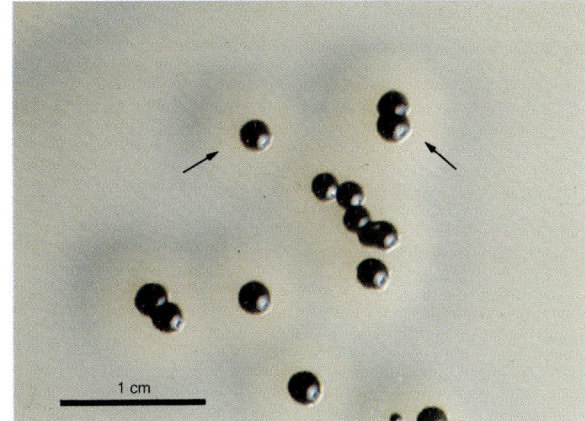

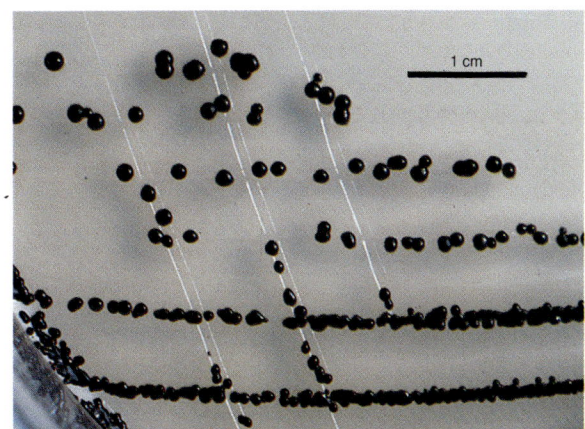

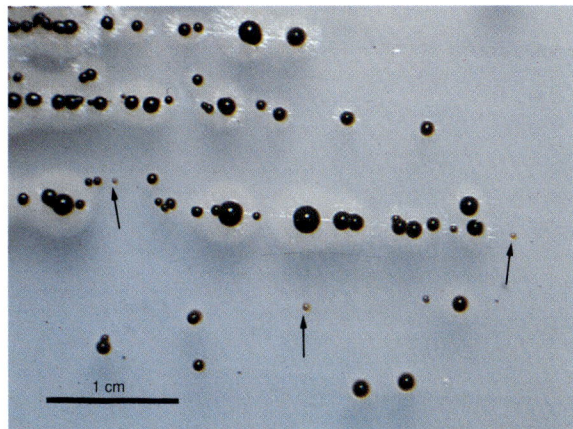

196–198. Isolation of *Staph. aureus* on Baird–Parker agar medium. Baird–Parker (BP) medium is a highly effective medium for the selective isolation of *Staph. aureus*. However, the appearance of colonies rarely conforms to the published description and there may be considerable variation.

In most cases, the shiny, jet black appearance is the most distinctive feature of *Staph. aureus* (**196**). There is often an accompanying opaque zone around colonies (arrowed), but this usually develops after 48 h incubation. Clearing of the medium and a white rim around colonies are often absent, and the colonies may appear simply as jet black without any other distinguishing feature (**197**).

In addition to variations in colonial appearance, colony size may also vary considerably. **198** illustrates number of strains of *Staph. aureus* growing on one streak plate; with the exception of the small, brown colonies (arrowed), which are a species of *Micrococcus*, all of the colonies are of a coagulase-positive *Staph. aureus*. (*Baird–Parker agar: 24h at 37°C.*)

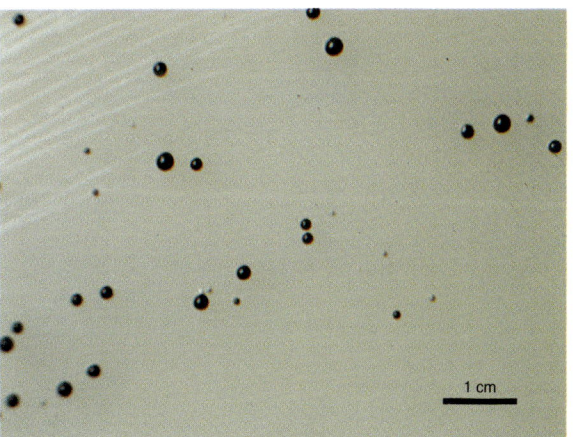

199 & 200. Selectivity of Baird–Parker medium. BP is a strongly selective medium and few other foodborne bacteria are able to develop on it. Some species of *Bacillus* grow well on BP, and while old colonies may be recognised by a brownish matt appearance, differences between young colonies and some strains of *Staph. aureus* may not be easy to discern. Young colonies of *Bacillus* (**199**) are, however, slightly less regular and many have a characteristic 'dimple'.

Species of *Proteus* may also be a problem, although with most foods this appears to be due less to swarming than to colonies being confused with those of atypical atrains of *Staph. aureus*. If necessary, the growth of *Proteus* may be suppressed by the addition of sulphamethazine.

Various other additives have been used to increase the selectivity of Baird–Parker medium including bacitracin, acriflavin and polymyxin (White *et al.*, 1988). (*Baird–Parker agar: 24h at 37°C.*)

Despite the advantages of RPF agar, many laboratories continue to use the unmodified Baird-Parker medium in preference. Initially, this was due to difficulties in preparing RPF agar, but the medium is now available in a complete, dehydrated form (Unipath©). In other cases, laboratories have complained of difficulties in recognising coagulase-positive colonies on crowded plates, while others prefer the egg yolk-tellurite diagnostic system. In some cases, at least preference for unmodified Baird-Parker medium stems from familiarity rather than logic.

An alternative approach to the incorporation of the coagulase reaction is the on-plate determination of thermonuclease activity (Lachica, 1980). Initial isolation is made on either unmodified Baird-Parker medium or RPF (Tatini et al., 1984). The isolation plate is then heated for 2 hours at 65°C, over-layered with toluidine blue-DNA agar and reincubated at 37°C. Pink zones in the agar are indicative of TNase activity (cf page 260).

Although highly specific, this technique is cumbersome and colonies for further examination must be picked off the isolation plate before heating.

A highly selective, two-stage procedure for the isolation of Staph. aureus from foods has been described by Isigidi et al. (1989). The first stage involves a 'solid repair method' in which samples are inoculated onto Baird-Parker base (without egg yolk and tellurite) and incubated for 1 hour at 37°C. After a brief drying at room temperature, the inoculated plate is over-layered with Baird-Parker medium containing a supplement of egg yolk, acriflavine, colistin, potassium tellurite and sodium sulphamezathine and incubated for 24 hours at 37°C. The selectivity of the method is such that any colony developing after 24 hours may be accepted as Staph. aureus irrespective of the appearance. A comparative study showed significantly improved recovery using the two-stage procedure in samples where large numbers of competitive micro-organisms were present.

Baird-Parker medium, or one of its derivatives, is considered to be the medium of choice for Staph. aureus by many microbiologists. However, a second egg yolk containing medium KRANEP agar (**Table 96**; Sinell and Baumgart, 1967) is widely used in continental Europe, especially Germany. Unlike Baird-Parker agar, KRANEP agar recovers most species of coagulase-negative staphylococci and is thus the medium of choice for 'total Staphylococcus'.

Differentiation of Staphylococcus from other genera

The simple criteria listed in **Table 91**, page 237, are usually adequate to differentiate Staphylococcus from other genera and in most cases differentiation between Micrococcus and Staphylococcus may be made on the basis of glucose fermentation alone. Few special precautions are required, although standard Hugh and Leifson agar is not suitable for the determination of an oxidative/fermentative attack on glucose and the modification of Baird-Parker (1966) should be used.

A glucose–peptone–yeast extract medium is used for the determination of the catalase reaction, and cultures should be incubated for 24 hours at 30°C (Baird-Parker, 1979).

In cases where differentiation between Micrococcus and Staphylococcus is difficult due to atypical oxidation/fermentation reactions, the combined use of lysostaphin sensitivity, lysozyme resistance and acid production from glycerol in the presence of erythromycin is recommended (Schleiffer and Kloos, 1975). However, for many purposes, the lysostaphin test alone is of sufficient specificity although ca 4% of Staph. aureus are insensitive (Bennett et al., 1986). Resistance to lysozyme and erythromycin is not suitable for use as sole criteria.

A number of methods have been proposed for determining lysostaphin sensitivity, but the broth method of Males et al. (1975) as modified by Baker (1984) is recommended: 0.1 ml of lysostaphin solution (50µg/ml lysostaphin in 0.1M Tris: 0.15M NaCl pH 7.5) is added to 0.4 ml culture in veal broth incubated overnight at 37°C. The culture is observed for clearing. The use of veal broth, that is of high glycine content but limited in serine, is essential and care must also be taken in the source of lysostaphin, that supplied by the Sigma Chemical Company being recommended (other brands may be unsuitable).

The lysostaphin test requires careful control, Staph. aureus ATCC 12600, NCIB 6571 and Micrococcus sp. ATCC 15306, NCIB 9278 are recommended.

Differentiation of Staph. aureus from other species

The coagulase test forms the basis of the differentiation of Staph. aureus from other species. Several approaches have been taken for the determination of coagulase, of these the slide test should not be used and either the tube test or the more recent co-agglutination tests are recommended. The latter are simpler to use than the tube test and usually highly sensitive and specific (**Figures 201–204**; Berke and Tilton, 1986).

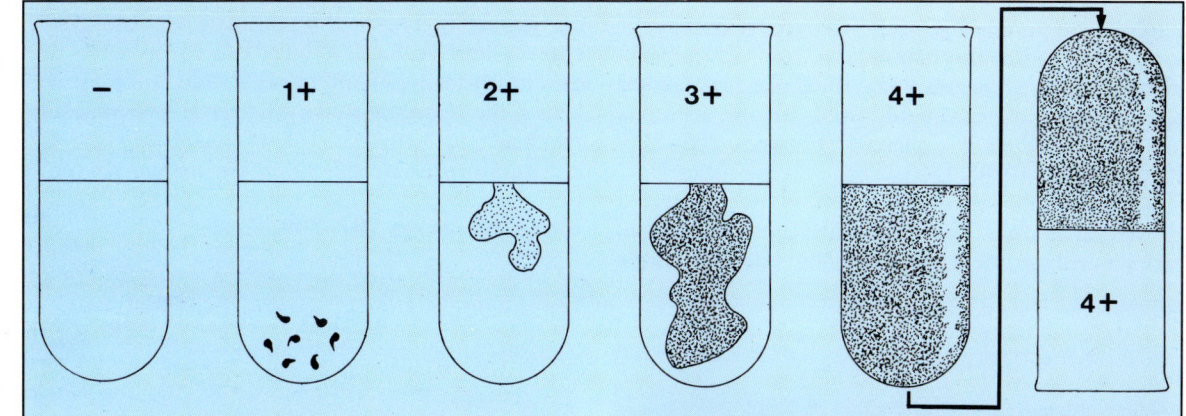

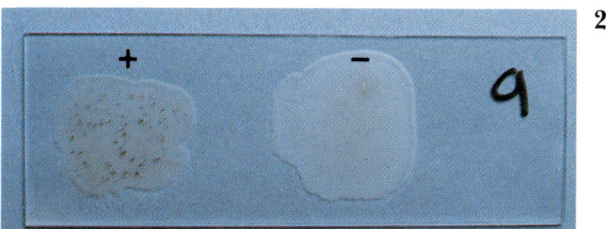

201–204. The coagulase test for the confirmation of identity of *Staph. aureus*. The tube coagulase test (**201**) is the definitive means of determining the coagulase reaction, and detects 'free coagulase'. Coagulase thombrin is the coaggulating principle which is the result of the the the interaction between coagulase and the reacting factor in plasma. Human plasma is recommended to differentiate between *Staph. aureus* and *Staph. hysicus* (Amstberg, 1979; Harvey and Gilmour, 1985), but commercially prepared rabbit plasma with EDTA is most convenient for general purposes. There may, however, be variation in clot formation from one brand to another, and it is thus essential that the test is carefully controlled. Visual interpretation of the strength of the reaction inevitably introduces subjectivity, and there is some disagreement over what constitutes a positive reaction. It is recommended that a reaction corresponding to 3+ or 4+ should be regarded as positive, and that weaker reactions, 1+ and 2+, should be supported by the thermonuclease test (see **205** & **206**; ICMSF, 1978).

To a large extent, the traditional coagulase test has been supplanted for routine use by commercially produced rapid agglutination tests, which use sensitised latex or red blood cells to detect the staphylococcal clumping factor ('bound coagulase') and/or protein A.

The 'Staphylase©' (**202**) detects the clumping factor. Coloured latex particles are used to aid recognition of positive (+) reactions and a negative control is incorporated (−).

The 'Staphaurex©' detects both the clumping factor and protein A. The test is made on a special card which permits easy recognition of both positive (**203**) and negative (**204**) reactions. Difficulties have been reported with latex kits due to a false-positive response by *Staph. saprophyticus* (Gregson *et al*., 1988), although as a urinary pathogen this species is unlikely to be encountered in foods in significant numbers.

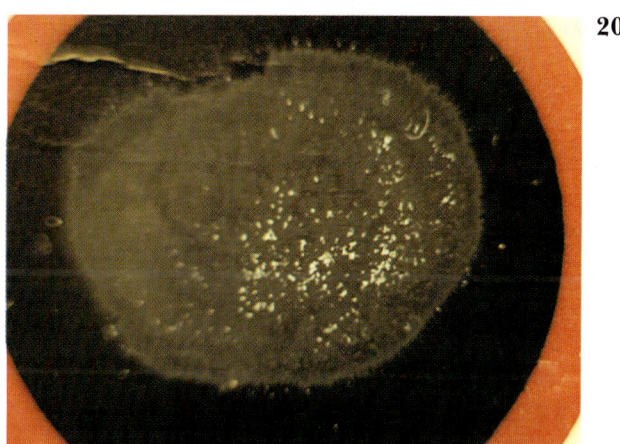

Where required, the coagulase reaction may be supported by the thermonuclease test. The microslide technique of Lachica *et al.*, (1972) using σ-toluidine blue agar is most commonly used but the antibody inhibition method is preferred (**Figures 205–206**; Lachica *et al.*, 1979).

Where a full characterisation is necessary, or species other than *Staph. aureus* are present, identification may be made according to the criteria in **Table 92**, page 238. The media and methods employed by Harvey and Gilmour (1985) should be used; those developed for other micro-organisms, such as members of the *Enterobacteriaceae*, are likely to give misleading results.

Miniaturised identification kits are an alternative to conventional methods, the use of two types is illustrated in **Figures 207–209**.

205

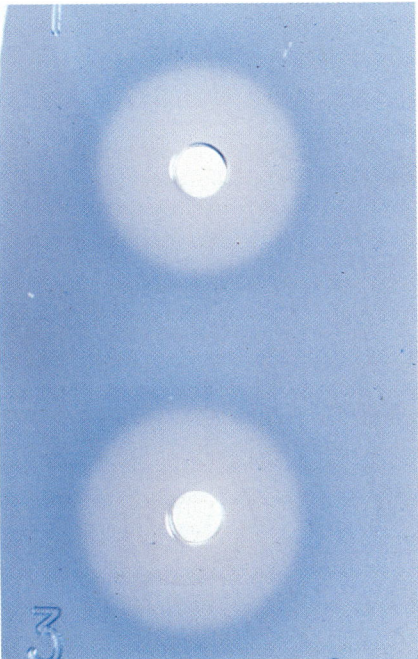

206

205 & 206. Detection of staphylococal thermonuclease using an antibody inhibition method. Staphylococcal thermonuclease is determined by gel diffusion in toluidine blue-DNA agar, the medium changing from blue to pink in the presence of thermonuclease. The specificity of the test is increased and the need to heat the cell suspension eliminated by the parallel use of substrate agar with and without a specific antibody to staphylococcal thermonuclease. **205** illustrates the commercial 'staphynuclease©' kit. The toluidine blue agar in the left-hand compartment (darker blue) contains no antibody and a positive reaction (**206**) is seen in two wells. In the right-hand compartment, an antibody to thermonuclease is present in the agar and the reaction is inhibited. Thermonucleases from bacteria other than *Staph. aureus* would react both in the presence and absence of an antibody.

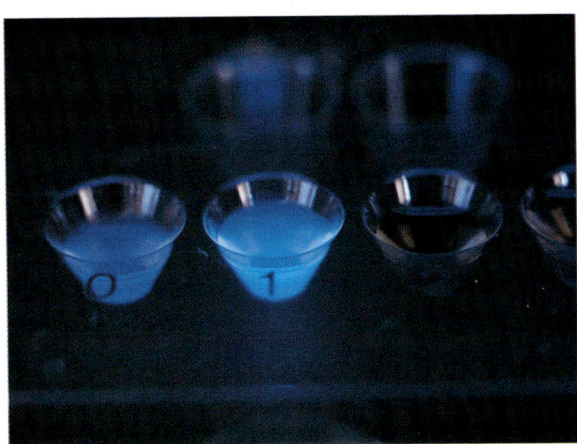

207

207–209. Use of miniaturised kits for the presumptive identification of *Staphylococcus* species. The Rapidec Staph© system is based on the detection of pre-formed enzymes and permits recognition of *Staph. aureus*, *Staph. intermedius*, *Staph. xylosus*, *Staph. epidermidis* and *Staph. saprophyticus* within 2 h. The first fluorogenic reaction (**207**) detects an enzyme involved in the coagulase reaction which is specific to *Staph. aureus*, while the second (alkaline phosphatase) and third (β-galactosidase) reactions are used to differentiate the other species (**208**).

In contrast to the Rapidec system, the ATB 32 Staph© provides a full identification of a large number of species of *Staphylococcus* and *Micrococcus*. The tests employed are a combination of conventional biochemical reactions and specific enzyme determinations. Incubation for 24 h is required. **209** illustrates the identification of a strain of *Staph. warneri* isolated from lemonade. (*Rapidec Staph: 2 h at 37°C; ATB 32 Staph: 24 h at 37°C.*)

Note: The long exposure required for photography of the fluorescent reaction (**207**) has lessened the difference between the control cupule (0) and the test cupule (1).

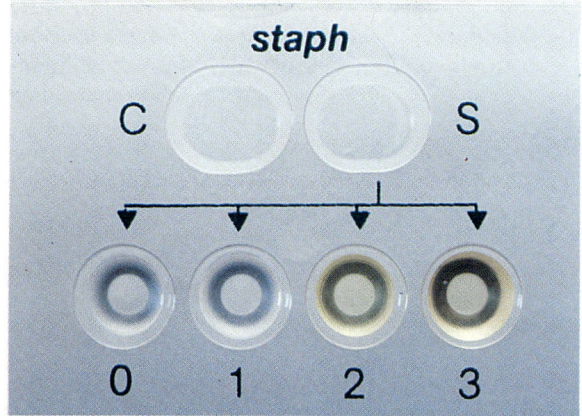

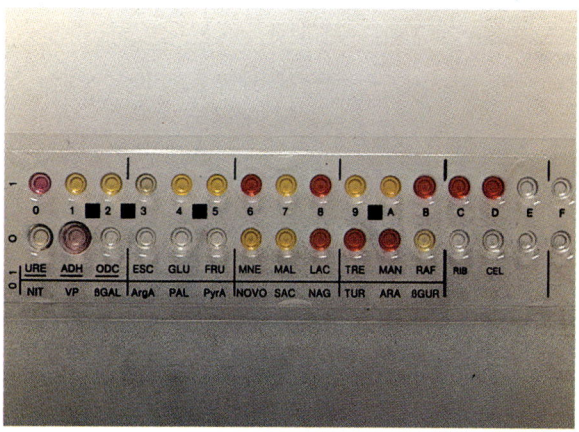

Typing of *Staph. aureus*

Phage typing

Staph. aureus is phage typed according to the generalised protocol described in page 41. Details of the specific procedure for the organism may be obtained from de Saxe *et al.* (1982). Strains of *Staph. aureus* are assigned to phage groups rather than phage types (**Figure 210**). The majority of food poisoning strains are of phage group III, although some also show patterns of reaction with Group I phages (group I/III).

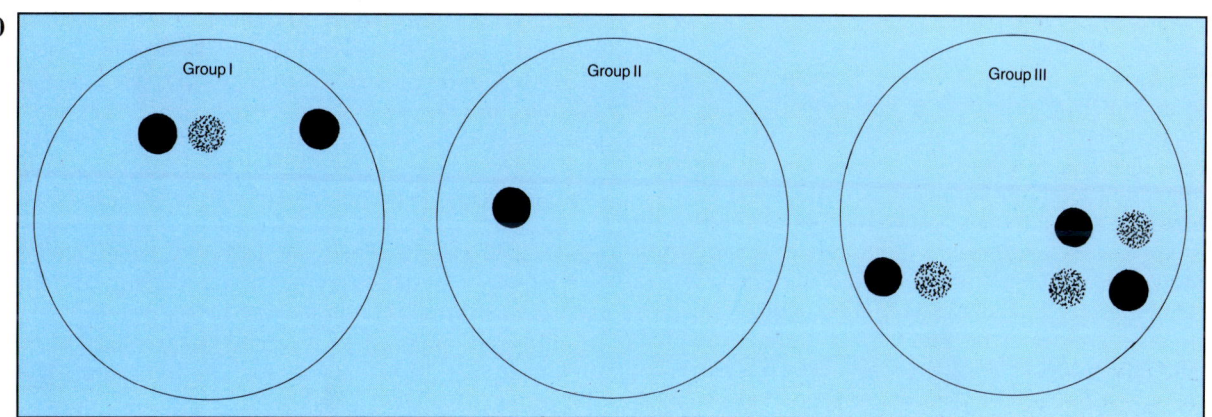

210. Phage-typing patterns of *Staph. aureus*. Phages are assigned to lytic groups depending on their host range. Examples of typing patterns for lytic groups I to III are illustrated.

Detection of staphylococcal enterotoxins

Although the detection of enterotoxins produced by pure cultures of micro-organisms growing under optimal conditions for toxogenesis is relatively straight-forward, the detection of enterotoxins in food samples continues to cause considerable problems. This largely stems from the fact that the very considerable advances in methods for the detection of enterotoxins have not been matched by similar advances in the extraction and concentration of enterotoxins from foods.

Production of enterotoxin by isolates in pure culture

Enterotoxin production varies according to the medium used. Brain-heart infusion medium (BHI) is widely used (Bennett, 1986), but other media including a 3% pancreatic digest of casein plus 1% yeast extract (Bergdoll, 1972), or columbia blood agar (de Saxe *et al.*, 1982), are also suitable. In all cases, it is necessary to be aware of the possibility of differences in medium performance, both from batch to batch and from manufacturer to manufacturer.

The simplest means of producing enterotoxins involves the growth of the culture in semisolid (0.7% agar) BHI at pH 5.3 (Casman and Bennett, 1963; Bennett and McLure, 1976). This method is suitable for use where facilities are limited, and has been adopted as an AOAC method (AOAC, 1984). The yield of toxin is, however, relatively low, and the method is only suitable where a sensitive detection system is available, or with strains which produce large quantities of enterotoxin.

In recent years, simultaneous growth-concentration methods have been widely used. The first of these, the sac-culture method (Donnelly et al., 1967), was used successfully for many years, but is now being replaced by the cellophane–over–agar technique (Robbins et al., 1974). A separate concentration stage is not usually required, but if necessary the use of a Minicon© concentrator (see below) is recommended.

It may be necessary to pass heat-stressed Staph. aureus through four to five subcultures to allow the organism to regain the ability to synthesise enterotoxins (Batish et al., 1989).

Extraction, purification and concentration of enterotoxins from foods

To some extent, failure to develop methods for extraction and concentration of enterotoxins from foods stems from improvements in the sensitivity of detection methods. It was widely thought that the improved detection methods would obviate the need for complicated extraction and concentration methods while maintaining a sufficiently high overall sensitivity to detect levels of enterotoxin capable of causing illness. However, the level of toxin required to produce illness is now known to be much lower than previously thought (page 243). At the same time, it is necessary to appreciate that samples of foods suspected of involvement in food poisoning may be small in size and returned for examination in a deteriorated condition. Thus, while a simple extraction method combined with highly sensitive detection may be satisfactory in many circumstances, techniques for, and expertise with, concentration and purification of enterotoxins should be available.

Extraction

Initial extraction usually involves homogenisation in a buffer or in distilled water, followed by low-speed centrifugation to remove particulate matter. Phosphate buffered saline (PBS; phosphate buffer, 0.07 mol/l containing 0.15 mol/l NaCl, pH 7.2) is commonly used both for extraction and as a diluent.

Additions to PBS may be made for specific purposes. These include the addition of Tween 20 to eliminate non-specific reactions in some ELISA assays (Notermans, 1982), or of sodium hexametaphosphate for a similar purpose when renneted dairy products are assayed using some types of reverse passive latex agglutination (Rose et al., 1989).

Where distilled water is used for extraction, it is necessary to adjust the pH value of the homogenate to 4.5 before centrifugation. The pH value is then readjusted to 7.5 (Reiser et al., 1974).

When examining dried foods, or foods which have become dehydrated prior to examination, the recovery of enterotoxin may be improved by allowing the homogenised material to stand for 4 hours at 4°C before centrifugation.

Purification and concentration

Two methods of purification have been recommended for routine use (Bergdoll and Bennett, 1984). The first involves column chromatography using a carboxymethylcellulose ion exchanger and the second batch adsorption of enterotoxin using a CG50 ion exchange resin (Reiser et al., 1974). The two methods are of similar efficiency, but the batch adsorption method is faster. More recently copper chelate sepharose© has been used in a batch process (Dickie and Akhtar, 1989) and holds promise as a simple and effective means of concentration. A novel method using dye ligand affinity chromatography has also been developed (Reynolds et al., 1988), which improves both yield and purity when used with pure cultures and which may be applicable to foods.

Defatting is necessary with some foods and is sometimes applied routinely. Chloroform is commonly used, and may conveniently be removed after defatting by filtration through water saturated Whatman 40 filters.

Concentration of the extracted enterotoxin is not always necessary and indeed may not be recommended. Where used there are three basic approaches: freeze drying, concentration by dialysis (usually against polyethylene glycol), or the use of a hollow fibre concentration device such as the Minicon© (Amicon Ltd). The latter method is convenient and the loss of enterotoxin during concentration is minimised. Loss is a significant problem with freeze drying and dialysis against polyethylene glycol (Windemann et al., 1989), and in the case of the extraction of SEC and SED from salami, concentration is not recommended.

Detection of enterotoxins

Animal methods

Although a number of animal methods have been proposed, only kittens, rhesus monkeys and chimpanzees are of sufficient sensitivity. Few laboratories have the facilities to handle these animals and with the exception of special purposes, the use of animals has been superseded by serological assays.

Serological methods

The production of specific antibodies to staphylococcal enterotoxins is relatively straightforward, and a wide variety of serological methods have been applied to their detection (**Table 97**). For a number of years gel diffusion techniques, such as the microslide method (Simkovicova and Gilbert, 1971), were the methods of choice. Such methods are technically straightforward but insensitive. The optimum sensitivity plate method (Robbins *et al.*, 1974) remains in use as a means of screening a large number of cultures for enterotoxin production, although a more recently developed single radial diffusion method (Sokari and Anozi, 1989) permits 50 or more isolates to be tested simultaneously and is likely to be of particular value in developing nations where sophisticated reagents are not readily available.

The relative insensitivity of the earlier serological detection methods was recognised and a number of proposals were made to increase their sensitivity. Radioimmunoassay was applied to the detection of staphylococcal enterotoxins at an early stage, and had the advantages of being both highly sensitive and rapid. As with all radiometric assays, the expensive reagents and equipment, coupled with the difficulties associated with handling radio-isotopes, severely restricted the application of the technique.

In recent years, interest has focused on two sensitive, but non-isotopic methods, co-agglutinations and ELISAs. Early work with agglutination techniques indicated that reverse passive haemagglutination (RPHA) was likely to be most suitable. However, the advantage of high sensitivity was negated by a number of disadvantages including a high level of non-specific cross-reactions, and so RPHA is considered too unreliable for routine use (Bennett, 1986).

An alternative agglutination technique, reverse passive latex agglutination (RPLA), has been developed (**Flow diagram 24**; Shingaki *et al.*, 1981), and is available as a commercial kit (Denka-Seiken; Unipath). A sensitivity in the order of 0.25 ng/ml was noted independently in two evaluations of the kit (Park and Szabo, 1986; Rose *et al.*, 1989), a sensitivity which, significantly, is greater than that claimed by the manufacturer. These results differed, from those of a third evaluation (Wieneke and Gilbert, 1987),

Table 97. Summary of serological methods for assay of staphylococcal enterotoxins.

Gel diffusion
Microslide[1]
Optimum sensitivity plate[2]
Single radial diffusion[3]

Gel precipitin
Capillary tube test[4]

Radioimmunoassay
Various assays[5, 6]

Co-agglutination
Reverse passive haemagglutination[7, 8]
Reverse passive latex agglutination[9]

Enzyme linked immunosorbent assay
Various assays[10, 11]

[1] Simkovicova and Gilbert (1971).
[2] Robbins *et al.* (1974).
[3] Meyer and Palmieri (1980).
[4] Fung (1973).
[5] Lindroth and Niskanen (1977).
[6] Miller *et al.* (1978).
[7] Silverman *et al.* (1968).
[8] Bergdoll *et al.* (1976).
[9] Shingaki *et al.* (1981).
[10] Kuo and Silverman (1980).
[11] Fey (1986).

Flow diagram 24. Detection of staphylococcal enterotoxins by reverse passive latex agglutination.

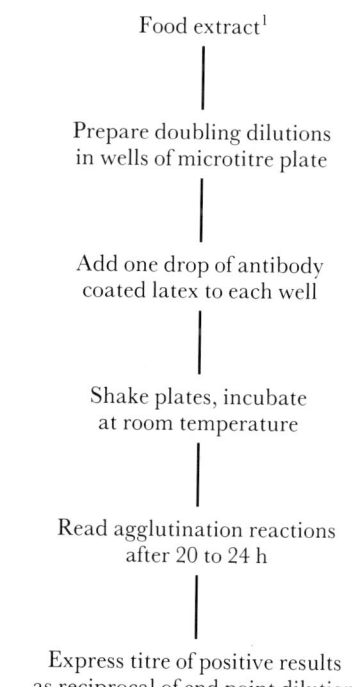

Food extract[1]

Prepare doubling dilutions
in wells of microtitre plate

Add one drop of antibody
coated latex to each well

Shake plates, incubate
at room temperature

Read agglutination reactions
after 20 to 24 h

Express titre of positive results
as reciprocal of end point dilution

[1] The extraction procedure of Reiser *et al.* (1974) is recommended.

which indicated that RPLA, while convenient to use, was not always of sufficient sensitivity.

Initial problems were encountered with false-positive reactions in renneted dairy products such as many types of cheese, but these were overcome by a simple modification to the extraction buffer (see page 262). Filtration through a 0.22μm filter is required to produce a clear extract.

RPLA conventionally requires 16 hours to obtain a result, but this may be reduced to 3 hours by using high density latex particles (Fujikowa and Igarashi, 1988).

Although the commercial availability of RPLA has resulted in wide usage, most development effort has been expended on ELISA and it is probably this technique, with its general advantages of high sensitivity, specificity and relatively simple methodology, that holds the greatest potential for the future.

A number of approaches have been taken to ELISA assays for staphylococcal enterotoxins, and both the competitive and sandwich configurations have been employed, general consensus now favouring the sandwich assay (Fey *et al.*, 1984; Bennett, 1986; Lapeyre *et al.*, 1988). Mention should also be made of the assay for SEB described by Morita and Woodburn (1978) which is one of the few examples in food microbiology of a homogeneous enzyme immune assay. The assay had a sensitivity of 5ng/ml, but was not widely adopted.

The original ELISAs for enterotoxins used polyclonal antibodies (PCAs), but subsequently a number of monoclonal antibodies have been developed (Thompson *et al.*, 1985). In a comparative study, it was concluded that, in a quantitative assay, MCA ELISA gave more reliable and sensitive results with a lower, non-specific background response than PCA ELISA (Schonwalder *et al.*, 1988). However, the recovery of enterotoxin from food extracts was less efficient with MCAs than with PCAs.

Most ELISAs for staphylococcal enterotoxins employ polystyrene microtitre plates as solid phase. An important exception is the assay method of Stiffler-Rosenberger and Fey (1978), which employs polystyrene balls individually coated with antibody. The sensitivity of this assay is considerably greater than that of the microtitre assays previously used, but it is prone to interference from food ingredients which necessitates the dilution of the extract and consequently their is a loss of sensitivity. A practical disadvantage is that the difficulty of manipulating the polystyrene balls mean that it is impractical to test a large number of samples (Wieneke and Gilbert, 1987).

A number of modifications have been made to microtitre ELISA to increase the sensitivity to that of the polystyrene ball method without the associated problems of interference from food particles. Enzyme amplification (Stanley *et al.*, 1985) has been incorporated into a microtitre assay (Windemann *et al.*, 1989), while Lapeyre *et al.* (1988) have devised an indirect double sandwich ELISA. This assay uses a specific MCA as a coating antibody and polyspecific PCAs as probing antibodies and has a sensitivity of 1 ng/g sample. This sensitivity is also achieved by use of high-efficiency, polyclonal capture antibodies in a commercial sandwich ELISA kit.

Two ELISA assays are available as commercial kits (**Flow diagram 25**), the polystyrene bead assay (Fey *et al.*, 1984; Diagnostic Laboratory, Dr W. Bommeli), and the microtitre plate assay mentioned above (TECRA©). The two tests are of equivalent sensitivity in use, but the microtitre plate method is technically simpler and the results may be read visually. The microtitre plate method does not distinguish between the different enterotoxins, which is a disadvantage in some circumstances.

Flow diagram 25. Detection of staphylococcal entero-toxins by enzyme linked immunosorbent assay.

A Polystyrene ball assay

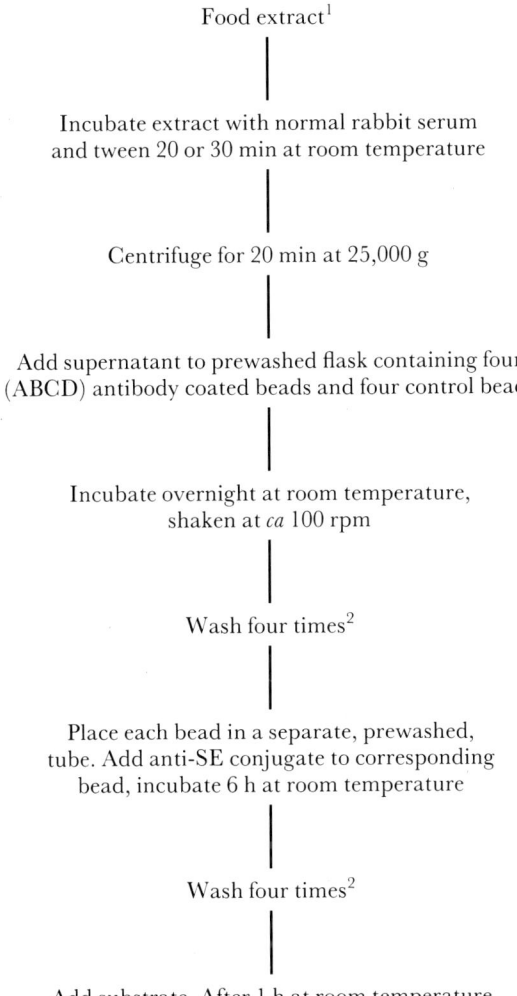

Food extract[1]

|

Incubate extract with normal rabbit serum and tween 20 or 30 min at room temperature

|

Centrifuge for 20 min at 25,000 g

|

Add supernatant to prewashed flask containing four (ABCD) antibody coated beads and four control beads

|

Incubate overnight at room temperature, shaken at *ca* 100 rpm

|

Wash four times[2]

|

Place each bead in a separate, prewashed, tube. Add anti-SE conjugate to corresponding bead, incubate 6 h at room temperature

|

Wash four times[2]

|

Add substrate. After 1 h at room temperature transfer 0.2 ml from each tube to separate wells of a microtitre plate. Read absorbance at 405 nm[3]

B Microtitre plate assay

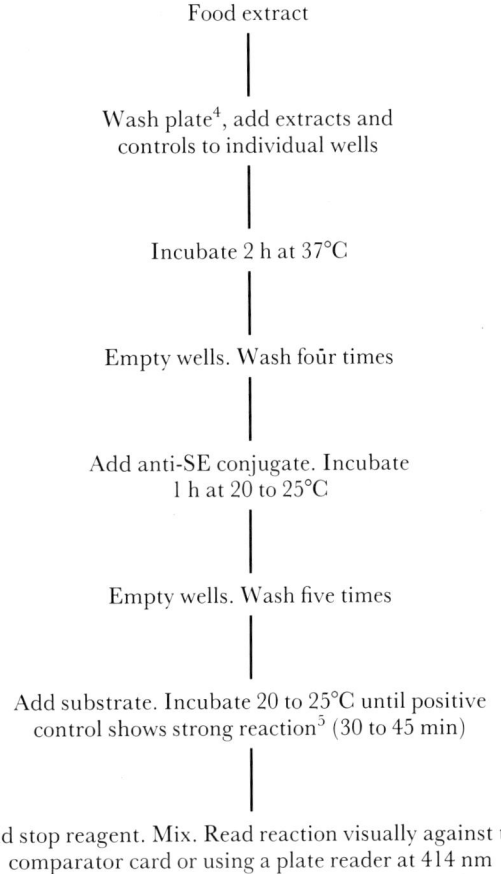

Food extract

|

Wash plate[4], add extracts and controls to individual wells

|

Incubate 2 h at 37°C

|

Empty wells. Wash four times

|

Add anti-SE conjugate. Incubate 1 h at 20 to 25°C

|

Empty wells. Wash five times

|

Add substrate. Incubate 20 to 25°C until positive control shows strong reaction[5] (30 to 45 min)

|

Add stop reagent. Mix. Read reaction visually against the comparator card or using a plate reader at 414 nm

[1] The extraction method of Reiser *et al.* (1974) is recommended.
[2] The manufacturer's instructions specify three washes but four are recommended.
[3] Where a large number of samples are to be assayed the reactions may be stopped by adding 2N NaOH to each tube before transfer to the microtitre plate.
[4] In the TECRA, assay plates are precoated with anti-SE.
[5] The required strength of the positive control reaction is defined by the manufacturer. If this is not attained within 45 minutes the test is invalid and must be repeated after corrective action.

13. *Bacillus*

Introduction

The genus *Bacillus* is a member of the family *Bacillaceae* and comprises Gram-positive, catalase positive, rod-shaped bacteria. A distinctive feature of the genus is the formation of heat resistant endospores, that are of considerable importance in the ecology and taxonomy of the organism (**Figures 211–213**).

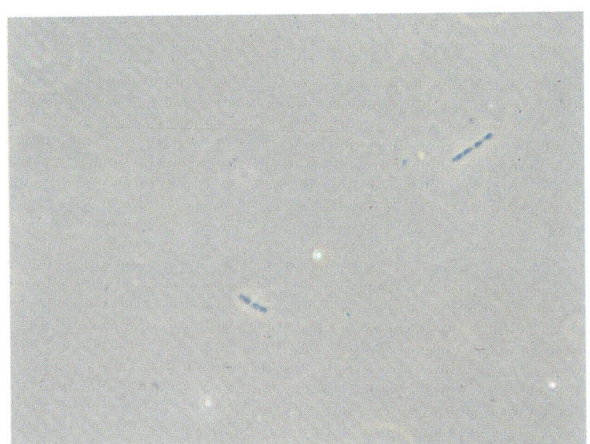

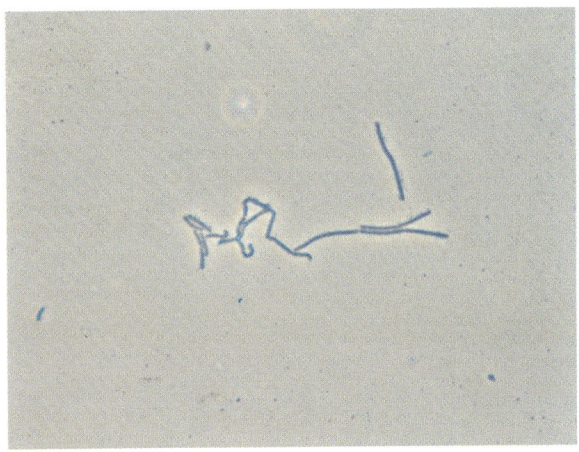

211 & 212. Germination and outgrowth of *B. cereus* endospores. Germination and outgrowth can be rapid under favourable conditions (see **213**). An endospore-containing culture of emetic *B. cereus* was photographed directly after inoculation into rice slurry (**211**) and after 4 h at 28°C (**212**). In this period, germination and outgrowth had occurred and vegetative growth was well established.

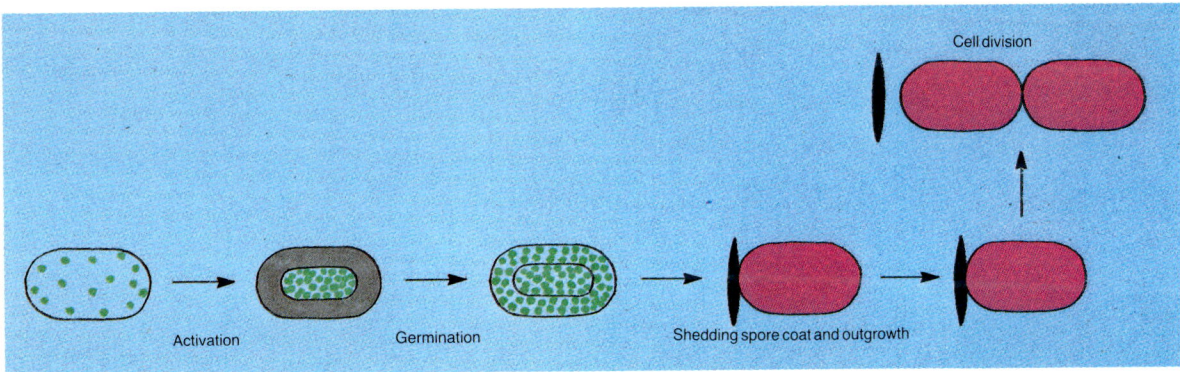

213. Sequence of events during activation, germination and outgrowth of endospores of *Bacillus*. Newly formed endospores remain largely dormant, even if conditions are favourable for germination. Dormancy can be broken by a number of treatments such as heat shocking, which are collectively referred to as activation. Activation also occurs during storage (in the absence of specific treatments), an increasing percentage of the population becoming capable of germination with increasing time. In contrast to activation by heat shocking, this activation is irreversible.

Germination of activated spores occurs when conditions are favourable and requires a chemical trigger such as L-alanine, ribosides or glucose. Germination is a rapid process and marked by a loss of refractility and a burst of metabolic activity which is due to pre-formed, but inactive, enzymes in the spore protoplast. During this time, some 30% of the dry weight of the endospore is lost as spore exudate, which consists largely of calcium dipicolinate derived from the protoplast, and peptidoglycan fragments derived from the disintegrating cortex. The loss of resistance to heat and other factors which accompanies the other changes is of considerable importance with respect to processing foods for safety.

Outgrowth follows germination providing that the conditions are suitable for vegetative growth. The endospore swells within the spore coat while at the same time the vegetative cell wall is synthesised. The newly-formed vegetative cell emerges from the spore coat and commences cell division.

The genus *Bacillus* embraces a large number of bacteria with a great diversity of properties (Parry *et al.*, 1983). There are currently over 30 species (Claus and Berkeley, 1986), as well as a number of unclassified strains. Unusually among bacteria, the colonial morphology is not only a valuable feature in recognition of the genus but, in the opinion of at least some authors, is a valuable aid in the differentiation of species (**Figures 214–219**).

Bacillus species are ubiquitous in nature and some species are of importance in the spoilage of heat-treated foods (see page 279). With the exception of *B. anthracis*, that is primarily a pathogen of cattle and sheep but that also affects man, the genus has not generally been considered to be a human pathogen.

However, five species, have now been recognised as causative organisms of food poisoning, and the genus has also been increasingly associated with various other types of human and animal disease (Gilbert *et al.*, 1981; Parry *et al.*, 1983; Logan, 1988).

Of the five species that have been associated with food poisoning, *B. cereus*, *B. brevis*, *B. subtilis*, *B. licheniformis* and *B. sphaericus*, *B. cereus* is of by far the greatest importance. This organism causes two distinct types of food poisoning, a diarrhoeal and an emetic type, the latter being commonly associated with the consumption of rice. In addition, food is an occasional route of infection for *B. anthracis*, while other members of the genus may also be associated with food poisoning.

214–217. Colonial morphology of *Bacillus* species on non-selective agar media. Most species of *Bacillus* have a distinctive colonial morphology which is easily recognised on non-selective media, even when large numbers of other bacteria are present. Although some workers feel that recognition is facilitated by supplementing media with blood, there is no difficulty in recognising species of *Bacillus* on the unsupplemented nutrient agars commonly used by food microbiologists.

214

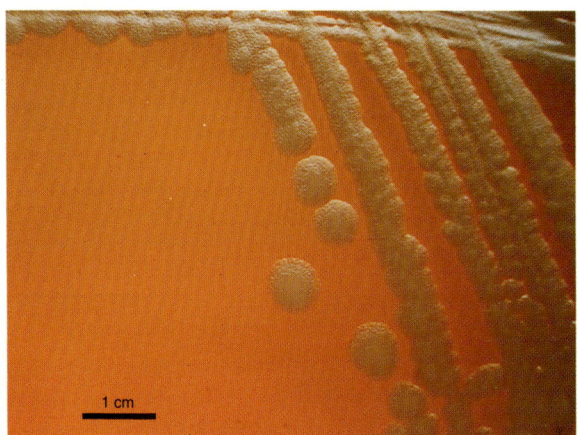

21

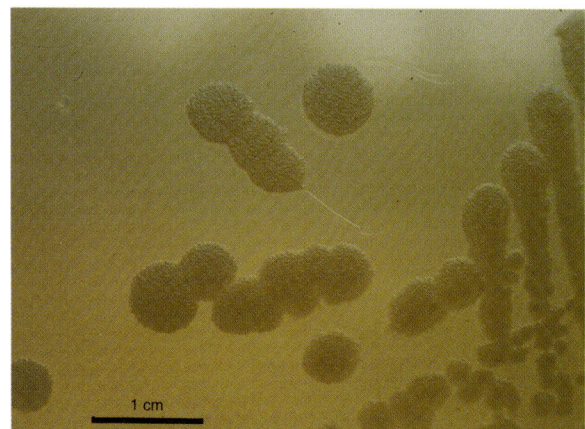

216

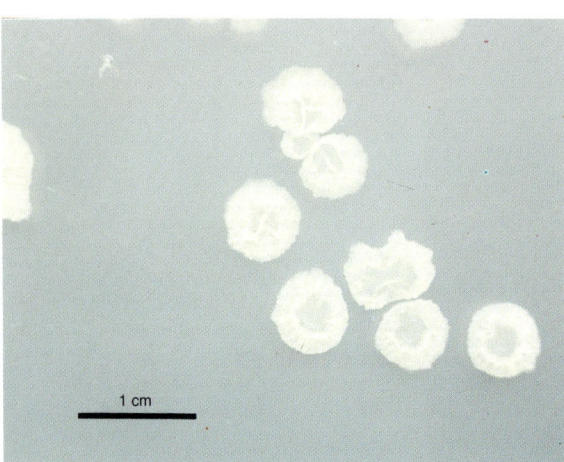

214 & 215. *B. cereus*. Colonial morphology of *B. cereus* is similar on blood (**214**) or on plate count agar (**215**). *B. cereus* has a remarkably wide range of colonial morphologies and yet are usually recognised as *B. cereus* by even the relatively inexperienced (Parry *et al.*, 1983). The colonies of *B. anthracis* and *B. thuringiensis* are, however, of very similar appearance. Note that the strain of *B. cereus* illustrated is non-haemolytic.

216. *B. subtilis*. A range of colony types may be found ranging from rough friable types to moist or mucoid types. The colonies illustrated are of the mucoid type, which appears to be most common in many foods. Dehydration of colonies of this type is very rapid and the typical mucoid appearance has been lost.

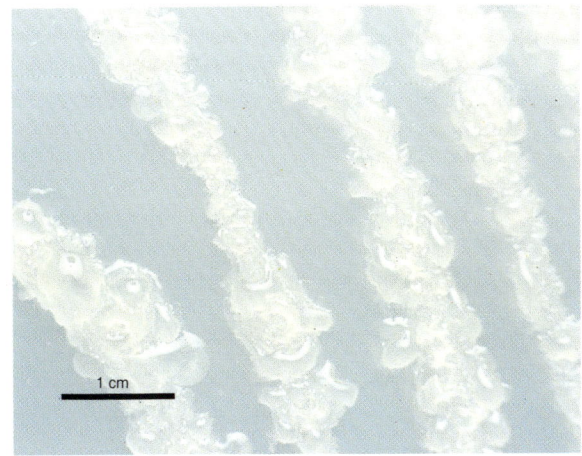

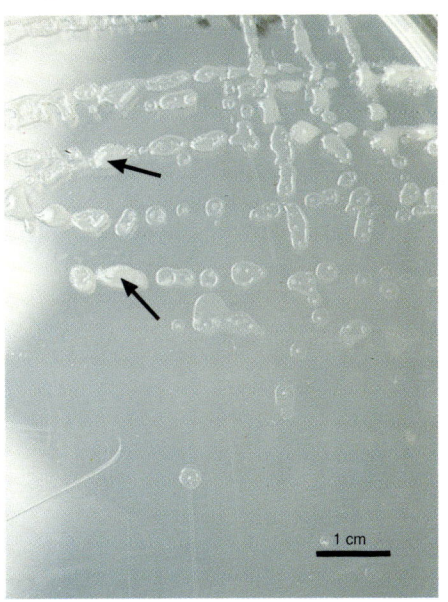

217. *B. licheniformis*. Although colonies of *B. licheniformis* may resemble those of the closely related *B. subtilis*, appearances may be more bizarre. The colonies illustrated are large and partially mucoid and have coalesced despite being originally well separated. (*Blood agar/plate count agar: 24 h at 37°C.*)

218 & 219. Development of *B. licheniformis* colonies. *B. licheniformis* grows rapidly at 37°C and visible colonies may develop within 12 h (**218**). Small amounts of mucous fluid are present (arrowed). After 24 h incubation, colonies are fully developed and the fluid has evaporated (**219**). (*Plate count agar: 12/24 h at 37°C.*)

Isolation of *Bacillus*

B. cereus is present in small numbers in many foods and although enrichment methods have been described they are rarely used. In some circumstances, *Bacillus* species are present in heat-treated foods in large numbers, and the use of selective plating media is unnecessary, the organism being recognised by its colonial morphology on blood or other media. Recovery from clinical samples is also often possible on blood agar.

Where samples contain significant numbers of other bacteria, the supplementation of blood agar with polymyxin B is an effective means of selecting for *Bacillus* species (Kramer *et al.*, 1982).

A number of selective and diagnostic media designed specifically for *B. cereus* have been described. These are based on similar principles and include the widely used polymyxin-pyruvate-egg yolk-mannitol-bromothymol blue agar (*B. cereus* selective medium; Holbrook and Anderson, 1980). All common species of *Bacillus* are recovered on these media but, with the exception of *B. cereus*, produce atypical diagnostic reactions. *The isolation of* Bacillus *is discussed in greater detail, with respect to choice of method and laboratory practice, in* **Laboratory Methods***, pages 283–285.*

Taxonomy of *Bacillus*

Differentiation from other genera

The genus *Bacillus* is unusual in that recognition is usually possible · on the basis of two characters, aerobic growth and endospore formation. The dependence on endospore formation to define the genus means that its members comprise a wide range of types. Molecular studies have suggested that the present genus would probably be divisible into five to six genera. Such a division is thought to be premature (Claus and Berkeley, 1986) and unfair to people working with the genus on a practical basis.

Differentiation within the genus

Current classification of *Bacillus* is based on the work of Smith *et al.* (1952), which was a largely successful attempt at bringing order to taxonomic chaos. This scheme included a primary subdivision on the basis of endospore and sporangium morphology (Morphological groups 1, 2 and 3; **Figures 220–222**), and the further subdivision to species level on the basis of biochemical tests (**Figures 223–225**).

220   **221**   **222**

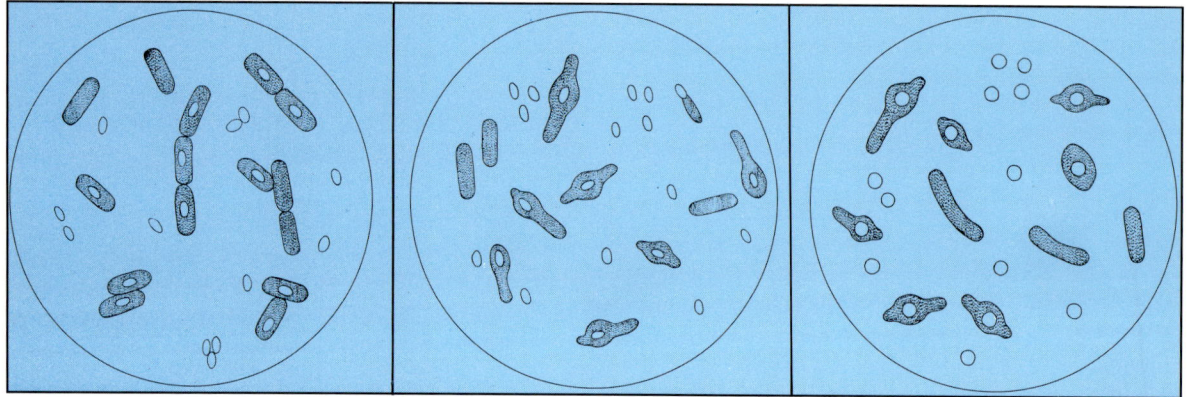

220–222. Morphological groups of *Bacillus*. Although the concept of morphological groups is now of reduced importance in the taxonomy of the genus, the defining characteristics of the groups provide valuable information to the experienced microbiologist. In conjunction with other easily obtained data such as colonial morphology and reaction on selective media, a presumptive identification may be obtained at an early stage. However, experience is required to correctly assess the defining characteristics.

220. Morphological group 1 contains the major foodborne pathogen, *B. cereus*, as well as *B. subtilis*, *B. licheniformis* and *B. anthracis*. Species important in food spoilage such as *B. coagulans* are also members. The defining characteristics are:
- Sporangia not swollen or, only very slightly swollen, by endospores.
- Endospores ellipsoidal/cylindrical, central/terminal.
- Gram-positive, occasional isolates may be gram-variable.

221. Morphological group 2 contains the occasional foodborne pathogen *B. brevis*, as well as the thermophilic spoilage organism *B. stearothermophilus*. The defining characteristics are:
- Sporangia definitely swollen by the endospores.
- Endospores oval, rarely cylindrical, central, subterminal or terminal.
- Gram-positive/negative/variable.

222. Morphological group 3 contains only two defined species, the occasional human pathogen *B. sphaericus*, and *B. pasteurii* which will not grow on normal laboratory media. The defining characteristics are:
- Endospores spherical, subterminal or terminal.
- Sporangium swollen.

Note: Defining characteristics based on *Parry et al.* (1983).

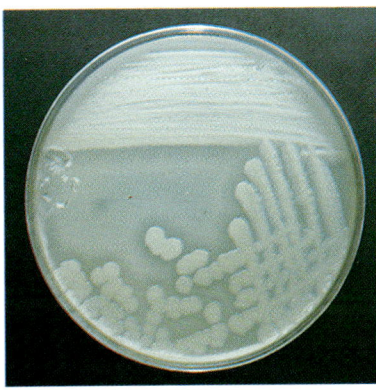 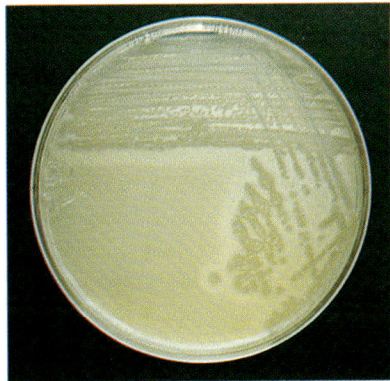 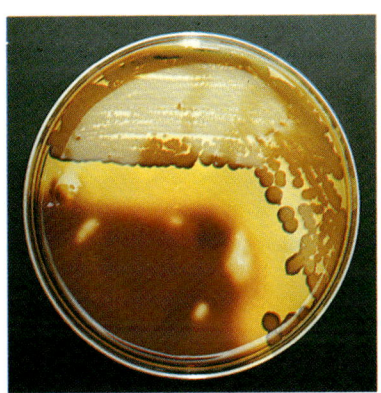

223 & 224. Lecithovitellin/lecithinase production (LV reaction). The LV reaction test is based on the hydrolysis of the phosphatid lecithin of egg yolk by the currently poorly defined action of phospholipase-C. The test is positive only for *B. cereus* and closely related species, and is an important diagnostic feature. *B. cereus*, streaked onto egg yolk agar shows a strong positive reaction (**223**). Some other species including *B. subtilis* show a weak or 'restricted' reaction (**224**) which, for diagnostic purposes, is recorded as negative. (*Egg yolk agar: 24h at 37°C.*)

225. Production of extracellular hydolytic enzymes by *Bacillus* species. A feature of *Bacillus* species is the production of extracellular enzymes which hydrolyse a range of compounds including starch, casein and gelatin. These reactions form the basis of diagnostic tests, although they are of less value than the LV test. **225** illustrates the hydrolysis of starch by *B. cereus*; the broad, pale coloured zone adjacent to the colonies indicates the extent of hydrolysis after flooding the plate with iodine solution. (*Starch agar: 24h at 37°C.*)

The original scheme of Smith *et al.* has been extended over the years in attempts to accommodate the diverse members of the genus (Gordon *et al.*, 1973). This process has continued to date (Claus and Berkeley, 1986), and while the primary subdivision into morphological groups has been superseded, endospore and sporangium morphology are still important differential features.

In practice, the number of species likely to be encountered in foods is relatively small, although the proportion of strains that cannot be assigned to any species may be high. Differential properties of *Bacillus* species known to be associated with foodborne disease are listed in **Table 98** (Kramer *et al.*, 1982; Parry *et al.*, 1983). Identification using this scheme involves a large number of tests, and a simplified scheme (**Table 99**) was devised by Deak and Timor (1988) using statistical techniques to select the minimum number of tests required for identification.

An attractive alternative means of identifying *Bacillus* species is based on the use of commercially produced API miniaturised identification kits originally developed for the *Enterobacteriaceae*. Identification is based on the use of the first 12 tests of the API 20E© kit, and all of the tests of the API 50CH© kit (Logan *et al.*, 1978, 1979; Logan and Berkeley, 1981, 1984). It must be stressed that use of the kits in this context represents an identification scheme *per se*, and does not merely involve the substitution of miniaturised methods for conventional methods.

The identification of Bacillus *is discussed in greater detail, with respect to choice of methods and laboratory practice, in* **Laboratory Methods**, *pages 285–287.*

Table 98. Differentiation of species of _Bacillus_: conventional scheme.

	Bacillus										
	cereus[1]	licheni-formis	brevis	subtilis	sphaericus	mega-terium	coagulans	macerans	pumilis	stearo-thermophilus	anthracis[2]
Motility	+	+	+	+	+	+	+	+	+	+	−
Morphological group	I	I	II	I	III	I	I/II	II	I	II	I
Anaerobic growth	+	+	−	−	−	−	+	+	−	−	−
Voges–Proskauer	+	+	−	+	−	−	+/−	−	+	−	+
pH of VP broth	4.3–5.6	5.0–6.5	8.0–8.6	5.0–8.0	>6.0	4.5–6.8	4.2–4.8	5.0–6.0	5.0–8.0	4.8–5.8	5.0–6.0
Lecithinase	+	−	−	−	−	−	−	−	−	−	+/−
Nitrate reduction	+	+	+/−	+	−	−/+	−/+	+	−	+/−	+/−
Growth at											
45°C	+/−	+	+/−	+		+	+	+/−	+	+	−
55°C	−	+/−	−	−		−	+	−	−	+	−
65°C	−	−	−	−		−	−	−	−	+	+
Growth in											
7% NaCl	+	+	−	+	−	+	−	−	+	−	+
0.001% lysozyme	+	−	−/+	−		−	−	−	+/−	−	+
0.02% azide	−	−	−	−		−	+	−	−	−	−
Acid from											
Glucose	+	+	+	+	−	+	+	+	+	+	+
Arabinose	−	+	−	+	−	+/−	+/−	+	+	−/+	−
Mannitol	−	+	−	+	−	+/−	+/−	+	+	−/+	−
Xylose	−	+	+/−	+	−	+/−	+/−	+	+	−/+	−
Hydrolysis of											
Starch	+	+	−	+	−	+	+	+	−	+	+
Gelatin	+	+	+	+	+	+	−	+	+	+/−	−
Tyrosine	+	−	+	−		+/−	−	−	−	−	−
Citrate utilisation	+	+	−/+	+	−	−/+	−/+	−/+	+	−	−/+
Propionate utilisation	+	+	−	−		−	−	−	−	−	+

[1] _B. cereus_ var. _mycoides_ (_B. mycoides_) is very similar to _B. cereus_ and is differentiated primarily on the basis of colony form and lack of motility.

[2] _B. anthracis_ is differentiated primarily on the basis of being non-haemolytic, non-motile and sensitive to penicillin. Although rare in foods, the possibility of isolating the organism can never be totally discounted. If there is any reason to believe an isolate is _B. anthracis_, work on the isolate should be the minimum possible and a Class I exhaust protective cabinet used.

Note: Based on Kramer _et al._ (1982); Parry _et al._ (1983).

Table 99. Differentiation of species of *Bacillus*: simplified scheme.

Scheme A	*Bacillus*			
	cereus[1]	*licheniformis*	*macerans*	*anthracis*[2]
Basic tests				
Anaerobic growth	+	+	+	+
Starch hydrolysis	+	+	+	+
Voges–Proskauer	+	+	+	+
Swollen sporangium	−	−	−	−
Acid from glucose	+	+	+	+
Gas from glucose	−	+/−	+	−
pH <6.0 in VP broth	+	+/−	+	+
Supplementary tests				
Growth in 7% NaCl	+	+	−	+
Growth at 50°C	−	+	+	−
Mannitol fermentation	−	+	+	−
Nitrate reduction	+	+	+	+/−
Casein hydrolysis	+	−	+	+

Other features

B. cereus: Contains poly-β-hydroxybutyrate within vegetative cells.

B. macerans: A terminal endospore.

B. anthracis: Sensitive to gamma phage; non-motile; non-haemolytic.

Table 99. *Continued.*

Scheme B	*Bacillus*						
	brevis	*subtilis*	*sphaericus*	*megaterium*	*coagulans*	*pumilis*	*stearothermophilus*
Basic tests							
Anaerobic growth	−	−	−	−	+	−	−
Starch hydrolysis	−	+	−	+	+	−	+
Voges–Proskauer	−	+	−	−	−	+	−
Swollen sporangium	+	−	+	−	+/−	−	+/−
Acid from glucose	+	+	−	+	+	+	+
Gas from glucose	−	−	−	−	−	−	−
pH <6.0 in VP broth	−	+/−	−	+/−	+	+	+
Supplementary tests							
Growth in 7% NaCl	−	+	+/−	+	−	+	−
Growth at 50°C	+	+	−	−	+	+	+
Nitrate reduction	+/−	+	−	+/−	+/−	+	+

Other features
B. sphaericus: A round endospore.

B. megaterium: Contains poly-β-hydroxybutyrate within vegetative cells.

[1] *B. cereus* var. *mycoides* (*B. mycoides*) is very similar to *B. cereus* and is differentiated primarily on the basis of colony form and lack of motility.

[2] *B. anthracis* is differentiated primarily on the basis of being non-haemolytic, non-motile and sensitive to penicillin. Although rare in foods the possibility of isolating the organism can never be totally discounted. If there is any reason to believe an isolate is *B. anthracis*, work on the culture should be the minimum possible and a Class I exhaust protective cabinet used.

Typing of *Bacillus*

Serotyping

Serotyping of *B. cereus* is based on the flagellar (H) antigens. This results from the work of Norris and Wolf (1961) who established that while the spore antigen possessed the highest species specifity, the H antigen was most highly strain specific.

The first serotyping scheme was that described by Lemille *et al.* (1969), but this scheme was based largely on isolates from insects and was of extremely limited value in studies of foods. A typing scheme of proven value in the epidemiological investigation of food poisoning was developed subsequently in the UK (Taylor and Gilbert, 1975; Gilbert and Parry, 1977; Kramer *et al.*, 1982). The scheme employs a basic set of more than 20 agglutinating sera and constant monitoring is required to assess the need for further expansion of the basic set. The basic UK scheme is likely to require modification when used in geographically distinct regions. Thus the serotyping scheme used in Japan is based on UK serovars 1 to 18 supplemented with 12 additional antisera to Japanese strains (Kramer *et al.*, 1982). Antisera are not commercially available, and it is necessary to refer strains for typing to the appropriate reference centre.

Serovars that have been associated with outbreaks of food poisoning are listed in **Table 100**. Some serovars (1, 8, 12, 19) have been associated with both emetic and diarrhoeal types. It appears that the two syndromes cannot be related to specific serovars, although 70% of all emetic cases reported have involved serovar 1. The recovery of multiple serovars from food involved in illness is not uncommon.

Table 100. Serovars of *Bacillus cereus* associated with food poisoning.

	Serovars
Diarrhoeal	1 2 6 8 9 10 12 19 non-typable
Emetic	1 3 4 5 8 12 19 non-typable

Serotyping schemes have not been developed for other species of *Bacillus* associated with food poisoning.

The serotyping of Bacillus *is discussed in greater detail, with respect to Methodology, in* **Laboratory Methods**, page 287.

Biotyping

No formal biotyping scheme exists for *B. cereus* or other species associated with food poisoning. However, there has been considerable interest, and some controversy, concerning biochemical differentiation between strains causing the emetic and diarrhoeal types of food poisoning.

Strains causing the emetic syndrome have been variously found to be unable to hydrolyse starch (Shinagawa *et al.*, 1985) or ferment salicin (Gilbert and Taylor, 1976; Raevuori *et al.*, 1977), although Gilbert *et al.*, (1981) subsequently stated that some emetic strains fermented salicin slowly. In studies using the API system, it was found possible to differentiate emetic strains on the basis of the failure to ferment dextrin, starch and glycogen (**Figures 226–228**; Logan and Berkeley, 1981), although diarrhoeal strains could not be differentiated from non-pathogenic strains. Some workers have found no consistent differences between diarrhoeal and emetic strains (Major *et al.*, 1979).

Plasmid profiling

Plasmid profiling has recently been applied in studies of *B. cereus* (De Bueno *et al.*, 1988; Ellison *et al.*, 1988, 1989). In an investigation of an outbreak among patients and staff in a nursing home, in which beef stew was the vehicle of infection, plasmid profiling was considered to be 'quite useful' where isolates could not be serotyped (De Bueno *et al.*, 1988).

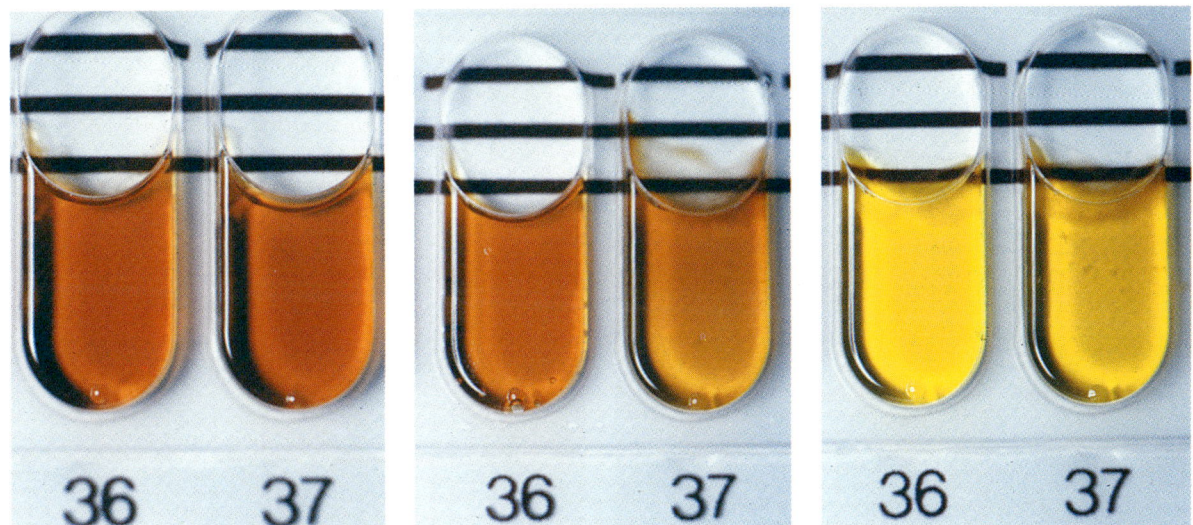

226–228. Biochemical differentiation between emetic and diarrhoeal strains of *B. cereus*. Emetic strains of *B. cereus* may be differentiated from diarrhoeal strains by biochemical tests, including acid production from starch and glycogen. Emetic strains are negative after 2 days' incubation (**226**), while diarrhoeal strains show weak acid production after 24h incubation (**227**), and give a full positive reaction after 2 days (**228**). Some workers, however, do not accept that differences between emetic and diarrhoeal strains are consistent (see above). (*API 50CHB©1 24/48h at 37°C*.)

Symptoms of *Bacillus* food poisoning

B. cereus

Symptoms of *B. cereus* food poisoning are of two distinct types – diarrhoeal and emetic. The two syndromes are the result of the activity of different toxins (see pages 277–278), and there are also epidemiological differences. *B. cereus* diarrhoeal syndrome is associated with a wide range of food including cooked meat dishes, soups, vegetable puddings and sauces, whereas the emetic syndrome is almost always associated with rice or pasta dishes.

Diarrhoeal syndrome is characterised by an incubation period of 8 to 16 hours, followed by abdominal cramps, profuse diarrhoea with watery stools and rectal tenesmus. Vomiting and fever are occasional symptoms. Recovery is usually within 24 hours and complications are rare. Two deaths are known to have occurred, the autopsy revealing toxic signs and lesions on the spleen, liver, kidneys, lungs and brain (Logan, 1988).

Emetic syndrome has a short incubation period of 1 to 6 hours. Initial symptoms are of nausea, followed by vomiting and malaise. Diarrhoea and fever are rare but not unknown. Recovery is usual within 24 hours and complications are very rare although peri-orbital oedema has been noted (Logan, 1988).

Differential diagnosis of *B. cereus* food poisoning can be difficult in the absence of bacteriological examination of stools, since symptoms of diarrhoeal syndrome closely resemble those of *Clostridium perfringens* food poisoning, while symptoms of emetic syndrome resemble those of staphylococcal intoxication. The situation is complicated by the fact that the epidemiology of *B. cereus* can be similar to that of the two other organisms and the organism may be present in conjunction with *Staphylococcus aureus* or *Cl. perfringens* in food associated with food poisoning.

Other food poisoning species

Food poisoning due to other species of *Bacillus* is less common than that due to *B. cereus*, and symptoms are often less well defined. Symptoms vary from species to species, but a common feature is the notably short incubation period.

Symptoms of *B. licheniformis* food poisoning resemble those of *B. cereus* diarrhoeal syndrome, but the incubation period is rather shorter usually extending between 4 to 15 hours.

Vomiting is the dominant symptom of *B. subtilis* food poisoning, although diarrhoea is relatively common. Fever and abdominal cramps may be present on rare occasions. Incubation periods as short as 15 minutes have been reported and symptoms invariably appear within 10 hours (Kramer *et al.*, 1982).

B. subtilis has also been implicated, although not conclusively, in a hospital outbreak in which elderly patients suffered intermittent and irregular diarrhoea over an extended period.

Both vomiting and diarrhoea are common symptoms of *B. brevis* food poisoning; reported incubation periods range from 1 to 9.5 hours.

Symptoms of food poisoning due to unidentified strains of *Bacillus* (**Figures 229** & **230**) have tended to resemble *B. cereus* emetic syndrome. The incubation period tends to be longer (8 to 18 hours) and actual vomiting is relatively rare, symptoms primarily involving nausea and malaise.

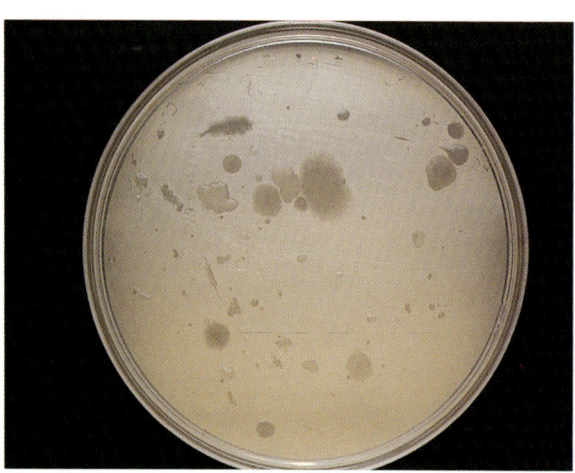

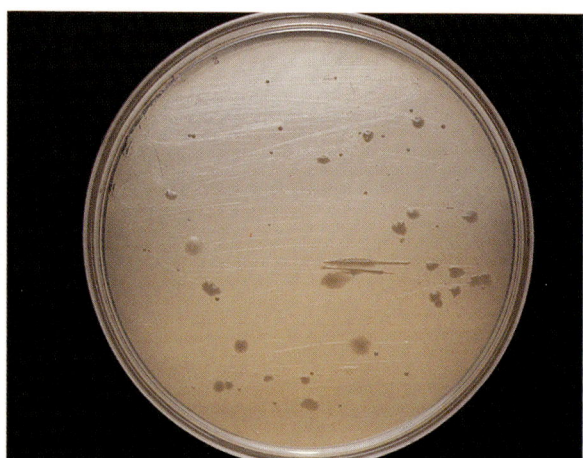

229 & 230. *Bacillus* species from foods. *Bacillus* species may be readily isolated from a wide range of foods including black pepper (**229**) and, more surprisingly, honey (**230**). Although the species present may not be those usually associated with food poisoning, any member of the genus should be treated with suspicion if allegations of food poisoning have been made. (*Plate count agar: 2 days at 30°C.*)

Persons at particular risk from *Bacillus* food poisoning

No clearly defined, high-risk people or groups exist with respect to food poisoning due to *B. cereus* or other species of *Bacillus*, although the young, the elderly and the debilitated might be expected to be more susceptible. However, *B. cereus* is most commonly associated with large-scale food preparation, and in the case of emetic syndrome there is a strong link with Chinese cookery and more particularly with traditional means of preparing rice. Emetic syndrome was commonly referred to as Chinese-restaurant syndrome, although the cause was earlier believed to be over-use of monosodium glutamate. Habitual consumers of Chinese cuisine would therefore seem to be at higher risk of emetic syndrome. Indeed, members of the 'public house subculture' who habitually follow the Saturday night consumption of numerous pints of beer with a Chinese take-away meal, appear to accept the occasional bout of vomiting as an inevitable part of their life style, and of no more consequence than the Sunday morning hangover.

People at risk from anthrax are predominantly workers with animals, or animal products such as hides and fleece, in areas where anthrax is endemic. Intestinal anthrax is primarily a risk where shortages and economic conditions encourage the consumption of meat from infected animals.

Treatment of *Bacillus* food poisoning

Food poisoning due to *Bacillus* species is almost always self-limiting and specific treatment is not required.

Intestinal anthrax is a life-threatening disease and requires antimicrobial therapy. *B. anthracis* is highly susceptible to penicillin and treatment is succesful if commenced as soon as symptoms appear. Supportive therapy may also be necessary.

Mechanisms of pathogenicity

B. cereus produces a number of toxins including two distinct toxins that are responsible for the diarrhoeal and emetic syndromes respectively (Melling *et al.*, 1976). For many years previously, it was thought that the factors involved with human pathogenicity were those responsible for *B. cereus* haemolytic, phospholipolytic and mouse lethal activities, but later work demonstrated a distinct diarrhoeal toxin (Spira, 1974; Spira and Goepfort, 1975), and subsequently the existence of a separate emetic toxin was confirmed (Melling *et al.*, 1976). The various toxins produced by *B. cereus* are summarised in **Table 101**.

Toxin production has not been demonstrated in *B. subtilis, B. licheniformis* or other species associated with food poisoning.

Table 101. Summary of toxins produced by *B. cereus*.

A Toxins definitely associated with food poisoning	
Diarrhoeal (Synonyms: Fluid accumulation factor, vascular permeability factor, dermonecrotic factor, intestino-necrotic toxin, mouse lethal factor I)	See below for description
Emetic	See below for description
B Toxins of no known role in food poisoning	
Haemolysin I (Synonym: cereolysin I)	MW 49 to 59,000, pI 6.3 to 6.7. Cytolytic thiol activated toxin related to streptolysin O and clostridial O toxin. Heat labile, trypsin sensitive (Coolbaugh and Wiliams, 1978)
Haemolysin II (Synonym: secondary haemolysin)	MW 29 to 34,000, pI 4.92. Heat labile, sensitive to pepsin, pronase and trypsin. Not thiol activated or related to strepolysin O. *In vivo* toxicity not proven
Phospholipase C Synonyms: lecithinase, egg yolk turbidity factor)	MW 23 to 29,000, pI 6.5 to 8.1. May be two enzymes. One is cytotoxic metallo-enzyme requiring Zn^{++}. Relatively stable, not trypsin sensitive. Not related to *Cl. perfringens* products. *In vitro* toxicity not proven
Mouse lethal factor II	Unstable, closely related or identical to haemolysin I
Ezepchuk and Fluer toxin	MW 57,000. Inactivated at 60°C for 20 minutes. May be related to diarrhoeal toxin

Nature of *B. cereus* toxins

Diarrhoeal toxin

B. cereus diarrhoeal toxin is a protein of MW *ca* 50,000 and has an isoelectric point of 4.9. It probably consists of several subunits which differ slightly in molecular weight and isoelectric point (Kramer, 1984). The toxin is heat labile, being inactivated by heating at 56°C for 5 minutes (Turnbull, 1976). It is also sensitive to trypsin and pronase (Goepfort *et al.*, 1973) and is unstable during storage at either 4°C or 25°C (Spira and Goepfort, 1975). The diarrhoeal toxin may be related to the toxin described by Ezepchuk and Fluer (1973).

Emetic toxin

B. cereus emetic toxin is of MW less than 5,000 and probably non-polar in nature (Johnson, 1984). The toxin is heat stable to 126°C for 90 minutes, and is not sensitive to trypsin or pepsin. The toxin is also stable during storage at 4°C for two months.

Production of *B. cereus* toxins

The production of both types of toxin is associated with the growth of the organism in foods, but while the diarrhoeal toxin is synthesised during exponential growth, production of the emetic toxin appears to occur during sporulation. The relationship between sporulation and toxin production is not clear (Turnbull, 1986).

Little is known of the genetic control of toxin production, and tools to study the genetics of *B. cereus* are limited (Glatz, 1986). Some years ago, it was suggested that lysogeny may be linked to toxin production but this possibility has not been further explored. More recently, it has been suggested that toxin production is mediated by episomes or plasmids (Gilbert and Kramer, 1984), and in this context it is of interest that toxin production has been linked to plasmids in both *B. anthracis* and *B. thuringiensis*.

Toxin production by *B. cereus* varies widely from strain to strain (Turnbull *et al.*, 1979a), and is also influenced by a number of extracellular factors. Many of these are of relevance to toxin production in foods and are discussed in **Factors affecting growth, survival and enterotoxin production**, page 281.

Action of *B. cereus* enterotoxins

The action of *B. cereus* diarrhoeal toxin is stimulation of the adenylate cyclase-cyclic AMP system (Turnbull, 1976; Terranova and Blake, 1978), and this may also be responsible for fluid accumulation in the intestines. This toxin is probably also responsible for both pyogenic and pyrogenic symptoms in extraintestinal

B. cereus infections. The mechanism of action of emetic toxin is not known.

Detection of *B. cereus* toxins

Diarrhoeal toxin

Until recently, detection of diarrhoeal toxin has involved two tests, the vascular permeability reaction (VPR) test and the ligated rabbit ileal loop test. A serological assay method based on reverse passive latex agglutination is now available as a commercial kit (Unipath: Denka Seiken), and should permit toxin determinations to be undertaken by a wider range of laboratories.

Emetic toxin

For many years the only means of detecting *B. cereus* emetic toxin has been monkey feeding with rice culture slurries (Melling *et al.*, 1976). This is expensive and inconvenient with very few laboratories equipped to handle primates. However, a tissue culture technique has been described (Hughes *et al.*, 1988), and offers obvious advantages.

The determination of B. cereus *enterotoxins is discussed in* **Laboratory Methods***, page 288.*

Human carriage of *Bacillus*

B. cereus may be isolated from the stools of 14 to 15% of healthy people (Ghosh, 1978; Shinagawa, 1979). This probably reflects dietary intake and is not surprising given the widespread occurrence of the organism in foods. In most cases, carriage is relatively short

lived, excretion persisting for less than two weeks.

Human carriage is not considered to be of any significance in food poisoning due to *B. cereus* or other species.

Bacillus in the environment

Bacillus species are extremely widespread in nature and may readily be isolated from soil, water, dust, air and many types of food (see below). The natural habitat of most species is soil, and direct contamination from soil is of importance with respect to some foods. For example, *B. cereus* accounts for some 10% of the microflora of soil in rice paddies (Asanuma *et al.*, 1979).

In some situations, care must be taken to ensure

that the isolation of *Bacillus* species is placed in its correct context. Endospores of the genus are extremely persistent and the isolation of the organism from a given habitat does not necessarily imply a major role in the ecology of that habitat. This is an important principle which may be extended to the isolation of small numbers of *B. cereus* and other potentially pathogenic species from foods.

Foods commonly associated with *Bacillus*

Meat and meat products

B. cereus and other species may be isolated from fresh meat, the incidence and numbers present reflecting the degree of contamination of the carcass with soil etc. In Japan, for example, an incidence of 6.6% has been reported in raw meat although numbers rarely exceeded 100/g. In a study in the UK, Sooltan *et al.*, (1987) found that 6.9% of raw chickens contained detectable numbers of *B. cereus*, although it was considered that the organism may have been under-recovered due to competition from other organisms. The incidence in meat products such as sausages is often higher, reflecting the contribution made from additives such as spices and fillers.

Endospores of *Bacillus* will survive the cooking process, but significant numbers are found only where the temperature has not been adequately controlled after cooking. High numbers are most common in products such as stews, pies, soups and roasts, and these are also the meat products most commonly associated with food poisoning due to *B. cereus* and other species. In the case of *B. cereus*, diarrhoeal syndrome is virtually always involved.

There has been a particularly strong association between *B. cereus* food poisoning and the central European goulash. This is probably a consequence of a high level of endospores derived from spices and of traditional cooking practices (Powers *et al.*, 1976).

Meat dishes are the common vehicles of infection in food poisoning due to *B. subtilis* and *B. licheniformis*. A range of products have been involved including pies, mince, curry, chicken and turkey.

Intestinal anthrax is a consequence of eating the meat of infected animals. Large outbreaks have occurred in developing nations where animal immunisation programmes have been interrupted by warfare. A large outbreak of intestinal anthrax also occurred in the Siberian city of Sverdlovsk during the 1980s. This outbreak was of note in that the United States government alleged that the source of infection was a biological warfare plant. Subsequent investigations have shown this to be unlikely, and the American scientific community at least, appear to accept that consumption of illegally butchered meat from infected cattle was responsible (Anon, 1988).

Milk and milk products

B. cereus has, for many years, been associated with the spoilage of fresh milk, classical spoilage patterns being bitty cream formation (**Figure 231**; Billing and Cuthbert, 1958), or sweet curdling (**Figure 232**; Overcast and Atmaram, 1974). The incidence of *B. cereus* spoilage has fallen in recent years, partly due to the widespread introduction of refrigeration for pasteurised dairy products, and partly because of a reduced incidence of contamination of incoming raw milk. This in turn reflects improved farm hygiene and changes in practice such as the substitution of bulk refrigerated milk tanks for churns. Despite this, *B. cereus* may be readily isolated from pasteurised milk (Christiansson, 1989) and related products such as cream. Viable cells of *B. cereus* have been isolated from ultra-heat treated (UHT) milk (Mostert *et al.*, 1979; Westhoff and Dougherty, 1981). Heat resistant strains of *B. cereus* may survive the UHT process if present in sufficient numbers initially (Rajkowski and Mikolajcik, 1987), and while there have been no reported cases of *B. cereus* food poisoning due to UHT milk, there are

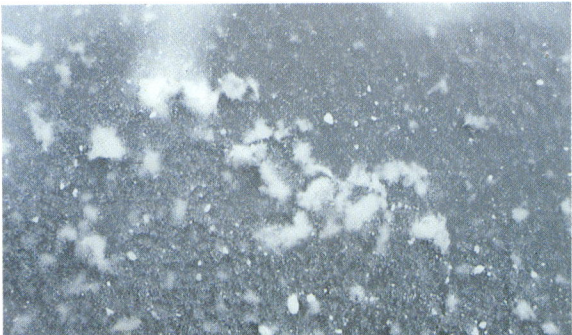

231

232

231 & 232. *B. cereus* as a spoilage organism in milk. Before the introduction of refrigeration for pasteurised milk, spoilage by *B. cereus* was a major problem during summer months. The classical spoilage is bitty cream (**231**), although in practice soft clotting (**232**) may be a more common fault.

important public health implications for the extension of UHT processing to meat soups etc.

B. cereus is a common contaminant of dried milk, although numbers present are usually small. There is disagreement over the significance of *B. cereus* in this product; Woes (1976), for example, did not consider that there was a major health risk, while Walthew and Luck (1978) reported that reconstituted dried skim milk must remain at room temperature for 78 hours before a hazard develops. Dried milk has been implicated as the source of *B. cereus* in outbreaks involving vanilla slices (Pinegar and Buxton, 1977) and macaroni cheese (Holmes *et al.*, 1981).

B. cereus has been implicated in the spoilage of yoghurt (Driessen and Stadhouders, 1980), and under some circumstances may survive during the manufacture of cheese.

Despite the presence of *B. cereus* in a wide range of dairy products, there have been few published reports of food poisoning apart from those involving dried milk as the source of the organism. There has been an unconfirmed report of an outbreak of emetic syndrome in which feta cheese was implicated (Schmitt *et al.*, 1976), and there have also been anecdotal accounts of diarrhoeal syndrome associated with the consumption of a milk-based caramel dessert and a fruit-flavoured soft-cheese dessert.

Various explanations have been suggested for the low level of *B. cereus* food poisoning incidents associated with milk and milk products (**Table 102**) but none are proven.

Table 102. Possible reasons for the low incidence of *Bacillus cereus* food poisoning from dairy products.

1 Visible spoilage occurs before numbers are sufficiently high to cause illness.
2 Dairy products are nutritionally deficient for toxin production.
3 'Dairy' strains are usually non-pathogenic (Chopra *et al.*, 1980).
4 Insufficient aeration in fluid dairy products for toxin production (Christiansson, 1989).

Cereal products

Species of *Bacillus*, including those associated with food poisoning, may be isolated from cereal grains and flour in high, but variable, numbers. The incidence of *B. cereus* in rice is high and the association of emetic syndrome with rice dishes is well known (Gilbert *et al.*, 1981; Kramer *et al.*, 1982). The traditional Chinese practice involves boiling a large quantity of rice which is then kept at room temperature before frying. There is resistance to the refrigeration of the boiled rice since this, allegedly, has an adverse effect on the finished product.

B. cereus serovar 1 is involved in 70% of incidents of emetic syndrome, but this is not matched by a predominance of that serovar in uncooked rice or the environment (Gilbert and Parry, 1977), selection for the more heat resistant serovar 1 strains occurring during cooking (Parry and Gilbert, 1980).

Chinese dishes other than rice have been implicated in *B. cereus* food poisoning (Schiemann, 1978; Sly and Ross, 1982), and in incidents involving Chinese food, it should not be assumed that the vehicle is rice.

B. cereus is less common in wheat flour, but pasta products including macaroni cheese (Holmes *et al.*, 1981) have also been implicated in emetic food poisoning.

Although most cases of *B. cereus* food poisoning associated with starchy cereal foods are of the emetic type, diarrhoeal syndrome has also been involved. The first definite account of *B. cereus* diarrhoeal food poisoning involved vanilla sauce, the organism being derived from the corn starch used as a thickener. Six hundred people were affected in four outbreaks (Hauge, 1955). Cooked rice, eaten cold, has also been implicated as a cause of diarrhoeal syndrome (Shinagawa *et al.*, 1979).

B. licheniformis and *B. subtilis* are common in wheat flour and may cause ropey spoilage of bread. *B. subtilis* and, possibly, *B. licheniformis* have also been implicated as the cause of food poisoning due to bread (Kramer *et al.*, 1982; Communicable Disease Surveillance Centre, 1988). Wholemeal bread is usually involved and rope is usually incipient or, less commonly, fully developed. Although still unusual, the number of cases of food poisoning associated with the growth of *Bacillus* in bread has increased since the addition of anti-microbial agents to bread has to a large extent been discontinued.

Dried products

In addition to dried milk, endospores of *B. cereus* and other species of *Bacillus* can be isolated from a wide range of dried foods including egg, potato, legumes, spices, formulated infant feeds, sauce mixes and soups. The presence of endospores is of no consequence unless reconstituted foods are stored at high temperatures – a particular hazard with respect to infant feeds.

The practice of soaking dried legumes before cooking has also been considered hazardous (Blakey and Priest, 1980).

B. cereus may also be isolated from live legume and other seeds used for sprouting. Seeds were the probable source of the organism in an outbreak involving vegetable sprouts (Portnoy *et al.*, 1973). Subsequent investigations showed that while many types of seed including wheat, alfalfa and mung bean were often contaminated with *B. cereus*, only wheat would support good growth.

Although the incidence of *B. cereus* in spices is high, numbers are usually small. However, numbers as high as 10^3 cfu/g may occur especially where the spices are dried on earth. Spices themselves present no risk, but are an important source of the organism in some dishes.

Miscellaneous foods

A number of different foods have been implicated as vehicles in outbreaks of *Bacillus* food poisoning. These are summarised in **Table 103**.

Table 103. Miscellaneous foods as vehicles of food poisoning due to species of *Bacillus*.

Food	Species
Mashed potato[1]	*B. cereus* diarrhoeal
Green bean salad[1]	*B. cereus* diarrhoeal
Honey[2]	*Bacillus* species[3]
Instant breakfast[1]	*B. cereus* mixed symptoms
Canned tuna[1,4]	*B. cereus* mixed symptoms
Pork sandwich[5]	*B. subtilis*
Egg mayonnaise sandwich[6]	*B. cereus* and *Staph. aureus*
Pizza[5]	*B. subtilis*

[1] Schmitt *et al.* (1976).
[2] Sutherland *et al.* (1986).
[3] Identification not possible using biochemical criteria.
[4] Probably post-process contamination.
[5] Kramer *et al.* (1982).
[6] Communicable Disease Surveillance Centre (1988).

Factors affecting the growth, survival and production of toxin by *B. cereus* and other species of *Bacillus* in foods

It is important to distinguish between the growth and survival of the vegetative cell, and the germination, and outgrowth, and the survival of the endospore. The distinction is usually clear when discussing resistance to heat and other sterilising agents, but confusion can occur in the case of resistance to inhibitors. In most cases, the effect of inhibitors on growth of the vegetative cell is of greatest importance.

Temperature

The upper growth limit of *B. cereus* is generally accepted as being 50 to 55°C (Doyle, 1988; Johnson *et al.*, 1983) although Gilbert *et al.*, (1974) noted a maximum growth temperature (in rice) of 43°C. Maximum temperature for the germination and outgrowth of endospores has been recorded as 50°C (Johnson *et al.*, 1983).

The growth of *B. cereus* is optimal at *ca* 37°C. However, there is disagreement over the temperature range for optimal growth, Johnson *et al.* (1983) finding an optimal range of 35 to 40°C, while Gilbert *et al.* (1974) considered the lower temperatures of 30 to 37°C to be optimal. The optimum temperature for endospore germination is 30°C (Johnson *et al.*, 1983).

Although a minimum growth temperature of 15°C has been claimed for *B. cereus*, there is strong evidence that many strains are able to grow at 7 to 8°C (Rajkowski and Mikolajcik, 1987; Christiansson *et al.*, 1989). Temperatures in this range have also been reported to be the minimum for endospore germination.

Toxin production by *B. cereus* closely correlates with exponential growth in the case of the diarrhoeal toxin, and sporulation in the case of the emetic toxin. Where other conditions are favourable, toxin production will take place at 8°C (Christiansson *et al.*, 1989).

Vegetative cells of *Bacillus* have no enhanced resistance to heat and are destroyed at time-temperature combinations which inactivate non-endospore forming bacteria.

Endospores of *B. cereus* show considerable variation in thermal resistance. In some cases, this is related to serovar and accounts for the predominance of serovar 1 strains in cooked rice (Gilbert and Parry, 1977). A study of eight strains showed the D_{95} values (in phosphate buffer) to range from 1.2 to 20.2 minutes with an average z value of 9.2°C (Johnson et al., 1982). These D values are of the same order as those of 2.5 to 36.2 minutes and 5.0 to 36.0 minutes determined by Parry and Gilbert (1980) and Gilbert et al. (1980) respectively, but markedly higher D_{95} values of 256.7 and 5,122.3 minutes were reported for two isolates from underprocessed canned soup (Bradshaw et al., 1975).

Non-linear survival curves for B. cereus have been noted at temperatures above 100°C (Johnson et al., 1987; Rajkowski and Mikolajcik, 1987). Non-linear survival ('tailing') is a common phenomenon with endospores, the basis of which has been fully discussed by Cerf (1977), and has important implications in canning and UHT processing.

The diarrhoeal and emetic toxins differ markedly in thermal stability (see page 277). The practical implications are that emetic toxin is able to withstand the reheating of foods such as rice, while the diarrhoeal toxin is not.

Irradiation

Vegetative cells of Bacillus species have a resistance to ionising radiation similar to that of non-endospore forming bacteria, and are inactivated by proposed treatments of 0.3 Kgy. The endospores are significantly more resistant and withstand treatments of 20 to 30 Kgy. Doses of this order bring about unacceptable organoleptic changes in foods, and ionising radiation is not considered to be a practical means of controlling B. cereus or other species of Bacillus.

pH value and organic acids

The minimum pH value for the growth of vegetative cells of B. cereus has been determined as 5.0 when hydrochloric acid is used as an acidulant. The limiting pH value in food may be lower, growth occurring to some extent in meat at pH 4.35 (Raevuori and Genigeorgis, 1975). The inhibitory effect is increased by the use of organic acids as an acidulant, limiting pH values in the presence of 0.1M acetic, formic and lactic acids being 6.1, 6.0 and 5.6 respectively (Wong and Chen, 1988).

A lower pH value, or higher concentration of organic acids, is required to inhibit endospore germination. At concentrations of 0.1M of acetic, formic and lactic acid, pH values of 4.4, 4.3 and 4.2 respectively were required to inhibit endosphere germination by 50%.

Water activity

Relatively little is known of the effect of a_w on the growth of Bacillus species in foods, although a limiting value for growth of 0.95 was reported for B. cereus. There is no evidence of an ability to grow, or for endospores to germinate, at levels significantly below this.

Oxidation-reduction potential

Of the Bacillus species associated with food poisoning, B. subtilis, B. brevis and B. sphaericus are obligate aerobes. In foods containing nitrate, however, limited anaerobic growth of B. subtilis and some strains of B. brevis is possible due to nitrate respiration. The E_h range permitting growth of B. subtilis has been determined as +135 to −100 mV (Jacob, 1970).

B. cereus and B. licheniformis are facultative anaerobes and are able to grow at E_h values below −200 mV. Significantly, B. cereus requires a high level of aeration for toxin production, and toxin was produced in aerated, but not in non-aerated, milk (Christiansson et al., 1989).

Preservatives

Recent work with B. cereus has been focused on combinations of anti-microbial agents at concentrations acceptable in foods. Examples include the use of 1.7% NaCl, 120 mg/l $NaNO_2$, 0.15% glucono-δ-lactone, 0.05% sodium erythorbate and 0.05% citric acid to control B. cereus in liver sausage (Asplund et al., 1988). Sodium erythorbate was effective only in the presence of other inhibitors. Other preservatives suggested for the control of B. cereus include a combination of sorbic acid and potassium sorbate (Raevuori, 1976) and nisin (Shehata and Hassen, 1981).

Chemicals added to foods for technological purposes may also be effective against both vegetative growth and endospore germination. The anti-oxidants, butylated hydroxyanisole and tertiary butylated hydroxyquinone, for example, are effective against *B. subtilis* at concentrations of 150 and 100 mg/l respectively (Al-Khayat *et al.*, 1987). However, only a certain portion of the endospore population is affected.

The sequestering agent ethylenediaminetetraacetic acid (EDTA) also has activity against the outgrowth of *Bacillus* endospores (Bulgarelli and Shelef, 1985).

Competition from other micro-organisms

High levels of other micro-organisms have been repeated to reduce the growth of *B. cereus* (Steele and Miles, 1981). Lactic acid bacteria, especially, *Lactococcus (Streptococcus)* were particularly inhibitory (Wong and Cheni, 1988).

Laboratory methods

Isolation of *Bacillus*

Enrichment

Enrichment is not usually necessary when examining foods for *Bacillus*. Non-selective enrichment in nutrient or lactose broths may be used, however, in the examination of foods specifically for *B. cereus* (Parry et al, 1983), although a selective enrichment broth, trypticase-soy-polymyxin broth (Harmon, 1980) is recommended where it is necessary to recover small numbers of *B. cereus* or for examination of dehydrated or starchy foods. Incubation for 18 to 24 hours at 37°C is usual, although 30°C may be preferred for dairy products.

Each of the enrichment media is also suitable for the recovery of species of *Bacillus* other than *B. cereus*.

Plating media

The use of non-selective blood agar has been proposed by Kramer *et al.*, (1982) for the recovery of *B. cereus* from foods implicated in food poisoning and from clinical specimens. Blood agar is simple and effective, and may also be used in the routine examination of cooked meat products as a means of determining not only the presence of food poisoning species of *Bacillus*, but also of indicating slow cooling after cooking.

With many food samples, the presence of other bacteria makes the use of a selective medium necessary. The simplest selective medium is blood agar supplemented with polymyxin. Columbia blood agar base, which contains starch, is particularly suitable but other types may be used. Blood agar-polymyxin provides a suitable level of selectivity, while remaining relatively simple to prepare. *B. cereus* and other species may usually be recognised by colonial morphology (**Figures 214–219**, pages 268–269), and

colony types are easily differentiated for serological testing (Kramer *et al.*, 1982).

A number of selective media have been devised for *B. cereus*. These include mannitol-egg yolk-polymyxin agar (Mossel *et al.*, 1967), KG agar (Kim and Goepfort, 1971) and polymymxin-pyruvate-egg yolk-mannitol-bromothymol blue agar (PEMBA, *B. cereus* selective medium; Holbrook and Anderson, 1980). The underlying principles of each medium is similar (**Table 104**).

Table 104. Principles of selective and diagnostic media for *Bacillus cereus*.

Component/characteristic	Function
Polymyxin B	Selective agent
Mannitol and Indicator[1]	Differentiation (*B. cereus* negative)
Egg yolk	Differentiation (*B. cereus* positive)
Low level of peptone/ absence of meat extract[2]	Encourage sporulation Inhibit lecithinase production by *B. polymyxa*

[1] Not present in KG medium.
[2] Levels normal in mannitol-egg yolk-polymyxin agar.

A number of comparisons of performance have been made (eg Peters *et al.*, 1985) but no medium appears to be consistently or significantly superior. Mannitol-egg yolk-polymyxin agar does not induce sporulation and permits lecithinase production by species such as *B. polymyxa*. Both KG agar and PEMBA overcome these problems, but the inclusion of pyruvate in PEMBA facilitates the detection of weak lecithinase producing strains and reduces colony size. This prevents problems which occur with KG agar due to spreading colonies (Stec and Burzynska,

1980). The substitution of bromocresol purple for bromothymol blue as the indicator in PEMBA is claimed to shorten incubation from 48 to 22 hours and to enhance general performance (Szabo *et al.*, 1984) but this modification has not been widely adopted.

PEMBA is considered to be the medium of choice

although it may not be sufficiently selective to permit the optimal recovery of *B. cereus* from samples containing large numbers of competing micro-organisms (Sooltan *et al.*, 1987). Its use is illustrated in **Figures 233–240**. The medium is available as a complete dehydrated formulation (Unipath).

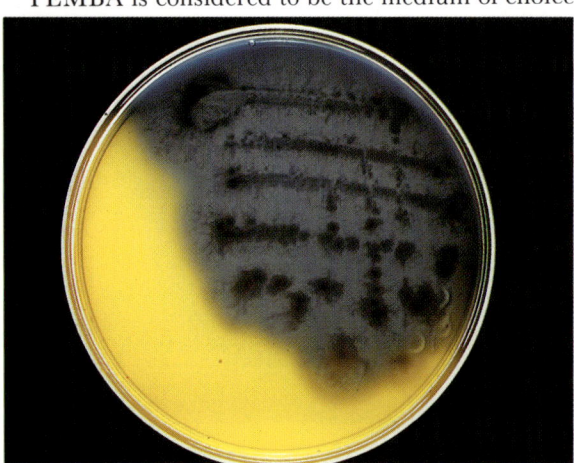

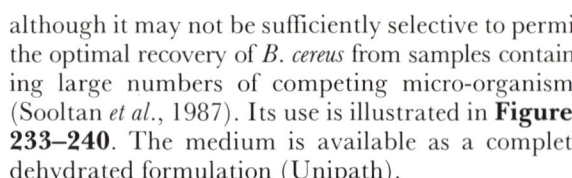
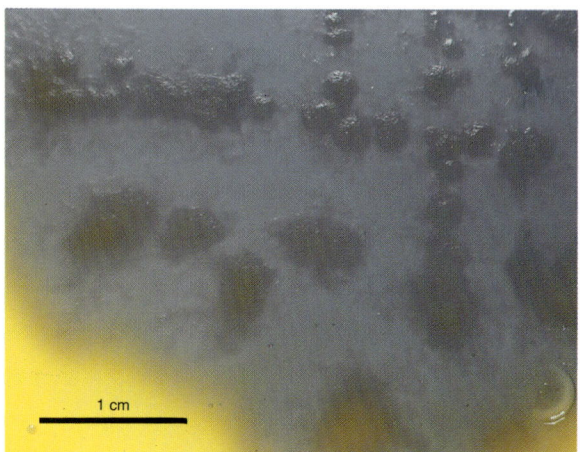

233 & 234. Isolation of *B. cereus* on Polymyxin-Pyruvate-Egg yolk-Mannitol-Bromothymol blue agar medium (PEMBA: *B. cereus* selective agar). The appearance of *B. cereus* on PEMBA is striking (**233**) and colonies are usually easily recognised. The zone of precipitation due to the LV reaction is coincident with the zone of alkalinity due to the inability of *B. cereus* to ferment mannitol (**234**). Although a highly successful medium,

difficulties may occur with LV-negative strains of *B. cereus* or, conversely, where zones of precipitation merge making recognition of LV-negative colonies difficult. Where *B. cereus* is growing in mixed culture with strong mannitol-fermenting bacteria, diffusion of acid may result in the loss of the characteristic blue colour associated with *B. cereus*. (*PEMBA medium: 24h at 37°C.*)

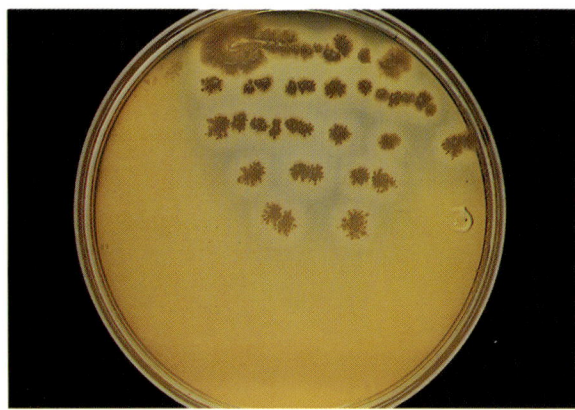

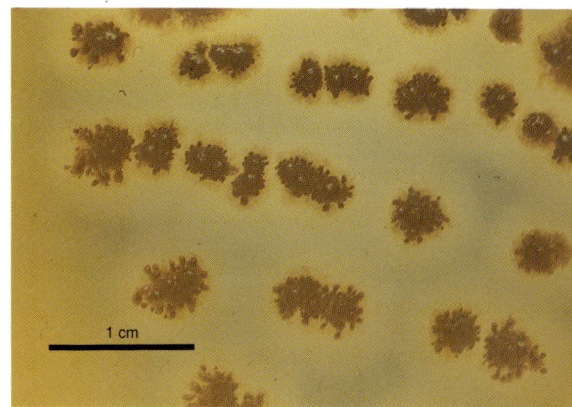

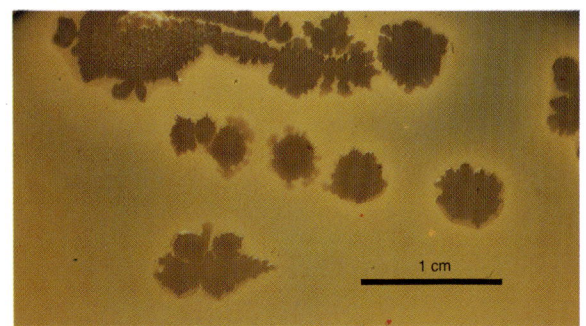

235–237. Isolation of other species of *Bacillus* on PEMBA medium. PEMBA serves as a selective isolation medium for all species of *Bacillus* with the exception of species with specific nutritional requirements. Colonies of species other than *B. cereus* cannot be differentiated from each other, although differentiation from colonies of genera other than *Bacillus* is straightforward.

In most cases, the differentiation between *B. cereus* and other *Bacillus* species such as *B. licheniformis* (**235**) or *B. subtilis* (**236**) presents no difficulty even to the inexperienced. Small zones of precipitation in the egg yolk (**237**) are of no significance and may be ignored. (*PEMBA medium: 24h at 37°C.*)

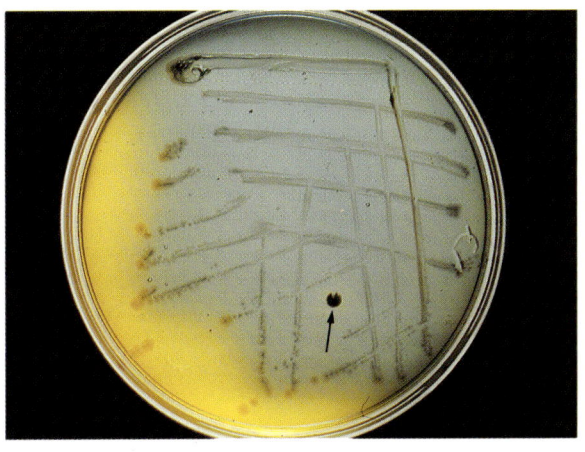

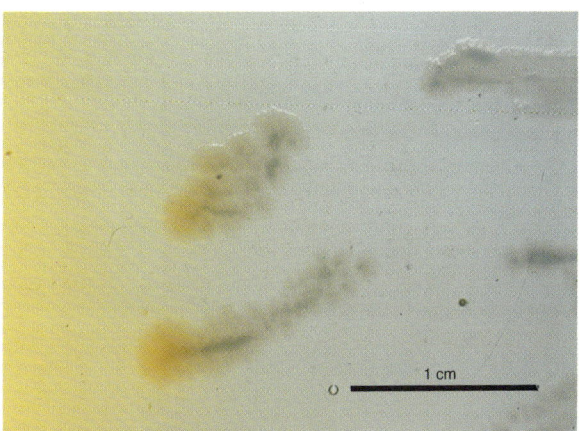

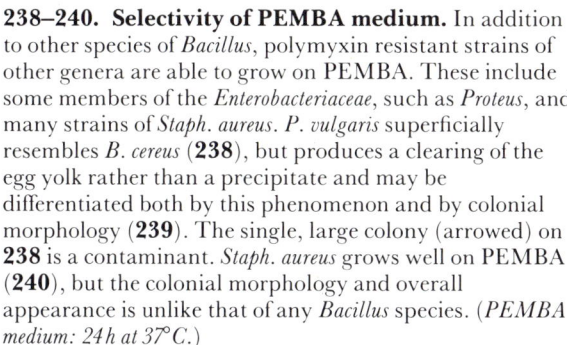

238–240. Selectivity of PEMBA medium. In addition to other species of *Bacillus*, polymyxin resistant strains of other genera are able to grow on PEMBA. These include some members of the *Enterobacteriaceae*, such as *Proteus*, and many strains of *Staph. aureus*. *P. vulgaris* superficially resembles *B. cereus* (**238**), but produces a clearing of the egg yolk rather than a precipitate and may be differentiated both by this phenomenon and by colonial morphology (**239**). The single, large colony (arrowed) on **238** is a contaminant. *Staph. aureus* grows well on PEMBA (**240**), but the colonial morphology and overall appearance is unlike that of any *Bacillus* species. (*PEMBA medium: 24h at 37°C.*)

Identification of *Bacillus*

Differentiation from other genera

In practice, differentiation of *Bacillus* from other genera rarely presents any difficulty. Identification is readily made on the basis of colonial and cellular morphology. However, inexperienced workers can be confused by the Gram-variable and even Gram-negative reaction on staining (**Figure 241**). This is often associated with older cultures and it is recommended that Gram-staining should be carried out on cultures no more than 24 hours old.

241. Loss of Gram-positivity by a culture of *B. cereus*. Cultures of many strains of *Bacillus* which have been incubated for more than 18 to 24h contain cells, or groups of cells, which have lost Gram-positivity. The culture illustrated was stained after 30h incubation at 30°C. Although some cells have lost Gram-positivity entirely, the vestiges of the purple colouration may be seen in the majority of cells. This, together with the size and overall appearance of the cells should prevent undue confusion. In situations where confusion does arise, it is important to appreciate that the remedy lies not only with staining cultures at an earlier stage, but in ensuring that staff are adequately trained and, above all, encouraged to actually think.

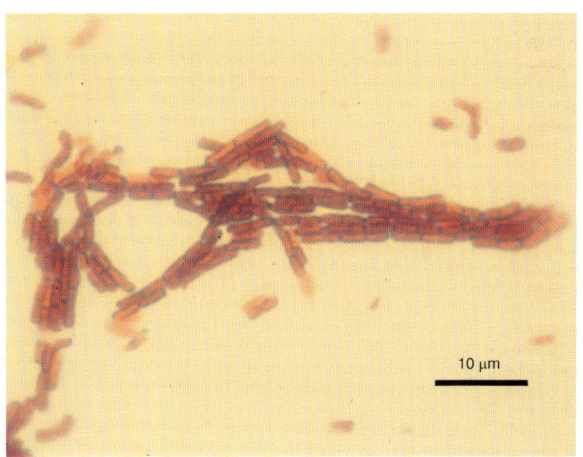

Differentiation between species of Bacillus

In the case of *B. cereus* growing on selective and diagnostic media, no further identification may be considered necessary. Over 95% of typical colonies on mannitol-egg yolk-polymyxin agar, for example, are identical with *B. cereus* (Bouwer-Hertzberger and Mossel, 1982). Colonial morphology on blood agar permits presumptive identification of *B. cereus* and some other species, but confirmatory testing is required.

Fermentation of mannitol and the egg yolk reaction may be conveniently determined using diagnostic BC medium (Gilbert and Taylor, 1976). This medium does *not* contain a selective agent and should not be confused with *B. cereus* selective agar. Additional tests are required for identification beyond the presumptive level, but in the case of *B. cereus* it is usually only necessary to determine a few key reactions (**Table 105**). It must be appreciated that *B. cereus* can be very variable in its biochemical properties; many strains are citrate-negative (Berkeley *et al.*, 1984) and some may fail to reduce nitrate. Acid production from sugars, particularly sucrose, can also be variable (Mosso *et al.*, 1989) and particular care is required when determining anaerobic growth (**Figures 242–244**). Biochemical tests may be replaced by

Table 105. Selection of tests for confirmation of identity of *Bacillus cereus*.

Bouwer-Hertzberger and Mossel (1982)	Harmon (1980)	Kramer *et al.* (1982)
Microscopic examination	Microscopic examination	Acid from: Glucose (+)
Glucose fermentation (acid not gas)	Glucose fermentation (acid not gas)	Mannitol (−) Xylose (−)
Gelatin hydrolysis (+)	Nitrate reduction (+)	Arabinose (−)
Acid from xylose (−)	Voges–Proskauer (+) Decomposition of L-tyrosine (+) Growth in 0.001% lysozyme (+)	

242

243

244

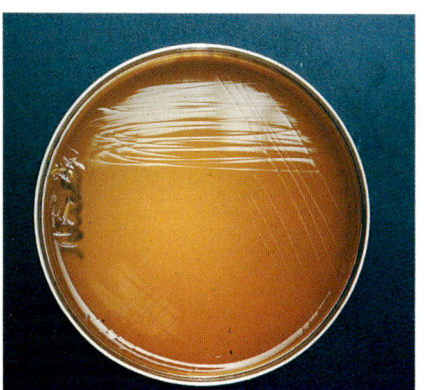

242–244. Anaerobic growth of *Bacillus* species.
Anaerobic growth on blood agar is a differentiating feature between *B. cereus*/*B. licheniformis* (+) and *B. subtilis* (−). Although the test is technically straightforward, anaerobic conditions must be rigorously maintained. *B. cereus* is readily detected (**242**) after growth on freshly prepared, pre-reduced blood agar in an anaerobic jar fitted with an active catalyst. Under the same conditions, *B. subtilis* is unable to grow (**243**). Where strict anaerobiosis is not maintained, in this example due to the failure to use freshly prepared and pre-reduced media, limited growth of *B. subtilis* is possible (**244**). In determining anaerobic growth, incubation should be for at least 2 days to avoid false negative results. (*Blood agar: 2 days at 37°C in 95% H_2/5% CO_2.*)

microscopic examination for lipid globules and endo-spores after the application of a combined staining technique (Holbrook and Anderson, 1980). This technique is not, however, widely used probably because of the difficulties associated with staining.

Where full characterisation of presumptive *B. cereus* is required, or where preliminary evidence suggests the involvement of other species, identification may be made according to one of the schemes on pages 272–273. Media and methods for conventional bio-chemical tests are described in Parry *et al.*, (1983) or, alternatively, a miniaturised kit such as API 20E©, API 50CHB© or Minitek© may be used to generate information (**Figures 245–247**).

245–247. The use of miniaturised test kits for the identification of food poisoning *Bacillus* species.
Some commercially available miniaturised test kits such as the API 20E© and API 50CHB© are suitable for the identification of *Bacillus* species. The API 20E kit, originally developed for the identification of members of the *Enterobacteriaccae*, may be used in either of two ways. As a direct substitute for conventional methodology with results being interpreted by reference to taxonomic schemes such as that summarised in **Table 98** or, in conjunction with the API 50CHB, in a taxonomic system based specifically on the use of the two API systems (see page 271).

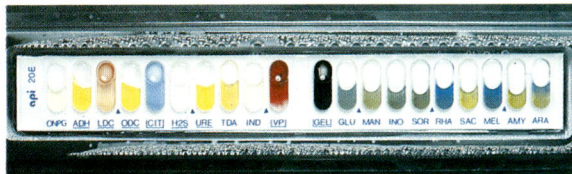

246

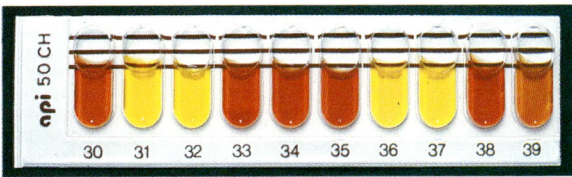

245. Use of API 20E© for the identification of *B. subtilis*. In this illustration, the kit is used as a direct substitute for conventional methods. Acid production from carbohydrates may be weak and care should be taken to avoid false-negative results. These tests are redundant when the kit is used with the 50CHB© and would not usually be inoculated. (*API 20E: 18h at 37°C.*)

 245

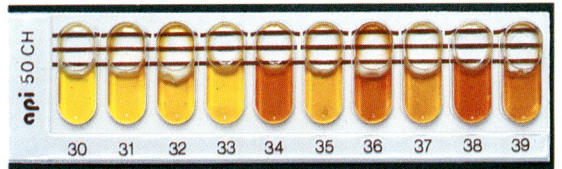

246 & 247. Use of API 50CHB© for the identification of *B. subtilis* and *B. cereus*. The 50CHB kit consists entirely of tests for carbohydrate fermentation. Although

247

further tests are required, clear differences in the fermentation pattern of *B. subtilis* (**246**) and *B. cereus* (**247**) can be seen for 10 of the tests. (*API 50CHB: 24h at 37°C.*)

Typing of *B. cereus*

Serotyping

B. cereus may exhibit considerable variation in colony form. Although colonial variants are usually of the same serovar, subcultures of all colony types must be examined. Where colony types are identical, at least three isolated colonies, chosen at random, should be typed.

It is essential to ensure that isolates are in pure culture. Blood-free medium such as nutrient agar is preferred for checking the purity of cultures.

It is necessary to passage cultures through two Craigie tubes at 30°C to ensure purity. Nutrient broth is seeded from the second Craigie tube and incubated shaken for 5 hours at 37°C. Formalin fixative is added and the culture allowed to stand for *ca* 1 hour. The culture may then be typed by agglutination (**Figure 248** Reproduced with permission from Kramer, J.M., Turnbull, P.C.B., Munshi, S. and Gilbert, R.J., *Isolation and Identification Methods for Food Poisoning Organisms.* © 1982 Society for Applied Bacteriology).

248

248. Serological typing of *B. cereus*. Doubling dilutions of antisera are prepared in WHO agglutination trays. Equal volumes of H antigen suspension are added, and the trays are incubated for 2 h at 50°C, or overnight at 37°C. Reactions are read using dark field illumination, the right-hand well showing a positive reaction (arrowed) and the left-hand a negative reaction.

Detection of *B. cereus* toxins

Diarrhoeal toxin

The recent introduction of a reverse passive latex agglutination test kit has simplified the detection of *B. cereus* diarrhoeal toxin and avoids the use of animals. The assay must be made on a culture filtrate (**Flow diagram 26**), the protocol of the assay being similar to that of other RPLA assays (*cf Staph. aureus*, page 263).

Flow diagram 26. Production of *B. cereus* diarrhoeal toxin (Kramer *et al.*, 1982).

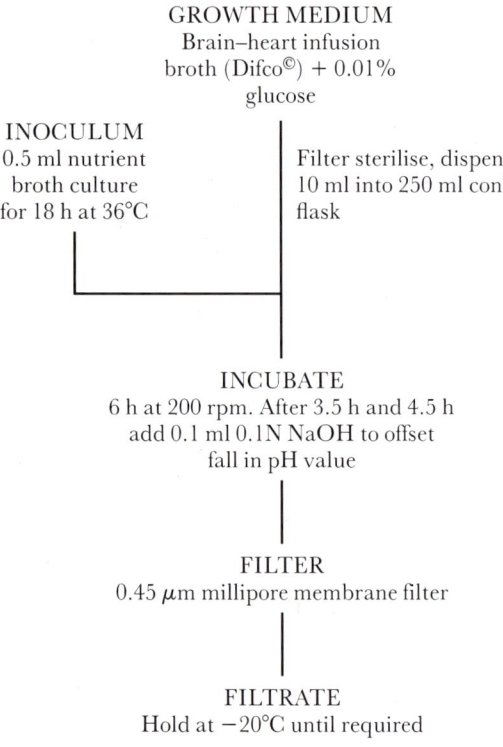

GROWTH MEDIUM
Brain–heart infusion broth (Difco©) + 0.01% glucose

INOCULUM
0.5 ml nutrient broth culture for 18 h at 36°C

Filter sterilise, dispense 10 ml into 250 ml conical flask

INCUBATE
6 h at 200 rpm. After 3.5 h and 4.5 h add 0.1 ml 0.1N NaOH to offset fall in pH value

FILTER
0.45 μm millipore membrane filter

FILTRATE
Hold at −20°C until required

Biological tests for diarrhoeal toxin remain in use. The ligated rabbit ileal loop test (**Figure 249**), which is based on that devised by De and Chatterjee (1953) for cholera toxin, is the definitive test. However, it is time consuming and for screening purposes the vascular permeability reaction test (VPR) is often preferred (Kramer *et al.*, 1982). Correlation between the two tests is usually good (Turnbull *et al.*, 1979b), but under some circumstances false-positive reactions may occur in the VPR due to cellular metabolites other than diarrhoeal toxin (Glatz *et al.*, 1974).

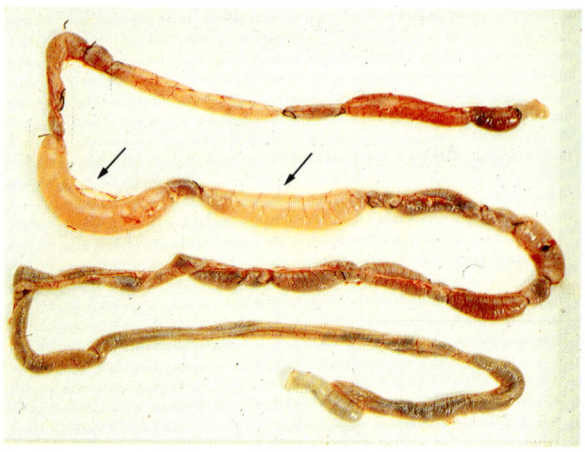

249. Ligated rabbit ileal loop test for detection of *B. cereus* diarrhoeal enterotoxin. Loops 1 to 9 show various degrees of response to different culture filtrates of *B. cereus*, loop 10 is a negative control. Loops 4 and 5 (arrowed) are considered to show a strong positive reaction. (Reproduced by courtesy of Dr J.M. Kramer, Central Public Health Laboratory, Colindale.)

Emetic toxin

The established method for the detection of *B. cereus* involves rhesus monkey feeding tests using rice culture slurries (Melling *et al.*, 1976). An alternative tissue culture test based on vacuole production in HEp-2 cells (**Figure 250**) has been proposed (Hughes *et al.*, 1988) but requires further evaluation.

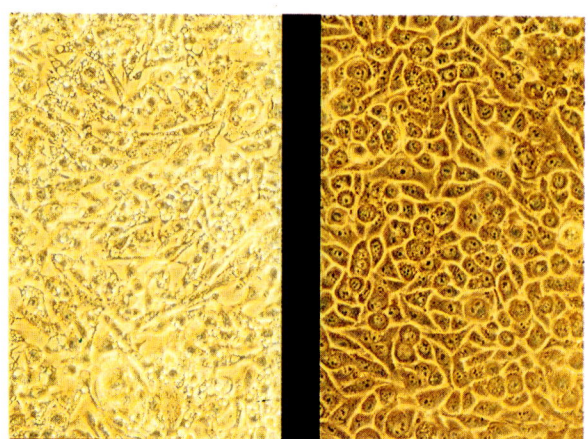

250. The response of cultured HEp-2 monolayers to a rice culture filtrate of an emetic strain of *B. cereus*. Vacuoles are seen in the cells on the left which have been exposed to emetic toxin; cells on the right are a negative control. (Reproduced with permission from Hughes, S., Bartholomew, B., Hardy, J.C. and Kramer, J.M. *FEMS Microbiology Letters* **52**, 7–12, © 1988 Elsevier Science Publishers BV [Biomedical Division].)

14. *Clostridium botulinum*

Introduction

The genus *Clostridium* comprises Gram-positive, catalase-negative, endospore-forming, rod-shaped bacteria which are obligate anaerobes (**Figures 251–253**).

The genus currently contains more than 80 species that are diverse in terms of metabolic activity, nutritional requirements and DNA relatedness. For these reasons, the genus as currently constituted may 'eventually' be split into at least two genera (Cato *et al.*, 1986).

Species of *Clostridium* are responsible for a number of diseases of man and animals, including food poisoning, tetanus and gas gangrene, wound and soft tissue infections, abscesses and pseudomembranous colitis.

Two species, *Cl. botulinum* and *Cl. perfringens* (type A) are primarily involved in food poisoning, although the production of botulinum toxin and consequent human disease by *Cl. baratii* and *Cl. butyricum* has been recorded (see below).

Although botulism is, fortunately, rare, the disease is severe with a high mortality. There are currently considered to be four categories (Pierson and Reddy, 1988):

1 Classical foodborne botulism due to ingestion of preformed toxin in food.
2 Wound botulism due to *in vivo* toxin production in an infected wound. This is the least common form of botulism.
3 Infant botulism due to the elaboration of toxin in the intestinal tract of infants following colonisation by *Cl. botulinum* (Pickett *et al.*, 1976). Over 600 cases have been reported since the initial report in 1976, and infant botulism may be one of the causes of infant sudden death syndrome (cot death; Thompson *et al.*, 1980). Infant botulism has also been caused by toxin producing strains of *Cl. baratii* (Hall *et al.*, 1985; Suen *et al.*, 1988a) and *Cl. butyricum* (Aureli *et al.*, 1986; McCroskey *et al.*, 1986; Suen *et al.*, 1988a).
4 Undetermined botulism is a classification for cases of botulism involving people older than one year in which no food or wound source is known. There is evidence that colonisation of the intestine in a manner analogous to that in infant botulism is involved (Morris and Hatheway, 1980).

The overwhelming majority of cases of botulism are of the classical foodborne type.

251   **252**   **253**

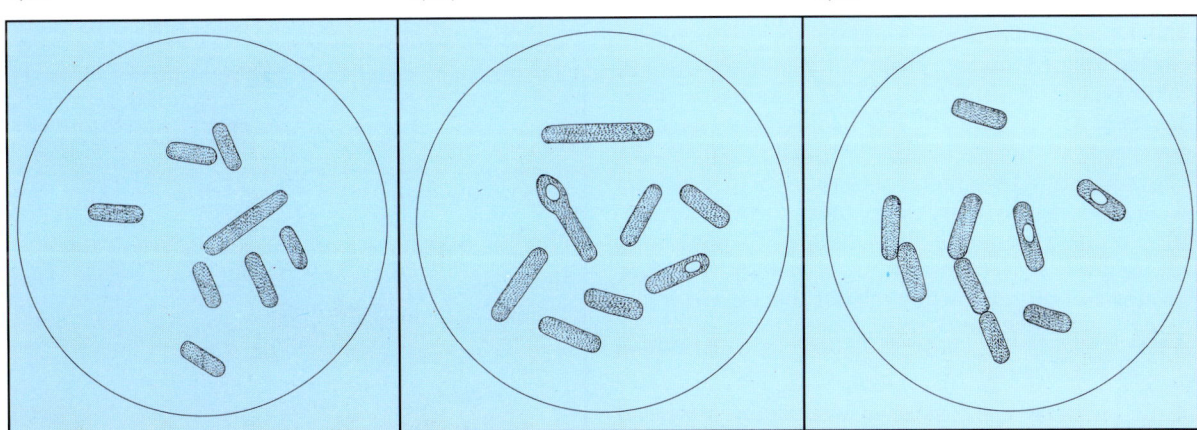

251–253. Morphology of *Cl. botulinum.* Although morphological groups are not used in the identification of clostridia (*cf. Bacillus*, page 270), the position and shape of the endospore, together with the size and shape of the vegetative cell, can be a useful guide to the experienced microbiologist. Non-sporulating cells of *Cl. botulinum* (**251**) are straight, or slightly curved, rods. Spores are oval, subterminal and usually, but not invariably, swell the sporangium (**252**). With type E and non-proteolytic strains of types B and F, swollen sporangia may not be present (**253**).

Isolation of *Cl. botulinum*

The isolation of *Cl. botulinum* from foods is generally considered to be of less significance than the detection of the toxin (Hobbs *et al.*, 1982), although methods for the isolation of the organism have been devised (Hobbs *et al.*, 1982; Kautter and Lynt, 1984; Foegeding, 1986), using a two-stage procedure. However, organisms identified with *Cl. botulinum* are biochemically diverse, and it should be appreciated that no single method will be equally successful in isolating all strains from all foods. Those involved in the isolation of *Cl. botulinum* should therefore have available a set of methods to be applied as experience and knowledge dictates.

Enrichment

Enrichment is almost invariably required, since *Cl. botulinum* is likely to be overgrown by other obligate or facultative anaerobes if direct plating is used. A two-fold approach is made, involving, where possible, self-enrichment in conjunction with conventional enrichment in liquid media. Non-selective enrichment media, such as Robertson's cooked meat broth or the more specialised botulinum enrichment medium, are most commonly used. Advantages have been claimed for the selective enrichment medium, sulphite-polymyxin broth (Mossel and DeWaart, 1968), but these have not necessarily been confirmed by other workers (Hobbs *et al.*, 1982).

Plating media

Enrichment broth and material (or swabs) from self-enrichment is streaked onto either non-selective, or selective, plating media. At this stage both enrichments are also tested directly for toxins by feeding mice or guinea pigs (see pages 309–310).

The use of non-selective plating media is most common practice, two media, horse blood agar and egg yolk agar being used in parallel. Selective media are considered to be over-selective with respect to non-proteolytic strains and are not widely used.

The isolation of Cl. botulinum *is discussed in greater detail, with respect to choice of media and laboratory practice, in* **Laboratory Methods**, *pages 307–308.*

Taxonomy of *Cl. botulinum*

Cl. botulinum consists of metabolically diverse bacteria which have, as a common property, the production of closely related toxins. Eight antigenically distinct toxin types – A, B, C_1, C_2, D, E, F and G – of *Cl. botulinum* are recognised, but distinctions based on toxin type are not necessarily reflected in phenotypic characteristics. Four metabolic groups of *Cl. botulinum* were described by Smith and Hobbs (1974), the validity of this grouping being subsequently confirmed on the basis of ribosomal RNA homology. The four groups and phenotypically related species are:

1 *Cl. botulinum* type A, proteolytic strains of types B and F: *Cl. sporogenes*.

2 *Cl. botulinum* type E, saccharolytic strains of types B and F.

3 *Cl. botulinum* types C and D: *Cl. novoyi* type A.

4 *Cl. botulinum* type G: *Cl. subterminale*.

From a purely taxonomic viewpoint, it is recognised that the situation concerning *Cl. botulinum* is unsatisfactory, and that in most circumstances, the different metabolic groups of *Cl. botulinum* would be assigned to different species. The present system is, however, likely to remain because of its utility to

Table 106. Differentiation of metabolic groups of *Clostridium botulinum*.

	Group			
	1	*2*	*3*	*4*
Gelatin hydrolysis	+	+	+	+
Motility	+/−	+	+/−	+
Indole	−	−	−/+	−
Lipase	+	+	+	−
Aesculin hydrolysis	+	−	−	−
Starch hydrolysis	−	+/−	−	−
Nitrate reduction	−	−	−	−
Acid from				
Amygdalin	−	−/+	−	−
Fructose	−/+	+	−/+	−
Inositol	−	−/+	+/−	−
Maltose	−/+	+	+/−	−
Melezitose	−	+/−	−	−
Sucrose	−	+	−	−
Trehalose	−	+	−	−

medical microbiologists and to avoid the possible confusion associated with a major change in taxonomy. Despite this a new, genetically homogeneous species, *Cl. argentinese*, has been proposed (Suen *et al.*, 1988b) which would include all strains of *Cl. botulinum* type G and some non-toxigenic strains previously described as *Cl. subterminale* or *Cl. hortiforme*. *Cl. botulinum* type G is, however, of only limited importance as a human pathogen.

Differentiating features of each metabolic group of *Cl. botulinum* and of closely related species, are listed in **Table 106**. However, it must be appreciated that recognition of *Cl. botulinum* lies ultimately with demonstration of toxin production and that biochemical tests are of only limited value. For example, it is not possible to reliably differentiate between *Cl. sporogenes* and *Cl. botulinum* type A or proteolytic strains of types B and F using either conventional or miniaturised methodology, or by gas chromatographic analysis of end-products (Holdemann *et al.*, 1977). The separation of phenotypically identical strains is, however, possible using electrophoretic protein patterns (Bom *et al.*, 1986).

The identification of Cl. botulinum *is discussed in greater detail, with respect to methodology, in* **Laboratory Methods**, page 311.

Typing of *Cl. botulinum*

Typing of *Cl. botulinum* is based on the serological differentiation of toxins. In the case of types B and F, further subdivision may be made on the basis of proteolytic activity.

Typing on this basis is satisfactory for most purposes and this, together with the difficulties of working with the organism, means that there has been little interest in developing more discriminatory typing systems.

However, the determination of electrophoretic protein patterns of vegetative cells does permits the recognition of individual strains of *Cl. botulinum* (Bom *et al.*, 1986). This is obviously extremely valuable in tracing routes of contamination, but the techniques used are not suitable for application on a routine basis.

Symptoms of botulism

Classical foodborne botulism

Symptoms usually appear within 12 to 72 hours, although periods of up to 6 days have been reported (Lecour *et al.*, 1988). A more serious condition is generally presented when symptoms appear in less than 24 hours (Schaffner, 1981). In some cases, symptoms develop gradually but progressively (Critchley *et al.*, 1989), and this may contribute to difficulties in diagnosis.

Symptoms reported include generalised muscular weakness, visual disturbances and difficulty with speech, headache, dizziness and extreme dry mouth, throat and skin. Intestinal manifestations are usually nausea, vomiting and severe constipation which may be preceded by transitory diarrhoea. Breathing difficulty leading to respiratory failure is a serious complication and may develop insidiously (Schmidt-Nowara *et al.*, 1983). Other complications include palpebral ptosis, dysphonia, urinary retention and postanal hypotension. Infectious complications with associated fever may also occur and in a review of 13 outbreaks, Lecour *et al.*, (1988) recorded infectious stomatitis, and pharyingitis, acute parotitis, oral candidiasis, pneumonia and urinary tract infection.

In many cases, the symptoms of foodborne botulism are severe and life threatening. Hospitalisation is commonly required and, despite modern therapy, the mortality rate is *ca* 10%. The severity of the disease may vary according to the host resistance, type and amount of toxin ingested and type of food (Pierson and Reddy, 1988). A study of 50 cases of type B botulism showed that in all the nature of the symptoms was identical, but the severity varied to the extent that 20 patients remained ambulant and were treated on an out-patient basis (Lecour *et al.*, 1988). Mild and non-specific symptoms with minimal neurological disturbances were also noted in a study of botulism in Alaskan native peoples (Wainwright *et al.*, 1988), the authors considering that 'mild botulism' requires further study.

The duration of classic botulism is usually extended and symptoms of life threatening severity are often present for 10 days or more. In non-fatal cases, this is followed by a slow recovery. Ultimate recovery is usually complete, but symptoms including slight fatigue, dry mouth, disturbed vision and constipation may persist for several months.

In areas where botulism is a relatively common problem, or where food history reveals consumption of a previously incriminated foodstuff, differential diagnosis may present little problem and is assisted, where botulism is suspected, by electromyography and the detection of toxin in faeces. In countries where botulism is uncommon, the symptoms may easily be confused with those of other diseases. This may be illustrated by reference to misdiagnoses reported in Portugal (**Table 107**) where the number of cases has increased, but physicians remain unfamiliar with the disease (Lecour, 1988). The situation in other countries is similar. In an outbreak in north-west England, the difficulties of diagnosis were complicated by the fact that patients presented singly over a relatively wide area and by atypical symptoms. These indicated segmental myelination and included sore throat, drowsiness and fever. There was widespread publicity over the epidemic, which occurred at a time when food poisoning was a topical political issue, and a number of patients themselves raised the possibility of botulism. Despite this, hospitalisation was delayed by up to 2 weeks by the refusal of practitioners to diagnose the disease (Critchley *et al.*, 1989).

In addition to clinical misdiagnosis, botulism in its early stages may be confused with drunkeness, both by the physician and the patient.

Table 107. Confusion of the symptoms of botulism with those of other syndromes.

Symptom	Misdiagnosis
Diarrhoea and vomiting	Acute gastroenteritis. Other (non-specific) food poisoning
Constipation, abdominal distention, reduced bowel sounds	Acute abdomen
Dry mouth, sore throat, redness of oral mucosa	Acute pharyngitis
Visual defects, other cranial nerve impairment and oropharyngeal signs	Diphtheric paralysis

Note: Based on data for cases in Portugal (Lecour *et al.*, 1988).

Infant botulism

Initial symptoms of infant botulism are constipation followed by poor feeding, weakness, lethargy and pharyngeal pooling. Loss of head control is a striking symptom and the cry is weak, or altered. Symptoms may be prolonged and persist for 6 weeks or longer. Occasional stool isolates of *Cl. botulinum* may be made from asymptomatic infants and a wide spectrum of severity may exist (Thompson *et al.*, 1980).

Infant botulism has been identified as one of the causes of infant sudden death syndrome (cot or crib death). *Cl. botulinum* is probably responsible for only a small percentage of cot deaths, the underlying causes of which are not fully understood.

Electrophysiological examination of children suffering infant botulism shows changes similar to those of classical botulism. Clinical suspicion must, however, be raised and thus initial diagnosis is of prime importance. Differential diagnosis is again likely to be most difficult in areas where the syndrome is rare and physicians have no prior experience of its symptoms. Under these circumstances, infant botulism may be confused with septicaemia, Guillain-Barré syndrome and paralytic poliomyelitis (Hinde *et al.*, 1987), although when symptoms are mild the diagnosis may be confined to viral syndrome or failure to thrive (Anon, 1987a).

Undetermined botulism

Cases of undetermined botulism have similar symptoms to those of classic foodborne botulism. A particular feature is the presence of the organism or toxin in clinical specimens for an extended period, and consequently a prolonged duration of symptoms. In one case, a woman survived for 240 days before succumbing to complications (Chia *et al.*, 1986).

Differential diagnosis poses the same problems as diagnosis of classic botulism. Further difficulties are due to the long period between infection and the lack of a demonstrable link with either a toxin-containing food or a wound.

Persons at particular risk of botulism

Classic foodborne botulism

Susceptibility

In addition to the general higher risk to those at extremes of age, or suffering an underlying illness, there is considerable person-to-person variation in susceptibility. However, evidence for this is circumstantial and based on the wide spectrum of severity of symptoms among individuals who have ingested similar amounts of toxin. The factors underlying differences in susceptibility are not understood (Lamanna and Carr, 1967; Lecour *et al.*, 1988).

Exposure

Certain cultural and ethnic groups are at enhanced risk of botulism due to a high level of exposure resulting from dietary habits and customs. The most striking example is that of the native Inuit and Indian population of Alaska, northern British Columbia and the Canadian Artic, where an average of 8 cases and 1.5 deaths per year result from the consumption of traditional foods such as muktak (fermented whale blubber with skin and meat) and fermented salmon eggs. Dietary custom can also be a risk factor among discrete social or religious groups in modern urban society: ribyetz (kapchunka), a whole, uneviscerated salted whitefish which is a delicacy in Jewish cuisine was responsible for a number of cases of type E botulism in New York (Anon, 1987b).

In addition to individuals and defined groups at particular risk, the general risk to entire populations varies according to geographical differences in the incidence of *Cl. botulinum* and variations in the type of toxin produced. Thus the low incidence of botulism in the UK is attributed to a corresponding low incidence of the *Cl. botulinum* in the environment. Equally, botulism in Europe is usually due to type B strains and less severe than that in the western USA where type A strains are dominant. Local variations in toxigenicity may occur over smaller areas and possibly account for the low mortality associated with the disease in France.

Infant botulism

Cases of infant botulism usually occur in the early stages of weaning when non-milk foods are first introduced, and for this reason the syndrome is rare in children over 6 months of age. Breast-fed children are at a markedly higher risk than those fed artificially (Anon, 1986).

In most cases, it has not been possible to relate cases of infant botulism to the consumption of a particular food, but honey has been identified as a risk factor in cases investigated in the western USA (Arnon, 1986). For this reason it is common practice to recommend that honey should not be fed to children aged less than 2 years.

Cases of infant botulism show a clear geographical distribution, and while the disease has been reported in Australia, Canada, Europe and South America, it is most common in the USA. A disproportionate number of cases occur in the state of California and account for *ca* 50% of the USA total. It has been suggested that this is an artefact due to the better diagnostic services available, but possible risk factors are the high incidence of *Cl. botulinum* endospores in the Californian environment and the feeding of honey to infants – a practice that is more common in California than elsewhere.

Undetermined botulism

Risk factors for undetermined botulism, at least where colonisation of the gut is involved, have been identified as recent gastrointestinal surgery and antibiotic therapy.

Treatment of botulism

Classical botulism

The underlying principles of therapy in cases of classic botulism are the elimination of toxin from the gastrointestinal tract and intensive respiratory care. Treatment usually involves the administration of botulinal antitoxin together with supportive therapy and symptomatic treatment as required. In severe cases, nasogastric intubation and bladder catheterisation may be necessary, in addition to respiratory assistance. Anti-microbial therapy is not recommended except in cases of infective complications.

Although antitoxin is often administered as a matter of course, there are doubts concerning its efficacy (Merson *et al.*, 1974; Bricaire, 1982), the rapid and irreversible binding of toxin at the neuromuscular junction being a possible impediment to the antitoxin (Lecour *et al.*, 1988). A more serious objection is the occurrence of hypersensitive reactions to the equine antitoxin used. These occur in up to 21% of patients and may in themselves be life threatening (Merson *et al.*, 1974). It is therefore considered that while antitoxin treatment is likely to be justified with the severe type A botulism (Tacket *et al.*, 1984) a cautious approach to its use should be adopted where other types of botulism are involved, particularly if symptoms are relatively mild (Sebald and Jougland, 1977; Tacket *et al.*, 1984).

Infant botulism

Treatment of infant botulism is based largely on supportive care including artificial ventilation where required. Neither anti-microbial therapy (Sakaguchi, 1979) nor administration of antitoxin are recommended (Arnon and Chin, 1979).

Undetermined botulism

Administration of antitoxin has been used in the treatment of botulism in adults where circumstances suggest gut colonisation (McCroskey and Hatheway, 1988). The long duration of symptoms means that supportive therapy and treatment of complications is often necessary.

Mechanisms of pathogenicity

Symptoms of all types of botulism are due entirely to the action of the toxin. Human botulism is normally caused by types A, B and E. Human botulism due to type F is rare and types C and D appear to be involved only in animal botulism. Toxin type G was also considered not to be active against man, but has been implicated in five sudden unexplained deaths (Sonnabend *et al.*, 1981).

Nature of botulinum toxin

Botulinum toxin is the term applied to eight different biological substances (toxins A, B, C_1, C_2, D, E, F and G). Each is antigenically distinct but may share structural and pharmacological properties (Sugiyama, 1980).

Botulinum toxins are proteins and consist of a single polypeptide chain of MW *ca* 150 Mdaltons, which is nicked by proteases when released from the cell to produce a dichain molecule. In the case of proteolytic strains of *Cl. botulinum*, the protease is produced by the bacterium itself, but with non-proteolytic strains an exogenous protease such as trypsin is involved. Activation occurs at the same time as nicking, but the two events may be unrelated.

With the exception of toxin C_2, the dichain molecule is composed of a light chain polypeptide A, of MW *ca* 55 Mdaltons, and a heavy chain polypeptide B, of MW *ca* 100 Mdaltons. The two components are linked by a disulphide bond. Type C_2 toxin consists of two components, I of MW *ca* 55 Mdaltons and II of MW *ca* 105 Mdaltons. The two components are separate molecules and are not covalently linked (Simpson, 1982).

Synthesis of botulinum toxin

Toxin plays no essential role in the growth and metabolism of *Cl. botulinum*. Relatively little toxin is present in filtrates of logarithmic phase cultures, but toxin is released from the cells during autolysis. The titre of toxin can therefore continue to increase even when cell growth has ceased (Simpson, 1981).

The genetic basis of the control of toxin production appears to differ according to the type of toxin. Bacteriophages are related to toxigenicity in types C and D in a system which closely resembles the classic prophage conversion of *Corynebacterium diptheriae* (Glatz, 1986). When toxigenic strains are cured of their resident prophages, the progeny are phage sensitive and do not produce toxins C_1 or D. Production of C_2, however, continues and does not require the presence of a prophage. With types C_1 and D more than one phage may be involved.

The relationship of the prophage to the bacterium has been termed pseudolysogeny since the phage can be lost on sporulation (Eklund and Poysky, 1973). This may explain the occurrence of non-toxigenic strains in nature and non-toxigenic variants in laboratory cultures (Glatz, 1986).

Prophage conversion has not been demonstrated with types A, B, E or F, and strains of types B and F have remained toxigenic after curing of one or two prophages. Thus prophage conversion either does not occur or other prophages are present (Eklund and Poysky, 1973).

Plasmids have been demonstrated in *Cl. botulinum* and it is possible to relate the plasmid pattern to the type of toxin produced (Strom *et al.*, 1984). In most cases, no related phenotypes have been found, and the plasmids remain cryptic (Scott and Duncan, 1978; Strom *et al.*, 1984). The first evidence for plasmid mediation of toxin production was obtained for type G (Eklund *et al.*, 1988a). It is not known if plasmids are involved in the production of other types of toxin.

It is often thought that individual strains of *Cl. botulinum* can synthesise and release only one toxin. This would seem to be correct in the majority of cases, but a strain which produced both type A and F toxins has been isolated (Gimminez and Ciccarelli, 1970;

Sugiyama *et al.*, 1972). Type F was the dominant toxin accounting for 90 to 99% of the total.

Action of botulinum toxin

Most botulinum toxins are recognised as neuropharmacological substances. There is, evidence that C_2 is not a neurotoxin and the action of both C toxins differs from that of other types.

Botulinum toxin is resistant to the acidic conditions in the stomach, resistance in some cases being due to an association with a haemagglutinin of MW *ca* 500 daltons. Unusually for a large molecule, botulinum toxin is absorbed intact from the gastrointestinal tract and enters the bloodstream. It is transported by the bloodstream and binds via gangliosides to nerve cells, the site of binding having been identified as the presynaptic cholinergic nerve ending. At this site, the toxin acts by blocking the release of the neurotransmitter, acetylcholine, which under normal circumstances triggers contractions of the skeletal muscle. A model has been developed for the action of botulinum toxin at the cholinergic nerve terminal (Simpson, 1980; 1981), which involves three stages – binding, translocation and lysis (**Figure 254**), the latter stage culminating in the blockage of transmitter release.

Type C_1 differs in producing degeneration of neuronal processes and rounding of neural somas in primary neuron cultures from foetal mouse brain (Kurokawa *et al.*, 1987). Electron microscopy demonstrated degenerated mitochondria, membranous dense bodies and vesicles which resembled *in vivo* Wallerian degradation. The mechanism of C_2 toxin has not been fully elucidated. *In vivo* the toxin causes hypotension, lung haemorrhage and fluid collection around the lungs and trachea, which may be due to changes in membrane permeability and/or changes in cellular secretion (Simpson, 1982).

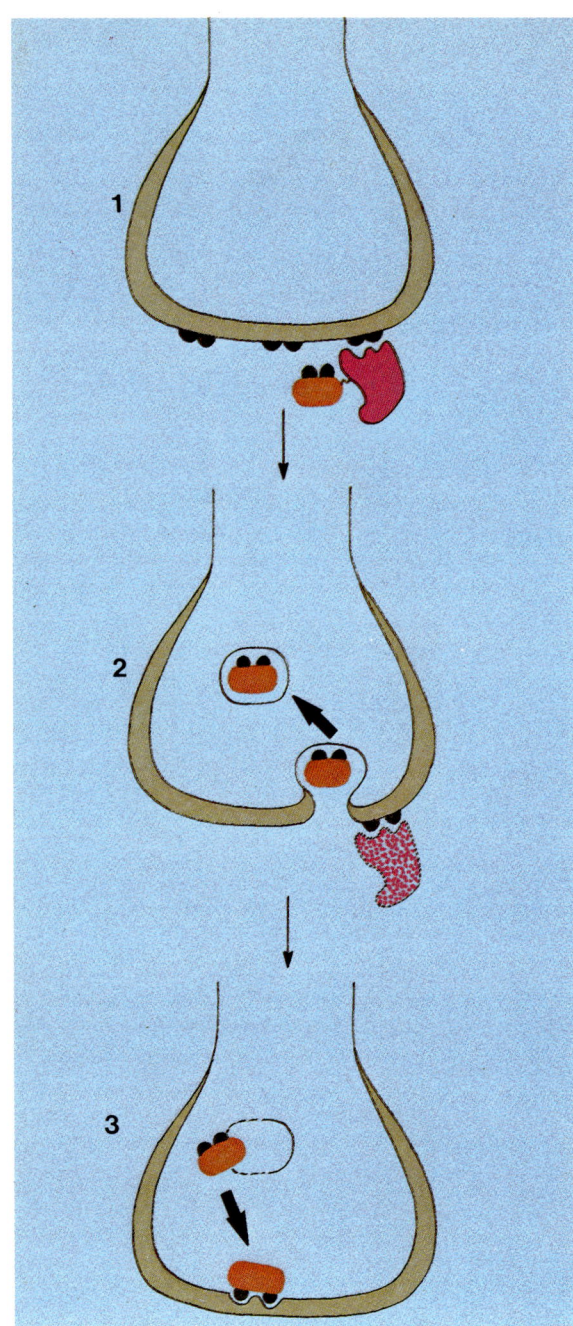

254. Three stage model for the interaction of botulinum toxin with the cholinergic nerve terminal.

1 Binding stage. During binding, the heavy chain of the toxin molecule binds to an unidentified receptor on the external surface of the nerve. This stage does not produce any obvious changes in nerve cell function.

2 Translocation stage. Like the binding stage, translocation produces no obvious change in nerve cell function and is non-toxic. Either the toxin molecule or, more probably, the light chain portion of it is translocated into the nerve interior by an active process which may involve pinocytosis.

3 Lytic stage. The lytic stage may, in reality, comprise a sequence of events which culminate in the blockade of transmitter release. This stage is slower than either binding or translocation and is nerve-activity dependent. The toxin may act at, or very close to, the transmitter release site. (Reproduced with permission from Simpson, L.L. *Journal of Pharmacology and Experimental Therapeutics* **212**: 16–21. © 1980 American Society for Pharmacology and Experimental Therapeutics.)

Toxin production by species other than *Cl. botulinum*

Production of botulinum toxin by *Cl. baratii* and *Cl. butyricum* has been demonstrated and it is possible that production occurs in other species. Production by other species has not been studied in detail, but it appears that the toxin is identical both structurally and pharmacologically with its counterpart produced by *Cl. botulinum*.

Colonisation of the gastrointestinal tract by *Cl. botulinum*

In contrast to classical foodborne botulism, infant botulism and some types of undetermined botulism in adults require colonisation of the intestinal tract. *Cl. botulinum* is not a normal inhabitant of the tract and thus lacks the mechanisms usually associated with colonisation. The organism usually becomes established only when the normal adult gut flora is not fully established, as in infant botulism or when it is disturbed by surgery or anti-microbial therapy in adult undetermined botulism. It is assumed that under these conditions, *Cl. botulinum* is able to become established in ecological niches usually occupied by members of the normal gut flora.

Human carriage of *Cl. botulinum*

Endospores of *Cl. botulinum* may occasionally be isolated from the faeces of healthy humans. This reflects dietary intake and passage is for a short period only. There are no implications in the transmission of botulism.

Cl. botulinum in the environment

Endospores of *Cl. botulinum* are widely distributed in the environment and may be isolated from cultivated and forest soils, shore and bottom deposits of rivers, lakes and coastal waters. In hot weather, the lower layers of lakes become anaerobic and, where the content of organic material is high, significant growth of *Cl. botulinum* may occur. The organism is also present in the digestive tract of animals and birds and the gills and viscera of shellfish. Corpses of wild animals and birds commonly contain large numbers of *Cl. botulinum* and may be a reservoir of the organism. In warm countries, virtually 100% of corpses are thought to be toxic and while occasional outbreaks of botulism occur involving animals which have eaten corpses, a much higher incidence might be expected among habitual carrion-eating animals.

Of the types usually associated with human disease, types A, B and F are usually of terrestial origin, while type E is most commonly associated with aquatic environments. In soils, types A and B have a contrasting pattern of geographical distribution. Type A is commonly found in the western USA, while type B is predominant in the eastern USA and Europe. However, type B strains from the USA are proteolytic, in contrast to those from Europe which are non-proteolytic. Type F is rare in the environment in comparison with types A, B and E, while until recently isolations of type G were restricted to grassland soils in Argentina.

Foods commonly associated with *Cl. botulinum*

Endospores of *Cl. botulinum* are present in small numbers in many foods. This is of no direct significance unless the food is to be consumed by infants or susceptible adults. Under suitable conditions, however, outgrowth and germination of endospores occurs followed by vegetative growth and toxin synthesis. These conditions are well defined (see below), and botulism has therefore tended to be associated with a limited range of food types of which canned foods is prominent. As the number of cases due to canned foods has fallen, other food types which less obviously meet the conditions for toxin production have become involved (MacDonald *et al.*, 1986). In consequence, it appears that no food can be automatically discounted as a source of botulinum toxin (Archer and Young, 1988).

Canned (fully heat sterilised) foods

Canned foods were for many years a major cause of botulism in the USA and elsewhere. In many cases, domestically canned, or bottled food, was involved, such products accounting for 72% of botulism outbreaks in the USA between 1899 and 1976 (Pierson and Reddy, 1988). As noted above, the number of botulism cases due to canned foods has followed a downward trend over a number of years. This may be attributed to a large extent to a decline in home canning, but also to improvements in commercial cannery practice. The occasional outbreaks of botulism which still occur due to commercially canned foods serve as drastic reminders that *Cl. botulinum* continues to present a major hazard, and that the highest standard of cannery operation is required for its control.

In canned foods, botulism results either from underheating of the cans and the subsequent survival of endospores of *Cl. botulinum*, or from entry of the organism into the cans after heating. Highly heat-resistant proteolytic strains of types A and B are normally involved (Sperber, 1982), where underprocessing is the cause. Outbreaks due to domestically canned foods invariably involve underprocessing and low acid home canned foods are considered inherently unsafe.

The majority of outbreaks of botulism involving commercially canned foods are also due to underprocessing. The publicity given to two outbreaks of botulism resulting from canned salmon, which was attributed to can leakage after processing (Ball *et al.*, 1979; Anon, 1982) has given rise to the misconception that such contamination is the major problem. This is not supported by the facts, and a survey of outbreaks of human botulism since 1940 showed that only four outbreaks involved can leakage (Anon, 1984).

A wide range of canned foods have been involved in outbreaks of botulism. Low acid foods are almost invariably involved, high acid foods only being implicated where the pH value has been raised due to other factors. The risk of botulism from proteolytic strains of *Cl. botulinum* is often considered to be greater in vegetable products than meat or fish due to the putrid odours associated with proteolysis serving to prevent consumption of the latter products if contaminated. Although this may be true to some extent, the likelihood of proteolytic spoilage cannot be a guarantee of safety: *'Thousands of deaths from botulism in the past . . . testify to the futility of relying on the consumer's sensory perception of atypical odour or other evidences of spoilage'* (Christiansen and Foster, 1965).

Meat products

A number of surveys have shown that the incidence of *Cl. botulinum* in meat and meat products to be low but variable (Greenberg *et al.*, 1966; Abrahamson and Riemann, 1971; Anon, 1971; Roberts and Smart, 1976; Hauschild and Hilsheimer, 1980). Only one isolation, a type C strain from chicken, was made from a total of 2,358 samples of raw pork, beef and chicken in the USA (Greenberg *et al.*, 1966), while in the UK a survey of dressed poultry showed a contamination rate of 0.32%. All strains isolated were type C, which has caused outbreaks among broiler chickens (Roberts *et al.*, 1973), but which has not been associated with cases in man (Roberts, 1982).

An exception among surveys was that of Roberts and Smart (1976), in which the incidence in one lot of bacon was 73%. In contrast to other surveys where direct plating was used, Roberts and Smart employed enrichment procedures. This may suggest that in many surveys the incidence of *Cl. botulinum* has been underestimated, although factors other than methodology may be involved.

Cl. botulinum, or its toxin, has been isolated from a number of other meat products including luncheon meat (Taclindo *et al.*, 1967), cooked ham (Abrahamson and Riemann, 1971), vacuum-packed frankfurters (Insalata *et al.*, 1969) and smoked turkey (Abrahamson and Riemann, 1971).

Botulism has only been associated with a small number of meat products. Cured meats are a potential risk product, but the record of commercial products is very good. However, botulism due to cured meats is

relatively common in countries where, due to custom or economic circumstance, home curing is widely practised. For example, in Portugal, 12 out of 13 recent outbreaks were due to home-cured meats, while in Poland, home curing accounts for the high prevailing incidence of botulism.

There has been some concern in recent years that changes in commercial curing practice may compromise the safety of the product. Earlier concern that the introduction of vacuum packaging of raw bacon would produce conditions favourable for the growth of *Cl. botulinum* has proved unfounded. However, reduction in the level of nitrite, due to fears of nitrosamine formation, and in the level of sodium chloride (or its partial substitution with other salts), due to dietary fashion, means that the capacity to inhibit *Cl. botulinum* in some cured meats is limited (see pages 303, 434, 442).

Both proteolytic and non-proteolytic strains of *Cl. botulinum* are able to withstand the cooking temperatures used in the manufacture of cooked (pasteurised) meat products, and outgrowth in the finished product is a potential hazard. In commercial practice, problems have been very few, an exception is poultry potpies which have been implicated in family outbreaks of botulism (Anon, 1983).

The *sous-vide* method of food preparation (see page 444), in which pre-prepared meals are vacuum-packed, pasteurised and stored under refrigeration until use, is considered to present a potential hazard. Until now, the method has been largely restricted to catering but an extension to the retail market has been proposed.

Milk products

Botulism in dairy products has been restricted to cheese and cheese spread. An outbreak of type B botulism involving Brie cheese was attributed to storage of the cheese at a high temperature for an excessive period. Approximately 45 people in Switzerland and 38 in France were affected (Anon, 1980).

Cheese spread was implicated in two outbreaks of botulism in the USA during 1951 (Kosikowski, 1977), and an outbreak in Argentina in 1974 involving commercially manufactured cheese spread with onion led to at least 6 cases and 3 deaths. Type A botulism was involved, contributory factors being temperature abuse of a relatively high pH value and a_w level food.

During 1989, a number of cases of botulism in north-west England were attributed to hazelnut-flavoured yoghurt. The toxin was derived from incorrectly processed hazelnut puree used as flavouring, and was not associated with the yoghurt *per se*. One of several noteworthy features of this outbreak (see also page 292) was that the lack of visible spoilage was taken to indicate the safety of individual cans, despite the fact that on an overall basis the number of 'blown' cans was high.

Fish and marine mammals

The incidence of *Cl. botulinum* in fish varies according to a number of factors including the type of fish, its habitat and its feeding habits. Particular attention has been paid to farmed fish where a combination of circumstances can lead to a high incidence of the organism. In a study of trout from Danish farms, for example, contamination rates varied from 5 to 100% in winter, to 85 to 100% in late summer (Huss *et al.*, 1974a). The intestinal tract and contents were the main reservoir, and the incidence was markedly lowered by gutting (Huss *et al.*, 1974b). The level of contamination may be further reduced by other management measures including starvation.

Although type E strains are most commonly associated with fish, types A and B may also be present and type F has also been isolated from salmon (Craig and Pilcher, 1966).

Fish has been the cause of botulism on a number of occasions including the highly publicised cases due to canned salmon (see above). With the exception of canned fish, all cases have involved fermented, smoked or salted products.

In general terms, the safety record of smoked fish is good with respect to botulism. In Europe, the exception is the traditional Norwegian delicacy rakefisk (fermented trout), which has been implicated in both type B (Skulberg, 1958) and type E (Norbo and Valland, 1967; Yndested, 1970) botulism. The product is usually home prepared and problems arise from a lack of care and knowledge (Erichsen, 1983).

Botulism due to fermented fish and marine mammals is a continuing problem among native peoples of northern Canada and Alaska (see also page 471). The foods are traditionally prepared and the hazard is well known. However, the incidence of botulism is increasing (Wainright *et al.*, 1988). This has been attributed to the introduction of novel, but unsophisticated, technology in the use of plastic containers and bags as fermentation vessels (Eisenberg and Bender, 1976), and the speeding up of the process by placing the ferment adjacent to a stove.

An ungutted, salted whitefish product, ribyetz or kapchunka, has been responsible for a number of outbreaks of type E botulism in recent years. The largest of these have been in New York and manufacture of the product is now banned in the USA (see also page 471; Anon, 1987c).

Smoked whitefish from the North American Great Lakes have also been responsible for botulism. The outbreak followed technological changes which markedly lowered the stability of the product with respect to microbial growth. At the same time, vacuum packaging was introduced which prevented the normal pattern of spoilage developing (Kautter, 1964).

Vegetables and other plant foods

In recent years, a number of vegetable products have been involved in outbreaks of botulism. Although these may appear to be unrelated incidents, there are common features from which lessons can be learnt and conclusions can be drawn.

In each case, the food had been treated in such a manner that the spoilage microflora was destroyed or inhibited, and conditions created that permitted the growth of *Cl. botulinum*. Contributory factors included changes in working practice and the creation of a new product.

Fried lotus rhizomes in mustard, a traditional Japanese delicacy, were implicated in a nation-wide outbreak of type A botulism which affected 36 people with 11 deaths (Otofuji *et al.*, 1987). The product was vacuum packed, which may have been a factor in promoting the growth of the organism, but which also inhibited the usual spoilage microflora. Although cases were distributed across a number of prefectures, none occurred in Kumamoto prefecture, where the product is manufactured. This may be attributed to differences in culinary habits, lotus rhizomes being reheated in Kumamoto but eaten cold elsewhere in Japan.

Baked potatoes that had been wrapped in foil and held for extended periods of time were implicated as the source of toxin in two outbreaks of botulism which followed consumption of potato salad and were the possible causative food in a third (Seals, *et al.*, 1981; Sugiyama, 1982). The foil wrapping was a significant factor in creating anaerobic conditions. Baked potatoes have become moderately popular as a fast-food in the UK and are sometimes retailed from warm cabinets in public houses as well as outlets unused to handling cooked foods, such as garages. The operating temperature of the cabinets may be no more than 30 to 35°C and baked potatoes may remain some considerable time at this temperature. The practice of foil wrapping to prevent dehydration appears to be coming more widespread and unless adequate controls are introduced, products of this nature must be seen as a potential hazard with respect to botulism.

Sauted onions were responsible for 28 cases of type A botulism (MacDonald *et al.*, 1985). The strain implicated was highly toxigenic (Solomon and Kautter, 1986), and it is thought that the coating of margarine provided the anaerobic conditions necessary for its growth and toxin production. This outbreak serves to illustrate the fine balance that may exist in foods between aerobic and anaerobic conditions.

Chopped garlic in soya oil has been responsible for a number of cases of botulism in recent years. The first of these outbreaks a restaurant outbreak in British Columbia, in which 37 people suffered type B botulism, was recognised retrospectively (St Louis *et al.*, 1988). Failure to store the product under refrigeration was originally thought to be a significant contributory factor, but subsequent investigation showed that non-proteolytic strains of *Cl. botulinum* was able to grow on the garlic at refrigeration temperatures. Following further cases in the USA, the Food and Drug Administration (USFDA) considered seizure of garlic in oil products which relied totally on refrigeration stating that such products; 'represent a special, immediate hazard' (Anon, 1989). Acidification to below pH 4.6 was recommended to ensure safety.

Before the outbreaks, both onion and garlic were considered to be unable to support the growth of *Cl. botulinum* (Solomon and Kautter, 1988). With hindsight there appears to be no reason for this assumption. Other plant products are known to support the growth of the organism in cans, and while both onions and garlic have anti-microbial properties, these are variable and selective in effect, possibly creating a selective advantage for those organisms which are not inhibited.

Coleslaw was the vector in an outbreak of type A botulism (Conner *et al.*, 1990). The source of the toxin was believed to be cabbage which had been packed in a modified atmosphere to extend storage life. Inoculation experiments showed that under these conditions, cabbage would support growth and toxigenesis by *Cl. botulinum* type A, but not type B.

The ever-present consumer demand for more convenience and the food industry demand for higher profit margins, may well lead to other developments of doubtful safety. Vacuum-packed cooked, or part cooked, potatoes or other vegetables, are considered to carry a high level of potential risk, for example.

Mushrooms may conveniently be discussed with plant foods. On some occasions at least mushrooms may be heavily contaminated with *Cl. botulinum*. Cases of botulism have resulted from underprocessing of home canned and commercially canned mushrooms (Lynt *et al.*, 1975) and from commercially manufactured marinated mushrooms (Todd *et al.*, 1974). Concern has also been expressed that the practice of packing mushrooms in polyvinylchloride film as a means of extending shelf life may, by depleting the concentration of oxygen and increasing the concentration of carbon dioxide, create conditions suitable for the growth of *Cl. botulinum* (Sugiyama and Yang, 1975). Puncturing the film was considered to be a means of reducing the possibility of growth (Sugiyama and Rutledge, 1978) and the USFDA has instructed packers as to the advisability of puncturing the packaging film.

The botulism risk in mushrooms has been reappraised by Notermans *et al.* (1989), who found a much lower incidence of *Cl. botulinum* than had been reported previously (Hauschild *et al.*, 1978). This was attributed to good hygiene during growing. Notermans *et al.* also noted that botulinum toxin was produced only when a large spore inoculum was introduced and that no toxin was detected in 1,078 packs of naturally contaminated mushrooms examined at the end of shelf life (Kautter *et al.*, 1978).

Infant foods

Concern over infant botulism has led to the examination of a number of infant formulations and other foods commonly used for infant feeding. Many of these foods are also consumed by older children and adults, but in these cases the presence of endospores of *Cl. botulinum* is usually of no significance.

Of the foods examined only honey is considered to be significant as a source of *Cl. botulinum*, and endospores have been isolated in a number of surveys (Midura *et al.*, 1979; Huhtanen *et al.*, 1981; Hauschild *et al.*, 1988). In the latter case, endospores were isolated from a sample of honey that had been implicated in a case of infant botulism (Anon, 1985). In most samples containing *Cl. botulinum*, only a single type is present which suggests build-up, possibly from a single endospore, rather than haphazard contamination. It has been suggested that build-up may occur in the bodies of dead bee larvae in the hive (Huhtanen *et al.*, 1981).

Endospores of *Cl. botulinum* have also been isolated from a single sample of rice cereal (Hauschild *et al.*, 1988), and from corn syrup (Kautter *et al.*, 1982). The risk from these products is not considered to be significant as the number of endospores is very low and there is no likelihood of growth during manufacture.

Factors affecting the growth and survival of *Cl. botulinum* in foods and the stability of its toxin

Temperature

Cl. botulinum falls into two groups with respect to its temperature relationships: proteolytic and non-proteolytic. Optimum growth and toxin production by proteolytic strains is *ca* 35°C, with very slow growth and toxigenesis continuing at both 12.5 and 50°C (Pierson and Reddy, 1988). In contrast, the optimum for growth and toxin production by non-proteolytic strains is ca 30°C, and the lower and upper limits 3.3 and 45°C. Refrigeration at temperatures above 3.3°C thus cannot be considered to be an absolute safeguard against any strain of *Cl. botulinum*, except those of type A.

Differences in growth:temperature relationships between proteolytic and non-proteolytic strains are reflected in differences in the heat resistance of endospores. The heat resistance of proteolytic strains is considerably greater, D_{121} values being in the range less than 0.1 to 0.25 minutes compared with D_{80} values of 0.6 to 3.3 minutes for non-proteolytic strains (Hobbs *et al.*, 1982). Thus inactivation of endospores of proteolytic strains requires a full sterilisation process, whereas a heavy pasteurisation process will inactivate endospores of non-proteolytic strains.

Processes which combine heavy pasteurisation to eliminate endospores of non-proteolytic strains with refrigeration to prevent the growth of surviving proteolytic have been proposed where full sterilisation would be unacceptable from the viewpoint of product quality. A process based on in-pack pasteurisation at 92.2°C for 65 minutes has been proposed for the inactivation of endospores of non-proteolytic types B and E in vacuum-packed, hot smoked fish of reduced NaCl content (Ecklund *et al.*, 1988b). Refrigeration is then required to prevent the growth of proteolytic strains. Similar principles are applied in *sous-vide* food preparation (Raffael, 1984; Light *et al.*, 1988). There are inevitable risks with vacuum-packed foods that rely solely on refrigeration as a means of controlling *Cl. botulinum*, especially where the vagaries of retail display and customer storage are involved. It is unfortunate that much published work has paid no attention to this hazard.

Tyndallisation has been proposed as a means of eliminating endospores of proteolytic strains of *Cl. botulinum*. The technique has been used for home canning low-acid foods, but more recently a two-stage process has been devised for commercial production of vacuum-packed cooked potatoes, or other vegetables, that can be stored for many months at ambient temperatures (Demuelemeester and Demuelemeester, 1982). The ability of *Cl. botulinum* to survive the process and produce toxin during storage at 25°C was demonstrated conclusively by Lund *et al.* (1988), and such products are considered inherently unsafe unless storage at temperatures below 4°C can be guaranteed, or a $F_0 3$ process is applied.

Sublethal heat treatments can be of importance in increasing the sensitivity of *Cl. botulinum* endospores to other inhibitory factors including nisin and NaCl.

In comparison with staphylococcal enterotoxins, botulinum toxin is heat labile and readily destroyed by boiling, or even warming, food. The toxin is totally inactivated by heating at 80°C for 10 minutes (Pierson and Reddy, 1988), and slow inactivation may take place at temperatures as low as 50°C (Bonventre and Kempe, 1959). At this temperature the synthesis and inactivation of toxin may proceed simultaneously.

Irradiation

Although vegetative cells of *Cl. botulinum* are likely to be inactivated by doses of ionising radiation in the order of 3 to 5 Kgy, endospores are likely to survive such doses and cannot be inactivated without unacceptable organoleptic changes.

Low-dose irradiation has been proposed to ensure the safety of low NaCl cured meats, but gamma radiation cannot totally compensate for NaCl reduction with respect to toxin production by *Cl. botulinum* (Barbut *et al.*, 1988).

Irradiation is not, therefore, seen as a practical means of controlling *Cl. botulinum* and indeed survival of the organism and its possible growth in the absence of spoilage micro-organisms is recognised as one of the major hazards associated with commercial scale irradiation.

Pre-formed botulinum toxin is unlikely to be inactivated at doses applied to foods (Rose *et al.*, 1988).

pH value and acidity

pH value and acidity are of major importance in the control of *Cl. botulinum*. For example, pH value is used to define the level of processing required for the safety of shelf-stable canned goods, while acidification is applied to control the growth of the organism in pasteurised foods (Notermans *et al.*, 1985).

The optimum pH value for the growth of all types of *Cl. botulinum* is 6.8 to 7.0, and the upper limiting value *ca* 8.5 (Hobbs *et al.*, 1982). There are, however, differences between proteolytic and non-proteolytic strains with respect to the lower limit for growth.

The minimum pH value for the growth of non-proteolytic strains is 5.0 to 5.2, and there is no evidence of growth at lower values. The situation with proteolytic strains is more complex. For many years it was accepted that *Cl. botulinum* would not germinate or grow at pH 4.8 or below (Townsend *et al.*, 1954) and, using a 'safety margin' of 0.2 pH units, a value of 4.6 has been used to define 'high-acid' and 'low-acid' foods when determining thermal processing parameters (Anon, 1987d).

Despite widespread acceptance of pH 4.8 as the limiting value for *Cl. botulinum*, there have been a number of reports of growth and toxin production below this. Growth at lower pH values has frequently been attributed to the existence of localised areas of higher pH value. The existence of such areas has been observed under mycelial mats in mould-spoiled tomato juice (Huhtanen *et al.*, 1976; Oldlaug and Pflug, 1979), and the existence of similar high pH areas in precipitated protein postulated without substantiating evidence (Tanaka, 1982). There have also been a number of instances where growth and toxin production by proteolytic *Cl. botulinum* has occurred at pH values as low as 4.0 (Raatjes and Smelt, 1979; Smelt *et al.*, 1982; Young-Perkins and Merson, 1987) and where this growth could not be explained by localised variations in pH value.

Growth at pH values below 4.6 was investigated in some detail by Young-Perkins and Merson (1987) who concluded that:

1 pH value alone is not a suitable index for prediction of toxicity endpoints in protein-containing material.
2 Titratable acidity (for a specific acidulant) may be a better means of defining acidification conditions that inhibit germination, outgrowth and toxin production by *Cl. botulinum*.
3 Three main parameters are necessary for the prediction of the likelihood of *Cl. botulinum* endospore germination; the acidity level of the system in terms of the acidulant, the redox potential and the protein concentration.

In a further study, Wong *et al.* (1988) confirmed that, under strictly anaerobic conditions (E_h −370 to −391 mV), *Cl. botulinum* was consistently able to germinate, grow and synthesise toxin at pH values below 4.6. Beef protein was a better substrate than soya protein under high acid conditions, while citric acid was a less effective inhibitor than hydrochloric acid at the same pH values. Differences in limiting pH value according to acidulant were also noted by Tsang *et al.* (1985), who recorded a minimum pH value for growth and toxin production of 4.2 with citric acid as the acidulant, but 5.0 with acetic acid. Care must be taken in defining the limiting pH value where citric acid is the acidulant since the inhibitory effect stems from chelation of divalent metal ions as well as an unfavourable pH value (Graham and Lund, 1986).

These findings have obvious implications for the safety of high-acid canned foods and other foods where safety is related to a low the pH value. It should, however, be appreciated that even if pH value alone will not prevent growth, it may contribute to inhibition by combination effects (Lund *et al.*, 1987a).

Botulinum toxins are stable at pH values of 7.0 and below but are unstable in alkaline conditions.

Redox potential and composition of atmosphere

Cl. botulinum is generally considered to grow optimally at E_h −350 mV (Smoot and Pierson, 1979), and to be unable to grow at E_h values above +150 mV (Roberts, 1982). The upper limit for growth may be modified by other factors such as pH value, while a substrate-linked effect occurs in fish (Huss *et al.*, 1979; 1980). In fresh herring, the maximum E_h value for growth was +100 mV, but in smoked herring growth continued up to +250 mV.

Controlled atmosphere packaging is now widely applied as a means of extending the shelf life of fresh foods, and there has been concern that such packaging may create conditions that favour the growth of *Cl. botulinum*. Work with pure cultures suggests that while hyperbaric carbon dioxide is likely to inhibit *Cl. botulinum*, the organism may be stimulated under atmospheric carbon dioxide (Sperber, 1982). In foods, results have been rather equivocal with *Cl. botulinum* type E growing poorly in a variety of sandwiches packed in nitrogen and stored under refrigeration (Kautter *et al.*, 1981). In fish, toxin production was the same whether stored under vacuum or under various CO_2:N_2 combinations (Post *et al.*, 1985). Growth and toxin production by *Cl. botulinum* type A in modified atmosphere packed cabbage is believed to have led to an outbreak of botulism in which coleslaw was vector (see page 300).

The inclusion of oxygen in gas mixtures used for controlled atmosphere packing has been claimed to inhibit *Cl. botulinum*. These claims should be treated with caution as experimental evidence is lacking, and since there is a 'paucity' of information concerning the level of oxygen inhibitory to growth (Wong *et al.*, 1988).

A mathematical model for predicting the safe storage life of modified atmosphere stored fish with respect to toxin production by *Cl. botulinum* has been devised (Baker and Genigeorgis, 1990). The model is based on the length of the lag phase, which can be significantly affected by temperature, inoculum level, fish type, pool of endospores and type of atmosphere.

Water activity

Reduction of water activity may be used as a means of controlling *Cl. botulinum*. With NaCl as the humectant, minimum a_w levels (at optimum temperatures) that permit the outgrowth of endospores and toxin synthesis by proteolytic strains are 0.95 for type A and 0.94 for type B. Non-proteolytic strains are more sensitive to reduction of a_w and type E strains will not grow below 0.97.

Limiting a_w values will vary according to the humectant, being lower if glycerol, for example, is used, and will also be modified by other growth controlling and inhibitory factors (Baird-Parker and Freame, 1967).

Curing ingredients

NaCl

Although NaCl is normally used in combination with nitrite in meats, it is the sole inhibitory ingredient in salt fish as well as in some types of meat product.

There is a marked difference in NaCl relationships between proteolytic and non-proteolytic strains. Proteolytic strains of A, B and F have a maximum NaCl concentration for growth of 10%, as opposed to a maximum for non-proteolytic strains of B, E and F of 5% (Hobbs *et al.*, 1982). Toxin production may be inhibited at slightly lower levels than growth, toxin production by *Cl. botulinum* type E being inhibited by 4% NaCl for over 35 days at 27°C (Cuppett *et al.*, 1987).

The continuing trend to lower NaCl concentrations in foods means that there is increasing reliance on a combination of NaCl and refrigeration. This combination may not be effective when non-proteolytic strains of *Cl. botulinum* are present, and it is somewhat ironic that, at a time when nitrite levels in meat products are being lowered, the addition of 220 mg/l nitrite to increase the safety of salt fish has been proposed (Cuppett *et al.*, 1987).

A further consequence of the desire for a low sodium diet has been the introduction of 'salt substitutes', which are usually potassium chloride, alone or in combination with other salts such as potassium sulphate and calcium glutamate. Use of some of these products have been proposed as a means of removing some, or all, NaCl from cured meats. Although relatively little information is available, there is evidence that the inhibitory effect of KCl and other substitute salts on *Cl. botulinum* is significantly less than that of NaCl. This factor must be considered when developing low-sodium cured meats.

Nitrite

Under the correct conditions, nitrite can exert a powerful inhibitory effect on *Cl. botulinum*. Inhibitory levels vary considerably depending on a combination of factors including NaCl, pH value, temperature of storage, presence of other inhibitors, and the initial level of contamination (Roberts *et al.*, 1981a). Some of these interactions are discussed below.

Nitrate

Although nitrate is a traditional curing ingredient, it is often omitted from modern cures and there has been considerable discussion over its role. Nitrate significantly reduced toxin formation, but not spoilage, by proteolytic strains of *Cl. botulinum* types A and B in pasteurised pork model systems (Roberts *et al.*, 1981a,b). However, the inhibitory effect was due to nitrite formed by reduction of the nitrate.

Curing adjuncts

Many modern curing formulations include adjuncts which are added primarily for a technological purpose, such as colour formation, but which contribute to the anti-microbial properties as a whole. The anti-microbial properties of curing adjuncts are likely to be of particular importance in formulations of reduced NaCl and/or nitrite levels.

Polyphosphates are added primarily to improve water retention, but some types are also inhibitory for *Cl. botulinum*. However, the situation is not straightforward and the polyphosphate curaphos 700 which, at a concentration of 0.3%, reduced toxin formation in a high pH (6.3 to 6.8) pasteurised cured meat model system slightly increased formation in the same system with low (5.5 to 6.3) pH meat (Roberts *et al.*, 1981a). In a study of reduced NaCl turkey frankfurters, Barbut *et al.* (1986) noted that tripolyphosphates enhanced toxin production, sodium hexametaphosphate had no effect and sodium acid pyrophosphate delayed toxin production.

Ascorbate and erythorbate (isoascorbate) are primarily added to cured meats as antioxidants and to enhance colour formation (ascorbate is permitted in the UK, erythorbate in the USA). There has been some controversy over the role of these compounds as inhibitors of *Cl. botulinum*, ascorbate being claimed both to act synergistically with nitrite to inhibit the organism, and to increase the possibility of its growing. However, erythorbate was found to reduce significantly toxin formation in both high and low pH model systems (Roberts *et al.*, 1981a,b).

The curing system as a whole

Although it is important to define the role of individual ingredients of cured meat, safety is determined by the inhibitory properties of the system as a whole. Challenge studies have provided much information on the role of different ingredients, but in the past have failed to provide adequate information on the likelihood of *Cl. botulinum* growth and toxin production in a given product (Hauschild, 1982). The situation has been adequately summarised by Ingram (1973): *'what we need . . . is not more incubated pack experiments but a rationale for interpreting them'*.

More recently, there has been considerable interest in the development of mathematical techniques to provide that rationale. These may be illustrated by two contrasting approaches which have been taken in the UK (Roberts and Jarvis, 1983). The first was by workers at the then Meat Research Laboratory, Bristol, who attempted to develop an overall database from a relatively enormous experiment which, if suitable, would lead to a single predictive equation for toxigenesis by *Cl. botulinum*. In reality, two formulae have been developed, for low (5.5 to 6.3) and high (6.3 to 6.7) pH systems (**Table 108**; Roberts *et al.*, 1981c).

Table 108. Formulae for estimating the probability of toxin production by *Clostridium botulinum* in pork slurry (Roberts *et al.*, 1981c).

A Meat of 'low' initial pH value (5.5 to 6.3)	
Probability of toxin production (P) $= 1/(1 + e^{-u})$, where u = 5.750	
$- (3.25 \times N)$	*where* N $=$ NaNO$_2$ µg/g $\times$ 100
$+ (0.04834 \times T)$	T $=$ Storage temperature °C
$- (1.854 \times S)$	S $=$ NaCl % w/v on water
$+ (1.846)$	*if* inoculum 1000 spores/bottle
$- (4.074) + (1.222 \times N) + (0.4861 \times S)$	nitrate added
$- (2.23) + (0.3617 \times N) + (0.3174 \times S)$	erythorbate added
$+ (2.557) - (0.3615 \times S) + (0.4233 \times N \times S)$	polyphosphate added
$+ (2.154) - (0.4510 \times T) - (0.2555 \times S)$	heat treatment low
$- (0.01424) - (0.0214 \times T) + (0.1256 \times S)$	heat treatment high
$- (0.532)$	heat low and erythorbate added
$+ (0.5193)$	heat high and erythorbate added
$+ (0.4618)$	heat low and polyphosphate added
$- (1.548)$	heat high and polyphosphate added
$+ (1.025)$	heat low and nitrate added
$- (1.048)$	heat high and nitrate added
B Meat of 'high' initial pH value (6.3 to 6.7)	
u $= 4.679$	
$- (1.47 \times N)$	*where* N $=$ NaNO$_2$ µg/g $\times$ 100
$- (1.104 \times S)$	S $=$ NaCl % w/v on water
$+ (0.129 \times T)$	T $=$ Storage temperature °C
$- (2.09) + (0.67 \times N)$	*if* µg/g nitrate added
$- (6.238) + (0.8264 \times S)$	1000 µg/g erythorbate added
$- (1.7049) + (0.8264 \times S)$	0.3% polyphosphate added
$- (1.771) + (0.3997 \times N)$	heat treatment high
$- (0.01937 \times N \times T) - (1.2824)$	nitrate and polyphosphate added
$+ (0.99)$	nitrate added and heat high

Note: Low heat = 80°C at 7 minutes; high heat = low + 70°C at 1 hour.

The second approach was that of the Leatherhead Food Research Association where studies were made on specific parts of an interactive system. Initially, a risk factor hypothesis for changes in preservative systems was developed (Jarvis *et al.*, 1979) and subsequently a number of predictive equations have been produced (**Table 109**). None of these is, in itself, complete, but all can be handled by computers to predict safety and stability (**Table 110**, Roberts and Jarvis, 1983).

It must be appreciated that while predictive systems are potentially extremely valuable tools other, and less easily quantified factors, may also be involved in determining safety. In investigations concerning the effect of pig breed, cut of meat and animal to animal variation on the behaviour of *Cl. botulinum*, it was shown that toxin production and spoilage was significantly greater in shoulder than in leg cuts, but that while there was considerable animal to animal variation, there was no systematic difference between breeds (Gibson *et al.*, 1982). Difference between cuts is likely to be a consequence of differences in pH value.

Table 109. Examples of predictive equations for the probability of toxin production by *Clostridium botulinum* (Roberts and Jarvis, 1983).

	NaCl (%)	NO$_2$ (µg/g)	Process	Storage temp. (°C)	Phosphate (%)
A	3.3–5.5	40–300	L	20, 25	0
B	3.5–5.5	40–300	H	20, 25	0
C	2.0–3.5	75–175	M	20	0–0.5

Probability of toxin production (P) = $1/(1 + e^{-u})$, where
A $u = 0.43T - 2.01S + 0.01N - 0.0013NT + 1.2$
B $u = 0.43T - 1.53S + 0.0064N - 0.001NT - 1.698$
C $u = 7.94 - 0.045N - 0.54S + 4.87P - 2.85SP$
where $S = NaCl$, $N = NO_2$, T = storage temperature, P = phosphate

Note: Processes: L = low (20 to 70°C in 7 minutes).
M = medium (20 to 70°C in 7 minutes plus 1 hour hold at 70°C).
H = high (20 to 70°C in 7 minutes plus 2 hour hold at 70°C).

Table 110. Risk factors for growth of *Clostridium botulinum* under different curing conditions.

A Average present-day practice
$$NaCl (3.5\%) + NO_2 (200\,mg/l) + P_{80} (6.7) + Storage (15°C) + pH (6.0)$$
$$R = 1 \times 1 \times 1 \times 1 \times 1 = 1$$

B Reduce NaCl and NO$_2$; increase storage temperature and pH
$$NaCl (2\%) + NO_2 (50\,mg/l) + Storage (20°C) + pH (6.4)$$
$$R = 4 \times 10 \times 5 \times 4 = 800$$

C Reduce NO$_2$ and process; increase NaCl and storage temperature
$$NaCl (4.5\%) + NO_2 (100\,mg/l) + Storage (20°C) + P_{80} (0.7)$$
$$R = 0.5 \times 3 \times 5 \times 3 = 22.5$$

Notes:
1 P_{80} = minutes at 80°C.
2 The higher the numerical value assigned to each parameter the greater the overall probability of toxin production.
3 From Jarvis and Patel (1979); Farber (1986).

Smoking

Smoking is usually employed as an additional process to curing or salting, although in a few traditional products smoking is the only means of preservation. Much of the inhibitory effect on *Cl. botulinum* of traditional smokes resulted from drying and consequent reduction of the a_w level. However, some of the compounds produced during smoking may have a specific inhibitory effect on *Cl. botulinum*. Type E, for example, is inhibited by 40 mg/l formaldehyde, a level attainable in smoked salmon (Nielsen and Pedersen, 1967).

Most modern smoking processes are very light and in many cases any inhibitory effect is likely to be minimal. Any claims made by manufacturers of liquid smokes with respect to activity against *Cl. botulinum* should be treated with considerable caution.

Other preservatives

Sorbic acid and sorbates

Sorbic acid and potassium sorbate is widely used as a preservative in foods, although its major function has been to inhibit the spoilage microflora, particularly yeasts and moulds. The clostridia in general have been considered to be relatively resistant to sorbic acid, which has been used as a selective agent in isolation media for the genus. Despite this, a role for these compounds as inhibitors of *Cl. botulinum* in foods has been suggested, a concentration of 250 mg/l undissociated sorbic acid being found to retard the growth of the organism from both endospores and vegetative cells over the pH range 5.5 to 7.0 in artificial media (Restaino *et al.*, 1981).

A systematic evaluation of the inhibitory effects of sorbic acid against *Cl. botulinum* types A and B was undertaken by Lund *et al.* (1987b). The inhibitory effect was confirmed and shown to be a function of undissociated sorbic acid. Using a mathematical model, it was shown that for a type A strain a total sorbic acid concentration of 1,000 mg/l would reduce the probability of growth at 30°C in 14 days by a factor of the order of 10^8 at pH 5.0. At lower pH values the factor would be significantly higher.

The use of sorbic acid, or sorbate, as a practical means of controlling *Cl. botulinum* in foods has been proposed in two main areas: processed cheese products, and as a partial substitute for nitrite in cured meat.

The possibility of using sorbate in processed cheese has been explored in Argentina following an outbreak of type A botulism (see page 298). A combined effect was noted between sorbate, a_w level and inoculum size. The former interaction was noteworthy in that it occurred over a very narrow range, 0.967 to 0.974. It was concluded that the use of 0.3% sorbate in a cheese with slightly reduced a_w level and pH value would significantly improve safety (Briozzo *et al.*, 1985).

A number of workers have evaluated the potential use of sorbate as a partial replacement for nitrite in cured meats, the overall conclusion being that combinations of a minimal nitrite level (40 mg/l) plus a sorbate/sorbic acid mixture is an effective means of controlling *Cl. botulinum*. However, allergic reactions in some people to sorbate in bacon led to the refusal of 'generally recognised as safe' (GRAS) status for potassium sorbate in bacon in the USA.

Nisin

The polypeptide antibiotic, nisin, has received much attention as an inhibitor of *Cl. botulinum* since the finding that the compound prevented clostridial spoilage of Swiss-type cheese (Hirsch *et al.*, 1951). Two main areas of application exist – canned products and processed cheese products. In the former case, it has been suggested that with the addition of nisin the heat treatment applied could be much reduced (Hurst, 1981). However, in current practice, the minimum heat treatment required to kill endospores of *Cl. botulinum* (botulinum cook) is still applied, nisin being used to control highly heat-resistant spoilage bacteria which otherwise require excessive heating.

Nisin has been used for many years in processed cheese products (Meyer, 1973), although its original role was control of spoilage. Nisin is now approved for use in the USA for control of *Cl. botulinum* in high moisture, low NaCl cheese spreads (Anon, 1988a,b).

Endospores of *Cl. botulinum* are more resistant to nisin than those of many other bacteria and inhibitory concentrations are up to one-hundred times greater (Sperber, 1982), a maximum concentration of 250 mg/l being permitted in the USA in processed cheese. The efficiency of nisin is affected by other factors, being more effective at pH 6.0 than at higher values and after higher heating temperatures (Scott and Taylor, 1981). The application in meat products may be limited by binding of the antibiotic to meat proteins (Somers and Taylor, 1981).

Parabens

Parabens (*p*-hydroxybenzoic acid esters) are broad-spectrum microbial inhibitors at pH values near neutral. In aqueous foods, the inhibitory effect of parabens on *Cl. botulinum* increases with increasing chain length, the methyl ester having a minimum inhibitory concentration of 1,000 µg/ml compared with 0.3 µg/ml for the undecyl ester (Dymicky and Huhtanen, 1979). In an oil-containing food, the inhibitory effect *lessens* with increasing chain length. This is attributed to the higher partition coefficient of the long chain parabens leading to a greater solubility in oil than in water (Sperber, 1982).

Plant extracts

Various 'natural ingredients' have been investigated as a means of controlling *Cl. botulinum*, including spice extracts (Hall and Maurer, 1986) and garlic or onion oil (DeWit *et al.*, 1979). Such substances may be attractive to the 'green' consumer and manufacturer alike, but cannot be seen as major defences against botulism, although some may contribute to the inhibitory activity of a preservative system. Both garlic and onions have been involved in recent outbreaks of botulism (see pages 299–300) and, in the case of garlic, there was no inhibitory effect on the causative type B strain (Solomon and Kautter, 1988).

Competition from other micro-organisms

Cl. botulinum is inhibited by acid-producing micro-organisms, and thus will not grow in the presence of active starter cultures. However, endospores will survive and outgrow if conditions are suitable at a later stage.

Modifications to the environment by other micro-organisms may create conditions which permit the growth of *Cl. botulinum*. Examples are lowering of the E_h by aerobic bacteria and raising of the pH value by moulds. Actively metabolising facultative anaerobes such as *Bacillus licheniformis*, may compromise the inhibitory effect of organic acids on *Cl. botulinum* in high-acid foods (Wong *et al.*, 1988).

Laboratory methods

Isolation

Enrichment

In most cases, it is possible to employ both self-enrichment and enrichment in artificial media. Self-enrichment involves quite simply vacuum-packing suspect food and incubating at 25 to 30°C for 6 to 7 days. The procedure is usually successful if the food will support good growth of the organism, but may fail if other anaerobes are capable of faster growth. Considerable quantities of gas may be produced during self-enrichment, and precautions should be taken against the possibility of the pack bursting or leaking.

Although advantages have been claimed for the selective sulphite-polymyxin broth in conventional enrichment procedures for *Cl. botulinum* (Mossel and DeWaart, 1968), the use of non-selective media is more common. A wide range of non-selective media continue to be used (**Table 111**). Of these the well

Table 111. Examples of enrichment media for *Clostridium botulinum*.

A Non-selective
Robertson's cooked meat medium[1]
Glucose-peptone broth
Fish infusion broth
Reinforced clostridial medium
Trypticase-peptone-glucose-yeast extract-trypsin broth
Corn extract medium
Botulism enrichment medium
B Selective
Sulphite-polymyxin broth

[1] Glucose may be added.

established Robertson's cooked meat broth remains the medium of choice in many laboratories. In some cases, 1% glucose is added, and although the value of this addition is not universally accepted, it can improve the isolation of non-proteolytic strains. A more significant increase in the isolation rate of these strains is achieved by the inclusion of trypsin in media such as trypticase-peptone-glucose-yeast extract-trypsin broth (Lilly *et al.*, 1971). Trypsin is considered to reduce the effect of inhibitory substances (Kautter *et al.*, 1966) and improve the endospore germination rate (Alderton *et al.*, 1974). The latter property has been exploited in botulinum enrichment medium that contains both trypsin and lysozyme (Holdemann *et al.*, 1977). This medium gave significantly better results than reinforced clostridial medium with naturally contaminated dried foods (Hobbs *et al.*, 1982).

Other compounds which promote germination of endospores also improve recovery when added to a basal enrichment medium. These include lactate plus L-alanine and bicarbonate (Crowther and Baird-Parker, 1984).

In the past, it was usual practice to heat enrichment cultures to 80°C for 10 minutes, or 70°C for 2 hours, before incubation to destroy commensal bacteria. This treatment is not recommended since non-proteolytic strains of types B, E and F may be killed. In most cases, pre-heating is not considered necessary, but if required a lower temperature of 60°C for 30 minutes is recommended. Alternatively, post-enrichment treatment with ethanol may be used, equal volumes of enrichment culture and absolute ethanol being mixed and held for *ca* 1 hour at 25°C. In either case, an untreated sample should be included for the recovery of vegetative cells in the absence of endospores.

Enrichment cultures for *Cl. botulinum* are normally incubated at 26 to 35°C for 5 to 10 days. With both self-enrichment and those made in artificial media, samples are removed for toxin testing at this stage, as well as for streaking onto plating media.

Non-selective plating media are usually preferred, the most widely used being fresh horse blood agar and egg yolk agar. Neither is selective, but typical reactions of slight haemolysis and a characteristic egg yolk reaction respectively aid in the differentiation of *Cl. botulinum* (**Figures 255–257**). The choice of basal medium for egg yolk agar is important, since the presence of a fermentable carbohydrate often leads to atypical reactions (Hobbs *et al.*, 1967). Freshly prepared media, such as meat infusion broth, are preferred but commercial dehydrated media such as brain-heart infusion agar are suitable. A fresh egg yolk agar is preferable to commercially prepared suspensions (Hobbs *et al.*, 1982). Both media should be incubated for 3 days at 30°C.

Antibiotics or other inhibitors may be added to either horse blood or egg yolk agars to impart selectivity but non-proteolytic strains tend to be inhibited. The most successful selective medium is probably CB1 agar (Dowell and Dezfulian, 1981), which contains cycloserine, sulphamethoxazole and trimethoprim in an egg yolk agar base. Some strains of *Cl. botulinum* type E are inhibited and type G strains are difficult to recognise. The medium is not fully selective, but usually permits the isolation of *Cl. botulinum* within 1 to 2 days.

An immunodiffusion technique for detection of *Cl. botulinum* types A, B, E and F on an isolation plate has been described (Lilly *et al.*, 1984). Enrichment cultures are streaked onto trypticase-glucose-yeast extract agar containing specific antitoxin and incubated anaerobically. Antitoxin:toxin precipitate, which is enhanced by staining with thiazine red R, forms around colonies of *Cl. botulinum*.

255

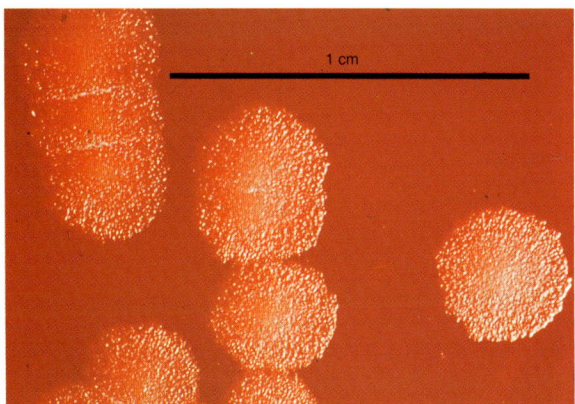

25

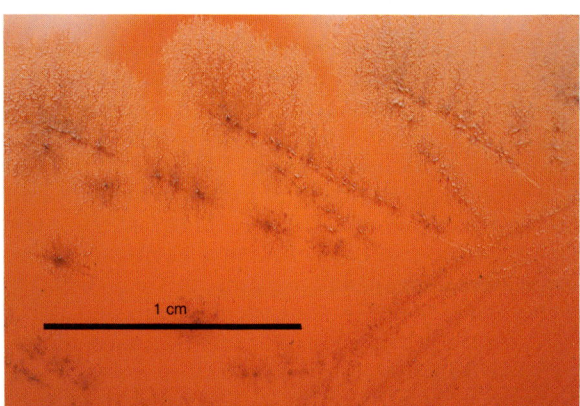

257

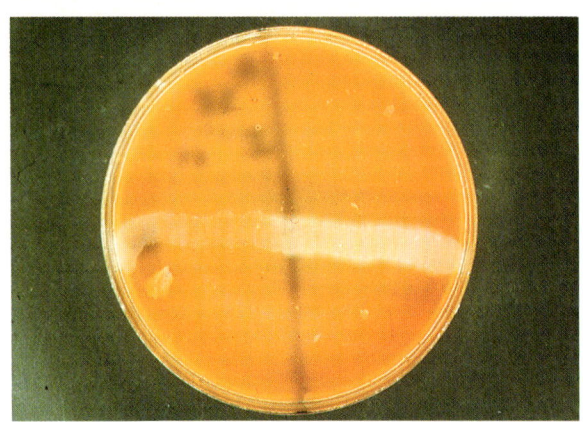

255–257. Isolation of *Cl. botulinum*. Horse blood agar and lactose-egg yolk agar are normally inoculated in parallel. Colonies of a proteolytic type B strain (**255**) have a typical appearance, although the characteristic weak haemolysis is lacking. Colonial appearance can vary considerably, however, a type A strain producing colonies with a rhizoidal appearance (**256**).

A strain of *Cl. botulinum* streaked onto lactose-egg yolk medium is illustrated in **257**. The precipitate under the growth is due to the release of fatty acids in the egg yolk by lipolysis, a property restricted to *Cl. botulinum* and a few other clostridia. (*Both media: 24h at 30°C in 95% H$_2$/5% CO$_2$.) (**255** and **257** Reproduced by courtesy of Dr T.A. Roberts, AFRC Institute of Food Research, Reading Laboratory.)

Detection of toxin

Initial confirmation of identity of *Cl. botulinum* isolates is by determination of toxin production. In investigation of outbreaks of botulism, it is usual to assay suspect foods, faeces etc and, as noted above, enrichment cultures for presence of toxin (**Flow diagram 27**).

Flow diagram 27. Detection of *Cl. botulinum* and its toxins in food (Hobbs *et al.*, 1982).

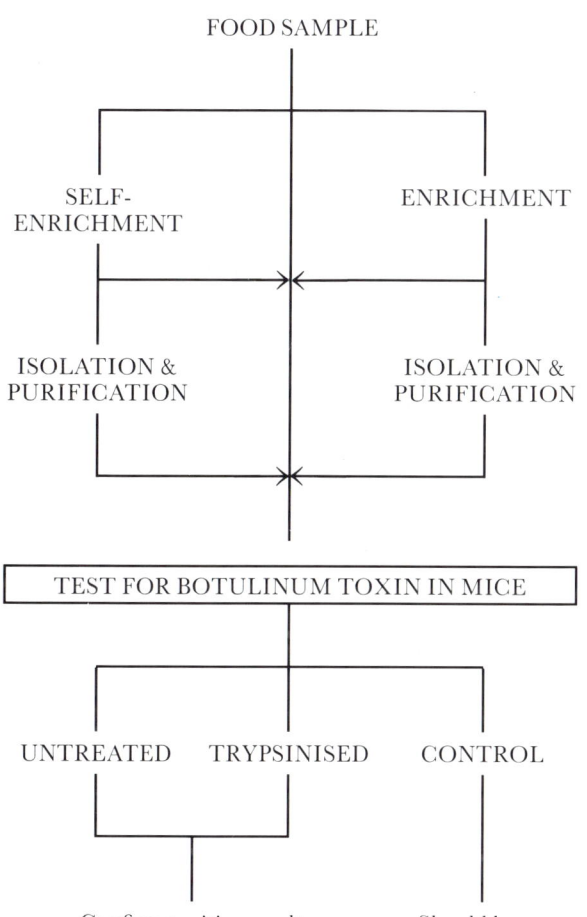

FOOD SAMPLE

SELF-ENRICHMENT   ENRICHMENT

ISOLATION & PURIFICATION   ISOLATION & PURIFICATION

TEST FOR BOTULINUM TOXIN IN MICE

UNTREATED   TRYPSINISED   CONTROL

Confirm positive result by mouse protection tests   Should be no symptoms

Preparation of toxin

Similar methods are applied to food, faeces and enrichment cultures. Solid samples of food and faeces are first homogenised in gelatin-phosphate buffer at pH 6.5. The ratio of sample to diluent should be as high as possible.

Colonies for testing should be picked off isolation plates and incubated for 3 days at 30°C in cooked meat broth. The purity of cultures which test positive should be carefully checked, taking account of precautions necessary when dealing with anaerobic bacteria (**Figures 258 & 259**). It is usually necessary to pick a large number of colonies, since many of those present on isolation plates are likely to be non-toxigenic.

Cultures, or sample extracts, are allowed to settle, or centrifuged at 1,200 g for 5 to 25 minutes (Hobbs *et al.*, 1982). The supernatant is divided into three, one portion remaining untreated, one heated to demonstrate the labile nature of the toxin, and one trypsinised to demonstrate the presence of inactive protoxin (**Flow diagram 28**).

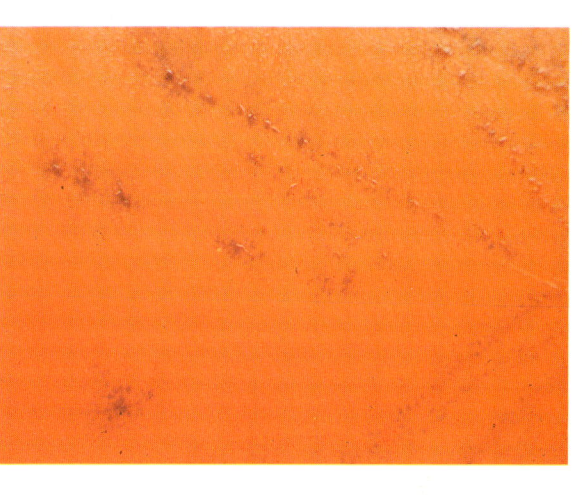

258

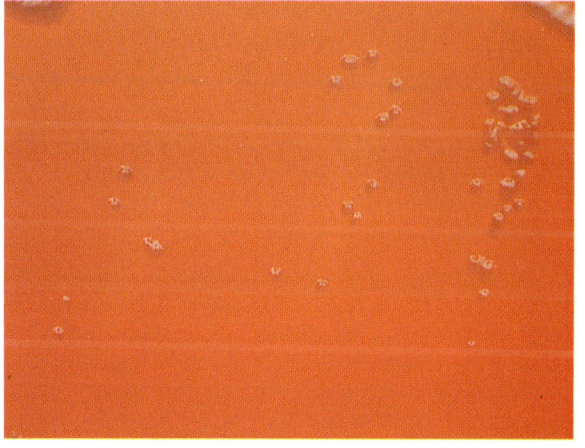

259

258 & 259. Ensuring the purity of cultures of *Clostridium* species. Culture under anaerobic conditions imposes selective pressures, and streak plates incubated under these conditions may not permit colony formation by aerobic contaminants. For this reason, it is necessary to ensure purity by streaking onto a second plate and incubating aerobically. In **258**, an apparently pure culture of *Cl. botulinum* (see **257**) was shown to contain an aerobic contaminant (**259**). (*Blood agar: 24 h at 30° C in 95% H₂-5% CO₂ /air.*)

Flow diagram 28. Characterisation of *Cl. botulinum* toxin.

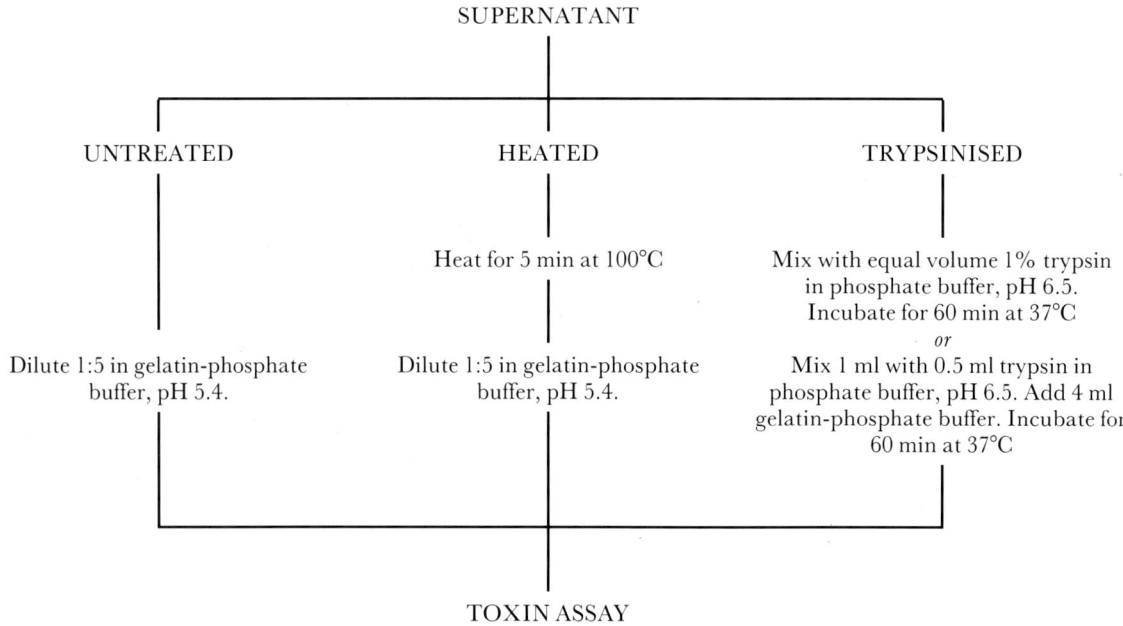

SUPERNATANT

UNTREATED	HEATED	TRYPSINISED
	Heat for 5 min at 100°C	Mix with equal volume 1% trypsin in phosphate buffer, pH 6.5. Incubate for 60 min at 37°C
		or
Dilute 1:5 in gelatin-phosphate buffer, pH 5.4.	Dilute 1:5 in gelatin-phosphate buffer, pH 5.4.	Mix 1 ml with 0.5 ml trypsin in phosphate buffer, pH 6.5. Add 4 ml gelatin-phosphate buffer. Incubate for 60 min at 37°C

TOXIN ASSAY

Note: Either method of trypsinisation is satisfactory, but the first is more effective when toxin levels are low (Hobbs *et al.*, 1982).

Mouse challenge test

Although a number of *in vitro* methods have been proposed (see below), the mouse challenge test remains the most sensitive and reliable. The test (**Flow diagram 29**) involves two stages:

1 Demonstration of the presence of a substance toxic to mice, preferably with typical symptoms.
2 Identification of serological toxin type by protection with specific antisera.

In addition to the general difficulties associated with animal tests, the mouse protection test suffers from specific problems, taking up to 3 days to obtain a result when toxin levels are low and being subject to interpretational difficulties due to non-specific death or atypical symptoms. The latter problems may usually be overcome either by diluting the sample 1:10 or 1:20, by diluting the sample and adding 10% mixed clostridial antiserum and either 200 mg/l chlortetracycline or 1,000 mg/l streptomycin sulphate (Hobbs *et al.*, 1982) or by treatment of the sample with bovine serum albumen (Solberg *et al.*, 1985).

Flow diagram 29. The mouse challenge test for detection of *Cl. botulinum* toxin.

A Detection of substance toxic to mice

Inject 0.5 ml of each test supernatant intraperitoneally into each of two 18 to 29 g mice

Observe over three day period for symptoms of botulism and death

B Identification of serological toxin type

Mix equal quantities of specific antitoxin with toxic test material. Stand for 60 min at room temperature

Inject a pair of 18 to 20 g mice with 0.5 ml mixture of antitoxin and test material and a further pair with a control mixture containing no antitoxin

Observe over three-day period. Protection of a single pair indicates the presence of a single type of toxin. Samples containing more than one toxin occasionally occur and necessitate retesting with mixtures of antitoxin

Note: The level of toxin can be determined by injecting dilutions of the test material made in gelatin-phosphate buffer. The reciprocal of the highest dilution giving typical symptoms and death is the minimum lethal dose (MLD).

In-vitro assays

Although a number of *in vitro* assays have been developed none, until recently, has approached the sensitivity and specificity required even for screening purposes.

In vitro assays are essentially of two types: serological and tissue culture. The former have advantages in being technically simpler and requiring no special expertise in tissue culture techniques. However, problems have arisen with the specificity of the antibodies, while a more fundamental objection to serological methods is that antigenicity is determined rather than toxicity.

Serological tests

A number of serological methods have been proposed, earlier methods having been briefly reviewed by Hobbs *et al.* (1982) and Foegeding (1986). Most of these methods were unsatisfactory with respect to both sensitivity and specificity. A highly sensitive (5 to 10 mld/ml) radioimmunoassay for toxins A and B has been developed (Boroff and Shu-Chen, 1973), but application has been severely restricted by the requirement for labelled toxin of high specific radioactivity as well as the general difficulties of handling radioactive material.

Of the non-radiometric assays, ELISA has shown most promise but with the exception of an assay for type G toxin (Lewis *et al.*, 1981) none of those based on polyclonal antibodies have approached the required sensitivity (Gibson *et al.*, 1987). More recently, sandwich ELISA assays have been developed using monoclonal antibodies and enzyme amplified detection for type A (Shone *et al.*, 1985) and type B (Modi *et al.*, 1986) toxins. The assay for type A toxin has been evaluated by Gibson *et al.* (1987), who concluded that assays of this type have considerable potential in screening tests and proposed the development of a commercial kit.

Tissue culture methods

A range of animal cell lines has been used in attempts to develop a method for the detection of *Cl. botulinum* toxin. These include human embryonic lung, HeLa, mouse muscle fibroblasts and rhesus monkey kidney (DeWaart *et al.*, 1972). Attempts have also been made to use protozoa and algae, but none of these assays have been of sufficient sensitivity. Some promise has been shown with tissue culture monolayers originating from mammalian (mouse, sheep, human) nervous tissue (Gibbs *et al.*, 1987). However, further work is needed before such a method could be adopted for practical use.

Biochemical characteristics

Where required, biochemical characteristics may be determined using either conventional methodology or miniaturised kits (**Figures 260–262**). Useful tests include glucose fermentation, indole production, casein hydrolysis, lecithinase, lipase and proteolytic activity, as well as the determination of volatile and non-volatile acids from peptone-yeast extract-glucose medium (Dowell and Dezfulian, 1981). The methods of Holdemann *et al.* (1977) are recommended for the determination of conventional test criteria.

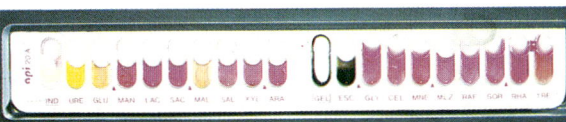

260

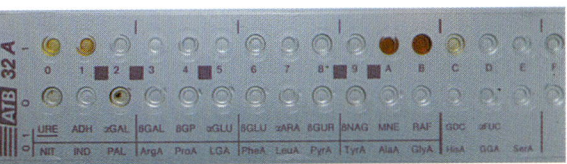

261

260 & 261. Use of miniaturised test kits for the identification of clostridia. The API 20A© (**260**) and API ATB 32A© kits use different technology, in that the former is based on the tests used in conventional identification systems and requires anaerobic incubation, whereas the latter is based on the determination of specific enzymes and may be incubated aerobically. The use of enzyme determinations also means that results are available after 4h incubation at 37°C, as opposed to 18 to 24h for the 20A.

Neither system is capable of reliably differentiating *Cl. botulinum* from phenotypically similar, but non-toxigenic bacteria, and a further problem is that many clostridia, including some types of *Cl. botulinum* and *Cl. sporogenes* (illustrated), are largely identified on negative criteria. (*API 20A: 24h at 37°C in 95% H₂/5% CO₂; API ATB 32A: 4h at 37°C in air.*)

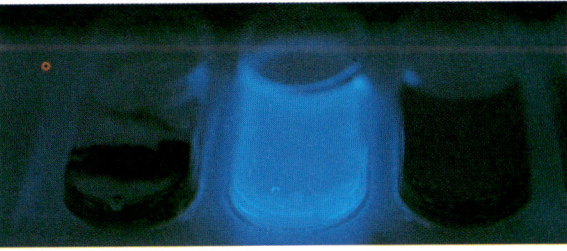

262

262. Misinterpretation of the aesculin hydrolysis test. Misinterpretation of the test for aesculin hydrolysis is common with both conventional and miniaturised methodology. The most common problem is false-positive reactions in which blackening is caused by the reduction of sulphite to sulphide. It is therefore necessary to check presumptive positive reactions under ultra-violet illumination. Fluorescence is indicative of the presence of aesculin and is thus a negative reaction.

15. *Clostridium perfringens*

Introduction

Clostridium perfringens is a Gram-positive, catalase-negative, endospore-forming, rod-shaped bacterium (**Figure 263**). It is a typical member of the genus *Clostridium* and requires anaerobic conditions for growth.

263

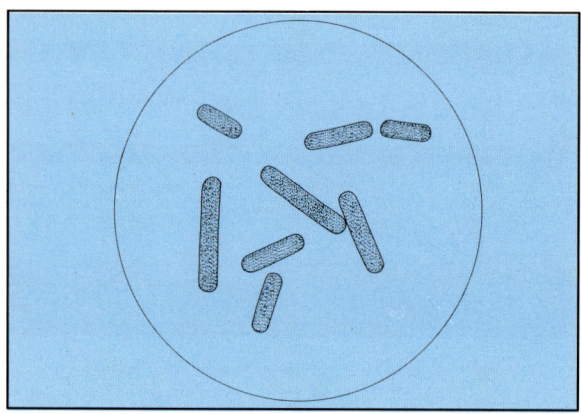

263. Morphology of *Cl. perfringens*. Cells of *Cl. perfringens* are straight, blunt-ended rods which occur singly, or in pairs. Endospores are only rarely seen *in vivo* or in common *in vitro* conditions. If present, the spores are large, oval and either central or subterminal, and the sporangium is distended.

There are five types of *Cl. perfringens*, A, B, C, D and E, which may be distinguished by their lethal toxins (**Table 112**; Sterne and Batty, 1975). These toxins are distinct from, and should not be confused with, the enterotoxin produced by food poisoning strains of *Cl. perfringens* type A (and some strains of C and D).

Enterotoxigenic *Cl. perfringens* type A is established as an important agent of foodborne disease, and has also been implicated as a possible cause of antibiotic associated diarrhoea (Borriello *et al.*, 1984). In such cases, the symptoms differ from those produced by classic *Cl. perfringens* food poisoning (see page 315), and person-to-person spread and environmental contamination are likely to be involved as vehicles

Table 112. Lethal toxins produced by different types of *Clostridium perfringens* (Cato *et al.*, 1986).

Type	Toxin			
	Alpha	*Beta*	*Epsilon*	*Iota*
A	+	–	–	–
B	+	+	+	–
C	+	+	–	–
D	+	–	+	–
E	+	–	–	+

(Borriello *et al.*, 1985; Larson and Borriello, 1988). *Cl. perfringens* type A may also be involved in some cases of infantile diarrhoea (Luzzi *et al.*, 1986), and a role in sudden infant death syndrome ('cot death') has been hypothesised (Murrel *et al.*, 1987).

Cl. perfringens type A is the cause of a number of extraintestinal diseases of man and animals, including gas gangrene (clostridial myonecrosis), intravascular haemolysis and bacteraemia. *Cl. perfringens* is also the most common species of *Clostridium* isolated from human infections, such as lower body abscesses, peritonitis and soft tissue infections. These infections are often polymicrobic and the organisms are probably derived from the colonic microflora (Finegold, 1977).

Other types of *Cl. perfringens* are agents of both human and animal disease. Type C is the cause of necrotic enteritis (pigbel), a food-associated disease of man, of importance in Papua, New Guinea and occasionally elsewhere, while *Cl. perfringens* type B and C cause necrotic enteritis and enterotoxaemia in a wide range of farm animals, as well as non-specific infections in domestic animals. *Cl. perfringens* is among the organisms which have been associated with necrotising enteritis in newborn infants, although a causative role is not definite (Borriello and Stephens, 1984).

Isolation of *Cl. perfringens*

Enrichment

Two-stage enrichment:selective plating techniques are rarely used in methodology for *Cl. perfringens* (Mead *et al.*, 1982). The usual approach is to use direct plating where numbers are likely to be high and most probable number techniques where numbers are low. However, if required enrichment may be made using either non-selective or selective media. Self-enrichment may also be employed with meats.

Direct plating

A number of solid media for the recovery and enumeration of *Cl. perfringens* from foods have been devised. Virtually all use one or more antibiotics as selective agents. Widely used media include tryptose-sulphite-cycloserine medium and its egg yolk-free variant, oleandomycin-polymyxin-sulphadiazine-perfringens medium and Shahidi Ferguson-perfringens medium.

Most probable number estimation

A large number of liquid media are available for the enumeration of small numbers of *Cl. perfringens*. These include differential reinforced clostridial medium (DRCM), LSUP medium, LS medium and iron-milk medium.

Lecithinase as an indicator of the prior growth of *Cl. perfringens*

A recognised problem during the investigation of food poisoning outbreaks where *Cl. perfringens* is suspected, is the possibility of a significant decline in the numbers of vegetative cells during the refrigerated storage of food samples. In an attempt to obviate the possibility of false low numbers of the organism being recovered, Harmon and Kautter (1970, 1974) proposed the determination of lecithinase (alpha toxin) as an indicator of the prior presence of large numbers of *Cl. perfringens*, and a standard method for the assay has been devised (Harmon and Duncan, 1984). Detectable levels of lecithinase are produced when numbers of *Cl. perfringens* reach 10^5 to 10^6 cfu/g — this number is significant as it corresponds with the population level commonly used as confirmation of an outbreak of food poisoning.

Despite the potential of the lecithinase test, its shortcomings are numerous (Foegeding, 1986), and include variation in the amount of lecithinase produced according to the strain of *Cl. perfringens* (Foegeding and Busta, 1980), food substrate and the time and temperature of growth (Park and Mikolajcik, 1979). It is possible to conclude, therefore, that while a detectable level of lecithinase is significant, the absence of detectable lecithinase does not exclude the possibility of the prior presence of large numbers of *Cl. perfringens*.

Cl. perfringens type C

Although *Cl. perfringens* type C is easily isolated from clinical cases, isolation from normal people is extremely difficult due to interference from the much more numerous type A organism. However, an affinity method using antibody coated silicate beads has been developed (Lawrence *et al.*, 1984).

Methods for the isolation of Cl. perfringens *are discussed in greater detail, with respect to choice of method and laboratory practice, in* **Laboratory Methods***, pages 322–325.*

Taxonomy of *Cl. perfringens*

Differentiation from other species

Differentiation from other species of *Clostridium* may be made on the basis of the biochemical criteria listed in **Table 113**. Conventional test methods and media or miniaturised methods such as API 20A©, API ATB 32A© or Minitek© may be used.

For a number of years, it has been common practice to use the Nagler reaction as a means of confirming the identity of *Cl. perfringens*. This reaction is based on the inhibition of lecithinase activity by type A antitoxin. A higher level of differentiation is obtained by combining the Nagler reaction with a test for fermentation of lactose (Hobbs *et al.*, 1971), but while these reactions are indicative of *Cl. perfringens* they are not specific to the genus. Culturally and biochemically similar species include *Cl. absonum*, *Cl. baratii* (*Cl. paraperfringens*), *Cl. perenne* and *Cl. sardiniensis* (Mead *et al.*, 1982). This technique also fails to detect lecithinase-negative strains of *Cl. perfringens* (Pinegar and Stringer, 1977). In some cases, at least lecithinase production may be demonstrated for apparently negative strains by using a medium modification (Serrano and Schneider, 1978). Alternatively, some workers have used the CAMP reaction with *Streptococcus agalactiae* as a means of presumptive identification (Hansen and Elliot, 1980).

Table 113. Differentiation of *Cl. perfringens* from related bacteria (Mead *et al.*, 1982).

	Clostridium						
	perfringens	*absonum*	*sordellii*[1]	*celatum*	*baratii*[4]	*perenne*	*sardiniensis*
Motility	−	−[2]	+	−	−	−	+[5]
Nitrate reduction	+	+	−	+	+	+	+[6]
Gelatin hydrolysis	+	+[3]	+	−	−	−	+[3]
Fermentation of							
Lactose	+	+	−	+	+	+	+
Raffinose	+	−	−	−	−	−	−
Salicin	−	+	−	+	+	+/−	+
Lecithinase	+	+	+	−	+	+/−	+
Inhibition of lecithinase by *Cl. perfringens* antitoxin	Total	Partial	Partial	N/A	Total	Total	Partial

[1] *Cl. bifermentans* gives identical reactions in these tests.
[2] Weak motility may be observed.
[3] Hydrolysis is slow.
[4] Synonymous with *Cl. paraperfringens*.

[5] Motility is weak.
[6] Nitrate reduction is weak with some strains.
Note: N/A = Not applicable.

Table 114. Differentiation of food poisoning and non-food poisoning strains of *Clostridium perfringens* by GLC (Kondo and Nagasui, 1988).

	End product		
	Acetic acid	*Butyric acid*	*Propionic acid*
Food poisoning	+	+	−
Non-food poisoning	−	−	+

Note: End products determined after growth in cooked meat medium containing 2% fructose.

Cl. perfringens is a relatively homogeneous species and, until recently, it has been considered that types cannot be reliably differentiated on the basis of cellular or colonial morphology, biochemical reactions or GLC analyses of fatty and organic acid metabolic endproducts (Cato *et al.*, 1986). More recently, however, Kondo and Nagasui (1988) have described a GLC technique which is claimed to differentiate food poisoning strains (**Table 114**).

The taxonomy of Cl. perfringens *is discussed in greater detail, with respect to methodology, in* **Laboratory Methods**, *page 325.*

Typing of *Cl. perfringens* type A

Serotyping

Almost all strains of *Cl. perfringens* that cause food poisoning are type A (Fruin, 1978), and it is not usually necessary to determine the lethal toxins produced. Serotyping is well established in epidemiological investigations of *Cl. perfringens* food poisoning, its value first being recognised by Hobbs *et al.* (1953). Serotyping is particularly useful where food samples are not available, or where *Cl. perfringens* cannot be isolated (Gross *et al.*, 1989) from implicated foods. It is recommended therefore that criteria for the confirmation of *Cl. perfringens* as a cause of food poisoning should include the presence of the same serovar in stool samples from most ill people, but not from the stools of people at risk but not ill (Hauschild, 1975; IAMFES, 1987).

The type-specific antigens of *Cl. perfringens* are located in the capsular polysaccharides (Cherniak and Henderson, 1972). Seventy-five antisera are in routine use at the Food Hygiene Laboratory, Colindale, UK, typing methods being a combination of slide and tube agglutination (Stringer *et al.*, 1982). Typing antisera are not commercially available.

In an investigation of outbreaks of *Cl. perfringens* food poisoning in the UK between 1970 and 1980, a causative serovar was established in 414 (64%) of 646 outbreaks. The use of an additional 68 American and Japanese antisera increased the typability by only 5%. This was attributed to differences in the source of strains used to raise antiserum (Stringer *et al.*, 1982). Serovars established as causative in outbreaks of food poisoning in the UK are listed in **Table 115**. In some outbreaks, two or more serovars were involved.

Serotyping of Cl. perfringens *is discussed in greater detail with respect to methodology in* **Laboratory Methods**, *page 326.*

Table 115. Serovars of *Clostridium perfringens* responsible for food poisoning in the UK, 1970–80 (Stringer *et al.*, 1982).

Serotype	Number of outbreaks
1	34
3, 4 and 3/4	125
11 and 11/13	27
29	31
41	37
Other types	232

Note: In a number of outbreaks two or more serotypes were involved.

Other typing methods

Bacteriocin typing

A bacteriocin typing scheme for *Cl. perfringens* has been described by Watson (1985). Fifty bacteriocins were used to type 802 isolates from food poisoning outbreaks and other sources. In most cases, the same serovar had the same pattern of bacteriocin sensitivity, but some isolates which could not be typed serologically were typable using bacteriocins.

A much higher proportion of food poisoning strains produced bacteriocins than isolates from human and animal infections or from the environment (Watson *et al.*, 1983).

Plasmid profiling

The possibility of employing plasmid profiling has been explored (Longden and Phillips-Jones, 1988), and is considered to be a potentially useful method in epidemiological studies. Further work is required to fully develop the technique.

Symptoms of *Cl. perfringens* infections

Food poisoning due to *Cl. perfringens* is characterised by abdominal pain, diarrhoea and nausea. In a few cases, nausea is severe and accompanied by vomiting. The incubation period is usually 8 to 22 hours (Pinegar and Suffield, 1982), but may be as short as 6 hours (Naik and Duncan, 1977) or, according to anecdotal accounts, less. The illness is of short duration and recovery within 1 to 2 days is normal.

Differential diagnosis is usually straightforward, although there may be confusion with the diarrhoeal form of *Bacillus cereus* food poisoning.

Antibiotic-associated diarrhoea due to *Cl. perfringens* resembles that due to *Cl. difficile* (see page 356). The onset of symptoms is usually explosive with severe diarrhoea and bloody faeces. The disease is of long duration, diarrhoea persisting for up to 11 days (Larson and Borriello, 1988).

Necrotic enteritis (pigbel) results in a severe, acute, colicky pain in the region of the umbilicus. The pain is precipitated by oral intake, a few sips of water being sufficient to produce symptoms. In the majority of cases, the patient is constipated but a few loose and bloody stools may be passed. Vomiting is an occasional symptom; fever, lethargy and a dry tongue are common but do not occur in all cases. The mortality is as high as 30 to 40%.

Persons at particular risk from *Cl. perfringens* food poisoning

As far as is known there are no predisposing medical conditions specific to *Cl. perfringens* food poisoning. From a cultural viewpoint, people at greatest risk are those who, by circumstance or preference, are exposed to institutional cooking. Hospitals and old peoples' homes are often implicated, but outbreaks are also common in schools, factory canteens and restaurants.

Antibiotic-associated diarrhoea due to *Cl. perfringens* occurs among elderly and hospitalised persons. The disease is more common in female patients than in male. Despite the terminology, cases of this type have been reported in people not receiving antibiotic therapy (Borriello *et al.*, 1984, 1985).

Necrotic enteritis, as its popular name 'pigbel' suggests, is often associated with the consumption of pork. It is not known whether the disease is due to *Cl. perfringens* derived from food, or to endogenous strains from the gut. Meat is not necessarily a risk factor, outbreaks having occurred among non-meat eaters in India. In all cases, necrotic enteritis occurs when feasting follows an extended period of poor nutrition, and is thus a disease of developing nations, although cases occurred in Germany (Darmbrand) during the food shortages which followed the Second World War (Zeissler and Rassfeld-Sternberg, 1949). A major predisposing factor is the low level of digestive proteases resulting from a low protein diet. In many cases, this is exacerbated by the presence of heat-stable trypsin inhibitors in dietary staples such as sweet potatoes.

Necrotic enteritis is more common among children than adults, the latter normally having a high level of immunity.

Treatment of *Cl. perfringens* food poisoning

Food poisoning due to *Cl. perfringens* is self-limiting and treatment is not usually necessary. Supportive measures may be required if symptoms are of particular severity or if the patient is very young, old or debilitated.

In the case of necrotic enteritis, broad-spectrum antibiotics are administered. Bowel rest is also necessary, and where prolonged treatment is required nutritional support should be administered. Surgical intervention may be required if there is delay in diagnosis.

Mechanisms of pathogenicity

Symptoms of *Cl. perfringens* are due to the action of an enterotoxin. This is usually, although not invariably, produced in the intestine and for this reason *Cl. perfringens* food poisoning is not an intoxication in the classic sense, as the ingestion of viable organisms is necessary (McDonel, 1986).

Nature of *Cl. perfringens* enterotoxin

Cl. perfringens enterotoxin consists of a single polypeptide of molecular weight *ca* 34 Kda (McDonel, 1986). The enterotoxin is situated either in the spore coat or trapped between core layers but, contrary to earlier thinking, is unlikely to be a major structural component of the coat (Ryu and Labbe, 1989). In its native form, the molecule comprises about 20% random coil structure and about 80% β-sheet (Granum, 1986). A 48 Kda *Cl. perfringens* enterotoxin-related protein is present in both enterotoxin-positive and enterotoxin-negative strains, and may be a precursor which can only be processed to enterotoxin by positive strains. Alternatively, a different regulator mechanism may be involved in positive and negative strains (Ryu and Labbe, 1989).

The enterotoxin is heat labile and biological activity is lost in 5 minutes at 60°C (McDonel, 1986).

Synthesis of *Cl. perfringens* enterotoxin

Ingestion of a large number of organisms is necessary both to initiate growth in the intestine, and to permit survival of a significant proportion of the population through the acid regions of the stomach (Sutton and Hobbs, 1971). Estimation of the numbers required vary from 10^6 cfu/g (Mead *et al.*, 1982) to 10^8 cfu/g (Foegeding, 1986).

The traditional view of *Cl. perfringens* enterotoxin production is that enterotoxin is produced during sporulation and released along with the mature endospore on cell lysis (Labbe, 1980; McDonel, 1986). Little is known of the genetic control of enterotoxin synthesis (Glatz, 1986), but it has been proposed that enterotoxin-producing strains have an altered regulation of synthesis, that leads to its production in quantities in excess of those required for spore coat development.

Although the traditional view of enterotoxin production by *Cl. perfringens* has been widely accepted, enterotoxin production has been observed by non-sporulating cultures (Goldner *et al.*, 1986), and some workers question the concept of production being sporulation specific (McLane *et al.*, 1988).

In addition to the usual situation of enterotoxin production in the intestine, production has also been shown to be possible in food (Naik and Duncan, 1977; Craven, 1980; Al-Obaidy *et al.*, 1985), and the former authors were able to demonstrate enterotoxin in a food sample actually involved in an outbreak. In the study of Al-Obaidy *et al.*, the enterotoxin was produced by vegetative cells. In general, the production of enterotoxin in food in significant quantities is con-

sidered to be rare. However, the use of more sensitive methods of toxin detection may reveal the phenomenon to be more widespread than was previously thought.

Mechanism of *Cl. perfringens* enterotoxin

Cl. perfringens enterotoxin is strongly cytotoxic and induced effects are observable within 15 to 60 minutes in either cultured cells or animal models. This may be contrasted with the delay of several hours before effects are observable with Shiga toxin (McLane *et al.*, 1988). In the rabbit, enterotoxin is most active in the ileum, mildly active in the jejunum and of very low activity in the duodenum and colon (McDonel, 1986).

Cl. perfringens enterotoxin superficially resembles cholera and related toxins in stimulating fluid loss from the intestinal mucosa. However, there are important differences from these toxins in that *Cl. perfringens* enterotoxin inhibits glucose absorption and produces histopathological damage to the wall of the small intestine (McDonel, 1986). Damage is characterised by desquamation of the villus tips of the intestinal wall (McDonel, 1980).

The action of the enterotoxin follows a sequence of events which are summarised in **Figure 264**. The

Time (minutes)	Action	
0	Binding Complex formation	
2	Insertion Ion permeability	Ca^{++} Non-dependent
5		
8	Macromolecular synthesis inhibition	
15	Morphological damage	
	Larger permeability alterations	
		Ca^{++} dependent
	Cell lysis Villi tip	
30	desquamation	
	Fluid loss	

264. Action of *Cl. perfringens* enterotoxin.
(Reproduced with permission from McLane, B.A., Hanna, P.C., Wnek, A.P. *Microbial Pathogenicity* **4:** 317–323. © 1988 Academic Press.)

initial event is the specific binding to one, or more, protein receptors followed by the insertion of the entire enterotoxin molecule into the membranes to form a complex of MW 150,000. Structural changes occur when the toxin molecule is exposed to the membrane environment, and amphiatic α-helical structures are exposed which can form membrane pores (Granum, 1990). A sudden change is initiated in ion fluxes, this change following immediately after (and indeed almost simultaneously with) insertion. The nature of events leading to changes in ion flux are not fully understood but are likely to involve the formation of membrane pores. Cellular metabolism is affected by changes in intracellular ion levels and macromolecular synthesis is inhibited. The influx of calcium ions is increased, resulting in morphological damage and permeability alterations for larger molecules. It is not known if these effects are linked.

Cell death occurs with the cumulative effects of the enterotoxin on cellular metabolism and histopathological damage appears. This appears to be responsible for the loss of normal intestinal function and the resulting secretion of fluids and electrolytes. The clinical manifestation of this effect is diarrhoea.

Detection of *Cl. perfringens* enterotoxin

Biological methods

A number of biological methods have been used for the detection of enterotoxin. The most widely used is the rabbit ligated ileal loop test (cf. *B. cereus*, page 288; Duncan *et al.*, 1968, Hauschild *et al.*, 1971). This and other earlier biological methods are summarised by Stringer *et al.* (1982).

Biological assays have largely been superseded by serological techniques, although in recent years there has been some interest in sensitive tissue culture assays using Vero cells (Mahoney *et al.*, 1989).

Serological methods

A number of serological methods have been devised for the detection of *Cl. perfringens* enterotoxin from culture filtrates (**Table 116**). In the past, immunoprecipitation methods such as double gel diffusion and counterimmunoelectrophoresis have been the

methods of choice (Stringer *et al.*, 1982, Foegeding, 1986), but a kit system based on reverse passive latex agglutination (RPLA) provides a straightforward and sensitive method of detection. RPLA may be used for the detection of enterotoxin in faeces (Birkhead *et al.*, 1988), as well as in culture filtrates. A simple DOT-ELISA method has also been developed, which is suitable for use by unsophisticated laboratories in developing nations (Mehta *et al.*, 1989)

Enterotoxin production may also be determined directly on colonies growing on an agar plate using a sensitive ELISA method (Stelma *et al.*, 1985).

Practical aspects of detection of Cl. perfringens *enterotoxin are discussed in greater detail in* **Laboratory Methods** *page 326.*

Table 116. Serological detection of *Clostridium perfringens* enterotoxin.

Double gel diffusion	Modification of Casman *et al.* (1969)
Electroimmunodiffusion	Duncan and Somers (1982)
Counterimmunoelectrophoresis	Naik and Duncan (1977)
Single gel diffusion	Genigeorgis *et al.*, (1973)
Reverse passive haemagglutination	Uemura *et al.* (1973)
Reverse passive latex agglutination	Commercial kit
Sandwich ELISA	Bartholomew *et al.* (1985)

Cl. perfringens type C

The symptoms of necrotic enteritis have been attributed to the action of the β-toxin. This changes the mucosa, reduces the mobility of the villi and enhances bacterial attachment. The absorbed toxin causes patchy necrosis of the intestinal wall below affected mucosa. Isolates from infants suffering necrotising enteritis have a high level of histidine decarboxylase, and it has been postulated that histamine may act as a promoting factor for a type I hypersensitivity reaction (Arseculeratne *et al.*, 1980) and that a similar mechanism may be involved in necrotic enteritis.

Human carriage of *Cl. perfringens*

Cl. perfringens type A is common in the intestine of man, and surveys have shown that 80 to 100% of the healthy population are excretors of the organism (Uemura *et al.*, 1974, Brant *et al.*, 1978).The main site of carriage is the distal gastrointestinal tract, although a small percentage carry *Cl. perfringens* in the normal oral or cervicovaginal microflora. *Cl. perfringens* may also be present in the urine (Cato *et al.*, 1986). Despite this level of carriage, it is doubtful whether food handlers as faecal excretors are of any significance with respect to *Cl. perfringens* food poisoning (Roberts, 1982).

It should be noted that when counts of *Cl. perfringens* endospores in faeces are made as a part of an epidemiological investigation, criteria for implication of *Cl. perfringens* as causative organism include a faecal spore count of greater than 10^5 cfu/g. Some care is needed in interpreting faecal spore counts in investigations of hospital outbreaks, since 'long stay' patients may carry high numbers of *Cl. perfringens* and yet remain free of diarrhoeal illness (Stringer *et al.*, 1983). Determination of faecal enterotoxin is recommended to obviate confusion of the aetiology in outbreaks where both ill and diarrhoea free people have similar faecal spore counts (Birkhead *et al.*, 1988).

Cl. perfringens in the environment

In addition to the intestine of man, *Cl. perfringens* is common in the intestinal microflora of other warm-blooded animals and in soil and dust. There is some variation in distribution according to organism type (Smith, 1975). Only type A is found as part of the microflora of both soil and intestinal tracts, this type having been isolated from a very wide range of animals. In contrast, types B, C, D and E appear to be obligate parasites. These types are usually isolated from animals, but occasionally are found in man.

The association with the gastrointestinal tract means that *Cl. perfringens* is invariably present in sewage, and enters water courses either as a result of sewage pollution or of contamination from surface run off. The organism is also present in marine sediments (Schroder and Schau, 1978).

In potable water, the presence of *Cl. perfringens* may be used as an indicator of long-term faecal pollution (Davis, 1981).

Foods commonly associated with *Cl. perfringens*

Meat and meat products

Cl. perfringens is present in small numbers in a wide range of meat products. This is of no direct significance, but in raw meats the level of contamination tends to reflect the standard of abattoir hygiene. The incidence recorded in surveys inevitably varies, but in raw meat, isolation rates as high as 73.1% for beef and veal, 68.8% for pork, 85% for lamb and 62.8% for poultry have been reported (Roberts, 1982). The organism is also common in cooked meats. Low numbers reflect the level of contamination of raw materials, but high numbers are a consequence of delayed or inadequate cooling. Growth on cooked products can be very rapid, and while spoilage of food may occur before dangerous levels are reached (Al-Obaidy *et al.*, 1985), this should not be relied on as a safety factor in the absence of correct practice.

Meat products are the predominant cause of *Cl. perfringens* food poisoning, being responsible in the USA, for over 70% of reported cases (Genigeorgis, 1986), for example. Cured products are only rarely involved (Labbe, 1988), and the classic dish is a meat stew where cooking has helped to create the anaerobic conditions necessary for the growth of *Cl. perfringens*. Under appropriate conditions, however, sufficiently anaerobic conditions exist in large joints and, in the case of poultry, within large birds. In a smaller number of cases, the prime causative factor is under-cooking rather than slow cooling or warm holding (Anon, 1984).

Examples of *Cl. perfringens* food poisoning due to meat products are listed in **Table 117**.

Table 117. Examples of *Clostridium perfringens* food poisoning due to meat products.

Product	Number affected	Circumstances
Pork	21	Pork cooked previous day and reheated
Turkey	4	Family Christmas dinner
Chicken liver pate	6	Customers in public house
Boiled chicken	65	Chicken held overnight at ambient temperature
Beef stew and dumplings	9	Not known
Meat gravy	100	Gravy made previous day and held overnight at ambient temperature
Roast beef	48	Meat cooked for *ca* 4 hours then held at 6°C for 14 hours
Shepherd's pie	30	Pie prepared on previous day
Turkey	49	Turkey inadequately cooked
Chicken	2 (1 died)	Prepared at home

Milk and milk products

Cl. perfringens, usually derived from faeces, is frequently present in small numbers in many raw and pasteurised milk products. This is unimportant with respect to food poisoning, although the organism may, on rare occasions, be responsible for spoilage of pasteurised milk. Storage at elevated temperatures is inevitably a major contributory factor to spoilage of this type.

Food poisoning due to *Cl. perfringens* is rare in milk products. An outbreak in the UK involving 6 people was attributed to the cheese sauce used for making cheeseburgers (Communicable Disease Surveillance Centre, 1987), the cheese for the sauce having been prepared on the same work surface as raw beef. A second, unpublished, outbreak concerned a milk pudding which had been kept 'warm' for an extended period, but in this case a full epidemiological investigation was not made.

Fish and fish products

Cl. perfringens may be isolated from fish, but there appears to be no systematic information concerning its incidence. According to Labbe *et al.* (1988), the body surfaces and alimentary canal of several species of fish harbour the organism, but ICMSF (1986) consider that *Cl. perfringens* gains access during processing or food service operations.

Fish is involved in infrequent outbreaks of *Cl. perfringens* food poisoning in the USA (Labbe, 1988), and the situation appears to be similar elsewhere. A large outbreak in the UK, which involved as many as 800 people in two incidents on consecutive days, was attributed to salmon mayonnaise (Communicable Disease Surveillance Centre, 1988). Salmon appears to present a greater risk than other fish, earlier large outbreaks having been reported by Hewitt *et al.* (1986). A common feature in each case was storage of the salmon for a long period between boiling and serving.

Miscellaneous products

The widespread distribution of *Cl. perfringens* and the persistence of its endospores means that the organism can be isolated from a wide range of foods. Salads and vegetables, for example, may contain *Cl. perfringens* as a result of contamination from soil (NRC, 1985). Isolations have also been reported from gums (tragacanth, locust bean, acacia; NRC, 1985), spices (Powers *et al.*, 1975) pepper (Pruthi, 1980), cereal products (NRC, 1985) and dried animal products such as blood, plasma and gelatin (ICMSF, 1986). The presence of *Cl. perfringens* in dried animal products is considered to be sufficiently hazardous to require sampling and laboratory examination on a regular basis (ICMSF, 1986), but in other cases the numbers present are very low and there has been no involvement with food poisoning.

The Authors are, however, aware of unconfirmed reports of *Cl. perfringens* food poisoning attributed to the vegetarian equivalent of meat dishes such as vegetable goulash. In addition, a confirmed outbreak occurred at a Caribbean special day in the UK when at least 40 people were ill after eating aubergine and coconut cream (Communicable Disease Surveillance Centre, 1988).

Factors affecting the growth and survival of *Cl. perfringens*

Temperature

Temperature is the single most important factor in the control of *Cl. perfringens*. The organism will grow at temperatures as low as 6°C, but growth is severely restricted below 15 to 20°C (Roberts, 1982), and thus cooling to below 10°C is usually an effective means of control. Growth does not occur during the first 2 hours after cooking, irrespective of temperature (Blankenship *et al.*, 1988), and thus a cooling period not exceeding 2 hours should ensure safety. However, growth is extremely rapid at optimal growth temperatures in the region of 45°C with a mean generation time of 10 to 12 minutes (Mead, 1969).

Cl. perfringens is able to grow at temperatures as high as 50°C, although the growth rate is reduced as the upper limit approaches. A minimum holding temperature of 60°C should therefore be used if food is to be kept hot before serving.

Vegetative cells of *Cl. perfringens* have no enhanced level of heat resistance, and heating food to 75°C ensures their destruction. This temperature should be attained when chilled food is reheated.

Cl. perfringens endospores fall into two categories with respect to heat resistance, with survival times at 100°C varying from a few minutes to several hours. Both types have been involved in food poisoning (Roberts, 1982). D_{90} values for heat resistant (non-haemolytic) strains and heat sensitive (haemolytic) strains are 15 to 145 minutes and 6 to 7 minutes respectively.

Sub-lethal heat treatment increases the sensitivity of *Cl. perfringens* endospores to curing ingredients (see below), and thus contributes to the safety of cooked, cured meat products (Chumney and Adams, 1980).

pH value

Cl. perfringens will grow over the pH range 5.0 to 9.0. At the lower end of the range, the limiting pH value depends on the acidulant and there are interactions with other factors such as temperature.

Water activity

With NaCl as humectant, the limiting water activity level for *Cl. perfringens* is 0.95 to 0.96. Limiting values depend on humectant, being lower, for example, with glycerol and higher with KCl. There is also strong interaction with pH value and temperature, the limiting a_w being significantly higher at sub-optimal incubation temperatures and pH values below 7.0 (Bartsch and Walker, 1982).

Redox potential (E_h)

Cl. perfringens is relatively tolerant of oxygen and will grow at E_h values as high as +350 mV depending on the substrate and the pH value. The redox potential of meat is lowered by cooking, and *Cl. perfringens* is often able to grow under what appear to be aerobic conditions.

Curing ingredients

Cl. perfringens is not a problem in raw, cured meats although it is able to grow at the concentrations of NaCl and $NaNO_2$ present in the modern product. A significant safety factor in cooked, cured meats is the sensitising effect of sub-lethal heat treatment to NaCl and $NaNO_2$ (Chumney and Adams, 1980). Damage occurred over the temperature range 70 to 90°C, and reduced the recovery of heat treated cells (relative to heat activated) by 94.7% in the presence of 200 mg/l $NaNO_2$ and by 99.6% in the presence of 3.5% NaCl.

Other inhibitors

Butylated hydroxyanisole totally inhibits *Cl. perfringens* at a concentration of 150 mg/l, and partially inhibits the organism at 100 mg/l. At the lower level, there is a strong interaction with pH value, the inhibitory effect being markedly increased at extremes of the pH range 5.5 to 8.5. Butylated hydroxyanisole is most effective when used in combination with nitrite, sorbic acid or parabens, but its effect is greatly reduced in the presence of lipids or surfactants (Klindworth *et al.*, 1979).

Competition from other micro-organisms

Cl. perfringens competes poorly with many spoilage micro-organisms. This is rarely of significance in cooked foods, although there are instances when the growth of the organism on large joints appears to have been inhibited by the growth of heat resistant species of *Enterococcus* and *Lactobacillus* (Authors' unpublished observations).

Laboratory methods

Isolation

Enrichment

Non-selective enrichment media are preferred in many cases, and a satisfactory method is to use cooked meat medium incubated at 43 to 45°C to introduce a degree of selectivity and to allow sufficiently rapid growth of *Cl. perfringens* to permit subculture in a working day (6 to 8 hours incubation) when the inoculum is large. Where a large number of competitive bacteria are present, selective enrichment media are likely to be more effective, and the medium described by Debevere (1979) is recommended. This comprises a basal medium of fluid thioglycollate without glucose, supplemented with cycloserine. Incubation at 46°C is recommended.

Self-enrichment is not widely used, at least partly because many laboratories lack vacuum packaging equipment. The technique is only effective where the product supports the growth of *Cl. perfringens*, and may fail with meats if a large number of competitive facultative anaerobes are present. Incubation is at 43 to 45°C and the swelling of the pack indicates the likely presence of *Cl. perfringens*. Precautions should be taken against the pack bursting and against spurting on opening.

Plating media

The various plating media are based on similar principles; antibiotics, singly or in combination, are used for selectivity and a differential reaction is incorporated to differentiate clostridia from non-clostridia. The active constituents of widely used plating media are summarised in **Table 118** and usage illustrated in **Figures 265–267**.

265

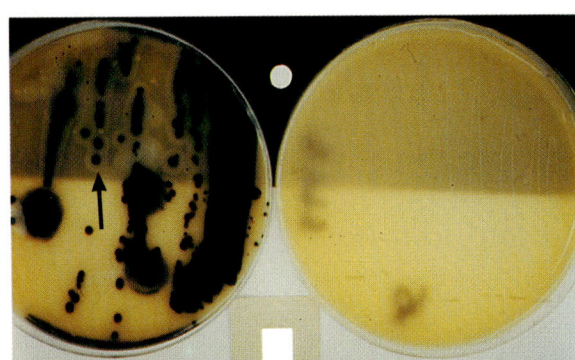

266

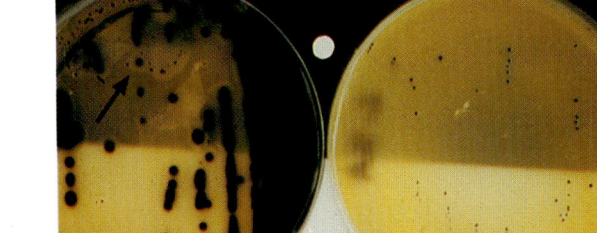

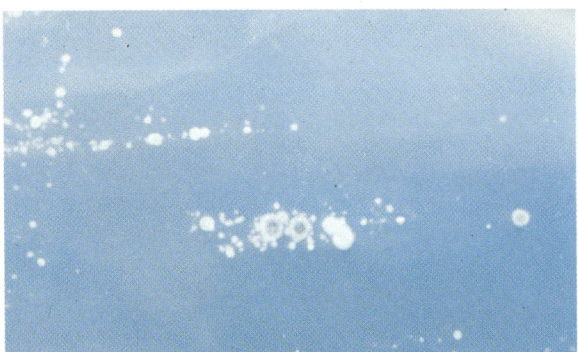

265–267. Isolation of *Cl. perfringens*. Presumptive recognition of *Cl. perfringens* on media containing egg yolk and sulphite, such as Shahidi Ferguson agar, is by intense blackening and a zone of precipitation around the colonies (arrowed; **265** & **266**(Reproduced by courtesy of Dr T.A. Roberts, AFRC Institute of Food Research, Reading Laboratory)). Blackening was absent, however, when a strain of *Cl. perfringens* was grown on SPS agar (**267**). This was probably due to conditions in the medium being insufficiently reducing as a result of using plates which had been stored for some time in air. (*Both media: 24 h at 37°C in 95% H$_2$/5% CO$_2$.*).

Table 118. Examples of media for isolation of Clostridium perfringens by direct plating.

Tryptose-sulphite-cycloserine[1]	
Selective agents	Cycloserine
Indicator system	Sulphide blackening: egg yolk
Egg yolk-free tryptose-sulphite-cycloserine[2]	
Selective agents	Cycloserine
Indicator system	Sulphide blackening[3]
Shahidi Ferguson-perfringens[4]	
Selective agents	Sulphadiazine: polymyxin: kanamycin
Indicator system	Sulphide blackening: egg yolk
Oleandomycin-polymyxin-sulphadiazine-perfringens[5]	
Selective agents	Oleandomycin: polymyxin: sulphadiazine
Indicator system	Sulphide blackening
Neomycin-blood[6]	
Selective agents	Neomycin
Indicator system	Haemolysis[7]
Sulphite-polymyxin-sulphadiazine[8]	
Selective agents	Polymyxin: sulphadiazine
Indicator system	Sulphide blackening

[1] Harmon *et al.* (1971).
[2] Hauschild and Hilsheimer (1974b).
[3] Egg yolk is added to base medium but not to the overlay.
[4] Shahidi and Ferguson (1971).
[5] Hanford (1974).
[6] Sutton and Hobbs (1968).
[7] Permits differentiation between heat-sensitive and heat-resistant strains of *Cl. perfringens*.
[8] Angelotti *et al.* (1962).

Of three widely-used media, oleandomycin-polymyxin-sulphadiazine-perfringens medium (OPSP), tryptose-sulphite-cycloserine medium (TSC) and Shahidi Ferguson-perfringens medium (SFP), the latter is least selective allowing the growth of several other genera including strains of *Serratia*, *Streptococcus lactis*, *Proteus*, *Enterobacter* and *Enterococcus*. Although differentiated by the sulphide-blackening reaction, these other genera may cause problems when present in large numbers. In contrast to SFP, TSC allows the growth of only *Serratia* and *Str. lactis*, while OPSP allows the growth only of occasional strains of *Enterococcus*. Incubation at 46°C increases the selectivity of all media, but the lower recovery compared with 37°C may be unacceptable. The addition of egg yolk to SFP and TSC to detect leci-thinase activity is not entirely successful, and black colonies should be considered to be presumptive *Cl. perfringens*, irrespective of the presence or absence of a reaction in egg yolk. In many cases, an egg yolk-free variant of TSC, which is easier to prepare but which gives comparable results, to TSC, is preferred.

Neomycin-blood agar is simple to prepare and permits differentiation of haemolytic heat-sensitive strains of *Cl. perfringens* and non-haemolytic heat-resistant strains. Recovery of heat-resistant strains may, however, be poor (Spencer, 1969), and the medium is not recommended where numbers of the organism are low.

The difficulties caused by sulphite reducing clostridia other than *Cl. perfringens* are potentially much greater, particularly if *Cl. bifermentans* or *Cl. butyricum* are present. These species produce large, black colonies which tend to spread and obscure the whole medium. *Cl. bifermentans* and *Cl. butyricum* are inhibited on OPSP medium, but grow on both TSC and SFP, *Cl. bifermentans* being partially inhibited on TSC. OPSP is the medium of choice with foods where the two other clostridia are commonly present.

Although SFP is the least selective of the three media, recovery of *Cl. perfringens* may be greater, particularly if cells have been stressed. This medium may therefore be preferred where the number of competing bacteria is not significant. However, TSC medium, or its egg yolk-free variant, is often considered to be the most satisfactory general purpose medium.

Pour plates are recommended when OPSP is used (Handford, 1974), but surface spread plates with an overlay of the same medium (without egg yolk where appropriate) may be used with TSC and SFP. Spread plates are generally considered to be more convenient, but recovery may be greatest if pour plates are used (Hauschild and Hilsheimer, 1973).

Although incubation in an atmosphere of hydrogen: carbon dioxide is often specified for the recovery of *Cl. perfringens*, Mead *et al.* (1982) stated that there is no evidence that a particular gas mixture is critical. However, reducing conditions are required for consistent blackening (Mead, 1969).

A number of deep culture methods have been proposed to avoid the use of anaerobic jars (Mead *et al.*, 1982), but have rarely been used in practice. A potential alternative to the use of anaerobic jars is the incorporation of an oxygen reducing membrane fraction that permits normal plate culture techniques (Hoskins and Davidson, 1988).

Most probable number techniques

A number of liquid media have been proposed for most probable number estimations of *Cl. perfringens* (**Table 119**), and usage is illustrated in **Figures 268 & 269**. Counts obtained using this method are often

very inaccurate, especially where media contain no selective agents and rely entirely on the sulphide blackening reaction to differentiate *Cl. perfringens*. The blackening reaction may be caused by many bacteria other than clostridia (or specifically *Cl. perfringens*), and dependence on this reaction is likely to lead to false high counts. Despite this differential reinforced clostridial medium, which is of this type, is popular in the UK food industry.

Pasteurisation of presumptive positive cultures followed by subculture into the same medium is commonly used as a means of confirming the presence of *Clostridium spp*, although not specifically of *Cl. perfringens*. The pasteurisation procedure is both cumbersome and unreliable, and may lead to an underestimate of the number of *Cl. perfringens* (and other clostridia) present (Roberts and Smart, 1976).

LSUP contains polymyxin as selective agent, this antibiotic does not, however, markedly increase the selectivity of the medium. Cycloserine may well be of greater value (Smart *et al.*, 1979).

Lactose fermentation is used as a differential character in a number of media including rapid perfringens medium, LS medium and iron-milk medium. *Cl. perfringens* is recognised in LS medium both by sulphide blackening and gas production from lactose, and confirmatory tests are not required when incubation is at 46°C. Gas formation occurs between 18 and 24 hours (Beerens *et al.*, 1982). The simple iron-milk medium has been reported to give good recovery and specificity (Abeyta, 1985), but in the Authors' experience gas formation can be variable.

An incubation temperature of 46°C is recommended for most probable number determination of *Cl. perfringens*, and results are obtained within 24 hours.

268 & 269. Liquid media for the enumeration of *Cl. perfringens* by the most probable number technique.
The most probable number technique is widely used for the examination of foods which contain small numbers of *Cl. perfringens*. 'Positive' reactions in differential reinforced clostridial medium (DRCM) are illustrated in **268**. In this case, sulphide blackening is the sole criterion for determining the presumptive presence of *Cl. perfringens*. When the same sample (meat extract) was inoculated into LS medium, however, the cultures were judged to be negative for *Cl. perfringens*, since despite the sulphide blackening (**269**) there was no evidence of gas production from lactose, the second differential feature in that medium. It should also be noted that the air space above the medium is too great and bottles should contain a greater volume of medium. (*DRCM/LS medium: 24h at 46°C.*)

Table 119. Examples of media for enumeration of *Clostridium perfringens* by most probable number methods.

LS[1]	
Selective agents	None
Indicator system	Sulphide blackening: lactose fermentation
Rapid perfringens[2]	
Selective agents	Neomycin: polymyxin
Indicator system	Lactose fermentation
Iron–milk[3]	
Selective agents	None
Indicator system	Lactose fermentation
Differential reinforced clostridial[4]	
Selective agents	None
Indicator system	Sulphide blackening

[1] Beerens *et al.* (1982).
[2] Ericksen and Deibel (1978).
[3] St John *et al.* (1982).
[4] Hirsch and Grinstead (1954).

Isolation of Cl. perfringens *type C from foods*

The affinity technique for the isolation of *Cl. perfringens* type C, when massively outnumbered by type A, is based on the antigenic dissimilarity of the two types (Lawrence *et al.*, 1984). The isolation involves two major stages:

1 Binding of type C cells to beads coated with specific type C antibody. Most type A cells fail to bind and are washed away.
2 Plating out of the beads, on non-selective medium to recover *Cl. perfringens* type C.

Rapid detection

DNA probes for the detection of enterotoxigenic strains of *Cl. perfringens* are now available, although not on a commercial basis (Van Damme-Jongsten *et al.*, 1990).

Identification of *Cl. perfringens*

Full characterisation of clostridia is a time consuming procedure, and for *Cl. perfringens* it is usual to restrict identification to a few key tests. Four reactions, lack of motility, nitrate reduction, gelatin liquefaction and lactose fermentation, are usually sufficient to distinguish *Cl. perfringens* from interfering micro-organisms. In some cases, additional tests may be required, including the fermentation of raffinose (+1) and salicin (−).

The four main tests may conveniently be made in the combined lactose-gelatin (Hauschild and Hilsheimer, 1974a) and nitrate-motility (Hauschild and Hilsheimer, 1974b) media. These media are not available as complete preparations, and in the case of lactose-gelatin medium, gelatin must be obtained from reputable sources to avoid inconsistent results (Hauschild *et al.*, 1977). Fermentation of raffinose and salicin may be determined using the medium and method of Harmon and Duncan (1984).

Miniaturised test kits such as the API 20A©, API ATB 32A© and Minitek© are a convenient alternative to conventional methods, and have the additional advantage of permitting a wider range of characters to be determined with little additional work. The API 20A and ATB 32A systems may also be used in conjunction with a computerised database.

Despite the shortcomings of the test (page 314), the Nagler reaction is widely used as a means of confirming the presence of *Cl. perfringens*. In some cases, a positive Nagler reaction may be taken as adequate confirmation without further testing, while in others it is considered to be presumptive evidence only, and further biochemical tests are carried out on Nagler positive cultures (**Figures 270** & **271**).

The possibility of the presence of lecithinase-negative strains of *Cl. perfringens* must always be considered even where the modified medium of Serrano and Schneider (1977) is used. It is recommended that when clostridia are isolated during the investigation of food poisoning outbreaks, biochemical confirmation of identity should be made irrespective of the results of the Nagler test.

The technique of the CAMP test is the same as that used with other organisms such as *Listeria monocytogenes* (see page 353). *Str. agalactiae* is used to induce synergistic haemolysis on blood agar plates, the two organisms being streaked at right angles (Hansen and Elliot, 1980). The CAMP test is not widely used with *Cl. perfringens* and is in any case applicable only to haemolytic strains. Interpretation of results may be difficult and should be made only by experienced people.

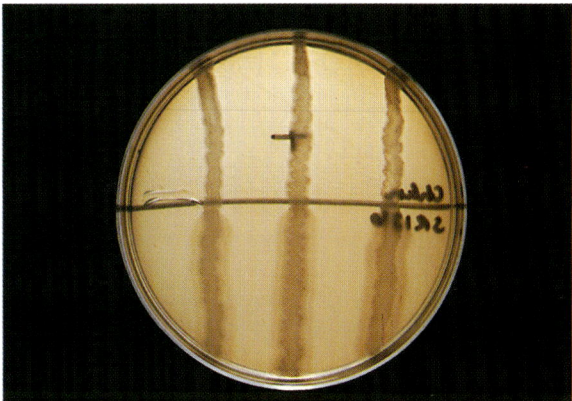

270

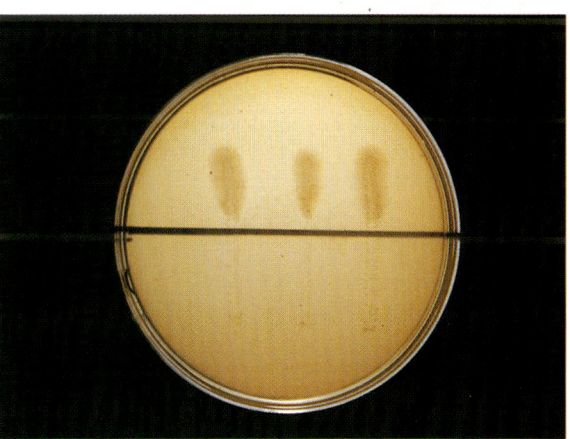

271

270 & 271. Inhibition of lecithinase by *Cl. perfringens* antitoxin (Nagler reaction). 'Half plates are used, one half containing egg yolk agar and antitoxin, the other containing egg yolk agar alone. Each isolate under test is inoculated by a single streak across the two halves. In a typical positive reaction, lecithinase production, but not growth, is inhibited in the presence of antitoxin (**270**). Growth, however, may also be inhibited (**271**). (*Egg yolk agar: 24 h at 37°C.*)

Typing of *Cl. perfringens*

Serotyping

Typing antisera for *Cl. perfringens* are not commercially available, and where serotyping is required, reference should be made to the appropriate national organisation. The description of methodology is therefore brief and is based on practice at the Central Public Health Laboratory, London, UK (Stringer *et al.*, 1982).

Smooth colonies are required for typing of *Cl. perfringens*, roughness being associated with loss of the capsule and thus type-specific antigens. Attempts to convert strains from rough to smooth have met only limited success.

Slide agglutination

Seventy-five antisera are used on a routine basis. These are arranged in a system of eight primary pools, each of which contain nine or ten antisera and which is divided into three secondary pools containing three or four antisera.

Colonies, growing on either blood agar or nutrient egg yolk medium, are emulsified in 0.85% physiological saline on a glass slide, mixed with a loopful of antiserum at a dilution of 1:5 and observed for agglutination. Agglutination with more than one antiserum is usually a consistent property, and such strains are designated multiple types.

Tube agglutination

Tube agglutination is used to determine homologous and cross-reacting titres. Doubling dilutions of antisera are made in 0.85% physiological saline to give a final volume of 0.5 ml in each well of an agglutination tray. Two drops (0.04 ml) of antigen suspension are added to each well, and the tray incubated at 37°C for 18 hours. The titre is calculated as the reciprocal of the highest dilution at which strong agglutination occurred.

Detection of *Cl. perfringens* enterotoxin

Sporulation medium

Although the concept of sporulation dependent enterotoxin production has been challenged (see page 317), it remains practice to use a sporulation medium to enhance production of enterotoxin. A number of sporulation media have been devised, but may not be suitable for all strains. However, good results have been obtained using the medium of Tsai *et al.* (1974). A means of increasing sporulation in a medium by using a pre-reduced anaerobic sterilisation technique or with carbon dioxide or carbonate has been described more recently (Craven, 1988), and would seem to be of considerable potential value.

Detection

In recent years, serological methods have largely superseded animal tests for the determination of *Cl. perfringens* toxin, and while tissue culture assays using Vero cells have also been developed, these are not suitable for use in most food industry laboratories.

A number of serological means of detection have been applied to *Cl. perfringens* enterotoxin (see **Table 116**, page 318 and Stringer *et al.*, 1982), but only two are to be discussed further, the reverse passive latex technique for culture filtrates and faeces, and the ELISA technique for direct detection of enterotoxin on an agar medium plate (Stelma *et al.*, 1985).

Reverse passive latex agglutination (RPLA)

A test kit based on RPLA has been developed (Denka-Seiken; Unipath©), and enables less sophisticated laboratories to assay for enterotoxin production in both culture supernatants and stool samples. An independent evaluation (Harmon and Kautter, 1986) showed the kit to be consistent and accurate both in identifying and quantifying *Cl. perfringens* enterotoxin. In the absence of interfering substances which may occur in faeces, the sensitivity is *ca* 3 ng/ml. Although less sensitive than either ELISA or counter-immunoelectrophoresis, RPLA is easier to use and requires less specialised reagents (Harmon and Kautter, 1986).

Colony ELISA

The ELISA method devised by Labbe (1980) employs a standard assay technique. Colonies must be blotted from the agar surface after growth and sporulation to remove cellular components, probably the spore coat protein, which non-specifically binds rabbit IgG. Enterotoxin remaining at the site of the colony may then be bound onto nitro-cellulose filter paper prior to the ELISA assay.

16. *Listeria monocytogenes*

Introduction

The genus *Listeria* is composed of small, Gram-positive, coccoid to rod-shaped, bacteria which exhibit a characteristic tumbling motility at room temperature. The organisms are catalase-positive, oxidase-negative and have a fermentative metabolism of glucose, producing acid but not gas. Endospores are not formed.

Seven species of *Listeria* are currently recognised: *L. monocytogenes*, *L. ivanovii*, *L. innocua*, *L. welshimeri*, *L. seeligeri*, *L. grayi*, *L. murrayi*. Only *L. monocytogenes* is associated consistently with disease in man, but two other haemolytic species, *L. ivanovii* and *L. seeligeri*, have been implicated on rare occasions (Seeliger, 1984; Kluge and Hof, 1986; Rocourt *et al.*, 1986). An eigthth species, *L. denitrificans*, is not considered to be a member of the genus, but its taxonomic position remains uncertain (Seeliger and Jones, 1986).

Before the 1980s, *L. monocytogenes* was of concern primarily as a cause of disease in animals. Ruminants are most commonly involved, the organism being associated with abortions and encephalitis in cattle and sheep. Subsequently, the role of food as a vehicle for *Listeria* infections in humans has been recognised, and the organism has been the subject not only of scientific but also of very considerable public concern. It is unfortunate that some of the more sensational aspects of lay reporting have entered the scientific debate, and that the term 'listeria hysteria' should be used so freely in relation to justified and systematic studies of the organism and its significance in public health.

Isolation and detection of *L. monocytogenes*

Conventional cultural techniques

Attempts to recover *L. monocytogenes* from foods by direct plating have been largely unsuccessful and a two-stage enrichment:selective plating technique is most commonly used. Resuscitation is employed only rarely. The sudden increase in interest in *L. monocytogenes* and the previous lack of generally accepted media and methods for its isolation, has led to the development of a vast number of new media which, in many ways, has served to confuse the practising microbiologist rather than to facilitate isolation of the organism.

Enrichment

There are two approaches to enrichment for *L. monocytogenes*, low-temperature enrichment normally using non-selective broth, and enrichment at higher temperatures using a selective broth.

A number of two-stage enrichments have also been proposed, which either combine low temperature and selective procedures (Hayes *et al.*, 1986), or use a less selective broth for the first stage of enrichment (Lee and McClain, 1986).

Selective plating media

A very large number of solid selective media for *L. monocytogenes* have been developed in recent years. Each of these media has a claim for its own special advantages (Cassiday and Brackett, 1989) but modified McBride agar (McBride and Girard, 1960; Francis *et al.*, 1984) has remained widely used. Later media which have been applied widely include GNT agar (Martin *et al.*, 1984), LPM agar (Lee and McLain, 1986) and Oxford agar (Curtis *et al.*, 1989).

Conventional cultural methods for the isolation of L. monocytogenes *are discussed in greater detail, with respect to choice of method and laboratory practice, in* **Laboratory Methods**, *pages 343–349.*

Rapid detection methods

A number of serological and genetic methods for the detection of *Listeria* species have been developed and commercial kits are available based on ELISA and DNA hybridisation respectively. Some of these methods are not specific for *L. monocytogenes*, but detect all species of *Listeria* and a similar restriction applies to the use of electrical techniques. However, species specific probes for *L. monocytogenes* are under development (Datta, 1988; Flamm *et al.*, 1989), and two serological methods which specifically detect *L. monocytogenes* are also available – flow cytometry, which is not suitable for routine application (Donnelly and Baigent, 1986), and a direct immunofluoresence test (McLauchlin and Pini, 1989) that offers promise of routine application.

Some aspects of rapid detection methods for L. *monocytogenes are discussed in greater detail, with respect to choice of method and laboratory practice, in* **Laboratory Methods**, *pages 348–351.*

Taxonomy of *L. monocytogenes*

Differentiation from similar genera

Listeria may be differentiated from similar genera using the criteria in **Table 120**. Conventional media may be used, but it is convenient to use the API 20 strep[©] system.

Table 120. Differentiation of *Listeria* from similar genera.

	Listeria	*Brochothrix*	*Lactobacillus*	*Kurthia*
Blue-green colonies[1]	+	−[2]	−	−
Facultative anaerobe	+	+	+	−
Motile at 25°C	+	−	+[3]	+
Growth at 35°C	+	−[4]	+[5]	+
Acid from glucose	+	+	+	−
Catalase	+	+	−[6]	+

[1] When viewed microscopically using oblique lighting on tryptose agar.
[2] There are unconfirmed reports that some strains of *Brochothrix* have a similar colonial appearance to *Listeria*.
[3] Strains of some species may be motile on high pH value, low carbohydrate media.
[4] Occasional strains of '*Brochothrix*-like' bacteria will grow at 35°C.
[5] Some strains, especially from meat, will not grow at 35°C.
[6] Strains of some species may give a positive reaction on a high pH value, low carbohydrate medium, or in the presence of haem.

Differentiation of *L. monocytogenes* from other species

L. monocytogenes may be differentiated from other species of *Listeria* according to the criteria in **Table 121**. The API 20 strep[©] system forms a useful means of determining the basic biochemical characteristics. The API 20 strep alone cannot, however, reliably differentiate between species, and this kit should be used in conjunction with the CAMP test. Alternatively, the Mast ID[©] or, less conveniently, the API 50 CH[©] systems may be used for species identification without additional tests (Kerr *et al.*, 1990). A technique for the direct identification of pathogenic species (but not specifically *L. monocytogenes*) may also be of value (Blanco *et al.*, 1989).

Serological methods are of no use for confirming the identity of *L. monocytogenes*, although serology may be used epidemiologically to support the classification of the isolate (Lovett, 1988; see below). Pathogenicity is still sometimes used as a means of confirming the classification of *L. monocytogenes*, but this is considered to be neither necessary nor confirmatory (Lovett, 1988).

Table 121. Differentiation of *Listeria monocytogenes* from other species of *Listeria*.

	Listeria							
	monocytogenes	*ivanovii*	*seeligeri*	*innocua*	*grayi*	*murrayi*	*welshimeri*	*denitrificans*[2]
ß-haemolysis	+	+	+	−	−	−	−	−
CAMP reaction								
Staph. aureus	+	−	+[1]	−	−	−	−	−
R. equi	−	+	−	−	−	−	−	−
Nitrate reduction	−	−	−	−	−	+	−	+
Voges–Proskauer	+	+	+	+	+	+	+	−
Acid from								
Mannitol	−	−	−	−	+	+	−	−
Rhamnose	+	−	−	+/−	−	+/−	+/−	−
Xylose	−	+	+	−	−	−	+	+
Hippurate hydrolysis	+	+		+	−	−		−

[1] Weak reaction only.
[2] *L. denitrificans* is not considered to be a member of the genus *Listeria*.

Typing of *L. monocytogenes*

Serotyping

L. monocytogenes may be typed on the basis of the agglutination reactions with highly absorbed rabbit antisera (Seeliger and Hohne, 1979). A number of O and H antigens have been identified and used to subdivide *L. monocytogenes* into 13 serovars; 1/2a, 1/2b, 1/2c, 3a, 3b, 3c, 4a, 4ab, 4b, 4c, 4d, 4e and 7. Surveys in a number of countries have shown that serotyping is of limited practical value since a large proportion of strains from both human and animal infections belong to one of a small number of serovars. For example, in the UK, 91% of all isolates from humans comprised only three serovars, the majority of which (59%) were serovar 4b (**Table 122**; McLauchlin, 1987).

Typing sera are not generally available on a commercial basis, although serovars 1 and 4 can be determined with material available in some countries. However, isolates from foods are often non-typable using commercial antisera (Lovett, 1988).

Table 122. Relative occurrence of serovars of *Listeria monocytogenes* in human illness in the UK (McLaughlin, 1987).

Serovar	Number of cases	
1/2a	130	(18%)
1/2b	99	(14%)
1/2c	29	(4%)
3	11	(1%)
4b	423	(59%)
Other 4	29	(4%)
Non-typable	1	
Total	722	

Phage typing

Lysogenic phages are present in all species of *Listeria*, and a typing system has been devised for *L. monocytogenes* (Audurier *et al.*, 1979). An evaluation of the scheme in the UK (McLauchlin *et al.*, 1986a), showed it to be highly reproducible and discriminatory, but that only 64% of strains were typable (37% for serogroup 1/2 and 82% for serogroup 4). The shortcomings of the current phage typing scheme are recognised, and attempts are being made, on an international collaborative basis, to develop a more 'useful' typing set (Rocourt *et al.*, 1985). Despite this it sems unlikely, that inherent problems with phage typing of *L. mono-*

cytogenes can be fully overcome (McLauchlin, 1987). However, phage typing has been succesfully applied to restricted situations, such as differentiation of listerias present as sporadic contaminants in dairy factories from factory-specific strains (Loessner and Busse, 1990).

Other typing schemes

A number of approaches have been made to the problem of devising a sensitive and suitably discriminatory typing scheme for *L. monocytogenes*. Attempts have been made to use plasmid-typing (Mottice *et al.*, 1987), but this method is inherently limited by the existence of non-plasmid bearing, but pathogenic, strains and by the fact that multiple plasmid types may occur in the same epidemic (Schlech, 1988).

Isoenzyme analysis shows some promise as a typing method (Bibb *et al.*, 1986), but the most effective means of typing may well be DNA fingerprinting. This technique, which reveals DNA restriction fragment length polymorphisms, has already been used in epidemiological investigations with encouraging results (Facinelli *et al.*, 1988; McLauchlin *et al.*, 1988), although further work is required to substantiate the technique.

Symptoms of *L. monocytogenes* infection

Ingestion of *L. monocytogenes* and subsequent invasion of macrophages may be marked by a 'flu-like' illness with symptoms of malaise, diarrhoea and a mild fever. Symptoms usually appear between one and several weeks after ingestion of infected food (WHO, 1988), although an incubation period as short as 4 days has been recorded (Kerr *et al.*, 1988). This 'enteric phase' may, however, be symptom-free or so mild as to go unnoticed (Lovett and Twedt, 1988).

In healthy people, listeriosis usually never develops beyond the 'enteric phase', but in susceptible people the organism gains access to all parts of the body. Fully developed human listeriosis is characterised either by the formation of miliary granulomas (listeriomas), masses of inflamed tissue comprising many small lesions, and focal necroses, or by suppuration in the infected tissue (Seeliger and Finger, 1976). The size, number and site of lesions vary according to the mode of infection, dose of organism and age and resistance of the host. The strain of *L. monocytogenes* involved may also be important in determining the site of infection, a strain involved in a Swiss common source outbreak having a distinct tropism for the brain stem of healthy adults (Malinverni *et al.*, 1986).

Listeriosis during pregnancy

The usual symptoms of infection in pregnant women are fever, chills, headache, backache and discoloured urine. Accompanying symptoms may include pharyngitis, diarrhoea and pyelitis, but meningitis is rare during pregnancy. In some cases, symptoms are very mild.

During the stage when symptoms are apparent, *L. monocytogenes* may be isolated from a wide range of tissues including blood, umbilical cord blood, vaginal mucus and placental tissues (Seeliger and Finger, 1976). Infection in the mother usually, but not invariably leads to infection of the foetus by either the transplacental route or during delivery. The interval between maternal and foetal infection is poorly defined and variable. In some cases, abortion or stillbirth occurs immediately after development of the maternal symptoms, while in others, these events are delayed for several weeks. Most cases of foetal listeriosis occur after the fifth month of pregnancy, although cases have been reported before the fourth month.

Although recovery by the mother is usually complete, *L. monocytogenes* may persist in the genital tract for an extended period, and cause problems during a subsequent pregnancy (Seeliger, 1961).

Listeriosis of the newborn (granulomatosis infantiseptica)

Infection of the foetus by the transplacental route usually results in a bacteraemia, which in itself may be responsible for dissemination of the organism through various organs and tissues (Seeliger, 1961). Alternatively, *L. monocytogenes* may be present in foetal urine which is discharged into, and thus infects, the amniotic fluid. The amniotic fluid is aspirated by the foetus which results in infection of the respiratory and gastrointestinal tracts (Marth, 1988).

Symptoms of listeriosis of the newborn may vary considerably, but common symptoms are respiratory distress, heart failure, cyanosis, refusal to drink, vomiting and convulsions. Symptoms are often accompanied by a soft whimpering, and there may be an early discharge of meconium and mucous stools. The presence of small cutaneous granulomas in the posterior pharyngeal wall is a common and characteristic symptom, and there may also be cutaneous nodules on the back and lumbar regions (Seeliger and Finger, 1976). Temperature may be either elevated or subnormal (Marth, 1988).

The mortality rate in listeriosis of the newborn is in the order of 30%, and while recovery is usually complete, there may be long-term medical difficulties, particularly where the disease was severe. Two out of eight survivors who were examined at 16 months of age appeared to be of normal intelligence but had, respectively, mild and moderate spastic diplegia (Evans *et al.*, 1984). In both cases, the children had been born prematurely with severe perinatal disease and had complications of the central nervous system immediately after birth.

Meningitis and meningoencephalitis

Meningitic listeriosis usually affects very young, or newborn, children, and people, usually male, aged over 50 years. The symptoms differ between the age groups, but cannot be distinguished from meningitis due to other bacteria. The onset of the disease is usually sudden and development of severe symptoms rapid, the mortality is *ca* 70% where treatment is not available or is delayed. Meningitis may also develop from other forms of listeriosis such as septicaemic or oculoglandular listeriosis (see below; Seeliger and Finger, 1976).

Symptoms of meningitic listeriosis in infants include shallow or rapid breathing, cyanosis, lethargy, fever and anorexia. These may be accompanied by convulsions or muscular twitching. As the illness progresses there are sudden interruptions in respiration with severe cyanosis and irritability. In fatal cases, the infant is usually lethargic or delirious as death approaches.

Symptoms in adults vary considerably particularly in the later stages of the illness. Initial symptoms are usually similar to those of influenza, followed by severe headache, pains and cramps in the limbs, chills, high temperature, stiff neck, nausea, vomiting and photophobia. The patient becomes somnolent and there are intermittent, but increasingly frequent bouts, of convulsion and delirium before coma and death.

Encephalitic meningitis has two distinct phases. In the first, which is usually of about 10 days duration, the patient suffers symptoms which include headache, backache, vomiting, conjunctivitis and rhinitis. The onset of the second phase is marked by a high fever followed by disturbance in the central nervous system. If treatment is not available, death follows in 2 to 3 days (Marth, 1988).

Complications may occur during recovery from meningitic and encephalitic listeriosis. Examples include hydrocephalus, ptosis, impaired vision and squint, difficulty in walking, failing memory, incoherent speech and palsy (Seeliger, 1961).

Septicaemic listeriosis

Although the primary symptom of this form of *L. monocytogenes* infection is fever, this is usually present in conjunction with severe pharyngitis and a leukocytosis accompanied by mononucleosis. Complete recovery is usual, but septicaemic listeriosis sometimes progresses to the meningitic form. This syndrome closely mimics infectious mononucleosis (Marth, 1988), and differential diagnosis may be difficult.

Oculoglandular listeriosis

This form may develop independently of other *L. monocytogenes* infections, or may accompany septicaemic listeriosis. Symptoms involve a localised conjunctivitis that can progress to a purulent meningitis which may be fatal (Seeliger and Finger, 1976).

Granulomatosis septica and pneumonic listeriosis

Listeriosis may take a septic course with various clinical manifestations. This form may accompany septicaemic listeriosis. In some cases, high fevers predominate and the symptoms resemble those of typhoid fever; symptoms of pneumonia or subacute endocarditis may also develop (Seeliger and Finger, 1976).

Other forms of listeriosis

Other forms of listeriosis occasionally occur. Cervicoglandular listeriosis is characterised by inflammation of lymph nodes which may become purulent and discharge pus. Surgery may be required to permit drainage. Cutaneous listeriosis involves the development of skin nodules which increase in size from that of a pin-head to that of a pea over 1 to 2 days. The centre of the nodule becomes pustular and the edge reddened. This form of listeriosis is contracted by direct contact with infected animal tissues. Focal infections also occur and may result in arthritis, osteomyelitis, peritonitis, spinal and brain abscesses and cholecystitis (Armstrong, 1985).

Persons at particular risk from *L. monocytogenes* infection

Susceptibility

Under the correct circumstances anyone can be infected with *L. monocytogenes*, but in most cases people remain symptom-free (Marth, 1988). However, there are fairly distinct sub-populations who, if infected, are likely to develop the full disease. These include pregnant women, neonates and infants and adults with compromised cell-mediated immunity. The latter group comprises people suffering from leukaemia or some other malignancy, undergoing long-term treatment with corticosteroids, cytotoxic drugs, or both, and people undergoing treatments such as haemodialysis. Alcoholism may also be a predisposing factor (Griffiths, 1989).

In contrast to expectations, infections with *L. monocytogenes* are relatively rare among people with acquired immunodeficiency syndrome (AIDS), although some cases have been reported (Harvey and Chandraesekar, 1988). A number of reasons have been postulated for the relatively low incidence in these circumstances, including the existence of an altered gut microflora, non-specific host defence mechanisms and a T cell subset other than T4, which provides anti-listeria activity. (Harvey and Chandraesekar, 1988).

All well-documented cases of listeriosis among AIDS patients have involved male homosexuals. It has been suggested that, like cryptosporidiosis, listeriosis occurs as an opportunistic pathogen in AIDS patients only when homosexual activity was the risk factor for acquisition of the human immunodeficiency virus (Harvey and Chandraesekar, 1988). It may be necessary to wait until the relationship between host factors and the pathogenic properties of *L. monocytogenes* is more fully elucidated before the situation with respect to AIDS can be explained.

Within the general pattern of susceptibility to *L. monocytogenes*, there are differences related to age and sex (in addition to the relationship between sex and listeriosis in pregnancy and age and listeriosis of the newborn). Meningitic listeriosis is primarily a disease of the newborn and of older people, usually males more than 50 years old, while septicaemic listeriosis and cervicoglandular listeriosis are more common in older adults than children.

Although there is no doubt of the greater susceptibility to listeriosis of the subsets of the population discussed above, their relative susceptibility to that of the 'normal' population must be placed in perspective. Outbreaks of listeriosis involving people without any known predisposing factors have been described on a number of occasions (Schech *et al.*, 1983; Malinverni *et al.*, 1986; Bannister, 1987; Azadian, 1989), and a survey among young people showed that over half the patients were previously well and had no detectable predisposing condition (Tim *et al.*, 1984).

Although some *L. monocytogenes* infections among the 'normal' population may be attributed to ingestion of unusually large numbers of the organism, or to exceptional virulence of the causative strain, there is also likely to be variation in resistance among individuals. By analogy with mice, it is considered that this variation may have a genetic basis (Marth, 1988). Alternative explanations involving extrinsic factors such as prior infection with *Salmonella* or another enteric pathogen have, however, been proposed (Cox, 1989) and while not tested have received some support (see below).

Exposure

Although *L. monocytogenes* has been isolated from a wide range of food, only a few types have been related to outbreaks of listeriosis (see pages 337–340). Soft cheese has been most widely implicated, and consumption of varieties such as brie, camembert and blue-vein types are considered to present an unacceptable risk of exposure to *L. monocytogenes* for susceptible people. In some countries, this is indicated at the point of sale, while in others such as the UK, advice to this effect is issued in literature produced by the Government and by other bodies. Some microbiologists believe that chicken and other foods should also be avoided, unless these can be shown to have been adequately cooked (Kerr and Lacey, 1988; Schwartz *et al.*, 1988).

L. monocytogenes is a cause of animal disease and exposure to domestic pets, farm animals or their housing may increase risk for susceptible people. An enhanced risk of occupational exposure exists among workers with animals and workers in the general agricultural environment. For example, cutaneous listeriosis is usually a disease of farmers and veterinarians who come into contact with the tissue of an infected animal such as a cow's placenta, while a fatal case of listeriosis is believed to have been acquired by inhaling contaminated dust in a sheep house (Seeliger, 1961).

Treatment of *L. monocytogenes* infections

Treatment of listeriosis involves antibiotic therapy. This should be commenced as early as possible and, in the case of meningitic listeriosis, an early lumbar puncture is recommended to obtain cerebro-spinal fluid for culture since blood cultures are often negative (Bannister, 1987).

L. monocytogenes is usually susceptible to a wide range of antibiotics when tested *in vitro*, including ampicillin, benzyl penicillin and erythromycin, which are sometimes recommended as the drugs of choice (Marth, 1988). These antibiotics are not always successful in treatment and chloramphenicol is considered by many to offer the best chance of a rapid response (Bannister, 1987). In the treatment of meningitis this efficiency is probably related to its penetration of the blood-brain barrier but a more valuable general property may be its ability to enter cells such as phagocytes which protect *L. monocytogenes* from many other antibiotics.

Wide-spectrum cephalosporins are becoming more widely used in the treatment of culture-negative meningitis, but these drugs are ineffective against *L. monocytogenes* (Hall *et al.*, 1985).

Mechanisms of pathogenicity

The mechanisms of pathogenicity of *L. monocytogenes* are not well understood and, indeed, only one virulence factor, a haemolysin (listeriolysin O), has been fully identified. The importance of host factors in determining susceptibility to infection has been discussed previously (see above), and makes the infective dose virtually impossible to determine. It has been considered (Cox, 1989) that the prime role played by the host, and possibly other factors unrelated to the properties of the organism itself, indicates that *L. monocytogenes* is a circumstantial rather than an opportunistic pathogen.

Genetic control of virulence

Although there has been some speculation concerning the possession by *L. monocytogenes* of a 'Yersinia-like' virulence plasmid, there is now evidence that virulence is under chromosomal control, the gene for listeriolysin production residing in the chromosome at a region crucial for virulence (Cossart, 1988).

Plasmids have been demonstrated in some strains of *L. monocytogenes*, but there is no difference in virulence between plasmid-containing and plasmid-free strains and it is concluded that plasmids have no role in determining virulence.

There have been a number of reports suggesting that *L. monocytogenes* is of greater virulence when grown at low temperatures. This suggested the possibility of temperature regulation of virulence genes (*cf. Y. enterocolitica*, page 139).

The effect of growth temperature on virulence has been investigated by a number of workers with conflicting results. Resistance to killing by human neutrophils increased as the growth temperature decreased, possibly as a result of increased resistance to hydrogen peroxide (Stecha *et al.*, 1989), but Czuprynski *et al.* (1989) found that growth at 4°C significantly increased the virulence of *L. monocytogenes* in intravenously inoculated, but not intragastrically inoculated, mice. It was suggested that isolates grown at 4°C may revert to the 37°C phenotype by the time the epithelial barrier is crossed or, alternatively, that competing microorganisms or other gut conditions compromise the potential virulence of listerias grown at 4°C. Subsequently, Brackett and Beuchat (1990) found growth at low temperature on crabmeat to have no significant effect on virulence, and it is clear that further work is required to resolve the situation.

Adherence and penetration

L. monocytogenes competes poorly with the normal gut microflora (Czuprynski and Balish, 1981), and thus is likely to be favoured by conditions which change the microbial ecology of the gut. Gastric acidity is also likely to play a protective role against the organism (Schlech *et al.*, 1986). The organism has no known factors which mediate mucosal adherence and penetration, although there are some differences in cell-wall composition between serovars which may be related to virulence (Mann, 1969).

It has been hypothesised that host conditions which permit penetration of the epithelium by *L. monocytogenes* are produced by simultaneous enteric infection with a pathogen such as *Salmonella* (Cox, 1989). Evidence of an association between listeriosis and extraintestinal *Salmonella* infections in three renal transplant patients was cited in support, but the hypothesis has otherwise not been tested.

Systemic spread

L. monocytogenes crosses the epithelial barrier at Peyer's patches and translocates to the mesenteric lymph nodes from where dissemination occurs. This initially involves the spleen and liver, the latter being of prime importance in the capture of the organism.

Survival and proliferation in host tissue

Resistance to phagocytosis

The response of mice to intravenous injection of a sub-lethal dose of *L. monocytogenes* has been studied in some detail (Kongshavn and Skamene, 1984), and it is supposed that there are similarities in response between mice and humans (Marth, 1988). There are essentially three phases of resistance to infection:

1 Within 10 minutes of injection, *ca* 90% of cells have been captured by the liver and most of the remainder by the spleen. Within 6 hours the number of cells in the liver has been reduced by 90% due to rapid killing by resident tissue macrophages sometimes known as Kupffer cells (Kongshavn and Skamene, 1984).
2 Remaining viable cells of *L. monocytogenes* commence rapid multiplication in the liver and spleen, and maximum populations occur 48 to 72 hours after the initiation of infection.
3 At the point of maximum population of *L. monocytogenes*, rapid inactivation by macrophages commences. This is initiated by acquired cellular resistance, the macrophages being activated by lymphokines of sensitised thymus-derived lymphocytes (North *et al.*, 1973). Recovery of the host then follows (Kongshavn and Skamene, 1984).

In mice and, possibly, if the analogy between the two species is valid, in humans, the resistance trait is genetically controlled, regulation being by one major gene. Genetic variation may well account for variations in susceptibility in 'normal' people since absence of the resistance trait leads to severe infection. Resistance is obviously much reduced by immuno-incompetence, and organism-specific virulence factors may also be involved since cell-wall extracts of *L. monocytogenes* increase susceptibility to infection and are immunosuppressant (Schlech, 1988).

Haemolysin plays a key role in the survival and proliferation of *L. monocytogenes* within macrophages, and the haemolysin gene is absolutely essential for intracellular multiplication (Cosart *et al.*, 1989). Work with Caco-2 cells has demonstrated that hae-molysin produced inside the phagolysosomal compartment damages the phagosome membrane, and thus allows the organism to multiply freely in the cytosol (Gaillard *et al.*, 1987, Cossart *et al.*, 1989).

Although haemolysin plays a key role, other factors probably contribute to the organism's resistance to macrophages, including superoxide dismutase which imparts resistance to toxic oxygen products. However, catalase does not have a critical role (Leblond-Francillard *et al.*, 1989).

Interaction with host cell microfilaments

Studies using advanced techniques, such as confocal microscopy, have shown that while listeriolysin mediates lysis of the phagocytic vacuole, interaction with host cell microfilaments is of major importance in intracellular and cell-to-cell spread (Mouncer *et al.*, 1990). Cells of listeria became coated with F-actin shortly after lysis of the phagosomal membrane, and within two hours of entering the host cell were observed following organised routes which corresponded to stress fibres. After a further two hours, random movement of some bacteria was indicated by trails of F-actin, this movement resulting in penetration of adjacent cells, which led in turn to the formation of vacuoles limited by a double membrane. After lysis of these vacuoles, bacteria were released into the cytoplasm where multiplication occurred, followed by the invasion of new cells.

Production of toxins

Nature of haemolysin

The haemolysin of *L. monocytogenes* is a 58 Kda protein (Geoffrey *et al.*, 1987). The haemolysin is one of a family produced by a variety of Gram-positive bacteria including *Streptococcus*, *Clostridium* and *Bacillus*. The prototype is streptolysin O, and members of the family share many properties including a similar molecular weight, binding and inactivation by cholesterol, serological cross reactions and thiol activation.

L. monocytogenes haemolysin possesses additional properties, the most significant of which is probably phospholipase activity.

Genetic and biochemical evidence has been presented for production of listeriolysin by *L. seeligeri* and *L. ivanovii*, as well as *L. monocytogenes* (Leimeister-Wachter and Chakraborty, 1989). However, Vasquez-Boland *et al.* (1989) have described the haemolysin produced by *L. ivanovii* (ivanolysin-O) as being distinct from listeriolysin.

Other toxins

Various other toxic activities have been attributed to *L. monocytogenes* but the role in pathogenicity, if any, is uncertain. An endotoxin with the same characteristics as endotoxins of Gram-negative bacteria has been described (Wexler and Oppenheim, 1979), but its presence or absence in avirulent listerias has not been determined. *L. monocytogenes* bacteraemia in pregnant women does, however, have some of the features of Gram-negative endotoxaemia (Schlech, 1988).

The diarrhoeal prodrome associated with some cases of listeriosis has led to speculation that a true enterotoxin is produced, but there is no evidence of this.

L. ivanovii produces a second cytolysin, sphingomyelinase C, but its pathogenic significance is not known (Vazquez-Boland *et al.*, 1989).

Determination of virulence

Virulence of *L. monocytogenes* is commonly demonstrated using Swiss white mice. The test is generally unsatisfactory since other organisms which resemble *L. monocytogenes*, including *L. denitrificans*, will kill mice under the test conditions. More recently, an immunocompromised mouse model that uses carrageenan to inactivate the mouse macrophages before intraperitoneal injection of the test organism, has been described (Stelma *et al.*, 1987). Pathogenic strains of *L. monocytogenes* show five or more logarithmic reductions in LD_{50} values for immunocompromised mice whereas there is little change in LD_{50} when nonpathogenic strains are injected.

Virulence in *Listeria* is usually associated with haemolysis, although variants which are haemolytic but avirulent have been described (Hof, 1984). Loss of virulence is usually accompanied by the development of a rough colony form, which contrasts with the smooth form which is usual for virulent strains.

Human carriage of *L. monocytogenes*

A carrier state may develop following infection with *L. monocytogenes*, and carriage rates ranging from 5 to 15% have been reported in asymptomatic people (Griffiths, 1989). The carriage of *L. monocytogenes* may be related to occupation (Manev *et al.*, 1975), but alternatively carriage may be sporadic and transient, and merely reflect the dietary load (Kwantes and Isaac, 1975). The organism is also shed in the faeces of people recovering from listeriosis for an indefinite time (Marth, 1988). Contamination of food by this route is an obvious possibility, but there is no known outbreak of foodborne listeriosis where the involvement of a carrier has been suspected, and the clinical significance of asymptomatic carriage of *L. monocytogenes* in faeces has been questioned (Kwantes and Isaac, 1975).

L. monocytogenes in the environment

L. monocytogenes is both a common environmental organism which may be isolated from a wide range of environmental sources, and an important cause of zoonoses. Indeed, there has been some debate as to whether the organism should be regarded as primarily zoonotic or primarily environmental. However, to some extent this debate is irrelevant to the fundamental understanding of the routes by which food may become contaminated with *L. monocytogenes* and the interaction between man, animals and the environment (see page 337).

Animals and birds

L. monocytogenes is pathogenic to a wide range of animals and may also be found in the intestinal tract of healthy animals and in animals with subclinical infections. There are well-substantiated reports of humans acquiring infection direct from diseased animals, or indirectly via contact with faeces. Ticks and flies may also be involved in the transmission of *L. monocytogenes* from diseased animals to man (Ralovich, 1984), and ants may be of some importance in contaminating foods (Surak and Barefoot, 1987).

Feed has long been recognised as a source of *L. monocytogenes* in infections among farm animals, especially ruminants. There has been a particular association with silage made from a wide range of crops, particularly when the material was not properly fermented. In this context, particular suspicion has attached to 'big bale' silage. Birds, particularly seagulls, have been considered to be of importance in introducing *L. monocytogenes* to silage (Fenlon, 1985). It is likely the vegetation itself is also a source of the organism (see below).

Animals from which *L. monocytogenes* has been isolated are listed in **Table 123**.

Table 123. Animals from which *Listeria monocytogenes* has been isolated (Seeliger, 1961; Brackett, 1988).

Mammals		
Cattle*	Sheep*	Goats*
Pigs*	Horses*	Dogs*
Cats*	Deer*	Raccoons*
Rats*	Mice*	Lemmings*
Rabbits*	Guinea pigs*	Chinchillas*
Skunks*	Mink*	Ferrets*
Foxes*	Voles	Moose*

Birds		
Canaries*	Chaffinches	Chickens*
Cranes	Doves	Ducks*
Eagles*	Geese*	Hawks
Lorikeets	Owls*	Parrots*
Partridges*	Pheasants	Pigeons*
Seagulls*	Turkeys*	Whitegrouse
Whitethroat	Woodgrouse	

Others		
Frogs	Fish	Crustaceans
Flies	Ticks	Ants
Snails		

* Indicates that cases of listeriosis are known.

Soil and vegetation

In addition to *L. monocytogenes* derived from animal faeces and other sources such as sewage sludge (Watkins and Sleath, 1981), soil itself may be an important reservoir of the organism. *L. monocytogenes* is believed to live a saprophytic life in close association with soil (Weis and Seeliger, 1975), and the organism has been isolated from both cultivated and uncultivated soils. Numbers are greatest in mud and moist soils, and the frequency of isolation is greater from surface soils and fallow fields than from cultivated fields. *L. monocytogenes* in dust may contaminate a wide range of materials including the food itself, packaging, transport vehicles and the food handling environment.

L. monocytogenes is also associated with plant material that may be a common source in infection (Welshimer and Donker-Voet, 1971). The organism has been isolated from a range of plants including shrubs, wild grasses and food plants such as corn, cereals and soya beans (Mitserlich and Marth, 1984). Although silage has been identified as a major source of *L. monocytogenes*, it seems likely that infection may also be acquired during normal grazing, especially if animals are allowed to feed on natural vegetation. In western Canada, Ponderosa pine needles appear to be a source of *L. monocytogenes* and lead to frequent abortions in cattle grazing them (Schlech, 1984).

Water

L. monocytogenes is present in the water cycle and is frequently present in river waters in greater numbers than *Salmonella* (Watkins and Sleath, 1981). *L. monocytogenes* has also been isolated from lakes and canals (Dijkstra, 1982). It is possible that some of the organisms are derived from soil and vegetation by runoff, but major focal sources are sewage and drainage from abbatoirs and poultry-processing plants (Watkins and Sleath, 1981). It may also be postulated that untreated drainage from such places as farmyards is a major source.

Under many conditions, *L. monocytogenes* is able to survive sewage treatment and is present in the final effluent and, in lesser numbers, in the sewage sludge cake (Watkins and Sleath, 1981; Al-Ghazali and Al-Azawi, 1988a). Survival in the sludge cake is of considerable importance where the sludge is used as a fertiliser, but solar dewatering, correctly applied, will destroy surviving cells of *L. monocytogenes* (Al-Ghazali and Al-Azawi, 1988b).

Drinking untreated water may contribute to human and animal listeriosis, and a risk of contamination exists where untreated river water is used for irrigation or washing crops.

The food processing environment

The ability of *L. monocytogenes* to colonise the food processing environment, including process equipment, is now well recognised (Coleman, 1986; Surak and Barefoot, 1987; Griffiths, 1989). The primary source in processing plants is considered to be floors and floor drains, a large survey isolating listerias in decreasing order of frequency from drains, floors, standing water, residues and food-contact surfaces (Cox *et al.*, 1989). Significantly, no samples from dry culinary food units contained listerias, and it is likely that dry conditions and the restriction of food residues contribute to the control of the organism.

L. monocytogenes may also be isolated from domestic kitchens (Finch *et al.*, 1978; Scott *et al.*, 1982), and while studies have been less extensive, it seems that dishcloths could be an important source of the organism in the home.

The ubiquitous nature of *L. monocytogenes*, together with specific properties such as its ability to attach to stainless steel under certain conditions (Herald and Zottola, 1987), mean that the organism is always likely to present a problem with respect to plant sanitation. It has been considered that control, through good management practice, should be no more difficult than the control of any other non-endospore forming bacteria (El-Kest and Marth, 1988). However, it seems likely that measures for the control of *Listeria* will have to be applied more rigorously and with more attention to detail than measures for the control of 'traditional' pathogens (Engel, 1988).

Foods commonly associated with *L. monocytogenes* infections

L. monocytogenes has been isolated from a wide range of foods. This reflects both the variety of means by which *L. monocytogenes* may enter food (**Flow diagram 30**) and the ability of the organism to grow in food at low temperatures. Despite this, a positive association between listeriosis and specific food products has been demonstrated in only a small number of outbreaks, and a high proportion of cases are of a sporadic nature. Epidemiological investigations may be inconclusive for a number of reasons, including the possibility of a long period between the consumption of infected food and the development of symptoms. It is probable that food has been the vehicle of infection in many of the sporadic cases of listeriosis. However, some caution is required and it is important that the possibility of other vehicles of transmission being involved should not be overlooked (McLaughlin *et al.*, 1986b).

Flow diagram 30. Possible cycles of infection for *L. monocytogenes* (Reproduced with permission from Brackett, R.E. *Food Technology* **42**(3): 162–164. © 1988 Institute of Food Technologists.)

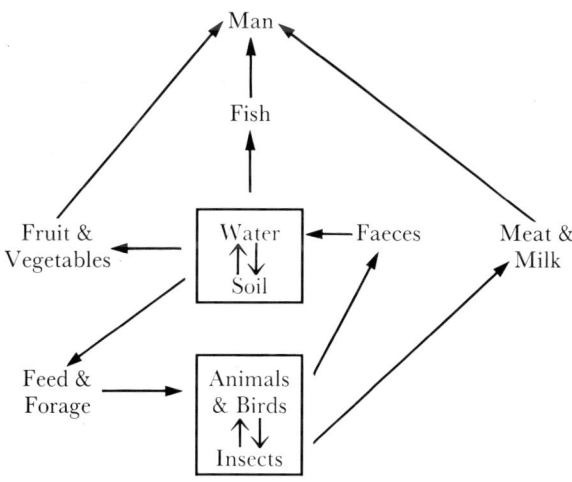

Meat and poultry products

Although results of surveys have, inevitably, differed in detail, the incidence of *L. monocytogenes* in raw meat and poultry appears to be high. The organism was recovered from 57% of the fresh and frozen poultry sampled by Kwantes and Isaac (1971), while Gitter (1976) isolated the organism from 15% of oven-ready poultry samples. In a more recent survey (Lee and McClain, 1987), isolation rates were ground beef, 70%; pork sausage, 43%; and poultry, 48%. The presence of *L. monocytogenes* in slaughterhouse effluents (Watkins and Sleath, 1981) is supportive evidence for a high incidence of the organism in raw meats.

L. monocytogenes has a higher level of heat resistance than most vegetative pathogens, and cooking processes for some meat products may be marginal for safety (Anon, 1989a; see page 341). Undercooked chicken and hot dogs have been implicated on epidemiological grounds as vehicles of sporadic listeriosis in the USA (Schwartz *et al.*, 1988), although the association is not universally accepted (Anon, 1988). However, turkey frankfurters have subsequently been implicated as a cause of human listeriosis in the USA. *L. monocytogenes* has been recovered from a wide range of cooked meats as diverse as roast beef and pâté (Morris and Ribeiro, 1989), although it is not known whether the presence of the organism was due to underprocessing or post-process contamination.

In the UK, there has been particular concern over the isolation of *L. monocytogenes* from 'ready-meals' (cook-chill recipe products usually meat based). Surveys have demonstrated contamination rates of 18 to 25%, despite the fact that many such products are labelled as 'ready to eat' (Kerr and Lacey, 1988; Gilbert *et al.*, 1989a; Lacey and Kerr, 1989). Three cases of listeriosis in the UK have been associated with cook-chill chickens (Gilbert *et al.*, 1989b), and fears concerning the safety of such products have been increased by evidence that microwave reheating is frequently unreliable. It has, however, been argued (Cox, 1989) that cook-chill products have, in various forms, existed for many years, and that the link between these products and listeriosis is not proven. In proposing this argument, the very long shelf lives given to these products in recent years has been overlooked and, while several factors may be involved in the increased incidence of *Listeria* infection, it is considered that the length of refrigerated storage is a major factor in enabling the organism to reach significant numbers. The importance of avoiding multiplication in foods is recognised in legislation in some countries such as the UK, which imposes a statutory maximum upper temperature limit for storage of this type of product. However, a maximum storage life of 3 days between cooking and consumption has been proposed (Lacey and Kerr, 1989), which contrasts with the current 10 to 20 days (Sheppard, 1988).

In the UK, cook-chill catering is being introduced on a wide scale into hospitals, where the number of people susceptible to listeriosis is likely to be high. However, surveys suggest that risk primarily lies with meals brought into hospitals, rather than those prepared on the premises (Armstrong et al., 1989; Gilbert et al., 1989a).

L. monocytogenes is relatively resistant to curing ingredients (see page 342). The organism has been isolated from salami, and this product has been identified as a risk factor for listeriosis in the USA. In early 1990, a major recall was made in the UK of vacuum-packed ham products which were contaminated at a high level by L. monocytogenes, and political controversy surrounding food poisoning was reawakened by allegations of unecessary delay by the Department of Health before warning notices were issued (Anon, 1990).

Milk and milk products

Dairy products currently have a stronger association with listeriosis than other types of food, with both raw and pasteurised milk and certain soft cheeses having been identified as causes of food poisoning outbreaks. The organism has also been isolated from a range of other dairy products, although listeriosis has not been associated with these products.

Milk

A number of surveys concerning the incidence of L. monocytogenes in raw milk have been made. Two surveys in the USA reported an average incidence of 4.7% (Lovett et al., 1987), and 12% (Hayes et al., 1986), while in the Netherlands an incidence of 7.7% has been reported (Beckers et al., 1987). In other European countries, the incidence varied from 4% in France (Skovgaard, 1988), to 45% in Spain (Dominguez Rodriguez et al., 1985). Listeria spp can be recovered from 12% of raw milk in the Ontario province of Canada (Farber, 1988a), but the incidence of L. monocytogenes was only 1.3%. In all cases, the numbers of L. monocytogenes present were low, the concentration usually being less than 1/ml (Lovett et al., 1987). L. monocytogenes is able to grow in raw milk and there is likely to be some increase in numbers during storage in farm bulk tanks and in silos.

In addition to geographical differences in the incidence of L. monocytogenes, there is strong evidence of seasonal variation. In Spain, the incidence is highest from October to March (Fernandez Garayzabal et al., 1987). This reflects the higher incidence of bovine listeriosis in winter months, which may in turn reflect the feeding of poor quality silage or the greater number of pregnancies. In Canada, the incidence of L. monocytogenes in raw milk is higher in summer than winter (Farber et al., 1988a). The difference between the two countries may probably be explained by different milk production methods (Griffiths, 1989).

The main source of L. monocytogenes in milk is probably faecal contamination, although the organism is a rare cause of mastitis (Gitter et al., 1980). Shedding of L. monocytogenes by infected animals is intermittent, and can continue for periods in excess of 12 months and thus into successive lactations (Hyslop, 1975, Slade et al., 1990).

Although cases of listeriosis have been attributed to raw milk (Griffiths, 1989), greatest concern was raised by an outbreak in the USA in which pasteurised milk caused some 49 cases (Fleming et al., 1985). Epidemiological investigation failed to fully resolve the circumstances leading to the outbreak, and the suggestion was made that the organism was of sufficient heat resistance to survive commercial pasteurisation. This possibility has now been largely discounted (see pages 340–341), although the safety margin is less for L. monocytogenes than for other organisms, and also less when high-temperature short-time processing is used than low-temperature long-time processing.

Cheese

The first outbreak of listeriosis associated with cheese involved Jalisco-brand Mexican-style cheese, 86 people being affected in the western US (James et al., 1985). In the aftermath of this outbreak, soft cheeses were the subject of a high level of surveillance, and isolation of L. monocytogenes was reported from Liederkranz, Morbier Rippoz, Touvre de Aubier, Fromages des Burans, Brie and several Swiss types (Ryser and Marth, 1987a). During early 1986 almost 60% of French Brie cheese on the US market was recalled because of the presence of L. monocytogenes, which resulted in economic hardship for producers.

Surveys in a number of countries have shown L. monocytogenes to be a common contaminant of soft cheese and that numbers may be as high as $10^6/g$ (Breer, 1986; Terplan et al., 1986a; Beckers et al., 1987; Farber et al., 1987a; Gilbert and Pini, 1988). In addition to isolated outbreaks of listeriosis associated with cheese (Bannister, 1987), a major outbreak involving more than 50 cases was associated with the Swiss Vacherin Mont d'Or cheese (Anon, 1987). Subsequently, a UK produced goat's milk cheese,

Anari, was the cause of a single case of meningitis (Communicable Disease Surveillance Centre, 1988).

Although the use of unpasteurised milk for cheese-making is the cause for greatest concern, *L. monocytogenes* has been isolated from pasteurised milk cheese (Breer, 1986; Terplan *et al.*, 1986b). Recontamination would generally seem to be of considerably greater importance than improper pasteurisation, and contamination of the finished product with raw milk was probably also the underlying cause of the outbreak involving Jalisco cheese (Prentice and Neaves, 1988).

Considerable attention has been paid to the behaviour of *L. monocytogenes* in various types of cheese (see page 457 for a more detailed discussion). The problem is greatest with cheeses of the Brie and Camembert type, where growth may occur on the surface of the cheese as a consequence of the increase in pH value during ripening (Ryser and Marth, 1987b). However, *L. monocytogenes*, if present in raw milk in numbers of *ca* 100/ml, can survive a number of cheese-making processes and can remain viable in the final product for a considerable length of time (Griffiths, 1989). In the case of soft cheese, the risk to susceptible people is recognised, and in many countries pregnant women and the immunosuppressed are advised to avoid consumption. Soft cheese may well play a major role in sporadic listeriosis, and it again seems likely that inherent problems are exacerbated by the long storage life given to such products by both retailer and consumer.

Other dairy products

Ice cream

L. monocytogenes survives ice cream manufacture without sustaining significant injury, and the organism has been isolated from ice cream in the USA (Rosenow and Marth, 1987a).

Cream and butter

Cream has been implicated on epidemiological grounds in an outbreak of food poisoning in the UK. *L. monocytogenes* will also survive the manufacturing process for cultured cream, and persist in the final product for at least 7 days, providing that the original milk contains 5×10^5 cfu/ml (Stajner *et al.*, 1979).

Many of the bacteria present in cream are removed with the butter milk during the manufacture of butter, and the population of *L. monocytogenes* is reduced by *ca* 92 to 97.5% (Olsen *et al.*, 1988). The organism is able to grow during the initial stages of storage and persists for more than 70 days.

Non-fat dried milk

Inoculation experiments have demonstrated that *L. monocytogenes* is reduced in number by 1 to 1.5 log cycles during drying (inlet temperature 165°C; outlet temperature 67°C) to a moisture content of 3.6 to 6.4% (Doyle *et al.*, 1985). The number of surviving cells decreased during storage at 25°C in moisture-proof packs, numbers being reduced by more than four log-cycles after 16 weeks.

Salads and vegetables

Salad and other vegetable crops are inevitably contaminated with soil and the presence of *L. monocytogenes* is unavoidable, especially if nightsoil or animal manure is applied or if polluted water is used for irrigation. The use of manure from a flock of sheep affected by ovine listeriosis (circling disease), was an important factor in an outbreak of listeriosis associated with coleslaw (Schlech *et al.*, 1983). In this case, the cabbage used in the manufacture of the coleslaw was infected with *L. monocytogenes* from the manure, the organism growing during subsequent cold storage and surviving in the finished mayonnaise. The outbreak involved 34 cases of perinatal listeriosis and seven of listeriosis in adults.

An association between consumption of lettuce and other raw vegetables and listeriosis among hospitalised, immunosuppressed patients was suggested by Ho *et al.* (1981), and a link between sporadic listeriosis in the community and salads was suspected for some years. Support for this possibility came from the isolation of *L. monocytogenes* from a large number of pre-packed salads sold in the UK.

The ability of *L. monocytogenes* to grow in cabbage at 5°C was established during the course of investigations following the outbreak involving coleslaw (Beuchat *et al.*, 1986), and the organism has subsequently been shown to be capable of growth on asparagus, broccoli and cauliflower (Berrans *et al.*, 1989). The risk of listeriosis is likely to be increased by the use of extended storage at low temperature, but potential problems are exacerbated by controlled atmosphere packaging which extends the storage life by maintaining an acceptable appearance, but has no effect on the growth of *L. monocytogenes* (Berrans *et al.*, 1989).

Fish and fish products

L. monocytogenes has been isolated from frozen seafood from nine countries (Weagant *et al.*, 1989), as well as from smoked salmon, imported surimi (seafood analogue) and shrimps in the USA. In some cases (Anon, 1988b), imported shrimps were threatened with detention because of contamination with ants which are becoming recognised as vectors of *L. monocytogenes*. Seafood has also been associated with a listeriosis out-break in New Zealand, but no conclusive evidence was available (Lennon *et al.*, 1984). In contrast to reports which imply that *L. monocytogenes* may be a major problem in fish and other seafood, no isolations were made from 35 tropical fish and fishery products in India (Fuchs and Surendran, 1989), although *L. innocua* was isolated from 14 samples.

Factors affecting growth and survival of *L. monocytogenes* in foods

Temperature

As with all micro-organisms, temperature is of prime importance in the control and inactivation of *L. monocytogenes*. However, there are continuing discrepancies in results concerning both the minimum temperature for growth and the thermal tolerance of the organism.

The maximum temperature for growth of *L. monocytogenes* is *ca* 45°C, and the optimum in the range 30 to 37°C (Seeliger, 1961; Wilkins *et al.*, 1972). Behaviour at low temperatures is less easy to define, and while there is agreement that *L. monocytogenes* is able to grow at low temperatures, a minimum of 1.1 ± 0.3°C being defined by Juntilla *et al.* (1988), there is some disagreement as to how well it will grow (Griffiths, 1989).

Growth dynamics of *L. monocytogenes* at temperatures below 5°C are characterised by a lag phase lasting several days, followed by slow growth of the organism. Cells of strains of *L. monocytogenes* previously grown in tryptose-phosphate broth and inoculated into sterilised milk, had a lag phase lasting 2 days at 4°C (Donnelly and Briggs, 1986). In contrast, four strains of *L. monocytogenes* previously grown in skim milk at 30°C had a lag phase of 5 days before growing in sterilised skim, full cream and chocolate milk, with an average generation time of *ca* 37.6 hours at 4°C and a range of 29.7 to 45.6 hours (Rosenow and Marth, 1987b). A similar generation time of 30 hours at 4°C was reported by Mossel *et al.* (1987).

Discrepancies over the behaviour of *L. monocytogenes* at low temperatures may be partly explained by strain variation, but other factors such as growth medium and incubation temperature of the inoculum may also be important (Walker and Stringer, 1987). The effect of growth medium was demonstrated by Rosenow and Marth (1987b,c), who found growth enhancement in chocolate milk and by cane sugar, cocoa powder and carrageen stabiliser. It was not possible to confirm the observation of Donnelly and Briggs (1986) that growth of *L. monocytogenes* was enhanced by milk fat.

Growth at a low temperature may enhance the virulence of *L. monocytogenes*. This is discussed in more detail in **Mechanisms of Pathogenicity**, page 333.

The ability of *L. monocytogenes* to survive heat treatments in the order of those applied during pasteurisation has been the subject of considerable controversy which has not yet been fully resolved. Should *L. monocytogenes* be able to survive pasteurisation processes, however, the implications for safety are obvious.

The possibility of survival of *L. monocytogenes* during pasteurisation of milk was first suggested by Bearns and Girard (1958), who claimed the organism could survive a treatment of 61.7°C for 35 minutes. However, subsequent work revealed technical flaws which accounted for the apparent survival (Donnelly *et al.*, 1987). More convincing evidence for survival of commercial heat treatments came from experiments using milk from cows inoculated with *L. monocytogenes*, which was processed using a small scale commercial heat exchanger (Doyle *et al.*, 1987). Under these conditions, *L. monocytogenes* could be recovered by enrichment following heating at 72.2°C for 16.4 seconds, a combination which may be used in commercial practice.

Survival of *L. monocytogenes* has been attributed to the protection conferred by their existence within bovine leukocytes. However, this theory was tested by Bunning *et al.* (1988), who found no difference in heat resistance between free and internalised cells when heated in sterile milk. Subsequently, the consensus of opinion supported the contention that, in contrast to the findings of Doyle *et al.* (1987), *L. monocytogenes* cannot survive adequate pasteurisation (Farber *et al.*, 1988b; Lovett *et al.*, 1988; Stroupe *et al.*, 1988). The situation was comprehensively reviewed by Mackey and Bratchell (1989), who concluded that while normal pasteurisation processes will inactivate *L. monocytogenes* in milk, the safety margin is considerably greater with a low-temperature long-time process (62.8°C/30 minutes: 39D reduction) than with the more commonly used high-temperature short-time process (71.7°C/15 seconds: 5.2D reduction). This latter value suggests that survival of *L. monocytogenes* is possible, if numbers of the organism are sufficiently high. This appears highly improbable since the level of contamination is usually in the order of 1 cell/ml (Lovett *et al.*, 1987). However, work in the USA suggested only a 3.7D reduction, and a minimum pasteurisation process of 77°C for not less than 16 seconds has subsequently been recommended (Coleman, 1986; Surak and Barefoot, 1987).

Concern over the possibility of survival of *L. monocytogenes* during pasteurisation has again increased, following the finding that cells may survive heating, but are not culturable on media commonly used. The use of strictly anaerobic incubation (Hungate technique) resulted in the recovery of otherwise non-culturable cells after heating in milk (Knabel *et al.*, 1990). This was attributed to inactivation of superoxide dismutase and catalase, and it was further suggested that any protection conferred by localisation within phagocytes may result from the presence of thermostable phagocytic superoxide dismutase, rather than from any heat shield effect. These workers considered that *L. monocytogenes* could survive both HTST and LTLT pasteurisation and that further research is required.

Heat resistance of micro-organisms can vary markedly according to the nature of the heating menstruum and claims have been made of 'exceptional' heat resistance of *L. monocytogenes* in meat and poultry products. The organism was isolated from grilled meatballs heated to an internal temperature of 78 to 85°C when the original inoculum was 10^3/g (Karaioannoglou and Xenos, 1980), and from chicken breasts cooked to a temperature of 82°C (Carpenter and Harrison, 1989). Such claims have not been supported elsewhere. Microwave cooking of poultry to an internal temperature of *ca* 70°C resulted in a 10^6-fold reduction in numbers (Lund *et al.*, 1989), while Mackey and Bratchell (1989) found no evidence of unusual heat resistance of *L. monocytogenes* during heating in beef. D_{60} values ranged from 5.02 to 8.32 minutes, and z values from 7.39 to 5.98 when determined accurately in thick slurries of autoclaved chicken or beefsteak (Gaze *et al.*, 1989). Such *D* values are not compatible with survival of the organism above 80°C (Mackey and Bratchell, 1989).

In most cases, it is likely that reports of unusual heat resistance result from uneven heating. The possibility of protection of the organism by fat or drying exists, however, (Mackey and Bratchell, 1989), and solutes have been shown to protect *L. monocytogenes* from thermal injury (Smith and Hunter, 1988). Heat resistance may also vary within a population, some cells surviving moist heat treatment of poultry to 73.9°C, irrespective of population size (Carpenter and Harrison, 1989). Significantly, survivors could re-establish themselves at population levels at least equal to those present before heating during storage at either 4 or 10°C.

L. monocytogenes has been reported to survive the boiling of shrimps (Anon, 1988c), three of seven samples being positive after 3 minutes boiling and one of seven after 5 and 10 minutes boiling. It is probable that survival is due to uneven heating, and a two-stage process involving boiling followed by freezing to −20°C has been proposed to overcome the problem (Anon, 1989b).

Information on the effect of food type on heat resistance of *L. monocytogenes* is generally lacking in other products, although the organism has been shown to be slightly more resistant when heated in liquid whole egg than in milk (Leasor and Foegeding, 1988), and considerably less resistant when heated in cabbage juice (Beuchat *et al.*, 1986). Information of the behaviour of the organism during heating in a wider range of foods is required.

An increase in the heat resistance of *L. monocytogenes* resulting from tempering at sub-lethal temperatures has been observed in both milk (Fedio and Jackson, 1989; Knabel *et al.*, 1990) and meat (Farber and Brown, 1990). This has obvious consequences for some manufacturing processes, the enhanced resistance persisting for at least 24 hours in cells transferred to storage at 4°C (*cf Salmonella*, page 70).

Irradiation

L. monocytogenes is inactivated by ionising radiation at similar dose levels to those required to inactivate *Salmonella* spp (Patterson, 1989), and the organism is therefore unlikely to present any unique problem in any radiation pasteurisation process.

L. monocytogenes is susceptible to short wave ultra-violet energy at a wavelength of 254 nm but not 365 nm (Yousef and Marth, 1988).

pH value and organic acids

In the past, *L. monocytogenes* has been considered to be relatively sensitive to low pH values, and a limiting value of 5.6 has been widely quoted (Seeliger, 1961). More recently, growth at values below pH 5.0 has been demonstrated (Conner *et al.*, 1986), while George *et al.* (1988) found the limiting pH value at 30°C to be between 4.42 and 5.05 depending on strain. There is a strong interaction with temperature, and at 20°C the lower limiting pH value is 4.95 to 5.2 (George *et al.*, 1988; McLure *et al.*, 1989). There are also strong interactions with NaCl concentration (see below).

Lower limiting pH values for growth are dependent on the acidulant, acetic acid being most effective and hydrochloric acid the least (Farber *et al.*, 1989). Similar results were obtained by Ahamad and Marth (1989).

L. monocytogenes is relatively little affected by increases in the pH value and is able to grow at pH 9.5 (Lovett and Twedt, 1988).

A_W level

Minimum a_w level for *L. monocytogenes* is 0.932 with glycol as humectant and 0.942 with sucrose or NaCl (Griffiths, 1989).

Composition of atmosphere

Relatively little attention has been paid to the effect of atmosphere on growth of *L. monocytogenes* in foods. However, work with cooked chicken loaf showed that neither 50 nor 80% carbon dioxide inhibited growth, although the rate of multiplication was reduced. Carbon dioxide, however, selected for *L. monocytogenes* over spoilage organisms such as *Pseudomonas fragi* (Ingham *et al.*, 1990)

Curing ingredients

NaCl

L. monocytogenes has been reported to be relatively tolerant to NaCl, but in many cases reports refer to survival in concentrations of NaCl in excess of those used in the vast majority of foods, and have little relevance to modern food manufacturing.

A multi-factorial approach using gradient plates and turbidometric techniques was made by McClure *et al.* (1989). The results of the work with gradient plates clearly illustrated that *L. monocytogenes* is capable of tolerating high NaCl concentrations within a clearly defined pH range, the organism being able to

grow in 10% NaCl at 35°C and at neutral pH value. At 25°C, growth continued to a NaCl concentration of 9.9% in the pH range 6.6 to 7.9, while at 20°C, 8.8% NaCl was required to limit growth between pH 6.6 and 7.6.

Nitrite

Nitrite is not inhibitory to *L. monocytogenes* at permitted levels unless there is an interaction with other inhibitory agents or conditions (Shahamat *et al.*, 1980). For example, nitrite is inhibitory at a concentration of 100 mg/l providing that the NaCl concentration is at least 3%, the pH value less than 5.5 and the incubation temperature 5°C. However, under similar conditions in Finnish fermented sausage, the organism was able to survive a 21 day fermentation during which numbers were reduced by *ca* 1 log cycle (Junttila *et al.*, 1989).

Other inhibitors

Sorbic and benzoic acids

L. monocytogenes is susceptible to both sorbic and benzoic acids. In each case there is interaction with pH value and temperature, inhibition being greatest at pH 5.0 and 4°C (El-Shenawy and Marth, 1988a,b; Ryser and Marth, 1988). Sodium benzoate is most effective when used in combination with organic acids (El-Shenawy and Marth, 1989).

Nisin

Although reports have been conflicting, *L. monocytogenes* appears to be relatively resistant to nisin. On exposure to the antibiotic there is initially a substantial reduction in numbers, followed by a lag and then multiplication (Somers and Taylor, 1987). *L. monocytogenes* strain Scott A is able to multiply in tryptose broth containing 2,000 iu/ml nisin and other strains in broth containing 500 iu/ml.

The lactoperoxidase system and hydrogen peroxide

A number of proposals have been made for stimulating the naturally occurring lactoperoxidase system (Reiter, 1981), either as a means of cold sterilisation or for specific purposes such as preparation of infant feeds in countries where pure water supplies are not available (Banks and Board, 1985). The effect of the lactoperoxidase system on *L. monocytogenes* is similar to that on other Gram-positive bacteria, in that growth is delayed but not prevented (Siragusa and Johnson, 1989). The organism is also relatively resistant to hydrogen peroxide (Dominguez Rodriguez *et al.*, 1987), the basis of the resistance probably being the high levels of catalase and superoxide dismutase present in *L. monocytogenes* (Dallmier and Martin, 1988).

Phenolic compounds

In an investigation of the activity of phenolic compounds against *L. monocytogenes*, the most inhibitory was the antioxidant, tertiary butylhydroquinone, and the most effective permitted compound, propyl paraben. At pH values close to neutrality, both were significantly more effective than potassium sorbate, when tested at 35°C. The activity of propyl paraben, in contrast to that of tertiary butylhydroquinone, was consistent and independent of the level of inoculum (Payne *et al.*, 1989).

Effect of other micro-organisms on the growth of *L. monocytogenes*

On the basis of growth dynamics, it has been predicted that *L. monocytogenes* would compete poorly with true psychrotrophs during storage of foods at low temperatures (Griffiths, 1989). However, a study of the growth of *L. monocytogenes* in minced beef showed that while growth was inhibited by *Lactobacillus plantarum*, multiplication occurred in the presence of *Pseudomonas fluorescens* (Gouet *et al.*, 1978). Although pseudomonads have little or no effect on the growth of *L. monocytogenes* when the organisms are co-inoculated into milk in approximately equal numbers, the prior growth of *Ps. fluorescens* and, to a lesser extent, *Ps. fragi*, significantly enhances the growth of *L. monocytogenes* (Marshall and Schmidt, 1988).

In addition to non-specific inhibition of *L. monocytogenes* by lactic acid bacteria, there is evidence that some strains of *Pediococcus acidilactici* used in meat starter cultures have a specific inhibitory effect on *L. monocytogenes*. This is probably due to production of a bacteriocin (pediocin), and it may be possible to exploit this property to increase the safety of fermented meats such as salamis. However, bacteriocin activity cannot guarantee the absence of *L. monocytogenes*, and in semidry sausages heat treatment is still required (Berry *et al.*, 1990).

Laboratory methods

Isolation of *L. monocytogenes*

Enrichment

Although a large number of enrichment broths and procedures have been devised, it is apparent that no single enrichment technique will recover all strains of *L. monocytogenes* under all circumstances. This situation is similar to that with other foodborne pathogens such as *Salmonella*, but the problem currently appears to be greater with *L. monocytogenes* than with other organisms.

The immediate decision concerning choice of enrichment procedure is whether to adopt low temperature enrichment in non-selective media, or selective enrichment at higher incubation temperatures. In many cases, the choice is dictated by the length of incubation for low temperature enrichment. In some cases, cold enrichment has been continued for weeks and even months, although there is evidence that the optimum period is 1 week at 4°C (Doyle and Schoeni,

1986; Pini and Gilbert, 1988). However, even this period may be unacceptably long.

Comparisons between low temperature and selective enrichment procedures have produced differing results. For example, a comparison of two selective enrichments and a low temperature enrichment for the isolation of *L. monocytogenes* from soft cheese, showed the low temperature procedure to be most effective (Doyle and Schoeni, 1987). In contrast, Pini and Gilbert (1988) found a selective procedure to be more effective with chicken samples, but little difference with soft cheese. However, neither method alone yielded all isolates. It appears that the relative efficiency of the two approaches to enrichment varies according to the nature of the sample and this may, in turn, reflect differences in the competing microflora and in the extent of sub-lethal damage suffered by *Listeria* during processing. Different methods may also yield different serovars from the same sample, and thus it is likely that the efficiency of enrichment procedures is also serovar dependent.

Low temperature enrichment procedures for *L.*

monocytogenes invariably employ non-selective broth media such as nutrient broth. The use of strongly buffered media may be considered advantageous, and nutrient broth may be buffered with morpholino-propane sulphonic acid buffer (MOPS) or, alternatively, a medium such as tryptose-phosphate broth may be used.

Self-enrichment may be employed for foods such as milk, which will support the growth of *L. monocytogenes*. This procedure is not used widely, although preliminary enrichment in milk diluted 1:5 was part of a procedure for investigating a milkborne outbreak of listeriosis (Hayes *et al.*, 1986). More recently, McCarthy *et al.* (1990) found that recovery of heat-stressed cells from shrimp was improved by holding the sample for 3 days at 4°C without the addition of broth.

Low temperature enrichments for *L. monocytogenes* are incubated at 4°C for up to 12 weeks. Subcultures should be made at weekly intervals and, as noted above, most isolations may be made after incubation for only 1 week. The possibility of using different time:temperature combinations (*cf Y. enterocolitica*, page 146) has not been explored.

Selective enrichment is now more widely used than low temperature enrichments. A wide range of selective agents are used in various combinations, commonly used broths being summarised in **Table 124**. There is considerable variation in the selectivity of different broths. That devised by Lovett *et al* (1987) is only moderately selective, while another widely used enrichment broth, the medium of Doyle and Schoeni (1986), is of significantly greater selectivity. However, all selective enrichment broths are to some extent inhibitory to *L. monocytogenes*. The extent of inhibition by the various selective agents is strain dependent, and the effect is most marked with stressed cells, although non-stressed cells may also be inhibited. Two-stage enrichment procedures have been devised to minimise the inhibition of *L. monocytogenes* during enrichment (**Flow diagram 31**) (see page 349), but the extra manipulations involved have restricted application. Alternatively, a non-selective

resuscitation stage may be introduced.

Additional procedures may be used with the intention of increasing the selective pressure on non-listerias while limiting the effect on *L. monocytogenes*. In the isolation technique of Doyle and Schoeni (1986), enrichment is made in a side-arm flask which is evacuated and refilled three times with a gas mixture of composition 5% O_2: 10% CO_2: 85% N_2 followed by incubation in an orbital incubator at 100 rpm. This procedure is effective in minimising competition from aerobic bacteria, but is time consuming and requires equipment which may not be available in food industry laboratories. An alternative means of increasing selectivity is post-enrichment treatment with alkali (KOH). This procedure may reduce the recovery of *L. monocytogenes* if used during isolations from samples with only a small population of non-listerias (Parish and Higgins, 1989), but improve recovery where the number of competing organisms is high (Pini and Gilbert, 1988).

Table 124. Commonly used media for the selective enrichment of *Listeria monocytogenes*.

Medium	Selective inhibitors
Doyle and Schoeni[1]	Polymyxin B, acriflavin, nalidixic acid[2]
FDA broth[3]	Acriflavin, nalidixic acid, cycloheximide
University of Vermont broth[4] (UVB)	Acriflavin[5], nalidixic acid
Hayes' enrichment broth[6]	Nalidixic acid, potassium thiocyanate[7]

[1] Doyle and Schoeni (1986).
[2] Additional selectivity is obtained by incubation under microaerobic conditions.
[3] Lovett *et al*. (1987).
[4] Lee and McClain (1986).
[5] The broth is used in a two-stage procedure, in the first the acriflavin concentration is 20 mg/l and in the second 25 mg/l.
[6] Hayes *et al*. (1986).
[7] Used after cold enrichment.

Selective plating media

The situation with selective plating media is similar to that with enrichment broths in that a wide variety of media utilising many different combinations of antimicrobials are available. The more commonly used are summarised in **Table 125**. Media vary in selectivity, and there is evidence that there is variation in relative efficiency according to the nature of the food sample. In a comparison of six media, GNT medium was found to be most suitable for recovery of the organism from pasteurised milk and chocolate ice cream, but RISA medium was preferred for recovery from Brie cheese (Golden *et al.*, 1988).

Variation in suitability of different selective agars according to the nature of the sample presents obvious problems in laboratories examining a wide range of sample types. On an overall basis, LPM medium has been found superior (Loessner *et al.*, 1988), and this medium is certainly effective for the examination of highly contaminated samples (Cassiday and Brackett, 1989). However, it may be inhibitory to some strains of *L. monocytogenes* (Swaminathan *et al.*, 1988) as well as to *L. grayi* and *L. murrayi*, and a less selective media such as modified McBride medium may be preferred where competitive pressures are less (**Figures 272–274**). The more recently developed Oxford medium is more selective than modified McBride medium of equal sensitivity (**Figures 275–279**). A further

Table 125. Commonly used selective plating media for *Listeria monocytogenes*.

Medium	Selective inhibitors
McBride *Listeria* agar[1] (MLA)	Phenylethanol, glycine anhydride, lithium chloride
Modified McBride agar[2]	As above but containing additionally cycloheximide[3]
Gum base-nalidixic acid-tryptone-soya agar[4] (GNT)	Nalidixic acid[5]
Lithium chloride-phenylethanol-moxalactam agar[6] (LPM)	Phenylethanol, glycine anhydride, lithium chloride, moxalactam
Acriflavin-ceftazidime agar[7] (AC)	Acriflavin, ceftazidime
Rodriguez isolation agar[8] (RISA)	Acriflavin, nalidixic acid
van Netten agar[9]	Phenylethanol, acriflavin, nalidixic acid
Oxford agar[10]	Acriflavin, lithium chloride, cycloheximide, colistin, cefotetan, fosfomycin
Lithium chloride-ceftazidime agar[11] (LCC)	Lithium chloride, ceftazidime, glycine, anhydride

[1] McBride and Girard (1960).
[2] Francis *et al.* (1984). There are many modifications of McBride's medium.
[3] This formulation further differs from the original in being blood-free.
[4] Martin *et al.* (1984).
[5] Gum base replaces agar as gelling agent.
[6] Lee and McClaine (1986).
[7] Bannerman and Bille (1988).
[8] Dominguez Rodriguez *et al.* (1985).
[9] van Netten *et al.* (1988).
[10] Curtis *et al.* (1989).
[11] Lachica (1990c).

recently developed medium, LCC medium is of interest in being specially developed for use in conjunction with rapid identification of *L. monocytogenes*, using modified conventional methodology (see page 352). This medium has also been recommended for recovery of stressed cells (Lachica, 1990a).

The use of supplements to enhance recovery of stressed cells of *L. monocytogenes* has been investigated by Smith and Buchanan (1990), who found that the addition of Tween 80, bovine foetal serum or egg yolk-tellurite increased 100-fold the ability of modified Vogel-Johnson medium to detect heat-stressed cells without any loss of selectivity.

Many of the selective plating media for *L. monocytogenes* contain no diagnostic system, the organism being recognised by its colonial appearance under indirect (Henry) illumination (**Figures 272 & 273**).

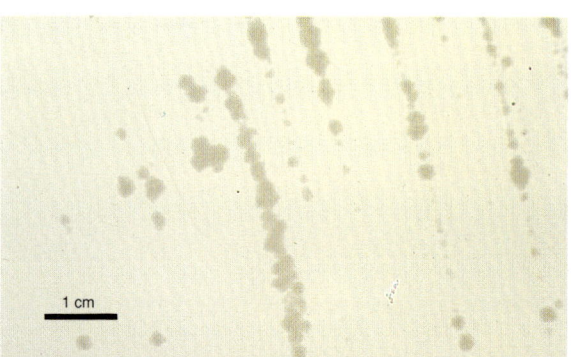

272 & 273. Isolation of *Listeria* spp. on modified McBride agar medium. Modified McBride agar is widely used for the isolation of listerias from foods. Typical colonies are irregular in shape and have a rather crystalline appearance (**272**), but their appearance can vary and colonies of some strains are smaller and more regular (**273**). *Listeria* is most easily recognised by using a binocular microscope with indirect illumination (Henry technique). Under these conditions colonies have a pale blue-grey or blue-green appearance. This is rather subtle and is best appreciated in photographic reproduction by comparing **272** and **273** with colonies of a species of *Bacillus* growing on modified McBride agar (see **274**), which has a distinct yellow colour under indirect illumination. Although the colour of *Listeria* colonies is difficult both to describe and illustrate, it is highly characteristic and, once seen, is often easily recognised on subsequent occasions. (*Modified McBride agar: 24h at 30°C.*)

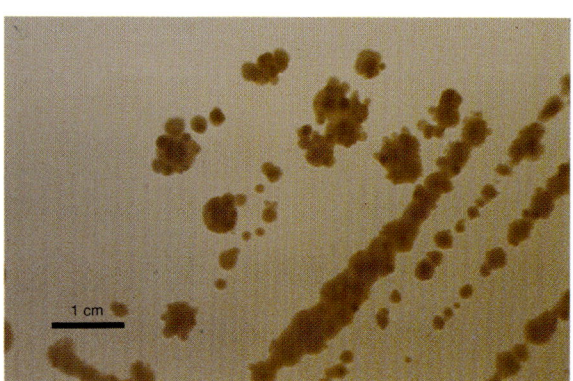

274. Selectivity of modified McBride agar. Modified McBride agar is only moderately selective and will support the growth of a number of Gram-positive bacteria. The greatest problem is likely to occur in foods with significant populations of *Bacillus*. Many species, such as the strain of *B. cereus* illustrated, superficially resemble some listerias when growing on this medium. As noted in the previous figure, differentiation is possible by examination under Henry illumination, but this technique is both time consuming and tedious where large numbers of samples are involved. (*Modified McBride agar: 24h at 30°C.*)

In this method, (Henry, 1933; Gray *et al.*, 1950), the colonies are viewed using a binocular microscope, the light source being reflected from a mirror positioned to reflect the light onto the base of the petri dish at a 45° angle. The technique produces better results when thinly poured plates (2 to 3 mm depth) are used (Pini and Gilbert, 1988), or where the agar component of the medium is replaced with a self-gelling colloid gum (Martin *et al.*, 1984). Although the Henry technique is widely used, it may not be effective under all circumstances and Hao *et al.* (1987) reported it to be of only limited help in distinguishing *L. monocytogenes* from non-listerias isolated from cabbage. A simplified modification has been described using high-intensity direct illumination (Lachica, 1990b).

More recently, attempts have been made to avoid the problems of earlier media by incorporating aesculin and ferric ammonium citrate in order to detect *Listeria* species by aesculin hydrolysis. Media of this type include AC medium (Bannerman and Bille, 1988), van Netten medium (van Netten *et al.*, 1988) and Oxford medium (**Figures 275–279**; Curtis *et al.*, 1989).

In some cases, the antimicrobial agents used may cause practical or logistic problems; the potency of moxalactam, for example, may vary considerably and thus a high level of quality control is required for media which contain this compound. A further problem, particularly in developing nations, is that some antibiotics such as ceftazidime may be very difficult to obtain.

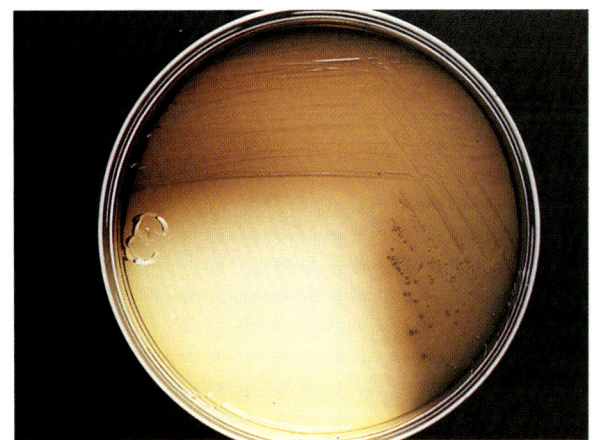

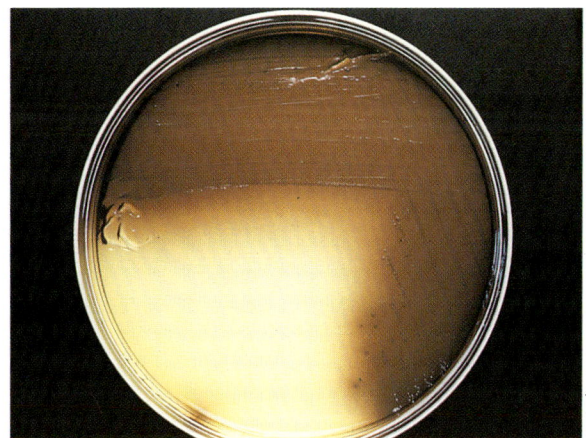

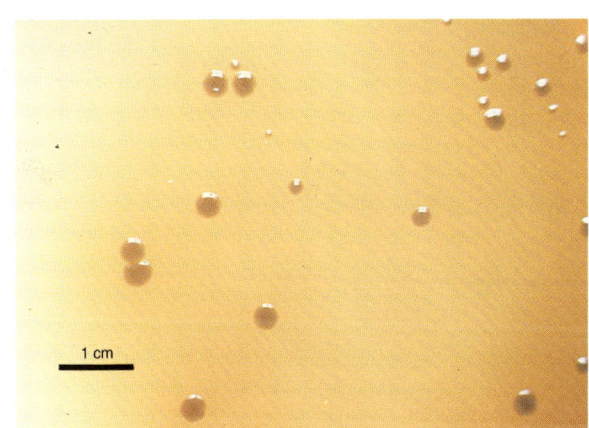

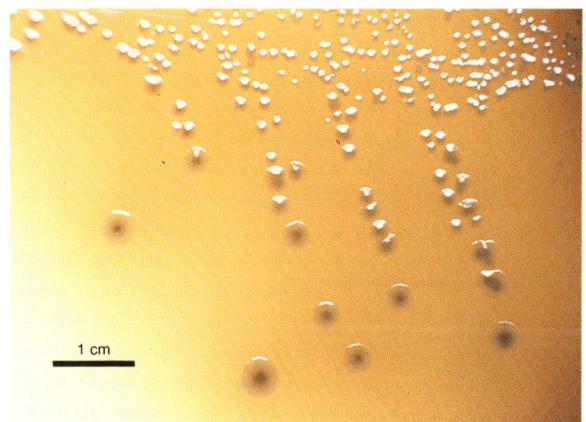

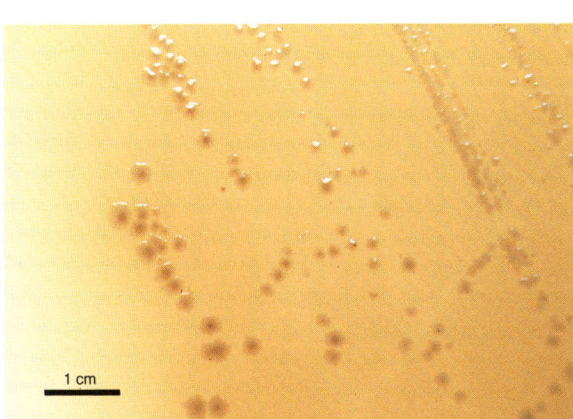

275–279. Isolation of *Listeria* spp. on Oxford agar media. Colonies of *Listeria* on Oxford agar are recognised by a black or dark brown colouration due to hydrolysis of aesculin. The appearance of *Listeria* is distinctive (**275**) and there is no need for special illumination. A minority of strains require incubation for 48 rather than 24 h, but in most cases this is not necessary and in the example illustrated there was little change during additional incubation (**276**).

All species of *Listeria* are recovered on Oxford agar including *L. monocytogenes* (**277**), *L. innoccua* (**278**) and *L. welshimerii* (**279**). Some strains of *L. seeligeri* may, however, be inhibited. There is some difference in colonial appearance, but this tends to be strain rather than species dependent. Oxford agar is more selective than modified McBride agar and bacteria such as *B. cereus* are unable to develop. (*Oxford agar: 24/48h at 37°C.*)

Complete protocol for isolation of L. monocytogenes

A large number of protocols for the isolation of *L. monocytogenes* have been published, but no single procedure has been universally adopted. In addition there are, a large number of *ad hoc* and unpublished methods devised by individual laboratories. Four protocols are described in **Flow diagram 31**, each of which illustrates a different approach to the isolation of *L. monocytogenes*. Reference should also be made to Martin *et al.* (1984); Dominguez Rodriguez (1984); Doyle and Schoeni, (1986); Beckers *et al.*, (1987); Slade and Collins-Thompson (1987), Bannerman and Bille, (1988); and van Netten *et al.* (1988).

Rapid detection of *L. monocytogenes*

The increased interest in *L. monocytogenes* has led to efforts in the development of rapid methods which have paralleled efforts in conventional methodology. The major lines of development have been the same as those for other pathogens; electrical, serological and genetic, and general observations made apply to the application of these methods to the detection of listerias.

Most of the methods currently in use detect all listerias rather than *L. monocytogenes* specifically. Manufacturers of instruments and detection kits present this as an advantage, since species other than *L. monocytogenes* can be pathogenic to man. However, *L. monocytogenes* is considered to be of overwhelming importance among the listerias, and under many circumstances the ability to detect *L. monocytogenes*, but not other members of the genus, would be beneficial.

Electrical methods

Protocols for the detection of *L. monocytogenes* have been developed for the major impediometric monitoring systems such as Bactometer© and Malthus©. All species of *Listeria* are detected and confirmatory testing is required.

Serological methods

Most attention in the development of serological methods has been paid to ELISA techniques, although there is also the prospect of the development of latex co-agglutination methods. Immunofluorescent microscopy may also find a widescale application.

Flow diagram 31. Protocols for the isolation of *L. monocytogenes*.

A Cold enrichment procedure for milk[1]

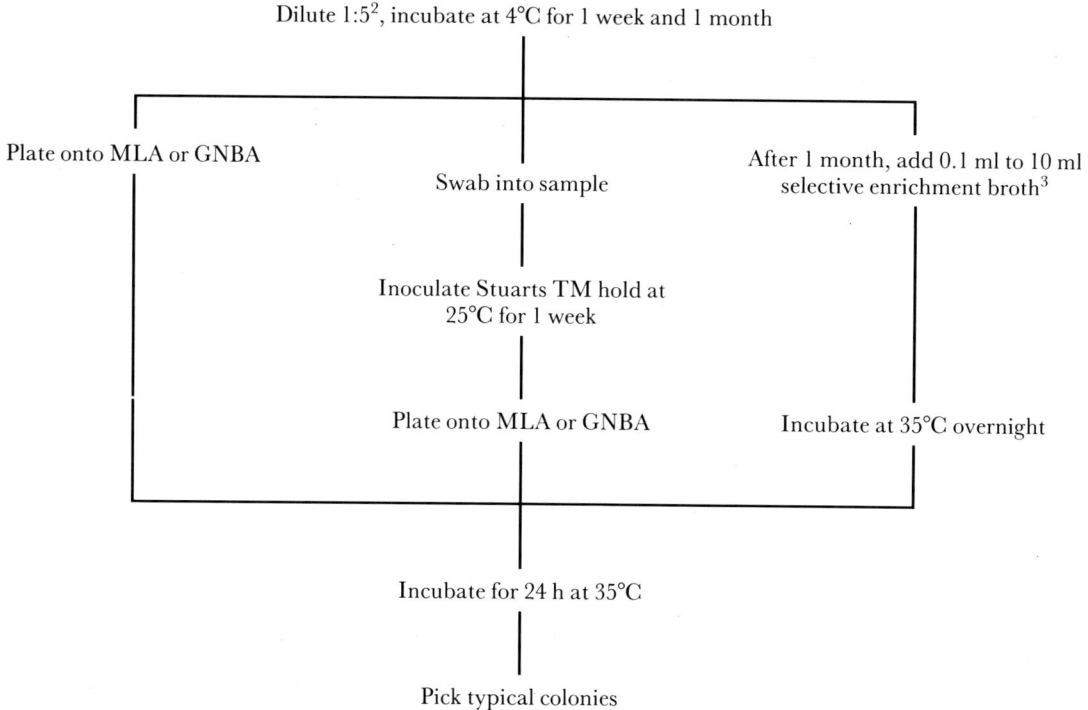

Dilute 1:5[2], incubate at 4°C for 1 week and 1 month

Plate onto MLA or GNBA

Swab into sample

After 1 month, add 0.1 ml to 10 ml selective enrichment broth[3]

Inoculate Stuarts TM hold at 25°C for 1 week

Plate onto MLA or GNBA

Incubate at 35°C overnight

Incubate for 24 h at 35°C

Pick typical colonies

348

B USFDA Procedure[4]

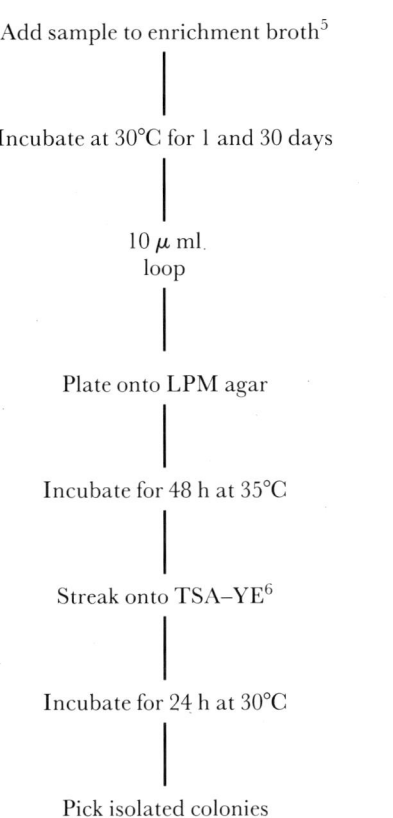

Add sample to enrichment broth[5]

|

Incubate at 30°C for 1 and 30 days

|

10 μ ml.
loop

|

Plate onto LPM agar

|

Incubate for 48 h at 35°C

|

Streak onto TSA–YE[6]

|

Incubate for 24 h at 30°C

|

Pick isolated colonies

C Method of Lee and McClane (1986)

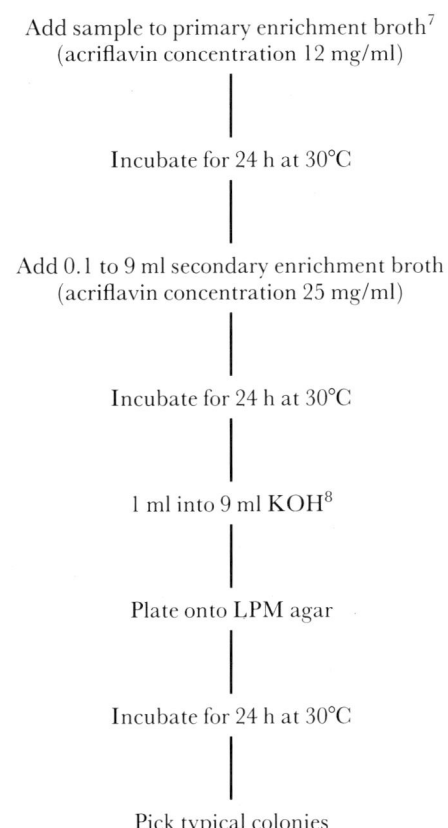

Add sample to primary enrichment broth[7]
(acriflavin concentration 12 mg/ml)

|

Incubate for 24 h at 30°C

|

Add 0.1 to 9 ml secondary enrichment broth
(acriflavin concentration 25 mg/ml)

|

Incubate for 24 h at 30°C

|

1 ml into 9 ml KOH[8]

|

Plate onto LPM agar

|

Incubate for 24 h at 30°C

|

Pick typical colonies

D General purpose protocol[9]

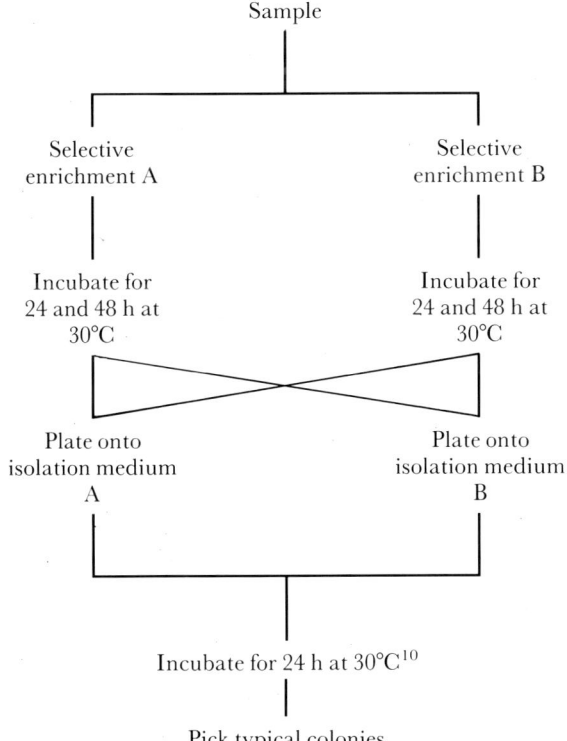

Sample

Selective enrichment A — Selective enrichment B

Incubate for 24 and 48 h at 30°C — Incubate for 24 and 48 h at 30°C

Plate onto isolation medium A — Plate onto isolation medium B

Incubate for 24 h at 30°C[10]

Pick typical colonies

[1] Hayes *et al.* (1986).
[2] Diluent: nutrient broth containing 0.1M MOPS buffer.
[3] Hayes' enrichment broth.
[4] Modified procedure of Lovett *et. al.* (1987). Original procedure employed alkali treatment of enrichment culture and modified McBride agar.
[5] FDA broth.
[6] Trypticase soy agar plus yeast extract.
[7] University of Vermont broth. In modifications selective agents may be omitted from the primary enrichment broth (Lammerding and Doyle, 1989).
[8] Primary enrichment broth may also be KOH treated and plated onto selective agar.
[9] This protocol may be adopted to suit individual requirements by selecting enrichment and selective media for specific properties.
[10] Incubation of negative plates of Oxford agar should continue to 48 hours.

ELISA techniques

Several approaches have been taken in the development of ELISA assays for *L. monocytogenes*. A number of colony assays have been devised (Farber and Speirs, 1987; Farber *et al.*, 1988c) using a monoclonal antibody. Milk samples were filtered through a nitrocellulose membrane that was incubated overlaid on a modification of McBride agar before removal of the membrane for the assay. A similar procedure is used for solid samples except that a colony blotting technique is used to transfer colonies from solid selective media onto nitrocellulose membranes. Where direct plating is used, the method is insensitive due to the lack of an enrichment stage, and in common with most other rapid detection methods species of *Listeria* other than *L. monocytogenes* are detected.

An alternative approach, which has been adopted for the development of a commercial kit (Organon Teknika©), is the use of two monoclonal antibodies, one for antigen capture and one for enzyme conjugation, to produce a highly specific and sensitive detection system (Butman *et al.*, 1988; Mattingley *et al.*, 1988). The assay is made on enrichment cultures, different protocols being used for meat and dairy products. However, this assay also lacks the ability to differentiate *L. monocytogenes* from other species of *Listeria*.

An independent evaluation of methods for the detection of *Listeria* showed that the commercial ELISA assay detected 68% of 44 positive vegetable samples compared with 75% detected by the conventional USFDA cultural method (Heisick *et al.*, 1989). The analytical procedure for the detection of *Listeria* by ELISA is similar to that for other bacteria and toxins (see pages 81–82, 264–265). Particular attention should be paid to washing of the wells to avoid false positives, and better results are obtained when seven washes are used instead of the six recommended.

Immunofluorescent microscopy

A direct immunofluoresence test (DIFT) has been developed which may be used on samples of soft cheese without culturing (McLauchlin and Pini, 1989). Suspensions prepared from soft cheese were fixed onto microscope slides, stained with an FITC/monoclonal antibody conjugate and examined using fluorescent microscopy. The method was used successfully for the detection of *L. monocytogenes* in rindless soft cheeses containing more than 10^2/g, but was unable to detect the organism in soft cheeses with rinds even at levels in excess of 10^4/g. Although further development is required, this method has promise as a means of detecting *L. monocytogenes* in food samples in less than 2 hours.

Genetic methods

The most recent developments in methodology for the detection of *L. monocytogenes* have involved gene probes. A hybridisation assay based on 16S rRNA sequences (Gene-Trak©) has been developed for the detection of *Listeria* species in dairy products and environmental samples (Klinger *et al.*, 1988). Hybridisation is carried out after a two-stage broth enrichment. According to Heisick *et al.* (1989), this procedure was less effective than ELISA, recovering 45% of positive samples from vegetables. A contrasting approach was made by Datta *et al.* (1988), who developed a probe specifically for *L. monocytogenes*. This probe detects the β-haemolysin gene and consists of an internal DNA fragment containing 500 base pairs of the haemolysin gene, cloned into plasmid pUC8 and isotopically labelled by nick translation. The probe is used in a colony assay whereby samples are plated directly onto selective agar and, after incubation, colonies blotted onto filter paper before hybridisation and detection of positive colonies by autoradiography.

Enrichment is not required unless samples contain less than 10 *L. monocytogenes*/g but, where direct plating is used, false-negative results may be obtained due to the inhibitory action of selective agents on stressed cells (Smith and Archer, 1988). A similar hybridisation procedure is used by the USFDA for confirming the identity of *L. monocytogenes* recovered using conventional methodology (see page 352). In either case, the use of isotopic labelling is likely to limit the application of colony hybridisation in the food industry.

A further species-specific probe for *L. monocytogenes* has been developed which detects a gene encoding a major secreted polypeptide. The nature of the polypeptide is not known, but it is probably not listeriolysin O. Under stringent hybridisation conditions, the DNA sequences were not detected in other *Listeria* species, *Bacillus* or *Streptococcus* (Flamm *et al.*, 1989).

Taxonomy of *L. monocytogenes*

Differentiation of Listeria *from other species*

Presumptive *Listeria* colonies on isolation media should be streaked onto a non-selective medium such as trypticase-soy agar supplemented with 0.6% yeast extract. Picking colonies from the isolation plate may be difficult, and a fine needle is often easier to use than a loop. After incubation for 24 hours at 30°C, colonies on the non-selective medium should be checked for purity using the Henry technique if necessary. Wet mounts should be made from isolated colonies and examined, preferably using phase contrast illumination. *Listeria* exhibits a typical gently tumbling and rotating motility. Colonies should then be tested for a positive catalase reaction and a Gram-stained smear prepared. *Listeria* cells are short, Gram-positive rods which occur either as straight pairs, V-forms and paired straight pairs (**Table 126**).

The use of these three screening tests will eliminate most non-listerias. Confirmation is on the basis of biochemical reactions and may be made, using conventional methodology, according to the criteria in **Table 120** (page 328). Alternatively, rapid test kits may be used. The API strep© has been widely applied to identification of *Listeria*, but identification is to genus level only, and it is usual to carry out further tests concurrently to identify *L. monocytogenes*.

Table 126. Cell morphology of *Listeria* species.

Straight pairs	Twin cells joined end to end
V-forms	Two straight pairs joined at various angles to produce a V-shape
Paired straight pairs	Straight pairs lying adjacent to each other. The pairs may be in contact or separated by a small gap

Note: Other arrangements may be present, but the above predominate.

Differentiation of species may be made according to the criteria in **Table 121** (page 329) or by means of the API 20 strep. kit together with the determination of the CAMP reaction using *Staphylococcus aureus* and *Rhodococcus equi* (**Figures 280–285**). Particular care must be taken in determination of haemolysis. *L. monocytogenes* is only weakly haemolytic and haemolysis may not be detectable where surface streaks are used. Stabbing sheep-blood agar rather than streaking is therefore recommended (Lovett, 1988). Determination of haemolysis by *L. seeligeri* is even more difficult, but there are no problems with *L. ivanovii*. For this reason, the use of the CAMP reaction is recommended with conventional methodology both to confirm and determine haemolysis.

L. monocytogenes may be differentiated from other species without additional tests using either the Mast ID© system or, less conveniently, the API 50CH© (Kerr *et al.*, 1990), but in many cases the use of the API strep© in combination with the CAMP reaction remains the method of choice.

An alternative to miniaturised kits is the use of accelerated conventional tests. Large colonies (greater than 2.5 mm diameter) are used to provide an inoculum for determination of the CAMP reaction, haemolysis and fermentation of rhamnose and xylose (Lachicha, 1990c). These tests are used in conjunction with examination by phase contrast microscopy, the catalase test and determination of gram reaction by viscoscity in 3% KOH, rather than by staining. This procedure provides same day identification colonies from selective media. Large colonies are required and LCC selective medium was developed for use in conjunction with this system. Other media which support large colonies such as Oxford medium, would, presumably, also be suitable.

Confirmation of identity is not possible using conventional serological techniques, but may be made using either the direct immunofluoresence technique or colony DNA hybridisation. In the former technique, haemolysis must be used as an additional test to permit reliable differentiation of *L. monocytogenes* and *L. innocua* (McLauchlin and Pini, 1989).

Colony DNA hybridisation techniques for confirmation of the identity of *L. monocytogenes* employ the same technology as that developed for the determination of the organism after direct plating onto a selective medium. In the procedure used by the USFDA, however, presumptive listerias recovered by a standard enrichment:isolation technique are picked and streaked onto a non-selective medium before blotting onto filters and hybridisation. The use of colony hybridisation in this way is highly effective, and in a survey of four methods gave the highest recovery (Heisick *et al.*, 1989).

280

28
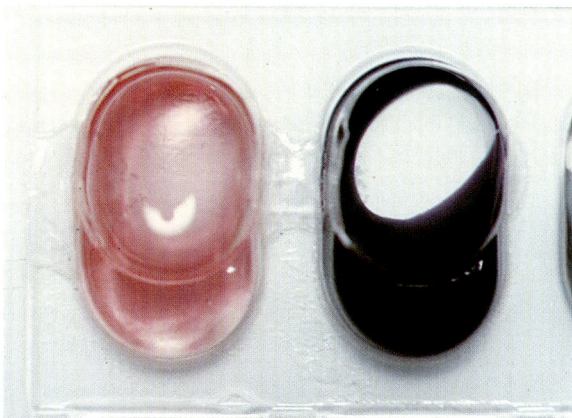

282
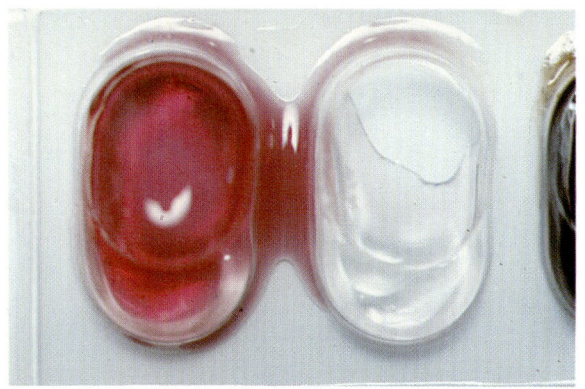

280–282. Identification of *Listeria* spp. using the API 20 strep© kit. The API 20 strep kit is a convenient means of confirming the identity of *Listeria* spp. (**280**). Differences between species occur in reactions such as hippurate hydrolysis (**281+, 282−**), but additional tests such as the CAMP reaction are required for reliable species differentiation. (*API 20 strep: 24h at 30°C.*)

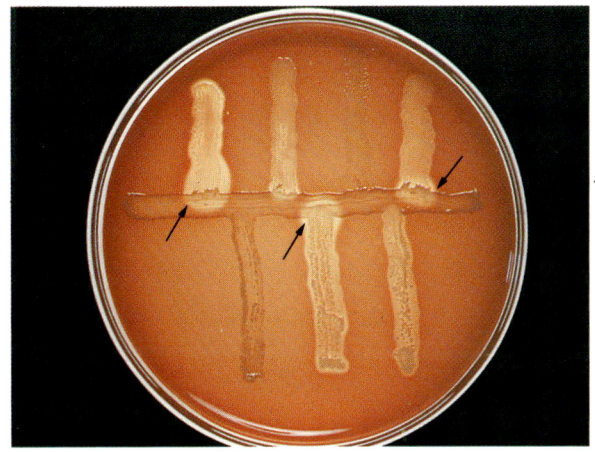

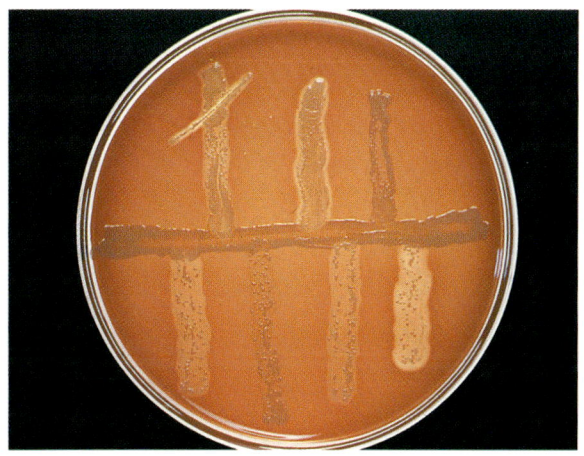

283–285. Use of the CAMP reaction in the differentiation of species of *Listeria*. The CAMP test is based on the enhancement of haemolysis of a culture under test by metabolites diffused into the medium from a second organism. Two organisms are used in the case of *Listeria*, *Staphylococcus aureus* and *Rhodococcus equi*. Each of these species is streaked vertically on a plate of blood agar overlying plate count agar, and isolates of *Listeria* are then streaked at right-angles. Haemolysis by *L. monocytogenes* is stimulated by *Staph. aureus* (**283**: arrowed), but not by *R. equi* (**284**). *L. monocytogenes* is not strongly haemolytic, and the degree of stimulation in the vicinity of the *Staph. aureus* streak is limited (**285**).

285

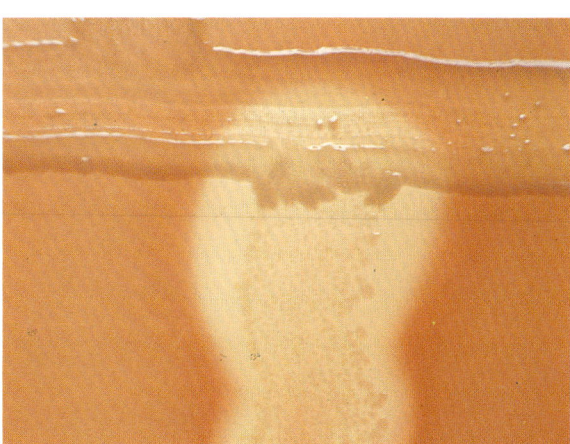

17. Other bacterial agents of foodborne disease

Introduction

Over the years a great many bacteria have been implicated as agents of foodborne disease. The organisms involved are diverse in nature, and the symptoms vary from mild diarrhoea to severe and life-threatening illnesses such as tuberculosis.

In many cases, the organisms involved may be placed in one of three categories. In the first place are organisms such as *Mycobacterium bovis*, which in the past were major causes of human disease. Many of these organisms were milkborne and in developed countries, the incidence has been drastically reduced by pasteurisation and eradication from the national herd. In other countries, however, these diseases may still be endemic. Second are those organisms which have long been associated with foodborne disease, but whose causative role remains controversial. The classic examples are *Enterococcus faecalis* and *Enterococcus faecium*. Third are putative pathogens which have been implicated in recent years and where a causative role in illness may not be proven. These include members of the *Enterobacteriaceae*, some of which produce an *Escherichia coli*-like heat stable toxin. Recognition of the causative organism of enteric illness may be difficult and in some cases bacteria have been incriminated on inadequate grounds. At the same time, it should be recognised that bacteria now considered to be important potential pathogens were previously thought to have been of no significance in illness.

Brucella

The genus *Brucella* comprises cocci, coccobacilli or short rods, arranged singly or, less commonly, in pairs, short chains or small groups. The organism is Gram-negative, catalase-positive and non-motile and has a respiratory metabolism. Carbon dioxide may be required for its growth, especially on primary isolation. Acid is not produced from carbohydrates (Corbel and Brinley-Morgan, 1984).

Three species, *Brucella abortus*, *Br. melitensis* and *Br. suis*, are most commonly responsible for the disease of man, brucellosis. Acute brucellosis has a sudden onset after an incubation period lasting from several days to several weeks. Symptoms of fever, drenching sweats, severe headache and pains in joints and back appear suddenly and gradually increase in severity.

The temperature swings between 37.7 and 39.4°C with a spike of 40.5 to 41°C. There is an accompanying loss of appetite, weakness and fatigue, and a sense of illness and depression. These symptoms reflect the bacteraemic phase of infection which is followed by localisation in the reproductive organs and reticuloendothelial system. If untreated, the symptoms gradually disappear and recovery is common within 14 days in the milder form of the illness caused by *Br. abortus*. Recurrence is common even after antibiotic treatment and the disease thus enters the subacute phase.

Subacute brucellosis (undulating fever) may follow on from the acute phase or may develop insidiously. Fatigue is the predominant symptom accompanied by backache and, in the case of infection by *Br. melitensis*, sciatica. The temperature is irregular rather than undulating, and there are occasional attacks of acute brucellosis. Overall subacute brucellosis is characterised by misery and dejection rather than by definable symptoms and acute signs. Subacute brucellosis usually persists for 12 to 16 months, but in about 20% of patients it becomes chronic and persists for many years. Death is uncommon but may occur especially in infections with *Br. melitensis* or *Br. suis*. Treatment is by administration of antibiotics such as gentamycin, tetracyclines or rifampicin.

Under natural conditions, *Brucella* behaves as a obligate parasite and has no existence independent of its hosts. Cattle are the most important hosts of *Br. abortus*, sheep and goats of *Br. melitensis* and pigs of *Br. suis*. In developed countries, brucellosis is largely an occupational disease affecting such people as veterinarians, farmers and slaughterhouse workers, but the disease may also be contracted via the milk of infected animals. The organism will also survive cheese making and will remain viable in hard cheese for many months (Chapman and Sharpe, 1981), and on rare occasions undercooked meat may be the vehicle of infection.

Foodborne brucellosis has largely been eliminated in developed nations due to the pasteurisation of milk and the eradication of *Br. abortus* from national herds. Occasional cases do occur due to consumption of raw cows' or goats' milk or raw milk cheese. The disease is still endemic in less developed nations.

Methods for the detection of *Brucella* in foods are insensitive and unsuitable for routine application, but serological techniques are available to detect infection in herds.

Clostridium difficile

Cl. difficile is a typical member of the genus (see page 289) and may be differentiated from other members on the basis of biochemical criteria. The species name may truly be seen as a caveat and reflects the difficulty encountered in its isolation and study (Cato *et al.*, 1986).

Cl. difficile is widely distributed in the environment and is present in meats and, probably, other foods as well as the faeces of *ca* 3% of asymptomatic persons. The carriage rate in neonates is much higher than in older children and adults, ranging from 15 to 70%. For a number of years, *Cl. difficile* was not associated with any specific disease, but is now known to cause enteric disease varying from mild antibiotic-associated diarrhoea, through colitis to severe and life-threatening pseudomembranous colitis. The classic symptoms of pseudomembranous colitis are watery diarrhoea, fever, abdominal pain and leucocytosis (Gebhard *et al.*, 1985). Occasionally, diarrhoea may be bloody following treatment with penicillin and ampicillin (Borriello, 1984), but in other cases occult blood is usually present on testing. Examination with a sigmoidoscope reveals the presence of elevated, yellow shining plaques, 2 mm to 2 cm in diameter on the surface of the colonic mucous membrane. Plaques consist of transudated fibrin with cellular debris and polymorphonuclear leukocytes.

Infection usually occurs following antibiotic therapy, but cases may also occur spontaneously and there is also an association with antineoplastic agents (Borriello, 1984). All antibiotics with the exception of vancomycin have been shown to be capable of initiating pseudomembranous colitis, and an unresolved paradox is that *Cl. difficile* is exquisitely sensitive to many of the antibiotics which induce its overgrowth. The most frequently implicated antibiotics are ampicillin, cephalosporins and clindamycin, although this probably reflects usage rather than colitis-producing potential. Pseudomembranous colitis usually develops within 4 to 9 days of commencing therapy, but the disease may appear after only a single dose of antibiotics or be delayed for up to 10 weeks after therapy has ceased (Tedesco, 1982). Females appear more prone to pseudomembranous colitis than males, and patients are typically over 50 years of age. Neonates and infants do not contract the disease even in the presence of faecal *Cl. difficile* cytotoxin (Griffin, 1989).

Mild antibiotic-associated diarrhoea may be treated successfully simply by withdrawal of the antibiotic. This is not always possible, and treatment with vancomycin, which is highly effective against all known strains of *Cl. difficile*, was widely adopted. Problems, however, emerged due to a relapse rate of 5 to 55% and also because of the high cost of treatment. Metronidazole and bacitracin are used as effective and relatively cheap alternatives to vancomycin, while ion-exchange resins such as cholestyramine have been used to bind *Cl. difficile* toxins. This treatment is not always effective and also reduces vancomycin activity by binding. Treatment after relapse is particularly difficult, and usually involves retreatment with the same antibiotic or substitution of an alternative drug. Restoration of a 'normal' gut flora using faecal enemata from controlled donors has been applied successfully (Bowden *et al.*, 1981), but is likely to be rejected by most patients on aesthetic grounds (Griffin, 1989).

During the first stage of *Cl. difficile* colitis, colonisation and overgrowth with the organism occurs in the colonic lumen. The organism may have originated by oral ingestion of endospores in food or from pre-existing carriage (George, 1986). Having become established, *Cl. difficile* synthesises and releases toxins causing fluid secretion, tissue damage and an inflammatory response. At least four toxins are synthesised by *Cl. difficile*: an 'enterotoxin', toxin A; a cytotoxin, toxin B; a motility enhancing factor; and a fluid accumulation toxin. Toxins A and B have been purified, but their mode of action is not known (Lyerly and Williams, 1984). Toxin A is not a classical enterotoxin and has strong cytotoxic activity, enterotoxigenicity probably resulting from disintegration of the brush border cells as a result of this activity (Tucker *et al.*, 1990). This toxin is believed to be primarily responsible for symptoms (Mitchell *et al.*, 1986).

A selective medium has been developed for the isolation of *Cl. difficile* from faecal samples, which consists of a sheep blood agar base containing fructose, with cycloserine and cefoxitin as selective agents. *Cl. difficile* is recognised on a presumptive basis by colonial morphology (after 48 hours anaerobic incubation at 37°C, colonies are 4 to 6 mm in diameter, irregular, raised, opaque, and grey-white in colour) and confirmation of identity is by biochemical testing. *Cl. difficile* is, however, very variable in biochemical activity (Borriello, 1984). Various typing systems are available for epidemiological studies including phage typing, biotyping (Borriello, 1984), plasmid profile typing and clindamycin susceptibility (Clabots *et al.*, 1988).

Cl. difficile cytotoxin may be detected in faeces by a tissue culture assay involving the neutralisation of activity by *Cl. sordellii* antitoxin. This test is tedious and time consuming, and various immunological tests for both toxins A and B have been devised. Their specificity is currently undetermined and controversial (Griffin, 1989).

Coxiella burnetii

Coxiella is a member of the tribe Rickettsieae, which comprises small, pleomorphic, Gram-negative micro-organisms which are obligate intracellular parasites. Coxiella grows preferentially in the vacuoles of the host cell, in contrast to Rickettsia which prefers the cytoplasm or nucleus (Weiss and Moulder, 1984).

Cox. burnetii is the causative organism of Q fever, symptoms of which are fever, severe headache and interstitial pneumonia. Symptomless infection is common. Death is rare, but other organs may become involved and a rare but highly fatal complication is Q fever endocarditis. Tetracyclines are the antibiotics of choice in the treatment of the disease.

Infection is usually acquired by the aerosol route from infected domestic or farm animals, but may be contracted from milk. Cox. burnetii has a high level of resistance to chemical and physical agents and survives at 63°C for 30 minutes. The high level of heat resistance has been attributed to the formation of endospore-like forms (McFaul and Williams, 1981). In many countries, such as the USA, pasteurisation processes have been based on time:temperature combinations necessary for the inactivation of the organism (Anon, 1978), and milkborne Q fever due to consumption of unpasteurised milk is rare. However, infection is still associated with consumption of unpasteurised milk and a survey in southern California showed an infection rate of 0.9% in pasteurised milk drinkers compared with 10.7% for unpasturised milk drinkers.

Epidemiologically, Q fever may be regarded as a 'place' infection, outbreaks occurring in localised areas which have undergone no change in exposure to the organism.

Cox. burnetii may be detected in milk by either guinea-pig inoculation or by agglutination tests (Ormsbee, 1980). The latter is a rapid and simple technique of use in determining the prevalence of the organism in dairy herds.

The Enterobacteriaceae

In addition to the well-established pathogens such as Salmonella, a number of members of the Enterobacteriaceae have been implicated in food poisoning. There has been speculation concerning the possibility of inter-generic transfer of virulence factors, and the production of an E. coli-like heat stable toxin by a strain of Citrobacter freundii is well established (Guarino et al., 1987). A number of the genera implicated in food poisoning are common in the environ-ment, and some strains are capable of growth at low temperature, growing on and, in some cases, being involved in the spoilage of refrigerated foods. It seems likely that pathogenicity is restricted to a small number of strains within each genus, but a fuller understanding is dependent on knowledge of the underlying genetic relationships in the family Enterobacteriaceae and other Gram-negative bacteria, and of the nature of virulence factors.

Citrobacter

Citrobacter is established as an opportunistic pathogen but, while previously considered to be an entero-pathogen, is now generally considered to be a normal inhabitant of the human intestine (Sakazaki, 1984). A role in diarrhoeal disease has been suggested for two species, Cit. freundii and Cit. diversus (Guerrant et al., 1976; Wadstrom et al., 1976; Finn, 1978; Guarino et al., 1987).

Citrobacter is common in the environment, and food and may be isolated using standard media such as violet red-bile-lactose agar. Identification kits are a convenient means of differentiating Citrobacter from other members of the Enterobacteriaceae, but additional tests may be required for differentiation of species.

Edwardsiella

Edwardsiella contains four species of which one, Ed. tarda, has been implicated as a cause of diarrhoea. Virtually all reports have been made in developing countries. The natural habitat of Ed. tarda appears to be the intestine of animals (Farmer and McWhorter, 1984), from which faeces may contaminate the environment and foods. The ability of the organism to grow in foods is not known.

Edwardsiella differs from other members of the Enterobacteriaceae in being more fastidious, producing smaller colonies on artificial media; being relatively inactive with respect to carbohydrate fermentation; and in being more sensitive to the Vibriostat O/129. The organism will grow on general purpose media for the Enterobacteriaceae, although the rate of growth is relatively slow. Edwardsiella is differentiated on the basis of a number of biochemical reactions rather than a single test, and while miniaturised kits employing a large number of tests are suitable for identification to species level, kits relying on a smaller number of tests may not differentiate Ed. tarda from other species.

Enterobacter

Enterobacter is a widely distributed organism in the environment, and may readily be isolated from human and animal faeces where its role is usually regarded as commensal. *Ent. cloacae* is known to be an opportunistic pathogen in extraintestinal infections, and has also been associated with diarrhoea in children. A causal role in diarrhoea is not well established, although strains of *Ent. cloacae* have been shown to produce a heat-stable toxin similar to that produced by *E. coli* (Klipstein and Engert, 1976).

Ent. sakazakii is rarely encountered in clinical specimens and is more common in environmental samples and food. The organism is an occasional pathogen causing neonatal meningitis and bacteraemia, and has been involved in a number of cases of meningitis in which the vehicle of infection is thought to have been dried milk (Biering *et al.*, 1989).

Environmental strains of *Enterobacter* are capable of growth in foods at refrigeration temperatures, and grow better at 20 to 30°C than at 37°C. *Enterobacter* grows well on the commonly used media for the *Enterobacteriaceae* (**Figure 286**), and may be differentiated to species level using identification kits. However, environmental strains, are likely to give atypical biochemical reactions if incubated at 37°C.

286

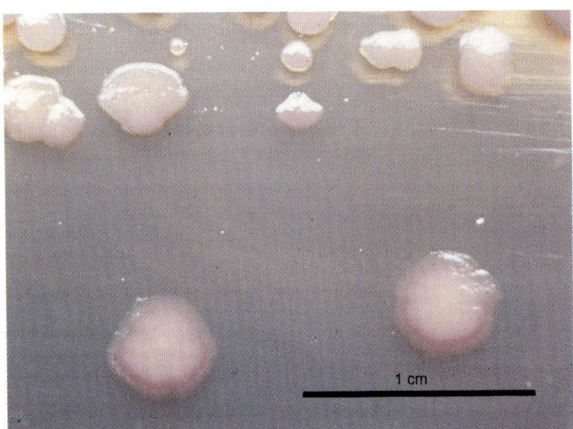

286. Isolation of *E. sakazakii* on violet red-bile agar (VRBA). *E. sakazakii* is a typical member of the *Enterobacteriaceae* and readily recovered on media such as VRBA. Various colony forms may be seen including smooth, mucoid rubbery and dry (Farmer *et al.*, 1980). However, it is not possible to distinguish *E. sakazakii* from other members of the *Enterobacteriaceae* on the basis of colonial appearance, and biochemical identification is required. (*VRBA: 20h at 37°C.*)

Klebsiella

Klebsiellas are recognised as being opportunistic pathogens and have become of increasing importance in nosocomial infections. This is probably a consequence of multiple antibiotic resistance. The gut is probably the main reservoir of opportunistic strains, and species of *Klebsiella* are present in small numbers in a high proportion of healthy individuals. One species, *K. pneumoniae*, has been implicated as a cause of diarrhoea and some strains are enterotoxigenic (Klipstein *et al.*, 1977).

K. pneumoniae has been connected with ankylosing spondylitis for several years, and victims of the disease usually carry larger numbers of *K. pneumoniae* in their intestines than do healthy individuals (Ebringer *et al.*, 1978).

K. pneumoniae and other klebsiellas have been isolated from foods and from environmental sources, although it is likely that many earlier isolations were of two more recently described species *K. terrigena* and *K. planticola*. *Klebsiella* may be isolated on standard media for the *Enterobacteriaceae*, although a specific medium, MacConkey-inositol-carbenicillin agar, has been devised (Bagley and Seidler, 1978). Identification to species level is possible using miniaturised kits.

Proteus, Providencia and Morganella

In the past *Proteus vulgaris*, *Pr. mirabilis* and *Morganella morganii* (*Pr. morganii*) have all been associated with diarrhoeal disease and food implicated as a vehicle of infection. However, considerable doubt has been cast on the validity of earlier reports and, in the case of *M. morganii*, a causative role seems unlikely. In the case of *Proteus*, a measure of support for a causative role in diarrhoea is the demonstration of enterotoxin production by the organism (Back *et al.*, 1980). However, some 25% of the healthy population are carriers of *Proteus* and it seems likely that, if the organism is the cause of diarrhoea, symptoms are produced only by certain strains or under certain uncommon conditions.

Providencia alcafaciens, particularly serovar O:3, has also been implicated as a cause of diarrhoea. The organism is not normally present in the stools of healthy people, but it has been isolated from diarrhoeal specimens on a number of occasions. However, it is not entirely clear if *Prov. alcafaciens* is the causative organism, or a commensal that is able to take advantage of conditions caused by infections with other micro-organisms.

Proteus, *Providencia* and *Morganella* may all be isolated on general purpose media for the *Enterobacteriaceae*, and identified to species level using identification kits. Swarming by strains of *Proteus* can make isolation of the genus difficult, or may cause problems during the isolation of other genera such as *Salmonella*.

Serratia

Serratia species are recognised opportunistic pathogens and *Ser. marcescens* (**Figure 287**) is particularly important as a cause of hospital infections. Both *Ser. marcescens* and *Ser. liquefaciens* have been associated with diarrhoea, although confirmation of a causative role is lacking. Species of *Serratia* are common in the environment and on plants, and are involved in food spoilage (Sutherland *et al.*, 1986). *Ser. liquefaciens* is capable of growth at 4°C.

Species of *Serratia* may be isolated on general purpose media for the *Enterobacteriaceae* and identified to species level using miniaturised kits.

287. Serratia marcescens. *Ser. marcescens* is best known for its production of the red pigment, prodigiosin. Pigmentation, however, is easily lost by mutation, while some strains of biogroup A4 produce a second pigment, pyrimine. On general purpose isolation media for the *Enterobacteriaceae*, *Ser. marcescens* appears as a typical lactose non-fermenting isolate and biochemical testing is required for confirmation of identity. Species of *Serratia* may also be isolated using a selective medium, caprylate-thallous agar, production of deoxyribonuclease being used as a confirmatory test (Grimont and Grimont, 1984). (*Plate count agar: 24 h at 30°C.*)

Enterocococcus

For many years, the role of *Enterococcus faecalis* and *Enterococcus faecium* (faecal streptococci, Group D streptococci) in food poisoning has been controversial. Evidence for a role has been based on the isolation of enterococci from suspected foods, and the results of volunteer studies have been conflicting (Roberts, 1982). Where illness has been produced by ingestion of enterococci, doses have been very large and similar numbers have been eaten in foods without ill effects. The situation has been reviewed by Bryan (1979), who suggested that only certain strains of some species are pathogenic.

Symptoms associated with *Enterococcus* appear within 2 to 36 hours of ingestion (average 6 to 12 hours) and are mild, typically involving nausea, abdominal pain and, sometimes, vomiting. Cooked meats have been frequently implicated in the past, but other foods implicated include turkey dressing cheese and evaporated milk (Jay, 1978).

It is noteworthy that there have been fewer allegations of enterococcal food poisoning in recent years. This may be attributed either to improved hygiene and the widespread application of refrigeration, or to a change in attitude towards the role of *Enterococcus* as an enteric pathogen. In contrast, its role as an opportunistic pathogen in serious extraintestinal disease including endocarditis is now fully recognised (George and Uttley, 1989). In both the UK and the USA its importance as a nosocomial pathogen is increasing in proportion to the usage of broad-spectrum β-lactam antibiotics.

Enterococci are common in the intestine of man and animals. They are resistant to many unfavourable conditions and may readily be isolated from foods. A role as an index organism for faecal pollution has been suggested, but in many cases their presence bears little relationship to the likely presence of estab-

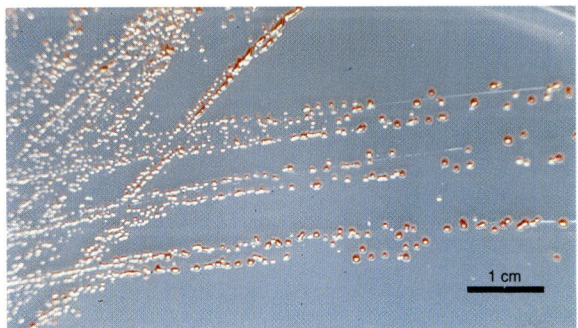

288

288. Isolation of *Enterococcus* spp. on Barnes' thallous acetate-tetrazolium-glucose medium (TTG). Barnes' TTG medium is one of the older media for *Enterococcus*, but is still widely used in the UK food industry. In many ways, the shortcomings of this medium typify general problems with selective media for enterococci; selectivity is poor and while non-enterococci usually produce white, or pale-pink colonies, overgrowth of *Enterococcus* may occur. At the same time, the medium is inhibitory to some strains of *Enterococcus*, which grow poorly, or not at all. (*Barnes' TTG medium: 24 h at 37°C.*)

lished pathogens such as *Salmonella*. Enterococci have a high level of heat resistance, and often survive in marginally underprocessed meat products.

An extremely large number of media have been devised for *Enterococcus*, but all have shortcomings relative to the degree of selectivity, quantitative recovery or differential ability (APHA, 1984). Packer's crystal violet-azide-blood agar and kanamycin-aesculin-azide agar are among the less unsatisfactory, while Barnes' TTG agar (**Figure 288**) remains widely used in the food industry in the UK. The API 20 strep.© kit is a convenient means of confirming identity.

Mycobacterium

The mycobacteria are slow-growing, Gram-positive, rod-shaped bacteria which are characteristically acid-fast when stained by the Ziehl-Neelson method. Cells are fairly regular but branched filaments may be present. Traditionally, the mycobacteria have been treated as a family, the *Mycobacteriaceae*, but the organisms are closely related to *Corynebacterium, Nocardia* and *Rhodococcus* (Wayne and Kubica, 1986).

The genus *Mycobacterium* is best known as the causative organism of two serious diseases of man, leprosy and tuberculosis. Tuberculosis is most commonly acquired by person-to-person transmission via inhalation, and under these circumstances the causative species is *M. tuberculosis*. Tuberculosis may also be acquired by consumption of milk, or less commonly meat, from animals infected with the closely related *M. bovis*, or by contact with infected animals.

Although *M. bovis* and *M. tuberculosis* are phenotypically distinct, they form a macrocluster along with *M. microti* and *M. africanis*, which is distinct from other members of the genus. Tuberculosis caused by *M. bovis* and *M. tuberculosis* cannot be distinguished clinically.

Milkborne *M. bovis* may enter the body by a number of routes. The organism may, for example, settle on the tonsils and pass to the cervical lymph nodes, or reach the intestines and cause intestinal tuberculosis, or be absorbed and cause tuberculosis in any part of the body, especially the bones and joints.

Tuberculosis generally responds to chemotherapy, although the older compounds used, isoniazid, streptomycin and para-aminosalicylic acid required treatment to be continued for as long as 18 months to 2 years. More modern antimicrobials such as rifampicin, pyrazinamide and ethambutal permit shorter courses of treatment, and may be used in combination with older compounds, a combination of isoniazid and rifampicin being particularly effective (Christie, 1987). Surgical removal of infected tissue may be required in cases of bone and joint tuberculosis.

In developed countries, the morbidity due to tuberculosis has been dramatically reduced by the introduction of milk pasteurisation and the elimination of *M. bovis* from cattle. Badgers have been identified as a reservoir of the organism from which cattle are affected, and in the UK this has resulted in a highly controversial campaign of destruction of badgers by gassing. The scale of achievement in reducing tuberculosis to very low levels should not be underestimated and in the UK and other developed countries, most cases are now imported. In the nations of the Indian subcontinent, Africa and South America, the disease remains endemic.

Cultural methods for the detection of *M. bovis* in milk are not suitable for routine use, although serological methods are widely used for testing cattle for infection. In some countries, suspected tubercular lesions found on carcasses at inspection are routinely cultured for the organism. Various media are available including acid-egg medium, Lowenstein-Jensen medium and Dorset egg medium.

Pseudomonas aeruginosa

Ps. aeruginosa is a common environmental organism and may also be isolated from the faeces of man, and animals associated with man. It is recognised as an opportunistic pathogen and is of particular concern when present in bottled non-carbonated mineral water. Bottled water is frequently used for vulnerable groups such as infants and invalids in the mistaken belief that it is inherently safer than piped water, and under these circumstances *Ps. aeruginosa* presents an obvious risk. In addition, an outbreak of *Ps. aeruginosa* infection occurred among infants in a nursery when a well providing drinking water was contaminated by seepage of sewage and the infiltration of stream water (Weber *et al.*, 1987).

A method for the isolation of *Ps. aeruginosa* from water has been described (Mossel *et al.*, 1976). This involves enrichment at 42°C in a broth containing nitrofurantoin followed by streaking onto King's media supplemented with nitrofurantoin. Alternative selective agents are a combination of cetrimide and nalidixic acid. In bottled water, the numbers of *Ps. aeruginosa* may be sufficiently high to permit detection by direct plating.

Pigment production is not a reliable means of identifying *Ps. aeruginosa*, and identity should be confirmed using polytropic tubes (Mossel *et al.*, 1976) or, more conveniently, an identification system such as API 20NE© (**Figure 289**).

289. Growth of *Ps. aeruginosa* using single organic substrates as the sole carbon and energy source. *Ps. aeruginosa*, like many other members of the genus, is able to grow using a wide range of simple organic compounds as the sole carbon and energy source. The range of compounds which may be used is species dependent, and are used as a means of identification using both conventional methodology and in some miniaturised kits such as the API 20NE© illustrated. There are wider implications in that the ability to utilise single organic substrates at a low concentration permits the growth of the organism in some types of bottled water. Note the

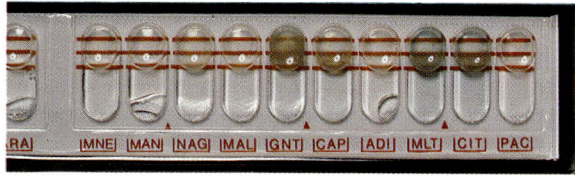

differential effect on pigment production; some substrates do not support production of any pigment, while in others the ratio of fluorescein to pyocyanin is obviously variable. (*API 20NE: 24 h at 30°C.*)

Streptococcus

Streptococcus pyogenes (group A, β-haemolytic) is the cause of scarlet fever and streptococcal pharyngitis (streptococcal sore throat). Pharyngitis may be followed by post streptococcal glomerulonephritis, a disease characterised by temporary kidney failure due to immune complexes becoming lodged in the glomeruli of kidneys. Pharyngitis may also be followed by rheumatic fever, a disease in which the heart is enlarged and there is temporary arthritis. Rheumatic fever, in turn, may be followed by the more serious rheumatic heart disease. The mechanism of disease is not known in the cases of rheumatic fever and rheumatic heart disease.

Str. pyogenes is also a cause of mastitis in cattle, and raw milk containing *Str. pyogenes* resulting from clinical, or subclinical, mastitis, was a common cause of infection in man before the introduction of widescale pasteurisation. Occasional outbreaks still occur due to the consumption of raw milk, and infections remain common in less developed countries.

Cases of *Str. pyogenes* infections associated with foods other than milk occur on rare occasions, and are caused by contamination of the food by infected handlers, including asymptomatic carriers. Salads and other foods consumed without further heating are most commonly involved. In most cases, carriage is in the throat but *Str. pyogenes* is among the streptococci involved in skin sepsis on the hands of meat handlers, and there have been anecdotal accounts of infection of food via this route. Personnel with skin lesions or sepsis, or suffering sore throats, should be excluded, but otherwise normal hygiene precautions are adequate to prevent the contamination of foods by handlers.

Str. zooepidemicus, a β-haemolytic species of Lancefield group C, has previously been rare in man, but in recent years has been involved in a small number of outbreaks of severe disease. Raw milk (Ghanheim and Cooke, 1980; Edwards *et al.*, 1988) and cheese made from raw milk (Report, 1983) have been implicated as vehicles of infection. Clinical features of the infection include septicaemia, meningitis and endocarditis, and the fatality rate may be high. In the case involving raw milk cheese in New Mexico, USA, there were 2 deaths out of 16 known cases, while in an outbreak in the north of England no less than 8 deaths occurred out of 12 known cases, although in no case was the infection totally responsible.

The English outbreak was fully investigated and the immediate source of infection was traced to a farm where the herd suffered intermittent attacks of mild mastitis. The full extent of the outbreak was not known since young and healthy people probably suffered no more than mild 'flu-like' symptoms. Most of the fatal cases involved the elderly, and one further fatal case was that of a premature, one-day-old infant (Edwards *et al.*, 1988).

Methods that are sufficiently selective and quantitative for routine examination of foods for either group A or group C β-haemolytic streptococci are not available.

Spoilage bacteria

The activities of spoilage bacteria are associated with scombroid poisoning in some types of fish, and histamine toxicity in some varieties of cheese. These are discussed fully in pages 458 and 465–466 respectively.

18. Foodborne viral infection

Introduction

Historically, the major foodborne viral disease has been hepatitis A infection, and the problem has been recognised world-wide for several years. Hepatitis A infection is primarily a disease of underdeveloped nations and is declining in developed countries. In the USA, the incidence remains relatively high among the urban poor, as well as among risk groups including sewer workers and homosexuals. Contaminated food and water can also be involved in the spread of hepatitis E (enterically transmitted non-A, non-B), but not hepatitis B.

Since the 1970s, there have been an increasing number of reports of foodborne viral gastroenteritis. Most of these have involved shellfish and small, round, structured viruses (SRSVs) are of overwhelming importance as causal agents.

Despite this, SRSVs account for only a small percentage of diarrhoeal viruses reported. In the UK, for example, rotaviruses consistently account for over 80% of cases, followed by adenoviruses, which account for *ca* 12% of cases (Anon, 1988a). These viruses are classically responsible for outbreaks of childhood diarrhoea that are transmitted by person-to-person contact rather than by foods. Adenoviruses have never been associated with foodborne transmission and rotaviruses only very rarely. In the case of adenoviruses, this may be explained satisfactorily by a very high level of immunity in older age groups, but this explanation is not entirely satisfactory in the case of rotavirus, since adults are susceptible in waterborne and institutional outbreaks. However, rotavirus C, one of a number of newly isolated rotaviruses, was implicated in an outbreak of foodborne gastroenteritis, which affected Japanese children and their teachers (Matsumato *et al.*, 1989), while food may be one of the vehicles of infection in the very large outbreaks of diarrhoea due to rotavirus B, which have occurred in recent years in the Republic of China.

Although SRSVs, like other diarrhoeal viruses, are capable of spread from person to person, it may be argued that the less common foodborne route is of greater social and commercial significance (Anon, 1988a). In an example quoted, several thousand people in the USA were infected with an SRSV following widespread distribution of cakes contaminated by a food handler. Immediate consequences were the disruption of medical and surgical activity in at least one hospital where staff were affected, and delays in the opening of several schools.

Despite the fact that viral foodborne gastroenteritis can no longer be described as a 'new' form of food poisoning, a number of misconceptions remain, not least the totally erroneous concept that viruses may multiply in some foods. It should also be appreciated that the full extent of the problem is not known due partly to under-reporting, and partly to the shortcomings of current viral methodology.

Foods are occasionally involved in the spread of other viruses such as poliovirus, but such transmission is of limited importance.

In the UK, and elsewhere in 1990, there was the start of considerable publicity given to the possibility of bovine spongiform encephalopathy (BSE), the 'mad cow' disease of popular newspapers, infecting humans via meat products containing nervous tissue. Spongiform encephalopathies are caused by agents distinct from conventional viruses, and are poorly understood. Available evidence was examined by a UK Government Working Party, which concluded that the risk of human infection is remote (Anon, 1989).

Mention should also be made of acquired immunodeficiency syndrome (AIDS), as there have inevitably been anxieties expressed concerning the possibility of foodborne transmission. Current opinion is that there is no known risk of transmission of the virus via foods, providing that basic sanitary precautions are observed.

Types of virus associated with foodborne disease

Small gastroenteritis viruses

Failure to propagate the majority of small viruses associated with gastroenteritis means that full characterisation has only rarely been possible. As a result, an unsatisfactory system of nomenclature based on morphological appearance or geographical origin has been evolved (Anon, 1988a). Some order has been introduced by the nomenclature scheme of Caul and Appleton (1982), although the interim nature of this scheme should be appreciated (**Table 127**).

Table 127. Nomenclature of foodborne diarrhoeal viruses (Caul and Appleton, 1982).

Structured particles	
Classical surface structure	
5 or 6 pointed star	Astrovirus
Stain-filled cups	Calicivirus
('Star of David')	
Amorphous structure lacking surface geometry SRSVs	Norwalk
	Hawaii
	Montgomery County
	Snow Mountain agent
	Angel Island agent
	Sapporo agent
	Taunton agent
	Amulree agent
Featureless particles	
No resolvable surface structure and a sharply delineated outer edge	Cockle agent
	Wollan agent
	Ditchling agent
	Parramata agent

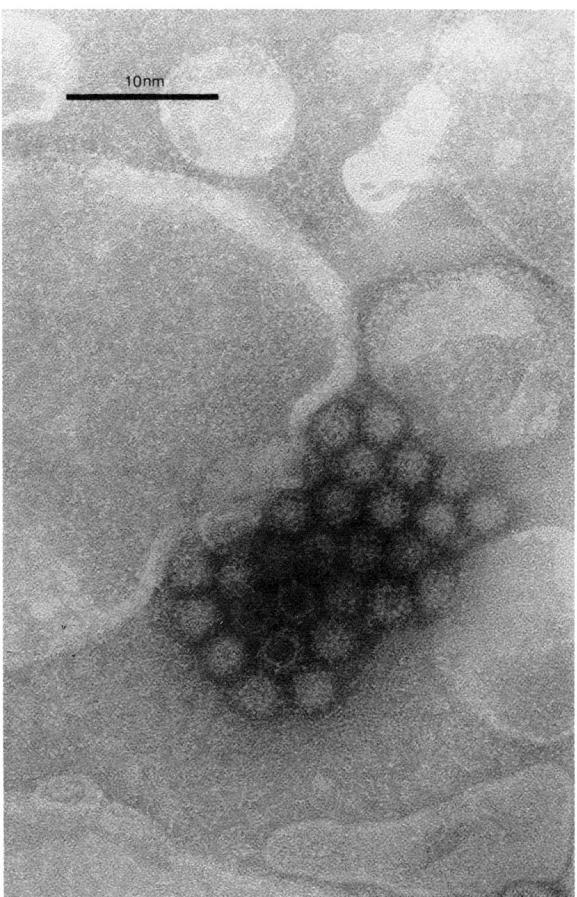

290. Transmission electron micrograph of a small, round, structured virus. (Reproduced by courtesy of Dr E.O. Caul, Public Health Laboratory Service, Bristol.)

Small, round, structured viruses

SRSVs (**Figure 290**) are of much the greatest importance in foodborne gastroenteritis. The first virus of this type to be described was Norwalk virus (Kapikian *et al.*, 1972), and the term 'Norwalk-like virus' has been used to describe those small, round viruses which are indistinguishable from the original Norwalk virus. In the UK, it is considered that the term Norwalk-like implies a degree of identity that has not been proven, and the Public Health Laboratory Service recommend that at the present time all viruses indistinguishable from Norwalk virus should be collectively identified as SRSVs (Anon, 1988a). Immunoelectron microscopy has shown that Norwalk virus is related to Montgomery County agent, but not to Snow Mountain agent and other SRSVs. However, radioimmunoassays of antibodies showed some relationship between Norwalk and Snow Mountain agent (Cubitt, 1989).

Norwalk virus may be a member of the Caliciviridae (Cubitt *et al.*, 1987), and there is both biophysical and antigenic similarity with calicivirus. Norwalk virus, like the Caliciviridae, has a single major structural polypeptide of MW 66,000, but cannot definitely be placed in the family until the nature of the nucleic acid has been determined (Cubitt, 1989).

Viruses with classical surface structure

There are two morphologically distinct types of virus with a classic surface structure, astrovirus and calicivirus. Identification is theoretically straightforward, but the characteristic surface structure is not always readily discernible (Anon, 1988a). Astrovirus is usually associated with gastroenteritis in infants aged less than 1 year, but has also been involved in foodborne illness, in co-infections with putative parvoviruses (Appleton, 1987).

Caliciviruses were originally observed in specimens from sporadic cases of gastroenteritis in babies, and later from outbreaks of gastroenteritis and winter vomiting among babies and young children (Halligan and Auty, 1988). There is one known foodborne outbreak (Appleton, 1987).

Featureless viruses

Small, featureless viruses have been recovered from foods associated with gastroenteritis on a number of occasions. These are thought to be parvoviruses, a view strengthened by preliminary evidence that they have a DNA genome. However, no virus of this type has conclusively been shown to cause gastroenteritis (Anon, 1988a).

Enteroviruses

The enteroviruses form a large group encompassing hepatitis A virus, poliovirus, echovirus and coxsackie virus. The most significant from the viewpoint of food safety is hepatitis A virus. Foodborne transmission of poliovirus has also been demonstrated, and foods may also be associated with echovirus infections (Cliver et al., 1984).

Hepatitis E

Hepatitis E is one of at least three types of hepatitis non-A, non-B. Hepatitis E is epidemic and usually transmitted by contaminated food or water. Person-to-person spread may also occur. Hepatitis E is prevalent in the Indian subcontinent and other areas where standards of hygiene are poor. The virus is an unenveloped single stranded RNA virus of 32–34 nm diameter. It is extremely labile and biochemical properties suggest a relationship with the caliciviruses (Zuckerman, 1990).

Symptoms of foodborne viral infections

Small, round, structured viruses

The incubation period of illness caused by Norwalk and other SRSVs is variable, and may range from 15 hours to 2 days, possibly depending on dose. The attack rate is usually very high (Eyles, 1986; Appleton, 1987). Symptoms are typically diarrhoea or diarrhoea and vomiting; diarrhoea alone being reported in 54% of cases, diarrhoea and vomiting in 38% and vomiting alone in only 8% of cases (Butcher et al., 1989). These symptoms may be 'spectacular' and involve both explosive diarrhoea and projectile vomiting. These symptoms are often accompanied by malaise, abdominal pain, pyrexia and nausea. The duration of the illness is short, and clinical symptoms usually persist for no more than 1 to 3 days. However, malabsorption of D-xylose and fat often occurs, and may persist for several days after the disappearance of clinical symptoms.

Symptoms are rarely of sufficient severity or duration for hospitalisation, or indeed for most patients to seek any medical assistance. Rehydration is required only on rare occasions and there is no specific treatment. Although mild in healthy people, infection with SRSVs has hastened death in debilitated patients (Kaplan et al., 1982).

Hepatitis A

The incubation period of hepatitis A is usually 3 to 6 weeks, although it may be as short as 2 weeks and as long as 8 weeks. The illness begins with a prodromal stage of non-specific malaise, symptoms commonly involving fatigue, fever, loss of appetite and nausea. The disease progresses to abdominal pain and vomiting, with jaundice and darkening of the urine developing later. Mortality is low but patients may be incapacitated for several weeks. In children and, occasionally, in adults, infection may be asymptomatic. Passive immunisation using immunoglobulin is effective and should be given to people at particular risk.

Hepatitis E

The incubation period of hepatitis E is rather longer than that of hepatitis A. In most patients, the disease involves a self-limiting hepatitis of moderate severity, symptoms being similar to those of hepatitis A, but with generalised pruritis being more common. The disease is of considerably greater severity among pregnant women, and in the third trimester of pregnancy the case fatality is 10 to 20%. With the exception of pregnancy, recovery is complete and there is no evidence of chronic liver disease resulting from the infection. Symptoms rarely develop in children.

Mechanisms of pathogenicity

Small, round, structured viruses

Attempts to find an animal model for SRSVs have been unsuccessful, and what little is known has been obtained from adult human volunteer studies in the USA. In affected patients, there is a broadening and blunting of the villi of the proximal small intestine. The epithelial cells remain intact, but the microvilli are reduced in length. Unknown genetic or physiological factors may predispose certain individuals, but not others, to gastroenteritis following exposure (Kapikian *et al.*, 1982).

Hepatitis A

The mechanisms of pathogenicity of hepatitis A virus are poorly understood. The likely course of events is that the virus enters the body by ingestion and first multiplies in the intestinal epithelium. This stage is followed by its spread, presumably via the bloodstream, to the liver where the virus infects the parenchymal cells. None of these events have been conclusively proven.

The mechanism of pathogenicity of hepatitis E virus is not known.

Human carriage and the role of the food handler

Small, round, structured viruses

There is no true carrier state associated with infection by SRSVs. Shedding of detectable virus continues for up to 4 days (Riordan, 1988), although the period is usually rather shorter and may be as little as 12 hours (Cubitt, 1989). It should be appreciated that these periods are determined using insensitive electron microscopic techniques, and that excretion at a low level may continue for a longer period. The need for exclusion of food handlers after suffering viral diarrhoea is widely accepted, but there has been considerable controversy over the length of the exclusion period. The current UK practice is to allow handlers to return to work 48 hours after diarrhoea has ceased (Anon, 1988a; Reid *et al.*, 1988), and this seems reasonable *in the current state of knowledge*, but should always be open to reassessment. Difficulties are caused by the fact that infection may be symptomless, or extremely mild, particularly after re-infection, as well as, inevitably, the failure to report illness (see below).

The role of the food handler in the spread of SRSV infection is of particular importance because of the extremely low infective dose and also because the nature of the symptoms. Diarrhoea is explosive and difficult to contain, possibly leading to contamination of washrooms, and placing a strain on all but the most rigorous personal hygiene. Vomiting may occur with little or no warning, is projectile and uncontrollable, leading to the possibility of contamination of the food preparation environment, or of the food itself. Infection with SRSVs may be acquired by inhalation via the nasopharynx to the gastrointestinal tract, and thus may spread among food handlers by aerosol inhalation as well as by the more familiar hand-mouth route. It seems likely that cycles of infection and re-infection can be established, and account must be taken of this even where staff remain free of overt symptoms (**Flow diagram 32**).

The point of contamination is usually during final preparation and the serving of food, and thus most cases involve the home or food service operations. To some extent, this may be due to the level of handling of food involved. Intensive hands-on activity has been

Flow diagram 32. Possible cycles of infection with small, round, structured viruses.

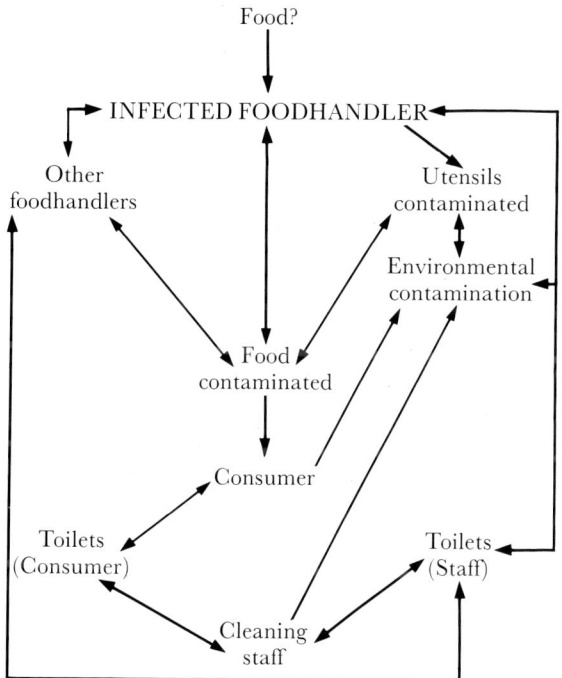

identified in a number of cases where a particular food has been implicated, examples including de-pipping grapes, garnishing melon and stirring cake-mix (Anon, 1988a).

The ability of rotavirus to survive on hands and be transferred to other surfaces has been conclusively demonstrated (Ansari et al., 1988), and there is no reason to believe the situation to be different with SRSVs. In the food service industry, however, it is also necessary to take account of poor standards of sanitation which often exist as well as poor standards of staff training.

In contrast to many types of bacterial food poisoning, SRSV infections are often, after the initial explosive symptoms, sufficiently mild to work through without undue discomfort and the common low wages and lack of job security in food service operations mean that the economic pressure to do so can be very strong. There has certainly been at least an instance in the UK where an outbreak of SRSV infection could be attributed to an infected food handler who continued to work because of fear of dismissal. The consequences of such actions may, of course, be very serious; a baker who worked during a diarrhoeal illness was responsible for more than 3,000 cases of illness in the USA.

People in closed communities, particularly when elderly, are often at particular risk of SRSV infection; person-to-person transmission is probably of major importance, although the outbreak may be initiated by contaminated food. Health care workers may be involved in the spread of infection, and should be excluded from work if suffering diarrhoeal illness (Butcher et al., 1989). Person-to-person spread may also be involved in incidents where point source infection has previously been suspected. Investigations of consecutive outbreaks of diarrhoea on a cruise ship (Ho et al., 1989) suggested that person-to-person transmission via aerosolised vomitus was largely responsible. A pattern of consecutive outbreaks of diarrhoea among new guests has also been observed in hotels and conference centres and a similar pattern of spread is likely. In a closed environment, illness among guests may also contribute to a cycle of infection involving food handlers, and food may be involved in either initiating or reinforcing the cycle (**Flow diagram 32**).

Hepatitis A

Person-to-person transmission is of greatest import-ance in the spread of Hepatitis A virus, but spread by food handlers may occur. There is no chronic carrier state and excretion is associated with clinical illness. Excretion of the virus is at a peak during the prodromal stage and before onset of the typical symptoms of hepatitis, and for this reason exclusion at the appro-priate time is unlikely. Shedding of the virus in the faeces is most common, but urinary carriage also occurs; in an outbreak on a US Navy ship, officers were allegedly infected after a disgruntled chef had urinated in a salad.

Many outbreaks of hepatitis due to infected food handlers occur in the home or in food service oper-ations, but a notable case in Scotland involved frozen raspberries, which were probably contaminated at picking and/or packing (Reid and Robinson, 1987). Raspberries and other fruits are often picked by itiner-ant labour who may have neither the inclination nor the facilities to pay sufficient attention to hygiene. Hepatitis A virus survives freezing and the risk with raspberries is increased by the fact that the fruit are not usually washed before processing.

Environmental sources of foodborne viruses

Foodborne viruses are shed with faeces and thus enter the sewage system. Viruses are not eliminated by sewage treatment and thus enter rivers, while estuarine and coastal waters, the growing grounds of shellfish, are contaminated either from the rivers or from sewage outfalls. A number of outbreaks of viral diarrhoeal disease in the USA have been attributed to municipal water supplies and swimming pools (Kaplan *et al.*, 1982; Cubitt, 1989) and, while no outbreaks have been reported, other recreational waters may be a source of viruses.

Soils may be contaminated from irrigation water or from the use of sewage, or sewage sludge, as fertiliser (Katznelson and Mills, 1984). Although contamin-ation has traditionally been thought of as involving the surface of plants, the possibility of intrinsic con-tamination of the plant by assimilation of the virus from the soil into the body of the plant via the root exists.

Foods associated with viral disease

Molluscan shellfish

Molluscan shellfish – oysters, clams, cockles, mussels and scallops – are the major known primary source of SRSV infections and hepatitis A virus. Shellfish are also vehicles of transmission of hepatitis E virus, ingestion of raw shellfish being associated with 12.5% of cases in hospitals in Baltimore, USA (Gerba, 1988), small featureless viruses, astrovirus and calicivirus. Contamination of shellfish is usually associated with sewage contamination of growing beds. The shellfish are filter feeders which concentrate the particulate content of the filtrate, including any bacteria and viruses present, up to a thousand-fold in their gut. The incidence of SRSVs cannot, at present, be deter-mined, but no shellfish can be assumed to be free of contamination. The incidence will vary according to the extent of pollution around growing beds. In the UK, for example, cockles landed at Leigh-on-Sea, Essex, have been associated with a particularly large number of outbreaks of diarrhoeal illness.

Cockles and, usually, mussels are heat processed before cooking, but a relatively mild process is used to prevent shrinkage and toughening of the meat and this, together with inadequately controlled heating, has meant the survival of SRSVs (see page 467). The heating applied to open shells of other fish is also inadequate to inactivate the virus. Oysters are eaten raw and must rely on depuration for purification, but it is now generally recognised that this process, as currently applied, cannot guarantee the removal of viruses.

Diarrhoeal disease associated with shellfish is a continuing and, indeed, growing problem in many countries. The practice of serving shellfish as part of dinners and banquets means that large numbers of victims may be involved, while a nation-wide out-break in Australia involved more than 2,000 people.

The high incidence of contamination of molluscan shellfish by SRSVs and, to a lesser extent, other viruses, has broader implications for food safety in that cross-contamination from shellfish to other foods can be a significant problem. Potential problems posed by raw poultry etc, with respect to cross-contamination are, in general, well recognised and precautionary measures are well established. This may not be the case with respect to shellfish. It should further be appreciated that different sanitation programmes may be necessary for employment against viruses than against bacteria (see page 370).

Fruit and vegetables

As discussed above, vegetables have the potential to act as a primary source of viral infection. However, while salad dishes account for about half the reports of foodborne viral infections in the USA, there is no clear evidence of primary contamination of vegetables, and contamination during preparation of the salads appears more likely (Anon, 1988a). The situation is likely to be different in developing countries where nightsoil is commonly used as fertiliser, and where there have been a number of reports of hepatitis A transmission by salads. It is noteworthy that survival of viruses on lettuce, the basis of most salad dishes, is greater than on other vegetables (Ward and Irving, 1987).

Although primary contamination of fruit is a possibility, contamination by a handler, or infected water, represents a greater risk. Melon was the vehicle for SRSV infection in an incident involving more than 200 people, while in another incident the vehicle was a fruit salad. Fruit can also be the vehicle of infection for hepatitis A virus, a recent outbreak involving raspberries (see above). Although all of the above incidents involved infection by food handlers, contaminated water was implicated as the cause of SRSV infection associated with the consumption of various fruits in a Spanish hotel.

Water and ice

Water is now well established as a vehicle of viral infection, and SRSVs may account for as much as 23% of acute, non-bacterial diarrhoea in the USA (Cliver, 1987). Hepatitis A virus and rotavirus have also been implicated in a small number of incidents in the USA (NRC, 1985), but water is of considerably greater importance as a vehicle of these viruses in developing nations. Consumption of contaminated water, or its use in food preparation, is also believed to be a major factor in epidemic hepatitis E outbreaks in the Indian subcontinent, Algeria, Burma and East Africa. Several outbreaks in East Africa have involved refugee camps where even basic hygiene is difficult to maintain.

Ice was responsible for an outbreak of viral diarrhoea which involved over 950 known cases in Pennsylvania and Delaware, USA. It was estimated that the total number of people affected may have exceeded 5,000. The cause of the outbreak was probably a SRSV, which had entered the wells supplying water to the ice manufacturer with floodwater from a local river. At the same time, an increase in diarhoeal illness was noted among residents in the area who drank water from flooded wells (Anon, 1988b).

Milk and milk products

Although milk products are subject to contamination by food handlers, their association with viral disease is concerned with poliovirus, unpasteurised or recontaminated milk being the most common food vehicles (Cliver et al., 1984).

Other foods

Other foods which have been associated with viral infection have all been contaminated, directly or indirectly, by food handlers. Contamination of any food is possible, but most vulnerable are those that receive no cooking before consumption and whose preparation involves a high level of handling. For example, sandwiches have been implicated on a number of occasions in both SRSV infections (Pether and Caul, 1983; Communicable Disease Surveillance Centre, 1989) and in hepatitis A (Eyles, 1986).

Factors affecting survival of viruses

Viruses do not grow in foods or in the environment, but the infective dose is very low and control must be directed firstly at prevention of contamination, and secondly at destruction. SRSVs cannot currently be cultured and thus the effect of heating, disinfectants etc cannot be readily determined. In situations, therefore, where processes have been designed to control SRSVs, the parameters have been based on those determined for hepatitis A virus, or other enteroviruses. Thus equipment installed for continuous flow steaming of shellfish at Leigh-on-Sea, UK, was designed according to results obtained for the survival of hepatitis A virus (Anon, 1988). However, it may be said that foodborne viruses of all types are robust, and capable of survival for prolonged periods in foods, or in the food handling environment.

Temperature

Hepatitis A virus in shellfish can be inactivated by heating the meat to a temperature of 90°C for 90 seconds (Millard *et al.*, 1987). A pilot plant based on these principles has been constructed (see above), and although it is not known if at this temperature treatment is sufficient to inactivate SRSVs (Anon, 1988a), survival under these conditions seems unlikely.

Storage of foods at refrigeration temperatures of *ca* 4°C offers no protection against foodborne viruses once contamination has occurred and, indeed, is likely to enhance survival. Freezing may cause a loss of definition of the virus (Cubitt *et al.*, 1987), but has no effect on viability.

Irradiation

Viruses are highly resistant to ionising irradiation (ICMSF, 1980), and would survive currently proposed pasteurisation processes designed for vegetative bacteria. Ultra-violet irradiation may have a limited application in areas such as depuration.

pH value and acidity

Hepatitis A virus is generally considered to be less stable in acidic conditions than other enteroviruses, and it has been suggested that in some foods such as fruit juices, the virus may be inactivated. Insufficient is known, however, to attempt to draw any conclusions with respect to other foods or to food safety.

Disinfectants

The importance of contamination in the food handling environment is such that disinfection programmes specifically targeted against viruses must be available where necessary. These are likely to differ from programmes devised for the control of bacteria. Chlorine-based disinfectants are recognised as being effective against viruses and their use is recommended.

Detection of foodborne viruses

The nature of viruses and their requirement for a living-host system for growth mean that the methods used for their detection are unfamiliar to most food microbiologists, and require equipment and expertise which is not available outside specialist laboratories. Furthermore, even where suitable expertise exists, methods currently available are involved and time consuming, and may not be suitable for application on a routine basis.

Small, round, structured viruses

Attempts to propagate SRSVs have so far been unsuccessful. This places a restriction on methodology available for the detection of the viruses, and makes meaningful determination of properties such as heat resistance difficult.

Methods currently used for the detection of SRSVs in faeces are based on electron microscopy, and involve a three-stage process of separation of virus particles from faecal material, concentration and microscopy. Procedures can vary considerably in detail, but the most common approach is to employ differential centrifugation for separation, and caesium chloride gradients for concentration. Following these stages, selected fractions are stained and examined by transmission electron microscopy. Negative staining is the most widely used technique, and has the advantage of providing the opportunity to detect any virus (Anon, 1988a).

However, electron microscopy requires expensive capital equipment and experienced personnel. There are also serious technical shortcomings in that there is an inherent insensitivity, and 1 to 10 million particles/g are required for detection. Furthermore, detection is possible only if morphologically intact particles are present. Techniques such as solid-phase immuno-electron microscopy and immunogold labelling have improved both the sensitivity of electron microscopy and the ability to classify viral particles of indistinct morphology (Oliver and Phillips, 1988). The application of these techniques, however, is limited by the requirement for reference sera, which is not routinely available.

As with other foodborne viruses such as hepatitis A virus, the detection of SRSVs in food is much more difficult than in faeces. This is due to the concentration usually being much lower in foods and also to the possibility of interference from food components. SRSVs have been successfully detected in foods using extraction and concentration methods based on those for other viruses (see below). However, there are at present no routine tests capable of detecting SRSVs in food.

Considerable effort is being expended in attempts to develop alternative methods for detecting SRSVs in both faeces and foods. These involve assays based on radioimmunoassay, enzyme immunoassays and nucleic acid probes. An enzyme immunoassay for the detection of Norwalk agent (Herrman et al., 1985), and a radioimmunoassay for Snow Mountain agent (Dolin et al., 1986) in faeces, have been developed but their application is limited by availability of sera.

Hepatitis A

Cell culture methods have been developed for hepatitis A virus (Provost and Hilleman, 1979) but growth of the virus is slow and primary isolation from foods and the environment is difficult. Electron microscopy has therefore been widely used, although improvements to tissue culture systems may permit fuller investigation of outbreaks of foodborne hepatitis (Anon, 1988).

A large number of methods have been devised for the recovery of hepatitis A and other enteroviruses from foods. In many cases, it is necessary to vary the procedure according to the nature of the food, but methodology commonly involves homogenisation, pH adjustment to detach the virus from food particles, or adsorb it to them, and centrifugation (**Flow diagram 33**; Larkin, 1986). More recently, a method involving polyethylene glycol precipitation has been developed (Lewis and Metcalf, 1988), which has been used successfully for the detection of a number of viruses including hepatitis A virus and rotavirus, and which may enhance recovery of SRSVs.

Techniques for the detection of hepatitis A virus are more advanced than for SRSVs, and progress has been made in the development of nucleic acid probes. A RNA probe is in use at the Central Public Health Laboratory, Colindale, London, for the detection of the virus in faeces, although refinements are required to obtain a suitable sensitivity for use with foods or environmental samples (Anon, 1988a). Dot-blot hybridisation has been applied successfully to the detection of hepatitis A virus in the environment by Margolin *et al.* (1986), and Metcalf and Jiang (1988). In the latter case, the probe did not detect the virus in samples of shellfish, but this was attributed to insufficient samples being examined. Dot-blot hybridisation techniques used currently are unable to distinguish between infective and non-infective hepatitis A virus.

Flow diagram 33. Some procedures for the recovery of viruses from food (Larkin, 1986).

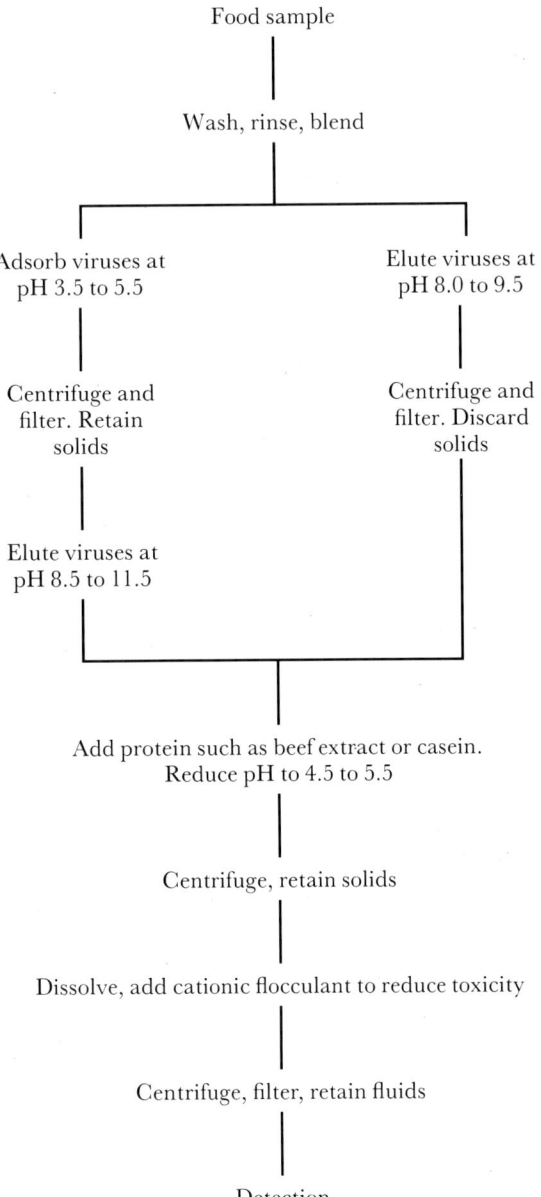

19. Protozoa

Introduction

Although *Entamoeba histolytica*, the causative organism of amoebiasis, has been recognised as a human pathogen since the late nineteenth century, the part played by other protozoa in disease has only recently been appreciated. There remain, however, many unresolved problems concerning both the organisms themselves and their control.

Protozoa are unable to replicate in food, but the cyst form of the organism can remain infectious in foods for extended periods and has a very low infective dose. Protozoa such as *Cryptosporidium parvum* and *Giardia lamblia*, which were previously associated with disease in male homosexuals, AIDS patients and other immunocompromised people are now recognised as causing diarrhoea in the general population.

Although food and, more significantly, water act as vehicles of protozoal infection, it should be appreciated that the epidemiology is complex and that the organisms may be acquired in a number of ways other than ingestion of food and water.

The upsurge in interest in protozoal infections has resulted in improved methodology, including the means of detecting the organisms. Investigation of foodborne protozoal disease, however, is still hampered by the relative insensitivity of detection methods and the lack of an equivalent to the enrichment culture used for the detection of bacterial pathogens. Methods used differ from those of bacteriology, and the necessary expertise is unlikely to be available in most food microbiology laboratories.

Cryptosporidium parvum

Properties of *Cryptosporidium parvum*

Cryptosporidium parvum (**Figures 291** & **292**) is a coccidian protozoan, but differs from other coccidia

in a number of ways (**Table 128**). *Cryptosporidium* is an obligate parasite of ubiquitous occurrence, which

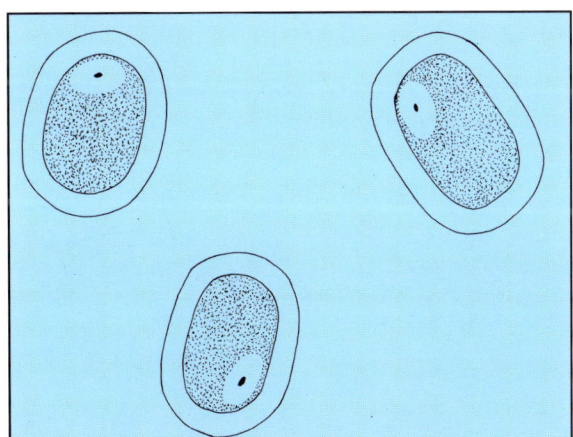
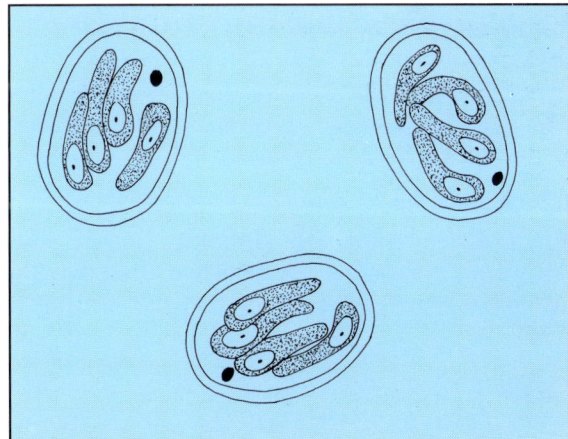

291 & **292.** *Cryptosporidium*. Diploid zygotes of *Cryptosporidium* (**291**) are produced following fertilisation of macrogametes by microgametes. Approximately 80% of zygotes form thick-walled oocysts which sporulate in the host cell producing four haploid infectious sporozoites (**292**). Thick-walled oocysts are responsible for the transmission of *Cryptosporidium* and almost all are passed

unaltered in the faeces. Remaining zygotes produce oocysts which do not form an oocyst wall and which are not passed with faeces. In this type of oocyst, the sporozoites are surrounded only by a single unit membrane that ruptures shortly after release from the host cell, the invasive sporozoites entering other host cells and thus reinitiating the endogenous infective cycle.

Table 128. Specific properties of *Cryptosporidium parvum* (Casemore, 1989).

1 Monoxenous: Life cycle completed within single host
2 Infection: Follows ingestion of environmentally resistant oocyst containing four naked, motile sporozoites. Sporozoites are released into lumen and are actively invasive
3 Development: Through six stages – excystment, merogeny (asexual), gamogony (sexual), zygote and oocyst formation, sporulation
4 Tissue stages: Intracellular but extracytoplasmic. Lacks both species and tissue specificity
5 Reinfection: Reinfection (autoinfection) results from recycling of asexual stages and the presence of thin-walled oocysts and permits very heavy infection to develop
6 Oocyst: Does not require a period of external 'ripening'

has been described in a large number of host species including man, farm and pet animals. The majority of infections in man and most farm animals are with *Cryptosporidium parvum*. It is possible that isolates of this species can be divided into host-adapted strains or antigenic types, and there appears to be antigenic variation which may reflect the host species (Casemore, 1988).

Although cryptosporidiosis is often considered to be a zoonotic disease, person-to-person transmission is now recognised to be common (see page 377), and provides evidence in support of the contention that the disease in man is not primarily a zoonosis (Casemore and Jackson, 1984). However, many human infections with *Cryptosporidium* are derived from farm animals, particularly cattle (Casemore, 1989), for which wild rodents may act as a reservoir (Klesius *et al.*, 1986).

Cryptosporidium infections in man

Cryptosporidium causes acute and often protracted gastroenteritis which, while self-limiting in the normal individual, is often chronic and life-threatening in the immunocompromised. There may also be extraintestinal complications.

It is not usually possible to define accurately the incubation period, but in most cases symptoms appear within 3 days to a week, or, occasionally longer. In immunocompetent people, symptoms involve diarrhoea varying in severity from mild to severe and lasting from several days to more than a month (Current *et al.*, 1983).

In immunodeficient people, especially those suffering from AIDS, cryptosporidiosis usually results in a prolonged and life-threatening illness which resembles cholera in symptoms. In many cases, diarrhoea becomes irreversible and fluid loss excessive. Passage of 3 to 6 litres per day of watery faeces is common, and as much as 17 litres has been reported. Extraintestinal symptoms may occur and both respiratory and biliary cryptosporidiosis have been reported. Finding cryptosporidiosis in the immunodeficient host often carries an ominous prognosis (Current and Owen, 1989).

In immunocompetent people, cryptosporidiosis is a self-limiting illness and antimicrobial therapy is not usually necessary. Supportive therapy includes fluid replacement and, in chronic cases, parenteral nutrition. No truly effective remedy is available for cryptosporidiosis in the immunocompromised, although a number of compounds show some promise including the macrolide antibiotic spiramycin. Immunomodulation therapy may also be of value, with transfer factor, recombinant interleukin-2 and hyperimmune bovine colostrum all being used successfully in small-scale trials.

The incidence of cryptosporidiosis varies from region to region. In the developed world, surveys have produced contrasting results, with an incidence of less than 1% shown in some areas, contrasting with others where *Cryptosporidium* can be the third or fourth most commonly identified pathogen and, at certain times, the most common. In north Wales, for example, detection rates can exceed 10% in children and *Cryptosporidium* is then the most common finding (Casemore, 1989). A survey conducted in the UK over a two-year period showed *Cryptosporidium* to be twice as common among children 1 to 5 years as *Salmonella* (Palmer and Biffin, 1987).

In developing countries the incidence of *Cryptosporidium* infection is higher, and there is evidence of hyperepidemicity. The organism is an important cause of travellers' diarrhoea (Jokipii *et al.*, 1985; Sterling *et al.*, 1986) and mixed infections, especially with *Giardia* are common. This may suggest a common epidemiology involving contaminated water or food. Polymicrobial infections involving *Campylobacter* and *Sh. sonnei* are also common in developing countries, and becoming more common in developed areas, particularly among AIDS patients. Unlike most causes of travellers' diarrhoea, travel-associated *Cryptosporidium* infections may be acquired by city-dwelling West Europeans on visits to country areas in their own, or neighbouring, countries (Palmer and Biffin, 1987).

In addition to differences in the geographical occurrence of *Cryptosporidium*, various seasonal or temporal peaks have been observed. For example, in north America the incidence of infection is greatest in spring or late summer, while in Central America and India peak infection occurs during the rainy season. Studies in the UK and Eire (Casemore *et al.*, 1986; Corbett-Feeney, 1987) have shown a peak of incidence in the spring and, on some occasions, a second peak in late autumn or early winter. These peaks may indirectly reflect rainfall and farming events and practices. Experience in the UK over a five-year period indicates outbreaks, or temporal clusters of apparently sporadic cases, in different parts of the country. These tend to occur at about the same time each year, but in different localities each year (Casemore, 1987a). The underlying reasons for this pattern are not known.

Age distribution of cryptosporidiosis tends to vary geographically, and in some Scandinavian countries such as Sweden, the disease is mainly among adults (Atterholm *et al.*, 1987). Elsewhere there appears to be a primary peak of infection among children aged less than 5 years and a secondary peak in adults aged 20 to 40 years, possibly reflecting family exposure to children or occupational exposure. In the UK, at least, cryptosporidiosis is rare in people over the age of 40 years (Casemore, 1989). Although there is agreement that the incidence is high among children aged 1 to 5 years, there are discrepancies concerning the disease in children aged less than 1 year. In the UK, some surveys have shown cryptosporidiosis to be less common in children aged less than 1 year, and rare in those under 6 months of age (Palmer and Biffin, 1987; Thomson *et al.*, 1987). This pattern is similar to that reported in the USA, rural Costa Rica, Liberia, Rwanda, Guinea-Bissau and Haiti (Casemore, 1989). In contrast, other surveys in the UK (Baxby and Hart, 1986), Eire (Corbett-Feeney, 1987), and other developing countries including Guatemala and India (Mathan *et al.*, 1985; Cruz *et al.*, 1988) have shown *Cryptosporidium* infections to be common in children aged less than 1 year. These differences may be related to differences in maternal immunity, exposure and weaning practices.

Mechanisms of pathogenicity

Little is known of the mechanisms of pathogenicity of *Cryptosporidium*. In immunocompetent people, the jejunum and ileum are most heavily infected, examination of biopsy specimens by light microscopy showing patchy involvement with partial villous atrophy and mild to moderate cellular infiltration of the *lamina propria* by plasma cells, lymphocytes and neutrophils. These changes may contribute to malabsorption due to a reduced absorptive area and decreased levels of membrane-bound enzymes (Crawford and Vermund, 1988). There is also evidence of depletion of host nutrients, the 'feeder organelle' being separated from the enterocyte cytoplasm by only a degenerating parasitophorous vacuole membrane and a sheet of microfilaments. The existence of an enterotoxin responsible for the symptoms of secretory diarrhoea has also been postulated, but not demonstrated (Crawford and Vermund, 1988).

Transmission of *Cryptosporidium*

Routes of transmission of *Cryptosporidium* to man are complex (**Flow diagram 34**), and probably vary with relative importance according to circumstance.

Flow diagram 34. Routes of transmission and reservoirs of *Cryptosporidium*.

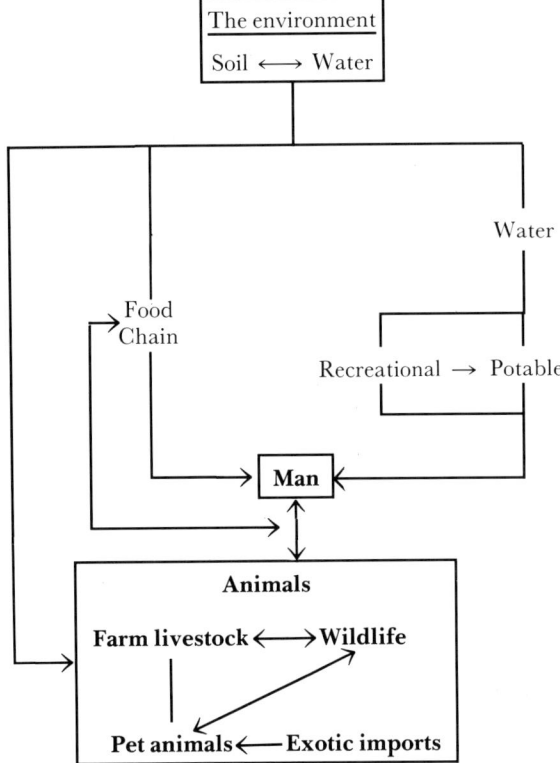

Note: Reservoirs are printed in **bold**.

Food and water

Direct incrimination of food in the transmission of *Cryptosporidium* is difficult due to the problem of detecting small numbers of oocysts. However, there is strong epidemiological evidence of the involvement of foods (Elsser *et al.*, 1985; Casemore 1989), and raw, or undercooked, meat, offal and unpasteurised milk have all been associated with outbreaks of cryptosporidiosis (Barrett, 1986, Casemore, 1988), probably as a result of contamination by faeces. A study in north Wales showed that people most at risk from raw milk were those who only rarely consumed it, or who had just commenced consumption (Casemore, 1987b). Regular consumers were thought to maintain sufficient immunity to prevent symptomatic infection.

Although fruit and vegetables have not been associated specifically with outbreaks of cryptosporidiosis, there appears to be a significant risk of contamination of these crops, either from animal manure of sewage sludge used as fertiliser, or from contaminated irrigation water.

During the acute diarrhoeal stage of infection, faeces may contain as many as 10^6 oocysts/ml and any handler attempting to work through an attack of cryptosporidiosis will obviously put food at risk. The duration of excretion after diarrhoea has ceased varies, and while the stools of some patients become oocyst-free shortly after cessation of diarrhoeal symptoms, excretion in others continues for prolonged periods. Excretion may be accompanied by continuing non-diarrhoeal symptoms. In immunologically normal people, excretion is normally thought to cease within 1 to 4 weeks after clinical symptoms disappear (Jokipii and Jokipii, 1986; Casemore, 1988), but when assessed using more sensitive immunofluorescence techniques, it appears that excretion may continue for longer periods (McLauchlin *et al.*, 1987). At present, exclusion of food handlers is not considered justified after diarrhoea has ceased, providing that a high standard of personal hygiene can be maintained.

There is limited evidence for asymptomatic carriage in immunocompromised people (Roberts *et al.*, 1987). Such cases may excrete only a small number of oocysts and yet still provide a reservoir of infection (Casemore, 1989). The significance with respect to food handling is not known.

Although oocysts are environmentally resistant they are susceptible to freezing (Sherwood *et al.*, 1982), heating above 45°C for 5 to 20 minutes (Anderson, 1985) and prolonged drying (Anderson, 1986). Behaviour in foods has not, in general, been determined. Oocysts have a high level of resistance to disinfectants including chlorine, but the use of hot solutions in sanitising equipment and utensils is an effective means of control.

Water is established as a vehicle of infection for *Cryptosporidium*, both by ingestion and contamination of foods and the food handling environment. Although some cases have involved consumption of surface water, piped supplies of treated water have also been implicated. The largest known waterborne outbreak involved *ca* 13,000 cases in a town in Georgia, USA (Anon, 1988); and confirmed outbreaks have occurred in various parts of the UK including Scotland (Smith *et al.*, 1988) and the Thames Valley (Casemore, 1989). Outbreaks have been caused by contamination of water after treatment or by inadequate water treatment. Inadequate treatment has usually involved missing or defective filtration or flocculation. As noted above, *Cryptosporidium* has a high level of resistance to chlorine and continuing problems of waterborne cryptosporidiosis are anticipated (Casemore, 1989). Recreational water has also been involved in the transmission of *Cryptosporidium*.

Other routes of infection

Cryptosporidium may be transmitted to man by a variety of routes unrelated to food or potable water. These, together with likely circumstances, are summarised in **Table 129**.

Table 129. Routes of transmission of *Cryptosporidium* under different circumstances.

Zoonotic
Livestock
    Occupational exposure
    Domestic exposure
    Educational farm visits
    Social and recreational exposure (Horse riding?)

Companion animals
    Domestic exposure
    Occupational exposure (Veterinary surgeons, etc)
    Social, recreational and educational exposure

Indirect zoonotic exposure
    Household contacts of animal handlers
    Transmission of excreta on shoes and clothing

Person-to-person
Urban or non-zoonotic[1]
    Within families
    In day-care entres, etc

Nosocomial infection
    Direct or, possibly, indirect via instruments

Sexual transmission
    Male homosexuals

Environmental
Land disposal of excreta
    Direct aerosol infection
    Recreation on contaminated land[2]
    Recreation in contaminated water[3]

[1] Transmission may be direct, via fomites, or via vomit or sputum.
[2] Walkers, cross-country runners, etc may be at risk where animal faeces have been spread across footpaths.
[3] Recreational use of water is increasing in some countries and a larger number of people are thus at risk. An outbreak associated with a swimming pool occurred in Doncaster, UK.

Entamoeba histolytica

Properties of *Entamoeba histolytica*

The genus *Entamoeba* (**Figure 293**) comprises several species of parasites, species designation being based on the number of nuclei in the mature cyst, which is four in the case of *Entamoeba histolytica*. This species is the only member of the genus which is capable of causing disease in humans, man being the sole reservoir and source of infection. *Entamoeba polecki*, a parasite in pigs, is occasionally found in humans but is non-pathogenic.

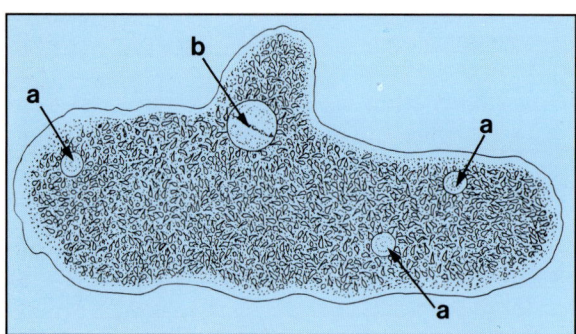

293

293. *Entamoeba histolytica*. The motile trophozoite of *E. histolytica* is illustrated. The remains of degraded tissue components may be seen in cytoplasmic vesicles (arrowed **a**) in pathogenic strains isolated from invasive lesions. With the exception of the nucleus (arrowed **b**) other membrane organelles are absent except under unusual conditions.

The complete life cycle of *Entamoeba histolytica* consists of three consecutive stages – trophozoite, cyst and metacyst. The trophozoite lives in the lumen of the large intestine and generally produces no signs or symptoms, but can act as a gut pathogen. Trophozoites are the motile forms of *Entamoeba histolytica*, and are highly dynamic and pleomorphic cells, both form and motility being extremely sensitive to the nature of the physicochemical environment (Martinez-Palomo *et al.*, 1989). Trophozoites lack a cell wall and are covered only by a 10 nm plasma membrane and a minimal surface coat composed mainly of glyco-proteins. The endoplasm is composed of vacuoles containing a variety of lysosomal enzymes. This structure is of importance in the pathogenesis of amoebiasis (see page 380).

Trophozoites multiply by binary fission and encyst, producing typical four-nucleated cysts after two successive nuclear divisions (Martinez-Palomo, 1989). Cysts never develop in tissues. Cysts remain viable in faeces for several days after excretion, depending on environmental temperature and humidity, and survive for weeks in water. When ingested they are able to survive passage through the stomach and excyst in the terminal ileum.

When viewed by light microscopy, cysts appear as round, or slightly oval, hyaline bodies, the cytoplasm being vacuolated and containing glycogen deposits. Cysts possess a rigid cell wall comprising a closed mesh of chitin-containing filamentous material. This wall provides protection during the period outside the intestine.

Non-pathogenic strains of *Entamoeba histolytica* can be differentiated from pathogenic strains by analysis of isoenzyme patterns, the latter clustering into eight different patterns.

Entamoeba histolytica infections in man

Intestinal invasive amoebiasis

Symptoms of intestinal invasive amoebiasis develop gradually, reflecting invasion and ulceration of the intestinal mucosa, and are of two types depending on whether the lesions are localised in the rectosigmoid or in the more proximal regions of the colon. The first type involves dysentery with bloody mucous stools, colicky pain and tenesmus, while the second type involves diarrhoea often with bloodstained stools. These syndromes account for 90% of cases of intestinal amoebiasis and constitute classic 'ambulatory dysentery' as opposed to bacterial dysentery (Martinez-Palomo, 1989). Fever and systemic manifestations are rare, and symptoms readily respond to treatment, or may disappear spontaneously. Treatment may be by orally administered metronidazole or, alternatively and painfully, by intramuscularly administered dehydroemetine.

There are three forms of intestinal invasive amoebiasis which involve symptoms of markedly greater severity, fulminating amoebic colitis, amoeboma of the colon, and amoebic appendicitis. These are life-threatening conditions and in the case of fulminating amoebic colitis, the mortality rate is 45% even among those fit for surgery.

Extraintestinal amoebiasis

Amoebic liver abscess is the most common form of extraintestinal amoebiasis. *Entamoeba histolytica* is thought to spread from intestinal lesions via lymphatic and blood vessels, and while only 9% of patients with amoebic liver abscess have detectable amoebic colitis, autopsy data suggests that only 38% of patients with amoebic liver disease are free of intestinal involvement (Martinez-Palomo, 1989). Liver abscesses usually occur singly and on the right face of the liver. The onset of symptoms is abrupt with an intense and constant pain, usually in the right upper quadrant. Intermittent fever of 38 to 40°C occurs in most cases, often accompanied by night sweats. Nausea and vomiting may be present, sometimes accompanied by diarrhoea and *ca* 50% of patients have a non-productive cough. Symptoms of mild jaundice are present in *ca* 33% of cases.

In some cases, the abscess heals spontaneously, the size being rapidly reduced leaving a small area of fibrous tissue. In other cases, the abscess may attain a large size before finally rupturing. Rupture may be in any direction but is usually into the bronchi, abdomen or, less commonly, into the pleural or pericardial cavity. The latter occurrence is often associated with a high mortality. The overall death rate resulting from amoebic liver abscess is *ca* 2%, although mortality is likely to be higher where hospital facilities are not available. Treatment usually involves chemotherapy using drugs such as metronidazole, surgical treatment should be reserved for people not responding to chemotherapy or where perforation has occurred.

Amoebic brain abscesses may occur concurrently or subsequently to liver abscesses, and are usually sited in the cerebrum. Such abscesses are usually fatal. Cutaneous abscesses are secondary to intestinal abscesses and arise from the erosion of the abdominal wall, while lesions on the penis result from what some authors describe, rather coyly, as 'unnatural sexual practices'.

Prevalence and distribution of Entamoeba histolytica infections

Although the distribution of *Entamoeba histolytica* is worldwide, human amoebiasis is prevalent only in certain geographical areas. The virulence of pathogenic strains varies, and the most virulent strains appear to be in areas where the disease is common, less virulent strains being of far wider distribution.

Endemic infection with *Entamoeba histolytica* is associated with inadequate sanitation and a generally low level of hygiene, and occurs in both developed countries, usually among the urban or semi-urban poor, and in developing countries. Although only *ca* 10% of those infected develop clinical symptoms, amoebiasis is a major problem in areas of Africa, Asia and Latin America, where a combination of inadequate sanitation and the presence of highly virulent strains of *Entamoeba histolytica* results in a high level of symptomatic infection (Martinez-Palomo, 1989). It was estimated (Walsh, 1986) that in 1984 some 500 million people were infected by the organism and 38 million developed some form of amoebiasis. Amoebiasis accounts for at least 40,000 deaths each year.

The incidence of amoebic liver abscess is strongly associated with age and sex. The disease rarely affects children and is most common in adults aged between 20 and 60 years. It is three times more frequent in males than females, and is of particular prevalence in the poor urban populations of some developing countries (Martinez-Palomo, 1989).

Mechanisms of pathogenicity

Entamoeba histolytica becomes established in the mucosal crypts of the intestine, penetrates the submucosa, and multiplies to form abscesses which rupture to the gut lumen and develop into ulcers. During the early stage of invasion, ulceration is superficial, focal, and there is only minimal cell infiltration at the site of invasion.

Ulcers are of two types, nodular and irregular, and both may be present in the same person. Nodular ulcers are 0.5 to 1.0 cm in size and appear as rounded, slightly raised areas of the mucosa with necrotic centres surrounded by a 'shaggy' rim of oedematous tissue. Irregular ulcers are larger – 3 to 5 cm in length – and may occur singly.

As the disease progresses, ulceration may become confluent, further advancement of lesions leading to the loss of the mucosa and submucosa over the muscle layer and to eventual rupture of the serosa.

Ulceration results from the cytolytic activity of virulent strains of *Entamoeba histolytica*. Cytolysis has been studied using cell cultures and the process may be divided into four stages: adhesion, contact-dependent cytolysis, phagocytosis and intracellular degradation.

On the basis of *in vitro* studies, adherence of *Entamoeba* to target cells appears to be an absolute prerequisite for lysis of the host cell (Martinez-Palomo, 1989). The basis of adherence is not known although a lectin may be involved. There is no difference, however, between levels of the lectin in strains of different virulence (Kobiler and Mirelman, 1981). *Entamoeba* does not remain attached after cytolysis, this process having been described as a 'hit-and-run' phenomenon (Martinez-Palomo, 1989), although 'kiss of death' might be rather more appropriate.

Lysis of host cells appears to be a very rapid process, and is probably mediated through direct damage to the plasma membrane (Martinez-Palomo *et al.*, 1985). *Entamoeba histolytica* produces a membrane-associated phospholipase which may be involved in damage to the host plasma membrane (Long–Krug *et al.*, 1985), while there is also evidence that calcium fluxes within the parasite and the entry of calcium into the target cell are involved in the cytopathic effect (Radvin and Guerrant, 1982).

A variety of substances that are contained within the vacuoles of *Entamoeba histolytica* and released on contact via the action of a vermiform microfilament, may be involved in cytolysis (**Table 130**). In no case, however, has an involvement in the pathogenesis of human amoebiasis been proven.

Following cytolysis, lysed cells are ingested by engulfment, the parasite also being capable of ingesting living cells.

Although the sequence of events in cytolysis is known, the factor or factors imparting virulence remain obscure. It has been suggested that the virulence of any individual strain of *Entamoeba histolytica* depends to a large extent on the reducing power of the trophozoites (Martinez-Palomo, 1989). This suggestion is supported by experimental evidence which showed that anaerobic conditions and the presence of ingested bacteria increased virulence apparently by favouring the lowering of the E_h within the parasite (Bracha and Mirelman, 1984). However, it should be appreciated, that *Entamoeba histolytica* is in no sense an obligate anaerobe, and its ability to consume oxygen may benefit the organism during growth in organs, that have an abundant blood supply, such as the liver.

Table 130. Putative cytolytic substances produced by *Entamoeba histolytica*.

Substance	Effect in vitro
Proteases	Round up cultured cells[1]
Pore-forming proteins	Insert into membranes and create an ionic imbalance[2]
Various enzymes	Degradation of collagen and oligosaccharides in extracellular matrix[3]
Neurotransmitter-like compounds	Induce water secretion in intestinal samples[4]

[1] Lushbaugh *et al.* (1984).
[2] Lynch *et al.* (1982).
[3] Munoz *et al.* (1982).
[4] McGowan *et al.* (1983).

Transmission of *Entamoeba histolytica*

The two most common routes of transmission are by food or water contaminated with cysts, or person-to-person transmission by the faecal-oral route. Contamination of foods may occur when human faeces are used as fertiliser for crops such as salads, or sewage polluted water is used either for irrigation or in the preparation of foods eaten without further cooking.

Food handlers are likely to be of considerable significance in the epidemiology of *Entamoeba histolytica* in countries where hygiene is poor. Asymptomatic cyst passers are common, especially in areas where amoebiasis is endemic and are an obvious risk where engaged in food handling. Treatment of carriers is

considered justified, both to reduce the spread of the parasite and as a precaution against the possibility that a luminal form of amoebiasis may be transformed into invasive amoebiasis (Martinez-Palomo, 1989).

Cysts of *Entamoeba histolytica* are sensitive to chlorine and thus the parasite is unlikely to be transmitted by adequately treated drinking water supplies. Domestic use of untreated, polluted water is a route of transmission in developing countries. In the past, sewage contamination of piped, chlorinated water supplies have led to outbreaks of amoebiasis in the USA and elsewhere.

Giardia lamblia

Properties of *Giardia lamblia*

Giardia lamblia (**Figures 294** & **295**) is a flagellate protozoan which contains three morphologically distinct species: *Giardia lamblia* (synonym *G. duodenalis*), *G. muris* and *G. agilis*. *G. lamblia* is now recognised as

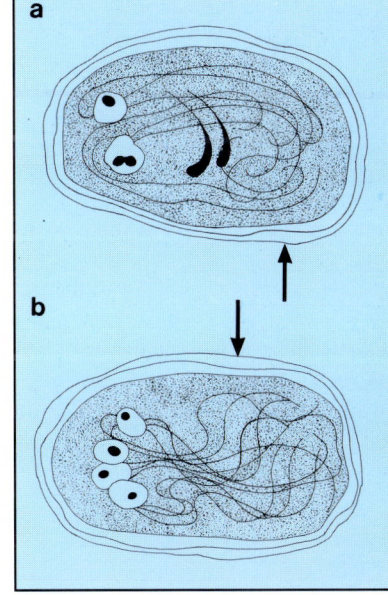

294 & **295.** ***Giardia lamblia.*** A motile trophozoite is illustrated in **294**. The dorsal surface (**a**) is smooth and convex, while the ventral surface (**b**) is concave and bears the distinctive ventral disc (arrowed). The paired flagella are structurally similar to those of eukaryotes and arise from basal bodies found at the anterior poles of both

nuclei. The median body of *G. lamblia* resembles a 'hammer claw' in appearance, and is distinct in shape from the median bodies of other species of *Giardia*. In **295**, both binucleate (**a**) and quadrinucleate (**b**) cysts have a thick chitin containing cell wall (arrowed) that imparts resistance to environmental hazards.

an ubiquitous enteropathogen causing diarrhoeal disease in both immunocompetent and immuno-deficient individuals. Members of the genus *Giardia* are widely distributed throughout the animal kingdom, although only *G. lamblia* is of importance as a parasite of man. Beavers and other wild animals may act as a reservoir of the parasite.

G. lamblia has a two-stage life cycle involving the trophozoite and the cyst. As with other protozoan parasites, the cyst is the infective form, but is not involved in colonisation, which cannot begin until excystation (Farthing, 1989). Excystation is believed to be triggered by exposure to low pH value and pancreatic exocrine secretions. Following excystation, trophozoites colonise the duodenum and proximal jejunum. The site of colonisation appears to be determined by availability of nutrients, especially bile. This, in turn, appears to be related to a requirement for pre-formed phospholipid, which is plentiful in bile and uptake of which is facilitated by the presence of bile salts (Farthing *et al.*, 1985). Encystation occurs in the small intestine (Farthing *et al.*, 1984) and possibly the large intestine. Both trophozoites and cysts are shed with faeces, but only cysts are of importance with respect to transmission of the parasite.

The motile trophozoite of *Giardia* is characterised by its symmetry, and has two nuclei and four pairs of flagella. The organism has a distinctive ventral disc which is a complexity of microtubules and microfilaments (Crossley *et al.*, 1986). The genus also has a unique median body which contains microtubules and contractile proteins, but is of undefined function. Differences in the median body are used to differentiate the three species of *Giardia* (Farthing, 1989). Mitochondria and Golgi bodies are notable by their absence in trophozoites, and in the presence of oxygen respiration is by an iron-sulphur protein and flavin electron-transport system (Lindmark, 1980).

Cysts of *Giardia* are oval in shape and have both a binucleate and a quadrinucleate stage. The cyst wall contains chitin and the system is resistant to most environmental conditions, surviving for up to 3 months in water when the temperature is less than 10°C (deRegnier *et al.*, 1989).

Giardia lamblia infections in man

Clinical symptoms are apparent in only a small percentage of giardiasis cases. Most infections involve apparently asymptomatic passage of cysts, although it is not known if the individuals involved suffered an initial clinical or subclinical illness. Symptoms, where present, appear after an incubation period of 1 to 3 weeks and are of two types, acute diarrhoea or chronic diarrhoea with intestinal malabsorption. Stools are often watery in the initial stages of illness, and there may be abdominal pain, but nausea and vomiting are rare. Symptoms may resolve in a matter of days but between 30 and 50% of those clinically affected by *G. lamblia* go on to develop a subacute or chronic illness. Up to 50% of these cases involve malabsorption with steatorrhoea and weight loss, which persists for many weeks in the absence of treatment (Farthing, 1989). Malabsorption of a variety of nutrients has been described, although results of different studies have varied considerably. Abnormal tests with respect to fat, D-xylose, lactose and, less commonly, vitamin B_{12} have been reported.

In addition to malabsorption, chronic diarrhoea due to *G. lamblia* is invariably associated with anorexia, voluminous flatulence and a bloated abdomen. Occasional allergic reactions, including urticaria, arthralgia and myalgia, may also occur (Farthing *et al.*, 1984).

Giardiasis is common among children and there has been considerable discussion concerning the possibility that *G. lamblia* has a marked effect on growth during infancy and early childhood. Definitive evidence is lacking, but it seems likely that symptomatic *G. lamblia* infections can reduce gain in both height and weight, particularly in children under 5 years of age.

Metronidazole and tinidazole are the drugs of choice for the treatment of *Giardia* infections in many countries, but are not approved in the USA because of concern over safety. Mepacrine is an effective treatment, but may be less well tolerated, while furazolidone is available as a suspension and used for the treatment of children in the USA. However, it is relatively ineffective and has a number of adverse effects. None of the drugs available for treatment of giardiasis can be considered safe for use in pregnancy (Farthing, 1989).

G. lamblia is a widely distributed pathogen and is endemic in most countries of the world. Prevalence rates vary from up to 10% in the USA and Europe, to greater than 20% in many parts of the developing world. *Giardia* is a cause of travellers' diarrhoea that may be acquired during visits to places other than those conventionally associated with the syndrome, including Leningrad in the USSR and Aspen, Colorado, in the USA. In these cases the parasite is usually waterborne (see page 384).

Mechanisms of pathogenicity

Attachment to the intestinal epithelium is an important factor in colonisation of the duodenum and proximal jejunum. The ventral disc is considered to be important in attachment, and a combination of mechanical and hydrodynamic forces may be involved. A proposed mechanism involving generation of a suction pressure under the ventral disc by the lateral flagella (Holberton, 1974) is considered unlikely following studies using a scanning electron microscope (Erlandsen and Feeley, 1984). However, it is probable that contractile proteins present in the disc are able to modify the disc's size and shape and thus permit close contact with the microvillus membrane of epithelial cells.

Attachment by the ventral disc requires correct orientation with respect to the intestinal mucosa, and lectin-mediated attachment has been proposed as an additional or alternative mechanism (Farthing *et al.*, 1986). A mannose-binding lectin has been shown to be associated with the surface membrane of *G. lamblia*. This lectin is distributed throughout the surface membrane and thus specific orientation is not required. It is possible that adherence involves a two-stage process, with lectin-mediated attachment being involved in the initial stages followed by more permanent attachment by means of the ventral disc.

Colonisation alone is not responsible for the development of clinical symptoms, and the mechanisms responsible for diarrhoea and malabsorption are not fully understood. Current hypotheses involve two mechanisms, small intestine mucosal injury and luminal factors (Farthing, 1989).

Mucosal injury

Examination of small intestine biopsy material by light microscopy reveals considerable variation in appearance from complete normality at one extreme, to subtotal villus atrophy at the other. However, transmission electron microscopy is able to detect damage to microvilli which appear normal by light microscopy. There appears to be a relationship between impairment of intestinal absorption and extent of mucosal damage (Duncombe *et al.*, 1978).

The mechanism by which mucosal damage occurs is not known. Invasion is a rare event and cannot account for the widely observed effects, and while there is some support for the direct effects of the parasite, this explanation is not fully satisfactory. According to Farthing (1989), the most plausible explanation is that mucosal inflammation causes both morphological changes in the villi and, possibly, functional alterations with respect to absorption. 'Sensitisation' to *Giardia* antigens may be involved in the inflammatory response.

Luminal factors

For many years it was believed that *Giardia* was either an effective competitor for host nutrients, or formed a physical barrier to absorption. Neither of these possibilities now seems likely. The most likely luminal factors involve bile salts. Deconjugation of bile salts may be a contributory factor to malabsorption of fats, and under some circumstances is associated with the bacterial overgrowth that occurs in the small intestine of some people suffering giardiasis (Tandon *et al.*, 1977). Deconjugation by *Giardia* itself may occur but this has not been confirmed (Inge *et al.*, 1985). A further possibility is that bile salt uptake by *Giardia* reduces the intraluminal bile salt concentration and depletes the bile salt pool, but this also remains unproven (Farthing, 1989). Bile salt is required for the activity of pancreatic lipase and uptake by *Giardia* may well be sufficient to account for the inhibition of this enzyme (Smith *et al.*, 1981).

A cholera toxin-like protein has been isolated from *G. lamblia* that exhibited partial immunological relatedness with cholera toxin, and activated adenylate cyclase in an identical manner (Archer and Young, 1988). It is not known how widespread production of this toxin is amongst different strains of *G. lamblia*, or its role, if any, in the induction of diarrhoea.

Transmission of *G. lamblia*

Transmission of *G. lamblia* may be by person-to-person spread, a route of particular importance in institutions serving infants or children, or by ingestion of contaminated food or water. Waterborne giardiasis is of particular importance in travellers' diarrhoea and is a major factor in maintaining the level of infection in areas of particular prevalence. In the USA 26% of waterborne disease is due to *G. lamblia*, and between 1965 and 1981 more than 20,000 individuals had been affected. Many outbreaks have been associated with drinking untreated surface waters and the likelihood of a reservoir of *Giardia* among creatures of the wild means that, as with *Campylobacter*, consumption of water from a pristine mountain stream may lead to infection. *G. lamblia* is accordingly a causative organism of 'trekkers' trots'.

In a population of cysts of *Giardia*, at least some are chlorine resistant, and giardiasis has been associated with supplies where chlorination is the only treatment or where there has been a failure of filtration. Correctly treated mains supplies have been found to have a residual cysticidal effect (deRegnier *et al.*, 1989), which may offer protection in the event of post-treatment contamination of supplies. Ice made from contaminated water has also been implicated as a cause of giardiasis, although cysts are usually considered to be inactivated by freezing (Bingham *et al.*, 1979).

Food is less commonly involved in giardiasis than water, and only a few outbreaks have been described. Cysts of *G. lamblia* are inactivated by boiling for 10 minutes, and thus foods involved have been those which are not cooked or fully reheated prior to consumption. Transmission by food handlers was involved in each of two outbreaks in the USA. In the first, the person involved had changed the nappy of a baby with asymptomatic giardiasis shortly before transferring home-canned salmon to serving containers (Osterholm *et al.*, 1981); while in the second, the person who had prepared a noodle salad, although asymptomatic at the time of preparation, developed clinical symptoms of giardiasis a day later (Petersen *et al.*, 1988). Multiple modes of transmission were involved in a further US outbreak at a nursing home (White *et al.*, 1989), there being evidence of both food-borne and person-to-person spread. Some of the person-to-person spread involved infection of elderly nursing-home residents by children from an attached day-care centre where the outbreak began, and rigorous controls are required where the socially desirable mixing of young children with the elderly takes place in institutions.

Food is probably a much more important vehicle of *Giardia* infection than is suggested by the small number of reported cases (Petersen *et al.*, 1988).

Other protozoa

Isopora belli

Human disease due to *Isopora belli* used to be considered a curiosity, but the organism is now recognised as an occasional cause of travellers' diarrhoea in the immunocompetent as well as diarrhoea among AIDS sufferers and other immunosuppressed people.

The cause of infection is probably sporozoites, which are ingested with food. Clinical manifestation of *I. belli* infection is limited to brief, self-limiting diarrhoea.

Toxoplasma gondii

Toxoplasma gondii resembles *I. belli* in being a parasite of the margin of the gut lumen. The parasite occurs in a wide range of mammalian and avian hosts, as crescent-shaped zoites. These invade host cells, primarily macrophages, and multiply until the whole cell is packed with zoites. This is often referred to as a pseudocyst. Rupture subsequently releases *Toxoplasma* to infect other cells. Proliferation occurs throughout the body, but primarily in the central nervous system, muscles and lungs.

The foetal nervous system is particularly vulnerable to infection, and thus prenatal toxoplasmosis is often severe. Cerebral calcification, chorio-retinitis, hydrocephalus or microcephalus and psychiatric disturbances are all common after pregnancies proceed to term, but infection frequently results in stillbirth. *Postpartum* infection is much less serious and typically involves mild lymphadenitis, which may be accompanied by a fever. Infection can persist for a prolonged period in the latent phase. Serological investigations have shown that *ca* 30% of the UK population has been infected, usually asymptomatically, by *T. gondii*. For a number of years, infection was thought to be acquired from raw, or undercooked meat, or from occupational exposure resulting from work as a meat handler etc. More recently, it has been found that the cat is the definitive host of *T. gondii*, and it is thought that a high percentage of infection arises through contact with a cat, usually in its role as companion animal. For this reason, pregnant women are advised to be cautious in their dealings with cats and to avoid handling feline faeces. Other domestic animals including dogs may also be of importance in the transmission of *T. gondii*.

Balantidium coli

Trophozoites of the ciliophore *Balantidium coli* usually exist as parasites of the gut lumen, where they graze on intestinal bacteria. Occasionally *Balantidium coli* invades the tissues of the gut epithelium, resulting in an illness which clinically resembles intestinal invasive amoebiasis. Ulceration resembling that of *Entamoeba histolytica* occurs, and stools may be bloody and mucoid. Balantidial peritonitis and vaginitis have been recorded but are rare. *Balantidium coli* is primarily associated with pigs, and many human patients have occupational contact with these animals. It is also an occasional cause of travellers' diarrhoea and food can be involved in its transmission.

Sarcocystes

Sarcocystes are primarily parasites of cattle (*Sarcocystis bovihominis*) and pigs (*S. suihominis*), the crescent-shaped trophozoites which resemble those of *Toxoplasma* being present in the muscle of lungs, oesophagus and diaphragm. In some areas, the incidence of infection particularly among cattle is high, but illness in man is very rare, apparently involving mild diarrhoea (Heydoorn, 1977).

Free-living pathogenic amoebae

Small, free-living amoebae of the genus *Acanthamoeba* or *Naegleria* are now recognised as potential pathogens of man and animals (Martinez, 1985). *Naegleria fowleri* causes primary amoebic meningitis and is a primary cause of death in humans, while *Acanthamoeba culbertsoni* is recognised as an opportunistic pathogen, primarily among the immunocompromised, which causes granulomatous amoebic encephalomyelitis. Several other species of *Acanthamoeba* have been recovered from the eyes of people suffering amoebic keratitis. In most cases, wear of contact lenses was involved (Anon, 1986). A third genus, *Vahlkampfia*, was originally isolated from a person with severe diarrhoea (T. K. Sawyer, personal communication), although a causal relationship was not established. Pathogenic species of free-living amoebae can be readily isolated from municipal sewage as well as from sewage, or thermally, polluted water and soil (DeJonckheere, 1981, Sawyer, 1989).

Less pathogenic species are almost universally present in the environment.

Although muddy water or dust are often considered to be the routes of infection of *Acanthamoeba* and other free-living genera, the involvement of food cannot be discounted in view of the widespread contamination of the environment. Trophozoites of three potentially pathogenic species of *Acanthamoeba* were detected during a survey of salad and other vegetables in the USA (Rude *et al.*, 1984), while the finding of pathogenic species in sewage polluted shellfish waters (Munson and Sawyer, 1987) suggests another possible vehicle of infection.

Although there is no direct evidence that foods are associated with free-living amoebae infections, the potential obviously exists, particularly for the immunosuppressed. There is an obvious need for further research in this area.

Detection of protozoa

Foods are not routinely examined for the various cysts of protozoan pathogens, and while the procedures used are not intrinsically of great difficulty, they require expertise which is not usually available in a food microbiology laboratory.

A major difficulty associated with the detection of protozoal cysts in foods is that the numbers present are usually small and the microscopic methods used insensitive even after concentration.

Various concentration and separation procedures are used including filtration, centrifugation and flotation. Methods chosen vary according to the food and the nature of the cyst, but flotation using zinc sulphate solutions may be particularly useful with foods.

Microscopic examinations are commonly made using bright field microscopy after fixing and staining with Lugol's iodine, modified Ziehl-Neelsen, or another method. Considerable experience is required both to differentiate cysts from other bodies such as fungal spores, which may have a similar microscopic appearance, and to correctly identify any cysts present; people involved in this work should have received specific training.

In some cases, identification of cysts is aided by the use of alternative microscopic techniques such as Nomarski interference, but, in the long term, serological techniques may offer the greatest promise in terms of detection and definitive identification. In the case of *Cryptosporidium*, for example, serological methods currently available include immunofluorescence microscopy using either polyclonal (Stibbs and Ongerth, 1986) or monoclonal antibodies (McLauchlin *et al.*, 1987), ELISA and latex coagglutination (Pohjola *et al.*, 1986; Casemore, 1989). Genetic techniques are also likely to be of future importance — a method for detection of *T. gondii* by the polymerase chain reaction has been developed to complement serology in diagnosis (Savva *et al.*, 1990), and the wider application of the technique to detection of protozoa may be anticipated.

20. Control of pathogenic micro-organisms in food: management aspects

One common sanitary law

The provision of food is a vast and complex operation conducted by organisations which range in size from multinational corporations employing thousands of people on a world-wide basis, to the sole trader working, perhaps, in a remote and underdeveloped region. All of these organisations have a basic responsibility to the consumer to ensure, as far as possible, that food is safe. Furthermore, it must be appreciated that safety cannot be considered to be an optional extra to be selected or rejected as whim or finance dictates. It must be an integral part of any food-based operation and take precedence, where necessary, above all other aspects of commerce.

Within the structure of the food industry, overall responsibility for safety lies with management. Defined policies for food safety are required in larger organisations and should be, but rarely are, included in corporate planning from its earliest stages.

At its simplest, management responsibility may be seen in terms of provision of resources – for ensuring that machinery and buildings are suitable and that sufficient trained staff are available, for example – as well as ensuring that company systems exist for the control of processing and distribution etc. Responsibility goes further, however, and includes the development of a corporate culture in which all staff are able and willing to accept their share of responsibility for safety. Safety must, indeed, come down from the top, and up from the bottom of the hierarchy. It is all too common to encounter on the one hand senior management blissfully unaware of actual conditions at the operating levels, and on the other operatives frustrated in their wish to proceed correctly by the ill-considered decisions of higher management.

Although most managements accept, at least in theory, their responsibility for food safety, a minority pay lip-service at best. It is often thought that this results from ignorance but, while this is true in some instances, in others the problem is more deep rooted and appears to stem from basic personality traits. It is not pertinent to speculate on reasons for negative attitudes to food safety, but it should be stressed that abdication of responsibility by management can represent a very serious threat to public health.

Management systems

All but the very smallest companies employ a number of management systems which are used to control all aspects of activity. In the food industry, the control of safety during the manufacture and handling of food is of prime importance, and is best achieved by means of the hazard analysis critical control point technique (HACCP). It is also necessary to have a means of responding rapidly in the event of processing faults which have gone undetected at plant level, and for this reason a complaints and withdrawal system is also required.

Hazard analysis critical control point technique (HACCP)

HACCP is best regarded as a systematic approach to hazard identification, assessment and control (ICMSF, 1989). The technique was first described in the USA at the 1971 National Conference on Food Protection (Anon, 1972), although Shapton (1988) contends that the first application of the concept, but not the name, was the use of pasteurisation to control the hazard of tuberculosis in milk. Since that time, the technique has been fully developed and, while originally devised for manufacturing operations (Kauffman and Schaffner, 1974), extended to food service (Bryan and McKinley, 1979; Wheelock, 1988), and to ensuring the safety of recipes (Recipe hazard analysis, Zottola and Wolf, 1981). In use, it is common practice to include considerations of spoilage as well as safety. It may also be applied to situations not involving micro-organisms, such as the elimination of pesticide residues from food (Anon, 1988a).

The exact approach to HACCP will vary according to individual circumstances and needs, but the underlying principles are the same in each case and are outlined below. The terms used are defined in the context of HACCP in **Table 131**.

Table 131. Definition of terms used in HACCP (NRC, 1985; Anon, 1987a; ICMSF, 1989).

Hazard:	The potential to cause harm to the consumer or damage to the product (safety) or damage to the product (spoilage).
Hazard analysis:	Any system used to analyse the significance of a hazard with respect to consumer safety (or product acceptability).
Critical control point[*]:	A location, stage or process which, if not properly controlled, provides a threat to consumer safety. A raw material is a critical control point *if* there is no processing stage that can be depended upon to eliminate pathogens in that material.
Monitoring:	The verification that the processing or handling procedure at each critical control point is correctly carried out.
Risk:	The probability that a hazard will be realised.

[*] In some systems two types of critical control are identified:
  CCP1 will assure control of a hazard.
  CCP2 will minimise but cannot assure control of a hazard.

Basis of HACCP

The basic approach to HACCP involves three considerations:

1 The determination of hazards. In the manufacturing context this embraces all stages from acquisition of raw materials to product sale and consumption, but in the broader sense also includes those hazards associated with production at farm level.
2 Identification of critical control points required to control those hazards.
3 Establishment of procedures for monitoring critical control points.

Hazard analysis

The hazard analysis is intended to:

1 Identify potential hazards in raw foods.
2 Identify specific sources and points of contamination over the whole of the food processing and handling procedure.
3 Determine the potential for survival and growth of over the whole of the food processing and handling procedure.

For an effective hazard analysis it is necessary to define both the nature of the hazard and its severity. However, care must be taken to avoid less severe hazards being 'written-off' as 'unimportant'.

Critical control points

The determination of critical control points is by a combined evaluation of the ingredients used in the product and the processing, if any, applied. In products which are not processed to eliminate pathogens, the ingredients themselves are critical control points. In other cases, stages designed either to eliminate, or control the growth of micro-organisms, are the control points. In some, but not all, systems differentiation is made between those points which *assure* control of a hazard and those which minimise risk but do *not assure* control. Critical control points are not necessarily specific processing stages such as heating, but may include more general factors such as sanitation and the food handlers themselves.

In addition to determining the critical control points, it is necessary to define the seriousness of failure of control at a given critical point. This may be referred to as level of concern, and is derived primarily from knowledge of hazards and the risk of their occurrence. Many users of HACCP define four levels of concern (**Table 132**).

Table 132. Critical control points: levels of concern.

1	High:	Life threatening risk in absence of control
2	Medium:	Risk of sufficient severity to require positive control
3	Low:	Little risk of food poisoning (or spoilage), but control may be advisable
4	None:	No risk of food poisoning

Monitoring

The purpose of monitoring is to ensure that the processing or handling at a critical control point is under control. The monitoring system should be designed to detect any loss of control sufficiently rapidly for corrective action to be taken before the product must be rejected.

Monitoring should always be made on a systematic basis and, in most cases, must be in accordance with a statistically based plan. The design of such plans requires specialist knowledge, and further information may be obtained from ICMSF (1986, 1989).

The decision to establish a HACCP-based system may be seen as a positive commitment to safety assurance. The implications are wide ranging and may result in major, and expensive, changes such as alterations to production routines or retraining operatives. To be successful, therefore, the support and co-operation of all levels of staff is necessary and, in the case of executive management, support must include the necessary finance and resources for the successful implementation of HACCP.

Good men and true: the composition of the HACCP committee

HACCP analysis is a complex operation, and while microbiological input is likely to be of paramount importance, a multidisciplinary approach is likely to be most effective. This is best achieved by a committee, or group comprising experts in the relevant fields (an example of a typical HACCP committee is shown in **Table 133**). Each member of the committee must have a good knowledge of the purpose and principles and, with the exception of the Chairman and Secretary, have sufficient expertise in their discipline to give authoritative advice to the Chairman and other members.

Each member of the committee should be recognised as an expert and treated on an equal basis irrespective of his or her position in the management hierarchy. Although it is desirable that the Chairman should be of sufficient seniority to ensure the necessary co-operation from people not directly involved and to ensure the availability of the necessary resources, he or she should not attempt to impose his or her own views on those of the expert members. Furthermore, although consensus is desirable, the expert in each discipline must be regarded as the ultimate authority in relevant matters. Disagreements should be resolved purely on the basis of technical evidence, and the temptation to proceed by committee vote should be avoided. The microbial world is not a democracy and bacteria have no more respect for majority vote than for management hierarchy.

In some cases, where in-house expertise is considered inadequate, an outside expert, the 'Godfather', may be introduced to assist with the preparation of the audit. Review of a completed audit by an outside expert is considered good practice in all circumstances, and may avoid errors due to over concern for detail at the expense of principle.

The HACCP audit

The audit is often considered to be the most important stage in the establishment of HACCP. The actual procedure can vary considerably and the stages described below represent a generalised approach rather than a definitive guide. The major points discussed are, however, those of key importance to the successful establishment of a HACCP system. Examples taken from actual situations are used to illustrate the points made.

Table 133. HACCP Committee: typical composition.

Chairman:	Convenes group, plans overall approach and ensures technique properly applied. Ensures Committee is run efficiently and with enthusiasm. Liaises with higher management.
Microbiologist:	Advises Chairman and Committee on hazards, risks and other facets of microbiology.
Other technical specialists:	Typically chemist and quality assurance manager. Provide expert opinion analogous to that of microbiologist.
Production specialist:	Prepares initial process flow diagram. Advises on production procedures and constraints and on technological objectives of the various stages. In conjunction with engineer, advises on technical limitations of equipment.
Engineer:	Advises on technical limitations of equipment and services. Liaises with outside specialists in areas such as instrumentation.
Other members:	A variety of other people may need to attend at least some meetings to provide advice on specific topics. Such people include packaging technologists, sales staff, training and personnel managers.
'Secretary':	A technically qualified person should act as 'secretary', drawing up documentation which records the results of audits and the overall progress of the Committee. This role may be undertaken by a member of the Committee, but it is not desirable to have such a person distracted by the more mundane tasks which are likely to befall its 'secretary'.

Audit: Stage 1.

The first stage involves the preparation of the process flow diagram and the list of ingredients for each product under consideration.

Preparation of the flow diagram requires a detailed knowledge not only of the actual process but of working practices, cleaning schedules etc (**Table 134**). The flow diagram is the prime responsibility of the production specialist and should be based on actual practices. Where processing varies to meet differing circumstances, alternative procedures should be included in the flow diagram, and the circumstances under which these are used clearly defined (**Flow diagram 35**). The first stage flow diagram for a ready meal is illustrated in **Flow diagram 36**.

The ingredient list (**Table 135**) is also prepared by the production specialist. All ingredients including water should be listed, even when prepared formulations are imported from outside suppliers. Where the ingredient list varies because of availability or any other reason, alternatives must be included.

After preparation of the flow diagram and ingredient list, copies are circulated to all members of the HACCP committee. Each should familiarise himself or herself with the product and process.

Table 134. Factors to be considered in the preparation of HACCP flow diagram.

Process related factors:	Operating parameters (time/ temperature, etc) Design and construction of equipment Plant layout Cleaning and maintenance procedures Health of staff
Production routine:	Shift arrangements including overtime etc Deployment of staff including arrangement for breaks Staff expertise
Factory management:	Design and layout Surfaces and flooring Provision of services

Table 135. HACCP audit: ingredients.

Major ingredient	Coating sauce[*]	Others
Chicken	**Stabilised yoghurt** (pasteurised yoghurt and gelatin) **Herb mix**	Courgettes **Dried mint**

[*] Formulation of coating sauce is that of the amended formulation.
Note: Ingredients listed in **bold** are CCPs.

Flow diagram 35. Preparation of HACCP flow diagram: alternative procedures.

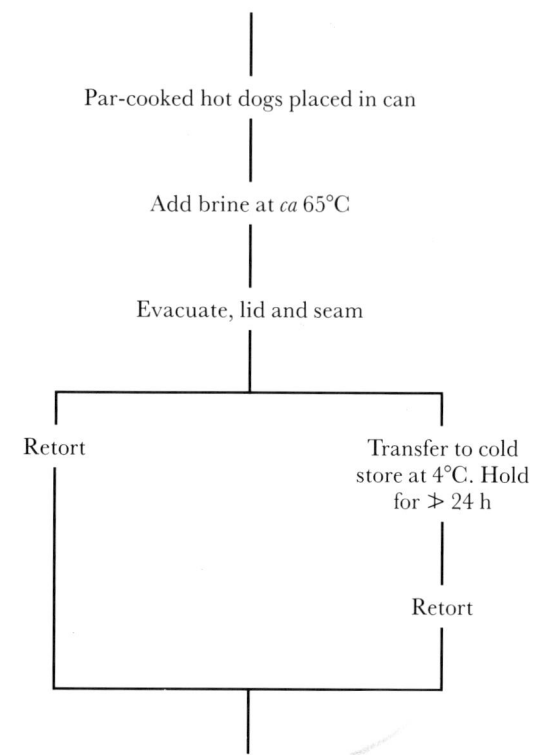

Note: The figure is a simplified flow diagram of a small-scale cannery procedure for hot dogs. The cans were hot filled and retorted immediately after seaming, taking precedence over cold-filled products to avoid delay. The process calculations assumed an initial temperature of not less than 50°C. On some occasions, problems with steam supplies meant that retorts could not be operated, and cans were transferred to a cold store for a maximum period of 24 hours. When removed from the cold store, the cans were retorted immediately with no adjustment made to process parameters for the significantly lower initial temperature.

Before the HACCP audit, no written record of the alternative procedure had been made and no consideration given either to consequences for safety or to remedying the long-term steam supply problem. It is fortunate that the process calculations allowed a massive, if unplanned, margin of safety.

Flow diagram 36. HACCP Audit procedure.

A Stage 1

Basic flow diagram

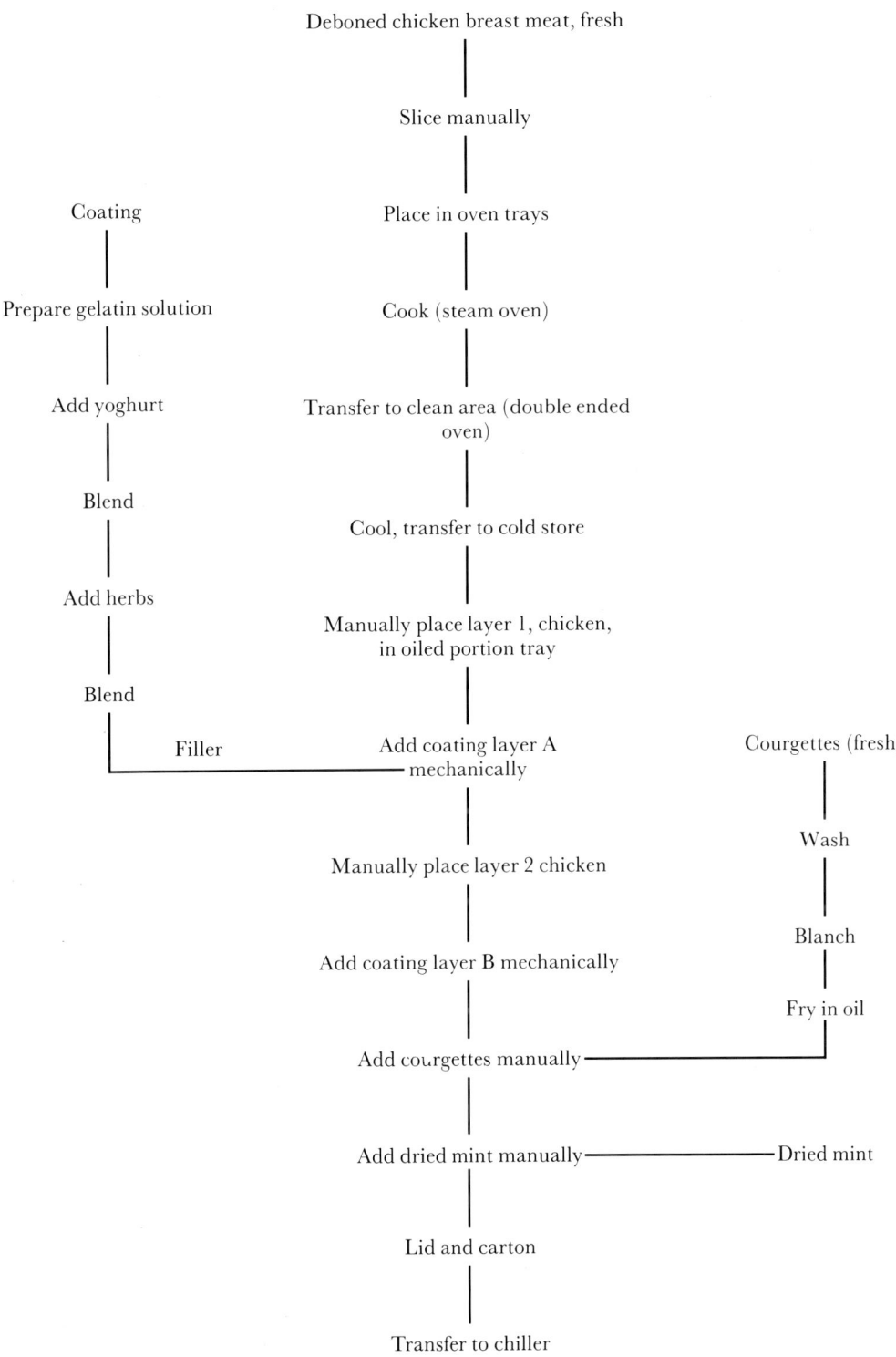

Note: For the sake of clarity the flow diagram has been simplified by omitting process parameters such as temperature and time of cooking, and the handling of the product between processing stages. In practice, these would be included and the flow diagram would also be applied to handling during transport and distribution.

Audit: Stage 2.

Stage 2 involves the first formal meeting of the HACCP committee. Although all members should have been briefed individually, the Chairman should reiterate the purpose of the Committee and outline how he or she intends to proceed. The production specialist then reviews the flow diagram and ingredient list, ensuring the technological objectives are fully understood and explaining background considerations. A visit to the production site may be made at this stage, or may be made directly before hazards are considered. The co-operation of production personnel is essential, and the Chairman should therefore ensure that all affected are aware of the purpose of the visit.

Table 136. HACCP: food product factors and hazard related parameters.

Food product factor	Hazard related parameter
Storage	Ambient temperature
	Chill (defined)
	Frozen
	Shelf life
Formulation	pH value (and effect of acidulant)
	Preservatives (including gas packing)
	Water activity
	Redox potential and oxygen tension
	Competitive inhibition
	Food structure
Packaging	Type
	Integral to product stability?
	Damage resistance
Retail practices	Sold as pre-pack?
	Contamination
	Temperature abuse
	Embellishment
	Out of code sales
Consumer practices	Temperature abuse
	Cooking method (if any)
	Reheating
	Defrosting
	Contamination
	Preparation
	Recipe enhancement
Target groups	Home use (normal)
	Home use (infants, geriatrics, compromised)
	Catering use (general)
	Catering use (institutional including infants, geriatrics, compromised)

Audit: Stage 3.

The third part of the audit is concerned with identifying the characteristics of the product and conditions of its usage (storage, distribution, retailing and customer handling), which in turn define its inherent hazards. It is again advisable to adopt a structured approach, first considering food-product factors and then hazard-related parameters (**Table 136**).

It will be appreciated that knowledge of risk may derive from one of two sources. First, epidemiological evidence based on problems known to have occurred with the actual product or with products of a similar nature. Second, assessment of technical information which indicates that a hazard may exist. The latter type of information is often considered a less satisfactory basis for decision making than the former, and likely to lead to over control (NRC, 1985). However, the use of historical epidemiological information also requires caution and the fact that a potential hazard has never been realised may be simply a matter of luck.

Audit: Stage 4.

During the fourth stage, process analysis, information and assessments concerning both the process and inherent hazards in the product are brought together in order that the critical control points (CCPs) in the process may be identified, and an assessment made of the degree of concern. Before commencing this stage, it is recommended that the Chairman should briefly review the activities to date to ensure that agreement exists and there have been no 'second thoughts' or 'revised opinions'. If a Committee visit to the processing floor has not been made earlier, one should be made at this stage. This is important even for those members who are routinely associated with processing, since the operations may be viewed in a different, and more revealing, context.

In determining CCPs, each stage of the process, including procurement of starting-point materials, is discussed. This involves discussion not only of distinct processing operations such as heating, but intermediate stages between processing operations. For example, there is often no stipulated maximum time before transfer of a finished product to a cold store, and delays may introduce a hazard to an otherwise well-controlled plant.

The three main criteria to be determined are:

1 Can micro-organisms be introduced and, if so, will there be growth or elaboration of toxin?

2 Will micro-organisms already present in the food be able to grow?

3 Are micro-organisms destroyed?

In many situations, of the likelihood of growth of micro-organisms is difficult to determine, and while much effort has been expended in determining the effects of physical conditions and preservatives on the growth of pathogens, it is often not possible to relate these studies to specific foods and processes. Predictive modelling techniques to determine the possibility, or probability, of growth or toxin production by microbial pathogens (**Table 137**) are likely to find increasing application in the future. At present, however, problems remain, particularly with foods that are of heterogeneous composition (**Figures 296 & 297**).

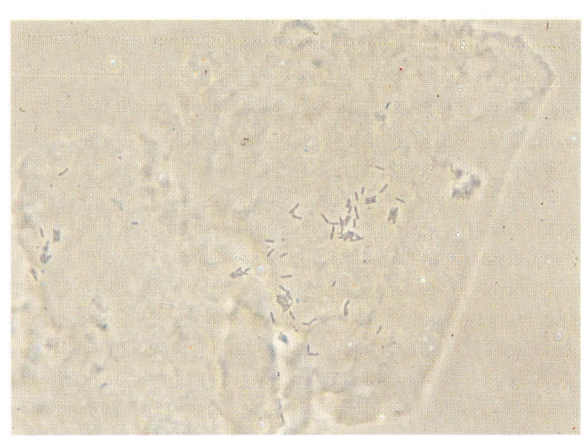

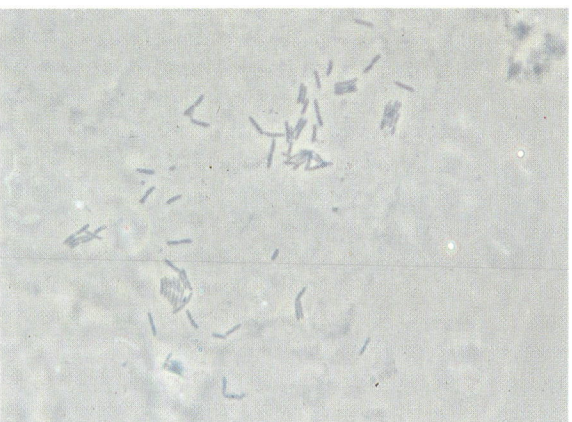

296 & 297. Localised growth of *Bacillus cereus* in a low pH value product. *B. cereus* is a recognised problem in rice dishes and in this product, a rice salad, the pH value of the sauce was designed to be below the minimum value for growth of the organism. The product, however, was heterogenous and localised areas of higher pH value around the rice (**296**) permitted growth of the organism during cooling (**297**) even though the measured pH value of the whole product was within satisfactory limits.

Table 137. Principles of predictive modelling. *'Mathematical modelling is a procedure leading to the description of a process of phenomena of interest by one or more mathematical equations'* (Rand, 1984).

Applications

1 Prediction of the relative microbial stability of a new food product, which may differ only slightly from an established product.

2 Aiding the assessment of the direct and interactive effects on microbial growth of a new combination of preservatives in a food.

3 Assessment of the safety of foods when subjected to temperature abuse during passage through the distribution chain from manufacturer to consumer.

Stages in the modelling procedure

1 Planning: Define, clearly and specifically, the problem. Decide on the parameters to be modelled.

2 Data collection.

3 Model fitting.

4 Model validation.

Limitations of modelling

1 Extrapolation should be avoided and no predictions made beyond the limits of the experiment.

2 Experiments done using pure cultures should not be used to predict the behaviour of micro-organisms in foods.

3 It is only possible to comment with certainty on the *relative* effects of different treatment combinations. *Definitive* statements concerning the ability to grow and/or produce toxins in foods cannot be made.

Note: Information from Farber (1986); Baird–Parker and Kilsby (1987).

Under most circumstances, level of concern may be determined according to a system employing four levels (**Table 132**, page 388). More complex systems inevitably involve long, unfruitful discussions and add little or nothing to the efficiency of the HACCP system. It should be appreciated that recognition of different degrees of risk is intended to ensure that effort is directed to the most critical areas, and not wasted on the relatively trivial. There is no benefit, therefore, in creating distinctions between the degree of risk represented by contamination with, for example, *Salmonella* and *Campylobacter*.

Three basic means exist by which level of concern may be assessed. The first of these is essentially empirical and involves expert judgement of risk. Where possible, two or more people with suitable qualifications and expertise should be involved in this process. Situations may arise where assessment of the risk is not possible and where the dilemma must be resolved by testing. This possibility must be accepted by the Chairman and other committee members, and no pressure should be placed on the 'experts' to make a decision which can only be based on guesswork.

The second means of determining level of concern depends on comparison with the same or related products that have a satisfactory record of safety. This approach is valid, providing that it is possible to ensure that products are genuinely comparable in terms of formulation, technology and conditions of manufacture, and that the satisfactory record truly reflects the safety status of the known product.

The third means is probabilistic and assesses the probability of failure. The concept of 'botulinum cook' (see page 480) is probably the best developed example of the probabilistic approach. A much broader example is that described by Mossel (1984). This defines an acceptable level of safety for foods processed for safety as 75% of the population being exposed to a pathogen at a frequency not exceeding about once in a hundred years when eating one portion daily.

A brief review of the advantages and disadvantages of the probabilistic approach suggested that to be effective, 'chance' must be accurately defined. It was also noted that calculations involving very low levels of probability usually depend on extrapolation of data, with the consequence that predictions are unlikely to be either accurate or precise. Where cumulative probabilities are used, 'chances' and, accordingly, errors are multiplied (Anon, 1987a). It should also be appreciated that probabilistic risk assessment is essentially applicable only to a 'normal' situation, and can have no role in predicting the abnormal situation which leads to a high proportion of food poisoning incidents. Naturally, probability can *never* be applied to predict the vagaries of human behaviour.

The means by which risk is assessed will vary according to circumstances. In some cases, a combination of methods is likely to be used. With a new product, for example, the risk from established pathogens such as *Salmonella* may be assessed on the basis of expert judgement, but the risk from less familiar pathogens such as *Listeria* on the basis of published work relating to products of a similar nature and technology.

The system(s) used for assessing level of concern should be recorded. It is also advisable that supplementary records should be kept concerning any assumptions that have been made, references to published material that have been used, and the basis for the more contentious decisions.

Audit: Stage 5.

The fifth stage comprises two related parts – first a review of process and product, and second determination of control options. During the first part the process and product are reviewed with the objective of identifying areas where alterations would either reduce the level of concern, or permit more effective control. Possible changes (**Table 138**) should be fully discussed with production personnel, food technologists etc, and should be agreed, at least in prin-

Table 138. Possible process/product modifications arising from process analysis.

1 Modification of recipe
2 Change in process parameters
3 Change of packaging method
4 Reorganisation of production
5 Reorganisation of working practices
6 Change in storage and distribution patterns
7 Change in storage temperature
8 Modification to instructions to consumers
9 Change in target consumer groups

Table 139. Examples of conflict between safety of food and perceived organoleptic quality.

Food	Process factor	Conflict
Roast beef	Cooking	Preference for rare meat vs survival of pathogens
Pâté	Cooking	Avoidance of 'rendering out' vs survival of pathogens
Cheese (raw milk)	Pasteurisation	Alleged better flavour/texture vs survival of pathogens
Egg dishes (raw egg)	Cooking	Exploitation of culinary properties of raw egg vs survival of pathogens
Cured meats	Curing ingredients	Consumer demand for 'additive-free' foods vs growth of pathogens

ciple, at this stage. Where a process or product is considered actually unsafe, immediate action would be taken. Whatever the circumstances, there may well be opposition to change if only because of inertia. Increased cost will always result in problems of acceptance, but a more insidious problem is that of a possible conflict between perceived quality and safety (**Table 139**). For example, with many products the organoleptic difference resulting from more stringent processing is minimal and often grossly overstated. In any case, safety *must* have precedence.

When changes are agreed, a modified flow diagram is prepared (**Flow diagram 37**) and the process analysis (Stage 4) repeated as appropriate.

Flow diagram 37. HACCP Audit procedure.

B Stage 5

Amended flow diagram

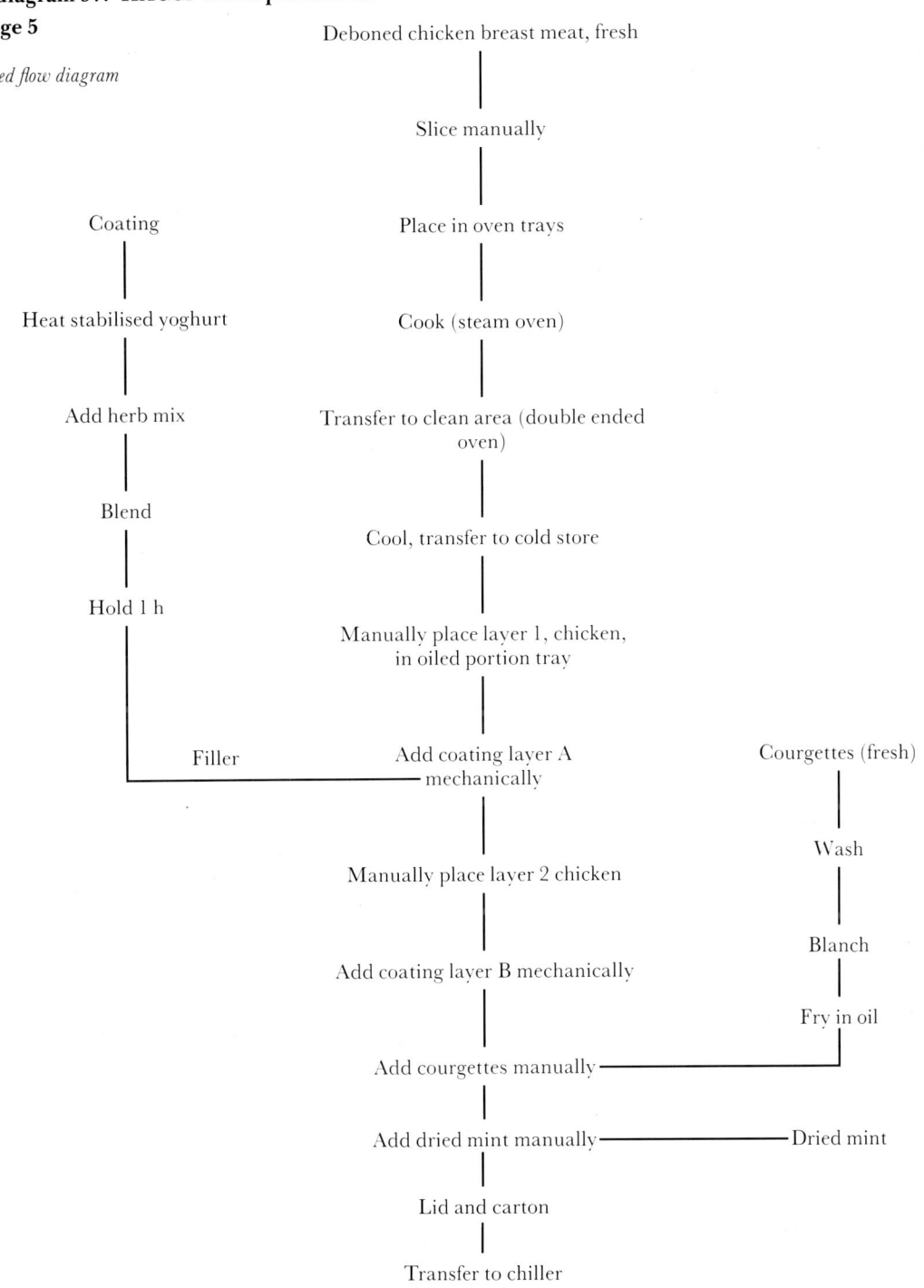

Note: Modification was necessary because the original means of preparing the coating sauce was unsatisfactory with respect to the use of unheated gelatin. Although the stabilised yoghurt imported into the plant is pasteurised, the product is reheated for technological reasons.

The second part of Stage 5 consists of determination of control options. This is primarily the function of the microbiologist, chemist, production specialist, quality assurance specialist and process engineer. Examples of control options for common food manufacturing operations are illustrated in **Table 140**.

It is often useful at this stage to review the overall philosophy of quality assurance. A traditionally minded management and workforce may find it difficult to accept the concept expressed by Board (1984) that: '. . . *everyone involved in the operation from management to line operatives should be encouraged, instructed and made responsible for as much of the quality (and safety)*

management as possible', yet acceptance is seen as a prerequisite for successful implementation of HACCP. As Board further notes, production staff are already stationed at control points and their involvement is not only logical but, in the vast majority of cases, beneficial.

Control options should be recorded as part of the documentation, and possibly combined with the process analysis.

The control and monitoring of the operation of pasteurisation, a critical control point in the production of pasteurised milk is illustrated, in the context of HACCP in (**Figures 298–304**).

Table 140. Examples of control options.

Incoming ingredients (catering)	Ensure organoleptically acceptable and from an approved supplier
Cooking	Define and ensure time:temperature requirements
Fermentation	Define and ensure pH value and acidity criteria to be achieved by given time
Cooling	Define and ensure cooling rate criteria and final temperature
Distribution (chilled products)	Define and ensure target and maximum permissible temperatures
Storage (dried products)	Define and ensure maximum humidity
Pre-service display (catering)	Define and ensure minimum temperature
Label instructions	Inform consumer of storage conditions and use of product

298

29

298–304. Pasteurisation of milk: control and monitoring. Pasteurisation is the most important critical control point in the processing of milk. Control is imposed by heating for a mandatory temperature/time combination, in most cases using the high temperature-short time process, in which the milk is heated in a plate heat exchanger (**298**). Temperature is controlled thermostatically and a flow diversion valve operates if the temperature of the milk leaving the pasteuriser falls below a predetermined lower limit. The operation of the plant should be controlled to prevent milk surging in the pasteuriser and causing small quantities to pass underheated without activating the flow diversion device.

Monitoring procedures for temperature control involve ensuring correct setting of the thermostat; ensuring the flow diversion device operates correctly; and by checking temperatures using both an accurate digital thermometer and a temperature recorder (**299**). Care should be taken to ensure that the probes of the measuring devices are correctly placed and that the instruments are correctly calibrated. Even where digital data logging devices are used to allow computer analysis of plant performance, a conventional temperature recorder should be fitted to permit simple visual assessment of operation by the operator.

The correct length of treatment is ensured by passing the heated milk through a holding tube (arrowed, **300**). The length and diameter of the holding tube are fixed and thus holding time depends on flow rate. This is determined by a positive action pump and should be monitored.

Protection of the pasteurised milk from recontamination is an integral part of the processing operation. Most pasteurising plant combines the heating and cooling functions in a single piece of equipment, outgoing heated milk first warming incoming raw milk and then being cooled by circulating cooling fluid.

During this time, the pasteurised milk is separated from potential contaminants by a system of stainless steel plates and rubber gaskets (**301**). Contamination may occur through pin-holes in the stainless steel or through worn gaskets, and while further protection is provided by operating the plant with higher pressure on the pasteurised milk side, this does not totally obviate the possibility of contamination. The condition of heat exchanger plates and gaskets should be carefully monitored, and replacements made at the first sign of wear. Pin holes may not always be visible by eye and special dye tests are required. Verification of the plant should therefore be made according to a regular schedule. Unless specialist staff are available verification should be made by the manufacturer.

Several outbreaks of food poisoning have resulted from contamination of pasteurised with raw milk via cross-connections. An important part of plant operation is to ensure that the parts handling pasteurised and raw milk are entirely separate, including the cleaning-in-place circuit. The pipework within a dairy can be extremely complex (**302**), and reference must be made to the plant plans. Verification is necessary after alterations to the plant are made to ensure that no cross-connections have been introduced.

300

301

302

Bottling (**303**) is a further stage at which contamination may occur, but is relatively unimportant with respect to risk from pathogens and adequate protection is provided by good general hygiene and simple protective screens. However, long-term contamination of pasteurised milk by *Yersinia* spp. was attributed to colonisation of the filler. Poor washing of returnable bottles was also considered to be a contributory factor and steam cleaning is recommended.

A simple chemical test, the phosphatase test, may be used as a means of verifying the adequacy of pasteurisation and should be supported by temperature records and other records relating to plant operation, such as the correct functioning of the flow diversion valve. It is also necessary to verify that cleaning schedules are adequate and that staff are suitably trained and competent.

It should be appreciated that plate heat exchangers are very widely used in the food industry, and that the level of control required for pasteurised milk may not be required. A small plate heat exchanger used for pasteurisation of a low-risk product, apple juice (**304**), for example, requires less sophisticated control, but is still operated according to good practice and the process monitored using a chart recorder.

Audit: Stage 6.

Stage 6, the design of monitoring procedures for control options, follows naturally from their determination (Stage 5), and in practice the two stages are often combined. Monitoring is seen as involving five types of procedure:

1 Documentation of control. A feature of HACCP-based control is a high level of documentation, designed not to create a bureaucracy, but to ensure that control measures are correctly applied. In plant operation, for example, operators should be required to affirm by signature that controls are correctly set and that the operating procedure is correctly followed. Similarly, where sensitive food products are handled, operators should be required to sign a positive declaration of health.

2 Monitoring of process parameters. Process parameters such as temperature should be recorded at intervals for retrospective assessment of control. Continuous recording by means of a chart recorder for heating/cooling cycles provides essential information that can be assessed very quickly. In some cases, such as canning, retention of process data for a given period is mandatory. However, manual retention of records is very cumbersome, and Sharpe (1986) foresees the on-line transmission of such data from the plant to a regulatory agency.

3 Validation of control equipment. The performance of equipment used in the control of food processes requires regular validation. This may require only very simple procedures such as ensuring visually that paper is correctly fitted to a chart recorder. In other cases, calibration of instrumentation is required, such as the calibration of process thermometers to National Physical Laboratory or other standards. Specialist assistance may be required with validation of automated control equipment, especially where they are computer based.

4 Analysis in the monitoring of control points. Analysis can be of value in the monitoring of control points. Acceptance/rejection testing, for example, may be used for sensitive raw materials that are used without further processing, and which thus comprise critical control points. Visual and organoleptic analysis is of value, as well as chemical or physical analysis, but microbiological analysis is limited by the time taken to obtain a result.

5 End product testing. End product testing has little application in safety assurance. However, it may be used as verification that the HACCP system is operating correctly (see page 401).

Failure mode effect analysis

Although the HACCP audit should take account of alternative working practices (page 390), it can only deal with known contingencies. It is therefore a useful supplementary process to test the HACCP process against all possible abnormal situations. This concept often emerges spontaneously during the course of Committee discussions, and may be dealt with informally. Alternatively, a more formal approach may be taken in which the consequences of all possible types of failure including human error, equipment and control systems are assessed, and the consequences for product safety determined. This procedure may be referred to as failure mode effect analysis.

Implementation of a HACCP system

It is true to say that while many organisations have considered HACCP, and some have proceeded to the audit, few have moved to formal implementation of a HACCP system. However, in some cases the main principles of HACCP have been incorporated into an *ad hoc* technical management system, and in others the system is used on an informal basis. It is likely, therefore, that the most difficult aspect of HACCP will be its implementation, and it is at this stage that the political skills of the Chairman and his knowledge of the corridors of power are needed most.

When the agreement of executive management is obtained, it is necessary to face the practical problems involved. This is likely to involve interaction with several different departments and the diplomatic skills of the Chairman are again likely to be necessary.

The first step is to ensure that the factory Code of Practice on which Good Manufacturing Practice is based, is in accordance with the control options and monitoring procedures identified during the HACCP audit. A new set of master manufacturing instructions should be prepared, which should incorporate the HACCP flow diagram annotated with hazards and control points (**Flow diagram 38**). Documentation for control and monitoring procedures should also be prepared at this time.

Following preparation of master manufacturing instructions and accompanying documents, it is necessary to consider related factors such as cleaning schedules and training needs (**Table 141**). The extent to which changes are required varies, but is usually minimal in a well-run organisation.

Although the decision to implement HACCP can only be made by executive management, it is production management and operatives who bear the responsibility for operating the system, and it is essential that these key groups are fully aware of developments throughout the audit. Although some management tend to use workforce intransigence as an excuse for their own inertia, it is usually found that new responsibilities are accepted with enthusiasm, providing that existing relations are good. It is also necessary to ensure that no member of the workforce suffers financially through the introduction of HACCP and, indeed, the recognition of greater responsibility through a wage increase seems reasonable where appropriate.

Table 141. Implementation of HACCP: secondary considerations.

Factor	Possible changes
Raw material procurement:	New supply sources and purchase agreements Handling procedures and storage Operator training
Process modification:	New or modified equipment Changes in layout Building modifications Operator training
Control and monitoring:	Instrument purchase Operator training
Cleaning procedures:	Change in schedules Operator training
Work pattern:	Changes in shift arrangements, meal breaks, bonus payments, etc
Medical requirements/personal hygiene:	Additional input from medical advisers Hygiene education Handwashing facilities Protective clothing

Flow diagram 38. HACCP Audit procedure.

C Implementation of HACCP

Annotated flow diagram

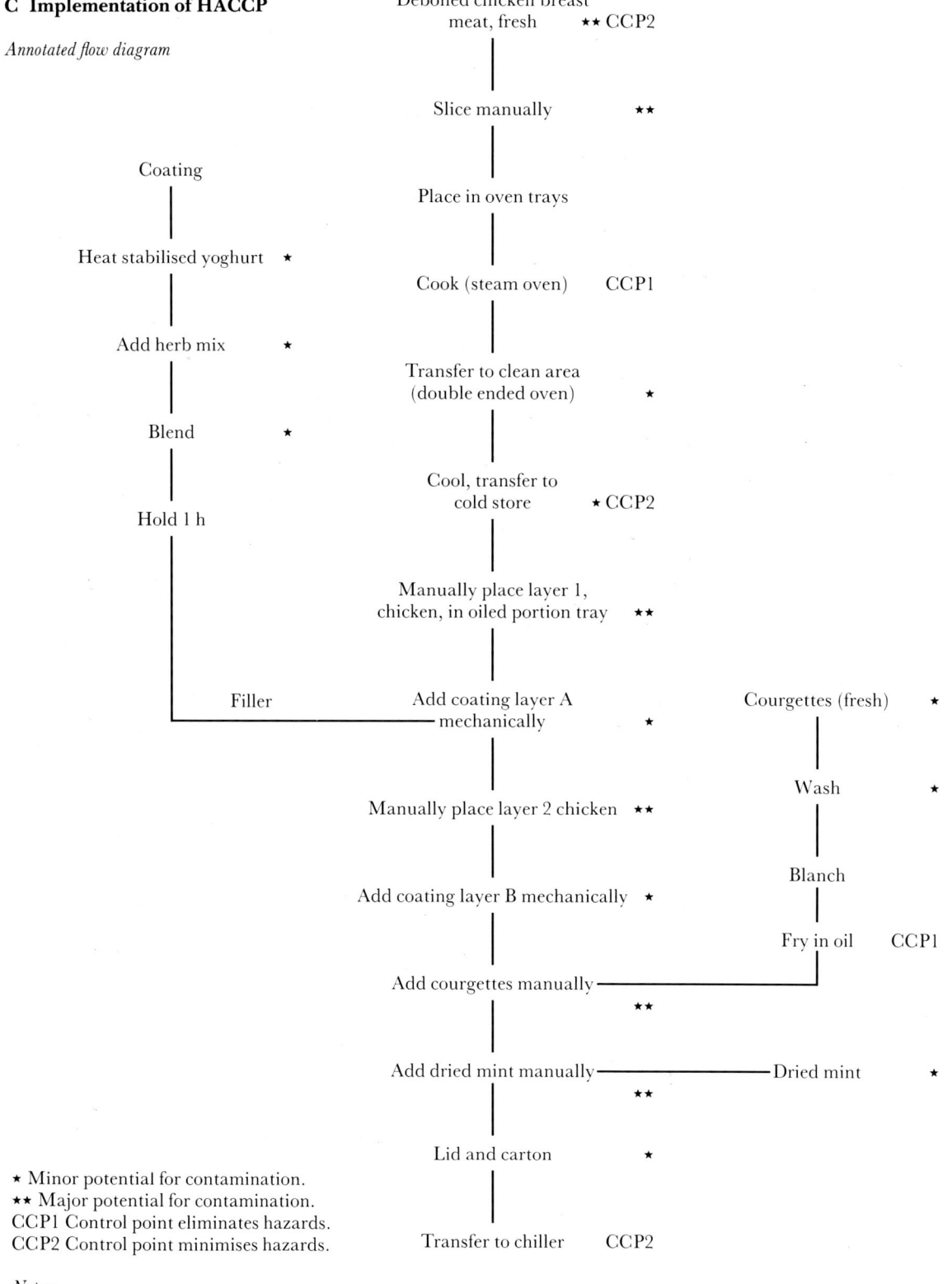

★ Minor potential for contamination.
★★ Major potential for contamination.
CCP1 Control point eliminates hazards.
CCP2 Control point minimises hazards.

Notes:

1 The heating of stabilised yoghurt is not a CCP as the process is for technological reasons and may not eliminate pathogens.

2 The combined process of blanching and frying courgettes is a CCP1.

Application of HACCP on a continuing basis

HACCP should not be regarded as a technique which is used only once, but as a management tool for use on a continuing basis. The HACCP committee should remain in being with two major functions: to review the performance of the HACCP-based control system and to incorporate into the system changes in process operations, product type etc.

Ultimately, the performance of any safety assurance system must be judged on its efficiency in preventing a potential hazard being realised. Verification usually involves information from a number of sources, including end product testing and storage life trials as well as feedback from customers. Complaints from customers should be treated seriously, and assessed for any indication that control and monitoring systems are inadequate. This requires a properly organised complaints department (see below).

Changes in product formulation, process operations etc should not be made without revision of the hazard analysis and introduction of new control and monitoring systems as required. The Chairman should be informed of even minor changes, and while there may be no need to involve the entire Committee, the appropriate expert should be consulted and the changes recorded. Totally new products should be fully assessed, although in many cases only parts of the process need be audited in detail.

Complaints and withdrawal system

The Complaints and Withdrawal system comprises two separate parts. That dealing with complaints essentially forms a part of everyday operations, while the withdrawals procedure is a key part of crisis strategy. The two parts are linked and often involve the same key personnel.

Complaints

At some time or another, all organisations dealing with food will receive complaints. In small organisations such as restaurants, these are usually dealt with informally and on a personal basis by the proprietor or a senior member of staff. This is not practical for manufacturers or multiple retailers, and in these cases a formal procedure is employed often involving input from both technical experts and customer (public) relations personnel.

The complaints procedure has, or should have, three objectives:

1 To maintain customer confidence and avoid adverse publicity. This is usually the most important function from the corporate viewpoint, but of no significance with respect to food safety.
2 To verify the operation of the safety and quality assurance system (**Figure 305**).
3 To provide, *under some circumstances*, an early warning that control has failed to such an extent that a public health or other serious hazard exists, and that product withdrawal should at least be considered.

Operation of the complaints procedure should be a specific management responsibility. The Complaints Manager is seen as having a number of important functions, and it is regrettable that in many organisations the key importance of the role is not recognised.

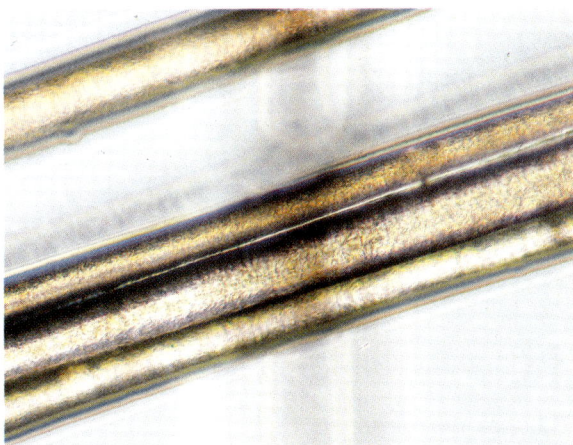

305. Early indications of control failures. People involved with a customer complaints operation should be aware that first evidence that control is failing may be unrelated to microbiology. A spate of complaints concerning human hair in a food, for example, might indicate generally falling standards of supervision, which might in turn compromise microbiological safety.

Withdrawal procedure

A procedure for the withdrawal of potentially hazardous products is an essential part of safety management, and requires careful planning and preparation. This requires the support of management of the highest level, who must be prepared not only to acknowledge the need for such a procedure, but be prepared to allocate the time and resources necessary for its establishment.

An effective withdrawal procedure consists of three parts:

1 Recognition that a hazardous, or potentially hazardous situation exists.
2 Assessment of the action required to control that situation and to decide whether or not withdrawal is required.
3 Where necessary to withdraw the product.

Information leading to recognition of a hazardous situation can come from a number of sources. These include the company complaints system, regulatory bodies, results of end-product testing, knowledge of problems with related products etc. In some cases, 'clues' may be obtained from several sources, and recognition of their significance depends on well-trained and experienced personnel.

Assessment of the action required should be made purely according to technical criteria, and thus should involve technical personnel. In the case of microbial food poisoning, final responsibility must lie with the microbiologist. It must be appreciated that disaster may strike at weekends, or during holidays, and a procedure must exist for contacting key per-

Table 142. Use of USFDA guidelines as a basis for a rational approach to product withdrawal.

USFDA Classification I

Problem
A situation in which there is a reasonable probability that the use of, or exposure to, a violative product will cause serious, adverse health consequences or death.

Examples
Botulism; acute poisons at high LD_{50} levels.

Depth of recall
Consumer; retail store; warehousing and transit.

Public communication
Urgent and immediate over whole distribution area of product lot involved.

Effectiveness of withdrawal
Objective is 100% return. Extensive audit is required at retail level.

USFDA Classification II

Problem
A situation in which the use of, or exposure to, a violative product may cause temporary or medically reversible adverse health consequences or where the probability of serious adverse consequences is remote.

Examples
Pathogens such as most salmonellas, *Staph. aureus*, etc; acute poisons at low LD_{50} level; widespread glass contamination.

Depth of recall
Retail store; warehousing and transit.

Public communication
Prompt press release, may be targeted at high-risk groups.

Effectiveness of withdrawal
As complete as possible; spot checks made over as wide an area as possible.

USFDA Classification III

Problem
A situation in which the use of, or exposure to, a violative product is not likely to cause adverse health consequences. Legal contraventions.

Examples
Adulteration with harmless substance; spoilage by non-pathogenic micro-organisms; major quality fault, widespread short weight.

Depth of recall
Carriers and direct deliveries.

Public communication
Usually none.

Effectiveness of withdrawal
Warehouse supervision.

---

sonnel or suitable deputies.

The decision to withdraw a product is a major decision and should be based on a rational approach. A suitable basis is the Guidelines of the US Food and Drug Administration (**Table 142**). It must be appreciated that, in some cases, assessment of the action required presents considerable difficulty. On some occasions it is possible to defer action until further information is available, but a fast response is of crucial importance in product withdrawal and it may be necessary to make decisions on a 'worse-possible situation' basis. The temptation to prevaricate and avoid commitment to a course of action is understandable, but must be avoided.

Once a decision to withdraw has been made it is necessary to assess the extent of withdrawal, that is, whether only a particular batch is affected or whether all product is suspect. It is usually only possible to restrict withdrawal to a single batch, or a few batches, when there is definite evidence that the fault occurred over a discrete and known period. If such evidence is not available, all product must be withdrawn.

The actual implementation of withdrawal involves a large number of people and requires efficient communications and effective co-ordination. This requires advance planning: *'There is no time to be arguing about who does what when disaster strikes. The forward-thinking manufacturer will have devised plans for dealing with emergencies'* (Goldie, 1988).

The most effective approach is for all aspects of the actual withdrawal to be controlled by a team led by a senior manager or director acting as co-ordinator (**Table 143**). The Co-ordinator should be chosen for his or her management skills and be of sufficient seniority to authorise the necessary actions. Each member of the team should be accepted as the expert in his given area, and all members should give absolute priority to the successful completion of the withdrawal.

Table 143. Generalised composition of product withdrawal team.

Co-ordinator (Director or Senior Manager)	Co-ordinates and monitors withdrawal procedures. Authorises action where necessary. Defines and re-defines objectives as necessary.
Deputy co-ordinator (Senior Manager)	In normal circumstances 'shadows' co-ordinator and is able to take over operation in an emergency. Is responsible for record keeping.
Quality Assurance Manager	Identifies extent of fault, product(s) involved and codes.
Microbiologist	Is not directly involved in logistics of withdrawal, but in cases of microbial food poisoning should be consulted for advice as required.
Distribution Manager	Advises on distribution chain. Organises recovery from depots and warehouses. Notifies carriers and stops further distribution.
Sales Manager (or equivalent)	Identifies final recipients and arranges contact. Instructs recipients on action required.
Public relations (or equivalent)	Liaises with media and arranges for placement of advertisements in press as required. Arranges for publicity material to be made available at retail outlets. Arranges for advice to be made available to consumers contacting company directly.
Office Manager	Not directly involved but ensure that facilities such as telephones, telex, etc are available.

In addition to the selection of a team for withdrawal, planning should involve preparation of a Manual and training for team members. The Manual should be as concise as possible, but should contain all necessary information including guidelines for withdrawal, withdrawal procedures, telephone numbers of key personnel and contacts, telephone, telex and fax numbers of wholesalers, shippers etc. Training is to ensure that everyone is aware of their role in an emergency and should include working through a mock emergency (**Table 144**).

The further the goods to be withdrawn have travelled down the distribution system, the greater the difficulties of securing effective withdrawal become. Where it is necessary to withdraw products from the consumer, the effective use of news media is essential.

Table 144. Training and preparation for withdrawal.

1 Preparation and issue of guidance documents (The Manual)

2 Establishment of withdrawal team

3 Theoretical work through of a recall using essential documentation.

4 Mock recall using real time and real data

5 Review of mock recall, identification of additional facilities and procedures required

6 Retrain and review every 6 to 12 months in a mock recall situation

Newspaper advertisements are an important means of informing consumers, and the Manual should contain the necessary contacts for the purchase of advertising space at short notice. Local and national radio and television news services are also of value as a means of contacting consumers, and multiple retailers have made effective use of warning notices in stores during the withdrawal of own-label brands. There will inevitably be telephone calls from consumers seeking information or reassurance, and provision of necessary advice is an important part of a successful withdrawal. This requires both physical facilities such as adequate telephones, and people confident at dealing with the public. Care should be taken to ensure that information is correct and unambiguous.

Food handlers and associated personnel

General aspects

Provision of adequate staff and the training necessary to enable them to perform efficiently, is an important management function in industries of all kind. In the food industry, food handling personnel are of particular importance since they must be aware not only of the mechanistic aspects of their immediate function, but of the basic principles of hygiene. Furthermore, it must be appreciated that the food handler is a potential source of micro-organisms that may compromise the safety of the final product.

It is important that management should have policies both for the training of food handling personnel, and for ensuring their health as handlers, and these policies are discussed in detail below. There are, however, more general matters of personnel policy which impinge directly or indirectly on food safety.

It is generally recognised that the most satisfactory performance is achieved by workforces which are reasonably content with wages, conditions and their treatment by management. Such a situation is considered to be of particular importance in industries producing 'sensitive' products such as the food industry. It is particularly unfortunate that in some parts of the food industry, the overall conditions of employment are such that it is virtually impossible to assemble a responsible and properly trained workforce. The situation in the catering industry is often particularly bad due to a combination of poor wages and a lack of job security, but management behaviour in the tradition of the worst of the nineteenth-century ironmasters is encountered in other areas, particularly where alternative employment is limited. The bullying of a workforce and a denial of basic workplace rights such as Trades Union representation or even a decent canteen, may do much for management machismo, but is disastrous in terms of encouraging a positive attitude towards hygiene and safety.

Training of food handlers and associated personnel

The training of food industry personnel varies in content according to different job requirements, as well as to differences in activity from one organisation to another. However, the underlying objectives are the same (**Table 145**).

Table 145. Training objectives for food industry personnel.

1 Basic principles of hygiene
2 Special hygiene training associated with specific job requirements
3 Specific training for plant/process operation. To include basic principles of technology and cleaning schedules
4 Continuing training, formal or informal as required
5 Specialist training: supervisors and technical managers

Table 146. Basic knowledge required by all food industry personnel.

Food poisoning
  Food poisoning as a disease
  Role of micro-organisms in food poisoning
  Basic means of controlling micro-organisms in food
  Circumstances leading to actual food poisoning

Personal responsibilities
  Personal hygiene and protective clothing
  Exclusion during sickness
  Procedures relating to sickness, infected lesions, etc

Job-related factors
  Importance of following laid down procedures
  Reporting abnormal situations

Note: The knowledge listed refers only to food poisoning. Actual training programmes would also be concerned with food spoilage.

Basic hygiene training

All employees in the food industry, including temporary or part time staff, should receive a basic training in the principles of hygiene and good practice in food handling (**Table 146**). Training should be given as soon as possible after commencing employment, and in the case of handlers before actually being involved in food production. In many cases, basic hygiene training may be given as part of the company induction programme, but sufficient time must be allowed for the subject matter to be fully presented and to reflect its importance. The material presented should be straightforward and factually correct, and should be designed to help operatives understand the special responsibilities of the food handler and the reasons for the special restrictions associated with employment in the food industry. The importance of micro-organisms should be understood (**Figures 306–310**) but it is important to remember that the training is intended for food handlers not for microbiologists.

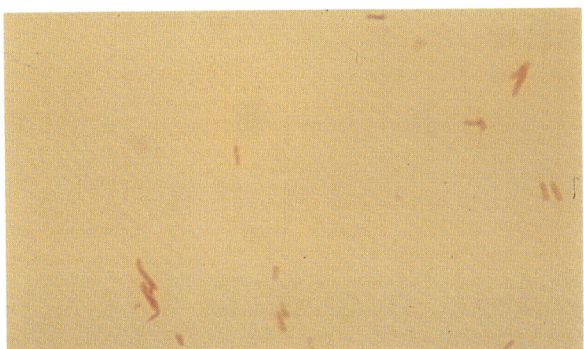

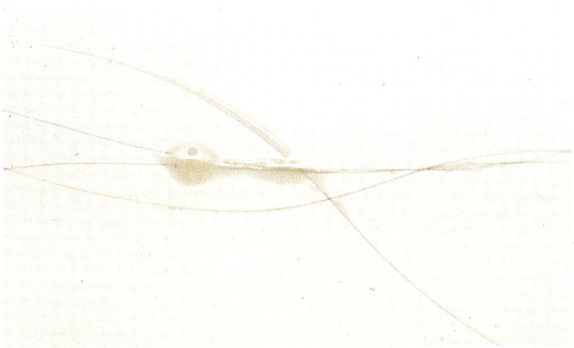

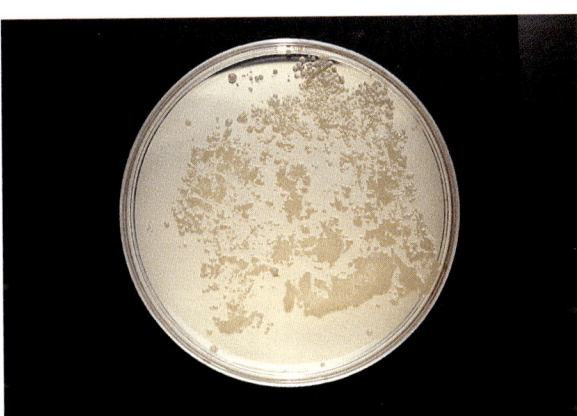

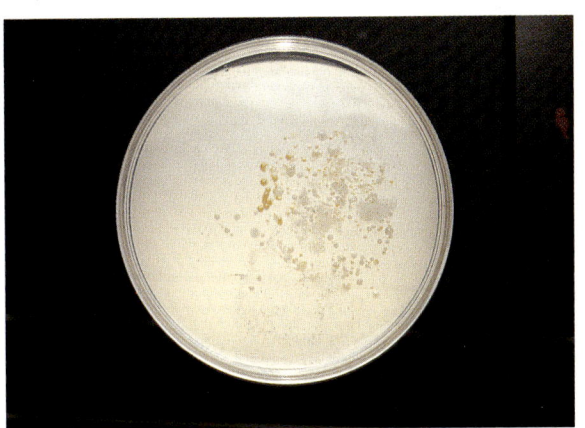

306–310. Use of demonstration plates in the education of food handlers. One of the most difficult aspects of training food handlers is that, while there may be familiarity with names such as *Salmonella* and *Listeria*, there is no concept of the nature of micro-organisms. Furthermore, at a more practical level, untrained people may find it difficult to accept the ubiquitous nature of micro-organisms and that apparently clean objects may be heavily contaminated. Although microphotographs of bacteria can be useful (**306**), demonstration plates often hold a degree of fascination when encountered for the first time and are an extremely effective aid during training. Effective examples include bacteria growing around a hair (**307**), derived from a dirty towel (**308**) and, for light relief, a cat's paw (**309**). Contact slides (see **318**) are also useful for the demonstration of contamination of surfaces and utensils such as a knife (**310**).

Precautions must be taken in the use of demonstration plates. The preparation of plates and their demonstration should be undertaken only by a qualified microbiologist, plates should be securely sealed and materials available to deal with any mishap. Exposure of colonies to chloroform vapour before sealing plates is a useful precaution, but cannot be guaranteed to kill all bacteria. Known

pathogens should not be demonstrated to untrained people, but it must be remembered that all bacteria are potentially pathogenic. Demonstration plates should not be taken into the food handling area and the precaution of handwashing should be observed.

A number of films, video and slide-tape presentations about food hygiene are available and, while both factual content and presentational quality vary, can be an effective means of introducing the subject. Where possible, they should be supported by an experienced person who is able to expand the points made and answer questions. A simple booklet for each member of staff which covers the points raised during the presentation is useful (**Figures 311–313**).

It is considered important that a basic training in hygiene should be given to members of staff who have no contact with the food handling process. This is seen as an important part of engendering a company-wide culture in which hygiene is given the highest priority.

Basic hygiene training for registered food handlers is due to become mandatory in the UK during 1991, and it is to be hoped that this will lead to an improvement in standards. However, some companies, particularly in catering and small-scale retailing, are reluctant to apply even basic training, often on the grounds that high staff turnover and widespread use of temporary staff mean that such training is not 'worthwhile'. The consequences of such attitudes have been suffered by many consumers (**Table 147**); the solution must lie in preventing high staff turnover rather than in abdication of responsibility for hygiene.

311

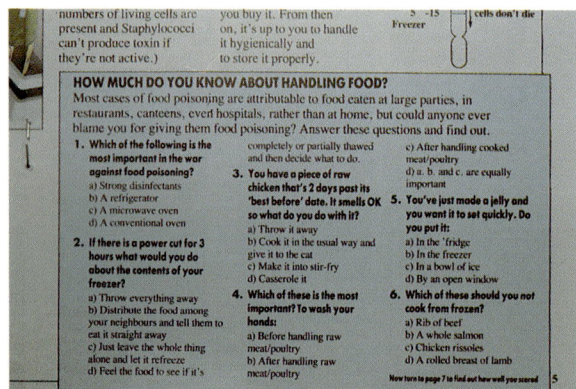

311–313. Hygiene education booklets. It is recognised that a high proportion of family outbreaks of food poisoning are due to mishandling in the home, and education of the public is seen as an important part of the initiative to reduce morbidity due to food poisoning. In recent years a major supermarket chain (**311**), food trade organisations (**312**) and the Government (**313**), have each produced booklets that, in general, were well designed, easily understood and informative.

Booklets are also widely used in the education of food handlers and as such are produced by companies and local authorities. It is essential that one booklet is available for each food handler. Language difficulties may negate the value of booklets and in areas of high immigrant populations in the UK, some local authorities have produced booklets in Chinese and other languages. (**311** Reproduced by courtesy of J. Sainsbury plc. **312** Reproduced by courtesy of the Food and Drink Federation. **313** Reproduced from *Food Safety: A Guide from HM Government* by permission of the Ministry of Agriculture, Fisheries and Food. © Crown copyright.)

312

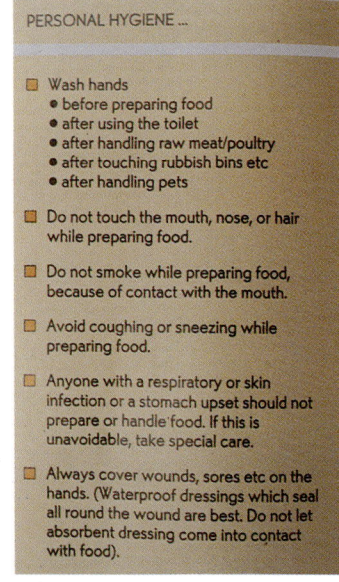

313

Table 147. Food poisoning: involvement of untrained personnel.

Organism
  Salmonella typhimurium

Number affected
  14

Circumstances
  Schoolgirl employed as a part-time waitress at a wedding reception contaminated cooked chicken by placing plates in the 'raw-meat' refrigerator. Plates of cooked chicken were then removed, visible drip wiped-off, and stored at room temperature for several hours.

Contributory factors
  Poor supervision; inadequate refrigeration facilities

Organism
  Staphylococcus aureus

Number affected
  Up to 50

Circumstances
  An apprentice at an engineering works was ordered, as a punishment, to assist in preparation of the works' Christmas dinner of roast pork and turkey. Several infected cuts were present on the hands of the apprentice, and these remained unprotected during his work of setting out portions. All people affected ate during the second sitting when the meat had been held at a warm ambient temperature for several hours before re-heating. The extent of the outbreak could not be accurately determined since a number of apprentices were drunk and the actual cause of vomiting was not known.

Contributory factors
  Poor supervision; inadequate refrigeration facilities

Organism
  Clostridium perfringens

Number affected
  10 known but probably more

Circumstances
  A girl assistant at a snack bar removed home-made beef and tomato soup from the hot-plate to prevent a skin forming. After several hours the 'warm' soup was discarded due to complaints of off-taste.

Contributory factors
  Poor supervision; preparation of an unnecessarily large quantity of soup too far in advance of serving.

Special training associated with specific job requirements

Although a basic knowledge of hygiene is required for all food industry personnel, a higher level of knowledge is required for certain positions. In the context of a supermarket, for example, only a basic knowledge is required by a check-out operator, but a rather higher level is necessary for an assistant working in an in-store bakery or on a delicatessen counter. Hands-on operations invariably require specific training, and special attention must be given to those handling particularly sensitive foods. People with even the minimum of supervisory responsibilities should be given extra training on their appointment. This should be directly task related and is of a much less detailed nature than the additional training required for high-level supervisors and technical management.

Special training for plant/process operatives

All employees operating plant or responsible for an entire process, require training directed to the task they have to perform. This may be straightforward, requiring only simple in-house training, but operators must be aware of the nature of the control if working at a CCP and know how to operate monitoring systems. The level of training required is higher for a complex plant such as pasteurisation equipment (see pages 396–398), or where an operator is responsible for an entire process. In these cases, in-house training may need to be supported by courses at technical or other colleges, or courses run by manufacturers of the equipment.

The need for special training is not restricted to process operatives, but is also necessary for cleaning teams and other ancillary workers.

Continuing training

Training of food operatives should be seen as a continuing process. The introduction of new products or new processes, for example, usually requires retraining during which basic hygiene should be re-emphasised.

In addition to retraining for specific purposes, it is desirable to continually reinforce the principles of hygiene and good food handling practice. This is achieved partly by good supervision in correcting faults and misconceptions. The use of posters, articles in company magazines etc, all have a role in maintaining an appreciation of the importance of hygiene. Other approaches have included bulletin boards carrying information relating to hygiene and food poisoning as well as, in some cases, quality-control results – simplified and annotated as necessary.

Specialist training: supervisors and technical management

According to circumstances, supervisors and technical management may require training of a more specialised nature, either to meet the immediate day-to-day requirements of their position or because of the needs of job progression. Such training can take a number of forms, including attendance at conferences and symposia, advanced courses offered by universities, research institutes etc, as well as full, or part-time, formal courses leading to an academic qualification. For higher level staff, mention should be made of the *Mastership in Food Control* administered by the United Kingdom Institute of Food Science and Technology.

Effectiveness of training

A common problem is assessing the effectiveness of hygiene training programmes. Obvious errors and misconceptions should be corrected by supervisors, who should inform the people responsible for training if misunderstandings are widespread. Management must recognise that a small percentage of food handlers will be totally recalcitrant with respect to hygiene and will refuse to accept the restrictions necessary. The most obvious example is smoking, but refusal to obey other hygiene-related regulations also occurs. The wilful and persistent offender can be a major threat to public health and cannot be permitted to continue work as a food handler. Disciplinary action is necessary, culminating where necessary in dismissal.

Medical policies

It is recognised that pathogenic micro-organisms can be transmitted from man to food. The prime means of control are the education of food handlers and strict supervision, and the implementation of personal hygiene and good food handling practices. However, it is necessary to have laid down medical policies that deal with the health of food handlers in relation to their employment, as well as normal occupational medicine. In the latter case, it is necessary to be aware that food handlers can be at risk from a wide range of zoonotic diseases (Bryan, 1979).

Medical examinations for food industry employees

The question of medical examinations for food handlers continues to raise controversy. In some countries, including certain EEC members, codes of public health practice stipulate that food handlers must undergo examination on commencing work and at yearly intervals thereafter. In other countries such as the UK, there is no legal requirement and practice varies from one organisation to another. In some cases it must be recognised the examinations proposed are absurd and can only distract from more important aspects of hygiene management.

Pre-employment medical examinations are employed by companies of all types to ensure the prospective employee's fitness to work. In the food industry it is common practice to extend these examinations to the specific health requirements of food handlers. Obtaining a medical history can be of value where there is evidence of a previous infectious disease such as typhoid fever, chronic diarrhoea or skin infection, while a physical examination is likely to detect a current skin infection and may detect signs of other acute or chronic infections. However, adequate physical examinations and the taking of medical histories are expensive, time consuming and require considerable follow-up (ICMSF, 1989). Furthermore, the examination is only valid at the time it is carried out.

It is not uncommon to find a physical examination supplemented by radiology, stool samples and even vaginal swabs and serological tests for sexually transmitted disease. Many of these tests such as chest x-ray examination may be beneficial to the worker, but are irrelevant to the prevention of foodborne disease (Cruickshank, 1990). Stool examinations are certainly capable of detecting a wide variety of pathogenic micro-organisms, but under most circumstances, laboratory examinations of stools are not cost effective (**Table 148**).

Table 148. Ineffectiveness of stool examination in the control of food poisoning.

1 State of health is transitory and screening cannot reliably detect people who have acquired infections.

2 With the exception of salmonellas such as *Salmonella typhi* which are host adapted to humans, and *Shigella*, man is only rarely the primary source of contamination. Exceptions occur primarily where the handler is ill and suffering clinical symptoms.

3 Except in areas where human-adapted salmonellas are highly endemic, the rate of carriage of these micro-organisms is low and screening for excreters is a massive and expensive operation which is rarely effective.

4 Large-scale screening for excreters diverts resources from more effective means of controlling food poisoning and can lead to false security when tests prove negative.

A report by the World Health Organisation (Anon, 1989) concluded that the general health of an employee is best assessed by a health questionnaire conducted by a suitably qualified nurse who could clarify the terminology where necessary. Physical examinations and other tests can then be targeted to situations where the questionnaire has revealed doubt over the employee's fitness to handle food. Completion of the questionnaire is also a suitable time to introduce personnel to their medical responsibilities as a food handler, and to explain procedures for reporting illness, exclusion etc.

Although the general application of stool screening is unwarranted, there may be justification for testing in special circumstances, such as the recruitment of staff from an area where enteric disease is endemic. Stool testing of food handlers is common practice during an outbreak of food poisoning but, while it is desirable to identify excreters, the finding of pathogenic micro-organisms does not necessarily imply that those excreting them were the source of the outbreak (Cruickshank, 1990).

Exclusion during sickness

No person should work as a food handler while suffering from the clinical symptoms of gastroenteritis. A policy of self-exclusion is most effective, and should also apply to high fever and severe sore throat. It is the responsibility of the employee to inform the employer immediately of the nature of the illness, and to accept advice concerning return to work etc. Where the illness is of sufficient severity to require medical attention, the physician should be informed of the patient's occupation as a food handler. Many companies issue a card for handlers to show to the doctor in the event of treatment being required.

Food handlers must report infected cuts, boils, abscesses etc, to supervisors before entering the workplace. The supervisor, with medical advice, should then decide the action to be taken. This usually involves total exclusion, advice to seek treatment or a transfer to employment not involving food handling. Handlers must similarly report mild illnesses including colds, influenza-like symptoms, sore throat etc. It should be appreciated that illness due to some enteric pathogens does not necessarily involve the classic symptoms associated with food poisoning. Dominant symptoms associated with *Yersinia enterocolitica* infections, for example, may involve pharyngitis (page 137), while there is also some evidence of non-enteric prodromal symptoms with other bacteria. It may be necessary to modify current responses to non-enteric illness among food handlers when these phenomena are better understood.

In cases of severe food poisoning, it is unlikely that even the most determined could work through the symptoms. For a number of reasons, food handlers may attempt to continue work when suffering mild symptoms, or return to work before symptoms have fully abated. Such action can present a significant risk to public health (**Table 149**). The most significant reason for not reporting an illness appears to be financial, and results from a fear (sometimes quite justified) of loss of wages and even of employment when unable to work. It is considered essential, therefore, that food handlers should not be financially penalised for reporting an illness. Inevitably, this brings accusations of malingering where absence is more than occasional. Recurring enteric illness in a food handler should in any case be fully investigated since it may indicate a chronic infection.

A second cause of failure to report illness is the 'indispensable man syndrome', which stems from an overdeveloped sense of self-importance. This syndrome tends to be more common among management, the very people who should know better.

Although a properly administered system of self-exclusion is highly effective, it must be remembered that supervisor and line management have a responsibility to look for signs of unreported illness among handlers. Examination of hands for cuts and other lesions is commonplace in hands-on operations with sensitive foods, and in some cases throats are also examined on a regular basis. The value of the latter practice is dubious and a more effective system for workers in sensitive areas is to require a positive declaration of health signed daily in the presence of a nurse competent to answer any queries. Outside

Table 149. Outbreaks of food poisoning in which working while ill was a major causative factor.

Organism
  Small, round, structured virus

Number affected
  Not known

Circumstances
  Inexperienced female catering assistant failed to report illness due to fear of dismissal.

Organism
  Shigella

Number affected
  12 known

Circumstances
  A male assistant at a college salad bar worked while still suffering diarrhoea. The assistant was under severe financial pressure and feared being unable to finance the completion of his own college course.

Organism
  Yersinia enterocolitica

Number affected
  None known but large number exposed to risk

Circumstances
  A female assistant on a supermarket delicatessen counter had suffered recurrent gastroenteritis over a period of several months. This had been reported but no medical examination given and the employee was warned for malingering. Following this warning she worked through several bouts of enteritis until she was finally hospitalised.

Organism
  Staphylococcus aureus

Number affected
  38 known

Circumstances
  A male chef with an infected cut on his finger, who had been warned by his doctor not to handle cooked food, insisted on carving turkey at a works Christmas dinner as 'no one else could do it properly'. Those affected were shift workers who ate turkey which had been held for several hours at room temperature.

Organism
  Salmonella agona

Number affected
  None known but a large number exposed to risk

Circumstances
  A male manager in a large supermarket continued to work packing cheese despite suffering severe diarrhoea and vomiting. The manager considered his action praiseworthy and to contrast favourably with that of similarly affected staff who, quite correctly, had excluded themselves from work.

Organism
  Salmonella typhimurium

Number affected
  8 known

Circumstances
  A chef-proprietor of a restaurant had insisted on preparing his 'special' dessert for a gourmet dinner while suffering clinically diagnosed salmonellosis. The chef had concealed his occupation from the doctor.

Organism
  Small, round, structured virus

Number affected
  More than 3,000

Circumstances
  A male baker involved in the preparation by hand of confectionery worked during a diarrhoeal illness. He was aware of the potential consequences and claimed to have washed his hands carefully after a visit to the toilet.

sensitive areas, supervisors should concentrate on detecting skin lesions and signs of illness such as jaundice or a large number of visits to the toilet.

An apparently innocuous practice that should be strongly discouraged is that of workers eating raw foods of animal origin while at work. The practice of giving raw milk to farm workers is an important means by which infection can be cycled at farm level, and in Scotland it is now illegal to give raw milk in lieu of wages. Large-scale infection of food handlers can also occur (**Table 150**).

Table 150. Food poisoning among food handlers attributed to the consumption of raw food.

Organism
  Salmonella

Number affected
  11 over a 7 month period, and then 13 in a single outbreak

Circumstances
  The consumption of raw pork sausage meat became a cult among some workers in a meat products factory, and was associated with sporadic salmonellosis among cult members. The sporadic outbreaks of illness were followed by a point source outbreak of *S. typhimurium* infection among workers in the cooked meat part of the plant, who were not members of the raw sausage cult. Although a proper epidemiological investigation was not made, the likely source of infection was a single meal prepared by a canteen worker who had suffered salmonellosis following the consumption of raw sausage meat as a 'treat'.

Organism
  Salmonella/Campylobacter

Number affected
  Not known

Circumstances
  Sporadic outbreaks of enteritis involving small numbers of cases had occurred among workers in a milk processing plant over a lengthy period. No investigation was made until the appointment of a new Medical Officer and in a subsequent outbreak, *S. heidelberg* and *C. jejuni* were isolated from different stool samples. It was discovered that on some occasions unpasteurised skim milk was provided to the canteen to make good milk shortages, while some workers were in the habit of drinking unpasteurised skim milk while working.

Exclusion after illness

Policies concerning exclusion after illness continue to cause confusion and controversy. The unthinking extrapolation from severe and highly infectious diseases such as typhoid to common types of food poisoning, has led to extended and unnecessary exclusion from employment and even dismissal from employment under circumstances that cannot possibly be justified on public health grounds. Essentially, this reflects the lay attitude which places undue importance on the role of the carrier in the spread of enteric disease, and it should be noted that this attitude is likely to be reflected in the judgements of law. In one widely quoted example (Bush, 1985), a judge upheld the dismissal of a *Salmonella* excreter even though it was agreed that his duties involved no contact whatsoever with food.

People recovering from enteric illness may continue to excrete the causative micro-organism for some time after the clinical symptoms. For example, approximately half of patients excrete detectable salmonellas 5 weeks after suffering salmonellosis, and the excretion of other pathogenic micro-organisms may be similarly prolonged. This is convalescent excretion and should not be confused with true, long-term carriage which is rare with organisms other than human-adapted salmonellas such as *S. typhi*.

The consequences of excretion with respect to food safety have been considered in detail, and it is generally agreed that, with the exception of *S. typhi* and *S. paratyphi*, an adult having a solid formed stool after recovery from diarrhoea, and who has good standards of personal hygiene, need not be excluded from any work including food handling unless they are in a special-risk category (**Table 151**). From the viewpoint of the food industry, only people who handle unwrapped foods to be consumed raw or without further cooking are of relevance (Cruickshank and Humphrey, 1987).

Although there is a high level of support for the exclusion of those in special-risk groups (Bahl, 1987; Bostock, 1987), some workers consider that even this restricted exclusion is unnecessary (Walker and Jones, 1987).

Table 151. Special risk groups in the spread of food poisoning (PHLS Salmonella Sub-Committee, 1983).

Group 1 Food handlers whose work involves handling unwrapped foods to be consumed raw or without further cooking.

Group 2 Health care, nursery or other staff who have direct contact, or contact through food, with highly susceptible people, in whom an intestinal infection would have particularly serious consequences.

Group 3 Children aged less than 5 years attending nurseries, playgroups, etc.

Group 4 Older children and adults with poor standards of personal hygiene. This includes the mentally ill or handicapped, or the infirm aged and those in other circumstances such as refugee camps where sanitation may be unreliable. In exceptional circumstances, children in infant schools may fall into this category.

All people other than those in these groups present a minimal risk of spreading gastrointestinal disease if they are well and have normal, firm stools.

Criteria for return to work after exclusion

Periods of exclusion and criteria for return to work vary according to the nature of the micro-organism, the severity of the disease, and the infective dose. Periods are summarised in **Table 152**. Lengthy exclusion is required only in the case of *S. typhi* and *S. paratyphi*, and nobody who has been infected by these organisms, or who is a carrier, should handle foods until properly cleared. Carriage may be prolonged and in some cases people will never be able to resume work as food handlers. The criteria for a return to work for food handlers involves 12 consecutive negative stool specimens. Stool specimens do not, however, give an infallible indication that *S. typhi* has been eliminated. For example, the urinary excreter who initiated the Aberystwyth typhoid epidemic of 1947 had been declared cleared after suffering the disease 9 years earlier (Rothwell, 1981).

Recommended periods of exclusion should be treated as guidelines only, and each case should be assessed individually before return to work is permitted. Factors which should be considered are summarised in **Table 153**.

Table 152. Criteria of clearance for food handlers after infectious disease (PHLS Salmonella Sub-Committee, 1983).

Salmonella typhi ⎱ *S. paratyphi* ⎰	12 consecutive negative stool samples over a 6 month period
Other salmonellas	None where personal hygiene standards are high, otherwise 3 consecutive negative stool specimens
Shigella	3 consecutive negative stool specimens
Vibrio cholerae (O1 and non-O1)	3 consecutive negative stool specimens
Escherichia coli (outbreaks only)	3 consecutive negative stool specimens
Staphylococcus aureus Vomiting	None
Skin lesions	Until lesions treated and healed
Small, round, structured viruses	2 days after diarrhoea ceases
Hepatitis A virus	7 days after appearance of jaundice
Rotavirus	7 days after complete recovery
Entamoeba histolytica	3 consecutive negative stool specimens
Threadworms and 'pork' tapeworm (*Taenia solium*)	Until treated

Table 153. Non-microbiological factors affecting clearance for return to work after illness.

1 Nature of work.
2 Final consumer of food (*ie* normal or high-risk people).
3 Personal cleanliness of handler.
4 Facilities for handwashing at workplace.
5 General hygiene standard of workplace.

Note: In developed countries, lack of handwashing facilities or poor general workplace hygiene is not acceptable as a reason for extended exclusion. If that situation exists, immediate remedial action is required.

Contact with people suffering gastroenteritis

In non-endemic situations, family or social contacts with people suffering gastroenteritis has often been an undue cause for concern, and innocent employees have suffered financial loss and the odium of exclusion on totally illogical grounds. In an extreme case, it was proposed that, because of the higher incidence of carriage of non-typhoid salmonellas among children, the parents of children aged less than 11 years should not be considered for employment by a supermarket chain – parenthood as a notifiable disease, perhaps.

There is no evidence that 'healthy' contact with those suffering gastroenteritis presents any risk and under most circumstances exclusion is not justified.

Exceptions are possible under specific circumstances, and it may be considered good practice for food handlers to notify management of illness among close contacts.

The situation with *S. typhi* and *S. paratyphi* is rather different, although in the latter case, household contacts of infected persons have been allowed to work. In an outbreak of paratyphoid fever in the UK associated with a fish-and-chip shop, there was evidence of household spread, despite good hygiene (Francis *et al.*, 1989), and it was recommended that food handlers sharing a household, or in intimate contact with cases or carriers of *S. paratyphi*, should be excluded from work until all close contacts cease excreting.

The food handler and personal hygiene

High standards of personal hygiene are of the utmost importance in the safe production and handling of food. Attaining and maintaining the necessary standard is an individual responsibility for each food-handler, but management have an equal responsibility in actively encouraging good hygienic practice and in the provision of facilities. Good habits develop most readily in a stable work situation, and where good hygienic practice is an integral part of the work ethos.

Hand washing and hand care

Hand washing is of major importance in preventing the spread of pathogenic micro-organisms, and a fundamental part of the safe handling of foods. In the USA hand washing is considered to be the single most important procedure for preventing nosocomial infection (Gomer and Favero, 1985), while the efficiency of a simple wash with soap and water in removing large numbers of salmonellas from intentionally contaminated hands has been conclusively demonstrated (Pether and Scott, 1982). It is unfortunate that in some situations more importance is attached to the wearing of jewellery or the angle of the headgear than to encouraging handwashing (**Figure 314**).

314. Supermarket in-store bakery: priorities for hygiene. This notice is prominently displayed at the entrance to the in-store bakery from the sales floor, which is widely used by bakery staff including people returning from the canteen and toilets. Despite the concern for jewellery and protective clothing there is no attempt to reinforce the importance of handwashing directly before commencing work. Furthermore, no handwashing sink is stationed at that entrance. Operations in the bakery include the production of sensitive items such as cream filled and custard confectionery.

314

315

315. **Handwashing reminder.** Although handwashing after using the toilet should be second nature to the experienced food handler, simple notices serve as a powerful visual reminder to the less experienced and a prompt to the conscience of the less hygiene minded.

The great majority of people are aware of the particular importance of hand washing in the context of food handling. From the start of employment, food handlers should be actively encouraged to acquire a hand washing habit. It is usual to stipulate occasions on which washing is mandatory; the commencement of each work period, for example, and after a visit to the toilet (**Figure 315**), but with experience washing on appropriate occasions should be second nature.

The food handler must also appreciate that a perfunctory splash in water is not adequate, and a correct routine should be demonstrated as part of training (**Figures 316 & 317**). Hands should be well washed with soap in a stream of running water, using a nail brush to remove material under fingernails, rinsed and dried. The importance of drying is often overlooked, but is of particular importance in the control of *Campylobacter*, which is particularly sensitive to desiccation (Coates *et al.*, 1987).

The importance of correct hand washing should also be stressed to engineering personnel and all others who have reason to visit the food handling area.

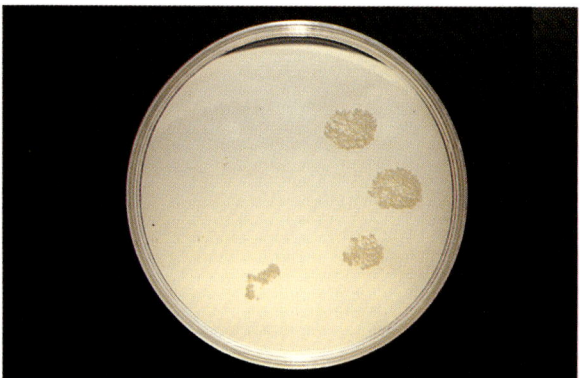

316

317

316 & 317. **Demonstration of correct handwashing technique.** In addition to their use in general hygiene training, nutrient agar culture plates can be very useful emphasising the need for correct handwashing. **316** shows the imprint of visually clean, but unwashed fingers, after holding a dirty cloth, while **317** shows the imprint of the same fingers after washing in soap and water.

It is a management responsibility to monitor hand washing and also to provide and maintain suitably equipped and situated facilities. It is not uncommon to find hand washing facilities, particularly in washrooms, poorly maintained even in generally well-managed factories. Maintaining standards can certainly be a problem where facilities are used by a large number of people in a short period, and management, with their own more salubrious facilities, are often totally unaware of conditions in washrooms.

Washrooms and toilets are of importance not only because of obvious physical requirements, but also to emphasise the importance of the food-production area (Craig, 1988). Cleaning and maintenance should be according to a properly planned schedule that reflects the patterns of heavy use. Monitoring that washrooms and toilets are clean and fully equipped with soap, towels etc, should be the responsibility of supervisors or line management, and incorporated in individual job descriptions. Those responsible for the condition of washrooms should also have the authority to order additional or remedial cleaning. The workforce should be encouraged to report shortages or defects. This often has the additional advantage of reducing vandalism.

Use of antiseptic soaps and creams

Soaps containing antiseptics are sometimes recommended for use by food handlers. It should be emphasised that physical removal of transient bacteria by soap and water is the basis for effective hand washing. Antiseptic soaps have a secondary role, but may distract attention from correct hand washing procedures.

Antiseptic creams or tinctures may be used in addition to washing, and although relatively expensive are more effective than antiseptic soaps. When used in restricted situations such as 'clean room' packing of sensitive foods, their value may largely lie in psychologically reinforcing the commitment to hygiene, but care must again be taken that their use does not detract from the need for effective hand washing.

Hypochlorite hand dips containing up to 50 mg/l available chlorine are sometimes used in processing plants, but there is little evidence they are effective. Problems may arise from their irritant action on the skin.

Hand care

In addition to hand washing, a positive approach should be adopted to overall care of hands. Nails should be kept clean and short, although with younger employees the potential conflict between fashion and hygiene may require careful handling. The handler should be encouraged to be actively concerned with the condition of the skin, and seek advice immediately any problem becomes apparent. Skin disorders may not be the result of infection, but may also arise from exposure to food or its ingredients, detergents and other materials, and the advice of a dermatologist should be sought where necessary.

Protective clothing

Provision of suitable protective clothing is an important aspect of hygienic practice. The basic requirements are simple (**Table 154**), but not always easily fulfilled. Major problems lie in two areas, first in design and second in laundering and cleanliness.

Table 154. Basic requirements of protective clothing.

1 To provide a suitable level of protection for the wearer.
2 To provide a suitable level of protection for the food.
3 Comfortable to wear and of acceptable appearance.
4 Are coloured such that contamination is easily seen.
5 Are easily cleaned and retain their essential properties through a large number of washes.

The basic problem in design is that clothing that is effective in protecting both the person and the food is uncomfortable to wear, while clothing which is comfortable to wear is ineffective in offering protection. This need not be an insuperable problem, but requires some thought, consultation with those who will actually have to wear the clothing, and willingness to spend more than the bare minimum. At all times, function must take precedence over fashion. Particular difficulties in this respect may occur in the case of clothing for people such as those who handle foods in public view. Here concern for appearance and corporate identity may result in a garb which is not only of little use with respect to hygiene, but is uncomfortable and contrives to make the wearer look ludicrous.

Problems concerning laundering and cleanliness probably stem largely from established custom and practice. Protective clothing inevitably becomes soiled and regular changing is required for both practical and aesthetic reasons. In this context, it is noteworthy that the practice in the USA is to change clothing daily, or at most every two days, while in the UK, reportedly the worst of the industrialised countries, change is made at weekly, or even longer intervals (Anon, 1987b). Alarmingly, a survey in the UK in 1988 showed that only 40% of employees considered their clothing to be usually clean, and 25% found it to be usually dirty or very dirty (Anon, 1988b).

Problems of cleanliness may be compounded by unsatisfactory laundering. In over 50% of UK food manufacturers, workers are expected to wash their own protective clothing (Anon, 1987b). This cannot be considered good practice and results in a wide variation in the standard of laundry. Furthermore, the use of low-temperature wash cycles permit the survival of micro-organisms and contamination from soiled children's clothing etc must be a possibility (Anon, 1987b). However, the apparent low priority given to the provision of clean clothing in the UK means that laundry is more likely to be regular if responsibility lies with the employee rather than the employer.

Footwear

The nature of many food-processing operations dictates the wearing of wellingtons or other protective footwear. The potential for the introduction of pathogens either to food via the hands or to the food-processing environment exists, but is often overlooked. Cleaning and sanitising footwear worn in sensitive areas of the plant should be part of the responsibilities of the individual, using facilities provided by the management. Footwear for use in sensitive areas should not leave the factory premises, and this ruling is obviously easier to enforce where they are provided by the employer.

Smoking, chewing and eating

These practices are not only aesthetically unacceptable, but carry a significant risk of transferring micro-organisms from mouth to food via the hand. Some individuals will persist in smoking at all costs, in which case disciplinary action, if necessary culmi-nating in dismissal, must be applied. Large-scale covert, or sometimes overt, smoking is evidence of a breakdown of discipline, and often results from serious weaknesses in supervision and management, or from widespread employee dissatisfaction. Whatever the reason, action is required to solve underlying problems.

Smoking regulations are enforced most easily when they are applied throughout the plant, including office areas. However, it is necessary to provide an area where the truly addicted may smoke during breaks.

Safe food handling in the context of plant operations

The role of the food handler in ensuring the safe handling of food cannot be treated in isolation, but must be seen in the context of plant operations. While there is an obvious and recognised need for high productivity, it is essential to ensure that the demands of production do not contradict good hygienic practice. For example, there is little point in inculcating the need for good personal hygiene if insufficient time is allowed for adequate hand washing. This is of particular importance with respect to the invidious and Dickensian practice of timing visits to the toilet. In a more general sense, it is difficult to maintain good practice where the workforce is over-dependent on bonus earnings, or where overtime working is at an exceptionally high level. Financial pressures may also apply in a more subtle manner. For example, profit-sharing schemes may lead to individual initiatives to increase profit at the expense of good practice. Despite their undoubted benefits, profit sharing and other incentive schemes do encourage greed, a vice which is usually incompatible with safe food handling.

Buildings and equipment

Design and construction of buildings

'Buildings should be located, designed, constructed, adapted and maintained to suit the operations carried out in them and to facilitate the protection of materials and products from contamination or deterioration' (Anon, 1987c).

Location

The dominant factors dictating location tend to be commercial. However, factors relating to food processing and hygiene should not be overlooked, the most obvious being sources of environmental contamination. Consideration at the outset must be given to the availability of services, including power and water supplies and drainage. These are specialist matters and advice should be obtained from suitably qualified people. The availability of a service may be particularly difficult to guarantee in remote areas, and product safety at a Norwegian frozen food plant was compromised by regular night time power interruptions. In such cases, stand-by facilities should be available.

The environment is an important source of pathogenic micro-organisms and proposed locations should be surveyed for potential hazards (Hayes, 1985). Account should be taken of the seasonal nature of agricultural activities such as manure spreading. Careful consideration should be given to site drainage and the proximity of watercourses, since flooding is a particular hazard. Seasonal variation should again be considered, the dry and innocuous ditch in summer can become a major source of contamination in winter. *S. dublin* is believed to have entered a cooked meat packing plant via contaminated footwear following flooding of an unofficial football pitch by a long defunct canal, the course of which could hardly be discerned in dry months.

Design

The requirements for safe food handling must be incorporated into every stage of plant design. This includes the physical construction of the plant, the necessary services and the layout. Layout is of particular importance in the segregation of 'clean' and 'unclean' operations, and it is necessary to decide whether physical separation is required or whether separation by distance is acceptable. Complete physical separation is required between raw and cooked products and this must be reflected in the design requirements.

Although design should allow for good hygienic practice in all parts of the plant, it must be appreciated that 'clean' and 'unclean' are not absolute terms, but that meanings vary according to the food handling context. Thus conditions perfectly acceptable in a slaughterhouse would be totally unacceptable in an aseptic filling room. It is counterproductive, therefore, to attempt to apply the same standards throughout the plant. The level of 'cleanliness' required in each area should be determined at the design stage, particular attention being paid to the provision of suitable facilities for particularly sensitive operations.

Construction

Building construction involves many different techniques and materials. These are the responsibility of the architect and the engineer and beyond the scope of the present discussion. However, the construction must meet several hygiene-related requirements.

Protection of ingredients and foods from external contamination

A basic function of the factory building is to protect stored ingredients and foods from external contamination. Such contamination may be introduced by birds, insects and animals such as rodents or by flooding and seepage or via dust. Precautions are necessary in the construction of both the factory buildings and the surrounding yards and roadways. There are often many ways by which rodents can enter buildings (Katsuyama and Strachan, 1980), and specific preventive measures must be taken. Flying insects can be adequately controlled by electric knock-down devices but ants, which are increasingly recognised as vectors of pathogenic micro-organisms, can be a major problem particularly in older buildings with cracked or shallow foundations. Air intakes for ventilation systems should be situated away from obvious sources of contamination, and air should be filtered where necessary.

417

Protection of foods and ingredients from sources of contamination within the building

Colonisation of the building itself by pathogenic micro-organisms such as *Listeria* is increasingly recognised to be a major hazard, while buildings may also harbour vermin, birds and insects. Construction should be such as to minimise these hazards and also to provide protection from contamination arising from activities within the building.

Construction of floors, walls and other surfaces.
As a general rule, surfaces should be constructed both to minimise the build-up of contamination and to permit easy cleaning. Floor and wall surfaces should be constructed with waterproof, impervious and washable material and may usefully be placed in one of three categories (**Table 155**) with respect to requirements for hygiene.

Particular problems with rodents and insects may result from build-up of food material in inaccessible areas. In new buildings, the creation of such areas should be avoided at the design stage, but in existing or converted buildings all voids must be located and sealed.

Table 155. Classification of surfaces within food handling plants (ICMSF, 1989).

1 Surfaces of materials that are intended to come into contact with food such as silos and storage bins: the risk of contamination is **high**.

2 Surfaces of materials such as walls which are not intended to come into contact with foods but which may do so by accident: the risk of contamination is **low**.

3 Surfaces of materials such as floors and ceilings that are not intended to come into contact with foods. The prime concern with these surfaces involves aesthetics and personnel safety, but they need to be cleanable and kept clean. The risk of contamination is usually **low** but can be **high** where bad practice such as the placing of food close to, or even on, the floor is permitted.

Water supply.
Water is of fundamental importance in the food industry, not only as an ingredient, for hand washing and cleaning, but in process operations such as heating, cooling and steam raising. In the UK, approximately 40% of water used is piped mains, the remainder being from other sources such as boreholes (Jacob, 1988). Potable water must always be used for recipe formulation, direct product contact and some cleaning and, where necessary, non-mains water must be suitably treated.

The water distribution system in the plant must be designed to avoid any possibility of contamination. Contamination of water tanks by bird droppings or other sources has led to outbreaks of gastroenteritis and where possible the use of storage tanks should be minimised. Such tanks, however, must be covered and protected from contamination. Fluctuating water pressures in different parts of large distribution systems may lead to back siphonage from effluent lines into the potable water system, and so systems must be constructed so that sufficient air breaks are present to avoid this occurring.

Potable and non-potable water should be distributed in separate colour-coded pipes with no cross-connections.

Drainage.
The drainage system of food plants should be of sufficient capacity to meet maximum demand. Floors should be sloped towards drains, open gulleys should be avoided. The system should be suitably vented to permit efficient operation, and traps should be employed to eliminate odours. Overall drainage flow should be away from 'clean' areas to minimise the consequences of blockages downstream of 'dirty' areas.

Ventilation and air conditioning.
Although, in most cases, the significance of airborne micro-organisms is overestimated, there are important exceptions. It is therefore necessary to prevent natural or mechanical airflow acting as a pathway from 'dirty' to 'clean' areas. Ventilation systems must be planned and operated in accordance with plant layout and processes (**Table 156**). Plants producing foods with a fine powder component, such as dried milk, pose particular problems, and particular care is needed to minimise airborne transfer of *Salmonella* (ICMSF, 1989).

Some operations such as aseptic filling require filtered air supplies, and flows may be directed to give further protection.

Air-conditioning systems may themselves be the source of micro-organisms, including respiratory pathogens such as *Legionella*. These pathogens are of no consequence with respect to food, but are of obvious concern to employees.

Table 156. Control measures: airborne contamination.

1	Eliminate, where possible, sources of contamination.
2	*Either:* Physically separate sources of contamination from the rest of the plant. *Or:* Arrange local ventilation at source of contamination to minimise spread.
3	Ensure air-flows (whether in open rooms or via ventilation ducting) do not permit the rapid transfer of contamination from one area to another. Arrange air-flows to protect sensitive operations.
4	Ensure incoming air does not carry contamination from external sources.

Maintenance

The maintenance of the building and its services is of considerable importance from the viewpoint of safe function as well as structural integrity. Maintenance requirements will obviously vary widely, but a three-fold approach is recommended. This should consist of unscheduled remedial work such as repair to damaged surfaces etc; planned short-cycle maintenance such as changing filters in ventilation systems which can be undertaken without significantly interrupting production; and planned long-cycle maintenance such as floor or wall resurfacing, which can only be undertaken during plant closures.

Equipment

Many factors will affect the choice of equipment, but most of these are a matter of commercial detail, and in the context of food safety the requirements for processing equipment is quite straightforward. In the specific case of equipment designed to eliminate, reduce or control microbial populations, it must be able to function effectively, while in the general case of all food handling equipment, it must be able to function without compromising the safety of food by introducing contaminants.

It is the responsibility of food technologists to ensure that equipment is suitable for its purpose, and shortcomings should be detected by control and monitoring systems. Despite this, problems have arisen in small companies due to the use of unsuitable types of heat exchanger, while inadequate performance by refrigeration equipment poses wider problems. Even in large companies, it is not uncommon to find cold stores over-filled, or stores that are essentially conservators being used to actually reduce temperatures. In retail stores, the performance of display cabinets is often grossly inadequate, with poor design being exacerbated by unsatisfactory practices such as overloading. In many cases, difficulties in maintaining correct temperatures have been compounded by the fitting of undershelf display lighting. It is also of note that in at least one major UK supermarket chain the responsibility for monitoring the temperature of display cabinets lies not with local management, but with engineers working from a central location who visit stores only occasionally. It is ironical that while check-out operators are entrusted with the operation of sophisticated laser scanners and computerised tills, management seem to be judged incapable of using a simple digital thermometer.

In the UK, a forthcoming Food Act will require display cabinets to maintain a temperature below 5°C when used for storage of ready meals, and some other sensitive products. This is an obvious advance, but there are likely to be problems of enforcement and, while the more responsible companies will apply the new standards to *all* products (Quarmby, 1989), in other cases substandard display cabinets will continue in use for products not covered by the Act.

Serious problems arise from the use in some catering establishments of domestic equipment, which simply cannot meet the demands of relatively large-scale production.

The ability of equipment to operate without introducing contaminants is dependent to a large extent on the cleanability of equipment, as well as on the application of correct cleaning schedules (see pages 422–423). It is unfortunate that cleanability has taken low priority in the design of equipment, and equally has often received little attention by the purchaser before operation. However, sanitary standards such as the 3-A sanitary standards of the International Association of Milk, Food and Environmental Sanitarians exist, and should be consulted as required.

Cleanability

Cleanability depends on a number of factors including overall equipment design, accessibility to contact surfaces and construction materials. Food contact surfaces should be smooth, continuous and non-porous to permit the easy removal of food residues and soil. The constructional material should be sufficiently robust and resistant to withstand normal use without becoming excessively roughened, scratched or corroded. Particular attention should be given to joins in material. Welds, for example, unless properly made may permit soils to become lodged, while precautions must also be taken to 'seal' the joint between different materials such as metal and plastic.

The above factors, together with cost, limit the range of materials suitable for the construction of equipment for food handling. Stainless steel is the material of choice in many applications, and is likely to remain so despite increased use of plastics. Glass and rubber also find specialised applications. Wood cannot be considered suitable for professional applications.

The level of polish of stainless steel is important and is usually specified in food equipment standards. However, microscopic examination of apparently smooth surfaces may reveal cracks, peaks and valleys, which trap soils and micro-organisms. Electropolishing is superior to conventional finishing (Freeman, 1988), especially when used for pipelines and other surfaces cleaned by cleaning-in-place procedures.

Maintenance of equipment

Effective maintenance is an essential part of the control of safety. An overall approach similar to building maintenance (page 419) should be adopted, with the overall philosophy being one of prevention. With specialist equipment, inspection by the manufacturer may be required.

During maintenance, attention must be paid not only to the physical function of equipment components such as pumps, but to the condition of food contact surfaces and the correct operation of control equipment and recording devices.

Where there are repairs to the fabric of the equipment, these should use either the same material as the original or a compatible material with the same properties and finished to the same standard. When components are replaced, the new part should be approved for use. Problems can arise from the substitution of non-approved components. In dairies, for example, there has been a tendency in recent years to replace ring-stem valves with the cheaper butterfly type. The latter type are not suitable for use in cleaning-in-place systems since daily disassembly is required (Karpinsky and Bradley, 1988).

Temporary repairs should be avoided and, if necessary, the nature of the defect and the repair effected must be assessed by a qualified person to ensure that product safety is not compromised. The nature of the repair should be noted and a maximum time stipulated before a permanent repair is made. 'Bodges' and 'string' engineering should be avoided at all times.

Cleaning and sanitisation

Cleaning and sanitisation of equipment and the food handling environment is an essential part of safe food handling. Cleaning schedules (**Table 157**) should be drawn up by qualified and experienced people, and each should take account of the nature of the equipment to be cleaned, the nature of the soil to be removed, the development of biofilms and any particular feature of the food or process which may effect cleaning efficiency.

Table 157. Basic requirements for cleaning schedules (Fuller, 1989).

1 What is to be cleaned?
2 Who is to clean it?
3 When is it to be cleaned?
4 How is it to be cleaned?
5 Which materials are to be used?
6 What precautions are to be taken?
7 What protective clothing is to be worn?
8 Who will monitor the operation?
9 How is the operation to be monitored?

Deposition and accumulation of soil

A number of factors affect deposition and accumulation of soil (**Table 158**), only some of which such as surface finish, porosity and hardness can be controlled.

Table 158. Factors affecting the deposition and accumulation of soil on food contact surfaces (Jennings, 1965; Dunsmore et al., 1981).

Substrate (nature of surface)
Surface finish
Porosity
Hardness
Wettability by liquid soil
Geometry
Reactivity with soil
Soil
Particle size
Viscosity
Surface tension
Mutual solubility of components
Reactivity with substrate

Long-term accumulation of soil is not (or certainly *should* not be) a continuous process, since most is removed by cleaning at the end of each production cycle. However, no routine system is completely effective and long-term accumulation of soil is inevitable. This may be overcome by using a more energetic periodic clean (Dunsmore *et al.* 1981).

Development of biofilms

In the past, micro-organisms on food contact surfaces were considered to be either associated with food material or physically entrapped in macroscopic or microscopic crevices etc. However, there has been increased awareness of the ability of foodborne bacteria, including pathogens, to adhere to food contact surfaces. Extensive colonisation and formation of biofilms may occur and act as a continuous focus of contamination. Colonisation has been recorded on both stainless steel (Lewis and Gilmour, 1987; Lewis *et al.*, 1987; Herald and Zottola, 1988a,b) and rubber (Lewis and Gilmour, 1987; Czechowski, 1988). Colonisation by *Yersinia enterocolitica* (Herald and Zottola, 1988a), *Listeria monocytogenes* (Herald and Zottola, 1988b) and *Staphylococcus aureus* (Dodd, 1988; Dodd *et al.*, 1988), has been observed, but the phenomenon seems unlikely to be restricted to these organisms.

Nature of cleaning systems

A cleaning system has three main components (Swartling, 1959):

1 Removal of food residues that serve as nutrients for bacteria remaining on the surface.
2 Destruction of any bacteria not killed or physically removed from the surface with food residues.
3 Storage of equipment under conditions which discourage or prevent growth of surviving micro-organisms during non-use periods.

From the toxicological viewpoint, a fourth component, removal of any residues which may contaminate the product, is desirable (Dunsmore *et al.*, 1981).

There is continuing discussion concerning the relative roles and importance of the detergent (removal) and sanitiser (killing) components. It is generally agreed that detergency is quantitatively of greatest importance (Dunsmore, 1980; Tebbutt, 1984), with up to 99.8% of both soil and bacteria being removed by cleaning without a sanitiser. However, it should be appreciated that to some extent the distinction between cleaning and sanitisation is arbitrary since all components of a cleaning system have, to a greater or lesser extent, roles both as cleaners and sanitisers.

The choice of both detergent and sanitisers depends on a number of factors (**Table 159**) of which the nature of the soil is likely to be most important. Commonly used detergents are summarised in **Table 160** and sanitisers in **Table 161**.

Although chemical sanitisers are very widely used, the use of heat should not be overlooked and, where application is possible, is often regarded as first choice.

Table 159. Factors affecting choice of detergents and sanitisers.

Primary
    Nature of soil
    Material to be cleaned
    Product requirement

Secondary
    Cost
    Availability of resources

Table 160. Choice of detergent for various soils (Elliot, 1980).

Soil	Detergent
High protein (meat, poultry, fish)	Chlorinated alkaline
High fat (fat meat, butter, oils)	Mildly alkaline (strongly alkaline if required)
High starch (fruits and vegetables)	Mildly alkaline
Sugars, organic acids, salt	Mildly alkaline
Stone-forming foods (milk products, beer)	Two types used in 5-day cycle: chlorinated or mildly alkaline detergent for 4 days; acid cleaner on day 5
Heat-precipitated water hardness	Acid cleaner

421

Table 161. Main types of chemical sanitisers.

Chlorine and chlorine compounds
Very widely used. Powerful biocides which are cheap and easy to use, but are rapidly inactivated by organic matter and are corrosive. Sodium hypochlorite is the most widely used inorganic compound. Organic chlorine compounds are stable as a powder and have a longer lasting activity. Chlorine gas only used for the treatment of process water.

Iodophors
Broad spectrum of activity. Most effective at low pH values. Relatively unaffected by organic matter at pH <4.0. Non-corrosive but may be adsorbed by rubber. Expensive and used mainly for cleaning-in-place of dairy plant.

Quaternary ammonium compounds
Most effective against Gram-positive micro-organisms and in slightly alkaline conditions. Activity limited below pH 5.0 and in the presence of anionic and some non-ionic detergents. Not significantly affected by organic matter but inactivated by some plastics. Non-corrosive and long lasting but may damage rubber. Relatively expensive but widely used for the disinfection of open surfaces.

Amphoteric compounds
Not widely used due to expense and foaming. Not affected by organic matter or water hardness, but inactivated by soap and anionic or non-ionic detergents.

Polymeric biguanides
Similar properties to quaternary ammonium compounds but greater activity against Gram-negative bacteria. Not compatible with anionic detergents. Unaffected by hard water but inactivated by milk residues.

Peroxygen compounds
Employed as hydrogen peroxide or peracetic acid. Very effective against micro-organisms including endospores. Peracetic acid is highly corrosive. Widely used to sterilise packaging used in aseptic filling and some types of aseptic filling equipment.

Application of cleaning systems

Simple contact with detergent and sanitiser solutions is not sufficient to remove most types of soil and some physical means of assisting removal is required. The simple scrubbing brush still finds application, but in large plants more sophisticated methods involving such means as the impact of sprays on surfaces or turbulent flow in a pipeline are now used widely. Suitable wet cleaning methods are cleaning-in-place, high pressure low volume, low pressure high volume and foam/gel. Each of these methods, and manual cleaning, has application according to the type of equipment and nature of soil (ICMSF, 1989). Cleaning-in-place is very effective for closed pipeline systems, and suitable for a high level of automation, but requires careful design and performance must be verified in each individual application. Outbreaks of salmonellosis have resulted from cross-contamination from raw milk to pasteurised via cleaning-in-place operations, and for this reason there must be no common circuitry, cleaning solution tanks etc.

The intercycle period when equipment is out of use is frequently overlooked when planning cleaning schedules, but is of particular importance with respect to the establishment of biofilms. Properly designed equipment which is self-draining goes some way to controlling the problem, but where feasible optimal results are obtained by leaving equipment filled with a solution of a long-acting sanitiser, or by leaving a sanitiser solution on an open surface. However, such procedures do require a rinse to prevent contamination of food with sanitiser residues, as do alternative procedures such as post-intercycle, pre-use sanitising rinses.

Frequency of cleaning is usually dictated by the nature of the food and the process. Continuous processes such as milk pasteurisation must usually depend on a thorough clean at the end of each production run, while in other cases, such as pre-packing cooked meats, cleaning can be at intervals through the day. A common procedure is 'interim cleaning' after each work period or, where appropriate, on change of product followed by a more rigorous clean at the end of each shift. In many situations, the frequency and efficiency of cleaning is more important than the use of a sanitiser (Gilbert and Maurer, 1968).

Use of combined detergent-sterilisers

Detergent-sterilisers are compatible combinations of a detergent with a separate sanitiser, and are available in a range of formulations to meet different requirements. The efficiency of detergent-sterilisers is lower than that of systems using separate components, but despite this have become popular particularly where soil is light and there is a need to avoid high temperature cleaning. A major reason for their popularity, is that the time taken to clean is less and a lower level of operator training required.

Dry cleaning

Although wet cleaning is necessary where many perishable foods including milk and meat products are handled, dry cleaning achieves better micro-biological control in plants processing dry foods and ingredients. Removal of soil is entirely by physical methods, and sanitising agents are not normally required.

Cleaning of floors and walls

Although the above discussion is mainly concerned with cleaning of equipment, similar principles apply to floors, walls and, on occasion, ceilings. Laid down procedures must be followed and facilities should be available at all times to deal with accidental spillages.

Care must be taken not to contaminate food contact surfaces with material from floors, drains etc. For this reason, vacuum-cleaning techniques are preferred and high-pressure hoses should be avoided. Cleaning primarily requires soil removal, and sanitisers are considered unnecessary for general applications even in hospitals (Watt, 1988). However, in view of concern over pathogens such as *L. monocytogenes* becoming established in the food handling environment, the use of a sanitiser as an *adjunct* to efficient soil removal is worthwhile in plants handling sensitive foods.

The periodic clean

The use of an occasional, more rigorous clean to remove accumulated soil is a well-established principle in all cleaning systems. The use of a sanitiser is essential to destroy any micro-organism that had previously been protected. The periodic clean may involve one of three basic measures (**Table 162**) or a combination of measures.

Table 162. Summary of periodic cleaning measures.

1 Cyclic use of a complementary cleaning agent.
2 Cleaning solutions of higher energy (usually more powerful detergents, higher detergent concentrations or higher application temperatures).
3 Increased mechanical removal (a common example is the periodic manual cleaning of plant usually cleaned by cleaning-in-place methods).

The emergency clean

It is always necessary to have facilities available to deal with contamination in an emergency, such as a food handler vomiting in the workplace. This is a particular hazard associated with infection with small, round, structured viruses and can result in a wide area being contaminated. The exact nature of emergencies cannot, of course, be predicted, but generalised guidelines should be available covering matters such as obtaining technical advice, stopping of production and quarantine of materials.

Health of cleaning staff

The same health regulations are necessary for people cleaning food contact surfaces as for food handlers. Particular care must be taken with people employed by outside contractors.

Sanitation of engineering materials

A minor risk to food safety is presented by materials such as conveyer lubricants and cooling fluids, which may be contaminated by pathogens. For example, inoculation experiments have shown a potential for colonisation of conveyer lubricants by *L. monocytogenes* unless a fast-acting biocide such as glutaraldehyde is incorporated (Rossmore, 1988). *Salmonella* has been isolated from a sweetwater cooling system and 'other potential' pathogens were detected in both sweetwater and glycol systems (Overdahl and Zottola, 1988). *L. monocytogenes* was also able to persist in a simulated milk cooling system and may be a hazard especially in sweetwater systems that have been contaminated with a small quantity of milk (Petron and Zottola, 1988).

The hazard presented by engineering materials should be assessed during preparation of the HACCP audit and control measures should be devised. Where sanitisation is required, procedures should be laid down as part of the overall operating schedule.

Monitoring and verification

Monitoring of cleaning is usually by inspection of cleaned equipment to ensure that it looks, feels and smells clean. The inspecting person must be highly experienced and be able to place in correct perspective the impact of cleaning and disinfection on the safety and quality of the food being processed (ICMSF, 1989). Even under these conditions, apparent cleanliness can be misleading and it is necessary to make microbiological examinations. The time required to obtain the results make these unsuitable for monitoring, but such tests can be useful in verification. Ideally, selective microbiological examinations should be used to support the results of inspections (Tebbutt and Southwell, 1989).

Several approaches may be taken to microbiological sampling including the use of swabs and various types of contact plates and slides (**Figures 318 & 319**) to sample surfaces, environmental samples and end-product testing of the food being produced. A further method which is useful with complex equipment is flow-sheet sampling, which consists of determining microbial numbers in food samples collected after each step in the production sequence (ICMSF, 1989).

Critical analysis of the standards of sanitation often reveal grave defects. There are several possible immediate causes (**Figures 320–325**) but major underlying problems result from failure to allow sufficient time for cleaning – a major potential problem particularly with multi-shift working, and difficulties with recruitment, training and the motivation of staff. Problems of staffing may themselves stem from the failure of management to appreciate the importance of cleaning, and to allocate it the priority and resources required.

In an increasing number of cases, cleaning is contracted out to 'specialist' cleaning companies. The standard of cleaning provided by these contractors varies considerable from the very good to the appalling. Too often the choice of contractor is made solely on the grounds of cheapness, and no consideration is made of likely performance. It is considered that the use of in-house personnel is always preferable due to the far greater degree of control available. Contract cleaners should be a reputable company who have a proven record and who are members of a genuine trade association. A detailed specification should be agreed including materials used, methods and the standards to be attained. Monitoring and verification must remain the responsibility of the food manufacturer.

318

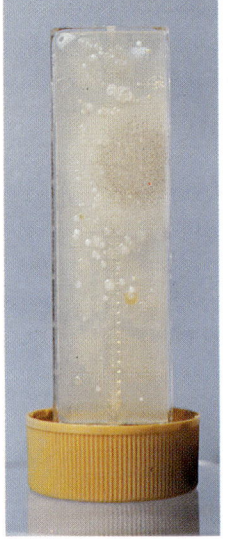

319

318 & 319. Surface sampling for microbiological contamination. A large number of methods are used for sampling surfaces. Contact slides (**318**), which consist of a thin film of agar on a plastic carrier are easy to use, and are available with several selective media in addition to the non-selective nutrient agar (Roche©) illustrated. An alternative approach is that of the Biomet© microstrips (**319**), in which a test pad containing dehydrated nutrients is bonded to a flexible plastic strip. After incubation, recognition of colonies is facilitated by the incorporation of a tetrazolium salt which is reduced by many bacteria to produce red colonies. The area sampled is smaller than with a contact slide, but the flexible nature enables inaccessible areas to be sampled more easily.

320

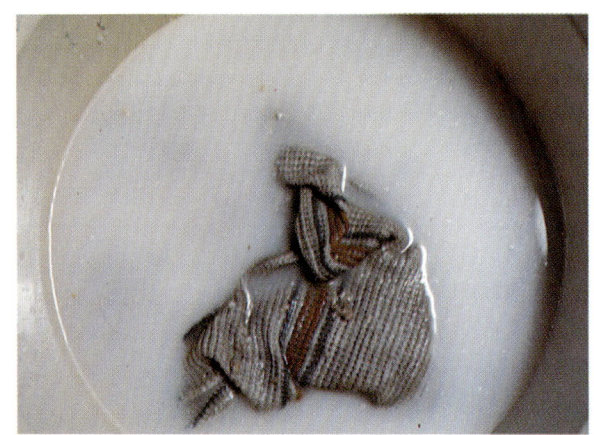

320–322. Poor cleaning practice in small scale catering. The standard of cleaning is often particularly poor in small scale catering operations where staff are untrained and unsuitable domestic materials employed. In the example illustrated, cleaning of a 'dirty' nylon cutting board (**320**) involved nothing more than a wiping with a disinfectant-soaked cloth. The disinfectant solution had been made up at the start of cleaning and was heavily contaminated with milk and other organic matter (**321**). Although wiping with the cloth removed visible dirt from the cutting board (**322**) the surface was left wet and there was effectively no sanitisation. The open weave cloth used entrapped food particles and was itself a potential source of micro-organisms, especially as surfaces used for handling cooked foods were cleaned after those used for handling uncooked. Examination of the 'cleaned' cutting board using a contact slide inevitably confirmed the totally ineffective nature of the 'cleaning'.

322

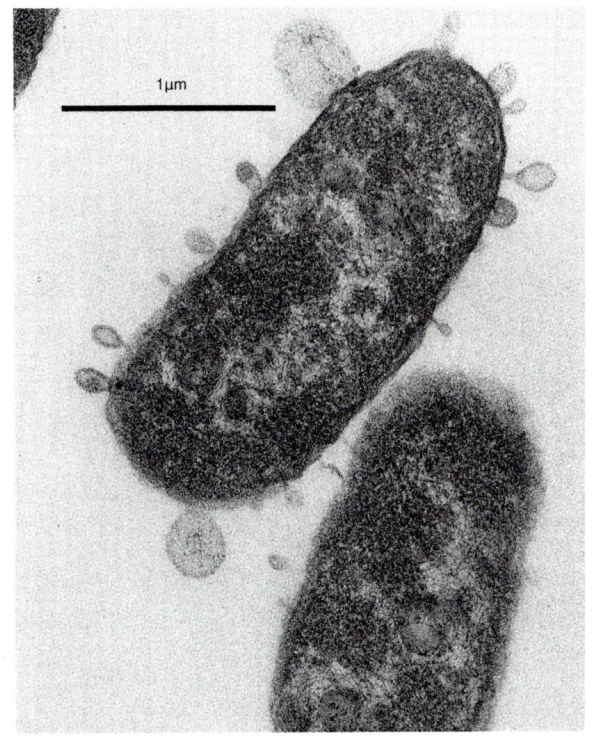

323

323. Effect of disinfectants on micro-organisms. The transmission electron micrograph illustrates the effect of an amphoteric disinfectant at a concentration of 10 mg/l on *Pseudomonas aeruginosa*. There is evidence of loss of integrity of the cell outer membrane and, at a higher concentration, complete lysis occurs. *Ps. aeruginosa* is highly resistant to disinfectants and can withstand concentrations which are effective against most other bacteria. Disinfectants are, however, relatively expensive and attempts to economise by making up under-strength solutions can lead to the survival of pathogens. (Reproduced by courtesy of Dr M.V. Jones, Unilever Research, Colworth House.)

324 & 325. Entrapment of soil in surface damage. Problems with cleaning can result from the condition of surfaces. Visual inspection of a cutting board after cleaning indicated a generally satisfactory appearance, but concern over a deep cut (arrowed, **324**) resulted in targeted microbiological sampling using Biomet© microstrips. A focus of contamination could clearly be seen after incubation (**325**).

324

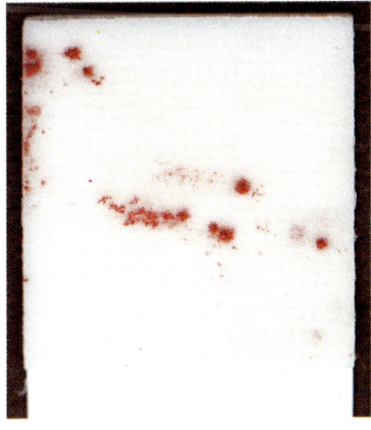

325

21 Control of pathogenic micro-organisms in meat

Control at farm level

Animals and birds utilised for meat are subject to infection by a wide range of micro-organisms, including some such as *Salmonella* and *Campylobacter*, which are important pathogens of man. With reason, recent publicity has focused on the role of modern intensive agricultural practices in the spread of foodborne infection, and there is no doubt that many current practices are undesirable. However, it must be appreciated that zoonotic disease is neither new nor associated exclusively with intensive agriculture: tuberculosis and brucellosis were endemic in the UK national herd which was managed almost entirely using traditional practices.

Farm animals may be infected via a number of routes (**Flow diagram 39**). The relative importance of these varies from species to species. Within a species, there is also considerable variation related to differences in agricultural practice. From the viewpoint of food safety, every stage in the rearing of an animal, from birth to arrival at the abattoir,

comprises a critical control point at which spread of infection should be controlled by attention to good practice. It should be stressed that there is no conflict between requirements for food safety and *good* husbandry, and that both producer and consumer benefit from the observation of basic precautions.

Animal feeds

Animal feeds are an obvious source of infection and many *Salmonella* serovars currently prevalent in the UK were derived from imported animal protein sources such as fishmeal. The practice of 'recycling' waste protein, such as feathers and animal parts which cannot be utilised as human food as feed for food animals is undoubtably at least partly responsible for the continuing high level of *Salmonella* infections in poultry and other farm animals. The young chick is particularly susceptible and can be infected with *Salmonella* within a few hours of hatching (Xu *et al.*, 1988). The scale of the problem could be very much reduced by ensuring all protein feed is adequately heat-treated and, indeed, in the UK, legislation is already available to make such a treatment mandatory. A further possibility with poultry feed is treatment with formic acid (Hinton and Linton, 1988; Humphrey and Lanning, 1988). However, it must be appreciated that the benefits of feed treatment are not fully sustained up to slaughter because of the acquisition of infection from other sources.

Incorporation of scrapie-infected sheep brain and nervous tissue into cattle food is believed to have been responsible for the appearance of bovine spongiform encephalopathy (see page 363).

Infection from feed is not restricted to processed protein foods. *Listeria monocytogenes* infections of ruminants, for example, have been associated with poorly made silage, especially that of the 'big bale' type, and there is also evidence of infection from naturally contaminated forage. Pasture may also be contaminated as a result of fertilisation with animal manure or sewage sludge, by irrigation with contaminated water or by flooding. The importance of infection by this route is difficult to assess. Feed also plays an important secondary role in other routes of infection (see below).

Flow diagram 39. Pathways of infection of farm animals.

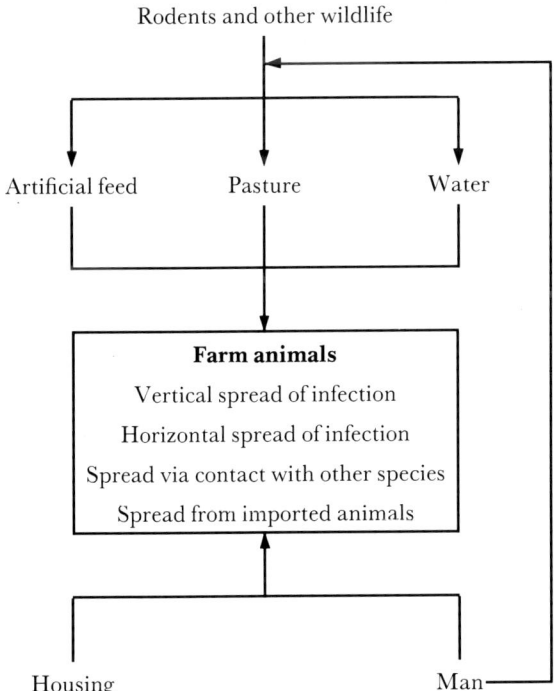

427

Water

Like feed, water plays an important secondary role in infection due to other primary sources. Water from contaminated supplies such as ditches and streams can also be a primary source of contamination, and may be of particular importance in infection of adult cattle with *Aeromonas* (Gray and Stickler, 1989) and *Campylobacter*. A long-term study is also being undertaken to reduce the incidence of *Campylobacter* in poultry based on the premise that the organism is introduced and maintained in water supplies. Initial results of a programme of cleaning and sanitising pipes and tanks is showing promising results (Skirrow, 1989). Water of potable quality should be supplied to all animals where possible, but it should be noted that there may be preference for water from natural, and contaminated sources.

Where borehole water is used, the quality should be monitored. Widespread *Salmonella* infection at a broiler rearing unit was attributed to contamination of a borehole from a remote source which was never positively identified.

Animal-to-animal spread of infection

Spread of infection from animal to animal is of particular importance when housing is intensive and hygiene poor (**Figures 326** & **327**). The faecal-oral route is of greatest importance, usually via litter, bedding, food or water. Spread of infection may be either vertical or horizontal, vertical spread often being of particular importance among poultry. Chickens may be infected before hatching either by trans-ovarian infection or by transmission through a porous or damaged egg.

It is particularly important to protect uninfected herds, or flocks, from infection from animals brought in from outside. The problem is well recognised in the poultry and parts of the pig industries where different stages of rearing are carried out at different sites. Precautions here should include quarantine of incoming stock and ensuring purchase is from a reputable source.

The more haphazard process of buying animals on the open market at auctions, stock sales etc, carries a potentially greater risk of introducing infection. Infections are readily acquired under the stressed and often unhygienic conditions of livestock markets, and there is usually no knowledge of the origin of the animals. Purchase of stock at open markets should be avoided but, if necessary, strict quarantine should be observed before contact with the herd is permitted.

Care must be taken to avoid the spread of infection between different species reared on the same farm. This requires strict segregation and good hygiene. Domestic cats and dogs have also been implicated as sources of infection, but such cases would appear to be rare.

326. Housing of pigs. Outdoor shelters, or 'arks', are often considered ideal for healthy pig rearing, offering protection from the elements while avoiding the opportunities for spread of infection offered by intensive indoor housing of both traditional and modern practice. Spread of infection between herd members is limited but disease may still be acquired from other sources.

327. Rearing of sheep. In many countries, sheep are housed indoors only under exceptional conditions. Sheep reared in remote regions often have a low incidence of infection with zoonotic disease. Elsewhere, however, the level of environmental contamination is such that levels of infection can be high, although often less than in cattle or pigs.

Infection by rodents

Rodents may be carriers of a number of zoonotic diseases and can be a source of infection of food animals. As far as possible, precautions should be taken to exclude rodents from animal housing and feed storage facilities.

Infection by other wild animals and birds

Infection of food animals can result from contamination of pasture or water supplies by the faeces of infected wild animals and birds. Evidence is usually circumstantial, but direct evidence exists for the role of seagulls in the transmission of salmonellas, and abortion in a cow was linked to *S. agona* infection in badgers (Humphrey and Bygrave, 1988). Grazing animals are considered to be at greatest risk, but precautions must also be taken against contamination of feed and water of all animals by birds.

Insects are probably involved in the transmission of disease among animals, but the relative importance is not known.

Infection by man

Where hygiene is poor, man can be important in the spread of infection among meat animals by passive transfer of contamination on hands, clothing and footwear. Occasional direct transmission from man to animals may occur, but such cases are rare and usually recorded for curiosity value.

Buildings

The fabric of buildings can become contaminated and serve as a vehicle of infection for succeeding batches of animals. Pathogens can persist for extended periods in contaminated buildings despite strenuous efforts at sanitisation. In one example, *S. typhimurium* persisted for several months in a premises of good construction despite depopulation, cleaning and disinfection (Twiddy *et al.*, 1988). In general, problems are considerably less in buildings constructed to modern standards.

Transport from farm to abattoir

Transportation to the abattoir can involve a major risk of infection both between animals in a given batch and from one batch to another. Risk should be minimised by strict attention to hygiene during transportation; vehicles, poultry crates etc should be cleaned and sanitised between different loads and the stress of the animals reduced to a minimum.

Competitive exclusion

Competitive exclusion (the Nurmi principle) was developed for the control of *Salmonella* in poultry (Nurmi and Rantala, 1973). The principle underlying competitive exclusion is to establish an adult gut microflora in chickens at the earliest opportunity, and thus provide the young bird with a degree of protection which is normally only available to the adult. This may be achieved by oral administration of either a saline suspension of the gut content of an adult *Salmonella*-free bird, or a culture of the major microorganisms present in such a suspension (Rantala and Nurmi, 1973). Although the effectiveness of this procedure has been confirmed on a number of occasions (Nurmi, 1985; Schleifer, 1985), applications have been limited. This has been due largely to concern that undefined material may transmit human or avian pathogens and, for this reason, more recent work has been concerned with the development of a defined gut flora treatment (Mead and Impey, 1986). To date, exclusion microflora developed are complex mixtures of obligate and facultatively anaerobic bacteria with up to 50 strains being required for optimal efficiency (Stavric *et al.*, 1985). This is probably due to interactive effects between the obligate and facultative strains (Goren *et al.*, 1984a).

While most defined exclusion microflora have been based on mixtures isolated from the gut of adult birds, an alternative approach was taken by Barrow *et al.* (1985), who searched for 'Salmonella-like organisms' which would occupy the ecological niches in the gut otherwise filled by *Salmonella* itself. A mixture of three strains of *Escherichia coli* provided some protection against *S. typhimurium*, but at a lower level that afforded by undefined gut cultures. To be effective in competitive exclusion, it was considered that a single strain must possess the colonisation characteristics of *Salmonella*, but not its virulence characteristics (Barrow *et al.*, 1987).

There is evidence that competitive exclusion is less effective in a commercial than in an experimental situation. This has been investigated by Impey *et al.* (1987) who, using the elegant feed challenge model of Hinton (1986), were able to suggest reasons for these discrepancies and to evaluate the practical implications. Results showed that protection of treated birds did not become effective until 24 or 48 hours after administration, and it was considered that to overcome this delay, protection should be given as early as possible, possibly by spraying the incubating eggs with protective microflora (Goren *et al.*, 1984b) rather than administering the protectant with drinking water. At the same time, the use of *Salmonella*-free feed would seem to be a prerequisite for the prevention of colonisation during the first 24 to 48 hours after contamination.

For unknown reasons, some birds do not acquire a protective microflora and are thus constantly susceptible to contamination (Impey *et al.*, 1987). Effective exclusion in commercial flocks therefore requires strict control of secondary challenges such as contaminated feed (Pivnick *et al.*, 1985; Impey *et al.*, 1987; Mead and Impey, 1986). The current inability to effectively control secondary challenges may be a limiting factor to the commercial application of competitive exclusion.

In its current form, competitive exclusion is applicable to the control of *Salmonella* only, and while the extension of the principle to other pathogens especially *Campylobacter* has been discussed, this may not be feasible.

There has recently been an upsurge of interest in the possibility of using immunisation to control host non-specific serovars of *Salmonella* in poultry (Mead and Barrow, 1990) but while there may be advantages over competitive exclusion, a considerable amount of further research is required before widescale application is possible.

Control in abattoir and packing plant

The slaughter and preliminary butchery of animals and birds inevitably involves contamination of the carcass. Contamination may occur at virtually all stages of handling, but major sources are the skins and feathers, especially if heavily contaminated with faeces, and the gastrointestinal tracts. Primary control is by good general hygiene, the physical or spatial separation of 'dirty' from 'clean' operations, and by good operator technique. In the latter context, it is noteworthy that while the superior facilities and conditions of the large slaughterhouse should, in theory, lead to improved standards of hygiene, in practice contamination is often less when animals are killed and dressed in a small-scale one-man, one-animal operation. Monitoring largely involves visual inspection and examination. Chilling of the dressed carcass, or of primal joints, is an important control point and time, temperature, airflow and relative humidity are all important factors in limiting microbial growth. Temperature and humidity should be measured and recorded on a continuous basis.

The scale and nature of operations in poultry packing plants leads to particular problems of cross-contamination. The process is summarised in **Flow diagram 40** and critical control points are listed in **Table 163**.

In most cases, the control of slaughter and butchery of meat animals and birds is limited to minimising contamination and controlling temperature. Various attempts have been made to positively reduce the numbers of micro-organisms on animal carcasses, including washing either with hot water or with various antimicrobial compounds including chlorine sprays (Kelly *et al.*, 1974), acetic acid and lactic acid (Snijders *et al.*, 1985). None of these has been fully successful, and thus washing cannot be regarded as a primary critical control point until effective and reliable procedures become commercially available (ICMSF, 1989).

Particular concern over levels of *Salmonella* and *Campylobacter* in poultry has led to the development of a large number of decontamination methods. Lactic and succinic acids have been used in various dip or spray treatments, and are sometimes combined with a freeze-thaw process (Stern *et al.*, 1985), while hydrogen peroxide has also been used in combination with lactic acid (Mulder *et al.*, 1987). Attempts to reduce contamination by raising (Humphrey *et al.*, 1981) or lowering (Okrend *et al.*, 1986) the pH value of the scald tank is effective in reducing the numbers of *Salmonella* and *Campylobacter* in the circulating water, but had no effect on the incidence of contamination on carcasses (Humphrey and Lanning, 1987).

The use of ionising radiation at doses of 3 Kgy has been proposed for several years to control *Salmonella* and other pathogens on poultry (Mossel, 1987), and is to be permitted in the UK. The safety of the process has been approved by expert panels (Anon, 1981; 1989), but considerable opposition remains among both the public (Webb and Lang, 1989) and among some food microbiologists. It is beyond the scope of this book to become involved in a detailed discussion of the merits or demerits of irradiation, but it is considered that to describe opposition to irradiation as 'emotionally rooted anxiety' (Mossel, 1989) is potentially misleading. Scientific opposition stems not from fears of induced radiation, but from concern that the use of irradiation is irrelevant to the real problems of hygiene and control facing the food industry and, furthermore, will encourage poor practice.

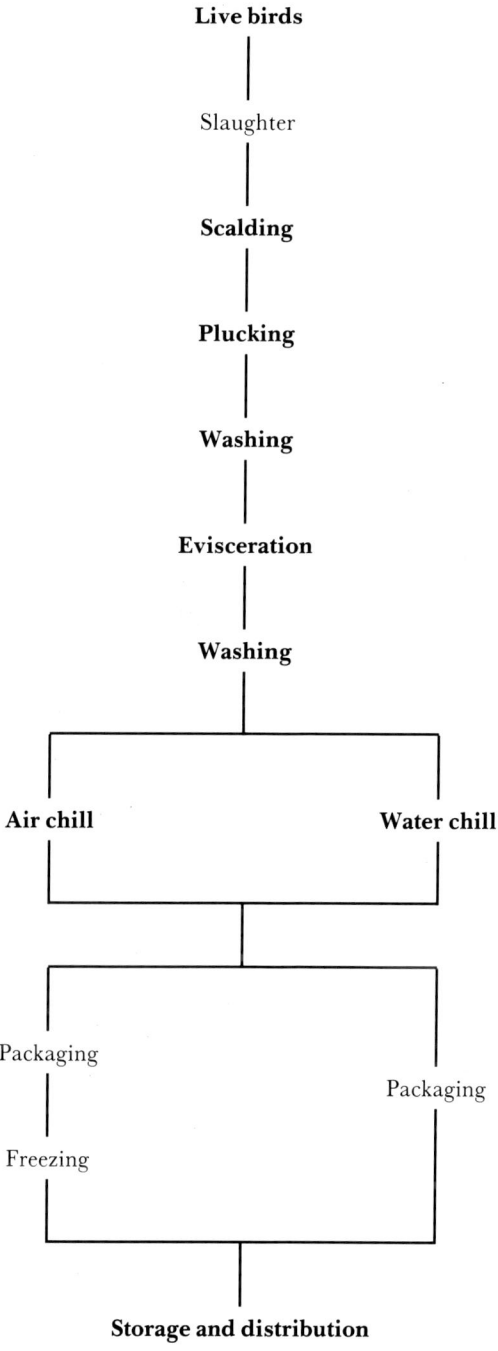

Flow diagram 40. Flow diagram: Poultry packing.

Notes:
1 Flow diagram is based on process for chickens.
2 Processes in **bold** are CCPs.

Table 163. Critical control points: poultry packing.

Live birds, CCP 2

Control

Achieved at farm level (see pages 427–430).

Monitoring

Ensure birds obtained from known and reputable supplier.

Scalding, CCP 2

Control

Limitation of spread of pathogens from bird to bird.

Build-up of pathogens in scald tank limited by maintaining correct temperature and by overflow and replacement of scald water.

Monitoring

Measurement of water temperature and overflow rate.

Comments

Birds for freezing are usually scalded at 60°C (hard scald) and those for chilling at 50°C (soft scald). Build-up of vegetative pathogens is a much greater problem when low-scald temperatures are used.

Plucking, CCP 2

Control

Limitation of spread of pathogens from bird to bird.

Spread is limited by applying rigorous cleaning procedures to the plucker fingers.

Monitoring

Ensure cleaning procedures correctly applied. Visual inspection of condition of plucker fingers.

Comments

Plucker fingers are readily colonised by *Staph. aureus* (see page 245) and this should be taken into account in the planning of cleaning procedures.

Verification may be made by the determination of *Staph. aureus* on plucker fingers (ICMSF, 1986).

Washing, CCP 2

Control

Removal of contaminating material including a proportion of surface microflora.

Monitoring

Ensure water pressure to washer is adequate and that spray nozzles are unblocked. Examine washed carcasses to ensure absence of soil.

Evisceration, CCP 2

Control

Prevention of contamination of birds by operatives.

Limitation of spread of pathogens from bird to bird.

Control requires use of skilled operators and correct application of technique.

Monitoring

Visual assessment of carcasses for contamination.

Where evisceration is manual, operators should be monitored to ensure techniques are correctly applied. The performance of mechanised evisceration machines should be monitored on a continuous basis and adjustments made as required.

Comments

Manual evisceration is increasingly being replaced by mechanical systems. Although this reduces the risk of contamination by the operator, the incidence of gut breakage is likely to be higher when evisceration is mechanical. The possibility of contamination with pathogens is therefore greater with mechanical systems.

Washing, CCP 2

Comments

General considerations of monitoring and control are the same as for the pre-evisceration washing. Special care is needed to ensure the body cavity is washed adequately.

Micro-organisms become more firmly attached to poultry skin with time and carcasses should be washed directly after evisceration.

Table 163. *Continued.*

Chilling, CCP 2 – Air
Control
Prevention of growth of pathogens.

Monitoring
Measurement of air temperature and flow rates.
Verification may be made by measurement of the deep muscle temperature at the end of the chilling process.

Chilling, CCP 2 – Water
Control
Prevention of growth of pathogens. Limitation of spread of pathogens from bird to bird and overall reduction in microbial contamination.
Counter-current immersion chilling is most effective and all water used should be chlorinated.

Monitoring
Measurement of water temperature, chlorine level and usage.
Verification may be made by measurement of the deep muscle temperature at the end of the chilling process.

Storage and distribution, CCP 2
Control
Maintenance of correct temperature.

Monitoring
Measurement of temperature at all stages.

Raw, uncured meat and poultry products

Raw meat and poultry and their products can inevitably contain pathogenic micro-organisms, as well as animal parasites such as tapeworms. None of the processing stages applied during the manufacture of meat products, such as fresh sausages and burgers, has any role in reducing the incidence of pathogens, while the additional handling and processes such as comminution may lead to the spread of contamination. In the UK, fresh sausages and burgers may contain sulphur dioxide as a preservative, while nitrite is added to some continental comminuted products as preservative, and to produce the distinctive appearance. (*NB* These products are not considered cured in the conventional sense.) The presence of these inhibitors has little, if any, effect on the persistence of pathogens, although where concentrations are sufficiently high the growth of pathogens will be prevented.

The use of vacuum- and controlled-atmosphere packaging for raw meats and raw meat products is for the extension of storage life and, despite the extravagant claims made by the representatives of some packaging companies, has no effect on the persistence of pathogens.

The safety of raw meat and raw meat products is ensured by cooking, although there is some conscious consumption of raw meat with its attendant risks (see also page 411). The physical size of the product can be an important factor in determining adequate cooking, and domestic ovens may be unable to cope with very large turkeys, particularly when stuffing is placed in the cavity. However, there are three particular areas of risk:

1 Undercooking of burger products, doner kebabs etc, in fast-food outlets at times of maximum demand. This problem may be exacerbated by the employment of inexperienced staff and, in the case of burgers, failure to defrost the product adequately.

2 Undercooking of burgers, sausages etc at barbecues. Large-scale barbecues are popular as social functions and in the gardens of inns during the summer months, and thus a large number of people can be at risk. Such cooked products may have a carbonised exterior and a barely warm interior. In parts of the USA, hog-bakes are popular and put people at risk not only through undercooking but through slow cooling (see page 319).

3 Inadequate or misleading instructions can lead to undercooking. For example, in the USA an outbreak of *E. coli* O157:H7 infection resulted from consumption of an undercooked burger product which had the *appearance* of having been fully cooked by the manufacturer, and was thus only warmed by the caterer preparing school meals. The nature of the product was not disclosed by the labelling instructions. Similarly, following the manufacturer's cooking instructions for a turkey burger with a cheese-sauce topping, which is retailed in the UK, results in the interior turkey meat being inadequately cooked. This fact is not appreciated until most of the burger has been consumed.

Raw cured meat products

Cured meats are prepared either by dry curing, in which the curing ingredients, NaCl, NaNO$_2$ and NaNO$_3$ are packed around the sides or joints or, more commonly, by brine curing, using a brine in which curing ingredients, together possibly with adjuncts such as polyphosphates, are dissolved. In the more traditional type of cure such as the Wiltshire, the meat is both injected with, and immersed in, brine but in more modern variants such as the sweetcure, brine is injected only, tumbling or massaging of the meat being employed to distribute the brine evenly. The Wiltshire process is outlined (**Flow diagram 41**) and critical control points summarised in **Table 164**.

The curing stage is a critical control point in the manufacture of all types of cured meat, although its importance with respect to safety varies according to the nature of the cured product.

Some traditionally cured products, particularly those produced in central Europe, contain sufficiently high levels of curing ingredients to inhibit the growth of NaCl tolerant pathogens such as *Staphylococcus aureus*, and proteolytic strains of *Clostridium botulinum*, as well as of most spoilage micro-organisms during storage at room temperature. However, such products are uncommon, and consumer preferences for a less salt product and toxicological fears concerning the possibility of carcinogenic nitrosamine formation have combined to effect a substantial reduction in the levels of both NaNO$_2$ and NaCl. Most modern cured meat products require refrigeration to prevent rapid spoilage, and pathogens such as *Staph. aureus* are able to grow in the absence of competition from other micro-organisms. Fears for the safety of vacuum-packed bacon where the aerobic spoilage microflora is inhibited have, however, proved unfounded due to the presence of a lactic acid bacteria microflora and, possibly, other unknown factors.

Reductions in the level of curing ingredients are continuing under the impetus of demands for 'healthier' food, and people involved in the development of such products must be aware of the possible consequences of such reductions. Some types of sweetcure product are of greatest concern, since the manufacturing process requires a heating stage which is designed to partially denature structural proteins, but which also markedly reduces the spoilage microflora (*cf.* cooked cured meats, pages 441–442).

It should be noted that the levels of curing ingredients in 'cured' poultry products are often very low, and make no contribution to either safety or shelf stability.

Although most raw, cured products are intended to be cooked before eating, a distinct type is eaten raw. This is typified by the Italian Parma ham. These products are usually dry cured, although a supplementary injection of curing brine is now more common. Following curing, the hams are matured at a cool room temperature, during which time a variable surface microflora develops. The curing process is essential to the safety of such products, and considerable skill is required to ensure an even distribution of curing ingredients. Despite the apparently haphazard approach of some manufacturers, the safety record of commercially produced hams of this type is very good, but consumption of home-cured varieties has led to a number of types of illness including botulism, staphylococcal intoxications and infection with animal parasites.

Moulds are commonly among the surface microflora which develops on Parma-type hams and contri-

Flow diagram 41. Flow diagram: Wiltshire bacon.

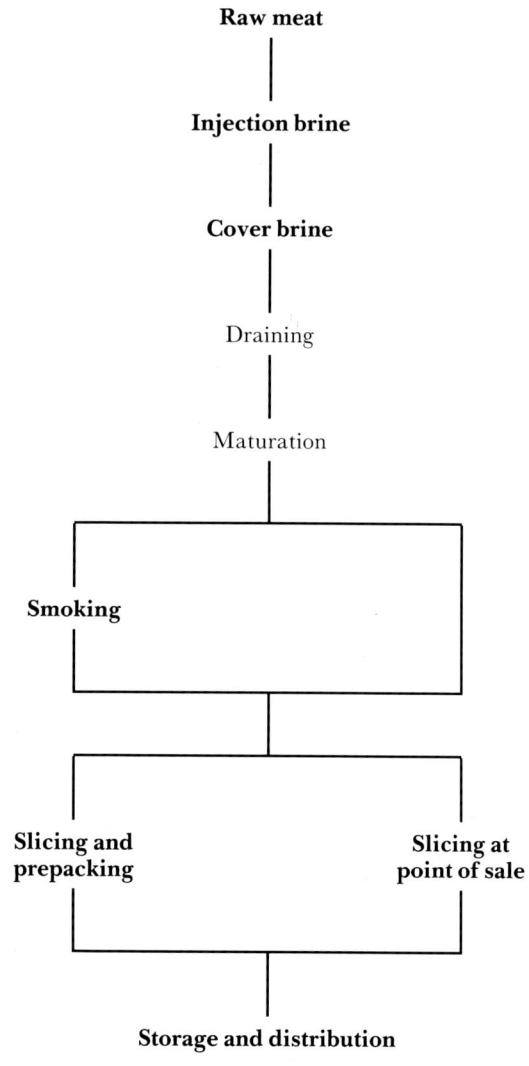

Raw meat

Injection brine

Cover brine

Draining

Maturation

Smoking

Slicing and prepacking

Slicing at point of sale

Storage and distribution

Note: Processes in **bold** are CCPs.

Table 164. Critical control points: Wiltshire bacon.

Raw meat, CCP 2

Control
Achieved at farm and abattoir level (see pages 427–430).

Monitoring
Ensure meat obtained from known and reputable supplier.

Injection brine, CCP 2

Control
Inhibits growth of pathogens.
Concentration of curing ingredients ($NaCl$, $NaNO_2$, $NaNO_3$) must be correct and brine must be correctly distributed in meat.

Monitoring
Curing ingredients must be accurately weighed into a known volume of water. Brine injector should be inspected for correct function before and during use. Brine usage should be measured.
Verification may be by chemical analysis of the brine, although this is rarely necessary in practice.

Cover brine, CCP 2

Control
Inhibits growth of pathogens.
Brine should be restored to full strength at the start of each curing cycle and in a stable condition. Meat should be fully immersed for correct period of time.
Temperature must be controlled.

Monitoring
$NaCl$, $NaNO_2$ and $NaNO_3$ should be assayed and the pH value determined before the start of each cycle and the constitution adjusted as required.
The loading of the tanks should be supervised by experienced personnel and a formal system should be used to ensure immersion for the correct period.
Temperature should be recorded on a continuous basis.
Verification of the stability of the brine may be made by recording changes in chemical composition and pH value over a number of curing cycles. This evidence may be supported by direct microscopic count and by specialist microbiological examination (see Gardner, 1982).

Comments
Injection brines are made up freshly before use and thus differ from cover brines, which are reused over extended periods. Pathogens such as *Salmonella* survive in brines for extended periods, but appear to be of little significance.

Smoking, CCP 2

Control
Inhibition of pathogens by deposition of phenolic compounds and surface drying.
Deposition of phenolics and surface drying must be even over the entire surface.

Monitoring
Monitoring of temperature and airflow within kiln. Visual assessment of correct functioning of kiln.
Verification is usually possible by visual assessment.

Comments
Smoking is a CCP *only* if the process is inhibitory to micro-organisms. In the case of liquid smoke processes, positive proof of inhibition is required.

Slicing and prepacking, CCP 2

Control
Prevention of contamination.

Monitoring
Strict control of personal hygiene and equipment cleaning.

Comments
The good safety record of prepacked (vacuum or gas packed) bacon is due, at least in part, to the inability of pathogens to grow in the pack. However, the slicing and packing process remains a CCP especially as continuing changes in levels of curing ingredients may lead to conditions which support the growth of pathogens.

Table 164. *Continued.*

Slicing at point of sale, CCP 2

Comments
General considerations of control and monitoring are the same as for slicing and prepacking. The risk of contamination is often higher due to the poorly controlled nature of operations.

Storage and distribution, CCP 2

Control
Maintenance of correct storage temperatures.

Monitoring
Measurement of temperature at all stages.

Comments
The importance of refrigeration during storage and distribution of Wiltshire bacon increases as the level of curing ingredients is reduced. Although control of spoilage is often of immediate concern, refrigeration is also of importance with respect to potential pathogens.

Note: Verification that the entire curing process has been satisfactory may be obtained by chemical analysis of the final product to ensure that levels of NaCl, $NaNO_2$ and $NaNO_3$ and pH value are within predetermined limits. Where doubt exists, specialist microbiological examination may be applied (see Gardner, 1982).

bute towards ripening. The moulds are not cultured, but are derived adventitiously from the cellars used for curing and maturation. Mycotoxin producing moulds such as *Aspergillus versicolor*, *A. ochraceus* and *Penicillium viridicatum*, have been isolated from maturing hams (Sutherland *et al.*, 1986), and while there is no direct link with human mycotoxicosis, an epidemiological link has been made between the consumption of mould-ripened hams and stomach cancer in Yugoslavia. It is recommended that only moulds which have been demonstrated not to produce mycotoxins should be used in the production of mould-ripened hams, and that ensuring the safety of the moulds should constitute a critical control point.

Smoking of cured meats

Traditional smoking procedures involved a considerable degree of surface drying with a corresponding raising of the surface concentration of curing ingredients and a lowering of the water activity together with heavy deposition of anti-microbial phenolic compounds. Such smokes were a major contribution to the stability and safety of cured meats.

The importance of smoking as a means of preserving meats is now much reduced, and most smoking serves no purpose other than to flavour the product. Traditional wood smokes have often been replaced by the use of liquid smoke flavouring, which is easier to use and which should be free from the potentially carcinogenic polycyclic aromatic hydrocarbons. There is some controversy over the anti-microbial activity of liquid smokes, which appears to be related to the type of wood used – Douglas fir extracts being most inhibitory against *A. hydrophila* and *Staph. aureus*, and lodge-pole pine and mesquite extracts being the least (Sofos *et al.*, 1988).

Where a product is smoked, the nature of the smoke and its technological role should be carefully evaluated. The process should be considered a critical control point with respect to safety, only if the numbers of any pathogens present are significantly reduced or if there is a significant inhibition of growth under feasible conditions of storage.

Raw fermented meat products

Raw fermented meats are typified by fully-dried products such as salamis, and semi-dried products such as cervelat. The processes are outlined in **Flow diagram 42** and critical control points summarised in **Table 165**.

The prime critical control point is the fermentation stage where *Staph. aureus* is controlled by a rapid fall in pH value. This should be achieved either by the use of a starter culture of *Pediococcus acidilactici*, or by acidification with glucono-δ-lactone which on hydrolysis yields gluconic acid. The older practice of 'back-slopping' fermented meat from the previous batch to act as a starter in the new, can give variable results and is not recommended, while reliance on a natural lactic fermentation being established under the selective conditions in the meat mix is not considered acceptable.

Control of *Trichinella* is necessary if the meat product contains pork. In fully dried products, this is attained by drying, but semi-dried products must be heated sufficiently to kill the parasite. In the USA a minimum temperature of 58.3°C is mandatory.

Although unable to grow in the finished product, pathogens such as *Salmonella* and *Listeria* may survive processing of both types of fermented meat, and may be present in meat up to the point of consumption. An outbreak of salmonellosis associated with a salami snack has occurred in the UK (see page 65), but otherwise problems are rare and, indeed, considerably less common than might be predicted. The significance of small numbers of *L. monocytogenes* in fermented meats has not been fully assessed. However, it has been observed that some strains of *P. acidilactici* produce bacteriocins active against *Listeria*, and the possibility of targeting starter cultures against that organism exists in the long term.

Fully dried salamis are stable for extended periods at ambient temperature. During this period moulds and, less commonly, yeasts may develop, and it is usual practice to remove these by wiping the skin with vegetable oil. Providing there has been no penetration of the skin, the product is used as normal. This practice is well established, but although there is no evidence of a mycotoxin hazard, it is seen as being undesirable and it is preferable that conditions of storage are such that moulds do not develop. Some fermented meat products are deliberately mould-ripened, in which case care must be taken to ensure that mycotoxigenic species are absent.

Flow diagram 42. Flow diagram: fermented sausages.

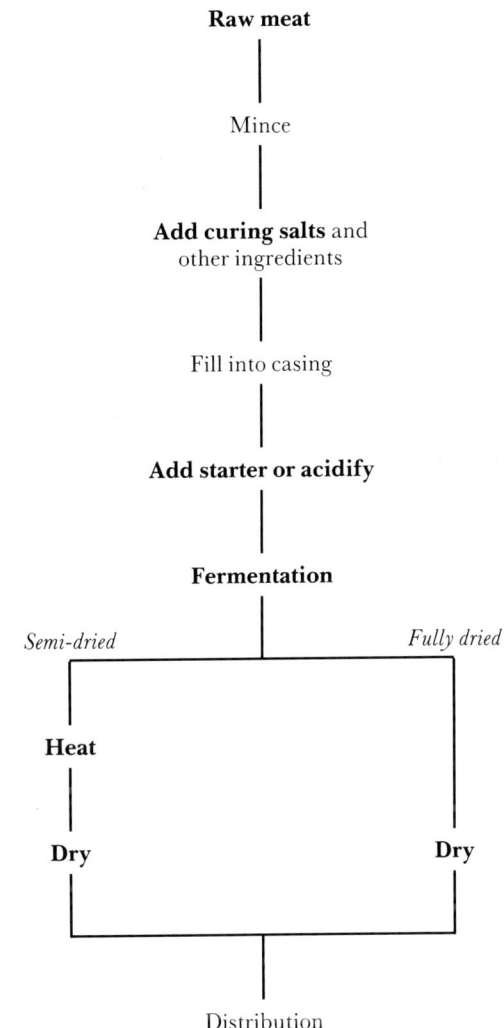

Notes:
1 Drying process is much shorter for semi-dried than for fully dried.
2 Processes in **bold** are CCPs.

Table 165. Critical control points: fermented sausages.

Raw meat, CCP 2
Control
Achieved at farm and abattoir level (see pages 427–430).

Monitoring
Ensure meat is obtained from known and reputable source.

Addition of curing ingredients, CCP 2
Control
Creates conditions favourable to rapid growth of starter culture.
Reduces risk of growth of some pathogens during production and storage.

Monitoring
Ensure accurate weighing of ingredients and correct distribution.
Verification may be made by chemical analysis of final product. Levels of nitrite (and nitrate) are likely to change during manufacture and storage and thus may not accurately reflect input levels.

Comments
In *most* cases, creation of a selective advantage for starter cultures is considered to be the main function of curing ingredients. The role in stability of the final product is likely to be greater in semi- than in fully dried products.

Add starter or acidify, CCP 2
Control
Prerequisite for correct fermentation

Monitoring
Ensure addition of correct inoculum of active starter culture *or* correct level of acidulant.

Fermentation, CCP 2
Control
Inhibition of pathogenic micro-organisms due to rapid fall in pH value.

Monitoring
Ensure correct reduction in pH value.
Verification of inhibition of *Staph. aureus* may be made by microbiological analysis (see pages 255–258). Numbers should be $<10^4$ cfu/g at end of fermentation. The thermonuclease test (see page 260) may be used to provide supportive evidence in cases of doubt.

Heat, CCP 2 (semi-dried only)
Control
Inactivation of *Trichinella*.

Monitoring
Ensure correct temperature/time combination applied.
Verification is by examination of thermograph records.

Comments
Heating is designed specifically for the control of *Trichinella* and may not be applied where no pork is included in the formulation. Any reduction in numbers of bacterial pathogens is incidental.

Drying, CCP 2
Control
Inactivates *Trichinella* in fully dried products.
Controls growth of bacterial pathogens.
Temperature and humidity of drying room must be controlled.

Monitoring
Ensure correct control of temperature and humidity.
Ensure product is dried for correct period.
Verification of correct drying of fully dried products may be made by determination of the a_w level of the finished product.

Comments
Drying is of no, or only limited, significance in the case of some semi-dried products.

Note: None of the critical control points in the production of fermented sausages are absolute and thus pathogenic bacteria derived from meat may be present in the final product.

Cooked meat and poultry products

Almost all meat and poultry is cooked at some stage before consumption, but in the present context 'cooked products' are those which are produced, usually as items of commerce, at a central facility for consumption remote from production. Commercially cooked meats may thus be distinguished from meats cooked in the home or in the course of catering operations, although meats prepared in central cook-chill operations may be regarded as being equivalent to commercial products. Naturally, hazards associated with the cooking process are the same irrespective of the place of cooking.

A very wide variety of cooked meat products exist, but many of these share a similar basic technology. Safety requirements are the same for all cooked meats and a generalised production process is outlined in **Flow diagram 43**, and critical control points summarised in **Table 166**. Specific requirements of individual product groups are discussed briefly below.

Flow diagram 43. Flow diagram: cooked meats.

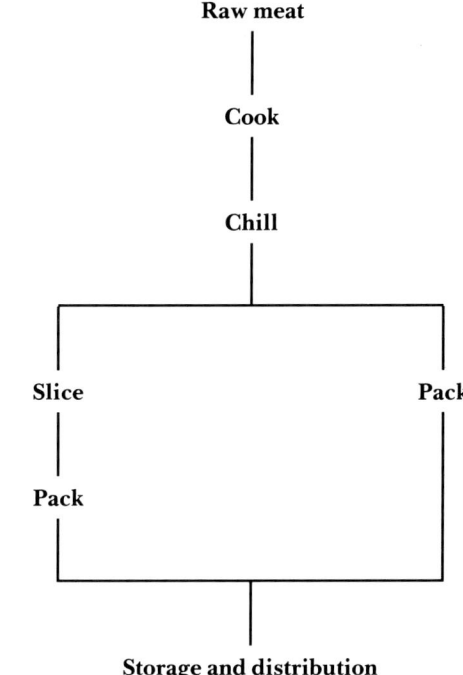

Note: All processing stages are CCPs.

Table 166. Critical control points: cooked meats.

Raw meat, CCP 2
Control
Achieved at farm and abattoir level (see pages 427–430).

Monitoring
Ensure meat is obtained from known and reputable supplier.

Cook, CCP 1
Control
Inactivation of vegetative pathogens by heating for predetermined temperature/time combinations.
Cooking should be controlled to ensure that all items receive the minimum safe process. Where items vary in size or physical characteristics, it is necessary to base control on those having the slowest heating profile. Similarly, where cold spots cannot be eliminated from ovens or heating baths, control should be based on temperatures attained in the coldest part.

Monitoring
Temperature of the heating medium (air, water, etc) should be monitored on a continuous basis.
Internal product temperatures should be measured and recorded throughout the cooking cycle.
Verification may be by enzyme inactivation tests (see **Figures 328** & **329**) together with examination of thermograph records. The importance of visual examination should not be underestimated.

Cool, CCP 1
Control
Rapid cooling to prevent germination and outgrowth of endospore forming pathogens and the growth of post-process contaminants.
Temperature/time of chilling should be controlled. Process should be based on items having the slowest cooling profile.

Table 166. *Continued.*

Cooling should commence as soon as possible after cooking is complete. If delays occur, the product should be maintained above 60°C until cooling is possible.

Care must be taken to ensure that post-process contaminants are not introduced during cooling. Where water is used this must be of potable quality and chlorinated. Air should be filtered.

Monitoring

Internal product temperatures should be measured and recorded throughout the cooling process.

The concentration of chlorine in cooling water should be measured regularly. Where spray cooling is used, equipment should be checked to ensure water pressure is adequate and that no spray nozzles are blocked. Where cooling is in tanks, the water should be changed on a regular basis and when visually dirty.

Verification of correct cooling may be made by examination of thermograph records. Microbiological analysis is desirable to verify the quality of cooling water and air.

Comments

With products such as large joints and pâtés it is not uncommon to find cooling procedures which are inadequate and marginal for safety, even when operated correctly. It is recommended that cooling procedures should be designed to permit the meat to pass through the temperature range 48.8 to 12.7°C within six hours.

Some products are sealed in impermeable plastic bags before cooking and are protected, providing seals are intact, from contamination during cooling. However, accumulation of juice within the cooking bag may necessitate rebagging. This involves handling of the cooked product and precautions are required. Rebagging is a *CCP 2*.

Additional ingredients, CCP 2
Control

Prevention of contamination of the cooked product by materials added after cooking.

Ensure products added after processing and the means of making the addition do not introduce pathogens.

Where necessary, process additional ingredients for safety (*cf.* gelatin, **Flow diagram 44**).

Monitoring

Ensure additional ingredients are obtained from known and reputable suppliers and have been correctly processed and handled.

Verification by microbiological analysis is likely to be necessary in some cases.

Comments

In addition to cooked products which have traditionally received additional ingredients after cooking, there is an increasing trend of adding value to basic products by embellishment at retail and food service establishments. Potentially, this practice carries a high level of risk.

Slice, CCP 2
Control

Prevention of contamination.

Strict control of personal hygiene and cleanliness of equipment and environment.

Monitoring

Monitoring hygiene of personnel. Visual inspection of cleanliness and condition of equipment.

Verification may be by microbiological analysis of equipment surfaces and environmental samples.

Packing, CCP 2
Control

Prevention of contamination. Protection of the product from contamination during subsequent handling. Reduction in growth rate of pathogens (see comments).

Strict control of personal hygiene and cleanliness of equipment and environment. Correct setting and operation of packaging machinery.

Monitoring

Monitoring hygiene of personnel. Visual inspection of cleanliness and condition of equipment. Examination of packs for correct formation, seal integrity, etc.

Verification of vacuum and gas packaging may be made by measuring the vacuum in vacuum packs and composition of the atmosphere in gas packs.

Comments

The role of packaging in controlling pathogenic micro-organisms is open to dispute and, indeed, the possibility always exists that packaging may create conditions, in combination with other factors, which favour the growth of *Cl. botulinum*.

Storage and distribution, CCP 2
Control

Prevention of growth of pathogens.

Ensure product is held under refrigeration at all times.

Monitoring

Measure and record temperature at all stages through storage and distribution.

Cooked, uncured meat and poultry

There is a large trade in roast meat and poultry. Meat may be sold as whole joints or sliced and pre-packed, while poultry is usually whole or half birds. Some types of cooked meat, and products such as turkey roll, are formed from pieces of meat rather than being a conventional joint.

It should be appreciated that from a microbiological viewpoint, the distinction between cured and uncured cooked meats is blurred. Many types of cooked ham contain only minimal levels of NaCl and nitrite, and should be treated as uncured with respect to safety. In these cases, the curing process is *not* a critical control point.

The cooking process is based upon attaining a 10^7-fold reduction in *Salmonella* numbers (Goodfellow and Brown, 1978). Examples of internal time/temperature combinations which meet these criteria are 37 minutes at 57.2°C, 12 minutes at 60°C and 5 minutes at 62.2°C. During the 1970s, a number of outbreaks in the USA of salmonellosis were attributed to roast beef (NRC, 1985). This appears to have been due to consumer desire for a 'rare appearance'. Some processes may be marginal with respect to *Listeria* and reevaluation of established processes is recommended (**Figures 328** & **329**).

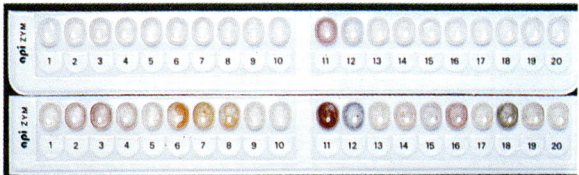

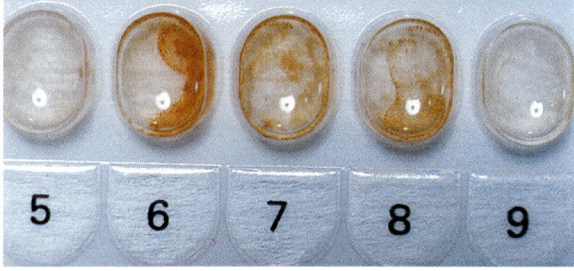

328 & **329. Use of the Api-zym© system to detect undercooking of meat.** Tests used in the past for verification of the adequate cooking of meat are cumbersome and cannot be applied at plant floor level. Attempts have been made to base verification on the determination of heat labile enzymes such as catalase, and these methods have had some success. The Api-zym system has also been found to be extremely useful. The system permits the simultaneous semi-quantitative determination of 19 enzymes and offers a clear discrimination between fully and undercooked meat (**328**). Leucine, valine and cystine arylamidase (**329**) appear to be particularly useful in detecting marginal undercooking of pork.

Outbreaks of salmonellosis (see page 65) and other food poisoning due to ingestion of cooked meats continue, but are largely due to post-cooking contamination. In some outbreaks involving small processors, no effective measures to separate raw and cooked product were taken, and the whole handling environment was contaminated. Interaction between retail economics and food safety may be demonstrated by reference to a number of incidents of *Salmonella* infection which occurred in north-west England during the 1980s. Local butchers, faced with increasing competition from supermarkets, installed rotisseries and supplied cooked chickens to market stalls and other outlets. Many of the butchers involved had no experience of handling cooked foods, the problem being compounded by the lack of refrigeration at the sales outlets.

Problems resulting from the inexperience of management and staff are not restricted to small companies. A number of poultry-packing operations which progressed to the production of cooked poultry, paid no real attention to the cardinal principle of separation of raw and cooked products, and in extreme cases the same, unwashed crates were employed for handling both.

Cooked meats are commonly sliced and pre-packed, vacuum packaging being common. Pre-packing operations offer considerable scope for contamination and this, together with long shelf lives and the opportunity for temperature abuse, suggest a high level of hazard. In practice, the safety record of vacuum-packed, sliced cooked meats is very good. This may be due to the reduced oxygen tension inhibiting staphylococcal enterotoxin production, or the inhibition of growth of pathogens by the lactic acid microflora which inevitably develops. Despite this, there is no doubt that food poisoning due to vacuum-packed, sliced cooked meats occurs even if only rarely reported. A far greater threat, however, is represented by catering operations such as buffets, where the meat may be sliced and laid out for several hours at room temperature before consumption.

Pasteurised canned hams are highly perishable products which contain only low levels of NaCl and NaNO$_2$. Refrigeration must be maintained throughout storage, and the provision of unambiguous labelling to this effect is a critical control point. The rigid container offers a high level of protection, but as with all cans, is vulnerable to leakage during cooling.

Cooked, cured meats

As noted above, this category is reserved for those products in which the curing stage is a critical control point. Such products are usually highly stable and, in the case of some traditionally made European products such as the Polish bozchek, have an extended storage life at ambient temperature. Refrigeration is required for less heavily cured products, and staphylococcal food poisoning has resulted from contamination at slicing followed by storage of the sliced product at room temperature.

An internal temperature of 70°C should be attained during the cooking of hams. Most modern processes involve a single cooking stage, but a two-stage process may still be used, the ham being manually derinded in the interval between the two cooks. The potential for contamination with *Staph. aureus* is high, and a high level of control is needed at this stage to prevent growth and toxin production.

NaCl and NaNO$_2$ are important safety determinants with respect to *Cl. botulinum*, and levels should not be reduced without a full appraisal of the consequences. Similar appraisals should be made where salts other than NaCl are used in an attempt to meet currently fashionable demands for low-sodium food, since there is evidence that these salts are markedly less effective than NaCl in the inhibition of *Cl. botulinum*.

Pies and puddings

Cooked meat pies and puddings are basically of two types. The first is represented by UK products such as steak and kidney pudding or shepherds' pie. In these, the meat filling is cooked separately before filling into a dish or a pastry mould and recooked. Such pies are usually eaten hot after further heating at the point of consumption. In the second type of pie, typified by the UK pork pie, the meat is filled into a pastry casing and the whole pie cooked in a single operation. The meat filling tends to shrink on cooking and a layer of gelatin is commonly injected between the meat and the pastry after cooling. Pies of this type are usually eaten cold.

Specific microbiological problems are rare with the first type of pie, although intoxications have resulted from the growth of *Staph. aureus* in the period between the two heating stages. It is unacceptable practice to depend on the second-stage cooking to eliminate pathogenic micro-organisms, and the intervening stage between the two cooks is considered to be a critical control point.

Botulism has been associated with frozen potpies following gross abuse by the customer.

Problems with cold eating pies such as pork pies usually result from the contamination of the gelatin injected into the pie after cooking. Gelatin is both a source of micro-organisms and a good medium for their growth, and mishandling has led directly to salmonellosis and staphylococcal intoxication. Virtu-

ally all aspects of the handling and use of gelatin are critical control points (**Flow diagram 44**). Gelatin and other materials may also be used as glaze on the crust of pies, and such material must also be treated as being potentially hazardous.

Although most problems with pork and similar pies relate to mishandling of gelatin, undercooking has led to food poisoning when large pies have been cooked in the same batch oven as sausage rolls. Control of the size of commodity is a critical control point in small factories where batch ovens are used for mixed loads.

Flow diagram 44. Handling of gelatin before addition to cooked meats.

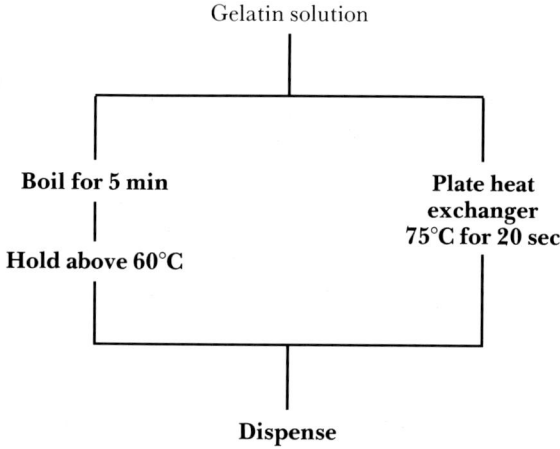

Notes:
1 Continuous heating using a plate heat exchanger is preferred, but is only practical where throughout is high.
2 A routine of continuous cleaning and sanitising equipment and utensils used for dispensing gelatin must be employed. The uncontrolled use of old teapots led to a large outbreak of salmonellosis from which the manufacturer was unable to financially recover.
3 All stages in handling are CCPs.

Comminuted products

Cooked comminuted products range in type from fine comminutes such as pâtés and liver sausage, to coarsely comminuted sausages such as the Spanish chorizos rosario and the Polish mysliwska. Most comminuted products contain either pre-cured meat, or curing ingredients, and while levels in pâtés and similar products are low, NaCl and NaNO$_2$ are important safety determinants in some types of European cooked sausage.

The cooking of pâtés requires careful monitoring, since measures to avoid problems of visual quality caused by the rendering out of fat, may mean that the cooking process is marginal for safety. In practice, most problems of food poisoning are related to post-process contamination. Some European manufacturers operate under conditions best equated with those of a cottage industry, where the environment for handling the cooked product is heavily contaminated. Gelatin used as a glaze or topping for pâtés is again a potential problem, and the various herbs etc added to some bowl pâtés as decoration, may also be sources of environmental pathogens such as *Listeria*.

Slow cooling is a problem associated with large liver sausages and bowl pâtés, especially when produced in small factories with inadequate cooling facilities. In some cases, cooling may be deliberately delayed for undefined and unnecessary organoleptic purposes. This practice is not acceptable.

Soups and stews

Meat soups and stews are notorious in catering operations for their association with *Cl. perfringens* food poisoning resulting from slow cooling. Soups and stews are also produced on a central basis, packaged in cardboard or plastic cartons and distributed under refrigeration. Cooling must be controlled carefully and a temperature of 10°C should be attained within 4 hours. If delays occur, the product should be held at a minimum temperature of 66°C before cooling commences.

Ultrahigh temperature (UHT) sterilisation techniques may now be applied to relatively viscous material containing small particles, and some soups are made by this process and packaged under aseptic conditions. The process is effectively the same as for UHT milk (pages 450–451), although it is possible that the existence of highly heat resistant endospores of pathogens, such as *Bacillus cereus* may be of greater consequence with meat products than with milk. UHT soups are stable at room temperature, but potential hazards have arisen from a similar package design being used for both UHT soups and conventionally prepared soups which require refrigeration. Package design of the two types should be distinct and, where necessary, the need for refrigeration clearly stipulated.

Cook-chill meat products (ready meals)

Cook-chill products are entire meals, or the major component of entire meals, and are prepared on a central basis, chilled and kept under refrigeration until heated directly before consumption. Cook-chill meals are used widely in large-scale catering including hospital and school meals, and are also produced for retail sale, where the term 'ready meal' is usually applied. Many types of ready meals are produced, most, but not all, being meat based and the flow diagram for the production of one such product is illustrated in detail with respect to HACCP in pages 390–400.

Ready meals present no more inherent problems than other cooked products and, indeed, traditional products such as the Cornish pasty may be considered to be early ready meals. Many modern ready meals are distinguished by the very high level of handling of cooked components, the highly perishable nature of the formulation, and the relatively long storage life which may exceed 14 days. A very high level of control is required during manufacture and subsequent handling, and this may be difficult to maintain in practice. Concern for the safety of these products has been heightened by isolations of *L. monocytogenes*. Legislation in the UK is to make storage at temperatures below 5°C mandatory, but it is doubted if this action alone is sufficient given the long storage lives increasingly applied. Limiting the storage life of ready meals to a safe period is considered to be a critical control point, and must be evaluated for each type of product individually. In this context, instructions to purchasers must be unambiguous and the use of 'best before' dating is considered to lack the necessary authority.

Conditions for the safe production and retailing of long-life ready meals have been discussed by Mossel (1989; **Table 167**). The practicality of some of these conditions must, however, be doubted.

Table 167. Conditions necessary for the sale of extended life cooked products (Mossel, 1989).

1   Strict microbiological quality assurance of raw materials to allow the intensity of primary processing by heat to be kept to a relatively low level, thus ensuring a product of maximum appeal.

2   Avoidance of bacteriological contamination from machinery and proliferation due to temperature abuse during formulation and processing.

3   Meticulous design and validation of heat processing procedures.

4   Rigorous avoidance, or else subsequent elimination, of post-process contamination; to be achieved by rapid post-process chilling and packaging under aseptic conditions or by post-packing heat treatment by, for example, microwave processing.

5   Pre-shipment storage as close to 0°C as possible.

6   Distribution and display at *food* temperatures guaranteed never to exceed 3°C. This is considered to be the most important factor of all.

7   Sales to be made only from stocks of items which have never been exposed to ambient temperatures or manipulation by potential customers.

8   Delivery to customers to be made exclusively in well-insulated shopping bags not accommodating other items.

9   Conspicuously clear ultimate date marking, both in words and the well known 'thermometer-logo' in red and green with 3°C as the cut-off temperature.

Sous-vide products

Sous-vide products were first developed for a French restaurant and are largely used in catering. Sous-vide products are par-cooked, vacuum-packed and then subjected to a heavy in-pack pasteurisation. The products are then refrigerated until use. The inherent hazards with respect to botulism are obvious, and the extension of sous-vide products to domestic use, where conditions of storage cannot be controlled, has been greeted with justified alarm.

Dried meat products

Substantial quantities of dried meat are produced for use in formulated foods such as powder soups, beef drinks etc, as well as for more specialist purposes such as hill-walking rations. Finely ground meat for soups etc, is dried in hot-air tunnels, the process being designed to reduce the water content to less than 15% in 1 to 2 hours. The process is outlined in **Flow diagram 45**, and critical control points are summarised in **Table 168**. Larger pieces of meat are freeze dried; this is an expensive process and normally only used for higher value products.

It is necessary to ensure that dried meat does not take up water during storage, and thus the choice of suitable packaging is a critical control point.

Flow diagram 45. Flow diagram: air dried meat.

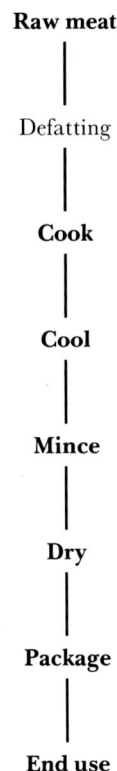

Raw meat

Defatting

Cook

Cool

Mince

Dry

Package

End use

Notes:

1 The flow diagram is for large-scale production of dried meat using hot air tunnels. Small quantities of sun dried meat such as biltong are also produced.

2 In some cases, the cooling stage is omitted and the meat is minced while still hot.

3 Processes in **bold** are CCPs.

Table 168. Critical control points: air dried meat.

Raw meat, CCP 2
Control
Achieved at farm and abattoir level (see pages 427–430).

Monitoring
Ensure meat obtained from known and reputable source.

Cook, CCP 1
Control
Inactivation of vegetative pathogens at correct temperature/time combinations.

Monitoring
Ensure cooking process correctly applied.
Verification is by examination of thermograph records. If doubt exists enzyme inactivation tests (see page 441) may be used.

Cool, CCP 1
Control
Prevention of germination and outgrowth of endospore forming pathogens.
Prevention of contamination during cooling.

Monitoring
Ensure cooling process correctly applied.
Ensure quality of water or air used for cooling.
Verification is by examination of thermograph records and by microbiological determination of the quality of the cooling medium. Where water is used chlorination should be verified by chemical analysis.

Comments
Where the process does not include a cooling stage, the meat must be maintained at 60°C or above until entering the dryer.

Mincing, CCP 1
Control
Prevention of contamination.
Reduction of meat to correct size for drying.

Monitoring
Visual examination to ensure cleaning schedules correctly applied.
Elimination of 'hands on' operations.
Visual assessment of correct operation.
Verification may be by microbiological sampling of surfaces. Size of particles leaving mincer should be determined at regular intervals.

Drying, CCP 1
Control
Reduction of water content to levels below those permitting growth of micro-organisms.

Monitoring
Monitoring of air temperature and, in continuous dryers, product flow rate.
Verification is by examination of thermograph and other records, and by determination of the moisture content of the final product.

Packaging, CCP 2
Control
Packaging material must be impermeable to moisture.

Monitoring
Visual assessment of pack and seal integrity.

End use, CCP 2
Control
Rehydrated product must not be stored at a temperature permitting the growth of pathogens.
Labelling instructions required for domestic consumers.

Comments
Dried meat of this type is normally used as an ingredient in dried formulated products such as soup mixes.

22 Control of pathogenic micro-organisms in milk

The vast majority of milk consumed in developed nations is cows' milk, although goats' milk is increasing in popularity in the UK, and significant quantities of sheep milk are consumed in Australia. In developing countries, sheep and goats may be of greater importance than cattle, while other sources of milk are camels and mares. The chemical composition of milk of different species varies considerably, but while this affects the organoleptic properties and, possibly, the development of the spoilage microflora, there is no evidence that the persistence, or growth, of pathogenic micro-organisms is significantly affected. The discussion in this Chapter refers largely to cows' milk but in general is applicable to that of other species.

Control at farm level

The general precautions against infection of dairy animals are the same as those for other stock described in Chapter 21. Infection with pathogens such as *Salmonella* may persist for extended periods, Giles *et al.* (1989) describing a *S. typhimurium* infection which persisted for 3½ years in a large herd, and which was associated with three incidents of human salmonellosis. However, persistent *Salmonella* infections normally involve *S. dublin*, carriage of other serovars being rare (Wray and Sojka, 1977).

Contamination of milk

Although milk may be contaminated by a number of routes (**Table 169**), the most important with respect to the major pathogens is contamination of milk from faeces, which usually occurs at milking.

In many countries, the conditions under which dairy cattle are housed and milked have improved markedly in recent years, although bad habits acquired in earlier years have not been entirely eliminated and problems can be exacerbated by poor hygiene. Modern machine milking installations are complex, and cleaning schedules must be rigidly adhered to. Even where conditions are very good, it is impossible to entirely eliminate the possibility of contamination of milk with faecal material, and pathogens may be present in the absence of faecal indicator bacteria. Milking staff should appreciate that even a seemingly minor incident, such as dropping a milking machine teat cluster, can have serious implications for safety, especially if subsequent storage conditions permit the growth of pathogens.

Udder excretion of various human pathogens may occur although, with the exception of common causative organisms of mastitis, the importance of this route is often overstated. *Salmonella* excretion from the udder, for example, is unusual (Wray and Sojka, 1977), although it has been reported in herds where acute outbreaks have occurred (Giles and King, 1987; Giles *et al.*, 1989). *Listeria monocytogenes* is a recognised cause of mastitis, although the number of reported cases is small (Gitter, 1989). Excretion of the organism for up to 3 years has been reported (Donker-Voet, 1965).

The relative importance of farm personnel as a source of contamination of milk is difficult to assess, but undoubtably occurs (Galbraith and Pusey, 1984). Where farm workers themselves drink unpasteurised milk, cycles of infection and re-infection may be established. It is desirable that people involved in handling milk at farm level should receive the same level of training as food handlers elsewhere, and that the same health rules should apply.

Table 169. Milk: possible sources of contamination.

Bovine faeces
Defaecation into milk
Contamination of udder via bedding etc
Contamination of milker's hands etc
Contamination of milking equipment after dropping on floor etc

Udder infections
Undetected sub-clinical mastitis or asymptomatic infection
Use of mastitic milk

Milking personnel
Direct transfer from unwell person to milk
Indirect transfer from unwell person via contamination of milking equipment

Environmental sources
Udder contamination from stream and pond water, infected pasture etc

Liquid milks

'It is the responsibility of all health professionals to see that the public and the policy makers are adequately informed about the scientific findings so that public policy on raw milk may be compatible with scientific knowledge and protective of the public's health' (Chin, 1982).

The epidemiology of milkborne disease has been the subject of numerous reviews (eg Bryan, 1983; Humphrey and Hart, 1988; D' Aoust, 1989), and among microbiologists there is consensus opinion that pasteurisation is an essential process in providing milk which is free of hazardous micro-organisms. This is illustrated most clearly by experiences in Scotland where legal controls on the sale of raw milk led to an immediate decline in the number of outbreaks of *Salmonella* and *Campylobacter* infections. The situation in Scotland may be contrasted with that in England and Wales, where controls are less restrictive, and where milkborne salmonellosis and campylobacteriosis continue to cause significant illness.

Despite the overwhelming microbiological and epidemiological evidence in support of milk pasteurisation, a small sector of the UK dairy industry campaigns vigorously for the continuing availability of raw milk. During 1989, the UK Government was persuaded to abandon its earlier plans to extend Scottish legislation on raw milk sales to the whole of the UK, although bottles will have to bear a 'health warning'.

Proponents of the continuing sale of raw milk tend to base arguments on an emotive blend of nostalgia for a lost rural Arcadia, spurious notions of individual freedom and unsubstantiated epidemiological and nutritional claims. A major fallacy lies in the contention that there is no risk from milk produced from healthy animals milked under good hygienic conditions, since raw milk produced under the most rigorous conditions has been associated with human infection (Werner *et al.*, 1984).

Pasteurisation processes are of two types; the batch low-temperature long-time process (LTLT) and the more commonly used high-temperature short-time process (HTST) which is a continuous process. In most countries, the temperature-time combination is defined by law; in the UK the minimum LTLT process is 62.8°C for 30 minutes, and the minimum HTST process 71.7°C for 15 seconds; similar processes are used in other countries. The emergence of *Listeria* as a significant pathogen has led to some processors voluntarily increasing the severity of the process (see page 341).

The operation of an HTST process with respect to application of HACCP has been discussed and illustrated in some detail on pages 396–398. The process is outlined in **Flow diagram 48**, and critical control points are summarised in **Table 170**.

Flow diagram 46. Flow diagram: pasteurised milk.

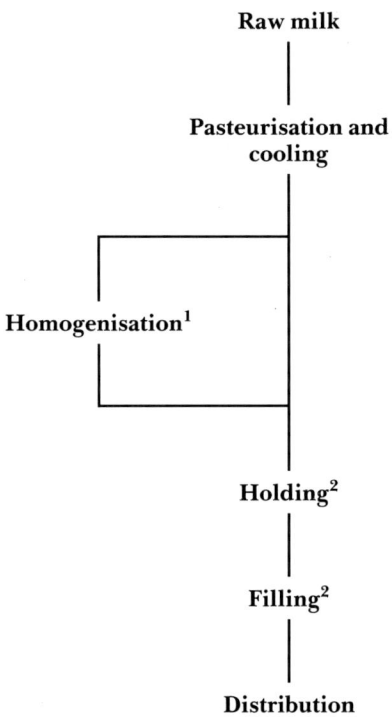

[1] Older types of homogeniser are notoriously difficult to clean and can be major sources of contamination.
[2] Care must be taken to prevent re-contamination of pasteurised milk at all stages after leaving the pasteuriser. Note: All processing stages are CCPs.

In recent years, a number of outbreaks of food poisoning have been associated with pasteurised milk. Inevitably these have been used as ammunition by the proponents of raw milk, and it must be stressed that while food poisoning due to raw milk reflects the inherently unsafe nature of that product, the pasteurised-milk incidents resulted from failures of control that can be remedied by attention to management and supervision. Infections have tended to result from poor plant design and maintenance, or from inadequate operator training, and there may have been an underlying tendency towards complacency. The proliferation of types of milk processed from straightforward, full-cream milk to skimmed, semi-skimmed, flavoured and vitamin-enriched milk also greatly increases the complexity of the plant and the possibility of cross-connections being made between raw milk and pasteurised.

In some types of flavoured milk, the flavouring is added after pasteurisation and is, therefore, a critical control point. Particular problems have been encountered with chocolate flavourings, and chocolate milk has been the vehicle in outbreaks of food poisoning due to both *Yersinia enterocolitica* and *Staphylococcus aureus*.

Mention should be made of Brainerd diarrhoea, a chronic diarrhoeal syndrome which was first recognised following a point-source outbreak associated with raw milk. It is likely that sporadic outbreaks had occurred on earlier occasions. Brainerd diarrhoea has been recognised as a distinct clinical condition with an infectious aetiology, but the causative agent remains unidentified (Archer and Young, 1988).

Table 170. Critical control points: pasteurised milk.

Raw milk, CCP 2
Control
Achieved at farm level (see pages 427–430).

Monitoring
General quality should be checked using the direct epifluorescent technique if possible.

Pasteurising and cooling, CCP 1
Control
Inactivation of vegetative pathogens by heating for approved temperature/time combinations.
Prevention of re-contamination in pasteurising plant and in subsequent handling.

Monitoring
Monitoring of temperature attained and ensuring the operation of under-temperature flow diversion devices.
Ensuring other plant operating parameters are normal.
Monitoring physical condition of plant.
Verification is by the phosphatase test and by examination of plant records.

Holding, CCP 2
Control
Control is by refrigeration and by ensuring that the period of holding is as short as possible and does not surpass pre-determined maxima.

Monitoring
Ensuring refrigeration is operated correctly.

Comments
Unused pasteurised milk may be returned to the raw milk side for re-processing. This practice is undesirable since the opportunity of contamination of pasteurised milk via the return line exists.

Filling, CCP 2
Control
Environmental hygiene in area of fillers and adequate sanitisation. Use of adequately cleaned bottles.

Monitoring
Environmental inspection supported by sampling if required. Monitoring of cleaning procedures and of bottle washing. Ensuring correct operation of bottle cappers and carton sealers.

Distribution, CCP 2
Control
Ensure refrigeration is applied as far down the chain as possible.

Monitoring
Checking and recording temperatures.

Comments
In countries such as the UK where doorstep delivery is still common, the temperature during delivery is uncontrolled and care must be taken in summer to avoid undue periods at high temperature. Temperature control of dairy display cabiniets is often particularly poor, and spot checking of retail outlets may be advisable during the summer months.

Ultrahigh temperature milk (UHT)

UHT milk is heated to a temperature sufficient to kill all vegetative bacteria, and all but the most heat-resistant endospores, and then packaged under aseptic conditions in pre-sterilised, plastic-coated cardboard. Unlike bottled pasteurised milk, UHT milk is stable for long periods at ambient temperatures. A process temperature of 132°C for not less than 1 second is employed, and there are two means of heating – direct by injecting live steam into the milk and indirect using plate, or tubular, heat exchangers. The production of UHT milk is outlined in **Flow diagram 47** and critical control points summarised in **Table 171**.

Production of UHT milk requires a high level of monitoring and provision of special facilities. The safety record, however, is very good, and microbiological problems are usually restricted to spoilage due to heat-stable proteolytic enzymes derived from psychrophilic bacteria growing in the raw milk.

Flow diagram 47. Flow diagram: UHT milk

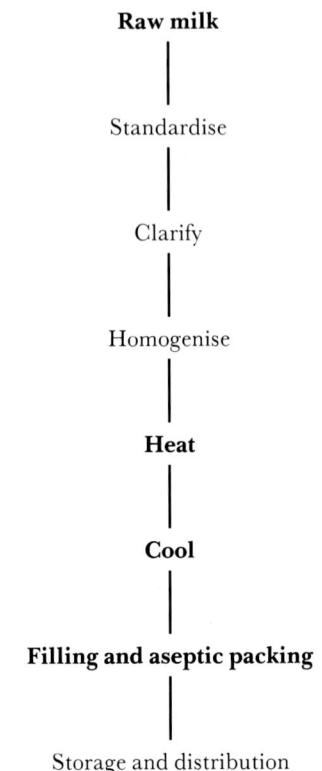

Raw milk

|

Standardise

|

Clarify

|

Homogenise

|

Heat

|

Cool

|

Filling and aseptic packing

|

Storage and distribution

[1] In some cases, homogenisation follows heating. This introduces a greater risk of contamination and in these circumstances homogenisation is a CCP.
Note: Processes in **bold** are CCPs.

Table 171. Critical control points: UHT milk.

Raw milk, CCP 2
Control
Achieved at farm level (see pages 427–430).

Monitoring
If possible, general quality should be checked using the direct epifluorescent technique.

Comments
Control of *spoilage* due to enzymes derived from psychrophilic bacteria is achieved by restricting pre-process storage periods.

Heating, CCP 1
Control
Inactivation of vegetative and endospore forming pathogens by heating for approved temperature/time combinations.
Prevention of re-contamination in cooling stage of plant.

Monitoring
Measurement of temperature, pressure and flow rates.
Ensuring correct operation of control and safety devices.
Monitoring physical condition of plant.
Verification is by examination of plant records.

Comments
Exact monitoring procedures vary according to whether *direct* or *indirect* heating is used.

Table 171. *Continued.*

Cooling, CCP 1
Control
Pre-sterilisation of plant.

Monitoring
Monitoring physical condition of plant.

Filling and aseptic packaging, CCP 1
Control
Pre-sterilisation of plant.
Installation of filler in 'clean' area with control of air flow and pressure.
Sterilisation of packaging material and of air or gases used for flushing pack.
Correct formation of pack and making of seals.

Monitoring
Monitoring physical condition of plant.
Ensuring air flow and pressure in packaging area is correct.
Ensuring correct operation of systems for sterilisation of packaging, air, etc
Ensuring correct formation of pack and integrity of seals.

Note: Verification of the entire process (from heating to aseptic packaging) may be made by incubation testing of finished products. With an established plant operating under normal conditions, 0.01% of each batch should be tested, but a higher rate is required under other circumstances. Non-destructive examination is sufficient for the detection of gas production and, using sonic testing, coagulation, but a proportion of packs should be opened and examined for abnormalities of appearance and smell and the pH value determined. Microbiological examination is required only in cases of doubt or for special purposes.

Cream and butter

Pasteurisation is also the key safety determinant in cream and butter and, with the exception of cream confections, problems are few with properly pasteurised products. Cream can undergo relatively complex processing after pasteurisation, including two-stage homogenisation and slow cooling, and stringent precautions against contamination are required. Cream may also be ultrahigh temperature treated, the process being identical to that for milk.

Cream-filled confections are a notorious cause of food poisoning, although the situation has improved with the general use of refrigerated display cabinets. Generally, the problem lies with poor standards of hygiene together with specific problems, including the high level of handling during the filling of cream cakes and the difficulty of cleaning equipment and utensils such as whipping machines and fillers.

Cream is also used as an ingredient in various other products such as desserts and fools. Manufacture requires a high level of handling with appropriate controls, and attention should also be paid to the handling of other sensitive ingredients such as gelatin. Ingredients which are used without heating are critical control points. The safety record of commercially produced desserts etc is good, but food poisoning has resulted from similar products prepared during catering operations. This has invariably resulted from contamination at the point of preparation followed by storage at a high temperature.

Butter made from pasteurised milk normally presents no specific safety problems. Staphylococcal food poisoning has been associated with whipped butter, but this would appear to be an exceptional circumstance.

Control of pathogenic micro-organisms in concentrated and dried milks

Concentrated milk

In its simplest form, milk is concentrated by heating in open pans, and while the resulting product is of poor quality, this process is still used in developing nations. The level of hygiene is usually low, and staphylococcal food poisoning may result from growth of the organism in the concentrated product.

Concentrated milk in developed nations is usually produced by heating under vacuum at relatively low temperatures. Milk is heated to a high temperature prior to entry into the evaporator and this, together with operating temperatures in the first stages of the evaporator, kills any vegetative pathogens present. Two types of concentrated milk are produced. Sweetened condensed milk contains sugar, usually sucrose, which is added to milk before concentration and which lowers the a_w level sufficiently to prevent growth of pathogenic organisms. Retail sale is usually in cans which do not undergo further heat treatment. Evaporated milk contains no sugar and will support the growth of *Staph. aureus*, and possibly other pathogens. Cans for retail sale are fully heat sterilised, and bulk evaporated milk must be kept under refrigeration. It is necessary to ensure that after leaving the evaporator, the concentrated milk is either cooled rapidly, or held at a temperature above 60°C.

A small quantity of concentrated milk is made by ultrafiltration. The lack of prolonged heating results in a product of higher quality, but considerable care is required in the sanitation of the ultrafiltration equipment.

Dried milk

Dried milk may be made by either the roller process in which concentrated milk is applied to steam-heated rollers, the dried milk forming a sheet which is removed by a scraping blade; or by the spray process in which concentrated milk is introduced into a tower drying chamber as atomised droplets, is dried by an upward stream of hot air, and falls as a powder to the bottom of the plant. The production of spray-dried milk is outlined in **Flow diagram 48** and critical control points summarised in **Table 172**.

In many countries, the roller process has been superseded by spray drying, with the exception of some special purposes or for animal feed. Spray-dried milk powder is of superior organoleptic quality, but

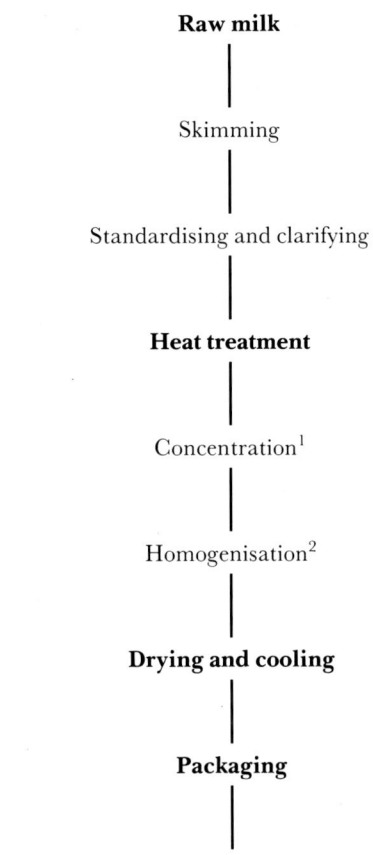

Flow diagram 48. Flow diagram: spray dried milk.

Raw milk
|
Skimming
|
Standardising and clarifying
|
Heat treatment
|
Concentration[1]
|
Homogenisation[2]
|
Drying and cooling
|
Packaging
|
Storage, distribution and sale

[1] The temperature in the first stages of the evaporator will inactivate many vegetable pathogens, but this is incidental and secondary to the control at the heat treatment stage.
[2] Older types of homogeniser are notoriously difficult to clean, and can be a major source of contamination unless high temperatures are maintained.

Note: Processes in **bold** are CCPs.

faces continuing problems of product safety which are related to the intrinsic nature of the process.

Historically, food poisoning problems with dried milk were associated with the growth of, and enterotoxin production by, *Staph. aureus* either in milk before concentration, or in the concentrated milk before drying. These hazards are now recognised and con-

Table 172. Critical control points: dried milk.

> **Raw milk, CCP 2**
> *Control*
> Achieved at farm level (see pages 427–430).
>
> *Monitoring*
> If possible, general quality should be checked using the direct epifluorescent technique.
> Standards at supplying farms should be monitored and milk not accepted from farms known to be infected with *Salmonella* or other possible milkborne pathogens.
>
> *Comments*
> 'Desperate' diseases require desperate remedies and consideration should be given to taking only heat-treated milk into the plant (see text).
>
> **Heat treatment, CCP 1**
> *Control*
> The level of control and monitoring should be the same as for pasteurised milk.
>
> **Drying and cooling, CCP 2**
> *Control*
> Before entering the dryer, concentrated milk should be held at a temperature sufficient to prevent the growth of *Staph. aureus*.
> Spatial separation is required between different areas of the plant. Maintenance and cleaning procedures should be designed to prevent the build up of contamination in the plant and its environment. Incoming air supplies should be filtered and the whole air handling system should be operated so that contamination of the environment with powder dust is avoided.
>
> *Monitoring*
> All plant operating parameters must be carefully monitored. Air movements should be monitored to ensure the correct function of the air supply and exhaust systems, and the efficiency of air filtration checked. The dryer chamber and ancillary equipment should be regularly checked for cracks and damage.
> Verification is by examination for *Salmonella* and possibly other pathogens in both powder samples and the environment.
>
> **Packing, CCP 2**
> *Control*
> The packing area should be separate and protected from environmental contamination.
> Impervious packaging should be used to prevent water uptake during storage.
>
> *Monitoring*
> End product testing is used including the presence of *Salmonella* and other pathogens and indicator organisms. A positive-release system should operate.

trolled, although the potential remains and occasional outbreaks are reported (see page 249). Since 1965, when contamination of dried milk with *S. newbrunswick* caused 29 cases of illness in the USA, the major hazard has been seen as being *Salmonella*. However, it should be noted that large outbreaks involving *Y. enterocolitica* and *Clostridium perfringens* have also occurred in recent years, while illness in infants due to contamination of milk powder with *Enterobacter sakazakii* suggests the possibility of an important role as a vehicle of opportunistic Gram-negative pathogens.

The detailed sequence of events leading to food poisoning obviously varies considerably, but a general pattern appears to be contamination of the environment, followed by contamination of the product stream, which in turn leads to growth of the pathogens in ecological niches, either in defects in the dryer wall or in the powder handling system. The powder itself is contaminated, sufficient escaping during handling to re-contaminate the environment. The fundamental protection lies in the prevention of contamination of the plant environment, and environmental sampling is of key importance in monitoring the operation. Additional product sampling is required if pathogens are detected in the environment (NRC, 1985), but once contamination is detected in the environment, the prognosis is poor for the future of the plant, since sooner or later the process itself is likely to become contaminated. The task of eliminating pathogens from a spray dryer and associated equipment is massive and may not be successful, even after extensive dismantling.

Prevention of environmental contamination is likely to require even more rigorous measures than those used at present. Following an outbreak of *S.*

ealing food poisoning in an infant formulation, where contamination was known to have entered the plant in raw milk, the recommendation was made that no raw milk should enter spray-drying plants and that pasteurisation should take place at a remote site. Attention is also needed in other aspects of the safety management of spray-drying plants, cracks and damage in the lining of the drying chamber, for example should be detected during routine inspections. Where a spray-drying plant has operated for many years past its design life, extreme care is necessary.

Fermented milk products

Fermented milk products represent an extremely old method of food preservation, and an enormous range is produced on a worldwide basis. In developed nations, the most important products are hard and soft cheeses and yoghurt, and discussion will be limited to these groups. However, it must be appreciated that the categories are not distinct and that many intermediate products exist. Furthermore, the wide variety of cheeses means that discussion is further restricted to representative types of cheeses which illustrate the problems of controlling pathogens. Cheese making is outlined in **Flow diagram 49** and critical control points are summarised in **Table 173**.

Notes:
1 The flow diagram depicts generalised practices only. There is considerable variation between varieties and the distinction between 'hard' and 'soft' varieties is not absolute. Further information on the manufacture of different types may be obtained from Chapman and Sharpe (1981).
2 Heating of 'soft' cheese may be before or after draining. Some types are not heated.
3 Processes in **bold** are CCPs.

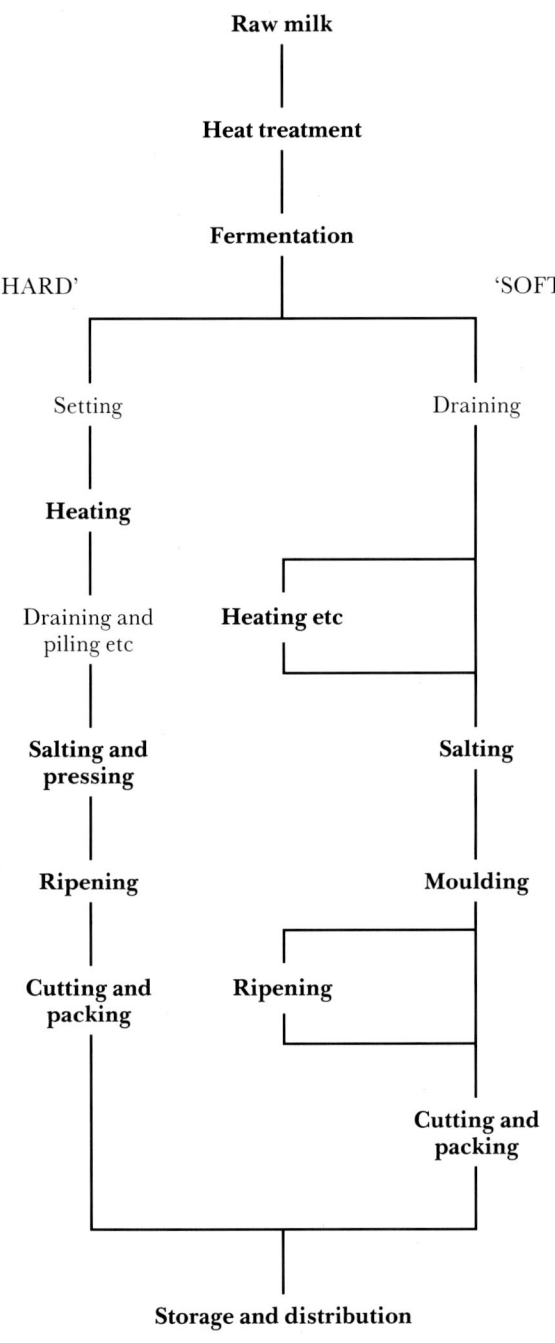

Flow diagram 49. Flow diagram: cheese.

Raw milk

Heat treatment

Fermentation

'HARD'   'SOFT'

Setting   Draining

Heating

Draining and piling etc   **Heating etc**

Salting and pressing   **Salting**

Ripening   **Moulding**

Cutting and packing   Ripening

Cutting and packing

Storage and distribution

Table 173. Critical control points: cheese.

Raw milk, CCP 2
Control
Achieved at farm level (see pages 427–430).

Monitoring
If possible, general quality should be checked using the direct epifluorescent technique.

Heat treatment, CCP 1
Control
The level of control and monitoring should be the same as for pasteurised milk.

Comments
Heat treatment of the milk is essential for the safety of cheese. However, on a number of occasions the treatment applied has failed to prevent food poisoning (see **Table 174**).

Fermentation, CCP 2
Control
Fall in pH value inhibits the growth of pathogens.
Active starter must be used, milk should be at the correct temperature and inhibitors should be absent (see **Table 175**).

Monitoring
Milk should be tested to ensure the absence of antibiotics.
Correct quantity of starter should be added according to the volume of milk in each vat.
Temperature should be monitored throughout process.
Equipment should be properly cleaned and all traces of disinfectant removed by rinsing.
Progress of fermentation should be monitored by determination of titratable acidity or pH value.

Heating, CCP 2
Control
Temperature controlled at <40°C to prevent the inactivation of a starter culture and to permit continuing acid production.

Monitoring
Measurement of temperature.
Continuing acid production monitored by the determination of titratable acidity or pH value.

Comments
Heating is applied to shrink the curd and release whey. It is not intended to inactivate pathogens. In soft cheese, release of whey may be achieved by pressing or simply by syneresis. These stages are not CCPs with respect to safety.

Salting, CCP 2
Control
Excess NaCl inhibits continuing acid production and favours the growth of *Staph. aureus*. In 'hard' varieties of cheese, too little NaCl may prolong the survival of pathogens.
NaCl level and distribution throughout curd must be within predetermined limits.

Monitoring
Required quantity of NaCl should be accurately weighed and distribution supervised by experienced personnel.

Pressing ('hard') cheese; moulding ('soft') cheese, CCP 2
Control
Prevention of contamination associated with high level of handling.
Strict hygiene of personnel and equipment required. 'Hands-on' operations should be minimised.

Monitoring
Health of personnel should be monitored on a positive basis. Efficiency of equipment cleaning should be assessed visually.

Ripening, CCP 2
Control
Time and temperature of ripening should be controlled to maximise inactivation of pathogens in 'hard' varieties and to minimise growth of pathogens in some 'soft' varieties. Consumers should be warned of risks associated with *L. monocytogenes* in Brie and similar varieties.

Table 173. *Continued.*

Monitoring

Temperature should be recorded throughout ripening and a formal system employed to ensure the period is correct. In some cases, monitoring of pH value is appropriate.

Verification that fermentation and later processing have been correct is by examination for *Staph. aureus* supported by the thermonuclease test where necessary. The thermonuclease test may be applied in the absence of detectable *Staph. aureus* if there is other evidence of starter failure. Where doubt exists concerning heat treatment, cheese should be examined for *Salmonella*.

Comments

Food poisoning associated with pathogenic strains of *E. coli* and, more recently, *L. monocytogenes* has led to suggestions that the safety of 'soft' cheese should be verified by examination for these bacteria. However, it is considered that current methodology is not adequate to permit routine use in this context.

Cutting and prepacking, CCP 2

Control

Prevention of contamination with potentially mycotoxigenic moulds and with other pathogens.

Packing unit should be separate from other activities and have filtered air supply under positive pressure.

Strict hygiene of personnel and equipment is required and hands-on operations should be minimised.

Monitoring

Air quality should be determined microbiologically. Health of personnel should be monitored on a positive basis. Cleaning of equipment should be assessed visually.

Where vacuum or gas packing is used to inhibit mould growth, regular examinations to ensure seam integrity are required.

Comments

Risk of bacterial pathogens is low with 'hard' cheese which will not support surface growth and which thus serves merely as a passive vehicle of infection. In the case of 'soft' cheese, pathogens are able to grow on the moist upper surface.

Risk of contamination is greatest in small-scale operations in supermarkets, etc, where there may be insufficient separation between cheese and other foods for prepacking such as meats. Conditions on service delicatessen counters where cheese is sliced at the point of sale may also be conducive to cross-contamination.

Storage and distribution, CCP 2

Control

Control of temperature and storage period to prevent the growth of pathogens.

Monitoring

Measurement of temperature at all stages of cold chain. Formal system of stock checking to ensure correct order of use.

Comments

Hazards are largely restricted to 'soft' cheese which will support the growth of pathogens. Problems of incorrect storage are often greatest at retail level, particularly in small stores.

Heat treatment of cheese milk

Heat treatment of cheese milk is seen by many, but not all, microbiologists as being of major importance in ensuring the safety of cheese. Opposition to heat treatment is based on claims that it is not possible to produce some varieties of cheese with heated milk due to the elimination of components of the raw milk microflora which contribute to the flavour of the product. There are also alleged textural problems due to partial denaturation of milk proteins during heating.

The alternative recommended procedure to ensure safety of hard cheese, holding for 60 days under refrigeration, is now recognised to be ineffective and both *Salmonella* and *Listeria* are known to survive beyond this period. Raw-milk cheese is known to have been associated with a large number of outbreaks of food poisoning involving not only well-known pathogens such as *Salmonella* and *Listeria*, but less common pathogens such as *Streptococcus zooepidemicus* (see page 361) which, in the USA, caused 16 cases of illness with 2 deaths following consumption of home-made Queso blanco cheese.

Despite the obvious risks associated with use of raw milk, it should be noted that a number of outbreaks have involved cheese made with heat-treated milk, and that some surveys have shown a similar incidence of *L. monocytogenes* in soft cheese made from raw and heat-treated milk. There are a number of reasons why heat treatment is less effective than expected (**Table**

174). Post-process contamination is probably of major importance, particularly in older premises where the degree of environmental contamination is likely to be high. Particular care is necessary where cheese is made on the farm, and where workers may handle both animals and cheeses.

Table 174. Heat treatment of cheese milk: failure to control pathogens.

1	Processes used are often less severe than full pasteurisation to minimise protein denaturation. Some are likely to permit the survival of *L. monocytogenes* and possibly other pathogens if growth has occurred in milk before treatment.
2	Heat treatment may be considered less critical for cheese milk and thus less well managed. Under temperature bypass devices may not be fitted or may be over-ridden manually. Process records may be inadequate.
3	Post-process contamination may occur. Several sources have been implicated in outbreaks including food handlers, starter culture and raw material, but in many cases environmental contamination is probably the most important source.

Control of fermentation

Control of fermentation is important to prevent the growth of pathogens. *Staph. aureus* is of greatest immediate importance, and food poisoning has resulted from the organism's growth and toxin production at this stage. However, growth of other pathogens increases the likelihood of their survival to consumption. Failure of fermentation is due to slow growth and acid production by the starter culture of lactic acid bacteria (**Table 175**). The use of active starter cultures is considered essential to the safety of cheese, and varieties made without starters are considered to be inherently hazardous.

In traditionally made yoghurt and some unripened soft cheeses, the high acidity of the final product is sufficient to inactivate many pathogens, and this contributes to the extremely good safety record of the products (Robinson and Tamime, 1981).

Table 175. Common reasons for starter failure.

1	Presence of antibiotics in milk.
2	Presence of sanitising agents such as quaternary ammonium compounds.
3	Infection of culture with bacteriophage.
4	Insufficient starter culture added.
5	Starter culture in poor physiological condition due to culturing in sub-optimal conditions.
6	Starter culture heavily contaminated by second micro-organism.
7	Temperature of fermentation too high or too low.
8	Excessive NaCl added after milling of curd (hard varieties only).

Note: Information on propagation and testing of starter cultures may be obtained from Sandine (1979) or Tamime (1981).

Behaviour of pathogenic bacteria during maturation

Pathogenic micro-organisms are unable to grow in hard cheese during ripening, and conditions of storage are intended to maximise the inactivation of any pathogens present. As noted above, however, inactivation cannot be guaranteed.

The situation in ripened soft cheeses is different in that a number of pathogens are able to grow in the later stages of ripening, particularly if the temperature is too high. In recent years, outbreaks of salmonellosis, shigellosis, botulism, listeriosis and infection with both enteropathogenic and enterotoxigenic *Escherichia coli* have been associated with consumption of Brie and similar varieties of cheese. The growth of the pathogens is associated with the rise in pH value due to proteolysis, although nutritional factors may also be involved. In the case of *Listeria*, growth appears to be limited to the crust of the cheese.

It is considered that control of pathogens in Brie-type cheeses must lie in heat treatment of milk, and the prevention of re-contamination. The problem may be reduced by strict control of temperature during maturation, but this is only likely to be effective against pathogens unable to grow at low temperatures such as *Salmonella*. People at high risk from listeriosis, including pregnant women, are advised not to consume Brie-type cheeses, and some people may consider the provision of correct warnings to be a critical control point.

E. coli is commonly present in ripened soft cheeses in large numbers, and may be a normal part of the microflora. The organism has no function as an index organism in cheese of this type, although diarrhoeagenic types should be absent.

457

Production of biogenic amines in cheese

Biogenic amines such as histamine and tyramine are produced in certain types of matured cheese by the decarboxylation of the amino acids histidine and tyrosine respectively. The presence of biogenic amines can cause a critical increase in blood pressure together with headache, flushing and sometimes rashes. In a large outbreak involving Stilton cheese, predominant symptoms were the sudden onset of vomiting, abdominal pain and diarrhoea in some cases.

Susceptibility to biogenic amines varies and people taking monoamine oxidase inhibiting (MAOI) drugs are particularly sensitive. Other factors may also be involved in determinig individual susceptibility.

In the past, Group D Streptococci (*Enterococcus*) has been associated with the production of histamine, but more recent work suggests that their role may be limited (Tham, 1988). *Lactobacillus buchneri* appeared to play a major role in outbreaks of histamine toxicity associated with the consumption of Swiss cheese (Sumner *et al.*, 1985), and some members of the *Enterobacteriaceae* may also be involved (*cf.* scombroid poisoning, pages 465–466). Free histidine is present only at low levels in cheese, and proteolysis is probably an initial step in the production of histamine (Sumner and Taylor, 1989). Histamine formation appears to occur more readily in some varieties than in others, but this may be an artefact due to inadequate recognition of this type of food poisoning.

Measures for controlling biogenic amine production are difficult to devise, partly because the process is not fully understood. The problem occurs only in aged cheese, and may thus be avoided by limiting maturation periods, but this is unacceptable to many consumers. Avoiding contamination with *L. buchhneri* has been suggested, but is likely to be difficult to implement since that species often forms part of the stable microflora of the plant (Chapman and Sharpe, 1981). Relatively straightforward methods are available for assay of histamine (AOAC, 1980, Lerke *et al.*, 1983), and methods have also been developed for the detection of histidine decarboxylating bacteria (Klausen and Huss, 1987; Sumner and Taylor, 1989). These may be used in the investigation of possible outbreaks of food poisoning associated with the production of histamine, or for verification of the status of suspect batches.

Control of value-added ingredients

A wide variety of flavourings and other ingredients may be added to fermented milk products, and the wider implications to safety management have been discusssed briefly on page 451. The need to monitor incoming ingredients must be stressed, since this aspect of control is often overlooked. There is particular complacency with respect to canned goods, and failure to adequately monitor the condition of hazelnut puree used to flavour yoghurt was a factor in an outbreak of botulism (see pages 298–299).

Processed cheese

Processed cheese is produced from a blend of hard cheeses and emulsifying salts, which is cooked in the molten state and re-solidified. The cooking is a critical control point with respect to the control of vegetative pathogens, but endospores survive and botulism has resulted from the growth of *Cl. botulinum* in the finished product (see page 298). Control of endospores is traditionally by low a_w level and low pH value, but these factors may not be sufficient and for that reason the antibiotic nisin is permitted in some countries. In some cases, the preservative system is only marginal for safe storage at ambient temperature and refrigeration is required. Care must be taken that the consumer does not confuse the less stable product with the more stable product, and suitable labelling is a critical control point. It is also necesary to ensure that hazards associated with added ingredients, such as ham pieces, are adequately controlled. This includes the possibility that such ingredients may locally destabilise the preservative system.

23 Control of pathogenic micro-organisms in eggs

Control at farm level

Under most circumstances, the interior contents of sound shell eggs are virtually free of micro-organisms and typically contain less than 10 organisms/g (Berquist *et al.*, 1984). However, pathogenic micro-organisms may enter eggs by two routes. The first route involves invasion from the exterior, the second route is by trans-ovarian infection during the development of the egg. Contamination of the contents of the egg by pathogens present on the outer shell surface when the egg is broken can also be a significant problem. In all cases, *Salmonella* is of greatest concern.

Infection of eggs from exterior contamination

Although the whole egg in the shell is largely self-protective, pathogens can enter if the shell and shell membranes are damaged, or if there is an excessive level of contamination at laying (**Figure 330**). The level of contamination may be reduced by providing clean nest material, by use of roll-away nests, and by frequent gathering of eggs. Egg trays should be washed and disinfected, and a high general standard of hygiene should be maintained.

Careful handling of eggs after gathering is also important and storage at high temperatures and high humidity should be avoided. Washing can remove gross contamination, but can predispose the egg to invasion if subsequent drying is inadequate. Furthermore, if the wash solution is lower in temperature than the egg, wash solution may be drawn into the egg through the cell pores when the contents contract (NRC, 1985).

Inspection is important in the reduction of risk due to external contamination of eggs, and heavily soiled and damaged eggs should not be used for human consumption.

Trans-ovarian infection of eggs

Until recently, external contamination of eggs has been of greatest significance with respect to food poisoning and, in hens eggs, trans-ovarian infection has been rare. The situation has changed in recent years, and trans-ovarian infection by *S. enteritidis* (phage type 4 in Europe, other phage types in the USA) is now a well-established problem (Anon, 1989; Humphrey *et al.*, 1989; Mawer *et al.*, 1989).

Although *S. enteritidis* PT4 is not host-adapted to chickens, the organism is invasive and is able to colonise extraintestinal sites such as the liver, spleen, oviducts and ovarian follicles (Anon, 1989). Unlike many other salmonellas, *S. enteritidis* PT4 infections are often associated with clinical symptoms. These include unevenness, weakness and septicaemia resulting in a high mortality and culling rate (Anon, 1988). However, while clinical symptoms are common among young broiler birds, *S. enteritidis* PT4 appears to behave differently in mature egg-laying hens. Such birds are susceptible to infection with as few as 100 cells of *S. enteritidis*, but overt symptoms rarely develop despite the organism being recoverable from viscera and egg contents (Humphrey, 1990).

Infection of flocks with *S. enteritidis* PT4 is widespread, but despite this the incidence of the organism in eggs is currently low. Studies of naturally infected flocks have shown that while *S. enteritidis* may be isolated from eggs during investigations of an outbreak, subsequent attempts at isolation are un-

330. Inspection of eggs. Inspection of eggs for soiling and damage is an important stage in the prevention of contamination from the exterior. It is virtually impossible, however, to produce eggs which are all totally free of soil, and small quantities such as that illustrated are normally of no concern. At the height of public concern over *S. enteritidis* PT4, eggs were rejected when very small amounts of soil were present on the exterior even though this offered no protection whatsoever from salmonellas which had entered by the trans-ovarian route.

successful despite continuing infection among the hens. Investigations using singly housed, naturally infected birds supported these observations, the production of contaminated eggs being clustered but intermittent (Humphrey, 1990).

Estimates of the incidence of contaminated eggs produced during outbreaks are in the order of 0.1 to 0.4%. Although this incidence is low, it must be appreciated that the number of eggs consumed is very large, and that in some egg dishes a single contaminated egg may result in the infection of a large number of people. The incidence may also be underestimated due to sampling problems resulting from the uneven distribution of salmonellas in infected eggs, and it is recommended that the entire contents *and* the shell should be examined during investigations.

In the long term control and prevention of problems associated with *S. enteritidis* in poultry can be resolved only by elimination of the organism from both layer and broiler flocks. In this context, the existence of trans-ovarian infection means that eradication depends on infection-free breeder flocks. The development of a sensitive test for the detection of *S. enteritidis* in flocks is a prerequisite for the elimination of the organism, and is currently considered to be a priority (Anon, 1989). Infected birds could then be culled and replacements bred from infection-free flocks, strict precautions being taken to prevent re-infection.

In the short term, some consumer protection can be afforded by the control of the temperature and storage period of eggs. The numbers of *S. enteritidis* present in freshly laid eggs are small (less than 10), and thus offer relatively little risk to healthy individuals, but multiplication is rapid during storage at room temperature. Multiplication is effectively controlled by storage at 8°C, while at 10°C there is a lag period of 5 days before multiplication commences (Humphrey, 1990). However, it is likely to be difficult to implement refrigerated storage of a product which traditionally has been stored at room temperature, and while producers may increasingly store at 8 to 10°C, room temperature storage, in the UK at least, continues to be very widespread at the retail level. It should further be appreciated that the five-day lag at 10°C may be of little relevance when subsequent storage is for extended periods.

Processing for safety

Adequate cooking and prevention of cross-contamination are the prime factors in the prevention of eggborne infection, irrespective of the route by which the egg was infected and the micro-organism involved. Despite repeated warnings to both the general public and to caterers concerning the need to ensure that eggs and egg dishes are adequately cooked, salmonellosis attributed to eggs has continued to be reported. Although this is at a lower level, the situation illustrates the difficulty of overcoming prejudices concerning foods perceived as safe.

It is unfortunate that raw eggs should be considered as 'virtuous' foods by some people, and that this concept should persist despite clear evidence as to the role of eggs as vehicles of infection. The hysterical reaction of other sectors of the community was equally unjustified but, unlike the cook or caterer who persisted in the use of raw eggs, put no one at risk. Mention should also be made of those with vested interests who continued to deny the existence of hazards associated with eggs as well as the confusion caused by cookery writers who, despite no true microbiological knowledge, chose to make lofty pronouncements on the subject and continued to describe raw egg recipes. Cases of food poisoning are known where the person preparing the food was aware of the potential hazards of raw eggs, but had been confused by conflicting information.

Most outbreaks of egg-associated *S. enteritidis* infections have involved either direct consumption of raw egg by, for example, body builders, or the preparation of dishes containing raw egg as an ingredient. Mayonnaise (see below) has been most commonly implicated, but other dishes involved include mousse, cassatta and home-made ice cream. The increased popularity of home ice cream making machines may have been a contributory factor to the increase in egg-associated infection. However, in some cases cooked egg dishes have also been involved.

Survival of *S. enteritidis* in cooked eggs

The involvement of hard-boiled egg sandwiches in early reported outbreaks of *S. enteritidis* PT4 infections led to suggestions that the organism is unusually heat resistant. However, subsequent investigations have shown that while *S. enteritidis* is less heat sensitive than some serovars, it does not have an unusual degree of heat resistance and is certainly less resistant than *S. senftenberg*. Experiments with artificially inoculated eggs showed that boiling for at least 7 minutes is required to enable the yolk of eggs to reach a sufficient temperature to kill the organism, and that survival is also likely in some fried and slowly scrambled eggs (Anon, 1989), and in lightly cooked omelettes (Perales and Audicana, 1988). The survival in hard-boiled egg may be attributed to the relatively light cooking used in food-service establishments both to permit rapid preparation and so that the egg retains sufficient liquid to cohere when used as a sandwich filling.

Survival of *S. enteritidis* and other micro-organisms is likely to be greatest in the yolk, due to its viscous nature slowing the penetration of heat and to the protective effect of fat (Humphrey, 1989).

Egg products

Liquid, frozen and dried eggs

In addition to trans-ovarian infection and contamination from the shell exterior, liquid, frozen or dried eggs are subject to contamination from personnel and from utensils and equipment. *Salmonella* is of greatest significance as a contaminant, but other micro-organisms including *Staphylococcus aureus* may be involved. Pasteurisation is necessary to assure the safety of egg products, and is mandatory in a number of countries. International trade in unpasteurised egg products represents a serious hazard as vehicles for the dissemination of salmonellas, and this trade should be discouraged (ICMSF, 1986).

Whole egg and egg yolks may be pasteurised at relatively high temperatures, a *minimum* treatment of 64.5°C for 2.5 minutes being recommended. Such temperatures cannot be applied to egg white (albumen); spray-dried albumen should be held at not less than 54.4°C for not less than 7 days, and pan-dried albumen at 51.7°C for not less than 5 days to ensure safety.

Tests for inactivation of the enzyme amylase may be used to verify the pasteurisation process for whole eggs and yolks, providing that the temperature of the process exceeds 64.5°C. The amylase test cannot be used to verify lower temperature treatments and freedom from *Salmonella* is verified by microbiological analysis according to a statistical sampling plan (USDA, 1980).

Other egg products

Egg is used as an ingredient in many products, but in most cases quantities are relatively small and the microbiology is largely that of the major ingredients, although the egg may be an important potential source of pathogens. However, there are a number of exceptions of which mayonnaise is most important.

Mayonnaise is an oil-in-water emulsion which contains egg, vegetable oil and an acidulant, usually acetic acid. Mayonnaise is used both directly as a salad dressing, and as a base for other products including tartare sauce.

The safety of mayonnaise made with raw egg relates directly to the pH value and the concentration and nature of the acidulant. In the past, a pH value of 4.1 or below and a concentration of acetic acid of at least 2.5% has been considered to be effective, not only in preventing the growth of such pathogens as *Bacillus cereus*, *Clostridium perfringens* and *Staph. aureus*, but in rapidly inactivating salmonellas during storage at room temperature (NRC, 1985). However, there is reason to doubt that such conditions do in fact ensure safety from salmonellas since *S. muenchen* has been shown to survive (Collins, 1985) and *S. enteritidis* PT4 to grow (Humphrey, 1989) at pH 4.0. Both serovars die rapidly at pH 3.0 to 3.5, but mayonnaise with this level of acidification is organoleptically unacceptable to many consumers. The possibility also exists that salmonellas may become acid habituated (Humphrey, 1989).

Acetic acid is the most effective acidulant, and mayonnaise made with citric or hydrochloric acid requires heat treatment of the eggs for safety.

Commercially prepared mayonnaise was responsible for major outbreaks of salmonellosis during the 1970s, the largest of which, in Denmark, involved approximately 10,000 cases. Following this outbreak, a procedure involving treatment of the egg yolk with vinegar for either 4 days at room temperature or for not less than 2 hours at 40°C was applied, and no further outbreaks have occurred. Pasteurisation of the egg is, however, considered to be the preferred treatment, although precautions must be taken against post-process contamination.

Although food poisoning due to mayonnaise normally involves *Salmonella*, other micro-organisms may be involved including *B. cereus* and *Staph. aureus*. As with *Salmonella*, home-made mayonnaise is most commonly involved.

A limited number of egg-based ready meals are available, including pre-prepared filled omelettes, and egg salad in aspic (Sutherland *et al.*, 1986).

Criteria for safety are adequate cooking of the egg and other ingredients, prevention of contamination after cooking, and control of temperature during storage. In many cases, a high degree of handling of the cooked product is necessary and may represent the major hazard.

24 Control of pathogenic micro-organisms in fish and shellfish

Fish

Fish are obtained either by capture or by aquaculture. Fish captured in seas and oceans are relatively free from human pathogenic micro-organisms. The major exceptions are *Clostridium botulinum*, which is a problem only in fish products, and species of *Vibrio* especially *V. parahaemolyticus*. The level of contamination reflects the geographical areas in which fish were captured and the feeding habits of the fish. For example, deep-sea fish are markedly less contaminated than those caught inshore, while bottom feeding fish usually have the highest level of contamination with *Cl. botulinum*. The method of fishing may also determine the level of contamination; fish caught in trawl nets are often more highly contaminated than those line caught due to the net stirring up bottom sediments.

Fish captured in lakes and rivers as well as in some estuarine and inshore situations often carry a wider range of pathogenic micro-organisms due to contamination of the water with sewage, land run-off, etc. Potential human pathogens such as *Aeromonas* and *Plesiomonas* are also common in freshwater and estuarine ecosystems, while *Salmonella* appears to be a natural inhabitant of some tropical and subtropical freshwater systems.

Aquaculture is now carried out on a large scale both in ponds at inland sites and at coastal situations. Specific problems of contamination exist which may be related to the practices employed (**Figures 331–334**).

Fish-associated food poisoning is not restricted to infections and intoxications due to pathogenic micro-organisms. The organs of some fish such as the pufferfish (tetradotoxin) in Japan, and some types of octopus and squid in the West Indies, are highly toxic. In other cases, toxins in fish are derived from dinoflagellates which form part of the food chain (see pages 464, 468–470). A third type of toxicity, scombrotoxic poisoning, is associated with the activity of the spoilage microflora (see pages 465–466). Consumption of raw fish is also increasingly associated with parasitic infections due to the nematode *Anisakis* (see below).

331

332

331–334. Rearing of trout by aquaculture. Although aquaculture has been practised on a small scale for many years, large-scale operations have developed relatively recently. A wide range of fish may be reared by this means, but aquaculture is only economically viable for high value species such as trout (**331**). Inland sites require a reliable supply of high-quality water which in this case is drawn from a chalk stream. Mud bottomed tanks (**332**) are widely used, but must be carefully managed to prevent excessive build up of endospores of *Cl. botulinum*. Concrete bottomed tanks (**333**) are readily cleaned when emptied (**334**).

Hazards associated with the consumption of raw fish

In the Western world, most fish is cooked directly before consumption, and the safety record is therefore good. However, raw fish has been eaten by tradition in Japan, south-east Asia and in parts of northern Europe, including Scandinavia, and there is an increasing trend to eating raw fish elsewhere, particularly in the USA. This stems from the increasing popularity of the Japanese cuisine and the conception of the raw fish dish sushi as a virtuous food.

Species of *Vibrio*, especially *V. parahaemolyticus*, present the greatest hazard of pathogenic microorganisms. Risk is significantly greater with fish caught in warm waters. In the USA, however, parasitic infections with *Anisakis* associated with the consumption of raw fish has been of major concern.

Anisakiasis is recognised to be increasing in incidence in the USA and the relatively small number of reported cases are believed to be a gross underestimate. The problem serves as an illustration of the way in which enthusiasm for a new cuisine may introduce new hazards, and of the difficulties that arise in informing the public of the degree of risk. It has been widely stated, for example, that an experienced sushi chef can detect *Anisakis* by careful inspection of the fish, but this is not possible if the parasite is lying parallel to the muscle of the fish and extremely difficult if present as a cyst. Furthermore, the allegation that *Anisakis* can be killed by thorough chewing has been totally dismissed by the USFDA: *'Chewing is not a recommended measure for prevention of parasitic diseases'* (Anon, 1989).

Anisakis is killed by normal cooking and may be inactivated in raw fish by freezing. However, a temperature of −20°C is required which can only be reliably obtained by commercial freezers. Home freezers and those employed in the food service do not reliably

reach this temperature. A holding period of 7 days is required to ensure inactivation.

Although anisakiasis is a new concern in the USA, the infection is widespread in many countries where raw fish consumption is common. Nematodes other than *Anisakis* may also infect man. These include *Terranova* spp and the common cod worm (*Phocanema* spp), although the latter is of low infectivity.

Dinoflagellate (ciguatera) poisoning

Ciguatera poisoning is a serious problem in tropical and subtropical regions, and is due to the consumption of fish containing ciguatoxin and possibly other closely related toxins. These are synthesised by the dinoflagellate *Gambierdiscus toxicus*, and possibly some other dinoflagellate species. These toxins enter the food chain as a result of dinoflagellates being eaten by small herbivorous fish, which in turn are eaten by larger and older carnivores. Toxicity is amplified by transfer through the food chain, and thus large carnivorous fish are usually most toxic. Numerous species may be involved including moray eel, amberjack and barracuda.

Symptoms of ciguatera poisoning may appear immediately after the consumption of toxic fish, or may be delayed by a few hours. Symptoms include severe nausea, vomiting and diarrhoea, and various neurosensory disturbances most commonly involving paraesthesia and dysaesthesia. Severe cases may result in shock, convulsions, muscular paralysis and death. Artificial respiration is required where symptoms involve paralysis. In fatal cases, death usually occurs within 24 hours and those surviving this period usually recover completely. However, convalescence may be prolonged for several months.

Factors responsible for the build up of dinoflagellate toxins are not well understood and, with the exception of ciguatoxin, the toxins involved are poorly defined. Assay methods exist but are of low precision and specificity. It is not therefore possible at present to apply surveillance programmes to the prevention of ciguatera poisoning. The problem is confined almost entirely to warm climates, and the importation of ciguatoxic fish is discouraged by many temperate countries.

Ciguatoxin is stable to heat and to high concentrations of NaCl etc, and protection is not offered by cooking or other processing. Some protection is afforded by avoiding unusually large reef fish and excluding the internal organs and the roe from the diet (Taylor, 1988).

Scombrotoxic poisoning

Scombrotoxic poisoning results from eating fish, usually of the families *Scombridae* and *Scombersocidae*, which have become toxic as a result of microbial contamination and growth. The symptoms of scombrotoxic poisoning appear within 10 minutes to 2 hours after consumption of toxic fish. A wide variety of symptoms can occur but usually only a few are experienced by individuals. The most common symptoms reported in the UK are a peppery or burning sensation in the mouth, flushing and a rash of the face and neck (Murray *et al.*, 1982). Flushing is often accompanied by a severe headache and cardiac palpitations may occur (Arnold and Brown, 1978). Death due to heart failure has been reported. Further symptoms include diarrhoea, nausea and vomiting, itching, dizziness and visual disturbances.

The symptoms of scombrotoxic food poisoning are essentially those of histamine toxicity, and it is considered that histamine produced by decarboxylation of histidine is primarily responsible for the illness. A number of micro-organisms are capable of decarboxylation of histidine including *Klebsiella pneumoniae* and *Morganella morganii*, which are involved in the spoilage of fish. It is unlikely, however, that histamine alone is responsible for scombrotoxic poisoning, since orally administered histamine is surprisingly non-toxic to humans (Bartholomew *et al.*, 1987). This paradox may be explained by the presence of histamine potentiators in the implicated fish (Taylor *et al.*, 1984).

A wide range of scombroid fish have been implicated in scombrotoxic food poisoning, including albacore, skipjack and yellow-fin tuna, mackerel, mahi-mahi, and bonito. Non-scombroid fish have also been implicated including sardines, pilchards and anchovies. Salmon has also been responsible for illness of a similar nature to scombrotoxic poisoning, although most of the fish contained only low levels of histamine and it is possible that the illness was caused by another toxin (Bartholomew *et al.*, 1987).

Outbreaks of scombrotoxic food poisoning are of two types. In the first, a cluster of cases occur which may be related to specific events, such as seasonal availability or increased popularity of a particular fish, leading to increased demand (**Flow diagram 50**). The second type are sporadic cases, involving only one person, which have increased in number in recent years. It is possible that sporadic cases are due to allergies to specific types of fish, but this seems unlikely and a more probable explanation is likely to lie either in the extremely localised distribution of histamine in the muscle of large fish such as tuna (Lerke *et al.*, 1978), or in the fact that only a small number of fish in a catch may contain significant quantities of histamine.

Scombrotoxic food poisoning is unique in being caused by the activities of spoilage micro-organisms, and as such it is wholly preventable. Control lies in rigorous application of refrigeration at all stages up to cooking, formation of histamine being very slow at 4°C and ceasing at 0°C. Modern, deep-sea fishing boats are equipped with mechanical refrigeration, but with line-caught fish such as some tuna, the first of the catch may be stored for excessive periods in relation to the efficiency of the refrigeration plant, resulting in a small number of fish becoming toxic. Inshore fishing boats often rely on ice which may be inadequate during hot weather, while day boats often have no means of refrigerating the catch whatsoever. Further opportunity for temperature abuse occur on landing, at auctions and during transport. Ideally, fish should be purchased direct from fishermen who are aware of the need for refrigeration and who have adequate on-board facilities. In cases of scombroid poisoning due to bluefish, a significant contributory factor was the purchase of fish from recreational fishermen who had no understanding of the need for refrigeration (Karolus *et al.*, 1985) and this practice should be avoided.

Histamine is unaffected by processing, including the heating involved in canning, and for this reason fish which is considered unacceptable for fresh consumption should not be used for any other purpose.

It is not always possible to monitor the handling of fish, and for this reason a means of determining the safety of scombroid and other sensitive fish is required. Visual inspection offers only a partial solution, since toxic fish may show no sign of spoilage. However, relatively simple methods of assaying for histamine are available (AOAC, 1980, Lerke *et al.*, 1983), and while uncertainty regarding the threshold toxic dose means that it is not possible to formulate regulations concerning the maximum permitted level of hista-mine, guidelines have been prepared (**Table 176**). However, it is necessary to be aware that the values quoted are only guidelines, and that scombrotoxic poisoning has been reported in fish containing only low levels of histamine.

Table 176. Levels of histamine in fish (Bartholomew et al., 1987).

<5 mg% histamine	Normal and safe for consumption
5 to 20 mg% histamine	Mishandled and possibly toxic
20 to 100 mg% histamine	Unsatisfactory and probably toxic
>100 mg% histamine	Toxic and unsafe for consumption

Flow diagram 50. Scombrotoxic food poisoning associated with smoked mackerel: contributory commercial factors.

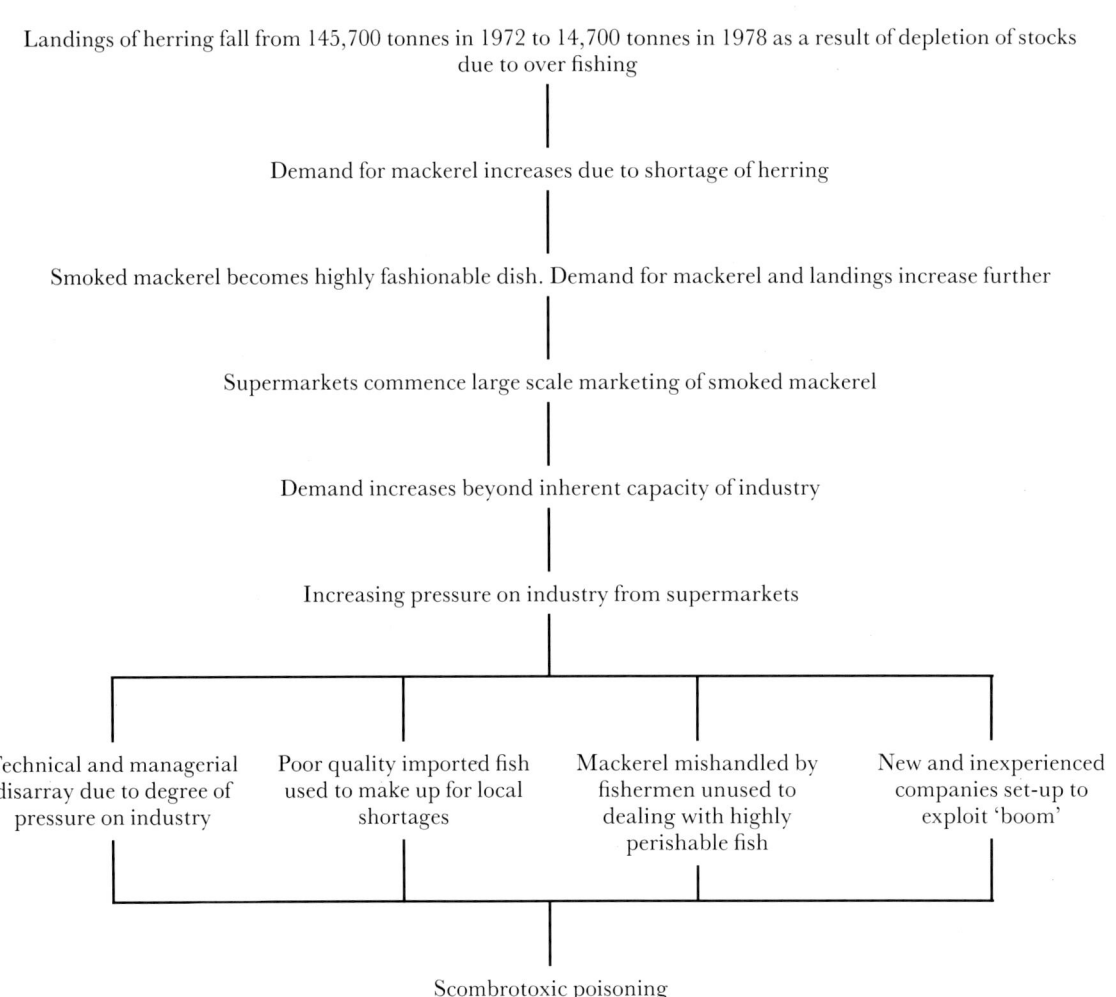

Landings of herring fall from 145,700 tonnes in 1972 to 14,700 tonnes in 1978 as a result of depletion of stocks due to over fishing

Demand for mackerel increases due to shortage of herring

Smoked mackerel becomes highly fashionable dish. Demand for mackerel and landings increase further

Supermarkets commence large scale marketing of smoked mackerel

Demand increases beyond inherent capacity of industry

Increasing pressure on industry from supermarkets

Technical and managerial disarray due to degree of pressure on industry

Poor quality imported fish used to make up for local shortages

Mackerel mishandled by fishermen unused to dealing with highly perishable fish

New and inexperienced companies set-up to exploit 'boom'

Scombrotoxic poisoning

Shellfish

Shellfish of commercial importance are of two major types, crustaceans which include crabs, lobsters, shrimps and prawns and molluscs which include oysters, mussels, cockles, clams and scallops.

Crustaceans

Crabs and lobsters are generally trapped and kept alive up to the point of cooking, while shrimps and prawns are captured by trawling and iced for transport to the cooking plant. Many crustaceans are caught at inshore sites, and are thus likely to be contaminated with pathogens derived from sewage as well as those derived from the marine ecosystem such as *V. parahaemolyticus* and other enteropathogenic vibrios.

Crustaceans are cooked before consumption and this, in theory, offers a considerable measure of protection against the pathogens present. However, the degree of processing is often determined by organoleptic considerations, and cooking is frequently uncontrolled. As a consequence, survival of many types of pathogens including enteropathogenic vibrios and *L. monocytogenes* have been reported. Further problems arise from the handling of the cooked fish, which is often of an unacceptably low standard (**Figure 335**).

Many of the problems with cooked crustaceans arise from the small scale, unstructured and archaic nature of the fishing industry. However, the production of cooked prawns is a major industry in the Indian subcontinent and parts of south-east Asia, substantial quantities being exported to Europe. Large-scale outbreaks of food poisoning including shigellosis and *Staphylococcus aureus* intoxications have resulted from recontamination of the cooked product at the processing plant, and irradiation treatment of imported cooked prawns has been suggested (Kayser and Mossel, 1984).

In the USA, there has been particular concern with respect to *V. vulnificus* infections among the immuno-compromised (see page 164), and it has been recommended that cooked seafood should be proscribed for AIDS patients and other immunocompromised people.

Molluscs

Most molluscs are harvested in estuarine waters and there is thus a possibility of contamination with sewage-derived pathogens, as well as those from the environment. Oysters and scallops are raked or trawled from the bottom, while clams and cockles are usually dug from the sand at low tide. Mussels are hand picked.

Molluscs are traditionally eaten raw or very lightly cooked. They are thus seen as being a high-risk food, and are widely associated with food poisoning probably as a consequence of increasing environmental pollution. Molluscs are major vehicles of hepatitis A virus and small, round, structured viruses. In the USA, some microbiologists consider sea-foodborne *V. cholerae* infections to pose a major threat to public health. Molluscs are also involved in various types of toxic poisoning (see below).

In the absence of an acceptable means of processing molluscan sea food for safety, attempts have been made to reduce hazards by control of the growing beds and by depuration or relaying. At best, these offer only partial solutions and may cynically be said only to reduce risk from very high to merely high.

335. Sale of cooked crabs from a seaside stall. The handling of cooked seafood at seaside stalls is often of an extremely low standard. Observation of operations at the stall illustrated showed crabs to be boiled for varying periods at an adjacent site, and cooled in water drawn by bucket from the harbour. The cooked crabs were then displayed for prolonged periods, and while a small quantity of ice was initially present this rapidly melted and was not replaced. The stall had no facilities for hand or utensil washing and no attempt was made to separate the handling of raw and cooked seafood.

Control of growing beds

In many countries, the harvesting of molluscs is restricted to areas which are free from direct sources of sewage pollution. Obvious sources of pollution may be detected by observation and a knowledge of local geography, but in other cases microbiological monitoring of the water is required. In the USA, the criterion for acceptability is the absence of 'coli-aerogenes' bacteria in 1 ml dilutions of growing area water (ICMSF, 1989). Verification may also be necessary for oysters based on both coliforms and aerobic plate count. It should be appreciated, however, that coliforms have no index function for *V. parahaemolyticus* and other enteropathogenic vibrios or for viruses. Any of these organisms may be present in the absence of coliforms.

Although control of growing beds is effective in controlling the activities of legitimate sea-food companies, fish from proscribed areas may reach the consumer through the activities of unlicensed people. In the UK, such activities are usually small scale and part time. In the USA, black-market trading in shellfish is a much greater problem, one dealer in South Carolina having sold over $1 million worth of illegally obtained shellfish over a one-year period (Anon, 1988a). The problem is contained only by active policing (Anon, 1986), people indicted facing a $20,000 fine and/or a five-year penitentiary sentence.

Depuration and relaying

Depuration involves placing shellfish from suspect growing beds into tanks of circulating seawater, which is continuously disinfected. Ultraviolet irradiation or chlorine is normally used to disinfect the water but ozone, although more expensive, may be more effective against vibrios (Anon, 1988b). Relaying may be used as an alternative to depuration for oysters and involves removal of the shellfish from suspect beds and their transfer to waters which are known to be unpolluted. Relaying is less effective than depuration, but involves no capital expenditure.

The underlying theory of both depuration and relaying is that any pathogens present will be shed into the water and flushed away. Properly controlled depuration will remove loosely attached bacterial cells, including those of sewage-derived pathogens; but a proportion of bacteria derived from the marine ecosystem such as *V. parahaemolyticus*, are capable of adhering more firmly to shellfish tissues, and are not readily removed and may even grow. Depuration is also of limited efficiency in the removal of viruses from the shellfish. This is possibly due to sequestration of the virus particles in the digestive gland or other tissues.

In the past, depuration has been verified by testing for the presence of coliform bacteria, *Escherichia coli* or enterococci. However, these organisms are not suitable as indicator organisms for vibrios or viruses, and alternative indicators are required.

As noted above, neither control of growing beds nor depuration can assure the safety of molluscan shellfish. These fish should be proscribed for the immunosuppressed while the healthy should be warned of hazards. In the long term, effective control may be possible by devising methods of cooking which inactivate pathogens without producing unacceptable organoleptic changes, particularly the development of an inedible rubbery texture. In the UK, a pilot plant for the heat treatment of cockles has been devised (see page 370). This plant is currently under evaluation and, if successful, it may be possible to extend the principle to other shellfish.

Toxic poisoning

Molluscan shellfish may become toxic due to the accumulation of various toxins acquired from dinoflagellates. In most cases, this happens only periodically when the dinoflagellates on which the shellfish feed 'bloom'. This occurs as a function of environmental conditions including water temperature, salinity, light and nutrient levels (Yentsch, 1984). The most important of these is temperature, and a minimum of 5 to 8°C is necessary for the development of a bloom. Such temperatures are usually attained only in shallow waters close to shore, and blooms thus often occur in the same location as the shellfish beds.

None of the toxins involved are heat sensitive, and thus no protection is offered by cooking or other processing.

Paralytic shellfish poisoning

Paralytic shellfish poisoning is a serious illness with a high mortality rate. The illness is a neurological disorder and symptoms, which usually appear within one hour of eating toxic fish, include tingling, burning and numbness of the lips and fingertips, ataxia, giddiness and staggering, drowsiness, dry throat and skin, incoherent speech, aphasia, rash, fever and peripheral and respiratory paralysis. In fatal cases, death usually occurs within 24 hours, and the prognosis is good if the patient survives this period; symptoms in non-fatal cases usually subside within a few days (Taylor, 1988). Artificial respiration is necessary in cases of paralysis, but no specific antidote is available.

Several dinoflagellates are involved in paralytic shellfish poisoning, including *Gonyaulax catenella* and *G. tamarensis*, the most important species in the USA. At least 12 related toxins known collectively as saxi-

336. Generalised structure of saxitoxins. (Hall and Reichardt, 1984).

	R1	R2	R3	R4	
1	H	H	H	H	Saxitoxin
2	H	H	H	SO_3^-	Saxitoxin B1
3	H	H	OSO_3^-	H	Gonyautoxin 2
4	H	H	OSO_3^-	SO_3^-	Saxitoxin C1
5	H	OSO_3^-	H	H	Gonyautoxin 3
6	H	OSO_3^-	H	SO_3^-	Saxitoxin C2
7	OH	H	H	H	Neosaxitoxin
8	OH	H	H	SO_3^-	Saxitoxin B2
9	OH	H	OSO_3^-	H	Gonyautoxin 1
10	OH	H	OSO_3^-	SO_3^-	Saxitoxin C3
11	OH	OSO_3^-	H	H	Gonyautoxin 4
12	OH	OSO_3^-	H	SO_3^-	Saxitoxin C4

toxins are produced (**Figure 336**), most toxic shellfish containing a mixture of several types. The toxins act by binding to the sodium channels in the nerve cell membranes and blocking nerve transmission.

Saxitoxins persist in shellfish for varying periods according to the fish and tissue involved. Some shellfish are toxic only for the actual period of the bloom, while others such as the Alaskan butter clam retains the toxin for many years and is never safe to consume (Schantz, 1984). In general, the muscle tissue (white meat) of shellfish contains less toxin than the internal organs (dark meat), and in some cases it is safe to consume the white, but not the dark meat.

Absolute control or prevention of paralytic shellfish poisoning is not possible. Prevention lies in the regular inspection of shellfish from the various beds, and analysis for saxitoxins using the mouse bioassay. A quarantine on affected beds is enacted when levels of saxitoxin reach 80 μg/kg (400 mouse units/100 g); commercial harvesting is stopped and signs erected to warn recreational harvesters. The quarantine is enforced until assay shows toxin levels to be within safe limits.

In the USA, the success of these preventative measures may be judged by the fall in the number of cases since quarantine restrictions were introduced. However, a significant number of cases do, still occur.

Neurotoxic shellfish poisoning

Neurotoxic shellfish poisoning is much less common than paralytic shellfish poisoning, and usually less severe. Symptoms include tingling and numbness of the lips, tongue, throat and area around the mouth, muscular aches, gastrointestinal disturbance and dizziness. The onset of symptoms varies from a few minutes to several hours, and the illness persists from a few hours to two to three days.

The dinoflagellate *Ptychodiscus brevis* (previously *Gymnodinium breve*) is associated with neurotoxic shellfish poisoning and is also responsible for the formation of red tides. At least two neurotoxins are produced and are referred to as brevetoxins B and C (Baden *et al.*, 1984). These toxins are polyether compounds which bind to nerve cells, but their mechanism of activity is otherwise unknown.

Prevention of neurotoxic shellfish poisoning is by monitoring for red tides, and the associated massive fish kills. Shellfish should then not be taken from the affected area for the duration of the tide (Taylor, 1988).

Diarrheic shellfish poisoning

Diarrheic shellfish poisoning has been described in Chile, the Netherlands and Japan (Kat, 1983; Yasumoto *et al.*, 1984). The organisms responsible are widespread, and diarrheic shellfish poisoning is likely to be a potential hazard elsewhere. The onset of the illness is between 30 minutes and several hours after the consumption of toxic fish, major symptoms being diarrhoea, abdominal pain, nausea and vomiting. The duration of symptoms is usually short, but may persist for several days in severe cases.

The toxins of diarrheic shellfish poisoning are produced by two species, *Dinophysis fortii* and *D. acuminata* and are polyether compounds. At least five related toxins are produced: two dinophysistotoxins, two pectenotoxins and okadaic acid. The mechanism of activity is not known.

The toxins are largely concentrated in the digestive organs, and the risk of diarrheic shellfish poisoning is reduced by avoiding the consumption of internal organs. A mouse bioassay is available for the toxins, and thus if necessary a system of surveillance and quarantine similar to that used for paralytic shellfish poisoning could be adopted.

Other toxins

A hitherto unknown toxin present in cultivated mussels from farms in Prince Edward Island,

Canada, caused more than 100 cases with three deaths in an outbreak during 1987 (Anon, 1987). This shellfish poisoning differed from other types in the long period between eating the toxic food, and the onset of symptoms – days rather than hours – and the bizarre nature of many of the symptoms suffered. These typically included burning sensations on the skin, memory loss, delirium and breathing difficulties.

Investigations showed the toxicity to be due to domoic acid produced by the chain forming phytoplankton *Nitzschia pungens*, a widely distributed organism which was not previously known to be toxigenic (Subba Rao *et al.*, 1988). Domoic acid (**Figure 337**) is an excitatory amino acid, which acts as an antagonist to glutamate in the central nervous system.

The conditions which led to the bloom of *N. pungens* are not fully understood, but the finding that such a widespread organism is toxigenic, has led to obvious concern. The wider question may also be posed as to whether still further species are toxigenic under certain conditions.

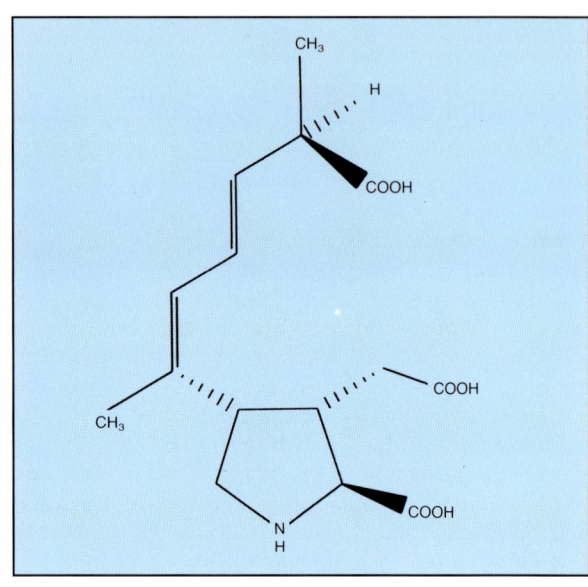

337. Structure of domoic acid. (Subba Rao *et al.*, 1988).

Fish products

Cooked fish products

A number of types of cooked fish products are manufactured, ranging from whole breaded or battered pieces of fish, through comminuted products such as fish cakes to complete dishes such as fish pie. In many cases, cooking is partial and must be completed by the end user. Many cooked fish products are frozen before distribution, although some are chilled only.

Fried, battered fish, usually eaten with chipped potatoes, is a traditional UK convenience food, and large quantities are produced despite competition from more modern convenience foods such as burgers and Indian or Chinese meals.

Critical control points for the safety of cooked fish products are essentially the same as for cooked meats (see pages 439–440). However, two particular points should be noted. In some cases fish are steamed or lightly cooked to facilitate the removal of the flesh from the bones before further processing. Removal of the flesh carries a relatively high risk of contamination, and strict hygiene of both equipment and personnel is required. Enterotoxin production by *Staph. aureus* is of particular concern, since the enterotoxin will survive subsequent cooking. Careful control of the batter used for coating fish prior to frying is also required since this will support the rapid growth of *Staph. aureus*, and cases of intoxication resulting from growth in the batter are known. Full hygienic precautions are required during the make up of the batter, which should be kept refrigerated until used. If this is not possible, small batches only must be made up. Left over batter should not be carried from one batch to another and all utensils must be fully sanitised between batches.

Salt, dried and smoked fish

Salt fish

Salt fish may be produced either by immersion in a brine or by packing in dry NaCl. Nitrite used in meat curing is not normally used with fish, although its addition has been suggested to improve the safety margin with respect to *Cl. botulinum*. The critical control point for the safety of salt fish is to ensure that NaCl, at an adequate concentration to inhibit the growth of pathogens, is distributed evenly throughout the fish.

Anisakis is sensitive to high NaCl concentrations, but inactivation is slow and may take up to 4 weeks.

Dried fish

Considerable quantities of sun-dried fish are produced in warm climates, especially in developing countries where refrigeration is not generally available. Small quantities are produced as delicacies elsewhere. Most dried fish is salted before the drying process, although small fish and thin strips of flesh are dried unsalted.

The microbiological safety of dried fish is dependent on the prevention of growth of pathogens during drying. In a hot, dry climate, small fish, and thin strips of flesh, dry sufficiently quickly to obviate the need for further precautions, but salting is required where larger fish are involved. Salting cannot guarantee the safety of uneviscerated fish, and the manufacture of products such as kapchunka has been proscribed in the USA following a number of outbreaks of botulism (see page 299).

Smoked fish

Fish are usually salt cured before smoking, and may also be fully or partly dried. Traditional smoked products were highly stable at room temperature, but modern processes are much less severe and in some cases have little antimicrobial role. This should be reflected in the handling and shelf life of the finished product. Where process changes result in changes in the inherent stability of the product, these should be understood by all people handling the fish, including retailers and consumers. This is of particular importance where a need for refrigeration is introduced and where vacuum packaging prevents the development of a spoilage microflora.

The safety of smoked fish depends on correct salting, adequate heating, drying and deposition of antimicrobial compounds during smoking and, in most cases, refrigeration of the final product. Smoking may be either by a hot or cold process, the temperature in the former often being sufficiently high to inactivate the endospores of non-proteolytic strains of *Cl. botulinum* which are capable of growth during refrigerated storage. Alternatively, the temperature may be raised at the end of the cold process to meet the same objective. Traditionally, constructed kilns are often notoriously difficult to control, and hazards arise from low temperatures permitting microbial growth during smoking and uneven drying, permitting growth in the finished product. These problems may be overcome by use of the more modern Torry kiln.

Anisakis is likely to survive both cold smoking and the relatively short curing process applied to smoked fish. Furthermore, in uneviscerated fish the rate of migration of *Anisakis* into the flesh may be increased by these processes. Commercially frozen fish should be used in cases where *Anisakis* will survive other processing.

Fermented and pickled fish

Fermentation of fish and marine mammals is a traditional means of food preservation which is important in many parts of the world today. Fermented fish products are essentially of two types: tropical and European.

Tropical fermented fish

A large variety of tropical fish products are made including various fish sauces, pastes and whole fish. The basic manufacturing process is the same; fish are packed with salt into earthenware pots and buried for several months while an anaerobic fermentation takes place (Erichsen, 1983). The final NaCl concentration of the product is 20 to 25%, and thus food poisoning is rare although even distribution of salt is necessary for safety.

European fermented fish

Although fermentation is less important as a means of preserving fish in Europe, large-scale production of some types continues. Details of processing varies according to type, but Scandinavian titbits, which are made from herring, are a typical example. During the fermentation, the fish are packed in wooden barrels with a mixture of NaCl and sucrose and allowed to mature in the brine formed for several months. At the end of this period, the NaCl concentration is *ca* 13% and the pH value 5.6 to 6.0. After maturation the fish is washed, filleted, cut into pieces and packed in cans in a marinade containing acetic and lactic acids, sucrose and benzoic acid. The final NaCl concentration is 8 to 9% and pH value *ca* 4.6. The cans receive no heat treatment and are not fully shelf stable, refrigeration being required (Erichsen, 1983).

Microbiological safety hazards in fermented fish are represented by *Staph. aureus* and *Cl. botulinum*. Safety during fermentation is assured by adding the correct quantity of NaCl and ensuring even distribution. Acid production from sucrose probably has a secondary role. Safety of the finished product is dependent on the marinade being correctly made up and on refrigeration.

Mention should be made of fermented fish and marine mammal products produced by native peoples in Alaska. These have a strong association with botulism (see page 299), and are considered to be intrinsically unsafe.

Pickled fish

Various types of unfermented fish pickled in a vinegar-based marinade are produced. In most cases, safety is assured by the low pH value and high concentration of acetic acid. Some marinades, however, are of relatively high pH value and low acetic acid concentration and require refrigeration. *Anisakis* may remain viable in products of this nature.

Surimi

Surimi is a fish muscle derivative that is increasingly widely used for the production of seafood analogues. In the production of surimi, the minced, washed flesh of high myosin content is ground with NaCl, the myosin solubilised, and the resulting fish muscle paste heat set to form a firm gel. Seafood analogues made from surimi closely resemble the actual fish, but are considerably cheaper and permit the economic use of underutilised species such as menhaden. The safety of surimi products is currently in doubt because of the high incidence of *Cl. botulinum* in surimi made from some types of fish. However, surimi made from pollock does not appear to have a significant problem with *Cl. botulinum*. The safety of imitation crabmeat made from surimi analogues may be assured by the control of *Cl. botulinum* type E and other non-proteolytic strains by high temperature pasteurisation, and by the presence of NaCl and other inhibitors. Refrigeration is necessary to control proteolytic strains of *Cl. botulinum*.

25 Control of pathogenic micro-organisms in miscellaneous foods and processes

Fresh vegetables and fruit

The greatest risk of contamination of vegetables and fruit is related to agricultural practices and involves the use of animal or human manure. The situation is most serious in less developed countries where the use of raw or partially treated sewage is commonplace. However, in developed regions, pathogenic micro-organisms may persist for extended periods in sewage sludge which is commonly used as fertiliser (Hyde, 1976; Watkins and Sleath, 1981). Risk is obviously greater with vegetables growing close to the ground, and thus subject to contamination from dust and rain splash. However, apples have been contaminated in orchards which have been manured with slurry or heavily grazed (Goverd *et al.*, 1979).

Contaminated irrigation water is a further potential source of pathogenic micro-organisms, and it should be appreciated that the liquid discharge from sewage treatment plants may contain significant numbers of pathogens.

Contamination of fruit and vegetables can occur during harvesting and subsequent handling, where, in many cases, precautions are directed at the prevention of damage to the produce rather than to the control of contamination. At farm level, precautions concerning the health of food handlers are rarely applied or even considered, while there are obvious, but not insurmountable, problems of supplying toilet and handwashing facilities. Problems are exacerbated by the employment of large numbers of casual and itinerant workers who may be housed under primitive conditions. It is difficult to assess morbidity due to contamination of vegetables or fruit by handlers, but an outbreak of hepatitis A was attributed to contamination during picking.

Refrigeration is widely used to extend the post-harvest storage life of some vegetables and fruit, but may permit the development of low-temperature pathogens such as *Listeria*. Refrigeration is also combined with modified atmosphere packing for display at retail level. This permits an extension of the shelf life by delaying wilting and other deteriorative changes in appearance, and by preventing mould growth. Refrigerated display cabinets for fruit and vegetables often operate at relatively high temperatures of 7 to 9°C, and growth of low-temperature pathogens is likely to be correspondingly relatively rapid. Growth of such pathogens is likely to be of particular significance with products which have been stored under refrigeration prior to packing.

Potential hazards are obviously greater with fruit and vegetables which are eaten without cooking. Salads may pose a particular hazard due to the difficulty of cleaning leafy components such as lettuce, and to the high degree of handling involved in preparation, and preprepared salad packs have come under particular scrutiny following the isolation of *L. monocytogenes* (see page 339). Salads also typify many of the general hazards associated with fruit and vegetables. Production of preprepared salads is outlined in **Flow diagram 51**, and critical control points are summarised in **Table 177**.

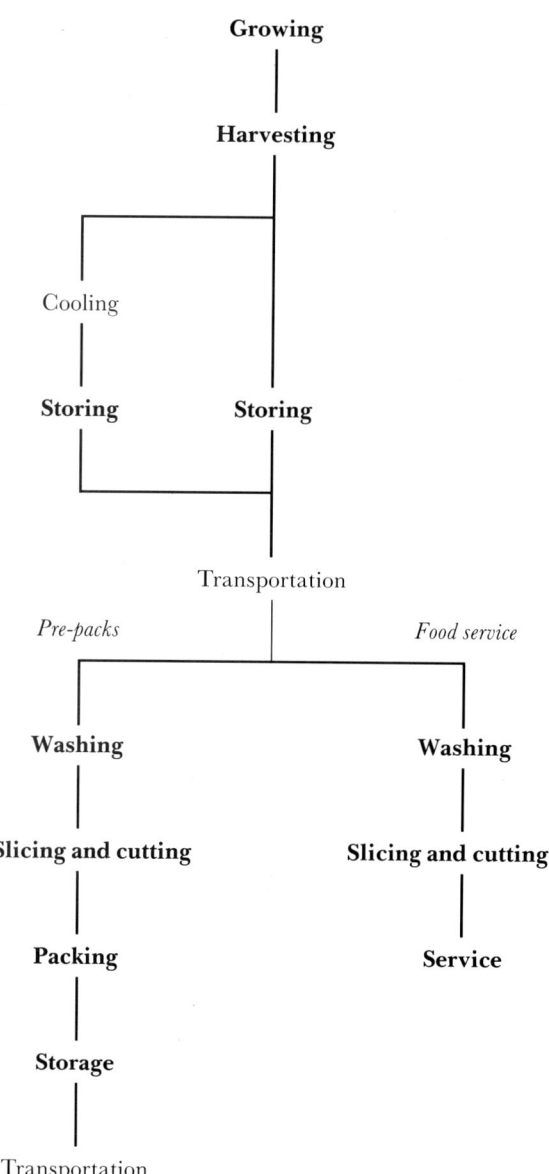

Flow diagram 51. Flow diagram: pre-prepared salad.

Growing

Harvesting

Cooling

Storing   **Storing**

Transportation

Pre-packs   *Food service*

Washing   **Washing**

Slicing and cutting   **Slicing and cutting**

Packing   **Service**

Storage

Transportation

Notes:
1 Some produce is washed and/or trimmed directly after harvesting.
2 Cooling is not a critical control point with respect to safety since the objective is to reduce the respiration rate and maintain quality.
3 Process in **bold** are CCPs.

Note: Examination of the end product for coliform bacteria is sometimes used to verify the product and process. The results obtained are likely to be highly misleading since members of the *Enterobacteriaceae* of no significance as index organisms are likely to be present in large numbers. Furthermore, although *E. coli* may be used as an index organism for pathogens of faecal origin, it is of no significance with respect to *L. monocytogenes* or other low-temperature pathogens.

Table 177. Critical control points: pre-prepared salads.

Growing, CCP 2
Control
Prevention of contamination of growing crop with pathogenic micro-organisms.
Only synthetic fertilisers should be used; sewage or sludge should not be used on ground used for raising salads or likely to be used for raising salads.
Irrigation water should not be drawn from sites downstream of sewage outfalls or close to other obvious sources of pollution.

Monitoring
Sources of fertiliser and irrigation water are monitored and any which are potential sources of contamination eliminated.
Verification is by microbiological examination of water and fertiliser for evidence of faecal pollution.

Harvesting, CCP 2
Control
Prevention of contamination by personnel or equipment.
Equipment should be cleaned on a regular basis. Handwashing and toilet facilities, portable if required, should be provided and personal hygiene controlled.

Monitoring
Visual examination of equipment. Observation of satisfactory personal hygiene.

Storage, CCP 2
Control
Prevention of contamination from environmental sources, wildlife, etc. In cold stores, limitation of growth of low-temperature pathogens.
Stores should be kept clean and protected from rodents, birds, etc. Temperature and maximum storage period should be controlled to limit build up of low-temperature pathogens.

Monitoring
Visual examination of cleanliness and condition of store. Measurement and recording of temperatures. Correct stock rotation.

Washing, CCP 2
Control
Removal of soil. Prevention of contamination.
Only potable water should be used and non-mains water should be chlorinated where necessary. Water for batch washing should be changed after each batch. Hygiene precautions should be observed.

Monitoring
Where in-house chlorination is used, chlorine levels should be measured and recorded. Supervision should ensure procedures are correctly followed.

Slicing and cutting, CCP 2
Control
Prevention of contamination.

Monitoring
Standards of equipment cleanliness and personal hygiene should be monitored.

Packing, CCP 2
Control
Prevention of contamination. Ensuring packaging is correctly formed to protect product from subsequent contamination.

Monitoring
Standards of equipment cleanliness and personal hygiene should be monitored.
Setting of packaging equipment should be regularly checked. Finished packs should be assessed visually for correct seals, etc.

Storage, CCP 2
Control
Prevention of growth of low-temperature pathogens.
Refrigeration temperature and storage period should be controlled to minimise risk of significant build up of low-temperature pathogens. Vegetables stored for extended periods in bulk should not be used.

Monitoring
Temperature should be measured and recorded. Proper stock rotation should be employed.

Problems associated with specific products

Watercress

Watercress has historically been of concern due to its association with liver fluke (*Fasciola hepatica*). This hazard may be controlled by preventing access to the growing beds by the primary hosts, cattle and sheep, and prevention of run-off from pastures (Anon, 1979). Watercress beds should also be protected from faecal pollution, and the quality of the water should be verified by microbiological examination.

Mushrooms

Prepacked mushrooms have been of concern with respect to *Clostridium botulinum*, although it is probable that the risk is small and may be controlled by attention to hygiene during growth and by perforating low-permeability packing films. However, some speciality mushrooms sold in the USA are vacuum-packed and while sold under refrigeration are considered to represent an unacceptable risk.

It should be appreciated that illness due to consumption of poisonous species is a continuing problem, and under most circumstances is the major cause of mushroom-associated illness.

Sprouting seeds

Sprouting seeds such as bean sprouts have been involved in a number of food poisoning incidents. Those involving *Bacillus cereus* were attributed to a particular type of domestic sprouter, but commercially produced sprouts have also been implicated in large outbreaks. These have been associated with contamination at two points, the incoming seed for sprouting and the water used for irrigation. Incoming seed should be of good microbiological quality and free from pathogens such as *Salmonella*; the microbiological quality may be verified by laboratory examination if required. Secondary protection lies in the high level chlorination of water used for soaking seeds. Water used for irrigation and other purposes should be of potable quality, and chlorination should be used where mains supplies have been stored or where water other than mains water is used. Commercial production of sprouts is largely automated, but precautions against contamination, by handlers should be taken. Peat or other growth medium is a further potential source of contamination and only sterilised materials should be used.

Fruit and vegetable products

Large quantities of fruit and vegetables are canned and frozen, but such processes do not involve the control of hazards specific to these commodities. Significant quantities are also dried, while smaller amounts of other products are manufactured.

Dried fruit and vegetables

Fruit and vegetables may be air-dried either in the sun or in artificially heated dryers, while considerable quantities of potato powder are produced by roller drying. The basic principles of control of pathogens are the same as for other dried commodities: prevention of contamination before drying, prevention of the growth of micro-organisms before drying, reduction of the a_w level to less than 0.7 and maintenance at that level. Several products are treated with sulphur dioxide, and while this is intended primarily to retard browning and maintain colour, it significantly reduces the initial microbial load.

Precautions against microbial growth during reconstitution should be taken, and where a lengthy period is required, as with dried legumes, refrigeration should be used. Reconstituted dried vegetables support the rapid growth of micro-organisms including pathogens, and precautions must be taken with products such as bean salads, which must be stored under refrigeration. The sale of reconstituted, pre-cooked peas from greengrocery outlets is considered unsafe since the need for refrigeration is often not appreciated, and since adequate refrigeration facilities are often not, in any case, available. Reconstituted legumes spoil extremely rapidly during storage at ambient temperature, and this may introduce a safety factor.

It is necessary to be aware of the intrinsic hazards due to a toxic haemagglutinin in red kidney beans (*Phaseolus vulgaris*). Symptoms appear within 1 to 3 hours and include nausea and vomiting followed by diarrhoea and, on some occasions, abdominal pain. The haemagglutinin is readily destroyed by cooking, such as boiling for at least 10 minutes. The increase in food poisoning due to red kidney beans during the 1970s and 1980s may be attributed to the movement towards 'natural foods', which extols the virtues of consuming raw foods, and to the unrelated increase in the use of low-temperature 'slow cookers' in which the temperature attained is too low to inactivate the haemagglutinin (Gilbert, 1983).

Fermented fruit and vegetables

Fruit such as green olives and vegetables such as cabbage (sauerkraut) and cucumbers may be preserved by fermentation. The fermentation takes place in the presence of NaCl added either dry or as a concentrated brine. The purpose of the NaCl is primarily to suppress spoilage bacteria until the fermentation is established, and the levels normally present do not totally inhibit the growth of halotolerant pathogens such as *Staphylococcus aureus* or proteolytic strains of *Cl. botulinum*. Safety is assured by the production of lactic acid by various species of lactic acid bacteria, and the fermentation should be monitored to ensure that the correct pH value is obtained. Fermentation is only suitable for use with raw materials which contain sufficient carbohydrate to support a lactic fermentation, and should not be applied to high-protein vegetables such as beans and peas.

Pickled vegetables

In the current context, pickled vegetables are defined as those which have undergone no fermentation, but where safety and stability are assured by the presence of acetic acid and other preservatives, such as lactic acid and NaCl, added to the pickle. Typical products are UK pickled onions and some types of pickled cucumber. Pathogens are unable to develop in a pickle of correct constitution and make up is a critical control point. Omission of lactic acid, for example, can permit the growth of *Staph. aureus* during the early stages of pickling. *Staph. aureus* also grows well on peeled onions before pickling, and delays before transfer to the pickle must be avoided.

Cooked vegetables

Small quantities of cooked vegetables, usually beetroot or potatoes, are produced for retail sale. In many cases, the production is small scale and carried out in primitive conditions with a high risk of post-process contamination and growth of micro-organisms during subsequent unrefrigerated display. Rapid spoilage probably provides a safety margin, although the safety of some cooked beetroot is assured by marinading in vinegar.

In some cases, vacuum packaging has been introduced to extend shelf life: such products, which are often retailed without refrigeration, are considered to be inherently unsafe.

Cereals and cereal products

Cereals may be contaminated by various human pathogens during growth, harvest and subsequent storage. Many of the pathogens are environmental organisms such as *B. cereus*, which is derived from soil, although faecal pathogens such as *Salmonella* may also be present as a result of contamination from manure. The a_w level of dried grain is too low to permit bacterial growth, and while organisms persist for extended periods, problems with finished products are few. However, food poisoning has been caused by *Bacillus* species derived from flour, and it is also common practice to specify the absence of *Salmonella* and other bacterial pathogens from flours used in infant formulations and special diets.

More significant hazards are presented by mycotoxins that may be produced by moulds, either while the cereal is growing or during storage if the a_w level is sufficiently high. With wheat, for example, infection with *Fusarium graminearum* during growth can lead to production of the trichothecenes divalenol or deoxydivalenol (vomitoxin) and zearaleone (Hagler *et al.*, 1984). Production is largely dependent on climatic conditions and occurs when the harvest months are cool and damp. These conditions also lead to heavy growth of mycotoxigenic moulds on other cereal crops, and also favour growth and ergotamine production by the sclerotia of *Claviceps purpurea* on rye. Although ergotism appears to be solely of historic interest, it remains a potential hazard, particularly under conditions of economic deprivation.

Grain should be inspected for the presence of mould directly after harvest to prevent infected cereal being mixed with uninfected cereal. Heavy infection can be detected by visual examination, some moulds producing characteristic signs such as the pink colouration of wheat kernels containing deoxydivalenol. Examination for florescence under ultraviolet illumination is also a useful indicator of the presence of mould, while chemical analysis is possible where required.

Mould growth on grain after harvest may be prevented by drying the grain to a final moisture content of 11 to 12%, and preventing moisture uptake during storage. Several species of mycotoxigenic moulds are able to grow if the moisture content is allowed to rise and re-inspection of grain is always necessary before milling or other processing.

Bakery goods

With the exception of cream-filled confections where hazards lie in the handling of the cream (see page 451), the safety record of bakery goods is extremely good. Vegetative pathogens are usually destroyed by baking, while the dry crust will not support the growth of contaminants.

Bread.

Endospores of *Bacillus* species, which may be present in flour in large numbers, survive baking and may outgrow in the finished loaf. Some species, notably *B. subtilis*, produce a particular type of spoilage, rope. Consumption of bread with incipient rope has led to mild food poisoning (see page 280). Although the number of cases are small, there has been an increase in recent years which corresponds to the removal of anti-microbial agents from bread.

It is not possible to produce flour which is free from endospores of *Bacillus*, and alternative control measures may be required if bread-associated food poisoning becomes more common. These include a shortening of the shelf life.

Sliced bread may be contaminated at the slicing stage, and contaminants may be able to grow during storage. *Staph. aureus* has been suspected of causing food poisoning by this route. Control lies in ensuring good standards of hygiene during slicing and subsequent handling.

Sandwiches

Commercially prepared sandwiches have been increasingly implicated as vehicles of food poisoning, although from a technical viewpoint control of hazards is straightforward (**Table 178**). In the UK, sales of pre-prepared sandwiches have increased significantly in recent years, and the number of organisations involved in making up and retailing sandwiches has increased accordingly. In many cases, problems have arisen from the actions of inexperienced management and personnel, as well as from the use of inadequately equipped premises. The problem has been exacerbated by vending practices including hawking and sale from non-traditional outlets, such as petrol filling stations. The magnitude of the problem of food handling in non-traditional outlets is illustrated by the finding that in filling stations, many employees thought it to be perfectly acceptable to refrigerate chocolate bars, but could not see the sense in applying refrigeration to sandwiches (Mills, 1989).

The situation with sandwiches illustrates the need for hygiene education to facilitate control of food poisoning. Compulsory registration of all food-handling premises to allow inspections to be made by local authorities etc, is also considered highly desirable.

Table 178. Control of hazards during the preparation and sale of sandwiches.

1 Ingredients should be properly cooked and care taken to separate raw materials from cooked. Pre-cooked ingredients should be obtained from known and reputable suppliers.
2 Sensitive ingredients should be stored under refrigeration until used. Stock control should be applied to avoid excessive storage periods.
3 Making up of sandwiches requires a high level of handling of ingredients. Strict control of personal hygiene and of the cleanliness of equipment is essential.
4 Sandwiches should be refrigerated during storage, distribution and display. Sandwiches should be date marked and sold on the day of make up.

Pasta products

Pasta products are manufactured from a dough prepared by mixing flour, usually wheat, with water at *ca* 30% prior to drying. Eggs or egg yolks are ingredients of some types of pasta, while others contain spinach or tomato as colouring and flavouring. The dough is then extruded, shaped, cut and dried. It must be appreciated that the dough is not heated sufficiently at any stage to inactivate vegetative microbial pathogens, and that it will support the growth of pathogens at all stages from preparation until the a_w is lowered sufficiently by drying. Control of hazards lies therefore in the selection of raw materials which are free of pathogens, strict control of personal hygiene and cleaning of equipment, and in minimising the opportunities for growth of pathogens. The drying stage, which involves reduction of the water content to less than 12%, may itself provide opportunities for rapid growth since the process is typically carried out at 35 to 40°C for periods which are usually in the order of 8 hours, but which may be as long as 18 hours (NRC, 1985). More modern processes such as one for macaroni which involves drying for 40 minutes at 71°C followed by 4 hours at 74°C eliminate the possibility of growth, but cannot be guaranteed to inactivate micro-organisms protected by evaporative cooling and a reducing water activity level.

Egg is the most important potential source of pathogens and is believed to have been the source of *Staph. aureus* in an international outbreak involving lasagne. Only pasteurised liquid egg, or dried egg made from pasteurised egg, should be used and precautions must be taken to prevent contamination of the egg before adding to the dough. However, it is misleading to think that only egg containing pasta is a high-risk product, and the same manufacturing precautions are required irrespective of the composition of the pasta.

End product testing is required to verify the safe processing of pasta. Examinations should be made as soon as possible after processing, since organisms will die out during storage and should determine both *Staph. aureus* and *Salmonella*. In equivocal cases, viable counts for *Staph. aureus* may be supported by the thermonuclease test or determinations of enterotoxins (see ICMSF, 1986).

Freshly prepared, undried pasta has become increasingly popular in recent years. Problems of growth during the drying stage are obviously avoided, but refrigeration must be maintained before cooking. A second type of undried pasta is vacuum-packed and subject to a heavy in-pack pasteurisation. In some types, meat or other filling is also present. Products of this nature are potentially hazardous and require refrigeration and carefully controlled storage lives. Some products of Italian origin sold in the UK carry only a small instruction indicating storage 'in a cool room', and are obviously open to temperature abuse and to storage for excessive periods.

Chocolate and sugar confectionery

Few types of confectionery pose problems related to microbial food poisoning, although in the long term the public health implications of dental decay and obesity resulting from excessive consumption are by no means negligible. Chocolate, however, is a major exception to the general rule, and has been associated with a number of outbreaks of salmonellosis (see page 67).

Chocolate

Chocolate is made by mixing cocoa butter, cocoa liquor and sugar. In the case of milk chocolate, dried milk is an additional ingredient (sweetened condensed milk may substitute for sugar and dried milk). The ingredients are mixed in a process referred to as conching, in which chocolate paste is kneaded under standardised conditions of temperature and aeration. Temperatures used range from 50 to 70°C and the process cannot therefore be relied upon to kill *Salmonella*, heat resistance being enhanced by the low a_w level (less than 0.50) of the chocolate paste. The safety of chocolate can therefore be assured only by control of ingredients and prevention of contamination during manufacture.

Cocoa butter and cocoa liquor are both derived from cocoa beans. Cocoa beans are known to be

sources of *Salmonella*, but require roasting at 150°C for 50 minutes. This is primarily undertaken for technological purposes, but also inactivates *Salmonella* and is a critical control point. The time and temperature of roasting should be measured and, if required, the process may be verified by microbiological analysis for *Enterobacteriaceae*.

Milk powder is a potential source of *Salmonella*, and each batch should be tested before use. The powder should also be obtained from a known and reputable supplier. Sweetened condensed milk does not normally present a hazard unless produced under primitive conditions (see page 452).

Control of the environment is critical to the prevention of contamination of chocolate and its ingredients during manufacture. Chocolate is in direct contact with the environment at several processing stages and clean room conditions are necessary. Environmental monitoring is an essential part of *Salmonella* control.

End product testing for *Salmonella* is required when the organism is detected in critical areas of the environment, and some manufacturers routinely monitor all batches on a positive release basis (ICMSF, 1989). This latter approach is recommended. Delays caused by conventional assays for salmonellas have led to the widescale adoption of impediometric monitoring, both for end-product testing, and for testing of sensitive ingredients. Testing must be made according to a statistical sampling scheme, a two-class plan being recommended

(ICMSF, 1986).

Chocolate is used as an ingredient in a range of other foods including ice cream, cakes, flavoured milk etc. Users of chocolate should be aware of possible hazards, and purchase only from known and reputable suppliers. Laboratory analysis to ensure the absence of salmonellas before use is recommended for large users.

Other ingredients of confectionery

A number of other ingredients which may be added both to chocolate and to other confectionery may be sources of pathogenic micro-organisms or toxins. These include mycotoxins from mouldy nuts and *Salmonella* from dried egg, coconut and gelatin. In most cases, there is no processing stage which will remove these hazards, and control is therefore dependent on obtaining safe raw materials.

Canned foods

Canning is a very widely used means of food preservation, and while to some extent the method has been replaced by freezing and other processes, production of canned foods is of very considerable economic importance.

The basis of the canning process is the use of thermal processing to achieve commercial sterility of the final product. Commercial sterility of a thermally processed canned product may be defined as the condition achieved by the application of heat, sufficient alone or in combination with other appropriate treatments, to render the food free from micro-organisms capable of growth under normal conditions of storage (ICMSF, 1989).

Although the principles of canning are the same for all foods, the severity of thermal processing varies according to a number of factors. The main processes are:

1 The 'botulinum cook' used for foods with an equilibrium pH value greater than 4.6 (low acid foods). The botulinum cook is a process capable of reducing the probability of survival of *Cl. botulinum* in a single container to 10^{-12} ($F_0 = 3$).

2 The lesser heat treatment applied to products containing effective quantities of curing ingredients.

3 The relatively mild heat treatment used for foods with an equilibrium pH value of 4.6 or below.

4 The mild heat treatment given to foods in which the water activity, or combination of the water activity with other inhibitory factors such as pH value, inhibits growth from endospores (ICMSF, 1986).

Although thermal processing is the most important stage in canning, a number of other stages involve procedures essential to safety. The production of canned foods is outlined in **Flow diagram 52** and critical control points summarised in **Table 179**.

It must be appreciated that the safety of canning operations may be compromised by apparently insignificant process changes. These include changes in product formulation and filling method, as well as in thermal processing parameters. For example, a change to vibrating fillers for mushrooms led to more mushrooms being packed into each can. This

Flow diagram 52. Flow diagram: canned foods.

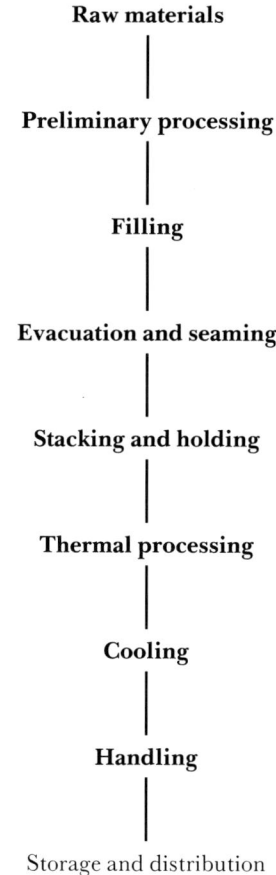

Raw materials

Preliminary processing

Filling

Evacuation and seaming

Stacking and holding

Thermal processing

Cooling

Handling

Storage and distribution

Note: Processes in **bold** are CCPs.

Table 179. Critical control points: canned foods.

Raw materials, CCP 2

Control
Avoidance of raw materials containing preformed toxins not destroyed by processing.

Monitoring
Ensure raw materials obtained from known and reliable sources. Monitor condition of incoming raw materials and ensure correct storage before processing.
Verification is possible in some cases by analysis.

Comments
Problems are usually restricted to certain types of food. Scombroid fish such as tuna, may contain high levels of histamine. Consideration must also be given to the possible presence of toxic chemicals such as heavy metals or pesticides.
Raw materials are not a CCP with respect to bacterial pathogens since these are destroyed by thermal processing.

Preliminary processing, CCP 2

Control
Ensure that preliminary processing produces a product of correct physical and chemical composition.
Ensure that preliminary processing introduces no hazard which is not eliminated during later stages.

Monitoring
Monitor all aspects of preliminary processing to ensure processing according to specifications.
Verification in some cases may be by analysis.

Comments
A wide variety of preliminary processing may be involved according to the nature of the product. This may range from simple chopping of vegetables to relatively complex formulation. Many aspects can have implications for safety: chopping to the wrong size, for example, may slow heat penetration, while incorrect formulation may lead to the pH value of a product being too high.
It is rare to introduce new hazards. However, cases have occurred of staphylococcal intoxication due to growth of *Staph. aureus* on vegetables after blanching.

Filling, CCP 2

Control
Ensure correct heat penetration during thermal processing.
Add correct weight of solid component and correct volume of liquid component.
Fill in correct order to avoid occlusion of air.
Where hot filling is used product must be at correct temperature.

Monitoring
Measure and record on a regular basis parameters relevant to each product, including weight of solids, volume of liquids, solid:liquid ratio, product density and headspace. Temperature should be measured with hot-filled products.

Evacuation and seaming, CCP 2

Control
Removal of residual air to reduce strains on seams during thermal processing and to create a vacuum in the final pack.
Prevention of contamination of product through faulty seams.
Correct maintenance and adjustment of seaming equipment by properly trained engineers who are aware of the safety implications of improper seaming.

Monitoring
Standard of can closure must be checked at regular intervals throughout operation. A formal visual examination should be made every 30 minutes and seam tear down measurements made at least once per shift and preferably more often. Seam measurements should be recorded in graphical form to enable pre-emptive action to be taken (see Sutherland *et al.*, 1986). Details of seam examination may be obtained from Anon (1973); Hersom and Hulland (1980).

Comments
The production of a vacuum in the finished pack is of no direct relevance to the growth of pathogens. The loss of vacuum *may*, however, indicate the growth of micro-organisms and offers some protection at all levels to consumption, although growth of pathogens is not invariably accompanied by any visible sign.

Table 179. *Continued*

Stacking and holding, CCP 2
Control
Prevention of damage to cans during stacking.
Elimination of undue delays.

Monitoring
Supervision of operations by experienced staff.

Thermal processing, CCP 1
Control
Inactivation of pathogenic micro-organisms.
Cans must be at correct temperature on entering retort and one can in each basket must be tagged with heat sensitive tape. Retort must be correctly loaded and fully vented.
Steam pressure and other operating parameters must be correctly set. Management systems must ensure that all cans are thermally processed.
Only retorts which have been correctly calibrated and for which the process is known to be valid should be used.
Retort operators should be fully trained and aware of the safety implications of improper thermal processing.

Monitoring
The initial product temperature is checked and the delay between filling and processing recorded. The loading of the retort and use of heat sensitive tape is checked.
The correct functioning of the retort, its controls and recording devices is checked visually or, in the case of computerised instrumentation, by test routines. Venting is monitored and, where appropriate, water levels and water temperature recorded.
Thermal processing should be timed from two points, the start of heating and the point at which the sterilisation temperature is reached. This should be determined using a properly calibrated and certified mercury in glass or platinum resistance thermometer. Other types are not suitable. Most retorts are fitted with temperature/time recorders which should be calibrated against the certified thermometer. These instruments are not sufficiently accurate to use for process control but provide a record of temperature profiles. Pressure should also be monitored during heating and cooling.
Other monitoring requirements vary according to retort design.
In addition to automatic recording instruments, the retort operator should keep full records of all aspects of operation.

Cooling, CCP 2
Control
Prevention of contamination by cooling medium.
Prevention of damage to can.
A high standard of hygiene must be maintained and cooling water should be chlorinated.
External pressure (where applied) and cooling rate should be controlled to prevent distortion of cans.

Monitoring
Free chlorine concentration should be measured *after* cooling water has been in contact with cans. Measurements should be made twice per shift. Overall performance of the cooling system should be monitored by determining residual chlorine, chlorine demand, suspended organic matter and pH value.
Verification may be made by microbiological analysis of cooling water.

Handling, CCP 2
Control
Prevention of contamination.
Manual handling of warm, damp cans must be avoided. Surfaces in contact with cans must be thoroughly cleaned and sanitised on a regular basis.
Can handling equipment should be designed to minimise mechanical shock.

Monitoring
Can handling should be supervised by a trained and responsible person. Cleanliness should be checked visually.
Verification of cleaning and sanitation of can contact surfaces may be made by microbiological examination.

Comments
The warm, damp can may readily be infected if exposed to excessive contamination in the area of the seams.
Supervision of handling of warm cans is often poor in comparison with that of other aspects of canning.

changed the heat penetration characteristics with the result that the thermal processing previously used was inadequate and a serious botulism hazard was created (NRC, 1985). It is essential that any technological changes should be notified to a product specialist and a microbiologist, and the implications properly evaluated. Where necessary, the safety of the modified process must be verified and the HACCP audit repeated. Any modifications to product or process should be recorded, and it is strongly recommended that management safeguards are devised to prevent modifications being made without consideration for safety.

Correct operation of equipment is crucial for safety and can require a detailed knowledge of complex plant. Operators should be well trained, responsible and aware of the basic microbiology of canning and of the importance of their work. Management should recognise that operators are not merely machine minders, but technical experts who should have the authority to halt production if equipment faults prevent safe processing.

Although most canned goods are stable at ambient temperatures, a few are pasteurised only and require refrigeration. This can lead to considerable problems during retail sale and domestic handling, since the metal can is generally associated with ambient temperature storage. Further confusion is likely to result from a blurring of the distinction between containers traditionally used for fully shelf-stable products, and those used for refrigerated products (Anon, 1988). Thus shelf-stable products may now be packed in pouches, semi-rigid bowls or trays, while pasteurised products may be packed in containers previously used for shelf-stable products. This situation can only be resolved by ensuring a high degree of product awareness among those handling the products during distribution and retail sale, and by clearly labelling the container with the requirement for refrigerated storage. In this context, it is essential that the clarity of storage instructions takes priority over the aesthetics of the package design.

The overall safety record of canned foods is very good. However, the extremely serious nature of the major hazard, botulism, means that production of canned foods must be under strict control and that all actions must be fully documented. Documentation is not a matter of bureaucracy but of safety and an essential part of operations. It must also be appreciated that the canning industry is not a place for assumptions and approximations. A salutary lesson should be drawn from experiences in the USA where a review of of process information which followed the introduction of new canning rules resulted in 'shelf stable' products being reclassified as requiring refrigeration. A pasteurised meat tortellini, for example, packaged in a modified atmosphere in a flexible pouch, purported to be shelf stable by virtue of low a_w level. Analysis, however, showed that the a_w level actually achieved was not sufficiently low to assure safety and refrigerated storage was introduced (Anon, 1988).

Frozen foods

The freezing process itself presents no microbiological hazards and, where problems arise, they are associated either with processing before freezing or with handling after freezing.

In many cases, the hazards associated with the pre-processing of foods for freezing are identical to those associated with the unfrozen equivalent. However, process modifications may be made to meet the specific requirements of freezing, and HACCP should be used to ensure process safety. The major concern is the two-stage cooking processes applied to some frozen meat products, which permit growth and enterotoxin production by *Staph. aureus*.

Problems resulting from the handling of frozen foods usually stem from a failure to maintain the correct storage temperature. In many cases, this is due to catastrophic failure of refrigeration. Defrosted food *must not* be refrozen, although food which has only partly defrosted and which retains ice crystals may be used for immediate consumption. In all cases, the judgement of experienced personnel is required.

In other cases, food may become defrosted and subsequently refrozen following delays in transportation or overloading of display cabinets. At least one incident of food poisoning is known where the temperature of 'frozen' peas rose sufficiently to permit the growth of *Staph. aureus*, although abuse to this extent is rare. It should be appreciated that overloading of display cabinets is not a fault restricted to small shops, but also occurs in large supermarkets. Monitoring the performance and correct loading of display cabinets should be a management responsibility.

Ice cream

Ice cream, ice lollipops etc, are unique in being the only products consumed in the frozen or part frozen state. Ice cream also retains the reputation of being an unsafe food.

Despite recent outbreaks of *S. enteritidis* infections associated with *home made* ice cream, the safety record of commercially made ice cream has been very good. In the UK, commercially made ice cream has not been associated with any known food poisoning outbreak since 1955 (Rothwell, 1981). This may be attributed to pasteurisation of the ice cream mix, and strict control of hygiene during subsequent operations. However, the isolation of *L. monocytogenes* from ice cream in the USA suggests that a re-evaluation of procedures with respect to this organism is necessary. Special attention should be paid to the efficiency of less-strict pasteurisation procedures, and to the control of *L. monocytogenes* in the environment.

Ice cream associated food poisoning continues to be reported in developing countries; this should be viewed in the context of generally low hygiene standards. Many problems result from contamination at point of sale, which is often the greatest risk in both industrialised and non-industrialised countries.

Handling of ice cream inevitably occurs during the dispensing of the product from a freezer-dispenser, or when scooping ice cream from bulk containers. Strict control of personal hygiene and of cleanliness of utensils is essential. Utensils should be kept in a stream of cold running water or, where this is not possible, in a container of fresh potable water which is changed at frequent intervals. Particular problems of maintaining hygiene standards exist among mobile vendors, who frequently operate from home on a part-time basis, and who often lack any knowledge of the necessary standards of personal hygiene and equipment cleanliness.

References

Chapter 1

Anon (1986), Consensus Development Conference Statement: Traveler's diarrhea. *Rev. Inf. Dis.*, **8** (Suppl.), S227–S223.

Archer, D.L. and Young, F.E. (1988), Contemporary issues: Diseases with a food vector. *Clin. Mic. Rev.*, **1**, 377–398.

Beebe, J.L. (1986), Atypical blood-borne bacteria as reflectors and indicators of cancer. *Lab. Manag.*, **24**(5), 43–50.

Blaser, M.J., Perez, G.P., Smith, P.F. *et al.* (1986), Extraintestinal *Campylobacter jejuni* and *Campylobacter coli* infections: host factors and strain characteristics. *J. Inf. Dis.*, **153**, 552–559.

Bodey, G.P. (1985), Overview of the problems of infections in the compromised host. *Am. J. Med.*, **79** (Suppl. 5B), 56–61.

Bryan, F.L. (1978), Impact of food-borne disease and methods of evaluation control programs. *J. Environ. Hlth.*, **40**, 315–326.

Burnett, J. (1966), *Plenty and Want*. Thomas Nelson: London.

Casemore, D.P. (1989), The epidemiology of human cryptosporidiosis. *PHLS Micro. Dig.*, **6**, 54–66.

Gamlin, L. (1989), Cooking up a storm. *New Scientist*, **July 8**, 45–49.

Gianella, R.A. (1986), Chronic diarrhea in travelers: Diagnostic and therapeutic considerations. *Rev. Imf. Dis.*, **8** (Suppl.), S223–S225.

Gilbert, R.J. (1983), Food-borne infections and intoxications — Recent trends and prospects for the future. In *Food Microbiology: Advances and Prospects*, Roberts, T.A. and Skinner, F.A. (Eds). Academic Press: London and New York.

Guerrant, R.L., Shields, D.S., Thorson, S.M. and Groschel, D.H.M. (1985), Evaluation and diagnosis of infectious diarrhea. *Am. J. Med.*, **78** (Suppl. 6B), 91–98.

Heydoorn, A.O. (1977), Sarkosporidieninfiziertes. Fleisch als mogliche krankheitsursach fur den menschen. *Arch. Lebensmittel.*, **28**, 27–32.

Hoskin, J.C. and Lior, H. (1985), Haemorrhagic colitis due to *Escherichia coli*, O157:H7. A rare disease? *Can. Med. Assoc. J.*, **134**, 25–26.

Kean, B.H. (1986), Travelers' diarrea: An overview. *Rev. Inf. Dis.*, **8** (Suppl.), S111–S116.

Jarvis, B., Chapman, W.B., Williams, A.P. *et al.*, (1982), Methods for the detection and identification of selected mycotoxins. In *Isolation and Identification Methods for Food Poisoning Microorganisms*, Corry, J.E.L, Roberts, D. and Skinner, F.A. (Eds). Academic Press: London and New York.

Noah, N.D., Bender, A.E., Reaidi, G.B. and Gilbert, R.J. (1980), Food poisoning from raw red kidney beans. *B. Med. J.*, **281**, 236–237.

Roberts, D. (1985), Food preparation — the faults that lead to foodborne disease. *Proceedings of the XIII International Congress of Nutrition*, Taylor, T.G. and Jenkins, N.G. (Eds). John Wiley: London.

Rosenberg, I.H., Beisel, W.R., Gordon, J.E. *et al.* (1974), Infant and child enteritis–malabsorption–malnutrition: the potential of limited studies with low-dose antibiotic feeding. *Clin. Nutr.*, **27**, 304–309.

Ross, G. (1981), A deadly oil. *B. Med. J.*, **283**, 424–425.

Ryan, C.A., Nickels, M.K., Hargrett-Bean, N.T. *et al.* (1987), Massive outbreak of anti-microbial resistant salmonellosis traced to pasteurised milk. *J. Am. Med. Assoc.*, **258**, 3269–3274.

Snyder, J.D. and Merson, M.H. (1982), The magnitude of the global problem of diarrheal disease: a review of active surveillance. *Bull. World Hlth. Org.*, **60**, 605–613.

Todd, E.C.D. (1983), Factors that contributed to food-borne disease in Canada, 1973–1977. *J. Fd. Prot.*, **46**, 737–745.

Turner, A. (1980), Food and public safety. *IFST Proc.*, **13**, 235–250.

Truswell, A.S., Asp, N.G., James, W.P.T. and McMahon, B. (1978), 'Conclusions'. *Proceedings of Marabou Symposium on Food and Cancer*. Caslon Press: Stockholm.

Van der Heijden, P.J., Stok, W. and Bianchi, A.T.J. (1987), Contribution of immunoglobulin-secreting cells in the murine small intestine to the total 'background' immunoglobulin production. *Immunology*, **62**, 551–555.

Chapter 2

Beckers, H.J., Daniels-Bosman, M.S.M., Ament, A. *et al.* (1985), Two outbreaks of salmonellosis caused by *Salmonella indiana*. A survey of the European summit outbreak and its consequences. *Int. J. Food. Mic.*, **2**, 185–195.

Cohen, D.R., Porter, I.A., Reid, T.M.S. *et al.* (1983), A cost benefit study of milk-borne salmonellosis. *J. Hyg.*, **91**, 17–23.

Coleman, T. (1965), *The Railway Navvies*. Hutchinson: London.

Howie, J.W. (1968), Typhoid in Aberdeen, 1964. *J. appl. Bact.*, **31**, 171–178.

Roberts, T. (1985), Microbial pathogens in raw pork, chicken, and beef: benefit estimates for control using irradiation. *Am. J. Agric. Econ.*, **67**, 957–965.

Roberts, J.A., Sockett, P.N. and Gill, O.N. (1989), Economic impact of a nationwide outbreak of salmonellosis: cost-benefit of early intervention. *B. Med. J.*, **298**, 1227–1230.

Rowe, B., Gilbert, R.J, Begg, N.T. *et al.* (1987), *Salmonella ealing* infections associated with consumption of infant dried milk. *Lancet*, **II**, 900–903.

Todd, E.C.D. (1987a), Legal liability and its economic impact on the food industry. *J. Fd. Prot.*, **50**, 1048–1057.

Todd, E.C.D. (1987b), Impact of spoilage and foodborne diseases on national and international economies. *Int. J. Fd. Mic.*, **4**, 83–100.

Todd, E.C.D. (1989), Preliminary estimates of costs of foodborne disease in the US. *J. Fd. Prot.*, **52**, 595–601.

Wheelock, J.V. (1988), Food-borne disease: the hidden hazard. Horton Publishing: Bradford.

Chapter 3

Ackland, M.R. and Manvell, P.M. (1987), A non-impedimetric electrical method for detecting and enumerating micro-organisms. *Food. Mic.*, **4**, 127–131.

Ackland, M.R., Manvell, P.M. and Bean, P.G. (1984), A rapid electrical method to detect microbial growth automatically. *Biotech. Lett.*, **6**, 137–142.

Anderson and Williams 1956 *cited by* Guinee, P.A.M. and van Leeuwen (1978), Phage typing of *Salmonella*. In *Methods in Microbiology*, Vol. 11, Bergan, T. and Norris, J.R. (Eds). Academic Press: London and New York.

Andrews, W.H. (1985), A review of culture methods and their relation to rapid methods for the detection of *Salmonella* in foods. *Food Tech.*, **39**(3), 77–82.

Archer, D.L., and Young, F.E. (1988), Contemporary issues: Diseases with a food vector, *Clin. Mic. Rev.*, **1**, 377–398.

Blaser, M.J., Berkowitz, I.D., LaForce, F. *et al.* (1979), *Campylobacter* enteritis: clinical and epidemiologic features. *Ann. Intern. Med.*, **91**, 179–185.

Brock, J.H. (1986), Iron and the outcome of infection. *B. Med. J.*, **293**, 518–520.

Brubaker, R.R. (1985), Mechanisms of bacterial virulence. *Ann. Rev. Micro.*, **39**, 21–50.

Bullock, R.D. and Frodsham, D. (1989), Rapid impedance detection of salmonellas in confectionery using modified LICNR broth. *J. appl. Bact.*, **66**, 385–392.

Christensen, G.D., Simpson, W.A. and Beachey, E.H. (1985), Microbial adherence in infection. In *Principles and practice of infectious diseases*, Mandell, G.L., Douglas, R.G. and Bennett, J.E. (Eds). John Wiley: London and New York.

Clark, C., Candlish, A.A.G. and Steell, W. (1989), Detection of *Salmonella* in foods using a novel coloured latex test. *Food Agric. Immunol.*, **1**, 3–9.

Costerton, J.W., Marrie, T.J. and Cheng, K-J. (1985), Phenomena of bacterial adhesion. In *Bacterial adhesion: mechanisms and physiological significance*, Savage, D.C. and Fletcher, M. (Eds). Plenum Press: New York.

Duguid, J.P. and Olds, D.C. (1980), Adhesive properties of Enterobacteriaceae. In *Bacterial adherence (Receptors and Recognition*, series B vol. 6), Beachey, E.H. (Ed). Chapman and Hall: London.

Dodd, C.E.R. (1988), Plasmid profiling: a new method of detection of critical control points. *Fd. Sci. Tech. Today*, **2**, 264–277.

Ebringer, A. (1983), The cross-tolerance hypothesis, HLA-B27 and ankylosing spondylitis. *B. J. Rheumatol.*, **22** (Suppl. 2), 53–66.

Edwards, P.R. and Ewing, W.H. (1972), *Identification of Enterobacteriaceae* 3rd ed. Burgess Publishing Co: Minneapolis.

Engvall, E. and Perlmann, P. (1971), Quantitative assay of immunoglobulin G. *Immunochem.*, **8**, 871–874.

Entis, P. (1986), Membrane filtration systems. In *Foodborne Microorganisms and their Toxins: Developing Methodology*, Pierson, M.D. and Stern, N.J. (Eds). Marcel Dekker: New York and Basel.

Essers, L. and Radebold, K. (1980), Rapid and reliable identification of *Staphylococcus aureus*, by a latex agglutination test. *J. Clin. Mic.*, **12**, 641–643.

Finkelstein, R.A., Sciortino, C.V. and McIntosh, M.A. (1983), Role of iron in microbe-host interactions. *Rev. Inf. Dis.* (Suppl.), S759–S777.

Firstenberg-Eden, R. (1986), Electrical impedance for detemining microbial quality of foods. In *Foodborne Microorganisms and their Toxins: Developing Methodology*, Pierson, M.D. and Stern, N.J. (Eds). Marcel Dekker: New York and Basel.

Firstenberg-Eden, R. and Eden, G. (1984), *Impedance Microbiology*. Research Studies Press: Letchworth.

Fitts, R.A. (1986), Detection of foodborne microorganisms by DNA hybridization. In *Foodborne Microorganisms and their Toxins: Developing Methodology*, Pierson, M.D. and Stern. N.J. (Eds). Marcel Dekker: New York and Basel.

Gotschlich, E.C. (1983), Thoughts on the evolution of strategies used by bacteria for evasion of host defenses. *Rev. Inf. Dis.*, **5** (Suppl.), S778–S792.

Harrigan, W.F. and McCance, M.E. (1976), *Laboratory Methods in Food and Dairy Microbiology*, Revised edition. Academic Press: London and New York.

Hinson, G., Knutton, S. and Lam-Po-Tang, M.K.-L. *et al.* (1987), Adherence to human colonocytes of an *Escherichia coli* strain isolated from severe infantile enteritis: molecular and ultrastructural studies of a fibrillar adhesin. *Inf. Immunol.*, **55**, 393–402.

Insalata, N.F., Mahnke, C.W. and Dunlap, W.G. (1973), Direct fluorescent-antibody technique for the microbiological examination of food and environmental swab samples for salmonellae. *Appl. Micro.*, **15**, 1145–1149.

Isenberg, H.D. (1988), Pathogenicity and virulence: another view. *Clin. Mic. Rev.*, **1**, 40–53.

Ishibashi, Y. and Arai, T. (1989), Susceptibility of *Salmonella typhimurium* and *Salmonella typhi* to oxygen metabolites. *FEMS Micro. Immunol.*, **47**, 279–284.

Jones, G.W., Abrams, G.D. and Freter, R. (1976), Adhesive properties of *Vibrio cholerae*: adhesion to isolated rabbit brush border membranes and haemagglutinating activity. *Immunol.*, **14**, 232–239.

Kochan, I. (1983), Neutralization of acquired antibacterial immunity with iron. In *Microbiology-1983*, Schlessinger, D. (Ed). American Society for Microbiology: Washington DC.

Kroll, R.G., Pinder, A.C., Purdy, P.W. and Rodrigues, U.M. (1989), A laser-light pulse counting method for automatic and sensitive counting of bacteria stained with acridine orange. *J. appl. Bact.*, **66**, 161–167.

Linggood, M.A., Roberts, M., Ford, M. *et al.* (1987), Incidence of the aerobactin iron uptake system among *Escherichia coli* isolates from infections of farm animals. *J. Gen. Mic.*, **133**, 835–842.

Luckey, T.D. (1972), An introduction to intestinal microecology. *Am. J. Clin. Nutr.*, **25**, 1292–1294.

Maurelli, A.T. (1989), Temperature regulation of virulence genes in pathogenic bacteria: a general strategy for human pathogens. *Micro. Path.*, **7**, 1–10.

Mims, C.A. (1982), *The pathogenesis of infectious disease* 2nd ed. Academic Press: London.

Minion, F.C., Abraham, S.N., Beachey, E.H. and Goguen, J.D. (1986), The genetic determinant of adhesive function in type 1 fimbriae of *Escherichia coli* is distinct from the gene encoding the fimbrial subunit. *J. Bact.*, **165**, 1033–1036.

Mossel, D.A.A. (1983), Essentials and perspectives of the microbial ecology of foods. In *Food Microbiology: Advances and Prospects*, Roberts, T.A. and Skinner, F.A. (Eds). Academic Press: London and New York.

Olds, R.J. (1975), *A Colour Atlas of Microbiology*. Wolfe Medical Publications: London.

Orskov, F. (1984), *Escherichia*. In *Bergey's Manual of Systematic Bacteriology*, Vol. 1, Krieg, N. and Holt, J.G. (Eds). Williams and Wilkins: Baltimore and London.

Pettipher, G.L. (1983), *The Direct Epifluorescent Filter Technique*. Research Studies Press: Letchworth.

Roberts, M., Parthasarathy, S., Lam-Po-Tang, M.K.L. and Williams, P.H. (1986), The aerobactin iron uptake system in enteropathogenic *Escherichia coli*: evidence for an extinct transposon. *FEMS Micro., Lett.*, **37**, 215–219.

Rodrigues, U.M. and Kroll, R.G. (1988), Rapid selective enumeration of bacteria in foods using a microcolony epifluoresence microscopy technique. *J. appl. Bact.*, **64**, 65–78.

Romick, T.L., Lindsay, J.A. and Busta, F.F. (1987), A visual DNA probe for detection of enterotoxigenic *Escherichia coli* by colony hybridisation. *Letts. Appl. Micro.*, **5**, 87–90.

Saiki, R.K., Schorf, S., Faloona, F. *et al.* (1985), Enzymatic amplification of ß-globin genomic sequences and restriction site analysis for diagnosis of sickle cell anaemia. *Science*, **230**, 1350–1354.

Saiki,R.K., Gelford, D.G., Stoffel, S. *et al.* (1988), Primer-directed enzymatic amplification of DNA with thermo-stable DNA polymerase. *Science*, **239**, 487–491.

Savage, D.C. (1980), Colonisation by and survival of pathogenic bacteria on intestinal mucosal surfaces. In *Adsorption of Organisms to Surfaces*, Bitton, G. and Marshall, K.C. (Eds). Wiley-Interscience: London.

Savage, D.C. (1983), Mechanisms by which indigenous microorganisms colonise gastrointestinal epithelial surfaces. *Prog. Food Nutr. Sci.*, **7**, 65–74.

Scotland, S.M. (1988), Toxins. *J. appl. Bact.*, **65** (Suppl.), 109S–129S.

Sharpe, A.N. (1986), Detection of foodborne micro-organisms and their toxins: the future. In *Foodborne Microorganisms and their Toxins: Developing Methodology*, Pierson, M.D. and Stern, N.J. (Eds). Marcel Dekker: New York and Basel.

Sharpe, A.N. and Michaud, G.L. (1974), Hydrophobic grid-membrane filters: new approach to microbiological enumeration. *Appl. Micro.*, **28**, 223–225.

Sharpe, A.N. and Michaud, G.L. (1975), Enumeration of high numbers of bacteria using hydrophobic-grid membrane filters. *Appl. Micro.*, **30**, 519–524.

Singleton, P. and Sainsbury, D. (1987), *Dictionary of Microbiology and Molecular Biology*. John Wiley: Chichester and New York.

Smith, H. (1983), The elusive determinants of bacterial interference with non-specific host defences. *Phil. Trans. Roy. Soc. Lond.* (B), **303**, 99–113.

Stannard, C.J. and Wood, J.M. (1983), The rapid estimation of microbial contamination of raw meat by measurement of adenosine triphosphate (ATP). *J. appl. Bact.*, **55**, 429–438.

Stendahl, O. (1983), The physicochemical basis of surface interaction between bacteria and phagocytic cells. In *Medical Microbiology*, Vol. 3, Easmon, C.S.F. and Jelzaszewicz, J. (Eds). Academic Press: London and New York.

Swaminathan, B. and Konger, R.L. (1986), Immunoassays for determining foodborne bacteria and microbial toxins. In *Foodborne Microorganisms and their Toxins: Developing Methodology*, Pierson, M.D. and Stern, N.J. (Eds). Marcel Dekker: New York and Basel.

Taylor, P.W. (1983), Bactericidal and bacteriolytic activity of serum against Gram-negative bacteria. *Mic. Rev.*, **47**, 46–83.

Thomason, B. (1981), Current status of immunofluorescent methodology for salmonellae. *J. Fd. Prot.*, **44**, 381–384.

van Leusden, F.M., van Schothorst, M. and Beckers, H.J. (1982), The standard *Salmonella* isolation method. In *Isolation and Identification Methods for Food Poisoning Organisms*, Corry, J.E.L., Roberts, T.A. and Skinner, F.A. (Eds). Academic Press: London and New York.

Waites, W.M. (1988), Recent advances in rapid detection and enumeration of microorganisms. *Food Sci. Tech. Today*, **2**, 256–258.

Wells, C.A., Maddaus, M.A. and Simmons, R.L. (1988), Proposed mechanisms for the translocation of intestinal bacteria. *Rev. Inf. Dis.*, **10**, 958–979.

Williams, P.H., Knutton, S., Brown, M.G.M. *et al.* (1984), Characterization of nonfimbrial mannose-resistant protein haemagglutinins of two *Escherichia coli* strains isolated from infants with enteritis. *Inf. Imm.*, **44**, 592–598.

Williams, P.H., Roberts, M. and Hinson, G. (1988), Stages in bacterial invasion. *J. appl. Bact.*, **65** (Suppl.), 131S–147S.

Chapter 4

Aguine, P.M., Cachi, J.B., Folgueira, L. *et al.* (1990), Rapid fluoresence method for screening Salmonella spp from enteric differential agars. *J. Clin. Mic.*, **28**, 148–149.

Alcock, S.J. (1985), Survival and growth of salmonellae at chill temperatures. *Proc. Chill Food Symp. May 7–9.* CFPRA: Stratford on Avon.

Alford, J.A. and Knight, N.L. (1969), Applicability of aeration and delayed addition of selenite to the isolation of salmonellae. *Appl. Micro.*, **18**, 1060–1064.

Anderson, E.S. (1964), The phage typing of *Salmonella* other than *S. typhi*. In *The World Problem of Salmonellosis*, van Oye, E. (Ed). Junk: The Hague.

Andrews, W.H. (1985), A review of culture methods and their relation to rapid methods for the detection of *Salmonella* in foods. *Fd. Tech-*, 39(3): 77–82.

Anon (1986), Salmonellosis outbreaks associated with commercial frozen pasta. *Morb. Mort. Wkly. Rep.*, **35**, 387.

Anon (1987), Antibiotic resistance linked to risk of infection in Illinois outbreak. *Fd. Chem. News*, **December 14**, 32–33.

Anon (1989), Memorandum of evidence to the agriculture committee inquiry on salmonella in eggs. *PHLS Micro. Dig.*, **6**, 1–9.

Archer, D.L. and Young, F.E. (1988), Contemporary issues: Diseases with a food vector. *Clin. Mic. Rev.*, **1**, 377–398.

Arnott, M.L., Gutteridge, C.S., Pugh, S.J. and Griffiths, J.L. (1988), Detection of salmonellas in confectionery products by conductance. *J. appl. Bact.*, **64**, 409–420.

Ashenafi, M. and Busse, M. (1989), Inhibitory effect of *Lactobacillus plantarum* on *Salmonella infantis*, *Enterobacter aerogenes* and *Escherichia coli* during tempeh fermentation. *J. Fd. Prot.*, **52**, 169–172.

Ayanwale, L. F., Kaneene, J.M.B., Sherman, D.M. and Robinson, R.A. (1980), Investigation of salmonella infection in goats fed corn silage grown on land fertilised with sewage sludge. *Appl. Environ. Micro.*, **41**, 285–286.

Barker, R.M., Old, D.C. and Tyc, Z. (1982), Differential typing of *Salmonella agona*, type divergence in a new serotype. *J. Hyg.*, **88**, 413–423.

Barrow, P.A., Lovell, M.A., Spitznagel, J.K. (1989), Functional homology of virulence plasmids in *Salmonella gallinarum*, *Salmonella pullorum* and *Salmonella typhimurium*. *Inf. Imm.*, **57**, 3136–3141.

Beckers, H.J., Tips, P.D, Delgou-van Asch, E. and Peters, R. (1986), Evaluation of an enzyme immunoassay technique for the detection of salmonellas in minced meat. *Lett. Appl. Mic.*, **2**, 53–56.

Beckers, H.J., Heide, J.V.D., Feniggen-Narucka, U and Peters, R. (1987), Fate of salmonellas and competing flora in meat sample enrichments in buffered peptone water and in Muller–Kauffman's tetrathionate medium. *J. appl. Bact.*, **62**, 97–104.

Beckers, H.J., Tips, P.D., Soentoro, P.S.S. *et al.* (1988), The efficacy of enzyme immunoassays for the detection of salmonellas. *Fd. Mic.*, **5**, 147–156.

Blackburn, C. de W., Stannard, J. and Gibbs, P.A. (1988), Evaluation of the TECRA Salmonella Visual Immunoassay for the detection of *Salmonella* in foods. *Tech. Note 65.* BFMIRA: Leatherhead.

Blaser, M.J. and Newman, L.S. (1982), A review of human salmonellosis: I. Infective dose. *Rev. Inf. Dis.*, **4**, 1096–1106.

Bockemuhl, J. (1972), Die lysosensibilitat von stammen der *Salmonella* Sub-genera I-IV gegenuben dem phagen 0–1. Ihre mogliche bedeutung fur die klassification der genus *Salmonella*. *Med. Mic. Immunol.*, **158**, 44–53.

Boring, J.R., Martin, W.T. and Elliot, L.M. (1971), Isolation of *Salmonella typhimurium* from municipal water, Riverside, California, 1965. *Am. J. Epid.*, **93**, 49–54.

Bradley, S.F. and Kauffman, C.A. (1988), Effect of protein malnutrition on salmonellosis and fever. *Inf. Imm.*, **56**, 1000–1002.

Bryan, F.L. (1976), Public Health Aspects of cream filled pastries. A review. *J. Milk Fd. Tech.*, **39**, 289–296.

Bryan, F.L. (1983), Epidemiology of milk-borne diseases. *J. Fd. Prot.*, **46**, 637–649.

Buchwald, D.S. and Blaser, M.J. (1984), A review of human salmonellosis: II. Duration of excretion following infection with nontyphi *Salmonella*. *Rev. Inf. Dis.*, **6**, 345–356.

Caprioli, A., D'Agnole, G., Falbo, V. *et al.* (1981), Isolation of *Salmonella wien* heat-labile enterotoxin. *Microbiology*, **5**, 1–7.

Cartwright, K.A.V. and Evans, B.G. (1988), Salmon as a food poisoning vehicle — two successive salmonella outbreaks. *Epid. Inf.*, **101**, 249–257.

Celum, C.L., Chaisson, R.E., Ratleiford, G.W. *et al.* (1987), Incidence of salmonellosis in patients with AIDS. *J. Inf. Dis.*, **156**, 998–1002.

Chambers, R.M. McAdam, P., de Sa, J.P.H. *et al.* (1987), A phage typing scheme for *Salmonella virchow*. *FEMS Micro. Lett.*, **40**, 155–157.

Chapman, H and Sharpe, M.E. (1981), Microbiology of Cheese. In *Dairy Microbiology, Vol 2. The Microbiology of Milk Products*, Robinson, R.K. (Ed). Applied Science Publishers: London and New Jersey.

Chart, H., Rowe, B., Threlfall, E.J. and Ward, L.R. (1989), Conversion of *Salmonella enteritidis* phage type 4 to phage type 7 involves loss of lipopolysaccharide with concommitant loss of virulence. *FEMS Micro. Lett.*, **60**, 37–40.

Clark, C., Candlish, A.A.G. and Steell, W. (1989), Detection of *Salmonella* in foods using a novel coloured latex test. *Fd. Agric. Immunol.*, **1**, 3–9.

Cohen, J.I., Bartlett, J.A. and Corey, G.R. (1987), Extraintestinal manifestations of salmonella infections. *Medicine*, **66**, 349–387.

Colonna, B., Nicoletti, M., Visca, P. *et al.*, (1989), Composite IS2 elements encoding hydroramate-mediated iron uptake in FIme plasmids from epidemic *Salmonella* spp. *J. Bact.*, **162**, 307–316.

Copal, L.R. and Deibel, R.H. (1975), Assay, characterization and localization of an enterotoxin produced by *Salmonella*. *Inf. Imm.*, **11**, 14–19.

Cowan, S.J. and Steel, K.J. (1974), *Manual for the identification of Medical Bacteria* 2nd ed. Cambridge University Press: Cambridge.

Cowden, J.M., O'Mahoney, M., Bartlett, G.G.N., *et al.* (1989), A national outbreak of *Salmonella typhimurium* DT126 caused by contaminated salami sticks. *Epid. Int*, **103**, 219–225.

Cox, N.A., Davis, B.H., Kendall, J.H., Watts, A.B. and Colver, A.R. (1972), *Salmonella* in the laying hen, 3. A comparison of various enrichment broths and plating media for the isolation of *Salmonella* from poultry faeces and poultry food products. *Poultry Sci.*, **51**, 1312–1316.

Craigie, J. and Yen, C.H. (1938), The demonstration of types of *B. typhosus* by means of preparation of type II Vi antigen. *Can. J. Pub. Hlth.*, **29**, 448–463; 484–496.

Craven, P.C., Mackel, D.C., Baine, W.B. *et al.* (1975), International outbreak of *Salmonella eastbourne* infection traced to contaminated chocolate. *Lancet*, **I**, 788–793.

Craven, S. (1978), *Salmonella* contamination of dried milk products, *Vic. Vet. Proc.*, **9**, 56–57.

D'Aoust, J.-Y. (1984), Effective enrichment-plating conditions for detection of *Salmonella* in foods. *J. Fd. Prot.*, **47**, 588–590.

D'Aoust, J.-Y. (1989), *Salmonella*. In *Foodborne Bacterial Pathogens*, Doyle, M.P. (Ed). Marcel Dekker: New York and Basel.

D'Aoust, J.-Y. and Sewell, A.M. (1986), Detection of *Salmonella* by enzyme immunoassay (EIA) technique. *J. Fd. Sci.*, **51**, 484–488.

D'Aoust, J.-Y. and Sewell, A.M. (1988), Reliability of the immunodiffusion 1–2 Test system for detection of *Salmonella* in foods. *J. Fd. Prot.*, **51**, 853–857.

D'Aoust, J.-Y., Aris, B.J., Thisdele, P. *et al.* (1975), *Salmonella eastbourne* outbreak associated with chocolate. *Can. Inst. Fd. Sci. Tech. J.*, **8**, 181–184.

D'Aoust, J.-Y., Emmary, D.B., McKellar, R., *et al.* (1987), Thermal inactivation of *Salmonella* species in fluid milk. *J. Fd. Prot.*, **50**, 494–501.

Davies, R.F. and Wahba, A.H. (1976), *Salmonella* infections in charter flight passengers. *Report on a visit to Spain (Canary Islands)*. **26 February–2 March 1976**. WHO Regional Office Europe: Copenhagen.

deSmedt, J.R. and Bolderdijk, R.F. (1987), Dynamics of *Salmonella* isolation with modified semi-solid Rappaport–Vassiliadis medium. *J. Fd. Prot.*, **50**, 658–663.

deSmedt, J.R., Bolderdijk, R.F., Rappold, R.F. and Lautenschlaeger, D. (1986), Rapid *Salmonella* detection in foods by motility enrichment in a modified semi-solid Rappaport–Vassiliadis medium. *J. Fd. Prot.*, **49**, 510–514.

Desmonts, C., Minet, J., Colwell, R. and Cormier, M. (1990), Fluorescent-antibody method for detecting viable but nonculturable *Salmonella* spp in chlorinated wastewater. *Appl. Environ. Micro.*, **56**, 1148–1152.

Dickson, J.S. (1989), Enumeration of Salmonellae by most-probable-number using the Salmonella 1–2 Test. *J. Fd. Prot.*, **52**, 388–391.

Duguid, J.P., Anderson, E.S., Alfredssen, G.A. *et al.* (1975), A new biotyping scheme for *Salmonella typhimurium* and its phylogenetic significance. *J. Med. Mic.*, **8**, 149–166.

Duguid, J.P., Morrison, B.P. and Swain, R.H.A. (1978), *Mackie and McCartney's Medical Microbiology*. Churchill Livingstone: Edinburgh.

Easter M.C. and Gibson, D.M. (1985), Rapid and automated detection of *Salmonella* by electrical measurements. *J. Hyg.*, **94**, 245–262.

Emswiler-Rose, B., Bennett, B. and Okrend, A. (1987), Comparison of cultural methods and the DNA hybridisation test for detection of Salmonellae in ground beef. *J. Fd. Sci.*, **52**, 1726–1727.

Entis, P. (1986), Membrane filtration systems. In *Foodborne Microorganisms and their toxins: Developing Methodology*, Pierson, M.D. and Stern, N.J. (Eds). Marcel Dekker: New York and Basel.

Erdman, E.E. (1974), ICMSF methods studies. IV International collaborative assay for the detection of *Salmonella* in raw meat. *Can. J. Mic.*, **20**, 715–720.

Ernst, R.K., Dombroski, D.M. and Merrick, T.M. (1990), Anaerobiosis, type 1 fimbriae and growth phase are factors that affect invasion of HEp-2 cells by *Salmonella typhimurium*. *Inf. Imm.*, **58**, 2014–2016.

Felix, A. and Callow, B.R. (1943), Typing of paratyphoid B bacilli by means of Vi bacteriophage. *B. Med. J.*, **2**, 127–131.

Fernandez-Beros, M.E., Gonzalez, C., McIntosh, M.A. and Cabello, F.C. (1989), Immune response to the iron-deprivation induced proteins of *Salmonella typhi* in typhoid fever. *Inf. Imm.*, **57**, 1271–1275.

Fields, P.I., Groisman, E.A. and Heffron, F. (1989), A *Salmonella* locus that controls resistance to microbicidal protein from phagocytic cells. *Science*, **243**, 1059–1062.

Finkelstein, R. and Markel, A. (1990), New and old drugs for treating typhoid fever. *J. Inf. Dis.*, **161**, 159.

Finlay, B.B. and Falkow, S. (1989), *Salmonella* as an intracellular parasite. *Mol. Micro.*, **3**, 1833–1841.

Fitts, R.A. (1985), Development of a DNA hybridisation test for the presence of *Salmonella* in foods. *Food Tech.*, **39** (3), 95–102.

Fitts, R.A. (1986), Detection of foodborne microorganisms by DNA hybridisation. In *Foodborne Microorganisms and their Toxins: Developing Methodology*. Pierson, M.D. and Stern N.J. (Eds). Morcel Dekker: New York.

Fitts, R.A., Diamond, M., Hamilton, C. and Neri, M. (1983), A DNA:DNA hybridization assay for *Salmonella* spp. in foods. *Appl. Environ. Micro*, **46**, 1146–1151.

Flowers, R.S., Chen, K.H., Robison, B.J. *et al.* (1987), Comparison of Salmonella Bio-Enzabead immunoassay and conventional culture procedure for detection of *Salmonella* in crustaceans. *J. Fd. Prot.*, **50**, 386–389.

Flowers, R.S., Klatt, M.J., Robison,B.J. and Mattingley, J.A. (1988a), Evaluation of abbreviated enzyme immunoassay method for detection of *Salmonella* in low-moisture foods. *J. Assoc. Off. Anal. Chem.*, **71**, 341–342.

Flowers, R.S., Klatt, M.J. and Keelan, S.L. (1988b), Visual immunoassay for detection of *Salmonella* in foods: Collaborative study. *J. Assoc. Off. Anal. Chem.*, **71**, 973–976.

Fontaine, R.G., Cohen, M.L., Martin, W.T. and Vernon, T.M. (1980), Epidemic salmonellosis from cheddar cheese: surveillance and prevention. *Am. J. Epid.*, **111**, 247–253.

Fricker, C.R. (1987), The isolation of salmonellas and campylobacters. *J. appl. Bact.*, **63**, 99–116.

Fricker, C.R., Quail, E. McGibbon, L. and Girdwood, R.W.A. (1985), An evaluation of commercially available dehydrated Rappaport–Vassiliadis medium for the isolation of salmonellae from poultry. *J. Hyg.*, **95**, 337–344.

Frost, J.A., Ward, L.R. and Rowe, B. (1989), Acquisition of a drug resistance plasmid converts *Salmonella enteritidis* phage type 4 to phage type 24. *Epid. Inf.*, **103**, 243–247.

Gastrin, B., Kampe, A, Nystrom, K. *et al.* (1972), An epidemic of *Salmonella dublin* caused by contaminated cocoa. *Lakartidningen*, **69**, 5335–5338.

Gibson, D.M. (1987), Some modification to the media for rapid automated detection of salmonellas by conductance measurement. *J. appl. Bact.*, **63**, 299–304.

Gill, O.N., Sockett, P.N., Bartlett, C.L.R. *et al.* (1983), Outbreak of *Salmonella napoli* infection caused by contaminated chocolate bars. *Lancet*, **I**, 574–577.

Gillies, R.R. (1978), Bacteriocin typing of *Enterobacteriaceae*. In *Methods in Microbiology*, Vol 11, Bergan, T. and Norris, J.R. (Eds). Academic Press: London and New York.

Glatz, B.A. (1986), Genetic regulation of toxin production by foodborne microbes. In *Food Microbiology, Vol 1. Concepts in Physiology and Metabolism*, Montville, T.J. (Ed). CRC Press: Boca Raton.

Goepfort, J.M. and Biggie, R.A. (1968), Heat resistance of *Salmonella typhimurium* and *Salmonella senftenberg* 775W in milk chocolate. *Appl. Micro.*, **16**, 1939–1940.

Goepfort, J.M., Olson, N.F. and Marth, E.H. (1968), Behaviour of *Salmonella typhimurium* during manufacture and curing of Cheddar cheese. *Appl. Micro.*, **16**, 862–866.

Gotuzzo, E. Morris, J.G., Benavente, L. *et al.* (1987), Association between specific plasmids and relapse in typhoid fever. *J. Clin. Mic.*, **25**, 1779–1781.

Greenwood, M.W. and Hooper, M.L. (1983), Chocolate bars contaminated with *Salmonella napoli* an infectivity study. *B. Med. J.*, **286**, 1394.

Greenwood, D.E. and Swaminathan, B. (1981), Rapid detection of salmonellae in mechanically deboned poultry meat. *Poultry Sci.*, **60**, 2253–2257.

Griffin, P.M. and Tauxe, R.V. (1988), Food counselling for patients with AIDS. *J. Inf. Dis.*, **158**, 668.

Gulig, R.A. (1990), Virulence plasmids of *Salmonella typhimurium* and other salmonellae. *Micro. Path.*, **8**, 3–11.

Hadfield, S.G., Lane, A. and McIllmurray, M.B. (1987), A novel coloured latex test for the identification of more than one antigen. *J. Immun. Met.*, **97**, 153–158.

Hargrove, R.E., McDonough, F.E. and Reamon, R.H. (1971), A selective medium and presumptive procedure for detecting *Salmonella* in dairy products. *J. Milk Food Tech.*, **34**, 6–11.

Hariharan, H., Booth, B.A., Brickman, T.J. *et al.* (1986), Competitive enzyme-linked immunosorbent assay for cholera-related enterotoxins in *Salmonella typhimurium*. *J. Clin. Mic.*, **24**, 298–300.

Harvey, R.W.S. and Price, T.H. (1980), *Salmonella* isolation with Rappaports medium after pre-enrichment in buffered peptone using a series of inoculum ratios. *J. Hyg.*, **85**, 125–128.

Harvey, R.W.S. and Price, T.H. (1982), *Salmonella* isolation and identification techniques. Alternative to the standard method. In *Isolation and Identification Methods for Food Poisoning Organisms*, Corry, J.E.L., Roberts, D. and Skinner, F.A. (Eds). Academic Press: London and New York.

Hess, C.B., Niesel, D.W. and Klimpel, G.R. (1989), The induction of interferon production in fibroblasts by invasive bacteria: a comparison of *Salmonella* and *Shigella* species. *Micro. Path.*, **7**, 111–120.

Holbrook, R., Anderson, J.M., Baird-Parker, A.C. *et al.* (1989), Rapid detection of *Salmonella* in foods — a convenient two day procedure. *Lett. Food Mic.*, **8**, 139–142.

Holmberg, S.D., Osterholm, M.T., Senger, K.A. and Cohen, M.L. (1984), Drug resistant *Salmonella* from animals fed antimicrobials. *New Eng. J. Med.*, **311**, 617–622.

Hsieh, F., Acott, K. and Labuza, T.P. (1976), Death kinetics of pathogens in a pasta product. *J. Fd. Sci.*, **41**, 516–519.

Hughes, D., Sutherland, P.S., Kelley, G. and Davey, G.R. (1987), Comparison of the TECRA Visual Immunoassay and standard cultural methods for the detection of salmonellae in foods. *Food Tech. Aust.*, **39**, 446–454.

Humbert, F., Salvat, G., Colin, P. *et al.* (1989), Rapid identification of *Salmonella* from poultry meat products by using the 'Mucap test'. *Int. J. Fd. Mic.*, **8**, 79–83.

Humphrey, T.J. and Lanning, D.G. (1988), The vertical transmission of salmonellas and formic acid treatment of chicken feed. A possible strategy for control. *Epid. Inf.*, **100**, 43–49.

Humphrey, T.J., Greenwood, M., Gilbert, R.J. *et al.* (1989), The survival of shell eggs cooked under simulated domestic conditions. *Epid. Inf.*, **103**, 35–45.

Hynes, M. (1942), The isolation of intestinal parasites by selective media. *J. Path. Bact.*, **59**, 193–207.

ICMSF (1980), *Microbial Ecology of Foods, Vol 1. Factors Affecting Life and Death of Microorganisms*. Academic Press: New York.

ICMSF (1986), *Microorganisms in Foods, 2. Sampling for Microbiological Analysis: Principles and Specific Applications*. Blackwell Scientific Publications: Oxford.

ICMSF (1989), *Microorganisms in Foods, 4. Applications of Hazard Analysis Critical Control Point (HACCP) System to Ensure Microbiological Safety and Quality*. Blackwell Scientific Publications: Oxford.

Islam, A., Butler, T., Kath, S.K. *et al.* (1988), Randomised treatment of patients with typhoid fever by using ceftriaxone or chloramphenicol. *J. Inf. Dis.*, **158**, 742–747.

Jayasheela, M., Singh, G., Sharma, N.C. and Saxena, N.C. (1987), A new scheme for phage typing *Salmonella bareilly* and characterization of typing phages. *J. appl. Bact.*, **62**, 429–432.

Jiminez, L., Muniz, I., Toranzos, G.G. and Hazen, T.C. (1989), Survival and activity of *Salmonella typhimurium* and *Escherichia coli* in tropical freshwater. *J. appl. Bact.*, **67**, 61–69.

Jiwa, S.F.H. (1984), Probing for enterotoxigenicity among the salmonellae: an evaluation of biological assays. *J. Clin. Mic.*, **14**, 463–470.

Jones, G.W., Richardson, L.A. and Uhlman, D. (1981), The invasion of Hela cells by *Salmonella typhimurium*: the contribution of mannose-sensitive and mannose-resistant haemagglutinating activities. *J. Gen. Mic.*, **127**, 361–370.

Kapperud, G., Lassen, J., Dommarsnes, K. *et al.* (1989), Comparison of epidemiological marker methods for identification of *Salmonella typhimurium* isolates from an outbreak caused by contaminated chocolate. *J. Clin. Mic.*, **27**, 2019–2024.

Kauffman, F. (1935), Weitere erfahrusen mit dem kombinierten anreicherungvergahlen fur *Salmonella* bazillen. *Zeitschrift fur Hyg. Infektion.*, **117**, 26–32.

Kauffman, F. (1966), *The Bacteriology of the Enterobacteriaceae*. Munksgaard: Copenhagen.

King, S. and Metzger, W.I. (1968), A new plating medium for the isolation of enteric pathogens. I Hektoen enteric agar. *Appl. Micro.*, **16**, 577–578.

Koo, F.C.W. and Peterson, J.W. (1982), Cell-free extracts of *Salmonella* inhibit protein synthesis and cause cytotoxicity in eukaryotic cells. *Toxicon*, **21**, 309–317.

Koo, F.C.W., Peterson, J.W., Houston, C.W. and Molina, N.C. (1984), Pathogenesis of experimental salmonellosis: inhibition of protein synthesis by cytotoxin. *Inf. Imm.*, **43**, 443–449.

Lanz, W.W. and Hartmann, P.A. (1976), Timed-release capsule method for coliform enumeration. *Appl. Environ. Micro.*, **32**, 716–722.

Lee, H.A., Wyatt, G.M., Branham, S. and Morgan, M.R.A. (1990), Enzyme-linked immunosorbent assay for *Salmonella typhimurium* in food: feasability of 1-day *Salmonella* detection. *Appl. Environ. Mic.*, **56**, 1541–1546.

Leifson, E. (1936), New selenite enrichment media for the isolation of typhoid and paratyphoid (*Salmonella*) bacilli. *Am. J. Hyg.*, **24**, 423–432.

Le Minor, L. (1984), *Salmonella*. In *Bergey's Manual of Systematic Bacteriology*, Vol. 1, Krieg, N.R. and Holt, J.G. (Eds). Williams and Wilkins: Baltimore and London.

Le Minor, L. and Popoff, M.Y. (1987), Designation of *Salmonella enterica* sp. nov., nom. rev., as the type and only species of the genus *Salmonella*. *Int. J. Syst. Bact.*, **37**, 465–468.

Le Minor, L., Pirchoid, M., Pichinoty, F. and Coynault, C. (1969), Etude par transduction sur les nitrate-tetrathionate et thiosulfate reductases de *S. typhimurium*. *Ann. Inst. Past.*, **117**, 637–644.

Ley, F.J. (1983), New interest in the use of irradiation in the food industry. In *Food Microbiology: Advances and Prospects*, Roberts, T.A. and Skinner, F.A. (Eds). Academic Press: London and New York.

Lin, F.R., Hsu, H.S. Mumaw, V.R. *et al.* (1989), Intracellular destruction of salmonellae in genetically resistant mice. *J. Med. Mic.*, **30**, 79–87.

Liu, S.-L., Miura, H., Matsui, K. and Yaabuchi, E. (1988), Intact motility as a *Salmonella typhi* invasion-related factor. *Inf. Imm.*, **56**, 1967–1973.

Lockman, H.A. and Curtis, R. (1990), *Salmonella typhimurium* lacking flagella or motility remain virulent in BALB/c mice. *Inf. Imm.*, **58**, 137–143.

Mackey, B.M. and Derrick, C.M. (1987a), The effect of heat shock on the thermoresistance of *Salmonella thompson* in foods. *Lett. Appl. Mic.*, **5**, 115–118.

Mackey, B.M. and Derrick, C.M. (1987b), Changes in the heat resistance of *Salmonella typhimurium* during heating at rising temperatures. *Lett. Appl. Mic.*, **4**, 13–16.

Madden, J.M., McCardell, B.A. and Archer, D.L. (1986), Virulence assessment of foodborne microbes. In *Foodborne Microorganisms and their Toxins: Developing Methodology*, Pierson, M.D. and Stern, N.J (Eds). Marcel Dekker: New York and Basel.

Mandal, B.K. and Brennan, J. (1988), Bacteraemia in salmonellosis: a 15 year retrospective study from a regional infectious diseases unit. *B. Med. J.*, **297**, 1242–1243.

McCormick, B.A., Stocker, B.A.D., Laux, D.C. and Cohen, P.S. (1988), Roles of motility, chemotaxis, and penetration through and growth in intestinal mucus in the ability of an avirulent strain of *Salmonella typhimurium* to colonize the large intestine of mice. *Inf. Imm.*, **56**, 2209–2217.

McCullough, N.B and Byrne, A.F. (1952), Relative efficiency of different culture mediums in isolation of the *Salmonella* group. *J. Inf. Dis.*, **90**, 71–75.

Mead, G.C. (1982), Microbiology of poultry and game birds. In *Meat Microbiology*, Brown, M.H. (Ed). Applied Science Publishers: London and New York.

Mims, C.A. (1987), *The Pathogenesis of Infectious Disease*, 3rd ed. Academic Press: London.

Minnich, S.A., Hartmann, P.A. and Heimsch, R.C. (1982), Enzyme assay for detection of salmonellas in foods. *Appl. Environ. Micro.*, **43**, 877–883.

Mitchell, E., O'Mahony, M., Lynch, D. *et al.* (1989), Large outbreak of food poisoning caused by definitive type 49 in mayonnaise. *B. Med. J.*, **298**, 99–101.

Morgan-Jones, S.C. (1982), A method for enumerating salmonellas from environments in the poultry industry. In *Isolation and Identification Methods for Food Poisoning Organisms*, Corry, J.E.L, Roberts, T.A. and Skinner, F.A. (Eds). Academic Press: London and New York.

Mountney, G.J and O'Malley, J. (1965), Acids as poultry meat preservatives. *Poultry Sci.*, **44**, 582–586.

Muotiala, A. and Makela, H.P. (1990), The role of IFN-G, in murine *Salmonella typhimurium* infection. *Micro. Path.*, **8**, 135–141.

Nath, E.J., Neidert, E. and Randall, C.J. (1989), Evaluation of enrichment protocols for the 1–2 test for *Salmonella* detection in naturally contaminated foods and faeces. *J. Fd. Prot.*, **52**, 489–499.

North, W.R. (1960), Use of crystal violet or brilliant green dyes for the determination of salmonellae in dried food products. *J. Bact.*, **80**, 861.

North, W.R. and Bartram, M.T. (1953), The efficiency of selenite broth of different compositions in the isolation of *Salmonella*. *Appl. Micro.*, **1**, 130–134.

NRCC (1969), An evaluation of the *Salmonella* problem. National Academy of Sciences: Washington DC.

NRC (1985), *An Evaluation of the Role of Microbiological Criteria for Foods and Food Ingredients*. National Academy Press: Washington DC.

Obafemi, A. and Davies, R. (1986), The destruction of *Salmonella typhimurium* in chicken exudate by different freeze-thaw treatments. *J. appl. Bact.*, **60**, 381–387.

Ogden, I.D. (1988), A conductance medium to distinguish between *Salmonella* and *Citrobacter* spp. *Int. J. Fd. Mic.*, **7**, 287–297.

Ogden, I.D. and Cann, D.C. (1987), A modified conductance medium for the detection of *Salmonella* spp. *J. appl. Bact.*, **63**, 459–464.

Oggel, J.J., Nundy, D.C. and Randall, C.J. (1990), Modified 1–2 test system as a rapid screening method for detection of *Salmonella* in foods and feeds. *J. Fd. Prot.*, **53**, 656–658.

Old, D.C. and Barker, R.M. (1989), Numerical index of the discriminating ability of biotyping for strains of *Salmonella typhimurium* and *Salmonella portatyphi* B. *Epid. Inf.*, **103**, 435–443.

Old, D.C., Munro, D.S., Reilly, W.J. and Sharp, J.C.M. (1985), Biotype discrimination of *Salmonella montevideo*. *Lett. Appl. Mic.*, **1**, 67–69.

Olson, V.M., Swaminathan, B., Pratt, D.E. and Stadelman, W.J. (1981), Effect of five cycle rapid freeze-thaw treatment in conjunction with various chemicals for the reduction of *Salmonella typhimurium*. *Poult Sci.* **60**, 1822–1826.

Ouwerkerk, T. (1983), Salmonella control in poultry through the use of gamma irradiation. *Proceedings of a Meeting on Combination Processes for Food Irradiation*. Colombo, Sri Lanka, IAEA Publication No STI/PUB/568. HMSO: London.

Panigrahi, D., Burks, M., Hariharan, H. and Finkelstein, R.A. (1987), Evaluation of the immuno–dot–blot assay for detection of cholera-related enterotoxin antigen in *Salmonella typhimurium*. *J. Clin. Mic.*, **25**, 702–705.

Patil, M.D. and Parhad, N.M. (1986), Growth of salmonellas in different enrichment media. *J. appl. Bact.*, **61**, 19–24.

Pavia, A.T., Shipman, L.D., Wells, J.G. *et al.* (1990), Epidemiologic evidence that prior antimicrobial exposure decreases resistance to antimicrobial-sensitive *Salmonella*. *J. Inf. Dis.*, **161**, 255–260.

Perales, I. and Audicana, A. (1989), The role of hens' eggs in outbreaks of salmonellosis in North Spain. *Int. J. Fd. Mic.*, **8**, 175–180.

Pinegar, A. and Suffield, A. (1982), The investigation of food poisoning outbreaks in England and Wales. In *Isolation and Identification Methods for Food Poisoning Organisms*, Corry, J.E.L., Roberts, T.A. and Skinner, F.A. (Eds). Academic Press: London and New York.

Price, T.H. (1983), Multiple plating from enrichment media as an aid to *Salmonella* enrichment. *J. Hyg.*, **90**, 435–440.

Prusak-Sochaczewski, E. and Luong, J.H.T. (1989), An improved ELISA method for the detection of *Salmonella typhimurium. J. Appl. Bact.*, **66**, 127–135.

Pugh, S.J., Griffiths, M.L., Arnott, M.L. and Gutteridge, C.S. (1988), A complete protocol using conductance for rapid detection of salmonellas in confectionery materials. *Lett. Appl. Mic.*, **7**, 23–27.

Rampling, A., Taylor, C.E.D. and Warren, R.E. (1987), Safety of pasteurised milk. *Lancet*, **II**, 1209.

Rappaport, F., Konforti, N. and Navon, B. (1956), A new enrichment medium for certain salmonellae. *J. Clin. Path.*, **9**, 261–266.

Rappold, H. Bolderdijk, R.F. and De Smedt, J.M. (1984), Rapid cultural method to detect *Salmonella*, in foods. *J. Fd. Prot.*, **47**, 46–48.

Ratnam, S. and March, S.B. (1986), Laboratory studies on *Salmonella*-contaminated cheese involved in a major outbreak of gastro-enteritis. *J. appl. Bact.*, **61**, 51–56.

Ratnam, S., Marsh. S.B. and Butler, R.W. (1984), A major outbreak of salmonellosis in Newfoundland traced to contaminated cheese: Laboratory aspects. *Abstract. Conjoint Meeting on Infectious Diseases PE 13.* Can. Assoc. Clin. Mic. Inf. Dis.: Vancouver.

Rhodes, M.W. and Kator, H. (1988), Survival of *Escherichia coli* and *Salmonella* spp. in estuarine environments. *Appl. Environ. Mic.*, **54**, 2902–2907.

Roberts, D. (1982), Bacteria of public health significance. In *Meat Microbiology*, Brown, M.H. (Ed). Applied Science Publishers: London and New York.

Robinson, R.K. and Tamime, A.Y. (1981), Microbiology of fermented milks. In *Dairy Microbiology, Vol. 2. The Microbiology of Milk Products.* Applied Science Publishers: London and New Jersey.

Rodrigues, U.M. and Kroll, R.G. (1990), Rapid detection of salmonellas in raw meats using a fluorescent antibody-microcolony technique. *J. appl. Bact.*, **68**, 213–223.

Rowe, B. Threlfall, E.J. and Hall, L.R. (1987a), Does chloramphenicol remain the drug of choice for typhoid? *Epid. Inf.*, **98**, 379–383.

Rowe, B., Hutchinson, D.N., Gilbert, R.J. *et al.* (1987b), *Salmonella ealing* infections associated with infant dried milk. *Lancet*, **II**, 900–903.

Rubin, H.E. (1985), Protective effect of casein towards *Salmonella typhimurium* in acid-milk. *J. appl. Bact.*, **58**, 251–255.

Ryan, C.A., Nickels, M.K., Hargrett-Bean, M.T. *et al.* (1987), Massive outbreak of antimicrobial-resistant salmonellosis traced to pasteurised milk. *J. Am. Med. Assoc.*, **258**, 3269–3274.

Ryan, C.A., Hargrett-Bean, N.T. and Blake, P.A. (1989), *Salmonella typhi* infections in the United States, 1975–1984: Increasing role of foreign travel. *Revs. Inf. Dis.*, **11**, 1–8.

Sandefur, P.D. and Peterson, J.W. (1976), Isolation of skin permeability factors from culture filtrates of *Salmonella typhimurium. Inf. Imm.*, **44**, 671–679.

Saxen, H. (1984), Mechanism of the protective action of anti-*Salmonella*, IgM in experimental mouse salmonellosis. *J. gen. Mic.*, **130**, 2277–2283.

Scholtens, R. (1962), A sub-division of *Salmonella typhimurium* into phage types based on the method of Craigie and Yen: phages adaptable to species of the B and D groups of *Salmonella*: phage adsorption as a diagnostic aid. *Antonie van Leeuwenhoek*, **28**, 373–381.

Scotland, S.M. (1988), Toxins. *J. appl. Bact.*, **65** (Suppl.), 109–129.

Sharp, J.C.M. (1986), Salmonella special: milk and dairy products. *PHLS Mic. Dig.*, **3**, 28–31.

Simonsen, B., Bryan, F.L., Christian, J.H.B. *et al.* (1987), Prevention and control of food-borne salmonellosis through application of Hazard Analysis Critical Control Point (HACCP). *Int. J. Fd. Mic.*, **4**, 227–247.

Small, P.C., Isberg, P.R. and Falkow, S. (1987), Comparison of the ability of enteroinvasive *Escherichia coli, Salmonella typhimurium, Yersinia pseudotuberculosis* and *Yersinia enterocolitica* to enter and multiply within HEp-2 cells. *Inf. Imm.*, **55**, 1674–1679.

Smith, H. (1984a), Bacterial subversion rather than suppression of immune defences. In *Bacterial and Viral Inhibition and Immunomodulation of Host Defences*, Falini, G., Compa, A., Smith, H. and Scott, G.M. (Eds). Academic Press: London.

Smith, H. (1984b), The biochemical challenge of microbial pathogenicity, *J. appl. Bact.*, **57**, 395–404.

Sperber, S.J. and Schleupner, C.J. (1987), Salmonellosis during infection with human immunodeficiency virus. *Rev. Inf. Dis.*, **9**, 925–934.

StClair, V.J. and Klenk, M.M. (1990), Performance of three methods for the rapid identification of *Salmonella* in naturally contaminated foods and feeds. *J. Fd. Prot.*, **53**, 961–964.

Stephen, J. and Pietrowski, R.A. (1981), *Bacterial Toxins.* Thomas Nelson: Walton on Thames.

Stevens, A., Joseph, C. and Bruce, J. (1989), A large outbreak of *Salmonella enteritidis* pt 4 associated with eggs from overseas. *Epid. Inf.*, **103**, 425–433.

Stinavage, P., Martin, L.E. and Spitznagel, J.K. (1989), O antigen and lipid A phosphoryl groups in resistance of *Salmonella typhimurium* LT-2 to nonoxidative killing in human polymorphonuclear neutrophils. *Inf. Imm.*, **57**, 3894–3900.

Stinavage, D.S., Martin, L.E. and Spitznagel, J.K. (1990), A 59 KDalton outer membrane protein of *Salmonella typhimurium* protects against oxidative intraleukocytic killing due to human neutrophils. *Mol. Mic.*, **4**, 283–293.

Sveum, W.H. and Kraft, A.A. (1981), Recovery of salmonellae from foods using a combined enrichment technique. *J. Fd. Sci.*, **46**, 94–99.

Swaminathan, B. and Konger, R.L. (1986), Immunoassay for detecting foodborne bacteria and microbial toxins. In *Foodborne Microorganisms and their Toxins: Developing Methodology*, Pierson, M.L. and Stern, N.J. (Eds).

Tamminga, S.K., Beumer, R.R., Kampelmacher, E.H. and van Leusden, F.M. (1975), Survival of *Salmonella eastbourne* and *Salmonella typhimurium* in milk chocolate prepared with artificially contaminated milk powder. *J. Hyg.*, **79**, 333–337.

Taplin, J. (1982), *Salmonella newport* outbreak—Victoria. *Comm. Dis. Int.*, **1**, 3–6.

Taylor, W.I. (1965), Isolation of shigellae. I. Xylose lysine agars; new media for isolation of enteric pathogens. *Am. J. Clin. Path.*, **44**, 471–475.

Taylor, D.N., Wachsmuth, I.K., Shangkuan, Y-H *et al.* (1982), Salmonellosis associated with marijuana: a multistate outbreak traced by plasmid fingerprinting. *New Eng. J. Med.*, **306**, 1249–1253.

Taylor-Robinson, S., Miles, R., Whitehead, A. and Dickinson, R.J. (1989), *Salmonella* infection and ulcerative colitis. *Lancet*, **I**, 1145.

492

Thomson, J.E., Cox, N.A. and Bailey, J.S. (1976), Chlorine, acid and heat treatments to eliminate *Salmonella* on broiler carcasses. *Poult. Sci.*, **55**, 1513–1517.

Threlfall, E.J., Hall, M.L.M. and Rowe, B. (1986a), *Salmonella gold-coast* from outbreaks of food-poisoning in the British Isles can be differentiated by plasmid profiles. *J. Hyg.*, **97**, 115–122.

Threlfall, E.J., Hall, M.L.M., Ward, L.R. and Rowe, B. (1986b), Use of plasmid profiling in investigating outbreaks of *Salmonella* food-poisoning in Britain. *IUMS XIV Int. Cong. Micro. Manchester: Abst. P.G5–60.*

Threlfall, E.J., Rowe, B. and Ward, L.R. (1989), Subdivision of *Salmonella enteritidis* phage types by plasmid profiles. *Epid. Inf.*, **102**, 459–465.

Threlfall, E.J., Rowe, B. and Ward, L.R. (1990), Plasmid profile typing can be used to subdivide phage-type 49 of *Salmonella typhimurium* in outbreak investigations. *Epid. Inf.*, **104**, 243–251.

Todd, L.S., Roberts, D., Bartholomew, B.A. and Gilbert, R.J. (1987), Assessment of an enzyme immunoassay for the detection of salmonellas in foods and animal feeding stuffs. *Epid. Inf.*, **100**, 186–192.

Truscott, R.B. and Lammerding, A.M. (1987), Millipore filtration and use of Rappaport–Vassiliadis medium for isolation of *Salmonella* from preenrichment broth. *J. Fd. Prot.*, **50**, 815–819.

Tufano, R., Ianniello, R. and Galdiero, M. (1989), Effect of *Salmonella typhimurium* porins on biological activities of human polymorphonuclear leukocytes. *Micro. Path.*, **7**, 337–341.

van Leusden, F.M., van Schothorst, M. and Beckers, H.J. (1982), The standard *Salmonella* isolation method. In *Isolation and Identification Methods for Food Poisoning Organisms*, Corry, J.E.L., Roberts, T.A. and Skinner, F.A. (Eds). Academic Press: London and New York.

van Schothorst, M. and Renauld, A. (1983), Dynamics of *Salmonella* isolation with modified Rappaport's medium (R10). *J. appl. Bact.*, **54**, 209–215.

Vassiliadis, P. (1983), The Rappaport–Vassiliadis (RV) enrichment for the isolation of salmonellas: An overview. *J. appl. Bact.*, **54**, 69–76.

Vassiliadis, P., Pateraki, E., Papaiconomou, N. *et al.* (1976), Nouveau procédé d'enrichissment de *Salmonella*. *Ann. Micro. (Inst. Pasteur)*, **127B**, 195–200.

Vassiliadis, P., Mavrommati, C.H., Kalopathaki, V. *et al.* (1985), *Salmonella* isolation with Rappaport–Vassiliadis enrichment medium seeded with different sized inocula of pre-enrichment cultures of meat products and sewage polluted water. *J. Hyg.*, **95**, 139–147.

Vieu, J.H., Hassan-Massoud, B., Klein, B. and Leherissey, M. (1981), Bacteriophage typing and biotyping of *Salmonella montevideo*. In *Salmonella and Salmonellosis*. Int. Symp. Eur. Fed. Mic. Soc. Istanbul.

Vladioanu, I.-R., Chang, H.R. and Pechere, J.-C. (1990), Expression of host resistance to *Salmonella typhi* and *Salmonella typhimurium*, bacterial survival within macrophages of murine and human origin. *Micro. Path.*, **8**, 83–90.

Wallis, T.S., Vaughan, A.T.M., Clark, G.J. *et al.* (1990), The role of leukocytes in the induction of fluid secretion by *Salmonella typhimurium*. *J. Med. Mic.*, **31**, 27–35.

Ward, L.A., de Saxe, J.D.H. and Rowe, B. (1987), A phage typing scheme for *Salmonella enteritidis*. *Epid. Inf.*, **99**, 291–294.

Watson, J. (1985), Another 'fast reactor' in Caithness. *Comm. Dis. (Scotland) Rep.*, 85/49.

Watson, C. and Walter, A.P. (1978), A modification of brilliant green agar for improved isolation of *Salmonella*. *J. appl. Bact.*, **45**, 195–204.

Wilkins, E.G.L. and Roberts, C. (1988), Extraintestinal salmonellosis. *Epid. Inf.*, **100**, 361–368

Williams P.H., Roberts, M. and Hinson, G. (1988), Stages in bacterial invasion. *J. appl. Bact.*, **65** (Suppl.), 131S–147S.

Wilson, W.J. and Blair, E.M.M. (1927), Use of a glucose bismuth sulphite iron medium for the isolation of *B. typhosus* and *B. proteus*. *J. Hyg.*, **26**, 374–391.

Worton, K.J., Candy, D.C.A., Wallis, T.S. *et al.* (1989), Studies on early association of *Salmonella typhimurium* with intestinal mucosa *in vivo* and *in vitro*. *J. Med. Mic.*, **29**, 283–294.

Chapter 5

Abbott, J.D. and Shannon, R. (1958), A method for typing *Shigella sonnei* using colicin production as a marker. *J. Clin. Path.*, **11**, 71–77.

Adler, B., Sasakawa, C., Tobe, T. *et al.* (1989), A dual transcriptional isolation system for the 230kb plasmid genes coding for virulence-associated antigens of *Shigella*. *Mol. Micro.*, **3**, 627–635.

Anon (1985), Gastroenteritis caused by *Escherichia coli* and *Shigella* retards the growth of children. *Nutr. Res.*, **43**, 169–170.

Anon (1988a), Multistate outbreak of *Shigella sonnei* gastroenteritis — United States. *J. Inf. Dis.*, **157**, 156–157.

Anon (1988b), Nationwide dissemination of multiply resistant *Shigella sonnei* following a common-source outbreak. *J. Fd. Prot.*, **51**, 157–158.

Aqua, M.F., Svinarich, D., Dhar, A. and Palchaudhuri, S. (1988), The stability of the O-antigen plasmid is determined by a chromosomal region of *Shigella dysenteriae* 1. *Can. J. Mic.*, **34**, 58–62.

Archer, D.L. (1985), Enteric microorganisms in rheumatoid diseases: causative agents and possible mechanisms. *J. Fd. Prot.*, **48**, 887–894.

Ashkenazi, S. and Cleary, T.G. (1989), Rapid method to detect Shiga toxin and Shiga-like toxin 1 based on binding to globiotriosyl ceramide (Gb$_3$), their natural receptor. *J. Clin. Mic.*, **27**, 1145–1150.

Ashkenazi, S., Cleary, K.R., Pickering, L.K. *et al.* (1990), The association of Shiga toxin and other cytotoxins with the neurologic manifestations of Shigellosis. *J. Inf. Dis.*, **161**, 961–965.

Bartlett, A.V., Prado, D., Cleary, T.G. and Pickering, L.K. (1986), Production of Shiga toxin and other cytotoxins by serogroups of *Shigella*. *J. Inf. Dis.*, **154**, 996–1002.

Bennish, M.L., Harris, J.R., Wojtyniak, B.J. and Struelens, M. (1990), Death in shigellosis: Incidence and risk factors in hospitalized patients. *J. Inf. Dis.*, **161**, 500–506.

Bergan, T. (1979), Bacteriophage typing of *Shigella*. In *Methods in Microbiology*, Vol. 13, Bergan, T. and Norris, J.R. (Eds). Academic Press, London and New York.

Berkow, R. (1982), Shigellosis. In *Merck Manual of Diagnosis and Therapy*. Merck, Sharpe and Dohme Research Laboratory: Rahway.

Bernadini, M.L., Mourier, J., d'Hauteville, H. *et al.* (1989), *ics*A, a plasmid locus of *Shigella flexneri*, governs bacterial intra- and intercellular spread through interaction with F-actin. *Proc. Nat. Acad. Sci.*, **80**, 3867–3871.

Bok, H.E., Holzapfel, W.H., Odendaal, E.S. and van der Linde, H.J. *et al.* (1986), Incidence of foodborne pathogens on retail broilers. *Int. J. Fd. Mic.*, **3**, 273–285.

Brenner, D.J. (1984), *Enterobacteriaceae*. In *Bergey's manual of Systematic Bacteriology*, Krieg, N.R. and Holt, J.G. (Eds). Williams and Wilkins: Baltimore and London.

Bryan, F.L. (1978), Factors that contribute to outbreaks of foodborne disease. *J. Fd. Prot.*, **41**, 816–822.

Bryan, F.L. (1979), Infections and intoxications caused by other bacteria. In *Foodborne Infections and Imtoxications*, 2nd ed., Riemann, H. and Bryan, F.L. (Eds). Academic Press: New York.

Butler, T., Islam, M.R., Azad, M.A.K. and Jones, D.K. (1987), Risk factors for the development of haemolytic uraemic syndrome during shigellosis. *J. Pediatr.*, **110**, 894–897.

Clerc, P.L. and Sansonetti, P.J. (1989), Evidence for clathrin mobilization during directed phagocytosis of *Shigella flexneri* by HEp-2 cells. *Micro. Path.*, **7**, 329–336.

Clerc, P.L., Ryter, A., Mounier, J. and Sansonetti, P.J. (1987), Plasmid-mediated early killing of eucaryotic cells by *Shigella flexneri* as studied by infection of J774 macrophages. *Inf. Imm.*, **55**, 521–527.

Cowan, S.T. and Steel, K.J. (1974), *Manual for the Identification of Medical Bacteria*, 2nd ed. Cambridge University Press: London.

Donohue-Rolfe, A. and Keusch, G.T. (1983), *Shigella dysenteriae* 1 cytotoxin: periplasmic protein releasable by polymyxin B and osmotic shock. *Inf. Imm.*, **39**, 270–274.

DuPont, J.L., Reves, R.R., Galindo, E. *et al.* (1982), Treatment of travelers' diarrhea with trimethoprim/sulfamethoxazole and with trimethoprim alone. *New England J. Med.*, **307**, 841–844.

Fehlhaber, K. (1981), Untersuchungen uber lebensmittelhygenisch bedeutsame Eigenschaften von *Shigella*. *Arch. Exp. Vet. Med.* (Leipzig), **35**, 955–964.

Fernandez, A., Sninsky, C.A., O'Brien, A.D. *et al.* (1984), Purified *Shiella* enterotoxin does not alter intestinal mobility. *Inf. Imm.*, **43**, 477–481.

Formal, S.B., Gemski, P., Gianella, R.A. and Austin, S. (1972), Mechanisms of *Shigella* pathogenesis. *Am. J. Clin. Path.*, **25**, 1427–1438.

Frorgren, V.C., Arondel, J. and Sansonetti, P.J. (1990), Contribution of superoxide dismutase and catalase activities to *Shigella flexneri* pathogenesis. *Inf. Imm.*, **58**, 529–535.

Goullet, P. and Picard, B. (1987), Differentiation of *Shigella* by esterase electrophoretic polymorphism. *J. gen. Mic.*, **133**, 1005–1007.

Gross, R.J., Rowe, B., Cheasty, T. and Thomas, L.V. (1981), Increase in drug resistance among *Shigella dysenteriae*, *Sh. flexneri* and *Sh. boydii*. *B. Med. J.*, **283**, 575.

Guerrant, R.L. (1985), Microbial toxins and diarrhoeal diseases: introduction and overview. In *Microbial Toxins and Diarrhoeal Diseases*, Evered, E. and Whelan, J. (Eds). CIBA Foundation Symposium No 112. Pitman Publishing Ltd: London.

Haider, K., Kay, B.A., Talukder, K.A. and Huq, M.I. (1988), Plasmid analysis of *Shigella dysenteriae* Type 1 isolates obtained from widely scattered geographical locations. *J. Clin. Mic.*, **26**, 2083–2086.

Hajna, A.A., (1955), A new enrichment broth medium for Gram-negative organisms of the intestinal group. *Pub. Hlth. Lab.*, **13**, 83–89.

Hale, T.L., Morris, R.E. and Bobventre, P.F. (1979), *Shigella* infection of Henle intestinal epithelial cells: role of the host cell. *Inf. Imm.*, **24**, 887–894.

Hale, T.L, Sansonetti, P.J., Shad, P.A. *et al.* (1983), Characterisation of virulence plasmids and plasmid-associated outer membrane proteins in *Shigella flexneri*, *Shigella sonnei* and *Escherichia coli*. *Inf. Imm.*, **40**, 340–350.

Hentges, D.J. (1967), Influence of pH on the inhibitory activity of formic and acetic acids for *Shigella*. *J. Bact.*, **93**, 2029–2030.

Hess, C.B., Niesel, D.W. and Holmgren, J. (1990), Interferon production by *Shigella flexneri*-infected fibroblasts depends upon intracellular bacterial metabolism. *Inf. Imm.*, **58**, 399–405.

Hyde, H.C. (1976), Utilization of wastewater sludge for agricultural soil enrichment. *J. Water Poll. Cont. Fed.*, **48**, 77–90.

Jacewicz, M., Feldman, H.A. and Donohue-Rolfe, A. (1989), Pathogenesis of *Shigella* diarrhoea XIV Analysis of Shiga toxin receptors on cloned HeLa cells. *J. Inf. Dis.*, **159**, 881–889.

Jackson, M.P. (1990), Structure-function analysis of Shiga toxin and the Shiga-like toxins. *Micro Path.*, **8**, 235–242.

Kayser, A. and Mossel, D.A.A. (1984), Intervention sensu Wilson: The only valid approach to microbiological safety of food. *Int. J. Fd. Mic.*, **1**, 1–4.

Ketyi, I. (1985), Toxins as virulence factors of bacterial enteric pathogens (a review). *Acta Microbiol. Hung.*, **32**, 279–304.

Keusch, G.T., Grady, G.F., Mata, L.J. and McIver, J. (1972), The pathogenesis of *Shigella* diarrhoea. I Enterotoxin production by *Shigella dysenteriae* 1. *J. Clin. Invest.*, **51**, 1212–1222.

Keusch, G.T., Donohue-Rolfe, A. and Jacewicz, M.C. (1985), *Shigella* toxin and the pathogenesis of shigellosis. In *Microbial toxins and diarrhoeal diseases*, Evered, E. and Whelan, J. (Eds). CIBA Foundation Symposium No 112. Pitman Publishing Ltd: London.

Kopecko, D.J., Baron, L.S. and Buysse, J. (1985), Genetic determinants of virulence in *Shigella* and dysenteric strains of *Escherichia coli*: their involvement in the pathogenesis of dysentery. *Curr. Top. Micro. Immun.*, **118**, 71–95.

Lampel, K.A., Jagow, J.A. and Truckess, M. (1990), Polymerase chain reaction for detection of invasive *Shigella flexneri* in food. *Appl. Environ. Micro.*, **56**, 1536–1540.

Madden, M.J., McCardell, B.A. and Archer, D.L. (1986), Virulence assessment of foodborne microbes. In *Foodborne Microorganisms and their Toxins: Developing Methodology*, Pierson, M.D. and Stern, N.J. (Eds). Marcel Dekker: New York and Basel.

Maurelli, A.T., Blackmon, B. and Curtiss, R. (1984), Temperature-dependent expression of virulence genes in *Shigella* species. *Inf. Imm.*, **43**, 195–201.

Mobassaleh, M., Donohue-Rolfe, A., Jacewicz, M. *et al.* (1988), Pathogenesis of *Shigella* diarrhoea: Evidence for a developmentally regulated glycolipid receptor for *Shigella* toxin involved in the fluid secretory response of rabbit small intestine. *J. Inf. Dis.*, **157**, 1023–1031.

Morris, G.K (1984), *Shigella*. In *Compendium of Methods for the Microbiological Examination of Foods*, 2nd ed. American Public Health Association: Washington DC.

Morris (1986), *Shigella*. In *Progress in Food Safety*, Cliver, D.O. and Cochrane, B.A. (Eds). Food Research Inst. Univ. of Wisconsin: Madison.

Mossel, D.A.A. (1982), *Microbiology of Foods*, 3rd ed. University of Utrecht: Utrecht.

Nakamura, M., Stone, R.L. and Krubsack, J.E. (1964), Survival of *Shigella* in sea water. *Nature*, **203**, 213–214.

Neill, R.J., Gemski, P., Formal, S.B. and Newland, J.W. (1988), Deletion of the Shiga toxin gene in a chlorate-resistant derivative of *Shigella dysenteriae* Type 1 that retains virulence. *J. Inf. Dis.*, **158**, 737–741.

Oaks, E.V., Wingfield, M.E. and Formal, S.B. (1985), Plaque formation by virulent *Shigella flexneri*. *Inf. Imm.*, **48**, 124–129.

O'Brien, A.D. and Holmes, R.K. (1987), Shiga and Shiga-like toxins. *Micro. Rev.*, **51**, 206–220.

Obrig, T,G., Del Vecchio, P.J., Brown, J.E. *et al.* (1988), Direct cytotoxic action of Shiga toxin on human vascular endothelial cells. *Inf. Imm.*, **56**, 2373–2378.

Ochiai, K., Yamanaka, T., Kimura, K. and Sowada, O. (1959), Studies on the inheritance of drug resistance between *Shigella* strains and *Escherichia coli* strains. *Nihon Iji Shimp.*, **1861**, 34–46.

Pai, T., Pacsa, A.S., Emody, L. *et al.* (1985), Modified enzyme-linked immunosorbent assay for detecting enteroinvasive *Escherichia coli* and virulent *Shigella* strains. *J. Clin. Mic.*, **21**, 415–418.

Parsonnet, J., Greene, K.D., Gerber, A.R. *et al.* (1989), *Shigella dysenteriae* Type 1 infection in US travellers to Mexico, 1988. *Lancet*, **I**, 543–545.

Payne, S.M. (1988), Iron and virulence in the family *Enterobacteriaceae*. *CRC Crit. Rev. Mic.*, **16**, 81–134.

Petrovskaya, V.G. and Bondarenko, V.M. (1977), Recommended corrections to the classification of *Shigella flexneri* on a genetic basis. *Int. J. Syst. Bact.*, **27**, 171–175.

Prizant, R. and Reed, W.P. (1980), Possible role of colonic content in the mucosal association of pathogenic *Shigella*. *Inf. Imm.*, **29**, 1197–1199.

Rothwell, J. (1981), Microbiology of ice cream and related products. In *Dairy Microbiology, Vol. 2. The Microbiology of Milk Products*, Robinson, R.K. (Ed). Applied Science Publishers: London and New Jersey.

Rowe, B. and Gross, R.J. (1984), *Shigella*. In *Bergey's Manual of Systematic Bacteriology*, Krieg, N.R. and Holt, J.G. (Eds). Williams and Wilkins: Baltimore and London.

Sansonetti, P.J., Kopecko, P.J. and Formal, S.B. (1981), *Shigella sonnei* plasmids: evidence that a large plasmid is necessary for virulence. *Inf. Imm.*, **34**, 75–83.

Sansonetti, P.J., D'Hauteville, H., Formal, S.B. and Toucas, M. (1982), Plasmid mediated invsiveness of 'shigella-like' *Escherichia coli*. *Ann. Micro. (Inst. Pasteur, Paris)*, **133A**, 351–355.

Sansonetti, P.J. D'Hauteville, H., Ecobichan, C. and Pourcel, C. (1983), Molecular comparison of virulence plasmids in *Shigella* and entero-invasive *Escherichia coli*. *Ann. Micro. (Inst. Pasteur, Paris)*, **134A**, 295–318.

Sansonetti, P.J., Ryter, A., Clerc, P. *et al.* (1986), Multiplication of *Shigella flexneri* within HeLa cells: lysis of the phagocytic vacuole and plasmid-mediated contact haemolysis. *Inf. Imm.*, **51**, 461–469.

Scotland, S.M. (1988), Toxins. *J. appl. Bact.*, **65** (Suppl.), 109–129.

Sethabutyr, O., Hanchalay, S., Echeverria, P. *et al.* (1985), A non- radioactive DNA probe to identify *Shigella* and enteroinvasive *Escherichia coli* in stools of children with diarrhea. *Lancet*, **II**, 1095–1097.

Smith, J.L. (1987), *Shigella* as a foodborne pathogen. *J. Fd. Prot.*, **48**, 887–894.

Smith, J.L. and Dell, B.J. (1990), Capability of selective media to detect heat-injured *Shigella flexneri*. *J. Fd. Prot.*, **53**, 141–144.

Stagard, C.E., Deskaleros, P.A. and Payne, S.M. (1989), A 101–kilodalton heme-binding protein associated with congo red binding and virulence of *Shigella flexneri* and enteroinvasive *Escherichia coli* strains. *Inf. Imm.*, **57**, 3534–3539.

Stephen, J. and Pietrowski, R.A. (1981), *Bacterial Toxins*. American Society for Microbiology: Washington DC.

Stoll, B.J., Glass, R.I., IIuq, M.I. *et al.* (1982), Epidemiologic and clinical features of patients infected with *Shigella* who attended a diarrheal disease hospital in Bangladesh. *J. Inf. Dis.*, **146**, 177–183.

Struelens, M.J., Patte, D., Kabir, D. *et al.* (1985), *Shigella* septicaemia: prevalence, presentation, risk factors and outcome. *J. Inf. Dis.*, **152**, 784–790.

Takeda, Y., Okamoto, K. and Miwitani, T. (1979), Mitomycin C stimulates production of a toxin in *Shigella* species that causes morphological changes in Chinese hamster ovary cells. *Inf. Imm.*, **23**, 178–180.

Thomas, M.E.M. and Tillett, H.E. (1973), Dysentery in general practice: a study of cases and their contacts in Enfield and an epidemiological comparison with salmonellosis. *J. Hyg.*, **71**, 373–389.

Tollison, S.B. and Johnson, M.G. (1985), Sensitivity to bile salts of *Shigella flexneri* sublethally heat stressed in buffer or broth. *Appl. Environ. Micro.*, **50**, 337–341.

Twedt, R.M. (1978), *Shigella*. In *FDA Div Microbiology, Bacteriological Analytical Manual*. Association of Official Analytical Chemists: Washington DC.

Venkatesan. M., Buysse, J.M., Vandendrien, E. and Kopecko, D.J. (1988), Development and testing of invasion-associated DNA probes for detection of *Shigella* species and entero-invasive *Escherichia coli*. *J. Clin. Mic.*, **26**, 261–266.

Venkatesan, M., Buysse, J.M. and Kopecko, D.J. (1989), Use of *Shigella flexneri* ipaC and ipaH gene sequences for the general identification of *Shigella* spp and entero-invasive *Escherichia coli*. *J. Clin. Mic.*, **27**, 2687–2891.

Wachsmuth, K. and Morris, G.K. (1989), *Shigella*. In *Foodborne Bacterial Pathogens*, Doyle, M.P (Ed). Marcel Dekker: New York.

Watanabe, H. and Timmis, K.N. (1984), A small plasmid in *Shigella dysenteriae* 1 specifies one or more functions essential for O-antigen production and bacterial virulence. *Inf. Imm.*, **46**, 55–63.

Weissman, J.B., Williams, S.V., Hinman, A.R. *et al.* (1974), Food-borne shigellosis at a county fair. *Am. J. Epid.*, **100**, 178–185.

Williams, P.H. and Roberts, M. (1985), Aerobactin-mediated iron uptake: a virulence determinant in enteropathogenic *Escherichia coli*? *Lancet*, **I**, 763.

Chapter 6

Abbar, F.M. (1988), Incidence of fecal coliforms and serovars of enteropathogenic *Escherichia coli* in naturally contaminated cheese. *J. Fd. Prot.*, **51**, 384–385.

Ahmed, R., Bopp, C., Borczyk, A. and Kasitaya, S. (1987), Phage-typing scheme for *Escherichia coli* O157:H7. *J. Inf. Dis.*, **155**, 806–809.

Anderson, J.M. and Baird-Parker, A.C. (1975), A rapid and direct plate method for enumerating *Eschericia coli* biotype 1 in food. *J. appl. Bact.*, **39**, 111–117.

Andrade, J.R.C., De Vaisa, V.F., De Santa Rosa, M.R. and Suassuna, I. (1989), An endocytic process in HEp-2 cells induced by enteropathogenic *Escherichia coli*. *J. Med. Mic.*, **28**, 49–57.

Anon (1987), Outbreak of gastrointestinal disease — Ontario. *J. Fd. Prot.*, **50**, 438–439.

Antai, S.P. and Anozie, S.O. (1987), Incidence of infantile diarrhoea due to enteropathogenic *Escherichia coli* in Port Harcourt metropolis. *J. appl. Bact.*, **62**, 227–229.

Back, E., Svennerholm, A.M., Holmgren, J. and Molby, R. (1979), Evaluation of a ganglioside immunosorbent assay for detection of *Escherichia coli* heat-labile enterotoxin. *J. Clin. Mic.*, **10**, 791–795.

Baldwin, T.J., Brooks, S.F., Knutton, S. *et al.* (1990), Protein phosphorylation by protein kinase C in HEp-2 cells infected with enteropathogenic *Escherichia coli*. *Inf. Imm.*, **58**, 761–765.

Basta, M., Karmali, M. and Lingwood, C. (1989), Sensitive receptor- specified enzyme-linked immunosorbent assay for *Escherichia coli* verocytotoxin. *J. Clin. Mic.*, **27**, 1617–1622.

Baudry, B., Savarinos, S.J., Vial, P. *et al.* (1990), A sensitive and specific DNA probe to identify enteroaggregative *Escherichia coli* a recently discovered diarrhoeal pathogen. *J. Inf. Dis.*, **161**, 1249–1251.

Beery, J.T., Doyle, M.P. and Schoeni, J.L. (1985), Colonization of chicken cecae by *Escherichia coli* associated with hemorrhagic colitis. *Appl. Environ. Micro.*, **49**, 310–315.

Beutin, L., Bode, L., Richter, T. *et al.* (1984), Rapid visual detection of *Escherichia coli* and *Vibrio cholerae* heat-labile enterotoxin by nitrocellulose enzyme-linked immunosorbent assay. *J. Clin. Mic.*, **19**, 371–375.

Beutin, L., Montenegro, M.A., Orskov, I. *et al.* (1989), Close association of verotoxin (Shiga-like toxin) production with enterohemolysin production in strains of *Escherichia coli*. *J. Clin. Mic.*, **27**, 2559–2564.

Bhan, M.K., Raij, P., Levine, M.M. *et al.* (1989), Enteropathogenic *Escherichia coli* associated with persistent diarrhoea in a cohort of rural children in India. *J. Inf. Dis.*, **159**, 1061–1064.

Bongaerts, G.D.A., Bruggeman-Ogle, K.M. and Mouton, R.P. (1985), Improvements in the microtitre G_{M1} ganglioside enzyme-linked immunosorbent assay for *Escherichia coli* heat-labile enterotoxin. *J. appl. Bact.*, **59**, 443–449.

Bowen, J.R., Congeni, B.L., Cleary, T.G. *et al.* (1989), *Escherichia coli* O114:non motile as a pathogen in an outbreak of severe diarrhoea associated with a day care center. *J. Inf. Dis.*, **160**, 243–247.

Brenner, D.J., Davis, B.R., Steigerwalt, A.G. *et al.* (1982a), Atypical biogroups of *Escherichia coli* found in clinical specimens and description of *Escherichia hermanii* sp. nov. *J. Clin. Mic.*, **15**, 703–713.

Brenner, D.J., McWhorter, A.C., Leete, Knutson, J.K. and Steigerwalt, A.G. (1982b), *Escherichia vulneris*: a new species of *Enterobacteriaceae* associated with human wounds. *J. Clin. Mic.*, **15**, 1133–1140.

Bryant, H.E., Athan, M.A. and Pai, C.H. (1989), Risk factors for *Escherichia coli* O157:H7 infections in an urban community. *J. Inf. Dis.*, **160**, 858–864.

Burgess, N.R.H., McDermott, S.N. and Whiting, J. (1973), Aerobic bacteria occurring in the hind-gut of the cockroach, *Blatta orientalis*. *J. Hyg.*, **71**, 1–7.

Chang, G.W., Brill, J. and Lum, R. (1989), Proportion of ß-D-glucuronidase negative *Escherichia coli* in human faecal samples. *Appl. Environ. Micro.*, **55**, 335–339

Clarke, R.C., McEwen, S.A., Gannan, V.P. *et al.* (1989), Isolation of verocytotoxin-producing *Escherichia coli* from milk filters in south-western Ontario. *Epid. Inf.*, **102**, 253–260.

Cleary, T.G. and Murray, B.E. (1988), Lack of Shiga-like cytotoxin production by enteroinvasive *Escherichia coli*. *J. Clin. Mic.*, **26**, 2177–2179

Cleary, T.G., Matthewson, J.J., Faris, E. and Pickering, L.K. (1985), Shiga-like cytotoxin produced by enteropathogenic *Escherichia coli* serogroups. *Inf. Imm.*, **47**, 335–337.

Cruickshank, R., Duguid, J.P., Marmion, B.P. and Swain, R.H.A. (1975), *Medical Microbiology Vol. II. The Practice of Medical Microbiology*. Churchill Livingstone: Edinburgh.

Czersinski, C. and Svennerholm, A.M. (1983), Ganglioside G_{M1} enzyme-linked immunospot assay for single identification of heat-labile enterotoxin producing *Escherichia coli*. *J. Clin. Mic.*, **17**, 965–969.

Danbara, H., Komare, K., Hiroshi, A. *et al.* (1988), Molecular analysis of enterotoxin plasmids of 14 different O serotypes. *Inf. Imm.*, **56**, 1513–1517.

Dickie, N, Speirs, J.I., Akhtor, M. *et al.* (1989), Purification of an *Escherichia coli* serogroup O157:H7 verotoxin and its detection in North American hemorrhagic colitis isolates. *J. Clin. Mic.*, **27**, 1973–1978.

Donnenberg, M.S., Donohue-Rolfe, A. and Keusch, G.T. (1989), Epithelial cell invasion: an overlooked property of enteropathogenic *Escherichia coli* (EPEC) associated with the EPEC adherence factor. *J. Inf. Dis.*, **160**, 452–459.

Donnenberg, M.S., Calderwood, S.B. and Donohue-Rolfe, A. (1990), Construction and analysis of *Tnpho* mutants of enteropathogenic *Escherichia coli* unable to invade HEp-2 cells. *Inf. Imm.*, **58**, 1565–1571.

Donta, S.I., Damiano-Burbach, P. and Poindexter, N.J. (1988), Modulation of enterotoxin binding and function *in vitro* and *in vivo*. *J. Inf. Dis.*, **157**, 557–564.

Dorn, C.R., Scotland, S.M., Smith, H.R. *et al.* (1989), Properties of verocytotoxin-producing *Escherichia coli* of human and animal origin belonging to serotypes other than O157:H7. *Epid. Inf.*, **193**, 83–95.

Doyle, M.P. and Schoeni, M.P. (1987), Isolation of *Escherichia coli* O157:H7 from retail fresh meats and poultry. *Appl. Environ. Micro.*, **53**, 2394–2396.

Doyle, M.P. and Padyhe, V.V. (1989), *Escherichia coli*. In *Foodborne Bacterial Pathogens*, Doyle, M.P. (Ed). Marcel Dekker: New York and Basel.

Duguid, J.P., Marmion, B.P. and Swain, R.H.A. (1978), *Mackie and McCartney Medical Microbiology*. Churchill Livingstone: Edinburgh.

Edelman, R., Karmali, M.A. and Fleming, P.A. (1988), Summary of the International Symposium and Workshop of Infections due to vero-cytotoxin (Shiga-like toxin)-producing *Escherichia coli. J. Inf. Dis.*, **157**, 1102–1104.

Ewing, W.H. (1972), *Differentiation of Enterobacteriaceae by Biocheical Reactions*. Center for Disease Control: Atlanta.

Eidels, L., Proia, R.L. and Hart, D.A. (1983), Membrane receptors for bacterial toxins. *Micro. Rev.*, **47**, 596–620.

Evans, D.J. and Evans, D.G. (1983), Classification of pathogenic *Escherichia coli* according to serotype and the production of virulence factors with special reference to colonization factor antigens. *Rev. Inf. Dis.*, **5** (Suppl. 4), S692–S701.

Evans, D.J., Evans, D.G. and Gorbach, S.L. (1974), Polymyxin B-induced release of low-molecular weight, heat-labile-enterotoxin from *Escherichia coli. Inf. Imm.*, **10**, 1010–1017.

Evans, D.J., Evans, D.G., DuPont, H.L. (1977), Patterns of loss of enterotoxigenicity by *Escherichia coli* isolated from adults with diarrhoea: Suggested evidence for an inter-relationship with serotype. *Inf. Imm.*, **17**, 105–111.

Farmer, J.J., Fanning, G.R., Davis, B.R. *et al.* (1985), *Escherichia fergusonii* and *Enterobacter taylorae* two new species of *Enterobacteriaceae* isolated from clinical specimens. *J. Clin. Mic.*, **21**, 77–81.

Feng, P.C.S. and Hartmann, P.A. (1982), Fluorogenic assays for immediate confirmation of *Escherichia coli. Appl. Environ. Micro.*, **43**, 1320–1329.

Fishbein, M., Mehlman, I.J., Chugg, L. and Olson, J.C. (1976), Coliforms, fecal coliforms, *E. coli* and enteropathogenic *E. coli*. In *Compendium of Methods for the Microbiological Examination of Foods*. American Public Health Association: Washington DC.

Fletcher, J.N., Saunders, J.R., Batt, R.M. *et al.* (1990), Attaching effacement of the rabbit enterocyte brush border is encoded on a single 96.5 kilobase-pair plasmid in an EPEC O111 strain. *Inf. Imm.*, **58**, 1316–1322.

Frampton, E.W., Restaino, L. and Blazko, N. (1988), Evaluation of the ß-glucuronidase substrate 5-bromo-4-chloro-3 Indolyl-ß-D-glucuronide (X-GLUC) in a 24 hour direct plating method for *Escherichia coli. J. Fd. Prot.*, **51**, 402–404.

Franke, V.G., Hahn, G and Tolle, A. (1984), *Zentralbl. Bakt. Hyg. A.*, **257**, 51–59.

Frost, J.A., Smith, H.R., Willshaw, G.A. *et al.* (1989), Phage-typing of vero-cytotoxin (VT) producing *Escherichia coli* O157 isolated in the United Kingdom. *Epid. Inf.*, **103**, 73–81.

Furrer, B., Candrian, U. and Luthy, J. (1990), Detection and identification of *E. coli* producing heat-labile enterotoxin type I by enzymatic modification of a specific DNA fragment. *Lett. Appl. Mic.*, **10**, 31–34.

Gibson, A.M. and Roberts, T.A. (1986), The effect of pH, water activity, sodium nitrite and storage temperature on the growth of enteropathogenic *Escherichia coli* and salmonellae in laboratory medium. *Int. J. Fd. Mic.*, **3**, 183–194.

Glatz, B.A. (1986), Genetic regulation of toxin production by foodborne microbes. In *Food Microbiology Vol. 1. Concepts in Physiology and Metabolism*, Montville, T.J. (Ed). CRC Press: Boca Raton.

Grandsen, W.R., Damm, M.A.S., Anderson, J.D. *et al.* (1985), Haemolytic cystitis and balanitis associated with verotoxin producing *Escherichia coli* O157:H7. *Lancet*, **II**, 150.

Greenhough, W.B., Glass, R.I., Holmgren, J. *et al.* (1988), Receptor blockade, cholera and *Escherichia coli* toxin. *J. Inf. Dis.*, **158**, 1147–1148.

Griffiths, F.L., Finkelstein, R.A. and Critchley, D.R. (1986), Characterization of the receptor for cholera toxin and *Escherichia coli* heat-labile toxin in rabbit intestinal brush borders. *Biochem. J.*, **238**, 313–322.

Gubash, S.M., Anand, C.M. and Stokman, M. (1988), Inhibition of *Escherichia coli* serotype O157:H7 by bromthymol blue. *J. Clin. Mic.*, **26**, 2248–2249.

Gunzburg, S.T., Burke, V. and Bettelheim, K.A. (1990), HEp-2 cell adherence and vero cell cytotoxin production by strains isolated from children with diarrhoea in New Zealand. *FEMS Micro. Lett.*, **69**, 181–186.

Haldane, D.J.M., Damm, M.A.S. and Anderson, J.S. (1986), Improved biochemical screening procedure for small clinical laboratories for Vero (Shiga-like) toxin producing strains of *Escherichia coli* O157:H7. *J. Clin. Mic.*, **24**, 24–26.

Hartman, P.A., Petzel, J.P. and Kaspar, C.W. (1986), New methods for indicator organisms. In *Foodborne Microorganisms and their Toxins: Developing Methodology*, Pierson, M.D. and Stern, N.J. (Eds). Marcel Dekker: New York and Basel.

Head, S.C., Karmali, M.A., Roscoe, M.E. *et al.* (1988), Serological differences between verocytotoxin 2 and Shiga like toxin II. *FEMS Micro. Lett.*, **51**, 211–216.

Hill, S.M., Phillips, A.D., Walker-Smith, J.A. *et al.* (1988), Antibiotics for *Escherichia coli* gastroenteritis. *Lancet*, **I**, 771–772.

Hiroyama, T., Ito, H. and Takeda, Y. (1989), Inhibition by the protein kinase inhibitors isoquinolinesulfonamides, of fluid secretion induced by *Escherichia coli* heat-stable enterotoxin, 8-bromo-cGMP and 8-bromo-cAMP in suckling mice. *Micro. Path.*, **7**, 255–261.

Hobbs, B.C., Rowe, B. and Kendall, M. (1976), *Escherichia coli* O27 in adult diarrhoea. *J. Hyg.*, **77**, 393–400.

Holbrook, R., Anderson, J.M. and Baird-Parker, A.C. (1980), Modified direct plate method for counting *Escherichia coli* in foods. *Food Tech. Aust.*, **32**, 78–83.

Honda, T., Taga, S., Takeda, Y. and Miwatani, T. (1981), Modified ELEK test for detection of heat-labile enterotoxigenic *Escherichia coli. J. Clin. Mic.*, **13**, 1–5.

ICMSF (1980), *Microbial Ecology of Foods Vol. 1. Factors Affecting Life and Death of Microorganisms*. Academic Press: London.

Jones, D. (1988), Composition and properties of the family *Enterobacteriaceae. J. appl. Bact.*, **65** (Suppl.), 1S–19S.

Kaper, J.B. (1987), Presentation to Society for General Microbiology: Durham.

Karch, H. and Meyer, T. (1989), Single primer pair for amplifying segments of distinct Shiga-like toxin genes by polymerase chain reactions. *J. Clin. Mic.*, **27**, 2751–2757.

Karch, H., Heesman, J., Laafs, R. *et al.* (1987), A plasmid of enterohaemorrhagic *Escherichia coli* O157:H7 is required for expression of a new fimbrial antigen and for adhesion to epithelial cells. *Inf. Imm.*, **55**, 455–461.

Knutton, S., Baldini, M.M., Kaper, J.B. and McNeish, A.S. (1987a), Role of plasmid-encoded virulence factors in adhesion of enteropathogenic *Escherichia coli* to HEp-2 cells. *Inf. Imm.*, **55**, 78–85.

Knutton, S., Lloyd. D.R. and McNeish, A.S. (1987b), Adhesion of enteropathogenic *Escherichia coli* to human intestinal enterocytes and cultured human intestinal mucosa. *Inf. Imm.*, **55**, 69–77.

Knutton, S., Baldwin, T., Williams, P.H. and McNeish, A.S. (1989), Actin accumulation at sites of bacterial adhesion to tissue culture cells: basis of a new diagnostic test for enteropathogenic and enterohaemorrhagic *Escherichia coli*. *Inf. Imm.*, **57**, 1290–1298.

Law, D. (1988), Virulence factors of enteropathogenic *Escherichia coli*. *J. Med. Mic.*, **26**, 1–10.

Levine, M.M. (1987), *Escherichia coli* that cause diarrhea: enterotoxigenic, enteropathogenic, enteroinvasive, enterohemorrhagic and enteroadherent. *J. Inf. Dis.*, **155**, 377–389.

Levine, M.M., Natara, J.P., Karch, H. *et al.* (1985), The diarrhoeal response of humans to some classic serotypes of *Escherichia coli* is dependent on enteroadhesiveness factor. *J. Inf. Dis.*, **152**, 550–559.

Levine, M.M., Prado, V., Robins-Browne, R. *et al.* (1988), Use of DNA probes and HEp-2 cell adherence assay to detect diarrheagenic *Escherichia coli*. *J. Inf. Dis.*, **158**, 224–229.

MacDonald, K.L., Eidson, M., Strohmeyer, C. *et al.* (1986), A multistate outbreak of gastrointestinal illness caused by enterotoxigenic *Escherichia coli* in imported semi-soft cheese. *J. Inf. Dis.*, **154**, 716–720.

March, S.B. and Ratnam, S. (1986), Sorbitol-MacConkey medium for detection of *Escherichia coli* associated with hemorrhagic colitis. *J. Clin. Mic.*, **23**, 869–872.

Marier, R., Wells, J.G., Swanson, R.C. *et al.* (1973), An outbreak of *Escherichia coli* foodborne disease traced to imported French cheese. *Lancet*, **II**, 1376–1378.

Martin, M.L., Shipman, L.D., Wells, J.G. *et al.* (1986), Isolation of *Escherichia coli* O157:H7 from dairy cattle associated with two cases of haemolytic uraemic syndrome. *Lancet*, **II**, 1043.

Matthewson, J.J. and Craviato, A. (1989), HEp-2 cell adherence as an assay for virulence amongst diarrheagenic *Escherichia coli*. *J. Inf. Dis.*, **159**, 1057–1060.

McGowan, K.L., Wickersham, E. and Strockbine, N.A. (1989), *Escherichia coli* O157:H7 from water. *Lancet*, **I**, 967–968.

Mehlman, I.J. and Romero, A. (1982), *Escherichia coli*: recovery from foods. *Fd. Tech.*, **36(3)**, 73–79.

Mehlman, I.J. and Lovett, J. (1984), *Escherichia coli*. In *FDA Bacteriological Analytical Manual*, 6th ed. Association of Official Analytical Chemists: Arlington.

Monsur, K.A., Clemens, J.D., Sack, D.A. *et al.* (1989a), The use of phage sensitivity patterns for tracing heat labile toxin-positive (LT+) *Eschericia coli*. *J. Med. Mic.*, **28**, 43–47.

Monsur, K.A., Begum, Y.A., Ahmed, Z.U. and Rahman, S. (1989b), Evidence of multiple infections in cases of diarrhea due to enterotoxigenic *Eschericia coli*. *J. Inf. Dis.*, **159**, 144–145.

Montenegro, M.A., Balte, M., Trumpf, T. and Aleksic, S. (1990), Detection and characterization of fecal verotoxin-producing *Escherichia coli* from healthy cattle. *J. Clin. Mic.*, **28**, 1417–1421.

Moseley, S.L., Huq, I., Alim, A.R.M.A. (1980), Detection of enterotoxigenic *Escherichia coli* by DNA hybridisation. *J. Inf. Dis.*, **142**, 897–898.

Morrison, D.M., Tyrell, D.L.J. and Jewell, L.D. (1986), Colonic biopsy in verotoxin-induced hemorrhagic colitis and thrombotic thrombocytopenic purpura (TTP). *Am. J. Clin. Path.*, **86**, 108, 112.

Mullaney, D.T. and Cantrey, J.R. (1984), Possible role of lipopolysaccharide in enteroadherent *Escherichia coli* diarrhoea. *Rev. Inf. Dis.*, **6**, 576–582.

Nataro, J.P., Baldini, M.M., Kaper, J.B. *et al.* (1985a), Detection of an adherence factor of enteropathogenic *Escherichia coli* in pig and rabbit intestines. *J. Inf. Dis.*, **152**, 560–565.

Nataro, J.P., Scaletsky, I.C.A., Kaper, J.B. (1985b), Plasmid-mediated factors conferring diffuse and localized adherence of enteropathogenic *Escherichia coli*. *Inf. Imm.*, **48**, 378–383.

Nicoletti, M., Superti, F., Conti, C. *et al.* (1988), Virulence factors of lactose-negative *Escherichia coli* strains isolated from children with diarrhoea in Somalia. *J. Clin. Mic.*, **26**, 524–529.

O'Brien, A.D. and LaVeck, G.D.C. (1983), Purification and characterization of a *Shigella dysenteriae* type I-like cytotoxin produced by *Escherichia coli*. *Inf. Imm.*, **40**, 675–683.

O'Brien, A.D. and Holmes, R.K. (1987), Shiga and Shiga-like toxins. *Micro. Revs.*, **51**, 206–220.

O'Brien, A.D. LaVeck, G.D., Thompson, M.R. and Formal, S.B. (1982), Production of *Shigella dysenteriae* type I-like cytotoxin by *Escherichia coli*. *J. Inf. Dis.*, **146**: 763–769.

Okrend, A.J.G., Rose, B.E. and Bennett, B. (1990), A screening method for the isolation of *Escherichia coli* O157:H7 from ground beef. *J. Fd. Prot.*, **53**, 249–252.

Oku, Y., Yutsudo, T., Hirayama, Y. *et al.* (1989), Purification and some properties of a vero toxin from a human strain of *Escherichia coli* that is immunologically related to Shiga-like toxin II (VT2). *Micro. Path.*, **6**, 113–122.

Orskov, F., Orskov, I. and Villon, J.A. (1987), Cattle as a reservoir of verotoxin-producing *Escherichia coli* O157: H7. *Lancet II*, 276.

Ostroff, S.M., Tarr, P.I., Neill, M.A. *et al.* (1989), Toxin genotypes and plasmid profiles as determinants of systemic sequelae in *Escherichia coli*, O157:H7 infections. *J. Inf. Dis.*, **160**, 994–998.

Padyhe, V.V., Kittel, F.B. and Doyle, M.P. (1986), Purification and physico-chemical properties of a unique vero cell cytotoxin from *Escherichia coli* O157:H7. *Biochem. Biophys. Res. Comm.*, **139**, 424–430.

Pai, C.H., Ahmed, N., Lior, H. *et al.* (1988), Epidemiology of sporadic diarrhea due to verocytotoxin-producing *Escherichia coli*: A two year prospective study. *J. Inf. Dis.*, **157**, 1054–1057.

Palumbo, S.A. (1986), Is refrigeration enough to restrain foodborne pathogens? *J. Fd. Prot.*, **49**, 1003–1009.

Park, C.E., El Derea, H.E. and Rayman, M.K. (1978), Evaluation of the staphylococcal thermonuclease (TNase) assay as a means of screening foods for growth of staphylococci and possible enterotoxin. *Can. J. Mic.*, **24**, 1135–1139.

Pavia, A.T., Nichols, C.R., Green, D. *et al.* (1988), Haemolytic uraemic syndrome (HUS) following an outbreak of *Escherichia coli* O157:H7 infection: leukocytosis as a predictor and sulfonamide exposure as a risk factor. *Abstracts 28th Interscience Symposium Antimicrobial Agents and Chemotherapy*, **369A**. American Society for Microbiology: Los Angeles.

Perry, M.B., Bundle, D.R., Gidney, M.A.J. and Lior, H. (1988), Identification of *Escherichia coli* serotype O157:H7 strains by using a monoclonal antibody. *J. Clin. Mic.*, **26**, 2391–2394.

Pickering, L.K., DuPont, H.L., Evans, D.G. *et al.* (1977), Isolation of enteric pathogens from asymptomatic students in the United States. *J. Inf. Dis.*, **135**, 1003–1005.

Pickering, L.K. (1979), Gastroenteritis due to enteropathogenic, enterotoxigenic and invasive *Escherichia coli*: A review. *Am. J. Med. Tech.*, **45**, 787–792.

Pieroni, P., Wanabec, E.A., Paranchych, W. and Armstrong, G.D. (1988), Identification of a human erythrocyte receptor for colonization factor antigen 1 pili expressed by HOO407 enterotoxigenic *Escherichia coli*. *Inf. Imm.*, **56**, 1334–1340.

Rasheed, J.K., Guzman-Verduzco, L.-M. and Kubersztoch, Y.M. (1990), Two precursors of the heat-stable enterotoxin of *Escherichia coli*: evidence of extracellular processing. *Mol. Mic.*, **4**, 265–273.

Ratnam, S., March, S.B., Ahmed, R. *et al.* (1988), Characterization of *Escherichia coli* serotype O157:H7. *J. Clin. Mic.*, **26**, 2006–2012.

Reis, C. (1980), Prevalence of enterotoxigenic *Escherichia coli* in some processed raw food from animal origin. *Appl. Environ. Micro.*, **39**, 270–271.

Riley, L.W. (1987), The epidemiological, clinical and microbiologic features of haemorrhagic colitis. *Ann. Rev. Mic.*, **41**, 383–408.

Riley, L.W. and Caffrey, C.J. (1990), Identification of enterotoxigenic *Escherichia coli* by colony hybridization with nonradioactive disoxigenin-labelled DNA probes. *J. Clin. Mic.*, **28**, 1465–1468.

Riley, L.W., Remis, R.S., Helgerson, S.D. *et al.* (1983), Haemorrhagic colitis associated with a rare *Escherichia coli* serotype. *New Eng. J. Med.*, **308**, 681–685.

Robins-Browne, R.M. (1987), Traditional enteropathogenic *Escherichia coli* of infantile diarrhoea. *Revs. Inf. Dis.*, **9**, 28–53.

Romick, T.L., Lindsay, J.A. and Busta, F.F. (1987), A visual DNA probe for detection of enterotoxigenic *Escherichia coli* by colony hybridization. *Lett. Appl. Micro.*, **5**, 87–90.

Romick, T.L., Lindsay, J.A. and Busta, F.F. (1989), Evaluation of a visual DNA probe for enterotoxigenic *Escherichia coli* detection in foods and wastewater by colony hybridisation. *J. Fd. Prot.*, **52**, 466–470.

Ryan, C.A., Tauxe, R.V., Hosek, G.W. *et al.* (1986), *Escherichia coli* O157:H7 in a nursing home: clinical, epidemiologic and pathological findings. *J. Inf. Dis.*, **154**, 631–638.

Sack, D.A., Merson, M.H., Wells, J.G. *et al.* (1975), Diarrhoea associated with heat-stable enterotoxin producing strains of *Escherichia coli*. *J. Inf. Dis.*, **131**, 631–638.

Sack, D.A., Huda, S., Neogi, P.K.B. *et al.* (1980), Microtiter ganglioside enzyme-linked immunosorbent assay for *Vibrio cholerae* and *Escherichia coli* heat-labile enterotoxins and antiserum. *J. Clin. Mic.*, **11**, 35–40.

Sakazaki, R., Tamura, K. and Saito, M. (1967), Enteropathogenic *Escherichia coli* associated with diarrhoea in children and adults. *Jap. J. Med. Sci. Biol.*, **20**, 387.

Salmon, R.L., Farrel, I.D., Hutchinson, J.G.P. *et al.* (1989), A christening party outbreak of haemorrhagic colitis and haemolytic uraemic syndrome associated with *Escherichia coli* O157:H7. *Epid. Inf.*, **103**, 249–254.

Samadpour, M., Liston, J., Ongerth, J.E. and Tarr, P.I. (1990), Evaluation of DNA probes for detection of Shiga-like-toxin-producing *Escherichia coli* in food and calf fecal samples. *Appl. Environ. Micro.*, **56**, 1212–1215.

Scaletsky, I.C.A., Silva, M.L.M. and Trabusi, L.R. (1984), Distinctive patterns of adherence of enteropathogenic *Escherichia coli* to HeLa cells. *Inf. Imm.*, **45**, 534–536.

Schlager, T.A., Wanke, C.A. and Guerrant, R.C. (1990), Net fluid secretion and impaired villous function induced by colonization of the small intestine by non-toxigenic colonizing *Escherichia coli*. *Inf. Imm.*, **58**, 1337–1343.

Scotland, S.M. (1988), Toxins. *J. Appl. Bact.*, **65** (Suppl.), 109–129.

Scotland, S.M. Smith, H.R., Willshaw, G.A. and Rowe, B. (1983), Vero cytotoxin production in strain of *Escherichia coli* is determined by genes carried on bacteriophage. *Lancet*, **II**, 216.

Scotland, S.M., Smith, H.R. and Rowe, B. (1985), Two distinct toxins active on vero cells from *Escherichia coli* O157. *Lancet*, **II**, 885–886.

Senerwa, D., Olsvik, O., Mutanda, L.N. *et al.* (1989), Colonization of neonates in a nursery ward with enteropathogenic *Escherichia coli* and correlation with the clinical history of the children. *J. Clin. Mic.*, **27**, 2539–2543.

Seriwatana, J., Echeverria, P., Taylor, D.N. *et al.* (1988), Type II heat-labile enterotoxin-producing *Escherichia coli* isolated from animals and humans. *Inf. Imm.*, **56**, 1158–1161.

Sherman, P.M. and Soni, R. (1988), Adherence of vero cytotoxin producing *Escherichia coli* of serotype O157:H7 to human epithelial cells in tissue culture: role of outer membranes as bacterial adhesins. *J. Med. Mic.*, **26**, 11–17.

Small, P.L.C. and Falkow, S. (1988), Identification of regions on a 230–K base plasmid from enteroinvasive *Escherichia coli* that are required for entry into HEp-2 cells. *Inf. Imm.*, **56**, 225–229.

Smith, H.R. Rowe, B., Gross, R.J. *et al.* (1987), Haemorrhagic colitis and vero-cytotoxin producing *Escherichia coli* in England and Wales. *Lancet*, **I**, 1062–1065.

Smith, H.R., Scotland, S.M., Willshaw, G.A. *et al.* (1988), Verocytotoxin and presence of VT genes in *Escherichia coli* strains of animal origin. *J. Gen. Mic.*, **134**, 829–834.

Stephen, J. and Pietrowski, R.A. (1981), *Bacterial Toxins*. American Society for Microbiology: Washington DC.

Strockbine, N.A., Marques, L.R.M., Newland, J.W. *et al.* (1986), Two toxin-converting phages from *Escherichia coli* O157:H7 strain 933 encode antigenically distinct toxins with similar biological activities. *Inf. Imm.*, **53**, 135–140.

Suez-Llorens, X., Guzman-Verduzco, L.M., Shelton, S. *et al.* (1989), Simultaneous detection of *Escherichia coli* heat-stable and heat-labile enterotoxin genes with a single RNA probe. *J. Clin. Mic.*, **27**, 1684–1688.

Svennerholm, A.M. and Holmgren, J. (1978), Identification of *Escherichia coli* heat-labile enterotoxin by means of a ganglioside immunosorbent assay (G_{M1}-ELISA) procedure. *Curr. Mic.*, **1**, 19–23.

Svennerholm, A.M. and Wiklund, G. (1983), Rapid G_{M1}-enzyme linked immunosorbent assay with visual reading for identification of *Escherichia coli* heat labile enterotoxin. *J. Clin. Mic.*, **17**, 596–600.

Svennerholm, A-M., Wikstrom, M., Lindblad, M. and Holmgren, J. (1986), Monoclonal antibodies to *Escherichia coli* heat-labile enterotoxins; neutralizing activity and differentiation of human and porcine LTs and cholera toxin. *Med. Biol.*, **64**, 23–30

Szabo, R.A., Todd, E.C.D. and Jean, A. (1986), Method to isolate *Escherichia coli* O157:H7 from food. *J. Fd. Prot.*, **49**, 768–782.

Tamura, K., Sakazaki, R., Kosako, Y. and Yoshizaki, E. (1986), *Leclercia adecarboxylata* gen. nov., comb. nov., formerly known as *Escherichia adecarboxylata*. *Curr. Mic.*, **13**, 179–184.

Tardelli Gomes, T.A., Midolli Vieira, M.A., Wachsmuth, I.K. *et al.* (1989), Serotype-specific prevalence of *Escherichia coli* strains with EPEC adherence factor genes in infants with and without diarrhea in Sao Paulo, Brazil. *J. Inf. Dis.*, **160**, 131–140.

Taylor, D.N., Echeverria, P., Sethabutr, O. *et al.* (1988), Clinical and microbiological features of *Shigella* and enteroinvasive *Escherichia coli* infections detected by DNA hybridisation. *J. Clin. Mic.*, **26**, 1362–1366.

Thoren, A., Wotde-Marion, T., Stinzing, G. *et al.* (1980), Antibiotics in the treatment of gastroenteritis caused by enteropathogenic *Escherichia coli*. *J. Inf. Dis.*, **141**, 27–31.

Todd, E.C.D., Szabo, R.A., Peterkin, P. *et al.* (1988), Rapid hydrophobic grid membrane filter-enzyme labelled antibody procedure for identification and enumeration of *Escherichia coli* O157 in foods. *Appl. Environ. Micro.*, **54**, 2536–2540.

Toth, I., Cohen, M.L., Rumschlag, H.S. *et al.* (1990), Influence of the 60–Megadalton plasmid on adherence of *Escherichia coli* O157:H7 and genetic derivatives. *Inf. Imm.*, **58**, 1223–1231.

Vadivelu, J., Dunn, J.T., Feachem, R.G. *et al.* (1987), Comparison of five assays for the heat-labile enterotoxin of *Escherichia coli*. *J. Med. Mic.*, **23**, 221–226.

Vial, P., Robins-Browne, R., Lior, H. *et al.* (1988), Characterization of enteroadherent-aggregative *Escherichia coli* a putative agent of diarrheal disease. *J. Inf. Dis.*, **158**, 70–78.

Vsai, L., Spezide, P. and Bozzini, S. (1990), Binding of collagens to an enterotoxigenic strain of *Escherichia coli*. *Inf. Imm.*, **58**, 449–455.

Wachsmuth, K. (1986), Molecular epidemiology of bacterial infection: Examples of methodology and an investigation of outbreaks. *Rev. Inf.*, **8**, 682–691.

Walmer-Touws, D. (1990), VTEC- is it a food-borne zoonosis? *J. Fd. Prot.*, **53**, 258–261.

Washington, J.A. and Timm, J.A. (1976), Unclassified citrate- positive member of the family *Enterobacteriaceae* resembling *Escherichia coli*. *J. Clin. Mic.*, **4**, 165–167.

Watkins, W.D., Rippey, S.R., Clavet, C.R. *et al.* (1988), Novel compound for identifying *Escherichia coli*. *Appl. Environ. Micro.*, **54**, 1874–1875.

Wells, J.G., Davis, B.R., Wachsmuth, K. *et al.* (1983), Laboratory investigations of hemolytic colitis outbreaks associated with a rare *Escherichia coli* serotype. *J. Clin. Mic.*, **18**, 512–520.

Yam, C.W., Lung, M.L. and Ng, M.H. (1988), Clonal origin, restricted natural distribution and conservation of virulence factors in isolates of enterotoxigenic *Escherichia coli* serogroup O126. *J. Clin. Mic.*, **26**, 1477–1481.

Yutsudo, T., Nakabayashi, N., Hirayama, T. and Takeda, Y. (1987), Purification and some properties of a vero toxin from *Escherichia coli* O157:H7 that is immunologically unrelated to Shiga toxin. *Micro. Path.*, **3**, 21–30.

Chapter 7

Aldova, E., Cerna, J., Janeckiova, M. and Pegrinkowi, J. (1975), *Yersinia enterocolitica* and its demonstration in food. *Czech. Hyg.*, **20**, 395–404.

Aleksic, S. and Bockemuhl, J. (1984), Proposed revision of the Wauters *et al.* antigenic scheme for serotyping of *Yersinia enterocolitica*. *J. Clin. Mic.*, **20**, 99–102.

Aleksic, S., Steigerwalt, A.G. and Bockemuhl, J. (1987), *Yersinia rohdei* sp. nov. isolated from human and dog feces and surface water. *Int. J. Syst. Bact.*, **37**, 327–332.

Anon (1977), *Yersinia enterocolitica* outbreak — New York. *Morb. Mort. Wkly. Rep.*, **26**, 53–54.

Anon (1987), Minutes: Judicial Commission of the International Committee on Systematic Bacteriology. *Int. J. Syst. Bact.*, **37**, 85–87.

Asakawa, Y., Akahane, S., Shizawa, K. and Honma, T. (1979), Investigations of source and route of *Yersinia enterocolitica* infection. In *Yersinia enterocolitica. Biology, Epidemiology and Pathology*, Carter, P.B., Lafleur, L. and Toma, S. (Eds). Karger: Basel.

Attwood, S.E.A., Cafferkey, M.T., West, A.B. *et al.* (1987), Yersinia infection and acute abdominal pain. *Lancet*, **I**, 529–530.

Aulisio, C.C.G. Mehlman, I.J. and Sanders, A.C. (1980), Alkali method for rapid recovery of *Yersinia enterocolitica* and *Yersinia pseudotuberculosis* from foods. *Appl. Environ. Micro.*, **39**, 135–140.

Aulisio, C.C.G., Hill, W.E., Stanfield, J.T. and Sellers, R.L. (1983), Evaluation of virulence factor testing and characteristics of pathogenicity in *Yersinia enterocolitica*. *Inf. Imm.*, **40**, 330–335.

Baker, T. (1983), *Plague and other Yersinia infections*. Plenum Press: New York.

Balligund, G., Laroche, Y. and Cornelis, G. (1985), Genetic analysis of a virulence plasmid from a serogroup 9 *Yersinia enterocolitica* strain: role of outer membrane protein P1 in resistance to human serum and auto-agglutination. *Inf. Imm.*, **48**, 282–286.

Barrett, N.J. (1986), Communicable disease associated with milk and dairy products in England and Wales. 1983–1984. *J. Inf.*, **12**, 265–272.

Bercovier, H. and Mollaret, H.H. (1984), *Yersinia*. In *Bergey's Manual of Systematic Bacteriology* Vol. 1, Krieg, N.R. and Holt, J.G. (Eds). Williams and Wilkins: Baltimore and London.

Bercovier, H., Alonso, J.M., Bentaiba, Z. *et al.* (1979), Contribution to the definition and taxonomy of *Yersinia enterocolitica*. *Cont. Mic., Imm.*, **5**, 12–22.

Bercovier, H., Brenner, D.J., Ursing, J. *et al.* (1980a), Characterization of *Yersinia enterocolitica sensu stricto*. *Curr. Mic.*, **4**, 201–206

Bercovier, H., Mollaret, H.H., Alonso, J.M. *et al.* (1980b), Intra- and inter-species relatedness of *Yersinia pestis* by deoxyribonucleic acid hybridization and its relationship to *Yersinia pseudotuberculosis*. *Curr. Mic.*, **4**, 225–230.

Bercovier, H., Steigerwalt, A.G., Guigoule, A. *et al.* (1984), *Yersinia aldovae* (formerly *Yersinia enterocolitica*-like organism — Group X2): A new species of Enterobacteriaceae isolated from aquatic ecosystems. *Int. J. Syst. Bact.*, **34**, 166–172.

Bisset, M.L., Powers, C., Abbott, S.L. and Janda, J.M. (1990), Epidemiologic investigations of *Yersinia enterocolitica* and related species: sources, frequency and serogroup distribution. *J. Clin. Mic.*, **28**, 910–912.

Black, R.G., Jackson, R.J., Tsai, M. (1978), Epidemic *Yersinia enterocolitica* infection due to contaminated chocolate milk. *New Eng. J. Med.*, **298**, 76–79.

Bolin, I., Portnoy, D.A. and Wolf-Watz, H. *et al.* (1985), Expression of the temperature-inducible outer membrane proteins of Yersinae. *Inf. Imm.*, **53**, 343–348.

Bolin, I., Forsberg, A., Norlander, L. *et al.* (1988), Identification and mapping of the temperature-inducible, plasmid encoded proteins of *Yersinia* spp. *Inf. Imm.*, **56**, 343–348.

Bottone, E.J. (1977), *Yersinia enterocolitica*: a panoramic view of a charismatic microorganism. *CRC Crit. Rev. Mic.*, **5**, 211–241.

Brackett, R.E. (1986), Growth and survival of *Yersinia enterocolitica* at acidic pH. *Int. J. Fd. Mic.*, **3**, 243–251.

Brackett, R.E. (1987), Effects of various acids on growth and survival of *Yersinia enterocolitica*. *J. Fd. Prot.*, **50**, 598–601.

Brenner, D.J., Bercovier, H., Ursing, J. *et al.* (1980a), *Yersinia intermedia*: a new species of Enterobacteriaceae composed of rhamnose-positive strains (formerly called *Yersinia enterocolitica* or *Yersinia enterocolitica*-like). *Curr. Mic.*, **4**, 207–212.

Brenner, D.J., Ursing, J., Bercovier, H. *et al.* (1980b), Deoxyribonucleic acid relatedness in *Yersinia enterocolitica* and *Yersinia enterocolitica*-like organisms. *Curr. Mic.*, **4**, 195–200.

Brubaker, R.R. (1979), Expression of virulence in *Yersinia*. In *Microbiology — 1979*, Schesinger, D. (Ed). American Society for Microbiology: Washington DC.

Brubaker, R.R. (1983), The Vwa$^+$ virulence factor of yersiniae: the molecular basis of the attendant nutritional requirement for calcium. *Rev. Inf. Dis.*, **5** (Suppl.), S748–S758

Buckeridge, S.A., Seaman, A. and Woodbine, M. (1980), Effects of temperature and carbohydrate on the growth and survival of *Yersinia enterocolitica*. In *Microbial Growth and Survival in Extremes of Environment*, Gould, G.W. and Corry, J.E.L. (Eds). Academic Press: London.

Carberry, W.L., Deffenbach, J.R., Schlender, G. *et al.* (1984), The Leavensworth, Washington case: environmental concerns with a water treatment plant failure. *J. Env. Hlth.*, **47**, 10–14.

CDC (1982), Multi-state outbreak of yersiniosis. *Morb. Mort. Wkly. Rep.*, **31**, 265–272.

Choo, W.-L., Ding, R.-J. and Chen. R.-S. (1988), Survival of *Yersinia enterocolitica* in the environment. *Can. J. Mic.*, **34**, 755–756.

Corniel, E., Mercerau-Puijalon, O. and Bonnefoy, S. (1989), The gene coding for the 190,000—Dalton iron-regulated protein of *Yersinia* species is present only in the highly pathogenic strains. *Inf. Imm.*, **57**, 1211–1217.

Cornelis, G., Bennett, P.M. and Grinsted, J. (1976), Properties of pGC1 a lac plasmid originating in *Yersinia enterocolitica* 842. *J. Bact.*, **127**, 1058–1062.

Cornelis, G., Laroche, Y., Balligund, G. *et al.* (1987), *Yersinia enterocolitica* as a primary model for bacterial invasiveness. *Rev. Inf. Dis.*, **9**, 64–87.

Daniels, J.J.H.M. (1965), Enteral infection with *Pasteurella pseudotuberculosis*. *B. Med. J.*, **2**, 997.

Davies, T.F. and Platzer, M. (1986), The T cell suppressor defect in autoimmune thyroiditis: evidence for a high set 'immunostat'. *Clin. Exp. Immunol.*, **63**, 73–79.

deBoer, E. and Seldom, W.M. (1987), Comparison of methods for the isolation of *Yersinia enterocolitica* from porcine tonsils and pork. *Int. J. Fd. Mic.*, **5**, 95–101.

Delmas, C.L. and Vidon, D.J.M. (1985), Isolation of *Yersinia enterocolitica* and related species from food in France. *Appl. Environ. Micro.*, **50**, 767–771.

Doyle, M.P. (1986), Detection and quantitation of food-borne pathogens and their toxins: Gram-negative bacterial pathogens. In *Foodborne Microorganisms and their Toxins: Developing Methodology*, Pierson, M.D. and Stern, N.J. (Eds). Marcel Dekker: New York and Basel.

Doyle, M.P. and Hughdahl, M.B. (1983), Improved procedure for the recovery of *Yersinia enterocolitica* from meats. *Appl. Environ. Micro.*, **45**, 127–135.

Eden, K.V., Rosenberg, M.L., Stoopler, M. *et al.* (1977), Waterborne gastroenteritis at a ski resort associated with isolation of *Yersinia enterocolitica*. *Pub. Hlth. Rep.*, **2**, 245–250.

El-Zawahry, Y. and Rowley, D.B. (1979), Radiation resistance and injury of *Yersinia enterocolitica*. *Appl. Environ. Micro.*, **37**, 50–54.

Feeley, J.C. and Schiemann, D.A. (1984), *Yersinia*. In *Compendium of Methods for the Microbiological Examination of Foods* 2nd ed., Speck, M.L. (Ed). American Public Health Association: Washington DC.

Feeley, J.C., Lee, W.H. and Morris, G.K. (1976), *Yersinia enterocolitica*. In *Compendium of Methods for the Microbiological Examination of Foods*, Speck, M.L. (Ed). American Public Health Association: Washington DC.

Francis, D.W., Spaulding, P.L. and Lovett, J. (1980), Enterotoxin and thermal resistance of *Yersinia enterocolitica* in milk. *Appl. Environ. Micro.*, **40**, 174–176.

Fukushima, H. (1985), Direct isolation of *Yersinia enterocolitica* and *Yersinia pseudotuberculosis* from meat. *Appl. Environ. Micro.*, **50**, 710–712.

Fukushima, H. (1987), New selective agar medium for isolation of virulent *Yersinia enterocolitica*. *J. Clin. Mic.*, **24**, 116–120.

Fukushima, H., Tsubokura, M., Otsuki, K. *et al.* (1985), Epidemiological study of *Yersinia enterocolitica* and *Yersinia pseudotuberculosis* infections in Shimone Prefecture, Japan. *Zentrallblat Bakt. Parasit. Infektion. Hyg. I. Abt. Orig. B*, **180**, 515–527.

Gemski, P., Lazere, J.R. and Casey, T. (1980), Plasmid associated with pathogenicity and calcium dependency of *Yersinia enterocolitica*. *Inf. Imm.*, **27**, 662–665.

Goguen, J.D., Walker, W.S., Hatch, T.P. and Yother, J. (1986), Plasmid-determined cytotoxicity in *Yersinia pestis* and *Yersinia pseudotuberculosis*. *Inf. Imm.*, **54**, 788–794.

Greenwood, M. and Hooper, W.L. (1987), Human carriage of *Yersinia* species. *J. Med. Mic.*, **23**, 345–348.

Greenwood, M. and Hooper, W.L. (1988), The use of alkalinity and incubation at 9°C for improved recovery of *Yersinia* spp. from faeces. *Epid. Inf.*, **101**, 53–58.

Gutman, L.T., Ottensen, E.A., Quan, T.J. *et al.* (1975), An interfamilial outbreak of *Yersinia enterocolitica* enteritis. *New Eng. J. Med.*, **288**, 1372–1377.

Hanna, M.O., Stewart, J.C., Zinc, D.L. *et al.* (1977a), Development of *Yersinia enterocolitica* in raw and cooked beef and pork at different temperatures. *J. Fd. Sci.*, **42**, 1180–1184.

Hanski, C., Kuschka, U., Schmoranza, H.P. *et al.* (1989), Immunohistochemical and electron microscopic study of interaction of *Yersinia enterocolitica* serotype O8 with intestinal mucosa during experimental enteritis. *Inf. Imm.*, **57**, 673–678.

Heim, F., Fehlhaber, K. and Scheibner, G. (1984), Studies into behaviour of *Yersinia enterocolitica* in different temperatures and curing salt concentrations. *Archiv. Exp. Vet.*, **38**, 729–734.

Heyma, P., Harrison, L.C. and Robins-Browne, R. (1986), Thyrotropin (TSH) binding sites on *Yersinia enterocolitica* recognised by immunoglobulins from humans with Grave's disease. *Clin. Exp. Immunol.*, **64**, 249–254.

Hughes, D. (1979), Isolation of *Yersinia enterocolitica* from milk and a dairy farm in Australia. *J. Appl. Bact.*, **46**, 125–130.

Inoue, M., Nakushima, H., Ueba, O. *et al.* (1984), Community outbreaks of *Yersinia pseudotuberculosis*. *Micro. Imm.*, **28**, 883–891.

Isberg, R. (1989a), Mammalian cell adhesion functions and cellular penetration of enteropathogenic *Yersinia* species. *Mol. Mic.*, **3**, 1449–1453.

Isberg, R. (1989b), Determinants for thermoinducible cell binding and plasmid encoded cellular penetration in the absence of the *Yersinia pseudotuberculosis* invasion protein. *Inf. Imm.*, **57**, 1998–2005.

Kanazawa, Y. and Ikemura, K. (1979), Isolation of *Yersinia enterocolitica* and *Yersinia pseudotuberculosis* from human specimens and their drug-resistance in the Niigata District of Japan. In *Yaersinia enterocolitica. Biology, Epidemiology and Pathology*, Carter, P.B., Lafleur, L. and Toma, S. (Eds). Karger: Basel.

Kandolo, K. and Wauters, G. (1985), Pyrazinamidase activity of *Yersinia enterocolitica* and related organisms. *J. Clin. Mic.*, **21**, 980–982.

Kaneko, S. and Maruyama, T. (1989), Evaluation of enzyme immunoassay for the detection of pathogenic *Yersinia enterocolitica* and *Yersinia pseudotuberculosis* strains. *J. Clin. Mic.*, **27**, 748–751.

Kapperud, G. (1982), Enterotoxin at 4°C, 22°C and 37°C among *Yersinia enterocolitica* and *Yersinia enterocolitica* like bacteria. *Acta Path. Microbiol. Scand.*, **B 90**, 185–189.

Kapperud, G. and Langeland, G. (1981), Enterotoxin production at refrigeration temperature by *Yersinia enterocolitica* and *Yersinia enterocolitica*-like bacteria. *Curr. Mic.*, **5**, 119–122.

Kapperud, G. and Bergan, T. (1984), Biochemical and serological characterization of *Yersinia enterocolitica*. In *Methods in Microbiology*, Vol 15, Bergan, T. (Ed). Academic Press: London.

Kapperud, G., Dommarsnes, K., Skurnik, M. and Hornes, E. (1990), A synthetic oligonucleotide probe and a cloned polynucleotide probe based on the yopA gene for detection and enumeration of virulent *Yersinia enterocolitica*. *Appl. Environ. Mic.*, **56**, 17–23.

Laird, W.J. and Cavanaugh, D.C. (1980), Correlation of autoagglutination and virulence of yersinae. *J. Clin. Mic.*, **11**, 430–432.

Larson, J.H. (1979), The spectrum of clinical manifestations of infections with *Y. enterocolitica* and their pathogenesis. In *Yersinia enterocolitica: Biology, Epidemiology and Pathology*, Carter, P.B., LaFleur, L. and Toma, S. (Eds). Karger: Basel.

Lassen, J. (1972), *Yersinia enterocolitica* in drinking water. *Scand. J. Dis.*, **4**, 125–127.

Leininger, H.V. (1974), Equipment, media, reagents, and stains. In *Compendium of Methods for the Microbiological Examination of Foods*, Speck, M.L.(Ed.), APHA: Washington DC.

Leistner, L., Hechelmann, H., Kashiwazaki, H. and Albertz, R. (1975), Nachweis von *Yersinia enterocolitica* in faces und fleisch von schweinen, rindern und geflagel. *Fleischwirt.*, **55**, 1599–1602.

Lovett, J., Bradshaw, J.G. and Peeler, J.T. (1982), Thermal inactivation of *Yersinia enterocolitica* in milk. *Appl. Environ. Micro.*, **44**, 5517–5519.

Marjai, E., Kalman, M., Kajary, I. *et al.* (1987), Isolation from food and characterization by virulence tests of *Yersinia enterocolitica* associated with an outbreak. *Acta. Mic. Hung.*, **34**, 97–109.

Martinez, R.J. (1983), Plasmid-mediated and temperature regulated surface properties of *Yersinia enterocolitica*. *Inf. Imm.*, **41**, 921–930.

Maurelli, A.T. (1989), Temperature regulation of virulence genes in pathogenic bacteria: a general strategy for human pathogens. *Micro. Path.*, **7**, 1–10.

Melby, K., Slandahl, S., Gutteberg, T.J. and Nordbrot, S.A. (1982), Septicaemia due to *Yersinia enterocolitica* after oral overdoses of iron. *B. Med. J.*, **285**, 467–468.

Miller, V.L. and Falkow, S. (1988), Evidence for two genetic loci in *Yersinia enterocolitica* that can promote invasion of epithelial cells. *Inf. Imm.*, **56**, 1242–1248.

Mingrone, M.G., Fantasia, M., Figura, N. and Guglielmetti, P. (1987), Characteristics of *Yersinia enterocolitica* isolated from children with diarrhoea in Italy. *J. Clin. Mic.*, **25**, 1301–1304.

Mehlman, I.J. and Aulisio, C.C.G. (1978), *Yersinia enterocolitica*, *Yersinia pseudotuberculosis* and related bacteria. In *Bacteriological Analytical Manual*. FDA: Washington DC.

Mehlman, I.J., Aulisio, C.C.G., Sanders, A.C. (1978), Problems in the recovery and identification of *Yersinia* from food. *J. Assoc. Off. Anal. Chem.*, **61**, 761–771.

Meliotis, M.D., Galen, J.E., Kaper, J.B. and Morris, J.G. (1989), Development and testing of a synthetic oligonucleotide probe for the detection of pathogenic *Yersinia* strains. *J. Clin. Mic.*, **27**, 1667–1670.

Mollaret, H.H., Bercovier, H. and Alonso, J.M. (1979), Summary of the data received at the World Health Organization Reference Centre for *Yersinia enterocolitica*. In *Yersinia enterocolitica: Biology, Epidemiology and Pathology*, Carter, P.B., LaFleur, L. and Toma, S. (Eds). Karger: Basel.

Moustafa, M.K., Ahmed, A.A.H. and Marth, E.H. (1983), Behaviour of virulent *Yersinia enterocolitica* during manufacture and storage of Colby-like cheese. *J. Fd. Prot.*, **46**, 318–320.

Neilsen, H.J.S. and Zeuthen, P. (1985), Influence of lactic acid bacteria and the overall flora of pathogenic bacteria in vacuum-packed cooked emulsion-style sausage. *J. Fd. Prot.*, **48**, 28–34.

Nesbakken, T. (1988), Enumeration of *Yersinia enterocolitica* O:3 and its occurrence on cut surfaces of pig carcasses and the environment of a slaughterhouse. *Int. J. Fd. Mic.*, **6**, 287–293.

Okamoto, K., Miyami, A., Takeda, T. *et al.* (1983), Cross-neutralization of heat-stable enterotoxin avtivity of *Escherichia coli* and *Yersinia enterocolitica*. *FEMS Micro. Lett.*, **16**, 85–87.

Pai, C.H., Mors, V. and Seemayer, T.A. (1980), Experimental *Yersinia enterocolitica* enteritis in rabbits. *Inf. Imm.*, **28**, 238–244.

Perry, R.D. and Brubaker, R.R. (1979), Accumulation of iron by yersiniae. *J. Bact.*, **137**, 1290–1298.

Perry, R.D. and Brubaker, R.R. (1983), Vwa$^+$ phenotype of *Yersinia enterocolitica*. *Inf. Imm.*, **40**, 166–171.

Pierson, D.E. and Falkow, S. (1990), Nonpathogenic *Yersinia enterocolitica* do not contain functional *inv*-homologous sequences. *Inf. Imm.*, **58**, 1059–1064.

Portnoy, D.A. and Falkow, S. (1981), Virulence associated plasmids from *Yersinia enterocolitica* and *Yersinia pseudotuberculosis*. *J. Bact.*, **148**, 877–883.

Portnoy, D.A., Moseley, S.L. and Falkow, S. (1981), Characteristics of plasmids and plasmid-associated determinants of *Yersinia enterocolitica* pathogenesis. *Inf. Imm.*, **31**, 775–782.

Prpic, J.K., Robins-Browne, R.M. and Davey, R.B. (1983), Differentiation between virulent and avirulent *Yersinia enterocolitica* isolates by using congo red agar. *J. Clin. Mic.*, **18**, 486–490. (*Erratum*, **19**, 446, 1984.)

Raccach, M. and Henningsen, E.C. (1984), Role of lactic acid bacteria, curing salts, spices and temperature in controlling the growth of *Yersinia enterocolitica*. *J. Fd. Prot.*, **47**, 354–358.

Restaino, L., Komatsu, K.K. and Syracuse, M.J. (1982), Effects of acids on potassium sorbate inhibition of food related microorganisms in culture media. *J. Fd. Sci.*, **47**, 134–138.

Robins-Browne, R.M., Rabson, A.R. and Koornhof, H.J. (1979), Generalized infection with *Yersinia enterocolitica* — the role of iron. In *Contributions to Microbiology and Immunology Vol 5. Yersinia, Pasteurella and Francisella*. Karger: Basel.

Robins-Browne, R.M. Milioti, M.D., Cianciosi, S. *et al.* (1989), Evaluation of DNA colony hybridisation and other techniques for detection of virulance in *Yersinia* species. *J. Clin. Mic.*, **27**, 644–650.

Rose, F.B., Camp, C.J. and Estes, E.J. (1987), Family outbreak of fatal *Yersinia enterocolitica* pharyngitis. *Am. J. Med.*, **82**, 636–637.

Rosqvist, R. Forsberg, A. and Rimpilainen, M. (1990), The cytotoxic protein Yop E of *Yersinia* obstructs the primary host defence. *Mol. Mic.*, **4**, 657–667.

Sanbe, K., Uchiyama, M., Koiwai, K. (1987), Community outbreak of *Yersinia pseudotuberculosis* occurred among primary schoolchildren. *J. Jpn. Assoc. Inf. Dis.*, **61**, 763–771.

Sato, K. (1987), *Yersinia pseudotuberculosis* infection. *Cont. Micro. Imm.*, **9**, 111–116.

Schiemann, D.A. (1983), Alkalotolerance of *Yersinia enterocolitica* as a basis for selective enrichment from food enrichments. *Appl. Environ. Micro.*, **46**, 22–27.

Schiemann, D.A. (1989), *Yersinia enterocolitica* and *Yersinia pseudotuberculosis*. In *Foodborne Bacterial Pathogens*, Doyle, M.P. (Ed). Marcel Dekker: New York.

Schiemann, D.A. and Olson, S.A. (1984), Antagonism by Gram-negative bacteria to growth of *Yersinia enterocolitica* in mixed cultures. *Appl. Environ. Micro.*, **48**, 539–544.

Scotland, S.M. (1988), Toxins *J. appl. Bact.*, **65** (Suppl.), 109–129.

Shiozawa, K., Akijama, M., Sabora, K. *et al.* (1987), Pathogenicity of *Yersinia enterocolitica* biotype 3b and 4, serotype O:3 isolates from pork samples and humans. *Cont. Mic. Imm.*, **2**, 190–195.

Shiozawa, K., Hayashi, M., Akiyama, M. *et al.* (1988), Virulence of *Yersinia pseudotuberculosis* isolated from pork and from the throats of swine. *Appl. Environ. Micro.*, **54**, 818–821.

Stern, N.J. (1982), *Yersinia enterocolitica*: recovery from foods and virulence characterisation. *Food Tech.*, **36**(3), 84–88.

Stern, N.J., Pierson, M.D. and Kotula, A.W. (1980), Effects of pH and sodium chloride on *Yersinia enterocolitica* growth at room and refrigeration temperature. *J. Fd. Sci.*, **45**, 64–67.

Svennerholm, A-M., Wikstrom, M., Winblad, M. and Holmgren, J. (1986), Monoclonal antibiodies against *Escherichia coli* heat-stable toxin (STa) and their use in a diagnostic ST ganglioside GM-1 enzyme linked immunosorbent assay. *J. Clin. Mic.*, **24**, 585–590

Swaminathan, B., Harmon, M.C. and Mehlman, I.J. (1982), A review. *Yersinia enterocolitica*. *J. appl. Bact.*, **52**, 151–183.

Tacket, C.O., Ballard, J., Harris, N. *et al.* (1985), An outbreak of *Yersinia enterocolitica* infection caused by contaminated tofu. *Am. J. Epid.*, **121**, 705–711.

Tauxe, R.V., Wauters, G. and Goosens, V. (1987), *Yersinia enterocolitica* infections and pork: the missing link. *Lancet*, **I**, 1129–1132.

Tsubokura, M., Otsuki, K., Sato, K. *et al.* (1989), Special features of distribution of *Yersinia pseudotubersulosis* in Japan. *J. Clin. Mic.*, **27**, 790–791.

Une, T. (1977), Studies on the pathogenicity of *Yersinia enterocolitica*. II Interaction with cultured cells *in vitro*. *Micro. Imm.*, **21**, 365–377.

Ursing, J., Brenner, D.J., Bercovier, H. *et al.* (1980), *Yersinia frederickensii*: a new species of Enterobacteriaceae composed of rhamnose-positive strains (formerly called atypical *Yersinia enterocolitica* or *Yersinia enterocolitica*-like). *Curr. Mic.*, **4**, 213–217.

Vandepitte, J. and Wauters, G. (1979), Epidemiological and clinical aspects of human *Yersinia enterocolitica* infections in elgium. In *Contributions to Microbiology and Immunology, Vol 5. Yersinia, Pasteurella and Francisella*. Karger: Basel.

Van Kooij, J.A. and deBoer, E. (1985), A survey of the microbiological quality of commercial tofu in the Netherlands. *Int. J. Fd. Mic.*, **2**, 349–354.

Velin, D. (1984), Enterotoxin production by *Yersinia entero — colitica* in food samples. *Acta Microbiol. Hung.*, **31**, 43–48.

Vesikari, T., Sundqvist, C. and Maki, M. (1983), Adherence and toxicity of *Yersinia enterocolitica* O:3 and O:9 virulence associated plasmids for various cultured cells. *Acta. Path. Mic. Immun. Scand.*, **Sect. B 91**, 121–127.

Vidon, D.J.M. and Delmas, C.L. (1981), Incidence of *Yersinia enterocolitica* in raw milk in Eastern France. *Appl. Environ. Micro.*, **41**, 355–359.

Walker, S.J. (1986), *Yersinia enterocolitica* and *Yersinia enterocolitica*-like bacteria in milk. Doctoral Thesis: Queen's University of Belfast.

Walker, S.J. (1987), *Yersinia enterocolitica: A review of a foodborne pathogen*, Technical Bulletin No 63. CFPRA: Chipping Campden.

Walker, S.J. and Gilmour, A. (1986a), A comparison of media and methods for the recovery of *Yersinia enterocolitica* and *Yersinia enterocolitica*-like bacteria from milk containing simulated raw milk microfloras. *J. appl. Bact.*, **60**, 175–183.

Walker, S.J. and Gilmour, A. (1986b), The incidence of *Yersinia enterocolitica* and *Yersinia enterocolitica*-like organisms in raw and pasteurised milk in Northern Ireland. *J. appl. Bact.*, **61**, 133–138.

Wauters, G. (1973), Improved methods for the isolation and recognition of *Yersinia enterocolitica*. In *Contributions to Microbiology and Immunology, Vol 2. Yersinia, Pasteurella and Francisella*. Karger: Basel.

Wauters, G. (1981), Antigens of *Yersinia enterocolitica*. In *Yersinia enterocolitica*, Bottone, E.J. (Ed). CRC Press: Boca Raton.

Wauters, G., Janssens, M., Steigerwalt, A.G. and Brenner, D.J. (1988), *Yersinia mollaretii* sp. nov. and *Yersinia bercovieri* sp. nov. Formerly called *Yersinia enterocolitica* biogroups 3A and 3B. *Int. J. Syst. Bact.*, **38**, 424–429.

Weissfield, A.S. and Sonnenwirth, A.C. (1980), *Yersinia enterocolitica* in adults with gastro-intestinal disturbances: need for cold enrichment. *J. Clin. Mic.*, **11**, 196–197.

Wenzel, B.E., Heeseman, J., Wenzel, K.W. and Scriba, D.C. (1988), Antibodies to plasmid-encoded proteins of enteropathogenic *Yersinia* in patients with autoimmune thyroid disease. *Lancet*, **I**, 56.

Williams, J.E. (1984), Proposal to reject the new combination *Yersinia pseudotuberculosis subspecies pestis* for violation of the International Code of Nomenclature of Bacteria: request for an opinion. *Int. J. Syst. Bact.*, **34**, 268–269.

Williams, P.H., Roberts, M. and Hinson, G. (1988), Stages in bacterial invasion. *J. appl. Bact.*, **65** (Suppl.), 131–147.

Wilson, H.D., McCormick, E. and Feeley, J.C. (1976), *Yersinia enterocolitica* infection in a 4-month-old infant associated with infections in household dogs. *J. Paediatr.*, **89**, 767–769.

Winblad, S. (1967), Studies on serological typing of *Yersinia enterocolitica*. *Acta. Path. Micro. Scand.*, **21** (Suppl.), 980–982.

Winblad, S. (1981), Erythema nodosum associated with infection with *Yersinia enterocolitica*. In *Yersinia enterocolitica*, Bottone, E.J. (Ed). CRC Press: Boca Raton.

Wormser, G.P. and Keusch, G.T. (1981), *Yersinia enterocolitica*: Clinical Observations. In *Yersinia enterocolitica*, Bottone, E.J. (Ed). CRC Press: Boca Raton.

Zinc, D.L., Feeley, J.C., Wells, J.G. *et al.* (1980), Plasmid-mediated tissue-invasiveness in *Yersinia enterocolitica*. *Nature*, **283**, 224–226.

Chapter 8

Abbott, S.L., Powers, C., Kaysner, C.A. *et al.* (1989), Emergence of a restricted bioserovar of *Vibrio parahaemolyticus* as the predominant cause of *Vibrio*-associated gastroenteritis on the West Coast of the United States and Mexico. *J. Clin. Mic.*, **27**, 2891–2893.

Allen, R.D. and Baumann, P. (1971), Structure and arrangement of flagella in species of the genus *Beneckea* and *Photobacterium fisheri*. *J. Bact.*, **107**, 295–302.

Amako, K., Shimodari, S., Imota, T. *et al.* (1987), Effects of chitin and its soluble derivatives on survival of *Vibrio cholerae* O1 at low temperature. *Appl. Environ. Micro.*, **53**, 603–605.

Andrews, C.R., Walter, M., Crosa, J.H. and Payne, S.M. (1983), Synthesis of siderophores by pathogenic *Vibrio* species. *Curr. Mic.*, **9**, 209–214.

Anon (1988), Ozone depuration reported effective against *Vibrio* in clams. *Fd. Chem. News.*, **May 30**, 11.

Ashan, C.R. and Ciznar, I. (1987), Release of endotoxin by toxigenic and non-toxigenic *Vibrio cholerae* O1. *J. Diarr. Dis.*, **5**, 7–15.

Bandekar, J.B., Chander, R. and Nerkar, D.P. (1987), Radiation control of *Vibrio parahaemolyticus* in shrimp. *J. Fd. Saf.*, **8**, 83–88.

Barker, W.H., Weaver, R.E., Morris, G.K. and Martin, W.T. (1975), Epidemiology of *Vibrio parahaemolyticus* infections in humans. In *Microbiology-1974*, Schessinger. D. (Ed). American Society for Microbiology: Washington DC.

Bartowski, C. J. and Manning, P.A. (1988), Molecular cloning of the plasmids of *Vibrio cholerae* O1 and the incidence of related plasmids in clinical isolates and other *Vibrio* species. *Inf. Imm.*, **56**, 1298–1304.

Basu, S. and Mukerjee, S. (1968), Bacteriophage typing of *Vibrio el tor*. *Experimentia*, **24**, 299–300.

Baumann, P. and Baumann, L. (1977), Biology of the marine enterobacteria *Beneckea* and *Photobacterium*. *Ann. Rev. Micro.*, **31**, 39–61.

Baumann, P. and Baumann, L. (1981), The marine Gram-negative eubacteria. In *The Prokaryotes, A Handbook on Habitats, Isolation and Identification of Bacteria*, Starr, M.P. *et al.* (Eds). Springer-Verlag: New York.

Baumann, P. and Schubert, R.H.W. (1984), *Vibrionaceae*. In *Bergey's Manual of Systematic Bacteriology*, Vol 1, Krieg, N.R. and Holt, J.G. (Eds). Williams and Wilkins: Baltimore and London.

Baumann, P., Baumann, L., Bang, S.S. and Woolkakis, M.J. (1980), Reevaluation of the taxonomy of *Vibrio*, *Beneckea* and *Photobacterium*: abolition of the genus *Beneckea*. *Curr. Mic.*, **4**, 127–132.

Baumann, P., Baumann, L. and Hall, B.G. (1981), Lactose utilization by *Vibrio vulnificus*. *Curr. Mic.*, **6**, 131–135.

Baumann, P., Furniss, A.L. and Lee, J.V. (1984), *Vibrio*. In *Bergey's Manual of Systematic Bacteriology*, Vol 1, Krieg, N.R. and Holt, J.G. (Eds). Williams and Wilkins: Baltimore and London.

Bennum, F.R., Roth, G.A., Monferren, C.B. and Cumar, F.A. (1989), Binding of cholera toxin to pig intestinal glycosphingolipids; relationship with the ABO blood group system. *Inf. Imm.*, **57**, 969–974.

Beuchat, L.R. (1977), Evaluation of enrichment broths for enumerating *Vibrio parahaemolyticus* in chilled and frozen crabmeat. *J. Fd. Prot.*, **40**, 592–595.

Blake, P.A. (1983), Vibrios on the half shell. What the walrus and the carpenter didn't know. *Ann. Int. Med.*, **99**, 558–559.

Blake, P.A., Rosenberg, M.L., Florencia, J. *et al.* (1977), Cholera in Portugal, 1974. II Transmission by bottled mineral water. *Am. J. Epid.*, **105**, 344–348.

Blake, P.A., Merson, M.H., Weaver, R.E. *et al.* (1979), Disease caused by a marine *Vibrio*: clinical characteristics and epidemiology. *New Eng. J. Med.*, **300**, 1–5.

Blake, P.A., Weaver, D.G. and Hollis, D.E. (1980a), Diseases of humans (other than cholera) caused by vibrios. *Ann. Rev. Mic.*, **34**, 341–367.

Blake, P.A., Allegra, D.T., Snyder, J.D. (1980b), Cholera — a possible epidemic focus in the United States. *New Eng. J. Med.*, **302**, 305–309.

Boutin, G., Bradshaw, J. and Stroup, W. (1982), Heat processing of oysters naturally contaminated with *Vibrio cholerae*. *J. Fd. Prot.*, **45**, 169–174.

Brandis, H. (1978), Vibriocin typing. In *Methods in Microbiology*, Vol 12, Bergan, T. and Norris, J.R. (Eds). Academic Press: London and New York.

Brayton, P.L., Tamplin, M.L., Huq, A. and Colwell, R.R. (1987), Enumeration of *Vibrio cholerae* O1 in Bangladesh by fluorescent-antibody direct viable count. *Appl. Environ. Micro.*, **53**, 2862–2865.

CDC (1986), Toxigenic *Vibrio cholerae* O1 infection — Louisiana and Florida. *Morb. Mort. Wkly. Rep.*, **35**, 606–607

Chart, H. and Griffiths, E. (1985), The availability of iron and the growth of *Vibrio cholerae* in the sera of patients with haemachromatosis. *FEMS Micro. Letts.*, **26**, 227–231.

Chatterjee, B.D., De, P.K. and Sen, T. (1977a), Sucrose teepol tellurite agar: a new selective indicator medium for isolation of *Vibrio* species. *J. Inf. Dis.*, **136**, 654–658.

Chatterjee, B.D., De, P.K. and Sen, T. (1977b), Erratum in Chatterjee *et al.* (1977a), *J. Inf. Dis.*, **136**, 716.

Chowdbury, M.A.R., Yamonako, H., Miyoshi, S.T. *et al.* (1989), Ecology of *Vibrio mimicus* in aquatic environments. *Appl. Environ. Micro.*, **55**, 2073–2078.

Colwell, R.R. (1984), *Vibrios in the Environment*. John Wiley: New York.

Colwell, R.R. (1986), *Vibrio cholerae* and related vibrios in the aquatic environment — an ecological paradigm. *J. appl. Bact.*, **61**, vii.

Datta-Roy, K., Dagupta, C. and Ghosh, A.C. (1989), Haemagglutination and intestinal adherence properties of clinical and environmental isolates of non-O1 *Vibrio cholerae*. *Appl. Environ. Micro.*, **55**, 2403–2406.

De, S.P., Sen, D, De, P.C. *et al.* (1977), A simple selective medium for isolation of vibrios with particular reference to *Vibrio parahaemolyticus*. *Ind. J. Med. Res.*, **66**, 398–399.

Delmore, R. and Crisley, P. (1979), Thermal resistance of *Vibrio parahaemolyticus* in clam homogenate. *J. Fd. Sci.*, **41**, 899–902.

DePaola, A., Motes, L. and McPhearson, R.M. (1988), Comparison of APHA and elevated temperature enrichment methods for recovery of *Vibrio cholerae* from oysters: collaborative study. *J.Assoc. Off. Anal. Chem.*, **71**, 584–589.

Desmarchelier, P.M. and Reichelt, J.L. (1982), Genetic relationships among clinical and environmental isolates of *Vibrio cholerae* from Australia. *Curr. Mic.*, **7**, 53–57.

Doyle, M.P. (1986), Detection and quantitation of foodborne pathogens and their toxins: Gram-negative bacterial pathogens. In *Foodborne Microorganisms and their Toxins: Developing Methodology*, Pierson, M.D. and Stern, N.J. (Eds). Marcel Dekker: New York and Basel.

Eaton, J.W., Brandt, P., Mahone, J.R. and Lee, J.T. (1982), Haptoglobin: a natural bacteriostat. *Science*, **215**, 691–693.

Ehara, M., Ishibashi, M., Watanabe, S. *et al.* (1986), Fimbriae of *Vibrio cholerae* O1: observation of fimbriae on the organism adherent to the intestinal epithelium and development of a new medium to enhance fimbrial production. *Trop. Med.*, **28**, 21–33.

Enomoto, K. and Gill, M.D. (1980), Cholera toxin activation of adenylate cyclase. *J. Biol. Chem.*, **255**, 1252–1258.

Entis, P. and Boleszczuk, P. (1983), Overnight enumeration of *Vibrio parahaemolyticus* in seafood by hydrophobic grid membrane filtration. *J. Fd. Prot.*, **46**, 783–786.

Eyles, M.J. and Davey, G.R. (1984), Microbiology of the commercial depuration of the Sydney Rock Oyster *Crasostrea commercialis*. *J. Fd. Prot.*, **47**, 703–706.

Eyles, M.J., Davey, G.R., Arnold, G. and Ware, H.M. (1985), Evaluation of methods for the enumeration and identification of *Vibrio parahaemolyticus* in oysters. *Fd. Tech. Aust.*, **37**, 302–304.

Field, M. (1979), Modes of action of enterotoxin from *Vibrio cholerae* and *Escherichia coli*. *Rev. Inf. Dis.*, **1**, 918–925.

Fleurat, A. (1986), Indigenous response to drought in sub-Saharan Africa. *Disasters*, **10**, 224–229.

Fujino, T., Sakaguchi, G., Sakazaki, R. and Takeda, Y. (1974), *International Symposium on Vibrio parahaemolyticus*. Saikon Publishing Company: Tokyo.

Galbraith, N.S., Barrett, N.J. and Sockett, D.N. (1987), The changing pattern of foodborne disease in England and Wales. *Pub. Hlth.*, **101**, 319–328.

Glatz, B.A. (1986), Genetic regulation of toxin production by foodborne microbes. In *Food Microbiology*, Vol 1, Montville, T.J. (Ed). CRC Press: Boca Raton.

Greenberg, E.P., Duboise, M., Palmofi, B. (1982), The survival of marine vibrios in *Mercinaria mercinaria*, the hardshell clam. *J. Fd. Safety*, **4**, 113–123.

Greenhough, W.B. (1985), *Principles and Practice of Infectious Diseases*. John Wiley: New York.

Greenhough, W.B., Glass, R.I., Holmgren, J. *et al.* (1988), Receptor blockage, cholera and *Escherichia coli* toxins. *J. Inf. Dis.*, **158**, 1147–1148.

Guerrant, R.L., Brunton, L.L., Schnaitman, T.C. *et al.* (1974), Cyclic adenosine monophosphate and alteration of Chinese hamster ovary cell morphology: a rapid sensitive *in vitro* assay for the enterotoxin of *Vibrio cholerae* and *Escherichia coli*. *Inf. Imm.*, **10**, 320–327.

Gunn, R.A., Kimball, A.M., Pollard, R.A. *et al.* (1979), Bottle feeding as a risk factor for cholera in infants. *Lancet*, **II**, 730–733.

Hackney, C.R. and Dicharry, A. (1988), Seafood-borne bacterial pathogens of marine origin. *Fd. Tech.*, **42**(3), 104–109.

Hackney, C.R., Roy, B. and Speck, M. (1980), Incidence of *Vibrio parahaemolyticus* in and the microbiological quality of seafoods in North Carolina. *J. Fd. Prot.*, **43**, 769–772.

Helms, S.D., Oliver, J.D. and Travis, J.C. (1984), Role of haem compounds and haptoglobin in *Vibrio vulnificus* pathogenicity. *Inf. Imm.*, **45**, 345–349.

Hoge, C.V., Watsky, D., Peeler, R.N. *et al.* (1989), Epidemiology and spectrum of *Vibrio* infection in a Chesapeke Bay community. *J. Inf. Dis.*, **160**, 985–993.

Hollis, D., Weaver, R., Baker, C. and Thornberry, C. (1976), Halophilic *Vibrio* species isolated from blood cultures. *J. Clin. Mic.*, **3**, 425–427.

Honda, T., Arita, M., Takeda, T. *et al.* (1985), Non-O1 *Vibrio cholerae* produces two newly identified toxins related to *Vibrio parahaemolyticus* haemolysin and *Escherichia coli* heat-stable enterotoxin. *Lancet*, **II**, 163–164.

Honda, S., Gato, I., Minematsu, I. *et al.* (1987), Gastro-enteritis due to Kanagawa negative *Vibrio parahaemolyticus*. *Lancet*, **I**, 331–332.

Honda, T., Ni, Y. and Miwatani, T. (1988a), Purification and characterization of a haemolysin produced by a classical isolate of Kanagawa phenomenon-negative *Vibrio parahaemolyticus* and related to the thermostable, direct haemolysin. *Inf. Imm.*, **56**, 961–965.

Honda, S.-I., Shimoiriso, K., Adachi, A. *et al.* (1988b), Clinical isolates of *Vibrio cholerae* O1 not producing cholera toxin. *Lancet*, **II**, 1486.

Honda, T., Kasemuksakal, K., Oguchi, T. *et al.* (1988c), Production and partial characterization of pili on non-O1 *Vibrio cholerae*. *J. Inf. Dis.*, **157**, 217–219.

Honda, T., Lertpocagombat, K. and Hato, A. (1989a), Purification and characterization of a protease produced by *Vibrio cholerae* non-O1 and comparison with a protease of *Vibrio cholerae* O1. *Inf. Imm.*, **57**, 2799–2803.

Honda, T., Ni, Y. and Miwatani, T. (1989b), Purification of a TDH-related hemolysin produced by a Kanagawa phenomenon-negative clinical isolate of *Vibrio parahaemolyticus* O6:K46. *FEMS Micro. Letts.*, **57**, 241–246.

Honda, T., Ni, Y., Yoh, M. and Miwatani, T. (1989c), Evidence of immunologic cross-reactivity between hemolysing of *Vibrio bullisue* and *Vibrio parahaemolyticas* demonstrated by monoclonal antibodies. *J. Inf. Dis.*, **160**, 1089–1090.

Hoover, D.G. (1985), Review of isolation and enumeration methods for *Vibrio* species of food safety significance. *J. Fd. Saf.*, **7**, 35–42.

Huq, A., Small, E.B., West, P.A. *et al.* (1983), Ecological relationships between *Vibrio cholerae* and planktonic crustacean copepods. *Appl. Environ. Mic.*, **45**, 275–283.

ICMSF (1980), *Microbial Ecology of Foods, Vol 1. Factors Affecting Life and Death of Microorganisms.* Academic Press: London and New York.

Johnson, J.M., Becker, St. and McFarland, L.M. (1985), *Vibrio vulniticus*: man and the sea. *J. Am. Med. Assoc*, **253**, 2580–2583.

Joseph, S.W., Colwell, R.R. and Kaper, J.B. (1982), *Vibrio parahaemolyticus* and related halophilic vibrios. *CRC Crit. Revs. Mic.*, **10**, 77–124.

Kaysner, C.A., Abeyta, C., Wekell, M.M. *et al.* (1987), Virulent strains of *Vibrio vulnificus* isolated from estuaries of the United States west coast. *Appl. Environ. Micro.*, **53**, 1349–1351.

Kaysner, C.A., Tamplin, M.L. and Wekell, M.M. (1989), Survival of *Vibrio vulnificus* in shellstock and shucked oysters (*Crassostrea gigans* and *Crassostrea virginica*) and effects of isolation ·medium on recovery. *Appl. Environ. Micro.*, **55**, 3073–3079.

Kaysner, C.A., Abeyta, C., Stott, R.F. *et al.* (1990), Incidence of urea-hydrolyzing *Vibrio parahaemolyticus* in Willopa Bay, Washington. *Appl. Environ. Micro.*, **56**, 904–907.

Kelly, M.T. (1982), Effect of temperature and salinity on *Vibrio* (*Beneckea*) *vulnificus* occurrence in Gulf Coast environment. *Appl. Environ. Micro.*, **44**, 820–827.

Kelly, M.T. and Stroh, M. (1989), Urease-positive, Kanagawa-negative *Vibrio parahaemolyticus* from patients and the environment in the Pacific Northwest. *J. Clin. Mic.*, **27**, 2820–2822.

Kilgen, M., Cole, M. and Gardner, R. (1987), Control of indicator and pathogenic bacteria in Louisiana shellstock oysters by ionizing radiation. Cited by Hackney and Dicharry, 1987.

Kobayashi, T., Enomoto, S., Sakazaki, R. and Kuwahara, S. (1963), A new selective medium for pathogenic vibrios: TCBS agar (modified Nakanishi's agar). *Jap. J. Bact.*, **18**, 387–392.

Kothary, M.H. and Kreger, A.S. (1985), Production and partial characterization of an elastolytic protease of *Vibrio vulnificus*. *Inf. Imm.*, **50**, 534–540.

Kushner, D.J. (1978), Life in high salt and solute concentrations: halophilic bacteria. In *Microbial Life in Extreme Environments*, Kushner, D.J. (Ed). Academic Press: London.

Lee, J.V. and Furniss, A.L. (1981), The phage typing of *V. cholerae* serovar O1. In *Acute Enteric Infection of Children*. Elsevier: Amsterdam

Lee, J.V., Shread, P., Furniss, A.L. and Bryant, T.N. (1981), Taxonomy and description of *Vibrio fluvialis* sp. nov. (synonym Group F vibrios, Group EFG). *J. appl. Bact.*, **50**, 73–94.

Levine, R.J. and Nalin, D.R. (1976), Cholera is primarily waterborne in Bangladesh. *Lancet*, **II**, 1305.

Levine, M.M., Kaper, J.B., Black, R.E. and Clements, M.L. (1983), New knowledge on pathogenesis of bacterial infections as applied to vaccine development. *Micro. Rev.*, **47**, 510–550.

Lowry, P.W., McFarland, L.M. and Threefoot, H.K. (1986), *Vibrio hollisae* septicaemia after consumption of catfish. *J. Inf. Dis.*, **154**, 730–731.

Madden, J., McCardell, B.A. and Reed, R. (1982), *Vibrio cholerae* in shellfish from US coastal waters. *Fd. Tech.*, **36**(3), 93–96.

Madden, J.M., McCardell, B.A. and Boutin, B.K. (1984), Isolation and identification of *Vibrio cholerae*. In *Bacteriological Analytical Manual*. Association of Official Analytical Chemists: Arlington.

Madden, J.M., McCardell, B.A. and Morris, J.G. (1989), *Vibrio cholerae*. In *Foodborne Bacterial Pathogens*, Doyle, M.P. (Ed). Marcel Dekker: New York.

Massad, G., Simpson, L.M. and Oliver, J.D. (1988), Isolation and characterisation of hemolysin mutants of *Vibrio vulnificus*. *FEMS Micro. Lett.*, **56**, 295–300.

Mhalu, F.S., Mtango, F.D.E. and Msengi, A.E. (1984), Hospital outbreaks of cholera transmitted through close person-to-person contact. *Lancet*, **II**, 82–84.

Mims, C.A. (1987), *The Pathogenesis of Infectious Disease*. Academic Press: London

Miwamoto, T., Miwa, H. and Hutano, S. (1990), Improved fluorogenic assay for rapid detection of *Vibrio parahaemolyticus* in foods. *Appl. Environ. Micro.*, **56**, 1480–1484.

Miyake, M., Honda, T. and Miwatani, T. (1988), Purification and characterisation of *Vibrio metschnikovii* cytolysin. *Inf. Imm.*, **56**, 954–960.

Morris, J.G. and Black, R.E. (1985), Cholera and other vibrios in the United States. *New Eng. J. Med.*, **312**, 343–350.

Morris, J.G., Wright, A.C., Roberts, D.M. *et al.* (1987), Identification of environmental *Vibrio vulnificus* isolates with a DNA probe for the cytotoxin hemolysin gene. *Appl. Environ. Micro.*, **53**, 193–195.

Morris, J.G., Takeda, T., Tall, B.D. *et al.* (1990), Experimental non-O group 1 *Vibrio cholerae* gastroenteritis in humans. *J. Clin. Invest.*, **85**, 697–705.

Mukerjee, S. (1978), Principles and practice of typing *V. cholerae*. In *Methods in Microbiology*, Vol 12, Bergan, T. and Norris, J.R. (Eds). Academic Press: London and New York.

Mukerjee, S., Guha, D.K. and Guha Roy, D.K. (1957), Studies on typing of cholera by bacteriophage, Part 1. Phage-typing of *V. cholerae* from Calcutta epidemics. *Ann. Biochem. Exp. Med.*, **17**, 161–176.

Nair, J.B., Oku, Y., Takeda, Y. *et al.* (1988), Toxin profiles of *Vibrio cholerae* non-O1 from environmental sources in Calcutta, India. *Appl. Environ. Micro.*, **54**, 3180–3182.

Nisibuchi, M. and Seidler, R.J. (1985), Rapid microimmunodiffusion method with species-specific antiserum raised to purified antigen for identification of *Vibrio vulnificus*. *J. Clin. Mic.*, **18**, 400–407.

Nisibuchi, M., Ishibashi, M., Takeda, Y. and Kaper, J.B. (1985), Detection of the thermostable direct hemolysin gene and related DNA sequences by the DNA colony hybridisation test. *Inf. Imm.*, **49**, 481–486.

NRC (1985), *An Evaluation of the Role of Microbiological Criteria for Foods and Food Ingredients*. National Academy Press: Washington DC.

O'Brien, A.D., Chen, M.E., Holmes, M.E. *et al.* (1984), Environmental and human isolates of *Vibrio cholerae* and *Vibrio parahaemolyticus* produce a *Shigella dysenteriae* 1 (Shiga)-like cytotoxin. *Lancet*, **I**, 77–78.

Okada, K., Miake, S., Moriya, T. *et al.* (1987a), Variability of haemolysin(s) produced by *Vibrio vulificus*. *J. Gen. Mic.*, **133**, 2853–2857.

Okada, K., Mitsuyama, M., Miake, S. and Amako, K. (1987b), Monoclonal antibodies against the haemolysin of *Vibrio vulnificus*. *J. Gen. Mic.*, **133**, 2279–2284.

Oliver, J.D. (1981), Lethal cold stress of *Vibrio vulniticus* in oysters. *Appl. Environ. Micro.*, **40**, 710–715.

Oliver, J.D. (1989), *Vibrio vulnificus*. In *Foodborne Bacterial Pathogens*, Doyle, M.P. (Ed). Marcel Dekker: New York.

Oliver, J.D. and Wanucho, D. (1990), Survival of *Vibrio vulnificus* at reduced temperature and elevated nutrient. *J. Fd. Safety*, **10**, 79–86.

Oliver, J.D., Warner, R.A. and Cleland, D.R. (1983), Distribution of *Vibrio vulnificus* and other lactose-fermenting vibrios in the marine environment. *Appl. Environ. Micro.*, **45**, 985–998.

Oscroft, C.A. (1987), Effects of freezing on the survival of *Vibrio parahaemolyticus* and the efficiency of enumeration methods to recover this organism from frozen and non-frozen prawn homogenates. *Tech. Memo.* 457. CFPRA: Chipping Campden.

Paile, D., Hackney, C., Riley, L *et al.* (1987), Seasonal variation in the fecal coliform population of Louisiana oysters and its relationship to microbiological quality. *J. Fd. Prot.*, **50**, 545–549.

Peres-Rosas, N. and Hazen, T.C. (1989), In situ survival of *Vibrio cholerae* and *Escherichia coli* in a tropical rain forest watershed. *Appl. Environ. Micro.*, **55**, 495–499.

Pinegar, J.A. and Suffield, A. (1982), The investigation of food poisoning outbreaks in England and Wales. In *Isolation and Identification Methods for Food Poisoning Organisms*, Corry, J.E.L., Roberts, D. and Skinner, F.A. (Eds). Academic Press: London and New York.

Pitrak, D.L. and Gindorf, J.D. (1989), Bacteremic cellulitis caused by non-serogroup O1 *Vibrio cholerae* acquired in a freshwater inland lake. *J. Clin. Mic.*, **27**, 2874–2876.

Pugsley, A.P. (1988), Report on Pasteur Centenary Symposium. Paris, 1987: Molecular Biology and Infectious Disease. *Micro. Sci.*, **5**, 84–86.

Rank, E.L., Smith, I.B. and Langer, M. (1988), Bacteraemia caused by *Vibrio hollisae*. *J. Clin. Mic.*, **26**, 375–376.

Rayman, M., Sack, D.A., Wadood, A *et al.* (1989), Rapid identification of *Vibrio cholerae* serotype O1 from primary isolation plates by coagglutination test. *J. Med. Mic.*, **28**, 39–41.

Reyes, A.L., Boutin, B.K., Peeler, J.T. and Twedt, R.M. (1985), Adherence and haemagglutination of mammalian cells by epidemiologically distinct strains of *Vibrio vulnificus*. *J. Fd. Prot.*, **48**, 783–785.

Richard, C. and Lhullier, M. (1977), *Vibrio parahaemolyticus* et vibrions halophiles, leur importance en pathologie humaine et dans l'environment marin. *Bull. Inst. Past. (Paris)*, **75**, 345–368

Riley, L.A. and Hackney, C.R. (1985), Survival of *Vibrio cholerae* during cold storage in artificially contaminated seafoods. *J. Fd. Sci.*, **50**, 838–839

Sack, D.A. and Sack, R.B. (1975), Test for enterotoxigenic *Escherichia coli* using Y1 adrenal cells in miniculture. *Inf. Imm.*, **11**, 334–336.

Saha, S. and Sanyal, S.C. (1988), Cholera toxin gene-positive *Vibrio cholerae* O1 Ogawa and Inaba strains produce the new cholera toxin. *FEMS Micro. Lett.*, **50**, 113–116.

Saha, S. and Sanyal, S.C. (1990), Immunobiological relationships of the enterotoxins produced by cholera-gene (CT^+) positive and — negative (CT^-) strains of *Vibrio cholerae* O1. *J. Med. Mic.*, **32**, 33–37.

Sakazaki, R. (1965), *Vibrio parahaemolyticus*. A non-choleragenic enteropathogenic Vibrio. In *Proceedings of the Second Cholera Research Symposium*, Bushnell, O.A. and Brookhyser, C.S. (Eds). US Dept. of Health and Welfare: Washington DC.

Sakazaki, R. and Balows, A. (1981), The genera *Vibrio, Plesiomonas* and *Aeromonas*. In *The Prokaryotes. A Handbook on Habitats, Isolation and Identification of Bacteria*, Vol 2, Stamm, N.D., Stolp, H., Trupen, H.G. *et al.* (Eds). Springer-Verlag: Berlin.

Salles, C.A. and Momen, H. (1981), A rapid visual test to characterize cholera vibrios. *J. appl. Bact.*, **51**, 433–437.

Sang, F.C., Hugh-Jones, M.E. and Hagstad, V. (1987), Viability of *Vibrio cholerae* O1 on frog legs under frozen and refrigerated conditions and low dose radiation treatment. *J. Fd. Prot.*, **50**, 662–664.

Sanyal, S.C., Alam, K., Neogi, P.K.B. *et al.* (1983), A new cholera toxin. *Lancet*, **I**, 1337–1338.

Sarkar, B.L., Kumar, R., De, S.P. and Pal, S.C. (1987), Hemolytic activity of and lethal toxin production by environmental strains of *Vibrio parahaemolyticus*. *Appl. Environ. Micro.*, **53**, 2696–2698.

Sengupta, D., Datta-Roy, K. and Ghose, A.C. (1989), Identification of some antigenically related outer-membrane-proteins of strains of *Vibrio cholerae* O1 and non-O1 serovars involved in intestinal adhesion and the protective role of antibodies to them. *J. Med. Mic.*, **29**, 33–39.

Setarunnahar, S. and Sanyal, S.C. (1989), Immunobiological relationships among new cholera toxins produced by CT gene-negative strains of *Vibrio cholerae* O1. *J. Med. Mic.*, **28**, 33–37.

Shultz, L., Rutledge, J., Grudner, R. and Biede, S. (1984), Determination of the thermal death time of *Vibrio cholerae* in blue crabs (*Callinectes sapidas*), *J. Fd. Prot.*, **47**:4.

Siddique, A.K. (1989), Simultaneous outbreaks of contrasting drug resistant classic and el tor *Vibrio cholerae* O1 in Bangladesh. *Lancet*, **II**, 396.

Simonson, J.G. and Siebeling, R.J. (1988), Coagglutination of *Vibrio cholerae, Vibrio mimicus* and *Vibrio vulnificus* with anti-flagellar monoclonal antibody. *J. Clin. Mic.*, **26**, 1962–1966.

Simpson, L.M. and Oliver, J.D. (1983), Siderophore production by *Vibrio vulnificus*. *Inf. Imm.*, **41**, 643–649.

Smith, H.L. (1970), A presumptive test for vibrios: the 'string' test. *Bull. World Hlth. Org.*, **42**, 817–818.

Smith, G.C. and Menkel, J.H. (1982), Collagenolytic activity of *Vibrio vulnificus*: potential contribution to its invasiveness. *Inf. Imm.*, **35**, 1155–1156.

Stelma, G.N., Spaulding, P.L., Reyes, A.C. and Johnson, C.H. (1988), Production of enterotoxin by *Vibrio vulnificus* isolates. *J. Fd. Prot.*, **51**, 192–196.

Tacket, C.O., Becker, S.F. and Blake, P.A. (1984), Clinical features and an epidemiological study of *Vibrio vulnificus* infections. *J. Inf. Dis.*, **149**, 558–561.

Takeda, Y. (1983), Thermostable direct haemolysin of *Vibrio parahaemolyticus*. *Pharm. Ther.*, **19**, 123–127.

Tamplin, M.L., Colwell, R.R. and Hall, S. (1987), Sodium-channel inhibitors produced by enteropathogenic *Vibrio cholerae* and *Aeromonas hydrophila*. *Lancet*, **I**, 975.

Tamura, K., Shimada, S. and Prescott, L.M. (1972), Vibrio agar: a new plating medium for isolation of *Vibrio cholerae*. *Jap. J. Med. Sci. Biol.*, **24**, 125–127.

Tassin, M.G., Siebling, R.J., Roberts, N.C. and Larson, A.D. (1983), Presumptive identification of *Vibrio* species with H antiserum. *J. Clin. Mic.*, **18**, 400–407.

Tauxe, R.V., Holmberg, S.D., Dodin, A. *et al.* (1988), Epidemic cholera in Mali: high mortality and multiple routes of transmission in a famine area. *Epid. Inf.*, **100**, 279–288.

Taylor, J.A., Miller, D.C., Barrow, G.I. *et al.* (1982), The isolation and identification of *Vibrio parahaemolyticus*. In *Isolation and Identification Methods for Food Poisoning Organisms*, Corry, J.E.L., Roberts, D. and Skinner, F.A. (Eds). Academic Press: London and New York.

Testa, J., Daniel, L.W. and Kreger, A.S. (1984), Extracellular phospholipase A_2 and lysophospholipase produced by *Vibrio vulnificus*. *Inf. Imm.*, **45**, 458–463.

Thompson, C.A. and Vanderzant, C. (1976), Relationship of *Vibrio parahaemolyticus* in oysters, water and sediment and bacteriological and environmental indices. *J. Fd. Sci.*, **41**, 117–120.

Thornley, M.J. (1960), The differentiation of *Pseudomonas* from other Gram-negative bacteria on the basis of arginine metabolism. *J. appl. Bact.*, **23**, 37–52.

Tilton, R.C. and Ryan, R.W. (1987), Clinical and ecological characteristics of *Vibrio vulnificus* in the northeastern United States. *Diag. Micro. Inf. Dis.*, **6**, 109–117.

Twedt, R.M. (1989), *Vibrio parahaemolyticus*. In *Foodborne Bacterial Pathogens*, Doyle, M.P. (Ed). Marcel Dekker: New York.

Twedt, R.M., Madden, J.M. and Colwell, R.R. (1984), In *Compendium of Methods for the Microbiological Examination of Foods*, 2nd ed, Speck, M.L. (Ed). American Public Health Association: Washington DC.

Van Heyningen, W.E. (1977), Cholera toxin. *Biol. Rev.*, **52**, 509–549.

Watkins, W.D., Thomas, C.D. and Cabelli, V.J. (1976), Membrane filter procedure for enumeration of *Vibrio parahaemolyticus*. *Appl. Env. Micro.*, **46**, 783–786.

West, P.A. (1989), The human pathogenic vibrios — a public health update with environmental perspectives. *Epid. Inf.*, **103**, 1–34.

West, P.A., Brayton, P.R., Bryant, T.N. and Colwell, R.R. (1986), Numerical taxonomy of *Vibrios* isolated from aquatic environments. *Int. J. Syst. Bact.*, **36**, 531–543.

Wright, A.C., Simpson, L.M. and Oliver, J.D. (1981), Role of iron in the pathogenesis of *Vibrio vulnificus* infections. *Inf. Imm.*, **34**, 503–507.

Wright, A.C., Simpson, L.M., Oliver, J.D. and Morris, J.G. (1990) Phenotypic evaluation of acapsular transposon mutants of *Vibrio vulnificus*. *Inf. Imm.*, **58**, 1769–1773.

Yamamoto, T. and Yokota, T. (1988), Production of cell associated hemagglutinins and *in vitro* adherence to mucus coat and epithelial surfaces of the villi and lymphoid follicles of human small intestines treated with formalin. *J. Clin. Mic.*, **26**, 2018– 2024.

Yamamoto, T. and Yokota, T. (1989), Adherence targets of *Vibrio parahaemolyticus* in human small intestines. *Inf. Imm.*, **57**, 2410–2419.

Yamanaka, H., Satoh, T., Katsu, T. and Shinoda, S. (1987), Mechanism of haemolysis by *Vibrio vulniticus*, haemolysin *J. Gen. Mic.*, **32**, 39–43.

Yamanaka, H., Sugiyama, K., Furuta, H. *et al.* (1990), Cytolytic action of *Vibrio vulnificus* hemolysin on mast cells from rat peritoneal cavity. *J. Med. Mic.*, **32**, 39–43.

Yoshida, S.-I., Ogawa, M. and Mizuguchi, Y. (1985), Relationship of capsular materials and colony opacity to virulence of *Vibrio vulnificus*. *Inf. Imm.*, **47**, 446–451.

Zakaria-Meehin, Z., Massad, G., Simpson, L.M. *et al.* (1988), Ability of *Vibrio vulnificus* to obtain iron from hemoglobin-haptoglobin complexes. *Inf. Imm.*, **56**, 275–277.

Chapter 9

Abeyta, C. and Wekell, M.M. (1988), Potential sources of *Aeromonas hydrophila*. *J. Fd. Safety*, **9**, 11–22.

Abeyta, C., Kaysner, C.A., Wekell, M.M. *et al.* (1986), Recovery of *Aeromonas hydrophila* from oysters implicated in an outbreak of foodborne illness. *J. Fd. Prot.*, **49**, 643–646.

Ablonglabor, D.E., Shonekar, R.A.O., Kazak, W.H. and Coker, A.O. (1982), *Aeromonas* food poisoning in Nigeria: a case report. *Cent. African J. Med.*, **28**, 36–38.

Agger, W.A. and Callister, S.M. (1987), Intestinal infection with *Aeromonas*. *Ann. Int. Med.*, **106**, 479.

Agger, W.A., McCormick, J.D. and Gurwith, M.J. (1985), Clinical and microbiological features of *Aeromonas hydrophila* associated disease. *J. Clin. Mic.*, **21**, 909–913.

Anon (1988), Evisceration, cold storage blamed for high *Aeromonas* counts on chickens. *Fd. Chem. News*, **June 6**, 37.

Archer, D.L. and Kvenberg, J.E. (1988), Regulatory significance of *Aeromonas* in foods. *J. Fd. Safety*, **9**, 53–58.

Asao, T., Kinoshita, Y., Kozaki, S. *et al.* (1984), Purification and some properties of *Aeromonas hydrophila* hemolysin. *Inf. Imm.*, **46**, 122–127.

Baumann, P. and Schubert, R.H.W. (1984), *Vibrionaceae*. In *Bergey's Manual of Systematic Bacteriology*, Vol 1, Krieg, N.R. and Holt, J.G. (Eds). Williams and Wilkins: Baltimore and London.

Beebe, J.L. (1986), Blood-borne bacteria as reflectors and indicators of cancer. *Lab. Man.*, **May**, 45–50.

Black, R.E. (1986), Pathogens that cause traveler's diarrhea in Latin America and Africa. *Rev. Inf. Dis.*, **8** (Suppl.), 131–135.

Borgcathi, S., Young, R., Olson, M.O.J. *et al.* (1989), Amonabactin, a novel tryptophan — or phenylalanine containing — phenolate siderophore in *Aeromonas hydrophila*. *J. Bact.*, **171**, 1811–1816.

Burke, V., Gracey, M., Robinson, J. *et al.* (1983), The microbiology of childhood gastroenteritis: *Aeromonas* species and other infective agents. *J. Inf. Dis.*, **148**, 68–73.

Burke, V., Robinson, J., Gracey, M. *et al.* (1984), Isolation of *Aeromonas hydrophila* from a metropolitan water supply: seasonal correlation with clinical isolates. *Appl. Environ. Micro.*, **48**, 361–366.

Callister, S.M. and Agger, W.A. (1987), Enumeration and characterization of *Aeromonas hydrophila* and *Aeromonas caviae* isolated from grocery store produce. *J. Fd. Prot.*, **50**, 317–323.

Campbell, J.D. and Houston, C.W. (1985), Effect of cultural conditions on the presence of a cholera toxin cross-reactive factor in culture filtrates of *Aeromonas hydrophila*. *Curr. Mic.*, **12**, 101–106.

Carnahan, A.M., Joseph, S.W. and Janda, J.M. (1989), Species identification of *Aeromonas* strains based on carbon substrate oxidation profiles. *J. Clin. Mic.*, **27**, 2128–2129.

Carnahan, A.M., Hammontree, L., Bourgeois, L. and Joseph, S.W. (1990), Pyrazinamidase activity as a phenotypic marker for several *Aeromonas* species isolated from clinical specimens. *J. Clin. Mic.*, **28**, 391–342.

Cattabiani, F. (1986), Sensibilita di disinfettani di *Aeromonas hydrophila* e di *Vibrio fluvialis*. *Arch. Vet. Ital.*, **37**, 65–72.

Chakraborty, T., Montenegro, M.A., Sanyal, S.E. *et al.* (1984), Cloning of enterotoxin gene from *Aeromonas hydrophila* provides conclusive evidence of production of a cytotonic enterotoxin. *Inf. Imm.*, **46**, 435–441.

Champsaur, H., Andremon, A., Mathieu, D. *et al.* (1982), Cholera like illness due to *Aeromonas sobria*. *J. Inf. Dis.*, **145**, 248–254.

Chopra, A.K. and Houston, C.W. (1989), Purification and partial characterisation of a cytotonic enterotoxin produced by *Aeromonas hydrophila*, *Con. J. Mic.*, **35**, 719–727.

Colwell, R.R., MacDonnell, M.T. and deLey, J. (1986), Proposal to recognize the family *Aeromonadaceae* form nov. *Int. J. Syst. Bact.*, **36**, 473–477.

Corrello, A., Solburn, K.A., Budder, J.R. and Chang, B.J. (1988), Adhesion of clinical and environmental *Aeromonas* isolates to HEp-2 cells. *J. Med. Mic.*, **26**, 392–393.

Cumbernatch, N., Gurwith, M.J., Langston, C. *et al.* (1979), Cytotonic enterotoxin produced by *Aeromonas hydrophila*: relationship of toxigenic isolates to diarrhoeal disease. *Inf. Imm.*, **23**, 829–837.

Dainty, R.H., Shaw, B.G. and Roberts, T.A. (1983), Microbial and chemical changes in chill-stored red meats. In *Food Microbiology: Advances and Prospects*, Roberts, T.A. and Skinner, F.A. (Eds). Academic Press: London and New York.

Davis, W.A., Kane, J.G. and Garagusi, V.F. (1978), Human *Aeromonas* infection: a review of the literature and a case report of endocarditis. *Medicine*, **57**, 267–277.

Dooley, J.S.G., McCubbin, W.D., Kay, C.M. and Trust, T.J. (1988), Isolation and biochemical characterisation of the S-layer protein from a pathogenic *Aeromonas hydrophila* strain. *J. Clin. Mic.*, **26**, 980–987.

Echeverria, P., Sack, R.B., Blacklow, N.R. *et al.* (1984), Prophylactic doxycycline for traveler's diarrhea in Thailand: further supportive evidence of *Aeromonas hydrophila* as an enteric pathogen. *Am. J. Epid.*, **120**, 912–921.

Edwards, P.R. and Ewing, W.H. (1972), *Identification of Enterobacteriaceae*, 3rd ed. Burgess Publishing Company: Minneapolis.

Elbashir, A.M. and Millership, S.E. (1989), Haemagglutinating activity of *Aeromonas* species from different sources, attempted use as a typing system. *Epid. Inf.*, **102**, 221–229.

Ellison, R.T. and Mostow, S.R. (1984), Pyogenic meningitis manifestations during therapy for *Aeromonas hydrophila* sepsis. *Arch. Inter. Med.*, **144**, 2078–2079.

Enfors, S.O., Molin, G. and Ternstrom, A. (1979), Effect of packaging under carbon dioxide, nitrogen or air on the microbial flora of pork stored at 4°C. *J. appl. Bact.*, **47**, 197–208.

FDA (1985), Pathogen surveillance sampling of *Aeromonas hydrophila* in foods. *Food Safety Compliance Program*, 7303–030. Washington DC.

Figura, N. and Gugliemetti, P. (1987), Differentiation of motile and mesophilic *Aeromonas* strains into species by testing for a CAMP-like factor. *J. Clin. Mic.*, **25**, 1341–1342.

Figura, N., Marvis, L., Vendioni, S. (1986), Prevalence, species differentiation and toxigenicity of *Aeromonas* strains in cases of childhood gastroenteritis and in controls. *J. Clin. Mic.*, **23**, 595–599.

Fricker, C.R. (1987), Serotyping of mesophilic *Aeromonas* spp. on the basis of lipopolysaccharide antigens. *Lett. Appl. Mic.*, **4**, 113–116.

Golden, D.A., Eylos, M.J. and Beuchat, L.N. (1989), Influence of modified atmosphere on the growth of uninjured and heat-injured *Aeromonas hydrophila*, *Appl. Environ. Micro.*, **55**, 3012–3015.

Gracey, N., Burke, V. and Robinson, J. (1982), *Aeromonas* associated gastroenteritis. *Lancet*, **II**, 1306.

Gracey, M., Burke, V. and Robinson, J. (1984), *Aeromonas* subspecies in traveller's diarrhoea. *B. Med. J.*, **289**, 658.

Gram, L., Troll, G. and Huss, H.H. (1987), Detection of specific spoilage bacteria from fish stored at low (0°C) and high (20°C) temperature. *Int. J. Fd. Mic.*, **4**, 65–72.

Grau, F.H., Eustace, I.J. and Bill, B.A. (1985), Microbial flora of lamb carcasses stored at 0°C in packs flushed with nitrogen and filled with carbon dioxide. *J. Fd. Sci.*, **50**, 482–485.

Gray, S.J. (1984), *Aeromonas hydrophila* in livestock: incidence, biochemical characteristics and antibiotic susceptibility. *J. Hyg.*, **92**, 365–375.

Harris, R.L., Fainstein, V. and Elting, L. (1985), Bacteremia caused by *Aeromonas* species in hospitalized cancer patients. *Rev. Inf. Dis.*, **7**, 314–320.

Herrington, T.L. (1984), Oyster-borne disease outbreaks in St Petersburg, Florida. *Memorandum* (1/31/84). Dept. Health and Human Services, FDA: Atlanta.

Heuschmann-Brunner, G. (1978), Aeromonads of the hydrophila-punctata group in fresh water fishes. *Arch. Hydrobiol.*, **83**, 99–125.

Hickman-Brenner, F.W, MacDonald, K.L., Steigerwalt, A.G. *et al.* (1987), *Aeromonas veronii* a new ornithine decarboxylase-positive species that may cause diarrhea. *J. Clin. Mic.*, **25**, 900–906.

Hickman-Brenner, F.W., Fanning, G.R., Arduino, M.J. *et al.* (1988), *Aeromonas schubertii*, a new mannitol-negative species found in human clinical specimens. *J. Clin. Mic.*, **26**, 1561–1564.

Holmberg, S.D. (1986), *Aeromonas* intestinal infections in the United States. *Ann. Intern. Med.*, **105**, 683–689.

Holmberg, S.D. and Farmer, J.J. (1984), *Aeromonas hydrophila* and *Plesiomonas shigelloides* as causes of intestinal infections. *Rev. Inf. Dis.*, **6**, 633–639.

Honda, T., Sato, M., Nishimura, T. *et al.* (1985), Demonstration of cholera toxin-related factor in cultures of *Aeromonas* species by enzyme-linked immunosorbent assay. *Inf. Imm.*, **50**, 322–323.

Howard, S.P., Garland, W.J., Green, M.J. and Bucklety, J.T. (1987), Nucleotide sequences of the gene for the hole-forming toxin aerolysin of *Aeromonas hydrophila*. *J. Bact.*, **169**, 2869–2871.

Hunter, P.R. and Burge, S.H. (1987), Isolation of *Aeromonas caviae* from ice-cream. *Lett. Appl. Mic.*, **4**, 45–46.

Janda, J.M. and Duffey, P.S. (1988), Mesophilic aeromonads in human disease: current taxonomy, laboratoty identification and disease spectrum. *Rev. Inf. Dis.*, **10**, 980–997.

Janda, J.M., Bottone, E.J. and Reitano, M. (1983), *Aeromonas* species in clinical medicine: significance, epidemiology and speciation. *Diag. Micro. Inf. Dis.*, **1**, 221–228.

Janda, J.M., Brendan, R. and Bottone, E.J. (1984), Differential susceptibility to human serum by *Aeromonas* species. *Curr. Mic.*, **11**, 325–328.

Janda, P.M., Oshiro, L.S., Abbott, S.L. and Duffey, P.S. (1987), Virulence markers of mesophilic aeromonads: association of the autoagglutination phenomenon with mouse pathogenicity and the presence of a peripheral cell-associated layer. *Inf. Imm.*, **55**, 3070–3077.

Janossy, G. and Torjan, V. (1980), Enterotoxigenicity of *Aeromonas* species on suckling mice. *Acta Mic. Acad. Sci. Hung.*, **23**, 63–69.

Johnson, W.M. and Lior, H. (1981), Cytotoxicity and suckling mouse reactivity of *Aeromonas hydrophila* isolated from human sources. *Can. J. Mic.*, **27**, 1019–1027.

Joseph, S.W., Jand, M. and Carnahan, A. (1988), Isolation, enumeration and identification of *Aeromonas* species. *J. Fd. Safety*, **9**, 23–36.

Kaper, J., Seidler, R.J., Lockman, H. and Colwell, R.R. (1979), A medium for the presumptive identification of *Aeromonas hydrophila* and the *Enterobacteriaceae*. *Appl. Environ. Micro.*, **38**, 1023–1026.

Kelly, M.T., Stroh, E.M.D. and Jessop, J. (1988), Comparison of blood agar, ampicillin blood agar, MacConkey-ampicillin-tween agar and modified cefsulodin-irgasan-novobiocin agar for isolation of *Aeromonas* species from stool specimens. *J. Clin. Mic.*, **26**, 1738–1740.

Kindschuch, M., Pickering, L.K., Cleary, T.G. and Ruiz-Palacios, G. (1987), Clinical and biochemical significance of toxin production by *Aeromonas hydrophila*. *J. Clin. Mic.*, **25**, 916–921.

Kirov, S.M., Rees, B., Wellock, R.C. *et al.* (1986), Virulence characteristics of *Aeromonas* spp. in relation to source and biotype. *J. Clin. Mic.*, **24**, 827–834.

Knochel, S. (1989), Effect of temperature on hemolysin production in *Aeromonas* species isolated from warm and cold environments. *Int. J. Fd. Mic.*, **8**, 149–154.

Knochel, S. (1990), Growth characteristics of motile *Aeromonas* species isolated from different environments. *Int. J. Fd. Mic.*, **10**, 253–254.

Krovakek, K., Peterz, M., Faris, A. and Mansson, I. (1989), Enterotoxigenicity and drug sensitivity of *Aeromonas hydrophila* isolated from well water in Sweden: a case study. *Int. J. Fd. Mic.*, **8**, 149–154

Kuijper, E.J., van Alphen, L., Leenders, E. and Zanen, H.C. (1989), Typing of *Aeromonas* strains by DNA restriction endonuclease analysis and polyacrylamide gel electrophoresis of cell envelopes, *J. Clin. Mic.*, **27**, 1280–1285.

LeChevalier, T., Evans, T.M., Seidler, R.J. *et al.* (1982), *Aeromonas sobria* in a chlorinated drinking water supply. *Micro. Ecol.*, **8**, 325–333.

Ljungh, A. and Wadstrom, T. (1983), Toxins of *Vibrio parahaemolyticus* and *Aeromonas hydrophila. J. Toxicol. Toxin Rev.*, **1**, 257–271.

Ljungh, A., Wretlind, B. and Molby, R. (1981), Separation and characterization of enterotoxin and two haemolysins from *Aeromonas hydrophila. Acta. Path. Micro. Scand.*, **B 89**, 387–391.

Ljungh, A., Enroth, P. and Wadstrom, T. (1982), Cytotonic enterotoxin from *Aeromonas hydrophila. Toxicon*, **20**, 787–794.

Marcus, L.C. (1971), Infectious diseases of reptiles. *J. Am. Vet. Med. Assoc.*, **159**, 1629–1631.

Megraud, F. (1986), Incidence and virulence of *Aeromonas* in faeces of children with diarrhoea. *Eur. J. Clin. Mic.*, **5**, 311–316.

Millership, S.E. and Want, S.V. (1989), Typing of *Aeromonas* species by protein fingerprinting: comparison of radiolabelling and silve staining for labelling proteins. *J. Med. Mic.*, **29**, 29–32.

Millership, S.E., Curnow, S.R. and Chattopadhyay, B. (1983), Faecal carriage rate of *Aeromonas hydrophila. J. Clin. Mic.*, **36**, 920–923.

Morgan, D.R. and Wood, L.V. (1988), Is *Aeromonas* species a foodborne pathogen? Review of the clinical data. *J. Fd. Safety*, **9**, 59–68.

Morse, J.W. and Hind, D.W. (1984), Bacteria isolated from lymph nodes of California slaughter swine. *Am. J. Vet. Res.*, **45**, 1648–1649.

Murray, R.G.E., Dooley, J.S., Whippey, P.W. and Trust, T.J. (1988), Structure of an S-layer on a pathogenic strain of *Aeromonas hydrophila. J. Bact.*, **170**, 2625–2630.

Namdari, H. and Bottone, E.J. (1989), Suicide phenomenon in mesophilic aeromonads as a basis for species identification. *J. Clin. Mic.*, **27**, 788–789.

Namdari, H. and Bottone, E.J. (1990), Microbiologic and clinical evidence supports the role of *Aeromonas caviae* as a pediatric enteric pathogen. *J. Clin. Mic.*, **28**, 837–840.

Namdari, H. and Cabelli, V.J. (1989), The suicide phenomenon in motile aeromonads. *Appl. Environ. Micro.*, **55**, 2615–2619.

Nishikawa, Y. and Kishi, T. (1987), Modification of bile salts brilliant green agar for isolation of motile *Aeromonas* from foods and environmental specimens. *Epid. Inf.*, **98**, 331–336.

Nishikawa, Y. and Kishi, T. (1988), Isolation and characterization of motile *Aeromonas* from human, food and environmental specimens. *Epid. Inf.*, **101**, 213–223.

Okrend, A.J.G., Rose, B.E. and Bennett, B. (1987), Incidence and toxigenicity of *Aeromonas* species in retail poultry, beef and pork. *J. Fd. Prot.*, **50**, 509–513.

Palumbo, S.A. and Buchanan, R.L. (1988), Factors affecting growth and survival of *Aeromonas hydrophila* in foods. *J. Fd. Safety*, **9**, 37–51.

Palumbo, S.A., Maxino, F., Williams, A.C. *et al.* (1985a), Starch-ampicillin agar for the qualitative detection of *Aeromonas hydrophila. Appl. Environ. Mivro.*, **50**, 1027–1030.

Palumbo, S.A., Morgan, D.R. and Buchanan, R.L. (1985b), Influence of temperature, NaCl and pH on the growth of *Aeromonas hydrophila. J. Fd. Sci.*, **50**, 1417–1421.

Palumbo, S.A., Jenkins, R.K., Buchanan, R.L. and Phillips, J.G. (1986), Determination of irradiation D-values of *Aeromonas hydrophila. J. Fd. Prot.*, **49**, 484–490.

Palumbo, S.A., Williams, A.C., Buchanan, R.L. and Phillips, J.G. (1987), Thermal resistance of *Aeromonas hydrophila. J. Fd. Prot.*, **50**, 761–764.

Pansone, A.C., Lewis, N.F. and Venugopal, V. (1986), Characterization of extracellular proteases of *Aeromonas hydrophila. Agric. Biol. Chem.*, **50**, 1743–1749.

Pitarangsi, C., Echeverria, P., Whitmore, R. *et al.* (1982), Enteropathogenicity of *Aeromonas hydrophila* and *Plesiomonas shigelloides*: prevalence among individuals with and without diarrhea in Thailand. *Inf. Imm.*, **35**, 666–673.

Popoff, M. (1984), *Aeromonas*. In *Bergey's Manual of Systematic Bacteriology*, Vol 1, Krieg, N.R. and Holt, J.G. (Eds). Williams and Wilkins: Baltimore and London.

Popoff, M. Cognault, C., Kinedjian, M. and Lemelin, M. (1981), Polynucleotide sequence relatedness among motile *Aeromonas* species. *Curr. Mic.*, **159**, 1629–1631.

Potomski, J., Burke, V., Robinson, J. *et al.* (1987a), *Aeromonas* cytotonic enterotoxin cross reactive with cholera toxin. *J. Med. Mic.*, **23**, 179–186.

Potomski, J., Burke, V., Watson, J. and Gracey, M. (1987b), Purification of cytotoxic enterotoxin of *Aeromonas spbria* by use of monoclonal antibodies. *J. Med. Mic.*, **23**, 171–177.

Rhamon, A.F.M.S. and Willoughby, J.M.T. (1980), Dysentery-like syndrome associated with *Aeromonas hydrophila. B. Med. J.*, **281**, 976.

Rippey, S.R. and Cabelli, V.J. (1980), Occurrence of *Aeromonas hydrophila* in limnetic environments: relationship of the organism to the trophic state. *Micro. Ecol.*, **6**, 45–54.

Rose, J.M., Chopra, A.K., Niesel, D.W. *et al.* (1987), Pathogenic mechanisms of *Aeromonas hydrophila*. Cited by Stelma, 1988.

Rose, J.M., Houston, C.W., Coppenhaver, O.H *et al.* (1989a), Purification and chemical characterization of a cholera toxin-cross-reactive cytolytic enterotoxin produced by a human isolate of *Aeromonas hydrophila. Inf. Imm.*, **57**, 1165–1169.

Rose, J.M., Houston, C.W. and Kurosky, A. (1989b), Bioactivity and immunological characterization of a cholera toxin-cross-reactive cytolytic enterotoxin from *Aeromonas hydrophila. Inf. Imm.*, **57**, 1170–1176.

Schultz, A.J. and McCardell, B.A. (1988), DNA homology and immunological cross-reactivity between *Aeromonas hydrophila* cytotonic toxin and cholera toxin. *J. Clin. Mic.*, **26**, 57–61.

Shattner, B., Jones, M.J. and George, W.L. (1988), Effect of incubation temperature on growth and soluble protein profiles of motile *Aeromonas* strains. *J. Clin. Mic.*, **26**, 392–393.

Shimoda, T., Sakazaki, R., Hariyame, K. *et al.* (1984), Production of a cholera-like enterotoxin by *Aeromonas hydrophila. Jap. J. Med. Sci. Biol.*, **37**, 141–144.

Shotts, E.B. and Rimler, R. (1973), Medium for the isolation of *Aeromonas hydrophila. Appl. Mic.*, **26**, 550–553.

Shotts, E.B., Gaines, J.L., Martin, C. and Prestwood, A.K. (1972), *Aeromonas* induced deaths among fish and reptile in a eutrophic inland lake. *J. Am. Vet. Med. Assoc.*, **20**, 1219–1228.

Shread, P., Donovan, T.J. and Lee, J.V. (1981), A survey of the incidence of *Aeromonas* in human faeces. *Soc. Gen. Mic Quartely*, **8**, 184.

Simard, R.E., Zee, J. and L'Heureux, L. (1984), Microbial growth in carcasses and boxed beef during storage. *J. Fd. Prot.*, **47**, 773–777.

Slade, P.J., Falah, M.A., and Al-Ghady, M.R. (1986), Isolation of *Aeromonas hydrophila* from bottled waters and domestic supplies in Saudi Arabia. *J. Fd. Prot.*, **49**, 471–476.

Stelma, G.N. (1988), Virulence factors associated with pathogenesis of *Aeromonas* isolates. *J. Fd. Safety*, **9**, 1–9.

Stelma, G.N. (1989) *Aeromonas hydrophila*. In *Foodborne Bacterial Pathogens*, Doyle, M.P. (Ed). Marcel Dekker: New York.

Stephenson, J.R., Millership, S.E. and Tabaqchali, S. (1987), Typing of *Aeromonas* species by polyacrylamide gel electrophoresis of radiolabelled cell proteins. *J. Med. Mic.*, **24**, 113–118.

Stern, N.J., Drazek, E.S. and Joseph, S.W. (1987), Low incidence of *Aeromonas* species in livestock faeces. *J. Fd. Prot.*, **50**, 66–69.

Tamplin, M.L., Colwell, R.R., Hall, S. *et al.* (1987), Sodium-channel inhibitors produced by enteropathogenic *Vibrio cholerae* and *Aeromonas hydrophila*. *Lancet*, **I**, 975.

Thomas, L.V., Gross, R.J., Cheasty, T. and Rowe, B. (1990), Extended serogrouping scheme for motile, mesophilic *Aeromonas*. *J. Clin. Mic.*, **208**, 980–984.

Timmis, K.N., Montenegro, M.A., Balling, E. *et al.* (1984), Genetics of toxin synthesis in pathogenic Gram-negative enteric bacteria. In *Bacterial Protein Toxins*, Aloufi, J., Freer, H., Fehrenhach, F. and Jeljaszewecz, J. (Eds). Academic Press: London.

Todd, L.S., Hardy, J.C., Stringer, M.F. and Bartholomew, B.A. (1989), Toxin production by strains of *Aeromonas hydrophila* grown in laboratory media and prawn puree. *Int. J. Fd. Mic.*, **9**, 145–156.

Travis, L.B. and Washington, J.A. (1986), The clinical significance of *Aeromonas*. *Am. J. Clin. Path.*, **85**, 330–336.

Trust, T.J. and Sparrow, R.A.H. (1974), The bacterial flora in the alimentary tract of freshwater salmonid fishes. *Can. J. Mic.*, **20**, 1219–1228.

Turnbull, P.C.B. Lee, J.V., Van der Walle, S. *et al.* (1984), Enterotoxin production in relation to taxonomic grouping and source of isolation of *Aeromonas* species. *J. Clin. Mic.*, **19**, 175–180.

van der Kooij, D. (1988), Properties of aeromonads and their occurrence and hygienic significance in drinking water. *Zbl. Bakt. Hyg.*, **B 187**, 1–17.

von Graevenitz, A. and Bucher, C. (1983), Evaluation of differential and selective media for isolation of *Aeromonas* and *Plesiomonas* species from human faeces. *J. Clin. Mic.*, **17**, 16–20.

Walker, S.J. and Stringer, M.F. (1987), Growth of *Listeria monocytogenes* and *Aeromonas hydrophila* at chill temperatures. *Tech. Memo.* No 462. CFPRA: Chipping Campden.

Watson, I.M., Robinson, J.O., Burke, V. and Gracey, M. (1985), Invasiveness of *Aeromonas* species in relation to biotype, virulence factors and clinical features. *J. Clin. Mic.*, **22**, 48–51.

Williams, L.A. and LaRoche, P.A. (1985), Temporal occurrence of *Vibrio* species and *Aeromonas hydrophila* in estuarine sediments. *Appl. Environ. Micro.*, **50**, 1490–1495.

Wohlegemuth, K., Pierce, R.L. and Kirkbridge, C.A. (1972), Bovine abortions associated with *Aeromonas hydrophila*. *J. Am. Vet. Med. Assoc.*, **160**, 1001–1002.

Wolff, R.L., Wiseman, S.L. and Kitchens, S.C. (1980), *Aeromonas hydrophila* bacteremia in ambulatory immuno-compromised hosts. *Am. J. Med.*, **68**, 238–242.

Chapter 10

Aldova, E. (1985), Experiences with serology of *Plesiomonas shigelloides*. I. O-antigen structure. *J. Hyg. Epid. Mic. Imm.*, **29**, 209–210.

Arai, T., Ikejima, N., Itoh, T. *et al.* (1980), A survey of *Plesiomonas shigelloides* from aquatic environments, domestic animals, pets and humans. *J. Hyg.*, **84**, 203–211.

Basu, S., Tharanathan, R.W., Kontroh, T. and Mayer, H. (1985), Chemical structure of the lipid A component of *Plesiomonas shigelloides* and its taxonomic significance. *FEMS Mic. Lett.*, **38**, 7–10.

Bimms, M.M., Vaughan, S., Sanyal, S.C. and Timmis, K.N. (1984), Invasive ability of *Plesiomonas shigelloides*. *Zentral. Bakt. Mik. Hyg.*, **A 257**, 345–347.

Brendan, R.A., Miller, M.A. and Janda, J.M. (1988), Clinical disease spectrum and pathogenic factors associated with *Plesiomonas shigelloides* in humans. *Rev. Inf. Dis.*, **10**, 313–316.

Canti, A.J., Lin, J.H. and Szabo, K. (1985), Overwhelming post-splenectomy infection with *Plesiomonas shigelloides* in a patient cured of Hodgkins disease. *Am. J. Clin. Path.*, **83**, 522–524.

Claesson, B.E.D., Holmlund, B.E.W., Lindhagen, C.A. and Matzsch, T.W. (1984), *Plesiomonas shigelloides* in acute cholecystitis: a case report. *J. Clin. Mic.*, **20**, 985–987.

Davis, W.A., Chretien, J.H., Garagusi, V.F. and Goldstein, M.A. (1978), Snake-to-human transmission of *Aeromonas* (PI) *shigelloides* resulting in gastroenteritis. *South. Med. J.*, **71**, 474–476.

Foster, B.G. and Rao, V.B. (1976), Isolation and characterization of *Aeromonas shigelloides* enterotoxin. *Texas J. Sci.*, **27**, 367–375.

Freund, S.M., Koburger, J.A. and Wei, C-I. (1988), Enhanced recovery of *Plesiomonas shigelloides* following an enrichment technique. *J. Fd. Prot.*, **51**, 110–112.

Gurwith, M.J. and Williams, T.W. (1977), Gastroenteritis in children: a two year review in Manitoba. I Etiology. *J. Inf. Dis.*, **136**, 239–247.

Habs, H. and Schubert, R.H.M. (1962), Uber die biochemischen merkinale und die taxonomische stellung von *Pseudomonas shigelloides* (Bader). *Zentralblatt. Bakt. Parasit. Inf. Hyg. Abt. I Orig.*, **B 186**, 316–327.

Hackney, C.R. and Dicharry, A. (1988), Seafood-borne bacterial pathogens of marine origin. *Food Tech.*, **42**(3), 104–109.

Holmberg, S.D. and Farmer, J.J. (1984), *Aeromonas hydrophila* and *Plesiomonas shigelloides* as causes of intestinal infection. *Rev. Inf. Dis.*, **6**, 633–639.

Holmberg, S.D., Wachsmuth, K., Hickman-Brenner, F.W. *et al.* (1980), *Plesiomonas* enteric infection in the United States. *Ann. Inter. Med.*, **105**, 690–694.

Hori, M., Hayaghi, K., Maeshima, K. *et al.* (1966), Food poisoning caused by *Aeromonas shigelloides* with an antigen common to *Shigella dysenteriae* 7. *J. Jap. Assoc. Infect. Dis.*, **39**, 433–444.

Humphreys, H.F., Keogh, B. and Kiane, C.T. (1986), Septicaemia and pleural infusion due to *Plesiomonas shigelloides*. *Postgrad. Med. J.*, **62**, 663–664.

Huq, M.I. and Islam, M.R. (1983), Microbiological and clinical studies in diarrhoea due to *Plesiomonas shigelloides*. *Ind. J. Med. Res.*, **77**, 793–797.

Ingram, C.W., Morrison, A.J. and Levitz, R.E. (1987), Gastroenteritis, sepsis and osteomyelitis caused by *Plesiomonas shigelloides* in an immunocompetent host: case report and review of the literature. *J. Clin. Mic.*, **25**, 1791–1793.

Janda, J.M. (1987), Effect of acidity and anti-microbial agent-like compounds on growth and survival of *Plesiomonas shigelloides*. *J. Clin. Mic.*, **25**, 1213–1215.

Johnson, W.W. and Lior, H. (1981), Cytotoxicity and suckling mouse reactivity of *Aeromonas hydrophila* isolated from human sources. *Can. J. Mic.*, **27**, 1019–1027.

Jones, D. (1988), Composition and properties of the family *Enterobacteriaceae*. *J. appl. Bact.*, **65** (Suppl.), 1–19.

Koburger, J.A. (1989), *Plesiomonas shigelloides*. In *Foodborne Bacterial Pathogens*, Doyle, M.P. (Ed). Marcel Dekker: New York.

Koburger, J.A. and Miller, M.L. (1984), *Plesiomonas shigelloides*: a new problem for the oyster industry? *Proc. 9th Ann. Trop. Subtrop. Fish Conf. Am.*, 337–342.

LeMinor, L., Chalon, A.M. and Veron, M. (1972), Recherches sur la presence de l'antigene commun des *Enterobacteriaceae* (Antigene Kunin) chez les *Yersinea*, *Levinea*, *Aeromonas*, et *Vibrio*. *Ann. Inst. Pasteur*, **123**, 761–774.

Lieh, S. (1983), Outbreak of shellfish associated gastroenteritis. *Bay County. Dept. of Health and Rehabilitation Services*. Tallahasee.

Ljungh, A. and Wadstrom, T. (1985), *Aeromonas* and *Plesiomonas* as possible sources of diarrhoea. *Infection*, **13**, 169–173.

MacDonell, M.T. and Colwell, R.R. (1985), Phylogeny of the Vibrionaceae and recommendation for two new genera, *Listonella* and *Shewanella*. *Syst. Appl. Micro.*, **6**, 171–182.

MacDonell, M.T., Swartz, D.G., Ortiz-Conde, B.A. *et al.* (1986), Ribosomal RNA phylogenies for the vibrioenteric group of eubacteria. *Micro. Sci.*, **3**, 172–178.

Martin, D.L. and Gustafsen, T.L. (1985), *Plesiomonas shigelloides* gastroenteritis in Texas. *J. Am. Med. Assoc.*, **253**, 3294–3295.

McNeeley, D., Ivy, P., Croft, J.C. and Cohen, I. (1984), *Plesiomonas*: biology of the organism and diseases in children. *Pediatr. Inf. Dis.*, **3**, 176–181.

Miller, M.L. and Koburger, J.A. (1985), *Plesiomonas shigelloides*: an opportunistic food and waterborne pathogen. *J. Fd. Prot.*, **48**, 449–457.

Miller, M.L. and Koburger, J.A. (1986a), Evaluation of inositol brilliant green bile salts and *Plesiomonas* agars for recovery of *Plesiomonas shigelloides* from aquatic samples in a seasonal survey of the Suwanee River estuary. *J. Fd. Prot.*, **49**, 274–277.

Miller, M.L. and Koburger, J.A. (1986b), Tolerance of *Plesiomonas shigelloides* to pH, sodium chloride and temperature. *J. Fd. Prot.*, **49**, 877–879.

Millership, S.E. and Chattopadhyaj, B. (1984), Methods for the isolation of *Aeromonas hydrophila* and *Plesiomonas shigelloides* from faeces. *J. Hyg.*, **92**, 145–152.

Newsome, R. and Gallais, C. (1982), Diarrheal disease caused by *Plesiomonas shigelloides*. *Clin. Mic. News*, **4**, 158–159.

Nolte, F.S., Poole, R.M., Murphy, G.W. *et al.* (1988), Proctitis and fatal septicaemia caused by *Plesiomonas shigelloides* in a bisexual man. *J. Clin. Mic.*, **26**, 388–391.

Olsvik, O., Wachsmuth, K., Kay, B.A. *et al.* (1985), Pathogenicity studies of clinical isolates of *Plesiomonas shigelloides*. *Abstract B 158*. American Society for Microbiology: Washington DC.

Pastian, M.R. and Bromel, M.C. (1984), Inclusion bodies in *Plesiomonas shigelloides*. *Appl. Environ. Mic.*, **47**, 216–218.

Paul, R., Siitonen, A. and Karkkainen, P. (1990), *Plesiomonas shigelloides* bacteremia in a healthy girl with mild gastroenteritis. *J. Clin. Mic.*, **28**, 1145–1146.

Pitarangsi, C., Echeverria, P., Whitmire, R. *et al.* (1982), Enteropathogenicity of *Aeromonas hydrophila* and *Plesiomonas shigelloides*: prevalence among individuals with and without diarrhoea in Thailand. *Inf. Imm.*, **35**, 666–673.

Rutala, W.A., Sarubbi, F.A., Finch, C.S. *et al.* (1982), Oyster-associated outbreak of diarrhoeal disease possibly caused by *Plesiomonas shigelloides*. *Lancet*, **I**, 739.

Sakata, T. and Todaka, K. (1987), Isolation of *Plesiomonas shigelloides* and its distribution in fresh water environments. *J. gen. Appl. Mic.*, **33**, 497–505.

Sanyal, S.C., Saraswathi, B. and Sharma, P. (1980), Enteropathogenicity of *Plesiomonas shigelloides*. *J. Med. Mic.*, **13**, 401–409.

Saraswathi, B., Agonwal, M.K. and Sanyal, S.C. (1983), Further studies on enteropathogenicity of *Plesiomonas shigelloides*. *Indian J. Med. Res.*, **78**, 12–18.

Sawle, G.V., Das, B.C., Acland, P.R. and Heath, D.A. (1986), Fatal infection with *Aeromonas sobria* and *Plesiomonas shigelloides*. *B. Med. J.*, **292**, 525–526.

Schubert, R.H.W. (1977), Uber den nachweiss von *Plesiomonas shigelloides* Habs und Schubert 1962 und ein elektivmedium den Inositol-Brilliantgrun-Galesalz agar. *E. Rodenwalt Arch.*, **4**, 97–103.

Schubert, R.H.W. (1981), Okologie von *Plesiomonas shigelloides*. *Zentrallblatt. Bakt. Parasiten. Infekt. Hyg. Abt. I Orig.* **B 172**, 528–533.

Schubert, R.H.W. (1984), *Plesiomonas*. In *Bergey's Manual of Systematic Bacteriology*, Vol 1, Krieg, N.R. and Holt, J.G. (Eds). Williams and Wilkins: Baltimore.

Shimada, T. and Sakazaki, R. (1978), Two epidemics of diarrheal disease possibly caused by *Plesiomonas shigelloides*. *J. Hyg.*, **80**, 275–280.

Vandepitte, J., van Damme, L., Fofana, Y. and Desmyter, J. (1980), *Edwardsiella tarda* et *Plesiomonas shigelloides*, Leur role comme agents de diarrhees et leur epidemiologie. *Bull. Soc. Path. Exat.*, **73**, 139–149.

van Loon, F.P.L., Rabin, Z., Chowbury, K.A. (1989), Case report of *Plesiomonas shigelloides* — persistent dysentery and pseudomembranous colitis. *J. Clin. Mic.*, **27**, 1913–1915.

513

von Graevenitz, A. (1980), *Aeromonas* and *Plesiomonas*. In *Manual of Clinical Microbiology*, 3rd ed., Lennette, E.H., Balows, A., Hausler, W.J. and Trust, T.-P. (Eds). American Society for Microbiology: Washington DC.

von Graevenitz, A. and Bucher, C. (1983), Evaluation of differential and selective media for the isolation of *Aeromonas hydrophila* and *Plesiomonas shigelloides* from human faeces. *J. Clin. Mic.*, **17**, 16–21.

Winson, D.K., Bloebaum, A.P. and Mathewson, J.J. (1981), Gram-negative, aerobic, enteric pathogens among intestinal microflora of wild turkey vultures (*Cathartis aura*) in West Central Texas. *Appl. Environ. Micro.*, **42**, 1123–1124.

Chapter 11

Abbott, J.D., Dale, B.A.S., Eldridge, J. *et al.* (1980), Serotyping of *Campylobacter jejuni/coli*. *J. Clin. Path.*, **33**, 762–766.

Abrams, R. (1985), Isoelectric focussing, a possible determination procedure for *Campylobacter* spp. In *Campylobacter* III, Pearson, A.D., Skirrow, M.B., Lior, H. and Rowe, B. (Eds). PHLS: London.

Altwegg, M., Bamers, A., Zollinger-Iten, J. and Penner, J.L. (1987), Problems in identification of *Campylobacter jejuni* associated with acquisition of resistance to nalidixic acid. *J. Clin. Mic.*, **25**, 1807–1808.

Anders, B.J., Paisley, J.W., Laurer, B.A. and Reller, L.B. (1982), Double-blind placebo controlled trial of erythromycin for treatment of Campylobacter enteritis. *Lancet*, **I**, 131–132.

Anon (1985), Outbreak of campylobacteriosis in a large educational institution — British Columbia. *J. Fd Prot.*, **48**, 458.

Anon (1989), *Fd. Chem. News*, **September 25**, 32–33.

Archer, D.L. and Young, F.E. (1988), Contemporary issues: diseases with a food vector. *Clin. Mic. Revs.*, **1**, 377–398.

Arimi, S.M., Fricker, C.R. and Park, R.W.A. (1986), The occurrence of Campylobacters in sewage and their removal by sewage treatment. *J. appl. Bact.*, **61**, xviii.

Arumugaswamy, R.K., Proudfoot, R.W. and Eyles, M.J. (1988), The response of *Campylobacter jejuni* and *Campylobacter coli* in the Sydney rock oyster (*Crassostrea commercialis*), during depuration and storage. *Int. J. Fd. Mic.*, **7**, 173–183.

Bar, W., Becker, K. and Hewel, C. (1989), Systemic spread of *Campylobacter jejuni* after intravenous infections. *FEMS Micro. Immunol.*, **47**, 263–270.

Benjamin, J., Leaper, S., Owen, R.J. and Skirrow, M.B. (1983), *Curr. Micro.*, **8**, 231–238.

Bernard, E., Rosen, P.M., Carles, D. *et al.* (1989), Diarrhea and *Campylobacter* infections in patients infected with human immunodeficiency virus. *J. Inf. Dis.*, **159**, 143–144.

Beuchat, L.R. (1987), Efficacy of some methods and media for detecting and enumerating *Campylobacter jejuni* in frozen chicken meat. *J. appl. Bact.*, **62**, 217–221.

Beumer, R.R., Cruysen, J.J.M. and Birtantie, I.R.K. (1988), The occurrence of *Campylobacter jejuni* in raw cows' milk. *J. Appl. Bact.*, **65**, 93–96.

Black, D. (1986), *The Plague Years: A Chronicle of AIDS — The Epidemic of Our Time*. Picador: London.

Blankenship, L.C. and Craven, S.E. (1982), *Campylobacter jejuni* survival in chicken meat as a function of temperature. *Appl. Environ. Micro.*, **44**, 88–92.

Blaser, M.J. (1985), *Campylobacter* enteritis. In *Principles and Practices of Infectious Diseases*, Mandell, G.L., Douglas, R.G. and Bennett, J.E. (Eds). John Wiley: New York.

Blaser, M.J., Berkowitz, D., LaForce, F.M. *et al.* (1979), Campylobacter enteritis: clinical and epidemiologic features. *Ann. Intern. Med.*, **91**, 179–185.

Blaser, M.J., Hardesty, H.L., Powers, B. and Wang, W.L. (1980), Survival of *Campylobacter fetus* subsp. *jejuni* in biological milieus. *J. Clin. Mic.*, **11**, 309–313.

Blaser, M.J., Wells, J.G., Feldman, R.A. *et al.* (1983), Campylobacter enteritis in the United States. *Ann. Inter. Med.*, **98**, 360–365.

Blaser, M.J. Feldman, R.A. and Wells, J.G. (1984), Epidemiology of *Campylobacter* infections. In *Campylobacter Infections in Man and Animals*, Butzler, J.P. (Ed). CRC Press: Boca Raton.

Blaser, M.J., Perez, G.P., Smith, P.F. *et al.* (1986), Extraintestinal *Campylobacter jejuni* and *Campylobacter coli* infections: host factors and strain characteristics. *J. Inf. Dis.*, **153**, 552–559.

Blaser, M.J., Smith, P.F., Repine, J.E. and Jainer, K.A. (1988), Pathogenesis of *Campylobacter fetus* infection. *J. Clin. Invest.*, **81**, 1434–1444.

Bolton, F.J. and Robertson, L. (1982), A selective medium for isolating *Campylobacter jejuni/coli*. *J. Clin. Path.*, **35**, 462–467.

Bolton, F.J. and Coates, D. (1983), A comparison of microaerobic systems for the culture of *Campylobacter jejuni* and *Campylobacter coli*. *Eur. J. Clin. Mic.*, **2**, 105–110.

Bolton, F.J., Coates, D., Hinchcliffe, P.M. and Robertson, L. (1983), Comparison of selective media for isolation of *Campylobacter jejuni/coli*. *J. Clin. Path.*, **36**, 78–83.

Bolton, F.J., Hutchinson, D.N. and Coates, D. (1984), Blood-free selective medium for isolation of *Campylobacter jejuni* from feces. *J. Clin. Mic.*, **19**, 169–171.

Bolton, F.J., Holt, V. and Hutchinson, C.N. (1985), Urease-positive thermophilic Campylobacters. *Lancet*, **I**, 1217–1218.

Bolton, F.J., Coates, D., Hutchinson, D.N. and Godfree, A.F. (1987), A study of thermophilic Campylobacters in a river system. *J. appl. Bact.*, **62**, 167–176.

Butzler, J.P. and Skirrow, M.A. (1979), *Campylobacter* enteritis. *Clin. Gast.*, **8**, 737–765.

CDC (1978), Waterborne *Campylobacter* gastroenteritis — Vermont. *Morb. Mort. Wkly. Rep.*, **27**, 207.

CDC (1983), Water-related disease outbreaks — Annual Summary. US Dept. Health and Human Services: Washington DC.

Carter, A.M., Pacha, R.E., Clark, G.W. and Williams, E.A. (1987), Seasonal occurrence of *Campylobacter* spp. in surface waters and their correlation with standard indicator bacteria. *Appl. Environ. Micro.*, **53**, 523–526.

Chevrier, D., Megraud, F., Larzul, D. and Guedson, J.-L. (1988), A new method for identifying *Campylobacter* species. *J. Inf. Dis.*, **157**, 1097–1098.

Coates, D., Hutchinson, D.N. and Bolton, F.J. (1987), Survival of thermophilic Campylobacters on fingertips and their elimination by washing and disinfection. *Epid. Inf.*, **99**, 265–274.

Cohen, D.I., Rouach, T.M. and Rogol, M. (1984), *Campylobacter* enteritis outbreak in a military base in Israel. *Israeli J. Med. Sci.*, **20**, 216–218.

Coid, C.R., O'Sullivan, A.M. and Dore, C.T. (1987), Variations in the virulence for pregnant guinea-pigs of *Campylobacter* isolated from man. *J. Med. Mic.*, **23**, 187–189.

Costas, M. and Owen, R.J. (1983), The classification and identification of Campylobacters using total protein profiles. In *Campylobacter II*, Pearson, A.D., Skirrow, M.B., Rowe, B. *et al.* (Eds). PHLS: London.

Cuk, Z., Annan-Prah, A., Janc, M. and Zajc-Satler, J. (1987), Yogurt: an unlikely source of *Campylobacter jejuni/coli. J. Appl. Bact.*, **63**, 201–205.

Dawkins, H.C., Bolton, F.J. and Hutchinson, D.N. (1984), A study of the spread of *Campylobacter jejuni* in four large kitchens. *J. Hyg.*, **92**, 357–364.

DeMélo, M.A. and Pechere, J.-C. (1990), Identification of *Campylobacter jejuni* surface proteins that bind to eukaryotic cells in vitro. *Inf. Imm.*, **58**, 1749–1756.

Doyle, M.P. and Roman, D.J. (1982), Recovery of *Campylobacter jejuni* and *Campylobacter coli* from inoculated foods by selective enrichment. *Appl. Environ. Micro.*, **43**, 1343–1353.

Dubreuil, J.P., Logan, S.M., Cubbage, S. *et al.* (1988), Structural and biochemical analysis of a surface array protein of *Campylobacter fetus. J. Bact.*, **170**, 4165–4173.

Edmonds, P., Patton, C.M., Griffen, P.M. *et al.* (1987), *Campylobacter hyointestinalis* associated with human disease in the United States. *J. Clin. Mic.*, **25**, 685–691.

Elharrif, Z and Megraud, F. (1986), Characterization of thermophilic *Campylobacter*. I. Carbon-substrate utilization tests. *Curr. Mic.*, **13**, 117–122.

Fennell, C.L., Totten, P.L., Quinn, T.C. *et al.* (1986), Isolation of '*Campylobacter hyointestinalis*' from an human. *J. Clin. Mic.*, **24**, 146–148.

Fernie, D.S. and Park, R.W.A. (1977), The isolation and nature of Campylobacters (microaerophilic Vibrios) from laboratory and wild rodents. *J. Med. Mic.*, **10**, 325–329.

Ferrero, R.L. and Lee, A. (1988), Motility of *Campylobacter jejuni* in a viscous environment: comparison with conventional rod-shaped bacteria. *J. Gen. Mic.*, **134**, 53–59.

Figura, N. and Guglielmetti, N. (1988), Clinical characteristics of *Campylobacter jejuni* and *Campylobacter coli*. *Lancet*, **I**, 942–943.

Fox, J.G., Taylor, N.S., Penner, J.L. *et al.* (1989), Investigation of zoonotically acquired *Campylobacter jejuni* enteritis with serotyping and restriction endonuclease DNA analysis. *J. Clin. Mic.*, **27**, 2423–2425.

Franco, D.A. (1988), *Campylobacter* species: considerations for controlling a foodborne pathogen. *J. Fd. Prot.*, **81**, 145–153.

Fricker, C.R. (1984), Procedures for the isolation of *Campylobacter jejuni* and *Campylobacter coli* from poultry. *Int. J. Fd. Mic.*, **1**, 149–154.

Fricker, C.R. (1985), A note on the effect of different storage procedures on the ability of Preston medium to recover campylobacters. *J. appl. Bact.*, **58**, 57–62.

Fricker, C.R. (1986), Campylobacter serotyping on shipped antigen extracts. *Lancet*, **I**, 554.

Fricker, C.R. (1987), The isolation of Salmonellas and Campylobacters: a review. *J. Appl. Bact.*, **63**, 99–116.

Fricker, C.R. and Park, R.W.A. (1987), Application of co-agglutination, passive haemagglutination and competitive ELISA to the study of Campylobacters. In *Immunological Techniques in Microbiology*, Grange, J.M. and Fox, A. (Eds). Blackwell Scientific Publishing: Oxford.

Fricker, C.R. and Park, R.W.A. (1989), A two-year study of the distribution of 'thermophilic' Campylobacters in human, environmental and food samples from the Reading area with particular reference to toxin production and serotype. *J. Appl. Bact.*, **66**, 477–490.

Fricker, C.R., Uradzinski, J., Alemohammad, M.M. *et al.* (1986), Serotyping of Campylobacters by coagglutination on the basis of heat-stable antigens. *J. Med. Mic.*, **21**, 83–86.

Fricker, C.R., Alemohammad, M.M. and Park, R.W.A. (1987), A study of factors affecting the sensitivity of the passive haemagglutination method for serotyping *Campylobacter jejuni* and *Campylobacter coli* and recommendations for a more rapid procedure. *Can. J. Mic.*, **33**, 33–39.

Goossens, H. and Butzler, J.-P. (1989), Letter to Editor: Isolation of *Campylobacter upsaliensis* from stool specimens. *J. Clin. Mic.*, **27**, 2143–2144.

Goossens, H., DeBoeck, M., Cosgnau, H. *et al.* (1986), Modified selective medium for isolation of *Campylobacter* spp. from faeces: comparison with Preston medium, a blood-free medium and a filtration system. *J. Clin. Mic.*, **24**, 840–843.

Grau, F.H. (1988), *Campylobacter jejuni* and *Campylobacter hyointestinalis* in the intestinal tract of calves and cattle. *J. Fd. Prot.*, **51**, 857–961.

Guerrant, R.L., Pennie, R.A., Barrett, L.J. and O'Brien, A. (1985), Studies of a cytotoxin for *Campylobacter jejuni*. In *Campylobacter* III, Pearson, A.D., Skirrow, M.B., Lior, H. and Rowe, B. (Eds). PHLS: London.

Hanninen, M.-L., Korkeala, H. and Pakkala, P. (1984), Effect of gas atmospheres on the growth and survival of *Campylobacter jejuni* on beef. *J. Appl. Bact.*, **57**, 89–94.

Hartog, B.J., DeWild, G.J.A. and DeBoer, E. (1983), Prevalence of *Campylobacter jejuni* in chicken wings. *Arch. Lebbensmittelhyg.*, **34**, 116–122.

Harvey, S.M. (1980), Hippurate hydrolysis by *Campylobacter fetus. J. Clin. Mic.*, **11**, 435–437.

Hebert, G.A., Hollis, D.G., Weaver, R.E. *et al.* (1982), 30 years of Campylobacters: biochemical features and a biotyping proposal for *Campylobacter jejuni*. *J. Clin. Mic.*, **15**, 1065–1073.

Hebert, G.A., Edmonds, P. and Brenner, D.J. (1984), DNA relatedness among strains of *Campylobacter jejuni* and *Campylobacter coli* with divergent serogroup and hippurate reactions. *J. Clin. Mic.*, **20**, 138–140.

Hodinka, R.L. and Gilligan, P.H. (1988), Evaluation of the Campyslide agglutination kit for confirmation of identification of selected *Campylobacter* species. *J. Clin. Mic.*, **26**, 47–49.

Hoffman, P., Krieg, N.R. and Smibert, R.M. (1979), Studies of the microaerophilic nature of *Campylobacter fetus* subsp. *jejuni*. I. Physiological aspects of advanced aerotolerance. *Can. J. Mic.*, **25**, 1–7.

Hudson, W.R. and Roberts, T.A. (1982), The occurrence of *Campylobacter jejuni* on commercial red meat carcases from an abattoir. In *Campylobacter: Epidemiology, Pathogenesis, Biochemistry*, Newell, D.G. (Ed). MTP Press: Lancaster.

Hughdahl, M.B. and Doyle, M.P. (1985), Chemotactic behaviour of *Campylobacter jejuni*. In *Campylobacter* III, Pearson, A.D., Skirrow, M.B., Lior, H. and Rowe, B. (Eds). PHLS: London.

Hughdahl, M.B., Beery, J.T. and Doyle, M.P. (1988), Chemotactic behaviour of *Campylobacter jejuni*. *Inf. Imm.*, **56**, 1560–1566.

Humphrey, T.J. (1986a), Techniques for the optimum recovery of cold injured *Campylobacter jejuni* from milk or water. *J. Appl. Bact.*, **61**, 125–137.

Humphrey, T.J. (1986b), Injury and recovery in freeze- or heat-damaged *Campylobacter jejuni*. *Lett. Appl. Mic.*, **3**, 81–84.

Humphrey, T.J. (1989), An appraisal of the efficacy of pre- enrichment for the isolation of *Campylobacter jejuni* from water and food. *J. Appl. Bact.*, **66**, 119–126.

Humphrey, T.J. and Cruickshank, J.G. (1985), Antibiotic and desoxycholate resistance in *Campylobacter jejuni* following freezing and heating. *J. Appl. Bact.*, **59**, 65–71.

Humphrey, T.J. and Beckett, P. (1987), *Campylobacter* in dairy cows and in raw milk. *Epid. Inf.*, **98**, 263–269.

Humphrey, T.J. and Muscat, I. (1989), Incubation temperature and the isolation of *Campylobacter jejuni* from water and food. *Lett. Appl. Mic.*, **9**, 137–139.

Hutchinson, D.N., Bolton, F.J., Jelley, W.C.N. *et al.* (1985a), *Campylobacter* enteritis associated with raw goat's milk. *Lancet*, **I**, 1037–1038.

Hutchinson, D.N., Bolton, F.J., Hinchcliffe, P.M. *et al.* (1985b), Evidence of udder excretion of *Campylobacter jejuni* as the cause of milkborne *Campylobacter* outbreak. *J. Hyg.*, **94**, 205–215.

Hutchinson, D.N., Bolton, F.J., Jones, D.M. *et al.* (1987), Application of three typing schemes (Penner, Lior, Preston) to strains of *Campylobacter* spp. isolated from three outbreaks. *Epid. Inf.*, **98**, 139–144.

Illingworth, D.S. and Fricker, C.R. (1987), Rapid serotyping of Campylobacters based on heat-stable antigens using a combined passive haemagglutination/co-agglutination technique. *Lett. Appl. Mic.*, **5**, 61–63.

Johnson, W.M. and Lior, H. (1984), Toxins produced by *Campylobacter jejuni* and *Campylobacter coli*. *Lancet*, **I**, 229–230.

Juven, B.J., Kanner, J., Weisslowicz, H. and Havel, S. (1988), Effect of ascorbic and isoascorbic acids on survival of *Campylobacter jejuni* in poultry meat. *J. Fd. Prot.*, **51**, 436–437.

Kakoyiannis, C.K., Winter, D.J. and Marshall, R.B. (1988), The relationship between intestinal *Campylobacter* species isolated from animals as determined by BRENDA. *Epid. Inf.*, **100**, 379–387.

Kaldor, J. and Speed, J.R. (1984), Gullain-Barre syndrome and *Campylobacter jejuni*: a serological study. *B. Med. J.*, **288**, 1867–1870.

Karmali M.A.and Fleming, P.C. (1979), Application of the Fortner principle to isolation of *Campylobacter* from stools. *J. Clin. Mic.*, **10**, 245–247.

Karmali, M.A., Penner, J.L., Fleming, P.C. *et al.* (1983), The serotype and biotype distribution of clinical isolates of *Campylobacter jejuni* and *Campylobacter coli* over a three year period. *J. Inf. Dis.*, **147**, 243–246.

Kielbauch, J.A., Albach, R.A., Baum, L.G. and Chang, K.P. (1985), Phagocytosis of *Campylobacter jejuni* and its intracellular survival in mononuclear phagocytes. *Inf. Imm.*, **48**, 446–451.

Klipstein, F.A., Engert, R.F., Shat, H. and Schenk, E.A. (1985), Pathogenic properties of *Campylobacter jejuni*: assay and correlation with clinical manifestation. *Inf. Imm.*, **50**, 43–49.

Koidis, P. and Doyle, M.P. (1984), Procedure for increased recovery of *Campylobacter jejuni* from inoculated unpasteurized milk. *Appl. Environ. Micro.*, **44**, 1150–1153.

Konkel, M.E. and Joens, L.A. (1989), Adhesion to and invasion of HEp-2 cells by *Campylobacter* spp. *Inf. Imm.*, **57**, 2984–2990.

Korolik, V., Colae, P.J. and Krishnapilla, V. (1988), A specific DNA probe for the identification of *Campylobacter jejuni*. *J. Gen. Mic.*, **13**, 521–529.

Kosunen, T.U., Danielsson, D. and Kjellander, J. (1980), Serology of *Campylobacter* ss *jejuni*, ('Related' Campylobacters). *Acta Path. Mic. Scand.*, **B 88**, 207–218.

Lanmerding, A.M., Garcia, M.M., Mann, E.D. *et al.* (1988), Prevalence of *Salmonella* and thermophilic *Campylobacter* in fresh pork, beef, veal and poultry in Canada. *J. Fd. Prot.*, **51**, 47–52.

Lastovica, A.J., LeRoux, E. and Penner, J.L. (1989), '*Campylobacter upsaliensis*' isolated from blood cultures of pediatric patients. *J. Clin. Mic.*, **27**, 657–659.

Laughan, B.E., Vernon, A.A., Druckmann, D.A. *et al.* (1988), Recovery of *Campylobacter* species from homosexual men. *J. Inf. Dis.*, **158**, 464–467.

Lauwers, S., Vloes, L. and Butzler, J.P. (1981), *Campylobacter* serotyping and epidemiology. *Lancet*, **I**, 158–159.

Lee, A., O'Rourke, J.L., Phillips, M. and Barrington, P.J. (1983), *Campylobacter jejuni* as a mucosa-associated organism; an ecological study. In *Campylobacter* II, Pearson, A.D., Skirrow, M.B., Rowe, B. *et al.* (Eds). PHLS: London.

Lee, A., O'Rourke, J.L., Barrington, P.J. and Trust, T.J. (1986), Mucus colonization as a determinant of pathogenicity in intestinal infection by *Campylobacter jejuni*: a mouse cecal model. *Inf. Imm.*, **51**, 536–546.

Lior, H. (1984a), Serotyping of *Campylobacter jejuni* and *Campylobacter coli* by slide agglutination based on heat-labile antigenic factors. In *Campylobacter Infection in Man and Animals*, Butzler, J.P. (Ed). CRC Press: Boca Raton.

Lior, H. (1984b), New extended biotyping scheme for *Campylobacter jejuni*, *Campylobacter coli* and '*Campylobacter laridis*'. *J. Clin. Mic.*, **15**, 761–768.

Lior, H. and Butzler, J.P. (1986), Serotyping *Campylobacter*. *Lancet*, 320.

Lior, H., Woodward, D.L., Edgar, J.A. *et al.* (1982), Serotyping of *Campylobacter jejuni* by slide agglutination based on heat-labile antigenic factors. *J. Clin. Mic.*, **15**, 761–768.

516

Luechtefeld, N.W., Wang, W.L., Blaser, M.J. and Reller, L.B. (1981), Evaluation of transport and storage techniques for isolation of *Campylobacter fetus* subsp. *jejuni* from turkey faecal specimens. *J. Clin. Mic.*, **13**, 438–443.

Malnick, H., Wiliams, K., Phil-Ebosie, J. and Levy, A.S. (1990), Description of a medium for isolating *Anaerobiospirillum* spp., a possible cause of zoonotic disease, from diarrheal feces and blood of humans and use of the medium in a survey of human, canine and feline feces. *J. Clin. Mic.*, **28**, 1380–1384.

Mancinelli, S., Palombi, L., Riccardi, F. and Morozzi, M.C. (1987), Serological study of *Campylobacter jejuni* infection in slaughterhouse workers. *J. Inf. Dis.*, **156**, 856.

Marinescu, M., Festy, B., Derimos, R. and Megraud, F. (1987), High frequency of isolation of *Campylobacter coli* from poultry meat in France. *Eur. J. Clin. Mic.*, **6**, 693–695.

Maurer, S.L. (1988a), Campylobacters in man and the environment in Hull and East Yorkshire. *Epid. Inf.*, **101**, 287–294.

Maurer, S.L. (1988b), The pathogenicity of environmental campylobacters — a human volunteer experiment. *Epid. Inf.*, **101**, 295–300.

McCardell, B.A. and Madden, J.M. (1985), Effect of iron concentration on toxin production by *Campylobacter jejuni* and *Campylobacter coli*. In *Campylobacter* III, Pearson, A.D., Skirrow, M.B., Lior, H. and Rowe, B. (Eds). PHLS: London.

McCardell, B.A., Madden, J.M. and Lee, E.C. (1984), *Campylobacter jejuni* and *Campylobacter coli* production of a cytotonic toxin immunologically similar to cholera toxin. *J. Fd. Prot.*, **47**, 943–949.

McManus, C. and Lanier, J.M. (1987), *Salmonella*, *Campylobacter jejuni* and *Yersinia enterocolitica* in raw milk. *J. Fd. Prot.*, **50**, 51–55.

McNulty, C.A.M. and Dent, J.C. (1987), Rapid identification of *Campylobacter pylori* (*C. pyloridis*) by pre-formed enzymes. *J. Clin. Mic.*, **25**, 1685–1686.

Megraud, F. (1987), Isolation of *Campylobacter* spp. from pigeon feces by a combined enrichment-filtration technique. *Appl. Environ. Mic.*, **53**, 1394–1395.

Megraud, F., Chevrier, D., Desplaces, N. *et al.* (1988), Urease-positive thermophilic *Campylobacter* (*Campylobacter laridis* variant) isolated from an appendix and from human faeces. *J. Clin. Mic.*, **26**, 1050–1051.

Metzing, L.O. (1981), Waterborne outbreaks of *Campylobacter* enteritis in Sweden. *Lancet*, **II**, 352–354.

Molbak, K., Hojlyng, N. and Gaarslev, K. (1988), High prevalence of Campylobacter excretors related to environmental contamination. *Epid. Inf.*, **100**, 227–237.

Morris, J.A. and Park, R.W.A. (1973), A comparison using gel electrophoresis of cell proteins of Campylobacters (Vibrios) associated with infertility, abortion and swine dysentery. *J. Gen. Mic.*, **78**, 165–178.

Moss, C.W., Lambert-Fair, A.M.A., Nicholson, M.A. and Guerant, G.O. (1990), Isoprenoid quinones of *Campylobacter cryoaerophila*, *C. cinaedi*, *C. hyointestalinis*, *C. pylori* and '*C. upsaliensis*'. *J. Clin. Mic.*, **28**, 395–397.

Munroe, D.L., Prescott, J.F. and Penner, J.L. (1983), *Campylobacter jejuni* and *Campylobacter coli* serotypes isolated from chickens, cattle and pigs. *J. Clin. Mic.*, **18**, 877–881.

Nachamkin, I., Stowell, C., Skalina, D. *et al.* (1984), *Campylobacter laridis* causing bacteremia in an immunocompromised host. *Ann. Intern. Med.*, **101**, 55–57.

Naess, V., Johannessen, A.C. and Hofstad, T. (1983), Adherence of *Campylobacter jejuni* to porcine brushborders. In *Campylobacter* II, Pearson, A.D., Skirrow, M.B., Rowe, B. *et al.* (Eds). PHLS: London.

Newell, D.G. and Pearson, A.D. (1984), The invasion of epithelial cell lines and the intestinal epithelium of infant mice by *Campylobacter jejuni/coli*. *J. Diarr. Dis. Res.*, **2**, 19–26.

Newell, D.G., McBride, H., Saunders, F. *et al.* (1985), The virulence of clinical and environmental isolates of *Campylobacter jejuni*. *J. Hyg.*, **94**, 45–54.

Ng, V.L., Hadley, W.K., Fennell, C.L. *et al.* (1987), Successive bacteraemias with '*Campylobacter cinaedi*' and '*Campylobacter fennelliae*' in a bisexual male. *J. Clin. Mic.*, **25**, 2008–2009.

Oosterom, J. (1983), *Campylobacter jejuni* in poultry processing and pig slaughtering. In *Campylobacter* II, Pearson, A.D., Skirrow, M.B., Rowe, B. *et al.* (Eds). PHLS: London.

Oosterom, J., den Uyl, C.H., Banffer, J.R.J. and Huisman, J. (1984), Epidemiological investigations on *Campylobacter jejuni* in households with a primary infection. *J. Hyg.*, **93**, 325–332.

Owen, R.J., Costas, M., Sloss, L. and Bolton, F.J. (1988), Numerical analysis of electrophoretic protein patterns of *Campylobacter laridis* and allied thermophilic campylobacters from the natural environment. *J. Appl. Bact.*, **65**, 69–78.

Owen, R.J., Costas, M. and Dawson, C. (1989), Application of different chromosomal DNA restriction digest fingerprints to specific and subspecific identification of *Campylobacter* isolates. *J. Clin. Mic.*, **27**, 2338–2343.

Pacha, R.E., Clark, G.W., Williams, E.A. *et al.* (1987), Small rodents and other mammals associated with mountain meadows as reservoirs of *Giardia* spp. and *Campylobacter* spp. *Appl. Environ. Micro.*, **53**, 1574–1579

Pai, C.H., Gillis, F., Toumanen, E. and Monks, M.I. (1983), Erythromycin treatment of Campylobacter enteritis in children. *Am. J. Inf. Dis. Child.*, **137**, 286–288.

Palmer, S.R., Gully, P.R., White, J.M. *et al.* (1983), Waterborne outbreak of *Campylobacter* gastroenteritis. *Lancet*, **I**, 287–290.

Pang, T., Wong, P.Y., Pathucheary, S.D. *et al.* (1987), *In vitro* and *in vivo* studies of a cytotoxin from *Campylobacter jejuni*. *J. Med. Mic.*, **23**, 193–198.

Park, C.E., Stankiewicz, Z.K., Lovett, J. *et al.* (1983), Effect of temperature, duration of incubation, and pH of enrichment culture on recovery of *Campylobacter jejuni* from eviscerated market chickens. *Can. J. Mic.*, **19**, 803–806.

Park, C.E., Smibert, R.M., Blaser, M.J. *et al.* (1984), *Campylobacter*. In *Compendium of Methods for the Microbiological Examination of Foods*, Speck, M.L. (Ed). American Public Health Association: Washington DC.

Pearson, A.D., Knott, J.R., Suckling, W.G. *et al.* (1983), A serodiagnostic study of *Campylobacter* infection in 450 children with suspected appendicitis. In *Campylobacter* II, Pearson, A.D., Skirrow, M.B., Rowe, B. *et al.* (Eds). PHLS: London.

Pearson, A.D., Hooper, W.L., Lior, H. *et al.* (1985), Why investigate sporadic cases? The significance of fresh and New York dressed chickens as a source of infection. In *Campylobacter* III, Pearson, A.D., Skirrow, M.B., Lior, H. and Rowe, B. (Eds). PHLS: London.

Pearson, A.D. (1986), Cited by Walker *et al.*, 1986.

Penner, J.L. (1988), The genus *Campylobacter*: a decade of progress. *Clin. Mic. Revs.*, **1**, 157–172.

Penner, J.L. and Hennessy, J.N. (1980), Passive hae-magglutination technique for serotyping *Campylobacter fetus* subsp. *jejuni* on the basis of heat-stable antigens. *J. Clin. Mic.*, **12**, 732–737.

Penner, J.L, Hennessy, J.N. and Congi, R.V. (1983), Serotyping of *Campylobacter jejuni* and *Campylobacter coli* on the basis of heat-stable antigens. *Eur. J. Clin. Mic.*, **2**, 378–380.

Perez-Perez, G.I., Cohn, D.L., Guerrant, R.L. *et al.* (1989), Clinical and immunologic significance of cholera-like toxin and cytotoxin production by *Campylobacter* species in patients with acute inflammatory diarrhoea in the United States. *J. Inf. Dis.*, **160**, 460–468.

Pignata, C., Gaundalini, S., Guarino, A. *et al.* (1984), Chronic diarrhea and failure to thrive in an infant with *Campylobacter jejuni*. *J. Ped. Gast. Nutr.*, **3**, 812–814.

Pinegar, J.A. and Suffield, A. (1982), The investigation of food poisoning outbreaks in England and Wales. In *Isolation and Identification Methods for Food Poisoning Organisms*, Corry, J.E.L., Roberts, D. and Skinner, F.A. (Eds). Academic Press: London and New York.

Popovic-Uroic, T. (1989), *Campylobacter jejuni* and *Campylobacter coli* diarrhoea in rural and urban populations in Yugoslavia. *Epid. Inf.*, **102**, 59–67.

Popovic-Uroic, T., Gmajnicki, B., Kalenic, S. and Vodopija, I. (1988), Clinical comparison of *Campylobacter jejuni* and *Campylobacter coli* diarrhoea. *Lancet*, **I**, 176–177.

Porter, I.A. and Reid, T.M.S. (1980), A milk-borne outbreak of *Campylobacter* infection. *J. Hyg.*, **84**, 415–419.

Preston, M.A. and Penner, J.L. (1989), Characterisation of cross-reacting serotypes of *Campylobacter jejuni*. *Can. J. Mic.*, **35**, 265–273.

Ray, B. and Johnson, C. (1984), Sensitivity of cold-stressed *Campylobacter jejuni* to solid and liquid enrichments. *Fd. Mic.*, **1**, 173–176.

Robinson, D.A. (1981), Infective dose of *Campylobacter jejuni* in milk. *B. Med. J.*, **282**, 1584.

Rogol, M. and Sechter, I. (1987), Serogroups of thermo-philic Campylobacters from human and from non-human sources, Israel 1982–1985. *Epid. Inf.*, **99**, 275–282.

Rogol, M., Sechter, Z., Greenberg, R. *et al.* (1985), Contamination of chicken meat and environment with various serogroups of *Campylobacter jejuni/coli*. *Int. J. Fd. Mic.*, **1**, 271–276.

Romero, S., Archer, J.R., Hamacher, M.E. *et al.* (1988), Case report of an unclassified microaerophilic bacterium associated with gastroenteritis. *J. Clin. Mic.*, **26**, 142–143.

Rosef, O. (1981), Isolation of *Campylobacter fetus* subsp. *jejuni* from the gallbladder of normal slaughter pigs using an enrichment procedure. *Acta Vet. Scand.*, **22**, 149–151.

Ruiz-Palacios, G.M., Torres, J., Torres, N.I. *et al.* (1983), Cholera-like enterotoxin produced by *Campylobacter jejuni*: characterization and clinical significance. *Lancet*, **II**, 250–251.

Saloma, S.M., Bolton, F.J. and Hutchinson, D.N. (1990), Application of a new phagetyping scheme for campylo-bacters isolated during outbreaks. *Epid. Inf.*, **104**, 405–411.

Sandstedt, K. (1982), *Campylobacter* transmission via food, especially milk. *Statens vet Anstalt*, 225–226.

Simmons, N.A. (1977), Isolation of Campylobacters. *B. Med. J.*, **II**, 707.

Simon, A.E, Karmali, M.A., Tabavji, T. and Roscoe, M. (1986), Abortion and perinatal sepsis associated with *Campylobacter* infections. *Rev. Inf. Dis.*, **8**, 397–402.

Sjogren, E., Ruiz-Palacios, G. and Kaijser, B. (1989), *Campylobacter jejuni* isolated from Mexican and Swedish patients with repeated symptomatic and/or asympto-matic diarrhoea episodes. *Epid. Inf.*, **102**, 45–57.

Skirrow, M.B. (1977), Campylobacter enteritis — a 'new' disease. *B. Med. J.*, **2**, 9–11.

Skirrow, M.B. (1984), *Campylobacter* infections of man. In *Medical Microbiology*, Vol 4, Easmonn, C.S.F. and Jeljaszewikz, J. (Eds). Academic Press: London.

Skirrow, M.B. (1987), A demographic survey of *Campylobacter*, *Salmonella* and *Shigella* infection in England. *Epid. Inf.*, **99**, 647–657.

Skirrow, M.B. (1989), Campylobacter perspectives. *PHLS Micro. Dig.*, **6**, 113–117.

Skirrow, M.B. and Benjamin, J. (1980), Differentiation of enteropathogenic Campylobacters. *J. Clin. Path.*, **33**, 1122.

Skirrow, M.B. and Benjamin, J. (1982), The classification of 'thermophilic' Campylobacters and their distribution in man and domestic animals. In *Campylobacter: Epidemi-ology, Pathogenesis and Biochemistry*, Newell, D.G. (Ed). PHLS Campylobacter Workshop. MTP Press: Lancaster.

Skirrow, M.B., Benjamin, J., Razi, M.H.H. and Waterman, S. (1982), Isolation , cultivation and identification of *Campylobacter jejuni* and *Campylobacter coli*. In *Isolation and Identification Methods for Food Poisoning Organisms*, Corry, J.E.L., Roberts, D. and Skinner, F.A. (Eds). Academic Press: London and New York.

Smibert, R.M. (1984), *Campylobacter*. In *Bergey's Manual of Systematic Bacteriology*, Vol 1, Krieg, N.R. and Holt, J.G. (Eds). Williams and Wilkins: Baltimore and London.

Sninsky, C.A., Ramphal, R., Gaskins, D.T. *et al.* (1985), Alterations of myoelectric activity associated with *Campylobacter jejuni* and its cell-free filtrate in the small intestine of rabbits. *Gastroenterology*, **89**, 337–344.

Sorqvist, S. (1989), Heat resistance of *Campylobacter* and *Yersinia* strains by three methods. *J. Appl. Bact.*, **67**, 543–549.

Steele, T.W. and McDermott, S. (1978), Campylobacter enteritis in South Australia. *Med. J. Aust.*, **2**, 404–406.

Steele, T.W. and McDermott, S.N. (1984), The use of membrane filters applied directly to the surface of agar plates for the isolation of *Campylobacter jejuni* from faeces. *Pathology*, **16**, 263–265.

Steele, T.W. and Owen, R.J. (1988), *Campylobacter doylei* subsp. *doylei* subsp. nov., a subspecies of nitrate-negative campylobacters isolated from human clinical specimens. *Int. J. Syst. Bact.*, **38**, 316–318.

Steele, T.W., Sangster, N. and Lanser, J.A. (1985), DNA relatedness and biochemical features of *Campylobacter* spp. isolated in Central and South Australia. *J. Clin. Mic.*, **23**, 71–74.

Stern, N.J. and Kotula, A.W. (1982), Survival of *Campylobacter jejuni* inoculated into ground beef. *Appl. Environ. Mic.*, **44**, 1150–1153.

Stern, N.J., Hernadez, M.P., Blankenship, L. *et al.* (1985), Prevalence and distribution of *Campylobacter jejuni* and *Campylobacter coli* in retail meats. *J. Fd. Prot.*, **48**, 595–599.

Tauxe, R.V. Deming, M.S. and Blake, P.A. (1985), *Campylobacter jejuni* infections on a college campus: a national survey. *Am. J. Pub. Hlth.*, **75**, 659–660.

Taylor, D.N., McDermott, K.T., Little, J.R. *et al.* (1983), *Campylobacter* enteritis from untreated water in the Rocky Mountains. *Ann. Intern. Med.*, **99**, 38–40.

Taylor, D.N., Echeverria, P., Pitarangsi, C. *et al.* (1988), Influence of strain characteristics and immunity on the epidemiology of *Campylobacter* infection in Thailand. *J. Clin. Mic.*, **25**, 150–151.

Tee, W., Dyall-Smith, M. and Dwyer, B. (1988), *Campylobacter cryoaerophila* isolated from a human. *J. Clin. Mic.*, **26**, 2469–2473.

Tenover, F.C., Conlon, L., Barbagallo, S. and Nachamkin, I. (1990), DNA probe culture confirmation assay for identification of thermophilic *Campylobacter* species. *J. Clin. Mic.*, **28**, 1284–1287.

Thompson, L.M., Smibert, R.M., Johnson, J.L. and Krieg, N.R. (1988), Phylogenetic study of the genus *Campylobacter. Int. J. Syst. Boct.*, **38**, 190–200.

Totten, P.A., Fennell, C.L., Tenover, F.C. *et al.* (1985), *Campylobacter cinaedi* (sp. nov.) and *Campylobacter fennelliae* (sp. nov.): two new *Campylobacter* species associated with enteric disease in homosexual men. *J. Inf. Dis.*, **151**, 131–139.

Totten, P.A., Nennell, C.L., Tenover, F.C. *et al.* (1987), Prevalence and characterisation of hippurate-negative *Campylobacter jejuni* in King County, Wsahington. *J. Clin. Mic.*, **25**, 1747–1752.

Vaira, D., D'Anastasio, C., Holton, J. *et al.* (1988), *Campylobacter pylori* in abattoir workers. Is it a zoonosis? *Lancet*, **II**, 725–726.

Vandamme, P., Falsen, E. and Pot, B. (1989), Identification of EF Group 22 campylobacters from gastroenteritis cases as *Campylobacter conciscus. J. Clin. Mic.*, **27**, 1775–1781.

Vandamme, P., Falsen, E. and Pot, B. (1990), Identification of *Campylobacter cinaedi* isolated from blood and faeces of children and adults. *J. Clin. Mic.*, **28**, 1016–1026.

Vogt, R.L., Little, A.A., Patton, C.M. *et al.* (1984), Serotyping and serology studies of Campylobacteriosis associated with consumption of raw milk. *J. Clin. Mic.*, **20**, 998–1000.

Walker, R.I., Caldwell, M.B., Lee, E.C. (1986), Pathophysiology of *Campylobacter* enteritis. *Micro. Res.*, **50**, 81–94.

Walsley, S.L. and Karmali, M.A. (1989), Direct isolation of atypical thermophilic *Campylobacter* species from human feces on selective agar medium. *J. Clin. Mic.*, **27**, 668–670.

Waterman, S.C. (1982), The heat sensitivity of *Campylobacter jejuni* in milk. *J. Hyg.*, **88**, 529–533.

Waterman, S,C., Park, R.W.A. and Bramley, A.J. (1984), A search for the source of *Campylobacter jejuni* in milk. *J. Hyg.*, **93**, 333–337.

Weinberg, E,D. (1984), Pregnancy-associated depression of cell-mediated immunity. *Rev. Inf. Dis.*, **6**, 814–831.

Whelan, C.D., Monaghan, P., Girdwood, R.W.A. and Fricker, C.R. (1988), The significance of wild birds (*Larus* sp.) in the epidemiology of *Campylobacter* infections in humans. *Epid. Inf.*, **101**, 259–267.

Wong, K.H., Skelton, S.K. and Feeley, J.C. (1986), Strain characterisation and grouping of *Campylobacter jejuni* and *Campylobacter coli* by interaction with lectins. *J. Clin. Mic.*, **23**, 407–410.

Yeen, W.P., Pathucheary, S.D. and Pang, T. (1983), Demonstration of a cytotoxin from *Campylobacter jejuni. J. Clin. Path.*, **36**, 1237–1240.

Zei, P. and Blaser, M.J. (1990), Pathogenesis of *Campylobacter fetus* infection. Role of surface array proteins in a mouse model. *J. Clin. Invest.*, **85**, 1036–1043.

Chapter 12

Attaie, R., Whalen, P.J., Shahani, K.M. and Amer, M.A. (1987), Inhibition of growth of *Staphylococcus aureus* during production of acidophilus yogurt. *J. Fd. Prot.*, **50**, 224–228.

Amstberg, G. (1979), Vergleichende biochemische und serologische untersuchungen an staphylokokken von schweinen und rindern unter besanderer berucksichtisung von *Staphylococcus hysicus* bzun *Staphylococcus epidermis* biotyp 2. *Zentralblatt fur Veterinamedizin*, **B 26**, 137–152.

Anderson, R. (1986), Inhibition of *Staphylococus aureus* and spheroplasts of Gram-negative bacteria by an antagonistic compound produced by a strain of *Lactobacillus plantarum. Int. J. Fd. Mic.*, **3**, 149–160.

Anderson, P.H.R. and Stone, D.M. (1955), Staphylococcus food poisoning associated with spray dried milk. *J. Hyg.*, **53**, 387–397.

Anon (1984), *Food-borne and water-borne disease in Canada, Annual Summary* 1978. Health Protection Branch, Health and Welfare Canada: Ottawa.

AOAC (1984), *Official Methods of Analysis*, 14th ed. Williams, S. (Ed). AOAC: Arlington.

Armigo, R., Henderson, D.A., Timothee, R. and Robinson, H.B. (1957), Food poisoning outbreaks associated with spray-dried milk. *Am. J. Pub. Hlth.*, **47**, 1093–1100.

Ayaz, M., Luedecke, L.O. and Branen, A.L. (1980), Antimicrobial effect of butylated hydroxytoluene on *Staphylococcus aureus. J. Fd. Prot.*, **43**, 4–6.

Baird-Parker, A.C. (1962), An improved diagnostic medium for isolating coagulase-positive staphylococci. *J. Appl. Bact.*, **25**, 12–69.

Baird-Parker, A.C. (1966), Methods for classifying staphylococci and micrococci. In *Identification Methods for Microbiologists*, Part A, Gibbs, B.M. and Skinner, F.A. (Eds). Academic Press: London and New York.

Baird-Parker, A.C. (1979), Methods for identifying staphylococci and micrococci In *Identification Methods for Microbiologists*, Skinner, F.A. and Lovelock, D.W. (Eds). Academic Press: London and New York.

Baker, J.S. (1984), Comparison of various methods for differentiation of staphylococci and micrococci. *J. Clin. Mic.*, **19**, 875–879

Barber, L.E. and Deibel, R.H. (1972), Effect of pH and oxygen tension on staphylococcal growth and enterotoxin production in fermented sausage. *Appl. Mic.*, **24**, 891–898.

Bashford, T.E., Gillespie, T.G. and Tomlinson, A.J.H. (1960), *Staphylococcal Food Poisoning Associated with Processed Peas*. CFPRA: Chipping Campden.

Batish, V.K., Nataraj, B. and Grover, S. (1989), Variation in the behaviour of enterotoxigenic *Staphylococcus aureus* after heat stress in milk. *J. Appl. Bact.*, **66**, 27–35.

Baumgartner, A., Nicolet, J. and Essimann, M. (1984), Plasmid profiles of *Staphylococcus aureus* causing bovine mastitis. *J. Appl. Bact.*, **56**, 159–163.

Bautista, L., Gaya, P., Medina, M. and Nunez, M. (1988), A quantitative study of enterotoxin production by sheep milk staphylococci. *Appl. Environ. Micro.*, **54**, 566–569.

Beckers, H.J., Leusden, F.M., Bindschleider, O. and Guerras, D. (1984), Evaluation of a rabbit plasma-bovine fibrinogen agar for the enumeration of *Staphylococcus aureus* in food. *Can. J. Mic.*, **30**, 470–474.

Bennett, R.W. (1986), Detection and quantitation of Gram-positive nonsporing pathogens and their toxins. In *Foodborne Microorganisms and their Toxins: Developing Methodology*, Pierson, M.D. and Stern, N.J. (Eds). Marcel Dekker: London.

Bennett, R.W. and McLure, F. (1976), Collaborative study of the serological identification of staphylococcal enterotoxins by microslide double gel diffusion test. *J. Assoc. Off. Anal. Chem.*, **59**, 594–601.

Bennett, R.W. and Amos, W.T. (1982), *Staphylococcus aureus* growth and toxin production in nitrogen-packed sandwiches. *J. Fd. Prot.*, **45**, 157–161.

Bennett, R.W., Yetevoin, M., Smith, W. *et al.* (1986), *Staphylococcus aureus* identification characters and enterotoxigenicity. *J. Fd. Sci.*, **51**, 1337–1339.

Bergdoll, M.S. (1972), The enterotoxins. In *The Staphylococci*, Cohen, J.O. (Ed). John Wiley and Sons: New York.

Bergdoll, M.S. (1979), Staphylococcal intoxications. In *Foodborne Infections and Intoxications*, Reimann, H. and Bryan, F.L. (Eds). Academic Press: New York.

Bergdoll, M.S. and Bennett, R.W. (1984), Staphylococcal enterotoxins. In *Compendium of Methods for the Microbiological Examination of Foods*, Speck, M.L. (Ed). American Public Health Association: Washington DC.

Bergdoll, M.S., Reiser, R. and Spitz, J. (1976), Staphylococcal enterotoxins — detection in food. *J. Fd. Tech.*, **30**, 80–84.

Berke, A. and Tilton, R.C. (1986), Evaluation of rapid coagulase method for the identification of *Staphylococcus aureus*. *J. Clin. Mic.*, **23**, 916–919.

Betley, M.J. and Mekalanos, J.A. (1988), Nucleotide sequence of the type A staphyloccal enterotoxin gene. *J. Bact.*, **170**, 34–41.

Board, R.G. (1983), *A Modern Introduction to Food Microbiology*. Blackwell Scientific: Oxford.

Bolton, K.J., Dodd, C.E.R. and Waites, W.M. (1988), Chlorine resistance of strains of *Staphylococcus aureus* isolated from poultry processing plants. *Lett. Appl. Mic.*, **6**, 31–34.

Bone, F.J. Bogie, D. and Morgan-Jones, S.C. (1989), Staphylococcal food poisoning from sheep's milk cheese. *Epid. Inf.*, **103**, 249–258.

Borrego, J.J., Florido, J.A., Mrocek, P.P. and Romero, P. (1987), Design and performance of a new medium for the quantitative recovery of *Staphylococcus aureus* from recreational waters. *J. Appl. Bact.*, **63**, 85–93.

Bouwer-Hertzberger, S.A., Mossel, D.A.A. and Bijker, P.G.J. (1982), Quantitative isolation and identification of *Staphylococcus aureus*. In *Isolation and Identification Methods for Food Poisoning Organisms*, Corry, J.E.L., Roberts, D. and Skinner, F.A. (Eds). Academic Press: London and New York.

Bramley, A.J., McKinnon, C.H., Staker, R.T. and Simkin, D.L. (1984), The effect of udder infections on the bacterial flora of the bulk milk of ten dairy herds. *J. Appl. Bact.*, **57**, 317–323.

Breckenridge, J.C. and Bergdoll, M.S. (1971), Outbreak of foodborne gastroenteritis due to a coagulase negative, enterotoxin producing *Staphylococcus*. *New. Eng. J. Med.*, **284**, 541–543.

Bryan, F.L. (1976), *Staphylococcus aureus*. In *Food Microbiology: Public Health and Spoilage Aspects*, de Figueiredo, M.P. and Splittstoesser, D.F. (Eds). AVI: Westport.

Bryan, F.L. (1980), Epidemiology of foodborne diseases transmitted by fish, shell fish and marine crustaceans in the United States, 1970–1978. *J. Fd. Prot.*, **43**, 859–876.

Bryant, R.G., Jarvis, J. and Guibert, G. (1988), Selective enterotoxin production by a *Staphylococcus aureus* strain implicated in a foodborne outbreak. *J. Fd. Prot.*, **51**, 130–131.

Casman, E.P. and Bennett, R.W. (1963), Culture medium for production of staphylococcal enterotoxin A. *J. Bact.*, **86**, 18–23.

Castro, R., Schaebitz, R., Mortes, L., Bergdoll, M.S. (1986), Enterotoxigenicity of *Staphylococcus aureus* strains isolated from cheese made with unpasteurised milk. *Fd. Sci. & Tech.*, **19**, 401–402.

CDC (1970), Staphylococcal food poisoning traced to butter — Alabama. *Morb. Mort. Wkly. Rep.*, **19**, 271.

CDC (1977), Presumed staphylococcal food poisoning asociated with whipped butter. *Morb. Mort. Wkly. Rep.*, **26**, 268.

Chapman, G.H. (1945), The significance of sodium chloride in studies of staphylococci. *J. Bact.*, **50**, 201–203.

Ciborowski, P. and Jeljaszewicz, J. (1985), Staphylococcal enzymes and virulence. In *Bacterial Enzymes and Virulence*, Holder, I.A. (Ed). CRC Press: Boca Raton.

Currier, R.W., Taylor, A. and Wolf, F.S. (1973), Fatal staphylococcal food poisoning. *South. Med. J.*, **66**, 703–705.

Dangerfield, H.G. (1973), Effects of enterotoxins after ingestion by humans. *Proceedings 73rd Ann. Meeting*. American Society for Microbiology: Miami Beach.

Daoud, S.M. and Debevere, J.M. (1985), The effect of *Bacillus subtilis* and *Streptococcus faecalis* var *liquefacium* on Staphylococcal enterotoxin A activity. *Int. J. Fd. Mic.*, **2**, 211–218.

Demchick, P.H., Palumbo, S.A. and Smith, J.L. (1982), Influence of pH on freeze-thaw lethality in *Staphylococcus aureus*. *J. Fd. Safety*, **4**, 185–189.

de Saxe, M., Coe, A.W. and Wieneke, (1982), The use of phage typing in the investigation of food poisoning caused by *Staphylococcus aureus* enterotoxins. In *Isolation and Identification Methods for Food Poisoning Organisms*, Corry, J.E.L., Roberts, D. and Skinner, F.A. (Eds). Academic Press: London and New York.

Devriese, L.A. (1984), A simplified system for biotyping *Staphylococcus aureus* strains isolated from different animal species. *J. Appl. Bact.*, **56**, 215–220.

Devriese, L.A., Yde, M., Godard, C. and Igisidi, B.K. (1985), Use of biotyping to trace the origin of *Staphylococcus aureus* in foods. *Int. J. Fd. Mic.*, **2**, 365–369.

Dickie, N. and Akhtar, M. (1989), Concentration of staphylococcal enterotoxin from food extract using copper chelate sepharose. *J. Fd. Prot.*, **52**, 903–904.

Dietrich, G.G., Watson, R.J. and Silverman, G.T. (1972), Effect of shaking speed on the secretion of enterotoxin B by *Staph. aureus*. *Appl. Mic.*, **24**, 561–566.

Dodd, C.E.R., Adams, B.W., Mead, G.C. and Waites, W.M. (1987), Use of plasmid profiles to detect changes in strains of *Staphylococcus aureus* during poultry processing. *J. Appl. Bact.*, **63**, 417–425.

Dodd, C.E.R., Chaffey, B.J. and Waites, W.M. (1988), Plasmid profiles as indicators of the source of contamination of *Staphylococcus aureus* within poultry processing plants. *Appl. Environ. Micro.*, **54**, 1541–1549.

Donnelly, C.B., Leslie, J.E., Black, L.A. and Lewis, K.H. (1967), Serological identification of enterotoxigenic staphylococci from cheese. *Appl. Micro.*, **15**, 1382–1387.

Donnelly, C.B., Leslie, J.E. and Black, L.A. (1968), Production of enterotoxin A in milk. *Appl. Micro.*, **16**, 917–924.

Duitschaever, C.L. and Irvine, D.M. (1971), A case study: effect of mould on growth of coagulase-positive staphylococci in cheddar cheese. *J. Milk. Fd. Tech.*, **34**, 583.

El-Dairouty, K.R. (1989), Staphylococcal intoxication traced to non-fat dried milk. *J. Fd. Prot.*, **52**, 901–902.

Elliot, P.H., Tomlins, R.I. and Gray, R.J.H. (1982), Inhibition and inactivation of *Staphylococcus aureus* in a sorbate/modified atmosphere combination system. *J. Fd. Prot.*, **45**, 1112–1116.

Evenson, M.L., Hinds, M.W., Bernstein, R.S. and Bergdoll, M.S. (1988), Estimation of human dose of staphylococcal enterotoxin A from a large outbreak of staphylococcal enterotoxin involving chocolate milk. *Int. J. Fd. Mic.*, **7**, 311–316.

Ewald, S. and Notermans, S. (1988), Effect of water activity on growth and enterotoxin production of *Staphylococcus aureus*. *Int. J. Fd. Mic.*, **6**, 25–30.

Fang, C.S., Post, L.S. and Solberd, M. (1985), Antimicrobial effect and disappearance of sodium nitrite in *Staphylococcus aureus*. *J. Fd. Sci.*, **50**, 1412–1416.

Fey, H., (1986), Staphylococcal enterotoxins. In *Methods of Enzymatic Analysis*, Bergmeyer, H.U. (Ed). VCH Verlagsgesselschaft: Weinheim.

Fey, H., Pfister, H. and Ruegg, O. (1984), Comparison of different enzyme-linked assay systems of the detection of staphylococcal enterotoxins A, B, C, and D. *J. Clin. Mic.*, **19**, 34–38.

Fisk, R.T. (1942), Studies on staphylococci. I Occurrence of bacteriophage carriers among strains of *Staphylococcus aureus*. *J. Inf. Dis.*, **71**, 153–160.

Firstenberg-Eden, R. (1986), Electrical impedance for determining microbial quality of foods. In *Foodborne Microorganisms and their Toxins: Developing Methodology*, Pierson, M.D. and Stern, N.J. (Eds). Marcel Dekker: New York and Basel.

Fujikowa, H. and Igarishi, H. (1988), Rapid latex agglutination test for detection of staphylococcal enterotoxins A to E that uses high density latex particles. *Appl. Environ. Micro.*, **54**, 2345–2348.

Fung, D.Y.C. (1973), Capillary agar tube system for the detection of staphylococcal enterotoxins in foods. In *Proceedings of the Conference on Staphylococci in Foods*. Pennsylvania State University: University Park.

Garcia, M.L., Francisco, J.J. and Moreno, B. (1986), Nasal carriage of *Staphylococcus* species by food handlers. *Int. J. Fd. Mic.*, **3**, 99–108.

Gecs, A., Sing,, J. and Tauber, M. (1983), Potential of lactic streptococci to produce bacteriocin. *Appl. Environ. Micro.*, **45**, 205–211.

Genigeorgis, C.A. (1976), Quality control for fermented meats. *J. Am. Vet. Med. Assoc.*, **169**, 1220–1228.

Genigeorgis, C.A. (1989), Present state of knowledge on staphylococcal intoxication. *Int. J. Fd. Mic.*, **9**, 327–360.

Genigeorgis, C.A., Riemann, H. and Sadler, W.W. (1969), Production of enterotoxin B in cured meats. *J. Fd. Sci.*, **34**, 62–68.

Gibbs, P.A., Patterson, J.T. and Harvey, J. (1978), Biochemical characteristics and enterotoxigenicity of *Staphylococcus aureus* strains isolated from poultry. *J. Appl. Bact.*, **52**, 251–258.

Gilbert, R.J. (1983), Food-borne infections and intoxications — recent trends and prospects for the future. In *Food Microbiology. Advances and Prospects*, Roberts, T.A. and Skinner, F.A. (Eds). Academic Press: London and New York.

Giolitti, C. and Cantoni, C. (1966), A medium for the isolation of staphylococci from foodstuffs. *J. Appl. Bact.*, **29**, 395–398.

Gockler, L., Notermans, S. and Kramer, J. (1988), Production of enterotoxins and thermonuclease by *Staphylococcus aureus* in cooked egg-noodles. *Int. J. Fd. Mic.*, **6**, 127–139.

Gower, J.C. (1966), Some distance properties of latent root and vector methods used in multivariate analysis. *Biometrika*, **53**, 325–328.

Gregson, D.B., Low, D.E., Skulnik, M. and Simon, A.E. (1988), Problems for rapid agglutination methods for identification of *Staphylococcus aureus* when *Staphylococcus saprophyticus* is being tested. *J. Clin. Mic.*, **26**, 1398–1399.

Hajek, V. (1978), Identification of enterotoxigenic staphylococci from sheep and sheep cheese. *Appl. Environ. Micro.*, **35**, 264–268.

Hajek, V. and Marsalek, E. (1973), The occurrence of enterotoxigenic *Staphylococcus aureus* strains in hosts of different animal species. *Zentralblatt. Bakt. Infekt. Hyg.*, I Abt. Orig. **A 223**, 63–68.

Halpin-Dohnalek, M.I. and Marth, E.H. (1989), *Staphylococcus aureus*: production of extracellular compounds and behaviour in foods — a review. *J. Fd. Prot.*, **52**, 267–282

Harbrecht, D.F. and Bergdoll, M.S. (1980), Staphylococcal enterotoxin B production in hard-boiled eggs. *J. Fd. Sci.*, **45**, 307–309.

Harvey, J. and Gilmour, A. (1985), Application of current methods for isolation and identification of staphylococci in raw bovine milk. *J. Appl. Bact.*, **59**, 207–221.

Harvey, J. and Gilmour, A. (1988), Isolation and characterisation of enterotoxigenic staphylococci from goats milk produced in Northern Ireland. *Lett. Appl. Mic.*, **7**, 79–82.

Harvey, J., Patterson, J.T. and Gibbs, P.A. (1982), Entero-toxigenicity of *Staphylococcus aureus* strains isolated from poultry: raw poultry carcasses as a potential food poisoning hazard. *J. Appl. Bact.*, **52**, 251–258.

Hauschild, A.H.W., Park, C.E. and Hilsheimer, R. (1979), A modified pork plasma agar for the enumeration of *Staphylococcus aureus* in foods. *Can. J. Mic.*, **25**, 1052–1057.

Herten, B., Board, R.G. and Mead, G.C. (1989), Conditions affecting growth and enterotoxin production by *Staphylococcus aureus* on temperature-abused chicken meat. *Lett. Appl. Mic.*, **9**, 145–148.

Hirooka, E.Y., Muller, E.E., Freitos, J.C. (1988), Entero-toxigenicity of *Staphylococcus intermedius* of canine origin. *Int. J. Fd. Mic.*, **7**, 185–191.

Holmberg, S.D. and Blake, P.A. (1984), Staphylococcal food poisoning in the United States. New facts and old misconceptions. *J. Am. Med. Assoc.*, **251**, 487–489.

Hoover, D.G., Tatini, S.R. and Maltais, J.B. (1983), Characterisation of Staphylococci. *Appl. Environ. Micro.*, **46**, 649–660.

Humphreys, H., Keane, C.T., Hone, R. *et al.* (1989), Enterotoxin production by *Staphylococcus aureus* isolates from cases of septicaemia and from healthy carriers. *J. Med. Mic.*, **28**, 163–172.

Ibrahim, G.F. and Baldock, A.K. (1981), Thermostable deoxyribonuclease content and enterotoxigenicity of Cheddar cheese made with sub-normal starter activity. *J. Fd. Prot.*, **44**, 655–660.

ICMSF (1978), *Microrganisms in Foods. I Their Significance and Methods of Enumeration*, 2nd ed. University of Toronto Press: Toronto.

ICMSF (1980), *Microbial Ecology of Foods, Vol 2. Food Commodities*. Academic Press: New York.

ICMSF (1986), *Microrganisms in Foods: Sampling for Micro-biological Analysis: Principles and Specific Applications*. Blackwell Scientific: Oxford.

Ikram, M. and Luedecke, L.O. (1977), Growth and enterotoxin A production by *Staphylococcus aureus* in fluid dairy products. *J. Fd. Sci.*, **40**, 769–771.

Isigidi, B.K., Devriese, L.A., Croegant, T. and van Hoof, J. (1989), A highly selective two-stage isolation method for the enumeration of *Staphylococcus aureus* in foods. *J. appl. Bact.*, **66**, 379–354.

Jarvis, A.W., Lawrence, R.C. and Pritchard, G.G. (1975), Glucose repression of enterotoxins A, B and C and other extracellular protein in staphylococci in broth and con-tinuous culture.

Jawetz, E., Melnick, J.L. and Adelberg, E.A. (1982), Pyogenic cocci. In *Review of Medical Microbiology*. Lange Medical Publications: Los Altos.

Kloos, W.E. and Schleifer, K.H. (1986), *Staphylococcus*. In *Bergey's Manual of Systematic Bacteriology*, Vol 2, Sneath, P.H.A., Mair, N.S. Sharpe, M.E. and Holt, J.G. (Eds). Williams and Wilkins: Baltimore and London.

Kokan, N.P. and Bergdoll, M.S. (1987), Detection of low-enterotoxin-producing *Staphylococcus aureus* strains. *Appl. Environ. Micro.*, **53**, 2675–2676

Kuo, J.K.S. and Silverman, G.J. (1980), Application of enzyme-linked immunosorbent assay for detection of staphylococcal enterotoxin in foods. *J. Fd. Prot.*, **43**, 404–407.

Lachica, R.V.F. (1980), Accelerated procedure for the enumeration and identification of food-borne *Staphylococcus aureus*. *Appl. Environ. Micro.*, **39**, 17–19.

Lachica, R.V.F., Hoeprich, P.D. and Genigeorgis, C. (1972), Metachromatic agar diffusion microslide tech-nique for detecting staphylococcal nuclease in foods. *Appl. Micro.*, **23**, 168–169.

Lachica, R.V.F., Jang, S.S. and Hoeprich. P.D. (1979), Thermonuclease seroinhibition test for distinguishing *Staphylococcus aureus* and other coagulase-positive staphy-lococci. *J. Clin. Mic.*, **9**, 141–143.

Lapeyere, C., Janin, F. and Kaveri, S.V. (1988), Indirect double sandwich ELISA using monoclonal antibodies for detection of staphylococcal enterotoxins A, B, C1 and D in food samples. *Fd. Mic.*, **5**, 25–31.

Lindroth, S. and Niskanen, A. (1977), Double antibody solid-phase radioimmunoassay for staphylococcal entero-toxin A. *Eur. J. Appl. Micro.*, **4**, 137–143.

Lombai, G., Janosi, L., Katona, F. *et al.* (1980), Properties and food hygiene importance of *Staphylococcus aureus* strains isolated from udders. *Arch. Lebensmittelhyg.*, **31**, 206–209.

Males, B.M., Rogers, W.A. and Parris, J.T. (1975), Virulence factors of biotypes of *Staphylococcus epidermis* from clinical sources. *J. Clin. Mic.*, **1**, 256–261.

Marcy, J.A., Kraft, A.A., Olson, D.G. *et al.* (1985), Fate of *Staphylococcus aureus* in reduced sodium fermented sausages. *J. Fd. Sci.*, **50**, 316–320.

McGowan, J.E., Terry, P.M. Huang, T.-S.R. *et al.* (1979), Nosocomial infections with gentamycin-resistant *Staphy-lococcus aureus*: plasmid analysis as an epidemiological tool. *J. Inf. Dis.*, **140**, 864–872.

McLean, R.A., Lilly, H.D. and Alford, J.A. (1968), Effects of meat curing salts and temperature on production of staphylococcal enterotoxin B. *J. Bact.*, **95**, 1203–1208.

Mead, G.C. and Adams, B.W. (1986), Chlorine resistance of *Staphylococcus aureus* from turkeys and turkey products. *Lett. Appl. Mic.*, **3**, 131–133.

Mead, G.C., Norris, A.P. and Bratchell, N. (1989), Differ-entiation of *Staphyloccus aureus* from freshly slaughtered poultry and strains 'endemic' to processing plants by biochemical and physiological tests. *J. Appl. Bact.*, **66**, 153–159.

Metaxopoulos, J. Genigeorgis, C.A., Fanelli, M.J. *et al.* (1981), Production of Italian dry salami. I. Initiation of staphylococcal growth in salami under commercial manufacturing conditions. *J. Fd. Prot.*, **44**, 347–352.

Meyer, R.F. and Palmieri, M.J. (1980), Single radial immunodiffusion method for screening staphylococcal isolates for enterotoxins. *Appl. Environ. Micro.*, **40**, 1080–1085.

Miller, B.A., Reiser, R.F. and Bergdoll, M.S. (1978), Detection of staphylococcal enterotoxins A, B, C, D, and E in foods by radioimmunoassay, using staphylo-coccal cells containing protein A as an immunoadsorbent. *Appl. Environ. Micro.*, **47**, 283–287.

Morita, T.N. and Woodburn, M.J. (1978), Homogeneous enzyme immune assay for staphylococcal enterotoxin B. *Inf. Imm.*, **21**, 666–668.

Morita, T.N., Patterson, J.E. and Woodburn, M.J. (1979), Magnesium and iron addition to casein hydrolysate for production of staphylococcal enterotoxin A, B and C. *Appl. Environ. Micro.*, **38**, 39–42.

Neill, R.J., Fanning, G.R., Delahoz, F. *et al.* (1990), Oligonucleotide probes for detection and differentiation of *Staphylococcus aureus* strains containing genes for entero-toxins A, B, and C and toxic shock syndrome toxin 1. *J. Clin. Mic.*, **28**, 1514–1518.

Newsome, R.L. (1988), *Staphylococus aureus* (Status summary). *Fd. Tech.*, **42** (4), 194–195.

Niskanen, A. and Nurmi, E. (1976), Effect of starter culture on staphylococcal enterotoxin and thermonuclease production in dry sausage. *Appl. Environ. Micro.*, **31**, 11–20.

Niven, C.F. (1958), Microbiological aspects of the radiation preservation of food. *Ann. Rev. Mic.*, **12**, 507–524.

Noble, W.C. (1981), *Microbiology of Human Skin*, 2nd ed., Lloyd-Luke Ltd: London.

Noleto, A.L.S., Malburg, L.M. and Bergdoll, M.S. (1987), Production of Staphylococcal enterotoxin in mixed cultures. *Appl. Environ. Mic.*, **53**, 2271–2274.

Notermans, S. (1982), Detection of staphylococcal enterotoxins (SE) with special reference to the enzyme linked immunosorbent assay (ELISA). In *Isolation and Identification Methods for Food Poisoning Organisms*, Corry, J.E.L., Roberts, D. and Skinner, F.A. (Eds). Academic Press: London and New York.

Notermans, S. and van Olterdijk, R.L.M. (1985), Production of enterotoxin A by *Staphylococccus aureus* in food. *Int. J. Fd. Mic.*, **2**, 145–149.

Notermans, S., Van Leeuwen, W.J. and Rost, J.A. (1983), *Staphylococcus aureus* indigenous to poultry processing plants, persistence, enterotoxigenicity and biochemical characteristics. In *Quality of Poultry Meat*, Lahellec, C., Ricard, F.H. and Colin, P. (Eds). Station Experimentale d'Aviculture.

Notermans, S., Heuvelman, K.J. and Wemars, K. (1988), Synthetic enterotoxin B DNA probes for detection of enterotoxigenic *Staphylococcus aureus*. *Appl. Environ. Micro.*, **54**, 531–533.

Nout, M.J.R., Notermans, S. and Rombouts, F.M. (1988), Effect of environmental conditions during soya-bean fermentation on the growth of *Staphylococcus aureus* and production and thermal stability of enterotoxins A and B. *Int. J. Fd. Mic.*, **7**, 299–309.

NRC (1985), *An Evaluation of the Role of Microbiological Criteria for Foods and Food Ingredients*. National Academy Press: Washington DC.

Otero, A., Garcia, M.C., Garcia, M.L. and Moreno, B. (1987), Production of staphylococcal enterotoxins C1 and C2 in ewe's milk. *Fd. Mic.*, **4**, 339–345.

Otero, A., Garcia, M.C., Garcia, M.L. and Moreno, B. (1988), Effect of growth of a commercial starter culture on growth of *Staphylococcus aureus* and thermonuclease and enterotoxin (C1 and C2) production in broth cultures. *Int. J. Fd. Mic.*, **6**, 107–114.

Parada, J.L., Chirife, L.J. and Magrini, R.C. (1982), Effect of BHA, BHT, and potassium sorbate on growth of *Staphylococcus aureus* in a model system and process cheese. *J. Fd. Prot.*, **45**, 1108–1111.

Park, C.E. and Szabo, R. (1986), Evaluation of the reversed passive latex agglutination (RPLA) test kit for detection of staphylococcal enterotoxin ABC and D in foods. *Can. J. Mic.*, **32**, 723–727.

Pierson, M.D., Smoot, L.A. and Stern, N.J. (1979), Effect of potassium sorbate on growth of *Staphylococcus aureus* in bacon. *J. Fd. Prot.*, **42**, 302–304.

Pinegar, J.A. and Suffield, A (1982), The investigation of food poisoning outbreaks in England and Wales. In *Isolation and Identification Methods for Food Poisoning Organisms*, Corry, J.E.L., Roberts, D. and Skinner, F.A. (Eds). Academic Press: London and New York.

Post, L.S., Dee, D.A., Solberg, M. *et al.* (1988), Development of staphylococcal toxin and sensory deterioration during storage of nitrogen and vacuum packaged nitrite-free bacon-like product. *J. Fd. Sci.*, **53**, 383–387.

Quinn, D.J., Anderson, A.W and Dyer, J.F. (1967), The inactivation of infection and intoxication microorganisms by irradiation in seafood. *Micro. Prob. Fd. Pres. Irrad. Panel Proc.*, STI/PUB/168. IAEA: Vienna.

Raccach, M. (1981), Control of *Staphylococcus aureus* in dry sausage by a newly developed meat-starter culture and phenolic-type antioxidants. *J. Fd. Prot.*, **44**, 665–669.

Rayman, K., Malik, N. and Jarvis, G. (1988), Performance of four selective media for enumerating *Staphylococcus aureus* in corned beef and in cheese. *J. Fd. Prot.*, **51**, 87–88.

Reichert, C.A. and Fung, D.Y.C. (1976), Thermal inactivation and subsequent reactivation of staphylococcal enterotoxin B in selected foods. *J. Milk Fd. Tech.*, **39**, 516–520.

Reiser, R.R., Conaway, D. and Bergdoll, M.S. (1974), Detection of staphylococcal enterotoxin in foods. *Appl. Mic.*, **27**, 83–85.

Reynolds, D., Tranter, H.S., Saye, R. and Hambleton, P. (1988), Novel method for purification of staphylococcal enterotoxin A. *Appl. Environ. Micro.*, **54**, 1761–1765.

Robbins, R., Gould, S. and Bergdoll, M.S. (1974), Detecting the enterotoxigenicity of *Staphylococcus aureus* strains. *Appl. Micro.*, **28**, 946–950.

Roberts, D. (1982), Bacteria of public health significance. In *Meat Microbiology*, Brown, M. (Ed). Applied Science Publishers: London and New York.

Rose, S.A., Bankes, P. and Stringer, M.F. (1989), Detection of staphylococcal enterotoxins in dairy products by the reversed passive latex agglutination (SET-RPLA) kit. *Int. J. Fd. Mic.*, **8**, 65–72.

Sawhney, D. (1986), The toxicity of potassium tellurite to *Staphylococcus aureus* in rabbit plasma fibrinogen agar. *J. Appl. Bact.*, **61**, 149–155.

Scheuber, P.H., Golecki, J.R., Kickhofen, B. *et al.* (1985), Skin reactivity of unsensitised monkeys upon challenge with staphylococcal enterotoxin B; a new approach for investigating the site of toxin action. *Inf. Imm.*, **50**, 869–876.

Scheuber, P.H., Denzlinger, C., Wilker, D. *et al.* (1987), Staphylococcal enterotoxin B as a nonimmunological stimulus in primates: the role of endogenous cysteinyl leukotrienes. *Int. Arch. Allergy Appl. Immun.*, **82**, 289–291.

Scheusner, D.L., Hood, L.L. and Harmon, L.G. (1973), Growth and enterotoxin production by *Staphylococcus aureus*. *J. Milk. Fd. Tech.*, **36**, 249–252.

Schleifer, K.H. and Kloos, W.E. (1975), A simple tube system for the separation of staphylococci from micrococci. *J. Clin. Mic.*, **1**, 337–338.

Schleifer, K.H. and Kramer, E. (1980), Selective medium for isolating staphylococci. *Zentrallblatt. Bakt. Parasiten. Infektions. Hyg. Abt 1. Orig.*, **C 1**, 270–280.

Schonwalder, H., Haaijman, J.J., Holbrook, R. *et al.* (1988), A collaborative study comparing three ELISA systems for detecting *Staphylococcus aureus* enterotoxin A in sausage extracts. *J. Fd. Prot.*, **51**, 680–684.

Seager, M.S., Banks, J.G., Blackburn, C.de W. and Board, R.G. (1986), A taxonomic study of *Staphylococcus* spp. isolated from fermented sausages. *J. Fd. Sci.*, **51**, 295–297.

Shingaki, M, Igarishi, H., Fujikouri, H. *et al.* (1981), Study on reversed passive latex agglutination for the detection of staphylococcal enterotoxins A–C. *Ann. Rep. Tokyo Metropolitan Res. Lab. Publ. Hlth.*, **32**, 128–131.

Silverman, G.J., Knott, A.R. and Howard, M. (1968), Rapid, sensitive assay for staphylococcal enterotoxin and a comparison of serological methods. *Appl. Micro.*, **16**, 1019–1023.

Simkovicova, M. and Gilbert, R.J. (1971), Serological detection of enterotoxin from food poisoning strains of *Staphylococcus aureus. J. Med. Mic.*, **4**, 19–30.

Sinell, H.J. and Baumgar, J. (1967), Selktivnahrboden mit Eigelb zur isolierung von pathogenen Staphylokokken aus lebensmitteln. *Zentralblatt. Bakt. Parasiten. Infektions. Hyg. Abt. I. Orig.*, **A 204**, 248–264.

Smith, J.L., Buchanan, R.L. and Palumbo, S.A. (1983), Effect of food components on staphylococcal enterotoxin synthesis: a review. *J. Fd. Prot.*, **46**, 545–555.

Sokari, T.G. and Anozie, S.O. (1989), Modified single radial immunodiffusion method for screening staphylococcal isolates for enterotoxin. *Fd. Mic.*, **6**, 45–48.

Sperber, W.H. and Tatini, S.R. (1975), Interpretation of the tube coagulase test for identification of *Staphylococcus aureus. Appl. Micro.*, **29**, 502–505.

Stadhouders, J., Hassing, F. and Van Aalst-Van Maren, N.O. (1976), A pour-plate method for the detection and enumeration of coagulase-positive *Staphylococcus aureus* in the Baird-Parker medium without egg yolk. *Neth. Milk Dairy J.*, **30**, 222–229.

Stanley, C.J., Johannsson, A. and Self, C.H. (1985), Enzyme amplification can enhance both the speed and the sensitivity of immunoassays. *J. Imm.*, **83**, 89–95.

Stephen, J. and Pietrowski, R.A. (1981), Membrane damaging toxins: primary event not known. In *Aspects of Microbiology: Bacterial Toxins*, 2nd ed. American Society for Microbiology: Washington DC.

Stersky, A.K., Szabo, R., Todd, E.C.D. *et al.* (1986), *Staphylococcus aureus* growth and thermostable nuclease and enterotoxin production in canned salmon and sardines. *J. Fd. Prot.*, **49**, 428–435.

Stiffler-Rosenberger, G. and Fey, H. (1978), Simple assay for staphylococcal enterotoxins A, B, and C: modification of enzyme linked immunosorbent assay. *J. Clin. Mic.*, **8**, 473–479.

Sugiyama, H., Chow, K.L. and Dragstedt, L.R. (1961), Study of emetic receptor sites for staphylococcal enterotoxins in monkeys. *Proc. Soc. Exp. Biol. Med.*, **108**, 92–95.

Tatini, S.R. (1973), Influence of food environments on growth of *Staphylococcus aureus* and production of various enterotoxins. *J. Milk. Fd. Tech.*, **36**, 559–563.

Tatini, S.R., Jezeski, J.J., Olson, J.C. and Casman, E.P. (1971), Factors affecting the production of staphylococcal enterotoxin A in milk. *J. Dairy Sci.*, **54**, 312–320.

Tatini, S.R., Hoover, D.G. and Lachica, R.V.F. (1984), Methods for the isolation and enumeration of *Staphylococcus aureus*. In *Compendium of Methods for the Microbiological Examination of Foods*, 2nd ed., Speck, M.L. (Ed). American Society for Microbiology: Washington DC.

Thompson, N.E., Bergdoll, M.S., Meyer, R.F. *et al.* (1985), Monoclonal antibodies to the enterotoxin and to the toxic shock syndrome toxin produced by *Staphylococcus aureus*. In *Monoclonal Antibodies*, Vol II, Macario, A.J.L. and Macario, E.C. (Eds). Academic Press: Orlando.

Tibora, A., Rayman, K., Akhtar, M. and Szabo, R. (1987), Thermal stability of staphylococcal enterotoxins A, B and C in a buffered system. *J. Fd. Prot.*, **50**, 239–242.

Todd, E., Szabo, R., Robern, H. *et al.* (1981), Variation in counts, enterotoxin levels and TNase in Swis-type cheese contaminated with *Staphylococcus aureus. J. Fd. Prot.*, **44**, 839–848.

Troller, J.A. (1971), Effect of water activity on enterotoxin B production and growth of *Staphylococcus aureus. Appl. Micro.*, **21**, 435–439.

Troller, J.A. and Stinson, J.V. (1978), Influence of water activity on the production of extracellular enzymes by *Staphylococcus aureus. Appl. Environ. Mic.*, **35**, 521–526.

Tuncan, E.U. and Martin, S.E. (1987), Lysostaphin lysis procedure for detection of *Staphylococcus aureus* by the firefly bioluminescent ATP method. *Appl. Environ. Micro.*, **53**, 88–91.

Tweten, R.K. and Iandolo, J.J. (1983), Transport and processing of staphylococcal enterotoxin B. *J. Bact.*, **153**, 297–303.

Umoh, V.J., Adesgan, A.A. and Gomwalk, N.E. (1988), Enterotoxin production by Staphylococcal isolates from Nigerian fermented milk products. *J. Fd. Prot.*, **51**, 534–537.

Upredi, G.C. and Hinsdill, R.D. (1975), Production and mode of action of lactocin 27: bacteriocin from a homofermentative *Lactobacillus. Antimic. Agents. Chemother.*, **7**, 139–145.

Vaamonde, G., Chirife, J. and Scorza, C.O. (1982), An examination of the minimal water activity for *Staphylococcus aureus* ATCC 6538P growth in laboratory media adjusted with less conventional solutes. *J. Fd. Sci.*, **47**, 1259–1262.

Valle, J., Gomez-Lucia, E., Piriz, S. *et al.* (1990), Enterotoxin production by staphylococci isolated from healthy goats. *Appl. Environ. Micro.*, **56**, 1323–1326.

Van Doorne, H., Baird, R.M., Hendricksz, D.T. *et al.* (1981), Liquid modification of Baird-Parker's medium for the selective enrichment of *Staphylococcus aureus. Antonie van Leeuwenhoek*, **47**, 267–268.

Varadaraj, M.C. and Nambudriped, V.K.N. (1986), Carryover of preformed staphylococcal enterotoxins and thermostable deoxyribonuclease from raw cow milk to Khou — a heat concentrated Indian milk product. *J. Dairy Sci.*, **69**, 340–343.

Waterman, S.H., Demarcus, T.A., Wells, J.G. and Blake, P.A. (1987), Staphylococcal food poisoning on a cruise ship. *Epid. Inf.*, **99**, 349–353.

White, D.E., Matos, J.S., Harmon, R.J. and Langlois, B.E. (1988), A comparison of six selective media for the enumeration and isolation of staphylococci. *J. Fd. Prot.*, **51**, 685–690.

Whiting, R.C., Benedict, R.C., Kussel, C.A. and Blalock, D. (1985), Growth of *Clostridium sporogenes* and *Staphylococcus aureus* at different temperatures in cooked corned beef with different levels of NaCl. *J. Fd. Sci.*, **50**, 304–307.

Wieneke, A.A. and Gilbert, R.J. (1987), Comparison of four methods for the detection of staphylococcal enterotoxins in foods from outbreaks of food poisoning. *Int. J. Fd. Mic.*, **4**, 135–143.

Windemann, H., Luthi, J. and Maurer, M. (1989), ELISA with enzyme amplification for sensitive detection of staphylococcal enterotoxins in food. *Int. J. Fd. Mic.*, **8**, 25–34.

Woolaway, M.C., Bartlett, C.L.R., Wieneke, A.A. *et al.* (1986), International outbreak of staphylococcal food poisoning caused by contaminated lasagne. *J. Hyg.*, **96**, 67–73.

Woodburn, M., Morita, T.N. and Venn, S.Z. (1973), Production of staphylococcal enterotoxins A, B and C in colloidal dispersion. *Appl. Micro.*, **25**, 825–833.

Wyatt, C.J. and Guy, V.H. (1981), The growth of *Salmonella typhimurium* and *Staphylococcus aureus* in retail pumpkin pie. *J. Fd. Prot.*, **44**, 418–421.

Yang, X., Board, R.G. and Mead, G.C. (1988), Influence of spoilage flora and temperature on growth of *Staphylococcus aureus* in turkey meat. *J. Fd. Prot.*, **51**, 303–309.

Chapter 13

Al-Khayat, M.A., Blank, G. and Biliadenis, C. (1987), Germination and outgrowth of *Bacillus subtilis* spores in the presence of selected antioxidants. *J. Fd. Prot.*, **50**, 206–211.

Anon (1988), Sverldlovsk: anthrax capital. *Science*, **240**, 383–385.

Asanuma, S., Tanaka, H. and Yatazawa, M. (1979), Rhizoplane microorganisms of rice seedlings as examined by scanning electron microscopy. *Soil Sci. Plant Nutr.*, **25**, 539–551.

Asplund, K., Nurmi, E., Hilli, P. and Hion, J. (1988), The inhibition of growth of *Bacillus cereus* in liver sausage. *Int. J. Fd. Mic.*, **7**, 349–352.

Berkeley, R.C., Logan, N.A., Shute, L.A. and Copley, A.G. (1984), Identification of *Bacillus* species. In *Methods in Microbiology*, Vol 16, Bergan, T. (Ed). Academic Press: London and New York.

Billing, E. and Cuthbert, W.A. (1958), 'Bitty' cream: the occurrence and significance of *Bacillus cereus* in raw milk supplies. *J. appl. Bact.*, **48**, 297–303.

Blakey, L.J. and Priest, F.G. (1980), The occurrence of *Bacillus cereus* in some dried foods including cereals and pulses. *J. appl. Bact.*, **21**, 65–78.

Bouwer-Hertzberger, S.A. and Mossel, D.A.A. (1982), Quantitative isolation and identification of *Bacillus cereus*. In *Isolation and Identification Methods for Food Poisoning Organisms*, Corry, J.E.L., Roberts, D. and Skinner, F.A. (Eds). Academic Press: London and New York.

Bradshaw, J.G., Peeler, J.T. and Twedt, R.M. (1975), Heat resistance of ileal loop reactive *Bacillus cereus* strains isolated from commercially canned food. *Appl. Micro.*, **30**, 943–945.

Bulgarelli, M.A. and Shelef, W.A. (1985), Effect of ethylenediaminetetraacetic acid (EDTA) on growth from spores of *Bacillus cereus*. *J. Fd. Sci.*, **50**, 661–664.

Cerf, O. (1977), Tailing of survival curves of bacterial spores. *J. appl. Bact.*, **42**, 1–19.

Chopra, P., Singh, A. and Kalra, M.S. (1980), Occurrence of *Bacillus cereus* in milk and milk products. *Indian J. Dairy. Sci.*, **33**, 248–252.

Christiansson, A., Naidu, A.S., Nilsson, I. *et al.* (1989), Toxin production by *Bacillus cereus* dairy isolates in milk at low temperatures. *Appl. Environ. Micro.*, **55**, 2595–2600.

Claus, D. and Berkeley, R.C.W. (1986), Genus *Bacillus*. In *Bergey's Manual of Systematic Bacteriology*, Vol 2, Sneath, P.H.A., Mair, N.S., Sharpe, M.E. and Holt, J.G. (Eds). Williams and Wilkins: Baltimore and London.

Coolbaugh, J.C. and Williams, R.P. (1978), Production and characterization of two toxins of *Bacillus cereas*. *Can. J. Mic.*, **24**, 1289–1295.

De, S.N. and Chatterjee, D.N. (1953), An experimental study of the action of *Vibrio cholerae* on the intestinal mucous membrane. *J. Path. Bact.*, **66**, 559–562.

Deak, T. and Timar, E. (1988), Simplified identification of aerobic spore-formers in the investigation of foods. *Int. J. Fd. Mic.*, **6**, 115–125.

DeBueno, B.A., Brondum, J., Kramer, J.M. *et al.* (1988), Plasmid, serotypic and enterotoxin analysis of *Bacillus cereus* in an outbreak setting. *J. Clin. Mic.*, **26**, 1571–1574.

Doyle, M.P. (1988), *Bacillus cereus. Fd. Tech.*, **42**(4), 199.

Driessen, F.M. and Stadhouders, I. (1980), A new defect in stirred yogurt. *Nordeurop. Mejeri-Tidsskr.*, **46**, 229–230.

Ezepchuk, Y.V. and Fluer, F.S. (1973), Der enterotoxisch effekt. *Mod. Medizin*, **3**, 20–25.

Ellison, A., Dodd, C.E.R. and Waites, W.M. (1988), The use of plasmid profiling to compare strains of *Bacillus cereus* isolated from liquid whole egg. *J. appl. Bact.*, **65**, xiv–xv.

Ellison, A., Dodd, C.E.R. and Waites, W.M. (1989), Use of plasmid profiles to differentiate between strains of *Bacillus cereus. Fd. Mic.*, **6**, 93–98.

Ghosh, A.C. (1978), Prevalence of *Bacillus cereus* in the faeces of healthy adults. *J. Hyg.*, **80**, 233–236.

Gilbert, R.J. and Taylor, A.J. (1976), *Bacillus cereus* food poisoning. In *Microbiology in Agriculture, Fisheries and Food*, Skinner, F.A. and Carr, J.C. (Eds). Academic Press: London.

Gilbert, R.J. and Parry, J.M. (1977), Serotypes of *Bacillus cereus* from outbreaks of food poisoning and from routine foods. *J. Hyg.*, **78**, 69–74.

Gilbert, R.J. and Kramer, J.M. (1984), *Bacillus cereus* enterotoxins: present status. *Biochem. Soc. Trans.*, **12**, 198–200.

Gilbert, R.J., Stringer, M.F. and Pearce, T.C. (1974), The survival and growth of *Bacillus cereus* in boiled and fried rice in relation to outbreaks of food poisoning. *J. Hyg.*, **73**, 433–444.

Gilbert, R.J., Turnbull, P.C.B., Parry, J.M. and Kramer, J.M. (1981), *Bacillus cereus* and other *Bacillus* species: their part in food poisoning and other clinical infection. In *The Aerobic Endospore-forming Bacteria*, Berkeley, R.C. W. and Goodfellow, M (Eds). Academic Press: London.

Glatz, B.A. (1986), Genetic regulation of toxin production by foodborne microbes. In *Food Microbiology: Vol 1. Concepts in Physiology and Metabolism*, Montville, T.J. (Ed). CRC Press: Boca Raton.

Glatz, B.A., Spira, W.M. and Goepfort, J.M. (1974), Alterations of vascular permeability in rabbits by culture filtrates of *Bacillus cereus* and related species. *Inf. Imm.*, **10**, 299–303.

Goepfort, J.M., Spira, W.M., Glatz, B.A. and Kim, H.U. (1973), Pathogenicity of *Bacillus cereus*. In *The Microbiological Safety of Foods*, Hobbs, B.C. and Christian, J.H.B. (Eds). Academic Press: London and New York.

Gordon, R.E., Haynes, W.C. and Pang C.H.-N. (1973), The Genus *Bacillus*. USDA: Washington DC.

Harmon, S.M. (1980), *Bacillus cereus*. In *Bacteriological Analytical Manual*, 5th ed. AOAC: Washington DC.

Hauge, S. (1955), Food poisoning caused by aerobic spore-forming bacilli. *J. Appl. Bact.*, **18**, 591–595.

Holbrook, R. and Anderson, J.M. (1980), An improved selective and diagnostic medium for the isolation and enumeration of *Bacillus cereus* in foods. *Can. J. Mic.*, **26**, 753–759.

Holmes, J.R., Plunkett, T., Pate, P. *et al.* (1981), Emetic food poisoning caused by *Bacillus cereus*. *Arch. Intern. Med.*, **141**, 766–767.

Hughes, S., Bartholomew, B., Hardy, J.C. and Kramer, J.M. (1988), Potential application of a HEp-2 cell assay in the investigation of *Bacillus cereus* emetic syndrome food poisoning. *FEMS Mic. Lett.*, **52**, 7–12

Jacob, H.E. (1970), Redox potential. In *Methods in Microbiology*, Vol 2, Norris, J.R. and Ribbons, D.W. (Eds). Academic Press: London and New York.

Johnson, K.M., Nelson, C.L. and Busta, F.F (1982), Germination and heat resistance of *Bacillus cereus* spores from strains associated with diarrhoeal and emetic foodborne illness *J. Fd. Sci.*, **47**, 1268–1271.

Johnson, K.M., Nelson, C.L. and Busta, F.F. (1983), Influence of temperature on germination and growth of spore of emetic and diarrhoeal strains of *Bacillus cereus* in a broth medium and in rice. *J. Fd. Sci.*, **48**, 286–287.

Johnson, K.M. (1984), *Bacillus cereus* foodborne illness — an update. *J. Fd. Prot.*, **47**, 145–153.

Kim, H.U. and Goepfort, J.M. (1971), Enumeration and identification of *Bacillus cereus* in foods. I. 24-hour presumptive test medium. *Appl. Micro.*, **22**, 581–587.

Kramer, J.M. (1984), *Bacillus cereus* exotoxins: production, isolation, detection and properties. In *Bacterial Protein Toxins*, Alouf, J.E., Freer, H., Fehrenbach, F.J. and Jeljaszewicz, J. (Eds).

Kramer, J.M., Turnbull, P.C.B., Munshi, G. and Gilbert, R.J. (1982), Identification and characterization of *Bacillus cereus* and other *Bacillus* species associated with food poisoning. In *Isolation and Identification Methods for Food Poisoning Organisms*, Corry, J.E.L., Roberts, D. and Skinner, F.A. (Eds). Academic Press: London and New York.

Lemille, F., Borjac, H. de and Bonnefoi, A. (1969), Etude serologique de *Bacillus cereus*. Mise en evidence de divers serotypes bases sur les antigenes flagellaires. *Ann. Inst. Past.*, **117**, 31.

Logan, N.A. (1988), *Bacillus* species of medical and veterinary importance. *J. Med. Mic.*, **25**, 317–324.

Logan, N.A. and Berkeley. R.C.W. (1981), Classification and identification of members of the genus *Bacillus* using API tests. In *The Aerobic Endospore-Forming Bacteria*, Berkeley, R.C.W. and Goodfellow, M. (Eds). Academic Press: London.

Logan, N.A. and Berkeley, R.C.W. (1984), Identification of *Bacillus* strains using the API system. *J. Gen. Mic.*, **130**, 1871–1892.

Logan, N.A., Berkeley, R.C.W. and Norris, J.R. (1978), Results of an international reproducibility trial using the API system applied to the genus *Bacillus*. *J. Appl. Bact.*, **45**, xviii–xxix.

Logan, N.A., Capel, B.J., Melling, J. and Berkeley, R.C.W. (1979), Distinction between emetic and other strains of *Bacillus cereus* using the API system and numerical methods *FEMS Micro. Lett.*, **5**, 373–375.

Major, P., Rimanoczi, I., Ormay, L. and Belteky, A. (1979), Characteristics of *Bacillus cereus* strains isolated from various foods. *Elem. Ipar.*, **33**, 314–315.

Melling, J., Capel, B.J., Turnbull, P.C.B. and Gilbert, R.J. (1976), Identification of a novel enterotoxigenic activity associated with *Bacillus cereus*. *J. Clin. Path.*, **29**, 938–940.

Mossel, D.A.A., Koopman, M.J. and Jongerius, E. (1967), Enumeration of *Bacillus cereus* in foods. *Appl. Micro.*, **15**, 650–653.

Mosso, A., Arribas, L.G., Cuera, J.A. and De la Rosa, C. (1989), Enumeration of *Bacillus* and *Bacillus cereus* spores from Spain. *J. Fd. Prot.*, **52**, 184–188.

Mostert, J.F., Luck, H. and Husmann, R.A. (1979), Isolation, identification and practical properties of *Bacillus* species from UHT and sterilized milk. *S. Afr. J. Dairy Technol.*, **11**, 125–132.

Norris, J.R. and Wolf, J. (1961), A study of the antigens of the aerobic spore-forming bacteria. *J. Appl. Bact.*, **24**, 42–56.

Overcast, W.W. and Atamaram, K. (1974), The role of *Bacillus cereus* in sweet curdling of fluid milk. *J. Milk. Fd. Tech.*, **37**, 233–236.

Parry, J.M. and Gilbert, R.J. (1980), Studies on the heat resistance of *Bacillus cereus* spores and growth of the organism in boiled rice. *J. Hyg.*, **84**, 77–82.

Parry, J.M., Turnbull, P.C.B. and Gibson, J.R. (1983), *A Colour Atlas of Bacillus Species*. Wolfe Medical Publications Ltd: London.

Peters, M., Wiberg, C. and Norberg, P. (1985), Comparison of media for isolation of *Bacillus cereus* from foods. *J. Fd. Prot.*, **48**, 969–970.

Pinegar, J.A. and Buxton, J.D. (1977), An investigation of the bacteriological quality of cream slices. *J. Hyg.*, **78**, 387–394.

Portnoy, B.L., Goepfort, J.M. and Harmon, S.M. (1973), An outbreak of *Bacillus cereus* food poisoning resulting from contaminated vegetable sprouts. *Am. J. Epid.*, **103**, 589–594.

Powers, E.M., Latt, T.G. and Brown, T. (1976), Incidence and levels of *Bacillus cereus* in processed spices. *J. Milk. Fd. Tech.*, **39**, 668–670.

Rajkowski, K.T. and Mikolajcik, E.M. (1987), Characteristics of selected strains of *Bacillus cereus*. *J. Fd. Prot.*, **50**, 199–205.

Raevuori, M.T. (1976), Effect of sorbic acid and potassium sorbate on growth of *Bacillus cereus* and *Bacillus subtilis* in rice filling of Karelian pastry. *Europ. J. Appl. Mic.*, **2**, 205–213.

Raevuori, M.T. and Genigeorgis, C.A. (1975), Effect of pH and sodium chloride on growth of *Bacillus cereus* in laboratory media and certain foods. *Appl. Micro.*, **29**, 68–73.

Raevuori, M.T., Kiutamo, T. and Niskanen, A. (1977), Comparative studies of *Bacillus cereus* isolated from various foods and food poisoning outbreaks. *Acta. Vet. Scand.*, **18**, 397–407

Schiemann, D.A. (1978), Occurrence of *Bacillus cereus* and the bacteriological quality of Chinese 'take-out' foods. *J. Fd. Prot.*, **41**, 450–454.

Schmitt, N., Bowmer, E.J. and Willoughby, B.A. (1976), Food poisoning outbreak attributed to *Bacillus cereus*. *Can. J. Pub. Hlth.*, **67**, 418–422.

Shehata, T.E. and Hassen, M.E. (1981), Effect of selected food additives on the growth of psychrophilic strains of *Bacillus*, I. Immediate action of chloramphenicol, EDTA, tylosin, and nisin on exponential growth of *Bacillus cereus* isolated from milk. *Res. Bull. Fac. Agr.*, Ain Shang University. No 1553.

Shinagawa, K., Kunita, N., Sasaki, Y. and Okamoto, A. (1979), Biochemical characteristics and heat tolerance of strains of *Bacillus cereus* isolated from uncooked and cooked rice after food poisoning outbreaks. *J. Fd. Hyg. Soc. Japan*, **20**, 431–436.

Shingawa, K., Matusaka, N., Konuma, H. and Kurata, H. (1985), The relationship between the diarrheal and other biological activities of *Bacillus cereus* involved in food poisoning outbreaks. *Jap. J. Vet. Sci.*, **47**, 557–565.

Sly, T. and Ross, E. (1982), Chinese foods: relationship between hygiene and bacterial flora. *J. Fd. Prot.*, **45**, 115–118.

Smith, N.R., Gordon, R.E. and Clark, F.E. (1952), *Aerobic Spore Forming Bacteria*, Monograph No 16. USDA: Washington DC.

Sooltan, J.R.A., Mead, G.C. and Norris, A.P. (1987), Incidence and growth potential of *Bacillus cereus* in poultry meat products. *Fd. Mic.*, **4**, 347–351.

Spira, W.M. (1974), Purification and characterization of *Bacillus cereus* enterotoxin. Doctoral Thesis: University of Wisconsin.

Spira, W.M. and Goepfort, J.M. (1975), Biological characterisation of an enterotoxin produced by *Bacillus cereus*. *Can. J. Mic.*, **21**, 1236–1246.

Stec, E. and Burzynska, H. (1980), Comparison of media for quantitative determination of *Bacillus cereus* in certain food products. *Roczn. Panst. Zak. Higieny*, **31**, 4007–412.

Steele, J.E. and Miles, M.E. (1981), Food poisoning potential of artificially contaminated vacuum packaged sliced ham in sandwiches. *J. Fd. Prot.*, **44**, 430–434.

Sutherland, J.P., Varnam, A.H. and Evans, M.G. (1986), *A Colour Atlas of Food Quality Control*. Wolfe Publishing Ltd: London.

Szabo, R.A., Todd, E.C.D. and Rayman, M.K. (1984), Twenty-four hour isolation and confirmation of *Bacillus cereus* in foods. *J. Fd. Prot.*, **47**, 856–862.

Taylor, A.J. and Gilbert, R.J. (1975), *Bacillus cereus* food poisoning: a provisional serotyping scheme. *J. Med. Mic.*, **8**, 543–550.

Terranova, W. and Blake, P.A. (1978), Current concepts: *Bacillus cereus* food poisoning. *New Eng. J. Med.*, **298**, 143–144.

Turnbull, P.C.B. (1976), Studies on the production of enterotoxin by *Bacillus cereus*. *J. Clin. Path.*, **29**, 941–948.

Turnbull, P.C.B. (1986), *Bacillus cereus* toxins. In *Pharmacology of Bacterial Toxins*. Pergamon Press: Oxford.

Turnbull, P.C.B., Kramer, J.M., Jorgensen, K. *et al.* (1979a), Properties and production characteristics of vomiting, diarrheal and necrotizing toxins of *Bacillus cereus*. *Am. J. Clin. Nutr.*, **32**, 219–228.

Turnbull, P.C.B., Jorgensen, K., Kramer, J.M. *et al.* (1979b), Severe clinical conditions associated with *Bacillus cereus* and the apparent involvement of enterotoxins. *J. Clin. Path.*, **32**, 289–293.

Walthew, J. and Luck, H. (1978), Incidence of *Bacillus cereus* in milk powder. *S. Afr. J. Dairy Res.*, **10**, 47–50.

Westhoff, D.C. and Dougherty, S.L. (1981), Characterization of *Bacillus* species isolated from spoiled ultra high temperature processed milk. *J. Dairy Sci.*, **64**, 572–580.

Woes, G. (1976), *Bacillus cereus* and *Clostridium perfringens* in dairy products. *Rev. Agr.*, **29**, 993–1005.

Wong, H.-C. and Chen, Y.-L. (1988), Effect of lactic acid bacteria and organic acids on growth and gemination of *Bacillus cereus*. *Appl. Environ. Micro.*, **54**, 2179–2184.

Chapter 14

Abrahamson, K. and Riemann, H. (1971), Prevalence of *Clostridium botulinum* in semipreserved meat products. *Appl. Micro.*, **21**, 543–544.

Alderton, G., Chen, J.K. and Ito, K.A. (1974), Effect of lysozyme on the recovery of heated *Clostridium botulinum* spores. *Appl. Mic.*, **27**, 613–615.

Anon (1971), Report on the incidence of *Clostridium botulinum* in dressed poultry. *Vet. Rec.*, **89**, 668–669.

Anon, (1980), *Behaviour of Pathogens in Cheese*, Document 122. International Dairy Federation: Brussels.

Anon (1982), Canned salmon recall/industry wide and international implications. *Annual Report*, National Food Packers Association Research Laboratories.

Anon (1983), Botulism and commercial pot pie, California. *Morb. Mort. Wkly. Rep.*, **32**, 39–40.

Anon (1984), Botulism risk from post-processing contamination of commercially canned foods in metal containers. *J. Fd. Prot.*, **47**, 801–816.

Anon (1985), A case of infant botulism — Quebec. *Comm. Dis. Wkly. Rep.*, **11**, 201–202.

Anon (1986), Infant botulism. *Lancet*, **II**, 1256–1257.

Anon (1987a), Infant botulism — California's 1985 and 1986 experience. *J. Fd. Prot.*, **50**, 807.

Anon (1987b), International outbreak of type E botulism associated with ungutted, salted whitefish. *Morb. Mort. Wkly. Rep.*, **36**, 633–634.

Anon (1987c), FDA may halt processing of uneviscerated fish. *Fd. Chem. News*, **Nov. 23**.

Anon (1987d), *Code of Federal Regulations. Acidified Foods.* Title 21, Part 114. USFDA: Washington DC.

Anon (1988a), Nisin preparation affirmed as GRAS for cheese spread. *Fd. Chem. News*, **April 11**.

Anon (1988b), Pasteurized process cheese spread standards amendment proposed. *Fd. Chem. News*, **April 11**.

Anon (1989), Garlic-in-oil products called immediate hazard. *Fd. Chem. News*, **Mar. 27**.

Archer, D.L. and Young, F.E. (1988), Contemporary issues: diseases with a food vector. *Clin. Mic. Rev.*, **1**, 377–398.

Arnon, S.S. (1986), Infant botulism: anticipating the second decade. *J. Inf. Dis.*, **154**, 201–206.

Arnon, S.S. and Chin, J. (1979), The clinical spectrum of infant botulism. *Rev. Inf. Dis.*, **1**, 614–621.

Aureli. P., Fenicia, B., Pasolini, M., *et al.* (1986), Two cases of type E infant botulism caused by neurotoxigenic *Clostridium butyricum* in Italy. *J. Inf. Dis.*, **154**, 207–211.

Baird-Parker, A.C. and Freame, B. (1967), Combined effect of water activity, pH and temperature on the growth of *Clostridium botulinum* from spore and vegetative cell inocula. *J. Appl. Bact.*, **30**, 420–429.

Baker, D.A. and Genigeorgis, C.A. (1990), Predicting the safe storage of fresh fish under modified atmospheres with respect to *Clostridium botulinum* toxigenesis by modelling the length of the lag phase of growth. *J. Fd. Prot.*, **53**, 131–140.

Ball, A.P., Hopkinson, R.B., Farrell, I.D. *et al.* (1979), Human botulism caused by *Clostridium botulinum* type E: the Birmingham outbreak. *Quarterly J. Med.*, **48**, 473–491.

Barbut, S., Tanaka, N., Gassens, R.G. and Maurer, A.J. (1986), Effects of NaCl reduction and polyphosphate addition on *Clostridium botulinum* toxin production in turkey frankfurters. *J. Fd. Sci.*, **51**, 1136–1138.

Barbut, S., Meske, L., Thayer, D.W. *et al.* (1988), Low dose gamma irradiation effects on *Clostridium botulinum* inoculated turkey frankfurters containing various sodium chloride levels. *Lancet*, **II**, 849–853.

Bom, I.J., Smely, J.P.P.M., Kersters, K. and Verrips, C.T. (1986), Identification and grouping of *Clostridium botulinum* strains by numerical analysis of their electrophoretic protein patterns. *J. Appl. Bact.*, **60**, 483–490.

Bonventre, P.F. and Kempe, L.L. (1959), Physiology of toxin production by *Clostridium botulinum* types A and B. III Effect of pH and temperature during incubation on growth, autolysis, and toxin production. *Appl. Micro.*, **7**, 374–381.

Boroff, D.A. and Shu-Chen, G. (1973), Radioimmunoassay for type A toxin of *Clostridium botulinum*. *Appl. Micro.*, **25**, 545–549.

Bricaire, F. (1982), Botulisme. *Presse Med.*, **11**, 1109–1110.

Briozzo, J., Amato de lagarde, A., Chirife, J. and Parada, J.L. (1985), Influence of potassium sorbate on toxin production by *Clostridium botulinum* type A in model systems of cheese spread. *J. Fd. Tech.*, **20**, 383–388.

Cato, E.P., George, W.L. and Finegold, S.M. (1986), *Clostridium*. In *Bergey's Manual of Systematic Bacteriology*, Vol 2, Sneath, P.H.A., Mair, N.S., Sharpe, M.E. and Holt, J.G. (Eds). Williams and Wilkins: Baltimore and London.

Chia, J.K., Clark, J.B., Ryan, C.A. and Pollock, M. (1986), Botulism in an adult associated with food-borne intestinal infection with *Clostridium botulinum*. *New Eng. J. Med.*, **315**, 239–241.

Christiansen, L.N. and Foster, E.M. (1965), Effect of vacuum packaging on growth of *Clostridium botulinum* and *Staphylococcus aureus* in cured meats. *Appl. Micro.*, **13**, 1023–1030.

Conner, D.E., Scott, V.N., Berned, D.T. and Kautter, D.A. (1990), Potential *Clostridium botulinum* hazard associated with extended shelf life refrigerated foods: A review. *J. Fd. Safety*, **10**, 131–153.

Craig, J.M. and Pilcher, K.S. (1966), *Clostridium botulinum* type F: isolation from salmon from the Columbus river. *Science*, **155**, 311.

Critchley, E.M.R., Hayes, P.J. and Isaacs, P.E.T. (1989), Outbreak of botulism in north west England and Wales, June 1989. *Fd. Mic.*, **5**, 1–7.

Crowther, J.S. and Baird-Parker, A.C. (1984), The pathogenic and toxigenic spore-forming bacteria. In *The Bacterial Spore*, Vol 2, Hurst, A. and Gould, G.W. (Eds). Academic Press: London.

Cuppett, S.L., Gray, J.I., Pestka, J.J. *et al.* (1987), Effect of salt level and nitrite on toxin production by *Clostridium botulinum* type E spores. *J. Fd. Prot.*, **50**, 212–217.

Demeulemeester, J.-R. and Demeulemeester, J.-M. (1982), UK Patent Application GB 2098850 A.

De Waart, J., Van Aken, F. and Pouw, H. (1972), Detection of orally toxic microbial metabolites in foods with bioassay systems. *Zentralbl. Bakt. Infekt. Hyg. Abt. I. Orig.*, **222**, 96–114.

DeWit, J.C., Notermans, S., Gorin, N. amd Kampelmacher, E.H. (1979), Effect of garlic oil or onion oil on toxin production by *Clostridium botulinum* in meat slurry. *J. Fd. Prot.*, **42**, 222–224.

Dowell, V.R. and Dezfulian, M. (1981), Physiological characterization of *Clostridium botulinum* and development of practical isolation and identification procedures. In *Biomedical Aspects of Botulism*, Lewis, G.C. (Ed). Academic Press: New York.

Dymicky, M. and Huhtanen, C.N. (1979), Inhibition of *Clostridium botulinum* by p-hydroxybenzoic acid n-alkyl esters. *Antimicrob. Agents Chemother.*, **15**, 798.

Eisenberg, M.S. and Bender, T.R. (1976), Plastic bags and botulism; a new twist to an old hazard of the north. *Alaska Med.*, **18**, 47–49.

Eklund, M.W. and Poysky, F.T. (1973), Bacteriophages and toxicity of *Clostridium botulinum*. In *The Microbiological Safety of Food*, Hobbs, B.C. and Christian, J.H.B. (Eds). Academic Press: London and New York.

Eklund, M.W., Poysky, F.T., Mseitif, L.M. and Strom, M.S. (1988a), Evidence for plasmid-mediated toxin and bacteriocin production in *Clostridium botulinum* type G. *Appl. Environ. Micro.*, **54**, 1405–1408.

Eklund, M.W., Peterson, M.E., Paranjpye, R. and Delroy, G.A. (1988b), Feasability of a heat-pasteurisation process for the inactivation of nonproteolytic *Clostridium botulinum* types B and E in vacuum packed, hot-process (smoked) fish. *J. Fd. Prot.*, **51**, 720–726.

Erichsen, I. (1983), Fermented fish and meat products: the present position and future possibilities. In *Food Microbiology: Advances and Prospects*, Roberts, T.A. and Skinner, F.A. (Eds). Academic Press: London and New York.

Farber, J.M. (1986), Predictive modelling of food deterioration and safety. In *Foodborne Microorganisms and their Toxins: Developing Methodology*, Pierson, M.D. and Stern, N.J. (Eds). Marcel Dekker: New York and Basel.

Foegeding, P.M. (1986), Detection and quantitation of sporeforming pathogens and their toxins. In *Foodborne Microorganisms and their Toxins: Developing Methodology*, Pierson, M.L. and Stern, N.J. (Eds). Marcel Dekker: New York and Basel.

Gibbs, P.A., Woods, L.F.J. and Neaves, P.A. (1987), *Improved Methods of Detecting Clostridium botulinum and Toxin: Initial Studies*. Research Report 606, BFMIRA: Leatherhead.

Gibson, A.M., Roberts, T.A. and Robinson, A. (1982), Factors controlling the growth of *Clostridium botulinum* types A and B in pasteurized cured meats. IV. The effect of pig breed, cut and batch of pork. *J. Fd. Tech.*, **17**, 471–482.

Gibson, A.M., Modi, N.K., Roberts, T.A. *et al.* (1987), Evaluation of a monoclonal antibody-based immunoasay for detecting type A *Clostridium botulinum* toxin produced in pure culture and an inoculated model cured meat system. *J. appl. Bact.*, **63**, 217–226.

Giminez, D.F. and Ciccarelli, A.S. (1970), Another type of *Clostridium botulinum*. *Zentralbl. Bakt. Parasit. Infekt. Hyg. Abt. 1. Orig.*, **215**, 221–224.

Glatz, B.A. (1986), Genetic regulation of toxin production by foodborne microbes. In *Food Microbiology, Vol 1. Concepts in Physiology and Metabolism*, Montville, T.J. (Ed). CRC Press: Boca Raton.

Graham, A.F. and Lund, B.M. (1986), The effect of citric acid on proteolytic strains of *Clostridium botulinum*. *J. Appl. Bact.*, **61**, 51–56.

Greenberg, R.A., Tompkin, R.B., Bladel, B.O. (1966), Incidence of mesophilic *Clostridium* spores in raw pork, beef and chicken in processing plants in the United States and Canada. *Appl. Micro.*, **14**, 789–793.

Hall, J.D., McCrosky, L.M., Pincomb, B.J. and Hatheway, C.L. (1985), Isolation of an organism resembling *Clostridium barati* which produces type F botulinal toxin from an infant with botulism. *J. Clin. Mic.*, **21**, 654–655.

Hall, M.A. and Maurer, A.J. (1986), Spice extracts, lauricidin and polypropylene glycol as inhibitors of *Clostridium botulinum* in turkey frankfurter slurries. *Poult. Sci.*, **65**, 1167–1171.

Hauschild, A.H.W. (1982), Assessment of botulism hazards from cured meat products. *Fd. Tech.*, **36** (12), 95–104.

Hauschild, A.H.W. and Hilsheimer, R. (1980), Incidence of *C. botulinum* in commercial bacon. *J. Fd. Prot.*, **43**, 564–571.

Hauschild, A.H.W., Aris, B. and Hilsheimer, R. (1978), *Clostridium botulinum* in marinated products. *Can. Inst. Fd. Sci. Tech. T.*, **8**, 84–87.

Hauschild, A.H.W., Hilsheimer, R., Weiss, K.F. and Burke, R.B. (1988), *Clostridium botulinum* in honey, syrup and dried infant cereals. *J. Fd. Prot.*, **51**, 892–894.

Hirsch, A., Grinsted, E., Chapman, H.R. and Mattick, A.T.R. (1951), A note on the inhibition of an anaerobic sporeformer in Swiss-type cheese by a nisin-producing streptococcus. *J. Dairy Res.*, **18**, 205–211.

Hobbs, G., Stiebrs, A. and Eklund, M.W. (1967), Egg-yolk reactions of *Clostridium botulinum* type E in different basal media. *J. Bact.*, **93**, 1192–1195.

Hobbs, G., Crowther, J.S. and Neaves, P. (1982), Detection and isolation of *Clostridium botulinum*. In *Isolation and Identification Methods for Food Poisoning Organisms*, Corry, J.E.L., Roberts, D. and Skinner, F.A. (Eds).

Holdemann, L.V., Cato, E.P. and Moore, W.E.C. (1977), *Anaerobe Laboratory Manual*, 4th ed. Virginia Polytechnic Institute: Blacksburg.

Huhtanen, C.N., Naghski, J., Custer, C.S. and Russell, R.W. (1976), Growth and toxin production by *Clostridium botulinum* in mouldy tomato juice. *Appl. Environ. Micro.*, **32**, 711–715.

Huhtanen, C.N., Knox, D. and Shimanaki, H. (1981), Incidence and origin of *Clostridium botulinum* spores in honey. *J. Fd. Prot.*, **44**, 812–814.

Hurst, A. (1981), Nisin. *Adv. Appl. Mic.*, **27**, 85–103.

Huss, H.H., Pedersen, A. and Cann, D.C. (1974a), The incidence of *Clostridium botulinum* in Danish trout farms. I. Distribution in fish and their environment. *J. Fd. Tech.*, **9**, 445–450.

Huss, H.H., Pedersen, A. and Cann, D.C. (1974b), The incidence of *Clostridium botulinum* in Danish trout farms. II. Measures to reduce contamination of the fish. *J. Fd. Tech.*, **9**, 451–458.

Huss, H.H., Schaeffer, I., Petersen, E.R. and Cann, D.C. (1979), Toxin production by *Clostridium botulinum* type E in fresh herring in relation to the measured oxidation reduction potential (Eh). *Nord. Vet.*, **31**, 81–87.

Huss, H.H., Schaeffer, I., Pedersen, A. and Jepsen, A. (1980), Toxin production by *Clostridium botulinum* type E in smoked fish in relation to the measured oxidation reduction potential, packaging method, and the associated microflora. *Adv. Fish Sci. Tech.*, **13**, 476–484.

Ingram, M. (1973), The microbiological effects of nitrite, *Proceedings 2nd Int. Symp. Nitrate, Meat Prod.*, Zeist, Pudoc: Wageningen.

Insalata, N.F., Witzeman, S.J., Fredericks, G.T. and Sunga, F.C.A. (1969), Incidence study of spores of *Clostridium botulinum* in convenience foods. *Appl. Micro.*, **17**, 542–544.

Jarvis, B. and Patel, M. (1979), *The Occurence and Control of Clostridium botulinum in foods*. Research Report 686, BFMIRA: Leatherhead.

Jarvis, B., Rhodes, A.C. and Patel, M. (1979), Microbiological safety of pasteurized, cured meats: inhibition of *Clostridium botulinum* by curing salts and other additives. In *Food Microbiology and Technology*, Jarvis, B., Christian, J.H.B. and Michener, H.A. (Eds). Medicina Viva: Parma.

Kautter, D.A. (1964), *Clostridium botulinum* type E in smoked fish. *J. Fd. Sci.*, **29**, 843–849.

Kautter, D.A. and Lynt, R.K. (1984), *Clostridium botulinum* (APHA Handbook). American Public Health Association: Washington DC.

Kautter,, D.A., Harmon, S.M., Lynt, R.K. and Lilly, T. (1966), Antagonistic effect on *Clostridium botulinum* type E by organisms resembling it. *Appl. Micro.*, **14**, 616–622.

Kautter, D.A., Lilly, T. and Lynk, R.K. (1978), Evaluation of the botulism hazard in fresh mushrooms wrapped in commercial polyvinylchloride film. *J. Fd. Prot.*, **41**, 120–121.

Kautter, D.A., Lilly, T., Lynt, R.K. and Solomon, H.M. (1981), Evaluation of the botulism hazard for nitrogen packed sandwiches. *J. Fd. Prot.*, **44**, 59–62.

Kautter, D.A., Lilly, T., Solomon, H.M. and Lynt, R.K. (1982), *Clostridium botulinum* species in infant foods: a survey. *J. Fd. Prot.*, **45**, 1028–1029.

Kosikowski, F.V. (1971), *Cheese and Fermented Milk Foods*, 2nd ed. Published by the author: Ann Arbor.

Kurokowa, Y., Oguma, K., Yokosawa, N. *et al.* (1987), Binding and cytotoxic effects of *Clostridium botulinum* Type A, C, and E toxins in primary neuron cultures from foetal mouse brains. *J. Gen. Mic.*, **133**, 2647–2657.

Lamanna, C. and Carr, C. (1967), The botulinal, tetanal and enterostaphylococcal toxins: a review. *Clin. Pharmacol. Ther.*, **8**, 286–332.

Lecour, H., Ramos, M.H., Almeida, B. and Barbosa, R. (1988), Food-borne botulism. A review of 13 outbreaks. *Arch. Intern. Med.*, **148**, 578–580.

Lewis, G.E., Kulinsky, S.S., Reichard, D.W. and Metzger, J.F. (1981), Detection of *Clostridium botulinum* type G toxin by enzyme linked immunosorbent assay. *Appl. Environ. Micro.*, **42**, 1018–1022.

Light, N.D., Williams, R., Barrett, J. *et al.* (1988), A pilot study on the use of vacuum cooking as a production system for high quality foods in catering and retail. *Int. J. Hosp. Man.*, **7**, 21–27.

Lilly, T., Harmon, S.M., Kautter, D.A *et al.* (1971), An improved medium for detection of *Clostridium botulinum* type E. *J. Milk Fd. Tech.*, **34**, 492–497.

Lilly, T., Kautter, D.A., Lynt, R.K. and Solomon, H.M. (1984), Immunodiffusion detection of *Clostridium botulinum* colonies. *J. Fd. Prot.*, **47**, 868–875.

Lund, B.M., Graham, A.F. and Franklin, J.G. (1987a), The effect of acid pH on the probability of growth of proteolytic strains of *Clostridium botulinum*. *Int. J. Fd. Mic.*, **4**, 215–226.

Lund, B.M., George, S.M. and Franklin, J.G. (1987b), Inhibition of type A and type B (proteolytic) *Clostridium botulinum* by sorbic acid. *Appl. Environ. Micro.*, **53**, 935–941.

Lund, B.M., Graham, A.F. and George, S.M. (1988), Growth and formation of toxin by *Clostridium botulinum* in peeled, inoculated, vacuum-packed potatoes after a double pasteurisation and storage at 25°C. *J. Appl. Bact.*, **64**, 241–246.

Lynt, R.K., Kautter, D.A. and Read, R.V.B. (1975), Botulism in commercially canned foods. *J. Milk Fd. Tech.*, **38**, 546–550.

MacDonald, K.L., Spengler, R.F., Hatheway, C.L. *et al.* (1985), Type A botulism from sauteed onions. *J. Am. Med. Assoc.*, **253**, 1275–1278.

MacDonald, K.L., Cohen, M.L. and Blake, P.A. (1986), The changing epidemiology of adult botulism in the United States. *Am. J. Epid.*, **124**, 794–799.

McCroskey, L.M. and Hatheway, C.L. (1988), Laboratory findings in four cases of adult botulism suggest colonization of the intestinal tract. *J. Clin. Mic.*, **26**, 1052–1054.

McCroskey, L.M., Hatheway, C.L., Fenicia, L. *et al.* (1986), Characterization of an organism that produces type E botulinum toxin which resembles *Clostridium butyricum* from an infant with type E botulism. *J. Clin. Mic.*, **23**, 201–262.

Merson, M., Hughes, J., Dowell, V. *et al.* (1974), current trends in botulism in the United States. *J. Am. Med. Assoc.*, **229**, 1305–1308.

Meyer, A. (1973), *Processed Cheese Manufacture*. Food Trade Press: London.

Midura, T.F., Snowden, S., Wood, R.M. and Arnon, S.S. (1979), Isolation of *Clostridium botulinum* from honey. *J. Clin. Mic.*, **9**, 282–285.

Modi, N.K., Shone, S.S., Hambleton, P. and Melling, J. (1986), Monoclonal antibody based amplified enzyme-linked immunosorbent assays for *Clostridium botulinum* toxin types A and B. *Proc. 2nd World Congress Foodborne Infections and Intoxications*, Vol II. Institute of Veterinary Medicine-Robert von Ostertag: Berlin.

Morris, J.G. and Hatheway, C.L. (1980), Botulism in the United States, 1979. *J. Inf. Dis.*, **142**, 302.

Mossel, D.A.A. and DeWaart, J. (1968), The enumeration of clostridia in foods and feeds. *Ann. Inst. Past. Lille*, **19**, 13–27.

Nielson, S.F. and Petersen, H.O. (1967), Studies on the occurrence and germination of *Clostridium botulinum* in smoked salmon. *Proc. 5th Int. Symp. Fd. Mic.*, Moscow.

Norbo, E. and Valland, M. (1967), Bacteriological and toxicological investigation on a case of human botulism type E by consumption of home-prepared fermented fish (rakefisk). *Nord. Vet. Med.*, **19**, 536–539.

Notermans, S., Dufrenne, J. and Keybets, M.J.H. (1985), Use of preservatives to delay toxin formation by *Clostridium botulinum* (type B, strain Okra) on vacuum-packed cooked potatoes. *J. Fd. Prot.*, **48**, 851–855.

Notermans, S., Dufrenne, J. and Gerrits, J.P.G. (1989), Natural occurrence of *Clostridium botulinum* on fresh mushrooms (*Agaricus bisporus*). *J. Fd. Prot.*, **52**, 733–736.

Oldlaug, T.E. and Pflug, I.J. (1979), *Clostridium botulinum* growth and toxin production in tomato juice containing *Aspergillus gracilis*. *Appl. Environ. Micro.*, **37**, 496–504.

Otofuji, T., Tokiwa, H. and Takahashi, K. (1987), A food-poisoning incident caused by *Clostridium botulinum* toxin A in Japan. *Epid. Inf.*, **99**, 167–172.

Pickett, J., Berg, B., Chaplin, E. and Brunstetter-Shafer, M. (1976), Syndrome of botulism in infancy: clinical and electrophysiologic study. *New Eng. J. Med.*, **295**, 770–772.

Pierson, M.D. and Reddy, N.R. (1988), *Clostridium botulinum* — Status summary. *Fd. Tech.*, **42**(4), 196–198.

Post, L.S., Lee, D.A., Solberg, M. *et al.* (1985), Development of botulinal toxin and sensory deterioration of vacuum and modified atmosphere packaged fish fillets. *J. Fd. Sci.*, **50**, 990–996.

Raatjes, G.J.M. and Smelt, J.P.P.M. (1979), *Clostridium botulinum* can grow and form toxin at pH values lower than 4.6. *Nature*, **281**, 398–399.

Raffael, M. (1984), Revolution in the kitchen. *Caterer and Hotelkeeper*. 16 Aug.

Restaino, L., Komatsu, K.K. and Syracuse, M.J. (1981), Effects of acids on potassium sorbate inhibition of food-related microorganisms in culture media. *J. Fd. Sci.*, **47**, 134–138.

Roberts, D. (1982), Bacteria of public health significance. In *Meat Microbiology*, Brown, M.H. (Ed). Applied Science Publishers: London and New York.

Roberts, T.A. and Smart, J.L. (1976), The occurrence and growth of *Clostridium* species in vacuum-packed bacon with particular reference to *Clostridium perfringens* (*welchii*) and *Clostridium botulinum*. *J. Fd. Tech.*, **11**, 239–244.

Roberts, T.A. and Jarvis, B. (1983), Predictive modelling of food safety with particular reference to *Clostridium botulinum* in model cured meat systems. In *Food Microbiology: Advances and Prospects*, Roberts, T.A. and Skinner, F.A. (Eds). Academic Press: London and New York.

Roberts, T.A., Thomas, A. and Gilbert, R.J. (1973), A third outbreak of botulism in broiler chickens. *Vet. Rec.*, **92**, 107–109.

Roberts, T.A., Gibson, A.M. and Robinson, A. (1981a), Factors controlling the growth of *Clostridium botulinum* types A and B in pasteurized, cured meats. I. Growth in pork slurries prepared from 'low' pH meat (pH range 5.5–6.3). *J. Fd. Tech.*, **16**, 239–266.

Roberts, T.A., Gibson, A.M. and Robinson, A. (1981b), Factors controlling the growth of *Clostridium botulinum* types A and B in pasteurized, cured meats. II. growth in pork slurries prepared from 'high' pH meat (pH range 6.3–6.8). *J. Fd. Tech.*, **16**, 267–282.

Roberts, T.A., Gibson, A.M. and Robinson, A. (1981c), Prediction of toxin production by *Clostridium botulinum* in pasteurized pork slurry. *J. Fd. Tech.*, **16**, 337–355.

Rose, S.A., Modi, N.K., Tranter, H.S. *et al.* (1988), Studies on the irradiation of toxins of *Clostridium botulinum* and *Staphylococcus aureus*. *J. Appl. Bact.*, **65**, 223–229.

Sakaguchi, G. (1979), Botulism. In *Foodborne Infections and Intoxications*, Riemann, H. and Bryan, F.L. (Eds). Academic Press: New York.

Schaffner, W. (1981), Botulism. In *Infectious Diseases*, Sanford, J. and Labs, J. (Eds). Grune and Stratton: New York.

Schmidt-Nowara, W. Samet, J. and Rosario, P. (1983), Early and late pulmonary complications of botulism. *Arch. Intern. Med.*, **143**, 451–456.

Scott, V.N. and Duncan, C. (1978), Cryptic plasmids in *Clostridium botulinum* and *Clostridium botulinum*-like organisms. *FEMS Mic. Lett.*, **4**, 55–58.

Scott, V.N. and Taylor, S.L. (1981), Temperature, pH and spare load effects on the ability of nisin to prevent the outgrowth of *Clostridium botulinum* spores. *J. Fd. Sci.*, **46**, 121–127.

Seals, J.E., Snyder, J.P., Edell, T.A. *et al.* (1981), Restaurant-associated type A botulism: transmission by potato salad. *Am. J. Epid.*, **113**, 436–444.

Sebald, M. and Jougland, J. (1977), Aspects actuels du botulism. *Rev. Pract.*, **27**, 173–176.

Shone, C.C., Wilton-Smith, P., Appleton, N. *et al.* (1985), Monoclonal antibody-based immunoassay for type A *Clostridium botulinum* toxin is comaparable to the mouse bioassay. *Appl. Environ. Micro.*, **50**, 63–67.

Simpson, L.L. (1980), Kinetic studies on the interaction between botulinum toxin type A and the cholinergic junction. *J. Pharm. Exp. Ther.*, **212**, 18–21.

Simpson, L.L. (1981), The origin, structure and pharmacological activity of botulinum toxin. *Pharm. Rev.*, **33**, 155–188.

Simpson, L.L. (1982), Comparison of the pharmacological properties of *Clostridium botulinum* type C_1 and C_2 toxins. *J. Pharm. Exp. Ther.*, **223**, 695–701.

Skulberg, A. (1958), A case of fish-borne type B Botulism. *Nord. Hyg. Tiddskrift*, **39**, 133–136.

Smelt, J.P.P.M., Raatjes, G.J.M., Crowther, J.S. and Verrips, C.T. (1982), Growth and toxin production by *Clostridium botulinum* at low pH values. *J. Appl. Bact.*, **52**, 75–82.

Smith, L.P.S. and Hobbs, G. (1974), Genus III *Clostridium* Prazmowski 1880, 23. In *Bergey's Manual of Determinative Bacteriology*, Buchanan, R.E. and Gibbons, N.E. (Eds). Williams and Wilkins: Baltimore and London.

Smoot, L.A. and Pierson, M.D. (1979), Effect of oxidation reduction potential on the outgrowth and chemical inhibition of *Clostridium botulinum* 10755A spores. *J. Fd. Sci.*, **44**, 706–709.

Solberg, M., Post, L.S., Furgang, D. and Graham, C. (1985), Bovine serum eliminates rapid nonspecific toxic reactions during bioassay of stored fish from *Clostridium botulinum* toxin. *Appl. Environ. Micro.*, **49**, 644–650.

Solomon, H.M. and Kautter, D.A. (1986), Growth and toxin production by *Clostridium botulinum* in sauteed onions. *J. Fd. Prot.*, **49**, 618–620.

Solomon, H.M. and Kautter, D.A. (1988), Outgrowth and toxin production by *Clostridium botulinum* in bottled chopped garlic. *J. Fd. Prot.*, **51**, 862–865.

Somers, E.B. and Taylor, S.L. (1981), Further studies on the antibotulinal effectiveness of nisin in acidic media. *J. Fd. Sci.*, **46**, 1972–1976.

Sonnabend, O., Sonnabend, W., Heinzl, R. *et al.* (1981), Isolation of *Clostridium botulinum* type G and identification of type G botulinal toxin in humans: report of five sudden unexpected deaths. *J. Inf. Dis.*, **143**, 22–27.

Sperber, W.H. (1982), Requirements of *Clostridium botulinum* for growth and toxin production. *Food. Tech.*, **36**(12), 89–93.

St. Louis, M.E., Peck, S.H.S., Bowering, D. *et al.* (1988), Botulism from chopped garlic: delayed recognition of a major outbreak. *Ann. Intern. Med.*, **108**, 363–368.

Strom, M.S., Eklund, M.W. and Poysky, F.T. (1984), Plasmids in *Clostridium botulinum* and related *Clostridium* species. *Appl. Environ. Micro.*, **48**, 956–962.

Suen, J.C., Hatheway, C.L., Steigerwalt, A.C. and Brenner, D.J. (1988a), Genetic confirmation of identities of neurotoxigenic *Clostridium boratii* and *Clostridium butyricum* implicated as agents of infant botulism. *J. Clin. Mic.*, **26**, 2191–2192.

Suen, J.C., Hatheway, C.L., Steigerwalt, A.G. and Brenner, D.J. (1988b), *Clostridium argentinese* sp. nov.: a genetically homogenous group composed of all strains of *Clostridium botulinum* toxin type G and some nontoxigenic strains previously identified as *Clostridium subterminale* or *Clostridium hortiforme*. *Int. J. Syst. Bact.*, **38**, 375–381.

Sugiyama, H. (1980), *Clostridium botulinum* neurotoxin. *Micro. Rev.*, **44**, 419–448.

Sugiyama, H. and Yang, K.H. (1975), Growth potential of *Clostridium botulinum* in fresh mushrooms packaged in semi-permeable plastic film. *Appl. Mic.*, **30**, 964–969.

Sugiyama, H. and Rutledge, K.S. (1978), Failure of *Clostridium botulinum* to grow in fresh mushrooms packaged in plastic film overwrap with holes. *J. Fd. Prot.*, **41**, 348–350.

Sugiyama, H., Mizutani, K. and Yang, K.H. (1972), Basis of type A and type F toxicities of *Clostridium botulinum* strain 84. *Proc. Soc. Exp. Biol. Med.*, **142**, 1063–1067.

Sugiyama, M.W. (1982), Botulism hazards from non-processed foods. *Fd. Tech.*, **36**(12), 113–118.

Tacket, C., Shandera, W., Mann, J. *et al.* (1984), Equine antitoxin use and factors that predict outcome in type A food-borne botulism. *Am. J. Med.*, **76**, 796–798.

Taclindo, C., Midura, T., Nygaard, G.S. and Bodily, H.L. (1987), Examination of prepared foods in plastic packages for *Clostridium botulinum*. *Appl. Mic.*, **15**, 426–430.

Tanaka, N. (1982), Toxin production by *Clostridium botulinum* in media at pH lower than 4.6. *J. Fd. Prot.*, **45**, 234–237.

Thompson, J.A., Glasgow, L.A., Worpinski, J.R. and Olson, C. (1980), Infant botulism: classical spectrum and epidemiology. *Pediatrics*, **66**, 936–942.

Todd, E., Pivnick, H. Chang, R.C. *et al.* (1974), *Clostridium botulinum* in commercially marinated mushrooms. *Can. J. Pub. Hlth.*, **65**, 63–64.

Townsend, C.T., Yee, L. and Mercer, W.A. (1954), Inhibition of the growth of *Clostridium botulinum* by acidification. *Fd. Res.*, **19**, 536–542.

Tsang, N., Post, L.S. and Solberg, M. (1985), Growth and toxin production by *Clostridium botulinum* in model acidified systems. *J. Fd. Sci.*, **50**, 961–965.

Wainright, R.B., Heywood, W.L., Middaugh, J.P. *et al.* (1988), Food-borne botulism in Alaska, 1947–1985: Epidemiology and clinical findings. *J. Inf. Dis.*, **157**, 1158–1162.

Wong, D.M., Young-Perkins, K.E. and Merson, R.L. (1988), Factors influencing *Clostridium botulinum* spore germination, outgrowth, and toxin formation in acidified media. *Appl. Environ. Micro.*, **54**, 1446–1450.

Yndested, M. (1970), A case of botulism, type E, after consumption of rakefisk. *Norsk Vet. Tiddkrift*, **82**, 321–323.

Young-Perkins, K.E. and Merson, R.L. (1987), *Clostridium botulinum* spore germination, outgrowth, and toxin production below 4.6: interactions between pH, total acidity, and buffering capacity. *J. Fd. Sci.*, **52**, 1084–1088.

Chapter 15

Abeyta, C., Michalovskis, A. and Wekell, M.M. (1985), Differentiation of *Clostridium perfringens* from related clostridia in iron milk medium. *J. Fd. Prot.*, **48**, 130–134.

Al-Obaidy, H.M., Khan, M.A., Blaschek, H.P. and Klen, B.P. (1985), Early detection of *Clostridium perfringens* enterotoxin and its relationship to sensory quality of cooked chicken. *J. Fd. Safety*, **7**, 43–45.

Angelotti, R., Hall, H.E., Foter, M.J. and Lewis, K.H. (1962), Quantitation of *Clostridium perfringens* in foods. *Appl. Micro.*, **10**, 193–199.

Anon (1984), *Food and Water-Borne Disease in Canada: Annual Summary 1978*. Health and Welfare Canada: Ottawa.

Arseculeratne, S.N., Panabotte, R.G. and Navaratnam, C. (1980), Pathogenesis of necrotising enteritis with special reference to intestinal hypersensitivity reactions. *Gut*, **21**, 265–268.

Bartholomew, B.A., Stringer, M.F., Watson, G.N. and Gilbert, R.J. (1985), Development and application of an enzyme linked immunosorbent assay for *Clostridium perfringens* type A. *J. Clin. Path.*, **38**, 222–228.

Bartsch, A.G. and Walker, H.W. (1982), Effect of temperature, solute and pH on the tolerance of *Clostridium perfringens* to reduced water activity. *J. Fd. Sci.*, **47**, 1754–1755.

Beerens, H., Romond, C., Lepage, C. and Criquelion, J. (1982), A liquid medium for the enumeration of *Clostridium perfringens* in food and faeces. In *Isolation and Identification Methods for Food Poisoning Organisms*, Corry, J.E.L., Roberts, D. and Skinner, F.A. (Eds). Academic Press: London and New York.

Birkhead, G., Vogt, R.L., Heun, E.M. *et al.* (1988), Characterization of an outbreak of *Clostridium perfringens* food poisoning by qualitative fecal culture and enterotoxin measurement. *J. Clin. Mic.*, **26**, 471–474.

Blankenship, L.C., Craven, S.E., Leffler, R.G. and Custer, C. (1988), Growth of *Clostridium perfringens* in cooked chilli during cooling. *Appl. Environ. Micro.*, **54**, 1104–1108.

Borriello, S.P. and Stephens, S. (1984), The development of the infant gut flora and the medical microbiology of infant botulism and necrotizing enterocolitis. In *Microbes and Infections of the Gut*, Goodwin, C.S. (Ed). Blackwell Scientific Publications: Oxford.

Borriello, S.P., Larson, H.E., Welch, A.R. *et al.* (1984), *Clostridium perfringens* — a possible cause of antibiotic associated diarrhoea. *Lancet*, **I**, 305–307.

Borriello, S.P., Barclay, E., Welch, A.R. (1985), Epidemiology of diarrhoea caused by enterotoxigenic *Clostridium perfringens*. *J. Med. Mic.*, **20**, 362–372.

Brant, P.C., Franti, C.E., Torres-Anjel, M.J. and Riemann, H.P. (1978), Prevalence of *Clostridium perfringens* type A in two groups of food handlers in Bello Horizonte, Minas Gerais, Brazil. *Rev. Latino-Americana Mic.*, **20**, 131–133.

Casman, E.P., Bennett, R.W., Dorsey, A.E. and Stone, J.E. (1969), The micro-slide gel double diffusion test for the detection and assay of staphylococcal enterotoxins. *Hlth. Lab. Sci.*, **6**, 185–198.

Cato, E.P., George, W.L. and Finegold, S.M. (1986), *Clostridium*. In *Bergey's Manual of Systematic Bacteriology*, Sneath, P.H.A., Mair, N.S., Sharpe, M.E. and Holt, J.G. (Eds). Williams and Wilkins: Baltimore and London.

Cherniak, R. and Henderson, B.G. (1972), Immunochemistry of capsular polysaccharides from *Clostridium perfringens*: selected Hobbs strains 1, 5, 9, and 10. *Inf. Imm.*, **6**, 32–37.

Chumney, R.K. and Adams, D.M. (1980), Relationship between the increased sensitivity of heat injured *Clostridium perfringens* spores to surface active antibiotics and to sodium chloride and sodium nitrite. *J. Appl. Bact.*, **49**, 55–63.

Craven, S.E. (1980), Growth and sporulation of *Clostridium perfringens* in foods. *Fd. Tech.*, **34**(4), 80–83.

Craven, S.E. (1988), Increased sporulation of *Clostridium perfringens* in a medium with the prereduced anaerobically sterilised technique or with carbon dioxide or with carbonate. *J. Fd. Prot.*, **51**, 704–706.

Davis, J.G. (1981), Microbiology of cream and dairy desserts. In *Dairy Microbiology, Vol 2. The Microbiology of Milk Products*, Robinson, R.K. (Ed). Applied Science Publishers: London and New Jersey.

Debevere, J.M. (1979), A simple method for the isolation and determination of *Clostridium perfringens*. *Eur. J. Appl. Mic. Biotech.*, **6**, 409–413.

Duncan, C.L. and Somers, E.B. (1972), Quantitation of *Clostridium perfringens* type A enterotoxin by electroimmunodiffusion. *Appl. Micro.*, **24**, 801–804.

Duncan, C.L., Sugiyama, H. and Strong, D.H. (1968), Rabbit ileal loop response to strains of *Clostridium perfringens*. *J. Bact.*, **95**, 1560–1566.

Erickson, J.E. and Deibel, R.H. (1978), New medium for rapid screening and enumeration of *Clostridium perfringens* in foods. *Appl. Environ. Micro.*, **36**, 567–572.

Finegold, S.M. (1977), *Anaerobic Bacteria in Human Disease*. Academic Press: London and New York.

Foegeding, P.M. (1986), Detection and quantitation of sporeforming pathogens and their toxins. In *Foodborne Microorganisms and their Toxins: Developing Methodology*, Pierson, M.D. and Stern, N.J. (Eds). Marcel Dekker: New York and Basel.

Foegeding, P.M. and Busta, F.F. (1980), Production of phospholipase C by nine strains of *Clostridium perfringens* at 37°C and at a constantly rising temperature. *J. Fd. Prot.*, **43**, 15–21.

Fruin, J.T. (1978), Types of *Clostridium perfringens* isolated from selected foods. *J. Fd. Prot.*, **41**, 768–769.

Genigeorgis, C. (1986), Problems associated with perishable processed meats. *Fd. Tech.*, **34**, 80–83.

Genigeorgis, C., Sakaguchi, G. and Riemann, H. (1973), Assay methods for *Clostridium perfringens* type A enterotoxin. *Appl. Micro.*, **26**, 111–115.

Glatz, B.A. (1986), Genetic regulation of toxin production by foodborne microbes. In *Food Microbiology, Vol 1. Concepts in Physiology and Metabolism*, Montville, T.J. (Ed). CRC Press: Boca Raton.

Goldner, S.B., Solberg, M., Jones, S. and Post, L.S. (1986), Enterotoxin synthesis by nonsporulating cultures of *Clostridium perfringens. Appl. Environ. Micro.*, **52**, 407–412.

Granum, P.E. (1986), Structure and mechanism of action of the enterotoxin from *Clostridium perfringens*. In *Proceedings 2nd European Workshop on Bacterial Protein Toxins*, Falmagne, P., Alouf, J., Fehrenbach, F. *et al.* (Eds). Gustav Fischer: Stuttgart.

Granum, P.E. (1990), *Clostridium perfringens* toxins involved in food poisoning. *Int. J. Fd. Mic.*, **10**, 101–112.

Gross, T.P., Kamara, L.B., Hatheway, C.L. *et al.* (1989), *Clostridium perfringens* food poisoning: use of serotyping in an outbreak setting. *J. Clin. Mic.*, **27**, 660–663.

Handford, P.M. (1974), A new medium for the detection and enumeration of *Clostridium perfringens* in foods. *J. Appl. Bact.*, **37**, 559–570.

Hanson, M.W. and Elliot, P.M. (1980), New presumptive identification test for *Clostridium perfringens*: reverse CAMP test. *J. Clin. Mic.*, **12**, 617–621.

Harmon, S.M. and Kautter, D.A. (1970), Method for estimating the presence of *Clostridium perfringens* in food. *Appl. Micro.*, **20**, 913–919.

Harmon, S.M. and Kautter, D.A. (1974), Collaborative study of the '- toxin method for estimating the presence of *Clostridium perfringens* in food. *J. Assoc. Off. Anal. Chem.*, **57**, 91–95.

Harmon, S.M. and Duncan, C.L. (1984), *Clostridium perfringens*. In *Compendium of Methods for the Microbiological Examination of Foods*, 2nd ed, Speck, M.L. (Ed). American Public Health Association: Washington DC.

Harmon, S.M. and Kautter, D.A. (1986), Evaluation of a reversed passive latex agglutination test kit for *Clostridium perfringens* enterotoxin. *J. Fd. Prot.*, **49**, 523–525.

Harmon, S.M., Kautter, D.A. and Peeler, J.T. (1971), Improved medium for isolation of *Clostridium perfringens. Appl. Micro.*, **22**, 688–692.

Hauschild, A.H.W. (1975), Criteria and procedures for implicating *Clostridium perfringens* in foodborne outbreaks. *Can. J. Publ. Hlth.*, **66**, 388–392.

Hauschild, A.H.W. and Hilsheimer, R. (1974a), Enumeration of food-borne *Clostridium perfringens* in egg yolk-free tryptose-sulphite-cycloserine agar. *Appl. Micro.*, **27**, 521–526.

Hauschild, A.H.W. and Hilsheimer, R. (1974b), Evaluation and modification of media for enumeration of *Clostridium perfringens. Appl. Micro.*, **27**, 78–82.

Hauschild, A.H.W., Hilsheimer, R. and Rogers, C.G. (1971), Rapid detection of *Clostridium perfringens* enterotoxin by a modified ligated intestinal loop technique in rabbits. *Can. J. Mic.*, **17**, 1475–1476.

Hauschild, A.H.W., Gilbert, R.J., Harmon, S.M. *et al.* (1977), Comparative study for the enumeration of *Clostridium perfringens* in foods. *Can. J. Mic.*, **23**, 884–892.

Hewitt, J.H., Begg, N., Hewish, J. *et al.* (1986), Large outbreaks of *Clostridium perfringens* food poisoning associated with the consumption of boiled salmon. *J. Hyg.*, **97**, 71–80.

Hirsch, A. and Grinsted, E. (1954), Methods for the growth and enumeration of anaerobic spore formers from cheese, with observations on the effect of nisin. *J. Dairy Res.*, **21**, 101–110.

Hobbs, B.C., Smith, M.E., Oakley, C.L. *et al.* (1953) *Clostridium welchii* Food poisoning. *J. Hyg.*, **51**, 75–101.

Hobbs, G., Williams, K. and Willis, A.T. (1971), Basic methods for the isolation of clostridia. In *Isolation of Anaerobes*, Shapton, D.A. and Board, R.G. (Eds). Academic Press: London and New York.

Hoskins, C.B. and Davidson, P.M. (1988), Recovery of *Clostridium perfringens* from food samples using an oxygen-reducing membrane fraction. *J. Fd. Prot.*, **51**, 187–191.

IAMFES (1987), *Procedures to Investigate Foodborne Illness*. IAMFES: Ames.

ICMSF (1986), *Microorganisms in Foods, Vol 2. Sampling for Microbiological Analysis: Principles and Specific Applications*. Blackwell Scientific Publications: Oxford.

Klindworth, K.J., Davidson, P.M., Brekke, C.J. and Brunen, A.L. (1979), Inhibition of *Clostridium perfringens* by butylated hydroxyanisole. *J. Fd. Sci.*, **45**, 564–567.

Kondo, F. and Nagasui, S. (1988), Gas-liquid chromatographic determination of organic acid production differentiates between the food poisoning and toxigenic-type strains of *Clostridium perfringens. J. Fd. Prot.*, **51**, 283–288.

Labbe, R.G. (1980), Relationship between sporulation and enterotoxin production in *Clostridium perfringens* type A. *Fd. Tech.*, **34**(4), 88–90.

Labbe, R.G. (1988), *Clostridium perfringens. Fd. Tech.*, **42**(4), 195–196.

Larson, H.E. and Borriello, S.P. (1988), Infectious diarrhoea due to *Clostridium perfringens. J. Inf. Dis.*, **157**, 390–391.

Lawrence, G., Brown, R., Bates, J. *et al.* (1984), An affinity technique for the isolation of *Clostridium perfringens* type C from man and pigs in Papua New Guinea. *J. Appl. Bact.*, **57**, 333–338.

Longden, M. and Phillips-Jones, M.K. (1988), The use of plasmid profiling to compare strains of *Clostridium perfringens. J. Appl. Bact.*, **65**, xv.

Luzzi, I., Caprioli, A., Falbo, V. *et al.* (1986), Detection of clostridial toxins in stools from children with diarrhoea. *J. Med. Mic.*, **22**, 29–31.

Mahoney, D.E., Gilliatt, E., Dawson, S. *et al.* (1989), Vero cell assay for rapid detection of *Clostridium perfringens* enterotoxin. *Appl. Environ. Micro.*, **55**, 2141–2143.

McClane, B.A. (1989), Characterization of calcium in the *Clostridium perfringens* type A enterotoxin-induced release of 3H-nucleotides from vero cells. *Mic. Path.*, **6**, 17–28.

McClane, B.A., Hanna, P.C. and Wnek, A.P. (1988), *Clostridium perfringens* enterotoxin. *Mic. Path.*, **4**, 317–323.

McDonel, J.L. (1980), Mechanism of action of *Clostridium perfringens* enterotoxin. *Fd. Tech.*, **34**(4), 91–95.

McDonel, J.L. (1986), Toxins of *Clostridium perfringens* type A, B, C, D and E. In *Pharmacology of Bacterial Toxins*, Dorner, F. and Drews, H. (Eds). Pergamon Press: Oxford.

Mead, G.C. (1969), The use of sulphite-containing media in the isolation of *Clostridium perfringens. J. Appl. Bact.*, **32**, 358–361.

Mead, G.C., Adams, B.W., Roberts, T.A. and Smart, J.L. (1982), Isolation and enumeration of *Clostridium perfringens*. In *Isolation and Identification Methods for Food Poisoning Organisms*, Corry, J.E.L., Roberts, D. and Skinner, F.A. (Eds). Academic Press: London and New York.

Mehta, M., Naraya, K.G. and Notermans, S. (1989), DOT-enzyme linked immunosorbent assay for detection of *Clostridium perfringens* type A enterotoxin. *Int. J. Fd. Mic.*, **9**, 45–50.

Mossel, D.A.A. and DeWaart, J. (1968), The enumeration of clostridia in foods and feeds. *Ann. Inst. Past.*, Lille, **19**, 13–27.

Murrell, T.G.C., Ingham, B.G., Moss, J.R. and Taylor, W.B. (1987), A hypothesis concerning *Clostridium perfringens* type A enterotoxin (CPE) and sudden infant death syndrome (SIDS). *Medical Hypotheses*, **22**, 401–403.

Naik, H.S. and Duncan, C.L. (1977), Enterotoxin formation in foods by *Clostridium perfringens* type A. *J. Fd. Safety*, **1**, 7–13.

NRC (1985), *An Evaluation of the Role of Microbiological Criteria for Foods and Food Ingredients*. National Academy Press: Washington DC.

Park, Y. and Mikolajcik, E.M. (1979), Effect of temperature on growth and alpha toxin production by *Clostridium perfringens*. *J. Fd. Prot.*, **42**, 533–539.

Pinegar, J.A. and Stringer, M.F. (1977), Outbreaks of food poisoning attributed to lecithinase-negative *Clostridium perfringens*. *J. Clin. Path.*, **30**, 491–492.

Pinegar, J.A. and Suffield, A. (1982), The investigation of food poisoning outbreaks in England and Wales. In *Isolation and Identification Methods for Food Poisoning Organisms*, Corry, J.E.L., Roberts, D. and Skinner, F.A. (Eds). Academic Press: London and New York.

Powers, E.M., Lawyer, R. and Masuoka, Y. (1975), Microbiology of processed spices. *J. Milk Fd. Tech.*, **38**, 683–687.

Pruthi, J.S. (1980), Spices and condiments: Chemistry, microbiology technology. *Adv. Fd. Res.*, (Suppl. 4). Academic Press: New York.

Roberts, D. (1982), Bacteria of public health significance. In *Meat Microbiology*, Brown, M.H. (Ed). Applied Science Publishers: London and New York.

Roberts, T.A. and Smart, J.L. (1976), The occurrence and growth of *Clostridium* spp. in vacuum packed bacon with particular reference to *Cl. perfringens*, (*welchii*) and *Cl. botulinum*. *J. Fd. Tech.*, **11**, 229–234.

Ryu, S. and Labbe, R.G. (1989), Coat and enterotoxin related proteins in *Clostridium perfringens* spores. *J. Gen. Mic.*, **135**, 3109–3118.

Schroder, G. and Schau, H.-P. (1978), Zur vorkommen pathogenen clostridien in der Antarktis. *Z. Gesante. Hyg.*, **24**, 704–708.

Serrano, A.M. and Schneider F.S. (1978), New modification of Willis and Hobbs' method for identification of *Clostridium perfringens*. *Appl. Environ. Micro.*, **35**, 809–810.

Shahidi, S.A. and Ferguson, A.R. (1971), New quantitative, qualitative and confirmatory media for rapid analysis of food for *Clostridium perfringens*. *Appl. Micro.*, **21**, 500–506.

Smart, J.L., Roberts, T.A., Stringer, M.C. and Shah, N. (1979), The incidence and serotypes of *Clostridium perfringens* on beef, pork and lamb carcasses. *J. Appl. Bact.*, **46**, 377–383.

Smith, L.D.S. (1975), Common mesophilic anaerobes including *Clostridium botulinum* and *Clostridium tetani*, in twenty-one soil samples. *Appl. Micro.*, **29**, 590–594.

Spencer, R. (1969), Neomycin-containing media for the isolation of *Clostridium botulinum* and food poisoning strains of *Clostridium welchii*, *J. Appl. Bact.*, **32**, 170–174.

Stelma, G.N., Johnson, C.H. and Shah, D.B. (1985), Detection of enterotoxin in colonies of *Clostridium perfringens* by a solid phase enzyme-linked immunosorbent assay. *J. Fd. Prot.*, **48**, 227–233.

Sterne, M. and Batty, I. (1975), *Pathogenic Clostridia*. Butterworths: London.

St John, W.D., Matches, J.R. and Wekell, M.M. (1982), Use of iron milk medium for enumeration of *Clostridium perfringens*. *J. Assoc. Off. Anal. Chem.*, **65**, 1129–1132.

Stringer, M.F., Watson, G.N. and Gilbert, R.J. (1982), *Clostridium perfringens* type A: serological typing and methods for detection of enterotoxin. In *Isolation and Identification Methods for Food Poisoning Organisms*, Corry, J.E.L., Roberts, D. and Skinner, F.A. (Eds). Academic Press: London and New York.

Stringer, M.F., Gilbert, R.J., Hassall, J.E. and Wallace, J.G. (1983), Carriage of *Clostridium perfringens* in different groups of the population. In *Food Microbiology: Advances and Prospects*, Roberts, T.A. and Skinner, F.A. (Eds). Academic Press: London and New York.

Sutton, R.G.A. and Hobbs, B.C. (1971), Resistance of vegetative cells of *Clostridium welchii* to low pH. *J. Med. Mic.*, **4**, 539–543.

Tsai, C.C., Torres-Anjel, M.J. and Riemann, H.P. (1974), Improved culture techniques and sporulation medium for enterotoxin production by *Clostridium perfringens* type A. *J. Formosan. Med. Assoc.*, **73**, 404–409.

Uemura, T., Genigeorgis, C., Riemann, H.P. and Franti, C.E. (1974), Antibody against *Clostridium perfringens* type A enterotoxin in human sera. *Inf. Imm.*, **9**, 470–471.

Van Damme-Jongsten, M., Rodhouse, J., Gilbert, R.J. and Notermans, S. (1990), DNA probes for detection of enterotoxigenic *Clostridium perfringens* strains isolated from outbreaks of food poisoning. *J. Clin. Mic.*, **28**, 131–133.

Watson, G.N. (1985), The assessment and application of a bacteriocin typing scheme for *Clostridium perfringens*. *J. Hyg.*, **42**, 144–149.

Watson, G.N., Stringer, M.F., Gilbert, R.J. and Mahoney, D.E. (1983), Development and application of a bacteriocin-typing scheme for *Clostridium perfringens*. In *Food Microbiology: Advances and Prospects*, Roberts, T.A. and Skinner, F.A. (Eds). Academic Press: London and New York.

Zeissler, J. and Rassfeld-Sternberg, L. (1949), Enteritis necroticans due to *Clostridium welchii* type F. *B. Med. J.*, **1**, 267–269.

Chapter 16

Ahmad, N. and Marth, E.H. (1989), Behaviour of *Listeria monocytogenes* at 7, 13, 21 and 35°C in tryptose broth acidified with acetic, citric or lactic acid. *J. Fd. Prot.*, **52**, 688–695.

Al-Ghazali, M.R. and Al-Azawi, S.K. (1988a), Effects of sewage treatment on the removal of *Listeria monocytogenes. J. Appl. Bact.*, **65**, 203–208.

Al-Ghazali, M.R. and Al-Azawi, S.K. (1988b), Storage effects of sewage sludge cake on the survival of *Listeria monocytogenes. J. Appl. Bact.*, **65**, 209–213.

Anon (1987), Swiss Vacherin Mont d'Or cheese, *Press Release* 87/419. DHSS: London.

Anon (1988a), Micro panel chairman Gilles indicates displeasure at CDC's choice of *Lancet. Fd. Chem. News*, **Oct 23**.

Anon (1988b), Imported shrimp may be detained because of ant contamination. *Fd. Chem. News*, **Feb 8**.

Anon (1988c), *Listeria* strain found to survive boiling of shrimp. *Fd. Chem. News*, **Mar 7**.

Anon (1989a), Current meat processing may not kill *Listeria*, study shows. *Fd. Chem. News*, **Feb 27**.

Anon (1989b), *Listeria* in shrimp killed by boiling plus freezing, FDA-er tells ASM. *Fd. Chem. News*, May 22.

Anon (1990), *The Guardian*, **Jan 18**.

Armstrong, D. (1985), *Listeria monocytogenes*. In *Principles and Practices of Infectious diseases*, 2nd ed, Mandell, G.L., Gouglas, R.G. and Bennett, J.E. (Eds). John Wiley: New York.

Armstrong, R., Aveyard, J. and Pinegar, J.A. (1989), *Listeria* in cook chill menu items. *Env. Hlth.*, **97**(2), 24–26.

Audurier, A., Chatelain, R., Chalons, F. and Picchaud, M. (1979), Lysotypie de 823 souches de *Listeria monocytogenes* isolees en France de 1958 a 1978. *Ann. Micro.*, **130B**, 179–189

Azadian, B.S., Finnerty, G.T. and Pearson, A.D. (1989), Cheese-borne *Listeria* meningitis in immunocompetent patient. *Lancet*, **I**, 322–323.

Banks, J.G. and Board, R.G. (1985), Preservation by the lactopcroxidase system (LP-S) of a contaminated infant milk formula. *Lett. Appl. Mic.*, **1**, 81–85

Bannerman, E.S. and Bille, J. (1988), A new selective medium for isolating *Listeria monocytogenes* from heavily contaminated material. *Appl. Environ. Micro.*, **54**, 165–167.

Bannister, B.A. (1987), *Listeria monocytogenes* meningitis associated with eating soft cheese. *J. Inf.*, **15**, 165–168.

Bearns, R.E. and Girard, K.F. (1958), The effect of pasteurisation on *Listeria monocytogenes. Can. J. Mic.*, **4**, 55–61.

Beckers, H.J., Soentoro, P.S.S. and Delfgou-van Asch, E.H.M. (1987), The occurrence of *Listeria monocytogenes* in soft cheeses and raw milk and its resistance to heat. *Int. J. Fd. Mic.*, **4**, 249–256.

Berrans, M.E., Brackett, R.E. and Beuchat, L.R. (1989), Growth of *Listeria monocytogenes* on fresh vegetables stored under controlled atmosphere. *J. Fd. Prot.*, **52**, 702–705.

Berry, E.D., Liewen, M.B., Mandigo, R.W. and Hutkins, R.W. (1990), Inhibition of *Listeria monocytogenes* by bacteriocin-producing *Pediococcus* during the manufacture of semidry sausage. *J. fd. Prota.*, **53**, 194–197.

Beuchat, L.R., Brackett, R.E., Hao, D.Y.-Y. and Connor, D.E. (1986), Growth and thermal inactivation of *Listeria monocytogenes* in cabbage and cabbage juice. *Can. J. Mic.*, **32**, 791–795.

Bibb, W.F., Kuffner, T.A. and Weaver, R.E. (1986), Typing of *Listeria monocytogenes* by isoenzyme analysis. *Abstracts American Society for Microbiology*, **C390**, 393.

Blanco, M., Fernandez-Garayzabel, J., Dominguez, L. *et al.* (1989), A technique for the direct identification of haemolytic-pathogenic listeria on selective plating media. *Lett. Appl. Mic.*, **9**, 125–128.

Brackett, R.E. (1988), Presence and persistence of *Listeria monocytogenes* in food and water. *Fd. Tech.*, **42**(2), 162–164.

Brackett, R.E. and Beuchat, L.R. (1990), Pathogenicity of *Listeria monocytogenes* grown on crabmeat. *Appl. Environ. Micro.*, **56**, 1216–1220.

Breer, C. (1986), The occurrence of *Listeria* spp. in cheese. *Proceedings 2nd World Congress Foodborne Infections and Intoxications*, **1**(1), 230–233.

Bunning, V.K., Donnelly, C.W., Peeler, J.T. *et al.* (1988), Thermal inactivation of *Listeria monocytogenes* within bovine milk macrophages. *Appl. Environ. Micro.*, **54**, 364–370.

Butman, B.T., Plank, M.C., Durham, R.J. and Mattingley, J.A. (1988), Monoclonal antibodies which identify a genus-specific *Listeria* antigen. *Appl. Environ. Micro.*, **54**, 1564–1569.

Carpenter, S.L. and Harrison, M.A. (1989), Fate of small populations of *Listeria monocytogenes* on poultry processed using moist heat. *J. Fd. Prot.*, **52**, 768–770.

Cassiday, P.K. and Brackett, R.E. (1989), Methods and media to isolate and enumerate *Listeria monocytogenes*: a review. *J. Fd. Prot.*, **52**, 207–214.

Coleman, G. (1988), Selective culture media may injure *Listeria* cells, reduce count. *Fd. Chem. News*, Oct 24.

Coleman, W.W. (1986), Controlling *Listeria* hysteria in your plant. *Dairy Fd. Sanit.*, **6**, 555–557.

Conner, D.E., Brackett, R.E. and Beuchat, L.R. (1986), Effect of temperature, NaCl and pH on growth of *Listeria monocytogenes* in cabbage juice. *Appl. Environ. Micro.*, **52**, 59–63.

Cossart, P. (1988), The listeriolysin O gene region: a chromosomal locus crucial for the virulence of *Listeria monocytogenes. Infection*, **16**(S2), 157–160.

Cossart, P., Vicente, M.F., Mengaud, J. *et al.* (1989), Listeriolysin O is essential for virulence of *Listeria monocytogenes*: direct evidence caused by gene complementation. *Inf. Imm.*, **57**, 3629–3630.

Cox, L.J. (1989), *Listeria* deserves a fair trial. *Fd. Mic.*, **6**, 63–67.

Cox, L.J., Kleiss, T., Cordier, J.L. *et al.* (1989), *Listeria* spp. in food processing, non-food and domestic environments. *Fd. Mic.*, **6**, 49–61.

Curtis, G.D.W., Mitchell, R.G., King, A.F. and Griffin, E.J. (1989), A selective differential medium for the isolation of *Listeria monocytogenes. Lett. Appl. Micro.*, **8**, 95–98.

Czuprynski, C.J. and Balish, E. (1981), Pathogenesis of *Listeria monocytogenes* for gnotobiotic rats. *Inf. Imm.*, **32**, 323–331.

Czuprynski, C.J., Brown, J.F. and Roll, J.T. (1989), growth at reduced temperatures reduces the virulence of *Listeria monocytogenes* for intravenously but not intragastrically inoculated mice. *Micro. Path.*, **7**, 213–223.

Dallmier, A.W. and Martin, C.E. (1988), Catalase and superoxide dismutase activity after heat injury of *Listeria monocytogenes. Appl. Env. Micro.*, **54**, 581–582.

Datta, A.R., Wentz, B.A. and Hill, W.E. (1988), Identification and enumeration of beta-hemolytic *Listeria monocytogenes* in naturally contaminated dairy products. *J. Ass. Off. Anal. Chem.*, **71**, 673–675.

Dijkstra, R.G. (1982), The occurrence of *Listeria monocytogenes* in surface water of canals and lakes, in ditches of one big polder and in the effluents of canals of a sewage treatment plant. *Zentralbl. Bakt. Parasit. Infekt. Hyg. Abt. I. Orig. B.*, **176**, 202–209.

Dominguez Rodriguez, L., Fernandez, G.S., Fernandez Garayzabal, J. and Ferri, E.R. (1984), New methodology for the isolation of *Listeria* microorganisms from heavily contaminated environments. *Appl. Environ. Micro.*, **47**, 1188–1190.

Dominguez Rodriguez, L., Fernandez Garayzabal, J., Boland, V.J.A. *et al.* (1985), Isolation of *Listeria* spp. from raw milk intended for human consumption. *Can. J. Mic.*, **31**, 938–941.

Dominguez Rodriguez, L., Fernandez Garayzabal, J., Ferri, E.R. *et al.* (1987), Viability of *Listeria monocytogenes* in milk treated with hydrogen peroxide. *J. Fd. Prot.*, **49**, 994–998.

Donnelly, C.W. and Baigent, G. (1986), Use of flow cytometry for the selective identification of *Listeria monocytogenes*. *Abstracts American Society for Microbiology*, **P-18**, 254.

Donnelly, C.W. and Briggs, E.H. (1986), Psychrotrophic growth and thermal inactivation of *Listeria monocytogenes* as a function of milk composition. *J. Fd. Prot.*, **50**, 14–17.

Donnelly, C.W., Briggs, E.H. and Donnelly, L.S. (1987), Comparison of heat resistance of *Listeria monocytogenes* in milk as determined by two methods. *J. Fd. Prot.*, **50**, 14–17.

Doyle, M.P. and Schoeni, J.L. (1987), Selective enrichment procedure for isolation of *Listeria monocytogenes* from fecal and biologic specimens. *Appl. Environ. Micro.*, **51**, 1127–1129.

Doyle, M.P., Meske, L.M. and Marth, E.H. (1985), Survival of *Listeria monocytogenes* during the manufacture and storage of non-fat dried milk. *J. Fd. Prot.*, **48**, 740–742.

Doyle, M.P., Glass, K.A., Beery, J.T. *et al.* (1987), Survival of *Listeria monocytogenes* in milk during high-temperature, short-time pasteurization. *Appl. Environ. Micro.*, **53**, 1433–1438.

El-Kest, S.E. and Marth, E.H. (1988), Inactivation of *Listeria monocytogenes* by chlorine. *J. Fd. Prot.*, **51**, 520–524.

El-shenawy, M.A. and Marth, E.H. (1988a), Sodium benzoate inhibits growth of or inactivates *Listeria monocytogenes*. *J. Fd. Prot.*, **51**, 525–530.

El-shenawy, M.A. and Marth, E.H. (1988b), Inhibition and inactivation of *Listeria monocytogenes* by sorbic acid. *J. Fd. Prot.*, **51**, 842–847.

El-shenawy, M.A. and Marth E.H. (1989), Inhibition or inactivation of *Listeria monocytogenes* by sodium benzoate together with some organic acids. *J. Fd. Prot.*, **52**, 771–776.

Engel, R. (1988), The prevalence and assessment of *Listeria monocytogenes*. WHO Informal Working Group on Foodborne Listeriosis, Geneva 15–19 February.

Evans, J.R., Allen, R.C., Bortolussi, R. *et al.* (1984), Follow-up study of survivors of foetal and early onset neonatal listeriosis. *Clin. Invest. Med.*, **7**, 329–333.

Facinelli, B., Varaldo, P.E., Casolari, C. and Fabio, U., (1988), Cross-infection with *Listeria monocytogenes* confirmed by DNA fingerprinting. *Lancet*, **II**, 1247–1248.

Farber, J.M. and Speirs, J.I. (1987), Monoclonal antibodies directed against the flagellar antigens of *Listeria* species and their potential in EIA-based methods. *J. Fd. Prot.*, **50**, 479–484.

Farber, J.M. and Brown, B.E. (1990), Effect of prior heat shock on heat resistance of *Listeria monocytogenes* in meat. *Appl. Environ. Micro.*, **56**, 1584–1587.

Farber, J.M., Johnston, M.A., Purvis, U. and Loit, A. (1987a), Surveillance of soft and semi-soft cheeses for the presence of *Listeria* spp. *Int. J. Fd. Mic.*, **5**, 157–163.

Farber, J.M., Sanders, G.W., Emmons, D.B. and McKellar, R.C. (1987b), Heat resistance of *Listeria monocytogenes* in artificially inoculated and naturally-contaminated raw milk. *J. Fd. Prot.*, **50**, 893.

Farber, J.M., Sanders, G.W. and Malcolm, S.A. (1988a), The presence of *Listeria* spp. in raw milk in Ontario. *Can. J. Mic.*, **34**, 95–100.

Farber, J.M., Sanders, G.W., Speirs, J.I. *et al.* (1988b), Thermal resistance of *Listeria monocytogenes* in inoculated and naturally contaminated raw milk. *Int. J. Fd. Mic.*, **7**, 277–286.

Farber, J.M., Sanders, G.W. and Speirs, J.I. (1988c), Methodology for isolation of *Listeria* from foods — a Canadian perspective. *J. Assoc. Off. Anal. Chem.*, **71**, 675–678.

Farber, J.M., Sanders, G.W., Dunfield, S. and Prescott, R. (1989), The effect of various acidulants on the growth of *Listeria monocytogenes*. *Lett. Appl. Mic.*, **9**, 181–183.

Fedio, W.M. and Jackson, H. (1989), Effect of tempering on the heat resistance of *Listeria monocytogenes*. *Lett. Appl. Mic.*, **9**, 157–160.

Fenlon, D.R. (1985), Wild birds and silage as resevoirs of *Listeria* in the agricultural environment. *J. Appl. Bact.*, **59**, 537–543.

Fernandez Garayzabal, J.F. Dominguez Rodriguez, L. and Vazquez-Boland, J.A. (1987), Occurrence of *Listeria monocytogenes* in raw milk. *Vet. Rec.*, **120**, 258–259.

Finch, J.E., Prince, J. and Hawksworth, M. (1978), A bacteriological survey of the domestic environment. *J. Appl. Bact.*, **45**, 357–364.

Flamm, R.K., Himrich, D.J. and Thomashaw, M.F. (1989), Cloning of a gene encoding a major secreted polypeptide of *Listeria monocytogenes* and its potential use as a species specific probe. *Appl. Environ. Micro.*, **55**, 2251–2256.

Fleming, D.W., Cochi, S.L., MacDonald, K.L. *et al.* (1985), Pasteurized milk as a vehicle of infection in an outbreak of listeriosis. *New Eng. J. Med.*, **312**, 406–407.

Francis, D.W., Hunt, J.M., Peeler, J. and Lovett, J. (1984), Distribution of *Listeria monocytogenes* into fractions of whole milk during seperation. *Abstracts Association of Official Analytical Chemists*, 19 (E147).

Fuchs, R.S. and Surendran, P.K. (1989), Incidence of *Listeria* in tropical fish and fishery products. *Lett. Appl. Mic.*, **9**, 49–51.

Gaillard, J.L., Berche, P., Mounier, J. *et al.* (1987), *In vitro* model of penetration and intracellular growth in the human enterocyte-like cell line Caco-2. *Inf. Imm.*, **55**, 2822–2829.

Gaze, J.E., Brown, G.D., Gaskell, D.E. and Banks, J.G. (1989), Heat resistance of *Listeria monocytogenes* in non-dairy menstrua. *Tech. Memo. 523.* CFPRA: Chipping Campden.

Geoffrey, C., Gaillard, J.L., Alouf, J. and Berche, P. (1987), Purification, characterisation and toxicity of the sulfhydryl-activated hemolysin listeriolysin O from *Listeria monocytogenes. Inf. Imm.*, **55**, 1641–1646.

George, S.M., Lund, B.M. and Brocklehurst, T.F. (1988), The effect of pH and temperature on initiation of growth of *Listeria monocytogenes. Lett. Appl. Mic.*, **6**, 153–156.

Gilbert, R.J. and Pini, P.N. (1988), Listeriosis and food-borne transmission. *Lancet*, **II**, 472–473.

Gilbert, R.J., Miller, K.L. and Roberts, D. (1989a), *Listeria monocytogenes* and cooked foods. *Lancet*, **I**, 383–384.

Gilbert, R.J., Hall, S.M. and Taylor, A.G. (1989b), Listeriosis update. *PHLS Micro. Dig.*, **6**, 33–37.

Gitter, M. (1976), *Listeria monocytogenes* in 'oven ready' poultry. *Vet. Rec.*, **118**, 240–242.

Gitter, M., Bradley, R. and Blampied, P.H. (1980), *Listeria monocytogenes* infection in bovine mastitis. *Vet. Rec.*, **107**, 390–393.

Golden, D.A., Beuchat, L.R. and Brackett, R.E. (1988), Evaluation of selective direct plating media for their suitability to recover uninjured, heat injured and freeze injured *Listeria monocytogenes* from foods. *Appl. Environ. Mic.*, **54**, 1451–1456.

Gouet, P., Labadie, J. and Serratorre, C. (1978), Development of *Listeria monocytogenes* in monoxenic and polyxenic beef minces. *Zentralbl. Bakt. Parasit. Infekt. Hyg. Abt. Orig.*, **B166**, 87–94.

Gray, M.L., Stafseth, H.I., Thorp, F. *et al.* (1948), A new technique for isolating Listerellae from bovine brain. *J. Bact.*, **55**, 471–474.

Gray, M.L., Stafseth, H.J. and Thorp, F. (1950), The use of potassium tellurite, sodium azide and acetic acid in a selective medium for the isolation of *Listeria monocytogenes. J. Bact.*, **59**, 443–444.

Griffiths, M.W. (1989), *Listeria monocytogenes*: its importance in the dairy industry. *J. Sci. Fd. Agric.*, **47**, 133–158.

Hall, C.J., Neider, C. and Melville, C.A.S. (1985), Listeriosis. *B. Med. J.*, **291**, 608.

Hao, D. Y.-Y., Beuchat, L.R. and Brackett, R.E. (1987), Comparison of media and methods for detecting and enumerating *Listeria monocytogenes* in refrigerated cabbage. *Appl. Environ. Micro.*, **53**, 955–957.

Harvey, R.L. and Chandraesekar, P.H. (1988), Chronic meningitis caused by *Listeria* in a patient infected with human immunodeficiency virus. *J. Inf. Dis.*, **157**, 1091–1092.

Hayes, P.S., Feeley, J.C., Graves, L.M. *et al.* (1986), Isolation of *Listeria monocytogenes* from raw milk. *Appl. Environ. Micro.*, **51**, 438–440.

Heisick, J.E., Harrell, F.M., Peterson, E.H. *et al.* (1989), Comparison of four procedures to detect *Listeria* spp. in foods. *J. Fd. Prot.*, **52**, 154–157.

Henry, B.S. (1933), Dissociation in the genus *Brucella. J. Inf. Dis.*, **52**, 374–402.

Herald, P.J. and Zottola, E.A. (1987), Attachment of *Listeria monocytogenes* and *Yersinia enterocolitica* to stainless steel at various temperatures and pH values. *J. Fd. Prot.*, **50**, 894–899.

Ho, J.L., Shands, K.N., Friedland, G. *et al.* (1981), A multihospital outbreak of Type 4b *Listeria monocytogenes* infection. *Proceedings 21st International Congress in Antimicrobial Agents and Chemotherapy*: Chicago.

Hof, H. (1984), Virulence of different strains of *Listeria monocytogenes* serovar 1/2a. *Med. Mic. Imm.*, **173**, 207–214.

Hyslop, N. St G. (1975), Epidemiologic and immunologic factors in listeriosis. In *Problems of Listeriosis*, Woodbine, M. (Ed). Leicester University Press: Leicester.

Ingham, S.C., Escade, J.M. and McCoon, P. (1990), Comparative growth rates of *Listeria monocytogenes* and *Pseudomonas fragi* in cooked chicken loaf stored under air and two modified atmospheres. *J. Fd. Prot.*, **53**, 289–291.

James, S.M., Fannin, S.L., Agree, B.A. *et al.* (1985), Listeriosis outbreak associated with Mexican-style cheese — California. *Morb. Mort. Wkly. Rep.*, **34**, 357–359.

Juntilla, J.R., Niemela, S.I. and Hirn, J. (1988), Maximum growth temperature of *Listeria monocytogenes* and non-haemolytic listeria. *J. Appl. Bact.*, **65**, 321–327.

Juntilla, J., Hirn, J., Hill, P. and Nurmi, E. (1989), Effect of different levels of nitrite and nitrate on the survival of *Listeria monocytogenes* during the manufacture of fermented sausage. *J. Fd. Prot.*, **52**, 158–161.

Karaioannoglou, P.G. and Xenos, G.C. (1980), Survival of *Listeria monocytogenes* in meatballs. *Hellenic Vet. Med. J.*, **23**, 117–118.

Kerr, K.G. and Lacey, R.W. (1988), *Listeria* in cook-chill food. *Lancet*, **II**, 37–38.

Kerr, K.G., Dealler, S.F. and Lacey, R.W. (1988), Materno-fetal listeriosis from cook-chill and refrigerated food. *Lancet*, **II**, 1133.

Kerr, K.G., Rotowa, N.A., Hawley, D.M. and Lacey, R.W. (1990), Evaluation of the Micro ID and API 50CH systems for identification of *Listeria* spp. *Appl. Environ. Micro.*, **56**, 658–660.

Klinger, J.D., Johnson, A., Croan, D. *et al.* (1988), Comparative studies of nucleic acid hybridisation for *Listeria* in foods. *J. Assoc. Off. Anal. Chem.*, **71**, 669–673.

Kluge, R. and Hof, H. (1986), Zur virulence von *Listeria welshimeri. Zentralbl. Bakt. Mik. Hyg. I. Abt. Orig.*, **A262**, 403–411.

Knabel, S.J., Walker, H.W., Hartmann, P.A. and Mendonca, A.F. (1990), Effect of growth temperature and strictly anaerobic recovery on the survival of *Listeria monocytogenes* during pasteurisation. *Appl. Environ. Micro.*, **56**, 370–376.

Kongshavn, P.A.L. and Skamene, E. (1984), The role of natural resistance in protection of the murine host from listeriosis. *Clin. Invest. Med.*, **7**, 253–263.

Kwantes, W. and Isaac, M. (1971), Listeriosis. *B. Med. J.*, **4**, 296.

Kwantes, W. and Isaacs, M. (1975), *Listeria* infection in West Glamorgan. In *Problems of Listeriosis*, Woodbine, M. (Ed). Leicester University Press: Leicester.

Lacey, R.W. and Kerr, K.G. (1989), Opinion: Listeriosis — the need for legislation. *Lett. Appl. Mic.*, **8**, 121–122.

Lachica, R.V. (1990a), Selective plating media for quantative recovery of food-borne *Listeria monocytogenes. Appl. Environ. Micro.*, **56**, 167–169.

Lachica, N.V. (1990b), Simplified Henry technique for initial recognition of *Listeria* colonies. *Appl. Environ Micro.*, **56**, 1164–1165.

Lachica, R.V. (1990c), Same day identification scheme for colonies of *Listeria monocytogenes. Appl. Environ. Micro.*, **56**, 1166–1168.

Lammerding, A.M. and Doyle, M.P. (1989), Evaluation of enrichment procedures for recovering *Listeria monocytogenes* from dairy products. *Int. J. Fd. Mic.*, **9**, 249–268.

Leasor, S.B. and Foegeding, P.M. (1988), Growth and inactivation of *Listeria monocytogenes* F5069 and Scott A in liquid whole egg. *Inst. Fd. Tech. Ann. Meet.*, Abst 168: New Orleans.

Leblond-Francillard, M., Gaillard, J.L. and Berche, P. (1989). Loss of catalase activity does not reduce growth of *Listeria monocytogenes* in vivo. *Inf. Imm.*, **57**, 2569–2573.

Lee, W.H. and McClain, D. (1986), Improved *Listeria monocytogenes* selective agar. *Appl. Environ. Micro.*, **52**, 1215–1217.

Lee, W.H. and McClain, D. (1987), Personal communication cited by Brackett (1988).

Leimeister-Wachter, M. and Chakraborty, T. (1989), Detection of listeriolysin, the thiul-dependent hemolysin in *Listeria monocytogenes*. *Listeria ivanovii* and *Listeria seeligeri*. *Inf. Imm.*, **57**, 7350–7357.

Lennon, D., Lewis, B., Mantell, B. *et al.* (1984), Epidemic perinatal listeriosis. *Ped. Inf. Dis.*, **3**, 30–34.

Loessner, M.J. and Busse, M. (1990), Bacteriophage typing of *Listeria* species. *Appl. Environ. Micro.*, **56**, 1912–1918.

Loessner, M.J., Bell, R.H., Jay, J.M. and Shelef, L.A. (1988), Comparison of seven plating media for enumeration of *Listeria* spp. *Appl. Environ. Micro.*, **54**, 3003–3007.

Lovett, J. (1988), Isolation and enumeration of *Listeria monocytogenes*. *Fd. Tech.*, **42**(2), 172–175.

Lovett, J. and Twedt, R.M. (1988), *Listeria. Fd. Tech.*, **42**(4), 188–191.

Lovett, J., Francis, J.W. and Hunt, J.M. (1987), *Listeria monocytogenes* in raw milk: detection, incidence and pathogenicity. *J. Fd. Prot.*, , 50, 188–193.

Lovett, J., Bradshaw, J.G., Francis, D.W. *et al.* (1988), Efficacy of high temperature short time pasteurisation for inactivation of *Listeria monocytogenes* in milk. *Abst. 75th Meeting IAMFES*: Tampa.

Lund, B., Knox, M.R. and Cole, M.B. (1989), Destruction of *Listeria monocytogenes* during microwave cooking of food. *Lancet*, **I**, 218.

Mackey, B.M. and Bratchell, N. (1989), A review: the heat resistance of *Listeria monocytogenes*. *Lett. Appl. Micro.*, **9**, 89–94.

Malinverni, R., Glauser, M.P., Bille, J. and Rocourt, J. (1986), Unusual clinical features of an epidemic of listeriosis associated with a particular phage type. *Eur. J. Clin. Mic.*, **5**, 169–171.

Manev, C., Yanakieva, M., Ivanova, E. *et al.* (1975), Healthy Bulgarians as *Listeria* carriers. In *Problems of Listeriosis*, Woodbine, M. (Ed). Leicester University Press: Leicester.

Mann, S. (1969), Uber die zelwandbansteine von *Listeria monocytogenes* und *Erysipelothrix rhusiopathi*. *Erst Abt. Orig.*, B., **209**, 510.

Marshal, D.L. and Schmidt, R.H. (1988), Growth of *Listeria monocytogenes* at 10°C in milk preincubated with selected pseudomonads. *J. Fd. Prot.*, **51**, 777–782.

Marth, E.H. (1988), Disease characteristics of *Listeria monocytogenes*. *Fd. Tech.*, **42**(2), 165–168.

Martin, R.S., Sumarah, R.K. and MacDonald, M.A. (1984), A synthetic based medium for the isolation of *Listeria monocytogenes*. *Clin. Invest. Med.*, **7**, 223–236.

Mattingley, J.A., Butman, B.T., Plank, M.C. and Durham, R.J. (1988), Rapid monoclonal antibody-base enzyme-linked immunosorbent assay for detection of *Listeria* in food products. *J. Assoc. Off. Anal. Chem.*, **71**, 679–681.

McBride, M.E. and Girard, K.F. (1960), A selective method for the isolation of *Listeria monocytogenes* from mixed bacterial populations. *J. Lab. Clin. Med.*, **55**, 153–157.

McCarthy, S.A., Motes, M.L. and McPharson, R.M. (1990), Recovery of heat stressed *Listeria monocytogenes* from experimentally and naturally contaminated shrimp. *J. Fd. Prot.*, **53**, 22–25.

McLauchlin, J. (1987), *Listeria monocytogenes*, recent advances in the taxonomy and epidemiology in humans. *J. Appl. Bact.*, **63**, 1–11.

McLauchlin, J. and Pini, P.N. (1989), The rapid demonstration and presumptive identification of *Listeria monocytogenes* in food using monoclonal antibodies in a direct immunofluorescence test (DIFT). *Lett. Appl. Mic.*, **8**, 25–27.

McLauchlin, J., Audurier, A. and Taylor, A.G. (1986a), The evaluation of a phage-typing system for *Listeria monocytogenes* for use in epidemiological studies. *J. Med. Mic.*, **22**, 367–377.

McLauchlin, J., Audurier, A. and Taylor, A.G. (1986b), Aspects of the epidemiology of human *Listeria monocytogenes* infections in Britain, 1967–1984; the use of serotyping and phage-typing. *J. Med. Mic.*, **22**, 367–377.

McLauchlin, J., Saunders, N.A., Ridley, A.M. and Taylor, A.G. (1988), Listeriosis and food-borne transmission. *Lancet*, **I**, 177–178.

McLure, P.J., Roberts, T.A. and Otto Oguru, P. (1989), Comparison of the effects of sodium chloride, pH and temperature on the growth of *Listeria monocytogenes* on gradient plates and in liquid medium. *Lett. Appl. Mic.*, **9**, 95–99.

Mitscherlich, E. and Marth, E.H. (1984), *Microbial Survival in the Environment: Bacteria and Rickettsia Important in Human and Animal Health*. Springer Verlag: Berlin.

Morris, I.J. and Ribeiro, C.D. (1989), *Listeria monocytogenes* and pate. *Lancet*, **I**, 1285–1286.

Mossel, D.A.A., van Netten, P. and Perales, I. (1987), Human listeriosis transmitted in food in a general medical-microbiological perspective. *J. Fd. Prot.*, **50**, 894–895.

Mottice, S., Robinson, W. and Perrier, L.G. (1987), Plasmid typing for the epidemiological characterization of *Listeria monocytogenes*. Cited by Schlech, 1988.

Mouncer, J., Ryter, A., Coquis-Rondon, M. and Sansonetti, P.J. (1990), Intracellular and cell-to-cell spread of *Listeria monocytogenes* involves interaction with F-action in the enterocyte like cell line Caco-2 *Inf. Imm.*, **58**, 1048–1058.

North, R.J. (1973), Cellular mediators of anti-Listeria immunity as an enlarged population of short-lived, replicating T cells. Kinetics of their production. *J. Exp. Med.*, **129**, 973–992.

Olsen, J.A., Yousef, A.E. and Marth, E.H. (1988), Growth and survival of *Listeria monocytogenes* during making and storage of butter. *Milchwissenschaft*, **43**, 487–489.

Parish, M.E. and Higgins, D.P. (1989), Extinction of *Listeria monocytogenes* in single strength orange juice: comparison of methods for detection in mixed populations. *J. Fd. Safety*, **9**, 267–277.

Patterson, M. (1989), Sensitivity of *Listeria monocytogenes* to irradiation on poutry meat and in phosphate-buffered saline. *Lett. Appl. Mic.*, **8**, 181–184.

Payne, K.D., Rico-Munoz, E. and Davidson, P.M. (1989), The antimicrobial activity of phenolic compounds against *Listeria monocytogenes* and their effectiveness in a model milk system. *J. Fd. Prot.*, **52**, 151–153.

Pini, P.N. and Gilbert, R.J. (1988), A comparison of two procedures for the isolation of *Listeria monocytogenes* from raw chickens and soft cheeses. *Int. J. Fd. Mic.*, **7**, 331–337.

Prentice, G.A. and Neaves, P. (1988), *Listeria monocytogenes* in food: its significance and methods for its detection. *Int. Dairy Fed. Bull.* No 223.

Ralovitch, B.(1984), *Listeriosis Research: present situation and Perspective*. Akademaii Kiado: Budapest.

Reiter, B. (1981), The impact of the lactoperoxidase syatem on the psychrotrophic microflora in milk. In *Psychrotrophic Microorganisms in Spoilage and Pathogenicity*, Roberts, T.A., Hobbs, G., Christian, J.H.B. and Skovgaard, N. (Eds). Academic Press: London and New York.

Rocourt, J., Audurier, A., Courtieu, A.L. *et al.* (1985), A multi-centre study on the phage typing of *Listeria monocytogenes*. *Zentralbl. Bakt. Mik. Hyg. Abt. 1, Orig. A. Med. Mik. Inf. Para.*, **259**, 489–497.

Rocourt, J., Hof, H., Schrettenbrunner, A. *et al.* (1986), Meningite purulente aigue a *Listeria seeligeri* chez un adulte immunocompetent. *Schweiz. Med. Wsch.*, **116**, 248–251.

Rosenow, E.M. and Marth, E.H. (1987a), *Listeria*, listeriosis and dairy foods: a review. *Cult. Dairy Prod. J.*, **22**, 13–17.

Rosenow, E.M. and Marth, E.H. (1987b), Growth of *Listeria monocytogenes* in skim, whole and chocolate milk, and in whipping cream during incubation at 4, 8, 13, 21 and 35°C. *J. Fd. Prot.*, **50**, 452–459.

Rosenow, E.M. and Marth, E.H. (1987c), Addition of cocoa powder, cane sugar, and carrageen to milk enhances growth of *Listeria monocytogenes*. *J. Fd. Prot.*, **50**, 726–729.

Ryser, E.T. and Marth, E.H. (1987a), Behaviour of *Listeria monocytogenes* during the manufacture and ripening of Cheddar cheese. *J. Fd. Prot.*, **50**, 7–13.

Ryser, E.T. and Marth, E.H. (1987b), Fate of *Listeria monocytogenes* during the manufacture and ripening of Camembert cheese. *J. Fd. Prot.*, **50**, 372–378.

Ryser, E.T. and Marth, E.H. (1988), Survival of *Listeria monocytogenes* in cold-pack cheese food during refrigerated storage. *J. Fd. Prot.*, **51**, 615–621.

Schlech, W.F. (1984), New perspectives on the gastro-intestinal mode of transmission in invasive *Listeria monocytogenes* infection. *Clin. Invest. Med.*, **7**, 321–325.

Schlech, W.F. (1988), Virulence characteristics of *Listeria monocytogenes*. *Fd. Tech.*, **42**(2), 176–178

Schlech, W.F., Lavine, P.M., Bortolussi, R.A. *et al.* (1983), Epidemic listeriosis evidence for transmission by food. *New Eng. J. Med.*, **308**, 203–206.

Schlech, W.F., Chase, D.P. and Badley, A. (1986), A rat model of *L. monocytogenes* infection via the oral route: 1. Development and effect of gastric acidity on infective dose. *Proceedings 26th International Congress on Antimicrobial Agents and Chemotherapy*. New Orleans.

Schwartz, B., Ciesielski, C.A., Broome, C.V. *et al.* (1988), Association of sporadic listeriosis with consumption of uncooked hot dogs and undercooked chicken. *Lancet*, **II**, 779–782.

Scott, E., Bloomfield, S.F. and Barlow, C.G. (1982), An investigation of microbial contamination in the home. *J. Hyg.*, **89**, 279–293.

Seeliger, H.P.R. (1961), *Listeriosis*. Hafner Publishing Co.: New York.

Seeliger, H.P.R. (1984), Modern taxonomy of the *Listeria* group: relationship to its pathogenicity. *Clin. Invest. Med.*, **7**, 217–221.

Seeliger, H.P.R. and Finger, H. (1976), *Listeriosis*. In *Infectious Diseases of the Fetus and Newborn Infant*, Remington, J.S. and Klein, J.O. (Eds). W.B. Saunders: Philadelphia.

Seeliger, H.P.R. and Hohne, K. (1979), Serotyping of *Listeria monocytogenes*. In *Methods in Microbiology*, Vol 13, Bergan, T. and Norris, J.R. (Eds). Academic Press: New York.

Seeliger, H.P.R. and Jones, D.M. (1986), *Listeria*. In *Bergey's Manual of Systematic Bacteriology*, Vol 2, Sneath, P.H.A., Mair, N.S., Sharpe, M.E. and Holt, J.G. (Eds). Williams and Wilkins: Baltimore and London.

Shahamat, M., Seaman, A. and Woodbine, M. (1980), Influence of sodium chloride, pH and temperature on the inhibitory avtivity of sodium nitrite on *Listeria monocytogenes*. In *Survival in Extremes of Environment*, Gould, G.W. and Corry, J.E.L. (Eds). Academic Press: London.

Sheppard, J. (1988), Cook-chill and beyond. Presented at the Symposium 'Cook chill and beyond'. London

Siragusa, G.R. and Johnson, M.G. (1989), Inhibition of *Listeria monocytogenes* growth by the lactoperoxidase-thiocyanate-H_2O_2 antimicrobial system. *Appl. Environ. Micro.*, **55**, 2802–2805.

Skovgaard, N. (1988), Listeriosis: a food-borne zoonotic disease. *Scand. D. Ind.*, **2**, 5–7.

Slade, P.J. and Collins-Thompson, D.L. (1987), Two-stage enrichment for isolating *Listeria monocytogenes* from raw milk. *J. Fd. Prot.*, **50**, 904–908.

Slade, P.J., Fistrovici, E.C. and Collins-Thompson, D.L. (1990), Persistence at source of *Listeria* spp in raw milk. *Int. J. Fd. Mic.*, **9**, 197–203.

Smith, J.L. and Archer, D.L. (1988), Heat induced injury in *Listeria monocytogenes*. *J. Ind. Mic.*, **3**, 105–110.

Smith, J.L. and Hunter, S.E. (1988), Heat injury in *Listeria monocytogenes* — prevention by solutes. *Fd. Sci. Tech.*, **21**, 307–311.

Smith, J.L. and Buchanan, R.L. (1990), Identification of supplements that enhance the recovery of *Listeria monocytogenes* on modified Vogel-Johnson agar. *J. Fd. Safety*, **10**, 155–163.

Somers, E. and Taylor, S.L. (1987), Personal communication cited by Doyle, 1988.

Stajner, B., Zacula, S., Kovinic, I. and Galic, M. (1979), Heat resistance of *Listeria monocytogenes* and its survival in raw milk products. *Vet. Glas.*, **33**, 109–112.

Stecha, P.F., Heyman, L.A., Roll, J.T. *et al.* (1990), Effects of growth temperature on the ingestion and killing of *Listeria monocytogenes* by human neutrophils *J. Clin. Mic.*, **27**, 1571–1576.

Stelma, G.N., Reyes, A.L., Peeler, J.T. *et al.* (1987), A pathogenicity test for *Listeria monocytogenes* using immuno-compromised mice. *J. Clin. Mic.*, **25**, 2085–2091.

Stroup, W.H., Prosser, J.T., Tierney, J.T. *et al.* (1988), Effect of high temperature short time pasteurisation on *Listeria monocytogenes. Proceedings 75th Meeting International Association of Milk and Environmental Sanitarians*. Tampa.

Surak, J.G. and Barefoot, S.F. (1987), Control of *Listeria* in the dairy plant. *Vet. Hum. Toxicol.*, **29**, 247–249.

Swaminathan, B., Hayes, P.S. and Przybyszewsli, V.A. (1988), Evaluation of enrichment and plating media for isolating *Listeria monocytogenes. J. Assoc. Off. Anal. Chem.*, **71**, 664–668.

Terplan, G., Schoen, R., Springmeyer, W. *et al.* (1986a), *Listeria monocytogenes* in milk and milk products. In *27 Arbeitstagung des Arbeitsgebietes 'Lebensmittelhygiene' Leitthema: 'Forschung und Praxis der Lebensmittelhygiene'*. Deutsch Veterinarmedizin Gesselschaft: Garnisch-Partenkirken.

Terplan, G., Schoen, R., Springmeyer, W. *et al.* (1986b), Occurrence behaviour and significance of *Listeria* in milk and dairy products *Arch. Lebensmittelhyg.*, **37**, 131–137.

Tim, M.W., Jackson, M.A., Shannon, K. *et al.* (1984), Non-neonatal infection due to *Listeria monocytogenes. Pediatr. Inf. Dis.*, **3**, 213–217.

van Netten, P., van der Ven, A., Perales, I. and Mossel, D.A.A. (1988), A selective and diagnostic medium for use in the enumeration of *Listeria spp* in foods. *Int. J. Fd. Mic.*, **6**, 187–188.

Vazquez-Boland, J.A., Dominguez, L., Rodriguez-Ferri, E.S. and Suarez, G. (1989), Purification and characterization of two *Listeria ivanovii* cytolysing, a sphingo-myelinase C and a thiol-activated toxin (Ivanolysin O). *Inf. Imm.*, **57**, 3928–3935.

Walker, S.J. and Stringer, M.F. (1987), Growth of *Listeria monocytogenes* and *Aeromonas hydrophila* at chill temperatures. *J. Appl. Bact.*, **63**, xx.

Watkins, J. and Sleath, K.P. (1981), Isolation and enumeration of *Listeria monocytogenes* from sewage, sewage sludge and river water. *J. Appl. Bact.*, **50**, 1–9.

Weagant, S.D., Sado, P.N., Colburn, K.G. *et al.* (1989), The incidence of *Listeria* species in frozen seafood products. *J. Fd. Prot.*, **51**, 655–657.

Weis, J. and Seeliger, H.P.R. (1975), Incidence of *Listeria monocytogenes* in nature. *Appl. Micro.*, **30**, 29–36.

Welshimer, H.J. and Donker-Voet, J. (1971), *Listeria monocytogenes* in nature. *Appl. Micro.*, **21**, 516–522.

Wexler, H. and Oppenheim, J.D. (1979), Isolation, characterization and biological properties of an endotoxin-like material from the Gram-positive organism *Listeria mono-cytogenes. Inf. Imm.*, **23**, 843–845.

WHO (1988), Working Group. Foodborne listeriosis. *Bull. WHO*, **66**, 421–428.

Wilkins, P.O., Bourgeois, R. and Murray, R.G.E. (1972), Psychrotrophic properties of *Listeria monocytogenes. Can. J. Mic.*, **18**, 543–549.

Yousef, A.E. and Marth, E.H. (1988), Inactivation of *Listeria monocytogenes* by ultraviolet energy. *J. Fd. Sci.*, **53**, 571–573.

Chapter 17

Anon (1978), Grade A pasteurized milk ordinance. (1978), recommendations. USPHS/FDA Publ. 229. US Government Printing Office: Washington DC.

APHA (1984), *Compendium of Methods for Microbiological Examination of Foods*, 2nd ed, Speck, M.L. (Ed). American Public Health Association: Washington DC.

Back, E., Molby., R., Kaijser, B. *et al.* (1980), Enterotoxigenic *E. coli* and other Gram-negative bacteria of infantile diarrhoea: surface antigens, hemagglutinins, colonization factor antigen and loss of enterotoxigenicity. *J. Inf. Dis.*, **142**, 318–327.

Bagley, S.T. and Seidler, R.J. (1978), Primary *Klebsiella* identification with MacConkey–inositol–carbenicillin agar. *Appl. Environ. Micro.*, **36**, 536–580.

Biering, G., Karlsson, S., Clark, N.C. *et al.* (1989), Three cases of neonatal meningitis caused by *Enterobacter sakazakii* in powdered milk. *J. Clin. Mic.*, **27**, 2054–2056.

Borriello, S.P. (1984), Typing *Clostridium difficile*. In *Antibiotic Associated Diarrhoeaand Colitis*. Martinus Nijhoff: Boston

Bowden, T.A., Mansberger, A.R. and Lykins, L.E. (1981), Pseudomembraneous enterocolitis: mechanism of restoring floral homeostasis. *Am. Surg.*, **47**, 178–183.

Bryan, F.L. (1979), Infections and intoxications caused by other bacteria. In *Food-borne Infections and Intoxications*, 2nd ed, Riemann, H. and Bryan, F.L. (Eds). Academic Press: London.

Cato, E.P., George, W.L. and Finegold, S.M. (1986), *Clostridium*. In *Bergey's Manual of Systematic Bacteriology*, Sneath, P.H.A., Mair, N.S., Sharpe, M.E. and Holt, J.G. (Eds). Williams and Wilkins: Baltimore and London.

Chapman, H.R.and Sharpe, M.E. (1981), Microbiology of Cheese. In *Dairy Microbiology, Vol 2. The Microbiology of Milk Products*, Robinson, R.K. (Ed). Applied Science Publishers: London and New Jersey.

Christie, A.B. (1987), *Infectious Disease*, Vol. 1, 4th ed., Churchill Livingstone: Edinburgh.

Clabots, R., Peterson, L.R. and Gerding, D.N. (1988), Characterization of a nosocomial *Clostridium difficile* outbreak by using plasmid profile typing and clindamycin susceptibility testing. *J. Inf. Dis.*, **158**, 731–736.

Corbel, M.J. and Brinley-Morgan, W.J. (1984), *Brucella*. In *Bergey's Manual of Systematic Bacteriology*, Vol 1, Krieg, N.R. and Holt, J.G. (Eds). Williams and Wilkins: Baltimore and London.

Ebringer, R.W., Cawdell, D.R., Cowling, P. and Ebringer, A. (1978), Sequential studies in ankylosing spondylitis: association of *Klebsiella pneumoniae* with active disease. *Ann. Rheum. Dis.*, **37**, 146–151.

Edwards, A.T., Roulson, M. and Ironside, M.J. (1988), A milk-borne outbreak of serious infection due to *Streptococcus zooepidemicus. Epid. Inf.*, **101**, 43–51.

Farmer, J.J. and McWhorter, A.C (1984), *Edwardsiella*. In *Bergey's Manual of Systematic Bacteriology*, Vol 1, Krieg, N.R. and Holt, J.G. (Eds). Williams and Wilkins: Baltimore and London.

Finn, V.G. (1978), The characteristics and clinical importance of bacteria of the genus *Citrobacter* isolated from patients with acute intestinal infections in the territory of Volgograd. *J. Hyg. Epid. Mic. Imm.*, **22**, 338–343.

Ganheim, A.T.M. and Cooke, E.M. (1980), Serious infection caused by Group C streptococci. *J. Clin. Path.*, **33**, 185–190.

Gebhard, R.L., Gerding, D.N., Olson, M.M. *et al.* (1985), Clinical and endoscopic findings in patients early in the course of *Clostridium difficile*-associated pseudomembranous colitis. *Am. J. Med.*, **78**, 45–48.

George, R.H. (1986), The carrier state: *Clostridium difficile*. *J. Antimicrob. Chemother.*, **18** (Suppl. A), 47–58.

George, R.C. and Uttley, A.H.C. (1989), Susceptibility of enterococci and epidemiology of enterococcal infections in the 1980s. *Epid. Inf.*, **102**, 403–441.

Griffin, G.E. (1989), *Clostridium difficile*. In *Enteric Infection*, Farthing, M.J.G. and Keusch, G.T. (Eds). Chapman and Hall: London.

Guarino, A., Capano, G., Malamisura, B. *et al.* (1987), Production of *Escherichia coli* STa-like heat-stable enterotoxin by *Citrobacter freundii* isolated from humans. *J. Clin. Mic.*, **25**, 110–114.

Guerrant, R.L., Dickens, M.D., Wenzel, R.P. and Kapikian, A.Z. (1976), Toxigenic bacterial diarrhea: nursery outbreak inVolving multiple bacterial strains. *J. Pediatr.*, **89**, 885–891.

Jay, J.M. (1978), *Modern Food Microbiology*, 2nd ed. D. Van Nostrand: New York.

Klipstein, F.A. and Engert, R.F. (1976), Partial purification of *Enterobacter cloacae* heat-stable enterotoxin. *Inf. Imm.*, **13**, 1307–1314.

Klipstein, F.A., Engert, R.A. and Short, H.B. (1977), Relative enterotoxigenicity of coliform bacteria. *J. Inf. Dis.*, **136**, 205–215.

Lyerly, D.M. and Wilkins, T.D. (1984), Characteristics of the toxins of *Clostridium difficile*. *Antibiotic Associated Diarrhoea and Colitis*. Martinus Nijhof: Boston.

McFaul, T.F. and Williams, J.C. (1981), Developmental cycle of *Coxiella burnetii*: structure and morphogenesis of vegetative and sporozoic differentiations. *J. Bact.*, **147**, 1063–1076.

Mitchell, T.J., Ketley, J.M., Haslam, S.C. *et al.* (1986), Effect of toxin A and B of *Clostridium difficile* on rabbit ileum and colon. *Gut*, **27**, 78–85.

Mossel, D.A.A., de Vor, H. and Eelderink, I. (1976), A further simplified procedure for the detection of *P. aeruginosa* in often contaminated aqueous substrata. *J. Appl. Bact.*, **41**, 307–309.

Ormsbee, R.A. (1980), Rickettsiae. In *Manual of Clinical Microbiology*. American Society for Microbiology: Washington DC.

Report (1983), Group C enterococcal infection associated with home made hard cheese. *Morb. Mort. Wkly. Rep.*, **32**, 510, 515–516.

Roberts, D. (1982), Bacteria of public health significance. In *Meat Microbiology*, Brown, M.H. (Ed). Applied Science Publishers: London and New York.

Sakazaki, R. (1984), *Citrobacter*. In *Bergey's Manual of Systematic Bacteriology*, Vol 1, Krieg, N.R. and Holt, J.G. (Eds). Wiliams and Wilkins: Baltimore and London.

Sutherland, J.P., Varnam. A.H. and Evans, M.G. (1986), *A Colour Atlas of Food Quality Control*. Wolfe Publishing Ltd: London.

Tedesco, F.J. (1982), Pseudomembranous colitis: pathogenesis and therapy. *Med. Clin. N. Am.*, **66**, 654–664.

Tucker, K.P., Canning, P.E. and Wilkins, T.D. (1990), Toxin A of *Clostridium difficile* is a potent cytotoxin. *J. Clin. Mic.*, **28**, 869–871.

Wadstrom, T.A., Aust-Kettis, D., Habte, D. *et al.* (1976), Enterotoxin-producing bacteria and parasites in stools of Ethiopian children with diarrheal disease. *Arch. Dis. Child.*, **51**, 865–870.

Wayne, L.G. and Kubica, G.P. (1986), The Mycobacteria. In *Bergey's Manual of Systematic Bacteriology*, Vol 2, Sneath, P.H.A., Mair, N.S., Sharpe, M.E. and Holt, J.G. (Eds). Williams and Wilkins: Baltimore and London.

Weber, G., Werner, H.B. and Matschnigg, H. (1977), *Pseudomonas aeruginosa* in trinkwasser als todersursache bei neugeborenen. *Zentralbl. Bakt. Parasiten. Infekt. Hyg. I. Abt. Orig.*, **216**, 210–214.

Weiss, E. and Moulder, J.W. (1984), *Coxiella*. In *Bergey's Manual of Systematic Bacteriology*, Vol 1, Krieg, N.R. and Holt, J.G. (Eds). Williams and Wikins: Baltimore and London.

Chapter 18

Anon (1988a), Foodborne viral gastroenteritis. *PHLS Micro. Dig.*, **5**, 69–75.

Anon (1988b), Outbreak of viral gastroenteritis — Pennsylvania and Delaware. *J. Fd. Prot.*, **51**, 158–159.

Anon (1989), *Report of the Working Party on Bovine Spongiform Encephalopathy*. Department of Health/Ministry of Agriculture, Fisheries and Food: London.

Ansari, S.A., Sattar, S.A., Springthorpe, V.S. *et al.* (1988), Rotavirus survival on human hands and transfer of infectious virus to animate and nonporous inanimate surfaces. *J. Clin. Mic.*, **26**, 1513–1518

Appleton, H. (1987), Small round viruses: classification and role in food-borne infections. In *Novel Diarrhoea Viruses* (Ciba Foundation Symposium 128). John Wiley: Chichester.

Butcher, I., Kudesia, G., Gordon, J, and Miller, J. (1989), Small round structured viruses and their spread. *Lancet*, **I**, 443.

Caul, E.O. and Appleton, H. (1982), The electron microscopical and physical characteristics of small round human fecal viruses: an interim scheme for classification. *J. Med. Vir.*, **9**, 257–265.

Cliver, D.O. Ellender, R.D. and Sobsey, M.D. (1984), Foodborne viruses. In *Compendium of Methods for the Microbiological Examination of Foods*, 2nd ed, Speck, M.L. (Ed). American Piblic Health Association: Washington DC.

Cliver, D.O. (1987), *Addendum to World Health Organisation Manual on Food Virology* (VPH/83.46). World Health Organisation: Geneva.

Cubitt, W.D. (1989), 'Norwalk' and small-round structured viruses. In *Enteric Infection*, Farthing, M.J.G. and Keusch, G.T (Eds). Chapman and Hall: London.

Cubitt, W.D., Blacklow, N.R., Hermann, J.E. *et al.* (1987), Antigenic relationships between human calicivirus and Norwalk virus. *J. Infect. Dis.*, **156**, 806–813.

Dolin, R., Roessner, K.D., Treanor, J.J. *et al.* (1986), Radioimmunoassay for detection of the Snow Mountain agent of viral gastroenteritis. *J. Med. Vir.*, **19**, 11–18.

Eyles, M.J. (1986), Transmission of viral disease by food: an update. *Fd. Tech. Aust.*, **38**(6), 239–242.

Gerba, C.P. (1988), Viral disease transmission by seafoods. *Fd. Tech.*, **42**(3), 99–103.

Halligan, A.C. and Auty, M.A.E. (1988), *Foodborne Viruses — A Survey of the Recent Literature*. Leatherhead Food R.A.: Leatherhead.

Herrman, J.E., Nowak, N.A. and Blacklow, N.R. (1985), Detection of Norwalk virus in stools by immunoassay. *J. Med. Vir.*, **17**, 127–133.

Ho, M.-S., Glass, R.I., Monroe, S.S. *et al.* (1989), Viral gastroenteritisaboard a cruise ship. *Lancet*, **II**, 961–965.

ICMSF (1980), *Microbial Ecology of Foods. Volume 1: Factors Affecting Life and Death of Microorganisms*. Academic Press: New York.

Kapikian, A.Z., Wyatt, R.G., Dolin, R. *et al.* (1972), Visualization by immune electron microscopy of a 27 nm particle associated with acute infectious non-bacterial gastroenteritis. *J. Virol.*, **10**, 1075–1081.

Kapikian, A.Z., Greenberg, H.B., Wyatt, R.G. *et al.* (1982), The Norwalk group of viruses, agents associated with viral gastroenteritis. In *Viral Infections of the Gastrointestinal Tract*, Tyrell, D.A. and Kapikian, A.Z. (Eds). Marcel Dekker: New York.

Kaplan, J.E., Gary, G.W., Baron, R.C. *et al.* (1982), Epidemiology of Norwalk gastroenteritis and the role of Norwalk virus in outbreaks of acute nonbacterial gastro-enteritis. *Ann. Int. Med.*, **96**, 756–761.

Katzenelson, E. and Mills, D. (1984), Contamination of vegetables with animal viruses via the roots. *Mongr. Virol. Kaiger*, **15**, 216–220.

Larkin, E.P. (1986), Detection, quantitation and public health significance of foodborne viruses. In *Foodborne Microorganisms and their Toxins: Developing Methodology*, Pierson, M.D and Stern, N.J. (Eds). Marcel Dekker: New York and Basel.

Lewis, D. and Metcalf, T.G. (1988), Removal of viruses in sewage treatment: assessment of feasibility. *Micro. Sci.*, **5**, 260–264.

Margolin, A.B., Hewlett, M.J. and Gerba, C.P. (1986), Use of a cDNA dot-blot hybridization technique for detection of enteroviruses in water. *Proc. Water Qual. Tech. Conf.* American Water Works Association: Denver.

Matsumato, K., Hataro, M., Kobayashi, K. *et al.* (1989), An outbreak of gastroenteritis associated with acute rotaviral infection in schoolchildren. *J. Inf. Dis.*, **160**, · 611–615.

Metcalf, T.G., and Jiang, X. (1988), Detection of hepatitis A virus in estuarine samples by gene probe assay. *Micro. Sci.*, **5**, 296–300.

Millard, J. Appleton, H. and Parry, J.V. (1987), Studies on heat inactivation of hepatitis A virus with special reference to shellfish. *Epid. Inf.*, **98**, 397–414.

NRC (1985), *An Evaluation of the Role of Microbiological Criteria for Foods and Food Ingredients*. National Academy Press: Washington DC.

Oliver, A.R. and Phillips, A.D. (1988), An electron microscopical examination of faecal small round viruses. *J. Med. Vir.*, **24**, 211–218.

Pether, J.V.S. and Caul, E.O. (1983), An outbreak of food-borne gastroenteritis in two hospitals associated with a Norwalk-like virus. *J. Hyg.*, **91**, 343–350.

Provost, J.P. and Hilleman, M.R. (1979), Propagation of human hepatitis A virus in cell culture in vitro. *Proc. Soc. Exp. Biol. Med.*, **160**, 213–217.

Reid, T.M.S. and Robinson, H.G. (1987), Frozen raspberries and hepatitis A. *Epid. Inf.*, **98**, 109–112.

Reid, J.A., Caul, E.O., White, D.G. and Palmer, S.R. (1988), Role of infected food handler in hotel outbreak of Noewalk-like viral gastroenteritis: implications for control. *Lancet*, **2**, 321–323.

Riordan, T. (1988), Investigation of outbreaks of winter vomiting disease. *Env. Hlth.*, **96**(8), 17–20.

Ward, B.K. and Irving, L.G. (1987), Virus survival on vegetables spray washed with waste water. *Water. Res.*, **21**(1), 57–63.

Zuckerman, A.J. (1990), Hepatitis E virus. THe main cause of enterically transmitted non-A, non B hepatitis. *B. Med. J.*, **300**, 1475–1476.

Chapter 19

Anderson, B.C. (1985), Moist heat inactivation of *Cryptosporidium* sp. *Am. J. Pub. Hlth.*, **75**, 1433–1434.

Anderson, B.C. (1986), Effect of drying on the infectivity of cryptosporidia-laden calf faeces for 3- to 7-day-old mice. *Am. J., Vet. Res.*, **47**, 2272–2273.

Anon (1986), *Ancanthamoeba* keratitis associated with contact lenses-United States. *Morb. Mort. Weekly Rep.*, **35**, 405–408.

Anon (1988), *Cryptosporidium*. *J. Am. Water Works Assn.*, **80**, 14–27.

Archer, D.L. and Young, F.E. (1988), Contemporary issues: diseases with a food vector. *Clin. Mic. Rev.*, **1**, 377–398.

Atterholm, I., Castor, B., Norlin, K. Cryptosporidiosis in southern Sweden. *Scand. J. Inf. Dis.*, **19**, 231–234.

Barret, N.J. Communicable disease associated with milk and dairy products in England and Wales: 1983–1984. *J. Inf. Dis.*, **12**, 265–272.

Baxby, D. and Hart, C.A. (1986), The incidence of cryptosporidiosis: a two-year prospective survey in a children's hospital. *J. Hyg.*, **96**, 107–111.

Bingham, A.K., Jarrall, E.L. and Meyer, E.A. (1979), *Giardia* sp.: physical factors of excystation *in vitro* and excystation vs eosin exclusion as determinants of viability. *Exp. Parasitol.*, **47**, 286–291.

Bracha, R. and Mirelman, D. (1984), Virulence of *Entamoeba histolytica* Grophozoites: effects of bacteria, microaerobic conditions and metronidazole. *J. Exp. Med.*, **160**, 353–368.

Casemore, D.P. (1987a), The antibody response to *Cryptosporidium*: development of a serological test and its use in a study of immunologically normal persons. *J. Inf.*, **14**, 125–134.

Casemore, D.P. (1987b), Cryptosporidiosis. *PHLS Micro. Dig.*, **4**, 1–5.

Casemore, D.P. (1988), Human cryptosporidiosis. In *Recent Advances in Infection*, 3, Reeves, D. and Geddes, A. (Eds). Churchill Livingstone: Edinburgh.

Casemore, D.P. (1989), The epidemiology of human cryptosporidiosis. *PHLS Micro. Dig.*, **6**, 54–66.

Casemore, D.P. and Jackson, F.B. (1984), Hypothesis: cryptosporidiosis in human beings is not primarily a zoonosis. *J. Inf.*, **9**, 153–156.

Casemore, D.P., Jessop, E.G., Douce, D. and Jackson, F.B. (1986), *Cryptosporidium* plus campylobacter: an outbreak in a semi-rural population. *J. Hyg.*, **96**, 95–105.

Corbett-Feeney, G. (1987), *Cryptosporidium* among children with acute diarrhoea in the west of Ireland. *J. Inf.*, **14**, 79–84.

Crawford F.G. and Vermund, S.H. (1988), Human cryptosporidiosis. *CRC Crit. Rev. Micro.*, **16**, 113–159.

Crossley, R., Marshall, J., Clark, J.T. and Holberton, D.V. (1986), Immunocytochemical differentiation of microtubules in the cytoskeleton of *Giardia lamblia* using monoclonal antibodies to α-tubulin and polyclonal antibodies to associated low molecular weight proteins. *J. Cell. Sci.*, **80**, 233–252.

Cruz, J.R., Cano, F., Caceres, P. *et al.* (1988), Infection and diarrhoea caused by *Cryptosporidium* sp. among Guatemalan infants. *J. Clin. Mic.*, **26**, 88–91.

Current, W.L. and Owen, R.L. (1989), Cryptosporidiosis and microsporidiosis. In *Enteric Infection*, Farthing, M.J.G. and Keusch, G.T. (Eds). Chapman and Hall: London.

Current, W.L., Reese, N.C., Ernst, J.V. *et al.* (1983), Human cryptosporidiosis in immunocompetent and immunodeficient persons: studies of an outbreak and experimental transmission. *New England J. Med.*, **308**, 1252–1257.

DeJonckheere, J.F. (1981), Pathogenic and non-pathogenic *Acanthamoeba* spp. in thermally polluted discharges and surface waters. *J. Protozool.*, **28**, 56–59.

de Regnier, D.P., Cole, L., Schapp, D.G. and Erlandsen, S.L. (1989), Viability of *Giardia* cysts suspended in lake, river and tap water. *Appl. Environ. Micro.*, **55**, 1223–1239.

Duncombe V.M., Bolin, T.D., Davis, A.E. *et al.* (1978), Histopathology in giardiasis: A correlation with diarrhoea. *Aust. N.Z. J. Med.*, **8**, 392–396.

Elsser, K.A., Moricz, M. and Proctor, E.M. (1985), *Cryptosporidium* infections: a laboratory survey. *Can. Med. Assoc. J.*, **135**, 211–213.

Erlandsen, S.L. and Feeley, D.E. (1984), Trophozoite motility and the method of attachment. In *Giardia and Giardiasis*, Erlandsen, S.L. and Meyer, E.A. (Eds). Plenum Press: New York.

Farthing, M.J.G. (1989), *Giardia lamblia*. In *Enteric Infection*, Farthing, M.J.G. and Keusch, G.T. (Eds). Chapman and Hall: London.

Farthing, M.J.G., Chong, S. and Walker-Smith, J.A. (1984), Acute allergic phenomena in giardiasis. *Lancet*, **II**, 1428.

Farthing, M.J.G., Keusch, G.T. and Carey, M.C. (1985), Effect of bile and bile salts on growth and membrane lipid uptake by *Giardia lamblia*: possible implications for pathogenesis of intestinal disease. *J. Clin. Invest.*, **76**, 1727–1732.

Farthing, M.J.G., Mata, L., Urrutia, J.J. and Kronmal, R.A. (1986), Giardiasis: impact on child growth. In *Diarrhoea and malnutrition in Childhood*, Walker-Smith, J.A. and McNeish, A.S. (Eds). Butterworth: London.

Heydoorn, A.O. (1977), Sarkosporidieninfiziertes. Fleisch als mogliche krankheitsursach fur den menschen. *Arch. Lebensmittelhyg.*, **28**, 27–32.

Holberton, D.V. (1974), Attachment of *Giardia* — a hydrodynamic model based on flagellar activity. *J. Exp. Biol.*, **60**, 207–221.

Inge, P.M.G., Webb, J.P.W. and Farthing, M.J.G. (1985), *Giardia* — bile salt interactions *in vitro*. Gut, **26**, A552.

Jokipii, L. and Jokipii, A.M.M. (1986), Timing of symptoms and oocyst excretion in human cryptosporidiosis. *New England J. Med.*, **315**, 1643–1647.

Jokipii, L., Pohjola, S. and Jokipii, A.M.M. (1985), Cryptosporidiosis associated with travelling and giardiasis. *Gastroenterology*, **89**, 838–842.

Klesius, P.H., Haynes, T.B. and Malo, L.K. (1986), Infectivity of *Cryptosporidium* sp. isolated from wild mice for calves and mice. *J. Am. Vet. Assn.*, **189**, 192–193.

Kobiler, D. and Mirelman, D. (1981), Adhesion of *Entamoeba histolytica* trophozoites to monolayers of human cells. *J. Inf. Dis.*, **144**, 539–546.

Lindmark, D.G. (1980), Energy metabolism of the anaerobic protozoan *Giardia lamblia*. *Mol. Biochem. Parasitol.*, **1**, 1–12.

Long-Krug, S.A., Fischer, H.J., Hysmith, R.M. and Ravdin, J.A. (1985), Phospholipase A enzymes of *Entamoeba histolytica*: description and subcellular localization. *J. Inf. Dis.*, **152**, 536–541.

Lushbaugh, W.B., Hofbauer, A.F. and Pittman, F.G. (1984), Proteinase activities of *Entamoeba histolytica* cytotoxin *Gastroenterology*, **87**, 17–27.

Lynch, E.C., Rosenberg, I.M. and Gitler, C. (1982), An ion-channel forming protein produced by *Entamoeba histolytica*. *E.M.B.O. J.*, **1**, 801–804.

Martinez, A.J. (1985), *Free Living Amebas: Natural History, Prevention, Diagnosis, Pathology, and Treatment of Disease.* CRC Press: Boca Raton.

Martinez-Palomo, A. (1989), *Entamoeba histolytica*. In *Enteric Infection*, Farthing, M.J.G. and Keutsch, G.T. (Eds). Chapman and Hall: London.

Martinez-Palomo, A., Gonzales-Robles, A., Chavez, B. *et al.* (1985), Structural bases of the cytolytic mechanisms of *Entamoeba histolytica*. *J. Protozool.*, **32**, 166–175.

Mathan, M., Venkatesan, S., George, R. *et al.* (1985), Cryptosporidium and diarrhoea in southern Indian children. *Lancet*, **II**, 1172–1175.

McGowan, K., Kane, A., Asarkof, W. *et al.* (1983), *Entamoeba histolytica* causes intestinal secretion: role of serotonin. *Science.* **221**, 262–264.

McLauchhin, J., Casemore, D.P., Harrison, T.G. *et al.* (1987), Identification of cryptosporidium oocsts by monoclonal antibody. *Lancet*, **I**, 51.

Munoz, M.L., Calderon, J. and Roskind, M. (1982), The role of collagenase in *Entamoeba histolytica*. *J. Exp. Med.*, **155**, 42–51.

Munson, D.A. and Sawyer, T.K. (1987), Distribution of *Acanthamoeba hatchetti* (Amoebida: Acanthamoebidae) in the Chester River, Maryland. *Trans. Am. Micro. Soc.*, **106**, 95–96.

Osterholm, M.T., Forgang, J.C., Ristemen, *et al.* (1981), An outbreak of foodborne giardiasis. *New Eng. J. Med.*, **304**, 24–38.

Palmer, S.R. and Biffin, A. (1987), Cryptosporidiosis. *PHLS Mic. Dig.*, **4**, 6–7.

Petersen, L.R., Cartter, M.L. and Hadler, J.L. (1988), A food-borne outbreak of *Giardia lamblia*. *J. Inf. Dis.*, **157**, 846–848.

Pohjola, S., Neuvonen, E., Niskanen, A. and Rantama, A. (1986), Rapid immunoassay for detection of cryptosporidium oocysts. *Acta Vet. Scand.*, **27**, 71–79.

Radvin, J.I. and Guerrant, R.L. (1982), A review of the parasite cellular mechanisms inVolved in the pathogenesis of amebiasis. *Rev. Inf. Dis.*, **4**, 1185–1207.

Roberts, W., Carr, M.F., Ma, J. *et al.* (1987), Prevalence of cryptosporidium in patients undergoing endoscopy — evidence for an asymptomatic carrier state. *Gastroenterology*, **92**, 1597.

Rude, R.A., Jackson, G.J., Bier, J.W. *et al.* (1984), Survey of fresh vegetables for nematodes, amoebae, and *Salmonella. J. Assoc. Off. Anal. Chem.*, **67**, 613–614.

Sawyer, T.K. (1989), Free-living pathogenic and non-pathogenic amoebae in Maryland soils. *Appl. Environ. Micro.*, **55**, 1074–1077.

Savva, D., Morris, J.C., Johnson, J.D. *et al.* (1990), Polymerase chain reaction for detection of *Toxoplasma gondii. J. Med. Mic.*, **32**, 25–31.

Sherwood, D., Angus, K.W., Snodgrass, D.R. and Tzipori, S. (1982), Experimental cryptosporidiosis in laboratory mice. *Inf. Imm.*, **38**, 471–475.

Smith, P.D., Horsburgh, C.R. and Brown, W.R. (1981), *In vitro* studies on bile acid deconjugation and lipolysis inhibition by *Giardia lamblia. Dig. Dis. Sci.*, **26**, 700–704.

Smith, H.V., Girdwood, R.W.A., Patterson, W.J. *et al.* (1988), Waterborne outbreak of cryptosporidiosis. *Lancet*, **II**, 1484.

Sterling, C., Seegar, K. and Sinclair, N.A. (1986), *Cryptosporidium* as a causative agent of traveller's diarrhoea. *J. Inf. Dis.*, **153**, 380–381.

Stibbs, H.H. and Ongerth, J.E. (1986), Immunofluorescence detection of *Cryptosporidium* oocysts in fecal smears. *J. Clin. Mic.*, **24**, 517–521.

Tandon, B.N., Tandon, R.K., Satpathy, B.K. and Shriniwas, A. (1977), Mechanism of malabsorption in giardiasis: a study of bacterial flora and bile salt deconjugation in upper jejunum. *Gut*, **18**, 176–181.

Thomson, M.A., Benson, J.W.T. and Wright, P.A. (1987), Two year study of cryptosporidium infection. *Arch. Dis. Child.*, **62**, 559–563.

Walsh, J. (1986), Problems in recognition and diagnosis of amebiasis: estimation of the global magnitude of morbidity and mortality. *Rev. Inf. Dis.*, **8**, 228–238.

White, K.E., Hedberg, C.V., Edmonson, L.M. *et al.* (1989), An outbreak of giardiasis in a nursing home with evidence for multiple modes of transmission. *J. Inf. Dis.*, **160**, 298–304.

Chapter 20

Anon (1972), *Proceedings of the 1971 National Conference on Food Protection*. United States Dept. of Health, Education and Welfare: Washington, DC.

Anon (1983), PHLS Salmonella Subcommittee. Notes on the control of human sources of gastrointestinal infections, infestations and bacterial intoxication in the United Kingdom. *Comm. Dis. Rep. Suppl. 1.* Communicable Disease Surveillance Centre: London.

Anon (1987a), *Guidelines to the Establishment of Hazard Analysis Critical Control Point (HACCP)*. CFPRA: Chipping Campden.

Anon (1987b), Hygiene and clothes: the role of workwear in the food industry. *B. Fd. J.*, **89**, 58–59.

Anon (1987c), *Food and Drink manufacture — Good Manufacturing Practice: a Guide to its Responsible Management*. Institute of Food Science and Technology (UK): London.

Anon (1988a), FMA urges Voluntary program and use of HACCP. *Food Chem. News*, Oct 24, 35–36.

Anon (1988b), *The Workwear Report. Part II*. British Workwear Rental Council: London.

Anon (1989), *Health surveillance and management procedures for food handling personnel. Report of a WHO consultation*. World Health Organisation: Geneva.

Bahl, M. (1987), Letter to Editor. *Lancet*, **II**, 865.

Baird-Parker, A.C. and Kilsby, D.C. (1987), Principles of predictive food microbiology. *J. Appl. Bact.*, **63** (Suppl.), 43S–49S.

Board, P.W. (1984), Quality assurance: everyone's responsibility. *Food Tech. Aust.*, **36**, 212–213.

Bostock, A.D. (1987), Letter to Editor. *Lancet*, **II**, 865.

Bryan, F.L. (1979), Infections and intoxications caused by bacteria. In *Foodborne Infections and Intoxications*, 2nd ed, Riemann, H. and Bryan, F.L. (Eds). Academic Press: New York.

Bryan, F.L. and McKinley, T.W. (1979), Hazard analysis and control of roast beef preparation in food-service establishments. *J. Fd. Prot.*, **42**, 4–18.

Bush, M.F.H. (1985), The symptomless salmonella excretor working in the food industry. *Comm. Med.*, **7**, 133–135.

Coates, D., Hutchinson, D.N. and Bolton, F.J. (1987), Survival of thermophilic campylobacters on fingertips and their elimination by washing and disinfection. *Epid. Inf.*, **99**, 265–274.

Craig, V. (1988), How clean is clean? *Food Sci. Tech. Today*, **2**, 267–268.

Cruickshank, J.G. (1990), Food handlers and food poisoning. *B. Med. J.*, **300**, 207–208.

Cruickshank, J.G. and Humphrey, T.J. (1987), The carrier food-handler and non-typhoid salmonellosis. *Epid. Inf.*, **98**, 223–232.

Czechowski, M.H. (1988), Bacterial attachment to Buna-N gaskets in milk processing equipment. *J. Fd. Prot.*, **51**, 824.

Dodd, C.E.R. (1988), Plasmid profiling: a new method of detection of critical control points. *Food Sci. Tech. Today*, **2**, 264–267.

Dodd, C.E.R., Chaffey, B.J. and Waites, W.M. (1988), Plasmid profiles as indicators of the source of contamination of *Staphylococcus aureus* endemic within poultry packing plants. *Appl. Environ. Micro.*, **54**, 1541–1549.

Dunsmore, D.G. (1980), bacteriological control of food equipment surfaces by cleaning systems. I Detergent effects. *J. Fd. Prot.*, **44**, 15–20.

Dunsmore, P.G., Twomey, A., Whittlestone, W.G. and Morgan, H.W. (1981), Design and performance of systems for cleaning product-contact surfaces of food equipment: a review. *J. Fd. Prot.*, **44**, 220–240.

Elliot, R.P. (1980), Cleaning and sanitizing. In *Principles of Food Sanitation*, Katsuyama. A and Strachan, J.P. (Eds). Food Processors' Institute: Washington DC.

Farber, J.M. (1986), Predictive modelling of food deterioration and safety. In *Foodborne Microorganisms and their Toxins: Developing Methodology*, Pierson, MD. and Stern, N.J. (Eds). Marcel Dekker: New York and Basel.

Francis, S., Rowland, J., Rattenbury, K. *et al.* (1989), An outbreak of paratyphoid fever in the UK associated with a fish-and-chip shop. *Epid. Inf.*, **103**, 445–448.

Freeman, L.K. (1988), Equipment and expertise are the keys to exploiting new products in a crowded market. *Food Eng. Int.*, **Aug, 33–37.**

Fuller, R. (1989), Red meat slaughter and hygiene. *Food Sci. Tech. Today*, **3**, 168–170.

Gilbert, R.J., and Maurer, I.M. (1968), The hygiene of slicing machines, cutting knives and can openers. *J. Hyg.*, **66**, 439–450.

Goldie, F.J. (1988), Thinking about the unthinkable: disaster management in the food industry. *B. Fd. J.*, **90**, 155–158.

Gomer, J.S. and Favero, M.S. (1985), *Guidelines for hand-washing and hospital environment control.* Centers for Disease Control: Atlanta.

Hayes, P.A. (1985), *Food Microbiology and Hygiene.* Elsevier: London and New York.

Herald, P.J. and Zottola, E.A. (1988a), Scanning electron microscopic examination of *Yersinia enterocolitica* attached to stainless steel at selected temperatures and pH values. *J. Fd. Prot.*, **51**, 445–448.

Herald, P.J. and Zottola, E.A. (1988b), Attachment of *Listeria monocytogenes* to stainles steel surfaces at various temperatures and pH values. *J. Fd. Sci.*, **53**, 1549–1552.

ICMSF (1986), *Microorganisms in Foods, 2. Sampling for Microbiological Analysis: Principles and Specifications.* Blackwell Scientific Publications: Oxford.

ICMSF (1989), *Microorganisms in Foods, 4. Application of hazard analysis critical control point (HACCP) system to ensure microbiological safety and quality.* Blackwell Scientific Publications: Oxford.

Jacob, M. (1988), Regulation of water quality for the food industry. *B. Fd. J.*, **90**, 114–116.

Jennings, W.G. (1965), Theory and practice of hard surface cleaning. In *Advances in Food Research*, Vol XIV, Mrak, E.M. and Stewart, G.F. (Eds). Academic Press: New York.

Karpinsky, J.L. and Bradley, R.L. (1988), Assessment of the cleanability of air-activated butterfly valves. *J. Fd. Prot.*, **51**, 364–368.

Katsuyama, A.M. and Strachan, J.P. (1980), *Principles of Food Processing.* Food Processors Institute: Washington DC.

Kauffman, E.L. and Schaffner, R.M. (1974), Hazard analysis, critical control points and good manufacturing practices regulations (sanitation) in food plant inspections. *Proc. IV Int. Cong, Food Sci. Tech.*, Chicago.

Lewis, A.J. and Gilmour, A. (1987), Microflora associated with the internal surfaces of rubber and stainless steel milk transfer pipeline. *J. Appl. Bact.*, **62**, 327–333.

Lewis, J.S., Gilmour, A., Frazer, T.W. and McCall, R.D. (1987), Scanning electron microscopy of soiled stainless steel inoculated with single bacterial cells. *Int. J. Fd. Mic.*, **4**, 279–284.

Mossel, D.A.A. (1984), Intervention as the rational approach to control diseases of microbial etiology transmitted by foods. *J. Fd. Safety*, **6**, 89–104.

NRC (1985), *An Evaluation of the Role of Microbiological Criteria for Foods and Food Ingredients.* National Academy Press: Washington DC.

Overdahl, B.J. and Zottola, E.A. (1988), Identification of microorganisms isolated from sweetwater and glycol cooling systems in dairy plants. *J. Fd. Prot.*, **51**, 824.

Pether, J.V.S. and Scott, R.J.D. (1982), *Salmonella* carriers: are they dangerous? A study to identify finger contamination with salmonellae by convalescent carriers. *J. Inf.*, **5**, 81–85.

Petron, R. and Zottola, E.A. (1988), Survival of *Listeria monocytogenes* in simulated milk cooling systems. *J. Fd. Prot.*, **51**, 172–175.

Quarmby, D.A. (1989), Food retailing: the challenge to food scientists and technologists. *Food Sci. Tech. Today*, **3**, 222–227.

Rand, W.M. (1984), Development and analysis of empirical mathematical kinetic models pertinent to food processing and storage. In *Computer-Aided Techniques in Food Technology*, Saguy, I. (Ed). Marcel Dekker: New York and Basel.

Rossmore, K. (1988), Evaluation and control of microbiological contamination of conveyor lubricants. *J. Fd. Prot.*, **51**, 826.

Rothwell, J. (1981), Microbiology of ice cream and related products. In *Dairy Microbiology, Vol 2. The Microbiology of Milk Products*, Robinson, R.K. (Ed). Applied Science Publishers: London and New Jersey.

Shapton, N. (1988), Hazard analysis applied to control of pathogens in the dairy industry. *J. Soc. Dairy Tech.*, **41**, 62–63.

Sharpe, A.N. (1986), Detection of foodborne microorganisms and their toxins. In *Foodborne Microorganisms and their Toxins: Developing Methodology*, Pierson, M.D. and Stern, N.J. (Eds). Marcel Dekker: New York and Basel.

Swartling, P. (1959), The influence of the use of detergents and sanitizers on the farm with regard to the quality of milk and milk products. *Dairy Sci. Absts.*, **21**, 1–10.

Tebbutt, G.M. (1984), A microbiological study of various food premises with an assessment of cleaning and disinfection practices. *J. Hyg.*, **66**, 439–450.

Tebbutt, G.M. and Southwell, J.M. (1989), Comparative study of visual inspections and microbiological sampling in premises manufacturing and retailing high risk foods. *Epid. Inf.*, **103**, 475–486.

Walker, D. and Jones, I.G. (1987), *Salmonella* and food handlers. *Lancet*, **II**, 1209–1210.

Watt, B. (1988), Disinfectants *vs* detergents. *Micro. Sci.*, **5**, 118.

Wheelock, V. (1988), *Food-borne disease — The Hidden Hazard.* Horton Publishing Ltd: Bradford.

Zottola, E.A. and Wolf, I.D. (1981), Recipe Hazard Analysis — RHAS — A systematic approach to analysing potential hazards in a recipe for food preperation/preservation. *J. Fd. Prot.*, **54**, 560–566.

Chapter 21

Anon (1981), Wholesomeness of irradiated food. *Tech. Rep. Ser. WHO No. 659.* World Health Organization: Geneva.

Anon (1989), *Food irradiation: a technique for preservation and improving the safety of food.* World Health Organization: Geneva.

Barrow, P.A., Smith, H.W. and Tucker, J.F. (1985), Reduction in salmonella infection in chickens by oral inoculation of Gram-negative facultatively anaerobic bacteria. *Proceedings Int. Symp. Salmonella, New Orleans*. American Association of Avian Pathologists: University of Pennsylvania.

Barrow, P.A., Tucker, J.F. and Simpson, J.M. (1987), Inhibition of colonization of the chicken alimentary tract with *Salmonella typhimurium* by Gram-negative facultatively anaerobic bacteria. *Epid. Inf.*, **98**, 311–322.

Gardner, G.A. (1982), Microbiology of processing: bacon and ham. In *Meat Microbiology*. Brown, M.H. (Ed.), Applied Science Publishers: London.

Goodfellow, S.J. and Brown, W.L. (1978), Fate of *Salmonella* inoculated into beef for cooking. *J. Fd Prot.*, **41**, 598–605.

Goren, E., de Jong, W.A., Doornenbal, P. *et al.* (1984a), Protection of chicks against *Salmonella infantis* infection induced by strict anaerobically cultured intestinal microflora. *Vet. Quarterly*, **6**, 22–26.

Goren, E., de Jong, W.A., Doornenbal, P. *et al.* (1984b), Protection of chicks against *Salmonella* infection induced by spray application of intestinal microflora in the hatchery. *Vet. Quarterly*, **6**, 73–79.

Gray, S.J. and Stickler, D.J. (1989), Some observations on the faecal carriage of mesophilic *Aeromonas* species in cows and pigs. *Epid. Inf.*, **103**, 523–537.

Hinton, M. (1986), The artificial contamination of poultry feed with Salmonella and its infectivity for young chickens. *Lett. Appl. Micro.*, **3**, 97–99.

Hinton, M. and Linton, A.H. (1988), Control of *Salmonella* infections in broiler chickens by the acid treatment of their feed. *Vet. Rec.*, **123**, 416–421.

Humphrey, T.J. and Bygrave, A. (1988), Abortion in a cow associated with *Salmonella* infections in badgers. *Vet. Rec.*, **123**, 160.

Humphrey, T.J. and Lanning, D.G. (1987), *Salmonella* and *Campylobacter* contamination of broiler chicken carcasses and scald tank water: the influence of water pH. *J. Appl. Bact.*, **63**, 21–26.

Humphrey, T.J. and Lanning, D,G. (1988), The vertical transmission of salmonellas and formic acid treatment of chicken feed: a possible strategy for control. *Epid. Inf.*, **100**, 43–49.

Humphrey, T.J., Lanning, D.G. and Beresford, D. (1981), The effect of pH adjustment on the microbiology of chicken-scald tank water with particular reference to the death rates of salmonellas. *J. Appl. Bact.*, **51**, 517–527.

ICMSF (1989), *Microorganisms in Foods, 4. Applications of the hazard analysis critical control point (HACCP) system to ensure microbiological safety and quality*. Blackwell Scientific Publications: Oxford.

Impey, C.S., Mead, G.C. and Hinton, M. (1987), Influence of continuous chllenge via the feed on competitive exclusion of salmonellas from broiler chicks. *J. Appl. Bact.*, **63**, 139–146.

Kelly, C.A., Dempster, J.F. and McLoughlin, A.J. (1974), The effects of washing lamb carcasses. *Proceedings 20th European Meeting of Meat Research Workers*: Dublin.

Mead, G.C. and Impey, C.S. (1986), Current progress in reducing *Salmonella* colonization of poultry by 'competitive exclusion' *J. Appl. Bact.*, **61** (Suppl.), 67–75.

Mead, G.C. and Barrow, P.A. (1990), *Salmonella* control in poultry by 'competitive exclusion' or immunization. *Lett. Appl. Mic.*, **10**, 221–227.

Mossel, D.A.A. (1987), Processing for safety of meat and poultry by radicidation. Progress, penury, prospects. In *Elimination of Pathogenic Organisms from Meat and Poultry*, Smulders, F.J.M. (Ed). Elsevier: Amsterdam.

Mossel, D.A.A. (1989), Adequate protection of the public against food-transmitted diseases of microbial aetiology. *Int. J. Fd. Mic.*, **9**, 271–294.

Mulder, R.W.A.W., van der Hulst, M.C. and Bolder, N.M. (1987), *Salmonella* decontamination of broiler carcasses with lactic acid, L-cysteine and hydrogen peroxide. *Poultry Sci.*, **66**, 1555–1557.

NRC (1985), *An Evaluation of the Role of Microbiological Criteria for Foods and Food Ingredients*. National Academy Press: Washington DC.

Nurmi, E. (1985), A review of competitive exclusion in poultry restricting intestinal colonization by *Salmonella* and some other enteric bacterial pathogens. *Proceedings Int. Symp. Salmonella, New Orleans*. American Society of Avian Pathologists: University of Pennsylvania.

Nurmi, E. and Rantala, M. (1973), New aspects of *Salmonella* infection in broiler production. *Nature*, **241**, 210–211.

Okrend, A.J., Johnston, R.W. and Moran, A.B. (1986), Effect of acetic acid on the death rates at 52°C of *Salmonella newport. Salmonella typhimurium* and *Campylobacter jejuni* in poultry scald water. *J. Fd. Prot.*, **49**, 500–503.

Pivnick, H., Barnum, D., Stavric, S. *et al.* (1985), Investigations on the use of competitive exclusion to control *Salmonella* in poultry. *Proceedings Int. Symp. Salmonella, New Orleans*. American Society of Avian Pathologists: University of Pennsylvania.

Rantala, M. and Nurmi, E. (1973), Prevention of the growth of *Salmonella infantis* in chicks by the flora of the alimentary tract of chickens. *B. Poult. Sci.*, **14**, 627–630.

Schleifer, J.H. (1985), A review of the efficacy and mechanism of competitive exclusion for the control of *Salmonella* in poultry. *World's Poult. Sci. J.*, **41**, 72–83.

Skirrow, M.B. (1989), *Campylobacter* perspectives. *PHLS Mic. Dig.*, **6**, 113–117.

Snijders, J.M.A., van Logtestijn, J.G., Mossel, D.A.A. and Smulders, F.J.M. (1985), Lactic acid as a decontaminant in slaughter and processing procedures. *Vet Quarterly*, **7**, 277–282.

Sofos, J.N., Maya, J.A. and Boyle, D.L. (1988), Effect of ether extracts from condensed wood smokes on the growth of *Aeromonas hydrophila* and *Staphylococcus aureus*. *J. Fd. Sci.*, **53**, 1840–1843.

Stavric, S., Gleeson, T.M., Blanchfield, B. and Pivnick, H. (1985), The effect of pure cultures of intestinal bacteria on competitive exclusion of *Salmonella* from chicks. *Proceedings Int. Symp. Salmonella, New Orleans*. American Society of Avian Pathologists: University of Pennsylvania.

Stern, N.J., Hernandez, M.P., Blankenship, L. *et al.* (1985), Prevalence and distribution of *Campylobacter jejuni* and *Campylobacter coli* in retail meats. *J. Fd. Prot.*, **48**, 595–599.

Sutherland, J.P., Varnam, A.H. and Evans, M.G. (1986), *A Colour Atlas of Food Quality Control*. Wolfe Publishing Ltd: London.

Twiddy, N., Hopper, D.W., Wray, C. and McLaren, I. (1988), Persistence of *Salmonella typhimurium* in calf rearing premises *Vet. Rec.*, **122**, 399.

Webb, T. and Lang, T. (1989), Food irradiation. *Lancet*, **I**, 498.

Xu, Y.M., Pearson, G.R. and Hinton, M. (1988), The colonization of the alimentary tract and visceral organs of chicks with Salmonellas following challenge via the feed: bacteriological findings. *B. Vet. J.*, **144**, 403–410.

Chapter 22

AOAC (1980), *Official Methods of Analysis*, 13th ed. AOAC: Washington DC.

Archer, D.L. and Young. F.E. (1988), Contemporary issues: diseases with a food vector. *Clin. Mic. Rev.*, **1**, 377–398.

Bryan, F.L. (1983), Epidemiology of milk-borne diseases. *J. Fd. Prot.*, **46**, 637–649.

Chapman, H.R. and Sharpe, M.E. (1981), Microbiology of Cheese. In *Dairy Microbiology, Vol. 2. The Microbiology of Milk Products*, Robinson, R.K. (Ed). Applied Science Publishers: London and New Jersey.

Chin, J. (1982), Raw milk: a continuing vehicle for the transmission of infectious disease agents in the United States. *J. Inf. Dis.*, **46**, 440–441.

D'Aoust, J.-Y. (1989), Manufacture of dairy products from unpasteurised milk: a safety asssessment. *J. Fd. Prot.*, **52**, 906–914.

Donker-Voet, J. (1965), Listeriosis in animals. *Bulletin de l'Office International des Epizooties*, **64**, 757–764.

Galbraith, N.S. and Pusey, J.J. (1984), Milkborne infectious disease in England and Wales 1938–1982. In *Health Hazards of Milk*, Freed, D.J.L. (Ed). Balliere Tindall: London.

Giles, N. and King, S.C. (1987), Excretion of *S. typhimurium* from a cow's udder. *Vet. Rec.*, **120**, 123.

Giles, N., Hopper, S.A. and Wray, C. (1989), Persistence of *S. typhimurium* in a large dairy herd. *Epid. Inf.*, **103**, 235–241,

Gitter, M. (1989), Veterinary aspects of listeriosis. *PHLS Mic. Dig.*, **6**, 38–42.

Humphrey, J.J. and Hart, R.J.C. (1988), *Campylobacter* and *Salmonella* contamination of unpasteurised cow's milk on sale to the public. *J. Appl. Bact.*, **65**, 463–467.

Klausen, N.K. and Huss, H.H. (1987), A rapid method for the detection of histamine-producing bacteria. *Int. J. Fd. Mic.*, **5**, 137–146.

Lerke, P.A., Parcana, M.N. and Chin. H.B. (1983), Screening test for histamine in fish. *J. Fd. Sci.*, **48**, 155–157.

NRC (1985), *An Evaluation of the Role of Microbiological Criteria for Foods and Food Ingredients*. National Academy Press: Washington DC.

Robinson, R.K. and Tamime, A.Y. (1981), Microbiology of fermented milks. In *Dairy Microbiology, Vol. 2. The Microbiology of Milk Products*. Applied Science Publishers: London and New Jersey.

Sumner, S.S. and Taylor, S.L. (1989), Detection method for histamine-producing, dairy-related bacteria using diamine-oxidase and leucocrystal violet. *J. Fd. Prot.*, **52**, 105–108.

Sumner, S.S., Speckhand, M.W., Somers, E.B. and Taylor, S.L. (1985), Isolation of histamine producing *Lactobacillus buchneri* from Swiss cheese implicated in a food poisoning outbreak. *Appl. Environ. Micro.*, **50**, 1094–1096.

Tham, W. (1988), Histamine formation by enterococci isolated from home-made goat cheeses. *Int. J. Fd. Mic.*, **7**, 103–108.

Werner, S.B., Morrison, F.R., Humphrey, G.L. *et al.* (1984), *Salmonella dublin* and raw milk consumption-California. *Morb. Mort. Weekly Rpt.*, **33**, 196–198.

Wray, C. and Sojka, W.J. (1977), Reviews in the progress of dairy science: bovine salmonellosis. *J. Dairy Sci.*, **44**, 383–425.

Chapter 23

Anon (1988), Poultry producers fall foul of enteritis. *Vet. Rec.*, **122**, 500.

Anon (1989), Memorandum of evidence to the Agriculture Committee Inquiry on *Salmonella* in eggs. *PHLS Micro. Dig.*, **6**, 1–9.

Berguist, D., Kraft, A., Cotteril, O. and Maguire, H. (1984), Eggs and egg products. In *Compendium of Methods for the Microbiological Examination of Foods*, Speck, M.L. (Ed). American Public Health Association: Washington DC.

Collins, M.A. (1985), Effect of pH and acidulant type on the survival of some food poisoning bacteria in mayonnaise. *Micro. — Aliment./Nutr.*, **3**, 215–221.

Humphrey, T.J. (1989), The effects of high or low temperature or acid on the survival of salmonellas in eggs, egg products or poultry meat. *Turkeys*, **37**, 13–14.

Humphrey, T.J. (1990), Public health implications of the infection of egg-laying hens with *Salmonella enteritidis* pt 4. *World's Poultry Sci. J.*, **46**, 5–13.

Humphrey, T.J., Cruickshank, J.G. and Rowe, B. (1989), *Salmonella enteritidis* PT4 and hens' eggs. *Lancet*, **I**, 280–281.

ICMSF (1986), *Microorganisms in Foods, 2. Sampling for Microbiological Analysis: Principles and Specific Applications*. Blackwell Scientific Publications: Oxford.

Mawer, S.L., Spain, G.E. and Rowe, B. (1989), *Salmonella enteritidis* PT4 and hens' eggs. *Lancet*, **I**, 280–281.

NRC (1985), *An Evaluation of the Role of Microbiological Criteria for Foods and Food Ingredients*. National Academy Press: Washington DC.

Perales, I. and Audicana, A. (1988), *Salmonella enteritidis* and eggs. *Lancet*, **II**, 1133.

Sutherland, J.P., Varnam, A.H. and Evans, M.G. (1986), *A Colour Atlas of Food Quality Control*. Wolfe Publishing Ltd: London.

USDA (1980), Regulations governing the inspection of eggs and egg products. *Code of Federal Regulations*. 7 CFR Part 2859.

Chapter 24

Anon (1986), The Cop on the boat. Tightening the net against unsafe shellfish. *FDA Consumer*, **Feb, 3–8.**

Anon (1987), Canadian researchers rush to find shellfish toxin. *New Scientist*, **Dec 24/31**, 8.

Anon (1988a), Arrests made after 3–state illegal shellfish harvesting crackdown. *Food Chem. News*, **Sept 5**, 44–45.

Anon (1988b), Ozone depuration reported effective against *Vibrio* in clams. *Food Chem. News*, **May 30**, 11.

Anon (1989), FDA-er corrects magazine article on sushi. *Food Chem. News*, **April 3**, 24–25.

AOAC (1980) *Official Methods of Analysis*. 13th ed. AOAC: Washington DC.

Arnold, S.H. and Brown, D.W. (1978), Histamine (?) toxicity from fish products. *Adv. Fd. Res.*, **24**, 113–154.

Baden, D.G., Mende, T.J., Poli, M.A. and Block, R.E. (1984), Toxins from Florida's red tide dinoflagellate *Ptychodiscus brevis*. In *Seafood Toxins*, Ragelis, E.P. (Ed). American Chemical Society: Washington DC.

Bartholomew, B.A., Berry, P.R., Rodhouse, J.C. and Gilbert, R.J. (1987), Scombrotoxic food poisoning in Britain: features of over 250 suspected incidents from 1976 to 1986. *Epid. Inf.*, **99**, 775–782.

Erichsen, I. (1983), Fermented fish and meat products: the present position and future possibilities. In *Food Microbiology: Advances and Prospects*, Roberts, T.A. and Skinner, F.A. (Eds). Academic Press: London and New York.

Hall, S. and Reichardt, P.B. (1984), Cryptic paralytic shellfish toxins. In *Seafood Toxins*, Ragelis, E.P. (Ed). American Chemical Society: Washington DC.

ICMSF (1989), *Microorganisms in Foods, 4. Application of the hazard analysis critical control point system to ensure microbiological safety and quality*. Blackwell Scientific Publications: Oxford.

Karolus, J.J., Leblanc, D.H., Marsh, A.J. *et al.* (1985), Presence of histamine in the bluefish *Domatomus saltatrix*. *J. Fd. Prot.*, **48**, 166–168.

Kat, M. (1983), *Dinophysis acuminata* blooms in the Dutch coastal area related to the diarrhetic mussel poisoning in the Dutch Waddensea. *Sarsia*, **68**, 81–87.

Kayser, A. and Mossel, D.A.A. (1984), Intervention *sensu* Wilson: The only valid approach to microbiological safety of food. *Int. J. Fd. Mic.*, **11**, 1–4.

Lerke, P.A., Werner, S.B., Taylor, S.L. and Guthertz, L.S. (1978), Scombrotoxic poisoning — report of an outbreak. *Western J. Med.*, **129**, 381–386.

Lerke, P.A., Parcana, M.N. and Chin, H.B. (1983), Screening test for histamine in fish. *J. Fd. Sci.*, **48**, 155–157.

Murray, C.K., Hobbs, G. and Gilbert, R.J. (1982), Scombrotoxin and scombrotoxin-like poisoning from canned fish. *J. Hyg.*, **88**, 215–220.

Schantz, E.J. (1984), Historical perspective on paralytic shellfish poison. In *Seafood Toxins*, Ragelis, E.P. (Ed). American Chemical Society: Washington DC.

Subba Rao, D.V., Quilliam, M.A. and Pocklington, R. (1988), Domoic acid — a neurotoxic amino acid produced by the marine diatom *Nitzschia pungens* in culture. *Can. J. Fish. Aquat. Sci.*, **45**, 2076–2079.

Taylor, S.L., Hui, J.Y. and Lyons, D.E. (1984), Toxicology of scombroid poisoning. In *Seafood Toxins*, Ragelis, E.P. (Ed). American Chemical Society: Washington DC.

Taylor, S.L. (1988), Marine toxins of microbial origin. *Fd. Tech.*, **42**(3), 94–98.

Yasumpto, T., Murata, M., Oshima, Y. *et al.* (1984), Diarretic shellfish poisoning. In *Seafood Toxins*, Ragelis, E.P. (Ed). American Chemical Society: Washington, DC.

Yentsch, C.M. (1984), Paralytic shellfish poisoning. In *Seafood Toxins*, Ragelis, E.P. (Ed). American Chemical Society: Washington DC.

Chapter 25

Anon (1973), *The Double Seam Manual*. The Metal Box Co. plc: Reading.

Anon (1979), *Approved Code of Practice for Watercress*. National Farmers' Union: St Albans.

Anon (1988), USDA prior approval for refrigerated foods seen insuring safety. *Food Chem. News*, **April 4**, 43–44.

Gilbert, R.J. (1983), Food-borne infections and intoxications — recent trends and prospects for the future. In *Food Microbiology: Advances and Prospects*, Roberts, T.A. and Skinner, F.A. (Eds). Academic Press: London and New York.

Goverd, K.A., Beech, F.W., Hobbs, R.P. and Shannon, R. (1979), The occurrence and survival of coliforms and salmonellas in apple juice and cider. *J. Appl. Bact.*, **46**, 521–530.

Hagler, W.M., Tyczcowska, K. and Hamilton, P.B. (1984), Simultaneous occurrence of deoxynivalenol, zearaleone, and aflatoxin in (1982), scabby wheat from the mid-western United States. *Appl. Environ. Microbiol.*, **47**, 151–154.

Hersom, A.C. and Hulland, E.D. (1980), *Canned Foods*. Churchill Livingstone: Edinburgh.

Hyde, H.C. (1976), Utilization of wastewater sludge for agricultural soil enrichment. *J. Water Poll. Con. Fed.*, **48**, 77–90.

ICMSF (1986), *Microorganisms in Foods, 2. Sampling for Microbiological Analysis: Principles and Specific Applications*. Blackwell Scientific Publications: Oxford.

ICMSF (1989), *Microorganisms in Foods, 4. Application of the hazard analysis critical control point (HACCP) system to ensure microbiological safety and quality*. Blackwell Scientific Publications: Oxford.

Mills, S. (1989), Bacteriological examination of sandwiches sold from petrol filling stations. *Brit. Fd. J.*, **91**(6), 28–36.

NRC (1985), *An Evaluation of the Role of Microbiological Criteria for Foods and Food Ingredients*. National Academy Press: Washington, DC.

Rothwell, J. (1981), Microbiology of ice cream and related products. In *Dairy Microbiology, Vol 2. The Microbiology of Milk Products*, Robinson, R.K (Ed). Applied Science Publishers: London and New Jersey.

Sutherland, J.P., Varnam, A.H. and Evans, M.G. (1986), *A Colour Atlas of Food Quality Control*. Wolfe Publishing Ltd: London.

Watkins, J. and Sleath, K.P. (1981), Isolation and enumeration of *Listeria monocytogenes* from sewage, sewage sludge and river water. *J. Appl. Bact.*, **50**, 1–9.

Further reading

Basic microbiology
Stanier, R.Y., Ingraham, J.L., Wheelis, M.L. and Painter, P.R. (1987), *General Microbiology*, 5th ed. Prentice Hall (US); Macmillan (UK): Englewood Cliffs, New Jersey; Basingstoke.

General food microbiology
Board, R.G. (1983), *A Modern Introduction to Food Microbiology*. Blackwell Scientific Publications: Oxford.

Jay, J.M. (1986), *Modern Food Microbiology*. Van Nostrand Rheinhold: New York.

Foodborne disease
Archer, D.L. and Young, F.E. (1988), Contemporary issues: Diseases with a food vector. *Clinical Microbiology Reviews*, **1**, 377–398 (Review).

Christie, A.B. (1987), *Infectious Disease, Epidemiology and Clinical Practice*, Vol. 1, 4th ed. Churchill Livingstone: Edinburgh. (This book is written from the viewpoint of clinical practice and while not exclusively concerned with foodborne disease, it offers a valuable insight into the clinician's approach to enteric pathogens.)

Concon, J.M. (1984), *Food Toxicology*, Parts A and B. Marcel Dekker: New York and Basel.

Leech, J.H., Sande, M.A. and Root, R.K. (1988), *Parasitic Infections*. Churchill Livingstone: Edinburgh.

Lund, B.M., Sussman, M., Jones, D. and Stringer, M.F. (1988), Enterobacteriaceae in the Environment and as Pathogens. Society for Applied Bacteriology Symposium Series 17, *Journal of Applied Bacteriology*, **65** (Suppl.).

Moss, M.O., Jarvis, B. and Skinner, F.A. (1989), Filamentous Fungi in Foods and Feeds. Society for Applied Bacteriology Symposium Series 18, *Journal of Applied Bacteriology*, **67**, (Suppl.).

Natari, S., Hashimoto, K. and Ueno, Y. (1989), *Mycotoxins and Phycotoxins '88*. Elsevier Science Publishers: Amsterdam.

Taylor, S.L. and Scanlan, R.A. (1989), *Food Toxicology: A Perspective of the Relative Risks*. Marcel Dekker: New York and Basel.

Tu, A.T (principal editor) (1983–1988), *Handbook of Natural Toxins*. Marcel Dekker: New York and Basel. (This handbook comprises a series of five volumes, of which Volume 1, *Plant and Fungal Toxins* and Volume 3, *Marine Toxins and Venoms*, are most useful.)

Pathogenesis
Mimms, C.A. (1987), *The Pathogenesis of Infectious Disease*. 3rd ed. Academic Press: London.

Investigation of foodborne illness
International Association of Milk, Food and Environmental Sanitarians (1990), *Procedures to Investigate Foodborne Illness*, 5th ed. IAMFES: Ames, Iowa.

Factors affecting growth and survival of pathogens
International Commission on Microbiological Specifications for Foods (1980), *Microbial Ecology of Foods. Volume 1: Factors Affecting Life and Death of Micro-organisms*. Academic Press: New York.

Processing for safety, hygiene and sanitation
Guthrie, R.K. (1988), *Food Sanitation*. Van Nostrand Rheinhold: New York.

Hayes, P.R. (1989), *Food Microbiology and Hygiene*. Elsevier Science Publishers: Amsterdam.

Institute of Food Science and Technology (UK) (1990), *Food and Drink Manufacture – Good Manufacturing Practice: A Guide to its Responsible Management*, 2nd ed. IFST: London.

International Commission on Microbiological Specifications for Foods (1989), *Micro-organisms in Foods 4. Application of the Hazard Analysis Critical Control Point (HACCP) System to Ensure Microbiological Safety and Quality*. Blackwell Scientific Publications: Oxford.

National Research Council (1985), *An Evaluation of the Role of Microbiological Criteria for Foods and Food Ingredients*. National Academy Press: Washington DC.

Laboratory methods
American Public Health Association (1976), *Compendium of Methods for the Microbiological Examination of Foods*. APHA: Washington DC.

Association of Official Analytical Chemists (1989), *FDA Bacteriological Analytical Manual*, 6th ed. AOAC: Arlington, Virginia.

Baird, R.M., Corry, J.E.L. and Curtis, G.D.W. (eds.) (1985), Pharmacopoeia of culture media for food microbiology, *International Journal of Food Microbiology*, **5**, 187–300.

Baird, R.M., Corry, J.E.L., Curtis, G.D.W., Mossel, D.A.A. and Skovgaard, N.P. (eds.) (1989), Pharmacopoeia of culture media for food microbiologists. *International Journal of Food Microbiology* (Additional monographs), **9**, 85–144. (The two *Pharmacopoeias* edited by Baird *et al.* provide authoritative formulae and methods for using and assessing the performance of selective media. Further information is available from the technical publications issued by media manufacturers such as API-bioMérieux, Becton-Dickinson, Difco Laboratories and Unipath (Oxoid).)

International Commission on Microbiological Specifications for Foods (1986), *Micro-organisms in Foods 2. Sampling for Microbiological Analysis: Principles and Specific Applications*, 2nd ed. ICMSF, Blackwell Scientific Publications: Oxford.

Pierson, M.D. and Stern, N.J. (1986), *Food-borne Micro-organisms and their Toxins: Developing Methodology*. Marcel Dekker: New York and Basel.

Threlfall, E.J. and Frost, J.A. (1990), The identification, typing and fingerprinting of *Salmonella*: laboratory aspects and epidemiological applications, *Journal of Applied Bacteriology*, **68** (Review), 5–16.

Walker, J. and Dougan, G. (1989), DNA probes: a new role in diagnostic microbiology, *Journal of Applied Bacteriology*, **67** (Review), 229–38.

Index

All numbers refer to page numbers. Those in **bold** refer to pages on which illustrations appear.

Abattoir
  control of micro-organisms, 430
  transport to, 428
Acanthamoeba sp, 386
Acidity
  effects on foodborne viruses, 370
  see also pH value
Acquired immunodeficiency syndrome
    (AIDS), 13, 363, 373
  Campylobacter and, 217
  cryptosporidiosis and, 374
  Isopora belli and, 385
  Listeria monocytogenes and, 217
  Salmonella and, 56
  shellfish risk for patients, 467
  virus, 11
Acute extraintestinal disease, 14
Adenosine triphosphate assay, 31-2, 237
Adenovirus, 363
Adherence, 45-6
  Aeromonas, 189
  Campylobacter, 218
  enterohaemorrhagic *Escherichia coli*, 115
  enteropathogenic *Escherichia coli*, **110-1**
  enterotoxigenic *Escherichia coli*, 113
  Listeria monocytogenes, 333
  Plesiomonas shigelloides, 205
  Salmonella, 61-2
  Shigella, 92
  Vibrio cholerae, 167-**8**
  Vibrio parahaemolyticus, 169-170
  Vibrio vulnificus, 171
  Yersinia, 140
Adhesins, 45-6, 49, 93, 110, 205
Adulteration of food, 10, 16
Aerobactin, 49, 63, 93, 140
Aerolysin *see* cholera toxin-cross-reactive
    cytolytic enterotoxin
Aeromonas, **185**-6
  control, **194**-5
  cytotonic enterotoxin, 190
  cytotoxic enterotoxin, 190-1
  in environment, 192
  epidemiology, 186
  foods associated with infection, **193**-4
  human carriage, 192
  isolation, 186-7, 195, **196-8**
  laboratory methods, 195, **196-200**
  pathogenicity mechanisms, 189-**91**
  people at particular risk from infection, 189
  species, **185**-6
  symptoms of infection, 188
  taxonomy, 187, **199-200**
  treatment of infection, 189
  typing, 188
Aflatoxins, 15-16
Airborne contamination, **246**, 418, 419
Algae, 14, 464-465, 468, **469-470**
Alkali tolerance, *Yersinia*, 146-**7**
Amines, biogenic, 458
Amoebae, free-living pathogenic, 13, 386
Amoebiasis, 373, 378-9
Amonobactin, 190
Animal feeds, 335, 427, 430
Animal products, dried and *Clostridium
    perfringens* infection, 320
Animals, association with foodborne
    pathogens, 65, 95, 142, 206, 206, 217, 221,
    245, 299, 319, 335-6, 337, 385
  see also meat animals
Anisakiasis, 464
Anisakis, 17, 463, 464

in smoked fish, 471
Ankylosing spondylitis, 49
  Klebsiella association, 358
Anthrax, intestinal, 276, 279
Antigenic heterogeneity, 14
Antigenic variation, 49
Aquaculture, **463-4**
Arsenic, 10
Aspergillus flavus, 15, 16
Aspergillus ochraceus, 16, 436
Aspergillus versicolor, 16, 436
Astrovirus, 13, 364
Atmosphere composition
  for *Campylobacter* cultivation, 225-6
  effects on *Aeromonas*, 194
  effects on *Campylobacter*, 224
  effects on *Clostridium botulinum*, 302-3
  effects on *Listeria monocytogenes*
  effects on *Staphylococcus aureus*, 252
  effects on *Vibrio*, 177
Audit for HACCP, 389-98, 400
  monitoring procedures, **396-8**
  predictive modelling, 393
  quality assurance, 396
  see also hazard analysis critical control point
    technique (HACCP)
Autochthonous micro-organisms, 44
Automated identification systems, 40

Bacillus, **267-9**
  in bread, 280, 478
  colonial morphology, **268-9**
  in environment, 278
  factors affecting growth, survival and toxin
    production, 281-3
  foods commonly associated with, **279-81**
  germination and outgrowth of endospores,
    267
  human carriage, 278
  identification, **285-7**
  isolation, 269, 283-**4**
  laboratory methods, 283, **284-8**
  morphological groups, **270**
  pathogenicity mechanisms, 277
  people at particular risk, 276
  species, **268-269**
  symptoms of food poisoning, 275-6
  taxonomy, **270-1**, 272-3
  treatment of food poisoning, 276
  typing, 274-**5**
Bacillus cereus, **268**, **275**
  in bean sprouts, 280, 476
  in cereals, 281, 477
  diarrhoeal toxin detection, **288**
  emetic toxin detection, **288**
  factors affecting growth, survival and toxin
    production, 281-3
  foods associated with, 279-81
  isolation, **284-5**
  in meat soups, 279, 280, 443
  toxins, 277-8
  typing, 274, **287**
Bacteraemia, 14, 15, 63, 190, 203
Bacterial restriction endonuclease analysis
    (*BRENDA*), 42
  Campylobacter, 214
Bactericidal effect of normal serum, 48, 140,
    171, 219
Bacteriocin typing, 41, 42
  Clostridium perfringens type A, 315
  Salmonella, 54
Bakery products, 249-50, 478

Balantidium coli, 385
Batter, 470
Bean sprouts, 476
Beetroot, cooked, 477
Benzoic acid effects on *Listeria monocytogenes*,
    342
Biochemical tests for identification of
    pathogens, 37-40
Biofilm development, 421
Biotyping, 42
  Bacillus, 274-**5**
  Campylobacter, 213, 234
  Salmonella, 54
  Staphylococcus aureus, 240
  Vibrio cholerae, 161
  Vibrio parahaemolyticus, 161-2
  Yersinia, 135
  Yersinia enterocolitica, 153
Birds, 417
  association with foodborne pathogens, 65,
    174, 206, 221, 223, 297, 335
  infection of livestock, 429
Botulinum toxin, 294-5
  action, 295, **296**
  species producing, 296
Botulism, 289
  Alaskan fermented fish and marine
    mammal products, 471
  classical foodborne, 289, 291-2, 293, 294
  dried fish, 471
  foods associated with, 297-300
  ham, 436
  hazard in canning, 297-8, 480
  infant, 289, 292, 293, 294
  people at particular risk, 293
  processed cheese, 297, 458
  sous-vide products, 298, 444
  symptoms, 291-2
  treatment, 294
  undetermined, 289, 292, 293, 294
  see also Clostridium botulinum
Bovine spongiform encephalopathy (BSE),
    11, 363
  infected animal feed, 427
Brain abscess, amoebic, 379
Brainerd diarrhoea, 449
Bread, 478
Brucella, 355
Brucellosis, 355
Buildings
  colonisation by pathogens, 418
  construction, 417-8
  design, 417
  livestock, 429
  location, 417
  maintenance, 419
  ventilation, 417, 418
  water supply, 418
Burgers, 433
Butchery, 430
Butter, 451

Caliciviridae, 364
Campylobacter, **209-10**
  complications, 215-6
  cultivation, 225-6
  detection, 210, 230
  diseases of animals and man, 209
  DNA hybridisation, 35, **232**
  ELISA assay, 34
  enteropathogenic species, 211-2
  in environment, 220-1

extraintestinal infections, 215-16
factors affecting survival, 223-4
foods associated with infection, 222-3
human carriage, 220
identification, 230, **231-2**
isolation, 210, 226-7, **228-9**, 230
laboratory methods, 225-7, **228-9**, 230,
   231-2, 233-4
in meat animals, 427
nutritional requirements, 225
pathogenicity mechanisms, 218-20
people at risk from infection, 216-7
species, 209, 211-2
symptoms of infections, 214-6
taxonomy, 210-2
treatment of infection, 217
typing, 213-4, 232, 233, 234
Campylobacter cinaedi, 212, 216
Campylobacter coli, 211-2, 214-6
Campylobacter conciscus, 212
Campylobacter cryoaerophila, 212
Campylobacter fennelliae, 212, 216
Campylobacter fetus, 14, 212, 216
Campylobacter hyointestinalis, 212, 216
Campylobacter jejuni, 14, 211-2, 214-6
Campylobacter laridis, 212, 216
Campylobacter upsaliensis, 212
Cancer, 14, 15
Canned foods, 480, 481-2, 483
   Clostridium botulinum in, 297-8
   Staphylococcus aureus in, 250
Cereals and cereal products
   Bacillus in 280-1, 477, 478
   bakery goods, 478
   control of pathogenic micro-organisms in,
     477-8
   mycotoxins, 477
   pasta products, 478-9
Cervelat, 437
Cervicoglandular listeriosis, 331
Cheese
   biogenic amines in, 458
   critical control points in production, 455-6
   heat treatment of milk, 456-7
   histamine toxicity, 361
   Listeria monocytogenes in, 339, 456, 457
   making, 454
   processed, 458
   ripening, 457
   Salmonella in, 66, 456, 457
   Staphylococcus aureus in, 249, 457
   Streptococcus in, 361, 456
Cheese, soft
   acidity, 457
   Escherichia coli in, 117-8, 457
   Listeria monocytogenes in, 332, 338-339, 457
   ripening, 457
Chemical food poisoning, 10, 16-7
Chemotaxis
   Campylobacter, 45, 218
   Salmonella, 61
   Vibrio cholerae, 45, **167**
Chicken
   see poultry and poultry products
Childhood diarrhoea, 11, 90, 108, 138, 189,
     204, 215, 216, 363, 365, 375, 382
Chocolate, 479-80
   Salmonella in, 67
Cholera, 157-8, 162
   toxin, 167, **168**, 169, **170**
   transmission, 175
Cholera toxin-cross-reactive cytolytic
     enterotoxin, 191
Chronic disease, 14
Ciguatoxin, 464-5
Citrobacter, 357
Claviceps purpurea, 15, 477
Cleaning, 420-3, **424-5**
   biofilm development, 421

deposition and accumulation of soil, 420-1,
   425
   dry, 423
   emergency, 423
   floors and walls, 423
   health of staff, 423
   monitoring, 423-**4**
   periodic, 423
   poor practice, **425**
   schedules, 420
   systems, 421, 422-3
   verification, 423-**4**
Clostridium botulinum, **289**
   biochemical characteristics, **311**
   colonisation of gastrointestinal tract, 296
   in environment, 297
   epidemiology, 289
   in fish, 299, **463-4**
   foods associated with, 297-300
   growth and survival in foods, 301-7
   human carriage, 297
   isolation, 290-1, 307-**8**
   laboratory methods, 307, **308-9**, 310, **311**
   and meat curing conditions, 303-5
   morphology, **289**
   in mushrooms, 300, 476
   pathogenicity mechanisms, 294-6
   people at particular risk, 293
   in processed cheese, 298, 458
   in smoked fish, 299, 471
   in surimi, 472
   symptoms of intoxication, 291-2
   taxonomy, 290-1
   toxin detection, 309, 310-1
   toxin preparation, 310
   toxin stability, 301
   typing, 291
   see also botulism
Clostridium difficile, 356
Clostridium perfringens, **312**
   in dried milk, 453
   factors affecting growth and survival, 321-2
   foods associated with, 319-20
   human carriage, 319
   identification **325**
   incidence in raw meat, 319
   isolation, 313, **322**, 323, **324**, 325
   laboratory methods, **322**, 323, **324-5**, 326
   lethal toxin production, 312
   nature of enterotoxin, 316
   pathogenicity mechanisms, 316, **317**, 318
   people at risk from food poisoning, 316
   symptoms of intoxication, 315-16
   taxonomy, 314
   treatment of food poisoning, 316
   type A, 312, 315, 325
   type C, 312, 313, 318, 325
   typing, 315, 326
Coconut, desiccated and *Salmonella*
     transmission, 68, 480
Cod worm, 17, 464
Colicin typing for *Shigella*, 90
Commercial consequences of food poisoning,
     22-3
Competition from other micro-organisms
   effects on *Bacillus*, 283
   effects on *Campylobacter*, 224
   effects on *Clostridium botulinum*, 307
   effects on *Clostridium perfringens*, 322
   effects on *Escherichia coli*, 120
   effects on *Listeria monocytogenes*, 343
   effects on *Salmonella*, 71
   effects on *Staphylococcus aureus*, 253, **254-5**
   effects on *Vibrio*, 177
Competitive exclusion, 429-30
Complaints system, 401-3
Complement cascade, 48
Concentrated foods, *Staphylococcus aureus* in,
     249

Confectionery
   ingredients, 480
   Salmonella transmission, 67-8
Contamination
   emergency cleaning, 423
   protection of foods, 417-18
Control of foodborne pathogens, principles,
     43
Cook-chill products, 443
   Listeria monocytogenes in, 337-338
Cost of foodborne disease, 21
   national, 23
   national economies, 24-26
   to company, 22
Coxiella burnetii, 357
Coxsackie virus, 365
Cream
   confections, 451
   Listeria monocytogenes in, 339
   pasteurisation, 451
Critical control points, 388
Cross tolerance hypothesis, 50
Crustaceans, 467
Cryptosporidiosis, 374-5
   age distribution, 375
Cryptosporidium parvum, 13, **373**-7
   detection, 386
   infections in man, 374-5
   mixed infections, 375
   pathogenicity mechanisms, 375
   properties, 373-4
   transmission, 376-7
Curing ingredients
   effects on *Clostridium botulinum*, 303-5
   effects on *Clostridium perfringens*, 321
   effects on *Escherichia coli*, 119
   effects on *Listeria monocytogenes*, 338, 342
   effects on *Salmonella*, **70**
   effects on *Shigella*, 97
   effects on *Staphylococcus aureus*, 252-253
   effects on *Yersinia*, 145
Curing of meat, 298, 434, 435, 436
Cutaneous listeriosis, 331
Cytolysis by *Entamoeba histolytica*, 380
Cytotoxin production
   Aeromonas, 190-191
   Campylobacter, 219-20
   Clostridium difficile, 356
   Salmonella, 64-5
   Shigella, 93-4
   Vibrio vulnificus, 172
   Yersinia enterocolitica, 141

Deoxydivalenol, 477
Detergent-sterilisers, 422
Detergents, 421
Diarrheic shellfish poisoning, 469
Diarrhoeal syndrome of *Bacillus*, 275
Diarrhoeal viruses, **364**-5
   food handlers, 366-8
   human carriage, 366-7
   pathogenicity mechanisms, 366
Dinoflagellates
   poisoning in fish, 464-5
   poisoning in shellfish, 468-70
Direct epifluorescence technique, 30-31
Disinfectants
   effect on micro-organisms, **425**
   effects on *Cryptosporidium parvum*, 376
   effects on foodborne viruses, 370
   see also sanitisers
Dissociation of colonies of *Yersinia*, **129**
Divalenol, 477
DNA:DNA hybridisation, for detection of
     micro-organisms, 35
   Salmonella, 35, 82-3
   Campylobacter, 35
   Clostridium perfringens, 325
   Escherichia coli, 35, 125, 126

Listeria monocytogenes, 35, 327, 350-1
Shigella, 98-9
Staphylococcus aureus, 237
Vibrio cholerae, 180
Vibrio vulnificus, 180
Yersinia enterocolitica, 152
DNA fingerprinting, *Listeria monocytogenes*, 330
Domoic acid, **470**
Drainage, 418
Dried foods
  Bacillus in, 281
  Staphylococcus aureus in, 249
Drying effects on *Campylobacter*, 224
Dysentery, 87, 90, 107, 163, 188, 214

Eating by food handlers, 416
Echovirus, 365
Edwardsiella, 357
Egg products, 461-2
Eggs, 10
  allergy, 17
  control at farm level, 459-60
  dried, 461
  exterior contamination, **459**
  frozen, 461
  liquid, 461
  pathogenic organisms in, 459-62
  pre-prepared meals, 461-2
  processing for safety, 460-1
  Salmonella in, 11, 51, 57, 67, 459, 460-1
  Staphylococcus aureus in, 248
  trans-ovarian infection, 459-60
  washing, 459
Elastase production, *Plesiomonas shigelloides*, 206
  Vibrio vulnificus, 172
Electrical techniques of detection, principles, 31
ELISA assay, principles, 32, **33**, 34-5
  for *Campylobacter* enterotoxin, 220
  for *Clostridium botulinum* toxin, 311
  for *Clostridium perfringens* enterotoxin, 326
  for *Cryptosporidium*, 386
  for enterohaemorrhagic *Escherichia coli* 125-6
  for *Escherichia coli* toxins 114, 116, 127-8
  for *Listeria monocytogenes* detection, 327, 350
  for Norwalk agent, 371
  for *Salmonella*, 81-**2**, 83
  for *Salmonella* toxin, 64
  for shiga toxin, 95
  for *Shigella*, 95, 98
  for *Staphylococcus aureus* enterotoxins, 263, 264-5
  for *Yersinia* pathogenicity, 141
Emetic syndrome of *Bacillus*, 275
Endotoxin
  enteropathogenic *Escherichia coli*, 112
  Listeria monocytogenes, 335
  Plesiomonas shigelloides, 205-6
  Salmonella, 64
  Vibrio cholerae, 169
Engineering materials sanitation, 423
Enrichment, 27-**8**
  Aeromonas, 186, 195
  Bacillus, 269, 283
  Campylobacter, 210, 226-7, 229-30
  Clostridium botulinum, 290, 307
  Clostridium perfringens, 313, 321-2
  Escherichia coli, 28, 103, 120
  Listeria monocytogenes, 327, 343-4
  Plesiomonas shigelloides, 202, 207-8
  Pseudomonas aeruginosa, 361
  Salmonella, 27-**8**, 51, 71-73
  Shigella, 88, 98
  Staphylococcus aureus, 237, 255
  Vibrio, 158, 178
  Yersinia, 130, 146, **147**
Entamoeba histolytica, 373, **378**-81

infections in man, 378-9
life cycle, 378
pathogenicity mechanisms, 380
prevalence and distribution of infections, 379
properties, 378
transmission, 381
Entamoeba polecki, 378
Enteric fever (typhoid) 51, 55, 63, 66, 412
  treatment, 57
Enteritis
  Campylobacter, 214-15
  necrotising, 312
  necrotic, *see* pigbel
  Salmonella, 55, 57
Enterobacter, **358**
  in dried milk, 453
Enterobacteriaceae, 357
  heat-stable enterotoxin production, 11, 357-8
  miniaturised identification systems, **38**, 39
  serotyping, **40**
Enterochelin, 63, 93, 140
Enterococcus, **359**-60
Enterococcus faecalis, 355, 359
Enterococcus faecium, 355, 359
Enterotoxin, 172
  Aeromonas 190-1
  Bacillus cereus, 277-8, **288**
  Campylobacter, 219
  Citrobacter, 357
  Clostridium perfringens, 316-8, 319, 326
  Plesiomonas shigelloides, 205
  Edwardsiella, 358
  Escherichia coli, 112-3, **114**
  Giardia lamblia, 383
  Salmonella, 64
  Shigella, 94
  Staphylococcus aureus, 236, 242-4, 250-4
  Staphylococcus spp., 236
  Vibrio cholerae, 167, **168**, 169, **170**
  Vibrio vulnificus, 172
  Yersinia, 141
Enzyme linked immunosorbent assay *see* ELISA assay
Equipment, 419
  canning, 483
  cleanability, 419-20
  cleaning, 420-3, **424-5**
  entrapment of soil in surface damage, **425**
  maintenance, 417, 420
Ergotamine, ergotism, 15
Escherichia coli, 101
  anaerogenic strain, **102**
  bacteriocin typing, 41
  citrate utilisation, **102**
  delayed lactose fermentation, **101**
  diarrhoeagenic strains, 103, 104-5
  enteroadherent-aggregative, 101, 108, 116, 125
  enterohaemorrhagic, 101, 105, 107-8, 115-6, 118-9, 121, 125
  enteroinvasive, 45, 101, 105, **114**-5, 118
  enteropathogenic, 101, 106, 107-9, **110-1**, 117
  enterotoxigenic, 101, 106-7, 112-4, 118, 125, 126, 127, **128**
  in environment, 117
  factors affecting growth and survival in foods, 119-20
  foods involved in transmission, 117-9
  human carriage, 116
  indole production, **102**
  isolation and detection, 31, 103, **122-4**, 125-6
  laboratory methods, 120-1, **122-4**, 125-7, **128**
  pathogenicity mechanisms, 109-116
  persons at particular risk, 108-9

serovar O157:H7, 11, 105, 106, 107, 115, 116, 117, 118, 119, 120, 121, **124**, 126, 127
  in soft cheese, 117-8, 457
  taxonomy, 104-5
  treatment of infections, 109
  typing, 106, 127
Exposure
  to *Aeromonas*, 189
  to *Balantidium coli* 385
  to *Brucella*, 355
  to *Campylobacter*, 217
  to *Clostridium botulinum*, 293
  to *Coxiella burnetii*, 357
  to *Cryptosporidium parvum*, 376-7
  to *Entamoeba histolytica*, 379
  to *Escherichia coli*, 108-9
  to *Giardia lamblia*, 384
  to *Listeria monocytogenes*, 332-3
  to *Plesiomonas shigelloides*, 204
  to *Pseudomonas aeruginosa*, 360
  to *Salmonella*, 56-7
  to *Shigella*, 90
  to *Staphylococcus aureus*, 241
  to *Streptococcus*, 361
  to *Toxoplasma gondii*, 385
  to *Vibrio*, 164-5
  to *Yersinia*, 138
Extraintestinal amoebiasis, 379

Facultative intracellular parasites, 47
Failure mode effect analysis in HACCP, 399
Farms, control of micro-organisms in meat, 427, **428**, 429-30
Fasciola hepatica, 17, 476
Featureless viruses, 365
Fermentation control in milk, 457
Fish
  Aeromonas in, 193-194, 463
  Aeromonas salmonicida, as piscine pathogen, 185
  Clostridium botulinum in, 299, **463-4**
  Clostridium perfringens infection, 320
  control of pathogenic organisms, 463-6
  dinoflagellate poisoning, 464-5
  freshwater, 463
  Listeria monocytogenes in, 340
  Plesiomonas shigelloides in, 206-7, 463
  raw, 463, 464
  Salmonella in, 67, 463
  scombroid, 12, 465
  scombrotoxic poisoning, 361, 465-6
  Staphylococcus aureus in, 248, 470
  toxins, 463, 464-5
  Vibrio in, 175, 463, 464
Fish products, 470-2
  Clostridium botulinum in, 299, 470, 471
  cooked, 470
  dried, 471
  fermented, 471-2
  Listeria monocytogenes in, 340
  pickled, 472
  salt, 470
  smoked, 471
Fluorescent antibody-microcolony technique for *Salmonella*, 82
Fluorescent antibody technique, 32
  Cryptosporidium, 386
  Listeria monocytogenes, 350
  Salmonella, **32**
Food handlers, 404
  eating and chewing, 416
  education, **405-6**
  exclusion after illness, 411-3
  hygiene training, 404-7
  medical examinations, 408-9
  medical policies, 408-13
  personal hygiene, 413-6
  plant operations, 416
  plant/process operatives, 407

protective clothing, 415-6
return to work after exclusion, 412-3
smoking, 416
training, 404, **405-6**, 407-8
training for specific job requirements, 407
untrained personnel, 407
Food intolerance, 16-7
Food poisoning
 agents, 9
 causes and effects, 12
 contributing factors, 18, 19
 importance of agents, 12
 industry wide repercussions, 23
 location of outbreaks, 17, 19
 nature of, 9
 poor practice, 19
 social impact, 26
 symptoms, 9, 13
 type of food and pathogenic
  micro-organism, 18, 19
Footwear, 416
Frozen foods, 483-4
Fruit
 control of pathogenic micro-organisms in,
  473
 Cryptosporidium parvum transmission, 376
 dried, 476
 fermented, 477
 foodborne viruses 368-9
 products, 476-7
Fungal toxins *see* mycotoxins
Fusarium graminearum, 477

Gambierdiscus toxicus, 464
Gastrointestinal tract, microbial ecology,
 44-5
Gelatin, 442, 443
Gene probes, *see* DNA:DNA hybridisation,
 for detection of micro-organisms
Genetic control of virulence, 44
 Aeromonas, 189
 enterohaemorrhagic *Escherichia coli*, 115
 enteropathogenic *Escherichia coli*, 109-10
 enterotoxigenic *Escherichia coli*, 112-3
 Listeria monocytogenes, 333
 Salmonella, 60
 Shigella, 91-2
 Vibrio cholerae, 166-7
 Vibrio parahaemolyticus, 170
 Vibrio vulnificus, 171
 Yersinia, 139
Genetic methods of typing, 42, 43
 Bacillus, 274
 Escherichia coli, 106
 Salmonella, 54-5
 Staphylococcus aureus, 240
Giardia lamblia, **373**, 381-4
 infections in man, 382-3
 life cycle, 381-2
 luminal factors, 383
 mixed infections, 375
 mucosal injury, 384
 pathogenicity mechanisms, 383
 toxin, 50, 383
 transmission, 383-4
Giardiasis, 382-3
Gonyaulax spp, 468
Granulomatosis infantiseptica *see* listeriosis of
 the newborn
Granulomatosis septica, 331

Haemagglutinin, toxic, 16, 476
Haemolysin
 of *Aeromonas*, 181
 of *Bacillus cereus*, 277
 of *Listeria ivanovii*, 335
 of *Listeria monocytogenes*, 335
 of *Vibrio parahaemolyticus*, 170-1
Haemolysin-cytotoxin, of *Vibrio vulnificus*, 172

Haemolytic uraemic syndrome (HUS), 55-6,
 90, 91, 107, 109, 137, 215-6
Haemorrhagic colitis, 11, 107
Ham
 canned, 441
 mycotoxins in, 436
Hand
 care, 415
 washing, **414**-5
Hazard analysis, 388
Hazard analysis critical control point
  technique (HACCP), 387-401
 application on continuing basis, 401
 audit, 389-98, 400
 basis, 388
 committee composition, 389
 critical control points, 388
 failure mode effect analysis, 399
 hazard analysis, 388
 implementation of system, 399
 monitoring, 388
 safety assurance system, 389-401
Heat labile toxin (LT) of enterotoxigenic
  Escherichia coli, 113, 114, 127-**8**
Heat stable toxin (ST) of enterotoxigenic
  Escherichia coli, 114, 128
 production by other micro-organisms, 11,
  357, 358

 detection, 372
 food handlers, 368
 human carriage, 368
 molluscs as vehicles, 368, 467
 pathogenicity mechanisms, 366
 symptoms, 365
 transmission on fruit and vegetables, 368,
  369
 water as vehicle, 369
Hepatitis E virus, 365
 symptoms, 366
Histamine
 in matured cheese, 361, 458
 in scombrotoxic poisoning, 465-6
 toxicity, 361
Honey, *Clostridium botulinum* in, 293, 300
Human carriers of *Salmonella*, 64, 411-413
Hydrodynamic flow, 45
Hydrogen peroxide effects on *Listeria
  monocytogenes*, 343
Hydrophobic grid membrane filter, 29-30
Hydrophobic grid membrane filtration
  (HGMF)
 for *Salmonella*, 30, 78
Hygiene
 mobile ice cream vendors, 484
 personal of food handlers, 413-16
Hyperferriaemia, 50

Ice, 369
Ice cream, 483-4
 home made, 483
 Listeria monocytogenes in, 339
 Salmonella in, 66
Identification, 37-40
 Aeromonas, 187, **199-200**
 Bacillus, **270-1**, 272-3, **285-7**
 Campylobacter, 210-2, 230, **231-2**
 Clostridium botulinum, **311**
 Clostridium perfringens, 314, **325**
 Enterobacteriaceae, 357-359
 Enterococcus, 360
 Escherichia coli, 104-5, 126-7
 Listeria monocytogenes, 328-9, 351, **352-3**
 Plesiomonas shigelloides, 202, 208
 Pseudomonas aeruginosa, **301**
 Salmonella, 52-3, 83-**4**
 Shigella, 88-**9**, 100
 Staphylococcus aureus, 237-9, 258, **259-61**
 Vibrio, **159**-60, 180, **181-2**

Yersinia, **131-2**, 133, **134**, 152-3
Illness of food handlers as causative factor in
 food poisoning, 410
Immunoimmobilisation methods for
 Salmonella, 30, **80**
Impedance measurement, for detection of
 micro-organisms, 31
 Escherichia coli, 31
 Listeria monocytogenes, 31
 Salmonella, 31, 81
 Staphylococcus aureus, 31
Incubation temperature effects
 Aeromonas, 199
 Vibrio, 159
 Yersinia, **131-132**
Indicator organism, *Staphylococcus aureus* as in
 foods, 246
Infant foods, *Clostridium botulinum* in, 300
Infective dose
 of *Campylobacter*, 218
 of *Clostridium perfringens*, 317
 of enteroinvasive *Escherichia coli*, 45
 of *Salmonella*, 58
 of *Shigella*, 45, 91
Insects, 335, 417, 418
Interaction with host cell microfilaments,
 Listeria monocytogenes, 334
Interferon, 63, 93
Intestinal invasive amoebiasis, 378-9
Intracellular proliferation, 46
Invasion
 Entamoeba histolytica, 380
 enteroinvasive *Escherichia coli*, **114**
 enteropathogenic *Escherichia coli*, 111-2
 of intestinal cells, 46
 Plesiomonas shigelloides, 205
 Salmonella, **58**, 61
 Shigella, 92
 Vibrio parahaemolyticus, 169
 Vibrio vulnificus, 171
Ionising radiation, and control of *Salmonella*
 in poultry, 430
Iron deprivation, 48-9
Iron deprivation effects on micro-organisms
 Aeromonas, 190
 Salmonella, 63
 Shigella, 93
 Vibrio vulnificus, 171
 Yersinia, 140-1
Irradiation effects
 on *Aeromonas*, 195
 on *Bacillus*, 282
 on *Campylobacter*, 224
 on *Clostridium botulinum*, 301
 on *Escherichia coli*, 119
 on foodborne viruses, 370
 on *Listeria monocytogenes*, 341
 on *Salmonella*, 69
 on *Shigella*, 97
 on *Staphylococcus aureus*, 251
 on *Vibrio*, 177
 on *Yersinia*, 145
Isolation of foodborne micro-organisms,
 principles, 27
 see also enrichment, resuscitation, selective
  plating
Isopora belli, 385

Kanagawa reaction, **183**
Kauffman-White serotyping scheme, 53, 54
Klebsiella, 358

Labelled bacteriophage detection, 36
 Listeria, 36
 Salmonella, 36, 83
Lactoferrin, 48
Lactoperoxidase system, effects on *Listeria
  monocytogenes*, 342-3
Latex co-agglutination tests, **32**

Bacillus cereus diarrhoeal toxin, 278
Campylobacter, 210, 230
Clostridium perfringens enterotoxin, 326
Escherichia coli labile-toxin, 127
Escherichia coli serovar O157:H7, **124**
Salmonella, 82-**3**
Shigella, 98
Staphylococcus enterotoxins 263-4
Laundry, of protective clothing, 415-6
Lecithinase as indicator of prior growth of
    Clostridium perfringens, 313
Legionella, 418
Listeria, 317
  colonisation of buildings, 418
  development in low temperature storage of
    fruit and
    vegetables, 473
  detection, 35, 36
  in fermented meat products, 437
  low temperature pathogen, 10
  in pâtés, 443
  in raw milk cheese, 456
  in salad vegetables, 473
  in soft cheese, 457
Listeria monocytogenes, 10, 327
  biofilm development, 421
  colonisation of conveyer lubricants, 423
  in cook-chill products, 337-8, 443
  detection, 327-8, 350-1
  ELISA assay, 34, 350
  in environment, 336
  epidemiology, 327
  factors affecting growth and survival, 340-3
  foods associated with infections, 337-40
  human carriage, 335
  in ice cream, 339, 484
  identification, 351, **352-3**
  infection cycles, 337
  isolation, 327, 343-5, **346-7**, 348-9
  laboratory methods, 343-5, **346-7**, 348-351,
    352-3
  pathogenicity mechanisms, 333-5
  people at particular risk, 332-3
  protocol for isolation, 348, 349
  serovar occurrence, 329
  in shellfish, 340, 467
  silage contamination, 335, 427
  species, 327
  survival during pasteurisation, 340-1
  symptoms of infection, 330-1
  taxonomy, 328, 329
  toxin production, 334-5
  treatment of infection, 333
  typing, 329-30
Listeriosis, 330-1
  of the newborn, 330-1
Liver abscess, amoebic, 379
Liver fluke, 17, 476
Liver sausage, 442
Lysozyme, 47, 48

Macrophages, 47-8
Management in the food industry, 387
Marine mammals
  Alaskan fermented products, 471-2
  Clostridium botulinum in, 299
Mastitis, 222, 246-7, 249, 338, 361, 447
Mayonnaise
  acetic acid in and safety, 70, 461
  Bacillus cereus in 460
  Clostridium perfringens in, 460
  Staphylococcus aureus in, 460
  pH value and safety, 461
  Salmonella enteriditis in, 68, 460, 461
  Salmonella transmission, 68, 460
Meat
  Aeromonas in, **193**
  air dried, 444-5
  Bacillus in, 279

Campylobacter in, 222-3
Clostridium botulinum in, 298
Clostridium perfringens in, 319
consumption of raw, 433
control of pathogenic organisms, 427-45
Cryptosporidium parvum transmission, 376
enterohaemorrhagic *Escherichia coli* in, 109,
  118-9
Listeria monocytogenes infection, 337-8
packaging, 433
raw uncured, 433-4
Salmonella in, 57, 65
Staphylococcus aureus in, 246
Yersinia in, 138, 143, 144
see also curing of meat
Meat animals
  animal-to-animal spread of infection, 428
  building contamination, 429
  butchery, 430
  carcass washing, 430
  competitive exclusion, 429-30
  infected feed, 427
  infection by man, 429
  slaughter, 430
  transport to abattoir, 429
  water quality, 428
Meat products
  Aeromonas in, 436
  Campylobacter in, 221, 430
  Clostridium botulinum in, 298
  Escherichia coli serovar O157:H7 in, 433
  Listeria monocytogenes in, 338, 437
  Staphylococcus aureus in, 244, 434, 436-7
  raw cured, 434, 435-6
  raw fermented, 437-8
Meat products, cooked, 439-45
  Bacillus cereus in, 443
  Clostridium botulinum in, 298
  Clostridium perfringens in, 319
  comminuted products, 442-3
  cook-chill products, 443
  critical control points, 439-40
  cured, 441-2
  dried, 444
  enterotoxigenic *Escherichia coli* in, 117
  extended life, 444
  gelatin in, 442, 443
  Listeria monocytogenes in, 337-8, 441, 443
  packaging, 441
  pies and puddings, 442
  Salmonella, 247-8, 442
  soups and stews, 443
  sous-vide products, 444
  Staphylococcus aureus, 247-8, 442
  uncured, 441
  undercooking, 441
Medical policies
  examinations for employees, 408-9
  exclusion after illness, 411-3
  exclusion during sickness, 409-411
  return to work after exclusion, 412-13
Membrane attack complex (MAC), 48-49
Membrane filter techniques, 29-30
Meningitis, 55-6, 330
Meningoencephalitis, 330
Microscopic techniques of detection,
  principles, 30-1
Milk
  Aeromonas in, 194
  Bacillus in, **279**-80
  Brainerd diarrhoea, 449
  Campylobacter in, 222, 448
  Clostridium perfringens in, 320
  concentrated, 452
  contamination, 11, 447
  control at farm level, 447
  control of pathogenic micro-organisms in,
    447-58
  Coxiella burnetii in, 357
  Cryptosporidium parvum in, 376

disease associated with, 10
fermented products, 454-8
flavouring, 448-9
foodborne viral infection, 369
heat treatment for cheese making, 456-7
Listeria monocytogenes in, 338-9, 447, 448, 456
pasteurisation, 448, 449
pasteurisation control and monitoring,
  396-8
powder, 479
raw, 448
Salmonella in, 11, 56, 57, 66, 447, 448, 456
Staphylococcus aureus in, 246-7, 248, 249, 449
Streptococcus in, 361, 449
survival of *Listeria monocytogenes* during
  pasteurisation, 340-1
tuberculosis and, 360
UHT, 443, 450, 451
Yersinia enterocolitica in, 143, 449
Milk, dried, 452-4
  Listeria monocytogenes in, 339
  Salmonella transmission, 66, 453-4
  Staphylococcus aureus in, 453
Milk products
  Clostridium botulinum in, 298-9, 458
  Clostridium perfringens in, 320, 453
  foodborne viral infection, 369
  Listeria monocytogenes in, 338-9, 457
Milking and contamination of milk, 447
Miniaturised identification systems, **38-9**, 40
Molecular mimicry, 14, 49
Molluscs, 467-70
  depuration, 468
  foodborne viral infection, 368-9
  growing beds, 467-8
  relaying, 468
  toxic poisoning, 468-70
Monocytes, 47
Mononuclear phagocytes, 63
Morganella, 358
Most probable number estimation
  Clostridium perfringens, 313, **324**
  Salmonella, 78, 80
  Staphylococcus aureus, 255
Motility as a virulence factor, 45
  Campylobacter, 45, 218
  Plesiomonas shigelloides, 205
  Salmonella, 61
  Shigella, 45, 92
  Vibrio cholerae, 45, **167**
Mouse challenge test for *Clostridium botulinum*
  toxin, 310
MUCAP test for *Salmonella*, 85
Mucus layer, 45
Multi-test media for identification of
  bacteria, 37
  Aeromonas, 199
  Plesiomonas, 208
  Salmonella, **38**, 83, 85
  Vibrio cholerae, 181
  Yersinia, 152, 153
Mushrooms, 476
  Clostridium botulinum in, 300, 476
Mycobacterium bovis, 355, 360
Mycobacterium tuberculosis, 360
Mycotoxins, 15, 16, 436, 477

NaCl
  and biochemical reactions of *Vibrio*, 159
  effect on *Aeromonas*, 195
  effect on *Plesiomonas shigelloides*, 207
  effect on *Vibrio*, 177
  in fermented fish, 471
  in fermented vegetables, 477
  in salt fish, 470
  see also curing ingredients
Naegleria sp, 386
Nematodes, 463, 464
Neoplasms, 15

Neurotoxic shellfish poisoning, 469
Neutrophils, 47-48
Nisin
  effects on *Clostridium botulinum*, 306
  effects on *Listeria monocytogenes*, 342
  in canned foods, 434
  in processed cheese, 306, 458
Nitrite
  as preservative, 433
  see also curing ingredients
Nitrosamines, 16, 298, 434
Nitzschia pungens, 470
Non-impediometric electrical techniques for
    detection of micro-organisms, 31
Norwalk virus, 364
  symptoms, 365
Nurmi principle *see* competitive exclusion

Oculoglandular listeriosis, 331
Olive oil adulteration, 16
Organic acids
  effects on *Bacillus*, 282
  effects on *Campylobacter*, 224
  effects on *Listeria monocytogenes*, 341-2
  effects on *Salmonella*, 70-1
Outer membrane proteins, 48
  Campylobacter, 218
  Salmonella, 63
  Shigella, 92
  Vibrio cholerae, 167
  Yersinia enterocolitica, 139, 140
Oxidation-reduction potential
  effects on *Bacillus*, 282
  effects on *Clostridium botulinum*, 302-3
  effects on *Clostridium perfringens*, 321
  effects on *Salmonella*, 70

Parabens
  effects on *Clostridium botulinum*, 305-6
  effects on *Clostridium perfringens*, 321
  effects on *Listeria monocytogenes*, 343
Paralytic shellfish poisoning, 468-9
Pasta
  products, 478-9
  Salmonella transmission, 68, 479
  Staphylococcus aureus, 249, 479
Pasteurisation
  butter, 451
  cream, 451
  of eggs, 461
  of milk, 340-1, 357, **396-8**, 448-9
Pasture contamination, 427
Pâtés, 442
Pathogenicity mechanisms, 44-50
Pediococcus acidilactici, 343, 437
Penetration
  Campylobacter, 218
  Listeria monocytogenes, 334
  Salmonella, **58-62**, 63
  Shigella, 92
  Vibrio parahaemolyticus, 169-70
  Vibrio vulnificus, 171
  Yersinia, 140
Personal impact of food poisoning, 26
pH value
  effects on *Aeromonas*, 194
  effects on *Bacillus*, 282
  effects on *Campylobacter*, 224
  effects on *Clostridium botulinum*, 301-2
  effects on *Clostridium perfringens*, 321
  effects on *Escherichia coli*, 119
  effects on foodborne viruses, 370
  effects on *Listeria monocytogenes*, 341-2
  effects on *Plesiomonas shigelloides*, 207
  effects on *Salmonella*, 69-70
  effects on *Shigella*, 97
  effects on *Staphylococcus aureus*, 252
  effects on *Vibrio*, 177
  effects on *Yersinia*, 145

fermentation of meat products, 437
fermented milk products, 457
pickled fish, 472
safety of canned foods, 301-7, 480
safety of mayonnaise, 461
safety of processed cheese, 298, 458
Phage typing, 41
  Campylobacter, 213
  Escherichia coli, 106
  Listeria monocytogenes, 329-30
  Salmonella, 54
  Shigella, 90
  Staphylococcus aureus, 239, **261**
  Vibrio cholerae, 161
  Yersinia, 136
Phagocytosis, 47-8
Phagocytosis, resistance of
  Campylobacter, 219
  Listeria monocytogenes, 334
  Salmonella, 63
  Shigella, 92-3
  Vibrio vulnificus, 171
  Yersinia, 140
Phase variation, 454-6, 49
Phenolic compound effects on *Listeria
    monocytogenes*, 343
Phocanema spp, 17, 464
Pigbel, 312, 316
Plague, 129
Plant
  extracts as preservatives and effects on
    Clostridium botulinum, 306
  material as reservoir of *Listeria monocytogenes*,
    336
Plasmid profiling, 42-3
  Bacillus, 274
  Clostridium perfringens type A, 315
  Escherichia coli, 106
  Listeria monocytogenes, 330
  Salmonella, 54, 55
  Shigella, 90
  Staphylococcus aureus, 240
  Vibrio cholerae, 161
Plesiomonas shigelloides, 201
  in environment, 206
  epidemiology, 201
  foods associated with, 206-7
  growth and survival, 207
  human carriage, 206
  incidence of infection, 204
  isolation, 202, 207
  laboratory methods, 207-8
  pathogenicity mechanisms, 205
  people at particular risk, 204
  symptoms of infection, 203
  taxonomy, 202
  treatment of infection, 204
  typing, 203
Pneumonic listeriosis, 331
Poliovirus, 365
  transmission in milk, 369
Polymerase chain reaction, 36
  Escherichia coli, labile-toxin producing, 125
  Escherichia coli, shiga-like toxin producing,
    116. 125
  Shigella, 99-100
  Toxoplasma gondii, 386
Potatoes, cooked, 477
Poultry
  Aeromonmas in, 195
  Bacillus in, 279
  Campylobacter in, 222, 427, 428, 430
  chemicals for killing salmonellas, 70-1
  Clostridium botulinum in, 298
  colonisation of packing plants by
    Staphylococcus aureus, **245-6**
  competitive exclusion of *Salmonella*, 337,
    429-30
  control of pathogens at farm level, 427-9

critical control points, 434-6
  Listeria monocytogenes infection, 341
  packing plants, 245, 246, 430, 431-3
  Plesiomonas shigelloides in, 207
  raw, 433-4
  Salmonella in, 65, 427, 428, 429, 430
  Staphylococcus aureus in, 246
Poultry products, cooked, 439-41, 443-4
  critical control points, 439-40
  Listeria monocytogenes in, 337-8
Pregnancy, listeriosis during, 331
Preservatives, effects on *Bacillus*, 282-3
Produce *see* vegetables
Product withdrawal, 401-4
Proliferation in host tissue, 47-9
  Campylobacter, 219
  Listeria monocytogenes, 334-5
  Salmonella, 63
  Shigella, 92-3
  Vibrio vulnificus, 171
  Yersinia, 140
Proteases, *Aeromonas*, 191
Protective clothing, 415-16
Protein kinase C activity of enteropathogenic
    Escherichia coli, 112
Protein profiles of *Campylobacter*, 214
Proteus, 358
Protozoa, 12, 13, 373
  detection, 386
Providencia, 358
Pseudomonas aeruginosa, 360, **361**
Pseudomonas fluorescens, 343
Pseudomonas fragi, 343
Pufferfish, 463

Q fever, 357

Rapid methods of detection of
    micro-organisms, 30, **31-3**, 34-6
  Campylobacter, 210, 230
  Clostridium perfringens, 325
  Escherichia coli, 125-6
  Listeria monocytogenes, 327-8, 350-1
  Salmonella, 52, **79-80**, 81, **82-3**
  Shigella, 88, 98-100
  Staphylococcus aureus, 237
  Vibrio, 159, 180
  Yersinia, 130, 152
Rapid visual test for *Vibrio cholerae*, 182
Receptor-mediated endocytosis, 46
  Salmonella, 59, 140
  Shigella, 92, 140
  Yersinia enterocolitica, 140
Red kidney beans, 16, 476
Red tides, 469
Resistance plasmid analysis for *Salmonella*, 55
Resuscitation, 27
  Campylobacter, 210, 226
  Escherichia coli, 103, 120
  Listeria monocytogenes, 327
  Salmonella, 27, 51, 71
  Shigella, 88, 97
Rickettsieae, 357
Rodents, 142, 221, 417, 418
  infection of livestock, 428
Rotavirus, 12, 363
  human carriage, 367

Salads
  Acanthamoeba in, 386
  Campylobacter in, 223
  Clostridium perfringens, 320
  control of pathogenic micro-organisms in,
    473-7
  enteroinvasive *Escherichia coli* in, 118
  foodborne viral infection, 368-9
  Listeria monocytogenes in, 339, 473
  Salmonella in, 68
  Shigella in, 96

Streptococcus pyogenes in, 361
Salami, 437
Salmonella, 51
  airborne transfer, 418
  in animal feeds, 427
  antibiotic resistance, 57
  in bean sprouts, 68, 476
  in cereals, 477
  in cocoa beans, 67, 479
  co-infection with *Listeria*, 332, 333
  competitive exclusion for control, 429-30
  cooling system contamination, 423
  detection, 30-2, 34-6, 52, **79-80**, 81, **82-3**
  DNA hybridisation detection, 35, 82-3
  in dried milk, 66, 453
  ELISA assay, 34, 35, 81-**2**
  in environment, 65
  factors affecting growth and survival in
    foods, **69-70**, 71
  farm building contamination, 429
  in fermented meat products, 65, 437
  foods associated with, 65-8
  human carriage, 19, 64, 411-3
  identification, 52-3, 83, **84**, 85
  impedance measurement, 31, 81
  in milk, 11, 66, 447
  isolation, 27, 28, 29, 30, 51-2, 71-2, **73-7**, 78
  labelled bacteriophage detection system, 36,
    83
  laboratory methods, 71-2, **73-7**, 78, **79-80**,
    81, **82-4**, 85
  latex co-agglutination assay, 34, 82-**3**
  in pasta, 68
  pathogenicity mechanisms, **58-62**, 63-4
  persons at particular risk of infection, 56-7
  in raw milk cheese, 66, 456
  subspecies, 53
  symptoms of infection, 55-6
  taxonomy, 52-3
  treatment of infections, 57
  typing, 40, 41, 54-5, 85
Salmonella dublin, 417
Salmonella enteriditis,
  home made ice cream, 483
  infection of eggs, 67, 459-60
  survival in cooked eggs, 67, 460-1
Salmonella Rapid Test, 78, **79**
Sandwiches, 478
Sanitisation, 420-**5**
Sanitisers, 421, 422, 423
Sarcocystes, 385
Sauerkraut, 477
Sausages, 433
Saxitoxins, 468-**9**
Scombroid
  fish, 12, 465
Scombrotoxic poisoning, 463, 465-6
Selective plating, 28-9
  Aeromonas, 186-7, 195, **196-8**
  Bacillus, 269, 283, **284-5**
  Campylobacter, 210, 227, **228-9**
  Clostridium botulinum, 290, **308**
  Clostridium perfringens, 313, **322-3**
  Enterobacter, **358**
  Enterococcus, **359**
  Escherichia coli, 103, 120-1, **122-4**, 125
  Listeria monocytogenes, 327, 345, **346-7**, 348
  Plesiomonas shigelloides, 202, 208
  Salmonella, 29, 52, **73-7**, 78
  Shigella, 88, **98-99**
  Staphylococcus aureus, 237, 255, **256-7**, 258
  Vibrio, 159, 178-9
  Yersinia, 130, 147, **148-51**, 152
Septicaemic listeriosis, 331
Serotyping, 40-1
  Aeromonas, 188
  Bacillus cereus, 41, 274, **287**
  Campylobacter, 213, 232-4

Clostridium botulinum, 291
Clostridium perfringens, 315, 326
Escherichia coli, 41, 106, 127
Listeria monocytogenes, 329
Plesiomonas shigelloides, 203
Salmonella, 40, 41, 54, 85
Shigella, 89, 100
Staphylococcus aureus, 239
Vibrio cholerae, 161
Vibrio parahaemolyticus, 161
Yersinia enterocolitica, 135-6, 153
Yersinia pseudotuberculosis, 136
Serratia, **358**
Sewage
  contamination of fishing site, 467
  foodborne micro-organisms in 65, 95, 117,
    174, 192, 221, 319, 336, 337, 368, 386
  mollusc harvesting, 467
  sludge as fertiliser, 473
Shellfish
  Aeromonas in, 193-4
  allergy, 17
  Campylobacter in, 223
  control of pathogenic micro-organisms,
    467-70
  diarrheic poisoning, 469
  foodborne viral infection, 368-9, 467-8
  Listeria monocytogenes in, 340, 341, 467
  neurotoxic poisoning, 469
  paralytic poisoning, 468-9
  Plesiomonas shigelloides in, 206-7
  Salmonella in, 67
  Shigella in, 96, 467
  Staphylococcus aureus in, 248, 467
  Vibrio in, 175
  Vibrio parahaemolyticus in, 467, 468
  Vibrio vulnificus in, 164, 467
  viral gastroenteritis, 363
Shiga toxin, 93, **94**
  detection, 95
Shiga-like toxins, 107, 112, 115-6, 169, 171
Shigella, 87
  in the environment, 95
  factors affecting growth and survival in
    foods, 97
  foods commonly involved in transmission,
    96
  human carriage, 95
  identification, 100
  infective dose, 45, 91
  isolation and detection, 88, 97, **98-9**, 100
  laboratory methods, 97, **98-9**, 100
  mechanisms of pathogenicity, 91-3, **94**, 95
  people at particular risk, 90-1
  species, 87
  symptoms of infection, 90
  taxonomy, **88**, 89
  toxin production, 93-4
  treatment of infection, 91
  typing, 41, 89-90, 100
Shigellosis from shellfish, 96, 467
Sickness, exclusion of workers during,
  409-411
Siderophores, 49
Silage, 335, 427
Small gastroenteritis viruses, 364
Small, round, structured viruses (SRSVs),
  363, **364**
  cycles of infection, 367
  detection, 371
  factors affecting survival, 370
  food handlers in spread, 366-7
  in fruit, 369
  in molluscan shellfish, 368, 369, 467-8
  symtoms, 365-6
  in water and ice, 369
Smoking
  of cured meats, 436
  effects on *Clostridium botulinum*, 305

Snow Mountain agent, 364
Soap, antiseptic, 415
Social impact of food poisoning, 26
Soil as reservoir of foodborne
  micro-organisms, 95, 278, 297, 319, 336,
  376,
Sorbate and sorbic acid,
  effects on *Bacillus cereus*, 282
  effects on *Clostridium botulinum*, 305-6
  effects on *Clostridium perfringens*, 321
  effects on *Listeria monocytogenes*, 343
  effects on *Staphylococcus aureus*, 253
  effects on *Yersinia*, 145
Soup, UHT, 443
Sous-vide products, 298, 444
Sprouting seeds, 476
Staphylococcus aureus, 235-6
  in bakery products, 249, 478
  biofilm development, 421
  cellular morphology, **235**
  coagulase production, 236, 238, 243-4, 256,
    258-9
  in concentrated and dried milk, 249, 452
  in cooked cured meats, 247, 442
  in environment, **245-6**
  enterotoxin detection, 237, 245, 261-5
  enterotoxin production, 236, **242-4**, 250-3,
    254-5
  factors affecting growth and survival, 250-3,
    254-5
  in fish products, 470-1
  foods associated with, 246-50
  in frozen peas, 483
  glucose fermentation, **235**
  human carriage, 19, 244
  identification, 259, **260-1**
  intoxication symptoms, 240-1
  isolation, 237, 255, **256-7**, 258
  laboratory methods, 255, **256-7**, 258,
    259-61, 262-5
  on onions, 249, 477
  in pasta, 249, 479
  pathogenicity mechanisms, **242-3**
  persons at risk from intoxication, 241
  pigmentation, **236**
  species, 236
  taxonomy, 237-8, 239
  thermonuclease production, 236, 238, 243-4,
    258, **260**
  treatment of intoxication, 241
  typing, 239-40, **261**
Starter cultures for fermented products, 343,
  437, 457
Stomach acidity, 45
Streptococcus, 361
String test for *Vibrio cholerae*, 182
Suicide phenomenon of *Aeromonas*, **185**, 199,
  200
Surface sampling for contamination, **424**
Surimi, 472
Survival in host tissue, 47-50
  Campylobacter, 219
  enteroinvasive *Escherichia coli*, **114**
  Listeria monocytogenes, 334
  Salmonella, **61-2**, 63
  Shigella, 92-3
  Yersinia, 140-1
Susceptibility
  to *Aeromonas*, 189
  to botulism, 293
  to *Campylobacter*, 216-7
  to *Escherichia coli*, 108
  to *Listeria monocytogenes*, 332
  to *Plesiomonas shigelloides*, 204
  to *Salmonella*, 56
  to *Shigella*, 90
  to *Staphylococcus aureus*, 241
  to *Vibrio*, 164
  to *Yersinia*, 138

Sushi, 17, 464
Systemic spread
  Aeromonas, 190
  Campylobacter, 219
  Listeria monocytogenes, 334
  Salmonella, 63
  Shigella, 92
  Vibrio parahaemolyticus, 170
  Yersinia, 140

Temperature effects
  on *Aeromonas*, **194**
  on *Bacillus*, 281-2
  on *Campylobacter*, 224
  on *Clostridium botulinum*, 301
  on *Clostridium perfringens*, 321
  on *Cryptosporidium parvum*, 376
  on *Escherichia coli*, 119
  on foodborne viruses, 370
  on *Listeria monocytogenes*, 340-2
  on *Plesiomonas shigelloides*, 207
  on *Salmonella*, **69**
  on *Shigella*, 97
  on *Staphylococcus aureus*, 251
  on *Vibrio*, 176-7
  on *Yersinia*, 145
Terranova spp, 464
Tetradotoxin, 463
Thermostable direct haemolysin production
  of *Vibrio parahaemolyticus*, 170-1
Tissue culture methods for *Clostridium*
  botulinum toxin, 311
Tofu, *Yersinia enterocolitica* transmission, 144
  Yersinia, 142
Toxin production by foodborne
  micro-organisms
  Aeromonas, 190-1
  Bacillus cereus, 50, 277-8, 281, **288**
  Campylobacter, 50, 219-20
  Citrobacter, 357
  Clostridium botulinum, 50, 289, 294-**6**, 297,
    301-5
  Clostridium perfringens, 316, **317**-8
  Enterobacter, 358
  Escherichia coli, 50, 112, 113-4, 115-6, 127-**8**
  Giardia lamblia, 50, 383
  Listeria monocytogenes, 334-5
  Plesiomonas shigelloides, 205-6
  Salmonella, 50, 63-4
  Shigella, 93-4
  Staphylococcus aureus, 50, 236, **242**-4, 250-3,
    254-5, 261-5
  Vibrio cholerae, 50, 167-9
  Vibrio parahaemolyticus, 170-1
  Vibrio vulnificus, 171-2
  Yersinia enterocolitica, 141
Toxoplasma gondii, 385, 386
Training of food handlers, 404, **405-6**, 407-8
  continuation, 408
  effectiveness, 408
  hygiene, 404, **405-6**, 407
  plant/process operatives, 407
  supervisors, 408
  technical management, 408
Transferrin, 49
Translocation, 46
Travellers' diarrhoea, 15, 24, 91, 108, 109,
  189, 204, 217, 375, 383, 384, 385
Trichinella, 437
Trichothecenes, 15
Tuberculosis, 360
Typhoid *see* enteric fever
Tyramine in matured cheese, 458

Udder excretion of pathogens, 447
Ulceration of gut

Balantidium coli, 385
Entamoeba histolytica, 380
Ulcerative colitis and *Salmonella*, 55
Ultrahigh temperature (UHT), 443
  milk, 450, 451
Undercooking hazards, 433, 439, 441
Undulating fever *see* brucellosis

V and W antigens of *Yersinia*, 140
Vahlkampfia sp, 386
Vegetable products, 476-7
Vegetables
  Aeromonas, 194
  Bacillus cereus in, 476
  Clostridium botulinum in, 299-300, 476
  control of pathogenic micro-organisms in,
    473, 474-5, 476-7
  Cryptosporidium parvum in, 376-7
  fermented, 477
  foodborne viral infection, 369
  Listeria monocytogenes infection, 339, 475
  pickled, 477
  Staphylococcus aureus in, 248, 477
Ventilation of buildings, 418-9
Vibrio, 157-8
  in environment, 174
  factors affecting growth and survival in
    foods, 176-7
  foods associated with infections, 175
  human carriage, 173
  identification, 180, **181-2**, 183
  isolation and detection, 158-9, 178, **179**, 180
  laboratory methods, 178, **179**, 180, **181-2**,
    183
  people at risk of infection, 164-5
  species, 157-8
  taxonomy, 159-60
  typing, 161-2
Vibrio cholerae
  in environment, 174
  exposure, 164-5
  foods associated with infections, 175
  heat resistance, 176
  pathogenicity mechanisms, 166, **167-8**, 169,
    170
  serological confirmation of identity, 182-3
  serotyping, 161
  susceptibility, 164
  symptoms of infections, 162-3
  treatment, 166
  virulence determination, 172
  see also cholera
Vibrio parahaemolyticus
  in environment, 174
  in fish, 463, 464
  foods associated with infections, 175
  heat resistance, 176
  pathogenicity mechanisms, 169-7
  in shellfish, 175, 467, 468
  susceptibility, 164
  symptoms of infections, 163
  treatment, 166
  virulence determination, 173, **183**
Vibrio vulnificus
  pathogenicity mechanisms, 171-2
  serological confirmation of identity, 182-3
  susceptibility, 164
  symptoms of infections, 163
  treatment, 166
  virulence determination, 173, 183
Viral infection, foodborne, 363
  diarrhoeal, 365-6
  environmental sources, 368
  factors affecting survival, 370
  foods associated with, 368-9
  pathogenicity mechanisms, 366

symptoms, 365-6
  types of virus, 364-5
Virulence determination
  Aeromonas, **191-2**
  Campylobacter, 220
  enteropathogenic *Escherichia coli*, 112
  Listeria monocytogenes, 335
  Shigella, 95, 100
  of *Vibrio*, 172-3, **183**
  of *Yersinia enterocolitica*, 141, 153, **154-5**
Viruses, 12, 363
  detection of foodborne, 370-1

Water
  flooding and location of buildings, 417
  foodborne micro-organisms in, 65, 95, 117,
    118, 142, 144, 174, 175, 192, 193, 206, 220,
    221, 223, 297, 319, 336-7, 360, 369, 376-7,
    381, 383, 384, 386
  irrigation, 417
  livestock drinking, 428
  supply for buildings, 418
Water activity effects
  on *Bacillus*, 282
  on *Campylobacter*, 224
  on *Clostridium botulinum*, 303
  on *Clostridium perfringens*, 321
  on *Escherichia coli*, 119
  on *Listeria monocytogenes*, 342
  on *Salmonella*, 70
  on *Staphylococcus aureus*, 252
Watercress, 476
Wheat, 477
Wiltshire bacon curing, 434, 435-6
Withdrawal procedure 401-3
  team composition, 403
  USFDA guidelines, 402

Yersinia, **129**-30
  dissociation of colonies, **129**
  in environment, 142
  factors affecting growth and survival in
    foods, 145
  foods commonly involved in transmission,
    143-4
  human carriage, 142
  identification, 152-3
  incubation temperature and biochemical
    reactions, **131-2**
  isolation, 130
  laboratory methods, 146, **147-51**, 152-3,
    154-5
  markers of virulence, 141-4, 153, **154-5**
  mechanisms of pathogenicity, 139-41
  people at particular risk from infections, 138
  species, 129-30
  symptoms of infection, 136-7, 409
  taxonomy, 131, 133, **134**
  treatment of infections, 138
  typing, **135**-6, 153
Yersinia enterocolitica, 136-7, 142, 143-4
  alkali tolerance, 146-7
  biofilm development, 421
  isolation and detection, 146, **147-51**, 152
  in milk, 143, 449
  typing, 153
  virulence determination, 153, **154-5**
Yersinia pseudotuberculosis, 137, 143, 144
  alkali tolerance, 146-7
  enrichment, 146
Yoghurt, 457
Yops, 139, 140

Zearaleone, 477